Electrodynamics

Dietmar Petrascheck • Franz Schwabl

Electrodynamics

 Springer

Dietmar Petrascheck
Institute for Theoretical Physics
Johannes Kepler University
Linz, Austria

Franz Schwabl
Physik Department
TU München
Garching, Bayern, Germany

This book is a translation of the original German edition "Elektrodynamik", 4th edition, by Dietmar Petrascheck and Franz Schwabl, published by Springer-Verlag GmbH, DE in 2023. The translation was done with the help of an artificial intelligence machine translation tool. A subsequent human revision was done primarily in terms of content, so that the book will read stylistically differently from a conventional translation. Springer Nature works continuously to further the development of tools for the production of books and on the related technologies to support the authors.

ISBN 978-3-662-71501-7   ISBN 978-3-662-71502-4   (eBook)
https://doi.org/10.1007/978-3-662-71502-4

Translation from the German language edition: "Elektrodynamik" by Dietmar Petrascheck and Franz Schwabl, © Der/die Herausgeber bzw. der/die Autor(en), exklusiv lizenziert an Springer-Verlag GmbH, DE, ein Teil von Springer Nature 2023 2024. Published by Springer Spektrum Berlin, Heidelberg. All Rights Reserved.

This Springer imprint is published by the registered company Springer-Verlag GmbH, DE, part of Springer Nature.
The registered company address is: Heidelberger Platz 3, 14197 Berlin, Germany

If disposing of this product, please recycle the paper.

Preface

Today, students learn electrodynamics, a classical field theory, based on the SI system. The representation of electrodynamics in the SI system with the constants for the permeability and permittivity in vacuum appears so excellent that many students are surprised when they are confronted with other representations in the further course of their studies.

Although most books contain a brief description of the other systems, this book on electrodynamics, a translation of the 4th edition of *Elektrodynamik* from German, has the special feature that all equations are written in a form that is equally valid for all systems.

Two constants, belonging to the Coulomb and Lorentz force are sufficient to specify the microscopic Maxwell equations and a third constant, the magnetic constant, is determined by Ampère's force law for SI. With these constants, the equations of electrodynamics for SI, Gauss, Heaviside-Lorentz, electrostatic or electromagnetic units are obtained by inserting the appropriate values.

The book covers (classical) electrodynamics including special relativity. Already in the first chapter the Maxwell equations are introduced together with the Lorentz force. Therefore electrostatics and magnetostatics are only special cases of Maxwell's equations.

Then the necessary approximations for the Maxwell equations in matter are made and electrostatics as well as magnetostatics are applied to matter. Concerning radiation, the dynamical theory of X-ray diffraction is discussed in detail.

Within the framework of the special theory of relativity, the Lagrange formalism is developed and the quantization of the electromagnetic field is carried out.

Linz, March 2025
Dietmar Petrascheck

Contents

1

Lorentz Force and Maxwell's Equations

Preliminary remark

Paragraphs appearing in the text in smaller font contain supplementary notes. However, if these paragraphs are, like the following paragraph, indented and separated from the rest of the text by horizontal lines, they supplement the topic being discussed:

Paragraphs like this contain intermediate calculations, formulas, supplements etc., which may be of interest in some cases.

In this chapter, we will be confronted quite abruptly with the Maxwell equations. To understand these, knowledge of vector analysis, or more precisely, of vector fields and the integral theorems of Gauss, Stokes, and Green are required. If this knowledge is lacking, it is advantageous to start with the appendix A.4.

Regarding notation, it should be noted that the so-called *Einstein's summation convention* is used throughout, which states that double indices are summed. The divergence of a tensor is thus

$$\nabla_j T_{ij} \equiv T_{ij,j} \equiv \sum_{j=1}^{3} \nabla_j T_{ij},$$

where the comma in $T_{ij,j}$ indicates the differentiation with respect to x_j.

1.1 Charges, Currents and Charge Conservation

Electrodynamics deals with the electric and magnetic fields generated by stationary and moving charges, the motion and interaction of charged particles under the influence of electromagnetic fields.

D. Petrascheck, F. Schwabl, *Electrodynamics*, https://doi.org/10.1007/978-3-662-71502-4_1

Experience shows that there are two types of charges, positive and negative. The assignment of the sign is pure convention[1]. As a result, protons carry a positive and electrons a negative charge.

The smallest electrical charge is referred to as the elementary charge e_0. The electron carries the charge $-e_0$, the proton the charge $+e_0$. Quarks with one and two thirds of the elementary charge are not included here, as these do not occur as free particles, but only in combinations with the charge 0, $\pm e_0$.

The charge carriers we will encounter here are stable. These are electrons, protons, atomic nuclei, which carry the charge Ze_0 with the atomic number Z and ions with the charge $\pm ze_0$, where z in electrolytes is equal to the valency.

In order to be able to specify the size of a charge, we must decide on a system of units. This not only determines the numerical values of electrical quantities, but also their dimensions and (almost) all relations between electromagnetic quantities. In the following, we strive to present electrodynamics, i.e. the Maxwell equations in general, without reference to a specific system of units. In the Gaussian system, the unit of charge 1 statC(oulomb) is defined in such a way that the force between two unit charges at a distance of 1 cm has a value of 1 dyn:

$$1\,\text{statC} = 1\,\text{cm}^{3/2}\,\text{g}^{1/2}\,\text{sec}^{-1}.$$

Now the Gaussian units are only of limited interest and for metrological applications one should use SI units (Système internationale; see p. 640). In these, the unit of charge is the Coulomb (1 C). This is the charge that flows through a conductor in one second at a current of 1 Ampere. The conversion to the Gaussian unit can be found in Tab. C.6, p. 644:

$$1\,\text{C} = 1\,\text{As} \approx 3\times10^9\,\text{statC}.$$

The elementary charge e_0 has the following values in these units

$$e_0^{\text{SI}} = 1.60218\times10^{-19}\,\text{C} \qquad \Leftrightarrow \qquad e_0 = 4.8032\times10^{-10}\,\text{statC}.$$

From this we can deduce that for a current of 1 microampere (μA) per microsecond (μs) the charges of 6×10^6 electrons are necessary. We conclude from this, that on the one hand the charge is discrete, i.e. it only appears as a multiple of the elementary charge, on the other hand the large number of elementary charges with an averaging over a small volume area allows the representation of ρ as a continuous charge density.

The first indications of the existence of an elementary charge were provided by the laws of electrolysis established by *Faraday*[2] in 1832. In today's formulation, the 1st Faraday's law states that in an electrolyte, the charge required for the deposition of one mole of a z-valued ion is given by $Q = zF$. Here, the Faraday constant $F = e_0 N_A = 96\,485\,\text{C}\,\text{mol}^{-1}$ indicates the amount of charge required to deposit 1 mol of a monovalent ion. $N_A = 6.022\times10^{23}\,\text{mol}^{-1}$

---

[1] If you rub a glass rod, it becomes positively charged; with hard rubber, the charge is negative.
[2] Michael Faraday, 1791–1867; English scientist; electromagnetic induction.

is the Avogadro constant. The existence of an elementary charge named *Electron* was proposed in 1874 by the Irish physicist *Stoney*[3] proposed. In 1881, *H. von Helmholtz*[4] presented the existence of an electric elementary charge as a consequence of the Faraday laws to the *Chemical Society* in London. The discovery of the electron in 1897 is independently attributed to *Emil Wiechert*[5] and *J. J. Thomson*[6] [Wiechert, 1897; J.J. Thomson, 1897].

The value of the elementary charge was experimentally determined in 1909 by *Millikan*[7] determined [Millikan, 1911].

Millikan experiment: Movement of oil droplets with the charge $e > 0$ in a gas. On the droplet, a sphere with the radius a, the electric field E (*Coulomb force*) acts and the gravitational field (reduced by the buoyancy of the sphere in the gas)

$$m\ddot{x} = eE - 6\pi\eta a\dot{x} - gm.$$

The first term on the right side indicates the acceleration of the droplet through the electric field, the second term is the (Stokes) friction in the gas with the viscosity η and the third term is the acceleration due to gravity.

Initially, the droplets have the charge $e = 0$. By ionizing the gas, the droplets acquire the charge $e = \pm Z e_0$. e is determined from their movement.

The electric charge is a fundamental quantity. In every interaction process of elementary particles, the total charge of the particles remains unchanged, regardless, whether this interaction involves the strong, weak, or electromagnetic interaction comes into play[8]. Therefore, a conservation law applies to charges:

In a closed system, the sum of all charges remains constant.

The extent of the charge in these particles varies. Electrons are point-like, i.e., finite dimensions are not detectable. Protons have a finite radius of $\sim 10^{-13}$ cm. But also the extent of the protons and nuclei can be neglected so that in the following all charge carriers are treated as point particles.

Note: Point particles also cause difficulties in classical electrodynamics. The self-energy of a finite charge diverges when the radius of the charge distribution tends to zero. Similarly, reducing the diameter of a wire while keeping the current constant leads to a divergence of the self-inductance.

From the dynamics of electrons, we learn that the assumption of a radius $a \lesssim r_e \approx 3 \times 10^{-13}$ cm of the charge distribution leads to incorrect results (see radiation reaction, section 8.5). Furthermore, at distances of the order of the Compton wavelength $\lambda_C \approx 4 \times 10^{-11}$ cm quantum mechanics would have to be taken into account. A solution is only offered by the QED (Quantum Electrodynamics).

---

[3] George Stoney, 1826–1911.

[4] Hermann von Helmholtz, 1821–1894, versatile scholar, among other things, vortex sets, wave equation. From 1870 professorship for physics in Berlin.

[5] Emil Wiechert, 1861–1928, from 1905 professor for geophysics in Göttingen.

[6] Joseph John Thomson, 1846–1940, Nobel Prize 1906 for el. conductivity of gases.

[7] Robert Andrews Millikan, 1868–1953; Nobel Prize 1923. Harvey Fletcher (unmentioned) was involved in the measurements.

[8] strong: $p + p \rightarrow \Sigma^0 + p + K^+$; weak: $n \rightarrow p + e^- + \bar{\nu}$; electromagnetically: $\pi^0 \rightarrow \gamma + \gamma$.

These rarely occurring difficulties can be circumvented, so that we stick to the concept of point particles despite the conceptual objections to them.

The total charge Q of an atomic nucleus is the sum of the charges of the protons

$$Q = \sum_p e_p = Z\, e_0 \,,$$

where Z is the atomic number. If you add the negative charge of the Z electrons, the total charge of the atom is zero. In other words, charges are additive, according to which the total charge Q is the sum of the partial charges q_i.

More important than the total charge Q is often the (spatial) charge density ρ, i.e. the charge ΔQ per volume element ΔV. Let there be n particles, numbered with i. They carry the charge e_i at location $\mathbf{x}_i(t)$ at time t. It is defined:

Microscopic charge density

$$\rho(\mathbf{x}, t) = \sum_{i=1}^{n} e_i\, \delta^{(3)}(\mathbf{x} - \mathbf{x}_i(t)). \tag{1.1.1}$$

Microscopic current density

$$\mathbf{j}(\mathbf{x}, t) = \sum_{i=1}^{n} e_i\, \delta^{(3)}(\mathbf{x} - \mathbf{x}_i(t))\, \dot{\mathbf{x}}_i. \tag{1.1.2}$$

Here, $\delta^{(3)}(\mathbf{x})$ is the three-dimensional *Dirac's delta function* (B.6.12). $\rho(\mathbf{x}, t)$ and $\mathbf{j}(\mathbf{x}, t)$ fulfill the *continuity equation*

$$\frac{\partial}{\partial t}\rho(\mathbf{x}, t) + \boldsymbol{\nabla} \cdot \mathbf{j}(\mathbf{x}, t) = 0, \tag{1.1.3}$$

which can be verified by differentiating the charge density with respect to t:

$$\frac{\partial}{\partial t}\rho(\mathbf{x}, t) = \sum_i e_i \frac{\partial}{\partial t}\delta^{(3)}(\mathbf{x} - \mathbf{x}_i(t)) = \sum_i e_i \left[\frac{\partial \delta(x - x_i)}{\partial x_i} \dot{x}_i\, \delta(y - y_i)\, \delta(z - z_i) \right.$$

$$\left. + \delta(x - x_i)\frac{\partial \delta(y - y_i)}{\partial y_i} \dot{y}_i\, \delta(z - z_i) + \delta(x - x_i)\, \delta(y - y_i)\frac{\partial \delta(z - z_i)}{\partial z_i} \dot{z}_i \right]$$

$$= -\sum_i e_i \boldsymbol{\nabla}\delta^{(3)}(\mathbf{x} - \mathbf{x}_i(t)) \cdot \dot{\mathbf{x}}_i.$$

Note: The continuity equation is valid for each individual particle i separately

$$\rho_i(\mathbf{x}, t) = q_i\, \delta^{(3)}(\mathbf{x} - \mathbf{x}_i(t)), \qquad\qquad \mathbf{j}_i(\mathbf{x}, t) = q_i \mathbf{v}_i\, \delta^{(3)}(\mathbf{x} - \mathbf{x}_i(t)).$$

Current through a surface

The charge within a fixed volume V is

$$Q(t) = \int_V \mathrm{d}^3x\, \rho(\mathbf{x}, t) = \sum_{\mathbf{x}_i(t) \in V} e_i\,. \qquad (1.1.4)$$

In many cases, one deals with charge carriers of one type (or a few types) so that the density can be brought to the form

$$\rho(\mathbf{x}, t) = q \sum_{i=1}^{n} \delta^{(3)}(\mathbf{x} - \mathbf{x}_i(t)) = q\, n(\mathbf{x}, t)\,. \qquad (1.1.5)$$

$n(\mathbf{x}, t)$ is the particle density. The current density $\mathbf{j}(\mathbf{x}, t)$ can be defined using an average velocity $\mathbf{v}(\mathbf{x}, t)$:

$$\mathbf{v}(\mathbf{x}, t) = \frac{1}{n} \sum_{i=1}^{n} \mathbf{v}_i \qquad \Rightarrow \qquad \mathbf{j}(\mathbf{x}, t) = \rho(\mathbf{x}, t)\, \mathbf{v}(\mathbf{x}, t)\,.$$

We now turn to Fig. 1.1 to determine the charge δQ flowing through the area da in time δt. The change in charge per unit of time results in the current through the area. $\mathbf{j} \cdot \mathbf{n}\, \mathrm{da}$ is the charge, which flows per unit of time through the unit area perpendicular to $\mathbf{j}$ ($[j] = \mathrm{A/m}^2$). During δt, the current through da leads to the loss of charge (see Fig. 1.1)

$$\delta Q = -\mathbf{j} \cdot \mathbf{n}\, \delta t \mathrm{da} = -\rho\, \mathbf{v} \cdot \mathbf{da}\, \delta t\,.$$

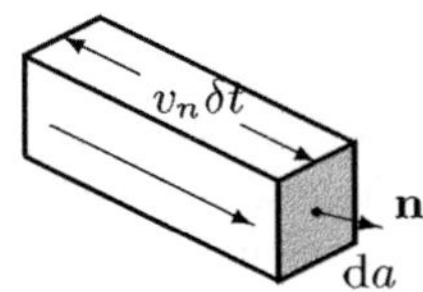

Fig. 1.1. $\mathbf{v}$ is the average speed of the charge carriers and $\mathbf{da} = \mathbf{n}\, \mathrm{da}$ the surface element through which in the time δt the particles located in the volume $\delta t\, \mathbf{v} \cdot \mathbf{n}\, \mathrm{da}$ flow

For a finite cross-section A one obtains

$$\frac{\delta Q}{\delta t} = -\iint_A \rho\, \mathbf{v} \cdot \mathbf{da}\,.$$

Continuous charge distributions

We recall that the distance between charge carriers in condensed matter (solid, liquid, plasma) is about 10^{-8} cm. For all macroscopic observations, distances of this magnitude are not resolvable. The particle structure of matter is therefore not visible, but one is dealing with an apparently continuous charge distribution which is formed by averaging the microscopic charge distribution in a volume $\Delta V(\mathbf{x})$.

According to *Lorentz*[9] it is obvious to combine all charges that are within a sphere $S_a = 4\pi a^3/3$ with the radius a around the considered point $\mathbf{x}$:

$$\Delta Q = \int_{S_a(\mathbf{x})} \mathrm{d}^3x' \sum_j e_j \delta^{(3)}\big(\mathbf{x}' - \mathbf{x}_j(t)\big) = \sum_{\mathbf{x}_j \in S_a(\mathbf{x})} e_j \,.$$

This gives the average density

$$\bar{\rho}(\mathbf{x},t) = \frac{\Delta Q}{S_a} = \int \mathrm{d}^3x' \, f(\mathbf{x}-\mathbf{x}')\rho(\mathbf{x}') \quad \text{with} \quad f(\mathbf{x}-\mathbf{x}') = \frac{\theta(a - |\mathbf{x}-\mathbf{x}'|)}{S_a},$$

where f is normalized

$$\int \mathrm{d}^3x' \, f(\mathbf{x}-\mathbf{x}') = 1$$

and $\theta(x)$ is the step function (Heaviside function) (B.6.17).

Note 1: The above distribution function f has the disadvantage of discontinuity on the sphere surface. We imagine f to be somewhat smoothed there, as sketched in Fig. 1.2, so that we obtain a continuous function, which can be achieved by

$$f(r) = \frac{\gamma}{\ln(e^{\gamma S_a} + 1)} \frac{1}{e^{\gamma(S_r - S_a)} + 1} \qquad \text{with} \qquad S_r = \frac{4\pi r^3}{3}. \qquad (1.1.6)$$

γ determines the smoothing of f. For $\gamma S_a = 50$ about 10% of the volume is in the transition area around $r=a$ and for $\gamma S_a \to \infty$ (1.1.6) transitions into the θ function.

In one dimension $(S_a = 2a)$ one obtains with

$$f(x) = \frac{\gamma}{\ln(e^{2\gamma a} + 1)} \frac{1}{e^{2\gamma(|x|-a)} + 1} \qquad\qquad (1.1.7)$$

an averaged *line charge density* $\bar{\lambda}(x)$. In Fig. 1.3 both the averaging of a linear charge distribution as well as (in the insert) the associated distribution function (1.1.7) is shown. If we had carried out the averaging with a rectangular distribution, small discontinuities would have become visible.

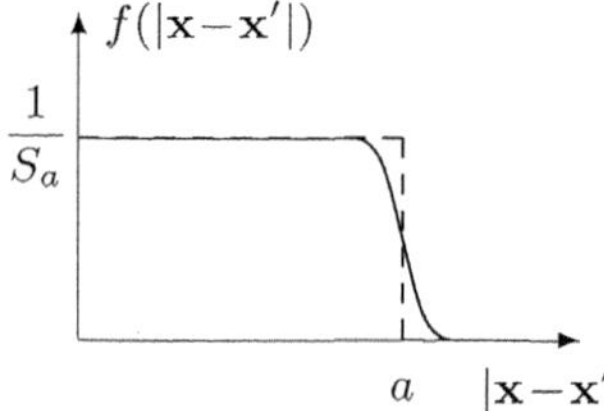

Fig. 1.2. Distribution function $f(|\mathbf{x} - \mathbf{x}'|)$, which, starting from a sphere S_a with radius a (dashed), is smoothed according to (1.1.6) at the edge

---

[9] Hendrik Antoon Lorentz, 1853–1928; Nobel Prize 1902 for Zeeman effect

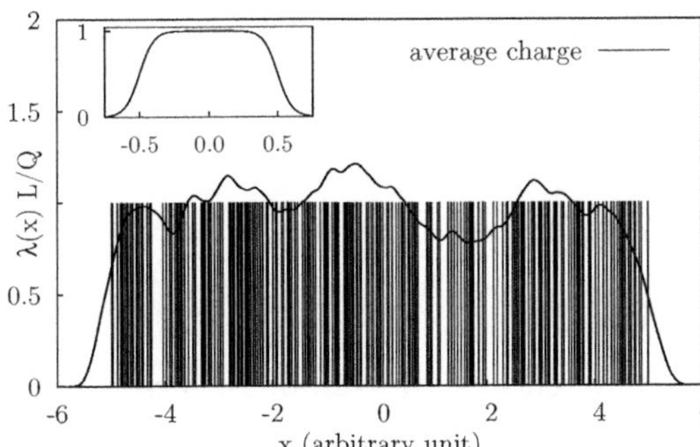

Fig. 1.3. Normalized average line charge density $\bar{\lambda}(x)/(Q/L)$; the vertical lines are point charges with the total charge Q, which are distributed over a length $L = 10$ (arbitrary length unit). The averaging was done with (1.1.7): $\gamma = 10$ and $a = 0.5$ (see insert) carried out

Note 2: In some contexts, an (additional) temporal averaging is useful. In a classical view, a point-like electron, which moves on a circular path with radius a and speed $\mathbf{v} = a\omega\mathbf{e}_\varphi$ around the nucleus generates the current

$$\mathbf{j}(\mathbf{x},t) = \rho(\mathbf{x},t)\,\mathbf{v}(\mathbf{x},t) = -e_0\delta(\varrho - a)\,\delta(\varphi - \omega t)\,\delta(z)\,\omega\,\mathbf{e}_\varphi\,.$$

The temporal averaging results in the ring current

$$\mathbf{j}(\mathbf{x}) = -\frac{e_0}{2\pi}\,\delta(\varrho - a)\,\delta(z)\omega\,\mathbf{e}_\varphi\,,$$

which generates a constant magnetic moment $\boldsymbol{\mu}$.

Note 3: The averaging mentioned here refers to charge carriers of one type and is intended to explain the use of continuous charge distributions. Not included are charge carriers (atoms or molecules) with charges of different signs, whose total charge disappears, but which have a finite (microscopic) dipole moment. We are dealing here with bound charges, as they occur in matter and which will only be discussed in section 5.2.

The average density is defined by

$$\bar{\rho}(\mathbf{x},t) \equiv \int \mathrm{d}^3x'\, f(\mathbf{x}-\mathbf{x}')\,\rho(\mathbf{x}',t) = e\,\bar{n}(\mathbf{x},t)\,. \tag{1.1.8}$$

With $\bar{n}(\mathbf{x},t)$ the particle density is denoted. For the current one gets

$$\bar{\mathbf{j}}(\mathbf{x},t) \equiv \int \mathrm{d}^3x'\, f(\mathbf{x}-\mathbf{x}')\,\mathbf{j}(\mathbf{x}',t) = \bar{\rho}(\mathbf{x},t)\,\mathbf{v}(\mathbf{x},t)\,. \tag{1.1.9}$$

The average speed $\mathbf{v}(\mathbf{x},t)$ is defined as

$$\mathbf{v}(\mathbf{x},t) = \frac{1}{\bar{n}(\mathbf{x},t)} \sum_{i=1}^{n} \int \mathrm{d}^3x'\, f(\mathbf{x}-\mathbf{x}')\,n_i(\mathbf{x}',t)\,\mathbf{v}_i\,.$$

It has to be shown that the averaged quantities also satisfy the continuity equation. This is the case if averaging and differentiation are interchangeable. Then the integrand satisfies the continuity equation and thus also the integral. For

$$\dot{\bar{\rho}}(\mathbf{x},t) = \int \mathrm{d}^3x'\, f(\mathbf{x}-\mathbf{x}')\dot{\rho}(\mathbf{x}',t) = \bar{\dot{\rho}}(\mathbf{x},t) \tag{1.1.10}$$

this is obvious and for

$$\boldsymbol{\nabla}\cdot\bar{\mathbf{j}}(\mathbf{x},t) = \boldsymbol{\nabla}\cdot\int_V \mathrm{d}^3x'\, f(\mathbf{x}-\mathbf{x}')\,\mathbf{j}(\mathbf{x}',t)$$

we use two "tricks" that will occur more often.

1. If a function depends only on the difference of two variables, then

$$\boldsymbol{\nabla} f(\mathbf{x}-\mathbf{x}') = -\boldsymbol{\nabla}' f(\mathbf{x}-\mathbf{x}'). \tag{1.1.11}$$

It follows

$$\boldsymbol{\nabla}\cdot\int_V \mathrm{d}^3x'\, f(\mathbf{x}-\mathbf{x}')\mathbf{j}(\mathbf{x}',t) = -\int_V \mathrm{d}^3x'\, \mathbf{j}(\mathbf{x}',t)\cdot\boldsymbol{\nabla}' f(\mathbf{x}-\mathbf{x}'). \tag{1.1.12}$$

2. *Integration by parts*: Supplement of the integrand to a complete divergence, on which then the Gauss's theorem (A.4.3) can be applied:

$$\int_V \mathrm{d}^3x'\, \mathbf{j}(\mathbf{x}',t)\cdot\boldsymbol{\nabla}' f(\mathbf{x}-\mathbf{x}') \tag{1.1.13}$$

$$= \int_V \mathrm{d}^3x'\, \left\{ \boldsymbol{\nabla}'\cdot\left[f(\mathbf{x}-\mathbf{x}')\,\mathbf{j}(\mathbf{x}',t) \right] - f(\mathbf{x}-\mathbf{x}')\boldsymbol{\nabla}'\cdot\mathbf{j}(\mathbf{x}',t) \right\}$$

$$= \oiint_{\partial V} \mathrm{d}\mathbf{a}'\cdot f(\mathbf{x}-\mathbf{x}')\,\mathbf{j}(\mathbf{x}',t) - \int_V \mathrm{d}^3x'\, f(\mathbf{x}-\mathbf{x}')\,\boldsymbol{\nabla}'\cdot\mathbf{j}(\mathbf{x}',t).$$

The boundary term here is a surface integral, which disappears because V is chosen large enough so that there are no more currents on the surface.

Thus, it holds

$$\boldsymbol{\nabla}\cdot\bar{\mathbf{j}}(\mathbf{x},t) = \overline{\boldsymbol{\nabla}\cdot\mathbf{j}}(\mathbf{x},t). \tag{1.1.14}$$

Averages and derivatives are therefore interchangeable and thus we obtain the continuity equation for the averaged charge and current density

$$\frac{\partial}{\partial t}\bar{\rho}(\mathbf{x},t) + \boldsymbol{\nabla}\cdot\bar{\mathbf{j}}(\mathbf{x},t) = 0. \tag{1.1.15}$$

A distinction between averaged, continuous and microscopic currents and densities is no longer necessary and is generally not made, so that for currents and charges, with the exception of special cases, only $\mathbf{j}$ and ρ are used.

We now integrate the 2nd term of the continuity equation over the volume V and apply Gauss's theorem (A.4.3):

$$I(t) = \int_V \mathrm{d}^3x\, \boldsymbol{\nabla}\cdot\mathbf{j}(\mathbf{x},t) = \oiint_{\partial V} \mathrm{d}\mathbf{a}\cdot\mathbf{j}(\mathbf{x},t). \tag{1.1.16}$$

∂V denotes the surface of V. With $\boldsymbol{\nabla}\cdot\mathbf{j}$ the sources (sinks) in V are denoted, which yield the total current I that passes through the surface ∂V. The temporal change of the total charge Q (see (1.1.4)) must yield the current I that passes through the surface enclosing the volume:

$$\dot{Q} + I = 0 \qquad \Leftrightarrow \qquad \frac{\mathrm{d}}{\mathrm{d}t}\int_V \mathrm{d}^3x\,\rho(\mathbf{x},t) + \oiint_{\partial V} \mathrm{d}\mathbf{a}\cdot\mathbf{j} = 0. \qquad (1.1.17)$$

Here we have the integral form of the continuity equation and can more precisely denote (1.1.15) as the differential form of the continuity equation. The same distinction between differential and integral forms is expressed in many places, especially in the Maxwell equations.

1.2 Lorentz Force

Electric charges are not defined per se, but through interaction forces they exert on each other. The force experienced by a charge moving in electric and magnetic fields is called the Lorentz force. In this case, exactly two constants, k_C and k_L, are freely selectable. Once they are set, the charges, currents, fields are also set and (with additional assumptions) the Maxwell equations.

1.2.1 Force on a Stationary Charge

According to Coulomb's law, two stationary electric charges exert the Coulomb force

$$\mathbf{F}_1 = q_1\,\mathbf{E}_2(\mathbf{x}_1) = k_C\,q_1 q_2\,\frac{\mathbf{x}_1 - \mathbf{x}_2}{|\mathbf{x}_1 - \mathbf{x}_2|^3} = -\mathbf{F}_2 \qquad (1.2.1)$$

on each other, where the factor k_C defines the charge. As a result, the current density $\mathbf{j}$ is determined with (1.1.2) and the electric field $\mathbf{E}$ with (1.2.2), as given in Tab. 1.1, p. 21.

$Coulomb$[10] discovered the law named after him in 1785 from the repulsion of two charges using a torsion balance. Priestley [1767, p. 732] has (on the advice of *Benjamin Franklin*) from the fact that no force acts on spheres that are in a charged metal vessel, concluded that the electrostatic force is $\sim 1/r^2$:

> *May we not infer from this experiment that the attraction of electricity is subject to the same laws with that of gravitation, and is therefore according to the squares of distances; since it is easily demonstrated, that where the earth in the form of a shell, a body in the inside of it would not be attracted to one side more than another?*

In a similar way, in 1773 *Cavendish*[11] determined the force law, but unfortunately did not publish it. It was left to *Maxwell*[12] to point out these mea-

---

[10] Charles Augustin de Coulomb, 1736–1806.

[11] Henry Cavendish, 1731–1810, English scientist.

[12] James Clerk Maxwell, 1831–1879; research on electrodynamics, kinetic gas theory, statistical mechanics.

surements [Maxwell, 1892, Art. 74a] and to make Cavendish's work known [Maxwell, 1879].

In the sketch Fig. 1.4, q_1 and q_2 have the same sign, i.e. the forces are repulsive. $\mathbf{E}_2(\mathbf{x}_1)$ is the electric field of the charge q_2, which the charge q_1

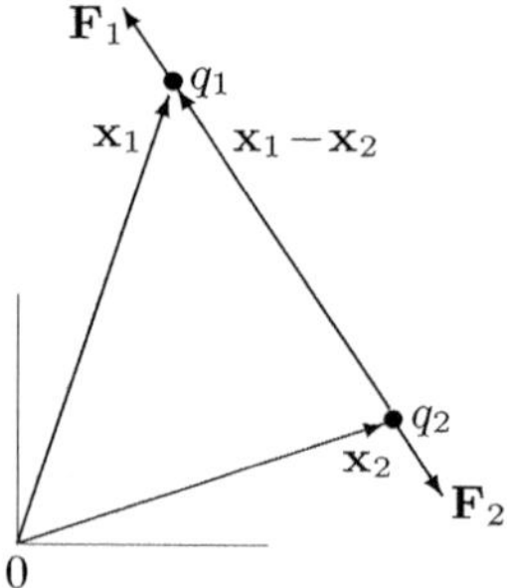

Fig. 1.4. The forces $\mathbf{F}_1$ and $\mathbf{F}_2$, which two stationary point charges q_1 and q_2 exert on each other, are anti/parallel to the connecting line $\mathbf{x}_1 - \mathbf{x}_2$. The forces for point charges of the same sign are shown

feels at location $\mathbf{x}_1$:

$$\mathbf{E}_2(\mathbf{x}_1) = k_C\, q_2 \frac{\mathbf{x}_1 - \mathbf{x}_2}{|\mathbf{x}_1 - \mathbf{x}_2|^3} = -k_C\, q_2 \boldsymbol{\nabla}_1 \frac{1}{|\mathbf{x}_1 - \mathbf{x}_2|}. \tag{1.2.2}$$

From this definition of force it follows that it can be derived from a potential, analogous to gravity:

$$\mathbf{F}_1 = -q_1 \boldsymbol{\nabla}_1 \phi(\mathbf{x}_1 - \mathbf{x}_2) \qquad \text{with} \qquad \phi(\mathbf{x}_1 - \mathbf{x}_2) = k_C \frac{q_2}{|\mathbf{x}_1 - \mathbf{x}_2|}. \tag{1.2.3}$$

ϕ is the scalar electrostatic potential, which in this form is only valid for stationary charges. The Coulomb force (1.2.1) is a central force, analogous to gravity.

Central forces always point to the center of force ($\mathbf{x}_1$, Fig. 1.4); the torque $\mathbf{N}$ vanishes, and the angular momentum $\mathbf{L}$ is constant over time

$$\mathbf{F}(\mathbf{x}) = F(|\mathbf{x} - \mathbf{x}_1|) \frac{\mathbf{x} - \mathbf{x}_1}{|\mathbf{x} - \mathbf{x}_1|}, \qquad\qquad \mathbf{N} = \dot{\mathbf{L}} = (\mathbf{x} - \mathbf{x}_1) \times \mathbf{F}(\mathbf{x}) = 0.$$

According to (1.2.2), it is

$$\mathbf{F}_1 = q_1 \mathbf{E}_2 = -q_2 \mathbf{E}_1. \tag{1.2.4}$$

1.2.2 Force on a Moving Charge

A magnetic field $\mathbf{B}$ exerts a force perpendicular to the direction of motion $\mathbf{v}$ of the charge. This force $\mathbf{F}_L$ must be added to the Coulomb force $\mathbf{F}_C$:

$$\mathbf{F} = \mathbf{F}_C + \mathbf{F}_L, \qquad \mathbf{F}_C = q\mathbf{E}, \qquad \mathbf{F}_L = k_L q \frac{\mathbf{v}}{c} \times \mathbf{B}. \tag{1.2.5}$$

The special choice of $\mathbf{v}/c$ in $\mathbf{F}_L$ with the speed of light c is based on the fact that the equivalence of inertial systems (systems that move uniformly without force) results in a universal speed c, as explained in section 12.1.1. Therefore, speeds appear in the ratio $\mathbf{v}/c$, as is taken into account in $\mathbf{F}_L$. This does not have to be considered, because k_L is a freely selectable constant that determines the magnetic field $\mathbf{B}$, but a choice with $k_L \neq 1$ interferes with the symmetry between electric field $\mathbf{E}$ and magnetic field $\mathbf{B}$.

Now all quantities that appear in the Maxwell equations (1.3.21), namely ρ, $\mathbf{j}$, $\mathbf{E}$ and $\mathbf{B}$, are determined by the two constants k_C and k_L. The system with $k_C = 1$ and $k_L = 1$ is called *Gaussian system*.

$\mathbf{B}$ is according to (1.3.1) the *magnetic flux density* and is (historically) called *magnetic induction*; as can be seen from (1.2.5), $\mathbf{B}$ is the physical field, the force effect of which on moving charges is measurable. To describe electromagnetic processes in matter, one introduces auxiliary fields, the dielectric displacement $\mathbf{D}$ and the magnetic field $\mathbf{H}$, which are defined by material equations (5.2.17). If one uses $\mathbf{B}$ as the designation for the magnetic field, one should, to be precise, refer to $\mathbf{H}$ as a *magnetic auxiliary field* [Griffiths, 2017, section 6.3] or the *magnetic excitation* [Sommerfeld, 1952, p. 10], which is rarely done. The force $\mathbf{F}_L$, which describes the deflection of moving particles in a magnetic field, is the part of the *Lorentz force* attributable to Lorentz. The literature is not consistent here, sometimes $\mathbf{F}_L$ is referred to as the Lorentz force [Griffiths, 2017, (5.1)], sometimes $\mathbf{F}$ [Jackson, 1998, (1.3)].

Work is defined as force times distance. The Lorentz component $\mathbf{F}_L$ does no work, as the force is perpendicular to the path $\delta\mathbf{s} = \mathbf{v}\delta t$:

$$\delta W = q\left(\mathbf{E} + \frac{k_L}{c}\mathbf{v}\times\mathbf{B}\right) \cdot \mathbf{v}\delta t = q\mathbf{E}\cdot\delta\mathbf{s}. \tag{1.2.6}$$

Notes: Maxwell [1892, Art. 599] refers to the field[13] $\mathbf{E}_1 = \mathbf{v}\times\mathbf{B}$ as the part of the electromotive force, which is to be assigned to the movement of a particle crossing magnetic field lines, (i.e. $\mathbf{F}_L = q\mathbf{E}_1$). $\mathbf{E}_1$ is a consequence of the induction law for moving conductors (1.3.4) and is given by Maxwell [already 1861, (77)]. One might think that $\mathbf{F}_L$ has been found with this.

It took about 30 years until Lorentz [1892, (61)] specifies the force on a charged particle moving in the field $\mathbf{B}$ with $\mathbf{F}_L$. At that time, an atomistic picture of matter with indications to electrons already existed. The direct experimental evidence through the deflection of cathode rays in electromagnetic fields is due to Lenard[14] and J.J. Thomson [1897]. The Hall voltage (discovered in 1879, section 5.3.3), which is caused by the deflection of electrons in a conductor, can be understood as indirect evidence.

---

[13] electrostatic units, $k_C = 1$, $k_L = c$, i.e. $\mathbf{E}_1 = (k_L/c)\mathbf{v}\times\mathbf{B} \to \mathbf{v}\times\mathbf{B}$.
[14] Philipp Lenard (1862–1947), Nobel Prize 1905.

1.3 Maxwell Equations

The equations that determine the behavior of charged particles in electro-magnetic fields, the Maxwell equations, can be stated in both differential and integral form. In the latter, there are integrals, especially over surfaces, volumes and lines, that enclose areas; this representation is usually better suited for the investigation of continuity conditions at interfaces.

First, it is necessary to define some terms:

$$
\begin{aligned}
&\text{magnetic flux through area } A: && \Phi_B = \iint_A da \cdot \mathbf{B}, \\[2mm]
&\text{electric flux through area } A: && \Phi_E = \iint_A da \cdot \mathbf{E}, \\[2mm]
&\text{circulation of the magnetic field: } && Z_B = \oint_{\partial A} d\mathbf{x} \cdot \mathbf{B}, \\[2mm]
&\text{circulation of the electric field: } && Z_E = \oint_{\partial A} d\mathbf{x} \cdot \mathbf{E}.
\end{aligned}
\tag{1.3.1}
$$

∂A is a closed curve that forms a right-handed helix as the boundary curve

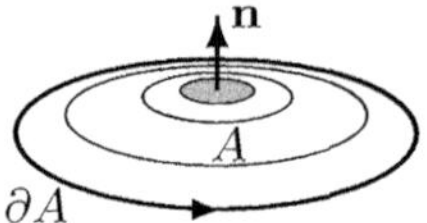

Fig. 1.5. Area A with boundary curve ∂A, which is traversed counterclockwise and with $\mathbf{da} = da\,\mathbf{n}$ forms a right-handed screw

of the surface A with its normal vector $\mathbf{n}$ as sketched in Fig. 1.5. The fields $\mathbf{B}$ or $\mathbf{E}$ must have vortices ($\mathbf{\nabla} \times \mathbf{E} \neq 0$) for a circulation to occur.

The electric (magnetic) flux is also referred to as electric (magnetic) permeation or as streamlines (force lines) number. $\mathfrak{E}$ or emf (electromotive force) are alternative expressions for the electric Z_E, where "force" stands in its older meaning for energy [Sommerfeld, 1952, p. 12].

The Maxwell equations, on which electrodynamics is based, are the result of experience with electrodynamic phenomena and thus roughly equivalent to the Newtonian axioms of mechanics. We anticipate that we will focus on two laws, the *Faraday's law of induction* and the *Ampère-Maxwell law*. The other two laws, Gauss's law and the divergence-free magnetic field, are additional axioms that follow from the assumption that $\mathbf{E}$ and $\mathbf{B}$ are vector fields with the electric charges as sources and sinks of $\mathbf{E}$.

1.3.1 Gauss's Law

Of interest is the flux Φ_E through the surface ∂V of the (simply connected) volume V, sketched in Fig. 1.6. Φ_E is proportional to the charge Q contained in V. This proportionality factor is $4\pi k_C$, as shown in (1.3.3').

Gauss's Law:

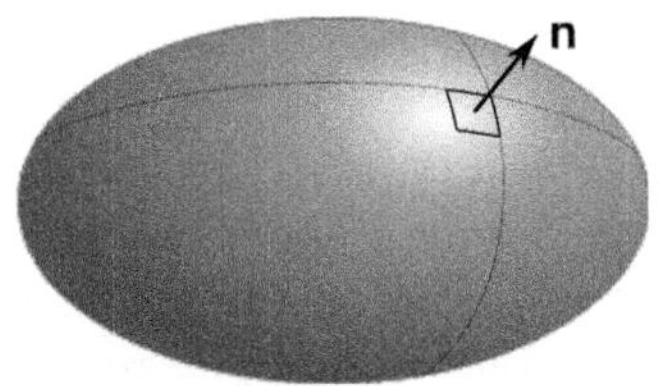

Fig. 1.6. Simply connected volume V with the surface ∂V; this is piecewise smooth (see p. 587), where $\mathbf{da} = \mathrm{d}a\,\mathbf{n}$ always points outwards

$$\oiint_{\partial V} \mathbf{da}\cdot\mathbf{E}(\mathbf{x},t) = 4\pi k_C Q(t). \tag{1.3.2}$$

If we convert the surface integral with Gauss's theorem (A.4.3) into a volume integral, we obtain Gauss's law in the form

$$\int_V \mathrm{d}^3 x \left(\boldsymbol{\nabla}\cdot\mathbf{E} - 4\pi k_C \rho(\mathbf{x},t)\right) = 0.$$

This equation is valid for arbitrary volumes V only if the integrand vanishes

$$\boldsymbol{\nabla}\cdot\mathbf{E}(\mathbf{x},t) = 4\pi k_C \rho(\mathbf{x},t), \tag{1.3.3}$$

with which we have derived the differential form of Gauss's law (1.3.2). If one substitutes the electric field (1.2.2) of the charge q_2 into (1.3.3), then one can verify the proportionality factor $4\pi k_C$:

$$\boldsymbol{\nabla}\cdot\mathbf{E}(\mathbf{x}) = -q_2 k_C \Delta\frac{1}{|\mathbf{x}-\mathbf{x}_2|} \overset{(2.1.7)}{=} 4\pi k_C q_2 \delta^{(3)}(\mathbf{x}-\mathbf{x}_2) \equiv 4\pi k_C \rho(\mathbf{x}). \tag{1.3.3'}$$

1.3.2 Faraday's Law of Induction

Faraday [1832] found that a current (field) is induced in a conductor loop when the magnetic flux Φ_B passing through the loop area A changes. This can happen through a neighboring loop in which the current is varied; but it can also be this loop (or a permanent magnet) moved relative to the conductor loop. In other words, the flux change is caused by a change in the magnetic field $\mathbf{B}$ and/or the area $A(t)$. With the flux change, a voltage is induced in the conductor loop, where $\mathbf{E}'$ is the field in the system of the loop:

$$\oint_{\partial A} \mathbf{ds}\cdot\mathbf{E}' = -k_3 \frac{\mathrm{d}}{\mathrm{d}t}\iint_A \mathbf{da}\cdot\mathbf{B}, \qquad\qquad k_3 \overset{(1.3.8)}{=} \frac{k_L}{c}. \tag{1.3.4}$$

k_3 is a yet undetermined proportionality factor, which in (1.3.8) is identified as $k_3 = k_L/c$. The negative sign anticipates that the magnetic field of the is opposed to the flux change generating the current (Lenz's rule). The direction of the path integral along ∂A is in the mathematically positive sense, i.e. counterclockwise, as indicated by the integral sign in (1.3.4) and sketched in Fig. 1.7. Using the definitions of the circulation Z_E and the magnetic flux Φ_B (1.3.1) the law of induction can be brought to the compact form

$$Z_E = -k_3 \dot{\Phi}_B. \tag{1.3.5}$$

In words according to Sommerfeld [1952, p. 13]:

Every change in the number of magnetic lines of force which traverse a given surface A produces in its boundary ∂A an electric loop tension (circulation of the electric field, i.e. electromotive force) which is numerically equal to the rate of change (multiplied with k_L/c), *but opposite in sign.*

The *Lenz's rule* indicates the direction of the current generated by the change of the magnetic flux passing through A. As sketched in Fig. 1.7(a), it is the "left-hand rule". If the boundary ∂A of the area A, as sketched in Fig. 1.7(b),

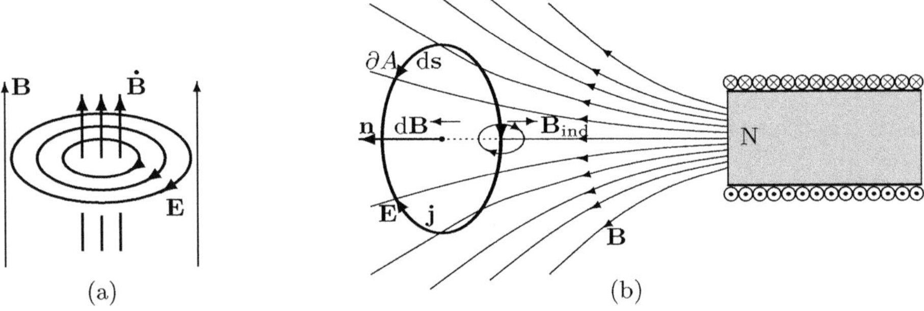

Fig. 1.7. (a) Induction law (left-hand rule); the magnetic field becomes stronger: $\dot{\mathbf{B}} \parallel \mathbf{B}$.
(b) (Stationary) conductor loop of the area A with the normal vector $\mathbf{n}$ and the boundary curve ∂A; if the field becomes stronger, then $\mathbf{n} \cdot \dot{\mathbf{B}} > 0$ and as a result $\dot{\Phi}_B > 0$. $\mathbf{E}$ and $\mathbf{j}$ are directed against the direction of the boundary curve ∂A. $\mathbf{j}$ induces a field $\mathbf{B}_{\mathrm{ind}}$, which is opposite to $\mathrm{d}\mathbf{B}$

represents a conductor loop, then the voltage induced in this causes a current, whose magnetic field $\mathbf{B}_{\mathrm{ind}}$ is directed against the field change $\mathrm{d}\mathbf{B}$.

Lenz's rule: The induced electric field is directed in such a way that the magnetic field induced by it opposes the causing flux change.

If $A(t)$ is moved, then in (1.3.4) $\mathbf{E}'$ is the field in the (current) rest system of the loop; here v/c should be sufficiently small so that higher orders are not to be considered; in the rest system of the loop, $A(t)$ is not deformed, which is why the derivative can be interchanged with the integral. To calculate $\frac{\mathrm{d}\mathbf{B}}{\mathrm{d}t}$ we use (A.2.35) for constant $\mathbf{a}$:

$$\boldsymbol{\nabla} \times (\mathbf{a} \times \mathbf{b}) = \mathbf{a}(\boldsymbol{\nabla} \cdot \mathbf{b}) - (\mathbf{a} \cdot \boldsymbol{\nabla})\mathbf{b}.$$

This gives the so-called *convective derivative*

$$\frac{\mathrm{d}\mathbf{B}}{\mathrm{d}t} = \frac{\partial \mathbf{B}}{\partial t} + (\mathbf{v} \cdot \boldsymbol{\nabla})\mathbf{B} = \dot{\mathbf{B}} - \boldsymbol{\nabla} \times (\mathbf{v} \times \mathbf{B}), \tag{1.3.6}$$

since the term $\mathbf{v}(\boldsymbol{\nabla}\cdot\mathbf{B})$ vanishes due to the divergenceless field $\mathbf{B}$ (1.3.19). Now apply the Stokes's theorem (A.4.13) to (1.3.4) and get:

$$\oint_{\partial A(t)} \mathrm{d}\mathbf{s}\cdot\mathbf{E}' = -k_3 \iint_A \mathrm{d}\mathbf{a}\cdot\dot{\mathbf{B}} + k_3 \oint_{\partial A(t)} \mathrm{d}\mathbf{s}\cdot\mathbf{v}\times\mathbf{B}. \tag{1.3.7}$$

We thus have the relationship between the field $\mathbf{E}'$ in the moving and $\mathbf{E}$ in the resting system:

$$\mathbf{E}' = \mathbf{E} + k_3\mathbf{v}\times\mathbf{B} \qquad \overset{(1.2.5)}{\Longrightarrow} \qquad k_3 = \frac{k_L}{c}. \tag{1.3.8}$$

Maxwell, who put the induction observed by Faraday into a mathematical form, derived this equation [Maxwell, 1861, (77)]. It describes the field that a moving charge q feels; hence k_3 can be given by comparison with the force (1.2.5) on a moving charge: $k_3 = k_L/c$. After applying Stokes' theorem (A.4.13), one obtains

$$\iint_A \mathrm{d}\mathbf{a}\cdot(\boldsymbol{\nabla}\times\mathbf{E} + \frac{k_L}{c}\dot{\mathbf{B}}) = 0. \tag{1.3.9}$$

Since the vanishing of the above integral does not depend on the area, the integrand must vanish and one obtains the induction law in differential form

$$\boldsymbol{\nabla}\times\mathbf{E} + \frac{k_L}{c}\dot{\mathbf{B}} = 0. \tag{1.3.10}$$

Notes: We have considered in (1.3.4) a current loop moving with $\mathbf{v}$ and only considered contributions linear in v [Jackson, 1998, sec. 5.15]. $\mathbf{E}$ here arises from the linear *Galilean transformation* (12.0.1) from $\mathbf{E}'$. Within the framework of the special theory of relativity (STR) we will see that the correct transformation, the *Lorentz transformation* in the moving system results in a contraction of the length parallel to $\mathbf{v}$, which is an effect of higher order in v; (13.1.29) is the transformation for $\mathbf{E}$. The differential Maxwell equations, like (1.3.10), are form-invariant under Lorentz transformations, which will be discussed in more detail in section 13.1.4.

1.3.3 Ampère-Maxwell Law

The *Ampère-Maxwell law* describes the construction of a magnetic field $\mathbf{B}$ around an electrical conductor:

$$\oint_{\partial A} \mathrm{d}\mathbf{s}\cdot\mathbf{B} = k_4\left[\frac{4\pi k_C}{c}\iint_A \mathrm{d}\mathbf{a}\cdot\mathbf{j} + \frac{1}{c}\frac{\mathrm{d}}{\mathrm{d}t}\iint_A \mathrm{d}\mathbf{a}\cdot\mathbf{E}\right], \qquad k_4 \overset{(1.3.15')}{=} \frac{1}{k_L}. \tag{1.3.11}$$

k_4 is a proportionality factor to be determined, which is identified in the following using the wave equation (1.3.15') as $1/k_L$. The correctness of the expression in the square bracket, which describes the total current $(4\pi k_C/c)I_{\text{tot}}$, is shown using the continuity equation.

To the (conduction) current density $\mathbf{j}$, the displacement current density $\mathbf{j}_d = \dot{\mathbf{E}}/4\pi k_C$ is added, thus forming the total current density. $\mathbf{j}_d$ (*displacement current*) is proportional to the temporal change of the electric flux density:

$$\mathbf{j}_{\text{tot}}(\mathbf{x}, t) = \mathbf{j}(\mathbf{x}, t) + \mathbf{j}_d(\mathbf{x}, t), \qquad \mathbf{j}_d(\mathbf{x}, t) = \frac{1}{4\pi k_C}\, \dot{\mathbf{E}}(\mathbf{x}, t). \tag{1.3.12}$$

The Ampère-Maxwell law thus takes the form of Ampère's (circuital) law

$$Z_B = k_4 \frac{4\pi k_C}{c}\, I_{\text{tot}} \quad \text{with} \quad k_4 \overset{(1.3.15')}{=} \frac{1}{k_L} \quad \text{and} \quad I_{\text{tot}} = \iint_A \mathbf{da} \cdot \mathbf{j}_{\text{tot}}, \tag{1.3.13}$$

where Z_B is the circulation (vortex strength) of the magnetic field. *Röntgen* experimentally demonstrated the displacement current through a dielectric moving in the electric field (*Röntgen current*) in 1888 [Röntgen, 1888], which Lorentz considered almost as important a discovery as that of the X-rays [Beier, 2013, p. 62].

For the Ampère-Maxwell law the following formulation of Sommerfeld [1952, p. 13] can be used if the displacement current density $\mathbf{j}_d$ is also included in the current density $\mathbf{j}$:

The number of electric current lines (multiplied with $4\pi k_C/ck_L$), which traverse an arbitrary surface A, is accompanied by a magnetic loop tension (magnetic circulation, magnetomotive force) in the bounding curve ∂A of A which is equal to it in both, magnitude and direction.

Of greater relevance is the transformation of the line integral into a surface integral with the vortex density $z_B = \mathbf{n} \cdot \mathbf{\nabla} \times \mathbf{B}$ by applying Stokes' theorem (A.4.13):

$$\iint_A \mathbf{da} \cdot \left\{ \mathbf{\nabla} \times \mathbf{B} - k_4 \left[\frac{1}{c} \dot{\mathbf{E}} + \frac{4\pi k_C}{c} \mathbf{j} \right] \right\} = 0. \tag{1.3.14}$$

Since the integral applies to every surface, the integrand must vanish, so that we get the differential form of the Ampère-Maxwell law:

$$\mathbf{\nabla} \times \mathbf{B} = k_4 \frac{4\pi k_C}{c} \left[\mathbf{j} + \frac{1}{4\pi k_C} \dot{\mathbf{E}} \right], \qquad k_4 \overset{(1.3.15')}{=} \frac{1}{k_L}. \tag{1.3.15}$$

The divergence is calculated using the continuity equation (1.1.15):

$$\mathbf{\nabla} \cdot \mathbf{\nabla} \times \mathbf{B} \overset{(1.3.3)}{=} k_4 k_C \frac{4\pi}{c} \left[\mathbf{\nabla} \cdot \mathbf{j} + \dot{\rho} \right] \overset{(1.1.15)}{=} 0.$$

Between $\mathbf{j}$ and $\mathbf{j}_d$ there can therefore be no further proportionality factor. To determine k_4 we take the Maxwell equations for the vacuum ($\mathbf{j} = \rho = 0$), derive the Ampère-Maxwell equation with respect to time and substitute for $\dot{\mathbf{B}}$ the induction equation:

$$\mathbf{\nabla} \times \dot{\mathbf{B}} = -\frac{c}{k_L} \mathbf{\nabla} \times (\mathbf{\nabla} \times \mathbf{E}) \overset{(A.2.38)}{=} \frac{c}{k_L} \Delta \mathbf{E} = \frac{k_4}{c} \ddot{\mathbf{E}}.$$

We have used in the calculation of $\nabla \times (\nabla \times \mathbf{E})$, that no sources are present ($\nabla \cdot \mathbf{E} = 0$) and so the wave equation for $\mathbf{E}$ is obtained:

$$\left(\frac{k_4 k_L}{c^2}\frac{\partial^2}{\partial t^2} - \Delta\right)\mathbf{E} = 0 \qquad\qquad \Longrightarrow \qquad\qquad k_4 = \frac{1}{k_L}. \qquad (1.3.15')$$

c is the propagation speed of electrical waves, which is why $k_4 = 1/k_L$.

Note: To (1.3.15) it should be added that around 1860 only *Ampère's circuital law*

$$\nabla \times \mathbf{B} = \frac{4\pi k_C}{ck_L}\mathbf{j} \qquad (1.3.16)$$

was known. Maxwell noticed that this could not be correct for time-dependent phenomena. It must be supplemented by a term, the displacement current $\mathbf{j}_d$, which ensures that $\nabla \cdot \nabla \times \mathbf{B} = 0$ and at the same time fulfills the charge conservation (continuity equation) as previously shown. Only with the displacement current $\mathbf{j}_d = \frac{1}{4\pi k_C}\dot{\mathbf{E}}$ are electromagnetic waves as solutions of the Maxwell equations in vacuum ($\rho = j = 0$) possible.

It was the merit of *Hertz*[15] around 1887 to prove the existence of electromagnetic waves and thus also to verify the Ampère-Maxwell equation, which was not experimentally secured at the time of its establishment.

We have already mentioned the Ampère-Maxwell law preceding *Ampère's law* for stationary currents, which in integral form is:

$$\oint_{\partial A} \mathrm{ds} \cdot \mathbf{B} = \frac{4\pi k_C}{ck_L}I, \qquad\qquad I = \iint_A \mathrm{da} \cdot \mathbf{j}. \qquad (1.3.17)$$

Even earlier (1820), Ørsted discovered that an electric current in a linear conductor generates a magnetic field, as shown in Fig. 1.8 and which can be calculated with the *Biot-Savart's law*, sometimes also referred to as *Ørsted's law*.

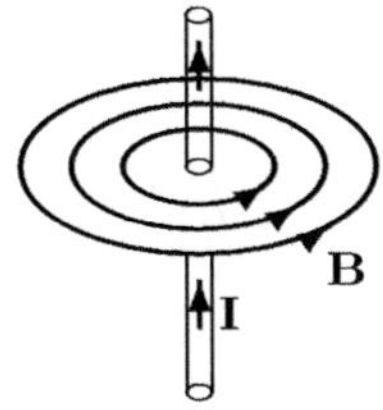

Fig. 1.8. A current I flows in the wire, which generates the magnetic field **B**. The right-hand rule is referred to as Ampère's rule when **I** points upwards and **B** is directed counterclockwise (right-hand screw)

The magnetic field is circular, and the magnetic circulation on a circle with radius R is given by

---

[15] Heinrich Hertz, 1857–1894.

$$Z_B = \oint_C \mathrm{ds}\cdot\mathbf{B} = 2\pi RB = \frac{4\pi k_C I}{ck_L}, \qquad \text{so that } B = \frac{k_C}{ck_L}\frac{2I}{R} \stackrel{\text{SI}}{=} \frac{\mu_0}{4\pi}\frac{2I}{R}.$$

This law, valid here for an infinitely long conductor (wire), follows both from the *Ampère's circuital law* and from the *Biot-Savart law* (1820). In a more general form, the wire is replaced by a current that spreads over the area A.

1.3.4 The Divergence–free Magnetic Field

In analogy to Gauss's law for the electric field, it also applies to $\mathbf{B}$ that $\nabla\cdot\mathbf{B}$ indicates the source density of the magnetic field. Due to the absence of magnetic charges (magnetic monopoles) the source density is zero and thus the magnetic flux through a closed surface ∂V also disappears

$$\oiint_{\partial V} \mathrm{da}\cdot\mathbf{B} = 0. \tag{1.3.18}$$

This (4th) Maxwell's law is analogous to Gauss's law for the electric field, with the already mentioned difference that so far no magnetic charges have been found. Therefore, the magnetic field is assumed to be source-free:

$$\nabla\cdot\mathbf{B}(\mathbf{x},t) = 0. \tag{1.3.19}$$

We have transformed the surface integral (1.3.18) into a volume integral with $\nabla\cdot\mathbf{B}$ as integrand, from which (1.3.19), the differential form of (1.3.18) follows. This completes the Maxwell equations.

1.3.5 Maxwell Equations in Integral Form

The term Maxwell equations includes Gauss's law (1.3.2), Faraday's law of induction (1.3.4), Ampère-Maxwell's law (1.3.12) and the divergence-free magnetic field (1.3.18). These are always addressed in the above order, whether in differential or integral form:

$$\text{(a)} \ \oiint_{\partial V}\mathrm{da}\cdot\mathbf{E} = 4\pi k_C\int_V \mathrm{d}^3x\,\rho, \qquad \text{(b)} \ \oint_{\partial A}\mathrm{ds}\cdot\mathbf{E} = -\frac{k_L}{c}\frac{\mathrm{d}}{\mathrm{d}t}\iint_A\mathrm{da}\cdot\mathbf{B},$$

$$\text{(c)} \ \oint_{\partial A}\mathrm{ds}\cdot\mathbf{B} = \frac{4\pi k_C}{ck_L}\iint_A\mathrm{da}\cdot(\mathbf{j}+\frac{\dot{\mathbf{E}}}{4\pi k_C}), \quad \text{(d)} \ \oiint_{\partial V}\mathrm{da}\cdot\mathbf{B}=0. \tag{1.3.20}$$

∂A is the boundary curve of the area A and ∂V the surface of the volume V.

Remarks: Sommerfeld [1952, p. 14] has shown that under certain conditions about the switch-off of the fields $\mathbf{E}$ and $\mathbf{B}$ the divergence-free magnetic field follows from the induction law (see (a)) and Gauss's law follows from Ampère-Maxwell's law (see (b)).

(a) If one takes the induction law and forms a closed surface around the curve C, where the second surface is oriented in the opposite direction, then the electric voltage disappears and one obtains

$$0 = -\frac{k_L}{c}\frac{\mathrm{d}}{\mathrm{d}t}\oiint \mathrm{d}\mathbf{a}\cdot\mathbf{B}.$$

The integration yields

$$\oiint \mathrm{d}\mathbf{a}\cdot\mathbf{B}(t) = \text{const} = \oiint \mathrm{d}\mathbf{a}\cdot\mathbf{B}(t_0).$$

The assumption that $\mathbf{B}(t_0)=0$, leads to const $= 0$ and thus to $\boldsymbol{\nabla}\cdot\mathbf{B} = 0$.

(b) In Ampère's law for a closed surface, the curve C must also additionally be traversed in the opposite direction. The magnetic circulation compensates itself on the two runs, and it remains

$$0 = 4\pi k_C \oiint \mathrm{d}\mathbf{a}\cdot\mathbf{j} + \frac{\mathrm{d}}{\mathrm{d}t}\oiint \mathrm{d}\mathbf{a}\cdot\mathbf{E}.$$

From this follows according to Gauss's theorem

$$0 = 4\pi k_C \int_V \mathrm{d}^3x\,\boldsymbol{\nabla}\cdot\mathbf{j} + \frac{\mathrm{d}}{\mathrm{d}t}\oiint \mathrm{d}\mathbf{a}\cdot\mathbf{E}.$$

In the 1st term on the right side, we insert the continuity equation. This gives

$$0 = -4\pi k_C \frac{\mathrm{d}}{\mathrm{d}t}\int_V \mathrm{d}^3x\,\rho(\mathbf{x},t) + \frac{\mathrm{d}}{\mathrm{d}t}\oiint \mathrm{d}\mathbf{a}\cdot\mathbf{E}.$$

The integration yields

$$\text{const} = -4\pi k_C \int \mathrm{d}^3x\,\rho + \oiint \mathrm{d}\mathbf{a}\cdot\mathbf{E}.$$

This is Gauss's law, when const$=0$ is assumed, which can be carried out analogously to the previous example.

1.3.6 Maxwell Equations in Differential Form

We have already pointed out in the case of Maxwell's equations in integral form that their differential form is obtained if it is established, as with Gauss' law or the divergence-free form of $\mathbf{B}$, that when the integral vanishes over any volume, the integrand must vanish (see (1.3.3) and (1.3.19)). We have represented Faraday's induction law and the Ampère-Maxwell law as integrals over arbitrary surfaces, whose integrands then also had to vanish (see (1.3.10) and (1.3.15)):

$$
\begin{aligned}
&\text{(a)} \quad \boldsymbol{\nabla}\cdot\mathbf{E} = 4\pi k_C\rho
&&\text{(b)} \quad \boldsymbol{\nabla}\times\mathbf{E} + \frac{k_L}{c}\dot{\mathbf{B}} = 0 \\[2mm]
&\text{(c)} \quad k_L\boldsymbol{\nabla}\times\mathbf{B} = \frac{4\pi k_C}{c}\mathbf{j} + \frac{1}{c}\dot{\mathbf{E}}
&&\text{(d)} \qquad\qquad \boldsymbol{\nabla}\cdot\mathbf{B} = 0.
\end{aligned}
\tag{1.3.21}
$$

The two inhomogeneous Maxwell equations are on the left side, the two homogeneous ones on the right side. These equations still require modification when applied in matter, in which the electrical and/or magnetic susceptibility play a role. Therefore, the equations (1.3.21) are sometimes more precisely referred to as microscopic Maxwell equations or Maxwell equations in vacuum to distinguish them from the Maxwell equations (5.2.16), which are mainly used in media.

We will not only limit ourselves to point particles, but also examine continuously distributed densities. We already know that by averaging over a resolution volume ΔV, continuous density distributions result from the microscopic densities. We will later see that from the microscopic Maxwell equations, which contain ρ and $\mathbf{j}$ as sources, by averaging similarly structured Maxwell equations for fields $\mathbf{E}$ and $\mathbf{B}$ averaged over ΔV are obtained, which $\bar{\rho}$ and $\bar{\mathbf{j}}$ are included as sources.

Here we have sequentially learned about Gauss's law, the (Faraday's) law of induction, the Ampère-Maxwell law and the divergence-free (source-free) magnetic flux density, which together form Maxwell's equations. While this order of laws is maintained within the book, there is no uniform sequence in the literature, so that the designation of the law of induction as the 2nd Maxwell's law is only of limited use. Therefore, we will mainly stick to the names of the laws, even though these primarily refer to the integral form, from us but are predominantly addressed in differential form.

1.3.7 Principle of Superposition

Initially, we found that to calculate the total charges, the point charges q_i are added like real numbers. From the Coulomb force (1.2.1) it follows immediately that then also the fields of individual charges add up to a total field:

$$\mathbf{E}(\mathbf{x}) = \sum_i \mathbf{E}^{(i)}(\mathbf{x}) = k_C \sum_i q_i \frac{\mathbf{x} - \mathbf{x}_i}{|\mathbf{x} - \mathbf{x}_i|^3} = -\boldsymbol{\nabla}\phi(\mathbf{x}).$$

This also applies to potentials (1.2.3) and charge distributions:

$$\phi(\mathbf{x}) = \sum_i \phi^{(i)}(\mathbf{x}) = k_C \sum_i \frac{q_i}{|\mathbf{x} - \mathbf{x}_i|},$$

$$\boldsymbol{\nabla}\cdot\mathbf{E} = -\sum_i \Delta\phi^{(i)} = 4\pi k_C \sum_i q_i \delta^{(3)}(\mathbf{x} - \mathbf{x}_i) = 4\pi k_C \rho(\mathbf{x}).$$

What has been shown here for point charges, of course, also applies to continuous charge distributions $\rho^{(i)}(\mathbf{x})$ and for current densities (1.1.1). This is evident from the continuity equation (1.1.3). Due to the linearity of Maxwell's equations, the principle of superposition also applies to the magnetic field $\mathbf{B}$, the later introduced vector potential $\mathbf{A}$ with $\mathbf{B} = \boldsymbol{\nabla} \times \mathbf{A}$ etc.

1.4 Notes on the Units

If you fix the factor k_C in the Coulomb force $\mathbf{F}_C$, the charge density ρ, the current density $\mathbf{j}$ and the electric field $\mathbf{E}$ are determined. Since k_L determines the magnetic field $\mathbf{B}$, all quantities that appear in the Maxwell equations are defined. With the help of Tab. 1.1, the Maxwell equations can thus be represented in any arbitrary system, whereby according to today's standards only three systems remain.

Tab. 1.1. Unit systems in electrodynamics: The Heaviside-Lorentz system and the SI system are called rational. For the electric and magnetic constants ϵ_0 and μ_0 applies (C.2.7): $\mu_0 = 1/k_L^2 \epsilon_0$.

System	k_C	k_L	$\rho' = \dfrac{1}{\sqrt{k_C}}\rho \qquad \mathbf{j}' = \dfrac{1}{\sqrt{k_C}}\mathbf{j}$	$\mathbf{E}' = \sqrt{k_C}\,\mathbf{E} \qquad \mathbf{B}' = \dfrac{\sqrt{k_C}}{k_L}\mathbf{B}$
Gauss	1	1	$\rho \qquad\qquad \mathbf{j}$	$\mathbf{E} \qquad\qquad \mathbf{B}$
Heaviside-Lorentz	$\dfrac{1}{4\pi}$	1	$\rho^{\text{H}} = \sqrt{4\pi}\,\rho \quad \mathbf{j}^{\text{H}} = \sqrt{4\pi}\,\mathbf{j}$	$\mathbf{E}^{\text{H}} = \dfrac{\mathbf{E}}{\sqrt{4\pi}} \quad \mathbf{B}^{\text{H}} = \dfrac{\mathbf{B}}{\sqrt{4\pi}}$
SI	$\dfrac{1}{4\pi\epsilon_0}$	c	$\rho^{\text{SI}} = \sqrt{4\pi\epsilon_0}\,\rho \quad \mathbf{j}^{\text{SI}} = \sqrt{4\pi\epsilon_0}\,\mathbf{j}$	$\mathbf{E}^{\text{SI}} = \dfrac{\mathbf{E}}{\sqrt{4\pi\epsilon_0}} \quad \mathbf{B}^{\text{SI}} = \sqrt{\dfrac{\mu_0}{4\pi}}\,\mathbf{B}$

1.4.1 Systems in Electrodynamics

The Gauss system

Systems with $k_L = 1$ are called *symmetric*. In these, the fields $\mathbf{E}$ and $\mathbf{B}$ have the same dimension. This applies to the systems named after Gauss and Heaviside-Lorentz. At the end of the 19th century, the Gauss units prevailed, but have since been replaced by the SI units. The Coulomb force is used with $k_C = 1$ to define the electric charge [Maxwell, 1892, Art. 41]:

The electrostatic unit of electricity is that quantity of positive electricity, which, when placed at unit of distance from an equal quantity, repels it with the unit of force.

In the Gauss system, the unit of force is 1 dyn and that of length is 1 cm. From (1.2.1) it follows that the charge q has the dimension (dyn=g cm s^{-2})

$$[q] = \sqrt{\text{dyn}}\,\text{cm} = \sqrt{\text{g cm}^3}\,\text{s}^{-1} = \text{statC(oulomb)}$$

has, which was already anticipated on p. 2. If you put $[q]$ into the Coulomb force (1.2.4), you get the dimension of $[E]$. The dimension of the current density $[j]$ follows from the continuity equation (1.1.15). With $k_L = 1$ the magnetic field $\mathbf{B}$ is determined, i.e. all quantities, which occur in the Maxwell equations, are defined. With the help of Tab. 1.1, the Maxwell equations can be represented in any system. The definitions of polarization and magnetization (5.2.17) are somewhat inconsistent, which has a negative effect in matter. Maxwell's equations:

$$
\text{G:}\quad
\begin{array}{ll}
\text{(a)} \quad \mathbf{\nabla}\cdot\mathbf{E} = 4\pi\rho, & \text{(b)} \quad \mathbf{\nabla}\times\mathbf{E} = -\dfrac{1}{c}\dot{\mathbf{B}}, \\[2ex]
\text{(c)} \quad \mathbf{\nabla}\times\mathbf{B} = \dfrac{4\pi}{c}\mathbf{j} + \dfrac{1}{c}\dot{\mathbf{E}}, & \text{(d)} \quad \mathbf{\nabla}\cdot\mathbf{B} = 0.
\end{array}
\qquad (1.3.21'')
$$

The Heaviside-Lorentz system

Systems in which the factor 4π appears in the denominator of Coulomb's law are called rational. In these, the Maxwell equations take a simple form. The Heaviside-Lorentz system is used in quantum electrodynamics, where natural units are usually assumed; but it would also be ideal in electrodynamics.

The SI system

If one wants to represent the charge as a characteristic base unit for electrodynamics, then in the Coulomb force (1.2.5) the factor k_C must be dimension-dependent so that F_C gets the dimension of a force. Now the force that two charges exert on each other in a dielectric is weakened by the polarization, i.e., by the displacement of the negative charges against the positive ones with $\propto 1/\epsilon_r$, where ϵ_r is the (relative) dielectric constant. In the SI system, such a constant (ϵ_0) is also introduced for the vacuum; hence the factor $k_C = 1/4\pi\epsilon_0$. However, the fourth base unit is the current and to determine the electric field constant ϵ_0, the force between two parallel currents (C.2.9) is measured. $\mu_0 = 1/k_L^2\epsilon_0$ defines the magnetic field constant.

The SI system has its roots in the MKS system proposed by Giorgi [1901] with a still undefined electrical unit. In 1948, the CPGM (Conférence Générale des Poids et Mésures) officially introduced the ampere as the fourth base unit (MKSA system) [PTB, 2007, p. 14] (see (C.2.9)). For physics, it would be more advantageous if, in the SI system, instead of the historically older variant with $k_L = c$ the symmetric ($k_L = 1$) one had been adopted. This would only affect magnetic quantities/units (μ_0, $\mathbf{B}$, ...), but the symmetry of the STR would be preserved. In Appendix C, the various systems are discussed in more detail. Maxwell's equations:

$$
\text{SI:}\quad
\begin{array}{ll}
\text{(a)} \quad \mathbf{\nabla}\cdot\mathbf{E} = \rho/\epsilon_0, & \text{(b)} \quad \mathbf{\nabla}\times\mathbf{E} = -\dot{\mathbf{B}}, \\[2ex]
\text{(c)} \quad \mathbf{\nabla}\times\mathbf{B} = \mu_0\mathbf{j} + \epsilon_0\mu_0\dot{\mathbf{E}}, & \text{(d)} \quad \mathbf{\nabla}\cdot\mathbf{B} = 0.
\end{array}
\qquad (1.3.21')
$$

On system independence

The formulas look unfamiliar in the notation valid for all systems, although only two quantities (k_C, k_L) are not specified. In matter, this impression is reinforced by the introduction of the field constants ϵ_0 and μ_0[16]. A not to be underestimated advantage is, however, that one immediately sees which formulas or quantities are the same in all systems.

The equations in matter have simpler factors in the SI system, not least because c is replaced in favor of ϵ_0 and μ_0. For this purpose, four different fields are also used in the vacuum, as can be seen from Tab. C.2, p. 634. For the electric field $\mathbf{E}$ $(\mathrm{V\,m^{-1}})$, the dielectric displacement $\mathbf{D}$ $(\mathrm{C\,m^{-2}})$, the magnetic fields $\mathbf{B}$ (T) and $\mathbf{H}$ $(\mathrm{A\,m^{-1}})$ have different dimensions, which could suggest four independent fields.

In our view, electrodynamics, which is a classical field theory, only has the fields $\mathbf{E}$ and $\mathbf{B}$, which are represented as tensor forces in the STR. $\mathbf{D} = \epsilon_r \epsilon_0 \mathbf{E}$ and $\mathbf{H} = (\mu_0 \mu_r)^{-1} \mathbf{B}$ are auxiliary fields (5.2.17), which arise when in matter the polarization originating from the bound charges and the magnetization originating from the bound currents are added to the fields $\mathbf{E}$ and $\mathbf{B}$. Therefore, $\mathbf{D}$ and $\mathbf{H}$ 'should' have the same dimensions as $\mathbf{E}$ and $\mathbf{B}$, i.e. $\epsilon_0 = \mu_0 = 1$ and the dielectric constant ϵ_r as well as the permeability μ_r are numbers.

The Lorentz transformation (LT) shows how $\mathbf{E}$ and $\mathbf{B}$ depend on each other: A stationary electron only has the field $\mathbf{E}$; in a system moving uniformly against the electron (boost) not only $\mathbf{E}$ is transformed, but the field $\mathbf{B}$ is also added, where the speed is parameterized with $\beta = \frac{v}{c} \leq 1$[17]. This is only clumsily represented in the SI system, which is why in the STR, i.e. in the covariant formulation of electrodynamics, the SI system is used less frequently.

Even in mechanics and quantum mechanics, the movement of particles in the electromagnetic field (Hamilton function or Schrödinger equation) is still predominantly represented on the basis of the Gauss system, which is particularly true for the textbooks *Statistical Mechanics, Quantum Mechanics* and *Advanced Quantum Mechanics* from Schwabl [2006, 2007, 2008]. However, the system-independent representation is equally valid for the SI system.

1.4.2 Electrodynamics and its Environment

We try to arrange the areas in which electrodynamics intervenes and which we will encounter in the following chapters in a scheme, the flow chart on p. 24. If we start here, at the end of the 1st chapter, with a classification of classical electrodynamics, we can do this on the basis of the knowledge of the Maxwell equations, to which the continuity equation and the Lorentz force

---

[16] In the Gaussian system all constants are equal to one: $k_C = k_L = \epsilon_0 = 1$

[17] With respect to the STR, it would be logical to define a reduced current density: $\mathbf{j}_r = \mathbf{j}/c = \rho\mathbf{v}/c$. c then only appears with the time derivatives, i.e., with the derivative according to $x_0 = ct$ one would have eliminated c in the Maxwell equations.

have been added. If you put an atom (molecule) in an electric field, it can move the electron shell against the nucleus. The atom is polarized, without the electrons losing their bond to the atomic nucleus; these electrons bound to the atom are to be treated differently than electrons freely moving in matter. The splitting of charges and currents into bound and free parts has in matter a division of the fields as a result and thus a restructuring of the Maxwell equations (section 5.1 and 5.2).

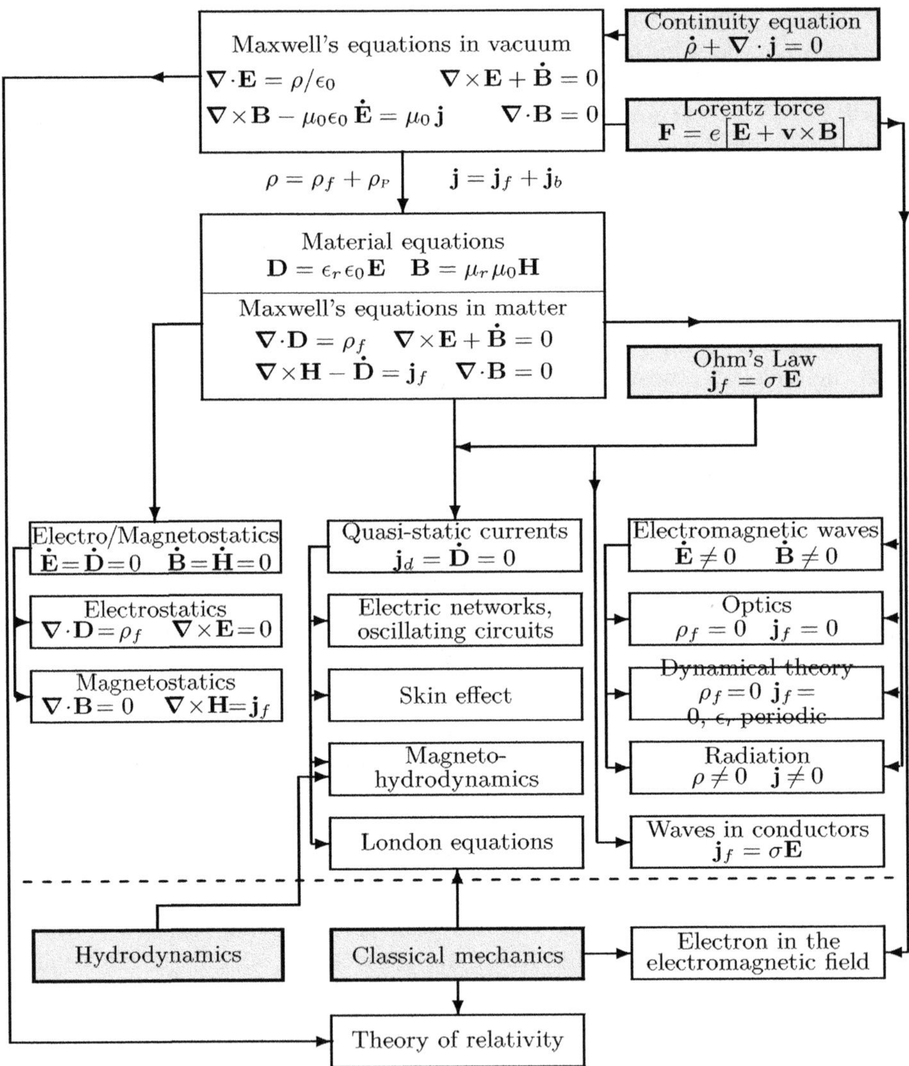

Fig. 1.9. Flowchart of the application areas of electrodynamics. Formulas are given in the SI system

Another factor for the classification and solution of the Maxwell equations is the time scale on which the processes take place and which are represented in the diagram in the form of three columns down to the dotted line:

1. Time-independent processes: The Maxwell equations decouple into electrostatic (Chap. 2, 3, 6) and magnetostatic (chap. 4, 7) components.
2. In slowly time-varying processes, the displacement current $\mathbf{j}_d$ is neglected, and it is $\nabla \cdot \mathbf{j}_f = 0$. The current is then source-free within the system and there is no radiation. This is used in electrical networks, in magnetohydrodynamics (chap. 9), for the London equations (section 5.4.3) or the skin effect (section 10.4.2). In all these systems, the (phenomenological) Ohm's law (section 5.3) also plays a role. In addition to magnetohydrodynamics, the hydrodynamics with the Navier-Stokes equations and the continuity equation for the density of the fluid are added. The London equations, which explain aspects of superconductivity, such as the Meissner effect, the non-penetration of the magnetic field into the superconductor, classically, use for the dynamics of the electrons equations of motion of mechanics, which take into account the inertia of conduction electrons.
3. Fast time-varying processes include the radiation of rapidly moving charges (Chap. 8), optics (Chap. 10) and the dynamical theory (Chap. 11). Many applications of optics and dynamic theory, such as the calculation of ray paths (laws of refraction), intensities etc. can however be carried out time-independently.

Outside of this scheme is the motion of the electron in the external electromagnetic field, as for this only the Lorentz force is introduced into the Euler-Lagrange equation of classical mechanics (section 5.4).

A larger space is devoted to the STR, where it is shown that the Maxwell equations are covariant under the Lorentz transformation (Chap. 13), but not the laws of classical mechanics, so that the latter must be transformed, in order to satisfy Lorentz covariance (Chap. 14). With the quantization of the electromagnetic field in section 14.2.4, QED is also briefly touched upon.

Problems for Chapter 1

1.1. *Coulomb force*: How large is the force with which the proton and electron attract each other at a distance $r_0 = a_B \approx 0.593$ Å?

1.2. *Sources and vortices of a vector field*: The vector field is given

$$\mathbf{v} = \lambda \frac{\mathbf{e}_z \times \mathbf{x}}{|\mathbf{e}_z \times \mathbf{x}|^2}.$$

Calculate the sources and vortices of $\mathbf{v}$, i.e. in particular $\nabla \cdot \mathbf{v}$ and $\nabla \times \mathbf{v}$. Additional question: What physical arrangement underlies $\mathbf{v}$, if you substitute $\lambda = 2I k_C / c k_L$ (see section 4.1.4)?

1.3. *Conservative vector field*: Let $\mathbf{E} = k_C f(r)\,\mathbf{x}$. Calculate $\nabla\!\cdot\!\mathbf{E}$, $\nabla\times\mathbf{E}$ and give the results for $\mathbf{E} = k_C\mathbf{x}/r^3$. Show in particular that $\nabla\times\mathbf{E} = 0$ also applies for $r = 0$, even if $\mathbf{E}$ is singular there.

1.4. *'Polarization potential'*: Let $\mathbf{Z} = \boldsymbol{\mu}/r$, where $\boldsymbol{\mu}$ should be a constant vector. Calculate $\phi = -\nabla\!\cdot\!\mathbf{Z}$, $\mathbf{A} = \nabla\times\mathbf{Z}$, $\mathbf{B} = \nabla\times\mathbf{A}$, $\mathbf{E} = \nabla\nabla\!\cdot\!\mathbf{Z}$ and $\mathbf{B}-\mathbf{E}$.

1.5. *Temporal change of the flow through an area $A(t)$*:

$$\frac{\mathrm{d}\Phi}{\mathrm{d}t} = \frac{\mathrm{d}}{\mathrm{d}t}\iint_A \mathrm{d}\mathbf{a}\cdot\mathbf{b}$$

indicates the flow of the vector field $\mathbf{a}$ through the area $A(t)$, which we calculated for (1.3.4) using the convective derivative. With the following "more direct" method, one obtains the relation

$$\frac{\mathrm{d}\Phi}{\mathrm{d}t} = \iint_{A(t)} \mathrm{d}\mathbf{a}\cdot\frac{\partial\mathbf{b}}{\partial t} + \iint_{A(t)} \mathrm{d}\mathbf{a}\cdot\mathbf{v}\,(\nabla\!\cdot\!\mathbf{b}) - \oint_{\partial A(t)} \mathrm{d}\mathbf{x}\cdot(\mathbf{v}\times\mathbf{b}), \tag{1.4.1}$$

whose validity you should demonstrate.

Hint: Form $\displaystyle\lim_{\delta\to 0}\frac{\Phi(t+\delta t)-\Phi(t)}{\delta t}$ and orient yourself to Fig. 1.10, which shows how the area integrals can be supplemented to a surface integral.

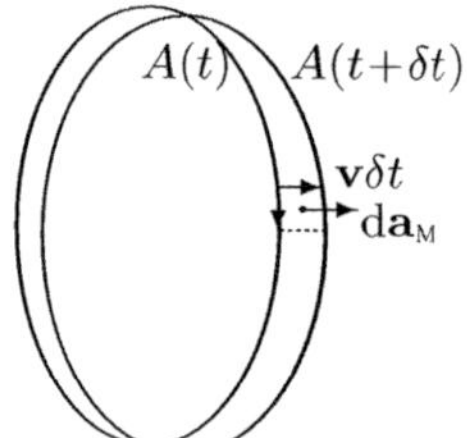

Fig. 1.10. $\mathrm{d}\mathbf{a}_M = \mathrm{d}\mathbf{x}\times\mathbf{v}\delta t$ is the vector of the surface element on the lateral area A_M. V denotes the volume enclosed by $A(t)$, $A(t+\delta t)$ and the lateral area A_M, whose surface is given by
$$\partial V = A(t+\delta t) + A(t) + A_M$$

1.6. *Induction in the earth's magnetic field*: The wings of an airplane (Airbus A380) have a wingspan of 80 m. The airplane flies at 900 km/h over Central Europe to the north, where the vertical component of the Earth's magnetic field is supposed to be $B = 45\times 10^{-6}$ T. Calculate the potential difference between the two ends of the wings and indicate which of the two is at a higher level.
Hint: The magnetic south pole is near the geographical North Pole.

1.7. *Poisson's equation*: Show that from Coulomb's law (1.2.1) the potential of a point charge and thus the Green's function of Poisson's equation can be derived. In addition, starting from a single point charge and using the principle of superposition, you obtain the Poisson equation for the microscopic charge density (1.1.1).

References

Beier W. *Wilhelm Conrad Röntgen*, Springer Berlin (2013)
Faraday M. *Experimental Researches in Electricity*, Phil. Trans. **122**, 125–162 (1832)

Giorgi G. *Unità Razionali di Elettromagnetismo*, Atti della Associazione Elettrotecnica Italiana **5**, 402, Torino (1901) and G. Giorgi, Nuovo Cimento (5) **4**, 11 (1902)

Griffiths D. *Introduction to Electrodynamics*, 4th ed., Cambridge University Press (2017)

Helmholtz H. von *On the Modern Development of Faraday's Conception of Electricity*, Science **2**, 182–185 (1881) or in J. Chem. Soc. **39**, 277-304 (1881)

Jackson J. D. *Classical Electrodynamics*, 3rd ed., John Wiley & Sons Inc. (1998)

Lenard P. *Ueber die magnetische Ablenkung der Kathodenstrahlen*, Ann. Physik. Neue Folge **52**, 23–33 (1894)

Lorentz H. A. *La théorie électromagnétique de Maxwell et son application aux corps mouvants*, E.J. Brill, Harlem (1892); in Archives Néerlandaises **XXV**, 363 (1892)

Maxwell J. C. *On Physical Lines of Force* Phil. Mag. **21**, part II, *A Dynamical Theory of the Electromagnetic Field*, 338–348 (1861)

Maxwell J. C. *A treatise on electricity and magnetism*, Vol I, 3rd ed. Clarendon Press Oxford (1892)

Maxwell J. C. *A treatise on electricity and magnetism*, Vol II, 3rd ed. Clarendon Press Oxford (1892)

Maxwell J. C. *The electrical research of the Honourable Henry Cavendish written between 1771 and 1781*, edited from the original manuscripts, Cambridge at the University Press (1879)

Millikan R. A. *The Isolation of an Ion, a Precision Measurement of its Charge, and the Correction of Stokes's Law*, Phys. Rev. (series 1) **32**, 349–397 (1911)

Priestley J. *The History and Present State of Electricity, with Original Experiments*, London (1767)

PTB Mitteilungen, Volume **117**, No. 2 (2007)

Röntgen W. C. *Über die durch Bewegung eines im homogenen elektrischen Felde befindlichen Dielektrikums hervorgerufene elektrodynamische Kraft*, Sitzungsberichte der Königlich Preussischen Akademie der Wissenschaften zu Berlin **7**, 23–29 (1888)

Schwabl F. *Statistical Mechanics*, 2nd ed. Springer Berlin (2006)

Schwabl F. *Quantum Mechanics*, 4th ed. Springer Berlin (2007)

Schwabl F. *Advanced Quantum Mechanics*, 4th ed. Springer Berlin (2008)

Sommerfeld A. *Electrodynamics*, Academic Press Inc., New York (1952)

Thomson J. J. *Cathode rays*, Phil. Mag. **44**, 293 (1897)

Wiechert E., Schriften der physikalisch-ökonomischen Gesellschaft zu Königsberg in Preussen, Volume **38**, No. 1. Sitzungsberichte 3–16 (1897)

2

Stationary Electric Charges and the Distribution of Electricity on Conductors

In the system-independent formulation of electrostatics, only the factor k_C attributable to the Coulomb force appears:

Gauss: $k_C = 1$	Heaviside: $k_C = \dfrac{1}{4\pi}$	SI: $k_C = \dfrac{1}{4\pi\epsilon_0}$.

2.1 Electrostatic Potential and Poisson Equation

In electrostatics, it is assumed that the charge density ρ is time-independent and the current density $\mathbf{j}$ vanishes. Then $\mathbf{E}$ and $\mathbf{B}$ are also time-independent and in the Maxwell equations (1.3.21") due to $\dot{\mathbf{E}} = \dot{\mathbf{B}} = 0$ decoupled:

$$(a) \quad \boldsymbol{\nabla}\cdot\mathbf{E} = 4\pi k_C \rho \;\Rightarrow\; \text{SI:}\, \boldsymbol{\nabla}\cdot\mathbf{E} = \rho/\epsilon_0 \qquad (b) \quad \boldsymbol{\nabla}\times\mathbf{E} = 0$$

$$(c) \quad \boldsymbol{\nabla}\times\mathbf{B} = 0 \qquad\qquad\qquad\qquad (d) \quad \boldsymbol{\nabla}\cdot\mathbf{B} = 0, \tag{2.1.1}$$

where for the phenomena of electrostatics only the first line of (2.1.1) is relevant. Gauss's law states that $\mathbf{E}$ has sources and sinks, and Faraday's law of induction tells us that in the presence of no or only static magnetic fields, the electric field is curl-free ($\boldsymbol{\nabla}\times\mathbf{E} = 0$). A curl-free field $\mathbf{E}$ can be represented as the gradient of a scalar potential ϕ:

$$\mathbf{E} = -\boldsymbol{\nabla}\phi, \tag{2.1.2}$$

since $\boldsymbol{\nabla}\times\mathbf{E} = -\boldsymbol{\nabla}\times\boldsymbol{\nabla}\phi = 0$. The integral

$$V = -\int_{\mathbf{x}_A}^{\mathbf{x}_B} \mathrm{d}\mathbf{x}\cdot\mathbf{E} = \int_{\mathbf{x}_A}^{\mathbf{x}_B}\left\{\frac{\partial\phi}{\partial x}\mathrm{d}x + \frac{\partial\phi}{\partial y}\mathrm{d}y + \frac{\partial\phi}{\partial z}\mathrm{d}z\right\} = \int_{\mathbf{x}_A}^{\mathbf{x}_B}\mathrm{d}\phi$$

$$= \phi(\mathbf{x}_B) - \phi(\mathbf{x}_A) \tag{2.1.3}$$

is a path independent line integral (see Fig. 2.1). The potential difference V

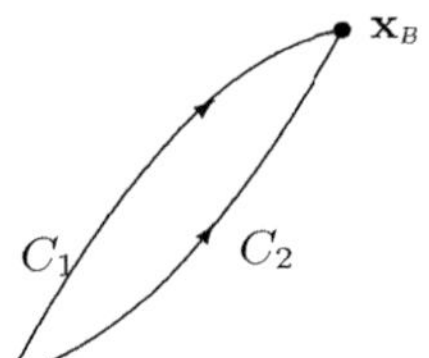

Fig. 2.1. The difference in the electric potential $\phi(\mathbf{x}_B)-\phi(\mathbf{x}_A)$ is the same for paths C_1 and C_2

is referred to as *voltage*. It is the potential difference from the endpoint to the starting point. If you go along a field line, the potential decreases.

The voltage is given in volts, the unit of the SI system, according to which $1\,\mathrm{Volt}=1\,\mathrm{J/(As)}$. The unit $1\,\mathrm{statV}\approx 300\,\mathrm{V}$ of the electrostatic or Gaussian system is rarely used.

If the path is closed, the loop voltage $\oint \mathrm{d}\mathbf{x}'\cdot\mathbf{E} = 0$, since the electrostatic field is irrotational. The force on a test particle of charge e is according to (2.1.3) $e\,\mathbf{E}$. To move this particle against the effect of the electric field from point A to point B one has to do the work

$$W_{AB} = -e \int_{\mathbf{x}_A}^{\mathbf{x}_B} \mathrm{d}\mathbf{x} \cdot \mathbf{E} = e\Big(\phi(\mathbf{x}_B) - \phi(\mathbf{x}_A)\Big) = eV. \tag{2.1.4}$$

Thus, $e\phi(\mathbf{x})$ can be considered as the potential energy of the test charge in the field. If $\mathbf{E} = -\boldsymbol{\nabla}\phi$ is inserted into the Gaussian law (2.1.1a), the Poisson equation is obtained

$$\Delta\phi(\mathbf{x}) = -4\pi k_C \rho(\mathbf{x}). \tag{2.1.5}$$

This can be integrated using the Green's function of the Poisson equation

$$G(\mathbf{x} - \mathbf{x}') \equiv G_0(\mathbf{x}, \mathbf{x}') = \frac{1}{|\mathbf{x} - \mathbf{x}'|}. \tag{2.1.6}$$

The notation G_0 is sometimes used for G. In potential theory, Green functions G_i with the stronger asymptotic decay $1/r'^{1+i}$ may be necessary to obtain convergent integrals of the form of (2.1.8) (see p. 594). G fulfills the differential equation

$$\Delta\frac{1}{|\mathbf{x} - \mathbf{x}'|} = -4\pi\delta^{(3)}(\mathbf{x} - \mathbf{x}'). \tag{2.1.7}$$

Multiplication with $\rho(\mathbf{x}')$ and integration $\int \mathrm{d}^3x'$ yields

$$\Delta \int \mathrm{d}^3x'\, \frac{\rho(\mathbf{x}')}{|\mathbf{x} - \mathbf{x}'|} = -4\pi\rho(\mathbf{x}),$$

from which the general solution of the Poisson equation

$$\phi(\mathbf{x}) = k_C \int \mathrm{d}^3x'\, \frac{\rho(\mathbf{x}')}{|\mathbf{x}-\mathbf{x}'|} \tag{2.1.8}$$

follows. It remains to be shown that $1/r$ is the Green function of the Poisson equation. With $r = |\mathbf{x}|$ one gets

$$\frac{\partial r}{\partial x_i} = \frac{x_i}{r}, \qquad \frac{\partial}{\partial x_i}\frac{1}{r} = -\frac{x_i}{r^3}, \qquad \frac{\partial^2}{\partial x_i^2}\frac{1}{r} = -\frac{1}{r^3} + 3\frac{x_i^2}{r^5},$$

$$\nabla\frac{1}{r} = -\frac{\mathbf{x}}{r^3} \quad \text{and} \quad \Delta\frac{1}{r} = -3\frac{1}{r^3} + 3\frac{r^2}{r^5} = 0 \qquad \text{for} \quad r \neq 0.$$

To determine $\Delta\frac{1}{r}$ at the zero point, calcultate the integral of a sphere S_ϵ with radius ϵ, where $\mathrm{da} = r^2\mathbf{e}_r\,\mathrm{d}\Omega$ and $\mathbf{e}_r = \mathbf{x}/r$:

$$\int_{S_\epsilon} \mathrm{d}^3x\,\nabla^2\frac{1}{r} = \oiint_{\partial S_\epsilon} \mathrm{da}\cdot\nabla\frac{1}{r} = -\oiint_{\partial S_\epsilon} \mathrm{da}\cdot\frac{\mathbf{x}}{r^3} = -\int \mathrm{d}\Omega = -4\pi. \qquad (2.1.9)$$

Thus, $\Delta\frac{1}{r} = -4\pi\delta^{(3)}(\mathbf{x})$, as claimed in (2.1.7). The Poisson equation is discussed in more detail in appendix A.4.4 and A.4.6.

2.2 Potential and Field for given Charge Distributions

Potentials and fields are now calculated for some simple and characteristic charge distributions for electrostatics. These can be used as building blocks for more complex charge distributions to calculate their potentials and fields by superposition.

2.2.1 Simple Arrangements of Charges

A spatially extended charge distribution generally has non-vanishing moments

$$M_{ij...} = \int \mathrm{d}^3x'\,\rho(\mathbf{x}')x_i'x_j'...$$

in any order. The zeroth order indicates the total charge, the first the dipole moment etc.. However, a point-like charge has no further moments apart from the zeroth one. For the point dipole, only the first moment does not vanish. Thus, one can study the angular distribution and the range of the field of the individual moments on simple arrangements of charges.

Point charge

At the position $\mathbf{x}_0$ there is a point charge e, whose charge density according to (1.1.1) is given by

$$\rho(\mathbf{x}) = e\,\delta^{(3)}(\mathbf{x}-\mathbf{x}_0).$$

The potential is determined by (2.1.8) as

$$\phi(\mathbf{x}) = k_C \frac{e}{|\mathbf{x}-\mathbf{x}_0|} \qquad \Longrightarrow \qquad \mathbf{E} = -\boldsymbol{\nabla}\phi = k_C e \frac{\mathbf{x}-\mathbf{x}_0}{|\mathbf{x}-\mathbf{x}_0|^3}. \qquad (2.2.1)$$

$\mathbf{E}$ points from positive to negative charge and is, as sketched in Fig. 2.2, around $\mathbf{x}_0$ radially symmetric, which is referred to as *Coulomb field.* The surfaces of equal potential, the *equipotential surfaces*, are spherical surfaces.

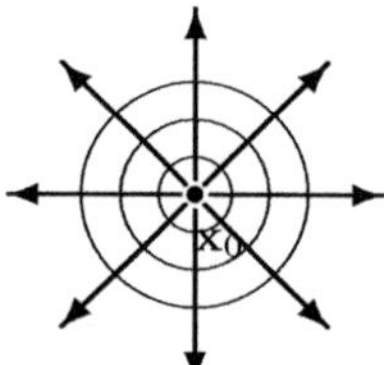

Fig. 2.2. Coulomb field around a positive point charge at the position $\mathbf{x}_0$ with equipotential lines (circles)

Electric Dipole

Given are two charges e and $-e$ at a distance d, with the negative one located at the origin, as sketched in Fig. 2.3. The charge density is then in first order of a Taylor expansion for $r \gg d$:

$$\rho(\mathbf{x}) = e\left[\delta^{(3)}(\mathbf{x}-\mathbf{d}) - \delta^{(3)}(\mathbf{x})\right] \approx -e\mathbf{d}\cdot\boldsymbol{\nabla}\delta^{(3)}(\mathbf{x})$$
$$= -\mathbf{p}\cdot\boldsymbol{\nabla}\delta^{(3)}(\mathbf{x}). \qquad (2.2.2)$$

$\mathbf{p} = e\mathbf{d}$ is the dipole moment of the charge distribution, where $\mathbf{p}$ points towards the positive charge. If the limit transition

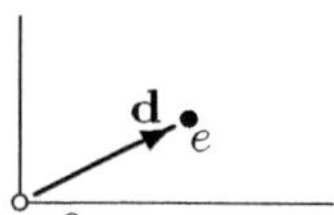

Fig. 2.3. The dipole $\mathbf{p} = e\mathbf{d}$ is directed towards the positive charge $e > 0$ (in contrast to the field $\mathbf{E}$, which points towards the lower potential)

$$\mathbf{p} = \lim_{d\to 0\ ;\ ed<\infty} e\mathbf{d} \qquad (2.2.3)$$

is carried out in such a way that $\mathbf{p}$ is finite, then a point dipole is present, whose charge density is given by (2.2.2). If one substitutes (2.2.2) into $\phi(\mathbf{x})$ (2.1.8), one obtains after integration by parts

$$\phi(\mathbf{x}) = -k_C \int \mathrm{d}^3x' \, \frac{\mathbf{p}\cdot\boldsymbol{\nabla}'\,\delta^{(3)}(\mathbf{x}')}{|\mathbf{x}-\mathbf{x}'|} = k_C \int \mathrm{d}^3x' \, \delta^{(3)}(\mathbf{x}') \left(\mathbf{p}\cdot\boldsymbol{\nabla}'\frac{1}{|\mathbf{x}-\mathbf{x}'|}\right).$$

The potential and field of a point dipole located at the origin are

$$\phi(\mathbf{x}) = -k_C \mathbf{p} \cdot \boldsymbol{\nabla} \frac{1}{r} = k_C \frac{\mathbf{p} \cdot \mathbf{x}}{r^3}, \tag{2.2.4}$$

$$\mathbf{E}(\mathbf{x}) = k_C \boldsymbol{\nabla}(\mathbf{p} \cdot \boldsymbol{\nabla} \frac{1}{r}) = -\frac{k_C}{r^3}\Big[\mathbf{p} - 3\frac{(\mathbf{p}\cdot\mathbf{x})\mathbf{x}}{r^2}\Big]. \tag{2.2.5}$$

Sometimes it is convenient not to equate the location of the dipole $\mathbf{x}_0$ with the location of the negative charge. In Fig. 2.4, $\mathbf{x}_0$ points to the middle of the connecting line of the two charges

$$\rho(\mathbf{x}) = e\Big[\delta^{(3)}\big(\mathbf{x}-(\mathbf{x}_0+\frac{\mathbf{d}}{2})\big) - \delta^{(3)}\big(\mathbf{x}-(\mathbf{x}_0-\frac{\mathbf{d}}{2})\big)\Big], \tag{2.2.6}$$

where the deviations of the density, the potential, and the field from the dipole contributions (2.2.7)-(2.2.9) are smallest

$$\rho(\mathbf{x}) = -\mathbf{p} \cdot \boldsymbol{\nabla}\, \delta^{(3)}(\mathbf{x} - \mathbf{x}_0), \tag{2.2.7}$$

$$\phi(\mathbf{x}) = -k_C\,\mathbf{p} \cdot \boldsymbol{\nabla}\frac{1}{|\mathbf{x} - \mathbf{x}_0|}, \tag{2.2.8}$$

$$\mathbf{E}(\mathbf{x}) = k_C \boldsymbol{\nabla}(\mathbf{p} \cdot \boldsymbol{\nabla}\frac{1}{|\mathbf{x} - \mathbf{x}_0|}). \tag{2.2.9}$$

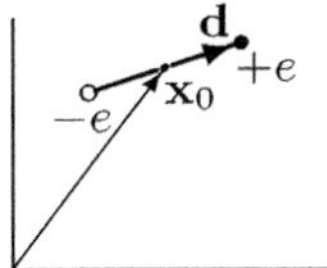

Fig. 2.4. The contributions of order $O(d^2)$ to ϕ and $\mathbf{E}$ are minimal when $\mathbf{x}_0$ is in the middle of the connecting line of the two charges

Electric quadrupole

We now consider two point charges each with the charges $e > 0$ and $-e$. The total charge q thus disappears. Additionally, it is required that the dipole moment also disappears. This is ensured by the approach

$$\rho(\mathbf{x}) = \sum_{i=1}^{4} \rho_i(\mathbf{x}) = e\big[\delta^{(3)}(\mathbf{x}-\mathbf{c}) - \delta^{(3)}(\mathbf{x}-\mathbf{a})\big] - e\big[\delta^{(3)}(\mathbf{x}+\mathbf{a}) + \delta^{(3)}(\mathbf{x}+\mathbf{c})\big].$$

Such a configuration is sketched in Fig. 2.5. At a great distance from the charge distribution localized around the origin, one obtains by Taylor expansion up to the 2nd order

$$\rho(\mathbf{x}) = e\Big[\delta^{(3)}(\mathbf{x}-\mathbf{c}) + \delta^{(3)}(\mathbf{x}+\mathbf{c}) - \delta^{(3)}(\mathbf{x}-\mathbf{a}) - \delta^{(3)}(\mathbf{x}+\mathbf{a})\Big]$$
$$\approx e\big(c_i c_j - a_i a_j\big)\nabla_i \nabla_j \delta^{(3)}(\mathbf{x}).$$

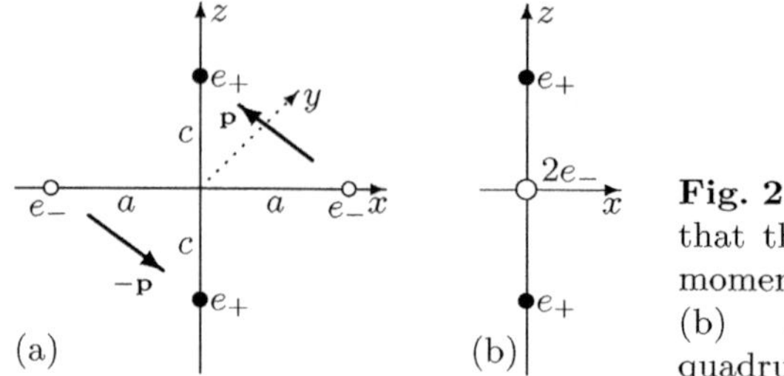

Fig. 2.5. (a) The charges are arranged so that the dipole moment and quadrupole moments Q_{ij} for $i \neq j$ disappear
(b) axially symmetric, elongated quadrupole as a limit $a \to 0$

The 2nd moments of this charge distribution are

$$M_{kl} = \int \mathrm{d}^3x\, x_k x_l\, \rho(\mathbf{x})$$

$$= e\big(c_i c_j - a_i a_j\big) \int \mathrm{d}^3x\, x_k x_l\, \nabla_i \nabla_j \delta^{(3)}(\mathbf{x}) \overset{\text{part. int}}{=} 2e\big(c_k c_l - a_k a_l\big).$$

Using

$$\rho(\mathbf{x}) \approx \frac{1}{2} M_{ij} \nabla_i \nabla_j \delta^{(3)}(\mathbf{x})$$

we now calculate the potential of the above point charges and obtain after twice integration by parts and rolling over the differentiation of $\nabla'_i \to -\nabla_i$

$$\phi_Q(\mathbf{x}) = k_C \frac{M_{ij}}{2} \int \frac{\mathrm{d}^3x'}{|\mathbf{x}-\mathbf{x}'|} \nabla'_i \nabla'_j \delta^{(3)}(\mathbf{x}') = k_C \frac{M_{ij}}{2} \int \mathrm{d}^3x'\, \delta^{(3)}(\mathbf{x}') \nabla'_i \nabla'_j \frac{1}{|\mathbf{x}-\mathbf{x}'|}$$

$$= k_C \frac{M_{ij}}{2} \nabla_i \nabla_j \frac{1}{r} = k_C \frac{M_{ij}}{2} \frac{3x_i x_j - \delta_{ij} r^2}{r^5}. \qquad (2.2.10)$$

Now the last term is transformed

$$M_{ij}\delta_{ij} r^2 = x_i x_j \delta_{ij} M \qquad \text{with} \qquad M = \operatorname{tr}\mathsf{M} = \sum_k M_{kk}.$$

We finally obtain

$$\phi_Q(\mathbf{x}) = \frac{k_C}{2} Q_{ij} \frac{x_i x_j}{r^5} \qquad \text{with} \qquad Q_{ij} = 3M_{ij} - \delta_{ij} \operatorname{Sp}\mathsf{M}. \qquad (2.2.11)$$

When brought to principal axis form, there are three diagonal quadrupole moments Q_i with $i = 1, 2, 3$. However, due to the tracelessness, only two independent moments remain. The charge distribution sketched in Fig. 2.5a has with a and c two parameters for two independent quadrupole moments, which is sufficient for the representation of any quadrupole in principal axis form. The point quadrupole is determined according to Fig. 2.5 by

$$Q_1 = -\lim_{\substack{a,c\to 0 \\ a^2 e<\infty;\, c^2 e<\infty}} e(4a^2 + 2c^2), \qquad Q_3 = \lim_{\substack{a,c\to 0 \\ a^2 e<\infty;\, c^2 e<\infty}} e(4c^2 + 2a^2). \qquad (2.2.12)$$

If the charge distribution is axially symmetric, $Q_1 = Q_2$, then there is only one quadrupole moment $Q = Q_3$. For $Q > 0$ the quadrupole is stretched, as sketched in Fig. 2.5b. The field of a quadrupole located at the origin is

$$E_i(\mathbf{x}) = -\nabla_i \phi_Q(\mathbf{x}) = k_C \Big[\frac{5}{2} Q_{jk} \frac{x_i x_j x_k}{r^7} - Q_{ij} \frac{x_j}{r^5} \Big]. \tag{2.2.13}$$

Line charge

The charge density ρ of a wire is homogeneous and its cross-section is so small that the wire can be treated as a line. On the line, there is the charge λ per unit length, the so-called line charge. $q = 2l\lambda$ is the total charge of a wire of length $2l$.

Straight line charge of infinite length
Fig. 2.6 shows a wire that lies in the z-axis where

$$\rho(\mathbf{x}) = \lambda \delta(x)\delta(y)\,\theta(l-|z|) \tag{2.2.14}$$

its charge density, as shown in Fig. 2.6, and

$$\phi(\mathbf{x}) = k_C \lambda \int_{-l}^{l} \frac{dz'}{\sqrt{\varrho^2 + (z-z')^2}} = -k_C \lambda \ln\Big[z - z' + \sqrt{\varrho^2 + (z-z')^2} \Big]\Big|_{-l}^{l} \tag{2.2.15}$$

are its corresponding potential. Let's assume that $l \to \infty$, then ϕ only depends

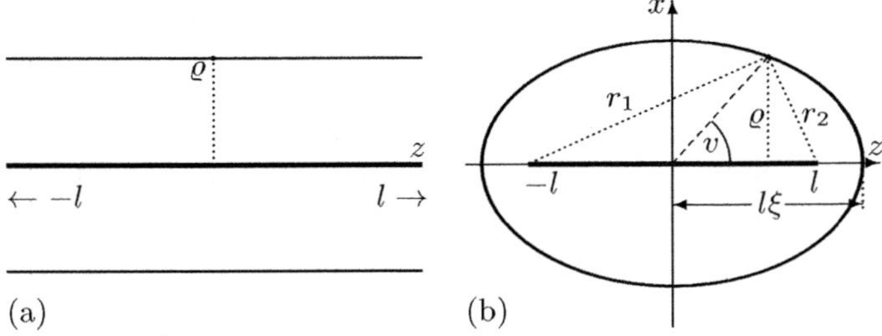

Fig. 2.6. (a) Line charge with $l \gg \varrho$; the equipotential line is a straight line   (b) Equipotential line (ellipse) of a line charge in the zx-plane. $r_{1,2}$ are the radii, l is the linear eccentricity and $l\xi$ is the semi-axis

on the distance $\varrho = \sqrt{x^2 + y^2}$ of the wire from the point of application, so we can set $z = 0$:

$$\phi(\varrho) = -k_C \lambda \ln \frac{-l + \sqrt{\varrho^2 + l^2}}{l + \sqrt{\varrho^2 + l^2}} = k_C \lambda \ln \frac{\big(l + \sqrt{\varrho^2 + l^2}\big)^2}{\varrho^2} \approx 2k_C \lambda \big[\ln 2l - \ln \varrho\big].$$

We have extended the numerator and denominator with $l + \sqrt{\varrho^2 + l^2}$ and developed for $l \gg \varrho$. In the limit $l \to \infty$ an infinite constant appears, which has no influence on the electrostatic field but can be avoided if the potential is only considered as a difference to any other point ϱ_0 (*regularization point*, see appendix p. 594 or problem 4.1) is determined. One obtains thus

$$\phi(\mathbf{x}, \mathbf{x}_0) = \lambda \lim_{l \to \infty} \left[\phi(\varrho) - \phi(\varrho_0)\right] = 2k_C\lambda\left(\ln \varrho_0 - \ln \varrho\right). \tag{2.2.16}$$

The dependence on ϱ_0 has no influence on $\mathbf{E}$. With the infinitely long wire, one has a two-dimensional problem. In two dimensions, it is typical that $|\phi(\varrho \to \infty)|$ does not decrease. For the field one obtains

$$\mathbf{E} = k_C(2\lambda/\varrho)\,\mathbf{e}_\varrho\,. \tag{2.2.17}$$

Straight line charge of finite length

When determining ϕ for a wire of finite length, as shown in Fig. 2.6b, one obtains from (2.2.15)

$$\phi(\mathbf{x}) = -k_C\lambda\ln\frac{z-l+r_2}{z+l+r_1} \qquad \text{with} \qquad r_{1,2} = \sqrt{\varrho^2+(z\pm l)^2}\,. \tag{2.2.18}$$

The electrostatic field is again obtained by gradient formation from the potential, where we have numerator and denominator extended with $-z-l+r_1$:

$$\frac{\partial \ln(z+l+r_1)}{\partial \varrho} = \frac{1}{z+l+r_1}\frac{\varrho}{r_1} = \frac{-z-l+r_1}{\varrho^2}\frac{\varrho}{r_1} = \frac{1}{\varrho} - \frac{z+l}{\varrho r_1}.$$

The derivative with respect to z can be read from (2.2.15), so that

$$\mathbf{E}(\mathbf{x}) = -\boldsymbol{\nabla}\phi(\mathbf{x}) = k_C\lambda\left[\left(\frac{z+l}{r_1} - \frac{z-l}{r_2}\right)\mathbf{e}_\varrho - \left(\frac{1}{r_1} - \frac{1}{r_2}\right)\mathbf{e}_z\right]. \tag{2.2.18'}$$

It remains to be shown that the equipotential surfaces are rotation ellipsoids with the focal points $z = \pm l$. First, we introduce new quantities:

$$\xi = \frac{1}{2l}(r_1+r_2), \qquad r_1 = l(\xi+\eta), \qquad l^2(\xi^2+\eta^2) = \frac{1}{2}(r_1^2+r_2^2) = \varrho^2+z^2+l^2,$$

$$\eta = \frac{1}{2l}(r_1-r_2), \qquad r_2 = l(\xi-\eta), \qquad \xi\eta = \frac{1}{4l^2}(r_1^2-r_2^2) = \frac{z}{l}\,.$$

This gives us

$$r_{1,2} + z \pm l = l[\xi \pm \eta + \xi\eta \pm 1] = l(\xi\pm 1)(1+\eta)$$

$$\phi(\mathbf{x}) = k_C\lambda\left[\ln(\xi+1) - \ln(\xi-1)\right]. \tag{2.2.19}$$

Thus, $\xi = (r_1+r_2)/2l = $ const represents an equipotential surface. Under this assumption, $r_{1,2}$ are the focal radii of the ellipse in Fig. 2.6, l is the linear eccentricity, $l\xi$ the semi-axis and the configuration is axially symmetric. The equipotential surfaces are thus stretched rotational ellipsoids and the field lines rotational hyperboloids.

To establish the connection with two-dimensional elliptical coordinates, we consider the zx plane. For an ellipse, $a = (r_1 + r_2)/2$. We find that

$$z = \xi\eta, \qquad\qquad x^2 = l^2(\xi^2 - 1)(1 - \eta^2),$$

which corresponds to elliptical coordinates. Here, $\eta = \cos v$ is the azimuthal angle in the zx plane with $0 \leq v < 2\pi$ and $1 \leq \xi < \infty$. If you set $\xi = \cosh u$, you get the zx representation of two-dimensional elliptical coordinates with

$$z = l\cosh u \,\cos v\,, \qquad\qquad x = l\sinh u \,\sin v\,.$$

If you look at the line charge from distances $r \gg l$, it appears as a point charge $q = 2l\lambda$ with a stretched axially symmetric quadrupole $Q_3 = \lambda 4l^3/3$.

Finite cross-section and curved lines

If the cross-section of the wire A is small compared to its length, it makes sense to treat the wire as a line (curve). Let s be the parametric representation for this curve, then

$$\int d^3x\, \rho(\mathbf{x}) = \int_C ds\, A(s)\, \rho = \int_C ds\, \lambda(s)$$

with the line charge density λ. The line integral extends over the curve C, as sketched in Fig. 2.7.

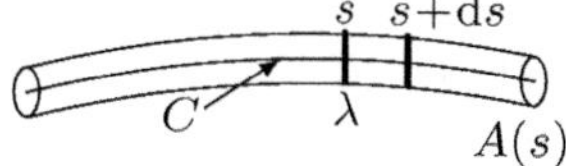

Fig. 2.7. Wire with the cross-section $A(s)$ and the charge λ per unit length

Starting from a line charge, local orthogonal coordinates $\boldsymbol{\xi}$ along C can be introduced. It then holds

$$\rho(\mathbf{x})\, d^3x = \rho(\xi)\, |\frac{\partial \mathbf{x}}{\partial \boldsymbol{\xi}}|\, d^3\xi = \lambda(\xi_3)\, \delta(\xi_1)\, \delta(\xi_2)\, d^3\xi,$$

where the integral over ξ_3 is to be calculated along the curve C.

Surface charge

Given is a plane surface uniformly covered with charges, as sketched in Fig. 2.8. It has the charge density

$$\rho(\mathbf{x}) = \sigma\, \delta(z). \qquad\qquad (2.2.20)$$

z is the coordinate normal to the charged surface, as can be seen from Fig. 2.8 and σ is the density of the charge applied to the surface. An attempt is made

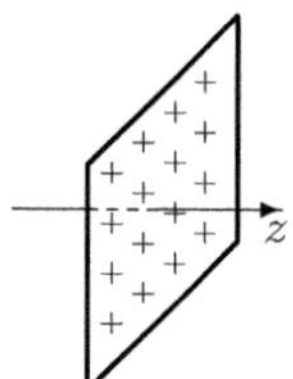

Fig. 2.8. Surface charge density σ in the plane with $z = 0$; it is $\rho(\mathbf{x}) = \sigma(\mathbf{x}_\parallel)\,\delta(z)$

to calculate the scalar potential ϕ using (2.1.8), where $x = y = 0$ can be set because ϕ should only depend on z. Since the integral over $\mathbb{R}^3$ diverges, we initially limit ourselves to a finite circle of radius R charged with charge:

$$\hat{\phi}(\mathbf{x}) = 2\pi k_C \sigma \int_0^R \mathrm{d}\varrho' \, \frac{\varrho'}{\sqrt{\varrho'^2 + z^2}} = 2\pi k_C \sigma \left(\sqrt{R^2 + z^2} - |z| \right).$$

The potential thus diverges with $R \to \infty$. This is not tragic, as only differences in potential, not its absolute value, are relevant. So one takes the potential $\hat{\phi}(\mathbf{x}_0)$ and forms the difference:

$$\phi(\mathbf{x}) = \lim_{R \to \infty} \left[\hat{\phi}(\mathbf{x}) - \hat{\phi}(\mathbf{x}_0) \right] = 2\pi k_C \sigma \left(|z_0| - |z| \right).$$

Alternative method: The Poisson equation simplifies due to $\phi(\mathbf{x}) = \phi(z)$ to

$$\Delta\phi = \frac{\mathrm{d}^2}{\mathrm{d}z^2}\,\phi = -4\pi k_C \sigma\,\delta(z). \tag{2.2.21}$$

From $\frac{\mathrm{d}}{\mathrm{d}x}\,\mathrm{sgn}\,x = 2\delta(x)$ follows $\frac{\mathrm{d}\phi}{\mathrm{d}z} = A - 2\pi k_C \sigma\,\mathrm{sgn}\,z$. Since $E_\perp = -\frac{\mathrm{d}\phi}{\mathrm{d}z}$ must be antisymmetric in these applications, it follows that $A = 0$. This results in

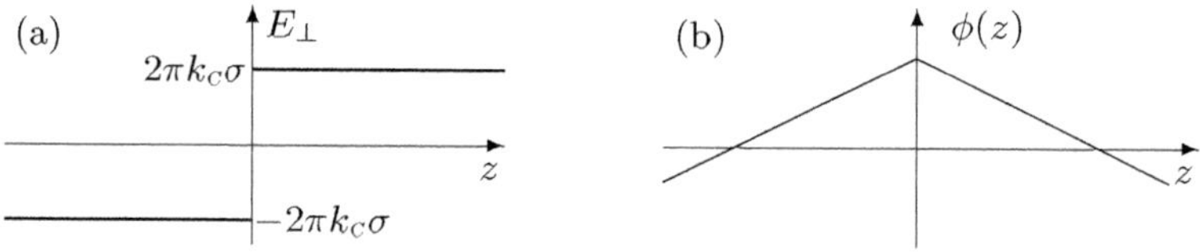

Fig. 2.9. (a) Jump of the electric field of a surface charge at the surface (b) Electrostatic potential ϕ of a surface charge in the plane $z = 0$

the field or potential sketched in Fig. 2.9

$$E_\perp = 2\pi k_C \sigma\,\mathrm{sgn}\,z, \qquad\qquad \phi = \phi_0 - 2\pi k_C \sigma |z|. \tag{2.2.22}$$

$E_\perp$ has a jump of $4\pi k_C \sigma$ at the charge layer. The absolute value of the potential is neither determinable nor physically relevant.

2.2.2 Field Lines

To represent the electrostatic potential ϕ, one can draw equipotential lines $\phi =$ const in analogy to the contour lines on geographical maps. Perpendicular to these equipotential lines are the field lines, formerly also referred to as force lines. These are used for the graphical characterization of the field. It is the

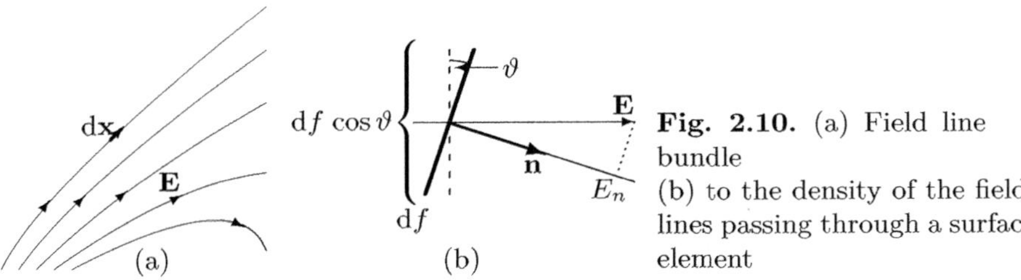

Fig. 2.10. (a) Field line bundle
(b) to the density of the field lines passing through a surface element

family of curves whose tangent at each point is parallel to the field strength $\mathbf{E}$ is $(\mathrm{d}\mathbf{x}\|\mathbf{E})$, as sketched in Fig. 2.10.

Properties of field lines

- The tangents of the field lines $\mathbf{x}(s)$ are along the path s parallel $\mathbf{E}(\mathbf{x}(s))$:

$$\frac{\mathrm{d}\mathbf{x}(s)}{\mathrm{d}s} = \frac{\mathbf{E}(\mathbf{x}(s))}{|\mathbf{E}(\mathbf{x}(s))|} \quad \overset{\mathrm{d}\mathbf{x}\|\mathbf{E}}{\text{or}} \quad \frac{\mathrm{d}x_i}{\mathrm{d}x_j} = \frac{E_i}{E_j} \quad \overset{i,j=1,2,3}{\Longleftrightarrow} \quad \frac{\mathrm{d}x_i}{E_i} = \frac{\mathrm{d}x_j}{E_j}. \tag{2.2.23}$$

- In electrostatics, only curl-free fields $(\nabla \times \mathbf{E} = 0)$ are known. In these, a field line begins at a higher potential (source, positive charge) and ends at a lower level (sink, negative charge).

- To determine the density of the field lines, a value Φ_0 is specified for the flux into which a field line falls. Through the area element $\Delta\mathbf{a}$ then go

$$\Delta\Phi_E/\Phi_0 = \Delta\mathbf{a}\cdot\mathbf{E}/\Phi_0$$

field lines (see Fig. 2.10b). Here is $\mathrm{d}\mathbf{a}\cdot\mathbf{E} = \mathrm{d}a\, E\cos{(\mathbf{n},\mathbf{E})}$.

Alternatively, we can freely choose the number of field lines emerging from the surface ∂V of a volume V with charge $Q > 0$ (proportional to Q). This number then remains unchanged in the charge-free space $(\Delta\phi = 0)$.

We consider a tube bounded by field lines (see Fig. 2.11):

$$\oiint_S \mathrm{d}\mathbf{a}\cdot\mathbf{E} = S_2\, E_2 - S_1\, E_1 = 0,$$

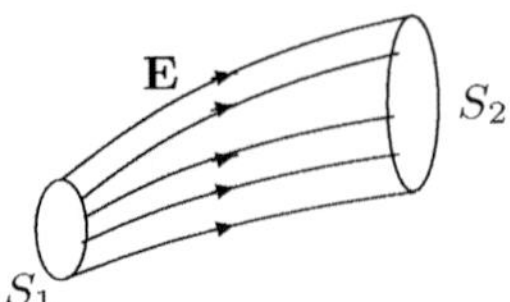

Fig. 2.11. Field lines on a tube

where S_1 and S_2 are perpendicular to the field lines. This implies that the number of field lines through S_2 is equal to the number of field lines through S_1 if no charges are present in the area.

If $\rho \neq 0$, is $4\pi k_c Q = \oiint \mathrm{d}\mathbf{f} \cdot \mathbf{E} = \oiint \mathrm{d}f \, E \, \cos(\mathbf{n}, \mathbf{E})$

$$= \text{Number of field lines through the surface.}$$

Analytical determination of the field lines for a dipole field

Given is a point dipole, which is located at the origin and points in the z direction, $\mathbf{p} = p\mathbf{e}_z$, as sketched in Fig. 2.12. According to (2.2.4), the field lines in the xz plane ($y = 0$)

$$\mathbf{E} = k_c \frac{p}{r^3}\left(3\frac{z\mathbf{x}}{r^2} - \mathbf{e}_z\right), \qquad \frac{\mathrm{d}z}{\mathrm{d}x} = \frac{E_z}{E_x} = \frac{2z^2 - x^2}{3zx} = \frac{2z}{3x} - \frac{x}{3z}. \qquad (2.2.24)$$

Although the vector field can be sketched with this, for a representation of the field lines the course of the field lines $z = f(x)$ must be determined from (2.2.24). With the coordinate transformation $z = ux$ it can be achieved that $g(u)\,\mathrm{d}u = f(x)\,\mathrm{d}x$:

$$\frac{\mathrm{d}z}{\mathrm{d}x} = x\frac{\mathrm{d}u}{\mathrm{d}x} + u = \frac{2u}{3} - \frac{1}{3u} \qquad \Rightarrow \qquad \frac{3u\,\mathrm{d}u}{u^2 + 1} = -\frac{\mathrm{d}x}{x}.$$

The integration and subsequent exponentiation yield

$$(3/2)\ln(u^2 + 1) = -\ln x + C \qquad \Rightarrow \qquad (u^2 + 1)^{3/2} = C/x.$$

A solution of the form $z(x)$ is sought, where x_0 with $z(x_0) = 0$ is the starting point on the x-axis. This determines the integration constant C:

$$z^2 = x^2\left[(C/x)^{2/3} - 1\right] \quad \overset{C = x_0}{\Longrightarrow} \quad z(x) = \pm x^{2/3}\sqrt{x_0^{2/3} - x^{2/3}}. \qquad (2.2.25)$$

For a given x_0, the corresponding field line, starting with $z = 0$, is determined. If you want to calculate the adjacent field lines, you have to determine the path piece 2δ on the x-axis through which a flux Φ_E of a given strength goes:

$$\Phi_E = \int_{x_0 - \delta}^{x_0 + \delta} \mathrm{d}x\, |E_z(z = 0)| = k_c p \int_{x_0 - \delta}^{x_0 + \delta} \mathrm{d}x\, x^{-3} = k_c \frac{2px_0\delta}{(x_0^2 - \delta^2)^2}.$$

The calculated field line density is related to a line, but not to an area ('flux tubes'). Fig. 2.12 shows the field lines of a point dipole.

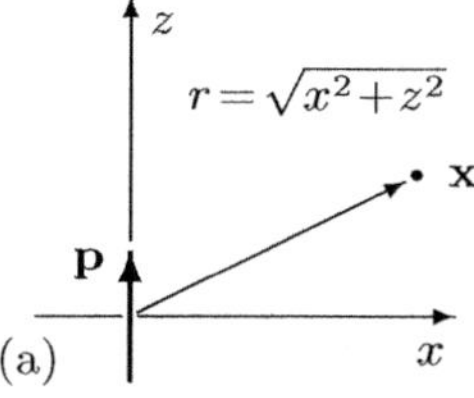

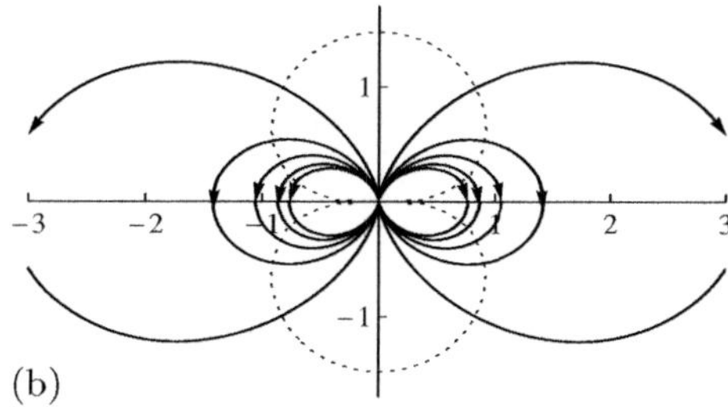

Fig. 2.12. (a) Point dipole (b) Field lines and equipotential lines (dashed) of the point dipole

Quadrupole field

In addition to the dipole field, we show without detailed calculation the field line image of four point charges in the arrangement of an axially symmetric quadrupole according to Fig. 2.13a. The resulting field line image Fig. 2.13b has in particular outside of the square the typical shape for a quadrupole.

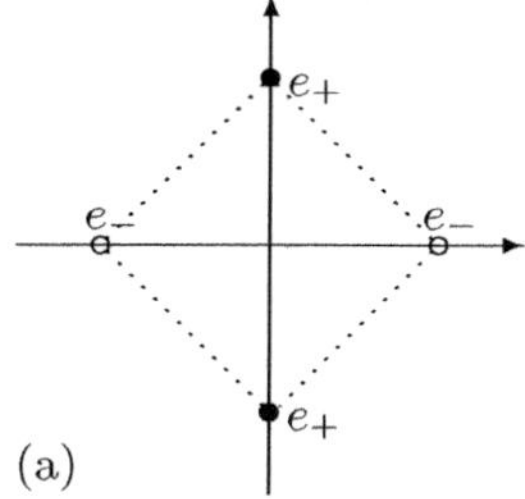

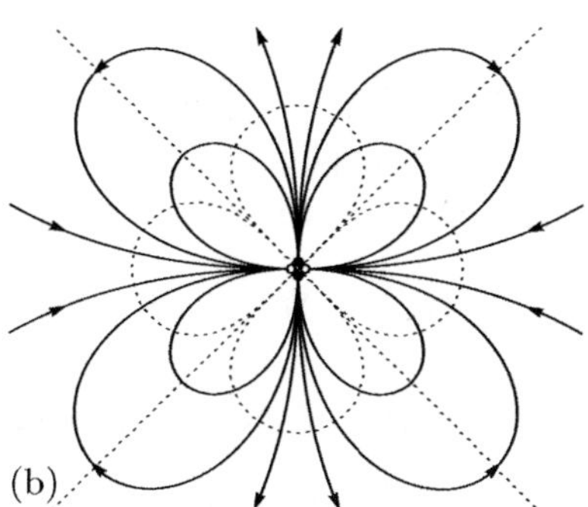

Fig. 2.13. (a) Arrangement of the quadrupole charges   (b) Associated field line image

2.2.3 Boundary Condition of the Electric Field at a Surface

Given is a surface with the surface charge σ. To determine are the boundary conditions for $\mathbf{E}$ at this interface, distinguishing between the normal- and the tangential components.

Jump in the normal component

When passing through a surface occupied with the surface charge σ, the normal component $\mathbf{E}_\perp$ makes a jump of $4\pi k_C \sigma$, as can be seen from (2.2.22). This jump in the normal component was derived from the potential of an infinitely extended, flat surface. Here it is shown that this result applies more generally.

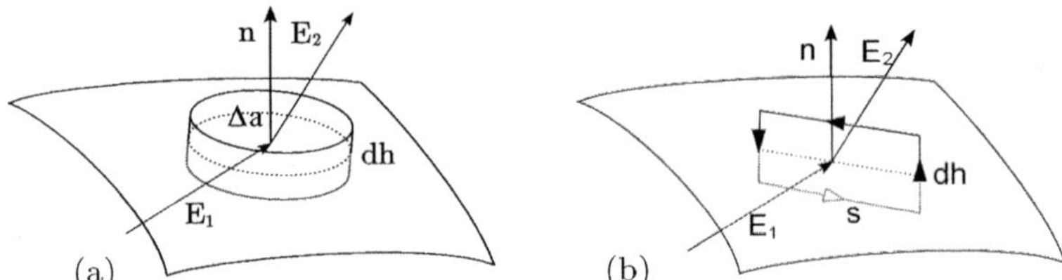

Fig. 2.14. (a) Jump in the normal component.   (b) Continuity of the tangential components

The interface is penetrated by a cylinder with radius Δr, the top surface $\Delta a = \Delta a \mathbf{n}$ and the infinitesimal height $\mathrm{d}h$, as sketched in Fig. 2.14a. Here, $\mathrm{d}h \ll \Delta r$ applies, so that the lateral surface $2\pi \mathrm{d}h \Delta r$ can be neglected compared to the top surfaces $\pi(\Delta r)^2$. The Gaussian law is applied to this cylinder:

$$\oiint \mathrm{d}\mathbf{a}\cdot\mathbf{E} = 4\pi k_C Q = 4\pi k_C \sigma \Delta a .$$

Now the surface integral is calculated. It applies $\Delta\mathbf{a}^{(2)} = -\Delta\mathbf{a}^{(1)} = \Delta a\mathbf{n}$.

$$\Delta\mathbf{a}^{(1)}\cdot\mathbf{E}^{(1)} + \Delta\mathbf{a}^{(2)}\cdot\mathbf{E}^{(2)} = \Delta a\mathbf{n}\cdot\left(\mathbf{E}^{(2)} - \mathbf{E}^{(1)}\right) = 4\pi k_C \sigma \Delta a .$$

From this it follows that the normal component of the electric field makes a jump when passing through the surface:

$$\mathbf{n}\cdot\left(\mathbf{E}^{(2)} - \mathbf{E}^{(1)}\right) = E_\perp^{(2)} - E_\perp^{(1)} = 4\pi k_C \sigma . \tag{2.2.26}$$

Continuity of the tangential components

When examining the tangential components at boundary surfaces, as sketched in Fig. 2.14b, a small rectangle perpendicular to the interface is taken and the flux through it is determined using Stokes' theorem. From $\nabla\times\mathbf{E} = 0$ it follows

$$\iint_A \mathrm{d}\mathbf{a}\cdot(\nabla\times\mathbf{E}) = \oint_{\partial A} \mathrm{d}\mathbf{s}\cdot\mathbf{E} = E_{tg}^{(2)}\Delta s + E_n\,\mathrm{d}h - E_{tg}^{(1)}\Delta s - E_n\,\mathrm{d}h$$
$$= (E_{tg}^{(2)} - E_{tg}^{(1)})\Delta s = 0 .$$

Thus, $E_{tg}^{(1)} = E_{tg}^{(2)}$, as long as $\nabla\times\mathbf{E} = 0$. This relation holds in the area, i.e., the two tangential components of the electric field $\mathbf{E}_\parallel$ are continuous:

$$\mathbf{E}_\parallel^{(1)} = \mathbf{E}_\parallel^{(2)} . \tag{2.2.27}$$

Note: The boundary conditions of a vector field $\mathbf{v}$ will be discussed on several occasions. The procedure is always the same: For the normal component, integrate over an infinitesimal cylinder V (*pillbox*) and for the tangential components over an infinitesimal rectangle A, as sketched in Fig. 2.14:

Tangential components $\qquad \oint_{\partial A} d\mathbf{x}\cdot\mathbf{v} \overset{\text{Stokes}}{=} \iint_A d\mathbf{a}\cdot\mathbf{\nabla}\times\mathbf{v},$

Normal component $\qquad \oiint_{\partial V} d\mathbf{a}\cdot\mathbf{v} \overset{\text{Gauss}}{=} \int_V d^3x\,\mathbf{\nabla}\cdot\mathbf{v}.$

2.2.4 Dipole Layer and Capacitor

Dipole layer on a flat surface

From two oppositely charged plates (see Fig. 2.15) for $d\to 0$ a dipole layer is formed when the limit $\sigma\to\infty$ is taken in such a way that the dipole surface density $D=\sigma d$ remains finite. The Taylor expansion of the charge density

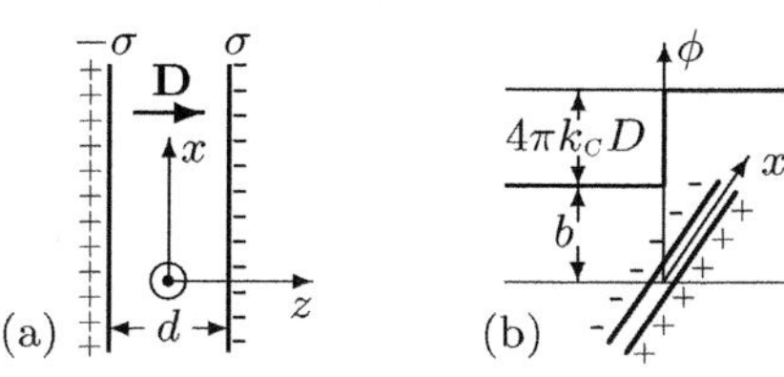

Fig. 2.15. (a) Dipole layer (b) Jump in the normal component when passing through the dipole layer; on the left let $\phi = b$ and on the right is $\phi = b + 4\pi k_C D$

results in

$$\rho(\mathbf{x}) = \sigma\left(\delta(z-\tfrac{d}{2}) - \delta(z+\tfrac{d}{2})\right) = -D\frac{\partial}{\partial z}\delta(z).$$

The dipole density $D=\sigma d$ is the dipole moment/area:

$$\frac{\partial^2}{\partial z^2}\phi = 4\pi k_C D\frac{\partial}{\partial z}\delta(z) \qquad\Rightarrow\qquad \frac{\partial}{\partial z}\phi = 4\pi k_C D\delta(z) + a.$$

For large distances $|z|\gg d$ the field must disappear, since the charges of the two layers compensate each other, from which $a=0$ follows:

$$\phi = 4\pi k_C D\theta(z) + b. \tag{2.2.28}$$

The potential jumps by $4\pi k_C D$.

Dipole layer on an arbitrary surface

The charge per unit area is $\sigma(\mathbf{x})$, where (see Fig. 2.16) the negative charge is on area 1 and the positive one is a distance $\mathbf{d}(\mathbf{x})$ on area 2.

Potential of 1st layer : $\quad \phi_1(\mathbf{x}) = -k_C \iint_A da'\,\dfrac{\sigma(\mathbf{x}')}{|\mathbf{x}-\mathbf{x}'|},$

Dipole density: $\qquad\qquad \mathbf{D}(\mathbf{x}) = \sigma(\mathbf{x})\,\mathbf{d}(\mathbf{x}),$

Potential of 2nd layer: $\quad \phi_2(\mathbf{x}) = k_C \iint_{A+d} da'\,\dfrac{\sigma(\mathbf{x}'-\mathbf{d}(\mathbf{x}'))}{|\mathbf{x}-\mathbf{x}'|}$

$$= k_C \iint_A da''\,\frac{\sigma(\mathbf{x}'')}{|\mathbf{x}-\mathbf{x}''-\mathbf{d}(\mathbf{x}'')|},$$

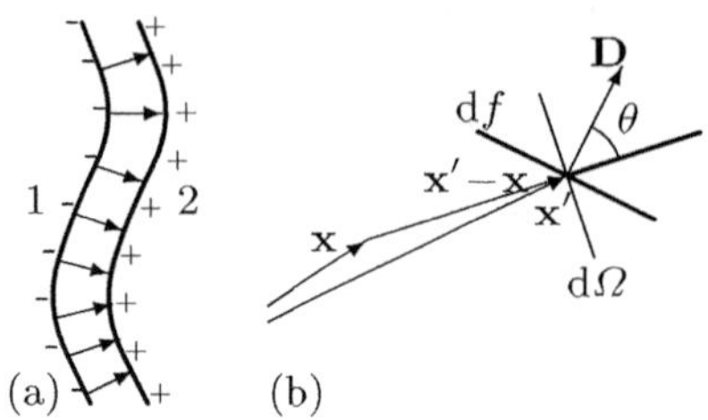

Fig. 2.16. (a) Dipole layer on any surface

(b) $\mathbf{x}$ is to the left of (behind) the layer, i.e. $\cos\theta \geq 0$

where $\mathbf{x}'' = \mathbf{x}' - \mathbf{d}(\mathbf{x}') \approx \mathbf{x}' - \mathbf{d}(\mathbf{x}'')$. This results in:

$$\phi(\mathbf{x}) = \phi_1 + \phi_2 \approx k_c \iint \mathrm{d}a'\, \sigma(\mathbf{x}')\, \mathbf{d}\cdot\boldsymbol{\nabla}'\frac{1}{|\mathbf{x}-\mathbf{x}'|} = k_c \iint \mathrm{d}a'\, \mathbf{D}(\mathbf{x}')\cdot\boldsymbol{\nabla}'\frac{1}{|\mathbf{x}-\mathbf{x}'|}$$

$$= -k_c \iint \mathrm{d}a'\, \mathbf{D}(\mathbf{x}')\cdot\frac{\mathbf{x}'-\mathbf{x}}{|\mathbf{x}'-\mathbf{x}|^3}\,.$$

According to the sketch in Fig. 2.16 is $\mathrm{d}a' = |\mathbf{x}'-\mathbf{x}|^2\,\mathrm{d}\Omega\,\dfrac{1}{\cos\theta}$.

If the angle θ between the dipole vector $\mathbf{D}$ and the surface normal $\mathbf{n}$ is less than $90°$, which is the case to the right of the double layer in the sketch 2.16, one obtains the positive sign in (2.2.29). The potential therefore again has a different sign to the left and right of the dipole layer:

$$\mathbf{D}(\mathbf{x}')\cdot\frac{\mathbf{x}'-\mathbf{x}}{|\mathbf{x}'-\mathbf{x}|^3} = \pm D\,\frac{\cos\theta}{|\mathbf{x}'-\mathbf{x}|^2}\quad \begin{cases} + & \text{left} \\ - & \text{right} \end{cases} \qquad (2.2.29)$$

$$\phi(\mathbf{x}) = \mp k_c \iint \mathrm{d}\Omega\, D \quad \begin{array}{l}\text{left}\\\text{right}.\end{array}$$

The potential at the location $\mathbf{x}$ has a positive or negative opening angle, on each side a maximum of 2π.

Plate and sphere capacitor

We are now making a preview of electrostatics in the presence of conductors, as the plates of a capacitor are conductors, on which the surface charge is applied. In a conductor, the charge carriers (electrons) are freely movable and arrange themselves so that no force acts on them. In equilibrium, the electric field therefore disappears thus the potential ϕ is constant. From the continuity conditions (see section 2.2.3) it follows that the tangential components of $\mathbf{E}$ in the exterior space at the conductor surface vanish and $\mathbf{E}$ thus stands perpendicular to the conductor surface.

Plate capacitor

We assume here that we have two layers in front of us, on one of which the surface charge $+\sigma$ and on the other the surface charge $-\sigma$ is applied, as shown in Fig. 2.17. ϕ is higher at the positive plate, as work must be done to bring

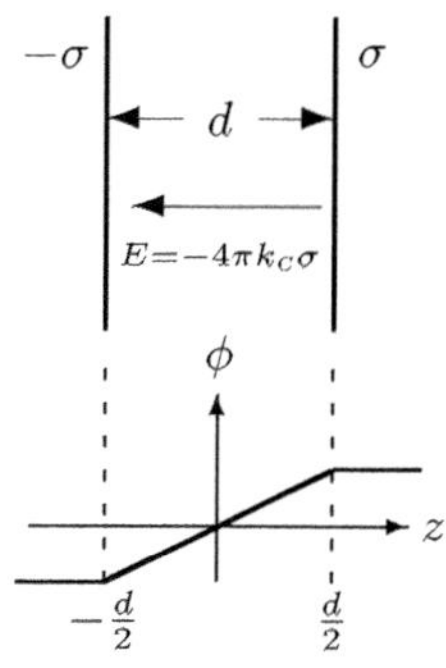

Fig. 2.17. Field E and potential ϕ. As can be seen from the sketch, the distance between the plates is d; we obtain the potential from (2.2.22)

$$\phi(z) = 2\pi k_C \sigma \left(|z + \frac{d}{2}| - |z - \frac{d}{2}| \right) = \begin{cases} -2\pi k_C \sigma d & z < -\frac{d}{2} \\ 4\pi k_C \sigma z & -\frac{d}{2} \le z \le \frac{d}{2} \\ 2\pi k_C \sigma d & z > \frac{d}{2}. \end{cases}$$

a positive charge there:

$$\mathbf{E}(\mathbf{x}) = \mathbf{e}_\perp 2\pi k_C \sigma \left[-\operatorname{sgn}(z + \frac{d}{2}) + \operatorname{sgn}(z - \frac{d}{2}) \right] = \begin{cases} -4\pi k_C \sigma \mathbf{e}_\perp & -\frac{d}{2} \le z \le \frac{d}{2} \\ 0 & \text{otherwise,} \end{cases}$$

$$\rho(\mathbf{x}) = \sigma \left[-\delta(z + \frac{d}{2}) + \delta(z - \frac{d}{2}) \right].$$

This result, valid for infinitely extended plates, can also be applied to finite ones, if their surface $A \gg d^2$. The total charge of a surface is $Q = \sigma A$. This gives us

$$\text{Voltage} \qquad V = -\int_{-d/2}^{d/2} \mathbf{ds} \cdot \mathbf{E} = 4\pi k_C \sigma d = 4\pi k_C \frac{Q}{A} d,$$

$$\text{Field strength} \qquad E = 4\pi k_C \frac{Q}{A},$$

$$\text{Capacitance} \qquad C \equiv \frac{Q}{V} = \frac{1}{4\pi k_C} \frac{A}{d}. \tag{2.2.30}$$

To calculate the energy stored in the capacitor, we must refer to (2.4.2):

$$U = \frac{1}{2} \int d^3 x \, \rho(\mathbf{x}) \phi(\mathbf{x}) = \frac{\sigma}{2} \iint_A dx dy \left[\phi(\frac{d}{2}) - \phi(-\frac{d}{2}) \right]$$

$$= \frac{1}{2} \sigma V A = \frac{1}{2} C V^2. \tag{2.2.31}$$

Actual field distribution: In Fig. 2.18(a) the field distribution of two semi-infinite plates of constant charge density $\pm\sigma$ at a distance d is according to

$$\mathbf{E}(\mathbf{x}) = k_C \sigma \left\{ \mathbf{e}_x \ln \frac{x^2 + (z+d/2)^2}{x^2 + (z-d/2)^2} \right. \tag{2.2.32}$$

$$\left. + \mathbf{e}_z \left[\pi \operatorname{sgn}(z - \frac{d}{2}) - 2\arctan \frac{x}{z - d/2} - \pi \operatorname{sgn}(z + \frac{d}{2}) + 2\arctan \frac{x}{z + d/2} \right] \right\}$$

outlined (see problem 2.5). Fig. 2.18(b) shows the field distribution when the plates are conductive (see problem 3.6). The field lines are perpendicular

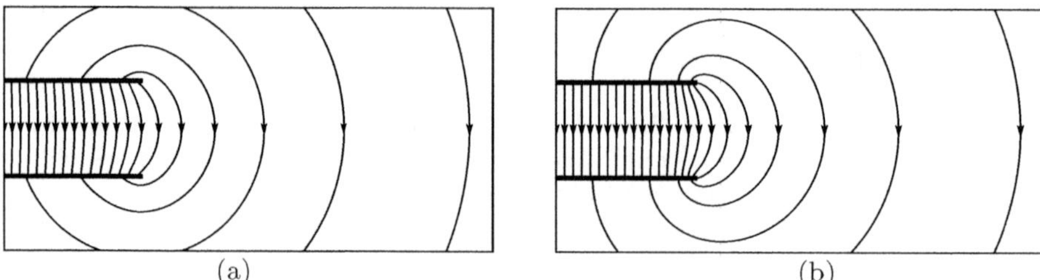

Fig. 2.18. Field lines of two oppositely charged infinite half-planes (plates) located at a distance d: (a) Constant charge densities $\pm\sigma$. (b) Metal plates (constant potential $\pm\phi$); the field lines are normal to the plates

to the plates and since the freely moving charge carriers in the metal repel each other, they are pushed to the edge. So overall the capacitance is larger. Calculations on the stray fields of the capacitor were made by *G. Kirchhoff*. For a circular plate capacitor with the radius R and the distance d Kirchhoff [1891, p. 109] obtained

$$C = \frac{1}{4\pi k_C}\left[\frac{R^2\pi}{d} + R\left(\ln\frac{16\pi R}{d} - 1\right)\right].$$

Spherical capacitor

The two spherical shells of the capacitor carry the charges $\pm Q$, which results in the surface charges

$$\sigma_{1,2} = \pm Q/(4\pi r_{1,2}^2).$$

The field points to the center of the sphere; so it is sufficient to consider the component E_r, which we get from Gauss's theorem (1.3.2) by integration over a sphere S_r with the radius $r_1 \leq r < r_2$:

$$\oiint_{S_r} \mathbf{da}\cdot\mathbf{E} = 4\pi r^2\, E_r = 4\pi k_C Q \qquad\qquad r_1 \leq r < r_2.$$

The field becomes stronger towards the inside, which is also indicated by the denser field lines in Fig. 2.19. The voltage applied to the capacitor plates is obtained from

$$V = -\int_{r_2}^{r_1} \mathrm{d}r\, E_r = k_C Q\left(\frac{1}{r_1} - \frac{1}{r_2}\right).$$

The capacitance of the spherical capacitor is equal to that of a plate capacitor of the area $A = 4\pi R^2$, where $R = \sqrt{r_1 r_2}$ is the geometric mean of the sphere radii:

$$C = \frac{Q}{V} = \frac{r_1 r_2}{k_C d} = \frac{1}{4\pi k_C}\frac{A}{d}. \tag{2.2.33}$$

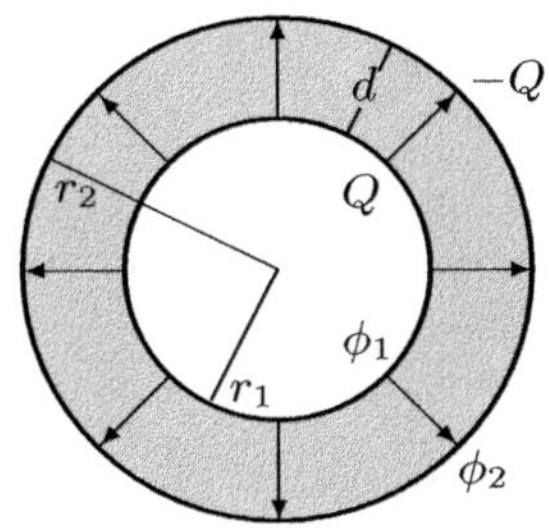

Fig. 2.19. Direction of the field in the spherical capacitor; the charges $\pm Q$ are applied, i.e., the surface charge density on the inner spherical shell is higher and the potential difference is $V = \phi_1 - \phi_2$

2.3 Fields of Stationary Charges in the Presence of Conductors

Conductors are media in which electrons can move freely. While the calculation of the field of fixed charges leads to an integral of the form of (2.1.8), one must consider in the presence of conductors that the electrons in conductors move freely and adapt to the given charges.

Electrical properties of conductors

1. *The field in the conductor is zero*: If a field were present at any point in time, the free moving electrons would redistribute until the field is compensated. Throughout the conductor, then $\mathbf{E} = 0$ and $\phi = \text{const}$.

2. *The charge density inside a conductor is zero*: Otherwise, due to $\nabla \cdot \mathbf{E} = 4\pi k_C \rho$, a field would be present inside.
   Note: Charges can occur on the surface.

3. *Tangential components* $\mathbf{E}_\parallel = 0$ *disappear at the metal surface*: From

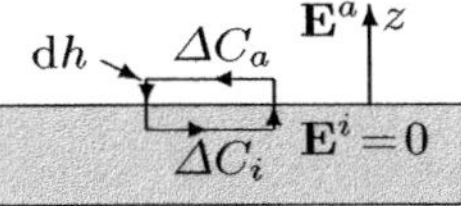

Fig. 2.20. $\mathbf{E}_\parallel$ disappears at the metal surface, and $\mathbf{E}^a$ is finite and is perpendicular to the surface

$\nabla \times \mathbf{E} = 0$ it follows (see Fig. 2.20), since in the metal $\mathbf{E}^i = 0$:

$$\oint_C \mathrm{d}s \cdot \mathbf{E} = E_\parallel^a \Delta C_a - E_z^a \mathrm{d}h + E_z^a \mathrm{d}h = E_\parallel^a \Delta C_a = 0.$$

The tangential components $\mathbf{E}_\parallel = 0$ disappear on the metal surface ∂L, which is why $\phi(\partial L)$ is constant there.

4. *Jump in the normal component*: A positive charge present in space induces a negative surface charge σ on a conductor. The normal component of the electric field experiences a jump when entering the metal from $4\pi k_C \sigma$, which is expected when passing through a surface charge. For this layer

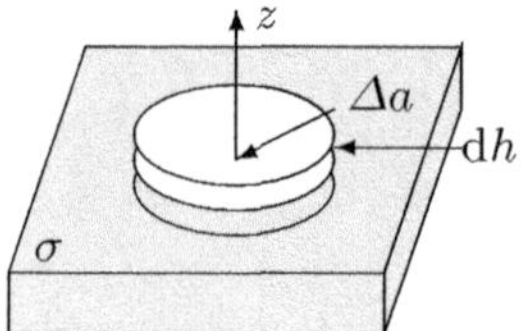

Fig. 2.21. Cylinder of volume V with height $2\mathrm{d}h$ and base Δa on metal surface

on the metal surface applies $\boldsymbol{\nabla}\cdot\mathbf{E} = 4\pi k_C \sigma\,\delta(z)$, if, as outlined in Fig. 2.21, the metal surface is characterized by $z = 0$. If you take as volume V the infinitesimal cylinder

$$\oiint_{\partial V} \mathrm{d}a \cdot \mathbf{E} = 4\pi\sigma k_C \Delta a,$$

thus follows by evaluating the surface integral

$$\Delta a E_n = 4\pi k_C \sigma \Delta a \rightarrow E_n = 4\pi k_C \sigma.$$

The field is perpendicular to the metal surface, which, as sketched in Fig. 2.23, causes the induced charges.

5. *Faraday cage*: This is, as sketched in Fig. 2.22, a cavity surrounded by a conductor. If there are no charges in it, the Laplace equation applies there, whose solutions are harmonic functions. For these, the maximum/minimum values are at the edge of the cavity (see appendix A.4.5). Since ϕ is constant there, this must apply to the entire cavity, i.e. $\mathbf{E} = 0$. However, this does

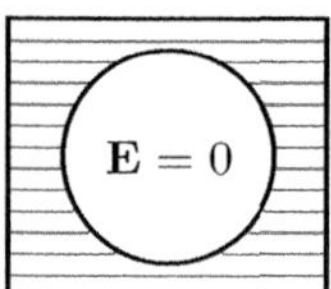

Fig. 2.22. In the cavity surrounded by a conductor, $\phi = \mathrm{const.}$, from which it follows that $\mathbf{E} = 0$

not mean that there can be no electromagnetic waves in the cavity, but these are solutions of the wave equation (see section 10.5).

From the first points it follows that a conductor can only be specified in one of the following two ways:

1. by the potential $\phi = \mathrm{const.}$ or

2. by the charge on the surface $\oiint_{\partial V} \mathrm{d}a \cdot \mathbf{E} = 4\pi k_C Q.$ (2.3.1)

Uniqueness theorem for the basic problem of electrostatics

In the basic problem of electrostatics, the solution of the Poisson equation $\Delta\phi(\mathbf{x}) = -4\pi k_C \rho$ for a given ρ, in the presence of conductors the additional conditions (2.3.1) must be met. Although there is no general method to solve this task, it can be shown that the found solution is unique [Becker, Sauter, 1973, §1.5].

We assume that we have two different solutions ϕ_1 and ϕ_2 for the same sources and boundary conditions. The difference $\phi_d = \phi_1 - \phi_2$ then satisfies $\Delta\phi_d = 0$, where in the conductor L_i according to (2.3.1) either the potential $\phi_{di} = 0$ or the charge $\oiint_{\partial L_i} \mathrm{d}\mathbf{f}_i \cdot \mathbf{E} = 0$ vanishes. We now use the 1st Green's theorem (A.4.19) with $\phi = \psi = \phi_d$ and then take into account that the second term on the right side vanishes due to the boundary conditions:

$$\int_{\mathbb{R}^3 \backslash \sum_i L_i} \mathrm{d}^3 x\, |\boldsymbol{\nabla}\phi_d(\mathbf{x})|^2 = \oiint_{r\to\infty} \mathrm{d}\mathbf{a} \cdot \phi_d \boldsymbol{\nabla}\phi_d - \sum_i \phi_{di} \oiint_{\partial L_i} \mathrm{d}\mathbf{a}_i \cdot \boldsymbol{\nabla}\phi_{di}$$

$$\overset{r\to\infty}{\sim}\; r^2 \frac{1}{r}\frac{1}{r^2} \to 0.$$

Assuming that charges and conductors are located in the finite region, $|\mathbf{E}|$ cannot asymptotically decrease slower than $1/r^2$. Thus the surface integral vanishes with at least $1/r$. Since the integrand $|\boldsymbol{\nabla}\phi_d|^2 \geq 0$, $\phi_d = 0$ must be, so that the integral vanishes. Thus, the solution is unique.

Remarks: The negative sign takes into account that the normal $\mathrm{d}\mathbf{a}_i$ points out of the volume L_i, thus it is directed opposite to $\mathbb{R}^3 \backslash \sum_i L_i$. The proof also holds for a finite volume, if ϕ is given at the boundary (*Dirichlet boundary condition*, see section 3.1). Then ϕ_d automatically vanishes at the boundary.

2.3.1 Method of Image Charges

In the presence of a conductor, such as a grounded infinite plate, as sketched in Fig. 2.23, negative charges from a great distance are attracted to it by the positive charge q. A negative charge layer forms, whose surface charge density σ is.

The field $\mathbf{E}$ is determined by applying *fictitious* (image) charges in the metal interior, i.e. outside the considered volume V, in such a way that these, together with the actual charge q, generate a field that satisfies the boundary conditions $\mathbf{E}_\parallel = 0$. Within V $\mathbf{E}$ then results from the superposition of the field of the charge with that of the image charges. The following boundary conditions are distinguished

1. Grounded metal body ($\phi = 0$): Find the image charge(s) q' for the charge q.
2. Ungrounded metal body with a given charge Q: Solution as under the first point, with the addition of the field of charge $Q - q'$, where $Q - q'$ is placed so that the field is perpendicular to the conductor surface (e.g. for a sphere in the center).
3. Metal body at fixed potential ϕ_0: The additional charge $Q - q'$ is to be replaced here by $C\,\phi_0$ ($C =$ capacitance of the conductor; (2.2.30)).

Point charge in front of a conductive plate

Given is an infinitely extended, grounded metal plate in front of which a point charge q is placed at a distance d ($\mathbf{d} = (0,0,d)$). This configuration is sketched in Fig. 2.23.

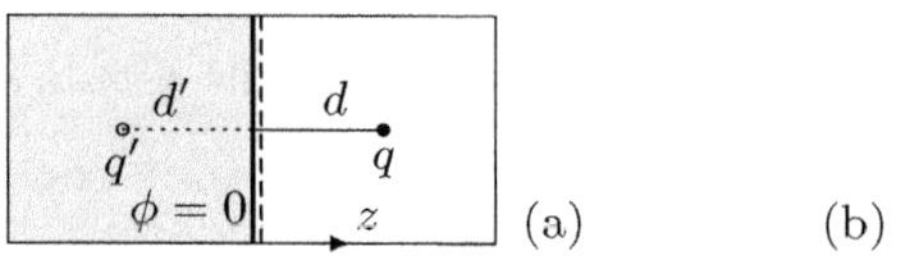

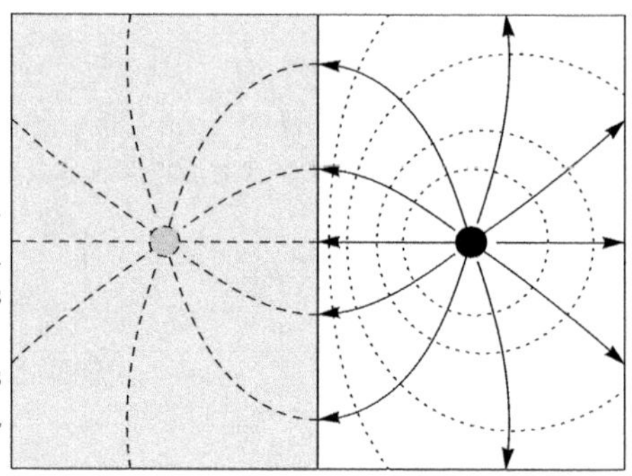

Fig. 2.23. (a) Charge $q > 0$ with image charge $q' = -q$ in front of a grounded metal plate; the induced surface charge is dashed; here is $d' = d$ and $q' = -q$
(b) Field and (dotted) equipotential lines of the positive point charge in front of a grounded metal plate

In the case of a single point charge q, an image charge $q' = -q$ at a distance $d' = -d$ from the plate is sufficient ($z = 0$), to ensure the disappearance of the tangential components $\mathbf{E}_{\parallel}$. It then applies

$$\phi(\mathbf{x}) = k_C q\Big(\frac{1}{|\mathbf{x} - d\mathbf{e}_z|} - \frac{1}{|\mathbf{x} + d\mathbf{e}_z|}\Big)$$

$$\mathbf{E}(\mathbf{x}) = k_C q\Big(\frac{\mathbf{x} - d\mathbf{e}_z}{|\mathbf{x} - d\mathbf{e}_z|^3} - \frac{\mathbf{x} + d\mathbf{e}_z}{|\mathbf{x} + d\mathbf{e}_z|^3}\Big)$$

for $z \geq 0$. $\qquad$ (2.3.2)

Field strength on the conductor surface $\mathbf{x} = (x, y, 0)$:

$$\mathbf{E}(x, y, 0) = \frac{-2k_C q d\mathbf{e}_z}{(\sqrt{x^2 + y^2 + d^2})^3} = E_n\mathbf{e}_z \qquad (\perp \text{ on surface}).$$

Surface charge density ($x^2 + y^2 = \varrho^2$):

$$\sigma = \frac{E_n}{4\pi k_C} = -\frac{qd}{2\pi(\sqrt{\varrho^2 + d^2})^3}. \qquad (2.3.3)$$

Finally, we show that the total charge induced on the surface is $-q$:

$$Q = \int d^3x' \, \sigma(\mathbf{x}') \, \delta(z') = -qd \int_0^\infty \frac{d\varrho' \, \varrho'}{\sqrt{\varrho'^2 + d^2}^3} = \frac{qd}{\sqrt{\varrho'^2 + d^2}}\Big|_0^\infty = -q.$$

The field of the image charge in the left half-space is the image of the real charge reflected at the $z = 0$ plane, except for the direction of the field lines, which in the first case are directed towards the image charge $q' < 0$. On the metal surface, they compensate so the tangential components of the two fields, while the normal components add up.

Inversion on conducting sphere

Given is a charge q, which is located at position $\mathbf{d}$ in front of a grounded sphere K with radius R. Together with a mirror charge q' at position $\mathbf{d}'$, we have the potential

$$\phi(\mathbf{x}) = k_C\left(\frac{q}{|\mathbf{x} - \mathbf{d}|} + \frac{q'}{|\mathbf{x} - \mathbf{d}'|}\right).$$

q' and $\mathbf{d}'$ are to be determined, so that ϕ vanishes on the sphere surface (see Fig. 2.24).

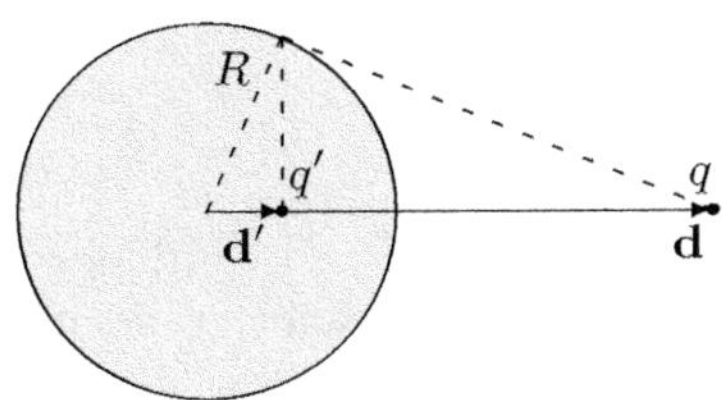

Fig. 2.24. Charge reflection of q on a sphere with radius R. Mirror charge q' at a distance $\mathbf{d}'$ from the center

---

Determination of the image charge and its position within the sphere:

$$\phi(\mathbf{R}) = 0 = k_C\left(\frac{q}{|\mathbf{R} - \mathbf{d}|} + \frac{q'}{|\mathbf{R} - \mathbf{d}'|}\right) \qquad \Rightarrow \qquad \frac{q}{|\mathbf{R} - \mathbf{d}|} = \frac{-q'}{|\mathbf{R} - \mathbf{d}'|}.$$

For symmetry reasons, $\mathbf{d}'\|\mathbf{d}$; then applies

$$q^2(R^2 + d'^2 - 2Rd'\cos\vartheta) = q'^2(R^2 + d^2 - 2Rd\cos\vartheta).$$

The terms dependent on ϑ are brought to the left side, considering that d' and q' must not depend on ϑ. The factor of $\cos\vartheta$ must therefore disappear:

$$(q'^2 d - q^2 d')2R\cos\vartheta = 0 = q'^2(R^2 + d^2) - q^2(R^2 + d'^2).$$

First, it follows that $q' = -q\sqrt{d'/d}$. Inserted into the right side of the above equation, it is obtained that $dd' = R^2$:

$$0 = d'(R^2 + d^2) - d(R^2 + d'^2) = (d' - d)(R^2 - dd').$$

So it applies

$$dd' = R^2 \qquad\qquad \text{and} \qquad\qquad q' = -q\sqrt{\frac{d'}{d}} = -q\,\frac{R}{d}. \qquad (2.3.4)$$

According to Fig. 2.24, $\mathbf{d}' = \frac{d'}{d}\mathbf{d}$. The potential of the image charge thus has the form

$$\phi_{q'}(\mathbf{x}) = \frac{k_C q'}{|\mathbf{x} - \mathbf{d}'|} = -\frac{R}{d}\,\frac{k_C q}{|\mathbf{x} - \frac{R^2}{d^2}\mathbf{d}|}. \qquad (2.3.5)$$

The induced charge $q' = -(R/d)q$ is distributed on the sphere surface according to:

$$\sigma^{\mathrm{ind}}(\vartheta) = \frac{E(\mathbf{R})}{4\pi k_C} = -\frac{q}{4\pi R}\frac{d^2 - R^2}{\sqrt{R^2 + d^2 - 2Rd\cos\vartheta}^{\,3}}. \tag{2.3.4'}$$

Conducting sphere at constant potential

In this case, the sphere is kept at the potential ϕ_0, which corresponds to a charge $Q = C\phi_0$ where the capacitance of the sphere $C = \frac{R}{k_C}$ is. This results in

$$\phi(\mathbf{x}) = k_C\left(\frac{q}{|\mathbf{x}-\mathbf{d}|} + \frac{q'}{|\mathbf{x}-\mathbf{d'}|}\right) + \frac{\phi_0 R}{r}, \qquad \sigma = \sigma^{\mathrm{ind}} + \frac{\phi_0}{4\pi k_C R}. \tag{2.3.5'}$$

Conducting sphere with charge Q

The sphere is now isolated, but already charged with $q' = -qR/d$ by influence, i.e., it still needs the charge $Q - q'$ to be supplied, so that it reaches the total charge Q:

$$\phi(\mathbf{x}) = k_C\left(\frac{q}{|\mathbf{x}-\mathbf{d}|} + \frac{q'}{|\mathbf{x}-\mathbf{d'}|} + \frac{Q + qR/d}{r}\right). \tag{2.3.5''}$$

The boundary condition of a constant potential on the sphere surface ∂S is fulfilled with q'. σ_P describes the displacement (polarization) of the surface charges by the external charge q:

$$\sigma = \frac{Q}{4\pi R^2} + \sigma_P, \qquad \sigma_P = \sigma^{\mathrm{ind}} + \frac{R}{d}\frac{q}{4\pi R^2}, \qquad \oiint_{\partial K} \mathrm{d}\Omega\,\sigma_P(\vartheta) = 0.$$

2.3.2 Maxwell's Stress Tensor

The Coulomb force that $\mathbf{E}$ exerts on a test charge e is according to (1.2.1) $\mathbf{F} = e\,\mathbf{E}$. The force density acting on $\rho(\mathbf{x})$ is accordingly

$$\mathbf{f}(\mathbf{x}) = \rho(\mathbf{x})\,\mathbf{E}(\mathbf{x}). \tag{2.3.6}$$

The force acting on the body is thus

$$\mathbf{F} = \int_V \mathrm{d}^3x\,\rho(\mathbf{x})\,\mathbf{E}(\mathbf{x}),$$

where V should enclose the entire charge. If there is no external field and/or no additional charge present, then it is expected that no force is exerted on the body:

$$\mathbf{F} = -\int_V \mathrm{d}^3x\,\rho(\mathbf{x})\,\boldsymbol{\nabla}\phi(\mathbf{x}) = k_C\int_V \mathrm{d}^3x\,\rho(\mathbf{x})\int \mathrm{d}^3x'\,\rho(\mathbf{x}')\frac{\mathbf{x}-\mathbf{x}'}{|\mathbf{x}-\mathbf{x}'|^3} = 0.$$

The integrand is antisymmetric under the exchange of $\mathbf{x} \leftrightharpoons \mathbf{x}'$, which is why the integral vanishes. With an external field or with a second body, this antisymmetry is broken $(\rho(\mathbf{x}') \to \rho(\mathbf{x}') + \rho_{\text{extern}}(\mathbf{x}'))$ and $\mathbf{F}$ indicates the force, that the external field exerts on the body, as sketched in Fig. 2.25.

According to the concept going back to Faraday and Maxwell, all forces are transmitted by the electromagnetic field in a continuous manner. So the forces transferred by $\mathbf{E}$ to the body must pass through a surface around the body, regardless of whether this surface is the surface of the body or a surface outside with $\rho = 0$. So the force $\mathbf{F}$ acting on the volume should be able to be replaced by a surface force.

From elasticity theory, we are quite familiar with the fact that volume forces are exerted on the interior of the body by a force (tension, pressure) acting on the surface. If a relation of the form

$$f_i(\mathbf{x}) = \nabla_j T_{ij}(\mathbf{x})$$

can be found, then the force density is given as the divergence of a tensor, the Maxwell stress tensor T_{ij}, and one can represent the force F_i through surface forces using Gauss's theorem

$$F_i = \int_V \mathrm{d}^3x \, T_{ij,j}(\mathbf{x}) = \oiint_{\partial V} \mathrm{d}a_j \, T_{ij} \tag{2.3.7}$$

where $T_{ik,k} \equiv \nabla_k T_{ik}$. The T_{ik} are therefore the stresses on the surface. For the

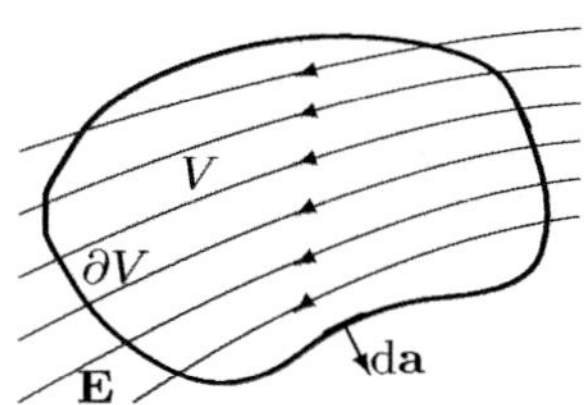

Fig. 2.25. Simply connected volume V with charge density $\rho(\mathbf{x})$ in the external field $\mathbf{E}(\mathbf{x})$

representation $f_i = \nabla_j T_{ij}$ we use Gauss's law $\nabla \cdot \mathbf{E} = 4\pi k_C \rho$ and the relation $\nabla_i E_j = -\nabla_i \nabla_j \phi = \nabla_j E_i$:

$$f_i = \frac{1}{4\pi k_C} E_i \nabla_j E_j = \frac{1}{4\pi k_C}\left[\nabla_j E_i E_j - E_j \nabla_j E_i\right] = \frac{1}{4\pi k_C}\left[\nabla_j E_i E_j - E_j \nabla_i E_j\right]$$
$$= \nabla_j \frac{1}{4\pi k_C}\left[E_i E_j - \frac{1}{2}\delta_{ij} E_l E_l\right].$$

The *Maxwell's stress tensor* is therefore

$$T_{ik} = \frac{1}{4\pi k_C}\left(E_i E_k - \frac{1}{2}\delta_{ik} E_l E_l\right) \tag{2.3.8}$$

with the trace

$$\mathrm{Sp}\,\mathsf{T} = \sum_i T_{ii} = \frac{1}{4\pi k_C} \sum_{i=1}^{3} \left(E_i\,E_i - \frac{1}{2} E^2 \right) = -\frac{1}{8\pi k_C} E^2 = -u(\mathbf{x}),$$

where u is the energy density (2.4.4). This representation is then useful when you have conductors in the volume V, whose charge distribution you do not need to calculate in order to determine the field strength $\mathbf{E}$.

Electrical tensions: The force acting on a body is according to (2.3.7)

$$F_i = \oiint_{\partial V} \mathrm{d}f\, T_{n\,i} \quad \text{with} \quad T_{n\,i} = T_{ij} n_j, \quad n_j = \mathbf{n}\cdot\mathbf{e}_j \quad \Leftrightarrow \quad \mathbf{F} = \oiint_{\partial V} \mathrm{d}a\, \mathbf{T}_n.$$

We take out a surface element $\mathrm{d}a$ and set the z-axis parallel to $\mathbf{n}$. The electric field vector $\mathbf{E}$ should, as in Fig. 2.26 drawn, lie in the xz plane: $E_x = E\sin\vartheta$, $E_y = 0$ and $E_z = E\cos\vartheta$. For the force acting on the surface element one obtains

$$\mathbf{T}_n = T_{xz}\mathbf{e}_x + T_{zz}\mathbf{e}_z = \frac{E^2}{8\pi k_C}\left(\sin(2\vartheta)\,\mathbf{e}_x + \cos(2\vartheta)\,\mathbf{e}_z \right).$$

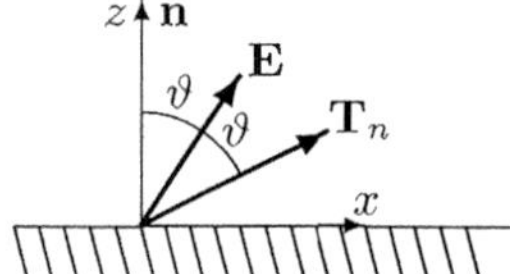

Fig. 2.26. Force on surface element $\mathrm{d}x\mathrm{d}y$, where $\mathbf{E}$ lies in the xz plane [Becker, Sauter, 1973, p. 76]

We see from this that for $\vartheta = 0$ the surface force $\mathbf{T}_n$ is parallel to $\mathbf{E}$, which corresponds to a tensile stress. If $\vartheta = \pi/4$, so lies $\mathbf{T}_n$ in the plane of the surface element, which represents a shear force (shear stress), and at $\vartheta = \pi/2$ $\mathbf{T}_n$ points perpendicularly into the interior of the body and therefore acts as pressure.

Force on a charge q in front of a metallic plate

We now consider the force with which a charge q is attracted by the grounded metal plate as sketched in Fig. 2.27. The stress tensor (2.3.8) of the metal

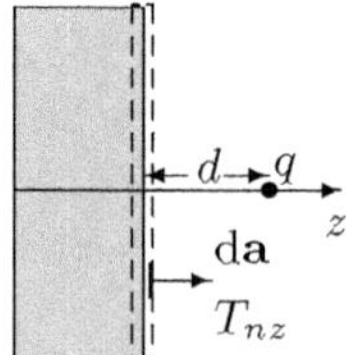

Fig. 2.27. Charge in front of a conductive surface: The dashed line sketches the volume V on whose surface element $\mathrm{d}a$ the surface force $T_{nz} = T_{zz}$ attacks.

plate is obtained from the electric field of the point charge at the boundary $z=0$:

$$\mathbf{E} = -\frac{2k_C q d \mathbf{e}_z}{(\sqrt{x^2+y^2+d^2})^3} = \mathbf{e}_z E_z,$$

$$T_{zz} = k_C \frac{q^2 d^2}{2\pi} \frac{1}{(x^2+y^2+d^2)^3}, \quad T_{xx}=T_{yy}=-T_{zz} \quad \text{and} \quad T_{ik}=0 \quad \forall\, i \neq k.$$

In the case outlined in Fig. 2.27, only the front side with $z = dz$ contributes from the volume V surrounding the metal plate with $z = 0$. Within the metal plate, the fields and thus also the T_{ik} disappear. This results in

$$F_i = \iint \mathrm{d}x\mathrm{d}y\, T_{iz} = k_C \frac{q^2 d^2}{2\pi} \delta_{iz} \iint \mathrm{d}x\mathrm{d}y \frac{1}{(x^2+y^2+d^2)^3}$$

$$= k_C \frac{q^2 d^2}{2\pi} \delta_{iz}\, 2\pi \int_0^\infty \mathrm{d}\varrho \frac{\varrho}{(\varrho^2+d^2)^3} = k_C \frac{q^2}{4d^2} \delta_{iz}.$$

Thus, the force $\mathbf{F}=k_C(q^2/4d^2)\mathbf{e}_z$ is exerted on the metal plate by q, regardless of the sign of its charge. The charge q induces a surface charge on the metal plate of opposite sign and these two charges attract each other.

2.3.3 Fields near Peaks

Fields are particularly strong near peaks. To put this into perspective, we consider two spheres, which are connected to each other. The potentials on the surface are:

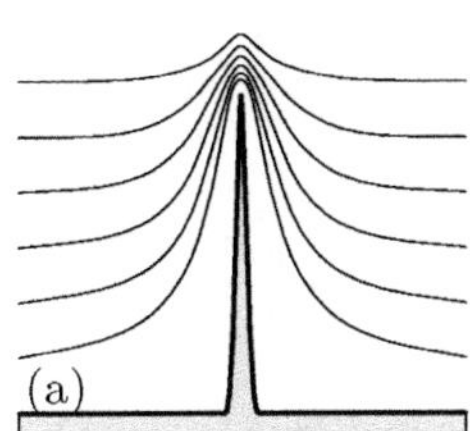

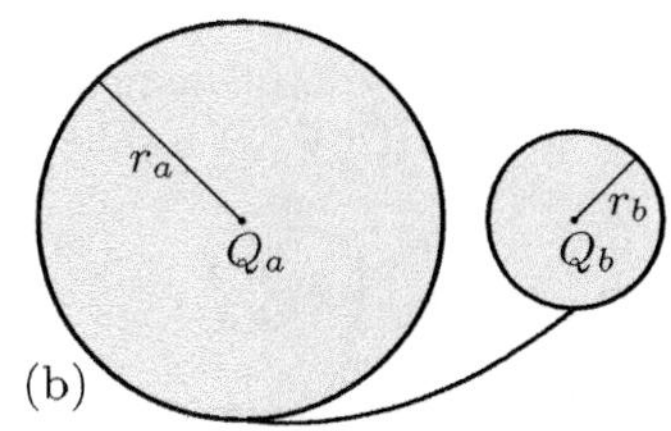

Fig. 2.28. (a) Equipotential lines around a tip (sketched)
(b) Two spheres at the same potential; the field around the smaller sphere is stronger by the factor r_a/r_b

$$\phi_a = k_C \frac{Q_a}{r_a}, \qquad \phi_b = k_C \frac{Q_b}{r_b} \qquad \Longrightarrow \qquad \frac{Q_a}{r_a} = \frac{Q_b}{r_b},$$

since the potentials must be equal. The interior is field-free, which is why the charges can only be on the surface. The normal component of the field strength is on the surface after (2.2.26) given by $E_n = 4\pi k_C \sigma$:

$$\sigma_a = \frac{Q_a}{4\pi r_a^2} \Rightarrow E_a = k_C \frac{Q_a}{r_a^2}, \quad \sigma_b = \frac{Q_b}{4\pi r_b^2} \Rightarrow E_b = k_C \frac{Q_b}{r_b^2}, \quad \frac{E_a}{E_b} = \frac{Q_a}{Q_b} \frac{r_b^2}{r_a^2} = \frac{r_b}{r_a}.$$

The field strength near the smaller sphere is greater. The effect of the spheres on each other is not taken into account here; for an estimate, however, the calculation is good enough. The field strength is greater, the smaller the radius of curvature is. If the field strength at a point $> 3 \times 10^6$ V, the air there is ionized and it leads to breakdown. The charge density is relatively high in the peaks because they are far away from the rest of the body and thereby the Coulomb energy becomes small. In the case of the *field electron microscope* and the *field ion microscope* the electrons emerge from the surface of the metal tip due to the strong field.

Fields at edges and corners

The field should only be specified near the edges, which in a two-dimensional model are peaks, as Fig. 2.29 for $\beta > \pi$ shows. In their vicinity, strong fields and thus high surface charges are expected, which the following calculation [Jackson, 1998, Section 2.11] confirms.

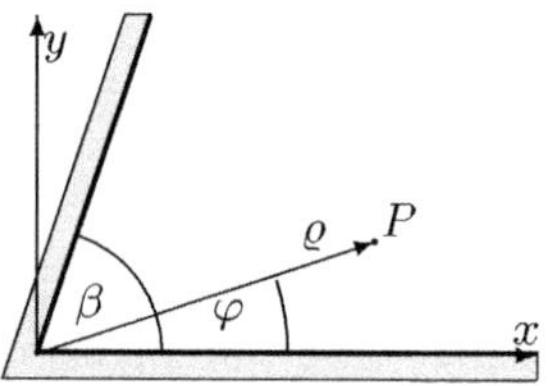

Fig. 2.29. Two conductive planes intersect at an angle β

In the exterior space, the two-dimensional Laplace equation is to be solved, preferably in plane polar coordinates, (A.3.20) and with a separation approach $\phi = k_c R(\varrho)\,\Psi(\varphi)$ results [Greiner, 1998, (2.28)]

$$\frac{1}{\varrho}\frac{\partial}{\partial\varrho}\left(\varrho\frac{\partial\phi}{\partial\varrho}\right) + \frac{1}{\varrho^2}\frac{\partial^2\phi}{\partial\varphi^2} = 0 \quad\Rightarrow\quad \frac{1}{R}\frac{\partial}{\partial\varrho}\left(\varrho\frac{\partial R}{\partial\varrho}\right) = \nu^2 = -\frac{1}{\Psi}\frac{1}{\varrho}\frac{\partial^2\Psi}{\partial\varphi^2}.$$

The solutions are

$$R(\varrho) = a_\nu\,\varrho^\nu + b_\nu\,\varrho^{-\nu} \qquad \text{and} \qquad \Psi(\varphi) = A_\nu\,\cos(\nu\varphi) + B_\nu\,\sin(\nu\varphi).$$

The case $\nu = 0$ must be treated separately:

$$R(\varrho) = a_0 + b_0\,\ln\varrho \qquad \text{and} \qquad \Psi(\varphi) = A_0 + B_0\,\varphi.$$

It is assumed that there are also no charges at the origin, i.e., that the Laplace equation applies there; thus all b_ν must vanish. The angular range of φ is limited to $0 \le \varphi \le \beta$ and the boundary condition states that $\phi(\varrho, 0) = \phi(\varrho, \beta)$. Then $B_0 = 0$ also applies and for ν the restriction $\nu = m\pi/\beta$ with $m \ge 1$ and integer. The general solution is therefore

$$\phi = \phi_0 + k_c \sum_{m=1}^{\infty} a_m\,\varrho^{m\pi/\beta}\,\sin\frac{m\pi\varphi}{\beta} \overset{\varrho\to 0}{\approx} \phi_0 + k_c a_1\,\varrho^{\pi/\beta}\,\sin\frac{\pi\varphi}{\beta}.$$

From this the fields result

$$E_\varrho = -\frac{\partial \phi}{\partial \varrho} \approx -k_C \frac{a_1 \pi}{\beta} \, \varrho^{\pi/\beta - 1} \, \sin \frac{\pi \varphi}{\beta},$$

$$E_\varphi = -\frac{1}{\varrho} \frac{\partial \phi}{\partial \varphi} \approx -k_C \frac{a_1 \pi}{\beta} \, \varrho^{\pi/\beta - 1} \, \cos \frac{\pi \varphi}{\beta}.$$

For $\beta \approx 2\pi$ one has in 2 dimensions a sharp tip with the fields

$$E_\varrho \approx -\frac{k_C a_1}{2\sqrt{\varrho}} \, \sin \frac{\varphi}{2}, \qquad\qquad E_\varphi \approx -\frac{k_C a_1}{2\sqrt{\varrho}} \, \cos \frac{\varphi}{2}.$$

The surface charge densities are the same on both surfaces, $\varphi = 0$ and $\varphi = \beta$

$$\sigma(\varrho) = \frac{E_\varphi(\varrho, 0)}{4\pi k_C} = -\frac{a_1}{4\beta} \, \varrho^{\pi/\beta - 1}.$$

2.4 Energy of the Electric Field

We determine the work that must be expended to create a configuration of finite extent with N point charges $(e_n, \mathbf{x}_n)$, where $n = 1, ..., N$. To do this, we bring the charges from infinity, where their interaction energy is zero, to the positions $\mathbf{x}_n$.

$$
\begin{array}{lll}
& \text{Work} & \text{Potential} \\
e_1 \to \mathbf{x}_1 & 0 & \phi_1(\mathbf{x}) \\
e_2 \to \mathbf{x}_2 & e_2\phi_1(\mathbf{x}_2) & \phi_1(\mathbf{x}) + \phi_2(\mathbf{x}) \\
e_N \to \mathbf{x}_N & e_N\left(\phi_1(\mathbf{x}_N) + \cdots + \phi_{N-1}(\mathbf{x}_N)\right) & \phi_1(\mathbf{x}) + \cdots + \phi_N(\mathbf{x})
\end{array}
$$

$\phi_n(\mathbf{x}) = k_C \dfrac{e_n}{|\mathbf{x} - \mathbf{x}_n|}$ is the Coulomb potential generated by n. The total work to be expended, i.e. the interaction energy of the configuration is

$$A = \sum_{n>m} e_n \phi_m(\mathbf{x}_n) = k_C \frac{1}{2} \sum_{n \neq m} \frac{e_n e_m}{|\mathbf{x}_n - \mathbf{x}_m|}. \tag{2.4.1}$$

This is the energy stored in this system. With point charges, whose charge density $\rho(\mathbf{x}) = \sum_n e_n \, \delta^{(3)}(\mathbf{x} - \mathbf{x}_n)$ is, we cannot write A in the form

$$A \stackrel{?}{=} \frac{1}{2} \int d^3x \, \rho(\mathbf{x})\phi(\mathbf{x}) = \frac{k_C}{2} \int d^3x \, d^3x' \, \frac{\rho(\mathbf{x})\rho(\mathbf{x}')}{|\mathbf{x} - \mathbf{x}'|}$$

because the self-energy of the point charges, which are the infinite $n=m$ terms, does not occur in (2.4.1). However, we try to represent A by continuous charge distributions.

We imagine the charges falling together in groups, within which the charge density can be replaced by a smeared one, as indicated in Fig. 2.30. Apparently, for the interaction energy of the charges belonging to different $\alpha \neq \beta$,

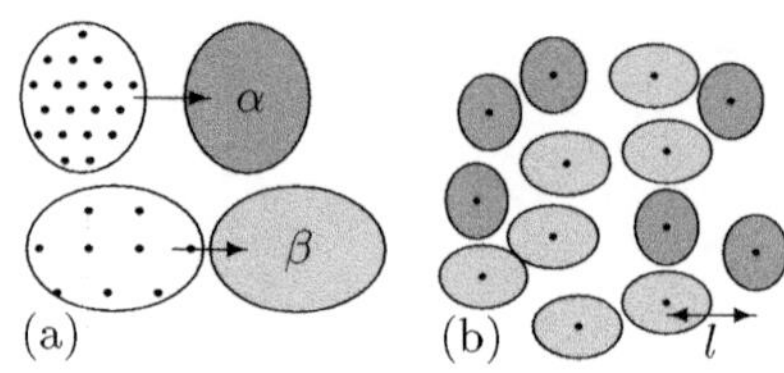

Fig. 2.30. (a) Point charges in two extended areas, α and β are grouped with smeared charge
(b) The average distance of the continuous charge distributions is l

$$A = \frac{1}{2} \sum_{\alpha,\beta\ \alpha\neq\beta} \int d^3x\, \rho_\alpha(\mathbf{x})\phi_\beta(\mathbf{x}).$$

For the interaction energy of the charges within such a charge cloud we can write $\frac{1}{2}\int d^3x\, \rho_\alpha(\mathbf{x})\phi_\alpha(\mathbf{x})$. However, we have made a mistake, since we count the self-interaction of a single smeared charge.

Estimation: The self-interaction part contributes about $N_\alpha\, e^2\, l^{-1}$ if l is the average extent of an area α.

The interaction energy of different areas $\alpha\neq\beta$ is indeed smaller due to the greater distance $l\, N_\alpha^{1/3}$, but the number with N_α^2 is significantly larger:

$$N_\alpha \frac{e^2}{l} \ll N_\alpha^2 \frac{e^2}{N_\alpha^{\frac{1}{3}} l} = \frac{N_\alpha^{\frac{5}{3}} e^2}{l}.$$

The self-interaction can therefore be neglected. Thus, the total energy, including the energy to build up charge distributions

$$U = \frac{1}{2}\int d^3x \sum_\alpha \rho_\alpha(\mathbf{x}) \sum_\beta \phi_\beta(\mathbf{x}) = \frac{1}{2}\int d^3x\, \rho(\mathbf{x})\phi(\mathbf{x})$$

$$\rho(\mathbf{x}) = \sum_\alpha \rho_\alpha(\mathbf{x}) \qquad \text{and} \qquad \phi(\mathbf{x}) = \sum_\beta \phi_\beta(\mathbf{x}).$$

Here, $\rho(\mathbf{x})$ contains all charges, including the induced ones:

$$U = \frac{k_C}{2}\int d^3x\, d^3x'\, \frac{\rho(\mathbf{x})\rho(\mathbf{x}')}{|\mathbf{x}-\mathbf{x}'|} = \frac{1}{2}\int d^3x\, \rho(\mathbf{x})\phi(\mathbf{x}). \tag{2.4.2}$$

The use of the Poisson equation yields

$$U = -\frac{1}{8\pi k_C}\int d^3x\,(\nabla^2\phi)\phi = -\frac{1}{8\pi k_C}\int d^3x\,\left[\nabla\cdot((\nabla\phi)\phi) - (\nabla\phi)^2\right].$$

The first term disappears, as can be seen by applying Gauss's theorem:

$$\oiint d\mathbf{a}\cdot(\nabla\phi)\phi \sim \int d\Omega\, R^2\, \frac{1}{R^2}\, \frac{1}{R} \to 0,$$

$$U = \frac{1}{8\pi k_C}\int d^3x\, \mathbf{E}^2(\mathbf{x}). \tag{2.4.3}$$

It is $\mathbf{E}^2 \geq 0$ and thus $U \geq 0$, as it contains the self-interaction of the charge distribution. The energy density is thus given by

$$u(\mathbf{x}) = \frac{1}{8\pi k_C} E^2(\mathbf{x}). \tag{2.4.4}$$

Note: In the first representations, the energy is represented as potential energy of the charges ρ_α in the potential ϕ_β of the others. This corresponds to the action-at-a-distance viewpoint. The last representation suggests the field-theoretical viewpoint, that the energy with the energy density $E^2/8\pi$ is stored in space. This is analogous to a stretched spring, to which the mass m is attached. The energy can be interpreted as higher potential energy of the mass m or as elastic energy of the stretched spring.

2.4.1 The Interaction Energy of two Charge Distributions

The energy of a charge distribution is given by (2.4.2) and (2.4.3) as

$$U = \frac{k_C}{2} \int d^3x\, d^3x' \frac{\rho(\mathbf{x})\,\rho(\mathbf{x}')}{|\mathbf{x} - \mathbf{x}'|} = \frac{1}{2} \int d^3x\, \rho(\mathbf{x})\,\phi(\mathbf{x}) = \frac{1}{8\pi k_C} \int d^3x\, E^2(\mathbf{x}).$$

If $\rho(\mathbf{x}) = \rho_1(\mathbf{x}) + \rho_2(\mathbf{x})$ consists of two (spatially separated) charge distributions, their potentials are given by $\phi(\mathbf{x}) = \phi_1(\mathbf{x}) + \phi_2(\mathbf{x})$. From this we obtain

$$U = \sum_{i,j=1}^{2} U_{ij} \qquad \text{with} \qquad U_{ij} = \frac{1}{2} \int d^3x\, \rho_i(\mathbf{x})\phi_j(\mathbf{x}). \tag{2.4.5}$$

This expression is divided into self-energy $U_s = U_{11} + U_{22}$ and interaction energy

$$U_w = U_{12} + U_{21} = \int d^3x\, \rho_1(\mathbf{x})\phi_2(\mathbf{x}). \tag{2.4.6}$$

$U_s > 0$ is positive, while U_w can also be negative. In (2.4.6) we used that $U_{12} = U_{21}$, which is sometimes referred to as *rReciprocity theorem* by Green. The proof is simple, as in (2.4.2) only $\mathbf{x}$ with $\mathbf{x}'$ has to be swapped.

The interaction energy of two point charges
The potential of 2 point charges is given by (2.2.1) through

$$\phi = \phi_1 + \phi_2 = \frac{k_C q_1}{|\mathbf{x} - \mathbf{x}_1|} + \frac{k_C q_2}{|\mathbf{x} - \mathbf{x}_2|}.$$

From this follows for the energy

$$U = \frac{1}{2} \int d^3x\, \rho(\mathbf{x})\,\phi(\mathbf{x})$$

$$= k_C \left\{ \frac{q_1^2}{2} \int d^3x\, \frac{\delta^{(3)}(\mathbf{x} - \mathbf{x}_1)}{|\mathbf{x} - \mathbf{x}_1|} + \frac{q_2^2}{2} \int d^3x\, \frac{\delta^{(3)}(\mathbf{x} - \mathbf{x}_2)}{|\mathbf{x} - \mathbf{x}_2|} + \frac{q_1 q_2}{|\mathbf{x}_1 - \mathbf{x}_2|} \right\}.$$

First, we notice the singular character of the self-energy of point charges, which is more evident in the alternative formulation of the energy

$$U = \frac{1}{8\pi k_C} \int d^3x\, E^2 \quad \text{with} \quad \mathbf{E} = -\boldsymbol{\nabla}\phi = k_C\left[q_1 \frac{\mathbf{x}-\mathbf{x}_1}{|\mathbf{x}-\mathbf{x}_1|^3} + q_2\frac{\mathbf{x}-\mathbf{x}_2}{|\mathbf{x}-\mathbf{x}_2|^3}\right].$$

For the self-energies, this results in

$$U_{ii} = \frac{q_1^2}{8\pi k_C}\int d^3x\, \frac{1}{|\mathbf{x}-\mathbf{x}_i|^4} = \lim_{\epsilon\to 0}\frac{q_i^2}{2k_C}\int_\epsilon^\infty dr\,\frac{1}{r^2} = \lim_{\epsilon\to 0}\frac{q_i^2}{2k_C\epsilon}.$$

They do not change upon approach of the two charges and correspond to the electrical part of the work that would have to be expended, to compress the charge q_i from an infinitely extended charge distribution to a sphere of radius ϵ, as can be seen from (2.4.8). The total energy U is positive definite because of the self-energies, while the interaction energy

$$U_w = 2U_{12} = \frac{k_C q_1 q_2}{|\mathbf{x}_1 - \mathbf{x}_2|}$$

is negative for charges with different signs, which is why then the two charges attract each other and the energy of the configuration is minimized upon approach. The result is not surprising in this form, since the force between two charges $\mathbf{F}_C = -\boldsymbol{\nabla}U_{12}$ is as expected the Coulomb force (1.2.1).

The calculation of the interaction energy from the field energy $E^2/8\pi k_C$ naturally brings the same result:

$$U_{12} = k_C\frac{q_1 q_2}{4\pi}\int d^3x\, \frac{(\mathbf{x}-\mathbf{x}_1)\cdot(\mathbf{x}-\mathbf{x}_2)}{|\mathbf{x}-\mathbf{x}_1|^3\,|\mathbf{x}-\mathbf{x}_2|^3} \qquad \begin{cases}\mathbf{x}' = \mathbf{x}-\mathbf{x}_1 \\ \mathbf{d} = \mathbf{x}_1 - \mathbf{x}_2\end{cases} \Rightarrow\ \mathbf{x}-\mathbf{x}_2 = \mathbf{x}'+\mathbf{d}$$

$$= k_C\frac{q_1 q_2}{4\pi}\int d^3x'\, \frac{\mathbf{x}'\cdot(\mathbf{x}'+\mathbf{d})}{r'^3\,|\mathbf{x}'+\mathbf{d}|^3} \qquad\qquad \text{Spherical coordinates}$$

$$= k_C\frac{q_1 q_2}{2}\int_{-1}^1 d\xi \int_0^\infty dr'\, \frac{r' + d\xi}{\left(r'^2 + d^2 + 2r'd\xi\right)^{3/2}}.$$

The integral is solvable, as the numerator contains the inner derivative of the denominator:

$$U_{12} = k_C\frac{q_1 q_2}{2}\int_{-1}^1 d\xi\, \frac{-1}{\left(r'^2 + d^2 + 2r'd\xi\right)^{1/2}}\Bigg|_0^\infty = k_C\frac{q_1 q_2}{d}.$$

2.4.2 The Self-Energy of a Homogeneously Charged Sphere

First, we calculate the field of a homogeneously charged sphere with radius R and the charge density ρ_0, which due to symmetry must be of the form

$\mathbf{E} = E\mathbf{e}_r$. $\mathbf{E}$ can then be determined using Gauss's law (1.3.2), where the integration volume should be a sphere S_r with radius r:

$$\oiint_{\partial S_r} \mathrm{d}\mathbf{a} \cdot \mathbf{E} = 4\pi r^2 E = 4\pi k_C \int_{S_r} \mathrm{d}^3 x\, \rho_0\, \theta(R-r) = 4\pi k_C \begin{cases} Q\, r^3/R^3 & r < R \\ Q & r \geq R. \end{cases}$$

$Q = \rho_0\, 4\pi R^3/3$ is the total charge of the sphere S_R. We distinguish between $\mathbf{E}^{(i)}$, the field inside, and $\mathbf{E}^{(a)}$, the field outside the sphere:

$$E = \begin{cases} E^{(i)} = k_C Q r/R^3 & r < R \\ E^{(a)} = k_C Q/r^2 & r \geq R. \end{cases}$$

By integration, we find the potentials

$$\phi^{(i)}(r) = k_C\left[-Q\frac{r^2}{2R^3} + C(R) \right] \qquad \text{and} \qquad \phi^{(a)}(r) = k_C Q\frac{1}{R}.$$

We set the integration constant in the outer space to zero and determine $C(R)$ from the continuity of the potentials on the sphere surface:

$$\phi^{(i)}(R) = \phi^{(a)}(R) \qquad \Rightarrow \qquad C(R) = \frac{3Q}{2R}.$$

For potential and field, this results in:

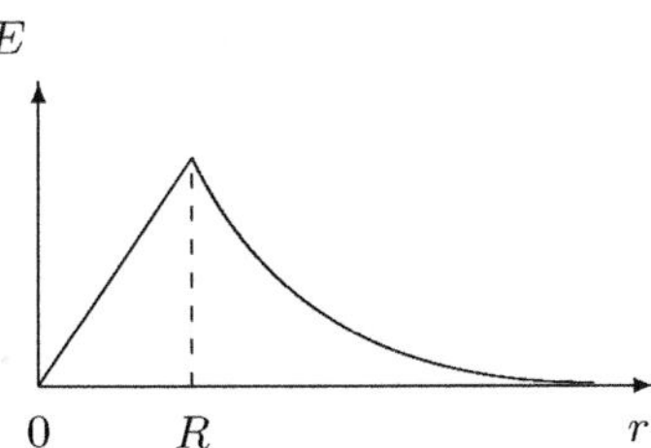

Fig. 2.31. Sketched field course for a homogeneous sphere with radius R

$$\phi(r) = k_C\left(\frac{3Q}{2R} - \frac{Qr^2}{2R^3} \right)\theta(R-r) + k_C\frac{Q}{r}\theta(r-R),$$

$$\mathbf{E} = k_C\left[\frac{Q}{R^3}\,\theta(R-r) + \frac{Q}{r^3}\,\theta(r-R) \right]\mathbf{x}, \qquad (2.4.7)$$

$$\nabla \cdot \mathbf{E} = k_C\frac{3Q}{R^3}\,\theta(R-r) = 4\pi k_C \rho_0\, \theta(R-r).$$

Fig. 2.31 shows the field course. For the energy inside and outside the sphere, one obtains from (2.4.7)

$$U^{(i)} = \frac{1}{8\pi k_C} \int_V \mathrm{d}^3x\, E^2 = \frac{k_C}{8\pi}\left(\frac{Q}{R^3}\right)^2 4\pi \int_0^R \mathrm{d}r\, r^4 = k_C \frac{Q^2}{2R^6}\frac{R^5}{5} = k_C\frac{Q^2}{10R},$$

$$U^{(a)} = k_C\frac{Q^2}{8\pi}\,4\pi \int_R^\infty \mathrm{d}r\,\frac{1}{r^2} = k_C\frac{Q^2}{2}\frac{1}{R}. \tag{2.4.8}$$

The total energy (self-energy) of the sphere is then

$$U = U^{(i)} + U^{(a)} = k_C\frac{Q^2}{R}\left(\frac{1}{10}+\frac{1}{2}\right) = k_C\frac{3Q^2}{5R}. \tag{2.4.9}$$

The self-energy of a sphere diverges with $R \to 0$. The proportion of the outer space is – regardless of the sphere radius – by a factor of 5 larger than that of the inside of the sphere. An internal structure would add contributions to the fields both inside and outside (multipole moments). However, there is no experimental evidence of a spatial structure of the charge in the electron. Since according to the Einstein formula $E = m_e c^2$, see (14.1.2), the electrostatic energy of a particle cannot become arbitrarily large, since this must be attributed to the mass, one can also make an estimate for the "radius of the electron" with

$$k_C\frac{3e_0^2}{5r_e} < k_C\frac{e_0^2}{r_e} = m_e c^2 \qquad \Longrightarrow \qquad r_e = k_C\frac{e_0^2}{m_e c^2} \approx 2.8\times 10^{-13}\,\mathrm{cm}.$$

r_e is the classical electron radius and m_e is the rest mass of the electron. r_e^2 gives the order of magnitude of the scattering cross section when light is scattered by electrons (see section 11.1.1). In classical models, the reverse path is usually taken by assigning to a radius a the mass corresponding to the electrostatic self-energy

$$m_{es} = U/c^2 = k_C e_0^2/2ac^2. \tag{2.4.10}$$

Here it was assumed that the charge is homogeneously distributed on the sphere surface, since then $U^{(i)} = 0$.

Note: The self-energy of the homogeneous sphere can of course also be calculated with

$$U = \frac{k_C}{2}\int \mathrm{d}^3x\,\mathrm{d}^3x'\,\frac{\rho(\mathbf{x})\rho(\mathbf{x}')}{|\mathbf{x}-\mathbf{x}'|} = \frac{1}{2}\int \mathrm{d}^3x\,\rho(\mathbf{x})\,\phi(\mathbf{x}),$$

where $\phi = \phi^{(i)}$. The calculation of $\phi^{(i)}$ results in

$$\phi^{(i)}(\mathbf{x}) = k_C\int_V \mathrm{d}^3x'\,\frac{\rho(\mathbf{x}')}{|\mathbf{x}-\mathbf{x}'|} \overset{\text{(B.5.23)}}{=} k_C\frac{3Q}{R^3}\left\{\frac{1}{r}\int_0^r \mathrm{d}r'\,r'^2 + \int_r^R \mathrm{d}r'\,r'\right\} = k_C\left[\frac{3Q}{2R} - \frac{Qr^2}{2R^3}\right].$$

2.4.3 Thomson's Theorem

Thomson's theorem[1] states that for n conductors, which are in a fixed (unchangeable) position and on which a certain total charge has been applied, the electrostatic energy of the system is an absolute minimum when the charges are distributed so that each surface represents an equipotential surface.

Physically, this is understandable if we imagine the charges on the conductors in any arrangement fixed. The moment the fixation is lifted, the charges within the conductors will distribute themselves so that they are force-free ($\mathbf{E}(\mathbf{x}) = 0$ for $\mathbf{x} \in L_i$), leading to a reduction of the field energy. Within each conductor, ϕ is then constant.

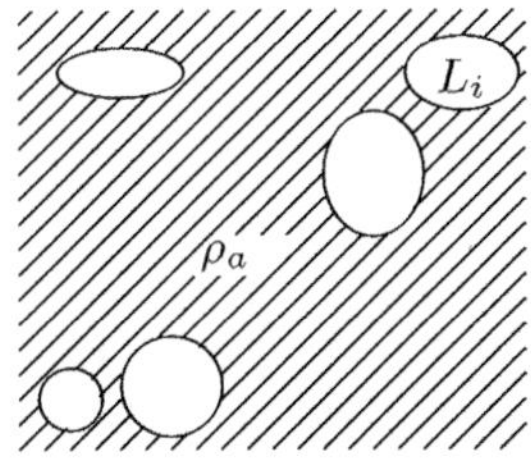

Fig. 2.32. n conductors with the volumes L_i and the total charges Q_i. Outside the conductors is the charge distribution ρ_a

There are n conductors with the volumes L_i. The configuration of the conductors and their charges Q_i are given, as indicated in Fig. 2.32. Outside of the conductor, there is still a solid external charge distribution with the potential ϕ_a. The energy of this arrangement is given by

$$U = \frac{k_C}{2} \sum_{i=1}^{n} \int_{L_i} \mathrm{d}^3 x_i\, \rho_i(\mathbf{x}_i) \left\{ \sum_{j=1}^{n} \int_{L_j} \mathrm{d}^3 x_j\, \frac{\rho_j(\mathbf{x}_j)}{|\mathbf{x}_i - \mathbf{x}_j|} + \phi_a(\mathbf{x}_i) \right\}. \qquad (2.4.11)$$

The 1st term is the energy of the charges on the surfaces of the conductors and the 2nd term is the interaction with the external potential ϕ_a.

The given total charges on the surfaces are

$$\int_{L_i} \mathrm{d}^3 x\, \rho_i(\mathbf{x}) = Q_i\,.$$

We now minimize the energy by varying the charge densities at fixed charges Q_i. ϕ_i are the Lagrange multipliers associated with this constraint:

$$\frac{\delta}{\delta \rho_k(\mathbf{x}_k)} \left\{ \frac{k_C}{2} \sum_{i=1}^{n} \int_{L_i} \mathrm{d}^3 x_i\, \rho_i(\mathbf{x}_i) \left\{ \sum_{j=1}^{n} \int_{L_j} \mathrm{d}^3 x_j\, \frac{\rho_j(\mathbf{x}_j)}{|\mathbf{x}_i - \mathbf{x}_j|} + \phi_a(\mathbf{x}_i) + \phi_i \right\} \right\} = 0.$$

This results, solved for ϕ_k

---

[1] W. Thomson, later Lord Kelvin, 1848; [see Maxwell, 1892, Art. 100]

$$\phi_k = \phi_a(\mathbf{x}_k) + \sum_{j=1}^{n} k_C \int_{L_j} \mathrm{d}^3 x_j \, \frac{\rho_j(\mathbf{x}_j)}{|\mathbf{x}_k - \mathbf{x}_j|} \qquad \mathbf{x}_k \in L_k \,. \qquad (2.4.12)$$

The Lagrange multiplier ϕ_k has the form of the potential on the conductor k. Since ϕ_k is now a constant for the conductor k regardless of $\mathbf{x}_k$, the potential is the same throughout the volume. The configuration with minimal (extremal) energy is therefore the one in which the charges distribute themselves so that on each conductor i the potential ϕ_i is constant.

Capacitance coefficients

If there are no charges outside the conductors, i.e. $\phi_a = 0$, and the charges Q_i on the conductors are given, then the energy (2.4.11) is

$$U = \frac{1}{2} \sum_{i=1}^{n} \int_{L_i} \mathrm{d}^3 x_i \, \rho_i(\mathbf{x}_i) \sum_{j=1}^{n} \phi_j(\mathbf{x}_i) = \frac{1}{2} \sum_{i=1}^{n} Q_i \, \phi_i \,.$$

We now calculate ϕ_i using the principle of superposition, by first calculating the potential of the configuration in which only the conductor k carries a charge:

$$\phi_i = \sum_{j=1}^{n} \phi_j(\mathbf{x}_i) \qquad \overset{Q_j = Q_k \delta_{jk}}{\Longrightarrow} \qquad \phi_i^{(k)} = \phi_k(\mathbf{x}_i) \qquad \mathbf{x}_i \in L_i \,.$$

If one now sums over all configurations $k = 1$ to n, one again obtains ϕ_i, but recognizes, that the individual summands are constant. Thus, the potentials and subsequently the energy can be brought into the form

$$\phi_i = \sum_{k=1}^{n} \phi_i^{(k)} = \sum_{k=1}^{n} \Gamma_{ik} Q_k \qquad \Longrightarrow \qquad U = \frac{1}{2} \sum_{i,k=1}^{n} Q_i \Gamma_{ik} Q_k. \qquad (2.4.13)$$

The *potential coefficients* Γ_{ik} will be discussed in more detail, where it is advantageous to introduce normalized particle densities $\hat{\rho}_j = \rho_j / Q_j$ and the corresponding potentials $\hat{\phi}_j = \phi_j / Q_j$

$$\hat{\phi}_j(\mathbf{x}_i) = k_C \int_{L_j} \mathrm{d}^3 x_j' \, \frac{\hat{\rho}_j(\mathbf{x}_j')}{|\mathbf{x}_i - \mathbf{x}_j'|} = \Gamma_{ij} \qquad \mathbf{x}_i \in L_i.$$

Of particular interest is the matrix C inverse to Γ:

$$U = \frac{1}{2} \sum_{i,j=1}^{n} \phi_i C_{ij} \phi_j \,. \qquad (2.4.14)$$

The C_{ij} are the so-called capacitance coefficients. Both the capacitance and the potential coefficients are symmetric, $C_{ij} = C_{ji}$. This is a consequence of the symmetry of the summands of (2.4.11):

$$U_{ij} = \frac{k_C}{2} \int d^3x\, d^3x'\, \frac{\rho_i(\mathbf{x})\rho_j(\mathbf{x}')}{|\mathbf{x}-\mathbf{x}'|} = \frac{1}{2} \int d^3x\, \rho_i(\mathbf{x})\phi_j(\mathbf{x}) = \frac{1}{2} \int d^3x\, \rho_j(\mathbf{x})\phi_i(\mathbf{x}).$$

The right side $(\rho_i\phi_j \rightleftharpoons \rho_j\phi_i)$ is the reciprocity theorem of Green.

2.5 Multipole Expansion

We start from a localized charge distribution and consider its far field. For

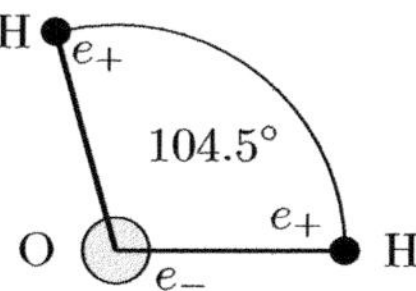

Fig. 2.33. Charge distribution in the water molecule. $e_+ \approx 0.4e_0$ and $e_- \approx -0.8e_0$

$\rho(\mathbf{x}')$ it should hold that $\rho(\mathbf{x}') = 0$ for $r' > R$. In atoms and molecules, the charge distributions are often not isotropic. The anisotropy is primarily manifested by an intrinsic dipole moment, as with the H_2O molecule (see Fig. 2.33). Due to the greater electronegativity of the O atom, a negative space charge forms around it and a positive one around the H atom. Molecules with ionic bonding always have an electric dipole moment, such as HCl[2]. The dipole moments of most molecules are smaller than 10 Debye.

In a systematic development, ϕ will be ordered according to the contributions of the moments of $\rho(\mathbf{x}')$ by inserting into

$$\phi(\mathbf{x}) = k_C \int d^3x'\, \frac{\rho(\mathbf{x}')}{|\mathbf{x}-\mathbf{x}'|} \tag{2.5.1}$$

a Taylor expansion of $1/|\mathbf{x}-\mathbf{x}'|$ for $r > R$. However, this is not the only way to define multipole moments. Especially when higher orders play a role, one expands $1/|\mathbf{x}-\mathbf{x}'|$ according to spherical functions. This expansion is only dealt with in section 3.3.3, where the necessary mathematical prerequisites are available. There, one has an expansion according to r'^l – as is ultimately also the Taylor expansion –, so that the moments of the same order can be transformed into each other by linear combinations (see (3.3.12)).

2.5.1 Multipole Moments of a Charge Distribution

The Taylor expansion of $1/|\mathbf{x}-\mathbf{x}'|$ for $r > R$, inserted into (2.5.1), results in

$$\phi(\mathbf{x}) = \sum_{n=0}^{\infty} \phi^{(n)}(\mathbf{x}) = k_C \frac{q}{r} + k_C \sum_{n=1}^{\infty} \frac{(-1)^n}{n!} M_{i_1\ldots i_n} \nabla_{i_1}\ldots\nabla_{i_n} \frac{1}{r}, \tag{2.5.2}$$

$$M_{i_1\ldots i_n} = \int d^3x'\, \rho(\mathbf{x}')\, x'_{i_1}\ldots x'_{i_n}\,. \tag{2.5.3}$$

---

[2] H_2O : $p = 1.84\,\text{Debye} = 6.14 \times 10^{-30}\,\text{Cm}$; HCl: $p = 1.03\,\text{Debye} = 3.44 \times 10^{-30}\,\text{Cm}$.

The entire potential is approximated by the superposition of the potentials of point charges, dipoles, etc. located at the origin.

Monopole

The zeroth moment is the total charge. Thus, one obtains for the first termof the Taylor expansion (2.5.2)

$$\phi^{(0)}(\mathbf{x}) = \phi_e(\mathbf{x}) = k_C \frac{q}{r}, \qquad\qquad q = \int d^3x'\, \rho(\mathbf{x}'),$$

$$E_i = -k_C q \nabla_i \frac{1}{r} = k_C q \frac{x_i}{r^3}. \tag{2.5.4}$$

Dipole

The second term of (2.5.2) determines the contribution of the dipole moment to the potential

$$\phi^{(1)} = \phi_p = -k_C p_j \nabla_j \frac{1}{r} \qquad\qquad p_i = \int d^3x'\, \rho(\mathbf{x}')\, x_i'$$

$$E_i = k_C p_j \nabla_i \nabla_j \frac{1}{r} = k_C \frac{p_j}{r^3}\left(3\frac{x_i x_j}{r^2} - \delta_{ij}\right), \tag{2.5.5}$$

which agrees with (2.2.5). The dipole field (2.5.5) is singular for $r=0$ and can be there treated separately. If a sphere S_a with radius a is placed around the origin, the total contribution of the dipole field within the sphere is given by

$$\int_{S_a} d^3x\, E_i(\mathbf{x}) = k_C p_j \int_{S_a} d^3x\, \nabla_i \nabla_j \frac{1}{r} = k_C \frac{p_i}{3} \int_{S_a} d^3x\, \nabla^2 \frac{1}{r}.$$

The same contribution independent of a is obtained by replacing $\mathbf{E}$ with:

$$\mathbf{E}(\mathbf{x}) = k_C \frac{\mathbf{p}}{3} \Delta \frac{1}{r} = -k_C \frac{4\pi}{3}\mathbf{p}\,\delta^{(3)}(\mathbf{x}).$$

The dipole field can thus be written as [see Jackson, 1998, (4.20)]:

$$\mathbf{E} = \lim_{a\to 0} k_C \left[3\frac{(\mathbf{p}\cdot\mathbf{x})\mathbf{x}}{r^5} - \frac{\mathbf{p}}{r^3}\right]\theta(r-a) - k_C \frac{4\pi}{3}\mathbf{p}\,\delta^{(3)}(\mathbf{x}). \tag{2.5.6a}$$

Notes: Naturally, $\mathbf{E}$ must also be curl-free in the form of (2.5.6a) and comply with Gauss's law. It is the contributions of the step function that ensure this, as is to be shown in problem 2.17.

The inner area does not have to be a sphere; it is sufficient to demand the inversion symmetry of the area so that the integrand $\nabla_i \nabla_j \frac{1}{r}$ for $i \neq j$ disappears. Instead of the step function, for the first term one can take the principal value P (see section B.1.2, p. 606) which excludes an infinitesimal sphere around the location of the dipole during integration [Fließbach, 2022, (12.33)].

$\mathbf{E}$ can be explained more vividly if one starts from a dipole consisting of two charges $\pm e$ fixed at a distance d as outlined in Fig. 2.4. A sphere with radius $a = d/2$ around the location of the dipole divides the field into an inner area with a strong field between the two charges and an outer area, which for $r > a$ in linear approximation gives a dipole field from $\mathbf{d}$. In this approximation, one obtains the (singular) field in the sphere [Brandt, Dahmen, 2004, Section 2.10.1]

$$\mathbf{E} = -k_C e \frac{\mathbf{d}}{a^3}\theta(a-r) \xrightarrow{a\to 0;\, ed=p} -k_C \frac{4\pi}{3}\mathbf{p}\delta^{(3)}(\mathbf{x}). \tag{2.5.6b}$$

Quadrupole

The next contribution to ϕ is provided by the quadrupole moment of $\rho(\mathbf{x})$

$$\phi^{(2)}(\mathbf{x}) = \phi_Q(\mathbf{x}) = \frac{k_C}{2}M_{ij}\nabla_i\nabla_j\frac{1}{r}, \tag{2.5.7}$$

$$M_{ij} = \int \mathrm{d}^3x'\, \rho(\mathbf{x}')\, x_i'x_j', \qquad\qquad M = \sum_i M_{ii}. \tag{2.5.8}$$

One can add diagonal terms $\delta_{ij} M$ to the M_{ij} without changing $\phi^{(2)}$ because

$$\delta\phi^{(2)} = k_C M\delta_{ij}\nabla_i\nabla_j\frac{1}{r} = k_C M\Delta\frac{1}{r} = 0 \qquad \text{for } r > R.$$

Instead of the M_{ij}, traceless quadrupole moments Q_{ij} are used:

$$\phi_Q = \frac{k_C}{6}Q_{kl}\nabla_k\nabla_l\frac{1}{r} = k_C\frac{Q_{kl}}{6}\frac{3x_kx_l - r^2\delta_{kl}}{r^5} = k_C\frac{Q_{kl}}{2}\frac{x_kx_l}{r^5},$$

$$Q_{kl} = 3M_{kl} - \delta_{kl}M = \int \mathrm{d}^3x'\, \rho(\mathbf{x}')\,(3x_k'x_l' - r'^2\,\delta_{kl}), \tag{2.5.9}$$

$$E_i = -\frac{k_C}{6}Q_{kl}\nabla_i\nabla_k\nabla_l\frac{1}{r} = k_C\frac{Q_{kl}}{2}\frac{5x_ix_kx_l - r^2(\delta_{kl}x_i+\delta_{il}x_k+\delta_{ik}x_l)}{r^7}$$

$$= k_C\frac{Q_{kl}}{2}\frac{5x_ix_kx_l}{r^7} - k_C Q_{ik}\frac{x_k}{r^5}.$$

In tensor notation, the field reads

$$\mathbf{E} = \frac{5k_C}{2r^7}(\mathbf{x}^T Q\mathbf{x})\mathbf{x} - \frac{k_C}{r^5}(Q\mathbf{x}). \tag{2.5.9'}$$

For charges of one sign (charge density of the atomic nucleus, mass density in gravitation) one can achieve by a suitable choice of the reference point that $\mathbf{p} = 0$. Then the quadrupole moment is a measure of the deviation from spherical symmetry.

When brought to diagonal form, the quadrupole tensor has only two independent elements due to $\sum_k Q_{kk} = 0$. If the charge density is additionally axially symmetric, only one element remains, the quadrupole moment

$Q = Q_{33}$. An axially symmetric, elongated, positive charge distribution has $Q = Q_{33} > 0$; for a flat distribution, $Q < 0$ would be.

 Examples of quadrupole moments of atomic nuclei

$$\frac{1}{e_0}Q_{\text{Deut}} = 2.87 \times 10^{-27} \text{cm}^2 \quad \frac{1}{e_0}Q_{\text{Lu}^{176}} = 8 \times 10^{-24} \text{cm}^2 \quad \frac{1}{e_0}Q_{\text{Bi}^{203}} = -4 \times 10^{-25} \text{cm}^2.$$

Higher multipole moments

The multipole moments resulting from the Taylor expansion are generally only used up to the quadrupole, because the number of terms to be summed increases much faster than the number of independent moments. One usually then takes the multipole moments defined by the expansion after Y_{lm}. "Traceless" octupole moments are

$$\phi^{(3)}(\mathbf{x}) = -\frac{k_C}{3!\,5}\, o_{ijk}\, \nabla_i \nabla_j \nabla_k \frac{1}{r}$$

$$= \frac{k_C}{10} o_{ijk}\left[5x_i x_j x_k - r^2(\delta_{ij}x_k + \delta_{ik}x_j + \delta_{jk}x_i)\right]\frac{1}{r^7} = \frac{k_C}{2} o_{ijk}\frac{x_i x_j x_k}{r^7}$$

$$o_{ijk} = 5M_{ijk} - (\delta_{ij}o_k + \delta_{ik}o_j + \delta_{jk}o_i) \tag{2.5.10}$$

$$= \int d^3x'\, \rho(\mathbf{x}')\left[5x_i' x_j' x_k' - r'^2(\delta_{ij}x_k' + \delta_{ik}x_j' + \delta_{jk}x_i')\right],$$

where $o_k = \sum_i M_{iik}$ and $\sum_i o_{iik} = 0$. The potential, expanden around the origin according to multipole moments up to the 2nd order, is

$$\phi(\mathbf{x}) = k_C\left\{\frac{q}{r} + \mathbf{p}\cdot\frac{\mathbf{x}}{r^3} + \frac{1}{2}Q_{kl}\frac{x_k x_l}{r^5} + \dots\right\} \tag{2.5.11}$$

Change of the spatial point around which it is expanded

If one moves the reference point of the expansion of $\rho(\mathbf{x})$ according to multipole moments from the origin to $\mathbf{x}_0$, then (2.5.1) becomes

$$\phi(\mathbf{x}) = k_C\int d^3x'\, \frac{\rho(\mathbf{x}')}{|\underbrace{\mathbf{x}-\mathbf{x}_0}_{\bar{\mathbf{x}}} - \underbrace{(\mathbf{x}'-\mathbf{x}_0)}_{\mathbf{x}''}|} = k_C\left\{\frac{q}{\bar r} - \bar{\mathbf{p}}\cdot\nabla\frac{1}{\bar r} + \frac{1}{6}\bar Q_{kl}\nabla_k\nabla_l\frac{1}{\bar r} + \dots\right\}$$

with the moments

$$\bar{\mathbf{p}} = \int d^3x'\, \rho(\mathbf{x})\,\mathbf{x}'' = \mathbf{p} + q\,\mathbf{x}_0,$$

$$\bar Q_{kl} = \int d^3x'\, \rho(\mathbf{x}')\,(3x_k'' x_l'' - r''^2)$$

$$= Q_{kl} + 3(p_k x_{0l} + p_l x_{0k} + q x_{0k} x_{0l}) - 2\mathbf{p}\cdot\mathbf{x}_0 - q r_0^2.$$

If the total charge ($q=0$) disappears, the dipole moment is independent of the reference point ($\bar{\mathbf{p}}=\mathbf{p}$). If $\mathbf{p}=0$ as well, then $\bar Q_{kl}=Q_{kl}$. It applies in general:

The lowest non-vanishing multipole moment is independent of the reference point.

In a charge distribution with $q \neq 0$ one can with $\bar{\mathbf{p}} = \mathbf{p} + q\mathbf{x}_0 = 0$ place the reference point so that the dipole moment disappears. This is analogous to the center of mass of the charge distribution. In the mass distribution, the quadrupole moment then indicates the deviation from spherical symmetry which in a charge distribution only applies to the excess of positive (negative) charge. In ions (Na^+) with a largely spherical symmetry, the centers of charge and mass will coincide, in others ($(OH)^-$) they will not. If one is interested in the electrical forces acting on an atom or molecule, the center of mass, which can differ from that of the charge, especially in molecules, is more suitable for the multipole expansion.

2.5.2 Energy of a Charge Distribution in an External Field

Given are a spatially limited charge distribution $\rho(\mathbf{x})$ and an external potential $\phi^e(\mathbf{x})$. Within the charge distribution, the change of $\phi^e(\mathbf{x})$ is so moderate that a Taylor expansion seems reasonable. The energy of the charge distribution in the external field is then [Jackson, 1998, (4.21)]

$$U = \int d^3x\, \rho(\mathbf{x})\, \phi^e(\mathbf{x}) = \int d^3x\, \rho(\mathbf{x}) \Big[\phi^e(0) + \mathbf{x}\cdot\boldsymbol{\nabla}\phi^e(0) + \frac{1}{2}x_i x_j \nabla_i \nabla_j \phi^e(0)\Big]$$

$$= q\,\phi^e(0) - \mathbf{p}\cdot\mathbf{E}^e(0) - \frac{1}{6}\,Q_{ij}\,E^e_{i,j}(0). \tag{2.5.12}$$

Q_{ij}, the second moment of the charge distribution, is called quadrupole moment, which couples to the gradient of the external field

$$E^e_{i,j} = \nabla_j E^e_i = E^e_{j,i} = -\phi^e_{,ij} = -\nabla_i \nabla_j \phi^e.$$

We obtain for the second order

$$Q_{ij}\, E^e_{i,j}(0) = \int d^3x\, \rho(\mathbf{x})\, \big(3x_i x_j - \delta_{ij} r^2\big) E^e_{i,j}(0).$$

The last term, which does not appear in (2.5.12), we have added. It does not contribute

$$-\delta_{ij}\, E^e_{i,j} = \Delta\phi^e = 0,$$

since there are no external charges in the volume V, and the field thus satisfies the Laplace equation, and trace-free quadrupole moments defined in this way are identical to (2.5.9).

Note: According to (2.4.2) one might think that in (2.5.12) the factor $1/2$ is missing. However, according to the rule there, it would be

$$U = \frac{1}{2}\int d^3x' \big[\rho(\mathbf{x}')\,\phi^e(\mathbf{x}') + \rho^e(\mathbf{x}')\,\phi(\mathbf{x}')\big],$$

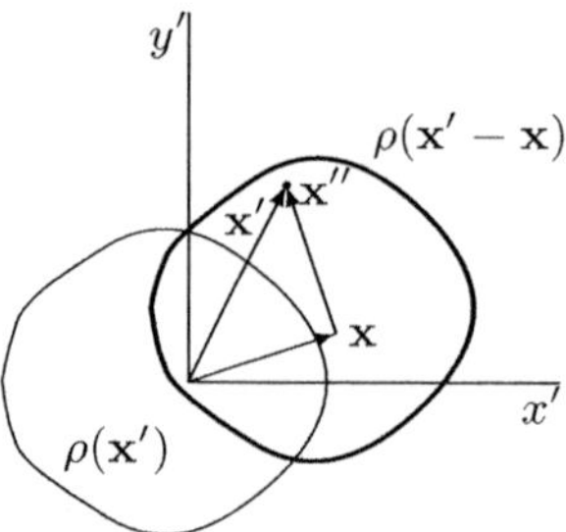

Fig. 2.34. The charge distribution $\rho(\mathbf{x}')$ is shifted by $\mathbf{x}$, where $\mathbf{x}'' = \mathbf{x}' - \mathbf{x}$

which exactly yields (2.5.12), as the two contributions are equal in size.

If the charge distribution concentrated around the origin, which we now denote by $\rho(\mathbf{x}')$, is shifted to the position $\mathbf{x}$ as outlined in Fig. 2.34, we have at $\mathbf{x}'$ the charge distribution $\rho(\mathbf{x}'-\mathbf{x})$. The energy of this charge distribution in the external potential is

$$U(\mathbf{x}) = \int \mathrm{d}^3 x' \, \rho(\mathbf{x}' - \mathbf{x}) \, \phi^e(\mathbf{x}') = q \, \phi^e(\mathbf{x}) - \mathbf{p} \cdot \mathbf{E}^e(\mathbf{x}) - \frac{1}{6} Q_{ij} \, E^e_{i,j}(\mathbf{x}), \quad (2.5.13)$$

where the calculation runs completely analogous to (2.5.12).

Interaction of two dipoles

As an example, the interaction of two dipoles $\mathbf{p}_1, \mathbf{x}_1$ and $\mathbf{p}_2, \mathbf{x}_2$ can be considered. Dipole 1 generates the field (2.2.4) at location $\mathbf{x}$

$$\mathbf{E}(\mathbf{x}) = \frac{k_C}{|\mathbf{x} - \mathbf{x}_1|^3} \left(-\mathbf{p}_1 + \frac{3 \, \mathbf{p}_1 \cdot (\mathbf{x} - \mathbf{x}_1)(\mathbf{x} - \mathbf{x}_1)}{|\mathbf{x} - \mathbf{x}_1|^2} \right).$$

The interaction energy of a dipole in the field $\mathbf{E}$ is (2.5.13)

$$U_{DD} = -\mathbf{E}(\mathbf{x}_2) \cdot \mathbf{p}_2 = k_C \left\{ \frac{\mathbf{p}_1 \cdot \mathbf{p}_2}{|\mathbf{x}_1 - \mathbf{x}_2|^3} - 3 \frac{\mathbf{p}_1 \cdot (\mathbf{x}_2 - \mathbf{x}_1) \, \mathbf{p}_2 \cdot (\mathbf{x}_2 - \mathbf{x}_1)}{|\mathbf{x}_1 - \mathbf{x}_2|^5} \right\}. \quad (2.5.14)$$

As can be seen from Tab.2.1, the dipole-dipole energy is smallest, when the two dipoles are parallel to each other and to $\mathbf{x}_1 - \mathbf{x}_2$. This applies to barium titanate ($\mathrm{Ba\,Ti\,O_3}$), a crystal with perovskite structure. Substances that have parallel aligned electric dipole moments are referred to as ferroelectrics. The force acting on a dipole $\mathbf{p}$ in the field $\mathbf{E}$ is[3]

$$\mathbf{F} = -\frac{\partial U}{\partial \mathbf{x}} = \boldsymbol{\nabla}(\mathbf{p} \cdot \mathbf{E}) = (\mathbf{p} \cdot \boldsymbol{\nabla})\mathbf{E} + \mathbf{p} \times (\boldsymbol{\nabla} \times \mathbf{E}), \quad (2.5.15)$$

---

[3] $\left(\mathbf{p} \times (\boldsymbol{\nabla} \times \mathbf{E})\right)_i = \epsilon_{ijk} p_j \, \epsilon_{klm} \nabla_l E_m = (\delta_{il}\delta_{jm} - \delta_{im}\delta_{jl}) p_j \nabla_l E_m = p_j \nabla_i E_j - p_j \nabla_j E_i.$
Written in vector form, this is $\mathbf{p} \times (\boldsymbol{\nabla} \times \mathbf{E}) = \boldsymbol{\nabla}(\mathbf{p} \cdot \mathbf{E}) - (\mathbf{p} \cdot \boldsymbol{\nabla}) \mathbf{E}.$

Tab. 2.1. Dipole-dipole energy for $\perp$ and $\parallel$ orientation of the dipoles on the connecting line

$\mathbf{p}_1$	$\mathbf{p}_2$	$\mathbf{x}_1 - \mathbf{x}_2$	U_{DD} measured in units : $U_d = p_1\,p_2/r_{12}^3$
$\uparrow$	$\uparrow$	$\rightarrow$	$U_d > 0$
$\uparrow$	$\downarrow$	$\rightarrow$	$-U_d < 0$
$\rightarrow$	$\rightarrow$	$\rightarrow$	$-2U_d < 0$ Ferroelectricity
$\rightarrow$	$\leftarrow$	$\rightarrow$	$+2U_d > 0$
$\uparrow$	$\rightarrow$	$\rightarrow$	$0 \rightarrow 0$

where the last term disappears because $\boldsymbol{\nabla}\times\mathbf{E} = 0$; alternatively, you can also use the symmetry resulting from the interchange of derivatives $E_{k,i} = E_{i,k} = -\phi_{,ik}$ for the transformation of (2.5.15). The torque acting on $\mathbf{p}$ is defined by

$$\mathbf{N} = \mathbf{p}\times\mathbf{E}. \tag{2.5.16}$$

Note: In mechanics, the torque $\mathbf{N} = \mathbf{x}\times\mathbf{F}$, where $\mathbf{F}$ is the force; if you replace $\mathbf{p} = q\mathbf{d}$, then $q\mathbf{E}$ is the (Coulomb) force that acts on the lever $\mathbf{d}$.

Force on a charge distribution

After determining the energy $U(\mathbf{x})$ (2.5.13) of a charge distribution in the external potential the next step is to calculate the force on this charge distribution

$$\mathbf{F}(\mathbf{x}) = -\boldsymbol{\nabla}U = -\int \mathrm{d}^3x' \,\boldsymbol{\nabla}\rho(\mathbf{x}'-\mathbf{x})\,\phi(\mathbf{x}') = \int \mathrm{d}^3x' \,(\boldsymbol{\nabla}'\rho(\mathbf{x}'-\mathbf{x}))\,\phi(\mathbf{x}')$$

$$= -\int \mathrm{d}^3x' \,\rho(\mathbf{x}'-\mathbf{x})\,\boldsymbol{\nabla}'\phi(\mathbf{x}').$$

During integration by parts, the surface term disappears. So it is

$$\mathbf{F}(\mathbf{x}) = \int \mathrm{d}^3x' \,\rho(\mathbf{x}'-\mathbf{x})\,\mathbf{E}(\mathbf{x}'). \tag{2.5.17}$$

With the Taylor expansion for the field $\mathbf{E}(\mathbf{x} + (\mathbf{x}'-\mathbf{x}))$ one obtains

$$F_i(\mathbf{x}) = \int \mathrm{d}^3x' \,\rho(\mathbf{x}'-\mathbf{x})\Big[E_i(\mathbf{x}) + (x_j'-x_j)E_{i,j}(\mathbf{x})$$

$$+\frac{1}{6}\big(3(x_j'-x_j)(x_k'-x_k) - \delta_{jk}(\mathbf{x}'-\mathbf{x})^2\big)E_{i,jk}(\mathbf{x}) + \dots\Big],$$

$$F_i(\mathbf{x}) = qE_i(\mathbf{x}) + p_j E_{i,j} + \frac{1}{6}Q_{jk}E_{i,jk} = -q\phi_{,i}(\mathbf{x}) - p_j\phi_{,ij} - \frac{1}{6}Q_{jk}\phi_{,ijk}. \tag{2.5.18}$$

The force on the charge is determined by the electric field and the on the dipole moment by the field gradients. For the force acting on the quadrupole moment, the 2nd derivatives of the electric field are responsible.

Torque

A charge distribution naturally also experiences a torque, which is analogous
to mechanics

$$\mathbf{N} = \int d^3x'\, \rho(\mathbf{x}'-\mathbf{x})\,(\mathbf{x}'-\mathbf{x})\times\mathbf{E}(\mathbf{x}').\tag{2.5.19}$$

The evaluation is easier for the individual components using the ϵ-tensor

$$\begin{aligned}
N_i &= \int d^3x'\, \rho(\mathbf{x}'-\mathbf{x})\,\epsilon_{ijk}(x'_j - x_j)E_k(\mathbf{x}')\\
&= \int d^3x'\, \rho(\mathbf{x}'-\mathbf{x})\,\epsilon_{ijk}(x'_j - x_j)\Big[E_k(\mathbf{x}) + (x'_l - x_l)E_{l,k}\\
&\quad + \frac{1}{6}\big(3(x'_l - x_l)(x'_m - x_m) - \delta_{lm}(\mathbf{x}'-\mathbf{x})^2\big)\,E_{l,mk}(\mathbf{x}) + \dots\Big].
\end{aligned}$$

The last term is of 3rd order in $(x'_i - x_i)$ and is neglected; the penultimate term
can be represented by the quadrupole moment, but has the prefactor $1/3$:

$$N_i = (\mathbf{p}\times\mathbf{E})_i - \frac{1}{3}\,\epsilon_{ijk}\,\phi_{,km}\,Q_{jm}\,.\tag{2.5.20}$$

The torque acting on a dipole is given by the electric field; the quadrupole
senses the electric field gradients.

Explicit determination of the components of the torque: We start with a field pointing
in the z direction and a dipole lying in the xz plane:

$$\mathbf{E}(0) = E_3\,\mathbf{e}_3\,,\qquad\qquad \mathbf{p} = \mathbf{e}_1\,p\,\sin\vartheta + \mathbf{e}_3\,p\,\cos\vartheta\,.$$

The z' axis is chosen parallel to the dipole moment, so that it also forms an angle
ϑ with the z axis, as shown in Fig. 2.35 $(\varphi = 0)$. The dipole moment $\mathbf{p}$ lies in
the z' axis: $\mathbf{p} = p(\sin\vartheta, 0, \cos\vartheta)$. The situation becomes somewhat more complex

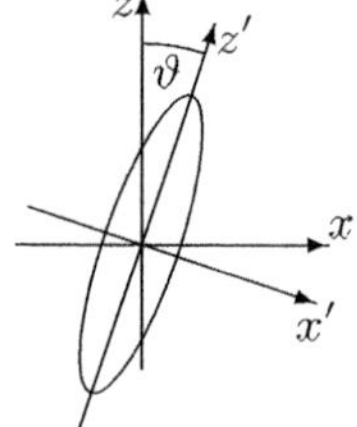

Fig. 2.35. Sketch of the position of the quadrupole moment; the
z'-axis is oriented so that it points in the direction of the dipole
$(\mathbf{p} = p\mathbf{e}_{z'})$

with the consideration of the quadrupole moment. The tensor of the field gradients
is diagonal with the simplifying assumption $E_3 = E_3(x_3)$ and naturally trace-free
$(\sum_i \phi_{,ii} = 0)$:

$$\begin{pmatrix} -\frac{1}{2}\phi_{,33} & 0 & 0 \\ 0 & -\frac{1}{2}\phi_{,33} & 0 \\ 0 & 0 & \phi_{,33} \end{pmatrix}\,.$$

The calculation of $\mathbf{N}$ according to (2.5.20) results in

$$N_1 = p_2 E_3 - p_3 E_2 + \frac{1}{3}\left(\epsilon_{123}\,\phi_{,22}\,Q_{32} + \epsilon_{132}\,\phi_{,33}\,Q_{23}\right) = 0\,,$$

$$N_2 = p_3 E_1 - p_1 E_3 + \frac{1}{3}\left(\epsilon_{231}\,\phi_{,33}\,Q_{13} + \epsilon_{213}\,\phi_{,11}\,Q_{31}\right) = -pE\,\sin\vartheta + \frac{1}{2}\phi_{,33}\,Q_{13}\,,$$

$$N_3 = p_1 E_2 - p_2 E_1 + \frac{1}{3}\left(\epsilon_{312}\,\phi_{,11}\,Q_{21} + \epsilon_{321}\,\phi_{,22}\,Q_{12}\right) = 0\,.$$

Of the torque, only N_2 is different from zero, i.e., it is perpendicular to the xz-plane and we only need the moment Q_{13}. We assume that the main axis of $\mathbf{Q}$ is also the z'-axis: $Q = Q_{3'3'}$. The calculation of the moments in the unprimed system is done in the following side calculation and results in $Q_{13} = Q_{31} = 3Q/2\,\sin\vartheta\,\cos\vartheta$. Thus we obtain the torque

$$\mathbf{N} = -\left(pE + \frac{3Q}{4}\,E_{3,3}\,\cos\vartheta\right)\sin\vartheta\,\mathbf{e}_2\,. \tag{2.5.21}$$

Dipole and quadrupole moment try to align parallel to $\mathbf{E} = (0,0,E)$.

Calculation of the quadrupole moments:

$$Q_{ik} = \begin{pmatrix} \cos\vartheta & 0 & \sin\vartheta \\ 0 & 1 & 0 \\ -\sin\vartheta & 0 & \cos\vartheta \end{pmatrix} \begin{pmatrix} -\frac{Q}{2} & 0 & 0 \\ 0 & -\frac{Q}{2} & 0 \\ 0 & 0 & Q \end{pmatrix} \begin{pmatrix} \cos\vartheta & 0 & -\sin\vartheta \\ 0 & 1 & 0 \\ \sin\vartheta & 0 & \cos\vartheta \end{pmatrix}$$

$$= Q \begin{pmatrix} -\frac{1}{2}\cos^2\vartheta + \sin^2\vartheta & 0 & \frac{3}{2}\sin\vartheta\,\cos\vartheta \\ 0 & -\frac{1}{2} & 0 \\ \frac{3}{2}\sin\vartheta\,\cos\vartheta & 0 & -\frac{1}{2}\sin^2\vartheta + \cos^2\vartheta \end{pmatrix},$$

$$Q_{13} = Q_{31} = \frac{3Q}{4}\sin 2\vartheta,\quad Q_{11} = \frac{Q}{4}(1 - 3\cos 2\vartheta),\quad Q_{22} = \frac{-Q}{2},\quad Q_{33} = \frac{Q}{4}(1 + 3\cos 2\vartheta).$$

Problems for Chapter 2

2.1. *Simple charge distributions*: Express the following charge distributions using δ-, θ-functions and volume, surface or line charge densities (ρ_0, σ_0 or λ), where all bodies should have the total charge Q.

Spherical coordinates
1. Homogeneously charged sphere, radius a.
2. Homogeneously charged spherical shell of infinitesimal thickness, radius a.
3. Point charge.

Cylindrical coordinates
4. Homogeneously charged cylinder, radius a and length $2l$.
5. Homogeneously charged hollow cylinder of infinitesimal thickness, radius a and length $2l$.
6. Homogeneous line charge, length $2l$.

2.2. *Dipole line*: Show that the field of two charges $\pm q$ at the locations $\mathbf{x}_{1,2}$ can be represented by the field of a line of dipoles, which are aligned parallel to the connecting line.

2.3. *Potential of a spherically symmetric charge distribution*: Show that the potential of a charge distribution $\rho(r)$ can be represented as

$$\phi(r) = 4\pi k_C \left[\frac{1}{r} \int_0^r \mathrm{d}r'\, r'^2\, \rho(r') + \int_r^\infty \mathrm{d}r'\, r'\, \rho(r') \right].$$

2.4. *Potential and field of the H-atom*: The charge density of the H-atom in the ground state is

$$\rho^{(H)}(\mathbf{x}) = e_0 \delta^{(3)}(\mathbf{x}) - (e_0/\pi a_B^3) e^{-2r/a_B} \qquad \text{with the Bohr radius } a_B \approx 0.529\,\text{Å}.$$

1. Calculate $\phi^{(H)}$, $\mathbf{E}^{(H)}$.
2. Sketch the course qualitatively.
3. Assume that the atomic nucleus has the radius $R = 1.5 \times 10^{-13}$ cm. Indicate the field strengths for $r = R$ and $r = a_B$.

2.5. *Field of a semi-infinite double layer*:

1. The electrostatic field of the charge density is to be calculated

$$\rho(\mathbf{x}) = \sigma \left[\theta(a+x) - \theta(x) \right] \theta(b-|y|) \delta(z).$$

2. In the limit $a, b \to \infty$ $\mathbf{E}$ diverges, but the difference to an (almost) arbitrary point $\mathbf{x}_0$: $\mathbf{E}(\mathbf{x}, \mathbf{x}_0) = \tilde{\mathbf{E}}(\mathbf{x}) - \tilde{\mathbf{E}}(\mathbf{x}_0)$ remains finite. Verify:

$$\tilde{\mathbf{E}}(\mathbf{x}) = k_C \sigma \left\{ -\mathbf{e}_x \ln(x^2 + z^2) + \mathbf{e}_z \left[\pi \operatorname{sgn} z - 2 \arctan \frac{x}{z} \right] \right\} + \text{const.} \qquad (2.5.22)$$

3. Now verify the field of the double layer (2.2.32):

$$\rho(\mathbf{x}) = \sigma \theta(-x) \left[\delta(z - \frac{d}{2}) - \delta(z + \frac{d}{2}) \right].$$
$$\int \frac{\mathrm{d}x}{(a^2 + x^2)\sqrt{a^2 + b^2 + x^2}} = \frac{1}{ab} \arctan \frac{bx}{a\sqrt{a^2 + b^2 + x^2}} \qquad \text{(auxiliary integral)}.$$

2.6. *Equipotential surfaces of two line charges*: Given is the charge density
$\rho(\mathbf{x}) = \lambda \delta(y) \left[\delta(x-a) - \delta(x+a) \right].$

1. Specify the potential of the line charges.
2. Determine potential surfaces for a given ϕ_0.
3. Sketch equipotential lines in the xy-plane.

Note: The negatively charged wire can also be understood as an image charge, so that the equipotential lines with $x > 0$ are the same as those of the wire in front of a grounded surface.

2.7. *Potential of a line charge in front of two metal plates*:

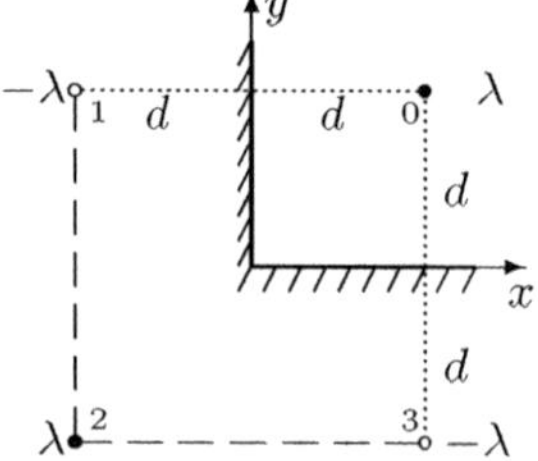

A line charge λ is at a distance d from two perpendicular grounded conductors, as indicated in the sketch. To fulfill the boundary conditions, three mirror charges are used.

Determine the potential ϕ, the field strength $\mathbf{E}$ and the charge induced on the conductor surface, σ.

2.8. *Point charge in front of conducting planes*

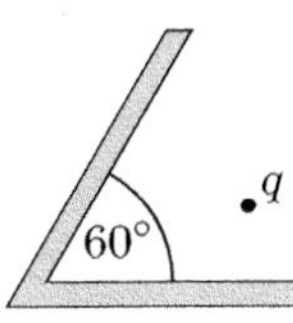

The point charge q is now in front of two infinitely extended metal plates inclined at an angle of $\alpha = 60°$ to each other (see sketch). q induces a charge on the metal surface which can be represented by the suitable placement of five mirror charges. Sketch the positions of the (mirror) charges and show why the boundary condition $\mathbf{E}_\parallel = 0$ is fulfilled on the metal surfaces.

2.9. *Dipole in front of a conducting plane*

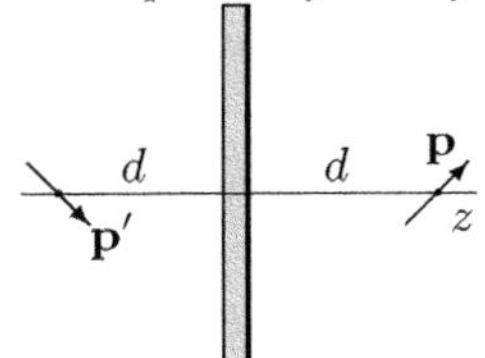

A point dipole is located at a distance d in front of a conductive plane.

1. Determine ϕ and $\mathbf{E}$ in the half-space $z \geq 0$.
2. Determine the induced surface charge.
3. Determine the configuration of minimal energy.

2.10. *Maxwell's stress tensor*:

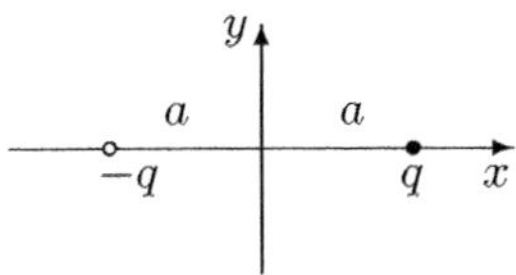

Given are a positive $(q > 0)$ and a negative charge at a distance $2a$, as shown in the adjacent sketch. Calculate the Maxwell's stress tensor for this arrangement and determine using the stress tensor the force on the negative charge.

2.11. *Interaction energy in the H-atom*: Determine the interaction energy between the nucleus and electron in the H-atom with the charge distribution

$$\rho_N(\mathbf{x}) = e_0 \delta^{(3)}(\mathbf{x}), \qquad\qquad \rho_e(\mathbf{x}) = -\frac{e_0}{\pi a_B^3}\, e^{-2r/a_B}\,.$$

2.12. *Interaction energy of two wire loops*

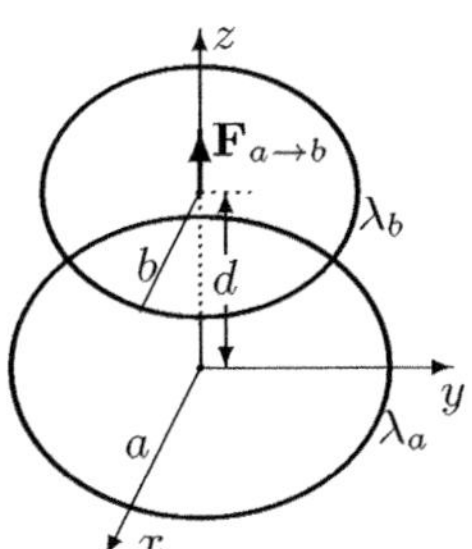

Determine the interaction energy and the force exerted by two concentric circular rings with radii a and b, whose centers are at a distance z from each other (see sketch 2.36).

Fig. 2.36. Two circular rings with line charges λ_a and λ_b

2.13. *On the energy of a dipole in an external field*: A dipole is located in a constant external field $\mathbf{E}^e$. Calculate the (interaction) energy of the dipole in the field $\mathbf{E}^e$. Show that

$$U = \frac{1}{4\pi k_C} \int \mathrm{d}^3x\, \mathbf{E}^e \cdot \mathbf{E}^p$$

does not lead to the correct result, where $\mathbf{E}^p$ is the field of the dipole. Does this also apply to a quadrupole?

2.14. *Self-energy of the spherical shell*: On a sphere of radius a, the charge Q is evenly distributed on the surface. Calculate the electrostatic self-energy and compare it with the full sphere, if Q is distributed homogeneously.

2.15. *Proof of Thomson's theorem*: There are n conductors with the volumes L_i and the charges Q_i, which are within a finite region. Outside the conductors, there is a charge distribution ρ_{ext}, also limited to a finite region.

In section 2.4.3, the charge distribution was varied at constant Q_i. In the energetically most favorable configuration, the field $\mathbf{E}=0$ vanishes in each conductor L_i. Here you should start from the solution for which $\mathbf{E}=0$ in the conductors, where

$$\oiint_{\partial L_i} \mathbf{df}_i \cdot \mathbf{E} = 4\pi Q_i \qquad \text{and} \qquad \boldsymbol{\nabla}\cdot\mathbf{E} = 4\pi\rho_{ext} \qquad (*)$$

are fulfilled. You should show that every field $\mathbf{E}'$, that fulfills the conditions $(*)$, leads to a higher energy if $\mathbf{E}' \neq 0$ is within the conductors.

2.16. *Capacitance matrix*: Determine the capacitance matrix of the following configurations

1. Two concentric spheres with radii $a < b$ and charges Q_a and Q_b.
2. Two spheres with radii a and b and charges Q_a and Q_b. Their centers are at a distance d from each other, where $d \gg a$ and $d \gg b$.

Consider for both configurations the case $Q_a = -Q_b = Q$ and determine the corresponding capacity.

2.17. *Divergence and curl of the dipole field*: If one separates the singularity from the dipole field according to (2.5.6a) $\mathbf{E}^s = -(4\pi k_C/3)\mathbf{p}\delta^{(3)}(\mathbf{x})$ at the location of the dipole from the 'rest' of the field, one should show that $\boldsymbol{\nabla}\cdot\mathbf{E} = 4\pi k_C\rho(\mathbf{x})$ and $\boldsymbol{\nabla}\times\mathbf{E} = 0$ are still fulfilled. Hint:

$$\lim_{a\to 0}\frac{3}{4\pi a^3}\theta(a-r)\widehat{=}\delta^{(3)}(\mathbf{x}). \tag{2.5.23}$$

References

Becker R., Sauter F. *Theorie der Elektrizität 1*, 21th ed. Teubner, Stuttgart (1973)
Brandt S., Dahmen H. D. *Elektrodynamik* 4th ed. Springer Berlin (2004)
Fließbach T. *Elektrodynamik*, 7th ed. Springer Spektrum Berlin (2022)
Greiner W *Classical Electrodynamics* Springer Berlin (1998)
Jackson J. D. *Classical Electrodynamics*, 3rd ed., John Wiley & Sons Inc. (1998)
Kirchhoff G. *Vorlesungen über mathematische Physik*, Vol. 3 *Electricität und Magnetismus*, Teubner Verlag, Leipzig (1891)
Maxwell J. C. *A Treatise on Electricity and Magnetism*, vol. I, 3rd ed. Oxford at the Clarendon Press (1892)

3

Boundary Value Problems in Electrostatics

Electrostatics is a potential theory, i.e. a theory for irrotational vector fields $(\nabla \times \mathbf{E} = 0)$. Within this, the vector field $\mathbf{E}$ can be derived from a scalar potential ϕ, which satisfies the (scalar) Poisson equation. First, conditions are sought that can be imposed on the boundary of the considered volume, so that the problem has a unique solution. In solving the Laplace or Poisson equation, spherical symmetry is often used. However, the more complicated cylindrical symmetry is also discussed. In some cases, when the configuration can be represented in two dimensions, complex analysis with the conformal mapping is a suitable solution method.

3.1 Solution of the Poisson Equation with Boundary Condition

A general method for solving the Poisson equation is not or only conditionally available. Therefore, it is all the more important to know which conditions can be imposed on the boundary surfaces and, if a solution has been found, that it is unique.

3.1.1 Uniqueness of the Solution of the Poisson Equation with Boundary Condition

We start from a local charge distribution on a simply connected region, as it is outlined in Fig. 3.1. To determine the potential ϕ, the Poisson equation must be solved, whereby on the surface ∂V the potential $\phi(\partial V)$ or the field $\mathbf{E}(\partial V)$ can be specified.

First, it is shown that when either the potential (or the parallel component of the field, both *Dirichlet boundary condition*) or the normal component of the field (*Neumann boundary condition*) is specified, the found solution is unique.

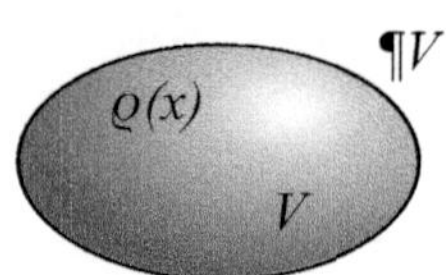

Fig. 3.1. Simply connected region of volume V, which is bounded by the surface ∂V. $\rho(\mathbf{x})$ is the charge density inside the region

Thus, the problem is generally over-determined when specifying potential and field. There are also so-called *mixed boundary conditions* with a specification of values of the potential and the field, which will not be discussed here.

Dirichlet boundary condition: $\phi(\partial V)$ specified.

Neumann boundary condition: $\dfrac{\partial \phi(\partial V)}{\partial n} = \mathbf{n} \cdot \boldsymbol{\nabla} \phi$ specified.

Typical Dirichlet boundary conditions for electrostatics are conductors that are held at different potentials, while predefined surface charge densities $\sigma = \mathbf{E} \cdot \mathbf{n}/4\pi k_C$ belong to the Neumann boundary conditions. The boundary conditions in the infinite $\mathbb{R}^3$ occupy a special position, where with $|\boldsymbol{\nabla}\phi(r \to \infty)| = 0$ all components of the field vanish, which is discussed in the Helmholtz decomposition theorem, section 7.1.2.

> *Theorem:* For Dirichlet or Neumann boundary conditions, the solution $\phi(\mathbf{x})$ of the Poisson equation $\Delta\phi = -4\pi k_C \rho$ is uniquely determined in this region.

Proof of the uniqueness of the solution

The proof of uniqueness is very similar to that of the basic problem of electrostatics, p. 48. We start from one of the two boundary conditions (Dirichlet or Neumann) and assume that there are two solutions $\phi_1(\mathbf{x})$ and $\phi_2(\mathbf{x})$ available. Then

$$\phi_d(\mathbf{x}) = \phi_1(\mathbf{x}) - \phi_2(\mathbf{x})$$

satisfies the Laplace equation $\Delta\phi_d = 0$ with $\phi_d(\partial V) = 0$ for Dirichlet,

$$\frac{\partial \phi_d(\partial V)}{\partial n} = 0 \text{ for Neumann.}$$

If one substitutes in the 1st Green's theorem (A.4.19)

$$\int_V \mathrm{d}^3x \left(\phi\nabla^2\psi + \boldsymbol{\nabla}\phi \cdot \boldsymbol{\nabla}\psi\right) = \oiint_{\partial V} \mathrm{d}\mathbf{a} \cdot \phi\boldsymbol{\nabla}\psi \tag{3.1.1}$$

$\phi = \psi = \phi_d$, then follows

$$\int_V \mathrm{d}^3x \left(\phi_d \underbrace{\nabla^2\phi_d}_{=0} + (\boldsymbol{\nabla}\phi_d)^2\right) = \oiint_{\partial V} \mathrm{d}\mathbf{a} \cdot \underbrace{\phi_d(\boldsymbol{\nabla}\phi_d)}_{=0} . \tag{3.1.2}$$

It should be noted that on the one hand $\nabla^2\phi_d = 0$, on the other hand on the surface either $\phi_d = 0$ or $\mathbf{n} \cdot \boldsymbol{\nabla}\phi_d = 0$. This results in

$$\int_V d^3x (\boldsymbol{\nabla}\phi_d)^2 = 0 \quad \Rightarrow \quad \boldsymbol{\nabla}\phi_d = 0, \quad \text{since} \quad (\boldsymbol{\nabla}\phi_d)^2 \geq 0 \quad \text{and} \quad \phi_d = \text{const.}$$

Dirichlet: $\phi_d = \text{const.} = 0$ on $\partial V \to \phi_1 = \phi_2$,
Neumann: $\phi_1 = \phi_2 + \text{const.}$

3.1.2 Solution of the Boundary Value Problem with Green's Functions

The 2nd Green's formula (A.4.20) reads

$$\int_V d^3x' (\psi \nabla'^2 \phi - \phi \nabla'^2 \psi) = \oiint_{\partial V} da' \cdot (\psi \boldsymbol{\nabla}'\phi - \phi \boldsymbol{\nabla}'\psi).$$

ϕ is the electrostatic potential of the problem and ψ is the Green function (2.1.6) or (A.4.22):

$$\nabla'^2 \phi(\mathbf{x}') = -4\pi k_C \rho(\mathbf{x}'),$$

$$\psi(\mathbf{x}, \mathbf{x}') = G(\mathbf{x}, \mathbf{x}') = \frac{1}{|\mathbf{x} - \mathbf{x}'|}, \qquad \nabla'^2 \psi = -4\pi \delta^{(3)}(\mathbf{x} - \mathbf{x}'). \qquad (3.1.3)$$

If you bring the 1st term of the Green's formula to the right side and divide by 4π, you get

$$\phi(\mathbf{x}) = k_C \int_V d^3x' \, \frac{\rho(\mathbf{x}')}{|\mathbf{x} - \mathbf{x}'|} + \frac{1}{4\pi} \oiint_{\partial V} da' \cdot \Big[\frac{\boldsymbol{\nabla}'\phi(\mathbf{x}')}{|\mathbf{x} - \mathbf{x}'|} - \phi(\mathbf{x}')\boldsymbol{\nabla}' \frac{1}{|\mathbf{x} - \mathbf{x}'|} \Big]. \quad (3.1.4)$$

Here $\phi(\mathbf{x}')$ and $\mathbf{n} \cdot \boldsymbol{\nabla}'\phi(\mathbf{x}')$ appear on the surface. The independent assignment of these two quantities is not allowed, as each uniquely determines the solution.

Green's function for the boundary value problem

The Green function for the boundary value problem

$$G_b(\mathbf{x}, \mathbf{x}') = G(\mathbf{x}, \mathbf{x}') + F(\mathbf{x}, \mathbf{x}'), \qquad\qquad G(\mathbf{x}, \mathbf{x}') = \frac{1}{|\mathbf{x} - \mathbf{x}'|}$$

is an extension of the Green function (2.1.6) with a solution of the Laplace equation $\Delta F(\mathbf{x}, \mathbf{x}') = 0$. Here the subscript b stands for D or N, depending on whether you have Dirichlet or Neumann boundary conditions. If you extend the Green function in (3.1.4) with a solution of the Laplace equation $\Delta F(\mathbf{x}, \mathbf{x}') = 0$, then ϕ, which is calculated with

$$G_b(\mathbf{x}, \mathbf{x}') = G(\mathbf{x}, \mathbf{x}') + F(\mathbf{x}, \mathbf{x}')$$

remains a solution of the boundary value problem, as one notices by applying Δ to ϕ in (3.1.4), since still

$$\Delta G_b(\mathbf{x}, \mathbf{x}') = -4\pi\delta^{(3)}(\mathbf{x}-\mathbf{x}')$$

remains unchanged:

$$\phi(\mathbf{x}) = k_C \int_V d^3x'\, G_b(\mathbf{x}, \mathbf{x}')\rho(\mathbf{x}') \qquad (3.1.5)$$

$$+ \frac{1}{4\pi} \oiint_{\partial V} d\mathbf{a}' \cdot \Big(G_b(\mathbf{x}, \mathbf{x}')\boldsymbol{\nabla}'\phi(\mathbf{x}') - \phi(\mathbf{x}')\boldsymbol{\nabla}'G_b(\mathbf{x}, \mathbf{x}')\Big).$$

Here too, ϕ and $\boldsymbol{\nabla}\phi$ appear on the surface. The Green function is now modified in (3.1.5) so that one of the two surface terms disappears and thus the boundary value problem is solvable.

(a) *Dirichlet boundary condition*

With the possibility to modify G_b so that

$$G_b(\mathbf{x}, \mathbf{x}') = G_D(\mathbf{x}, \mathbf{x}')|_{\mathbf{x}' \text{ on } \partial V} = 0$$

the boundary term with $\boldsymbol{\nabla}'\phi(\mathbf{x}')$ disappears. The configuration thus has a solution, which must be unique:

$$\phi(\mathbf{x}) = k_C \int_V d^3x'\, G_D(\mathbf{x}, \mathbf{x}')\rho(\mathbf{x}') - \frac{1}{4\pi} \oiint_{\partial V} d\mathbf{a}' \cdot \phi(\mathbf{x}')\boldsymbol{\nabla}'G_D(\mathbf{x}, \mathbf{x}'). \quad (3.1.6)$$

(b) *Neumann boundary condition*

It would be obvious to choose $\boldsymbol{\nabla}G_N = 0$ on ∂V. However, this is not permissible, because

$$\int_V d^3x'\, \nabla'^2 G_N(\mathbf{x}, \mathbf{x}') = -\int_V d^3x'\, 4\pi\delta^{(3)}(\mathbf{x} - \mathbf{x}')\,.$$

The application of Gauss's theorem yields

$$\oiint_{\partial V} d\mathbf{a}' \cdot \boldsymbol{\nabla}'G_N(\mathbf{x}, \mathbf{x}') = -4\pi\,.$$

The simplest boundary condition for $G_N(\mathbf{x}, \mathbf{x}')$, which is consistent with this surface integral is

$$\mathbf{n} \cdot \boldsymbol{\nabla}'G_N(\mathbf{x}, \mathbf{x}')\big|_{\mathbf{x}' \in \partial V} = -\frac{4\pi}{\partial V}\,,$$

where ∂V is the size of the surface. This implies

$$\phi(\mathbf{x}) = k_C \int_V d^3x'\, G_N(\mathbf{x}, \mathbf{x}')\rho(\mathbf{x}') + \frac{1}{4\pi} \oiint_{\partial V} d\mathbf{a}' \cdot G_N(\mathbf{x}, \mathbf{x}')\,\boldsymbol{\nabla}'\phi + \langle\phi\rangle_{\partial V}$$

$$\langle\phi\rangle_{\partial V} = \frac{1}{\partial V} \oiint_{\partial V} d a'\, \phi(\mathbf{x}') \qquad \text{with} \quad d\mathbf{a} = d a\,\mathbf{n}\,. \qquad (3.1.7)$$

$\langle\phi\rangle_{\partial V}$ is the average of ϕ over all boundary surfaces, which can be included as a constant in the definition of ϕ. For $\partial V \to \infty$, $\langle\phi\rangle_{\partial V} \to 0$.

Remarks

1. The Green's functions $G_b(\mathbf{x}, \mathbf{x}')$ do not depend on the boundary condition imposed on ϕ or $\mathbf{n} \cdot \nabla \phi$, but only on the shape of the surface. Once the Green's function has been found, any boundary value problem with the surface ∂V can be solved by integration.

2. $k_C G_b(\mathbf{x}, \mathbf{x}_0)$ gives the potential at the point $\mathbf{x}$ if there is a unit charge at the location $\mathbf{x}_0$ and the potential on the surface $(\phi(\partial V) = 0)$ vanishes (Dirichlet boundary condition):

$$G_b(\mathbf{x}, \mathbf{x}_0) = \frac{1}{|\mathbf{x} - \mathbf{x}_0|} + F(\mathbf{x}, \mathbf{x}_0).$$

Since $F(\mathbf{x}, \mathbf{x}_0)$ satisfies the Laplace equation in the interior with $\Delta F = 0$, this potential arises from the induced surface charges or from equivalent image charges. In the so-called external problem, with Neumann boundary conditions, one has a surface that lies at infinity, so that $\mathbf{n} \cdot \nabla \phi(\partial V) = -4\pi/\partial V \overset{\partial V \to \infty}{\longrightarrow} 0$.

3. *Reciprocity*: The Green's function for the Dirichlet problem is symmetric with respect to the source point $\mathbf{x}_0$ and the point of application (observation point) $\mathbf{x}$: $G_D(\mathbf{x}, \mathbf{x}_0) = G_D(\mathbf{x}_0, \mathbf{x})$.

   Proof: Substitute in 2. Green's theorem (A.4.20) $\phi(\mathbf{x}') = G_D(\mathbf{x}_0, \mathbf{x}')$ and
   $$\psi(\mathbf{x}') = G_D(\mathbf{x}, \mathbf{x}').$$

   First the surface terms (right side) of (A.4.20):

$$\oiint_{\partial V} \mathrm{d}\mathbf{a}' \cdot \left(G_D(\mathbf{x}_0, \mathbf{x}') \nabla' G_D(\mathbf{x}, \mathbf{x}') - G_D(\mathbf{x}, \mathbf{x}') \nabla' G_D(\mathbf{x}_0, \mathbf{x}') \right) = 0$$

$$\text{for Dirichlet applies}: \ G_D(\mathbf{x}_0, \mathbf{x}') = G_D(\mathbf{x}, \mathbf{x}')\Big|_{\mathbf{x}' \in \partial V} = 0.$$

Thus, the surface term vanishes; what remains is

$$0 = \int_V \mathrm{d}^3 x' \left(G_D(\mathbf{x}_0, \mathbf{x}') \nabla'^2 G_D(\mathbf{x}, \mathbf{x}') - G_D(\mathbf{x}, \mathbf{x}') \nabla'^2 G_D(\mathbf{x}_0, \mathbf{x}') \right)$$

$$= -4\pi \int_V \mathrm{d}^3 x' \left(G_D(\mathbf{x}_0, \mathbf{x}') \delta^{(3)}(\mathbf{x} - \mathbf{x}') - G_D(\mathbf{x}, \mathbf{x}') \delta^{(3)}(\mathbf{x}_0 - \mathbf{x}') \right)$$

$$= -4\pi \left(G_D(\mathbf{x}_0, \mathbf{x}) - G_D(\mathbf{x}, \mathbf{x}_0) \right) = 0 \quad \text{q.e.d.}$$

Point charge in front of a conductive plate

Potential and field of a point charge q in front of a metal plate, as sketched in Fig. 3.2, we have already calculated in section 2.3.1 using an image charge (or *mirror charge*) $-q$ at the same distance behind the conductive surface. Then ϕ disappears on the metal surface, so that $\phi(\mathbf{x}, \mathbf{d})$ fulfills the Dirichlet condition $G_D(\mathbf{x}, \mathbf{d}) = 0$ on the surface ∂V.

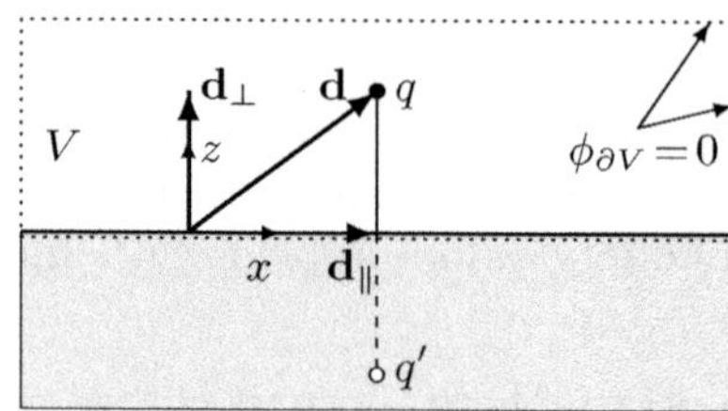

Fig. 3.2. Point charge in front of a metal plate in the half-space V with $z > 0$. The image charge $q' = -q$ ensures that $\phi(z=0) = 0$; also, $\phi(r \to \infty) = 0$, i.e. $\phi_{\partial V} = 0$

1. *Dirichlet-Green function*

If one takes (see section 2.3.1) instead of the point charge q a unit charge and replaces $\mathbf{d}$ with $\mathbf{x}'$, then $k_C\, G(\mathbf{x}, \mathbf{x}')$ represents the potential of the actual charge and the potential of the image charge the additional term

$$F(\mathbf{x}, \mathbf{x}') = \frac{-1}{|\mathbf{x} - \mathbf{x}'_\parallel + \mathbf{x}'_\perp|} = \frac{-1}{|\mathbf{x} - \mathsf{M}\mathbf{x}'|} \quad \text{with} \quad \mathsf{M} = \begin{pmatrix} 1 & 0 & 0 \\ 0 & 1 & 0 \\ 0 & 0 & -1 \end{pmatrix},$$

which satisfies the Laplace equation for $z > 0$. So we have

$$G_D(\mathbf{x}, \mathbf{x}') = \frac{1}{|\mathbf{x} - \mathbf{x}'_\parallel - \mathbf{x}'_\perp|} - \frac{1}{|\mathbf{x} - \mathbf{x}'_\parallel + \mathbf{x}'_\perp|}. \tag{3.1.8}$$

It is obvious that both the symmetry $G_D(\mathbf{x}', \mathbf{x}) = G_D(\mathbf{x}, \mathbf{x}')$ and the boundary condition $G_D(z=0, \mathbf{x}') = 0$ are fulfilled. The potential is determined by (3.1.6):

$$\phi(\mathbf{x}) = k_C \int_V \mathrm{d}^3x'\, G_D(\mathbf{x}, \mathbf{x}')\rho(\mathbf{x}') = k_C q \Big[\frac{1}{|\mathbf{x}-\mathbf{d}_\parallel-\mathbf{d}_\perp|} - \frac{1}{|\mathbf{x}-\mathbf{d}_\parallel+\mathbf{d}_\perp|} \Big],$$

where $\rho(\mathbf{x}') = q\, \delta^{(3)}(\mathbf{x}'-\mathbf{d})$. The fields are again given by (2.3.2).

2. *Constant potential ϕ_0 on conductive plane*

Now we modify the example by raising the potential on the ladder surface to $\phi(z=0) = \phi_0$. We expect that in the entire half-space $z \geq 0$ $\phi(\mathbf{x}) \to \phi(\mathbf{x})+\phi_0$. This contribution can only come from the surface term in (3.1.6):

$$\phi_{\partial V} = \frac{-1}{4\pi} \oiint_{\partial V} \mathrm{d}\mathbf{a}' \cdot \phi(\mathbf{x}') \boldsymbol{\nabla}' G_D(\mathbf{x}, \mathbf{x}') = \frac{\phi_0}{4\pi} \iint \mathrm{d}x'\,\mathrm{d}y'\, \frac{\partial}{\partial z'} G_D(\mathbf{x}, \mathbf{x}'). \tag{3.1.9}$$

Since $\phi_{\partial V}(\mathbf{x})$ should only depend on the distance z from the plane, we choose $x=y=0$ and after transforming to plane polar coordinates we get

$$\phi_{\partial V} = \frac{\phi_0}{4\pi} \iint \mathrm{d}x'\,\mathrm{d}y'\, \frac{2z}{|\mathbf{x}-\mathbf{x}'_\parallel|^3} = z\phi_0 \int_0^\infty \frac{\mathrm{d}\varrho'\, \varrho'}{\sqrt{\varrho'^2+z^2}^3} = \phi_0,$$

as it should be. Thus, we have

$$\phi(\mathbf{x}) = k_C q\Big(\frac{1}{|\mathbf{x}-\mathbf{d}|} - \frac{1}{|\mathbf{x}+\mathbf{d}|}\Big) + \phi_0 \, .$$

3. Different potentials on both half-planes

The two conductor plates are now, as sketched in Fig. 3.3, held at different potentials $\phi_1 \neq \phi_2$. G_D and ρ remain unchanged compared to the two previous cases, as the spatial configuration has not changed. Only the surface contribution $\phi_{\partial V}$ needs to be recalculated, which is covered in problem 3.1. Potential and fields are

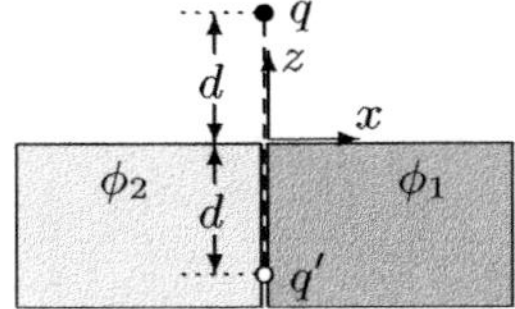

Fig. 3.3. Charge q at a distance d in front of a metal plate, which is divided and whose parts are kept at the potentials ϕ_1 and ϕ_2

$$\phi(\mathbf{x}) = k_C q\Big[\frac{1}{|\mathbf{x}-\mathbf{d}|} - \frac{1}{|\mathbf{x}+\mathbf{d}|}\Big] + \frac{\phi_1+\phi_2}{2} + \frac{\phi_1-\phi_2}{\pi}\arctan\frac{x}{z},$$

$$E_x = k_C q\Big[\frac{x}{|\mathbf{x}-\mathbf{d}|^3} - \frac{x}{|\mathbf{x}+\mathbf{d}|^3}\Big] - \frac{\phi_1-\phi_2}{\pi}\frac{z}{x^2+z^2},$$

$$E_z = k_C q\Big[\frac{z-d}{|\mathbf{x}-\mathbf{d}|^3} - \frac{z+d}{|\mathbf{x}+\mathbf{d}|^3}\Big] + \frac{\phi_1-\phi_2}{\pi}\frac{x}{x^2+z^2},$$

and the surface charge is determined by

$$\sigma = \frac{E_n(z=0)}{4\pi k_C} = \frac{1}{4\pi k_C}\Big[\frac{-2k_C dq}{\sqrt{x^2+d^2}^3} + \frac{\phi_1-\phi_2}{\pi}\frac{1}{x}\Big].$$

Fig. 3.4a shows the surface charge induced by the charge q and the metal surfaces $\phi_1 = -\phi_2$ induced surface charge. The singularity at $x = 0$ is caused by $\phi_2 = -\phi_1$. Since the potential of the right plate is equal to that of the point charge, the induced charge is rather small there. For larger values of x, the influence of the point charge diminishes, and the field lines approach semicircles. In Fig. 3.4b, field and equipotential lines for a point charge $q = 1$ ($\mathbf{d} = d\mathbf{e}_z$) in front of a divided metal plate with $\phi_{1,2} = \pm 1$ are drawn. The potential ϕ_1 is larger than the point charge, so no field lines from q go to the plate with $x > 0$.

Inversion on a sphere

Given is $\rho(\mathbf{x})$ outside a sphere with radius R, as in Fig. 2.24, p. 51 for a point charge or in Fig. 3.5 for a point dipole sketched. V is the volume outside the sphere with the boundary surface ∂V and S_R is the sphere volume with the fictitious charges.

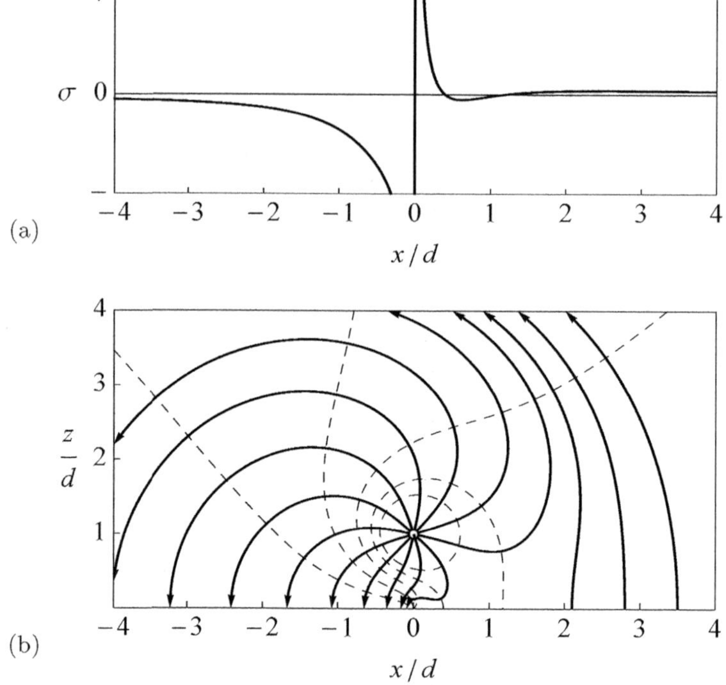

Fig. 3.4. (a) Surface charge, induced by point charge $q = 1$ and $\phi_1 = -\phi_2 = 1$
(b) Field and equipotential lines of a point charge $q = 1$ in front of a metal plate
with $\phi_1 = -\phi_2 = 1$

1. Dirichlet-Green Function

Using the image charge method, $\phi(\mathbf{x})$ of a point charge in front of a conducting sphere, when $\phi_{\partial V} = 0$ at the surface, has already been calculated. If we set $q = 1$ and replace $\mathbf{d}$ with $\mathbf{x}'$, we have the Green function G in the considered volume V and with the potential of the image charge the additional term $F(\mathbf{x}, \mathbf{x}')$, which in V fulfills the Laplace equation. We take (2.3.5)

$$F(\mathbf{x}, \mathbf{x}') = \frac{-R}{r'} \frac{1}{\left|\mathbf{x} - \frac{R^2}{r'^2}\mathbf{x}'\right|}. \tag{3.1.10}$$

The Green function of the (Dirichlet's) problem is thus

$$G_D(\mathbf{x}, \mathbf{x}') = \begin{cases} \dfrac{1}{|\mathbf{x} - \mathbf{x}'|} - \dfrac{R}{r'} \dfrac{1}{\left|\mathbf{x} - \frac{R^2}{r'^2}\mathbf{x}'\right|} & \text{for} \quad r, r' > R \\ 0 & \text{otherwise.} \end{cases} \tag{3.1.11}$$

Note: Due to the symmetry $G_D(\mathbf{x}, \mathbf{x}') = G_D(\mathbf{x}', \mathbf{x})$, $F(\mathbf{x}, \mathbf{x}')$ must also be symmetric under the exchange of $\mathbf{x}$ with $\mathbf{x}'$. This can be seen when transitioning to spherical coordinates $(\mathbf{x} \cdot \mathbf{x}' = rr' \cos\theta)$:

$$F(\mathbf{x}, \mathbf{x}') = \frac{-R}{r'\sqrt{r^2 + \dfrac{R^4}{r'^2} - 2\dfrac{R^2 r}{r'}\cos\theta}} = \frac{-1}{\sqrt{\dfrac{r^2 r'^2}{R^2} + R^2 - 2rr'\cos\theta}} . \tag{3.1.12}$$

We know that $\Delta F = 0$ in V, but $\Delta F \neq 0$ in the exterior space S_R, which will still be used. Now follows from (3.1.10)

$$\Delta F(\mathbf{x}, \mathbf{x}') = -\frac{R}{r'}\Delta\frac{1}{|\mathbf{x} - \frac{R^2}{r'^2}\mathbf{x}'|} = \frac{4\pi R}{r'}\delta^{(3)}\left(\mathbf{x} - \frac{R^2}{r'^2}\mathbf{x}'\right)$$

$$= \Delta' F(\mathbf{x}, \mathbf{x}') = \frac{4\pi R}{r}\delta^{(3)}\left(\mathbf{x}' - \frac{R^2}{r^2}\mathbf{x}\right) . \tag{3.1.13}$$

The symmetry $\mathbf{x} \rightleftharpoons \mathbf{x}'$ was exploited.

2. *Point charge in front of the sphere with* $\phi(\partial S_R) = 0$

   The surface term of (3.1.6) disappears. Potential and field are given by

$$\phi^{(0)}(\mathbf{x}) = k_C \int d^3x'\, G_D(\mathbf{x}, \mathbf{x}')\, q\delta^{(3)}(\mathbf{x}' - \mathbf{d}) = k_C\left[\frac{q}{|\mathbf{x} - \mathbf{d}|} - \frac{q\frac{R}{d}}{|\mathbf{x} - \frac{R^2}{d^2}\mathbf{d}|}\right]$$

$$\mathbf{E}^{(0)} = -\boldsymbol{\nabla}\phi^{(0)}(\mathbf{x}) = k_C q\left[\frac{\mathbf{x} - \mathbf{d}}{|\mathbf{x} - \mathbf{d}|^3} - \frac{R}{d}\frac{\mathbf{x} - \frac{R^2}{d^2}\mathbf{d}}{|\mathbf{x} - \frac{R^2}{d^2}\mathbf{d}|^3}\right]. \tag{3.1.14}$$

3. *Point charge in front of the sphere with* $\phi(\partial S_R) = \phi_0$

   The contribution of the surface ∂V comes solely from the sphere surface ∂S_R, since at infinity $\phi = 0$. The constant ϕ_0 comes before the integral and $\phi^{(0)}$ is given by (3.1.14):

$$\phi(\mathbf{x}) = \phi^{(0)}(\mathbf{x}) + \frac{\phi_0}{4\pi}\oiint_{\partial S_R} d\mathbf{a}' \cdot \boldsymbol{\nabla}' G_D(\mathbf{x}, \mathbf{x}') .$$

The normal vector of ∂V is $\mathbf{n}' = -\mathbf{e}_r$, which is why the contribution of the surface has a positive sign. With Gauss's theorem and (3.1.13) one obtains

$$\phi(\mathbf{x}) = \phi^{(0)}(\mathbf{x}) + \frac{\phi_0}{4\pi}\int_{S_R} d^3x'\, \Delta' F(\mathbf{x}, \mathbf{x}') = \phi^{(0)}(\mathbf{x}) + \frac{\phi_0 R}{r} . \tag{2.3.5'}$$

4. *Point charge in front of isolated conducting sphere with the total charge* Q

   We start with a grounded conducting sphere, in front of which at location $\mathbf{d}$ $(d > R)$ the point charge q is located. With the image charge q' at location $\mathbf{d}'$, the potential on the sphere surface vanishes.

If the grounding is now disconnected, the induced charge q' is located on the sphere surface. If the sphere surface is to have the total charge Q, this must be supplied with the charge $Q - q'$, what can be represented by an image charge at the center of the sphere:

$$\phi(\mathbf{x}) = \phi^{(0)}(\mathbf{x}) + k_C \frac{Q + \frac{qR}{d}}{r} \qquad \text{for} \quad r > R, \qquad (2.3.5")$$

where $\phi^{(0)}$ is given by (3.1.14). We have used the superposition principle.

Point dipole in front of conducting sphere

The next simplest configuration is the attachment of a point dipole in front of the conducting sphere, as sketched in Fig. 3.5, and its charge distribution is given by (2.2.7)

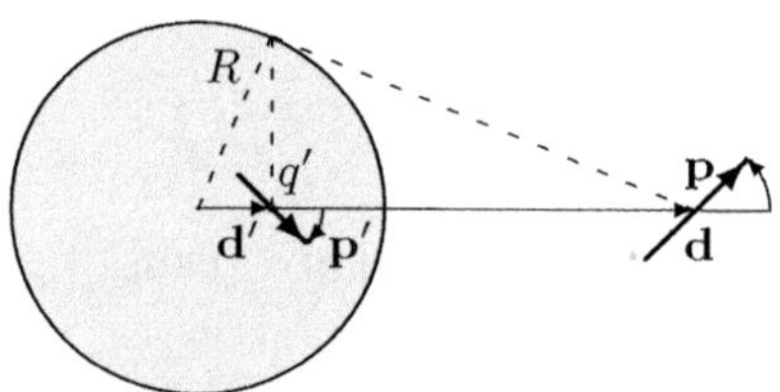

Fig. 3.5. Point dipole $\mathbf{p}$ in front of a conducting sphere with radius R. An additional mirror charge q' needs to be added to the mirror dipole $\mathbf{p}'$, which is negative when $\mathbf{p}$ is directed outward. When $\mathbf{p}$ is rotated counterclockwise, $\mathbf{p}'$ rotates clockwise

$$\rho(\mathbf{x}) = -\mathbf{p} \cdot \boldsymbol{\nabla} \delta^{(3)}(\mathbf{x} - \mathbf{d}).$$

The Green's function for this topological arrangement has already been determined with (3.1.11). The potential is thus given by

$$\phi(\mathbf{x}) = k_C \int d^3 x' \, G_D(\mathbf{x}, \mathbf{x}') \, \rho(\mathbf{x}')$$

$$= -k_C \int d^3 x' \left[\frac{1}{|\mathbf{x} - \mathbf{x}'|} - \frac{R}{r'} \frac{1}{|\mathbf{x} - \frac{R^2}{r'^2}\mathbf{x}'|} \right] \mathbf{p} \cdot \boldsymbol{\nabla}' \delta^{(3)}(\mathbf{x}' - \mathbf{d}).$$

This potential is only valid outside the sphere. Partial integration yields

$$\phi(\mathbf{x}) = k_C \int d^3 x' \, \delta^{(3)}(\mathbf{x}' - \mathbf{d}) \, \mathbf{p} \cdot \boldsymbol{\nabla}' \left[\frac{1}{|\mathbf{x} - \mathbf{x}'|} - \frac{\frac{R}{r'}}{|\mathbf{x} - \frac{R^2}{r'^2}\mathbf{x}'|} \right]$$

$$\overset{\mathbf{d}' = \mathbf{d} R^2/d^2}{=} k_C \left[\frac{\mathbf{p} \cdot (\mathbf{x} - \mathbf{d})}{|\mathbf{x} - \mathbf{d}|^3} + \frac{R}{d^3} \frac{\mathbf{p} \cdot \mathbf{d}}{|\mathbf{x} - \mathbf{d}'|} + \frac{R^3}{d^3} \frac{(\mathbf{x} - \mathbf{d}')}{|\mathbf{x} - \mathbf{d}'|^3} \cdot \left(2\frac{\mathbf{p} \cdot \mathbf{d}}{d^2}\mathbf{d} - \mathbf{p} \right) \right].$$

The first term is the potential of the point dipole $\mathbf{p}$ at location $\mathbf{d}$, the second is that of an image charge $q' = \frac{R}{d^3}\mathbf{p} \cdot \mathbf{d}$ at location $\mathbf{d}'$. This disappears when $\mathbf{p}$ is perpendicular to $\mathbf{d}$. If $\mathbf{d}$ is parallel to $\mathbf{p}$, then the charge is positive, when in an antiparallel position is negative. The dipole

$$\mathbf{p}' = \frac{R^3}{d^3}\Big(2\frac{\mathbf{p}\cdot\mathbf{d}}{d^2}\mathbf{d} - \mathbf{p}\Big)$$

has the strength $p' = \frac{R^3}{d^3}\,p$ and is parallel to $\mathbf{p}$, when this is oriented along $\mathbf{d}$ and antiparallel when $\mathbf{p}$ is perpendicular to $\mathbf{d}$.

3.2 Laplace's Equation in Spherical Coordinates

3.2.1 Separation Ansatz for Spherical Coordinates

The homogeneous Poisson equation $\Delta\phi = 0$ is called Laplace's equation. It has in Cartesian coordinates the form

$$\left(\frac{\partial^2}{\partial x^2} + \frac{\partial^2}{\partial y^2} + \frac{\partial^2}{\partial z^2}\right)\phi(\mathbf{x}) = 0 \tag{3.2.1}$$

and in spherical coordinates (see (A.3.36))

$$\left(\frac{1}{r^2}\frac{\partial}{\partial r}r^2\frac{\partial}{\partial r} + \frac{1}{r^2\sin\vartheta}\frac{\partial}{\partial\vartheta}\sin\vartheta\frac{\partial}{\partial\vartheta} + \frac{1}{r^2\sin^2\vartheta}\frac{\partial^2}{\partial\varphi^2}\right)\phi(r,\vartheta,\varphi) = 0. \tag{3.2.2}$$

The Laplacian can be divided into a radial and an angle-dependent part:

$$\Delta = \Delta_r - \frac{1}{r^2}\,\hat{\mathbf{L}}^2\,,$$

$$\Delta_r = \frac{1}{r^2}\frac{\partial}{\partial r}r^2\frac{\partial}{\partial r} \overset{(A.3.36')}{=} \frac{1}{r}\frac{\partial^2}{\partial r^2}r,$$

$$\hat{\mathbf{L}}^2 = -\frac{1}{\sin\vartheta}\frac{\partial}{\partial\vartheta}\sin\vartheta\frac{\partial}{\partial\vartheta} - \frac{1}{\sin^2\vartheta}\frac{\partial^2}{\partial\varphi^2}. \tag{3.2.3}$$

$\hat{\mathbf{L}}^2$ is the square of the (dimensionless) angular momentum operator

$$\hat{\mathbf{L}} = -\mathrm{i}\,\mathbf{x}\times\boldsymbol{\nabla}. \tag{3.2.4}$$

For the solution ϕ one makes the product ansatz of a radial function $R(r)$ with an angle-dependent function $Y(\vartheta,\varphi)$

$$\phi(r,\vartheta,\rho) = R(r)\,Y(\vartheta,\varphi) \tag{3.2.5}$$

and obtains the Laplace equation

$$Y\,\Delta_r R - R\,\frac{1}{r^2}\hat{\mathbf{L}}^2 Y = 0\,.$$

The radial part can be separated from the angle-dependent part if one multiplies on the left with $r^2/(Y R)$ and then brings the terms only dependent on ϑ,φ to the right side:

$$\frac{r^2}{R}\,\Delta_r\,R = \frac{1}{Y}\,\hat{\mathbf{L}}^2 Y = l(l+1).\tag{3.2.6}$$

The right side does not depend on r, the left not on ϑ and φ. Thus, both expressions can only be equal to a constant. Since both $\hat{\mathbf{L}}^2$ and Δ_r are positive semidefinite operators, this constant must be ≥ 0, where the choice $l(l+1)$ becomes comprehensible when solving the polar part. The parts are evaluated separately.

3.2.2 Radial Part

For the radial part one obtains

$$\frac{r^2}{R}\Delta_r = \frac{r}{R}\frac{\partial^2}{\partial r^2}\,rR = l(l+1).$$

With the further substitution $R = \dfrac{u}{r}$, the radial part results in

$$\frac{\mathrm{d}^2 u}{\mathrm{d}r^2} - \frac{l(l+1)}{r^2}u = 0.\tag{3.2.7}$$

The solution to this differential equation is

$$u = \alpha\,r^{l+1} + \beta\,r^{-l} \qquad \Longrightarrow \qquad R = \alpha\,r^l + \beta\,r^{-l-1}.\tag{3.2.8}$$

It is sufficient to restrict to $l \geq 0$ as $l < 0$ can be replaced by $l' = -l-1 \geq 0$. Then it is

$$R = \alpha\,r^l + \beta\,r^{-l-1} \to \alpha\,r^{-l'-1} + \beta\,r^{l'}.$$

A restriction of the l to whole numbers only comes through the polar part. If $l \geq 0$, the second term of R diverges for $r \to 0$, while the first term for $r \to \infty$ at least does not vanish.

3.2.3 Azimuthal Part

In the angle-dependent part (3.2.6)

$$\left(\hat{\mathbf{L}}^2 - l(l+1)\right)Y = 0\tag{3.2.9}$$

one inserts (3.2.3), whereby the dependence on φ is expressed by the differential operator

$$\hat{L}_z = -\mathrm{i}\frac{\partial}{\partial\varphi}.\tag{3.2.10}$$

$$\left[\frac{1}{\sin\vartheta}\frac{\partial}{\partial\vartheta}\sin\vartheta\frac{\partial}{\partial\vartheta} + l(l+1)\right]Y(\vartheta,\varphi) = \frac{1}{\sin^2\vartheta}\,\hat{L}_z^2\,Y(\vartheta,\varphi).\tag{3.2.11}$$

By means of a new product ansatz

$$Y(\vartheta, \varphi) = \Theta(\vartheta)\, \Phi(\varphi)$$

and the separation constant m^2 results in the differential equation for the azimuthal angle

$$\left(\hat{L}_z^2 - m^2\right)\Phi = -\left(\frac{\mathrm{d}^2}{\mathrm{d}\varphi^2} + m^2\right)\Phi = 0 \quad \text{with} \quad \Phi(\varphi) = \frac{1}{\sqrt{2\pi}}\, \mathrm{e}^{\pm \mathrm{i}m\varphi}\,. \tag{3.2.12}$$

From the requirement of a unique solution $\Phi(\varphi + 2\pi) = \Phi(\varphi)$ it follows

$$\mathrm{e}^{\pm 2\pi \mathrm{i}m} = 1 \qquad \text{with} \quad m = 0, \pm 1, \pm 2, \ldots$$

m must therefore be an integer. The solution

$$\Phi_m(\varphi) = \frac{1}{\sqrt{2\pi}}\, \mathrm{e}^{\pm \mathrm{i}m\varphi}\,, \qquad m \quad \text{integer}, \tag{3.2.13}$$

are plane waves with the base $[0, 2\pi]$. They are orthonormal

$$(\Phi_m, \Phi_{m'}) = \frac{1}{2\pi} \int_0^{2\pi} \mathrm{d}\varphi\, \mathrm{e}^{\mathrm{i}(m'-m)\varphi} = \delta_{mm'} \tag{3.2.14}$$

and complete

$$\sum_{m=-\infty}^{\infty} \Phi_m(\varphi)\, \Phi_m^*(\varphi') = \frac{1}{2\pi} \sum_m \mathrm{e}^{\mathrm{i}m(\varphi-\varphi')} = \delta(\varphi - \varphi')\,. \tag{3.2.15}$$

The validity of the orthonormality is evident. To prove the completeness (3.2.15) we develop a function $f(\varphi)$ according to the $\Phi_m(\varphi)$:

$$f(\varphi) = \int_0^{2\pi} \mathrm{d}\varphi'\, \delta(\varphi-\varphi') f(\varphi') = \sum_m \Phi_m(\varphi) \int_0^{2\pi} \mathrm{d}\varphi'\, \Phi_m^*(\varphi')\, f(\varphi')$$

$$= \sum_m \Phi_m(\varphi)\, (\Phi_m, f)\,.$$

We verify this development by multiplying from the left with $\displaystyle\int_0^{2\pi} \mathrm{d}\varphi\, \Phi_n^*(\varphi)$ and use the orthonormality $(\Phi_n, f) = \displaystyle\sum_m \delta_{nm}\, (\Phi_m, f)$.

3.2.4 Polar Part

For the polar angle ϑ one obtains the differential equation

$$\left[\frac{1}{\sin\vartheta} \frac{\partial}{\partial\vartheta}\left(\sin\vartheta \frac{\partial}{\partial\vartheta}\right) - \frac{m^2}{\sin^2\vartheta} + l(l+1)\right]\Theta(\vartheta) = 0\,. \tag{3.2.16}$$

This is the differential equation for the associated Legendre polynomials, which is usually specified in the variable $\xi = \cos\vartheta$.

For the transformation to $\xi = \cos\vartheta$ with $-1 \leq \xi \leq 1$ one needs

$$\frac{\mathrm{d}\xi}{\mathrm{d}\vartheta} = -\sin\vartheta = -\sqrt{1-\xi^2}, \qquad \frac{\mathrm{d}}{\mathrm{d}\vartheta} = \frac{\mathrm{d}\xi}{\mathrm{d}\vartheta}\frac{\mathrm{d}}{\mathrm{d}\xi} = -\sqrt{1-\xi^2}\,\frac{\mathrm{d}}{\mathrm{d}\xi},$$

$$\frac{1}{\sin\vartheta}\frac{\partial}{\partial\vartheta}\left(\sin\vartheta\frac{\partial}{\partial\vartheta}\right) = \frac{\mathrm{d}}{\mathrm{d}\xi}(1-\xi^2)\frac{\mathrm{d}}{\mathrm{d}\xi}.$$

The solutions of (3.2.16) $\Theta(\vartheta)$ are denoted with $P_l^m(\xi)$. Thus one gets the differential equation

$$\left[\frac{\mathrm{d}}{\mathrm{d}\xi}(1-\xi^2)\frac{\mathrm{d}}{\mathrm{d}\xi} - \frac{m^2}{1-\xi^2} + l(l+1)\right]P_l^m(\xi) = 0, \tag{3.2.17}$$

whose solution are the associated Legendre polynomials. To determine the solutions for the P_l^m, one starts from the case $m = 0$. Once you have the solutions P_l, they can be easily used to determine P_l^m, as will be shown later. Therefore, we will briefly discuss the Legendre polynomials $P_l(\xi)$ below.

Legendre's differential equation

For $m = 0$, one obtains from (3.2.17) the *Legendre's differential equation*

$$\left[(1-\xi^2)\frac{\mathrm{d}^2}{\mathrm{d}\xi^2} - 2\xi\frac{\mathrm{d}}{\mathrm{d}\xi} + l(l+1)\right]P_l(\xi) = 0 \tag{3.2.18}$$

with the domain $[-1, 1]$. Regular solutions are the *Legendre polynomials* $P_l(\xi)$, whose degree is denoted by l.

Legendre polynomials

Here, the P_l are derived in a "classical" way, which implies a power series approach with recursion conditions for the coefficients. With their help, the polynomial can then be determined.

The differential operator is invariant under $\xi \to -\xi$, so the solutions are even or odd functions of ξ. Furthermore, the differential operator is not singular for $|\xi| < 1$, which is why the solution as a power series

$$P(\xi) = \sum_{n=0}^{\infty} a_n\, \xi^n$$

can be represented. By inserting into (3.2.18)

$$\sum_{n\geq 2} n(n-1)a_n\, \xi^{n-2} - \sum_{n\geq 0}\left[n(n+1) - l(l+1)\right]a_n\, \xi^n = 0.$$

Since each power of ξ must vanish separately, the comparison of coefficients yields

$$\xi^n: \quad (n+2)(n+1)a_{n+2} - \big[n(n+1) - l(l+1)\big]a_n = 0$$

the recursion relation

$$a_{n+2} = \frac{n(n+1) - l(l+1)}{(n+1)(n+2)}a_n = -\frac{(l+n+1)(l-n)}{(n+1)(n+2)}a_n. \tag{3.2.19}$$

The series therefore contains only even or odd powers in ξ, depending on whether the solution is an even or odd function in ξ. For large n, the series behaves like $a_{n+2}/a_n \to n/(n+2)$ and diverges for $\xi = \pm 1$ like

$$\ln \frac{1+\xi}{1-\xi} = 2\Big(\xi + \frac{1}{3}\xi^3 + \frac{1}{5}\xi^5 + \cdots\Big).$$

For the solution to be finite at $\xi = \pm 1$, the series must terminate, which is the case according to (3.2.19) when $n = l$. From the termination condition it follows that l is an integer. The solution is therefore polynomials $P_l(\xi)$.

The P_l can be calculated using the recursion relation (3.2.19) (see problem B.1). One finally obtains

$$P_l(\xi) = \sum_n^l \frac{(-1)^{\frac{l-n}{2}}(l+n)!\,\xi^n}{2^l\left(\frac{l+n}{2}\right)!\left(\frac{l-n}{2}\right)!\,n!} \quad \begin{cases} n=0,2,...,l & \text{for even } l \\ n=1,3,...,l & \text{for odd } l. \end{cases} \tag{3.2.20}$$

P_l is therefore a polynomial of degree l in the base domain $[-1, 1]$. The simplest polynomials P_0 to P_3 are defined and illustrated in Fig. 3.6. In the appendix

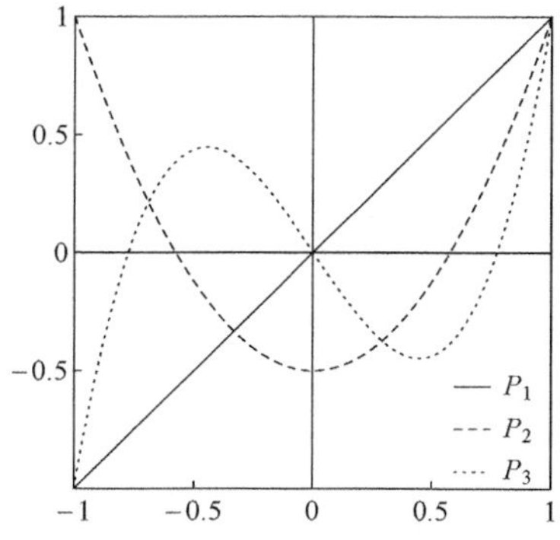

Fig. 3.6. First kind Legendre polynomials: $P_0 = 1,$ $P_1 = \xi,$ $P_2 = (3\xi^2 - 1)/2,$ $P_3 = (5\xi^3 - 3\xi)/2$. The even polynomials are symmetric, the odd ones antisymmetric

B.2 not only a more elegant way is chosen using the Rodrigues formula to derive the Legendre polynomials, but also some important relations associated with them are presented (which, however, go beyond the scope required here).

Symmetry property

The P_l are even or odd functions:

$$P_l(-\xi) = (-1)^l P_l(\xi) \qquad\qquad \text{symmetry.} \quad (3.2.21)$$

Orthogonality and completeness

The P_l form a complete and orthogonal system.

$$\int_{-1}^{1} d\xi\, P_l(\xi)\, P_{l'}(\xi) = \frac{2}{2l+1}\, \delta_{ll'} \qquad\qquad \text{orthogonality.} \quad (3.2.22)$$

$$\sum_{l=0}^{\infty} \frac{2l+1}{2}\, P_l(\xi)\, P_l(\xi') = \delta(\xi - \xi') \qquad\qquad \text{completeness.} \quad (3.2.23)$$

The proofs for the validity of the two relations are in the appendix ((B.2.13) and (B.2.15)).

Associated Legendre polynomials

We now want to determine the solution functions for the polar part (3.2.17) of the Laplace equation, the associated (associated) Legendre polynomials P_l^m. First, an approach of the form

$$P_l^m(\xi) = (1 - \xi^2)^{\frac{m}{2}} P_l^{(m)}(\xi) \qquad m = 0, 1, ..., l \qquad\qquad (3.2.24)$$

is made and the differential equation for the $P_l^{(m)}$ is derived:

$$\frac{d}{d\xi}(1-\xi^2)\frac{d}{d\xi}\left[(1-\xi^2)^{\frac{m}{2}} P_l^{(m)}\right] = \frac{d}{d\xi}\left[-\xi m(1-\xi^2)^{\frac{m}{2}} + (1-\xi^2)^{\frac{m}{2}+1}\frac{d}{d\xi}\right]P_l^{(m)}$$

$$= \left[-m(1-\xi^2)^{\frac{m}{2}} + \xi^2 m^2(1-\xi^2)^{\frac{m}{2}-1} - \xi m(1-\xi^2)^{\frac{m}{2}}\frac{d}{d\xi}\right.$$

$$\left. -\xi(m+2)(1-\xi^2)^{\frac{m}{2}}\frac{d}{d\xi} + (1-\xi^2)^{\frac{m}{2}+1}\frac{d^2}{d\xi^2}\right]P_l^{(m)}$$

$$= (1-\xi^2)^{\frac{m}{2}}\left[(1-\xi^2)\frac{d^2}{d\xi^2} - 2\xi(m+1)\frac{d}{d\xi} - m(m+1) + m^2(1-\xi^2)^{-1}\right]P_l^{(m)}.$$

This expression in (3.2.17) inserted, results in (3.2.25).

$$\left[(1-\xi^2)\frac{d^2}{d\xi^2} - 2\xi(m+1)\frac{d}{d\xi} + \left[l(l+1) - m(m+1)\right]\right]P_l^{(m)} = 0. \qquad (3.2.25)$$

The benefit of the approach (3.2.24) becomes apparent when one differentiates the Legendre differential equation (3.2.18) m times:

$$\frac{d^m}{d\xi^m}\left[(1-\xi^2)\frac{d^2}{d\xi^2} - 2\xi\frac{d}{d\xi} + l(l+1)\right]P_l(\xi) = 0. \qquad (3.2.26)$$

One obtains

$$\left\{(1-\xi^2)\frac{\mathrm{d}^2}{\mathrm{d}\xi^2} - 2\xi(m+1)\frac{\mathrm{d}}{\mathrm{d}\xi} + \left[l(l+1)-m(m+1)\right]\right\}\frac{\mathrm{d}^m}{\mathrm{d}\xi^m}P_l = 0\,. \quad (3.2.27)$$

This is exactly (3.2.25) for $P_l^{(m)}$, with which we can show that

$$P_l^{(m)}(\xi) = \frac{\mathrm{d}^m}{\mathrm{d}\xi^m}P_l$$

are determined by the m-th derivative of the P_l.

For the differentiation in (3.2.26) we use the auxiliary formula

$$\frac{\mathrm{d}^m(f\,g)}{\mathrm{d}\xi^m} = \sum_{k=0}^{m}\binom{m}{k}\frac{\mathrm{d}^k f}{\mathrm{d}\xi^k}\frac{\mathrm{d}^{m-k}g}{\mathrm{d}\xi^{m-k}}\,.$$

$$\frac{\mathrm{d}^m}{\mathrm{d}\xi^m}(1-\xi^2)\frac{\mathrm{d}^2}{\mathrm{d}\xi^2} = (1-\xi^2)\frac{\mathrm{d}^{m+2}}{\mathrm{d}\xi^{m+2}} - 2\xi m\,\frac{\mathrm{d}^{m+1}}{\mathrm{d}\xi^{m+1}} - m(m-1)\,\frac{\mathrm{d}^m}{\mathrm{d}\xi^m}\,.$$

$$-\frac{\mathrm{d}^m}{\mathrm{d}\xi^m}2\xi\frac{\mathrm{d}}{\mathrm{d}\xi} = -2\xi\frac{\mathrm{d}^{m+1}}{\mathrm{d}\xi^{m+1}} - 2m\frac{\mathrm{d}^m}{\mathrm{d}\xi^m}\,.$$

Thus, the $P_l^{(m)}$ are determined and consequently also the associated Legendre polynomials:

$$P_l^m(\xi) = (1-\xi^2)^{m/2}\frac{\mathrm{d}^m}{\mathrm{d}\xi^m}P_l(\xi) \quad\quad\quad (3.2.28)$$

$$= \frac{1}{2^l\,l!}(1-\xi^2)^{m/2}\frac{\mathrm{d}^{l+m}}{\mathrm{d}\xi^{l+m}}(\xi^2-1)^l \quad\quad -l\leq m\leq l.$$

For the P_l the Rodrigues formula (B.2.2) was used. The resulting formula is no longer limited to $0\geq m\geq 0$, but covers the entire range $-l\leq m\leq l$. The identity is shown in appendix B.3

$$P_l^{-m}(\xi) = (-1)^m\frac{(l-m)!}{(l+m)!}\,P_l^m(\xi)\,. \quad\quad\quad (3.2.29)$$

The simplest polynomials are

$$P_l^0 = P_l\,, \quad\quad\quad\quad\quad\quad P_l^l = (2l-1)!!\,(1-\xi^2)^{\frac{l}{2}}\,.$$

Orthogonality

The associated Legendre polynomials with the same m are orthogonal with respect to l, which is not shown here:

$$\int_{-1}^{1}\mathrm{d}\xi P_l^m(\xi)\,P_{l'}^m(\xi) = \frac{2}{2l+1}\frac{(l+m)!}{(l-m)!}\,\delta_{ll'}\,. \quad\quad\quad (3.2.30)$$

Note: The frequently used definition

$$P_l^m(\xi) = (-1)^m\,(1-\xi^2)^{m/2}\frac{\mathrm{d}^m}{\mathrm{d}\xi^m}P_l(\xi)$$

differs from (3.2.28) by the factor $(-1)^m$; this is e.g. to be considered in recursion relations.

Spherical harmonics

The solution of the angle-dependent part $Y(\vartheta, \varphi)$ of the Laplace equation are spherical harmonics

$$Y(\vartheta, \varphi) = \Theta(\vartheta)\,\Phi(\varphi) \rightarrow Y_{lm}(\vartheta, \varphi) = A_{lm}\,P_l^m(\cos\vartheta)\,\Phi_m(\varphi)\,.$$

The associated Legendre polynomials P_l^m are, unlike the Φ_m, not normalized. The normalization factor A_{lm} is obtained from the orthogonality relation (3.2.30):

$$A_{lm} = (-1)^m \sqrt{\frac{2l+1}{2}\frac{(l-m)!}{(l+m)!}}\,, \tag{3.2.31}$$

so that $\Theta_{lm} = A_{lm}\,P_l^m$ is the normalized associated Legendre polynomial. It is then

$$Y_{lm}(\vartheta, \varphi) = (-1)^m \left[\frac{2l+1}{4\pi}\frac{(l-m)!}{(l+m)!}\right]^{\frac{1}{2}} P_l^m(\cos\vartheta)\,\mathrm{e}^{im\varphi} \tag{3.2.32}$$

$$= \frac{(-1)^{m+l}}{2^l l!}\left[\frac{2l+1}{4\pi}\frac{(l-m)!}{(l+m)!}\right]^{\frac{1}{2}} \sin^m\vartheta\,\frac{\mathrm{d}^{l+m}\sin^{2l}\vartheta}{\mathrm{d}\cos\vartheta^{l+m}}\,\mathrm{e}^{im\varphi}\,.$$

Symmetry properties

The Y_{lm} fulfill the symmetry

$$Y_{l\,-m}(\vartheta, \varphi) = (-1)^m\,Y_{lm}^*(\vartheta, \varphi)\,, \tag{3.2.33}$$

which follows from (3.2.29) and $\Phi_{-m} = \Phi_m^*$.

In case of inversion $(\mathbf{x} \rightarrow -\mathbf{x})$ applies $(r, \vartheta, \varphi) \rightarrow (r, \pi - \vartheta, \pi + \varphi)$, since $\cos(\pi - \vartheta) = -\cos\vartheta$ and $\sin(\pi - \vartheta) = \sin\vartheta$. Thus, according to (3.2.32)

$$Y_{lm}(\pi - \vartheta, \varphi + \pi) = (-1)^l\,Y_{lm}(\vartheta, \varphi)\,. \tag{3.2.34}$$

Note: The spherical functions Y_{lm} are consistently defined and their relationship with the Legendre polynomials is given by

$$P_l(\cos\vartheta) = \sqrt{\frac{4\pi}{2l+1}}\,Y_{l0}(\vartheta, \varphi)\,. \tag{3.2.35}$$

Orthogonality

$\Theta_{lm} = A_{lm}\,P_l^m$ are orthonormal, as are Φ_m, from which it follows that $Y_{lm} = \Theta_{lm}\,\Phi_m$ must also be orthonormal:

$$\int_0^\pi \mathrm{d}\vartheta\,\sin\vartheta\int_0^{2\pi}\mathrm{d}\varphi\,Y_{lm}^*(\vartheta, \varphi)\,Y_{l'm'}(\vartheta, \varphi) = \delta_{ll'}\,\delta_{mm'}\,. \tag{3.2.36}$$

Completeness

$$\sum_{l=0}^{\infty}\sum_{m=-l}^{l} Y_{lm}(\vartheta, \varphi)\,Y_{lm}^*(\vartheta', \varphi') = (\sin\vartheta)^{-1}\,\delta(\vartheta - \vartheta')\,\delta(\varphi - \varphi')\,. \tag{3.2.37}$$

3.2.5 Solution of the Laplace Equation

The general solution of the Laplace equation (3.2.2), the multiplication of the radial part (3.2.8) with the angle-dependent part (3.2.32) results in

$$\phi(r, \vartheta, \varphi) = \sum_{l=0}^{\infty} \sum_{m=-l}^{l} (\alpha_{lm} r^l + \beta_{lm} r^{-l-1}) Y_{lm}(\vartheta, \varphi). \tag{3.2.38}$$

This is the expansion of the electrostatic potential according to spherical harmonics. This expansion presupposes a charge-free region in which the potential satisfies the Laplace equation. The solution is regular for finite r.

In a region that $r = 0$ contains, all $\beta_{lm} = 0$ must vanish, and in a region that $r = \infty$ contains, all $\alpha_{lm} = 0$ for $l > 1$.

A homogeneous field is taken into account, in which the potential $\phi \sim r$ is. The solutions of the Laplace equation are harmonic functions.

Definition: A function ϕ is harmonic in a region G, if it is at least $2\times$ continuously differentiable and satisfies the Laplace equation.

Properties of harmonic functions:

1. *Mean value property*: Let S be a sphere entirely in G with radius R, so the function value at the center of the sphere is equal to the (arithmetic) mean of the function values of ϕ on the sphere surface (see appendix B.1.2, page 607)

$$\phi(\mathbf{x}) = \frac{1}{4\pi R^2} \oiint_{\partial S} da' \, \phi(\mathbf{x}'). \tag{3.2.39}$$

Proof: In the 2nd Green's theorem (A.4.20) is inserted $\psi = \dfrac{1}{|\mathbf{x}' - \mathbf{x}|}$.

$$\oiint_{\partial S} d\mathbf{a}' \cdot \left(\frac{1}{|\mathbf{x}' - \mathbf{x}|} \boldsymbol{\nabla}' \phi(\mathbf{x}') - \phi(\mathbf{x}') \boldsymbol{\nabla}' \frac{1}{|\mathbf{x}' - \mathbf{x}|} \right)$$

$$= \int_S d^3 x' \left(\frac{1}{|\mathbf{x}' - \mathbf{x}|} \Delta' \phi(\mathbf{x}') - \phi(\mathbf{x}') \Delta' \frac{1}{|\mathbf{x}' - \mathbf{x}|} \right) = 4\pi \phi(\mathbf{x}).$$

In the left surface integral, $|\mathbf{x}' - \mathbf{x}| = R$ is inserted and the surface integral is transformed into a volume integral using Gauss's theorem, which vanishes due to $\Delta \phi = 0$

$$\oiint_{\partial S} d\mathbf{a}' \cdot \frac{1}{R} \boldsymbol{\nabla}' \phi(\mathbf{x}') = \frac{1}{R} \int_S d^3 x' \, \Delta' \phi(\mathbf{x}') = 0.$$

To calculate the surface terms, we perform the further transformation $\mathbf{x}'' = \mathbf{x}' - \mathbf{x}$ and take into account the surface element $d\mathbf{a}' = da'' \, \mathbf{e}_{r''}$

$$\phi(\mathbf{x}) = \frac{1}{4\pi} \oiint_{\partial S} da'' \, \mathbf{e}_{r''} \cdot \phi(\mathbf{x}'' + \mathbf{x}) \frac{\mathbf{e}_{r''}}{r''^2} = \frac{1}{4\pi R^2} \oiint_{\partial S} da'' \, \phi(\mathbf{x}'' + \mathbf{x}) \qquad \text{q.e.d.}$$

---

2. *Maximum and minimum principle*: A harmonic function ϕ in the domain G always has its maxima and minima at the boundary ∂G of the domain.

3. If a maximum/minimum is inside G, then ϕ is constant.

If one can expect due to the symmetry that ϕ is axially symmetric, one can select the xz-plane, where $\varphi = 0$. From (3.2.38) only terms with $m = 0$, which are proportional to the Legendre polynomials remain

$$\phi(r, \vartheta) = \sum_{l=0}^{\infty} (a_l r^l + b_l r^{-l-1}) P_l(\cos \vartheta) \,. \tag{3.2.40}$$

This expansion is easier to handle because of the simpler coefficients of the P_l.

Theorem of Earnshaw

The Earnshaw theorem states that no charged body can be held in a stable equilibrium solely under the influence of electrostatic forces.

For a body to be held in a stable equilibrium, the potential must have a minimum. If $\mathbf{x}_0$ is the location of the minimum, then in a neighborhood $\Delta\phi = 0$, since there are no charges there. The maximum and minimum principle states that there are function values of ϕ in a sphere around the minimum that are greater or smaller than $\phi(x_0)$ (it is a property of harmonic functions that they do not have maxima/minima, but only saddle points).

Note: The theorem also applies in magnetostatics, but only in regions that are charge- and current-free. In diamagnetic substances, currents are induced by the external magnetic field B whose magnetic field is opposed to the external field (see section 7.3.1). With very strong fields bodies can thus be kept in suspension (diamagnetic levitation).

3.3 Spherically Symmetric Problems

3.3.1 Properties of the Spherical Harmonics

In boundary value problems of electrostatics with spherical symmetry the spherical harmonics Y_{lm} play a central role. It is therefore necessary to go into more detail on these.

Expansion according to spherical harmonics

Since the Y_{lm} are complete, any function $f(\vartheta, \varphi)$ can be expanded according to them. One multiplies the completeness relation (3.2.37) from the left with
$\int d\vartheta' \sin \vartheta' \, d\varphi' \, f(\vartheta', \varphi')$:

$$f(\vartheta, \varphi) = \sum_{l=0}^{\infty} \sum_{m=-l}^{l} Y_{lm}(\vartheta, \varphi) \, (Y_{lm}, f),$$

$$(Y_{lm}, f) \equiv f_{lm} = \int_0^{\pi} d\vartheta' \sin \vartheta' \int_0^{2\pi} d\varphi' \, Y_{lm}^*(\vartheta', \varphi') \, f(\vartheta', \varphi') \tag{3.3.1}$$

and has obtained the expansion of $f(\vartheta, \varphi)$ according to spherical harmonics. In the expansion of $f(\vartheta = 0, \varphi)$ one has to consider that

$$Y_{lm}(0, \varphi) = \delta_{m0}\, Y_{l0}(0, \varphi) = \delta_{m0}\sqrt{\frac{2l+1}{4\pi}}\, P_l(1)\,,$$

where $P_l(1) = 1$. This results in

$$f(\vartheta = 0, \varphi) = \sum_{l=0}^{\infty} \sqrt{\frac{2l+1}{4\pi}}\,(Y_{l0}, f) = \sum_{l=0}^{\infty} \frac{2l+1}{4\pi}\,(P_l, f)\,,$$

$$(P_l, f) = \int_0^{\pi} d\vartheta'\, \sin\vartheta' \int_0^{2\pi} d\varphi'\, P_l(\cos\vartheta')\, f(\vartheta', \varphi')\,. \tag{3.3.2}$$

Spherical harmonics up to 2nd order

The following are the lowest polynomials Y_{lm}, it should be noted that in many cases, the orders up to $l = 2$ are sufficient. The polynomials themselves can be calculated using recursion formulas, which are listed in appendix B.3:

$$Y_{00} = \frac{1}{\sqrt{4\pi}}\,, \tag{3.3.3}$$

$$Y_{10} = \sqrt{\frac{3}{4\pi}} \begin{cases} \cos\vartheta \\ \dfrac{z}{r}\,, \end{cases} \qquad Y_{1\pm1} = \mp\sqrt{\frac{3}{8\pi}} \begin{cases} e^{\pm i\varphi}\sin\vartheta \\ \dfrac{x \pm iy}{r}\,, \end{cases}$$

$$Y_{20} = \sqrt{\frac{5}{16\pi}} \begin{cases} 3\cos^2\vartheta - 1 \\ \dfrac{3z^2 - r^2}{r^2}\,, \end{cases} \quad Y_{2\pm1} = \mp\sqrt{\frac{15}{32\pi}} \begin{cases} e^{\pm i\varphi}\sin 2\vartheta \\ \dfrac{(x \pm iy)z}{r^2}\,, \end{cases} \quad Y_{2\pm2} = \sqrt{\frac{15}{32\pi}} \begin{cases} e^{\pm 2i\varphi}\sin^2\vartheta \\ \dfrac{(x \pm ix)^2}{r^2}\,. \end{cases}$$

Addition theorem for spherical harmonics

A relationship that is important because of its wide range of applications is the so-called *addition theorem for spherical harmonics*

$$\sum_{m=-l}^{l} Y_{lm}(\vartheta, \varphi)\, Y_{lm}^*(\vartheta', \varphi') = \frac{2l+1}{4\pi} P_l(\cos\theta)\,. \tag{3.3.4}$$

θ is the angle enclosed by the vectors $\mathbf{x}$ and $\mathbf{x}'$, as sketched in Fig. 3.7. A special case of the addition theorem (3.3.4) for $l = 1$ is the spherical cosine theorem

$$\cos\theta = \cos\vartheta\,\cos\vartheta' + \sin\vartheta\,\sin\vartheta'\,\cos(\varphi - \varphi') \tag{3.3.5}$$

$$= \frac{4\pi}{3} \sum_{m=-1}^{1} Y_{1m}(\vartheta, \varphi)\, Y_{1m}^*(\vartheta', \varphi')\,,$$

and a proof of the addition theorem is in the appendix B.3.

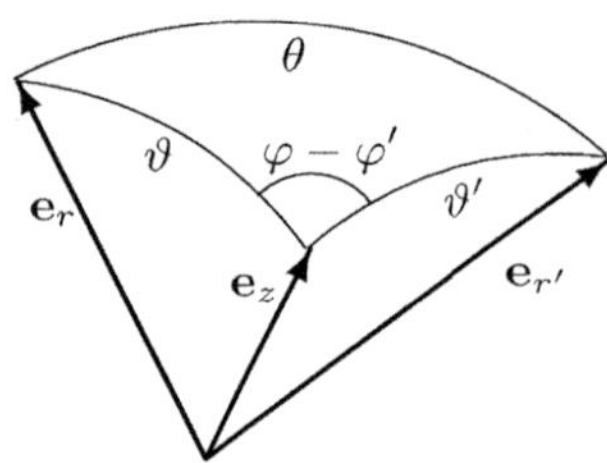

Fig. 3.7. Position of the vectors $\mathbf{x}$ and $\mathbf{x}'$ in relation to the $\mathbf{e}_z$-axis

3.3.2 Expansion of $|\mathbf{x}-\mathbf{x}'|^{-1}$ with Spherical Harmonics

The starting point is the generating function of the Legendre polynomials (B.2.6)

$$(1 - 2\xi t + t^2)^{-\frac{1}{2}} = \sum_{l=0}^{\infty} P_l(\xi)\, t^l \qquad\qquad |t| < 1. \tag{3.3.6}$$

Its derivation is explicitly carried out in the appendix. The function

$$\frac{1}{|\mathbf{x} - \mathbf{x}'|} = \frac{1}{\sqrt{r^2 + r'^2 - 2rr'\cos\theta}} = \frac{1}{r\sqrt{1 + \frac{r'^2}{r^2} - 2\frac{r'}{r}\cos\theta}}$$

can be directly expanded in powers of $\frac{r'}{r}$, as long as $r' < r$ or of $\frac{r}{r'}$, when $r < r'$. So $t = \frac{r}{r'}$ or $t = \frac{r'}{r}$, depending on whether $r < r'$ or $r > r'$ is to be inserted in (3.3.6). θ denotes the angle between $\mathbf{x}$ and $\mathbf{x}'$, where $\xi = \cos\theta$.

$$\frac{1}{|\mathbf{x}-\mathbf{x}'|} = \frac{1}{\sqrt{r^2 - 2rr'\cos\theta + r'^2}} = \begin{cases} \dfrac{1}{r} \displaystyle\sum_l \left(\dfrac{r'}{r}\right)^l P_l(\cos\theta) & r > r' \\[3mm] \dfrac{1}{r'} \displaystyle\sum_l \left(\dfrac{r}{r'}\right)^l P_l(\cos\theta) & r < r'. \end{cases} \tag{3.3.7}$$

This relationship is useful in calculating the potential in spherically symmetric problems. Using the addition theorem (3.3.4) we get

$$\frac{1}{|\mathbf{x}-\mathbf{x}'|} = \begin{cases} \dfrac{1}{r} \displaystyle\sum_{l=0}^{\infty} \dfrac{4\pi}{2l+1} \left(\dfrac{r'}{r}\right)^l \displaystyle\sum_{m=-l}^{l} Y_{lm}(\vartheta,\varphi) Y_{lm}^{*}(\vartheta',\varphi') & r > r' \\[4mm] \dfrac{1}{r'} \displaystyle\sum_{l=0}^{\infty} \dfrac{4\pi}{2l+1} \left(\dfrac{r}{r'}\right)^l \displaystyle\sum_{m=-l}^{l} Y_{lm}(\vartheta,\varphi) Y_{lm}^{*}(\vartheta',\varphi') & r < r'. \end{cases} \tag{3.3.8}$$

3.3.3 Multipole Expansion with Spherical Harmonics

The potential of the charge distribution limited to $r' < r$ is

$$\phi(r, \vartheta, \varphi) = k_c \sum_{l=0}^{\infty} r^{-l-1} \int \mathrm{d}^3 x' \, \rho(\mathbf{x}') \, r'^l \, P_l(\cos\theta), \tag{3.3.9}$$

when $\dfrac{1}{|\mathbf{x} - \mathbf{x}'|}$ is expanded according to (3.3.7) by Legendre polynomials. Using the addition theorem (3.3.4) we get

$$\phi(r, \vartheta, \varphi) = k_c \sum_{l=0}^{\infty} \sum_{m=-l}^{l} r^{-l-1} Y_{lm}(\vartheta, \varphi) \, Q_{lm} \, \frac{4\pi}{2l+1} \tag{3.3.10}$$

with

$$Q_{lm} = \int \mathrm{d}^3 x' \, \rho(\mathbf{x}') \, r'^l \, Y_{lm}^*(\vartheta', \varphi') = (-1)^m \, Q_{lm}^* \,. \tag{3.3.11}$$

The individual summands of (3.3.10) are the contributions of the 2^l-poles to the potential. These are proportional to $1/r^{l+1}$, and the Q_{lm} are the spherical multipole moments of the 2^l-pole (see in Tab. 3.1).

Tab. 3.1. 2^l-poles

l	2^l	
0	1	Monopole
1	2	Dipole
2	4	Quadrupole
3	8	Octupole
4	16	Hexadecapole

Dipole and quadrupole moments

The relationship between the dipole and quadrupole moments, which come from the Taylor expansion, can be established with the corresponding spherical multipole moments:

$$Q_{10} = \sqrt{\frac{3}{4\pi}} p_z, \qquad Q_{1\pm1} = \sqrt{\frac{3}{8\pi}}(\mp p_x + \mathrm{i} p_y), \tag{3.3.12}$$

$$Q_{20} = \sqrt{\frac{5}{16\pi}} Q_{zz}, \quad Q_{2\pm1} = \sqrt{\frac{5}{24\pi}}(\mp Q_{xz} + \mathrm{i} Q_{yz}), \quad Q_{2\pm2} = \sqrt{\frac{5}{96\pi}}(Q_{xx} - Q_{yy} \mp 2\mathrm{i} Q_{yx}).$$

Average electric field

The electric field of a given charge distribution ρ is uniquely determined by this. Exceptions are points where the field is singular. By integrating $\mathbf{E}$ over a sphere S_a with radius a one can define an average field at these points. Around the origin, one obtains

$$S_a \, \bar{\mathbf{E}}(0) = \int_{S_a} \mathrm{d}^3 x \, \mathbf{E}(\mathbf{x}) = -\mathbf{e}_i \int_{S_a} \mathrm{d}^3 x \, \boldsymbol{\nabla}\phi(\mathbf{x}) \cdot \mathbf{e}_i \tag{3.3.13}$$

$$= -\mathbf{e}_i \oiint_{\partial S_a} \mathrm{d}\mathbf{a} \cdot \mathbf{e}_i \, \phi(\mathbf{x}) = -k_c \int \mathrm{d}^3 x' \, \rho(\mathbf{x}') \oiint_{\partial S_a} \frac{\mathrm{d}\mathbf{a}}{|\mathbf{x} - \mathbf{x}'|} \,.$$

First, we used the identity $\mathbf{E} = \mathbf{e}_i(\mathbf{e}_i \cdot \mathbf{E})$, where $\mathbf{E} = -\boldsymbol{\nabla}\phi$. Subsequently, we transformed the volume integral into a surface integral using Gauss's theorem and substituted ϕ with (2.1.8). The factor $1/|\mathbf{x}-\mathbf{x}'|$ is represented by (3.3.8)

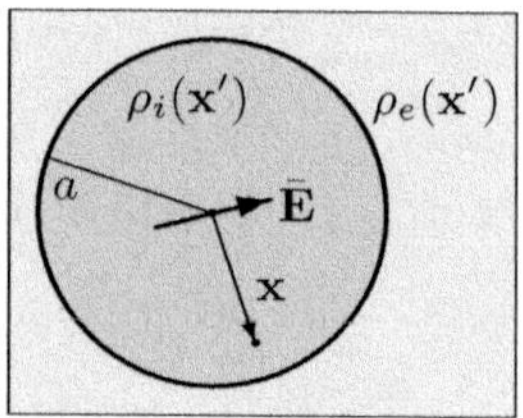

Fig. 3.8. Electric field $\bar{\mathbf{E}} = \bar{\mathbf{E}}_i + \bar{\mathbf{E}}_e$ formed by averaging over a sphere with radius a; ρ_i denotes the charge within and ρ_e the charge outside the sphere

through Y_{lm}, where the cases $r > r'$ and $r < r'$ are distinguished. $\bar{\mathbf{E}} = \bar{\mathbf{E}}_i + \bar{\mathbf{E}}_e$ thus consists of two contributions, one from the density $\rho_i(\mathbf{x}')$ within the sphere $r' < a$ and one from $\rho_e(\mathbf{x}')$ outside the sphere $r' > a$, as sketched in Fig. 3.8.

Contribution to the average field from the charges within the sphere
In (3.3.13) one sets (3.3.8) for $r' < r$ and $\mathbf{da} = a^2 \mathrm{d}\Omega\, \mathbf{e}_r$:

$$S_a\, \bar{\mathbf{E}}_i = -a^2 k_C \sum_{l,m} \frac{4\pi}{2l+1} \int \mathrm{d}^3x'\, \rho_i(\mathbf{x}')\, \frac{r'^l}{a^{l+1}}\, Y_{lm}(\Omega') \oiint \mathrm{d}\Omega\, Y_{lm}^*(\Omega)\, \mathbf{e}_r\,.$$

Now is

$$\mathbf{e}_r = \sin\vartheta(\cos\varphi\, \mathbf{e}_x + \sin\varphi\, \mathbf{e}_y) + \cos\vartheta\, \mathbf{e}_z$$

a linear combination of the $Y_{1m'}(\Omega)$. Due to the orthogonality one obtains

$$Y_{lm}(\Omega') \oiint \mathrm{d}\Omega\, Y_{lm}^*(\Omega)\, Y_{1m'}(\Omega) = \delta_{l1}\, \delta_{mm'}\, Y_{1m'}(\Omega')\,.$$

The linear combination of the $Y_{1m'}$ remains unchanged, only Ω is replaced by Ω', i.e. $\mathbf{e}_r$ by $\mathbf{e}_{r'}$:

$$S_a\, \bar{\mathbf{E}}_i = -\frac{4\pi k_C}{3} \int_{S_a} \mathrm{d}^3x'\, \rho_i(\mathbf{x}')\, r'\, \mathbf{e}_{r'} = -\frac{4\pi k_C}{3}\, \mathbf{p}_i\,. \tag{3.3.14}$$

So, only the dipole moment of the sphere centered around $\mathbf{x}$ with radius a contributes to $\bar{\mathbf{E}}_i(\mathbf{x})$. For a dipole field (2.5.6a) this contribution comes solely from the location $\mathbf{x}$ of the dipole (δ-term) [Jackson, 1998, (4.20)].

Contribution to the average field from the charges outside the sphere
The interior of the sphere S_a is now charge-free and thus a solution to the Laplace equation. For these harmonic solutions, the mean value property applies, according to which the average value of the field on any spherical surface

$r' \leq a$ is equal to the value $\mathbf{E}_e(\mathbf{x})$ at the center of the sphere. Therefore, the average field $\bar{\mathbf{E}}_e(\mathbf{x})$ is equal to the (not averaged) field $\mathbf{E}_e(\mathbf{x})$ of the charge distribution $\rho_e(\mathbf{x}')$.

If $r' > r$, then using (3.3.8) and a procedure analogous to ρ_i (application of the orthonormality of the spherical harmonics)

$$\bar{\mathbf{E}}_e = -k_C \frac{4\pi a^3}{3 S_a} \int_{r'>a} \mathrm{d}^3 x' \, \rho_e(\mathbf{x}') \frac{1}{r'^2} \mathbf{e}_{r'} = -k_C \boldsymbol{\nabla} \int_{r'>a} \mathrm{d}^3 x' \, \frac{\rho_e(\mathbf{x}')}{|\mathbf{x}-\mathbf{x}'|}\bigg|_{\mathbf{x}=0} = \mathbf{E}_e(0).$$

3.3.4 Conductive Sphere in a Homogeneous Field

The potential for a homogeneous field $\mathbf{E}_0 = (0, 0, E_0)$ is

$$\phi(\mathbf{x}) = -E_0 \, z = -E_0 r \, \cos\vartheta = -E_0 r \, P_1(\cos\vartheta). \tag{3.3.15}$$

At $\mathbf{x}=0$ there is a conductive sphere with radius R, as sketched in Fig. 3.9. It has the total charge Q. Then (3.3.15) only applies in the limit $r \to \infty$. If

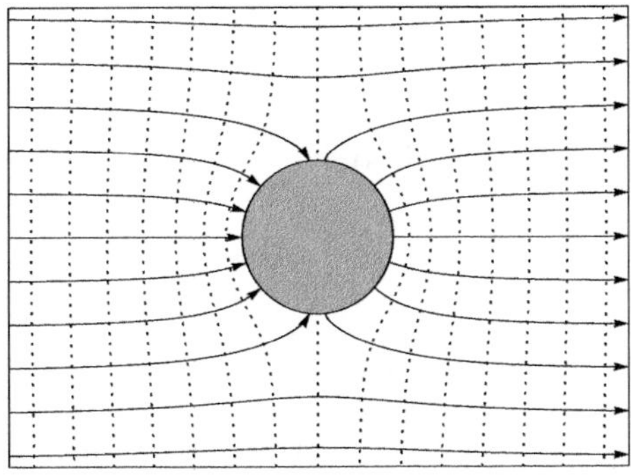

Fig. 3.9. Field lines of a conductive sphere with radius R and total charge $Q = 0$

we place the coordinate origin in the center of the sphere, so we can use the approach valid for axial symmetry (3.2.40)

$$\phi(\mathbf{x}) = \sum_l (a_l \, r^l + b_l \, r^{-l-1}) P_l(\cos\vartheta). \tag{3.3.16}$$

From (3.3.15) for $r \to \infty$ we get

$$a_1 = -E_0, \qquad\qquad a_l = 0 \qquad\qquad \text{for} \quad l \neq 1.$$

On the surface of the sphere, the potential is constant, i.e. the potential of the external field must be compensated by induced charges. This results in the approach for the sphere surface $r = R$

$$\phi(R) = \sum_{l=0}^{\infty} P_l(\cos\vartheta)\big(\delta_{l1} a_l R + b_l R^{-l-1}\big) = \text{const.}$$

The homogeneous field can only be compensated on the surface of the sphere by a term with the angular dependence $P_1(\cos\vartheta)$. A constant potential on the surface can only be added by $P_0(\cos\vartheta) = 1$. All other contributions must disappear:

$$P_0\, b_0\, R^{-1} = \text{const}, \quad P_1\big(a_1\, R + b_1\, R^{-2}\big) = 0\,, \quad P_l\, b_l\, R^{-l-1} = 0 \qquad \text{for } l > 1\,.$$

Potential and field thus have the form

$$\phi(\mathbf{x}) = b_0\, P_0\frac{1}{r} + a_1\Big(r - \frac{R^3}{r^2}\Big)P_1(\cos\vartheta) = b_0\,\frac{1}{r} - E_0\Big(1 - \frac{R^3}{r^3}\Big)z\,,$$

$$\mathbf{E}(R) = b_0\,\frac{\mathbf{e}_r}{R^2} + 3E_0\,\cos\vartheta\,\mathbf{e}_r\,.$$

To determine the constant b_0, we calculate

$$k_C Q = \frac{1}{4\pi}\oiint_{\partial S_R} \mathbf{da}\cdot\mathbf{E} = b_0 + 3E_0\, R^2\,\frac{1}{2}\int_0^\pi d\vartheta\,\sin\vartheta\,\cos\vartheta = b_0\,.$$

Thus, the potential is determined by

$$\phi(\mathbf{x}) = k_C\frac{Q}{r} - E_0\Big(1 - \frac{R^3}{r^3}\Big)z\,. \tag{3.3.17}$$

For vanishing total charge $Q = 0$ ($z = r\cos\vartheta$) one obtains

$$E_r = -\frac{\partial\phi}{\partial r} = E_0\Big(1 + 2\,\frac{R^3}{r^3}\Big)\cos\vartheta\,,$$

$$E_\vartheta = -\frac{1}{r}\frac{\partial\phi}{\partial\vartheta} = -E_0\Big(1 - \frac{R^3}{r^3}\Big)\sin\vartheta\,.$$

The resulting field line image is shown in Fig. 3.9. On the sphere surface, $E_\vartheta(R) = 0$ and $E_r = 3E_0\cos\vartheta$. The induced surface charge density is ($E_\perp = E_r$)

$$\sigma = \frac{E_r}{4\pi k_C} = \frac{3}{4\pi k_C}\,E_0\,\cos\vartheta\,. \tag{3.3.18}$$

The charge induced on the sphere has an induced dipole moment $\mathbf{p} \propto \mathbf{E}_0$. The potential can then also be written as

$$\phi(\mathbf{x}) = -\mathbf{E}_0\cdot\mathbf{x} + k_C\frac{\mathbf{p}\cdot\mathbf{x}}{r^3} \tag{3.3.19}$$

with the induced dipole moment

$$\mathbf{p} = R^3/k_C\,\mathbf{E}_0 = \alpha\,\mathbf{E}_0\,.$$

Here, $\alpha = R^3/k_C$ is the polarizability of the sphere.

3.4 Cylindrically Symmetric Problems

The solution of the Laplace equation in cylindrical coordinates leads us to Bessel functions, and the expansion of the Green function (2.1.6) in terms of Bessel functions proves to be more complicated than the one in terms of Legendre polynomials or spherical harmonics. The following section is therefore very formal and only of interest, if a given electrostatic configuration suggests an expansion after cylindrical functions.

3.4.1 Laplace Equation in Cylindrical Coordinates

In cylindrical coordinates, the Laplace equation (see (A.3.20)) takes the form

$$\left(\frac{1}{\varrho}\frac{\partial}{\partial\varrho}\varrho\frac{\partial}{\partial\varrho} + \frac{1}{\varrho^2}\frac{\partial^2}{\partial\varphi^2} + \frac{\partial^2}{\partial z^2}\right)\phi(\varrho,\varphi,z) = 0\,. \tag{3.4.1}$$

With the separation ansatz

$$\phi(\varrho,\varphi,z) = R(\varrho)\,\Phi(\varphi)\,Z(z) \tag{3.4.2}$$

one obtains, when multiplying from the left with $1/\phi$,

$$\frac{1}{R}\left(\frac{\partial^2 R}{\partial\varrho^2} + \frac{1}{\varrho}\frac{\partial R}{\partial\varrho}\right) + \frac{1}{\varrho^2\,\Phi}\frac{\partial^2\Phi}{\partial\varphi^2} + \frac{1}{Z}\frac{\partial^2 Z}{\partial z^2} = 0\,.$$

Separation of the azimuth angle
The separation constant $n^2 \geq 0$ for the azimuth angle must be positive to ensure the required periodicity $\Phi(\varphi) = \Phi(\varphi+2\pi)$. This results in

$$\varrho^2\left[\frac{1}{R}\left(\frac{\partial^2 R}{\partial\varrho^2} + \frac{1}{\varrho}\frac{\partial R}{\partial\varrho}\right) + \frac{1}{Z}\frac{\partial^2 Z}{\partial z^2}\right] = n^2 = -\frac{1}{\Phi}\frac{\partial^2\Phi}{\partial\varphi^2}$$

with

$$\left(\frac{\partial^2}{\partial\varphi^2}+n^2\right)\Phi(\varphi) = 0\,, \quad \Phi(\varphi) = A\cos(n\varphi)+B\sin(n\varphi), \quad n \text{ integer.} \tag{3.4.3}$$

Separation of the variable z
With the separation constant c one obtains

$$\frac{1}{R}\left(\frac{\partial^2}{\partial\varrho^2} + \frac{1}{\varrho}\frac{\partial}{\partial\varrho}\right)R(\varrho) - \frac{n^2}{\varrho^2} = c = -\frac{1}{Z}\frac{\partial^2 Z}{\partial z^2}\,.$$

We now distinguish the following cases:

1. *Separation constant $c = -k^2 < 0$*

   Easy to solve is

   $$\left(\frac{\partial^2}{\partial z^2} - k^2\right) Z = 0, \qquad\qquad Z(z) = \alpha e^{kz} + \beta e^{-kz}. \tag{3.4.4}$$

   For the radial function, we only note for now that from

   $$\left(\frac{\partial^2}{\partial \varrho^2} + \frac{1}{\varrho}\frac{\partial}{\partial \varrho} + k^2 - \frac{n^2}{\varrho^2}\right) R(\varrho) = 0 \quad R(\varrho) = C J_n(k\varrho) + D N_n(k\varrho) \tag{3.4.5}$$

   by means of the transformation $x = k\varrho$ the Bessel's differential equation (B.4.1)

   $$\left(\frac{d^2}{dx^2} + \frac{1}{x}\frac{d}{dx} + 1 - \frac{n^2}{x^2}\right) R(x) = 0 \tag{3.4.6}$$

   is obtained, whose solutions are the Bessel functions of the 1st kind $J_n(x)$ and 2nd kind, the Neumann functions $N_n(x)$.

2. *Separation constant $c = 0$*

   $$\frac{\partial^2 Z}{\partial z^2} = 0, \qquad\qquad Z(z) = \alpha + \beta z,$$

   $$\left(\frac{\partial^2}{\partial \varrho^2} + \frac{1}{\varrho}\frac{\partial}{\partial \varrho} - \frac{n^2}{\varrho^2}\right) R(\varrho) = 0, \qquad\qquad R(\varrho) = C \varrho^n + D \varrho^{-n}.$$

   The solution of $Z(z)$ requires no explanation. For R we make the power series approach $R(\varrho) = \sum_j a_j \varrho^j$, which we insert into the radial equation:

   $$(j(j-1) + j - n^2) a_j \, \varrho^{j-2} = 0 \qquad\qquad \Rightarrow \qquad\qquad a_j = \delta_{j \pm n}$$

   and take into account that the coefficients of all powers of ϱ must vanish separately.

3. *Separation constant $c = k^2 > 0$*

   $$\left(\frac{\partial^2}{\partial z^2} + k^2\right) Z = 0, \quad Z(z) = \alpha \cos(kz) + \beta \sin(kz),$$

   $$\left(\frac{\partial^2}{\partial \varrho^2} + \frac{1}{\varrho}\frac{\partial}{\partial \varrho} - k^2 - \frac{n^2}{\varrho^2}\right) R(\varrho) = 0, \quad R(\varrho) = C I_n(k\varrho) + D K_n(k\varrho).$$

   If one again switches to $x = k\varrho$, one obtains the modified Bessel's differential equation (B.4.5)

   $$\left(\frac{d^2}{dx^2} + \frac{1}{x}\frac{d}{dx} - 1 - \frac{n^2}{x^2}\right) R(x) = 0. \tag{3.4.7}$$

   If one replaces $x \to ix$, one returns to (3.4.6) and thus to the solutions $J_n(ik\varrho)$ and $N_n(ik\varrho)$. However, instead of these, the modified Bessel functions are used $I_\nu(x)$ and $K_\nu(x)$ (B.4.6).

3.4.2 Fourier-Bessel Expansion

With the Bessel functions J_ν and $\nu \geq -1$, one can define functions on the finite interval $[0, a]$, which are orthogonal and complete. Therefore, they can be used to expand functions $f(\varrho)$ into a series. The *Fourier-Bessel series* is a generalization of the (trigonometric) Fourier series [Oberhettinger, 1973]. In potential theory, it is applied to problems with cylindrical symmetry, but in electrodynamics with a focus on physics, it is rather rarely addressed [Rebhan, 2007, sec. 4.7.4], [Jackson, 1998, sec. 3.7]. For the orthogonality of functions, it is essential that each function $f_k(\varrho) \neq f_l(\varrho)$ inside the interval has a different number of zeros $k \neq l$ so that the integral can be zero. Let $f_{l-1}(\varrho) = J_n(\frac{\varrho}{a}x_{nl})$, where x_{nl} is the l-th zero of $J_n(x_{nl}) = 0$. f_{l-1} thus has $l-1$ zeros within the interval. $w(\varrho) = \varrho$ is the weight function of this expansion[1].

Orthogonality of the Bessel functions of the 1st kind

Let x_{nl} and x_{nk} be zeros of J_n with $J_n(x_{nl}) = J_n(x_{nk}) = 0$ and $0 \leq \varrho \leq a$, so the J_n are orthogonal with respect to different l and k for the same $n \geq 0$:

$$\int_0^a \mathrm{d}\varrho\varrho\, J_n(\frac{\varrho}{a}x_{nk})\, J_n(\frac{\varrho}{a}x_{nl}) = \delta_{lk}\, \frac{a^2}{2}\, J_{n+1}^2(x_{nl})\,. \tag{3.4.8}$$

Proof: We multiply the radial equation (3.4.5) from the left with ϱ and $J_n(q\varrho)$, the solution of (3.4.5) for q. We use the differential operator in the form of (3.4.1). Then we exchange $k \rightleftharpoons q$ and subtract the two equations:

$$J_n(q\varrho)\, \frac{\mathrm{d}}{\mathrm{d}\varrho}\Big(\varrho\, \frac{\mathrm{d}J_n(k\varrho)}{\mathrm{d}\varrho}\Big) - J_n(k\varrho)\, \frac{\mathrm{d}}{\mathrm{d}\varrho}\Big(\varrho\, \frac{\mathrm{d}J_n(q\varrho)}{\mathrm{d}\varrho}\Big) + \varrho\,(k^2 - q^2)\, J_n(q\varrho)\, J_n(k\varrho) = 0.$$

The first two terms can be combined into a complete differential:

$$\frac{\mathrm{d}}{\mathrm{d}\varrho}\, \varrho\, \Big(J_n(q\varrho)\frac{\mathrm{d}J_n(k\varrho)}{\mathrm{d}\varrho} - J_n(k\varrho)\frac{\mathrm{d}J_n(q\varrho)}{\mathrm{d}\varrho}\Big) + \varrho\,(k^2 - q^2)\, J_n(q\varrho)\, J_n(k\varrho) = 0\,.$$

Now we integrate over ϱ and get

$$(q^2 - k^2)\int_0^a \mathrm{d}\varrho\, \varrho\, J_n(q\varrho)\, J_n(k\varrho) = \varrho\,\Big\{J_n(q\varrho)\frac{\mathrm{d}J_n(k\varrho)}{\mathrm{d}\varrho} - J_n(k\varrho)\frac{\mathrm{d}J_n(q\varrho)}{\mathrm{d}\varrho}\Big\}\Big|_0^a. \tag{3.4.9}$$

Now we substitute for $k = x_{nk}/a$ and for $q = x_{nl}/a$ with $k \neq l$ in (3.4.9), the right side disappears. Then the integral on the left side must also disappear, with which (3.4.8) for $k \neq l$ is shown.

To calculate the normalization, we substitute for $q = k + \epsilon$ in (3.4.9) and develop up to the 1st order[2] in ϵ

$$2\epsilon k\int_0^a \mathrm{d}\varrho\, \varrho\, J_n(\varrho k)\, J_n(\varrho k) = \epsilon a k a\Big[J_n'(ka)J_n'(ka) - J_n(ka)\, J_n''(ka)\Big] + O(\epsilon^2).$$

---

[1] This is equivalent to the expansion with $f_{l-1}(\varrho) = \sqrt{\varrho}\, J_n(\varrho x_{nl}/a)$ and $w(\varrho) = 1$.
[2] $J_n'(ka) = k\, \mathrm{d}J_n(k\varrho)/\mathrm{d}\varrho\big|_{\varrho=a}$.

Now we substitute for $ka = x_{nk}$ and use the recursion relation $x\,J_n'(x) = n\,J_n(x) - x\,J_{n+1}(x)$, so all terms on the right side disappear up to $J_{n+1}^2(x_{nk})$, so that

$$\int_0^a \mathrm{d}\varrho\,\varrho\,J_n^2\!\left(\frac{\varrho}{a}x_{nk}\right) = \frac{a^2}{2}\,J_{n+1}^2(x_{nk}).$$

Fourier-Bessel series

In order to be able to develop according to Bessel functions with the weight function $w(\varrho) = \varrho$, we still require their completeness. The development of a given function $f(\varrho)$ according to Bessel functions of the first kind is

$$f(\varrho) = \sum_{k=1}^{\infty} c_{nk}\,J_n\!\left(\frac{\varrho}{a}x_{nk}\right), \tag{3.4.10}$$

$$c_{nk} = \frac{2}{a^2\,J_{n+1}^2(x_{nk})}(J_{nk}, f) \quad \text{and} \quad (J_{nk}, f) = \int_0^a \mathrm{d}\varrho\,\varrho\,J_n\!\left(\frac{\varrho}{a}x_{nk}\right) f(\varrho).$$

Completeness of the Bessel functions of the first kind

$$\sum_{l=1}^{\infty} \frac{2\,J_n\!\left(\frac{\varrho}{a}x_{nl}\right) J_n\!\left(\frac{\varrho'}{a}x_{nl}\right)}{a^2\,J_{n+1}^2(x_{nl})} = \frac{1}{\varrho}\,\delta(\varrho - \varrho'). \tag{3.4.11}$$

The completeness of the development is verified by substituting for the δ-function (3.4.11) insert:

$$f(\varrho) = \int_0^a \mathrm{d}\varrho'\,\delta(\varrho'-\varrho)\,f(\varrho') = \int_0^a \mathrm{d}\varrho'\,\varrho' \sum_{l=1}^{\infty} \frac{2\,J_n\!\left(\frac{\varrho}{a}x_{nl}\right) J_n\!\left(\frac{\varrho'}{a}x_{nl}\right)}{a^2\,J_{n+1}^2(x_{nl})}\,f(\varrho')$$

$$= \sum_{l=1}^{\infty} c_{nl}\,J_n\!\left(\frac{\varrho}{a}x_{nl}\right).$$

Fourier-Bessel transformation

For $k \gg n$ the zeros of J_n are according to (B.4.9)

$$x_{nk} = k\pi + \frac{n\pi}{2} - \frac{\pi}{4}.$$

At these values, J_{n+1} has the maximum or minimum values: $x_{nk} - \frac{\pi}{2}(n+\frac{3}{2}) = \pi(k-1)$ and one can use the asymptotic approximation for the Bessel function (see Tab. B.2, p. 622):

$$J_{n+1}(x_{nk}) = \sqrt{\frac{2}{\pi x_{nk}}}\,(-1)^{k-1}.$$

For the Fourier-Bessel series (3.4.10) this results in

$$f(\varrho) = \sum_{k=1}^{\infty} J_n\!\left(\frac{\varrho x_{nk}}{a}\right) (J_{nk}, f)\, \frac{x_{nk}}{a}\, \Delta k, \qquad\qquad \Delta k = \frac{x_{nk+1} - x_{nk}}{a} = \frac{\pi}{a}\,.$$

For $a \to \infty$ one goes to the continuous variable $k = x_{nk}/a$, noting that the range of k, in which the asymptotic approximation does not apply, shrinks to zero and the sum becomes an integral:

$$f(\varrho) = \int_0^{\infty} dk\, k\, c_n(k)\, J_n(k\varrho)\,, \tag{3.4.12}$$

$$c_n(k) = \int_0^{\infty} d\varrho\, \varrho\, f(\varrho)\, J_n(k\varrho)\,.$$

The transition from the Fourier-Bessel series to the Fourier-Bessel transformation is similar to that from the Fourier series to the Fourier transformation. However, the term *Hankel transformation* is more commonly used for (3.4.12).

Orthogonality and completeness are obtained from (3.4.8) and (3.4.11) in the limit $a \to \infty$:

$$\int_0^{\infty} d\varrho\, \varrho\, J_n(k\varrho)\, J_n(q\varrho) = \frac{1}{k}\, \delta(k - q) \qquad\qquad \text{orthogonality} \qquad (3.4.13)$$

$$\int_0^{\infty} dk\, k\, J_n(k\varrho)\, J_n(k\varrho') = \frac{1}{\varrho}\, \delta(\varrho - \varrho') \qquad\qquad \text{completeness.} \qquad (3.4.14)$$

3.4.3 Expansion of the Green's Function with Cylindrical Functions

Analogous to the expansion of $1/|\mathbf{x} - \mathbf{x}'|$ according to spherical harmonics (3.3.8) for boundary value problems with spherical symmetry, such a development can be carried out according to Bessel functions for boundary value problems with cylindrical symmetry, where the Green's function $1/|\mathbf{x} - \mathbf{x}'|$ is the potential $\phi(\mathbf{x})$ of a unit point charge ($q = 1$) at the location $\mathbf{x}'$ and satisfies the Laplace equation everywhere in space except at the location of the point charge.

We have distinguished the solutions of the Laplace equation according to the separation constant c, where for $c = -k^2$ these were the Bessel functions of the first and second kind and for $c = k^2$ the modified Bessel functions.

Expansion according to Bessel functions of the first kind

For $c = -k^2$ the partial solutions (3.4.4)

$$Z_k(z) = \alpha e^{kz} + \beta e^{-kz}$$

become singular for $z \rightarrow \pm\infty$. $\mathrm{e}^{-k|z-z'|}$ with $k > 0$ is a regular function, that only for $z' = z$ does not satisfy the Laplace equation and is used for the expansion of G. If one excludes the Neumann functions that are singular at the origin one obtains the approach

$$G(\mathbf{x}, \mathbf{x}') = \int_0^\infty \mathrm{d}k\, \mathrm{e}^{-k|z-z'|} \sum_{n=-\infty}^\infty \mathrm{e}^{\mathrm{i}n\varphi} J_n(k\varrho) A_{nk}(\varrho', \varphi'). \qquad (3.4.15)$$

We determine the coefficients A_{nk} from the Poisson equation [3]:

$$\Delta G(\mathbf{x}, \mathbf{x}') = -\boldsymbol{\nabla} \cdot \mathbf{E} = -\frac{4\pi}{\varrho}\, \delta(z - z')\, \delta(\varphi - \varphi')\, \delta(\varrho - \varrho').$$

The discontinuity of $\partial G/\partial z = -E_z$ is

$$\delta E_z = \int_{z'-\epsilon}^{z'+\epsilon} \mathrm{d}z\, \boldsymbol{\nabla} \cdot \mathbf{E} = E_z(z'+\epsilon) - E_z(z'-\epsilon) = \frac{4\pi}{\varrho}\, \delta(\varphi - \varphi')\, \delta(\varrho - \varrho').$$

It is taken into account here that the components E_ϱ and E_φ are continuous for $z = z'$. We calculate δE_z using (3.4.15):

$$\delta E_z = 2 \int_0^\infty \mathrm{d}k\, k \sum_{n=-\infty}^\infty \mathrm{e}^{\mathrm{i}n\varphi} J_n(k\varrho) A_{nk}.$$

In the next step, we multiply by $(1/4\pi) \int_0^{2\pi} \mathrm{d}\varphi\, \mathrm{e}^{-\mathrm{i}m\varphi}$ and use the Orthogonality of the plane waves:

$$\frac{1}{2\pi} \int_0^{2\pi} \mathrm{d}\varphi\, \mathrm{e}^{\mathrm{i}(n-m)\varphi} = \delta_{nm}, \qquad (3.4.16)$$

$$\int_0^\infty \mathrm{d}k\, k\, J_m(k\varrho) A_{mk} = \frac{1}{\varrho}\, \mathrm{e}^{-\mathrm{i}m\varphi'}\, \delta(\varrho - \varrho').$$

Now we multiply from the left with $\int_0^\infty \mathrm{d}\varrho\, \varrho\, J_m(q\varrho)$ and apply the orthogonality relations of the Fourier-Bessel transformation (3.4.13):

$$A_{mq}(\varrho', \varphi') = \mathrm{e}^{-\mathrm{i}m\varphi'} J_m(q\varrho').$$

Inserted into (3.4.15) we obtain for the expansion of G in terms of Bessel functions

$$G(\mathbf{x}, \mathbf{x}') = \frac{1}{|\mathbf{x} - \mathbf{x}'|} = \int_0^\infty \mathrm{d}k\, \mathrm{e}^{-k|z-z'|} \sum_{n=-\infty}^\infty \mathrm{e}^{\mathrm{i}n(\varphi - \varphi')} J_n(k\varrho) J_n(k\varrho'). \qquad (3.4.17)$$

---

[3] here $\boldsymbol{\nabla} \cdot \mathbf{E} = 4\pi \delta^{(3)}(\mathbf{x} - \mathbf{x}')$, i.e. $qk_c = 1$.

Expansion in terms of modified Bessel functions

In the expansion of $1/|\mathbf{x}-\mathbf{x}'|$ in terms of modified Bessel functions we now start from the partial solutions (3.4.7), for simplicity we write both the expansion in plane waves in the z direction ($Z_k = \mathrm{e}^{\mathrm{i}kz}$) and the expansion into a Fourier series ($\Phi_n = \mathrm{e}^{\mathrm{i}\varphi n}$) in complex form. The (3.4.15) corresponding approach is

$$G(\mathbf{x}, \mathbf{x}') = \int_{-\infty}^{\infty} \mathrm{d}k\, \mathrm{e}^{\mathrm{i}k(z-z')} \sum_{n=-\infty}^{\infty} \mathrm{e}^{\mathrm{i}n(\varphi-\varphi')}\, g_{nk}(\varrho, \varrho'),$$

$$g_{nk} = \alpha_{nk} \begin{cases} I_n(k\varrho)\, K_n(k\varrho')) & \text{for} \quad \varrho' > \varrho \\ K_n(k\varrho)\, I_n(k\varrho') & \text{for} \quad \varrho' < \varrho. \end{cases} \tag{3.4.18}$$

The distinction between $\varrho' > \varrho$ and $\varrho' < \varrho$ follows from the asymptotic behavior of the Bessel functions of the 2nd kind: $I_n(k\varrho)$ are regular at the origin and diverge for $\varrho \to \infty$. $K_n(k\varrho)$ are singular at the origin and vanish for $\varrho \to \infty$. The approach also reflects the symmetry $\mathbf{x} \leftrightarrows \mathbf{x}'$, since G is ultimately real.

The justification of the approach will be confirmed in the next steps: First, $\Delta G = -(4\pi/\varrho)\delta^{(3)}(\mathbf{x} - \mathbf{x}')$ is multiplied from the left with $\int_{-\infty}^{\infty} \frac{\mathrm{d}z}{2\pi}\, \mathrm{e}^{-\mathrm{i}qz}$:

$$\sum_{n=-\infty}^{\infty} \mathrm{e}^{\mathrm{i}n(\varphi-\varphi')}\left(\frac{1}{\varrho}\frac{\partial}{\partial\varrho}\varrho\frac{\partial}{\partial\varrho} - q^2 - \frac{n^2}{\varrho^2}\right)g_{nq}(\varrho, \varrho') = -\frac{2}{\varrho}\delta(\varphi-\varphi')\,\delta(\varrho-\varrho').$$

Then it is multiplied with $\int_0^{2\pi} \frac{\mathrm{d}\varphi}{2\pi}\, \mathrm{e}^{-\mathrm{i}m\varphi}$ and the orthogonality of the Fourier series (3.4.16) ist taken into account:

$$\left(\frac{1}{\varrho}\frac{\partial}{\partial\varrho}\varrho\frac{\partial}{\partial\varrho} - q^2 - \frac{m^2}{\varrho^2}\right)g_{mq}(\varrho, \varrho') = -\frac{1}{\pi\varrho}\delta(\varrho-\varrho'). \tag{3.4.19}$$

At the point $\varrho = \varrho'$, the solutions of (3.4.19) are continuous. We now verify the discontinuity of $\mathrm{d}g_{mq}/\mathrm{d}\varrho$ by integrating (3.4.19) around this:

$$\int_{\varrho'-\epsilon}^{\varrho'+\epsilon} \mathrm{d}\varrho \left(\frac{\mathrm{d}}{\mathrm{d}\varrho}\varrho\frac{\mathrm{d}}{\mathrm{d}\varrho} - \varrho\left(k^2+\frac{n^2}{\varrho^2}\right)\right)g(n, k; \varrho, \varrho') = -\frac{1}{\pi}.$$

This implies

$$\frac{\mathrm{d}g_{mq}(\varrho, \varrho')}{\mathrm{d}\varrho}\bigg|_{\varrho'+\epsilon} - \frac{\mathrm{d}g_{mq}(\varrho, \varrho')}{\mathrm{d}\varrho}\bigg|_{\varrho'-\epsilon} =$$

$$= \alpha_{mq}q\big(K'_m(q\varrho')\, I_m(q\varrho') - I_m(q\varrho')\, K_m(q\varrho')\big) = -\frac{1}{\pi\varrho'}. \tag{3.4.20}$$

The value of the constant $\alpha_{mq} = 1/\pi$ is determined with the asymptotic representations of I and K (see Tab. B.2, p. 622).

Note: The differential operator of (3.4.19)

$$L = \frac{\partial}{\partial x} p(x) \frac{\partial}{\partial x} + q(x)$$

is the Sturm-Liouville operator. If ψ_1 and ψ_2 are solutions of $L\psi = 0$, then (3.4.20) is the Wronskian determinant $W = \begin{vmatrix} \psi_1 & \psi_2 \\ \psi_1' & \psi_2' \end{vmatrix}$, which is proportional to $\frac{1}{p(x)}$.

Thus, for the Green's function

$$\frac{1}{|\mathbf{x} - \mathbf{x}'|} = \frac{1}{\pi} \int\limits_{-\infty}^{\infty} dk\, e^{ik(z-z')} \sum_{n=-\infty}^{\infty} e^{in(\varphi-\varphi')} I_n(k\varrho') K_n(k\varrho). \tag{3.4.21}$$

Here it is assumed that $\varrho' < \varrho$. If this is not the case, $\varrho \rightleftharpoons \varrho'$ must be swapped.

Dirichlet problem for the cylindrical surface

If $\phi(a, \varphi, z)$ is given on a cylindrical surface and the potential inside it is to be determined, one first determines the Dirichlet's Green function G_D for the interior of the infinitely long hollow cylinder with radius a (see Fig. 3.10):

$$G_D(\mathbf{x}, \mathbf{x}') = G(\mathbf{x}, \mathbf{x}') + F(\mathbf{x}, \mathbf{x}') \qquad \text{with}$$

$$\Delta F(\mathbf{x}, \mathbf{x}') = 0 \qquad \text{and} \qquad F(\mathbf{x}, \mathbf{x}')\Big|_{\varrho'=a} = -\frac{1}{|\mathbf{x} - \mathbf{x}'|}\Big|_{\varrho'=a}. \tag{3.4.22}$$

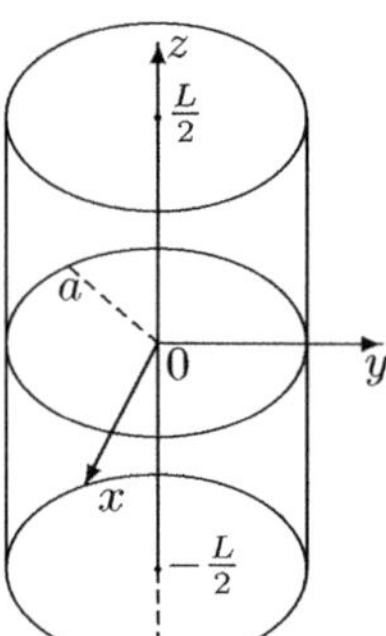

Fig. 3.10. Hollow cylinder with radius a and length $L \to \infty$; $G_D(\varrho, \varphi, z; a, \varphi', z') = 0$ on the lateral surface

For the expansion of F, only the modified Bessel functions I_n are considered, since due to the infinite length of the cylinder the solutions (3.4.4): $Z(z) = \alpha e^{kz} + \beta e^{-kz}$ diverge and the $K_n(k\varrho)$ are singular at the origin:

$$F(\mathbf{x}, \mathbf{x}') = \int\limits_{-\infty}^{\infty} dk z\, e^{ik(z-z')} \sum_{n=-\infty}^{\infty} e^{in(\varphi-\varphi')} \alpha_{nk} I_n(k\varrho) I_n(k\varrho'). \tag{3.4.23}$$

On the cylinder mantle, $F(\mathbf{x}, \mathbf{x}') = -G(\mathbf{x}, \mathbf{x}')$, where G is one of the two functions (3.4.17) or (3.4.21). If the two functions on the cylinder mantle (except for the sign) are equal, then all Fourier coefficients must also be equal. This results in (3.4.17)

$$\int_{-\infty}^{\infty} dk z\, e^{ik(z-z')}\, \alpha_{nk}\, I_n(k\varrho)\, I_n(ka) = -\int_0^{\infty} dk\, e^{-k|z-z'|}\, J_n(k\varrho)\, J_n(ka)\,.$$

To determine the α_{nk}, we multiply with $\int_{-\infty}^{\infty} dz\, e^{-iq(z-z')}$:

$$2\pi\, \alpha_{nq}\, I_n(q\varrho)\, I_n(qa) = -\int_0^{\infty} dk\, J_n(k\varrho)\, J_n(ka) \int_{-\infty}^{\infty} dz\, e^{-iq(z-z')}\, e^{-k|z-z'|}$$

$$= -\int_0^{\infty} dk\, \frac{2k}{k^2 + q^2}\, J_n(k\varrho)\, J_n(ka)\,.$$

Thus, $G_D(\mathbf{x}, \mathbf{x}') = G(\mathbf{x}, \mathbf{x}') + F(\mathbf{x}, \mathbf{x}')$ with

$$F(\mathbf{x}, \mathbf{x}') = \frac{-1}{\pi} \int_{-\infty}^{\infty} dq\, e^{iq(z-z')} \sum_{n=-\infty}^{\infty} e^{in(\varphi-\varphi')} \frac{1}{I_n(qa)} \int_0^{\infty} \frac{dk\, k}{k^2+q^2}\, J_n(k\varrho)\, J_n(ka).$$

$$(3.4.24)$$

3.5 Problems in two Dimensions

3.5.1 Potential Theory

Potentials, whose charge distributions are homogeneous in the z direction, can be treated two-dimensionally, as is the case with the infinitely extended line charge (see section 2.2.1):

$$\Delta\phi(\mathbf{x}) \equiv \left(\frac{\partial^2}{\partial x^2} + \frac{\partial^2}{\partial y^2}\right)\phi(x, y) = -4\pi k_c\lambda\delta(x)\delta(y).$$

We do not start here from a two-dimensional charge distribution $\sigma(x, y)$ with $\rho(\mathbf{x}) = \sigma(x, y)\,\delta(z)$, but from a charge distribution $\rho(x, y)$, which does not depend on z and satisfies

$$\Delta\phi(x, y) = -4\pi k_c\rho(x, y). \tag{3.5.1}$$

Green function for the two-dimensional Laplace operator

The Green function $G(x, y)$ for the two-dimensional Laplace operator is defined by[4]

$$\Delta G(\mathbf{x}-\mathbf{x}') = 2\pi\delta(x-x')\delta(y-y'). \tag{3.5.2}$$

---

[4] In three dimensions we had $\Delta G(\mathbf{x}) = -4\pi\delta^{(3)}(\mathbf{x} - \mathbf{x}')$.

We will prove that

$$G(\mathbf{x}-\mathbf{x}') = \ln|\mathbf{x}-\mathbf{x}'| \tag{3.5.3}$$

fulfills (3.5.2).

$$\frac{\partial G}{\partial x} = \frac{x-x'}{|\mathbf{x}-\mathbf{x}'|^2} \qquad\Rightarrow\qquad \frac{\partial^2 G}{\partial x^2} = \frac{1}{|\mathbf{x}-\mathbf{x}'|^2} - 2\frac{(x-x')^2}{|\mathbf{x}-\mathbf{x}'|^4}\,.$$

Therefore,

$$\Delta G = 0 \qquad \text{for}\quad \mathbf{x}\neq\mathbf{x}'\,.$$

The contribution for $\mathbf{x}=\mathbf{x}'$ still needs to be calculated, for which, as in three dimensions, the divergence theorem (3.5.4) for the area A is used:

$$\iint_A \mathrm{d}x\mathrm{d}y\,\Delta G(\mathbf{x}-\mathbf{x}') = \oint_{\partial A} \mathrm{d}\mathbf{n}\cdot\boldsymbol{\nabla} G(\mathbf{x}-\mathbf{x}')$$

$$= 2\pi \iint_A \mathrm{d}x\mathrm{d}y\,\delta(x-x')\delta(y-y') = 2\pi.$$

$A = A_{\varrho_0}$ is a circular area of radius ϱ_0 around $\mathbf{x}'$, i.e. $\mathbf{n} = \dfrac{\mathbf{x}-\mathbf{x}'}{|\mathbf{x}-\mathbf{x}'|}$ and $|\mathbf{x}-\mathbf{x}'| = \varrho_0$. Along the circle C, $\mathrm{d}\mathbf{n} = \varrho_0\mathrm{d}\varphi_0\,\mathbf{n}$. Thus,

$$\oint_{\partial A_{\varrho_0}} \mathrm{d}\mathbf{n}\cdot\boldsymbol{\nabla} G(\mathbf{x}-\mathbf{x}') = \int_0^{2\pi} \mathrm{d}\varphi_0 = 2\pi\,.$$

The divergence theorem in two dimensions

A suitable formulation for Gauss's theorem in two dimensions is sought. The starting point is Stokes's theorem

$$\iint_A \mathrm{d}\mathbf{a}\cdot(\boldsymbol{\nabla}\times\mathbf{v}) = \oint_{\partial A} \mathrm{d}\mathbf{s}\cdot\mathbf{v}\,.$$

Now let A be a flat area in the xy-plane and also $\mathbf{v}$ be a (two-dimensional) vector in the xy-plane:

$$\iint_A \mathrm{d}x\mathrm{d}y\,\Big[\frac{\partial}{\partial x}v_y - \frac{\partial}{\partial y}v_x\Big] = \oint_{\partial A} \mathrm{d}\mathbf{s}\cdot\mathbf{v}\,.$$

The (outward directed) normal vector $\mathrm{d}\mathbf{n}$ is perpendicular to ∂A, i.e. perpendicular to the vector $\mathrm{d}\mathbf{s}$, which is directed parallel to ∂A.

$$\begin{aligned}\mathrm{d}\mathbf{n} &= \mathrm{d}y\mathbf{e}_x - \mathrm{d}x\mathbf{e}_y \\ \mathrm{d}\mathbf{s} &= \mathrm{d}x\mathbf{e}_x + \mathrm{d}y\mathbf{e}_y\end{aligned} \qquad\Rightarrow\qquad \mathrm{d}\mathbf{n}\cdot\mathrm{d}\mathbf{s} = \mathrm{d}y\mathrm{d}x - \mathrm{d}x\mathrm{d}y = 0 \qquad \text{orthogonality.}$$

With the transformation $w_x = v_y$ and $w_y = -v_x$ one gets

$$\iint_A \mathrm{d}x\mathrm{d}y\,\Big(\frac{\partial}{\partial x}w_x + \frac{\partial}{\partial y}w_y\Big) = \iint_A \mathrm{d}a\,\boldsymbol{\nabla}\cdot\mathbf{w} = \oint_{\partial A} \mathrm{d}\mathbf{n}\cdot\mathbf{w}\,. \tag{3.5.4}$$

Thus, we have Gauss's theorem in two dimensions.

Electrostatic potential

The solution function of the two-dimensional Poisson equation (3.5.1), the electrostatic potential, is (in analogy to three dimensions)

$$\phi(\mathbf{x}) = -2k_C \int d^2x'\, G(\mathbf{x}-\mathbf{x}')\,\rho(\mathbf{x}') \qquad \text{with} \qquad G(\mathbf{x}) = \ln \varrho. \qquad (3.5.5)$$

Multipole expansion in two dimensions

The potential of a localized charge distribution $\rho(x,y)$ can be developed outside the same according to the moments of ρ:

$$\phi(\mathbf{x}) = -2k_C M_0^{(c)} \ln \varrho + 2k_C \sum_{n=1}^{\infty} \frac{1}{n\rho^n} \left[M_n^{(c)} \cos(n\varphi) + M_n^{(s)} \sin(n\varphi) \right]. \qquad (3.5.6)$$

The moments are

$$M_n^{(c)} = \int d^2x'\, \rho(\mathbf{x}')\,\varrho'^n \cos(n\varphi'), \qquad M_n^{(s)} = \int d^2x'\, \rho(\mathbf{x}')\,\varrho'^n \sin(n\varphi'). \qquad (3.5.7)$$

Derivation of (3.5.6) and (3.5.7) using the auxiliary formula [Gradshteyn, Ryzhik, 1965, No. 1514]

$$\ln(1 + t^2 - 2t\cos\alpha) = -2\sum_{n=1}^{\infty} \frac{t^n}{n}\cos(n\alpha) \qquad |t| < 1.$$

According to (3.5.5) applies

$$\phi(x,y) = -2k_C \int d^2x'\, \rho(x',y')\, \ln \sqrt{\varrho^2 + \varrho'^2 - 2\cos(\varphi-\varphi')}$$

$$= -k_C \int d^2x'\, \rho(\mathbf{x}') \left(2\ln\varrho + \ln\left(1 + \frac{\varrho'^2}{\varrho^2} - 2\frac{\varrho'}{\varrho}\cos(\varphi-\varphi')\right) \right)$$

$$= -2k_C \left\{ M_0^{(c)} \ln\varrho + \sum_{n=1}^{\infty} \frac{1}{n\varrho^n} \int d^2x'\, \rho(\mathbf{x}')\, \varrho'^n \cos\left(n(\varphi-\varphi')\right) \right\}.$$

If we substitute $\cos(n\varphi - n\varphi') = \cos(n\varphi)\cos(n\varphi') + \sin(n\varphi)\sin(n\varphi')$ into the above equation we obtain with the definitions (3.5.7) for the moments the expansion to be proven (3.5.6).

3.5.2 Complex Analysis

The methods of complex analysis are often very useful in two-dimensional problems. Some potentials, which are uniformly infinite in one dimension, can be treated as a two-dimensional problem more simply than in three dimensions. This is particularly true for systems with cylindrical symmetry, such as for a metal cylinder in an otherwise charge-free space. The z-coordinate is parallel to the cylinder axis, and the two-dimensional Laplace equation $\Delta\phi(x_1, x_2) = 0$ needs to be solved, for which complex-valued functions

$$f(z) = \phi(x, y) + \mathrm{i}\psi(x, y) \qquad \text{with} \quad z = x + \mathrm{i}y \tag{3.5.8}$$

are used. ϕ and ψ are real functions of the real variables x, y. If $f(z)$ is differentiable in an open area $G \subset \mathbb{C}$, then $f(z)$ is analytic. Properties of analytic (holomorphic) functions are listed in the appendix B.1.1, page 603.

According to Liouville's theorem, a function $f(z)$ that is analytic, unique and bounded in the entire z-plane, is a constant. $f(z)$ therefore has poles and these represent the sources; if the Laplace equation were fulfilled in the entire plane, then would the associated potential be constant (zero).

From the Cauchy-Riemann differential equations (B.1.3) follows

$$\frac{\partial\phi}{\partial x} = \frac{\partial\psi}{\partial y}, \quad \frac{\partial\phi}{\partial y} = -\frac{\partial\psi}{\partial x} \quad \Rightarrow \quad (\boldsymbol{\nabla}\phi)\cdot(\boldsymbol{\nabla}\psi) = \frac{\partial\phi}{\partial x}\frac{\partial\psi}{\partial x} + \frac{\partial\phi}{\partial y}\frac{\partial\psi}{\partial y} = 0\,.$$

The gradients of ϕ and ψ are therefore orthogonal to each other at every point. If ϕ_i=const are equipotential lines, then the lines ψ_i=const indicate the field course, as shown in Fig. 3.11. A suitable function $f(z)$ thus describes the

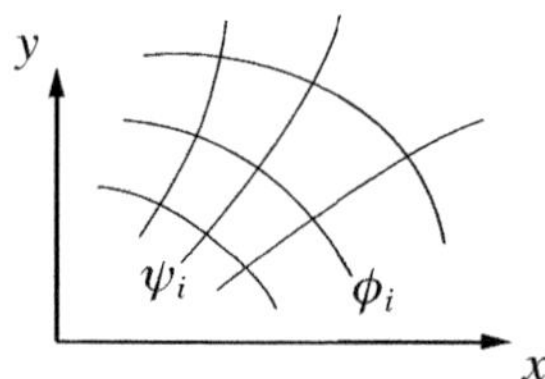

Fig. 3.11. The field lines ψ_i are perpendicular to the equipotential lines with constant ϕ_i; from hydrodynamics comes the name *stream function*

equipotential and field lines of a configuration.

Application of conformal mapping in potential theory

An analytical function provides at each point where its derivative does not vanish, a conformal, i.e., an angle and length preserving mapping (see page 607).

1. *Angle preserving*: The intersection angle α of two curves is not changed by the conformal mapping $z = g(\zeta)$:

$$z = z_0 + \delta z = z_0 + |\delta z|\mathrm{e}^{\mathrm{i}\alpha} = g(\zeta_0) + g'(\zeta_0)\,\delta\zeta \qquad \Rightarrow \qquad |\delta z|\mathrm{e}^{\mathrm{i}\alpha} = |g'(\zeta_0)|\mathrm{e}^{\mathrm{i}\beta'}\,|\delta\zeta|\mathrm{e}^{\mathrm{i}\beta}.$$

   For a δz of the same length, but different direction, $\delta z = |\delta z|\mathrm{e}^{\mathrm{i}(\alpha+\gamma)}$, only the angle from β to $\beta + \gamma$ changes in the conformal mapping. The intersection angle α of the two lines thus remains unchanged.
2. *Length preserving*: It states that the stretch ratio $|\delta z|/|\delta\zeta| = |g'(\zeta_0)|$ depends only on ζ_0 (i.e. z_0), but not on the direction.

If one has a representation with $f(z)$ that solves a given problem, one can determine the boundary conditions for another configuration by a conformal transformation $\zeta = g(z)$, where $g(z)$ is an analytic function. This provides the possibility to solve more complex configurations, by transforming to a known, usually simpler system, establishing the connection to the given problem.

Potential of a homogeneous field

We start from a homogeneous field that points in the x direction and is described by the potential $\phi = -Ex$. The complex potential is

$$f = -Ez = -E(x + iy) = \phi + i\psi.$$

The streamlines (field lines) are then given by $\psi = \text{const}$. These are lines parallel to the x-axis.

Line charge

A straight wire of infinitesimal thickness with the line charge density λ pierces the xy-plane at a right angle at the origin. The electric field $\mathbf{E} = E\mathbf{e}_\varrho$ is radial and is calculated with Gauss's law (1.3.2), where integration is done over a cylinder of length L with radius ϱ $(q = \lambda L)$:

$$\oiint \mathrm{da} \cdot \mathbf{E} = 2\pi\varrho L E = 4\pi k_C \lambda L.$$

Thus, it is

$$\mathbf{E} = k_C \frac{2\lambda \mathbf{e}_\varrho}{\varrho}$$

with $\varrho = \sqrt{x^2 + y^2}$. The corresponding electrostatic potential is

$$\phi = -2k_C \lambda \ln \varrho. \tag{3.5.9}$$

In two dimensions, the potential for $\varrho \to \infty$ does not drop to zero, which is seen as a consequence of the infinite extension of the line charge perpendicular to the xy plane. If the line charge is now placed at the point z_λ, the distance ϱ from the charge is to be replaced by

$$\varrho \to \sqrt{(x-x_\lambda)^2 + (y-y_\lambda)^2} = \sqrt{(z-z_\lambda)(z^* - z_\lambda^*)}. \tag{3.5.10}$$

This suggests the approach for the complex potential

$$f(z) = -2k_C \lambda \ln(z - z_\lambda). \tag{3.5.11}$$

This results in

$$\phi = \frac{f(z) + f^*(z)}{2} = -2k_C\lambda \ln \sqrt{(z - z_\lambda)(z^* - z_\lambda^*)}\,.$$

When switching to plane polar coordinates $z = \varrho e^{i\varphi}$, it follows that

$$\phi = -k_C\lambda \ln\left[\varrho^2 + \varrho_\lambda^2 - 2\varrho\varrho_\lambda\cos(\varphi - \varphi_\lambda)\right]$$

and ($\ln z = \ln \varrho + i\varphi$)

$$\psi = \frac{f(z) - f^*(z)}{2i} = ik_C\lambda \ln \frac{z - z_\lambda}{z^* - z_\lambda^*} = -2k_C\lambda\arctan\frac{y - y_\lambda}{x - x_\lambda}\,.$$

In plane polar coordinates, $\varphi = \arctan\dfrac{y}{x}$, i.e. if $\psi = \psi_0$ is constant, then $\varphi = \varphi_0$ is. The field lines are radial straight lines at fixed φ_0; if $z_\lambda \neq 0$, the lines originate from the point z_λ.

Cylindrical capacitor

An arrangement that is easy to treat using the methods of complex analysis is the cylindrical capacitor [Becker, Sauter, 1973, §1.6c]. It consists of 2 coaxial conductive circular cylinders with the radii $0 < \varrho_1 < \varrho_2$, and with the surface charge density $\sigma_1 > 0$ for the inner cylinder.

The height of the cylinder L is much greater than ϱ_2, so that edge effects can be neglected. The total charge of the inner cylinder is $q = \lambda L > 0$ and that of the outer $-q$. Here, λ is the charge per unit length and L is the length of the cylinder with $L \gg \varrho_2$. The field $\mathbf{E} = E\mathbf{e}_\varrho$ between the two metal plates is radially directed outward, as sketched in Fig. 3.12a. One obtains from Gauss's

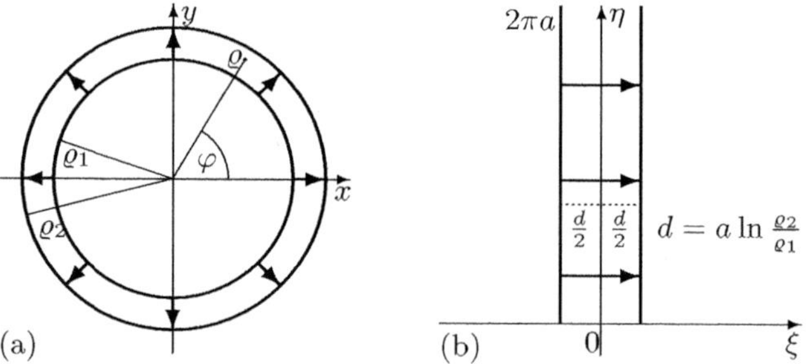

Fig. 3.12. Conformal mapping of a cylindrical capacitor into a plate capacitor of the same capacity
(a) Cylindrical capacitor: The inner cylinder is positively charged, as indicated by the field direction
(b) Plate capacitor: a is a scale factor for the lengths

law (1.3.2), by integrating over a cylinder with $\varrho_1 < \varrho < \varrho_2$ of length L:

$$\oiint da \cdot \mathbf{E} = 2\pi \varrho L E = 4\pi k_C (\sigma_1 2\pi \varrho_1 L) \,.$$

Expressed through the line charge λ, one obtains the electric field

$$\mathbf{E} = 2k_C \lambda \frac{\mathbf{x}}{\varrho^2}, \qquad\qquad \lambda = 2\pi \varrho_1 \sigma_1 = -2\pi \varrho_2 \sigma_2,$$

and from this the electrostatic potential for the area $\varrho_1 \leq \varrho \leq \varrho_2$

$$\phi = -2k_C \lambda (\ln \varrho - \ln \varrho_m).$$

$\varrho_m = \sqrt{\varrho_1 \varrho_2}$ is chosen so that $\phi(\varrho_2) = -\phi(\varrho_1)$. The voltage is applied to the capacitor

$$V = \phi(\varrho_1) - \phi(\varrho_2) = -2k_C \lambda \ln \frac{\varrho_1}{\varrho_2} \,.$$

The capacitance C of the capacitor is determined by

$$C = \frac{q}{V} = \frac{L}{2k_C \ln(\varrho_2/\varrho_1)} \,. \tag{3.5.12}$$

ψ results from the Cauchy-Riemann differential equations (B.1.3)

$$\frac{\partial \psi}{\partial x} = -\frac{\partial \phi}{\partial y} = 2k_C \lambda \frac{y}{\varrho} \quad\text{and}\quad \frac{\partial \psi}{\partial y} = \frac{\partial \phi}{\partial x} = -2k_C \lambda \frac{y}{\varrho} \,.$$

From this it follows initially

$$\psi = -2k_C \lambda \arctan \frac{y}{x} + \text{const.} = -2k_C \lambda (\varphi - \varphi_m).$$

The complex potential is then given by

$$f(z) = -2k_C \lambda \ln \frac{z}{z_m} \,. \tag{3.5.13}$$

Mapping of the cylindrical capacitor to a plate capacitor

With the method of complex analysis, the found solution can be used for another geometric arrangement through a conformal mapping $z = g(\zeta)$ with $\zeta = \xi + i\eta$.

Initially, one looks for the transformation that maps the circle with ϱ_m to the positive η-axis:

$$z = \varrho_m e^{\zeta/a} = \varrho_m e^{\xi/a} \, e^{i\eta/a} \,,$$

where $a > 0$ is a real scaling factor. The circle with $z = \varrho_m e^{i\varphi}$ is thus mapped to the straight line with $\xi = 0$ and $0 \leq \eta < 2\pi a$. Similarly, one obtains for the inner cylinder

$$z = \varrho_1 e^{i\varphi} = \varrho_m e^{\xi/a}\, e^{i\eta/a},$$

from which $\xi = -d/2 = a\ln(\varrho_1/\varrho_m)$ and $0 \le \eta < 2\pi a$ follows. The outer cylinder is mapped to the straight line of the same length at a distance $\xi = d/2$ from the η-axis, which results in the plate capacitor sketched in Fig. 3.12b.

For the complex potential $F(\zeta) = \Phi(\xi,\eta) + i\Psi(\xi,\eta)$ it holds that

$$F(\zeta) = f(g(\zeta)) = f(z).$$

Thus, the potentials on the conductor surfaces $z_{1,2} = \varrho_{1,2}\, e^{i\varphi}$ remain unchanged:

$$\Phi(\mp\frac{d}{2},\eta) = \phi(\varrho_{1,2}) \qquad \text{and} \qquad \Psi(\mp\frac{d}{2},\eta) = \psi(\varphi).$$

Since the charge q also remains unchanged, the capacitance of the plate capacitor is equal to that of the cylindrical capacitor (3.5.12). If you also set the distance of the plates $d = a\ln(\varrho_2/\varrho_1)$, you get the form known from the plate capacitor (2.2.30)

$$C = \frac{L}{2k_c \ln(\varrho_2/\varrho_1)} = \frac{aL}{2k_c d} = \frac{A}{4\pi k_c d} \qquad \text{with} \quad A = 2\pi aL.$$

Problems for Chapter 3

3.1. *Point charge in front of a split metal plate*: A point charge q is located in front of a metal plate ($z = 0$), which is divided along the line $x = 0$, as sketched in Fig. 3.3 on page 83. You can assume G_D for this configuration as known (3.1.8). Assume that the conductor for $x > 0$ is kept at the constant potential ϕ_1 and for $x < 0$ at ϕ_2. Calculate $\phi(\mathbf{x})$. Determine $\mathbf{E}$ for large r.
Auxiliary integrals: (B.5.10) and (B.5.17)

3.2. *Force between metal sphere and point charge*: In front of a metal sphere (radius R, charge $Q > 0$) there is a point charge q at a distance $d > R$. Determine the force between the metal sphere and the point charge and state the condition that you must impose on $q > 0$ so that the two positive charges attract each other.

3.3. *Spherical cosine theorem*: Prove the spherical cosine theorem without using the addition theorem for spherical functions to use.

3.4. *Earnshaw's theorem*: The validity of Earnshaw's theorem was shown on the basis of the principle of maximum/minimum or the mean value theorem. Here, derive the validity of the theorem using Gauss's theorem.

3.5. *Multipole expansion in two dimensions*: Derive the two-dimensional multipole expansion (3.5.6) with the moments (3.5.7), using the auxiliary formula

$$\ln(1 + t^2 - 2t\cos\alpha) = -2\sum_{n=1}^{\infty} \frac{t^n}{n}\cos(n\cos\alpha) \qquad |t| < 1$$

[Gradshteyn, Ryzhik, 1965, No. 1514] can be used.

3.6. *Edge correction of the plate capacitor according to Kirchhoff*: The scattered field of a plate capacitor, which extends as indicated in Fig. 3.13, from $x = -\infty$ to $x = 0$ [Sommerfeld, 1952, p. 329, problem II.4]. Show that the scattered field can be described by

$$z = g(\zeta) \quad \text{with} \quad g(\zeta) = \frac{d}{2\pi}\Big(1 + \frac{2\pi i\zeta}{V} + e^{\frac{2\pi i\zeta}{V}}\Big) \quad \text{and} \quad \begin{cases} z &= x + iy \\ \zeta &= \xi + i\eta \end{cases}$$

can be described, where $\xi \equiv \phi$ and $\eta \equiv \psi$. Show in particular that the field line

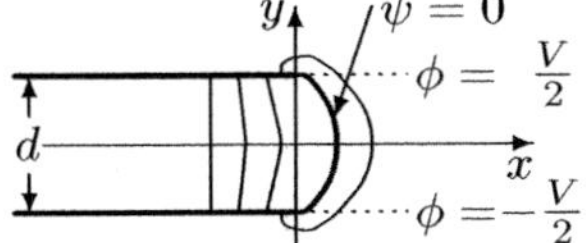

Fig. 3.13. Plate capacitor $(-\infty < x \leq 0)$ with the voltage difference V

$\psi = 0$, which connects the points $(0, d/2)$ with $(0, -d/2)$ (and is drawn in Fig. 3.13), is a cycloid:

$$x = \frac{d}{2\pi}\Big(1 + \cos(\frac{2\pi\phi}{V})\Big), \qquad\qquad y = \frac{d}{2\pi}\Big(\frac{2\pi\phi}{V} + \sin(\frac{2\pi\phi}{V})\Big).$$

3.7. *Boundary values given on rectangle*

1. Show using the separation approach $\phi(x,y) = f(x)g(y)$, that

$$f(x) = a\sin(\nu x) + b\cos(\nu x), \qquad g(y) = c\sinh(\nu y) + d\cosh(\nu y) \qquad (3.5.14)$$

   is a solution of the Laplace equation. a to d are integration constants and ν^2 the separation constant.
2. The potential ϕ of a rectangle with the side lengths l_x and l_y is only different from zero on one side, the upper edge $\phi(x, l_y) = \psi(x)$. Determine the potential $\phi(x,y)$.

References

Becker R., Sauter F. *Theorie der Elektrizität 1*, 21th ed. Teubner, Stuttgart (1973)
Gradshteyn I. S., Ryzhik I. M., *Table of Integrals, Series, and Products*, Academic Press N.Y. (1965)
Jackson J. D. *Classical Electrodynamics*, 3rd ed., John Wiley & Sons Inc. (1998)
Oberhettinger F. *Fourier Expansions*, Academic Press N.Y. (1973)
Rebhan E. *Theoretische Physik: Elektrodynamik*, Spektrum München (2007)
Sommerfeld A. *Electrodynamics*, Academic Press Inc., New York (1952)

Magnetostatics in Vacuum

The system-independent notation leads to unwieldy expressions in magnetostatics even in a vacuum. Therefore, some of the occurring combinations of constants are listed in Tab. 4.1.

Tab. 4.1. Unit systems in electrodynamics; k_C is the prefactor to Coulomb's law (1.2.1), k_L to Lorentz force (1.2.5), k_M to the magnetostatic force law (7.1.42) and k_r to rational systems. ϵ_0 and μ_0 are the electric and magnetic constants (C.2.6) and (C.2.10).

System	k_C	k_L	ϵ_0	$\mu_0 = \dfrac{1}{k_L^2 \epsilon_0}$	$k_r = k_C \epsilon_0$	$k_M = \dfrac{k_C}{k_L^2}$	$\dfrac{k_C}{ck_L}$	$\dfrac{4\pi k_C}{ck_L}$
Gauss	1	1	1	1	1	1	$\dfrac{1}{c}$	$\dfrac{4\pi}{c}$
Heaviside-Lorentz	$\dfrac{1}{4\pi}$	1	1	1	$\dfrac{1}{4\pi}$	$\dfrac{1}{4\pi}$	$\dfrac{1}{4\pi c}$	$\dfrac{1}{c}$
SI	$\dfrac{1}{4\pi\epsilon_0}$	c	ϵ_0	$\mu_0 = \dfrac{1}{c^2 \epsilon_0}$	$\dfrac{1}{4\pi}$	$\dfrac{\mu_0}{4\pi}$	$\dfrac{\mu_0}{4\pi}$	μ_0

4.1 Basic Equations of Magnetostatics

4.1.1 Maxwell's Equations

The basic equations of magnetostatics only concern the magnetic field $\mathbf{B}$. There are no electric charges and no electric fields, but only the (electric) current density $\mathbf{j}$, which is however time-independent. From the Maxwell equations (1.3.21) therefore remains the Ampère-Maxwell equation, but without the displacement current $\mathbf{j}_d = \dfrac{1}{4\pi k_C}\dot{\mathbf{E}}$, added by Maxwell, which does not occur in time-independent questions. Only the divergence-free magnetic field $\mathbf{B}$

D. Petrascheck, F. Schwabl, *Electrodynamics*, https://doi.org/10.1007/978-3-662-71502-4_4

is added:

$$\boldsymbol{\nabla}\times\mathbf{B} = \frac{4\pi k_C}{ck_L}\,\mathbf{j} \;\overset{\mathrm{SI}}{=}\; \mu_0\mathbf{j}, \qquad\qquad \boldsymbol{\nabla}\cdot\mathbf{B} = 0. \qquad\qquad (4.1.1\mathrm{a\text{--}b})$$

From $\boldsymbol{\nabla}\cdot\mathbf{B} = 0$ it follows, since $\boldsymbol{\nabla}\cdot(\boldsymbol{\nabla}\times\mathbf{A}) = 0$, that $\mathbf{B}$ can be represented by

$$\mathbf{B} = \boldsymbol{\nabla}\times\mathbf{A}. \qquad\qquad (4.1.2)$$

From Ampère's law follows by means of (A.2.38)

$$\boldsymbol{\nabla}\times\mathbf{B} = \boldsymbol{\nabla}\times(\boldsymbol{\nabla}\times\mathbf{A}) = \boldsymbol{\nabla}(\boldsymbol{\nabla}\cdot\mathbf{A}) - \Delta\mathbf{A} = -\Delta\mathbf{A} = \frac{4\pi k_C}{ck_L}\,\mathbf{j}$$

the vector-Poisson equation

$$\Delta\mathbf{A} = -\frac{4\pi k_C}{ck_L}\,\mathbf{j} \qquad\qquad \text{with} \qquad\qquad \boldsymbol{\nabla}\cdot\mathbf{A} = 0. \qquad (4.1.3)$$

These are the (scalar) Poisson equations for each component of $\mathbf{A}$, analogous to electrostatics (2.1.8) for the scalar potential. For $\mathbf{A}$ in (2.1.8) only ρ has to be replaced by $\mathbf{j}/ck_L$:

$$\mathbf{A}(\mathbf{x}) = \frac{k_C}{ck_L}\int \mathrm{d}^3 x'\, \frac{\mathbf{j}(\mathbf{x}')}{|\mathbf{x}-\mathbf{x}'|} = \frac{1}{4\pi}\int \mathrm{d}^3 x'\, \frac{\boldsymbol{\nabla}\times'\mathbf{B}(\mathbf{x}')}{|\mathbf{x}-\mathbf{x}'|}. \qquad (4.1.4)$$

The right side is obtained by substituting the Ampère's law (4.1.1a–b) into (4.1.4).

Note: Actually, the right side of (4.1.4) already implies that $\boldsymbol{\nabla}\cdot\mathbf{A} = 0$. We show it explicitly, by replacing $\boldsymbol{\nabla}\to -\boldsymbol{\nabla}'$ and integrating by parts:

$$\boldsymbol{\nabla}\cdot\mathbf{A} = \frac{k_C}{ck_L}\int \mathrm{d}^3 x'\, \mathbf{j}(\mathbf{x}')\cdot\boldsymbol{\nabla}\frac{1}{|\mathbf{x}-\mathbf{x}'|} = \frac{k_C}{ck_L}\int \mathrm{d}^3 x'\, \frac{1}{|\mathbf{x}-\mathbf{x}'|}\,\boldsymbol{\nabla}'\cdot\mathbf{j}(\mathbf{x}') = 0,$$

from which it becomes clear that this is not self-evident in the time-dependent case $\dot{\rho} \neq 0$. However, one can, if $\boldsymbol{\nabla}\cdot\mathbf{A}' \neq 0$, always add to $\mathbf{A}'$ the gradient of a scalar function χ without changing $\boldsymbol{\nabla}\times\mathbf{A}'$, i.e. without changing the field $\mathbf{B}$

$$\mathbf{A} = \mathbf{A}' + \boldsymbol{\nabla}\chi \qquad \text{with} \qquad \Delta\chi = -\boldsymbol{\nabla}\cdot\mathbf{A}' \qquad \Rightarrow \qquad \boldsymbol{\nabla}\cdot\mathbf{A} = 0.$$

4.1.2 Ampère's Law

Ampère's circuital law is the integral form of the Ampère-Maxwell equation in time-independent form (4.1.1a–b), applied to an area A with the boundary ∂A, as sketched in Fig. 4.1. This results in

$$\iint_A \mathrm{d}\mathbf{a}\cdot(\boldsymbol{\nabla}\cdot\mathbf{B}) = \frac{4\pi k_C}{ck_L}\iint_A \mathrm{d}\mathbf{a}\cdot\mathbf{j}$$

and applies the Stokes' theorem (A.4.13), from which the Ampère's law directly follows:

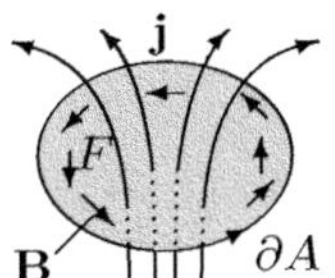

Fig. 4.1. Flow through the (open) surface A, bordered by the curve ∂A

$$Z_B = \oint_{\partial A} \mathrm{d}\mathbf{s} \cdot \mathbf{B} = \frac{4\pi k_C}{ck_L} I \stackrel{\mathrm{SI}}{=} \mu_0 I. \tag{4.1.5}$$

To be more precise, (4.1.5) should be referred to as the *Ampère's circuital law*, as *Ampère's law* [Sommerfeld, 1952, §3] or as *Ørsted's law* [Becker, Sauter, 1973, (5.4.1)]. In this case,

$$I = \iint_A \mathrm{d}\mathbf{a} \cdot \mathbf{j} \tag{4.1.6}$$

is the current through the area A and Z_B is the circulation of magnetic field **B**. Ampère's law is particularly useful in solving simple symmetrical problems, similar to Gauss's law in electrostatics (1.3.21").

4.1.3 Biot-Savart Law

The magnetic field **B** is obtained using (4.1.4) from the vector potential **A**:

$$\mathbf{B} = \boldsymbol{\nabla} \times \mathbf{A} = \frac{k_C}{ck_L} \int \mathrm{d}^3 x' \, \boldsymbol{\nabla} \times \frac{\mathbf{j}(\mathbf{x}')}{|\mathbf{x}-\mathbf{x}'|} = \frac{k_C}{ck_L} \int \mathrm{d}^3 x' \, \frac{\mathbf{j}(\mathbf{x}') \times (\mathbf{x}-\mathbf{x}')}{|\mathbf{x}-\mathbf{x}'|^3} \tag{4.1.7}$$

in a form that can be referred to as a generally held *Biot-Savart law* [Becker, Sauter, 1973, (5.4.14)].

Thin wires

In many cases, the current does not distribute over larger ranges of the volume, as (4.1.7) provides, but flows only within thin wires, as sketched in Fig. 4.2. The

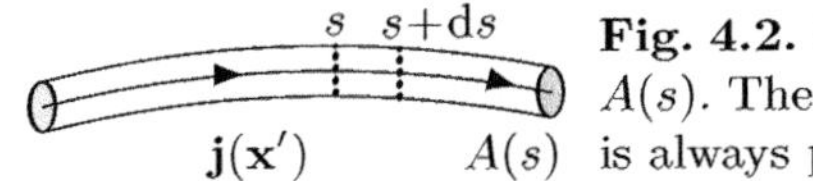

Fig. 4.2. Current-carrying wire with the cross-section $A(s)$. The direction $\hat{s}$ of the line of wire center points is always parallel to $\mathbf{j}(\mathbf{x}'(s)) = j(\mathbf{x}'(s))\hat{s}$

curve of the wire center points is given by $\mathbf{x}'(s)$, where s is the arc length. To integrate are expressions of the form

$$\int \mathrm{d}^3 x' \, \mathbf{j}(\mathbf{x}') \times \mathbf{v}(\mathbf{x}, \mathbf{x}'),$$

where $\mathbf{v}(\mathbf{x}, \mathbf{x}')$ should be approximately constant over the wire cross-section. We switch to a local, Cartesian coordinate system $(\mathbf{s}_\perp, s)$ in which the s-axis $\hat{s}$ is parallel to $\mathbf{j}(\mathbf{x}'(s))$. The functional determinant (Jacobi determinant) of the transformation $J = \det\left[\frac{\partial(\mathbf{x}')}{\partial(\mathbf{s}_\perp, s)}\right] = 1$. The current I is obtained as follows

$$\int \mathrm{d}^3x' \, \mathbf{j}(\mathbf{x}')... = \int \mathrm{d}s \, \mathrm{d}^2s_\perp \, \mathbf{j}(\mathbf{x}'(\mathbf{s}_\perp,s))..., = I \int_C \mathrm{d}\mathbf{s}... \tag{4.1.8}$$

$$I = \int \mathrm{d}^2s_\perp \, j(\mathbf{x}'(\mathbf{s}_\perp,s)) = j(s)\, A(s).$$

$j(s)$ denotes the current density averaged over the cross section. The wire is thus replaced by an infinitely thin line. A bit more loosely, we write

$$\mathrm{d}^3x' \, \mathbf{j}(\mathbf{x}') = I\delta^{(2)}(\mathbf{s}_\perp)\, \mathrm{d}^2s_\perp \, \mathrm{d}s.$$

The neglect of the finite cross section of the wire will not be justified in some cases, especially when calculating self-inductances (see section 7.2.2).

If you have a current-carrying wire that generates the field $\mathbf{B}$ according to (4.1.7), you get with the help of (4.1.8) the *Biot-Savart's law* [Abraham, Becker, 1930, (117)]

$$\mathbf{B}(\mathbf{x}) = \frac{k_C I}{ck_L} \int_C \frac{\mathrm{d}\mathbf{s}\times (\mathbf{x}-\mathbf{x}'(s))}{|\mathbf{x}-\mathbf{x}'(s)|^3}, \qquad \text{SI:} \; \frac{k_C}{ck_L} = \frac{\mu_0}{4\pi}. \tag{4.1.9}$$

For completeness, it should be added that Sommerfeld [1952, §15.A] refers to the differential form $\mathrm{d}\mathbf{B} = (k_C I/ck_L)\mathrm{d}\mathbf{s}\times\mathbf{x}/r^3$ as Biot-Savart's law, while Jackson [1998, (5.6)] gives (4.1.11) this name.

4.1.4 Magnetic Field of an Infinitely Long Wire

In this simple example, we will test four different methods, to determine the magnetic field $\mathbf{B}$ of an infinitely long wire. The wire is located, as sketched in Fig. 4.3, on the z-axis and thus $\mathbf{B}=B\,\mathbf{e}_\varphi$.

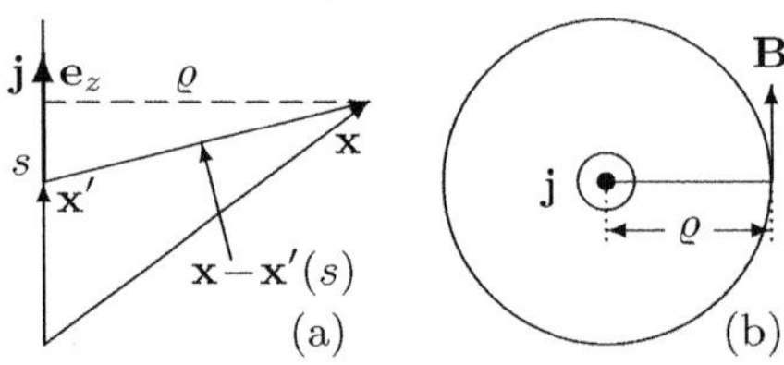

Fig. 4.3. (a) $\mathbf{j}\times(\mathbf{x}-\mathbf{x}'(s))$ is perpendicular to the paper plane and thus also $\mathbf{B}$. $\varrho = |\mathbf{e}_z\times(\mathbf{x}-\mathbf{x}')|$ (b) $\mathbf{B} = B\,\mathbf{e}_\varphi$

1. *Determination of B_φ using Ampère's law*

   From Ampère's law, we obtain the circulation by integrating over a circle S_ϱ and from it $\mathbf{B}$:

$$\oint_{S_\varrho} \mathrm{d}\mathbf{s}\cdot\mathbf{B} = \frac{4\pi k_C}{ck_L}I = 2\pi\varrho B_\varphi \quad \Longrightarrow \quad B_\varphi = \frac{k_C}{ck_L}\frac{2I}{\varrho} \underset{\text{SI}}{=} \frac{\mu_0 I}{2\pi\varrho}. \tag{4.1.10}$$

2. *Determination of B_φ using Biot-Savart's law*

   We substitute into Biot-Savart's law (4.1.9)

   $$\mathbf{ds} \times \big(\mathbf{x} - \mathbf{x}'(s)\big) = \mathrm{d}s\, \mathbf{e}_z \times (\boldsymbol{\varrho} + \mathbf{z} - s\mathbf{e}_z) = \mathrm{d}s\, \varrho\, \mathbf{e}_\varphi.$$

   This results in $(\varrho = \sqrt{x^2 + y^2})$

   $$\mathbf{B}(\mathbf{x}) = \frac{k_C}{ck_L} I \varrho\, \mathbf{e}_\varphi \int_{-\infty}^{\infty} \frac{\mathrm{d}s}{\sqrt{\varrho^2 + (z-s)^2}^{\,3}} \overset{v\varrho = s-z}{=} \frac{k_C}{ck_L} \frac{2I\mathbf{e}_\varphi}{\varrho} \int_0^{\infty} \frac{\mathrm{d}v}{\sqrt{1+v^2}^{\,3}}$$

   $$\overset{\text{(B.5.16)}}{=} \frac{k_C}{ck_L} \frac{2I}{\varrho} \mathbf{e}_\varphi \frac{v}{\sqrt{1+v^2}}\bigg|_0^{\infty} = \frac{k_C}{ck_L} \frac{2I}{\varrho} \mathbf{e}_\varphi. \tag{4.1.11}$$

   From this follows for the magnetic field of an infinitely long wire the same result as in (4.1.10).

3. *Calculation of $\mathbf{B}$ using the vector potential $\mathbf{A}$*

   We first determine $\mathbf{A}$ using (4.1.4), i.e. the solution of the vector Poisson equation for the straight wire of length $2l$:

   $$\mathbf{j}(\mathbf{x}') = I\delta(x')\delta(y')\theta(l-|z'|)\,\mathbf{e}_z \;\Rightarrow\; \mathbf{A}(\mathbf{x}) = \mathbf{e}_z \frac{k_C I}{ck_L} \int_{-l}^{l} \frac{\mathrm{d}z'}{\sqrt{\varrho^2 + (z-z')^2}}.$$

   The expression for A_z is, when replacing I/ck_L with λ, equal to the scalar potential ϕ (2.2.15) of a line charge. We orient ourselves to the calculations carried out there, whereby with regard to $l \to \infty$, $z = 0$ may be set:

   $$A_z = -\frac{k_C}{ck_L} I \ln\big(z - z' + \sqrt{\varrho^2 + (z-z')^2}\big)\bigg|_{-l}^{l} \overset{z=0}{=} -\frac{k_C}{ck_L} I \ln \frac{-l + \sqrt{\varrho^2 + l^2}}{l + \sqrt{\varrho^2 + l^2}}$$

   $$= -\frac{k_C}{ck_L} I \lim_{l \to \infty} \ln \frac{\varrho^2}{\big(l + \sqrt{\varrho^2 + l^2}\big)^2} = \frac{k_C}{ck_L} 2I \lim_{l \to \infty} \big[\ln(2l) - \ln \varrho\big]. \tag{4.1.12}$$

   The constant, although infinite, is irrelevant for the fields, which is why it is usually not taken into account in $\mathbf{A}$. The vector potential is better defined if it is defined as the difference to the potential of a regularization point $\mathbf{x}_0$, as already shown for the line charge (2.2.16):

   $$\mathbf{A}(\mathbf{x}, \mathbf{x}_0) = \frac{k_C}{ck_L} 2I(\varrho_0 - \varrho)\mathbf{e}_z$$

   $$\mathbf{B}(\mathbf{x}) = \boldsymbol{\nabla} \times \mathbf{A} = -\frac{k_C}{ck_L} \frac{2I}{\varrho} \mathbf{e}_\varrho \times \mathbf{e}_z = \frac{k_C}{ck_L} \frac{2I}{\varrho} \mathbf{e}_\varphi. \tag{4.1.13}$$

   Once again, we have obtained the well-known result for the $\mathbf{B}$-field of the infinite wire, but with the option to specify the field of the finite wire.

4. *Calculation of* **B** *via the two-dimensional Poisson equation*

j points in the z-axis, so that only the Poisson equation for A_z is inhomogeneous. Due to the infinite length of the wire, $A_z = A_z(x, y)$ is independent of z (and because of the axial symmetry also of φ):

$$\Delta A_z(x, y) = \left[\frac{\partial^2}{\partial x^2} + \frac{\partial^2}{\partial y^2} \right] A_z = -\frac{k_C}{ck_L} 2I \, 2\pi \, \delta(x)\, \delta(y).$$

A_z satisfies the Poisson equation (3.5.2) of the two-dimensional Green function $G(x, y)$, from which it follows:

$$\left[\frac{\partial^2}{\partial x^2} + \frac{\partial^2}{\partial y^2} \right] G(x, y) = 2\pi \, \delta(x)\, \delta(y) \quad \stackrel{(3.5.3)}{\Longrightarrow} \quad G(x, y) = \ln \varrho,$$

$$A_z(\varrho) = -\frac{k_C}{ck_L} 2I \ln \varrho, \qquad\qquad \text{SI:} \ \frac{k_C}{ck_L} = \frac{\mu_0}{4\pi}. \qquad (4.1.14)$$

By forming the curl of $\mathbf{A} = -\frac{k_C}{ck_L} 2I \ln \varrho \, \mathbf{e}_z$ we get again $\mathbf{B} = \frac{k_C}{ck_L} \frac{2I}{\varrho} \, \mathbf{e}_\varphi$.

4.2 Magnetic Dipole

Instead of a systematic expansion according to magnetic multipoles, the definition of the magnetic dipole moment **m** of a (finite) current distribution is introduced:

$$\mathbf{m} = \frac{k_L}{2c} \int \mathrm{d}^3 x' \, \mathbf{x}' \times \mathbf{j}(\mathbf{x}'). \qquad (4.2.1)$$

In section 12.1.1 it is shown that the equivalence of two inertial systems implies a universal speed, which is c. Velocities therefore appear parameterized in the form $\mathbf{v}/c$ or $\mathbf{j}/c$. This is broken in some systems and so there is a dependence on $(k_L/c)\mathbf{j}$ in magnetic moments. **m** is independent of the reference point. A translation by **a** gives the additional contribution

$$\mathbf{a} \times \int \mathrm{d}^3 x' \, \mathbf{j}(\mathbf{x}') = \mathbf{a} \times \dot{\mathbf{p}} = 0,$$

which vanishes due to (4.2.3). In a multipole expansion, the lowest non-vanishing multipole moment is always independent of the location. Since no magnetic monopoles have been found, it is the dipole moment.

4.2.1 Calculation of Moments of a Current Distribution

We consider a localized current distribution $\mathbf{j}(\mathbf{x}')$, which is only finite for $r' < R$. If we develop $\frac{1}{|\mathbf{x} - \mathbf{x}'|}$ in (4.1.4), we get

$$\mathbf{A}(\mathbf{x}) = \frac{k_C}{ck_L} \int \mathrm{d}^3 x' \, \mathbf{j}(\mathbf{x}') \left(\frac{1}{r} + \frac{\mathbf{x} \cdot \mathbf{x}'}{r^3} + \ldots \right) = \frac{k_C}{ck_L} \left(\frac{\mathbf{a}_1}{r} + \frac{\mathbf{a}_2}{r^3} + \ldots \right). \quad (4.2.2)$$

At the moment, we also allow time-dependent currents or charge densities and disregard the prefactors of the individual terms.

If you set for $\mathbf{j}$ the identity $\mathbf{j} = (\mathbf{j} \cdot \boldsymbol{\nabla}) \, \mathbf{x}$ in (4.2.2), then the 1st term results after integration by parts (Gauss's theorem) and application of the continuity equation (1.1.15):

$$\mathbf{a}_1 = \int \mathrm{d}^3 x' \, \mathbf{j} = \int \mathrm{d}^3 x' \, (\mathbf{j} \cdot \boldsymbol{\nabla}') \mathbf{x}' = -\int \mathrm{d}^3 x' \, (\boldsymbol{\nabla}' \cdot \mathbf{j}) \mathbf{x}' = \int \mathrm{d}^3 x' \, \dot{\rho} \, \mathbf{x}' = \dot{\mathbf{p}}. \quad (4.2.3)$$

The calculation of the 2nd term of (4.2.2) is somewhat more complex, but in principle similar, only that now the Grassmann identity (A.1.60) $\mathbf{a} \times (\mathbf{b} \times \mathbf{c}) = (\mathbf{a} \cdot \mathbf{c}) \mathbf{b} - (\mathbf{a} \cdot \mathbf{b}) \mathbf{c}$ and the definition of the dyadic (tensor) product (A.1.15) $\mathbf{a}(\mathbf{b} \cdot \mathbf{c}) = (\mathbf{a} \circ \mathbf{b}) \mathbf{c}$ are added

$$\begin{aligned} \mathbf{a}_2 \ &= \ \int \mathrm{d}^3 x' \, (\mathbf{x} \cdot \mathbf{x}') \, \mathbf{j} = \frac{1}{2} \int \mathrm{d}^3 x' \, (\mathbf{x} \cdot \mathbf{x}') \big[\mathbf{j} + (\mathbf{j} \cdot \boldsymbol{\nabla}') \, \mathbf{x}' \big] \\ &\overset{\text{Gauss}}{=} \ \frac{1}{2} \left\{ \mathbf{a}_2 - \int \mathrm{d}^3 x' \, \mathbf{x}' \left[(\mathbf{j} \cdot \boldsymbol{\nabla}')(\mathbf{x} \cdot \mathbf{x}') + (\boldsymbol{\nabla}' \cdot \mathbf{j}) \, (\mathbf{x} \cdot \mathbf{x}') \right] \right\} \\ &\overset{(1.1.15)}{=} \ \frac{1}{2} \left\{ \int \mathrm{d}^3 x' \, \left[(\mathbf{x} \cdot \mathbf{x}') \, \mathbf{j} - (\mathbf{x} \cdot \mathbf{j}) \, \mathbf{x}' \right] + \int \mathrm{d}^3 x' \, \dot{\rho} \, \mathbf{x}' (\mathbf{x}' \cdot \mathbf{x}) \right\} \\ &= \ \frac{1}{2} \int \mathrm{d}^3 x' \, \mathbf{x} \times (\mathbf{j} \times \mathbf{x}') + \frac{1}{2} \int \mathrm{d}^3 x' \, \dot{\rho} \, (\mathbf{x}' \circ \mathbf{x}') \, \mathbf{x}. \end{aligned}$$

In the first term, we can insert the magnetic dipole moment and in the other term the electric moments M_{ij} (2.5.3). We thus obtain

$$\mathbf{a}_2 = \int \mathrm{d}^3 x' \, (\mathbf{x} \cdot \mathbf{x}') \, \mathbf{j} = \frac{c}{k_L} \mathbf{m} \times \mathbf{x} + \frac{1}{2} \dot{\mathsf{M}} \, \mathbf{x}. \quad (4.2.4)$$

In later applications, such as multipole radiation, we will switch from M to the quadrupole tensor. We now insert (4.2.3) and (4.2.4) into (4.2.2) and obtain the vector potential of a magnetic dipole:

$$\mathbf{A}(\mathbf{x}) = \frac{k_C}{k_L^2} \frac{\mathbf{m} \times \mathbf{x}}{r^3} \qquad \Rightarrow \qquad \text{SI: } \mathbf{A}(\mathbf{x}) = \frac{\mu_0}{4\pi} \frac{\mathbf{m} \times \mathbf{x}}{r^3}. \quad (4.2.5)$$

Supplement: Method for calculating higher moments

We have calculated $\mathbf{A}$ using $\mathbf{a}_1$ and $\mathbf{a}_2$ directly. For higher moments, the process becomes more complex and it is worth deriving a general formula.

Let $f(\mathbf{x})$ and $g(\mathbf{x})$ be nonsingular functions within a volume V. If V is a source-free area on whose surface ∂V no current $\mathbf{j}(\mathbf{x})$ flows, then the continuity equation applies and subsequent integration by parts:

$$\int_V \mathrm{d}^3 x \, \dot{\rho} \, f g = -\int_V \mathrm{d}^3 x \, (\boldsymbol{\nabla} \cdot \mathbf{j}) \, f g = \int_V \mathrm{d}^3 x \left[(\mathbf{j} \cdot \boldsymbol{\nabla} f) g + f (\mathbf{j} \cdot \boldsymbol{\nabla} g) \right]. \quad (4.2.6)$$

In the case of a stationary current distribution, $\nabla \cdot \mathbf{j} = 0$:

$$\int_V \mathrm{d}^3 x \left[(\mathbf{j} \cdot \nabla f) g + f (\mathbf{j} \cdot \nabla g) \right] = 0. \tag{4.2.7}$$

Using (4.2.7) with $f = x_i$ and $g = 1$ for the zeroth moment of $\mathbf{j}$

$$\int \mathrm{d}^3 x \, j_i(\mathbf{x}) = \int \mathrm{d}^3 x \, \dot{\rho} x_i = \dot{p}_i = 0 . \tag{4.2.8}$$

For the next simplest terms, the first moments of $\mathbf{j}$, set $f = x_i$ and $g = x_k$ and get the auxiliary formula

$$\int \mathrm{d}^3 x \left(j_i \, x_k + j_k \, x_i \right) = \int \mathrm{d}^3 x \, \dot{\rho} \, x_i \, x_k = 0 , \tag{4.2.9}$$

which will be used several times. The right sides of (4.2.8) and (4.2.9) are electric dipole and quadrupole terms, which we will return to when discussing radiation of moving charge distributions. From (4.2.9) it follows that in integrals of stationary current distributions, whose integrands depend linearly only on $\mathbf{x}$ and $\mathbf{j}$, the quantities $\mathbf{j} \leftrightharpoons -\mathbf{x}$ can be interchanged.

4.2.2 Magnetic Dipole Field

The starting point is the magnetic dipole potential (4.2.5)

$$\mathbf{A} = \frac{k_C}{k_L^2} \frac{\mathbf{m} \times \mathbf{x}}{r^3} = -\frac{k_C}{k_L^2} \mathbf{m} \times \nabla \frac{1}{r}, \tag{4.2.10}$$

which is used to calculate the field

$$\mathbf{B} = \nabla \times \mathbf{A} = -\frac{k_C}{k_L^2} \nabla \times \left(\mathbf{m} \times \nabla \frac{1}{r} \right) = -\frac{k_C}{k_L^2} \left[\mathbf{m} \left(\nabla \cdot \nabla \frac{1}{r} \right) - (\mathbf{m} \cdot \nabla) \nabla \frac{1}{r} \right]$$

is used. Now $\Delta \frac{1}{r} = -4\pi \delta^{(3)}(\mathbf{x})$, so that

$$B_i = -\frac{k_C}{k_L^2} \left[m_j \nabla_i \nabla_j \frac{1}{r} - 4\pi m_i \delta^{(3)}(\mathbf{x}) \right] \tag{4.2.11}$$

is obtained. One can modify the 1st term by calculating the singular contribution for $r \to 0$ by integrating over a sphere S_ϵ of radius ϵ:

$$m_j \int_{S_\epsilon} \mathrm{d}^3 x \, \nabla_j \nabla_i \frac{1}{r} = m_i \int_{S_\epsilon} \mathrm{d}^3 x \, \frac{1}{3} \nabla^2 \frac{1}{r} = -\frac{4\pi}{3} m_i$$

and separately indicates

$$(\mathbf{m} \cdot \nabla) \frac{\mathbf{x}}{r^3} = -(\mathbf{m} \cdot \nabla) \nabla \frac{1}{r} = \frac{\mathbf{m}}{r^3} - 3 \frac{(\mathbf{m} \cdot \mathbf{x})\mathbf{x}}{r^5} + \frac{4\pi}{3} \mathbf{m} \, \delta^{(3)}(\mathbf{x}) .$$

The magnetic dipole field of a localized current is therefore

$$\mathbf{B} = \frac{k_C}{k_L^2}\left[P\Big(\frac{3(\mathbf{m}\cdot\mathbf{x})\mathbf{x}}{r^5} - \frac{\mathbf{m}}{r^3}\Big) + \frac{8\pi}{3}\,\mathbf{m}\,\delta^{(3)}(\mathbf{x})\right], \tag{4.2.12}$$

where P (principal value) means that an infinitesimal sphere $r \le \epsilon$ is to be excluded during integration. The situation is similar to that of the electric dipole (2.2.4) with $E_i = p_j \nabla_j \nabla_i \frac{1}{r}$. The singular term of the magnetic dipole contributes to the hyperfine structure of atomic s-states [Jackson, 1998, §5.7]. For $r > 0$ the $\mathbf{B}$-field (4.2.12) is completely identical to that of the electric point dipole (see Fig. 2.12). In Tab. 4.2 the properties of the potentials and fields of electric and magnetic dipoles are given. The magnetic analogue to

Tab. 4.2. Electric and magnetic point dipoles ($k_M = k_C/k_L^2$).

$\phi = -k_C\,\mathbf{p}\cdot\boldsymbol{\nabla}\dfrac{1}{r} = k_C\dfrac{\mathbf{p}\cdot\mathbf{x}}{r^3}$	$\mathbf{A} = -k_M\,\mathbf{m}\times\boldsymbol{\nabla}\dfrac{1}{r} = k_M\dfrac{\mathbf{m}\times\mathbf{x}}{r^3}$
$\mathbf{E} = -\boldsymbol{\nabla}\phi = k_C\boldsymbol{\nabla}(\mathbf{p}\cdot\boldsymbol{\nabla})\dfrac{1}{r}$ $\quad= k_C\Big[\dfrac{3(\mathbf{x}\cdot\mathbf{p})\mathbf{x}}{r^5} - \dfrac{\mathbf{p}}{r^3}\Big]$	$\mathbf{B} = \boldsymbol{\nabla}\times\mathbf{A} = k_M\Big[\boldsymbol{\nabla}(\mathbf{m}\cdot\boldsymbol{\nabla})\dfrac{1}{r} - \mathbf{m}\Delta\dfrac{1}{r}\Big]$ $\quad= k_M\Big[\dfrac{3(\mathbf{x}\cdot\mathbf{m})\mathbf{x}}{r^5} - \dfrac{\mathbf{m}}{r^3} + 4\pi\mathbf{m}\delta^{(3)}(\mathbf{x})\Big]$
$\boldsymbol{\nabla}\cdot\mathbf{E} = k_C\Delta(\mathbf{p}\cdot\boldsymbol{\nabla})\dfrac{1}{r} = 4\pi k_C\rho(\mathbf{x})$	$\boldsymbol{\nabla}\cdot\mathbf{B} = 0$
$\boldsymbol{\nabla}\times\mathbf{E} = 0$	$\boldsymbol{\nabla}\times\mathbf{B} = k_M\,\mathbf{m}\times\boldsymbol{\nabla}\Delta\dfrac{1}{r} = 4\pi k_M\dfrac{k_L}{c}\mathbf{j}(\mathbf{x})$
$\rho(\mathbf{x}) = -\mathbf{p}\cdot\boldsymbol{\nabla}\delta^{(3)}(\mathbf{x})$	$\mathbf{j}(\mathbf{x}) = -\dfrac{c}{k_L}\mathbf{m}\times\boldsymbol{\nabla}\delta^{(3)}(\mathbf{x})$
SI: $k_C = 1/4\pi\epsilon_0$	SI: $k_L = c,\qquad k_M = \mu_0/4\pi$

the dipole field of two opposing electric charges (Fig. 4.4a) is the $\mathbf{B}$-field of the circular current (Fig. 4.4b), which will be calculated in the following.

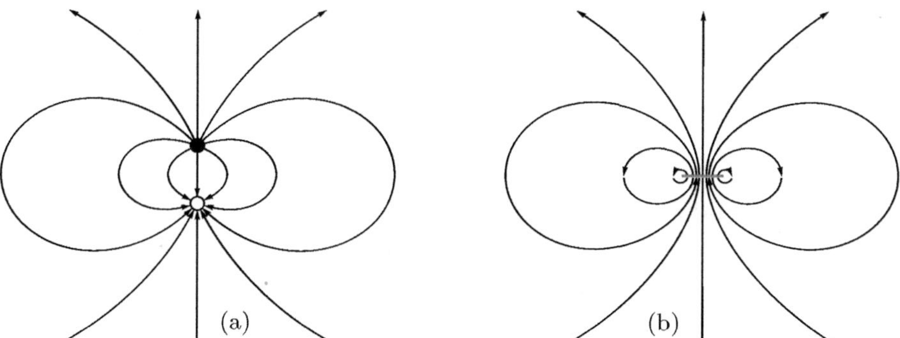

Fig. 4.4. (a) Field lines of an electric dipole; they are directed from the positive (upper) charge to the negative, lower charge.
(b) Field lines of a magnetic dipole, generated by a current loop; the current runs counterclockwise, and within the loop the field lines are directed upwards

4.2.3 Dipole Moment of a Current Loop

The magnetic moment **m** (4.2.1) of a closed, rigid circuit with constant current strength I is also referred to as magnetic dipole moment. Fig. 4.5 shows a section of a current loop consisting of a wire with the cross section $A(s)$, where s parametrizes the path along the loop. In (4.2.1) we insert (4.1.8):

$$\int d^3x\,\mathbf{j} = I\int_C ds \text{ and obtain}$$

$$\mathbf{m} = \frac{k_L}{2c}\int d^3x\,\mathbf{x}\times\mathbf{j}(\mathbf{x}) = \frac{k_L}{c}\frac{I}{2}\oint \mathbf{x}(s)\times d\mathbf{s}. \tag{4.2.13}$$

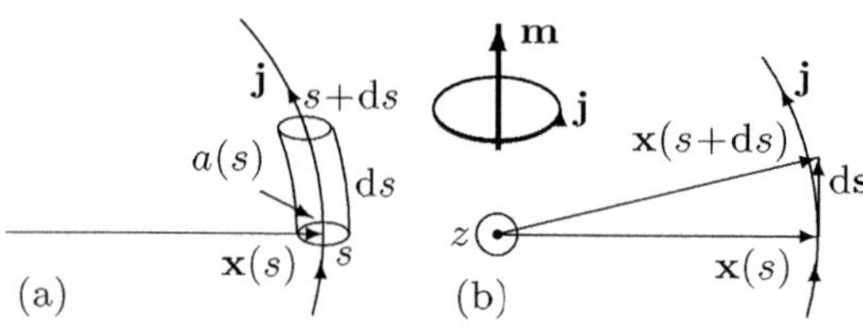

Fig. 4.5. (a) Wire with cross section $a(s)$, from which the current $I = j(s)\,a(s)$ results. (b) Sketch with planar current loop in the xy-plane, where $\mathbf{x}(s)$, the area A sweeps over

Plane current loop

As a special case of (4.2.13), we consider a flat current loop. The vectors $\mathbf{x}(s)$, $d\mathbf{s}$ and $\mathbf{x}(s+ds)$ form a triangle with the area

$$da = \frac{1}{2}|\mathbf{x}(s)\times d\mathbf{s}|,$$

as sketched in Fig. 4.5b. If you follow the entire closed curve along the path $C = \partial A$ according to (4.2.13), then you get

$$\oint_{\partial A} \mathbf{x}(s)\times d\mathbf{s} = 2A\mathbf{n} \qquad \Rightarrow \qquad \mathbf{m} = \frac{k_L}{c}IA\mathbf{n}. \tag{4.2.14}$$

Here I is the current, A the area of the current loop and $\mathbf{n}$ the normal vector to the surface.

4.2.4 Potential and Field of a Circular Current Loop

The fields of coils and thus also of circular loops, from which one can imagine a coil composed, occupy an important place in magnetostatics.

One either starts from **A** (4.1.4) the solution of the vector Poisson equation, and calculates $\mathbf{B} = \nabla\times\mathbf{A}$, or one takes the Biot-Savart equation (4.1.7) and thus obtains **B** directly; this is often faster.

For simple loops or coils, one can usually specify **B** in closed form, which includes elliptical integrals of the 1st to 3rd kind. Approximations are often more intuitive, with the development used in the following being good both inside and outside the loop/coil, except in the immediate vicinity of the wire.

Vector potential of a circular current loop

The circular current loop in the xy-plane, sketched in Fig. 4.6, is, as far as the magnetic field is concerned, the counterpart to the electric field of two opposite charges on the z-axis. Not only for this reason, but also because the circular loop is the basic element of a coil, there is a fundamental interest to $\mathbf{A}$ and $\mathbf{B}$, although we do not get the expressions in a form that allows a simple reduction to the potential $\mathbf{A}$ (4.2.10) or to the field $\mathbf{B}$ (4.2.12) of a magnetic dipole. Starting from the current density

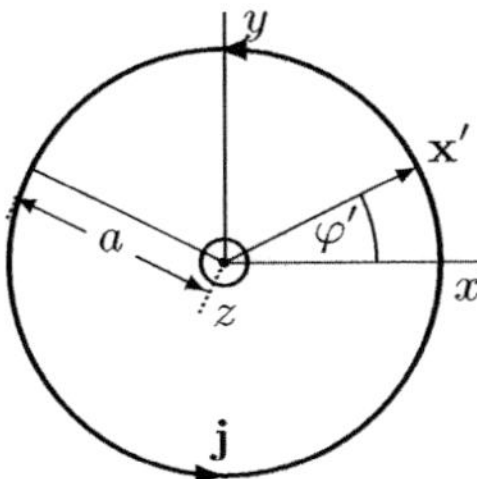

Fig. 4.6. Circular current loop of radius a; in the wire flows $\mathbf{j}(\mathbf{x}') = I\,\delta(a - \varrho')\,\delta(z')\,\mathbf{e}_{\varphi'}$; the magnetic moment of the loop is $\mathbf{m} = (k_L/c)\,I\,a^2\pi\,\mathbf{e}_z$

$$\mathbf{j}(\mathbf{x}) = I\delta(\varrho - a)\,\delta(z)\,\mathbf{e}_\varphi \tag{4.2.15}$$

the vector potential (4.1.4) in cylindrical coordinates is

$$\mathbf{A} = \frac{k_C}{ck_L}\int_0^{2\pi}\mathrm{d}\varphi'\,\frac{Ia\mathbf{e}_{\varphi'}}{\sqrt{z^2 + \varrho^2 + a^2 - 2a\varrho\cos(\varphi' - \varphi)}}. \tag{4.2.16}$$

In order to evaluate the integral, $\mathbf{e}_{\varphi'}$ must be replaced by a vector independent of φ' which is done in two steps by replacing φ' with $\varphi'' = \varphi' - \varphi$ and then replacing $\mathbf{e}_{\varphi'}$ with

$$\mathbf{e}_{\varphi''+\varphi} = -\sin(\varphi'' + \varphi)\,\mathbf{e}_x + \cos(\varphi'' + \varphi)\,\mathbf{e}_y = \cos\varphi''\,\mathbf{e}_\varphi - \sin\varphi''\,\mathbf{e}_\varrho\,.$$

In the integrand of (4.2.16), the term with $\sin\varphi''$ disappears for reasons of symmetry, so that

$$\mathbf{A} = \frac{k_C}{ck_L}Ia\mathbf{e}_\varphi K_0(\alpha,\beta)\quad\text{with}\quad K_0(\alpha,\beta) = \int_0^{2\pi}\mathrm{d}\varphi''\,\frac{\cos\varphi''}{\sqrt{\alpha - \beta\cos\varphi''}}. \tag{4.2.17}$$

Here, $\alpha = \varrho^2 + a^2 + z^2$ and $\beta = 2a\varrho$.

The exact solution

$K_0(\alpha,\beta)$ is an elliptic integral of the first kind (4.2.25) which is evaluated on page 133. So we can directly state the exact result:

$$\mathbf{A} = \mathbf{e}_\varphi\frac{k_C}{ck_L}\frac{4Ia}{k\sqrt{a\varrho}}\Big[(1 - \frac{k^2}{2})K(k) - E(k)\Big]\quad\text{with}\quad k^2 = \frac{4a\varrho}{(a + \varrho)^2 + z^2}. \tag{4.2.18}$$

K_0 diverges logarithmically when $k \lesssim 1$, which is equivalent to $z = 0$ and $\varrho = a$. Then one is in the immediate vicinity of the line wire with its singular current density.

Approximation method

Better insight than the exact solution provides the following "dipole approximation". It is not only applied to the circular loop, but also on coils and everywhere where elliptical integrals occur. It is obvious to develop the integrand of (4.2.17) according to

$$\kappa = \frac{\beta}{\alpha} = \frac{2\varrho a}{r^2 + a^2}. \tag{4.2.19}$$

κ is symmetric in ϱ and has its maximum value $\kappa = 1$, when $r = \varrho = a$; there, i.e. in the immediate vicinity of the wire, the approximation is worst. On the z-axis with $\varrho = 0$ the result becomes exact. It is now

$$K_0(\alpha, \beta) = \int_0^{2\pi} d\varphi \, \frac{\cos\varphi}{\sqrt{\alpha - \beta\cos\varphi}} = \frac{1}{\sqrt{\alpha}} \int_0^{2\pi} d\varphi \, \frac{\cos\varphi}{\sqrt{1 - \kappa\cos\varphi}}$$

$$\approx \frac{1}{\sqrt{\alpha}} \int_0^{2\pi} d\varphi \, \cos\varphi \left\{1 + \frac{1}{2}\kappa\cos\varphi + \ldots\right\} = \frac{2\pi}{\sqrt{\alpha}} \frac{\kappa}{4}. \tag{4.2.20}$$

If we now set $\mathbf{m} = \mathbf{e}_z(k_L/c)I\pi a^2$ and $\mathbf{e}_\varphi = \mathbf{e}_z \times \mathbf{e}_\varrho$, then

$$\mathbf{A} = \frac{k_C}{k_L^2} \frac{\mathbf{m} \times \boldsymbol{\varrho}}{\sqrt{r^2 + a^2}^3} \approx \frac{k_C}{k_L^2} \frac{\mathbf{m} \times \mathbf{x}}{r^3}. \tag{4.2.21}$$

Since $\mathbf{x} = \boldsymbol{\varrho} + z\mathbf{e}_z$, we could replace $\mathbf{m} \times \boldsymbol{\varrho}$ by $\mathbf{m} \times \mathbf{x}$ and for $r \gg a$ we obtained the vector potential (4.2.10) of a dipole. With our approximation for $\mathbf{A}$ we could establish a direct relationship to the dipole potential and also see, that (4.2.21) is a good approximation when not exactly $\varrho \approx a$.

The parameter κ is also used for the calculation of the field $\mathbf{B}$ of the circular loop and for $\mathbf{A}$ and $\mathbf{B}$ of coils, where it will be shown that in most cases the expansion can be truncated with the linear term in κ.

Elliptical integrals for loops and coils

We limit ourselves to circular loops and coils of radius a and coils of length $2l$. The current is linear for the loop (4.2.15) and for the coil it is surface-shaped (4.2.36). The occurring integrals each depend on two parameters, which are given as

$$\alpha = \varrho^2 + a^2 + (l \mp z)^2, \qquad\qquad \beta = 2a\varrho, \qquad\qquad 0 < \alpha \le \beta.$$

For the loop, $l = 0$, and if you have a coil, then $z = \pm l$ are the base surfaces. We start with $(\cos x = \cos(\pi + 2\varphi) = -\cos(2\varphi) = 2\sin^2\varphi - 1)$

$$K_1(\alpha, \beta) = \int_0^{2\pi} \frac{\mathrm{d}x}{\sqrt{\alpha - \beta \cos x}} \overset{2\varphi = x - \pi}{=} \int_{-\pi/2}^{\pi/2} \frac{2\,\mathrm{d}\varphi}{\sqrt{\alpha + \beta - 2\beta \sin^2 \varphi}}.$$

The above equation already has the form of an elliptic integral, if we pull $\alpha + \beta$ out of the root:

$$K_1(\alpha, \beta) = \frac{4}{\sqrt{\alpha + \beta}} \int_0^{\pi/2} \frac{\mathrm{d}\varphi}{\sqrt{1 - k^2 \sin^2 \varphi}} = \frac{4K(k)}{\sqrt{\alpha + \beta}}. \tag{4.2.22}$$

$K(k)$ is the complete elliptic integral of the first kind (B.5.4) with the modulus

$$k^2 = \frac{2\beta}{\alpha + \beta} = \frac{2a\varrho}{(a + \varrho)^2 + (z \mp l)^2} \leq 1. \tag{4.2.23}$$

Using the same transformations, we obtain

$$K_2(\alpha, \beta) = \int_0^{2\pi} \mathrm{d}x\,\sqrt{\alpha - \beta \cos x} = 4\sqrt{\alpha + \beta}\,E(k), \tag{4.2.24}$$

where $E(k)$ is the complete elliptic integral of the second kind (B.5.5). From K_1 and K_2 we get

$$
\begin{aligned}
K_0(\alpha, \beta) &= \int_0^{2\pi} \mathrm{d}x\,\frac{\cos x}{\sqrt{\alpha - \beta \cos x}} = \frac{1}{\beta} \int_0^{2\pi} \mathrm{d}x\,\frac{\alpha - \alpha + \beta \cos x}{\sqrt{\alpha - \beta \cos x}} = \frac{\alpha K_1 - K_2}{\beta} \\
&= \frac{4\sqrt{\alpha + \beta}}{\beta}\left[\frac{\alpha K(k)}{\alpha + \beta} - E(k)\right] = \frac{8}{k\sqrt{2\beta}}\left[\left(1 - \frac{k^2}{2}\right) K(k) - E(k)\right]
\end{aligned} \tag{4.2.25}
$$

together. The integrals that occur in the calculation of the fields are

$$
\begin{aligned}
K_3(\alpha, \beta) &= \int_0^{2\pi} \frac{\mathrm{d}x}{\sqrt{\alpha - \beta \cos x}^3} = \frac{4}{\sqrt{\alpha + \beta}^3} \int_0^{\pi/2} \frac{\mathrm{d}\varphi}{(1 - k^2 \sin^2 \varphi)\sqrt{1 - k^2 \sin^2 \varphi}} \\
&= \frac{4}{\sqrt{\alpha + \beta}^3}\,\Pi(-k^2, k) \overset{(B.5.9)}{=} \frac{4k^3}{\sqrt{2\beta}^3}\,\frac{E(k)}{1 - k^2},
\end{aligned} \tag{4.2.26}
$$

where $\Pi(q, k)$ is the complete elliptic integral of the 3rd kind (B.5.6), and

$$
\begin{aligned}
K_4(\alpha, \beta) &= \int_0^{2\pi} \frac{\cos x\,\mathrm{d}x}{\sqrt{\alpha - \beta \cos x}^3} = \frac{1}{\beta} \int_0^{2\pi} \mathrm{d}x\,\frac{\alpha - \alpha + \beta \cos x}{\sqrt{\alpha - \beta \cos x}^3} = \frac{\alpha K_3 - K_1}{\beta} \\
&= \frac{4k}{\beta\sqrt{2\beta}}\left[\frac{\alpha k^2}{2\beta}\,\frac{E(k)}{1 - k^2} - K(k)\right].
\end{aligned} \tag{4.2.27}
$$

Magnetic field of a circular current loop

We will not calculate $\mathbf{B}$ from $\boldsymbol{\nabla} \times \mathbf{A}$ (4.2.17), but will directly refer to the Biot-Savart law (4.1.7):

$$\mathbf{B} = \frac{k_C}{ck_L}\,Ia \int_0^{2\pi} \mathrm{d}\varphi'\,\frac{\mathbf{e}_{\varphi'} \times (\mathbf{x} - \mathbf{x}')}{|\mathbf{x} - \mathbf{x}'|^3}. \tag{4.2.28}$$

Auxiliary calculation: With the help of (A.3.7) we get ($\varphi'' = \varphi' - \varphi$):

$$\mathbf{e}_{\varrho'} = \cos\varphi''\,\mathbf{e}_\varrho + \sin\varphi''\,\mathbf{e}_\varphi\,, \qquad\qquad \mathbf{e}_{\varphi'} = -\sin\varphi''\,\mathbf{e}_\varrho + \cos\varphi''\,\mathbf{e}_\varphi\,.$$

For the numerator and denominator of the integrand of (4.2.28) this results in:

$$\mathbf{e}_{\varphi'} \times (\mathbf{x}-\mathbf{x}') = (\cos\varphi''\,\mathbf{e}_\varphi - \sin\varphi''\,\mathbf{e}_\varrho) \times \big[(\varrho - a\cos\varphi'')\mathbf{e}_\varrho - a\sin\varphi''\,\mathbf{e}_\varphi + (z-z')\mathbf{e}_z\big]$$

$$= (a - \varrho\cos\varphi'')\,\mathbf{e}_z + (z-z')\big[\cos\varphi''\,\mathbf{e}_\varrho + \sin\varphi''\,\mathbf{e}_\varphi\big], \qquad (4.2.29)$$

$$|\mathbf{x}-\mathbf{x}'|^2 = \varrho^2 - 2\varrho a\,\mathbf{e}_\varrho \cdot \mathbf{e}_{\varrho'} + a^2 + (z-z')^2 = \alpha - \beta\cos\varphi''.$$

Again, $\alpha = \varrho^2 + a^2 + z^2$ and $\beta = 2a\varrho$, since for the loop $z' = 0$.

B_φ vanishes for reasons of symmetry, so that

$$B_\varrho = \frac{k_C}{ck_L}\,IazK_4(\alpha,\beta), \qquad\qquad (4.2.30)$$

$$B_z = \frac{k_C}{ck_L}\,Ia\big[aK_3(\alpha,\beta) - \varrho K_4(\alpha,\beta)\big]. \qquad\qquad (4.2.31)$$

The exact field

The integrals K_3 and K_4 are expressed in (4.2.26) and (4.2.27) through elliptic integrals and can be inserted into the equations for B_ϱ and B_z, (4.2.30) and (4.2.31):

$$B_\varrho = \frac{k_C}{ck_L}\frac{Iz}{2\varrho}\frac{k}{\sqrt{a\varrho}}\Big[\frac{2-k^2}{1-k^2}E(k) - K(k)\Big], \qquad\qquad k^2 = \frac{4a\varrho}{(a+\varrho)^2 + z^2},$$

$$B_z = \frac{k_C}{ck_L}\frac{I}{c}\frac{k}{\sqrt{a\varrho}}\Big\{\frac{ak^2}{\varrho}\frac{E(k)}{1-k^2} - \Big[\frac{2-k^2}{1-k^2}E(k) - K(k)\Big]\Big\}.$$

In the $z=0$ plane, B_ϱ vanishes and B_z diverges near the line wire like $1/(a-\varrho)$, which would not be the case with a finite cross-section of the wire.

Approximate calculation of the field

If the exact calculation overestimates the field strength in the immediate vicinity, it is obvious to use the "dipole approximation" already applied for the calculation of $\mathbf{A}$ also for the field $\mathbf{B}$, especially since no singularity occurs here. This time we will consider the next correction to the dipole field and for this we have to take into account the next terms in the expansion of the integrands of K_3 and K_4:

$$\frac{1}{\sqrt{1-\epsilon}^3} \approx 1 + \frac{3}{2}\epsilon + \frac{3\cdot 5}{2\cdot 4}\epsilon^2 + \frac{3\cdot 5\cdot 7}{2\cdot 4\cdot 6}\epsilon^3 + \dots \qquad\qquad \text{binomial series}$$

$$\epsilon = \kappa\cos\varphi, \qquad\qquad\qquad\qquad \kappa = \frac{\beta}{\alpha} = \frac{2a\varrho}{r^2 + a^2}.$$

Substituted into K_3 and K_4 one obtains

$$K_3(\alpha, \beta) = \int_0^{2\pi} d\varphi \, \frac{1}{\sqrt{\alpha - \beta \cos \varphi}^{\,3}} \approx \frac{1}{\sqrt{\alpha}^{\,3}} \int_0^{2\pi} d\varphi \left(1 + \frac{15}{8} \kappa^2 \cos^2 \varphi\right)$$

$$= \frac{2\pi}{\sqrt{\alpha}^{\,3}} \left(1 + \frac{15}{16} \kappa^2\right), \tag{4.2.32}$$

$$K_4(\alpha, \beta) = \int_0^{2\pi} d\varphi \, \frac{\cos \varphi}{\sqrt{\alpha - \beta \cos \varphi}^{\,3}} \approx \frac{1}{\sqrt{\alpha}^{\,3}} \int_0^{2\pi} d\varphi \left(\frac{3}{2} \kappa \cos^2 \varphi + \frac{35}{16} \kappa^3 \cos^4 \varphi\right)$$

$$= \frac{2\pi}{\sqrt{\alpha}^{\,3}} \frac{3\kappa}{4} \left(1 + \frac{35}{12} \kappa^2 \frac{3}{8}\right). \tag{4.2.33}$$

In this approximation, the field lines do not qualitatively differ from the exactly calculated ones. The dipole field in Fig. 4.4 is based on (4.2.32). Now we only take the leading terms and set $\mathbf{m} = \mathbf{e}_z I \pi a^2 k_L / c$:

$$B_\varrho = \frac{k_C}{c k_L} I a \, \frac{z}{\sqrt{r^2 + a^2}^{\,3}} \frac{3\pi a \varrho}{r^2 + a^2} = \frac{k_C}{k_L^2} m \, \frac{3 z \varrho}{\sqrt{r^2 + a^2}^{\,5}},$$

$$B_z = \frac{k_C}{c k_L} I a \, \frac{2\pi}{\sqrt{r^2 + a^2}^{\,3}} - \frac{\varrho}{z} B_\varrho = \frac{k_C}{k_L^2} m \, \frac{2}{\sqrt{r^2 + a^2}^{\,3}} \left[1 - \frac{3\varrho^2}{2(r^2 + a^2)}\right].$$

In the approximation $r \gg a$ we obtain

$$r^5 \mathbf{B} \propto 3m(r^2 - \varrho^2)\mathbf{e}_z - m r^2 \mathbf{e}_z + 3 z \varrho \mathbf{e}_\varrho = 3(\mathbf{m} \cdot \mathbf{x}) z \mathbf{e}_z - r^2 \mathbf{m} + 3(\mathbf{m} \cdot \mathbf{x}) \varrho \mathbf{e}_\varrho.$$

This results, as expected, in the dipole field for $r > 0$:

$$\mathbf{B} = \frac{k_C}{k_L^2} \left[3 \frac{(\mathbf{m} \cdot \mathbf{x}) \, \mathbf{x}}{r^5} - \frac{\mathbf{m}}{r^3}\right].$$

Scalar magnetic potential of a current loop

Let there be a closed, local circuit, where the steady current should only flow within a small area, such as a wire. In the space outside the circuit, $\mathbf{j} = 0$ and thus also $\boldsymbol{\nabla} \times \mathbf{B} = 0$ (Ampère's circuital law in differential form (4.1.1a–b)). Thus, the fields there can be represented by a scalar potential ϕ_M [see Becker, Sauter, 1973, p. 111] representable, which admittedly is somewhat exotic. According to Biot-Savart's law (4.1.9) is

$$\mathbf{B}(\mathbf{x}) = \frac{k_C}{c k_L} I \oint_{\partial A} d\mathbf{x}' \times \boldsymbol{\nabla}' \frac{1}{|\mathbf{x} - \mathbf{x}'|}. \tag{4.2.34}$$

If we multiply (4.2.34) scalarly with any constant vector $\mathbf{a}$, cyclically exchange in the integrand and then apply Stokes's theorem, then

$$\mathbf{a}\cdot\mathbf{B} \propto \oint_{\partial A} d\mathbf{x}'\cdot\mathbf{\nabla}'\times\mathbf{a}\,\frac{1}{|\mathbf{x}-\mathbf{x}'|} = \iint_A d\mathbf{a}'\cdot\mathbf{\nabla}'\times\left(\mathbf{\nabla}'\times\frac{\mathbf{a}}{|\mathbf{x}-\mathbf{x}'|}\right)$$

$$\overset{(A.2.38)}{=} \iint_A d\mathbf{a}'\cdot\left[\mathbf{\nabla}'\left(\mathbf{\nabla}'\cdot\frac{\mathbf{a}}{|\mathbf{x}-\mathbf{x}'|}\right) - \Delta'\frac{\mathbf{a}}{|\mathbf{x}-\mathbf{x}'|}\right].$$

The 2nd term vanishes everywhere outside the wire, so that only the 1st term contributes, unless one calculates $\mathbf{B}$ along a path C that encloses the wire, so that it must cross A and so the 2nd term contributes. In the 1st term, one replaces a $\mathbf{\nabla}'$ with $-\mathbf{\nabla}$ and thus obtains

$$\mathbf{B} = -\mathbf{\nabla}\phi_M, \tag{4.2.35}$$

$$\phi_M = \frac{k_C}{ck_L}I\iint_A d\mathbf{a}'\cdot\frac{\mathbf{x}-\mathbf{x}'}{|\mathbf{x}-\mathbf{x}'|^3} \overset{r\gg r'}{=} \frac{k_C}{k_L^2}\mathbf{m}\cdot\frac{\mathbf{x}}{r^3}, \qquad \mathbf{m} = \frac{k_L}{c}I\iint_A d\mathbf{a}'.$$

$\mathbf{m}$ is the magnetic moment of the loop and ϕ_M is the scalar potential of an electric dipole. If one determines the circulation (Ampère's law) $\oint d\mathbf{s}\cdot\mathbf{B}$, it is $4\pi k_C I/ck_L \overset{\text{SI}}{=} \mu_0 I$, if the wire is enclosed by the path, otherwise it is zero. The circulation of a scalar potential always vanishes.

4.2.5 Potentials and Fields of Coils

The magnetic fields $\mathbf{B}$ of the following configurations are consistently calculated on the basis of Biot-Savart's law (4.1.7). For the field of the infinite straight coil, this is an unnecessary effort, since this can be determined most easily by means of the Ampère's law (4.1.5), as can be seen from Fig. 7.8, p. 232. Even for the finite coil, it is useful to refer back to the Biot-Savart law.

The straight coil

In applications, the exact knowledge of the field $\mathbf{B}$ in coils of finite length and finite cross-section is certainly in the foreground. Here we will only provide the basics for approximations to be made and for possible more precise (numerical) calculations. It is assumed that the density n of the windings is sufficiently

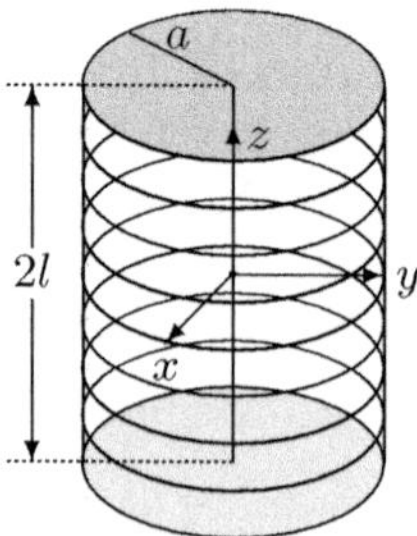

Fig. 4.7. Coil (solenoid) with the radius a and the length $2l$; the windings are so dense that the pitch of a winding can be neglected and one can thus assume a circular current

high so that the current I flowing in the wire can be represented by a uniform circular current $I_n = nI$ per unit of length on a cylindrical surface:

$$\mathbf{j} = I_n\delta(\varrho - a)\,\mathbf{e}_\varphi\,\theta(l - |z|)\,. \tag{4.2.36}$$

We substitute this current into the Biot-Savart law (4.1.7), whereby we refer back to the vector product on (4.2.29):

$$
\begin{aligned}
\mathbf{B}(\mathbf{x}) &= \frac{k_C}{ck_L}I_n a\int_0^{2\pi}\mathrm{d}\varphi'\int_{-l}^{l}\mathrm{d}z'\,\frac{\mathbf{e}_{\varphi'}\times(\mathbf{x}-\mathbf{x}')}{|\mathbf{x}-\mathbf{x}'|^3}\\
&\overset{u=z'-z}{=}\frac{k_C}{ck_L}I_n a\int_0^{2\pi}\mathrm{d}\varphi''\int_{-l-z}^{l-z}\mathrm{d}u\,\frac{(a-\varrho\cos\varphi'')\mathbf{e}_z-u(\cos\varphi''\mathbf{e}_\varrho+\sin\varphi''\mathbf{e}_\varphi)}{\sqrt{\varrho^2-2\varrho a\cos\varphi''+a^2+u^2}^{\,3}}\,.
\end{aligned}
$$

The integration can be performed using (B.5.16). For reasons of symmetry, B_φ disappears; the integration with respect to u yields

$$B_\varrho(\mathbf{x})=\frac{k_C}{ck_L}I_n a\int_0^{2\pi}\mathrm{d}\varphi'\,\frac{\cos\varphi'}{\sqrt{(\varrho^2-2a\varrho\cos\varphi'+a^2)+u^2}}\Bigg|_{-l-z}^{l-z}, \tag{4.2.37}$$

$$B_z(\mathbf{x})=\frac{k_C}{ck_L}I_n a\int_0^{2\pi}\mathrm{d}\varphi'\,\frac{a-\varrho\cos\varphi'}{\varrho^2-2a\varrho\cos\varphi'+a^2}\,\frac{u}{\sqrt{(\varrho^2-2a\varrho\cos\varphi'+a^2)+u^2}}\Bigg|_{-l-z}^{l-z}.$$

Exact field

$\mathbf{B}$ can be represented without approximation using complete elliptical integrals (B.5.4) – (B.5.6). So B_ϱ is given by K_0 (4.2.25) when we substitute α and β (4.2.41). This implies

$$B_\varrho = \frac{k_C}{ck_L}I_n a\frac{4}{\sqrt{a\varrho}}\sum_{j=1,2}(-1)^{j+1}\frac{1}{k_j}\left[\left(1-\frac{k_j^2}{2}\right)K(k_j)-E(k_j)\right] \tag{4.2.38}$$

with $k_{1,2}^2 = \dfrac{4a\varrho}{(a+\varrho)^2+z_{1,2}^2}$ and $z_{1,2}=l\mp z$.

The expression (4.2.37) for B_z is somewhat more complicated; the 1st term yields K_1 (4.2.22), and the 2nd term a complete elliptic integral of the 3rd kind:

$$
\begin{aligned}
B_z &= \frac{k_C}{ck_L}\frac{I_n}{2}\sum_{j=1,2}z_j\int_0^{2\pi}\mathrm{d}\varphi'\left[1-\frac{\varrho^2-a^2}{\varrho^2+a^2-2a\varrho\cos\varphi'}\right]\frac{1}{\sqrt{\varrho^2+a^2+z_j^2-2a\varrho\cos\varphi'}}\\
&= \frac{k_C}{ck_L}I_n\sum_{j=1,2}\frac{z_jk_j}{\sqrt{a\varrho}}\int_0^{\pi/2}\mathrm{d}\varphi''\left[1-\frac{\varrho-a}{\varrho+a}\frac{1}{1+q\sin^2\varphi''}\right]\frac{1}{\sqrt{1-k_j^2\sin^2\varphi''}}.\tag{4.2.39}
\end{aligned}
$$

Here, $\Pi(q,k_j)$ with $q = \dfrac{-4a\varrho}{(a+\varrho)^2}$ is the elliptic integral of the 3rd kind (B.5.6):

$$B_z = \frac{k_C}{ck_L}I_n\sum_{j=1,2}\frac{k_j}{\sqrt{a\varrho}}z_j\left[K(k_j)-\frac{\varrho-a}{\varrho+a}\Pi(q,k_j)\right],\quad q=\frac{-4a\varrho}{(a+\varrho)^2}. \tag{4.2.40}$$

The field lines in Fig. 4.8 were calculated with the "exact" fields (4.2.38) and

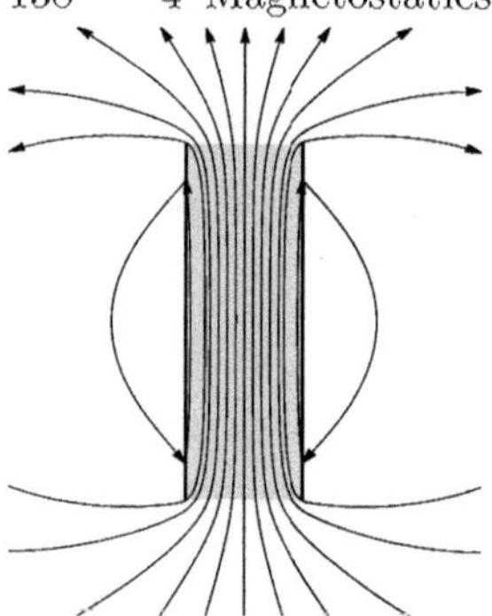

Fig. 4.8. Magnetic field **B** of a coil with radius a and length $2l = 6a$. The field lines are pushed from the edge towards the center. Especially near the ends of the coil one has a stray field

(4.2.40). B_z can be given analytically on the z-axis (7.1.33) and is at the ends of the coil about half as strong as in the middle.

Approximate calculation of the field

First, the same approximation is made for (4.2.37) as for the circular loop, except that now two parameters $\kappa_{1,2}$ are advantageous, which weight the contributions to **B** from the base and top surface of the coil differently on the z-axis:

$$\kappa_{1,2} = \frac{\beta}{\alpha_{1,2}} = \frac{2a\varrho}{\varrho^2 + a^2 + z_{1,2}^2} \qquad \text{with} \quad z_{1,2} = l \mp z. \qquad (4.2.41)$$

However, one must distinguish for which area the approximation is made, whether in the center of the coil, near the base or top surface of the cylindrical coil or asymptotically in the outer space. Corresponding calculations are set as problem 4.3. The above approximation results in the lowest order

$$B_\varrho(\mathbf{x}) = \frac{k_C}{ck_L} I_n a \, \frac{\pi}{2\sqrt{2a\varrho}} \left(\sqrt{\kappa_1}^3 - \sqrt{\kappa_2}^3 \right), \qquad (4.2.42)$$

$$B_z(\mathbf{x}) = \frac{k_C}{ck_L} I_n a \, \frac{1}{\sqrt{2a\varrho}} \sum_{j=1,2} z_j \sqrt{\kappa_j} \left[\frac{2\pi}{a} \, \theta(a-\varrho) + \frac{\kappa_j}{2} \begin{cases} -\pi/\varrho & a < \varrho \\ \pi\varrho/a^2 & a > \varrho \end{cases} \right].$$

The field of the coil is equal to that of a homogeneously magnetized rod magnet, as shown in magnetostatics in matter (see (7.1.4)). On the z-axis, the field can be given exactly ($\mathbf{z} = (0,0,z)$):

$$\mathbf{B(z)} = \frac{k_C}{ck_L} 2\pi I_n \left[\frac{l-z}{\sqrt{a^2+(l-z)^2}} + \frac{l+z}{\sqrt{a^2+(l+z)^2}} \right] \mathbf{e}_z \overset{|z|\to\infty}{=} \frac{k_C}{k_L^2} \frac{2\mathbf{m}}{|z|^3}. \qquad (4.2.43)$$

It is plotted in Fig. 7.6. In the asymptotic region $|z| \to \infty$ we have the field of a dipole, as shown in the problem 7.5.

The infinitely long coil

We take the integral obtained from the Biot-Savart law (4.2.37) for **B** of a coil of radius a and length $2l$, to determine the field of the infinitely long coil in the limit $l \to \infty$:

$$\mathbf{B}(\mathbf{x}) = \frac{2k_C}{ck_L} I_n a \mathbf{e}_z \int_0^{2\pi} d\varphi' \frac{a - \varrho \cos\varphi'}{\varrho^2 - 2a\varrho \cos\varphi' + a^2} \overset{(\text{B.5.21})}{=} \frac{4\pi k_C}{ck_L} I_n \theta(a - \varrho) \mathbf{e}_z.$$

$$(4.2.44)$$

Here we have returned to the designation of the circular current I_n per unit length of the coil to the current flowing in the wire $I_n = In$, where n is the number of windings per unit length. We will return to this result in section 7.1.4 of magnetostatics in matter.

4.2.6 The Semi-Infinite Coil and the Monopole

Of particular interest is the semi-infinite, infinitesimally thin coil, sketched in Fig. 4.9. The "upper" end of the coil represents a point source from which magnetic field lines $\mathbf{B}$ emanate, quite analogous to those of the electric field $\mathbf{E}$ of a point charge. Dirac [1931, 1948] took this configuration as the basis for the model of a magnetic monopole.

Note: Assuming the existence of monopoles, the Maxwell equations must be adapted, whereby a magnetic current $\mathbf{j}_m$ is to be introduced:

$$\boldsymbol{\nabla}\cdot\mathbf{E} = 4\pi k_C \rho_e, \qquad\qquad \boldsymbol{\nabla}\times\mathbf{E} = -\frac{k_L}{c}\left(\dot{\mathbf{B}} + 4\pi k_M \mathbf{j}_m\right), \qquad (4.2.45)$$

$$\boldsymbol{\nabla}\times\mathbf{B} = \frac{1}{ck_L}\left(\dot{\mathbf{E}} + 4\pi k_C \mathbf{j}_e\right), \qquad \boldsymbol{\nabla}\cdot\mathbf{B} = 4\pi k_M \rho_m, \qquad k_M = \frac{k_C}{k_L^2}.$$

This representation is not entirely consistent. Jackson [1998, (6.150)] defines current and charge as $\mathbf{j}'_m = \mu_0 \mathbf{j}_m$ and $\rho'_m = \mu_0 \rho_m$. As a result, the magnetic pole strength p_m [Sommerfeld, 1952, (7.10)] at Jackson [1998, (6.157), $g \equiv p_B$] also has to be multiplied by μ_0:

$$\rho_m = p_m \delta^{(3)}(\mathbf{x} - \mathbf{x}_m), \qquad p_B = \mu_0 p_m, \qquad \mathbf{B} = k_M p_m \frac{\mathbf{x} - \mathbf{x}_m}{|\mathbf{x} - \mathbf{x}_m|^3}. \qquad (4.2.46')$$

Continuity equations apply to both currents: $\dot{\rho}_{e,m} + \boldsymbol{\nabla}\cdot\mathbf{j}_{e,m} = 0$.

We will not go into the physical implications here, such as the quantization of charge, the observability of the dipole chain or differences in path in different dipole chains will not be discussed. We will only calculate $\mathbf{A}$ and $\mathbf{B}$ of this configuration, including their vortices and sources and we will ensure consistency with classical magnetostatics.

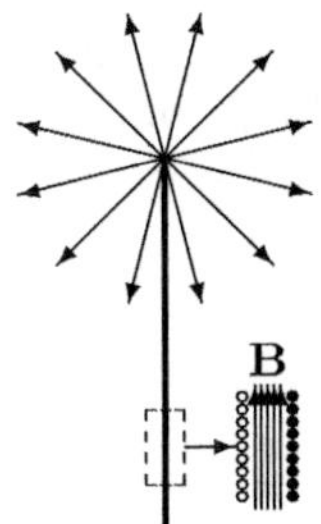

Fig. 4.9. A thin coil (solenoid), which goes from $z = -\infty$ to $z = 0$, has a (scattering) field at the origin, which is equivalent to that of a point charge.
The coil can be replaced by a chain of magnetic dipoles

Magnetic flux density of a semi-infinite coil

The current density, as already with the finite coil, has the form

$$\mathbf{j}(\mathbf{x}) = I_n \delta(\varrho - a)\, \theta(-z)\, \mathbf{e}_\varphi .$$

$I_n = nI$ is the current on the cylinder surface unit length, if n is the number of windings per unit length and I is the current in the wire. As with the finite coil, we calculate $\mathbf{B}$ directly using the Biot-Savart law (4.1.7) (and not over $\mathbf{A}$):

$$\mathbf{B}(\mathbf{x}) = \frac{k_C}{ck_L} I_n a \int_0^{2\pi} d\varphi' \int_{-\infty}^0 dz' \frac{\mathbf{e}_{\varphi'} \times (\mathbf{x} - \mathbf{x}')}{|\mathbf{x} - \mathbf{x}'|^3} .$$

We have already calculated the above vector product (4.2.29) for the wire loop ($\varphi'' = \varphi' - \varphi$):

$$\mathbf{e}_{\varphi'} \times (\mathbf{x} - \mathbf{x}') = (a - \varrho \cos\varphi'')\, \mathbf{e}_z + (z - z')\big[\cos\varphi''\, \mathbf{e}_\varrho + \sin\varphi''\, \mathbf{e}_\varphi\big],$$

$$|\mathbf{x} - \mathbf{x}'|^2 = \varrho^2 - 2\varrho a \cos\varphi'' + a^2 + (z - z')^2 .$$

Substituted into the Biot-Savart law, one obtains the integrals already known from the finite coil (4.2.37), where only the limits $-l < z' < l$ have to be changed to $-\infty < z' < 0$. For reasons of symmetry, the integral of $\mathbf{e}_\varphi$ vanishes:

$$\mathbf{B}(\mathbf{x}) \overset{u=z'-z}{=} \frac{k_C}{ck_L} I_n a \int_0^{2\pi} d\varphi'' \int_{-\infty}^{-z} du \frac{-u \cos\varphi''\, \mathbf{e}_\varrho + (a - \varrho \cos\varphi'')\mathbf{e}_z}{\sqrt{\varrho^2 - 2\varrho a \cos\varphi'' + a^2 + u^2}^{\,3}} .$$

The first integral is trivial and the second is listed in the appendix (B.5.16) p. 624:

$$\int dx \frac{x}{\sqrt{a^2 + x^2}^{\,3}} = \frac{-1}{\sqrt{a^2 + x^2}} , \qquad \int dx \frac{1}{\sqrt{a^2 + x^2}^{\,3}} = \frac{1}{a^2} \frac{x}{\sqrt{a^2 + x^2}} .$$

$$B_\varrho(\mathbf{x}) = \frac{k_C}{ck_L} I_n a \int_0^{2\pi} d\varphi'' \frac{\cos\varphi''}{\sqrt{\varrho^2 - 2a\varrho \cos\varphi'' + a^2 + z^2}} ,$$

$$B_z(\mathbf{x}) = \frac{k_C}{ck_L} I_n a \int_0^{2\pi} d\varphi'' \frac{(a - \varrho \cos\varphi'')}{\varrho^2 - 2a\varrho \cos\varphi'' + a^2} \left[1 - \frac{z}{\sqrt{\varrho^2 - 2a\varrho \cos\varphi'' + a^2 + z^2}}\right].$$

Ultimately, we are interested in the limit of finite $B\,\mathbf{e}_z$ at $a \to 0$ and pursue this with approximations that we have applied to the current loop, but also to the finite coil, i.e. we expand the root according to $\kappa = 2a\varrho/(r^2 + a^2)$

$$B_\varrho \approx \frac{k_C}{ck_L} I_n a \frac{1}{\sqrt{r^2 + a^2}} \int_0^{2\pi} d\varphi' \cos\varphi' \left(1 + \frac{\kappa}{2} \cos\varphi'\right)$$

$$= \frac{k_C}{ck_L} I_n a \frac{2\pi}{\sqrt{r^2 + a^2}} \frac{1}{4} \frac{2a\varrho}{r^2 + a^2} .$$

The 1st term of B_z results in a θ-contribution (see (B.5.21), p. 624) and we develop the root again according to κ

$$B_z = \frac{k_C}{ck_L} I_n a \left\{ \frac{2\pi}{a}\theta(a-\varrho) - \int_0^{2\pi} d\varphi' \frac{a - \varrho\cos\varphi'}{\varrho^2 + a^2 - 2\varrho a\cos\varphi'} \frac{z}{\sqrt{r^2+a^2}}\left(1 + \frac{\kappa}{2}\cos\varphi'\right)\right\}.$$

The remaining integral can be solved exactly (see (B.5.22)-problem 4.5)

$$B_z = \frac{k_C}{ck_L} 2\pi I_n \left\{ \theta(a-\varrho)\left(1 - \frac{z}{\sqrt{r^2+a^2}}\right) + \frac{a^2}{2}\frac{z}{\sqrt{r^2+a^2}^3}\left[1 - \theta(a-\varrho)\frac{a^2+\varrho^2}{a^2}\right]\right\}.$$

Now the flux Φ_B through the coil should be kept constant while reducing the radius $a \to 0$ ($z/\sqrt{r^2+a^2} \to -1$):

$$\Phi_B = \iint_{S_a} d\mathbf{a}\cdot\mathbf{B} = B\,a^2\pi = \frac{4\pi k_C}{ck_L} I_n a^2\pi = 4\pi k_M \frac{k_L}{c} a^2\pi, \qquad \text{SI: } k_M = \frac{\mu_0}{4\pi}.$$

The strength p_B of the point source is therefore [Jackson, 1998, $p_B\widehat{=}g$, (6.157)]

$$p_B = \mu_0 p_m, \qquad\qquad\qquad p_m = \frac{k_L}{c} I_n a^2\pi. \qquad\qquad (4.2.46)$$

Now we set $\lim\limits_{a\to 0} \dfrac{\theta(a-\varrho)}{a^2\pi}\widehat{=}\delta(x)\delta(y)$ and, neglecting the last term of B_z, we obtain

$$\mathbf{B} = k_M p_m \left[4\pi\,\delta(x)\,\delta(y)\,\theta(-z)\,\mathbf{e}_z + \frac{\mathbf{x}}{r^3}\right] = \mathbf{B}^s + \mathbf{B}^p. \qquad (4.2.47)$$

The 1st term $\mathbf{B}^s$ is the field of the coil and the 2nd term $\mathbf{B}^p$ that of a point source. As mentioned in the legend of Fig. 4.9, the coil can be replaced by a chain of magnetic dipoles. The Maxwell equations ($k_M = k_C/k_L^2$)

$$\boldsymbol{\nabla}\cdot\mathbf{B} = 0, \quad \boldsymbol{\nabla}\cdot\mathbf{B}^p = -\boldsymbol{\nabla}\cdot\mathbf{B}^s = 4\pi k_M p_m \delta^{(3)}(\mathbf{x}), \quad \text{SI: } \boldsymbol{\nabla}\cdot\mathbf{B}^p = \mu_0 p_m \delta^{(3)}(\mathbf{x}),$$

$$\boldsymbol{\nabla}\times\mathbf{B} = \boldsymbol{\nabla}\times\mathbf{B}^s = 4\pi k_M \frac{k_L}{c}\mathbf{j}, \qquad\qquad\qquad \text{SI: } \boldsymbol{\nabla}\times\mathbf{B} = \mu_0\mathbf{j}(\mathbf{x})$$

are fulfilled, which is to be shown in problem 4.6. $\mathbf{A}$ is determined solely by the coil with $\mathbf{B}^s$, since due to $\boldsymbol{\nabla}\times\mathbf{B}^p = 0$ the field of the point source does not contribute:

$$\mathbf{A} = \frac{1}{4\pi}\int d^3x' \frac{\boldsymbol{\nabla}'\times\mathbf{B}}{|\mathbf{x}-\mathbf{x}'|} = \frac{1}{4\pi}\int d^3x' \frac{\mathbf{B}\times(\mathbf{x}-\mathbf{x}')}{|\mathbf{x}-\mathbf{x}'|^3} = k_M p_m \int_C \frac{d\mathbf{x}'\times(\mathbf{x}-\mathbf{x}')}{|\mathbf{x}-\mathbf{x}'|^3},$$

where C is the path $-\infty < z' \le 0$. One obtains (problem 4.4)

$$\mathbf{A} = k_M p_m \frac{1-\cos\vartheta}{r\sin\vartheta}\mathbf{e}_\varphi \quad \overset{\text{arbitrary direction}}{\Longrightarrow} \quad \mathbf{A} = k_M p_m \frac{\mathbf{n}\times\mathbf{x}}{r(r+\mathbf{n}\cdot\mathbf{x})}. \qquad (4.2.48)$$

The expression on the right is for a straight path C (*Dirac-string*) of any direction, which is given by the unit vector $\mathbf{n}$.

4.3 Angular Momentum, Force and Torque

4.3.1 Angular Momentum and Magnetic Moment

The angular momentum of n particles, which are located at the positions $\mathbf{x}_n$ with the momentum $\boldsymbol{P}_n$, is given as

$$\mathbf{L} = \sum_n \mathbf{x}_n \times \boldsymbol{P}_n . \tag{4.3.1}$$

If one sets in (4.2.13) the current density $\mathbf{j}(\mathbf{x}) = \sum_n e\,\mathbf{v}_n\,\delta^{(3)}(\mathbf{x}-\mathbf{x}_n)$, then

$$\mathbf{m} = \frac{k_L}{2c}\int \mathrm{d}^3x\,\mathbf{x}\times\mathbf{j}(\mathbf{x}) = \frac{k_L}{2c}\sum_n e\,\mathbf{x}_n\times\mathbf{v}_n .$$

Without external field, $\boldsymbol{P}_n = m\,\mathbf{v}_n$. Thus, $\mathbf{m}$ can be expressed through the orbital angular momentum

$$\mathbf{m} = \frac{k_L}{c}\frac{e}{2m}\,\mathbf{L} . \tag{4.3.2}$$

The ratio of the magnetic moment to the angular momentum, here $k_L e/2mc$, is referred to as *gyromagnetic ratio*. In addition, there can be the *gyromagnetic factor g*. For the spin of the electrons $\mathbf{m}_s = (g k_L e/2mc)\mathbf{S}$ is $g = 2$ (almost). If an electromagnetic field is present, the canonical momentum (see (5.4.10)) is related to the velocity:

$$\boldsymbol{P}_n = m\,\mathbf{v}_n + \frac{k_L}{c}e\,\mathbf{A}(\mathbf{x}_n),$$

$$\mathbf{m} = \frac{k_L}{c}\frac{e}{2m}\,\mathbf{L} - \frac{k_L^2}{c^2}\frac{e^2}{2m}\sum_n \mathbf{x}_n\times\mathbf{A}(\mathbf{x}_n)$$

$$= \frac{k_L}{c}\frac{e}{2m}\,\mathbf{L} - \frac{k_L^2}{c^2}\frac{e^2}{2m}(x^2+y^2)\mathbf{B} \quad \text{for } \mathbf{B} = B\,\mathbf{e}_z \quad \text{and} \quad \mathbf{A} = \frac{1}{2}\mathbf{B}\times\mathbf{x}.$$

The natural unit for angular momentum is Planck's constant $\hbar$. We can thus express $\mathbf{m}$ in the form

$$\mathbf{m} = \frac{k_L}{c}\frac{e\hbar}{2m}\frac{\mathbf{L}}{\hbar} = -\mu_B\hat{\mathbf{L}}, \qquad \mu_B = \frac{k_L}{c}\frac{e_0\hbar}{2m_e}, \qquad \hat{\mathbf{L}} = \frac{\mathbf{L}}{\hbar}$$

specify. It is assumed here that the particle is an electron with the charge $e=-e_0$ and the mass $m=m_e$. μ_B is the Bohr magneton (see Tab. C.7, p. 645).

4.3.2 Force and Torque on a Current Loop

We now consider a current loop in a magnetic field $\mathbf{B}(\mathbf{x})$ and want to examine the effect of this field on the current loop.

Force on a current loop

Starting from the Lorentz force on a single one of the particles forming the current, we find the total force

$$\mathbf{F} = \frac{k_{\mathrm{L}}}{c}\sum_n e\,\mathbf{v}_n\times\mathbf{B}(\mathbf{x}_n) = \int \mathrm{d}^3x \sum_n \frac{e}{c}\mathbf{v}_n\times\mathbf{B}(\mathbf{x}_n)\delta^{(3)}(\mathbf{x}-\mathbf{x}_n)$$

$$= \frac{k_{\mathrm{L}}}{c}\int \mathrm{d}^3x\,\mathbf{j}(\mathbf{x})\times\mathbf{B}(\mathbf{x})\,. \tag{4.3.3}$$

Note: We assume that $\mathbf{B}(\mathbf{x})$ varies only slightly compared to the distance of the particles carrying the current. Then one can write the averaged current instead of the above microscopic current – and only for this do the used stationarity properties apply.

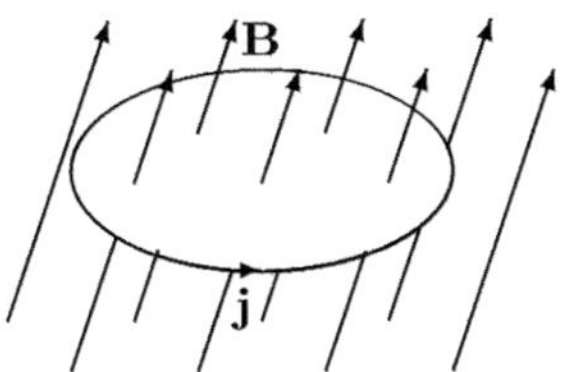

Fig. 4.10. Sketch of a current loop in the field $\mathbf{B}(\mathbf{x})$. The field should not vary too much in the area around the loop

For a closed current loop, as outlined in Fig. 4.10, the force can be expressed through its magnetic moment. We assume that the current loop is located near the origin of coordinates and develop

$$\mathbf{B}(\mathbf{x}) = \mathbf{B}(0) + (\mathbf{x}\cdot\boldsymbol{\nabla})\mathbf{B}|_{\mathbf{x}=0} + \cdots$$
$$B_k(\mathbf{x}) = B_k(0) + x_l\,B_{k,l}(0) + \cdots$$

The first term results in $\int \mathrm{d}^3x\,\mathbf{j} = 0$. Then we swap the terms until, from $(\mathbf{j}\times(\mathbf{x}\cdot\boldsymbol{\nabla})\mathbf{B})$ we get something with $\mathbf{x}\times\mathbf{j}$, i.e. with $\mathbf{m}$.

$$F_i = \frac{k_{\mathrm{L}}}{c}\int \mathrm{d}^3x\,\epsilon_{ijk}\,j_j(\mathbf{x})\,x_l\,B_{k,l}(0) = \frac{k_{\mathrm{L}}}{2c}\int \mathrm{d}^3x\,\epsilon_{ijk}\left(j_j\,x_l - j_l\,x_j\right)B_{k,l}(0),$$

where we have inserted the auxiliary formula (4.2.9). Now we replace still

$$x_l\,j_j - j_l\,x_j = (\delta_{ja}\,\delta_{lb} - \delta_{jb}\,\delta_{la})\,j_a x_b = \epsilon_{\mu jl}\,\epsilon_{\mu ab}\,j_a x_b = \epsilon_{\mu jl}\left[\mathbf{j}\times\mathbf{x}\right]_\mu\,.$$

The far right term already points to $\mathbf{m}$. What needs to be evaluated now is

$$\epsilon_{ijk}\left(j_j x_l - j_l x_j\right)B_{k,l} = \epsilon_{ijk}\,\epsilon_{\mu jl}\left[\mathbf{j}\times\mathbf{x}\right]_\mu B_{k,l} = (\delta_{i\mu}\delta_{kl} - \delta_{il}\delta_{k\mu})\left[\mathbf{j}\times\mathbf{x}\right]_\mu B_{k,l}$$
$$= \left[\mathbf{j}\times\mathbf{x}\right]_\mu B_{k,k} - \left[\mathbf{j}\times\mathbf{x}\right]_k B_{k,i}\,.$$

Now $B_{k,k} = \boldsymbol{\nabla}\cdot\mathbf{B} = 0$. Thus one obtains

$$F_i = \frac{k_{\mathrm{L}}}{2c}\int \mathrm{d}^3x\left(\mathbf{x}\times\mathbf{j}\right)_k B_{k,i} \overset{(4.2.13)}{=} m_k B_{k,i} = \nabla_i(\mathbf{m}\cdot\mathbf{B})$$

Using (A.2.33) one obtains

$$\mathbf{F} = \boldsymbol{\nabla}(\mathbf{m}\cdot\mathbf{B}) = (\mathbf{m}\cdot\boldsymbol{\nabla})\mathbf{B} + \mathbf{m}\times(\boldsymbol{\nabla}\times\mathbf{B})\,. \tag{4.3.4}$$

Potential energy of a dipole in the external field

It is reasonable to assume that the energy of a magnetic dipole in the field $\mathbf{B}$ is similar to that of the electric dipole in the field $\mathbf{E}$ (2.5.12): $U_e = -\mathbf{p}\cdot\mathbf{E}$. We can calculate this directly from the force (4.3.4) acting on a dipole in the field $\mathbf{B}$:

$$U = -\int_{-\infty}^{\mathbf{x}} \mathrm{d}\mathbf{x}'\cdot\boldsymbol{\nabla}'(\mathbf{m}\cdot\mathbf{B}(\mathbf{x}')) = -\mathbf{m}\cdot\int_0^{\mathbf{B}} \mathrm{d}\mathbf{B} = -\mathbf{m}\cdot\mathbf{B}. \qquad (4.3.5)$$

U does not include the energy required to maintain the current density $\mathbf{j}$, to keep $\mathbf{m}$ constant (see (7.2.9)). If one sets for $\mathbf{B}$ the field of a magnetic dipole (4.2.12), one obtains the interaction energy of two magnetic dipoles at the locations $\mathbf{x}_1 \neq \mathbf{x}_2$:

$$U = \frac{k_C}{k_L^2}\frac{1}{|\mathbf{x}_2-\mathbf{x}_1|^3}\left[\mathbf{m}_1\cdot\mathbf{m}_2 - 3\frac{(\mathbf{m}_1\cdot(\mathbf{x}_2-\mathbf{x}_1))(\mathbf{m}_2\cdot(\mathbf{x}_2-\mathbf{x}_1))}{|\mathbf{x}_1-\mathbf{x}_2|^2}\right]. \qquad (4.3.6)$$

The expression is analogous to the electric dipole-dipole interaction (2.5.14).

Torque

For the torque, similar transformations as for the force have to be carried out. Again the vector product $\mathbf{j}\times\mathbf{B}$ has to be transformed in such a way that an expression with the magnetic moment $(k_L/2c)\int \mathrm{d}^3x\,\mathbf{x}\times\mathbf{j}$ results:

$$\mathbf{N} = e\sum_n \mathbf{x}_n\times\mathbf{F}_n = \frac{k_L}{c}\sum_n \mathbf{x}_n\times(e\mathbf{v}_n\times\mathbf{B}(\mathbf{x}_n)) = \frac{k_L}{c}\int \mathrm{d}^3x\,\mathbf{x}\times(\mathbf{j}\times\mathbf{B}). \qquad (4.3.7)$$

In the first approximation, $\mathbf{B}(\mathbf{x}) = \mathbf{B}(0)$. Then, according to (4.2.9) $\mathbf{j} \leftrightharpoons -\mathbf{x}$ can be interchanged:

$$\mathbf{N} = \frac{k_L}{2c}\int \mathrm{d}^3x\left\{\mathbf{x}\times[\mathbf{j}(\mathbf{x})\times\mathbf{B}(0)] - \mathbf{j}(\mathbf{x})\times[\mathbf{x}\times\mathbf{B}(0)]\right\}.$$

Let's use the Jacobi identity (A.1.59): $\mathbf{x}\times(\mathbf{j}\times\mathbf{B}) = -\mathbf{j}\times(\mathbf{B}\times\mathbf{x}) - \mathbf{B}\times(\mathbf{x}\times\mathbf{j})$, so we can further transform (4.3.7):

$$\mathbf{N} = -\frac{k_L}{2c}\int \mathrm{d}^3x\,\mathbf{B}\times(\mathbf{x}\times\mathbf{j}(\mathbf{x})) = \mathbf{m}\times\mathbf{B}. \qquad (4.3.8)$$

Potential energy, force and torque of a dipole in the external field:

$$U \overset{(2.5.12)}{=} -\mathbf{p}\cdot\mathbf{E}, \qquad \mathbf{F} \overset{(2.5.15)}{=} \boldsymbol{\nabla}(\mathbf{p}\cdot\mathbf{E}), \qquad \mathbf{N} \overset{(2.5.16)}{=} \mathbf{p}\times\mathbf{E},$$

$$U \overset{(4.3.5)}{=} -\mathbf{m}\cdot\mathbf{B}, \qquad \mathbf{F} \overset{(4.3.4)}{=} \boldsymbol{\nabla}(\mathbf{m}\cdot\mathbf{B}), \qquad \mathbf{N} \overset{(4.3.8)}{=} \mathbf{m}\times\mathbf{B}.$$

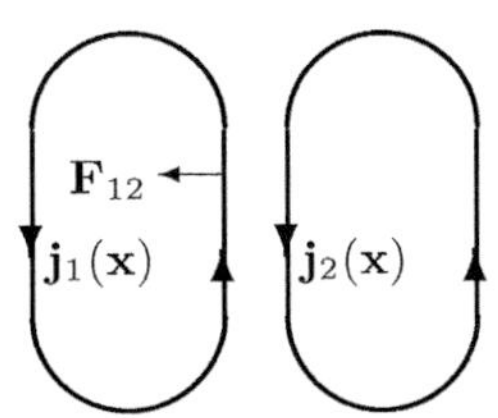

Fig. 4.11. Two current loops $\mathbf{j}_{1,2}$ repel each other, when the currents are antiparallel

4.3.3 Ampère's Force Law

The Biot-Savart law, to which we refer in the form of (4.1.7), informs us about the magnetic field of a current distribution. The force acting on a current loop located in a magnetic field is given in (4.3.3). With this we can determine the force that two current loops, as sketched in Fig. 4.11, exert on each other.

$$\mathbf{B}_2(\mathbf{x}_1) = \frac{k_C}{ck_L} \int \mathrm{d}^3 x_2 \, \frac{\mathbf{j}_2(\mathbf{x}_2) \times (\mathbf{x}_1 - \mathbf{x}_2)}{|\mathbf{x}_1 - \mathbf{x}_2|^3}.$$

The current $\mathbf{j}_2(\mathbf{x})$ of loop 2 generates the field $\mathbf{B}_2(\mathbf{x})$. Due to this field, the force acting on loop 1 is[1]

$$\mathbf{F}_{12} = \frac{k_L}{c} \int \mathrm{d}^3 x_1 \, \mathbf{j}_1(\mathbf{x}_1) \times \mathbf{B}_2(\mathbf{x}_1)$$

$$= \frac{k_C}{c^2} \int \mathrm{d}^3 x_1 \int \mathrm{d}^3 x_2 \, \frac{\mathbf{j}_1(\mathbf{x}_1) \times \left[\mathbf{j}_2(\mathbf{x}_2) \times (\mathbf{x}_1 - \mathbf{x}_2) \right]}{|\mathbf{x}_1 - \mathbf{x}_2|^3} = \frac{k_C}{c^2} \int \mathrm{d}^3 x_1 \int \mathrm{d}^3 x_2$$

$$\frac{\left(\mathbf{j}_1(\mathbf{x}_1) \cdot (\mathbf{x}_1 - \mathbf{x}_2) \right) \mathbf{j}_2(\mathbf{x}_2) - \left(\mathbf{j}_1(\mathbf{x}_1) \cdot \mathbf{j}_2(\mathbf{x}_2) \right) (\mathbf{x}_1 - \mathbf{x}_2)}{|\mathbf{x}_1 - \mathbf{x}_2|^3}.$$

The 2nd term already has the expected symmetry, i.e. only the sign of the force changes when loops 1 and 2 are swapped. Only the 1st term needs to be considered. This can be written in the following form:

$$-\frac{k_C}{c^2} \int \mathrm{d}^3 x_1 \int \mathrm{d}^3 x_2 \left(\mathbf{j}_1(\mathbf{x}_1) \cdot \boldsymbol{\nabla}_1 \frac{1}{|\mathbf{x}_1 - \mathbf{x}_2|} \right) \mathbf{j}_2(\mathbf{x}_2)$$

$$= \frac{k_C}{c^2} \int \mathrm{d}^3 x_1 \int \mathrm{d}^3 x_2 \left(\left(\boldsymbol{\nabla}_1 \cdot \mathbf{j}_1(\mathbf{x}_1) \right) \frac{1}{|\mathbf{x}_1 - \mathbf{x}_2|} \right) \mathbf{j}_2(\mathbf{x}_2) = 0,$$

since $\boldsymbol{\nabla}_1 \cdot \mathbf{j}_1 = 0$. This gives the force

$$\mathbf{F}_{12} = -\frac{k_C}{c^2} \int \mathrm{d}^3 x_1 \int \mathrm{d}^3 x_2 \, \mathbf{j}_1(\mathbf{x}_1) \cdot \mathbf{j}_2(\mathbf{x}_2) \frac{(\mathbf{x}_1 - \mathbf{x}_2)}{|\mathbf{x}_1 - \mathbf{x}_2|^3} \qquad (4.3.9)$$

in the form of Ampere's force law for the forces between 2 current loops, according to which parallel currents attract, antiparallel repel.

Force between two parallel wires

Given are two parallel wires of length $L \to \infty$, which have a distance d from each other (see Fig. 4.12). We calculate the integral analogously to section

---

[1] $\mathbf{a} \times (\mathbf{b} \times \mathbf{c}) = (\mathbf{a} \cdot \mathbf{c})\mathbf{b} - (\mathbf{a} \cdot \mathbf{b})\mathbf{c}$ (Grassmann identity, (A.1.60))

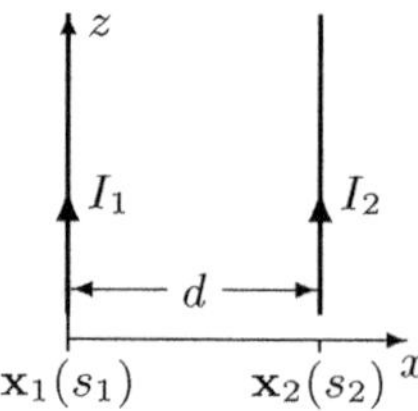

Fig. 4.12. Sketch with 2 wires at a distance d and $\mathbf{x}_{1,2}(s_{1,2})$ and parallel currents.

4.1.3 (Biot-Savart). The curve of the wire center points is given by $\mathbf{x}_i(s_i) = z_i\,\mathbf{e}_z$. For parallel currents, it is

$$\mathbf{j}_1(\mathbf{x}) = I_1\,\mathbf{e}_z\,\delta(x)\delta(y)\,, \qquad\qquad \mathbf{j}_2(\mathbf{x}) = I_2\,\mathbf{e}_z\,\delta(x-d)\delta(y)\,.$$

The wire cross-section is $F_i(s_i)$, so that $I_i = F_i(s_i)\,j_i(x_i)$.

$$\mathbf{x}_1(s_1) = \mathbf{e}_z s_1\,, \quad \mathbf{x}_2(s_2) = \mathbf{e}_x d + \mathbf{e}_z s_2 \quad \Rightarrow \quad \mathbf{x}_1 - \mathbf{x}_2 = -d\mathbf{e}_x + (s_1 - s_2)\mathbf{e}_z\,.$$

This gives the force acting on loop 1, if you set $s = s_1 - s_2$ and $L = \int_2 ds_2$:

$$\mathbf{F}_{1,2} = -\frac{k_C}{c^2}I_1 I_2 \int_1 ds_1 \int_2 ds_2 \frac{-d\mathbf{e}_x + (s_1-s_2)\mathbf{e}_z}{\sqrt{d^2 + (s_1-s_2)^2}^{\,3}} = -\frac{k_C}{c^2}I_1 I_2 L \int_{-\infty}^{\infty} ds \frac{-d\mathbf{e}_x + s\mathbf{e}_z}{\sqrt{d^2 + s^2}^{\,3}}\,.$$

The force per unit length $|\mathbf{F}_{1,2}/L|$, which wire 2 exerts on wire 1, does not depend on the location s_1 if wire 2 is infinitely long. The component F_z disappears for reasons of symmetry. So only F_x remains, i.e. $\mathbf{F}_{1,2} = F\mathbf{e}_x$, where it makes sense to only specify the force per unit length:

$$\frac{F}{L} = \frac{k_C}{c^2}I_1 I_2 \int_{-\infty}^{\infty} ds \frac{d}{\sqrt{d^2 + s^2}^{\,3}} = \frac{k_C}{c^2}\frac{I_1 I_2}{d} \left.\frac{s}{\sqrt{d^2 + s^2}}\right|_{-\infty}^{\infty} = \frac{k_C}{c^2}\frac{2I_1 I_2}{d}\,. \qquad (4.3.10)$$

The force per unit length on wire 1 is thus attractive for parallel currents and repulsive for antiparallel currents. Until 2019, (4.3.10) was used to determine the unit of current, the ampere (C.2.9). Now the ampere (C.2.8) is defined by the elementary charge (and the second). Thus, (4.3.10) determines the magnetic constant $\mu_0 = 1/c^2\epsilon_0$ independently of the ampere.

4.4 Magnetic Multipole Expansion

4.4.1 Moments of the Scalar Potential

First, using $\mathbf{B} = \boldsymbol{\nabla}\times\mathbf{A}$, (4.1.4), the field of local currents is calculated:

$$\mathbf{B} = \frac{k_C}{k_L c}\int_V d^3x'\,\boldsymbol{\nabla}\times\frac{\mathbf{j}(\mathbf{x}')}{|\mathbf{x}-\mathbf{x}'|} \overset{\boldsymbol{\nabla}\to-\boldsymbol{\nabla}'}{=} \frac{-k_C}{k_L c}\int_V d^3x'\left[\boldsymbol{\nabla}'\times\frac{\mathbf{j}(\mathbf{x}')}{|\mathbf{x}-\mathbf{x}'|} - \frac{\boldsymbol{\nabla}'\times\mathbf{j}(\mathbf{x}')}{|\mathbf{x}-\mathbf{x}'|}\right]$$

$$\overset{(A.4.5)}{=} \frac{-k_C}{k_L c}\left\{\oiint_{\partial V} d\mathbf{S}'\times\frac{\mathbf{j}(\mathbf{x}')}{|\mathbf{x}-\mathbf{x}'|} - \int_V d^3x'\,\frac{\boldsymbol{\nabla}'\times\mathbf{j}(\mathbf{x}')}{|\mathbf{x}-\mathbf{x}'|}\right\}, \qquad \text{SI: } \frac{k_C}{ck_L} = \frac{\mu_0}{4\pi}\,.$$

The first term disappears because there are no currents on the surface of V. We are only interested in the fields outside the local current distribution, where $\boldsymbol{\nabla} \times \mathbf{B} = 0$. There, a scalar potential ϕ_M (see p. 135) can be defined, which satisfies the Laplace equation $\Delta \phi_M = 0$. For this applies

$$
\begin{aligned}
\mathbf{x} \cdot \mathbf{B} &= -\mathbf{x} \cdot \boldsymbol{\nabla} \phi_M = -r \frac{\partial}{\partial r} \phi_M \\
&= \frac{k_C}{k_L c} \int_V \mathrm{d}^3 x' \, \frac{\mathbf{x} \cdot (\boldsymbol{\nabla}' \times \mathbf{j})}{|\mathbf{x} - \mathbf{x}'|} = \frac{k_C}{k_L c} \int_V \mathrm{d}^3 x' \, \frac{\boldsymbol{\nabla}' \cdot (\mathbf{j} \times \mathbf{x})}{|\mathbf{x} - \mathbf{x}'|}.
\end{aligned} \tag{4.4.1a}
$$

The following integral is zero, as each of the two terms on the right side disappears: The first after applying the Gauss's theorem, since the currents are local and the second term, since the scalar triple product has two parallel vectors:

$$
\int_V \mathrm{d}^3 x' \, \frac{\boldsymbol{\nabla}' \cdot \left[\mathbf{j} \times (\mathbf{x} - \mathbf{x}') \right]}{|\mathbf{x} - \mathbf{x}'|} = \int_V \mathrm{d}^3 x' \left\{ \boldsymbol{\nabla}' \cdot \frac{\mathbf{j} \times (\mathbf{x} - \mathbf{x}')}{|\mathbf{x} - \mathbf{x}'|} + \left[\mathbf{j} \times (\mathbf{x} - \mathbf{x}') \right] \cdot \boldsymbol{\nabla}' \frac{1}{|\mathbf{x} - \mathbf{x}'|} \right\} = 0.
$$

Thus, in the vector product of (4.4.1a), $\mathbf{x}$ can be replaced by $\mathbf{x}'$. Now $1/|\mathbf{x} - \mathbf{x}'|$ is expanded for $r > r'$ in powers of r'/r:

$$
\mathbf{x} \cdot \mathbf{B} = \frac{k_C}{k_L c} \int_V \mathrm{d}^3 x' \, \frac{\boldsymbol{\nabla}' \cdot (\mathbf{j} \times \mathbf{x}')}{|\mathbf{x} - \mathbf{x}'|} \tag{4.4.1b}
$$

$$
\overset{(3.3.8)}{=} \sum_{l=0}^{\infty} \frac{4\pi}{2l+1} \frac{1}{r^{l+1}} \sum_{m=-l}^{l} Y_{lm}(\vartheta, \varphi) \frac{k_C}{k_L c} \int_V \mathrm{d}^3 x' \, r'^l Y_{lm}^*(\vartheta', \varphi') \boldsymbol{\nabla}' \cdot (\mathbf{j} \times \mathbf{x}').
$$

(4.4.1b) can be represented with the help of

$$
\frac{1}{r^{l+1}} = -\frac{r}{l+1} \frac{\partial}{\partial r} \frac{1}{r^{l+1}}
$$

as an expansion of ϕ_M in spherical multipole moments

$$
\phi_M(\mathbf{x}) = \frac{k_C}{k_L^2} \sum_{l=0}^{\infty} \frac{4\pi}{2l+1} \frac{1}{r^{l+1}} \sum_{m=-l}^{l} Y_{lm}(\vartheta, \varphi) \, Q_{lm}^{(M)}, \tag{4.4.2}
$$

$$
Q_{lm}^{(M)} = \frac{k_L}{c(l+1)} \int_V \mathrm{d}^3 x' \, r'^l Y_{lm}^*(\vartheta', \varphi') \boldsymbol{\nabla}' \cdot (\mathbf{j} \times \mathbf{x}').
$$

We reshape the moments by integrating by parts:

$$
Q_{lm}^{(M)} = \frac{-k_L}{c(l+1)} \int_V \mathrm{d}^3 x' \, (\mathbf{j} \times \mathbf{x}') \cdot \boldsymbol{\nabla}' r'^l Y_{lm}^*(\vartheta', \varphi').
$$

Now the (dimensionless) angular momentum operator is defined by

$$
\hat{\mathbf{L}} = \mathbf{x} \times \hat{\mathbf{p}} = -i\mathbf{x} \times \boldsymbol{\nabla} = i\mathbf{e}_\vartheta \frac{1}{\sin \vartheta} \frac{\partial}{\partial \varphi} - i\mathbf{e}_\varphi \frac{\partial}{\partial \vartheta}, \qquad \hat{\mathbf{p}} = -i\boldsymbol{\nabla}. \tag{4.4.3}
$$

After cyclic permutation, we obtain the spherical multipole moments

$$Q_{lm}^{(M)} = \frac{-\mathrm{i}k_L}{c(l+1)} \int_V \mathrm{d}^3 x' \, r'^l \, \mathbf{j} \cdot \hat{\mathbf{L}}' Y_{lm}^*(\vartheta', \varphi'). \tag{4.4.4}$$

These can also be expressed by *vector spherical harmonics* (4.4.6) ([Hill, 1954] or [Jackson, 1998, (9.119)]) as follows:

$$Q_{lm}^{(M)} = \mathrm{i}\sqrt{\frac{l}{l+1}}\,\frac{k_L}{c}\int_V \mathrm{d}^3 x' \, r'^l \, \mathbf{j} \cdot \mathbf{X}_{lm}^*(\vartheta', \varphi'), \quad \mathbf{X}_{lm}(\vartheta, \varphi) = \frac{\hat{\mathbf{L}} Y_{lm}(\vartheta, \varphi)}{\sqrt{l(l+1)}}. \tag{4.4.5}$$

4.4.2 Vector Spherical Harmonics

To develop a vector field according to vector spherical harmonics, one constructs three vector functions, which should only depend on $\Omega = (\vartheta, \varphi)$:

$$\mathbf{Y}_{lm}(\Omega) = \mathbf{e}_r \, Y_{lm}(\Omega),$$

$$\mathbf{X}_{lm}(\Omega) = \frac{\hat{\mathbf{L}} Y_{lm}(\Omega)}{\sqrt{l(l+1)}} = \mathrm{i}\frac{r\,\boldsymbol{\nabla} \times \mathbf{Y}_{lm}(\Omega)}{\sqrt{l(l+1)}}, \tag{4.4.6}$$

$$\mathbf{Z}_{lm}(\Omega) = \mathbf{e}_r \times \mathbf{X}_{lm}(\Omega) = \frac{-r\hat{\mathbf{p}} Y_{lm}(\Omega)}{\sqrt{l(l+1)}}.$$

The thus defined vector spherical harmonics are orthonormal:

$$\begin{aligned}
(\mathbf{Y}_{l'm'}, \mathbf{Y}_{lm}) &= (\mathbf{X}_{l'm'}, \mathbf{X}_{lm}) = (\mathbf{Z}_{l'm'}, \mathbf{Z}_{lm}) = \delta_{ll'} \, \delta_{mm'}, \\
(\mathbf{Y}_{l'm'}, \mathbf{X}_{lm}) &= (\mathbf{Y}_{l'm'}, \mathbf{Z}_{lm}) = (\mathbf{X}_{l'm'}, \mathbf{Z}_{lm}) = 0.
\end{aligned} \tag{4.4.7}$$

The scalar products are integrals over the spherical surface $\mathrm{d}\Omega = \mathrm{d}\vartheta\mathrm{d}\varphi \sin\vartheta$.

Note: The definitions (4.4.6) are not "canonical", but are closely related to the definitions by Barrera et al. [1985]:

$$\mathbf{Y}_{lm}(\Omega) = \mathbf{e}_r \, Y_{lm}(\Omega),$$

$$\boldsymbol{\Phi}_{lm}(\Omega) = \mathbf{x} \times \boldsymbol{\nabla} Y_{lm}(\Omega) = \mathrm{i}\hat{\mathbf{L}} Y_{lm}(\Omega) = \mathrm{i}\sqrt{l(l+1)}\,\mathbf{X}_{lm}(\Omega), \tag{4.4.8}$$

$$\boldsymbol{\Psi}_{lm}(\Omega) = r\boldsymbol{\nabla} Y_{lm}(\Omega) = \mathrm{i}r\hat{\mathbf{p}} Y_{lm}(\Omega) = -\mathrm{i}\sqrt{l(l+1)}\,\mathbf{Z}_{lm}(\Omega).$$

Verification of orthogonality

It must be shown that the vector functions (4.4.6) are orthonormal, i.e., that they fulfill the conditions (4.4.7). It is immediately apparent that

$$\mathbf{Y}_{l'm'} \cdot \mathbf{X}_{lm} = \mathbf{Y}_{l'm'} \cdot \mathbf{Z}_{lm} = 0,$$

since $\mathbf{e}_r \cdot \hat{\mathbf{L}} = 0$ or $\mathbf{e}_r \cdot (\mathbf{e}_r \times \mathbf{X}_{lm}) = 0$. It is less trivial to show the orthogonality of

$$\mathbf{X}_{l'm'} \cdot \mathbf{Z}_{lm} = \mathbf{X}_{l'm'} \cdot (\mathbf{e}_r \times \mathbf{X}_{lm}) = (\mathbf{X}_{lm} \times \mathbf{X}_{l'm'}) \cdot \mathbf{e}_r,$$

which is done in the following auxiliary calculation:

Auxiliary calculation: $\mathbf{X}_{l'm'}\cdot\mathbf{Z}_{lm}=0$. For this we calculate

$$\mathbf{X}_{lm}\times\mathbf{X}_{l'm'}=\frac{-1}{l(l+1)}\left[\left(\mathbf{e}_\vartheta\frac{1}{\sin\vartheta}\frac{\partial}{\partial\varphi}-\mathbf{e}_\varphi\frac{\partial}{\partial\vartheta}\right)Y_{lm}\right]\times\left[\left(\mathbf{e}_\vartheta\frac{1}{\sin\vartheta}\frac{\partial}{\partial\varphi}-\mathbf{e}_\varphi\frac{\partial}{\partial\vartheta}\right)Y_{l'm'}\right]$$

$$=\frac{1}{l(l+1)}\frac{1}{\sin\vartheta}\left[(\mathbf{e}_\vartheta\times\mathbf{e}_\varphi)\frac{\partial Y_{lm}}{\partial\varphi}\frac{\partial Y_{l'm'}}{\partial\vartheta}+(\mathbf{e}_\varphi\times\mathbf{e}_\vartheta)\frac{\partial Y_{lm}}{\partial\vartheta}\frac{\partial Y_{l'm'}}{\partial\varphi}\right].$$

We thus obtain

$$\mathbf{X}_{l'm'}\cdot\mathbf{Z}_{lm}=\frac{1}{l(l+1)}\frac{1}{\sin\vartheta}\left[\frac{\partial Y_{lm}}{\partial\varphi}\frac{\partial Y_{l'm'}}{\partial\vartheta}-\frac{\partial Y_{lm}}{\partial\vartheta}\frac{\partial Y_{l'm'}}{\partial\varphi}\right].$$

For the scalar product of the vector functions, this results in

$$(\mathbf{X}_{l'm'},\mathbf{Z}_{lm})=\int\mathrm{d}\Omega\,\mathbf{X}^*_{l'm'}\cdot\mathbf{Z}_{lm}=\frac{1}{l(l+1)}\int\frac{\mathrm{d}\Omega}{\sin\vartheta}\left[\frac{\partial Y_{lm}}{\partial\vartheta}\frac{\partial Y^*_{l'm'}}{\partial\varphi}-\frac{\partial Y_{lm}}{\partial\varphi}\frac{\partial Y^*_{l'm'}}{\partial\vartheta}\right].$$

Now the 1st term with respect to ϑ and the 2nd term with respect to φ are partially integrated:

$$(\mathbf{X}_{l'm'},\mathbf{Z}_{lm})=\frac{1}{l(l+1)}\left\{\int_0^{2\pi}\mathrm{d}\varphi\,Y_{lm}\frac{\partial Y^*_{l'm'}}{\partial\varphi}\bigg|_{\vartheta=0}^{\vartheta=\pi}-\int\frac{\mathrm{d}\Omega}{\sin\vartheta}Y_{lm}\frac{\partial^2 Y^*_{l'm'}}{\partial\vartheta\partial\varphi}\right.$$

$$\left.-\int_0^\pi\mathrm{d}\vartheta\,Y_{lm}\frac{\partial Y^*_{l'm'}}{\partial\vartheta}\bigg|_{\varphi=0}^{\varphi=2\pi}+\int\frac{\mathrm{d}\Omega}{\sin\vartheta}Y_{lm}\frac{\partial^2 Y^*_{l'm'}}{\partial\vartheta\partial\varphi}\right\}.$$

Only the boundary terms remain, where the 2nd boundary term vanishes due to $Y_{lm}(\vartheta,\varphi)=Y_{lm}(\vartheta,\varphi+2\pi)$. This also applies to the 1st boundary term, when the integral over φ is performed:

$$\int_0^{2\pi}\mathrm{d}\varphi\,Y_{lm}(\vartheta,\varphi)\frac{\partial Y^*_{l'm'}(\vartheta,\varphi)}{\partial\varphi}=\frac{-m'}{m-m'}Y_{lm}(\vartheta,\varphi)\,Y^*_{l'm'}(\vartheta,\varphi)\bigg|_0^{2\pi}=0,\quad m\neq m'.$$

For $m=m'$ the integrand is independent of φ, and we get

$$\int_0^{2\pi}\mathrm{d}\varphi\,Y_{lm}(\vartheta,\varphi)\frac{\partial Y^*_{l'm}(\vartheta,\varphi)}{\partial\varphi}\bigg|_{\vartheta=0}^{\vartheta=\pi}=2\pi\mathrm{i}m\big[Y_{lm}(0,\varphi)Y^*_{l'm}(0,\varphi)-Y_{lm}(\pi,\varphi)Y^*_{l'm}(\pi,\varphi)\big].$$

Since the products are independent of φ, we set in the 1st term $\varphi\to\varphi+\pi$ and use the symmetry $Y_{lm}(\vartheta-\pi,\varphi+\pi)=(-1)^l\,Y_{lm}(\vartheta,\varphi)$. This shows that the three vector functions are orthogonal (2nd line of (4.4.7)).

It remains to show that the individual vectors are orthonormal, which follows directly from the orthonormality (3.2.36) of the Y_{lm}:

$$(\mathbf{Y}_{l'm'},\mathbf{Y}_{lm})=\int\mathrm{d}\Omega\,Y^*_{l'm'}(\Omega)\,Y_{lm}(\Omega)=\delta_{ll'}\delta_{mm'},$$

$$(\mathbf{X}_{l'm'},\mathbf{X}_{lm})=\frac{-1}{l(l+1)}\int\mathrm{d}\Omega\,(\hat{\mathbf{L}}Y^*_{l'm'})\cdot\hat{\mathbf{L}}Y_{lm},$$

$$=\frac{-1}{l(l+1)}\int\mathrm{d}\Omega\left[\frac{1}{\sin^2\vartheta}\frac{\partial Y^*_{l'm'}}{\partial\varphi}\frac{\partial Y_{lm}}{\partial\varphi}+\frac{\partial Y^*_{l'm'}}{\partial\vartheta}\frac{\partial Y_{lm}}{\partial\vartheta}\right].$$

After integration by parts one obtains

$$(\mathbf{X}_{l'm'}, \mathbf{X}_{lm}) = \frac{-1}{l(l+1)} \left\{ \int_0^\pi \frac{d\vartheta}{\sin\vartheta}\, Y_{l'm'}^* \left.\frac{\partial Y_{lm}}{\partial\varphi}\right|_{\varphi=0}^{\varphi=2\pi} + \int_0^{2\pi} d\varphi\, Y_{l'm'}^* \left.\frac{\partial Y_{lm}}{\partial\vartheta}\right|_{\vartheta=0}^{\vartheta=\pi} \right.$$
$$\left. - \int d\Omega\, Y_{l'm'}^*\, \hat{\mathbf{L}}^2 Y_{lm} \right\}.$$

The two boundary terms vanish for the same reasons that we have stated in the derivation of $(\mathbf{X}_{l'm'}, \mathbf{Z}_{lm})$ have stated. Y_{lm} are the eigenfunctions of $\hat{L}^2$ with the eigenvalues $l(l+1)$ (see (3.2.9)). With these we obtain the desired orthonormality condition

$$(\mathbf{X}_{l'm'}, \mathbf{X}_{lm}) = \delta_{ll'}\, \delta_{mm'}\,.$$

The orthonormality for $\mathbf{Z}_{lm}$ follows directly from

$$\mathbf{Z}_{l'm'}^* \cdot \mathbf{Z}_{lm} = (\mathbf{e}_r \times \mathbf{X}_{l'm'}^*) \cdot (\mathbf{e}_r \times \mathbf{X}_{lm}) = \mathbf{X}_{l'm'}^* \cdot \mathbf{X}_{lm}\,,$$

since $\mathbf{X}_{lm} \cdot \mathbf{e}_r = 0$. Thus, (4.4.7) is verified.

The expansion according to vector spherical harmonics

With the basis vectors, the vector spherical harmonics $\mathbf{X}_{lm}$, $\mathbf{Y}_{lm}$ and $\mathbf{Z}_{lm}$, an arbitrary vector field $\mathbf{v}$, like the magnetic vector potential $\mathbf{A}$, can be expanded [Barrera et al., 1985, (3.22)]:

$$\mathbf{v} = \sum_{l=0}^\infty \sum_{m=-l}^l \left\{ Q_{lm}^Y \mathbf{Y}_{lm} + Q_{lm}^X \mathbf{X}_{lm} + Q_{lm}^Z \mathbf{Z}_{lm} \right\}. \tag{4.4.9}$$

Here, the Q_{lm} are the spherical multipole moments of $\mathbf{v}$

$$Q_{lm}^Y(r) = (\mathbf{Y}_{lm}, \mathbf{v}), \quad Q_{lm}^X(r) = (\mathbf{X}_{lm}, \mathbf{v}), \quad Q_{lm}^Z(r) = (\mathbf{Z}_{lm}, \mathbf{v}), \tag{4.4.10}$$

which are integrals over the sphere surface. The Helmholtz decomposition theorem, Sect. 7.1.2, states that a continuously differentiable vector field $\mathbf{v}$, which falls off stronger than $1/r$ for $r \to \infty$, can be divided into a vortex-free ($\boldsymbol{\nabla} \times \mathbf{v}_l = 0$) and a source-free ($\boldsymbol{\nabla} \cdot \mathbf{v}_t = 0$) vector field:

$$\mathbf{v}(\mathbf{x}) = \mathbf{v}_l(\mathbf{x}) + \mathbf{v}_t(\mathbf{x}) = -\boldsymbol{\nabla}\phi(\mathbf{x}) + \boldsymbol{\nabla} \times \mathbf{A}(\mathbf{x}), \tag{7.1.15}$$

Vortex-free vector fields

The expansion of a vortex-free field $\mathbf{v}_l = -\boldsymbol{\nabla}\phi$ in vector spherical harmonics is given by (3.3.1):

$$\phi(\mathbf{x}) = \sum_{l=0}^{\infty} \sum_{m=-l}^{l} \phi_{lm}(r) Y_{lm}(\Omega), \quad \phi_{lm}(r) = \int d\Omega\, Y_{lm}^{*}(\Omega)\phi(\mathbf{x}), \quad (4.4.11)$$

$$\boldsymbol{\nabla}[\phi_{lm} Y_{lm}] = \frac{\partial \phi_{lm}}{\partial r}\mathbf{e}_r Y_{lm} + \frac{\phi_{lm}}{r}(r\boldsymbol{\nabla}Y_{lm}) \stackrel{(4.4.6)}{=} \frac{d\phi_{lm}}{dr}\mathbf{Y}_{lm} - i\sqrt{l(l+1)}\frac{\phi_{lm}}{r}\mathbf{Z}_{lm}.$$

From this follows for the coefficients of the expansion:

$$\mathbf{v}_l = \sum_{l=0}^{\infty} \sum_{m=-l}^{l} \left[-\frac{d\phi_{lm}}{dr}\mathbf{Y}_{lm} + i\sqrt{l(l+1)}\frac{\phi_{lm}}{r}\mathbf{Z}_{lm}\right], \quad\quad (4.4.12)$$

$$Q_{lm}^{Y} = (\mathbf{Y}_{lm}, \mathbf{v}_l) = -\frac{d\phi_{lm}}{dr}, \quad Q_{lm}^{X} = 0, \quad Q_{lm}^{Z} = (\mathbf{Z}_{lm}, \mathbf{v}_l) = i\sqrt{l(l+1)}\frac{\phi_{lm}}{r}.$$

This can be verified by direct calculation of the $Q_{lm}^{Y,X,Z}$.

Source-free vector fields – Debye's decomposition theorem

A source-free vector field $\mathbf{v}_t$ can be represented by two scalar fields χ and ψ as (*Debye's decomposition theorem*)

$$\mathbf{v}_t = \mathbf{x}\times\boldsymbol{\nabla}\psi + \boldsymbol{\nabla}\times(\mathbf{x}\times\boldsymbol{\nabla})\chi = i\hat{\mathbf{L}}\psi - \hat{\mathbf{p}}\times(\hat{\mathbf{L}}\chi). \quad\quad (4.4.13)$$

The source-freeness of the above approach is easy to verify:

$$\boldsymbol{\nabla}\cdot\mathbf{v}_t = -\hat{\mathbf{p}}\cdot\hat{\mathbf{L}}\psi + i\hat{\mathbf{p}}\cdot(\hat{\mathbf{p}}\times\hat{\mathbf{L}}\psi) = 0.$$

The scalar fields ψ and χ are now expanded according to (4.4.11) after $Y_{lm}(\Omega)$

$$\mathbf{v}_t = \sum_{l=0}^{\infty} \sum_{m=-l}^{l} \left[\psi_{lm}\, i\hat{\mathbf{L}}Y_{lm} - (\hat{\mathbf{p}}\chi_{lm})\times\hat{\mathbf{L}}Y_{lm} - \chi_{lm}\hat{\mathbf{p}}\times\hat{\mathbf{L}}Y_{lm}\right]$$

Auxiliary calculation: For the sake of abbreviation, the indices lm are omitted:

$$(\hat{\mathbf{p}}\chi)\times\hat{\mathbf{L}}Y = -\frac{1}{r}\frac{d\chi}{dr}\mathbf{x}\times(\mathbf{x}\times\boldsymbol{\nabla})Y = \frac{d\chi}{dr}(r\boldsymbol{\nabla}Y) = \frac{d\chi}{dr}\boldsymbol{\Psi}_i,$$

$$[\hat{\mathbf{p}}\times\hat{\mathbf{L}}Y]_i = -[\boldsymbol{\nabla}\times(\mathbf{x}\times\boldsymbol{\nabla})]_i Y = -\epsilon_{ijk}\epsilon_{klm}\nabla_j x_l \nabla_m Y = [2\nabla_i - x_i\boldsymbol{\nabla}^2 + (\mathbf{x}\cdot\boldsymbol{\nabla})\nabla_i]Y$$

$$= [2\nabla_i - x_i\boldsymbol{\nabla}^2 + (\mathbf{x}\cdot\boldsymbol{\nabla}\frac{1}{r})(r\nabla_i)]Y = [\nabla_i - x_i\boldsymbol{\nabla}^2]Y = \frac{1}{r}\boldsymbol{\Psi}_i + \frac{l(l+1)}{r}Y_i.$$

Here, $r^2\boldsymbol{\nabla}^2 Y_{lm} = -l(l+1)Y_{lm}$ and $r\boldsymbol{\nabla}Y_{lm} = \boldsymbol{\Psi}_{lm}$ and $\mathbf{e}_r Y_{lm} = \mathbf{Y}_{lm}$.

This gives us the source-free vector field:

$$\mathbf{v}_t = \sum_{l=0}^{\infty} \sum_{m=-l}^{l} \left\{\sqrt{l(l+1)}\left[\psi_{lm}\mathbf{X}_{lm} + i\frac{d(r\chi_{lm})}{r\,dr}\mathbf{Z}_{lm}\right] - l(l+1)\frac{\chi_{lm}}{r}\mathbf{Y}_{lm}\right\}. \quad (4.4.14)$$

The Debye potentials χ and ψ from $\mathbf{v}_t$ need to be determined:

$$X(\mathbf{x}) = \mathbf{x}\cdot\mathbf{v}_t = i\mathbf{x}\cdot\hat{\mathbf{L}}\,\psi - \mathbf{x}\cdot(\hat{\mathbf{p}}\times\hat{\mathbf{L}})\chi = -\hat{\mathbf{L}}^2\chi,$$

$$\Psi(\mathbf{x}) = \mathbf{x}\cdot(\boldsymbol{\nabla}\times\mathbf{v}_t) = i\hat{\mathbf{L}}\cdot\mathbf{v}_t = -\hat{\mathbf{L}}^2\psi$$

The fields X and Ψ are developed according to (4.4.11) after Y_{lm} and compared with χ and ψ. Now, $\hat{\mathbf{L}}^2 Y_{lm} = l(l+1)Y_{lm}$, from which the coefficients follow

$$X_{lm} = -l(l+1)\chi_{lm}, \qquad\qquad \Psi_{lm} = -l(l+1)\psi_{lm}.$$

Thus, the Debye potentials are determined, where $\chi_{00}(r) = \psi_{00}(r) = 0$ are set. This also implies that radially symmetric fields $\bar{\chi}(r)$ and $\bar{\psi}(r)$ can be added to the solutions without changing $\mathbf{v}_t$.

Problems for Chapter 4

4.1. *Infinite straight wire*: Calculate the regular potential

1. ϕ of a line charge λ
2. $\mathbf{A}$ of a line current I

using the Green function $\;G_1(\mathbf{x}-\mathbf{x}_0, \mathbf{x}'-\mathbf{x}_0) = \dfrac{1}{|\mathbf{x}'-\mathbf{x}|} - \dfrac{1}{|\mathbf{x}'-\mathbf{x}_0|},$
where the wire lies on the z-axis and $\mathbf{x}_0 \neq 0$ is a freely selectable convergence point in the xy-plane.

4.2. *Magnetic field of a rotating disk*: An infinitely thin circular disk of radius a is homogeneously charged (σ). The disk lies in the xy-plane with its center at the origin and rotates with the angular velocity $\boldsymbol{\omega}$ around the z-axis.

1. Calculate the magnetic moment $\mathbf{m}$ of the disk.
2. Calculate $\mathbf{B}$ approximately analogous to the circular loop and
3. exactly on the z-axis and compare with the dipole field for $|z| \gg a$.

4.3. *Field of a finite coil*: The field $\mathbf{B}$ of a coil of length $2l$ (radius a) is to be calculated in the first order of $\kappa = 2\varrho a/(\rho^2 + a^2 + (z\pm l)^2)$. For $r \gg l$ and $r \gg a$ the approximation should give a dipole field. Compare the approximation especially on the z-axis with the exact result.

4.4. *Vector potential of a semi-infinite dipole line.* To be calculated is

$$\mathbf{A} = k_M p_m \int_C \mathrm{d}\mathbf{x}' \times \frac{\mathbf{x} - \mathbf{x}'}{|\mathbf{x} - \mathbf{x}'|^3} \,.$$

The path C is the straight line: $\mathbf{x}' = s\mathbf{n}$ with $-\infty < s \leq 0$, where $\mathbf{n}$ is an arbitrarily oriented unit vector.

1. Calculate the vector potential (see (4.2.48)).

$$\int \mathrm{d}x\,\frac{1}{\sqrt{ax^2 + bx + c}^{\,3}} = \frac{4ax + 2b}{4ac - b^2}\,\frac{1}{\sqrt{ax^2 + bx + c}} \qquad \text{auxiliary integral}$$

2. Assume that $\mathbf{n} = \mathbf{e}_z$. The path C, the so-called *Dirac-string* is then the negative z-axis. Specify $\mathbf{A}$ in spherical coordinates (also see (4.2.48)) and calculate $\mathbf{B}^p = \boldsymbol{\nabla} \times \mathbf{A}$ in spherical coordinates for $0 \leq \vartheta < \pi$ and $r > 0$.

3. 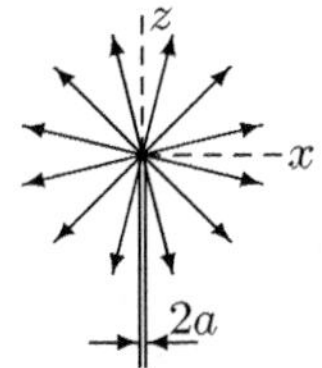 Using $\mathbf{B}^p$, determine the flow Φ_{B} through a circle S_ϱ of radius ϱ (see adjacent sketch). Show that Ampere's law

$$\oint_{\partial S_\varrho} d\mathbf{x} \cdot \mathbf{A} = \iint_{S_\varrho} d\mathbf{a} \cdot \mathbf{B}^p + 2\pi k_M p_m (1 - \operatorname{sgn} z)$$

is not fulfilled for $\vartheta > \pi/2$, as the flux of the dipole line (solenoid) from $\mathbf{B}^p$ is not captured.

4.5. *Field of a semi-infinite coil.* In a cylindrical, semi-infinite coil (solenoid) with radius a and winding density n the current I flows. n is high enough so that the slope per turn can be neglected and the current can thus be considered as a continuous circular current, as one would have with a charged rotating cylinder.

Fig. 4.13. Semi-infinite solenoid (coil) with the radius $a \to 0$; the field $\mathbf{B}$, which emerges at the end of the solenoid, is that of a monopole. The coil extends along the z-axis from $-\infty < z \leq 0$

1. Calculate $\mathbf{B}$ inside and outside the coil (Biot-Savart) and assume, that the coil is thin: $\mathbf{B} = \mathbf{B}^s + \mathbf{B}^p$

   Hint: If you calculate $\mathbf{B}$ without approximations, you will get the solution in the form of elliptic integrals of the 1st, 2nd and 3rd kind; since this is not very illustrative, you should make the approximations, which are very similar to those of $\mathbf{B}$ of the finite coil (see (4.2.41)); integrals can be found in section B.5.2.

2. Make the limit $a \to 0$ and specify $\mathbf{B}$ for this case: How does the current increase and how is the strength of the monopole defined?

4.6. *Monopole again*: The starting point is again the solenoid Fig. 4.13. For a general orientation $\mathbf{n}$, (4.2.48) applies

$$\mathbf{A} = k_M p_m \frac{\mathbf{n} \times \mathbf{x}}{r(r + \mathbf{n} \cdot \mathbf{x})}.$$

1. Verify

$$\mathbf{A} = \boldsymbol{\nabla} \times \mathbf{a} = k_M p_m \frac{\mathbf{n} \times \mathbf{x}}{r(r + \mathbf{n} \cdot \mathbf{x})}, \qquad \mathbf{a} = -k_M p_m \ln(r + \mathbf{n} \cdot \mathbf{x})\, \mathbf{n}.$$

2. You can use this approach for $\mathbf{A}$ to show that for $\mathbf{n} = \mathbf{e}_z$

$$\mathbf{B} = \boldsymbol{\nabla} \times \mathbf{A} = k_M p_m \frac{\mathbf{x}}{r^3} + 4\pi k_M p_m\, \mathbf{e}_z\, \delta(x)\delta(y)\theta(-z)$$

   is equal to that of a point source with a singular line.

   Hint: Using the representation $\mathbf{A} = \boldsymbol{\nabla} \times \mathbf{a}$ the field

$$\mathbf{B} = \boldsymbol{\nabla} \times \boldsymbol{\nabla} \times \mathbf{a} = \boldsymbol{\nabla}\boldsymbol{\nabla} \cdot \mathbf{a} - \Delta \mathbf{a} = \mathbf{B}^p + \mathbf{B}^s$$

   can be divided into two parts, which can be calculated individually. To determine in particular the singular contribution of $\mathbf{B}^s$, integrate over a sphere (cylinder) that encloses the negative z-axis.

3. Verify that $\nabla \cdot \mathbf{B}^s = -\nabla \cdot \mathbf{B}^p$ and $\nabla \cdot \mathbf{A} = 0$.
4. Show that $\mathbf{B}$ fulfills Ampere's (circuital) law, i.e., calculate $\nabla \times \mathbf{B}$.
   Hint: Assume that $\mathbf{B}^s$ is the limit case of the field in a coil with finite radius a, analogous to the finite wire that we could replace with a line.

4.7. *Magnetic field of a rotating sphere*: A homogeneously charged sphere with radius R and charge density ρ_0 rotates with the constant angular velocity $\boldsymbol{\omega}$.

1. Calculate the magnetic moment $\mathbf{m}$ of the sphere.
2. Calculate the vector potential outside the sphere, where you should express this through $\mathbf{m}$, and also calculate $\mathbf{B}$.

   Hint: The integration over ϑ' can with the help of the expansion of $\dfrac{1}{|\mathbf{x}-\mathbf{x}'|}$ after Legendre polynomials (3.3.7) be carried out relatively elegantly. From physical considerations, you know that $\mathbf{B}$ must be the field of a dipole.

4.8. *Force between current loops*: The force per unit length (4.3.10) that two infinitely long and parallel wires exert on each other was used to define the Ampere.

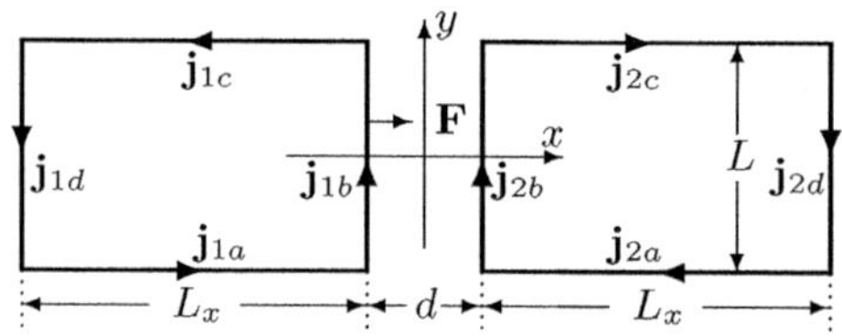

For finite length, we expect corrections, which here are to be calculated for two identical rectangular current loops. These are arranged in the xy-plane according to the sketch. Show that

$$\mathbf{F}_{1b\,2b} = \frac{2I_1 I_2}{dc^2}\left(\sqrt{L^2 + d^2} - d\right)\mathbf{e}_x$$

is the force that the two partial currents $\mathbf{j}_{1b}$ and $\mathbf{j}_{2b}$ exert on each other. Furthermore, calculate the force that acts between $\mathbf{j}_{1a}$ and $\mathbf{j}_{2a}$ and $\mathbf{j}_{1c}$ and $\mathbf{j}_{2c}$. Is this attractive or repulsive?

Result:

$$\mathbf{F}_{1a,2a} + \mathbf{F}_{1c,2c} = \frac{2I_1 I_2}{c^2}\ln\frac{d(d + 2L_x)}{(d + L_x)^2}\,\mathbf{e}_x\,.$$

4.9. *Interaction energy of magnetic dipoles.*

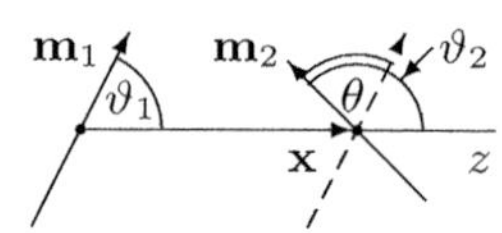

Given are two magnetic moments $\mathbf{m}_1$ and $\mathbf{m}_2$.
$\mathbf{m}_1$ is located at the origin and $\mathbf{x}$ points from $\mathbf{m}_1$ to $\mathbf{m}_2$, which does not necessarily lie on the z-axis, as the sketch suggests. $\mathbf{m}_1 \cdot \mathbf{m}_2 = m_1 m_2 \cos\theta$, $\mathbf{m}_1 \cdot \mathbf{x} = m_1 r \cos\vartheta_1$ and $\mathbf{m}_2 \cdot \mathbf{x} = m_2 r \cos\vartheta_2$. Auxiliary formula:
$\cos\theta = \cos\vartheta_1 \cos\vartheta_2 + \sin\vartheta_1 \sin\vartheta_2 \cos(\varphi_1 - \varphi_2)$.

1. Determine the interaction energy of the two dipoles and the force, that the two dipoles exert on each other.
2. How do the dipoles align when they are freely suspended, and what force do the two dipoles exert on each other in the configuration of minimal energy?

4.10. *Earth's magnetic field*: This can be approximated by the field of a magnetic point dipole at the center of the Earth's sphere. Furthermore, we assume that the geographical North Pole and the magnetic South Pole coincide, so that the southern magnetic latitudes (< 0) coincide with the northern geographical latitudes (> 0) (no declination). The field is thus parallel to the Earth's surface at the equator and points into the Earth on the northern hemisphere; this angle is called inclination.

1. Assume that on the 48th latitude the horizontal component $\mathbf{H}_{\|} = 18\,\mathrm{A/m}$ and calculate from this the magnetic moment of the Earth ($\mathbf{m} = 8.7\,\mathrm{Oe\,cm^2}$).
2. Give the relationship between geographical latitude and the inclination and show that this is about 66 degrees on the 48th latitude.

Hint: Take for the average radius of the Earth $R = 6\,370\,\mathrm{km}$.

4.11. *Sum rule for vector spherical harmonics*: Show that

$$\sum_{m=-l}^{l} |\mathbf{X}_{lm}(\vartheta, \varphi)|^2 = \sum_{m=-l}^{l} |\mathbf{Z}_{lm}(\vartheta, \varphi)|^2 = \frac{2l+1}{4\pi}.$$

References

Abraham M., Becker R. *Theorie der Elektrizität I*, 8th ed. Teubner Leipzig (1930)

Barrera R. G., Estévez G. A. and Giraldo J. *Vector spherical harmonics and their application to magnetostatics*, Eur. J. Phys. **6**, 287–294 (1985)

Becker R., Sauter F. *Theorie der Elektrizität 1*, 21th ed. Teubner, Stuttgart (1973)

Dirac P. A. M. *Quantised Singularities in the Electromagnetic Field*, Proc. R. Soc. **A133**, 60-72 (1931)

Dirac P. A. M. *The theory of Magnetic Poles*, Phys. Rev. **74**, 817 (1948)

Hill E. L. *The Theory of Vector Spherical Harmonics*, Am. J. Phys. **22**, 211–214 (1954)

Jackson J. D. *Classical Electrodynamics*, 3rd ed., John Wiley & Sons Inc. (1998)

Nowotny H. *Elektrodynamik und Relativitätstheorie*, script at the TU-Wien, http://tph.tuwien.ac.at/~rebhana/ED-Skriptum/ (2006)

Sommerfeld A. *Electrodynamics*, Academic Press Inc., New York (1952)

5

Electromagnetic Processes in Matter

The Maxwell equations (1.3.21") describe the electromagnetic processes in the presence of electric charges, currents and electromagnetic fields.

All processes considered so far were time-independent. Given charge distributions $\rho(\mathbf{x})$ generate electric fields and given currents $\mathbf{j}(\mathbf{x})$ magnetic fields. The forces and moments that the charge distributions exert on each other could be determined. As boundary conditions, areas of constant potential (conductors) have occurred. For the description, we used the Poisson equation with the corresponding (mostly Dirichlet's) boundary conditions.

In matter (solid bodies), the situation is more complex in that atoms or molecules are not freely movable, but are bound to their (lattice) places. They can be electrically neutral, carry a charge (ions) and/or have permanent dipole moments (multipole moments).

Under the influence of an electric field, the charges, such as the nucleus and electron shell, will shift against each other, resulting in an induced dipole moment. Of course, all higher multipole moments also occur, but their influence, especially at larger distances, is small in comparison to the dipole. These charges, which are bound to the molecule and/or the nucleus and are responsible for dielectric properties, are referred to as *bound* charges. These are distinguished from the *free* charges, the sources of the electric displacement density $\mathbf{D}$ (5.2.17).

The distinction between free and bound charges is obvious, as for the description of macroscopic properties, such as the polarizability of a medium, the use of microscopic fields is not suitable. The fields described by the Maxwell equations in vacuum are referred to in the following sections as microscopic fields $\mathbf{e}$ and $\mathbf{b}$. For macroscopic processes, it is neither possible nor necessary to calculate the fields of the bound charges from the microscopic Maxwell equations. Instead, one introduces average fields, an average charge distribution and an average current distribution.

D. Petrascheck, F. Schwabl, *Electrodynamics*, https://doi.org/10.1007/978-3-662-71502-4_5

5.1 The Microscopic Equations

The starting point is the Maxwell equations in vacuum (1.3.21), where here the microscopic fields are written in lowercase ($\mathbf{E} \rightarrow \mathbf{e}$ and $\mathbf{B} \rightarrow \mathbf{b}$), to make a clear distinction from the slowly varying macroscopic fields:

$$\text{(a)} \qquad \boldsymbol{\nabla} \cdot \mathbf{e} = 4\pi k_C \rho, \qquad\qquad \text{(b)} \quad \boldsymbol{\nabla} \times \mathbf{e} + \frac{k_L}{c} \dot{\mathbf{b}} = 0,$$

$$\text{(c)} \quad k_L \boldsymbol{\nabla} \times \mathbf{b} - \frac{1}{c} \dot{\mathbf{e}} = \frac{4\pi k_C}{c} \mathbf{j}, \qquad \text{(d)} \qquad\qquad \boldsymbol{\nabla} \cdot \mathbf{b} = 0. \tag{5.1.1}$$

We divide charges and currents into the contributions of free and bound charges and currents:

$$\begin{aligned} \rho(\mathbf{x},t) &= \rho_f(\mathbf{x},t) + \rho_b(\mathbf{x},t), \\ \mathbf{j}(\mathbf{x},t) &= \mathbf{j}_f(\mathbf{x},t) + \mathbf{j}_b(\mathbf{x},t). \end{aligned} \tag{5.1.2}$$

It is shown that the bound charge density ρ_b is responsible for the polarization $\mathbf{P}$ of the medium, and $\mathbf{j}_b$ for the magnetization $\mathbf{M}$ and the displacement current density $\dot{\mathbf{P}}/4\pi k_C \epsilon_0$ are responsible.

Bound charges

In matter, ρ contains, in addition to any free charges, primarily in atoms, ions or molecules, bound charges:

$$\rho_b(\mathbf{x},t) = \sum_n \rho_n(\mathbf{x} - \mathbf{x}_n(t)) \approx \sum_n \big[q_n - \mathbf{p}_n(t) \cdot \boldsymbol{\nabla} \big] \delta^{(3)}(\mathbf{x} - \mathbf{x}_n). \tag{5.1.3}$$

Here, the sum is over all molecules; $\mathbf{x}_n = \mathbf{x}_n(t)$ is the current position of molecule n. We know from the multipole expansion that even a complicated charge distribution for the calculation of the field at greater distance can be represented by a small number of moments. A simple estimate shows that it is sufficient to stop at the dipole term. q_n is the charge of the molecule and $\mathbf{p}_n$ its dipole moment (that can be permanent or induced).

Induced dipole moment: The atom n is electrically neutral. If e_n is the positive core charge, then the electron shell has the charge $-e_n$. In the field $\mathbf{E}$, the light shell moves mainly against the field direction by $\mathbf{d}$ and an electric dipole is created (see Fig. 5.1):

$$\rho_n(\mathbf{x}) = e_n \big[\delta^{(3)}(\mathbf{x}_n - \mathbf{x}) - \delta^{(3)}(\mathbf{x}_{n_-} - \mathbf{x}) \big].$$

In the center of gravity of the negative shell, the dipole moment of the negative shell disappears and the dipole is (antiparallel to $\mathbf{d}$) directed from there to the core: $\mathbf{x}_{n_-} = \mathbf{x}_n + \mathbf{d}_n$.

$$\rho_n(\mathbf{x}) \approx -e_n \mathbf{d}_n \cdot \boldsymbol{\nabla}_n \delta^{(3)}(\mathbf{x}_n - \mathbf{x}) = -\mathbf{p}_n \cdot \boldsymbol{\nabla} \delta^{(3)}(\mathbf{x} - \mathbf{x}_n).$$

The position of the core $\mathbf{x}_n$ is assumed to be time-independent, so that the entire time dependence is limited to the dipole moment $\mathbf{p}_n(t)$. If one now takes an ion instead of a neutral atom with the charge q_n, one arrives at (5.1.3) .

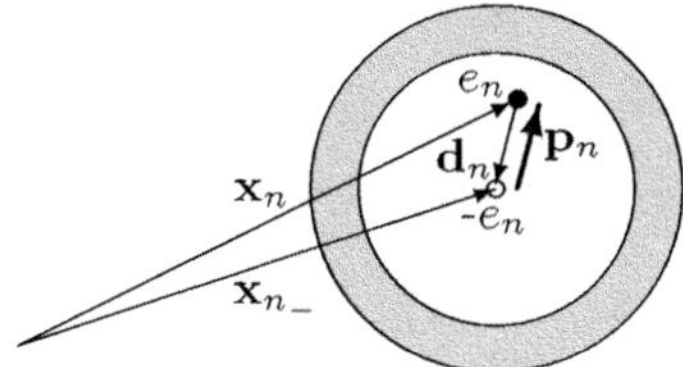

Fig. 5.1. Induced dipole moment of a neutral atom with the nuclear charge e_n. $\mathbf{x}_n$ is the location of the atomic nucleus and $\mathbf{x}_{n_-}$ the center of gravity of the electronic shell, which is shifted by $-\mathbf{d}_n$; it is $\mathbf{p}_n = -e_n\,\mathbf{d}_n$

Due to charge neutrality, we expect that only the dipole contribution

$$\rho_{\mathrm{p}}(\mathbf{x},t) = -\boldsymbol{\nabla}\cdot\sum_n \mathbf{p}_n(t)\,\delta^{(3)}(\mathbf{x}-\mathbf{x}_n) = -\boldsymbol{\nabla}\cdot\mathbf{p}(\mathbf{x},t) \tag{5.1.4}$$

occurs, where the dipole moments can be both, induced and permanent.

Bound currents

We consider the atom/molecule n, which is supposed to be at the origin $(\mathbf{x}_n = 0)$. Its charge distribution $\rho_n(\mathbf{x},t)$ is responsible for the polarizability and $\mathbf{j}_n(\mathbf{x},t)$ for the magnetic moment $\mathbf{m}_n$, which can be induced or permanent. Scalar and vector potential are given by

$$\phi_n(\mathbf{x},t) = k_C \int \mathrm{d}^3x' \frac{\rho_n(\mathbf{x}',t)}{|\mathbf{x}-\mathbf{x}'|}, \quad \mathbf{A}_n(\mathbf{x},t) = \frac{k_C}{ck_L}\int \mathrm{d}^3x' \frac{\mathbf{j}_n(\mathbf{x}',t)}{|\mathbf{x}-\mathbf{x}'|}. \tag{5.1.5}$$

These potentials are called *quasi-static*, as they are the static potentials of the charge and current distribution at time t, i.e. $\rho_n(\mathbf{x}',t)$ or $\mathbf{j}_n(\mathbf{x}',t)$ contribute instantaneously to the potentials. In the Ampère-Maxwell equation

$$\boldsymbol{\nabla}\times(\boldsymbol{\nabla}\times\mathbf{A}_n) \stackrel{(\mathrm{A.2.38})}{=} \boldsymbol{\nabla}(\boldsymbol{\nabla}\cdot\mathbf{A}_n) - \Delta\mathbf{A}_n = \frac{4\pi k_C}{ck_L}\,\mathbf{j}_n(\mathbf{x},t) - \frac{1}{ck_L}\,\boldsymbol{\nabla}\dot{\phi}_n$$

the term $\boldsymbol{\nabla}(\boldsymbol{\nabla}\cdot\mathbf{A}_n)$ is compensated by the displacement current $-\boldsymbol{\nabla}\dot{\phi}_n/4\pi k_C$, which can be verified using the continuity equation. This also means that the condition $\boldsymbol{\nabla}\cdot\mathbf{A}_n = 0$ valid in magnetostatics is violated.

We consider in the very good approximation in the atomic range only contributions of the electric and magnetic dipole moments $\mathbf{p}_n$ and $\mathbf{m}_n$, which according to (4.2.2) – (4.2.4) results in

$$\mathbf{A}_n(\mathbf{x},t) = \frac{k_C}{k_L^2}\left\{\frac{k_L}{cr}\,\dot{\mathbf{p}}_n + \frac{\mathbf{m}_n\times\mathbf{x}}{r^3} + \dots\right\}. \tag{5.1.6}$$

From (5.1.5) and (5.1.6) it follows

$$\Delta\mathbf{A}_n = \frac{k_C}{k_Lc}\int\mathrm{d}^3x'\,\Delta\frac{\mathbf{j}_n(\mathbf{x}',t)}{|\mathbf{x}-\mathbf{x}'|} = -\frac{4\pi k_C}{k_Lc}\mathbf{j}_n(\mathbf{x},t) = \frac{k_C}{k_Lc}\dot{\mathbf{p}}_n\Delta\frac{1}{r} - \frac{k_C}{k_L^2}\Delta\mathbf{m}_n\times\boldsymbol{\nabla}\frac{1}{r}.$$

The current $\mathbf{j}_n = \mathbf{j}_{np} + \mathbf{j}_{nm}$ consists of the polarization component $\mathbf{j}_{np}$ and the magnetic component $\mathbf{j}_{nm}$:

$$\mathbf{j}_{np} = \dot{\mathbf{p}}_n\,\delta^{(3)}(\mathbf{x}), \quad \mathbf{j}_{nm} = \frac{c}{4\pi k_L}\Delta\mathbf{m}_n\times\boldsymbol{\nabla}\frac{1}{r} = -\frac{c}{k_L}\mathbf{m}_n\times\boldsymbol{\nabla}\delta^{(3)}(\mathbf{x}). \quad (5.1.7)$$

Summed over all particles n follows

$$\mathbf{j}_{\mathrm{b}} = \mathbf{j}_p + \mathbf{j}_m = \sum_n\left(\dot{\mathbf{p}}_n(t) + \frac{c}{k_L}\boldsymbol{\nabla}\times\mathbf{m}_n(t)\right)\delta^{(3)}(\mathbf{x}-\mathbf{x}_n). \qquad (5.1.8)$$

Note 1: In the derivation, we neglected the retardation (see $\mathbf{A}$ in Coulomb gauge (8.2.52)). The quasi-static potentials (5.1.5) are only valid for small distances (*near field*) and suffice

$$\boldsymbol{\nabla}\cdot\mathbf{A}_n = \frac{k_C}{k_L^2}\left\{\frac{k_L}{c}\dot{\mathbf{p}}_n\cdot\boldsymbol{\nabla}\frac{1}{|\mathbf{x}-\mathbf{x}_n|} - \boldsymbol{\nabla}\cdot(\mathbf{m}_n\times\boldsymbol{\nabla}\frac{1}{|\mathbf{x}-\mathbf{x}_n|})\right\} \neq 0,$$

because on the right side only the 2nd term disappears.

Note 2: We considered the nuclei as static $\mathbf{x}_n(t) = \mathbf{x}_n(0)$, because the lighter electrons react faster to field changes; under this assumption we calculated both $\dot{\rho}_{np}$ and $\mathbf{A}_n$, so that the continuity equation

$$\dot{\rho}_n(\mathbf{x},t) = -\boldsymbol{\nabla}\cdot\mathbf{j}_n(\mathbf{x},t) = -\boldsymbol{\nabla}\cdot\mathbf{j}_{np}(\mathbf{x},t)$$

is consistent with the calculation; that also the nucleus contributes to the current density, is physically clear. The circular currents forming the magnetic moment $\mathbf{j}_m$ contribute nothing, because

$$\boldsymbol{\nabla}\cdot\mathbf{j}_{nm} = -(c/k_L)\boldsymbol{\nabla}\cdot\left(\mathbf{m}_n\times\boldsymbol{\nabla}\delta^{(3)}(\mathbf{x}-\mathbf{x}_n)\right) = 0.$$

The magnetic moments of the atoms do not have to result from circular currents, but can also be caused by the intrinsic torques of the electrons, the spin $\mathbf{S}$.

5.2 The Averaging of Microscopic Quantities

We initially assumed point-like charge carriers and in section 1.1 introduced continuous charge and current densities $\bar{\rho}$ and $\bar{\mathbf{j}}$ by averaging ρ and $\mathbf{j}$ over a small volume ΔV.

Temporal and spatial derivatives interchange with the averaging process, as we have shown in the derivation of the continuity equation for averaged charge distributions (1.1.15). Therefore, it makes no difference in the physical laws, whether our densities are either continuous or discrete, which is why we have dropped the distinction between averaged and continuous densities.

Here come the bound charges (currents) present in a medium, whose average density is represented as divergence (curl) of a vector field, which is ultimately attributed to the electric (magnetic) field and thus leads to a change in the Maxwell equations.

5.2.1 The Average Charge Distribution

We have averaged the charge density in section 1.1 with a normalized distribution

$$\int_V d^3x'\, f(\mathbf{x}-\mathbf{x}') = 1$$

realized with a range $\Delta V^{1/3}$, as sketched in Fig. 1.2. We will apply this averaging here to ρ_{b}, where for the average distance a between two atoms the inequalities

$$a \ll \Delta V^{1/3} \ll V^{1/3}$$

must be maintained. For the 1st term of ρ_{b} (5.1.3) due to charge neutrality

$$\overline{\sum_n q_n\, \delta^{(3)}(\mathbf{x}-\mathbf{x}_n)} = \int d^3x'\, f(\mathbf{x}-\mathbf{x}')\sum_n q_n \delta^{(3)}(\mathbf{x}'-\mathbf{x}_n) = \frac{1}{\Delta V}\sum_{\mathbf{x}_i \in \Delta V} q_i = 0.$$

The dipole contribution (5.1.4) is integrated by parts and exchanged $\boldsymbol{\nabla}' \leftrightharpoons -\boldsymbol{\nabla}$

$$\bar{\rho}_{\mathrm{p}}(\mathbf{x},t) = -\int d^3x'\, f(\mathbf{x}-\mathbf{x}')\boldsymbol{\nabla}'\cdot\sum_n \mathbf{p}_n\, \delta^{(3)}(\mathbf{x}'-\mathbf{x}_n)$$

$$= -\boldsymbol{\nabla}\cdot\sum_n \int d^3x'\, \mathbf{p}_n f(\mathbf{x}-\mathbf{x}')\, \delta^{(3)}(\mathbf{x}'-\mathbf{x}_n) = -\boldsymbol{\nabla}\cdot\mathbf{P}(\mathbf{x},t).$$

The dipole moment per unit volume is referred to as polarization $\mathbf{P}$:

$$\mathbf{P}(\mathbf{x},t) = \overline{\sum_n \mathbf{p}_n\delta(\mathbf{x}-\mathbf{x}_n)} = \sum_n \mathbf{p}_n \int d^3x'\, f(\mathbf{x}-\mathbf{x}')\delta^{(3)}(\mathbf{x}'-\mathbf{x}_n)$$

$$= \frac{1}{\Delta V}\sum_{\mathbf{x}_n \in \Delta V(\mathbf{x})} \mathbf{p}_n\,.$$

$$(5.2.1)$$

The averaged density is thus given by

$$\bar{\rho}(\mathbf{x},t) = \rho_f(\mathbf{x},t) + \rho_P(\mathbf{x},t) \qquad \text{with} \qquad \rho_P(\mathbf{x},t) = -\boldsymbol{\nabla}\cdot\mathbf{P}(\mathbf{x},t). \qquad (5.2.2)$$

For quantities such as polarization, which is defined as the sum of atomic (molecular) moments according to (5.2.1), the averaging can always be interchanged with the derivative:

$$\overline{\boldsymbol{\nabla}\cdot\sum_n \mathbf{p}_n(t)} = \boldsymbol{\nabla}\cdot\overline{\sum_n \mathbf{p}_n} = \boldsymbol{\nabla}\cdot\mathbf{P}.$$

This also applies to the time derivative, as was already shown in section 1.1.

5.2.2 The Average Current Density

According to (1.1.9), the average current density is defined by

$$\bar{\mathbf{j}}(\mathbf{x},t) \equiv \int \mathrm{d}^3x'\, f(\mathbf{x}-\mathbf{x}')\,\mathbf{j}(\mathbf{x}',t),$$

where in $\mathbf{j}$ the microscopic current density (5.1.8) is to be inserted

$$\mathbf{j}(\mathbf{x},t) = \mathbf{j}_f(\mathbf{x},t) + \sum_n \left[\dot{\mathbf{p}}_n(t) + \frac{c}{k_L}\boldsymbol{\nabla}\times\mathbf{m}_n(t)\right]\delta^{(3)}(\mathbf{x}-\mathbf{x}_n).$$

The averaging of the microscopic current yields

$$\bar{\mathbf{j}}(\mathbf{x},t) = \mathbf{j}_f(\mathbf{x},t) + \dot{\mathbf{P}}(\mathbf{x},t) + \frac{c}{k_L}\boldsymbol{\nabla}\times\mathbf{M}(\mathbf{x},t)\,, \tag{5.2.3}$$

where we have used (1.1.14), according to which the averaging with the partial derivative is interchanged may be:

$$\mathbf{j}_P = \overline{\sum_n \dot{\mathbf{p}}_n\delta^{(3)}(\mathbf{x}-\mathbf{x}_n)} = \dot{\mathbf{P}},$$

$$\mathbf{j}_M = \frac{c}{k_L}\overline{\sum_n \boldsymbol{\nabla}\times\mathbf{m}_n\delta^{(3)}(\mathbf{x}-\mathbf{x}_n)} = \frac{c}{k_L}\boldsymbol{\nabla}\times\overline{\sum_n \mathbf{m}_n\delta^{(3)}(\mathbf{x}-\mathbf{x}_n)} = \frac{c}{k_L}\boldsymbol{\nabla}\times\mathbf{M}.$$

Here $\mathbf{P}$ is the polarization (5.2.1) and $\mathbf{M}$ is the magnetization

$$\mathbf{M} = \overline{\sum_n \mathbf{m}_n\delta^{(3)}(\mathbf{x}-\mathbf{x}_n)} = \frac{1}{\Delta V}\sum_{\mathbf{x}_n\in\Delta V(\mathbf{x})}\mathbf{m}_n \tag{5.2.4}$$
$$= \text{magnetic dipole moment/volume unit.}$$

Since $\boldsymbol{\nabla}\cdot\mathbf{j}_M = 0$, the continuity equation is

$$\dot{\rho}_P(\mathbf{x},t) = -\boldsymbol{\nabla}\cdot\big(\mathbf{j}_P(\mathbf{x},t)+\mathbf{j}_M(\mathbf{x},t)\big) = -\boldsymbol{\nabla}\cdot\mathbf{j}_P(\mathbf{x},t)\,. \tag{5.2.5}$$

We can identify $-\rho_P = \boldsymbol{\nabla}\cdot\mathbf{P}$ with the divergence of $\mathbf{P}$ according to (5.2.2) and $\mathbf{j}_P$ with $\dot{\mathbf{P}}$ according to (5.2.3). Thus, the continuity equation is fulfilled. Referring back to (5.1.6), the vector potential

$$\mathbf{A} = \frac{k_C}{k_L^2}\left\{\frac{k_L}{cr}\dot{\mathbf{P}} - \mathbf{M}\times\boldsymbol{\nabla}\left(\frac{1}{r}\right)\right\}$$

does not satisfy the gauge condition $\boldsymbol{\nabla}\cdot\mathbf{A} = 0$.

5.2.3 Averaging of the Fields

The microscopic electric field $\mathbf{e}$ is composed according to the superposition principle from the sum of the fields of the individual charges, just as the magnetic field $\mathbf{b}$ can be represented as the sum of the contributions of the individual magnetic moments. Now, since the derivatives commute with the averaging at the charges and currents, this must also apply to the fields:

$$\boldsymbol{\nabla}\cdot\mathbf{E} = \boldsymbol{\nabla}\cdot\bar{\mathbf{e}} = \overline{\boldsymbol{\nabla}\cdot\mathbf{e}}.$$

The corresponding equations also apply to all other fields and their derivatives.

5.2.4 The Macroscopic Maxwell Equations

(a) Gauss's law $\boldsymbol{\nabla}\cdot\mathbf{e}=4\pi k_C(\rho_f+\rho_P)$.
The averaging results in

$$\boldsymbol{\nabla}\cdot\mathbf{E}=4\pi k_C\big(\rho_f+\rho_P\big), \qquad\qquad \rho_P=-\boldsymbol{\nabla}\cdot\mathbf{P}. \qquad (5.2.6)$$

A new macroscopic auxiliary field $\mathbf{D}$ is introduced, which takes into account the polarizability of the medium, the so-called *electric displacement*:

$$\mathbf{D}=\epsilon_0\mathbf{E}+4\pi k_r\mathbf{P}. \qquad (5.2.7)$$

Gauss's law now reads $(k_r=k_C\epsilon_0)$

$$\boldsymbol{\nabla}\cdot\mathbf{D}=4\pi k_r\rho_f.$$

(b) Faraday's law of induction $\boldsymbol{\nabla}\times\mathbf{e}+\dfrac{k_L}{c}\,\dot{\mathbf{b}}=0$.
The averaging does not change the form of the homogeneous Maxwell equations:

$$\boldsymbol{\nabla}\times\mathbf{E}+\frac{k_L}{c}\,\dot{\mathbf{B}}=0\,.$$

(c) Ampère-Maxwell equation $k_L\boldsymbol{\nabla}\times\mathbf{b}-\dfrac{1}{c}\dot{\mathbf{e}}=\dfrac{4\pi k_C}{c}\big(\mathbf{j}_f+\mathbf{j}_p+\mathbf{j}_m\big)$.
The averaging results in

$$k_L\boldsymbol{\nabla}\times\mathbf{B}-\frac{1}{c}\dot{\mathbf{E}}=\frac{4\pi k_C}{c}\big(\mathbf{j}_f+\mathbf{j}_P+\mathbf{j}_M\big).$$

Now $\mathbf{j}_P=\dot{\mathbf{P}}$. This contribution comes to $\dot{\mathbf{E}}$, which becomes $\epsilon_0\dot{\mathbf{E}}+4\pi k_r\dot{\mathbf{P}}=\dot{\mathbf{D}}$. The contribution of the magnetic dipoles to the current is

$$\mathbf{j}_M=\frac{c}{k_L}\boldsymbol{\nabla}\times\mathbf{M}.$$

Here, a new auxiliary field $\mathbf{H}$ is introduced by

$$\mathbf{B}=\mu_0(\mathbf{H}+4\pi k_r\mathbf{M}). \qquad (5.2.8)$$

This results in the Ampère-Maxwell equation in the form

$$\boldsymbol{\nabla}\times\mathbf{H}=\frac{1}{\mu_0\epsilon_0 k_L^2}\frac{k_L}{c}\big(4\pi k_r\mathbf{j}_f+\dot{\mathbf{D}}\big), \qquad\qquad \mu_0=\frac{1}{k_L^2\epsilon_0}. \qquad (5.2.8')$$

We have defined the still undetermined constant μ_0 in a way that makes sense for the systems in question here. In the appendix, (C.2.7) refers to this based on the field energy.

(d) Divergence-free magnetic field $\boldsymbol{\nabla}\cdot\mathbf{b}=0 \qquad\Longrightarrow\qquad \boldsymbol{\nabla}\cdot\mathbf{B}=0$.

Material equations

The Maxwell equations only form a closed system, when the material equations (5.2.7) for $\mathbf{D}$ and (5.2.8) for $\mathbf{H}$ are included. They are therefore also referred to as linking equations or less commonly as constitutive equations.

For the polarization $\mathbf{P}$, it applies that – with the exception of ferroelectric dielectrics – in the absence of a field $\mathbf{E}$ either no dipoles are present at all, or if there are spontaneous dipole moments, they are disordered. In any case, $\mathbf{P}=0$ if $\mathbf{E}=0$. For not too strong fields, $\mathbf{P}$ is proportional to $\mathbf{E}$:

$$\mathbf{P} = \epsilon_0 \chi_e \mathbf{E}. \tag{5.2.9}$$

χ_e is the electric susceptibility and

$$\epsilon_r = 1 + 4\pi k_r \chi_e \tag{5.2.10}$$

is the relative permittivity (dielectric constant). ϵ_0 is the permittivity of free space (electric constant) and the permittivity $\epsilon_0 \epsilon_r$ defines the (electric) displacement field:

$$\mathbf{D} = \epsilon_0 \epsilon_r \mathbf{E} = \epsilon_0 \mathbf{E} + 4\pi k_r \mathbf{P}. \tag{5.2.11}$$

Notes:

1. The linear relationship $\mathbf{D} = \epsilon_r \epsilon_0 \mathbf{E}$ applies except for Ferroelectrics (barium titanate, Seignette salt) up to very high fields. The usual fields produced in the laboratory $\lesssim 20\,\mathrm{kV\,cm^{-1}}$ are small compared to the interatomic fields $\sim 10^6\,\mathrm{kV\,cm^{-1}}$.

2. The scalar relationship applies to gases, liquids, isotropic and cubic solids. In materials with lower symmetry, ϵ_r is a tensor of 2nd order.

3. The (static) electric susceptibility $\chi_e \geq 0$. Thus is $\epsilon_r \geq 1$.

4. The susceptibilities determine the linear response of a system to an external disturbance; χ_e for example is defined as a tensorial quantity by

$$\chi_{ij}^e = \frac{\partial P_i}{\epsilon_0 \partial E_j}.$$

   When the anisotropy of the dielectric becomes noticeable, it expresses itself, as previously noted, in the dielectric constant.

As in electrostatics, we need a material equation that links $\mathbf{B}$ and $\mathbf{H}$. In analogy to electrostatics, the definition of the magnetic susceptibility χ_B through the linear relationship

$$\mathbf{M} = \frac{1}{\mu_0} \chi_B \mathbf{B} \tag{5.2.12}$$

follows according to (5.2.8)

$$\mathbf{H} = \frac{1}{\mu_0}\mathbf{B} - 4\pi k_r \mathbf{M} = \frac{1}{\mu_0}\left(1 - 4\pi k_r \chi_B\right)\mathbf{B} = \frac{1}{\mu_r \mu_0}\mathbf{B}.$$

However, $\mathbf{H}$ is usually not expressed through $\mathbf{B}$, but vice versa. So it is written

$$\mathbf{B} = \mu_r \mu_0 \mathbf{H} = \mu_0\left(\mathbf{H} + 4\pi k_r \mathbf{M}\right) = \mu_0(1 + 4\pi k_r \chi_m)\mathbf{H}. \tag{5.2.13}$$

μ_r is the relative (magnetic) permeability and χ_m, defined by

$$\mathbf{M} = \chi_m \mathbf{H}, \qquad\qquad \chi_m = \frac{\chi_B}{1 - 4\pi k_r \chi_B}, \tag{5.2.14}$$

is the magnetic susceptibility in the usual definition, where the right equation establishes the connection with the previously defined susceptibility χ_B. The connection to permeability is given by

$$\mu_r = \frac{1}{1 - 4\pi k_r \chi_B} = 1 + 4\pi k_r \chi_m. \tag{5.2.15}$$

In the usual formulation, magnetostatic problems with $\mathbf{j}_f = 0$ are directly comparable to electrostatic problems. One difference is that in magnetostatics for materials, which are called diamagnetic, $\mu_r < 1$ is. For $\mu_r > 1$ the material is paramagnetic.

The magnetic susceptibility χ_m is – analogous to χ_e – defined by

$$\chi_{ij}^m = \frac{\partial M_i}{\partial H_j},$$

i.e., that we also have to deal with a tensor in the most general case. This applies to some crystals of non-cubic structure. Thus, a few diamagnetic (graphite), paramagnetic (olivine) and especially ferromagnetic materials are anisotropic. With ferromagnets, it is added that $\mathbf{M}$ is not a linear function of $\mathbf{H}$. One must therefore additionally specify how χ_m or μ_r is defined.

If in matter $\epsilon_r(\mathbf{x})$ and/or $\mu_r(\mathbf{x})$ are not functions of $\mathbf{E}$ or $\mathbf{H}$, then the matter is called *linear* (*linear medium*).

Maxwell's equations in matter

It is irrelevant for the macroscopic Maxwell equations whether the free charges ρ_f (currents $\mathbf{j}_f$) represent a continuous (averaged) charge density (current density) or are discrete:

$$\text{(a)} \qquad \boldsymbol{\nabla}\cdot\mathbf{D} = 4\pi k_r \rho_f, \qquad\qquad \text{(b)} \quad \boldsymbol{\nabla}\times\mathbf{E} + \frac{k_L}{c}\dot{\mathbf{B}} = 0,$$
$$\text{(c)} \quad \boldsymbol{\nabla}\times\mathbf{H} - \frac{k_L}{c}\dot{\mathbf{D}} = 4\pi k_r \frac{k_L}{c}\mathbf{j}_f, \qquad \text{(d)} \qquad\qquad \boldsymbol{\nabla}\cdot\mathbf{B} = 0. \tag{5.2.16}$$

In the SI system $k_L = c$, $k_r = \frac{1}{4\pi}$ and in the Gaussian system $k_L = k_r = 1$

SI:
$$\text{(a)} \qquad \boldsymbol{\nabla}\cdot\mathbf{D} = \rho_f, \qquad\qquad \text{(b)} \;\; \boldsymbol{\nabla}\times\mathbf{E} + \dot{\mathbf{B}} = 0,$$
$$\text{(c)} \;\; \boldsymbol{\nabla}\times\mathbf{H} - \dot{\mathbf{D}} = \mathbf{j}_f, \qquad\qquad \text{(d)} \qquad\qquad \boldsymbol{\nabla}\cdot\mathbf{B} = 0. \tag{5.2.16'}$$

G:
$$\text{(a)} \qquad \boldsymbol{\nabla}\cdot\mathbf{D} = 4\pi\rho_f, \qquad\qquad \text{(b)} \;\; \boldsymbol{\nabla}\times\mathbf{E} + \frac{1}{c}\dot{\mathbf{B}} = 0,$$
$$\text{(c)} \;\; \boldsymbol{\nabla}\times\mathbf{H} - \frac{1}{c}\dot{\mathbf{D}} = \frac{4\pi}{c}\mathbf{j}_f, \qquad\qquad \text{(d)} \qquad\qquad \boldsymbol{\nabla}\cdot\mathbf{B} = 0, \tag{5.2.16''}$$

The averaging applies to bound charges (currents) in matter, where they establish the connection to the (auxiliary) fields in matter in the form of material equations

$$\mathbf{B} = \mu_r\mu_0\mathbf{H} = (1 + 4\pi k_r \chi_m)\mu_0\mathbf{H} = \mu_0(\mathbf{H} + 4\pi k_r\mathbf{M}),$$
$$\mathbf{D} = \epsilon_r\epsilon_0\mathbf{E} = (1 + 4\pi k_r\chi_e)\epsilon_0\mathbf{E} = \epsilon_0\mathbf{E} + 4\pi k_r\mathbf{P}. \tag{5.2.17}$$

SI:
$$\mathbf{B} = \mu_r\mu_0\mathbf{H} = (1+\chi_m)\mu_0\mathbf{H} = \mu_0(\mathbf{H}+\mathbf{M}),$$
$$\mathbf{D} = \epsilon_r\epsilon_0\mathbf{E} = (1+\chi_e)\epsilon_0\mathbf{E} = \epsilon_0\mathbf{E}+\mathbf{P}. \tag{5.2.17'}$$

G:
$$\mathbf{B} = \mu_r\mathbf{H} = (1+4\pi\chi_m)\mathbf{H} = \mathbf{H}+4\pi\mathbf{M},$$
$$\mathbf{D} = \epsilon_r\mathbf{E} = (1+4\pi\chi_e)\mathbf{E} = \mathbf{E}+4\pi\mathbf{P}, \tag{5.2.17''}$$

$\mathbf{E}$ electric field	$\mathbf{H}$ magnetic field in matter
$\mathbf{D}$ electric displacement field	$\mathbf{B}$ magnetic field/magnetic flux density
$\mathbf{P}$ polarization	$\mathbf{M}$ magnetization
ϵ_r (relative) permittivity	μ_r (relative) permeability
χ_e electric susceptibility	χ_m magnetic susceptibility ($\mathbf{H}$)

Although in some respects χ_B (5.2.12) would be the equivalent to χ_e, in the context of magnetism χ_m is used.

5.2.5 Boundary Conditions at the Interfaces of two Media

The integral form of the Maxwell equations is obtained from (1.3.20), by replacing the fields $\mathbf{E}$ and $\mathbf{B}$ according to (5.2.16) in the inhomogeneous equations with $\mathbf{D}$ and $\mathbf{H}$ and taking into account that only free charges and free currents contribute:

$$\text{(a)} \;\; \oiint_{\partial V} \mathrm{df}\cdot\mathbf{D} = 4\pi k_r \int_V \mathrm{d}^3x\, \rho_f, \qquad\qquad \text{(b)} \;\; \oint_{\partial A} \mathrm{dx}\cdot\mathbf{E} = -\frac{k_L}{c}\iint_A \mathrm{da}\cdot\dot{\mathbf{B}},$$
$$\text{(c)} \;\; \oint_{\partial A} \mathrm{dx}\cdot\mathbf{H} = \frac{k_L}{c}\iint_A \mathrm{da}\cdot(4\pi k_r\mathbf{j}_f + \dot{\mathbf{D}}), \qquad \text{(d)} \;\; \oiint_{\partial V} \mathrm{da}\cdot\mathbf{B} = 0. \tag{5.2.18}$$

From the integral form (5.2.18) one can derive the continuity conditions for the fields when transitioning from one medium to another if one considers infinitesimal volumes (cylinders for the normal component) or surfaces (rectangles for tangential components) at the interfaces.

From the Maxwell equations (a) and (d) the conditions for the normal components follow of $\mathbf{D}$ and $\mathbf{B}$ and with the material equations those of $\mathbf{E}$ and $\mathbf{H}$. The Maxwell equations (b) and (c) establish the transition conditions for the tangential components of $\mathbf{E}$ and $\mathbf{H}$. Again, the conditions for the missing fields, this time $\mathbf{D}$ and $\mathbf{B}$ are obtained from the material equations.

Divergence-free magnetic field

The starting point is the Maxwell equation (5.2.18d), where the volume ΔV is a cylinder of infinitesimal height $\mathrm{d}h$ with the base area Δa (see Fig. 5.2). The base area is located in medium 1, the top area in medium 2. The situation is thus completely analogous to the previously considered boundary conditions of the normal component E_n between vacuum and conductor or a surface charge:

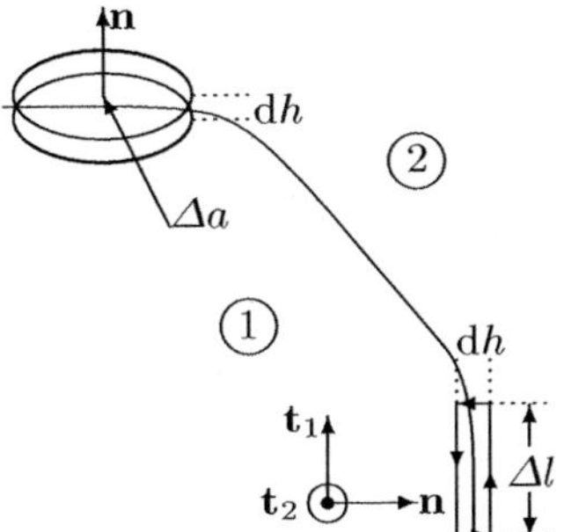

Fig. 5.2. Boundary conditions between media 1 and 2. Top left a cylinder with the volume $\Delta V = \Delta a\,\mathrm{d}h$. Bottom right a rectangle with the area $\Delta A = \Delta l\,\mathrm{d}h$. The tangential components $\mathbf{t}_1$, $\mathbf{t}_2$ and the normal vector $\mathbf{n}$ form a right-handed CS (coordinate system)

$$\oiint_{\partial \Delta V} \mathrm{d}a \cdot \mathbf{B} = (\mathbf{B}_2 - \mathbf{B}_1) \cdot \mathbf{n}\Delta a.$$

The contributions of the lateral surface disappear in each medium separately. So only the normal components remain, which, since there are no magnetic monopoles, are continuous:

$$B_{n1} = B_{n2}. \tag{5.2.19}$$

In magnetic media, the normal component of the magnetic field $\mathbf{H}$ has a discontinuity, which is caused by the magnetization:

$$\mu_2\,H_{n1} = \mu_1\,H_{n2}. \tag{5.2.20}$$

Gauss's law

The inhomogeneous Maxwell equation (5.2.18a) is Gauss's law, which applied to a cylindrical disk at the interface, as sketched in Fig. 5.2, becomes:

$$\oiint_{\partial \Delta V} \mathrm{da} \cdot \mathbf{D} = 4\pi k_r \int_{\Delta V} \mathrm{d}^3 x \, \rho_f = 4\pi k_r Q = 4\pi k_r \sigma \Delta a = (\mathbf{D}_2 - \mathbf{D}_1) \cdot \mathbf{n} \Delta a.$$

The contribution of the lateral surface disappears in both media separately, so that only the base and top surface remain. If there is a surface charge on the interface, then Q remains finite even as the height of the cylinder becomes smaller and the normal component of the electric displacement is discontinuous:

$$D_{n2} - D_{n1} = 4\pi k_r \sigma. \tag{5.2.21}$$

However, the normal component of the displacement field can also have a jump at the interface in the absence of free charges, which is caused by the polarization charges of the dielectric:

$$\epsilon_2 E_{n2} = \epsilon_1 E_{n1} + 4\pi k_c \sigma. \tag{5.2.22}$$

Faraday's law of induction

ΔA is a small rectangle with an infinitesimal side $\mathrm{d}h$ perpendicular to the interface between both media, as shown in Fig. 5.2. With $\mathrm{d}h \to 0$, also $\Delta A \to 0$. This implies in particular that also $\Delta A \mathbf{B} \cdot \mathbf{t}_2 \to 0$, since $\mathbf{B}$ (and the time derivative) are always finite and therefore their contribution with $\Delta A \to 0$ also vanishes:

$$\oint_{\partial \Delta A} \mathrm{ds} \cdot \mathbf{E} = -\frac{k_L}{c} \iint_{\Delta A} \mathrm{da} \cdot \dot{\mathbf{B}} = 0 = \mathbf{t}_1 \cdot (\mathbf{E}_2 - \mathbf{E}_1) \Delta l$$
$$= (\mathbf{t}_2 \times \mathbf{n}) \cdot (\mathbf{E}_2 - \mathbf{E}_1) \Delta l = (\mathbf{n} \times (\mathbf{E}_2 - \mathbf{E}_1)) \cdot \mathbf{t}_2 \, \Delta l.$$

It is taken into account that the contributions of the normal components $(i = 1, 2)$

$$\frac{\mathrm{d}h}{2} \left[\mathbf{E}_i(\mathbf{x}) - \mathbf{E}_i(\mathbf{x} + \Delta l \mathbf{t}_1) \right] \cdot \mathbf{n} = 0$$

cancel each other out. The above scalar product vanishes for every vector of the tangential plane

$$(\mathbf{n} \times (\mathbf{E}_2 - \mathbf{E}_1)) \cdot (\alpha \mathbf{t}_1 + \beta \mathbf{t}_2) = 0.$$

The continuity of the tangential components is thus represented by

$$\mathbf{n} \times (\mathbf{E}_2 - \mathbf{E}_1) = 0. \tag{5.2.23}$$

For the tangential components of the electric displacement, this results in

$$\epsilon_2 \mathbf{n} \times \mathbf{D}_1 = \epsilon_1 \mathbf{n} \times \mathbf{D}_2. \tag{5.2.24}$$

Ampère-Maxwell law

Again, a rectangle is considered that goes through the boundary surface of the two media, as shown in Fig. 5.2. The Ampère-Maxwell law for this is

$$\oint_{\partial \Delta A} d\mathbf{s} \cdot \mathbf{H} = \frac{k_L}{c} \iint_{\Delta A} d\mathbf{a} \cdot \left(4\pi k_r \mathbf{j}_f + \dot{\mathbf{D}} \right) = \frac{k_L}{c} 4\pi k_r \, \mathbf{K} \cdot \mathbf{t}_2 \, \Delta l .$$

with the surface current density [Jackson, 1998, sect. I.5]

$$\mathbf{K} = \int dh \left(\mathbf{j}_f + \frac{1}{4\pi k_r} \dot{\mathbf{D}} \right), \qquad \boxed{\text{SI: } [\mathbf{K}] = \text{A/m}}$$

which should be finite along the boundary surface: All components that are not parallel to the surface, disappear with $dh \to 0$, so that only the δ–part of the surface remains. So it is

$$\oint_{\partial \Delta A} d\mathbf{s} \cdot \mathbf{H} = \mathbf{t}_1 \cdot (\mathbf{H}_2 - \mathbf{H}_1) \Delta l = \frac{k_L}{c} 4\pi k_r \mathbf{K} \cdot \mathbf{t}_2 \, \Delta l .$$

The normal components of $\mathbf{H}$ do not contribute to the path integral. One substitutes again $\mathbf{t}_1 = \mathbf{t}_2 \times \mathbf{n}$ and cyclically exchanges:

$$\mathbf{n} \times (\mathbf{H}_2 - \mathbf{H}_1) = \frac{k_L}{c} 4\pi k_r \, \mathbf{K} = \mathbf{n} \times \mu_0 (\mu_2 \mathbf{B}_2 - \mu_1 \mathbf{B}_1),$$

$$\boxed{\text{SI: } \mathbf{n} \times (\mathbf{H}_2 - \mathbf{H}_1) = \mathbf{K} = \mathbf{n} \times \mu_0 (\mu_2 \mathbf{B}_2 - \mu_1 \mathbf{B}_1).}$$

$$(5.2.25)$$

In the presence of a surface current at the boundary of two media the tangential components of $\mathbf{H}$ are discontinuous, which of course also changes the jump of $\mathbf{B}$.

Summary of the boundary conditions

1. If the fields $(\mathbf{D}, \mathbf{B})$ at the boundary surfaces are source-free, i.e. if $\nabla \cdot \mathbf{D} = 0$ or $\nabla \cdot \mathbf{B} = 0$, then the associated normal component is continuous. This is shown using an infinitesimal cylinder (pillbox).
   The jump $(\mathbf{D}_2 - \mathbf{D}_1) \cdot \mathbf{n} = 4\pi k_r \sigma$ occurs when free charges are present on the surface.
2. If the fields $(\mathbf{E}, \mathbf{H})$ are free of vortices at the boundary surfaces, i.e. if $\nabla \times \mathbf{E} = 0$ or $\nabla \times \mathbf{H} = 0$, then the corresponding tangential components are continuous. This is represented by an infinitesimal rectangle passing through the boundary surface.
   If there is a surface current $\mathbf{K}$ at the boundary surface, a discontinuity of $\mathbf{n} \times (\mathbf{H}_2 - \mathbf{H}_1) = (k_L/c) 4\pi k_r \mathbf{K}$ occurs.

5.3 Ohm's Law

5.3.1 Drude Model of Electrical Conduction

Free electrons obey the equation of motion

$$m_e \dot{\mathbf{v}} = -e_0 \mathbf{E},$$

as the influence of the magnetic field can be neglected at low speeds. It is a remarkable consequence of wave mechanics that the periodic lattice, sketched in Fig. 5.3, does not disturb the motion of the electrons. The only influence is that the mass m_e is not the free mass, but an effective mass.

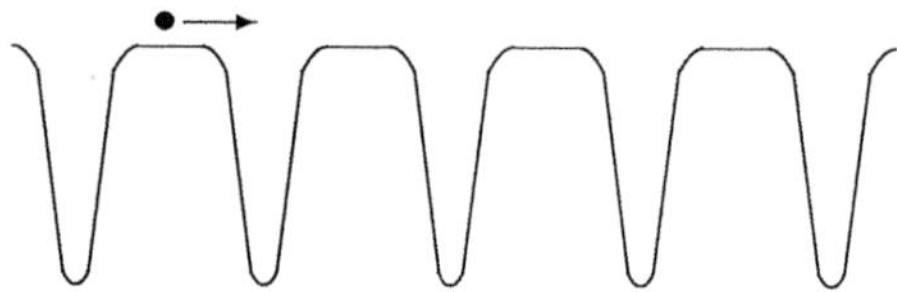

Fig. 5.3. Electron in one-dimensional periodic potential

In every lattice there are irregularities, and the collisions at such defects slow down the movement of the electrons. Defects play a role especially at low temperatures. However, as the temperature increases, there are interactions with the lattice vibrations, the phonons.

In the so-called *Drude model* for metals, one assumes free electrons that do not interact with each other, but only with the atoms, i.e. the phonons. Through a collision, the electron is on average slowed down to $\langle \mathbf{v}_0 \rangle = 0$ and after a time t, at which it collides again with the lattice, the speed $\mathbf{v}(t) = -e_0 t \mathbf{E}/m$. The times t are random, as sketched in Fig. 5.4, and therefore follow a Poisson distribution with the collision time τ. This gives the drift velocity

$$\mathbf{v}_D = \frac{1}{\tau} \int_0^\infty dt\, \frac{e\mathbf{E}t}{m_e} e^{-t/\tau} = -\frac{e_0 \mathbf{E}\tau}{m_e}. \tag{5.3.1}$$

The current density is given by (1.1.9)

$$\mathbf{j}_f = -e_0 n\, \mathbf{v}_D = (ne_0^2\tau/m_e)\,\mathbf{E} = \sigma\,\mathbf{E}, \tag{5.3.2}$$

where n indicates the density of the electrons. This linear relationship is the well-known Ohm's Law with the (direct current) conductivity

$$\sigma = ne_0^2\tau/m_e. \tag{5.3.3}$$

(5.3.2) may not have the familiar form associated with Ohm's law. If you take a homogeneous conductor (wire) of length l with the cross section A and apply a voltage $V = El$ to it, a current $I = jA$ flows, where the resistance is equal to $R = l/\sigma A$:

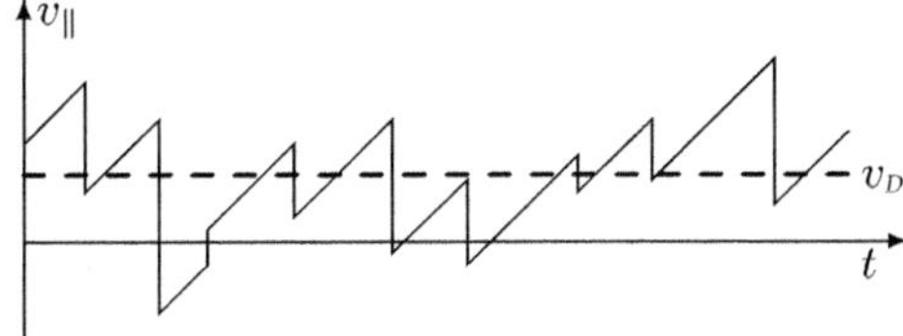

Fig. 5.4. Component of the velocity parallel to **E**; between the collisions the electrons are accelerated, resulting in a drift velocity v_D

$$(jA) = (El)(\sigma A/l) \qquad \Rightarrow \qquad I = V/R. \qquad (5.3.2')$$

Note: All these material relationships are valid as instantaneous relations only for slow temporal changes. However, Ohm's law can be formulated in a more general form for anisotropic, non-local and time-dependent interactions, where **j** also includes bound currents [Bruus, Flensberg, 2004, (6.15)]

$$j_i(\mathbf{x},t) = \int \mathrm{d}^3x'\,\mathrm{d}t'\,\sigma_{ij}(\mathbf{x},\mathbf{x}',t-t')\,E_j(\mathbf{x}',t'). \qquad (5.3.2'')$$

5.3.2 Joule's Heat

The energy that the particle gives off when it hits the lattice is attributable to *Joule's heat*. Therefore, the average kinetic energy that the electron gains in the electric field and through collisions with the medium is to be determined:

$$\mathrm{E}^e_{\mathrm{kin}}(\tau) = \frac{1}{\tau}\int_0^\infty \mathrm{d}t\,\frac{e_0^2 E^2 t^2}{2m_e}\mathrm{e}^{-t/\tau} = \frac{e_0^2\tau^2 E^2}{m_e}.$$

The heat produced per unit of time is for n electrons per unit volume (in the following, $\mathbf{j}_f$ will be replaced by $\mathbf{j}$ again)

$$\dot{\mathrm{E}}_{\mathrm{kin}} = \frac{n\mathrm{E}^e_{\mathrm{kin}}}{\tau} = \sigma E^2 = \mathbf{j}\cdot\mathbf{E}.$$

It is the increase in mechanical energy per unit volume $\dot{u}_{\mathrm{mech}}$. Thus, the total Joule's heat per unit of time is

$$\dot{U}_{\mathrm{mech}} = \int \mathrm{d}^3x\,\mathbf{j}(\mathbf{x})\cdot\mathbf{E}(\mathbf{x}). \qquad (5.3.4)$$

So far, we have assumed isotropic systems or systems with cubic symmetry. In anisotropic systems, σ is a tensor. In the presence of a magnetic field the symmetry relation $\sigma_{ik}(\mathbf{B}) = \sigma_{ki}(-\mathbf{B})$ applies. It follows from the symmetry properties of the *kinetic coefficients* (*Onsager reciprocity relationship*), which will not be discussed in more detail here [Ashcroft, Mermin, 1976, (13.90)]. We decompose σ into a symmetric and an antisymmetric part:

$$\sigma_{ik}(\mathbf{B}) = s_{ik}(\mathbf{B}) + a_{ik}(\mathbf{B}) \qquad \begin{cases} s_{ik}(\mathbf{B}) & = s_{ki}(\mathbf{B}) = s_{ki}(-\mathbf{B}) \\ a_{ik}(\mathbf{B}) & = -a_{ki}(\mathbf{B}) = a_{ki}(-\mathbf{B})\,. \end{cases} \qquad (5.3.5)$$

The Joule's heat generated per unit of time

$$\dot{u}_{\mathrm{mech}}(\mathbf{x}, t) = \mathbf{j} \cdot \mathbf{E} = s_{ik} E_i E_k \tag{5.3.6}$$

is determined solely by the symmetric part of the conductivity tensor, since $a_{ik} E_i E_k = 0$, as one can see immediately by swapping i with k. The s_{ik} depend on $\mathbf{B}$ only in the 2nd order, so that sufficiently small $\mathbf{B}$ do not contribute to the "mechanical" energy losses. This description has weaknesses. The mean free path length is more accessible to measurement. It is defined by $\tau = \bar{l}/v$.

The velocity $\mathbf{v}$ is composed of a velocity v_{th} and v_D. v_{th} is thought of as the thermal velocity when no field is present:

$$v_{th} = \sqrt{3 k_B T / m_e}\,.$$

In general, $v_{th} \gg v_D$ will be the case, so that v_D can be neglected. The conductivity is then $\sigma = n e_0^2 \bar{l} / m_e v_{th}$.

However, measurements show that $\bar{l}$ is larger than would be expected from the above formula. One reason is that mainly electrons near the Fermi energy contribute to the current and so v_{th} is to be replaced by $v_F = \sqrt{2 \mathrm{E}_F / m_e}$. For metals, we are in the range of $\bar{l} \sim 10^2 - 10^3$ Å.

Alternative descriptions of energy loss

1. One can compare the movement of charge carriers in a conductor with that in a viscous medium with large frictional resistance, which is more applicable to ion conductors than to metals. According to this, the charge carriers have a velocity proportional to the force $e\mathbf{E}$

$$\mathbf{v} = \mu_e\, e\mathbf{E}\,,$$

   where μ_e is the mobility. The current is obtained by multiplication with $\rho = en$:

$$\mathbf{j} = \sigma \mathbf{E} \qquad \text{with } \sigma = n e^2 \mu_e\,.$$

   If one replaces the mobility with the collision time $\tau = \mu_e m_e$, one again arrives at the direct current conductivity (5.3.3).

2. Due to the collisions, an electron, when viewed over a longer period of time, follow the field only with a delay. We can bring the equation of motion for the electron into the form:

$$m_e \mathbf{v}(t{+}\mathrm{d}t) = m_e \mathbf{v}(t) + e\mathbf{E}\mathrm{d}t + O(\mathrm{d}t^2)\,.$$

   The influence of the collisions is taken into account by the fact that in the interval $[t, t{+}\mathrm{d}t]$ the fraction $\mathrm{d}t/\tau$ experiences a collision. The equation of motion must therefore be modified, with terms of order $O(\mathrm{d}t^2)$ being neglected:

$$m_e \mathbf{v}(t{+}\mathrm{d}t) = \left(1 - \mathrm{d}t/\tau\right)\left(m_e \mathbf{v}(t) + e\mathbf{E}\mathrm{d}t\right)$$
$$= m_e \mathbf{v}(t) - \left[(m_e/\tau)\mathbf{v}(t) - e\mathbf{E}\right]\mathrm{d}t\,.$$

   This is a well-known equation of motion from mechanics

$$m_e \dot{\mathbf{v}} = -e_0 \mathbf{E} - \frac{m_e}{\tau}\,\mathbf{v} \tag{5.3.7}$$

for particles with a damping caused by friction $1/\mu_e$. We also arrive at Ohm's law here by multiplying (5.3.7) with $\mathbf{v}$ ($\mathbf{E}(\mathbf{x}) = \mathbf{E}$):

$$\frac{\mathrm{d}}{\mathrm{d}t}\left(\frac{m_e \dot{\mathbf{x}}^2}{2} + e_0 \mathbf{E}\cdot\mathbf{x}\right) = -\frac{m_e \dot{\mathbf{x}}^2}{\tau}\,.$$

The right side indicates the dissipation per unit of time. In the stationary case, i.e., when averaging over a longer time, $\dot{\mathbf{v}} = 0$ and thus

$$n\,e_0 \mathbf{E}\cdot\dot{\mathbf{x}} = -\mathbf{j}\cdot\mathbf{E} = -n\,\frac{m_e\dot{\mathbf{x}}^2}{\tau}\,, \quad \text{from which it follows} \quad \mathbf{j} = \frac{n e_0^2 \tau}{m_e}\,\mathbf{E} = \sigma \mathbf{E}\,.$$

3. Another, quite general approach is obtained by the temporal change of the kinetic energy of the particles. Starting from n particles, one can specify the change in the total kinetic energy with

$$\dot{\mathrm{E}}_{\mathrm{kin}} = \frac{\mathrm{d}}{\mathrm{d}t}\sum_n \frac{m_e \dot{\mathbf{x}}_n^2}{2} = \sum_n m_e \dot{\mathbf{x}}_n\cdot\ddot{\mathbf{x}}_n = \sum_n \dot{\mathbf{x}}_n\cdot\left[e\mathbf{E}(\mathbf{x}_n) + \frac{e}{c}\dot{\mathbf{x}}_n\times\mathbf{B}\right]$$

$$= \sum_n \dot{\mathbf{x}}_n\cdot e\mathbf{E}(\mathbf{x}_n) = \int \mathrm{d}^3 x \sum_n e\delta^{(3)}(\mathbf{x}-\mathbf{x}_n)\dot{\mathbf{x}}\cdot\mathbf{E}(\mathbf{x}_n) = \int \mathrm{d}^3 x\,\mathbf{j}(\mathbf{x})\cdot\mathbf{E}(\mathbf{x})\,.$$

Resistance of a wire

In Fig. 5.5 a wire of length L with the cross section $A(\mathbf{s})$, at whose ends the points A and B are located, is sketched. Starting from Ohm's law $\mathbf{j}(\mathbf{x}) = \sigma\,\mathbf{E}(\mathbf{x})$ we determine the potential difference between the endpoints A and B:

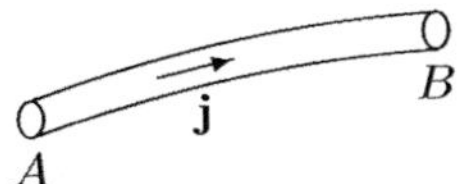

Fig. 5.5. Sketch with wire, ends A and B, cross section $A(s)$ and length L; $\phi_A > \phi_B$

$$\phi(\mathbf{x}_A) - \phi(\mathbf{x}_B) = \int_A^B \mathrm{d}\mathbf{s}\cdot\mathbf{E} = \int \mathrm{d}\mathbf{s}\cdot\frac{\mathbf{j}A(s)}{\sigma A(s)} = IR\,.$$

$I = j\,A(s)$ is the current flowing in the wire and R is the (total) ohmic resistance of the wire:

$$R = \int_A^B \mathrm{d}s\,\frac{1}{\sigma A(s)}\,. \tag{5.3.8}$$

Special case: If the conductivity σ and the cross-section A are constant, then $R = L/\sigma A$, where L is the length of the wire.

5.3.3 Hall Effect

If a current $\mathbf{j}$ flows in a conductor and the conductor is in a magnetic field $\mathbf{B}$, then the Coulomb and Lorentz forces (1.2.5) act on the electrons:

$$\mathbf{f}_C + \mathbf{f}_L = m_e \frac{d\mathbf{v}}{dt} = -e_0\left(\mathbf{E} + \frac{k_L}{c}\,\mathbf{v}\times\mathbf{B}\right).$$

We expect that the electrons will be deflected laterally towards the edge. At the same time, a voltage transverse to the wire builds up, as sketched in Fig. 5.6, until the restoring field $\mathbf{E}_H$ compensates the Lorentz force. In this stationary state, the electrons move again along the conductor in z-direction

$$\mathbf{E}_H = -\frac{k_L}{c}\,\mathbf{v}\times\mathbf{B} = -\frac{k_L}{nec}\,\mathbf{j}\times\mathbf{B}.$$

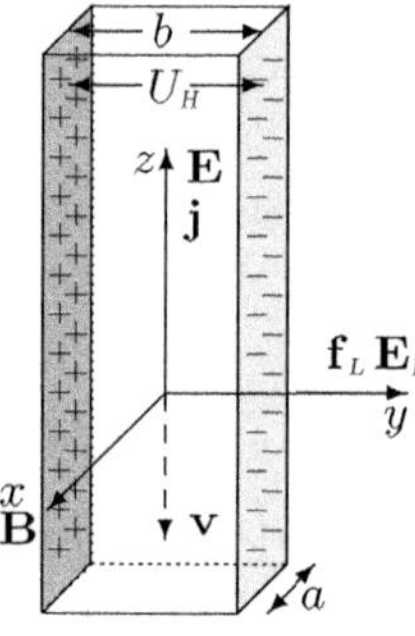

Fig. 5.6. Hall effect in a metallic conductor; the Lorentz force $\mathbf{f}_L$ drives the electrons to the edge, where the voltage U_H is built up, whose field $\mathbf{E}_H$ counteracts the force f_L

The Hall coefficient (for free charge carriers) is [Kittel, 2005, 6-(55)]

$$R_H = k_L/nec. \tag{5.3.9}$$

For electrons with $e = -e_0$ is $R_H < 0$. U_H, whose sign provides information about the charge carriers, can be easily demonstrated. For one type of charge carriers (electrons), the Hall voltage ($ab\mathbf{j} = -I\mathbf{e}_z$) results

$$U_H = R_H \int_0^b \mathbf{ds}\cdot(\mathbf{j}\times\mathbf{B}) = -R_H IB/a.$$

The effect was demonstrated in 1879 by Hall[1] at a time when neither the existence of electrons was confirmed, nor the electrical conduction understood was.

In deriving the Hall coefficient (5.3.9), we assumed isotropic systems (or systems with cubic symmetry). In systems with lower symmetry, we replaced σ

---

[1] Edwin Hall, American physicist 1855–1938.

with a tensor whose inverse R $R_{ik}(\mathbf{B}) = R_{ki}(-\mathbf{B})$ also fulfills the symmetry (5.3.5). Developed according to $\mathbf{B}^2$, one obtains

$$E_i = \left(R_{ik}^{(0)} + R_{ik;l}^{(1)} B_l + \frac{1}{2} R_{ik;lm}^{(2)} B_l B_m + \dots \right) j_k \, .$$

Here are

$$R_{ik}^{(0)} = R_{ki}^{(0)} \, , \qquad\qquad R_{ik;l}^{(1)} = -R_{ki;l}^{(1)} \, , \qquad\qquad R_{ik;lm}^{(2)} = R_{ki;lm}^{(2)} \, .$$

The Hall coefficients $R_{ik;l}^{(1)} = \epsilon_{ikj} R_{jl}^{H}$ can be represented by the nine components of a 2nd order tensor:

$$E_i = R_{ik}^{(0)} j_k + \epsilon_{ikj} R_{jl}^{H} B_l \, j_k \, .$$

In cubic (isotropic) systems, $R_{ik}^{(0)} = R_0$ and $R_{jl}^{H} = R_H \delta_{jl}$, from which follows

$$\mathbf{E} = R_0 \mathbf{j} + R_H \mathbf{j} \times \mathbf{B} \, .$$

Thomson Effect: If $\mathbf{B}$ is parallel to $\mathbf{j}$, the (transverse) Hall effect disappears and there is only a longitudinal resistance change in 2nd order in $\mathbf{B}$.

Hall Angle: If the transverse current (j_x) does not disappear, then the *Hall Angle* is defined by

$$\tan \vartheta_H = \left| \frac{E_H}{E} \right| = \frac{k_L e_0 B \tau}{m_e c} = \omega_c \tau \, .$$

$\mathbf{E}$, as sketched in Fig. 5.6, is parallel to the current. The total field is $\mathbf{E}_{\text{tot}} = \mathbf{E} + \mathbf{E}_H$.

$$\omega_c = k_L e_0 B / m_e c \qquad\qquad (5.3.10)$$

is the *cyclotron frequency* of the electron.

Magnetoresistance: Resistance changes $R(B)$ can in many cases be explained by the phenomenological *Kohler's rule*

$$(R(B) - R(0))/R(0) = f(B/R(0))$$

can be described. f is a function characteristic for the sample and $R(0) = R_0$ is the specific resistance for $\mathbf{B} = 0$. If the transverse current j_x disappears, i.e., if $\mathbf{E}_H$ compensates the Lorentz force, then the resistance R (in the Drude model) is independent of $\mathbf{B}$.

5.4 The Electron in the Electromagnetic Field

Somewhat out of the scope of this chapter, i.e. electrodynamics in matter, are the equations of motion for the electron in the electromagnetic field in vacuum. However, after the following two subsections, we are back with the London equations in matter.

---

[2] Of course, the linearity of $\mathbf{E}$ and $\mathbf{j}$ is not strictly valid.

5.4.1 Lagrange and Hamilton Function of the Electron

Principle of least action or Hamilton's principle

The action ([J s] or [erg s]) has in classical mechanics the form

$$S = \int_{t_1}^{t_2} dt\, L(q, \dot{q}, t)\,, \qquad (5.4.1)$$

where q and $\dot{q}$ are the generalized coordinates (location: $x_i, r, \vartheta, \varphi, \ldots$ and velocity: $\dot{x}_i, \dot{r}, \dot{\vartheta}, \dot{\varphi}, \ldots$) are. The Euler-Lagrange equations of motion are obtained by minimizing of S by variation of q and $\dot{q}$, while keeping the endpoints fixed, as sketched in Fig. 5.7:

$$\delta S = \int_{t_1}^{t_2} dt\, \left[L(q + \delta q, \dot{q} + \delta\dot{q}, t) - L(q, \dot{q}, t) \right] = \int_{t_1}^{t_2} dt\, \left[\frac{\partial L}{\partial q}\,\delta q + \frac{\partial L}{\partial \dot{q}}\,\delta\dot{q} \right] = 0\,.$$

In the second term we use the relation $\delta\dot{q} = d\delta q/dt$ and integrate by parts,

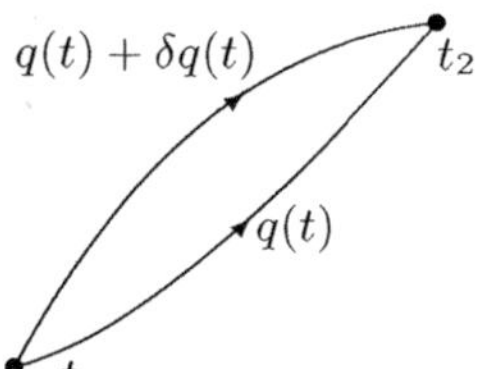

Fig. 5.7. Possible paths between the starting point $q(t_1)$ and the end point $q(t_2)$

where the boundary term due to $\delta q(t_1) = \delta q(t_2) = 0$ does not contribute:

$$\delta S = \frac{\partial L}{\partial \dot{q}}\,\delta q \Big|_{t_1}^{t_2} + \int_{t_1}^{t_2} dt\, \left[\frac{\partial L}{\partial q} - \frac{d}{dt}\left(\frac{\partial L}{\partial \dot{q}} \right) \right] \delta q = 0\,.$$

The integral vanishes for arbitrary δq only if

$$\frac{\partial L}{\partial q} - \frac{d}{dt}\left(\frac{\partial L}{\partial \dot{q}} \right) = 0\,. \qquad (5.4.2)$$

These are the *Euler-Lagrange equations*, which determine the path that the particle chooses between fixed points. The *generalized momentum* is defined by

$$p = \frac{\partial L}{\partial \dot{q}}\,. \qquad (5.4.3)$$

Now we consider paths that start again at time t_1 from location q_1 and go to a fixed location q_2, which they reach at different times t_2. If we replace $t_2 \to t$, we get the *Hamilton-Jacobi differential equation*

$$\frac{\mathrm{dS}}{\mathrm{d}t} = \mathrm{L} = \frac{\partial \mathrm{S}}{\partial q}\frac{\mathrm{d}q}{\mathrm{d}t} + \frac{\partial \mathrm{S}}{\partial t} \qquad \Rightarrow \qquad \frac{\partial \mathrm{S}}{\partial t} = \mathrm{L} - \mathrm{p}\,\dot{q} = -\mathrm{E}.$$

These paths satisfy the Euler-Lagrange equations, so that the integrand of δS vanishes, i.e. $\delta \mathrm{S} = \frac{\partial \mathrm{L}}{\partial \dot{q}}\delta q$ or $\frac{\partial S}{\partial q} = \mathrm{p}$, which we have inserted into the previous equation.

The *Lagrangian* for a point particle of mass m in potential $V(\mathbf{x})$ is [Schwabl, 2008, sect. 12.2.1], [Iro, 2015, (9.60)]

$$\mathrm{L}(\mathbf{x}, \mathbf{v}) = \frac{mv^2}{2} - V(\mathbf{x}) \tag{5.4.4}$$

with the Euler-Lagrange equations

$$\frac{\partial \mathrm{L}}{\partial \mathbf{x}} = \frac{\mathrm{d}}{\mathrm{d}t}\left(\frac{\partial \mathrm{L}}{\partial \mathbf{v}}\right) \qquad \Leftrightarrow \qquad \dot{\boldsymbol{p}} = m\ddot{\mathbf{x}} = -\boldsymbol{\nabla}V(\mathbf{x}). \tag{5.4.5}$$

Derivation of the Lagrangian from the Lorentz equation

By the *Lorentz equation* we understand the equation of motion for a charged particle (electron) under the influence of the Lorentz force (1.2.5). This is the Euler-Lagrange equation for a charged particle in the electromagnetic field

$$m\frac{\mathrm{d}\mathbf{v}}{\mathrm{d}t} = e\Big(\mathbf{E} + \frac{k_{\scriptscriptstyle L}}{c}\mathbf{v}\times\mathbf{B}\Big). \tag{5.4.6}$$

For this, the associated Lagrangian is to be determined. If $\mathbf{B}$ is represented by a vector field $\mathbf{A}$ (see (4.1.2))

$$\mathbf{B} = \boldsymbol{\nabla}\times\mathbf{A},$$

then $\boldsymbol{\nabla}\cdot\mathbf{B} = 0$ is automatically fulfilled. This follows from Faraday's law of induction (1.3.21c)

$$\boldsymbol{\nabla}\times\Big(\mathbf{E} + \frac{k_{\scriptscriptstyle L}}{c}\dot{\mathbf{A}}\Big) = 0\,.$$

If the curl of a vector vanishes, it can be derived from a scalar potential

$$\mathbf{E} + \frac{k_{\scriptscriptstyle L}}{c}\dot{\mathbf{A}} = -\boldsymbol{\nabla}\phi \qquad \Rightarrow \qquad \mathbf{E} = -\boldsymbol{\nabla}\phi - \frac{k_{\scriptscriptstyle L}}{c}\dot{\mathbf{A}}. \tag{5.4.7}$$

Substituted into (5.4.6) one obtains[3]

$$\frac{\mathrm{d}}{\mathrm{d}t}mv_i = -e\nabla_i\phi - \frac{k_{\scriptscriptstyle L}e}{c}\dot{A}_i + \frac{k_{\scriptscriptstyle L}e}{c}\big[\mathbf{v}\cdot(\nabla_i\mathbf{A}) - (\mathbf{v}\cdot\boldsymbol{\nabla})A_i\big].$$

Taking into account

---

[3] Auxiliary formula: $[\mathbf{a}\times(\boldsymbol{\nabla}\times\mathbf{b})]_i = \mathbf{a}\cdot\nabla_i\mathbf{b} - \mathbf{a}\cdot\boldsymbol{\nabla}\,b_i$

$$\frac{\mathrm{d}A_i}{\mathrm{d}t} = \dot{A}_i + \mathbf{v} \cdot \boldsymbol{\nabla} A_i \,,$$

(5.4.6) reads

$$\frac{\mathrm{d}}{\mathrm{d}t}\left(m\mathbf{v} + \frac{k_L e}{c}\mathbf{A}\right) = -e\boldsymbol{\nabla}\phi + \frac{k_L e}{c}\boldsymbol{\nabla}(\mathbf{v}\cdot\mathbf{A}). \tag{5.4.8}$$

It is easy to verify that (5.4.8) can be derived from the Lagrangian

$$\mathrm{L} = \frac{m\,v^2}{2} + \frac{k_L e}{c}\mathbf{v}\cdot\mathbf{A} - e\,\phi \tag{5.4.9}$$

can be derived. According to (5.4.3) is

$$\boldsymbol{p} = \frac{\partial \mathrm{L}}{\partial \mathbf{v}} = m\mathbf{v} + \frac{k_L e}{c}\mathbf{A} \tag{5.4.10}$$

the generalized momentum or *canonical momentum.*

$$m\mathbf{v} = \boldsymbol{p} - k_L e\mathbf{A}/c$$

is the *kinetic momentum.* The associated *Hamiltonian function* is defined by

$$\mathcal{H}(\boldsymbol{p},\mathbf{x}) = \boldsymbol{p}\cdot\mathbf{x} - \mathrm{L}(\mathbf{x},\mathbf{v}) = \frac{mv^2}{2} + e\phi.$$

One replaces $\mathbf{v}$ with $\boldsymbol{p}$ and obtains, when one again includes $V(\mathbf{x},t)$:

$$\mathcal{H}(\boldsymbol{p},\mathbf{x},t) = \frac{1}{2m}\left[\boldsymbol{p} - \frac{k_L e}{c}\mathbf{A}(\mathbf{x},t)\right]^2 + e\,\phi(\mathbf{x},t) + V(\mathbf{x},t)\,. \tag{5.4.11}$$

The consideration of the electromagnetic field in the Hamiltonian $\mathcal{H} = \boldsymbol{p}^2/2m + V(\mathbf{x})$ occurs by replacing $\boldsymbol{p} = m\mathbf{v} \to \boldsymbol{p} - k_L e\mathbf{A}/c$ and $\mathcal{H} \to \mathcal{H} + e\phi$, which is called *minimal substitution.*

5.4.2 Motion of a Particle in an External Field

Homogeneous magnetic field

The equation of motion for a charged particle is given by the Lorentz force (5.4.6) where $\mathbf{B}$ is the homogeneous external magnetic field. Without an electric field, one obtains

$$\frac{\mathrm{d}\mathbf{v}}{\mathrm{d}t} = \mathbf{v} \times \boldsymbol{\omega}_C \qquad \text{with} \qquad \boldsymbol{\omega}_C = \frac{k_L e}{mc}\mathbf{B}. \tag{5.4.12}$$

If one decomposes $\mathbf{v} = \mathbf{v}_\parallel + \mathbf{v}_\perp$ into its components parallel and perpendicular to $\mathbf{B}$, one obtains

$$\dot{\mathbf{v}}_\perp = \mathbf{v}_\perp \times \boldsymbol{\omega}_C \qquad \text{and} \qquad \dot{\mathbf{v}}_\parallel = 0\,. \tag{5.4.13}$$

One now fixes the coordinate system so that $\mathbf{B} = B\mathbf{e}_3$ and sets

$$\mathbf{v}_\perp(t) = \frac{v_\perp}{\sqrt{2}}(\mathbf{e}_1 - \mathrm{i}\mathbf{e}_2)\mathrm{e}^{-\mathrm{i}\omega_c t} \quad \text{with} \quad \omega_c = \frac{k_L e B}{mc} = |\omega_c|\,\mathrm{sgn}\,e, \qquad (5.4.14)$$

where $|\omega_c|$ is the cyclotron frequency (5.3.10) already encountered in the Hall effect. It is easy to convince oneself that (5.4.14) is a solution of (5.4.13):

$$\dot{\mathbf{v}}_\perp = \frac{v_\perp}{\sqrt{2}}(\mathbf{e}_1 - \mathrm{i}\mathbf{e}_2) \times \mathbf{e}_3 \omega_c \mathrm{e}^{-\mathrm{i}\omega_c t} = \frac{v_\perp}{\sqrt{2}}(-\mathbf{e}_2 - \mathrm{i}\mathbf{e}_1)\omega_c \mathrm{e}^{-\mathrm{i}\omega_c t} = -\mathrm{i}\omega_c \frac{\mathbf{v}_\perp}{\sqrt{2}}.$$

The particle moves according to

$$\mathbf{v}(t) = \mathbf{v}_{0\parallel} + \frac{v_\perp}{\sqrt{2}}(\mathbf{e}_1 - \mathrm{i}\mathbf{e}_2)\mathrm{e}^{-\mathrm{i}\omega_c t} \qquad (5.4.15)$$

with constant velocity $\mathbf{v}_\parallel$ spiraling along the magnetic field. The direction of rotation depends on the sign of the charge and is oriented clockwise for $e > 0$. This results in the following for the position vector

$$\mathbf{x}(t) = \mathbf{x}_0 + \mathbf{v}_{0\parallel}\, t - \mathrm{i}\,\mathrm{sgn}\,e\,\frac{v_\perp}{\omega_c\sqrt{2}}(\mathbf{e}_1 - \mathrm{i}\mathbf{e}_2)\,\mathrm{e}^{-\mathrm{i}\omega_c t}\,. \qquad (5.4.16)$$

The radius of the circular motion, the so-called *gyroradius* or *Larmor radius*, is

$$a = v_\perp/|\omega_c|\,. \qquad (5.4.17)$$

The energy of the particle $\mathrm{E} = mv^2/2$ remains constant, which we can expect when the force always acts perpendicular to the direction of motion:

$$\frac{\mathrm{dE}}{\mathrm{d}t} = m\mathbf{v}\cdot\dot{\mathbf{v}} = 0\,.$$

Notes: In (5.4.6) we have substituted for the momentum $p = m\mathbf{v}$, which only applies for velocities with $v \ll c$. The exact relation would be $p = \gamma m\mathbf{v}$ with $\gamma = 1/\sqrt{1-v^2/c^2}$ (see (14.2.14)). Then $\omega_c = k_L e B/(\gamma mc)$ becomes smaller and the radius corresponding to the screw motion larger.

In a more systematic solution method of (5.4.6) – see problem 5.1 – one defines

$$\dot{v}_i = \frac{k_L e B}{mc}\,\epsilon_{ijk} v_j \hat{B}_k = \omega_c\, R_{ij}\, v_j \qquad \text{with} \qquad R_{ij} = \epsilon_{ijk} B_k/B$$

the tensor R and has with

$$\mathbf{v}(t) = \mathrm{e}^{\omega_c \mathrm{R}t}\,\mathbf{v}_0$$

a solution that fulfills the initial condition, where $\omega_c = |\omega_c|\,\mathrm{sgn}\,e$.

Crossed magnetic and electric field

In addition to the homogeneous, static field $\mathbf{B} = B\mathbf{e}_z$, there is also a homogeneous and static field $\mathbf{E} = E_\perp \mathbf{e}_y + E_\| \mathbf{e}_z$. In component notation, (5.4.6) is

$$\begin{aligned} \dot{v}_x &= \omega_c v_y \\ \dot{v}_y &= \frac{e}{m}E_\perp - \omega_c v_x \\ \dot{v}_z &= \frac{e}{m}E_\| \, . \end{aligned} \qquad \Longrightarrow \qquad \dot{v}_x + i\dot{v}_y = \frac{ie}{m}E_\perp - i\omega_c(v_x + iv_y), \qquad (5.4.18)$$

$$(5.4.19)$$

The homogeneous solution, for which $E_\perp - 0$, we have already derived:

$$v_x + iv_y = v_{0\perp}\mathrm{e}^{-i\omega t + \varphi_0} \, .$$

For positive charges, this is the clockwise circular motion derived in (5.4.15). A special solution of the inhomogeneous equation is

$$v_x + iv_y = \frac{eE_\perp}{m\omega_c} = \frac{cE_\perp}{k_L B},$$

from which the solution follows:

$$\begin{aligned} v_x + iv_y &= v_{0\perp}\mathrm{e}^{-i\omega t + \varphi_0} + \frac{cE_\perp}{B} \\ v_z &= v_{0\|} + \frac{eE_\| t}{m}. \end{aligned} \qquad (5.4.20)$$

With $\mathbf{E}_\|$, the helical motion around $\mathbf{B}$ is maintained, only the pitch changes.

If one considers many particles, for which $v_{0\perp}$ vanishes on average at the beginning, one sees that in the plane perpendicular to $\mathbf{B}$, a mean drift velocity $\langle v_x + iv_y \rangle = cE_\perp/B$ is added in the x-direction, which is also independent of the sign of the charge. This drift is perpendicular to $\mathbf{E}_\perp$ and $\mathbf{B}$ and can be represented as

$$\langle \mathbf{v}_\perp \rangle = \frac{c}{B}\frac{\mathbf{E}_\perp}{B} \times \hat{\mathbf{B}}. \qquad (5.4.21)$$

We started from the non-relativistic approximation $\boldsymbol{p} = m\mathbf{v}$ and therefore have to limit the validity of (5.4.21) to $|\langle \mathbf{v}_\perp \rangle| \ll c$, from which $E_\perp \ll B$ follows. The further integration with the initial conditions $\mathbf{x}_0 = 0$ and $\varphi = 0$ results in

$$x = \frac{v_{0\perp}}{\omega_c}\sin(\omega_c t) + \frac{cE_\perp t}{B}, \qquad (5.4.22)$$

$$y = \frac{v_{0\perp}}{\omega_c}\big(\cos(\omega_c t) - 1\big). \qquad (5.4.23)$$

For a small $E_\perp$, as long as $\langle |\mathbf{v}_\perp| \rangle = cE_\perp/B < v_{0\perp}$, the circular motion, as sketched in Fig. 5.8a, becomes an elongated cycloid. For $E_\perp/B = v_{0\perp}B/c$, the path of the particle is an ordinary cycloid[4] (Fig. 5.8b) and for $E_\perp/B > v_{0\perp}/c$ a shortened cycloid (Fig. 5.8c).

---

[4] $x = v\tau - a\sin\tau$ and $y = v - a\cos\tau$ is for $a = v$ the ordinary, for $v < a$ the elongated and for $v > a$ the shortened cycloid.

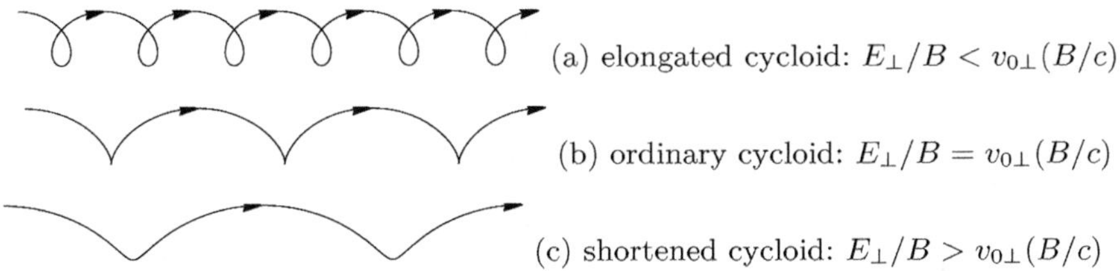

Fig. 5.8. Path of a charged particle in the xy-plane

Penning trap

In a region where $\Delta\phi = 0$, a particle cannot be brought into any stable equilibrium position, because according to Earnshaw's theorem (see p. 96) the potential ϕ has no minimum (maximum) there.

However, with an additional magnetic field $\mathbf{B}$, particles can be captured, as demonstrated by the Penning trap. Brown and Gabrielse [1986] mention in their review, among other things, precision measurements of the gyromagnetic factors g, which were carried out using such a trap [Van Dyck, Schwinberg, Dehmelt, 1984].

We now take a harmonic potential that confines positive ions (charge $q > 0$, mass M) in the z-direction to a small area and repels in the xy-plane:

$$\phi = \frac{V_0}{2d^2}\left(z^2 - \frac{\varrho^2}{2}\right), \tag{5.4.24}$$

where d is explained in Fig. 5.9. The equipotential surfaces of ϕ are circular

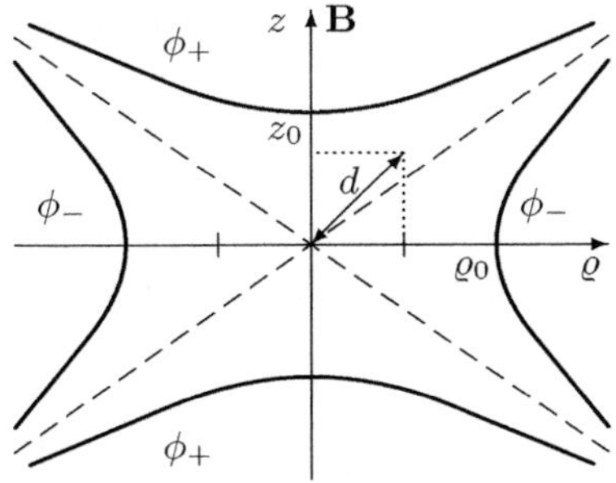

Fig. 5.9. The two electrodes with ϕ_+ are the circular hyperboloids $z^2 = z_0^2 + \varrho^2/2$, and the ring electrode with ϕ_- is the circular hyperboloid $z^2 = (\varrho^2 - \varrho_0^2)/2$.
Characteristic length of the trap:
$d = \sqrt{\varrho_0^2/4 + z_0^2/2}$

hyperboloids, which is also reflected in the form of the electrodes, as sketched in Fig. 5.9. They are equipotential surfaces of the potential (5.4.24), where $z_0^2 = \varrho_0^2/2$ is chosen so that $\phi_+ = -\phi_-$. The resulting fields are those of a quadrupole.

However, if we have a strong magnetic field $B\mathbf{e}_z$, the ions will perform circular motions around $\mathbf{B}$ and thus cannot escape. The equations of motion separate into the motion parallel and perpendicular to $\mathbf{B}$:

$$M\ddot{z} = qE_z, \qquad\qquad M\ddot{\boldsymbol{\varrho}} = q\mathbf{E}_\perp + \frac{k_L qB}{c}\,\dot{\boldsymbol{\varrho}} \times \mathbf{e}_z\,.$$

In the z-direction, this is a harmonic oscillation with $\omega_z = \sqrt{\dfrac{qV_0}{Md^2}}\,.$

If the field $\mathbf{e}_\perp = \frac{\omega_z^2}{2}\,\boldsymbol{\varrho}$ tends to zero, the ion performs circular motions with the cyclotron frequency $\omega_c = k_L qB/cM$. With a small field $E_\perp$, if this is constant, the drift (5.4.21) is added. Here, this is expressed in the two strongly different frequencies

$$\omega_\pm = \frac{\omega_c}{2}\left(\omega_c \pm \sqrt{\omega_c^2 - 2\omega_z^2}\right)$$

in the plane perpendicular to $\mathbf{B}$. For the particles to be captured, $\omega_c \geq \sqrt{2}\omega_z$ must be. The details of the calculation are part of the problem 5.3. With a sufficiently strong field we can assume that

$$\omega_- \ll \omega_z \ll \omega_+,$$

especially for electrons. The drift movement at constant field has become a slow circular motion with ω_-, superimposed by fast circles with $\omega_+ \sim \omega_c$.

5.4.3 London Equations

Conduction electrons show a mass inertia against field changes, which can be taken into account by (5.3.7):

$$m_e\dot{\mathbf{v}} - \frac{m_e}{\tau}\,\mathbf{v} = e\mathbf{E}\,.$$

τ has the meaning of a collision or relaxation time. If we now go from speed to current $\mathbf{j}$, we get

$$\tau\frac{\partial \mathbf{j}}{\partial t} + \mathbf{j} = \frac{ne^2\tau}{m_e}\,\mathbf{E} = \sigma_0\,\mathbf{E}\,. \tag{5.4.25}$$

σ_0 is the direct current conductivity. For harmonic oscillations of $\mathbf{j}$ and $\mathbf{E}(\mathbf{x}, t) = \mathbf{E}(\mathbf{x})\,e^{-i\omega t}$ follows

$$\mathbf{j} = \sigma\,\mathbf{E} \qquad\qquad \text{with} \qquad\qquad \sigma = \frac{\sigma_0}{1 - i\omega\tau}\,. \tag{5.4.26}$$

This is a modification of Ohm's law, where σ_0 is replaced by the *dynamic conductivity* σ. As long as $\omega \ll \tau^{-1}$, the resistance remains approximately the same (and real). The current $\mathbf{j}$, which for low frequencies is in phase with $\mathbf{E}$, becomes almost in phase with $\partial\mathbf{j}/\partial t$, which corresponds to an imaginary conductivity.

In superconductors, the direct current resistance disappears below a temperature T_c. We can phenomenologically describe this by $\tau \to \infty$, so that (5.4.25) takes the form

$$\frac{\partial \mathbf{j}}{\partial t} = \frac{n_s e^2}{m_e}\,\mathbf{E} \tag{5.4.27}$$

assumes. n_s is the particle density of the electrons involved in the superconductivity. (5.4.27) is referred to as the 1st *London equation*. We now form the curl and substitute for $\boldsymbol{\nabla}\times\mathbf{E} = -k_L\dot{\mathbf{B}}/c$ (Faraday's law of induction):

$$\frac{\partial}{\partial t}\Big(\boldsymbol{\nabla}\times\mathbf{j} + \frac{k_L n_s e^2}{m_e c}\,\mathbf{B}\Big) = 0\,. \tag{5.4.28}$$

If the time dependence of $\mathbf{j}$ is harmonic, (5.4.28) simplifies to

$$\boldsymbol{\nabla}\times\mathbf{j} = -\frac{k_L n_s e^2}{m_e c}\,\mathbf{B}\,. \tag{5.4.29}$$

(5.4.29) is referred to as the 2nd London equation. In a quasi-stationary approximation, the displacement current is neglected in the Ampère-Maxwell equation (5.2.16c). Assuming $\mu_r = 1$, we have

$$\boldsymbol{\nabla}\times\mathbf{B} = \frac{4\pi k_C}{k_L c}\,\mathbf{j} \stackrel{\text{SI}}{=} \mu_0\mathbf{j}. \tag{5.4.30}$$

Note: Using (5.4.30), a current $\mathbf{j}$ is determined for each $\mathbf{B}$. In the static case, (5.4.28) is fulfilled for all $\mathbf{B}$ and $\mathbf{j}$. This is not consistent with the experimental findings. F. and H. London found in 1935 with (5.4.30) the necessary restriction of the solutions.

Substituted into (5.4.29), using $\boldsymbol{\nabla}\times(\boldsymbol{\nabla}\times\mathbf{B}) = -\Delta\mathbf{B}$, we get

$$\Delta\mathbf{B} - \frac{4\pi k_C n_s e^2}{m_e c^2}\,\mathbf{B} = 0\,. \tag{5.4.31}$$

From (5.4.30) follows $\boldsymbol{\nabla}\cdot\mathbf{j} = 0$. If we form the curl of (5.4.29), we get

$$\Delta\mathbf{j} - \frac{4\pi k_C n_s e^2}{m_e c^2}\,\mathbf{j} = 0\,. \tag{5.4.32}$$

Meissner-Ochsenfeld effect

A sphere with $\mu_r = 1$ made of a material that is superconducting below T_c is located in a homogeneous, external field $\mathbf{B}_e$. Above T_c, the sphere shown in Fig. 5.10b is normally conducting and $\mathbf{B}$ crosses the sphere unhindered. As the temperature is lowered, the sphere becomes superconducting for $T < T_c$ and $\mathbf{B}$ is pushed away from the sphere, as sketched in Fig. 5.10a. This effect is named after its discoverers *Meissner-Ochsenfeld effect*[5]. Inside the superconductor is $\mathbf{B} = 0$, and due to $\boldsymbol{\nabla}\cdot\mathbf{B} = 0$ the normal component of $\mathbf{B}$ is continuous and thus also disappears at the surface of the superconductor.

---

[5] Walther Meissner, 1882–1974 and Robert Ochsenfeld, 1901–1993

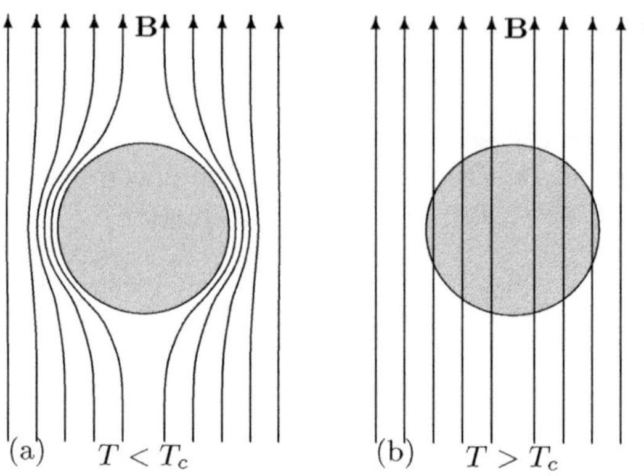

Fig. 5.10. (a) If a material is superconducting below T_c, then **B** does not penetrate into the superconducting area (b) Above T_c the field **B** is unchanged, since $\mu_r = 1$

The half-space with $z < 0$ is superconducting and is in an external field $\mathbf{B}_e = B_0 \mathbf{e}_x$. Small deviations from $\mu_r = 1$ are neglected, so that the tangential components of **B** at the interface are considered continuous. (5.4.31) then has the solution in the superconductor

$$\mathbf{B} = B_0\, e^{-|z|/\Lambda}\, \mathbf{e}_x \qquad \text{with} \qquad \Lambda = \sqrt{\frac{m_e c^2}{4\pi k_C n_s e^2}}. \tag{5.4.33}$$

Λ is the London penetration depth of the magnetic field into the superconductor:

$$\Lambda = r_s\sqrt{\frac{1}{3}\frac{n}{n_s}\frac{r_s}{r_e}} \qquad \text{with} \qquad r_s^3 = \frac{3}{4\pi n} \qquad \text{and} \qquad r_e = \frac{k_C e^2}{m_e c^2}. \tag{5.4.34}$$

r_s is the radius of the spherical volume occupied by a conduction electron which in most metals has values between $1\,\text{Å}$ and $2\,\text{Å}$. r_e is the classical electron radius ($2.8 \times 10^{-5}\,\text{Å}$, see Tab. C.7, p. 645). Thus, $\Lambda \gtrsim 100$ Å (see Tab. 5.1). We obtain the superconducting current from (5.4.30), Ampère's

Tab. 5.1. London penetration depth Λ

Substance		Al	Sn	Pb	Nb	Cd
Penetrationdepth[a]	Λ	160 Å	340 Å	370 Å	390 Å	1100 Å

[a] R. Meservey and B. B. Schwartz, in *Superconductivity*, edited by R. D. Parks Marcel Dekker Inc., New York (1969), p. 174

law ($z<0$):

$$\mathbf{j} = \frac{k_L c}{4\pi k_C}\,\boldsymbol{\nabla}\times\mathbf{B} = \frac{k_L c B_0}{4\pi k_C \Lambda}\, e^{-|z|/\Lambda}\,\mathbf{e}_y. \tag{5.4.35}$$

5.5 Dielectric Properties

5.5.1 Atomic Polarizability

Atoms and molecules are polarized under the influence of a local electric field
E. If we denote α as the polarizability of an atom, then its dipole moment is

$$p_i(\omega) = \alpha_{ij}(\omega)\, E_j(\omega)\,.$$

We will not address frequencies $\omega \neq 0$ or possible anisotropies of polarizability
here, so that

$$\mathbf{P} = n\,\alpha\,\mathbf{E} \qquad \text{with} \qquad n = N/V\,. \tag{5.5.1}$$

Note that α is system-dependent: $\alpha_{SI,H} = \alpha_G/k_C$. We distinguish between
electronic, ionic and orientation polarizability, which will be explained in the
following [Ashcroft, Mermin, 1976, p. 542].

Electronic polarizability

For atomic polarizability, we have a phenomenological model in mind, in which
the electrons are connected to the atomic nucleus by harmonic spring forces,
as indicated in Fig. 5.11. The electron shell is bound to the nucleus by the

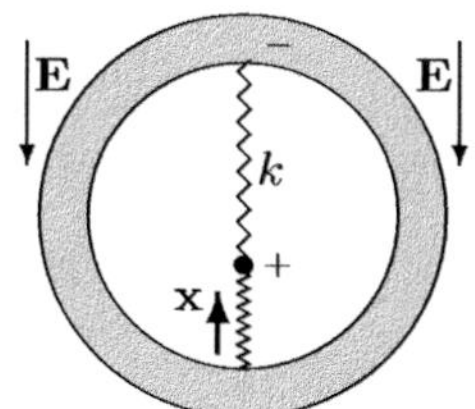

Fig. 5.11. The charge distribution of the electrons (electron shell) is connected to the atomic nucleus with an elastic spring k. Through a constant field **E**, the electron shell is displaced by **x**. The charge of the shell is $-Ze_0$ and its mass is Zm_e

spring force k, so that the harmonic restoring force is given by

$$\mathbf{F}_h = -m_e Z\,\omega_0^2\,\mathbf{x} \qquad \text{with} \quad \omega_0^2 = k/Zm_e\,.$$

The Coulomb force of the external field acts on the electron shell

$$\mathbf{F}_{\text{ext}} = -Ze_0\,\mathbf{E}$$

and is thus shifted by the electric field **E** against the nucleus. Thus, the
Coulomb force $\mathbf{F}_{\text{ext}}$ and the restoring harmonic spring $\mathbf{F}_h$ are parallel to **E**,
so that the problem can be treated in one dimension:

$$m\ddot{x} = -m\omega_0^2 x - Ze_0 E \quad \overset{\text{equilibrium}}{\Longrightarrow} \quad F_h + F_{\text{ext}} = -m\omega_0^2 x - Ze_0 E = 0.$$

The equilibrium position in the static limit case ($\omega \to 0$) is thus

$$x = -\frac{e_0}{m\omega_0^2}E, \qquad\qquad p = -Ze_0 x = \frac{Ze_0^2}{m\omega_0^2}E = \alpha E \qquad\qquad (5.5.2)$$

In order to make statements about the order of magnitude of α, assumptions about ω_0 must be made. In the H-atom, the electron is at a distance $a_B = \hbar^2/k_c m_e e_0^2 = 0.53\text{Å}$ (Bohr radius) in equilibrium, so that there the Coulomb force determines the strength of the spring:

$$\mathbf{F}_C = -\frac{\partial V}{\partial \mathbf{x}} = -k_c \frac{e_0^2 \mathbf{x}}{r^3} = -k\,\mathbf{x}\bigg|_{r=a_B}.$$

From this follows $k = m_e \omega_0^2 = k_c e_0^2/a_B^3$ and the estimate for the polarizability results in

$$\alpha = e_0^2/m_e\omega_0^2 = a_B^3/k_C.$$

Ionic polarizability

In substances with ionic bonding, the displacement of the ions in the electric field creates dipoles, whose density is the displacement polarization $\mathbf{P}$ of the induced dipoles. Here, two ions with the masses M_+ and M_- are coupled by a harmonic force $k_h = M\omega_0^2$. Their equations of motion are for the harmonic force k_h [Ashcroft, Mermin, 1976, (27.47)]:

$$M_+\ddot{\mathbf{u}}_+ = -k_h(\mathbf{u}_+ - \mathbf{u}_-) + e\mathbf{E},$$
$$M_-\ddot{\mathbf{u}}_- = -k_h(\mathbf{u}_- - \mathbf{u}_+) - e\mathbf{E}.$$

From difference and sum follows

$$M\,\ddot{\mathbf{u}} + M\,\omega_0^2\,\mathbf{u} = e\mathbf{E} \quad \text{with} \quad M = \frac{M_+ M_-}{M_+ + M_-} \quad \text{and} \quad \mathbf{u} = \mathbf{u}_+ - \mathbf{u}_-.$$

$\mathbf{u}$ is the relative displacement of the ions from the equilibrium positions, $\pm e$ their charge and $\mathbf{E}$ the local field. The frequency ω_0 is a typical mode in the crystal. At frequencies $\omega \gtrsim \omega_D$, the Debye frequency, the ions can no longer follow the field and the displacement polarization of the ions becomes small. In the limit case $\omega \to 0$ the equilibrium position is $\mathbf{u} = (e/M\omega_0^2)\mathbf{E}$. From this follows

$$\mathbf{p} = e\mathbf{u} = \alpha_{\text{ion}}\,\mathbf{E} \qquad\qquad \text{with} \qquad\qquad \alpha_{\text{ion}} = e^2/M\omega_0^2. \qquad (5.5.3)$$

The electronic polarizability $\alpha_{\text{el}}^{\pm}$ of the two ions is different, with negative ions usually being more polarizable. In the static limit case, the polarizabilities are added

$$\alpha = \alpha_{\mathrm{el}}^{+} + \alpha_{\mathrm{el}}^{-} + \alpha_{\mathrm{ion}} \, . \tag{5.5.4}$$

If n is the number of ions per unit volume, the susceptibility is obtained from the polarization, the dipole moment per unit volume (see (5.2.9))

$$\mathbf{P} = n\mathbf{p} = \chi\epsilon_0\mathbf{E}, \qquad\qquad \chi = n\alpha/\epsilon_0 .$$

The independent calculation of the individual components of the susceptibilities is only justified if it can be assumed that the movement of the (rigid) ions does not lead to any deformation in the shells.

Orientation polarizability

The molecule has a spontaneous dipole moment $\mathbf{p}_0$. At high temperatures, the $\mathbf{p}$ are unordered. The field $\mathbf{E}$ orients the moments. Given is a field $\mathbf{E}=(0 \ 0 \ z)$. Then at $T = 0$ all dipole moments are aligned parallel to the z axis. With increasing temperature, the average moment pointing in the z-direction becomes smaller according to the Boltzmann statistics:

$$\langle\mathbf{p}\rangle = \frac{\int \mathrm{d}\Omega\, \mathrm{e}^{-\beta H}\, \mathbf{p}}{\int \mathrm{d}\Omega\, \mathrm{e}^{-\beta H}} \approx \frac{1}{3}\frac{p_0^2}{k_B T}\mathbf{E} \quad \text{with} \quad \mathcal{H} = -\mathbf{p}\cdot\mathbf{E} \quad \text{and} \quad \beta = \frac{1}{k_B T}.$$

H is the Hamiltonian and k_B is the Boltzmann constant (see Tab. C.7, p. 645).

Calculation of the average dipole moment: $\mathbf{E} = E\,\mathbf{e}_z$

$$Z = \int \mathrm{d}\Omega\, \mathrm{e}^{\beta\mathbf{p}\cdot\mathbf{E}} = \int_0^{2\pi} \mathrm{d}\varphi \int_0^{\pi} \mathrm{d}\vartheta\, \sin\vartheta\, \mathrm{e}^{\beta E p_0 \cos\vartheta} = 2\pi \int_{-1}^{1} \mathrm{d}\xi\, \mathrm{e}^{\beta E p_0 \xi}$$

$$= \frac{4\pi}{\beta E p_0}\, \sinh(\beta E p_0)\, .$$

By differentiation one obtains ($\mathbf{p}_0 = p_0\,\mathbf{e}_z$):

$$\langle\mathbf{p}\rangle = \frac{1}{Z}\frac{\partial Z}{\beta\partial\mathbf{E}} \overset{u=\beta E p_0}{=} \frac{4\pi}{Z}p_0\,\mathbf{e}_z\,\frac{\partial}{\partial u}\frac{\sinh u}{u} = \mathbf{p}_0\,\frac{u}{\sinh u}\left(\frac{\cosh u}{u} - \frac{\sinh(u)}{u^2}\right)$$

$$= \mathbf{p}_0\left(\coth u - \frac{1}{u}\right) = \mathbf{p}_0\,L(u) \overset{u\to 0}{=} \mathbf{p}_0\frac{1}{u}\left(1 + \frac{u^2}{3} - ... - 1\right) = \mathbf{p}_0\,\frac{u}{3}\, .$$

$L(u)$ is the Langevin function. $u \to 0$ is the limit of high temperatures and/or small fields.

The different contributions to the polarizability, outlined in Fig. 5.12, can be determined separately, if an oscillating field is applied. In principle, the moments of lighter particles can follow the field at higher frequencies than the moments of heavier particles.

When ω becomes larger than the reorientation rate of the dipole $\mathbf{p}_0$, only α_{ion} and α_{el} remain until finally the ion movement also becomes too slow.

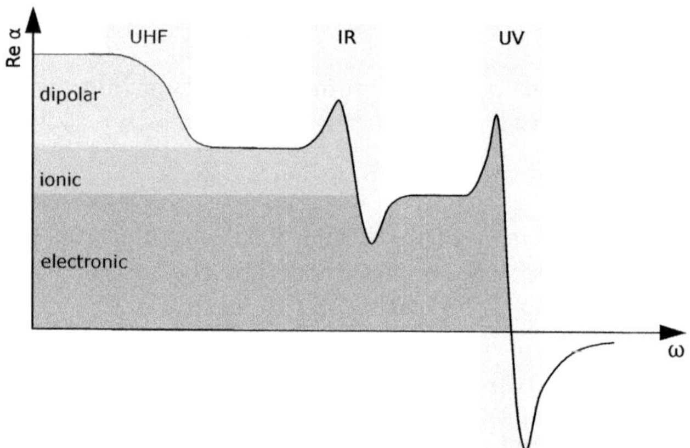

Fig. 5.12. Real part of the polarizability[a] separated into electronic, ionic and dipole contributions plotted against ω. The dipoles follow the field up to wavelengths λ in the millimeter range; (radio = UHF, microwave). For ions, the λ range goes up to infrared (IR) and for electrons up to ultraviolet (UV)

---

[a] adapted from Kittel [2005, Sect. 16, Fig. 8] with kind permission from © John Wiley & Sons 2005.

5.5.2 Dielectric Function

Drude-Lorentz model

The oscillator model used for static polarizability can also be used for dynamic polarizability with modifications. It is sometimes referred to as the *Lorentz model.*

In a time-varying local field $\mathbf{E} = \mathbf{E}_0\,e^{-i\omega t}$, the electron shell will instantaneously follow the field, $\mathbf{x} = \mathbf{x}_0\,e^{-i\omega t}$, but it should be considered that the electrons of the inner shells are deflected less strongly than the outer ones.

The individual electrons thus get different springs and the absorption is accounted for with a damping term. The spring k in Fig. 5.11 now connects the nucleus with a single electron:

$$m_e\big(\ddot{\mathbf{x}}_j + \gamma_j\dot{\mathbf{x}}_j + \omega_j^2\mathbf{x}_j\big) = -e_0\,\mathbf{E}_0 e^{-i\omega t} \tag{5.5.5}$$

and the local field $\mathbf{E}$ is periodic in time. We thus obtain the contribution of the j-th electron to polarizability

$$\mathbf{P}_j = -e_0\,\mathbf{x}_j = \frac{e_0^2}{m_e}\,\frac{1}{\omega_j^2 - \omega^2 - i\gamma_j\omega}\,\mathbf{E} = \alpha_j(\omega)\,\mathbf{E}\,.$$

An atom and/or molecule has Z electrons, which contribute to the polarizability with different oscillator strengths f contribute:

$$\alpha_{\mathrm{at}}(\omega) = \frac{e_0^2}{m_e}\sum_j \frac{f_j}{\omega_j^2 - \omega^2 - i\gamma_j\omega} \qquad \text{with} \qquad \sum_j f_j = Z\,. \tag{5.5.6}$$

The polarization, the dipole moment per unit volume, is given for n atoms per unit volume by

$$\mathbf{P} = n\,\alpha_{\mathrm{at}}\,\mathbf{E} = \chi_e\,\epsilon_0\,\mathbf{E}.$$

The relationship with the dielectric constant ϵ_r is known to be

$$\mathbf{D} = \epsilon_r\,\epsilon_0\,\mathbf{E} = \epsilon_0(\mathbf{E} + 4\pi k_C\mathbf{P}) = \epsilon_0(1 + 4\pi k_C n\alpha_{\mathrm{at}})\mathbf{E}\,.$$

This gives the dielectric function in the oscillator model, which is sometimes also referred to as the *Drude's formula* is referred to; The assignment of names is not uniform here. Well characterized is (5.5.7) as the dielectric function in the oscillator model:

$$\epsilon(\omega) = 1 + 4\pi k_C n \frac{e_0^2}{m_e} \sum_k \frac{f_k}{\omega_k^2 - \omega^2 - \mathrm{i}\gamma_k\omega}. \tag{5.5.7}$$

An analogous equation is also obtained in quantum mechanics, where $\hbar\omega_k$ are the energy differences of electron states and f_k are the corresponding matrix elements.

Approximations for small and large frequencies

If ω is far from the resonance frequencies ω_j, it makes sense to replace the ω_j and γ_j with averaged quantities. For small frequencies we obtain the static polarizability (5.5.2) and for frequencies that are far above the highest resonance ω_j, $\alpha_{at} \to 0$, where the negative Value implies that $\epsilon(\omega) \lesssim 1$:

$$\alpha_{\mathrm{at}}(0) = \frac{Ze_0^2}{m_e\omega_0^2}, \qquad \frac{1}{\omega_0^2} = \frac{1}{Z}\sum_j \frac{f_j}{\omega_j^2} \qquad \omega \ll \omega_j\;\forall j\,,$$

$$\alpha_{\mathrm{at}}(\omega) \approx -\frac{Ze_0^2}{m_e\omega^2}\left(1 - \mathrm{i}\frac{\gamma}{\omega}\right), \qquad \gamma = \frac{1}{Z}\sum_j f_j\gamma_j \qquad \omega \gg \omega_j\;\forall j. \tag{5.5.8}$$

Here, the electrons follow the electric field $\mathbf{E}$ as quasi-free particles. This applies to X-rays, whose energies $\sim 10\,\mathrm{keV}$ are much larger than the binding energies $\sim 10\,\mathrm{eV}$.

Conductivity

Metals have electrons that are not bound to atoms and can thus move in the medium. Let's assume that the i-th electron of the atom is not bound, so in the oscillator model (5.5.7) the frequency $\omega_i = 0$

$$\epsilon(\omega) = 1 + 4\pi k_C n \frac{e_0^2}{m_e}\left(\sum_{k\neq i} \frac{f_k}{\omega_k^2 - \omega^2 - \mathrm{i}\gamma_k\omega} - \frac{f_i}{\omega(\omega + \mathrm{i}\gamma_i)}\right)$$

$$= \epsilon_b(\omega) + \frac{4\pi k_C \mathrm{i}\sigma}{\omega}. \tag{5.5.9}$$

We have divided $\epsilon(\omega)$ into ϵ_b, in which the contributions of the electrons bound to the atom are summed up, and into the contribution of the free electron

$$\sigma = \frac{ne_0^2 f_i}{m_e(\gamma_i - i\omega)}.$$ (5.5.10)

We now insert (5.5.9) into the Ampère-Maxwell equation (5.2.16), where $\mathbf{E} = \mathbf{E}_0(\mathbf{x})e^{-i\omega t}$ and get

$$\boldsymbol{\nabla} \times \mathbf{H} + \frac{i\omega k_L}{c}\epsilon_b\epsilon_0 \mathbf{E} = \frac{4\pi k_C}{k_L c \mu_0}(\mathbf{j}_f + \sigma\mathbf{E}) \stackrel{\text{SI}}{=} \mathbf{j}_f + \sigma\mathbf{E},$$ (5.5.11)

where $\mathbf{j}_f$ takes into account any other existing charge carriers. If a field is applied to the medium, the free electrons generate the contribution determined by $\sigma\mathbf{E}$ to the current density, which is Ohm's law. We thus see that the transition from the dielectric to the conductor is smooth. We can identify the parameter $\gamma = \tau^{-1}$ according to (5.3.2) with the mean collision time. The static conductivity is (with the experimental value for Cu for $0°\mathrm{C}$)

$$\text{G:}\ \sigma(0) = \frac{ne_0^2}{m_e\gamma} \simeq 6 \times 10^{17}\,\mathrm{s}^{-1}, \qquad\qquad \gamma \simeq 4 \times 10^{13}\,\mathrm{s}^{-1}.$$

Kramers-Kronig dispersion relations

For dielectrics with only one resonance energy ω_0 applies

$$\epsilon(\omega) = 1 + \frac{\omega_p^2}{\omega_0^2 - \omega^2 - i\omega\gamma} = 1 + 4\pi k_C \chi_e, \qquad \omega_p^2 = \frac{4\pi k_C ne_0^2}{m_e},$$

where ω_p is the so-called plasma frequency. Fig. 5.13 shows the typical characteristics for susceptibilities $\chi_e' + i\chi_e''$. The imaginary part, which is responsible for the absorption (dissipation), is almost symmetric around the resonance frequency ω_0. The real part changes sign, which means that the oscillation for $\omega > \omega_0$ is out of phase. The damping $\gamma > 0$ is always positive. Both poles of

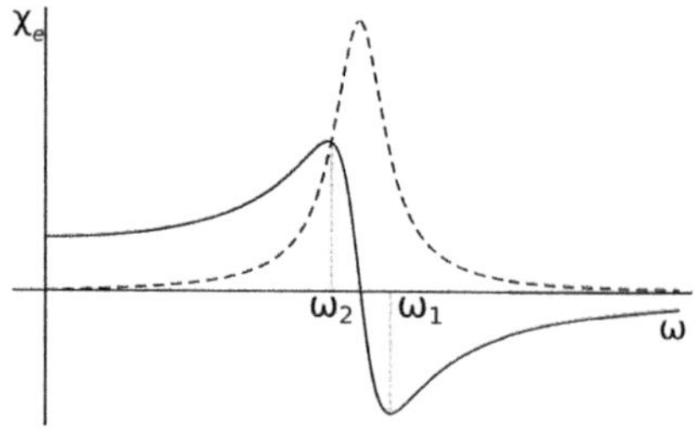

Fig. 5.13. Resonance ω_0 in the susceptibility χ_e or the polarizability. The imaginary part (dashed) determines the absorption, the real part the refraction

$\epsilon(\omega)$ are with

$$\omega = -\frac{i\gamma}{2} \pm \sqrt{\omega_0^2 - \gamma^2/2}$$

in the lower half-plane, i.e., ϵ is analytic in the upper half-plane and $\epsilon - 1$ decreases sufficiently fast for $|\omega| \to \infty$. Therefore, the Kramers-Kronig dispersion relations (B.1.13) apply, which connect the real part of the function χ_e that is analytic in the upper ω-half-plane with the imaginary part. Instead of χ, we give the dispersion relations (B.1.13) for $\epsilon = \epsilon' + i\epsilon''$:

$$\epsilon'(\omega) = 1 + \frac{1}{\pi} P \int_{-\infty}^{\infty} d\omega' \, \frac{\epsilon''(\omega')}{\omega' - \omega} = 1 + \frac{2}{\pi} P \int_0^{\infty} d\omega' \, \frac{\omega' \, \epsilon''(\omega')}{\omega'^2 - \omega^2}, \qquad (5.5.12)$$

$$\epsilon''(\omega) = \frac{-1}{\pi} P \int_{-\infty}^{\infty} d\omega' \, \frac{\epsilon'(\omega') - 1}{\omega' - \omega} = \frac{-2\omega}{\pi} P \int_0^{\infty} d\omega' \, \frac{\epsilon'(\omega') - 1}{\omega'^2 - \omega^2}. \qquad (5.5.13)$$

P indicates that the (Cauchy's) principal value of the integral is to be taken. Both the imaginary part and the real part contain all the information about the dielectric constant. Therefore, the measurement of absorption is sufficient to also get all the information about refraction. For the oscillator model, one obtains

$$\epsilon'(\omega) = 1 + \omega_p^2 \frac{\omega_0^2 - \omega^2}{(\omega_0^2 - \omega^2)^2 + \gamma^2 \omega^2}, \qquad (5.5.14)$$

$$\epsilon''(\omega) = \omega_p^2 \frac{\gamma\omega}{(\omega_0^2 - \omega^2)^2 + \gamma^2 \omega^2}. \qquad (5.5.15)$$

Therefore, $\epsilon'(-\omega) = \epsilon'(\omega)$ and $\epsilon''(-\omega) = -\epsilon''(\omega)$. These symmetry properties shown here for the Lorentz model apply generally.

The frequencies $\omega_{1,2} = -\gamma/2 + \sqrt{\omega_0^2 + \gamma^2/4}$ limit the range $\omega_1 \leq \omega \leq \omega_2$, in which the slope of the real part of χ_e is negative (and thus of the real part of the refractive index). This means that light of higher frequency is refracted less. This is then referred to as anomalous dispersion. Generally, the refractive index increases with frequency, which is referred to as normal dispersion.

5.6 Energy and Momentum Balance

5.6.1 Energy Balance

Lastly, we calculated the Joule's heat produced per unit of time. What is missing is a balance equation that relates the produced heat, which is the "mechanical" portion of the energy, to the field energy and the energy flow through the given volume V.

The starting point is (5.6.1), i.e. the Joule's heat $\dot{u}_{\text{mech}}$, produced per unit of time and volume, into which the Ampère-Maxwell equation (5.2.16) is inserted for $\mathbf{j}_f$ (SI: $4\pi k_r = 1$, $k_L = c$):

$$\dot{u}_{\text{mech}}(\mathbf{x}, t) = \mathbf{j}_f(\mathbf{x}, t) \cdot \mathbf{E}(\mathbf{x}, t) = \mathbf{E} \cdot \frac{1}{4\pi k_r} \left(\frac{c}{k_L} \boldsymbol{\nabla} \times \mathbf{H} - \dot{\mathbf{D}} \right). \qquad (5.6.1)$$

By means of $\nabla \cdot (\mathbf{E} \times \mathbf{H}) = \mathbf{H} \cdot (\nabla \times \mathbf{E}) - \mathbf{E} \cdot (\nabla \times \mathbf{H})$ and the law of induction $\nabla \times \mathbf{E} = -(k_L/c)\dot{\mathbf{B}}$ we obtain

$$\mathbf{E} \cdot (\nabla \times \mathbf{H}) = -\frac{k_L}{c} \mathbf{H} \cdot \dot{\mathbf{B}} - \nabla \cdot (\mathbf{E} \times \mathbf{H}). \tag{5.6.2}$$

Now we define the Poynting vector

$$\mathbf{S} = \frac{c}{k_L} \frac{1}{4\pi k_r} \mathbf{E} \times \mathbf{H} \quad \Rightarrow \quad \text{G: } \mathbf{S} = \frac{c}{4\pi} \mathbf{E} \times \mathbf{H}, \quad \text{SI: } \mathbf{S} = \mathbf{E} \times \mathbf{H} \tag{5.6.3}$$

and we obtain under the assumption that permittivity $\epsilon_r = \epsilon_r(\mathbf{x})$ and permeability $\mu_r = \mu_r(\mathbf{x})$ do not depend on time:

$$\dot{u}_{\text{mech}}(\mathbf{x}, t) + \nabla \cdot \mathbf{S} = -\frac{1}{4\pi k_r} \left[\mathbf{E} \cdot \dot{\mathbf{D}} + \mathbf{H} \cdot \dot{\mathbf{B}} \right]$$
$$= \frac{-1}{8\pi k_r} \frac{\partial}{\partial t} \left(\mathbf{H} \cdot \mathbf{B} + \mathbf{E} \cdot \mathbf{D} \right) = -\dot{u}(\mathbf{x}, t).$$

The energy density of the electromagnetic field is thus given by

$$u(\mathbf{x}, t) \equiv u_{\text{field}}(\mathbf{x}, t) = \frac{1}{8\pi k_r} (\mathbf{E} \cdot \mathbf{D} + \mathbf{H} \cdot \mathbf{B}) \tag{5.6.4}$$

and thus we get the balance equation for the energy, which has the form of a continuity equation for the total energy density u_g

$$\frac{\partial u_{\text{g}}(\mathbf{x}, t)}{\partial t} + \nabla \cdot \mathbf{S} = 0 \qquad \Leftrightarrow \qquad \dot{u}_{\text{field}} + \nabla \cdot \mathbf{S} = -\mathbf{j}_f \cdot \mathbf{E}, \tag{5.6.5}$$
$$u_{\text{g}}(\mathbf{x}, t) = u_{\text{mech}}(\mathbf{x}, t) + u_{\text{field}}(\mathbf{x}, t)$$

and represents the *Poynting's theorem*. In integral form, we obtain after applying Gauss's theorem

$$\frac{\mathrm{d}}{\mathrm{d}t} \left(U_{\text{mech}} + U_{\text{field}} \right) = \int_V \mathrm{d}^3 x \left(\mathbf{j}_f \cdot \mathbf{E} + \dot{u}_{\text{field}}(\mathbf{x}, t) \right) = -\oiint_{\partial V} \mathrm{d}\mathbf{a} \cdot \mathbf{S}. \tag{5.6.6}$$

The terms are, starting from the left

1. the work done on the free charges in V (= Joule's heat),
2. the change in field energy per unit of time in V and on the far right
3. the energy flow through the surface $\mathbf{S} = \dfrac{c}{k_L} \dfrac{1}{4\pi k_r} \mathbf{E} \times \mathbf{H}$.

The Poynting vector thus describes the energy flux through the surface. In (5.6.6) the field energy

$$U_{\text{field}} = \int \mathrm{d}^3 x \, u(\mathbf{x}, t) = \frac{1}{8\pi k_r} \int \mathrm{d}^3 x \, (\mathbf{E} \cdot \mathbf{D} + \mathbf{H} \cdot \mathbf{B}) \tag{5.6.7}$$

is used, where we usually use u instead of u_{field} for the energy density, see (5.6.7). The energy balance (5.6.6) also applies in this form to the microscopic fields.

5.6.2 Momentum Balance and Stress Tensor

The conservation of momentum for the macroscopic fields is obtained when starting from the Lorentz force for a particle:

$$\frac{\mathrm{d}\boldsymbol{p}}{\mathrm{d}t} = e\Big(\mathbf{E} + \frac{k_L}{c}\,\mathbf{v}\times\mathbf{B}\Big)\,.$$

If one transitions from a particle to continuous charge and current density, one obtains

$$\frac{\mathrm{d}\boldsymbol{P}_{\mathrm{mech}}}{\mathrm{d}t} = \int_V \mathrm{d}^3 x\,\dot{\boldsymbol{p}}_{\mathrm{mech}}(\mathbf{x},t) \quad\text{with}\quad \dot{\boldsymbol{p}}_{\mathrm{mech}} = \rho_f\mathbf{E} + \frac{k_L}{c}\,\mathbf{j}_f\times\mathbf{B}. \qquad (5.6.8)$$

Now according to (5.2.16)

$$\rho_f = \frac{1}{4\pi k_r}\,\boldsymbol{\nabla}\!\cdot\!\mathbf{D}, \qquad\qquad \mathbf{j}_f = \frac{c}{k_L}\frac{1}{4\pi k_r}\Big[\boldsymbol{\nabla}\times\mathbf{H} - \frac{k_L}{c}\dot{\mathbf{D}}\Big].$$

Thus, for (5.6.8) we obtain

$$\dot{\boldsymbol{p}}_{\mathrm{mech}} = \frac{1}{4\pi k_r}\Big[(\boldsymbol{\nabla}\!\cdot\!\mathbf{D})\mathbf{E} + (\boldsymbol{\nabla}\times\mathbf{H})\times\mathbf{B} - \frac{k_L}{c}\dot{\mathbf{D}}\times\mathbf{B}\Big]. \qquad (5.6.9)$$

The right side is transformed in the following auxiliary calculation so that physical quantities (field momentum, stress tensor) can be assigned to the individual terms.

First, we replace in the last term of (5.6.9)

$$\frac{k_L}{c}\dot{\mathbf{D}}\times\mathbf{B} = \frac{k_L}{c}\frac{\partial}{\partial t}(\mathbf{D}\times\mathbf{B}) - \frac{k_L}{c}\mathbf{D}\times\dot{\mathbf{B}} = \frac{k_L}{c}\frac{\partial}{\partial t}(\mathbf{D}\times\mathbf{B}) + \mathbf{D}\times(\boldsymbol{\nabla}\times\mathbf{E})\,,$$

where we have substituted for $\dot{\mathbf{B}} = -(c/k_L)\boldsymbol{\nabla}\times\mathbf{E}$ (induction equation). Thus we obtain

$$\begin{aligned}
\dot{\boldsymbol{p}}_{\mathrm{mech}} = \frac{1}{4\pi k_r}\Big\{&\big[(\boldsymbol{\nabla}\!\cdot\!\mathbf{D})\mathbf{E} - \mathbf{D}\times(\boldsymbol{\nabla}\times\mathbf{E})\big] + \big[(\boldsymbol{\nabla}\cdot\mathbf{B})\,\mathbf{H} - \mathbf{B}\times(\boldsymbol{\nabla}\times\mathbf{H})\big]\\
&- \frac{k_L}{c}\frac{\partial}{\partial t}(\mathbf{D}\times\mathbf{B})\Big\}.
\end{aligned}$$

We have also added the term $(\boldsymbol{\nabla}\!\cdot\!\mathbf{B})\mathbf{H}=0$ to establish symmetry between $\mathbf{E}\rightleftharpoons\mathbf{H}$ and $\mathbf{D}\rightleftharpoons\mathbf{B}$ and calculate with the auxiliary formula $\big[\mathbf{a}\times(\boldsymbol{\nabla}\times\mathbf{c})\big]_i = \mathbf{a}\!\cdot\!\nabla_i\mathbf{c} - \mathbf{a}\!\cdot\!\boldsymbol{\nabla}c_i$ the following expression, where we use the homogeneity $\epsilon_r=$const and $\mu_r=$const

$$\begin{aligned}
\big[(\boldsymbol{\nabla}\cdot\mathbf{D})\mathbf{E} - \mathbf{D}\times(\boldsymbol{\nabla}\times\mathbf{E})\big]_i &= E_i(\nabla_j D_j) - D_j(\nabla_i E_j) + D_j(\nabla_j E_i)\\
&\overset{\epsilon_r=\text{const}}{=} \nabla_j(E_i D_j) - \frac{1}{2}\nabla_i(E_j D_j)\,.
\end{aligned}$$

For the magnetic part, the corresponding term results, so that

$$\dot{\mathrm{p}}_{\mathrm{mech}\,i} = \frac{1}{4\pi k_r}\Big\{\big[\nabla_j(E_i D_j + H_i B_j) - \frac{1}{2}\nabla_i(\mathbf{E}\!\cdot\!\mathbf{D} + \mathbf{B}\!\cdot\!\mathbf{H})\big] - \frac{k_L}{c}\frac{\partial}{\partial t}(\mathbf{D}\times\mathbf{B})_i\Big\}. \qquad (5.6.10)$$

If we define the Maxwell stress tensor with

$$T_{ij} = \frac{1}{4\pi k_r}\left[E_i D_j + H_i B_j - \frac{1}{2}(\mathbf{E}\cdot\mathbf{D}+\mathbf{H}\cdot\mathbf{B})\delta_{ij}\right], \tag{5.6.11}$$

it is immediately apparent that the integrand of the 1st term of (5.6.10) has the form of a divergence $\nabla_j T_{ij} = T_{ij,j}$. The balance equation for the momentum densities (differential form of the momentum balance/of the momentum theorem) is thus

$$\dot{p}_{\mathrm{mech}\,i} + \dot{p}_{\mathrm{field}\,i} = T_{ij,j} \quad \text{with} \quad \dot{p}_{\mathrm{field}\,i} = \frac{1}{4\pi k_r}\frac{k_L}{c}\frac{\partial}{\partial t}(\mathbf{D}\times\mathbf{B})_i. \tag{5.6.12}$$

The integral form of the momentum balance is therefore [Jackson, 1998, (6.122)]

$$\left(\dot{\boldsymbol{P}}_{\mathrm{mech}} + \dot{\boldsymbol{P}}_{\mathrm{field}}\right)_i = \int_V \mathrm{d}^3 x\, T_{ij,j} \overset{(A.4.6)}{=} \oiint_{\partial V} \mathrm{d}a_k\, T_{ik}(\mathbf{x},t), \tag{5.6.13}$$

where the Gauss's theorem was applied to the volume forces $T_{ik,k}$. Here is

$$\boldsymbol{P}_{\mathrm{field}} = \frac{1}{4\pi k_r}\frac{k_L}{c}\int_V \mathrm{d}^3 x\,\mathbf{D}\times\mathbf{B} \tag{5.6.14}$$

the contribution of the electromagnetic field to the momentum. The surface force $\mathbf{T}^{(n)}$ outlined in Fig. 5.14 is defined by

$$\dot{\boldsymbol{P}}_{\mathrm{mech}} + \dot{\boldsymbol{P}}_{\mathrm{Feld}} = \oiint_{\partial V} \mathrm{d}a\,\mathbf{T}^{(n)} \quad \text{with} \quad \mathrm{d}a_k\,T_{ik} = \mathrm{d}a\,T_i^{(n)}.$$

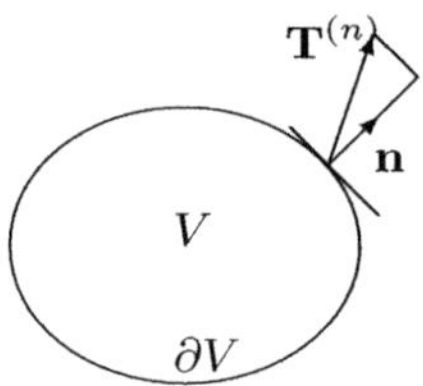

Fig. 5.14. Sketch of the tensions on the surface ∂V of the volume V. $\mathbf{T}^{(n)}\mathrm{d}a$ is the force acting on the surface element $\mathrm{d}a$

The force acts parallel to $\mathbf{n}$ on the element $\mathrm{d}a$ of the surface ∂V $\mathbf{T}^{(n)}\cdot\mathbf{n} = n_i T_{ik} n_k$. Of interest is the relationship of the field momentum density to the power density of the electromagnetic field:

$$\boldsymbol{p}_{\mathrm{Feld}} = \frac{1}{4\pi k_r}\frac{k_L}{c}\mathbf{D}\times\mathbf{B} = \frac{k_L^2}{c^2}\epsilon_r\epsilon_0\mu_r\mu_0\mathbf{S} = \frac{\epsilon_r\mu_r}{c^2}\mathbf{S}, \tag{5.6.15}$$

$$\text{G: }\boldsymbol{p}_{\mathrm{Field}} = \frac{1}{4\pi c}\mathbf{D}\times\mathbf{B}, \qquad\qquad \text{SI: }\boldsymbol{p}_{\mathrm{Field}} = \mathbf{D}\times\mathbf{B}. \tag{5.6.16}$$

It is clear that the field momentum density, when expressed by the Poynting vector, should not depend on electromagnetic units.

Problems for Chapter 5

5.1. *Electron in a homogeneous magnetic field*: A particle of charge e and mass m moves in a homogeneous static magnetic field $\mathbf{B}$. Show that

$$\mathbf{v} = e^{\omega_c t \mathsf{R}}\,\mathbf{v}_0 = \mathbf{v}_{0\parallel} + \big(\mathsf{R}\sin(\omega_c t) + \cos(\omega_c t)\big)\mathbf{v}_{0\perp}$$

with

$$\mathsf{R}_{ij} = \epsilon_{ijk}\hat{B}_k, \qquad\qquad \hat{\mathbf{B}} = \mathbf{B}/B, \qquad\qquad \omega_c = k_L e B/mc.$$

5.2. *Experiment by K.H. Nichols*: In a rapidly rotating metal disc, the density of conduction electrons due to the centrifugal force at the edge of the disc will be greater than in the center. This creates a field $\mathbf{E}$, which compensates the centrifugal force (impressed field). Calculate the voltage V^e between the center and the edge for a disc with a radius of $5\,\mathrm{cm}$ and a rotation speed of $10\,000$ revolutions per minute.

5.3. *Penning trap*: With the arrangement sketched on page 181, Fig. 5.9 of the electrodes, an electric field is to be generated that restricts the movement of a positively charged ion (charge q, mass M) in the z direction. With the homogeneous field $B\,\mathbf{e}_z$, the movement in the xy-plane is then limited, so that the movement of the ion is restricted to a finite volume.
The two positive end electrodes are the circular hyperboloids $z^2 = z_0^2 + \varrho^2/2$. The negative ring electrode is the circular hyperboloid $z^2 = (\varrho^2 - \varrho_0^2)/2$.

1. Show that

   $$\phi(\mathbf{x}) = a(z^2 - \varrho^2/2) + b$$

   describes the field generated by the electrodes with a suitable choice of parameters, where we restrict to the two cases $\phi_+ = -\phi_-$ and $\phi_- = 0$.
2. Solve the equation of motion for the ion and specify, given a, the minimum strength of B.

5.4. *Oscillator model*: (5.5.6) lists the contributions of individual electrons to $\alpha_{\mathrm{at}}(\omega)$. Here, the polarizability is isotropic when the electron is in a homogeneous alternating field $\mathbf{E} = \mathbf{E}_0 e^{-i\omega t}$. However, if the substance is exposed to an additional homogeneous, static magnetic field $\mathbf{B}$, the Lorentz force causes, that the electrons no longer exactly follow $\mathbf{E}$. Specify the tensor $\alpha_{ij}(\omega)$ for a single electron of the atom that is in an alternating field $\mathbf{E}$ and a static field $\mathbf{B}$.

5.5. *Dielectric tensor*: Continuing from problem 5.4, determine the tensor of the dielectric function $\epsilon_{ij}(\omega)$ in the limit of high frequencies (see (5.5.8)).

5.6. *Rotating wire loop*:

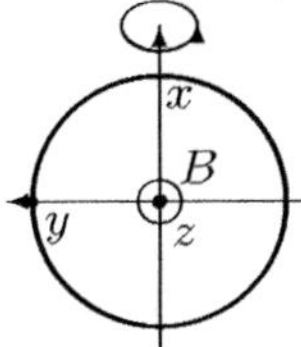

A wire loop has the radius a and the resistance R. The horizontal component of a magnetic field is $\mathbf{B} = B\,\mathbf{e}_z$. The loop rotates with ω in the field $\mathbf{B} = B\,\mathbf{e}_z$ around a vertical axis (the vertical component B_x contributes nothing and $B_y = 0$). We now assume that the wire is made of copper with a cross-section of $1\,\mathrm{mm}^2$. The Wire loop has a radius $a = 20\,\mathrm{cm}$ and rotates at 1000 revolutions per minute.

1. Determine the current in the wire loop and
2. the average Joule heat released to the wire per second.

Hint: Assume for the horizontal component of the (earth) magnetic field $H = 20\,\text{A/m}$. The conductivity of the wire (copper) is 58×10^6 Siemens/m.

5.7. *Induced voltage*: A straight wire of length l is moved at speed $\mathbf{v}$ in a homogeneous magnetic field $\mathbf{B}$. The circuit is closed via a voltmeter.

1. Show that the voltage (EMF) can be represented by $\mathbf{l} \cdot (\mathbf{v} \times \mathbf{B})/c$.
2. Assume that the rails of a railway track are insulated and connected with a voltmeter together. How high is the voltage that arises between the rails when the train travels at $120\,\text{km/h}$ (rail distance is $1.435\,\text{m}$); the strength of the earth's field is $30\,\text{A/m}$.

5.8. *Eddy current brake*: A flat metal plate, which is large enough that we can imagine it in the xy-plane as infinitely extended, is moved at the constant speed $\mathbf{v} = v\mathbf{e}_x$. The thickness of the plate is d $(\theta(d/2 - |z|))$. The plate crosses the magnetic field $\mathbf{B}(\mathbf{x}) = B\,\theta(\varrho_0 - \varrho)\,\mathbf{e}_z$, where $\varrho_0 \gg d$.

1. Calculate the (space-fixed) current density $\mathbf{j}(\varrho)$ in the plate.
   Hint: Decompose $\mathbf{E}$ into the components $\mathbf{E}(\varrho > \varrho_0)$ and $\mathbf{E}(\varrho < \varrho_0)$.
2. Determine the friction force $\mathbf{F}$ acting on the plate and the in the plate Joule heat developed per unit of time.

5.9. *Field momentum*: Verify for time-independent local charge and current densities, starting from (5.6.14):

$$\mathbf{P}_{\text{Field}} = \frac{k_L}{4\pi c k_C}\int \mathrm{d}^3x\,\mathbf{E} \times \mathbf{B} = \frac{1}{c^2}\int \mathrm{d}^3x\,\phi\,\mathbf{j} = \frac{1}{c^2}\int \mathrm{d}^3x\,(\mathbf{j}\cdot\mathbf{E})\,\mathbf{x} = \frac{k_L}{c}\int \mathrm{d}^3x\,\rho\,\mathbf{A}.$$

References

Ashcroft N., Mermin D. *Solid State Physics*, Holt, Rinehart and Winston, N.Y. (1976)

Brown L. S. and Gabrielse G. *Geonium theory: Physics of a single electron or ion in a Penning trap*, Rev. Mod Phys. **58**, 233 (1986)

Bruus H., Flensberg K. *Many Body Quantum Theory in Condensed Matter Physics: An Introduction*, Oxford University Press, Oxford (2004)

Griffiths D. *Introduction to Electrodynamics*, 4th ed., Cambridge University Press (2017)

Iro H. *A Modern Approach to Classical Mechanics*, 2nd ed., World Scientific (2015)

Jackson J. D. *Classical Electrodynamics*, 3rd ed., John Wiley & Sons Inc. (1998)

Kittel Ch. *Introduction to Solid State Physics*, 8th ed., John Wiley & Sons, Inc. (2005)

Nolting W. *Electrodynamics*, 1st, ed. Springer Berlin (2016)

Schwabl F. *Advanced Quantum Mechanics*, 4th ed. Springer Berlin (2008)

Van Dyck R. S., Schwinberg P. B., and Dehmelt H. G. in *Atomic Physics* **9**, ed. by R. S. Van Dyck and E. N. Fortson (World Scientific, Singapore) (1984)

Electrostatics in Matter

In a dielectric, under the influence of an electric field $\mathbf{E}$, dipoles are induced by displacement of the electron shells against the nuclei or permanent dipoles are aligned in the direction of the field. The polarization density is then proportional to $\mathbf{E}$ and the dielectric is a linear medium. The electrostatics of such linear media are discussed in more detail here.

Tab. 6.1. Parameters used in electrostatics.

System	k_C	k_L	ϵ_0	$k_r = k_C \epsilon_0$
Gauss	1	1	1	1
Heaviside-Lorentz	$1/4\pi$	1	1	$1/4\pi$
SI	$1/4\pi\epsilon_0$	c	ϵ_0	$1/4\pi$

6.1 Basic Equations and Continuity Conditions

Maxwell and material equations

The Maxwell equations are coupled via the time derivatives $\dot{\mathbf{D}}$ and $\dot{\mathbf{B}}$. If the fields are time-independent, the Maxwell equations in matter (5.2.16) also decouple into two independent equations for electrostatics and magnetostatics. Thus, from (5.2.16) we get the basic equations of electrostatics in dielectrics

$$\text{(a)} \quad \operatorname{div}\mathbf{D} = 4\pi k_r \rho_f(\mathbf{x}), \qquad\qquad \text{(b)} \quad \operatorname{rot}\mathbf{E} = 0 \qquad\qquad (6.1.1)$$

together with the material equations

$$\mathbf{P} = \chi_e \epsilon_0 \mathbf{E}, \qquad\qquad \boldsymbol{\nabla}\cdot\mathbf{P} = -\rho_P. \qquad\qquad (6.1.2)$$

This results in

$$\mathbf{D} = \epsilon_r \epsilon_0 \mathbf{E} = (1 + 4\pi k_r \chi_e)\epsilon_0 \mathbf{E} \qquad \text{with} \qquad \epsilon_r = 1 + 4\pi k_r \chi_e. \qquad (6.1.3)$$

From $\mathbf{\nabla}\times\mathbf{E}=0$ it follows that also in the dielectric

$$\mathbf{E}=-\mathbf{\nabla}\phi, \qquad\qquad \mathbf{\nabla}\cdot\mathbf{E}=4\pi k_C\rho=4\pi k_C(\rho_f+\rho_P). \qquad (6.1.4)$$

Continuity conditions on dielectrics

In vacuum, the continuity conditions concern the interfaces to conductors and surface charges. Now interfaces to dielectrics are added, so to interfaces where there are no free charges. Nevertheless, the situation is similar to that in electrostatics in vacuum.

1. $\mathbf{\nabla}\times\mathbf{E}=0$: From this follows (unchanged) the continuity of the two tangential components (Stokes's theorem).
2. $\mathbf{\nabla}\cdot\mathbf{D}=4\pi k_r\rho_f$: On the surface are free charges, induced charges on the metal surface or surface charges; $\rho_f=\sigma\delta(x_\perp)$. The jump in the normal component is then $D_n=4\pi k_r\sigma$ (see (2.2.26)).
3. $\mathbf{\nabla}\cdot\mathbf{D}=0$: An interface between two dielectrics was not available in the vacuum. $\mathbf{D}_\perp$ is here at the interface continuous, but the normal component of $\mathbf{E}$ still has a jump when crossing the boundary surface, which is caused by the polarization at the surface (see Fig. 6.1):

Fig. 6.1. Continuity of the normal component of $\mathbf{D}$ ($\rho_f=0$) and continuity of the tangential components of $\mathbf{E}$

$$
\begin{aligned}
\mathbf{\nabla}\times\mathbf{E}=0 \quad\rightsquigarrow\quad & \oint_C \mathrm{d}\mathbf{s}\cdot\mathbf{E}=0 \implies \quad \mathbf{E}_\parallel^{(1)}=\mathbf{E}_\parallel^{(2)}, \\
\mathbf{\nabla}\cdot\mathbf{D}=0 \quad\rightsquigarrow\quad & \oiint_{\partial V} \mathrm{d}\mathbf{a}\cdot\mathbf{D}=0 \implies \quad D_n^{(1)}=D_n^{(2)}.
\end{aligned}
\qquad (6.1.5)
$$

The second equation can also be written as $\epsilon_1 E_n^{(1)}=\epsilon_2 E_n^{(2)}$ written.

On the continuity of the electrostatic potential

When applying the continuity conditions at boundaries, the continuity of the electrostatic potential is often used. We would like to add that the continuity of ϕ at boundaries is equivalent to the continuity of the tangential components of $\mathbf{E}$. In Fig. 6.2, the interface $S(\mathbf{x})$ of two media is sketched by a line, on which the points $\mathbf{x}_a$ and $\mathbf{x}_b$ are located. If you integrate $\mathbf{E}$ from $\mathbf{x}_a$ to $\mathbf{x}_b$ along any path on S in medium 1, you get according to (2.1.3)

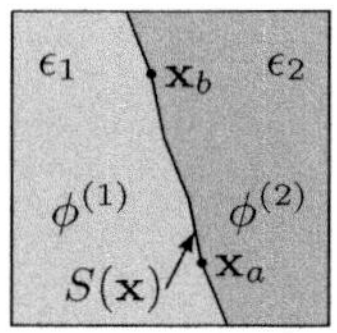

Fig. 6.2. The interface of two dielectric media with the permittivities (dielectric constants) ϵ_1 and ϵ_2 is $S(\mathbf{x})$

$$\phi^{(1)}(\mathbf{x}_b) - \phi^{(1)}(\mathbf{x}_a) = -\int_{\mathbf{x}_a}^{\mathbf{x}_b} d\mathbf{x}_\| \cdot \mathbf{E}_\|^{(1)}\,.$$

Now integrate on the same path in medium 2 and form the difference:

$$\phi^{(2)}(\mathbf{x}_b) - \phi^{(1)}(\mathbf{x}_b) = \phi^{(2)}(\mathbf{x}_a) - \phi^{(1)}(\mathbf{x}_a) - \int_{\mathbf{x}_a}^{\mathbf{x}_b} d\mathbf{x}_\| \cdot \left[\mathbf{E}_\|^{(2)} - \mathbf{E}_\|^{(1)}\right].$$

1. If the potential is continuous, i.e. $\phi^{(1)}(\mathbf{x}) = \phi^{(2)}(\mathbf{x})\ \forall \mathbf{x} \in S$, then $\mathbf{E}_\|^{(1)}(\mathbf{x}) = \mathbf{E}_\|^{(2)}(\mathbf{x})$ must hold, so that the integral for all $\mathbf{x}_a$ and $\mathbf{x}_b$ vanishes.

2. If the tangential components are continuous, i.e. $\mathbf{E}_\|^{(1)}(\mathbf{x}) = \mathbf{E}_\|^{(2)}(\mathbf{x})\ \forall \mathbf{x} \in S$, then $\phi^{(1)}(\mathbf{x})$ and $\phi^{(2)}(\mathbf{x})\ \forall \mathbf{x} \in S$ can at best differ by a constant. A potential that is discontinuous on the interface S would have a singular normal component there and thus a singular surface charge, which is not the case in dielectrics.

6.2 Application of Continuity Conditions

6.2.1 Configurations with Dielectrics, Conductors and Charges

Here, the basic equations of electrostatics are given for quite generally held configurations.

Two dielectric half-spaces

The separating surface of the two half-spaces is the xy-plane, as shown in Fig. 6.3. The dielectrics are homogeneous and $\epsilon_1 > \epsilon_2$. The field vector

$$\mathbf{E} = \theta(-z)\,\mathbf{E}^{(1)} + \theta(z)\,\mathbf{E}^{(2)}$$

lies in the xz-plane and is thus subject to the boundary conditions

$$\text{Tangential component}: \quad E_x^{(1)} = E_x^{(2)}, \qquad\qquad D_x^{(1)} = \frac{\epsilon_1}{\epsilon_2} D_x^{(2)} > D_x^{(2)},$$

$$\text{Normal component}: \quad E_z^{(1)} = \frac{\epsilon_2}{\epsilon_1} E_z^{(2)} < E_z^{(2)}, \quad D_z^{(1)} = D_z^{(2)}.$$

The continuity conditions for $\mathbf{E}$ form a "refraction law" for the field lines:

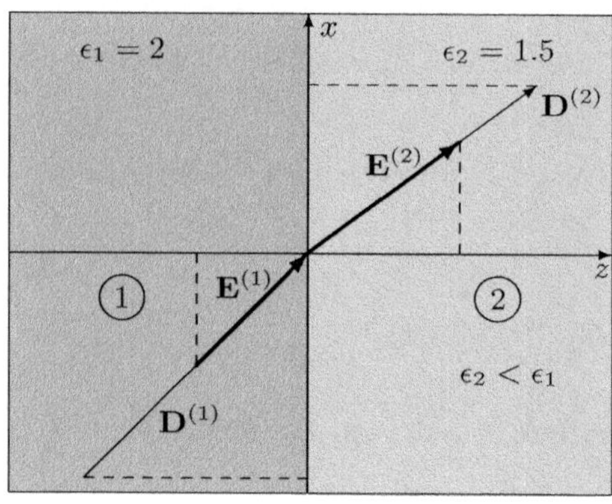

Fig. 6.3. The sketch shows the refraction between two media with the xy-plane as the separating surface; since $\epsilon_2 < \epsilon_1$, both, $\mathbf{E}$ and $\mathbf{D}$, are refracted towards the perpendicular. The dashed lines are of equal length, indicating continuity.

When transitioning to a medium with a larger dielectric constant, it is refracted from the perpendicular; this also applies to $\mathbf{D}$. According to Gauss's law, since $\rho_f = 0$,

$$\boldsymbol{\nabla} \cdot \mathbf{D} = \epsilon_0(\boldsymbol{\nabla} \cdot \mathbf{E} + 4\pi k_C \boldsymbol{\nabla} \cdot \mathbf{P}) = 0 \quad \Rightarrow \quad \boldsymbol{\nabla} \cdot \mathbf{E} = -4\pi k_C \boldsymbol{\nabla} \cdot \mathbf{P} = 4\pi k_C \rho_P.$$

Thus, because $\boldsymbol{\nabla} \cdot \mathbf{D}^{(1)} = \boldsymbol{\nabla} \cdot \mathbf{D}^{(2)} = 0$

$$\rho_P(\mathbf{x}) = \frac{1}{4\pi k_C} \boldsymbol{\nabla} \cdot \mathbf{E} = \frac{-1}{4\pi k_C} \delta(z)\left(E_z^{(1)} - E_z^{(2)}\right) = \frac{1}{4\pi k_C}\left(\frac{\epsilon_1}{\epsilon_2} - 1\right)E_z^{(1)}\,\delta(z).$$

This results in the surface charge at the interface of the two dielectrics:

$$\rho_P(\mathbf{x}) = \left(\sigma_{P_2} - \sigma_{P_1}\right)\delta(z) = \sigma_P\,\delta(z) \quad \text{with} \quad \sigma_P = \frac{1}{4\pi k_C}\left(\frac{\epsilon_1}{\epsilon_2} - 1\right)E_z^{(1)} > 0\,.$$

In an electric field, a force acts on the positive electric charges in the direction of the field. For $\epsilon_1 > \epsilon_2$ in a field directed to the right $E_z > 0$ the bound charges are shifted so that a positive charge layer is formed.

This fact can be interpreted microscopically as follows: In the medium with the larger dielectric constant, the polarization is greater, so the positive and negative charges are shifted more against each other, as sketched in Fig. 6.4. Therefore, there remains an excess of positive charge at the interface (and negative charge behind it), which is proportional to the applied field.

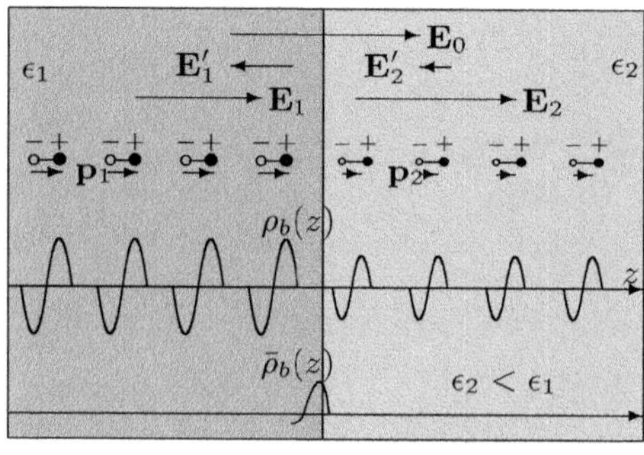

Fig. 6.4. Sketch of the displacement of the $\pm$ charges in the electric field; the medium 1 is more polarizable and therefore at the interface the positive charge from 1 is not completely compensated by the negative charge from 2. $\mathbf{E}_{1,2}' = 4\pi k_C \mathbf{P}_{1,2}$ are the induced fields

Charges and conductors in a homogeneous dielectric

ρ_f are the charges and $\epsilon_r(\mathbf{x}) = \epsilon_r$ is the location-independent dielectric constant. The following applies the Maxwell equations (6.1.1) of electrostatics:

$$\operatorname{div}\mathbf{E} = 4\pi k_C \rho_f/\epsilon_r, \qquad\qquad \operatorname{rot}\mathbf{E} = 0. \tag{6.2.1}$$

Compared to the vacuum, the field $\mathbf{E}$ generated by ρ_f is reduced by the factor $1/\epsilon_r$. If, as sketched in Fig. 6.5, conductors and free charges are embedded in a

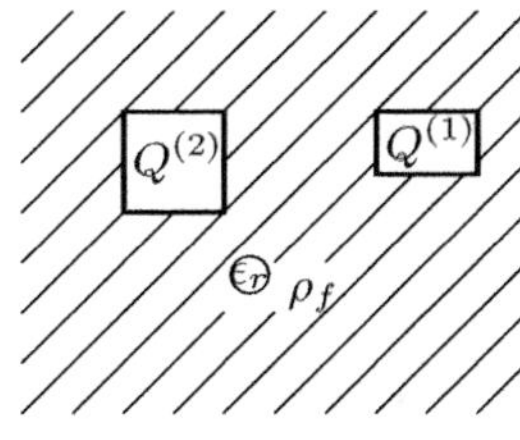

Fig. 6.5. Conductor with the charges $Q^{(i)}$ (potential $\phi^{(i)}$), embedded in a homogeneous dielectric, in which free charges ρ_f (potential ϕ_f) can also be found

dielectric, charges σ^{ind} are induced on the surfaces of the conductors, as they were calculated in (2.3.4') for a charge in front of a grounded sphere. Surface charges are for the dielectric external (free) charges:

$$D_n^{(i)} = 4\pi k_r \sigma^{(i)}, \qquad \mathbf{E}_\parallel^{(i)} = 0, \qquad E_n^{(i)} = 4\pi k_C \sigma^{(i)}/\epsilon_r. \tag{6.2.2}$$

We remember that in the conductor $\phi = \text{const.}$ and $\mathbf{E} = 0$, and that the normal component of the field has a jump at the boundary to the vacuum/dielectric. Their outwardly directed normal components are $s_n^{(i)}$. The field in the dielectric is determined from

$$\boldsymbol{\nabla}\cdot\mathbf{D} = 4\pi k_r \Big[\rho_f + \sum_i \sigma^{(i)}\,\delta(s_n^{(i)})\Big], \qquad\qquad \boldsymbol{\nabla}\times\mathbf{E} = 0. \tag{6.2.3}$$

Within the dielectric, therefore, applies

$$\boldsymbol{\nabla}\cdot\mathbf{E} = 4\pi k_C \frac{\rho_f}{\epsilon_r}, \qquad \mathbf{E} = -\boldsymbol{\nabla}\phi \qquad \Rightarrow \qquad \Delta\phi = -4\pi k_C \frac{\rho_f}{\epsilon_r}. \tag{6.2.4}$$

Imagine our configuration in Fig. 6.5 without dielectric, applies to isolated conductors

$$\phi_{\text{vac}} = \phi_{\text{vac}}^f + \sum_i \phi_{\text{vac}}^{(i)}, \qquad\qquad Q^{(i)} = \oiint_{\partial V^{(i)}} \mathrm{d}f\,\sigma^{(i)}.$$

1. *Charges on the conductors given*

$$\phi(\mathbf{x}) = \frac{1}{\epsilon_r}\Big(\phi_{\text{vac}}^f(\mathbf{x}) + \sum_i \phi_{\text{vac}}^{(i)}(\mathbf{x})\Big).$$

The polarization in the dielectric weakens $\mathbf{E} = \mathbf{E}_{\text{vac}}/\epsilon_r$, whereas the $Q^{(i)}$ remain unchanged.

2. Potentials on the conductors given

$$\phi(\mathbf{x}) = \frac{1}{\epsilon_r}\phi_{\mathrm{vac}}^{f}(\mathbf{x}) + \sum_i \phi_{\mathrm{vac}}^{(i)}(\mathbf{x}).$$

If we keep the potentials $\phi^{(i)}$ on the conductors constant, the charge there will be increased by the factor ϵ_r enlarged. Thus, $\mathbf{E}$ remains unchanged if $\rho_f = 0$. The weakening, however, also affects the (image) charges induced by ρ_f, which are included here in ϕ_{vac}^{f} are included.

6.2.2 Dielectric in the Plate Capacitor

In section 2.2.4 the essential features of the plate capacitor are given, but without dielectric. This deficiency is to be remedied here, by also addressing the capacitor partially filled with dielectric. Fig. 6.6b shows a capacitor provided with a dielectric in the ratio $\alpha = a_1/a$. If the separating surface is

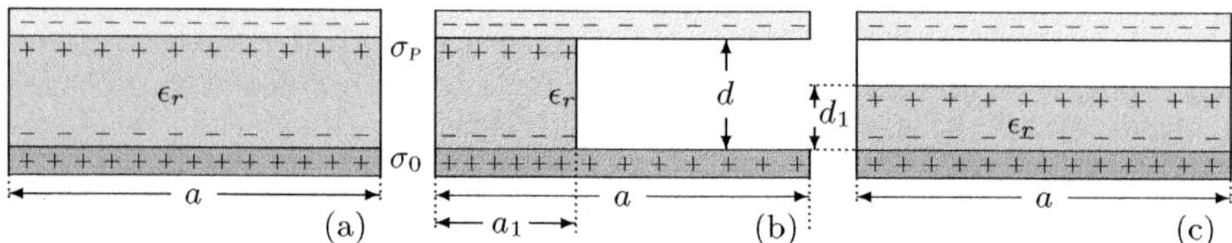

Fig. 6.6. (Isolated) Plate capacitor of area A and charge $Q = \sigma_0 A > 0$. (a) There is a dielectric between the plates. (b) The dielectric occupies only a part of the volume $\alpha = a_1/a$. (c) The dielectric is parallel to the plates and occupies the part $\alpha = d_1/d$ of the volume

perpendicular to the capacitor plate, the electric field is the same throughout the capacitor due to the continuity of the tangential components. If the charge $Q > 0$ of the capacitor plates is taken as given, the voltage V between the plates decreases as α increases. However, to keep $\mathbf{E}$ constant over the entire surface, the surface charge σ in the dielectric part must be larger by the factor ϵ_r than in the rest (problem 6.1).

In Tab. 6.2 the sizes given in (2.2.30) for the plate capacitor in the dielectric are listed, where $C_\perp \geq C_\parallel$. As for the energy of the capacitor, (2.2.31) $U = CV^2/2$ is also valid in the dielectric (see (6.3.1)). Regarding the displacement, we note that this, with the same charge Q, in the capacitor with or without dielectric is the same. If V is kept constant, then in the dielectric part D is larger.

6.2.3 Image Charges in Dielectrics

In homogeneous dielectrics, the method of image charges can be used for boundary value problems. The classic problem is the calculation of potentials

Tab. 6.2. Plate capacitor of area A and distance d between the plates, which are covered with $\pm Q = \pm\sigma_0 A$. $\alpha = a_1/a$ or $\alpha = d_1/d$ indicates the proportion of the filling with the dielectric. $V = Ed$

	empty $\epsilon_r=1$	(a): full $\epsilon_r>1$	(b): vertical $\alpha=a_1/a$ $\epsilon_r>1;\ \alpha$	$\epsilon_r=1;\ 1-\alpha$	(c): horizontal $\alpha=d_1/d$ $\epsilon_r>1;\ \alpha$	$\epsilon_r=1;\ 1-\alpha$
			Configuration of the dielectric in the capacitor			
Surface charge	$\sigma_0 = \dfrac{Q}{A}$	$\sigma_\epsilon = \sigma_0$	$\sigma_1 = \sigma_2 \epsilon_r$	$\sigma_2 = \dfrac{\sigma_0}{\epsilon_r \alpha + 1 - \alpha}$	σ_0	
Field strength	$E_0 = 4\pi k_C \sigma_0$	$E_\epsilon = \dfrac{E_0}{\epsilon_r}$	$E = \dfrac{E_0}{\epsilon_r \alpha + 1 - \alpha}$		$E_1 = \dfrac{E_0}{\epsilon_r}$	$E_2 = E_0$
Capacitance	$C_0 \equiv \dfrac{Q}{V_0}$	$C_\epsilon = \epsilon_r C_0$	$C_\perp = C_0\big[\epsilon_r \alpha + 1 - \alpha\big]$		$C_\parallel = C_0 \dfrac{\epsilon_r}{\alpha + \epsilon_r(1-\alpha)}$	

and fields for a charge q in front of the interface of two dielectrics, as outlined in Fig. 6.7. At the interface $z = 0$, the dielectric, compared to the metal, has the additional condition of continuity of the normal component $D_\perp$:

Metal-dielectric$_1$: $\mathbf{E}_{2\parallel} = \mathbf{E}_{1\parallel} = 0$ $\Rightarrow$ $\sigma = D_{1\perp}/4\pi k_r$,

Dielectric$_2$-dielectric$_1$: $\mathbf{E}_{2\parallel} = \mathbf{E}_{1\parallel}$ and $D_{2\perp} = D_{1\perp}$.

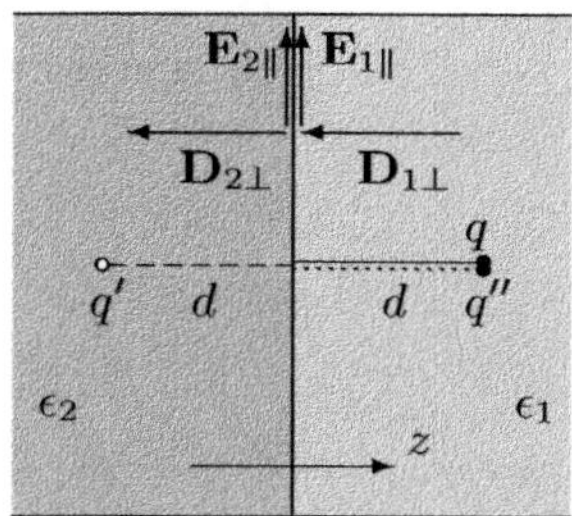

Fig. 6.7. Point charge in front of the interface of two homogeneous dielectrics; q and the image charge q' determine the potential ϕ_1 in the half-space with $z > 0$ and the image charge q'' the potential ϕ_2 in the half-space $z < 0$

The potential ϕ_1 in half-space 1 with $z > 0$ is determined by the charge q along with an image charge q' to be determined. Unlike in metals, $\mathbf{E}$ does not vanish in half space 2 with $z < 0$; due to the additional condition of the continuity of the normal component of $\mathbf{D}$, a further image charge q'' is necessary, which is responsible for the field $\mathbf{E}_2$. With the help of the boundary conditions at the interface $z = 0$

$$\frac{\partial \phi_1(\mathbf{x}_\parallel, 0)}{\partial \mathbf{x}_\parallel} = \frac{\partial \phi_2(\mathbf{x}_\parallel, 0)}{\partial \mathbf{x}_\parallel} \qquad \Leftrightarrow \qquad \phi_1(\mathbf{x}_\parallel, 0) = \phi_2(\mathbf{x}_\parallel, 0)$$

$$\epsilon_1 \frac{\partial \phi_1(\mathbf{x}_\parallel, 0)}{\partial z} = \epsilon_2 \frac{\partial \phi_2(\mathbf{x}_\parallel, 0)}{\partial z}$$

q' and q'' can be determined and with the approach $(\mathbf{d} = (0,0,d))$

$$\phi_1(\mathbf{x}) = k_C\Big[\frac{q}{\epsilon_1|\mathbf{x}-\mathbf{d}|} + \frac{q'}{\epsilon_1|\mathbf{x}+\mathbf{d}|}\Big], \qquad \phi_2(\mathbf{x}) = k_C \frac{q''}{\epsilon_2|\mathbf{x}-\mathbf{d}|}$$

one obtains

$$\frac{q+q'}{\epsilon_1} = \frac{q''}{\epsilon_2}, \qquad q-q' = q'' \qquad \Rightarrow \qquad q' = \frac{\epsilon_1-\epsilon_2}{\epsilon_1+\epsilon_2}q, \qquad q'' = \frac{2\epsilon_2}{\epsilon_1+\epsilon_2}q.$$

6.2.4 Dielectric Sphere in the External Field $\mathbf{E}_0$

A dielectric sphere ($\epsilon_r > 1$) with radius R is placed in a homogeneous external field $\mathbf{E}_0 = E_0\,\mathbf{e}_z$. The environment of the sphere has the dielectric constant $\epsilon_r = 1$. Since there are no charges, it is $\Delta\phi = 0$ inside ($r < R$) and outside

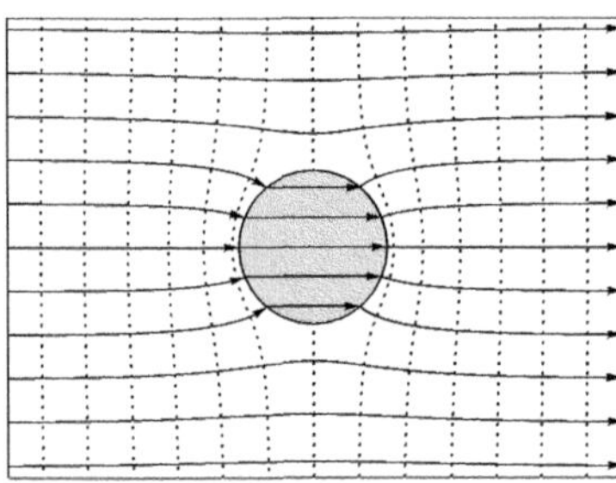

Fig. 6.8. Field lines around a sphere with radius R and $\epsilon_r = 5$; the field inside the sphere is weaker, which is not reflected in the field lines of the sketch, but at best can be read from the greater distance of the dashed equipotential lines in the area of the sphere

($r > R$) the sphere. Due to the symmetry, ϕ is independent of φ, therefore the approach (3.2.40) with Legendre polynomials:

$$\phi_i = \sum_{l=0}^{\infty} a_l\,r^l\,P_l(\cos\vartheta), \qquad \phi_a = \sum_{l=0}^{\infty} \left\{\alpha_l\,r^l + \beta_l\,r^{-l-1}\right\}P_l(\cos\vartheta).$$

Boundary condition for $r \to \infty$:

$$\phi_a(r \to \infty) = -E_0\,z = -E_0\,r\,\cos\vartheta \qquad \Rightarrow \qquad \alpha_l = -E_0\,\delta_{l1}.$$

Continuity of the tangential component $E_{i\vartheta}(R) = E_{a\vartheta}(R)$ on the surface of the sphere:

$$\frac{\partial\phi_i}{\partial\vartheta} = \sum_{l=1}^{\infty} a_l r^l \frac{\partial P_l(\cos\vartheta)}{\partial\vartheta}, \qquad \frac{\partial\phi_a}{\partial\vartheta} = \alpha_1 \frac{\partial P_1(\cos\vartheta)}{\partial\vartheta} + \sum_{l=1}^{\infty} \frac{\beta_l}{r^{l+1}} \frac{\partial P_l(\cos\vartheta)}{\partial\vartheta},$$

$$\frac{\partial\phi_i}{\partial\vartheta} = \frac{\partial\phi_a}{\partial\vartheta}, \qquad\qquad a_l = -E_0\,\delta_{l1} + \frac{\beta_l}{R^{2l+1}}.$$

For the normal component, $\epsilon_r E_{ir}(R) = E_{ar}(R)$ applies:

$$\frac{\partial\phi_i}{\partial r} = \sum_{l=0}^{\infty} l a_l r^{l-1}P_l(\cos\vartheta), \qquad \frac{\partial\phi_a}{\partial r} = \alpha_1\,\cos\vartheta - \sum_{l=0}^{\infty}(l+1)\frac{\beta_l}{r^{l+2}}\,P_l(\cos\vartheta),$$

$$\epsilon_r \frac{\partial\phi_i}{\partial r} = \frac{\partial\phi_a}{\partial r}, \qquad\qquad \epsilon_r l a_l = -E_0\,\delta_{l1} - (l+1)\frac{\beta_l}{R^{2l+1}}.$$

For $l \neq 1$ the equations only have the solution $a_l = \beta_l = 0$. So only a_1 and β_1 remain to be determined:

$$a_1 = -\frac{3}{\epsilon_r + 2} E_0, \qquad\qquad \beta_1 = \frac{\epsilon_r - 1}{\epsilon_r + 2} R^3 E_0.$$

Potential and field inside the sphere are

$$
\begin{aligned}
\phi_i &= -\frac{3}{2 + \epsilon_r} E_0 z = -\frac{3}{2 + \epsilon_r} \mathbf{E}_0 \cdot \mathbf{x}, \\
\mathbf{E}_i &= \frac{3}{2 + \epsilon_r} \mathbf{E}_0 = \mathbf{E}_0 - \frac{\epsilon_r - 1}{\epsilon_r + 2} \mathbf{E}_0.
\end{aligned}
\tag{6.2.5}
$$

The polarization, the field inside the sphere, expressed by $\mathbf{P}$ and the dipole moment of the sphere are

$$\mathbf{P} = \frac{\epsilon_r - 1}{4\pi k_C} \mathbf{E}_i = \frac{3}{4\pi k_C} \frac{\epsilon_r - 1}{\epsilon_r + 2} \mathbf{E}_0 \qquad\qquad r < R, \tag{6.2.6}$$

$$\mathbf{E}_i = \mathbf{E}_0 - \frac{4\pi k_C}{3} \mathbf{P} \qquad\qquad r < R, \tag{6.2.7}$$

$$\mathbf{p} = \frac{4\pi R^3}{3} \mathbf{P} = \frac{1}{k_C} \frac{\epsilon_r - 1}{\epsilon_r + 2} R^3 \mathbf{E}_0. \tag{6.2.8}$$

The polarization of the sphere is parallel to the external field $\mathbf{E}_0$. The macroscopic field inside the sphere $\mathbf{E}_i$ is also uniform and parallel to $\mathbf{E}_0$ with $E_i < E_0$.

In the outer space

$$\phi_a = -E_0 z + E_0 \frac{\epsilon_r - 1}{\epsilon_r + 2} R^3 \frac{\cos \vartheta}{r^2} = -\mathbf{E}_0 \cdot \mathbf{x} + k_C \frac{\mathbf{p} \cdot \mathbf{x}}{r^3} \tag{6.2.9}$$

is composed of the potential of the homogeneous field $\mathbf{E}_0$ and the field of the dipole $\mathbf{p}$ of the sphere; the potential is thus comparable to that of the conducting sphere (3.3.19), apart from the different dipole moment

$$\mathbf{E}_a = \mathbf{E}_0 - k_C \boldsymbol{\nabla} \frac{\mathbf{p} \cdot \mathbf{x}}{r^3}. \tag{6.2.10}$$

The field in outer space is thus the field $\mathbf{E}_0$ and that of the induced dipole $\mathbf{p}$. Finally, we calculate the charge density (surface charge) associated with $\mathbf{P}$, according to (6.2.6):

$$\rho_P(\mathbf{x}) = -\boldsymbol{\nabla} \cdot \mathbf{P}\, \theta(R - r) = \mathbf{P} \cdot \mathbf{e}_r\, \delta(R - r) = \frac{3}{4\pi k_C} \frac{\epsilon_r - 1}{\epsilon_r + 2} E_0\, \delta(R - r) \frac{z}{r}. \tag{6.2.11}$$

The field generated by ρ_P is opposite to $\mathbf{E}_0$ inside the sphere, as sketched in Fig. 6.9. $\mathbf{E}$ is a source field and $\mathbf{D}$ is a vortex field with the sources and vortices on the sphere surface:

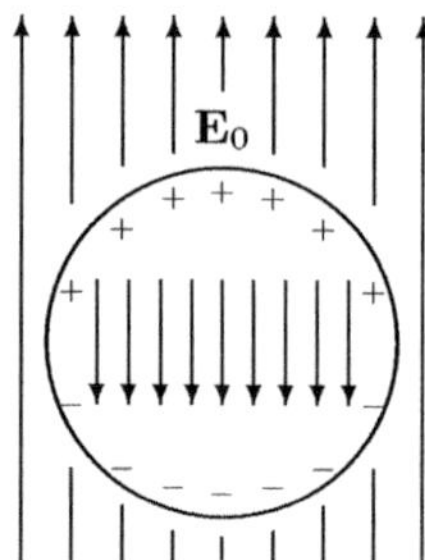

Fig. 6.9. The polarization creates a field inside the dielectric sphere that is directed opposite to the external one

$$\boldsymbol{\nabla}\cdot\mathbf{E} = 4\pi k_C\,\rho_P, \qquad\qquad \boldsymbol{\nabla}\times\mathbf{E} = 0.$$
$$\boldsymbol{\nabla}\cdot\mathbf{D} = 0, \qquad\qquad \boldsymbol{\nabla}\times\mathbf{D} = 4\pi k_r\mathbf{P}\times\mathbf{e}_r\,\delta(R-r), \qquad\qquad (6.2.11')$$

Note: We see in this example that for $\epsilon_r \to \infty$ the solution for a conductive sphere results. This is generally true in electrostatics. For $\epsilon_r \to \infty$ $E_i^n = 0$, so $\frac{\partial}{\partial n}\phi_i = 0$ on the entire surface of the conductor. Then $\phi_i = $ const. is the only solution to the Laplace equation and $\mathbf{E}_i = 0$.

Alternative calculation by displacement of two homogeneous charged spheres

For the dielectric sphere in the homogeneous field $\mathbf{E}_0$ we have calculated polarization, potential, and field. The polarization (6.2.6) of the sphere is homogeneous.

It can be shown that a small displacement of two overlapping and oppositely charged spheres also results in a homogeneous polarization which is equal to the homogeneous dielectric sphere, as sketched in Fig. 6.10.

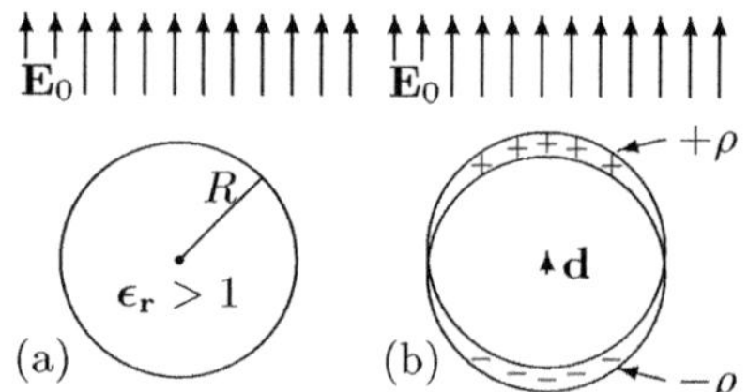

Fig. 6.10. (a) Dielectric sphere in a homogeneous field: For $r \leq R$ the field is homogeneous, for $r > R$ the sphere has the field of a point dipole   (b) The dielectric sphere can be represented by two spheres shifted by $\mathbf{d}$ with the charge densities $\pm\rho$

The potential of a homogeneous sphere with the charge density ρ and the total charge $Q = \rho\,4\pi R^3/3$ is

$$\phi(r) = \phi_i(r) + \phi_a(r) = k_C\Big[-\frac{Qr^2}{2R^3}\theta(R-r) + \frac{Q}{r}\theta(r-R)\Big].$$

The two spheres with $\pm\rho$ are shifted by $\pm\mathbf{d}/2$:

$$\rho^{(1)}(\mathbf{x}) = \rho\,\theta\big(R - |\mathbf{x} - \tfrac{\mathbf{d}}{2}|\big), \qquad\qquad \rho^{(2)}(\mathbf{x}) = -\rho\,\theta\big(R - |\mathbf{x} + \tfrac{\mathbf{d}}{2}|\big).$$

The potential is the sum of the potentials of both spheres. Inside the spheres, in the first order, one obtains

$$\phi_i = -\mathbf{E}_0\cdot\mathbf{x} + \frac{k_C}{2}\frac{Q}{R^3}\Big[\big(\mathbf{x} + \tfrac{\mathbf{d}}{2}\big)^2 - \big(\mathbf{x} - \tfrac{\mathbf{d}}{2}\big)^2\Big] \approx -\mathbf{E}_0\cdot\mathbf{x} + \frac{k_C}{2}\frac{Q}{R^3}\,2\mathbf{x}\cdot\mathbf{d},$$

$$\phi_a = -\mathbf{E}_0\cdot\mathbf{x} + \frac{k_C Q}{|\mathbf{x} - \tfrac{\mathbf{d}}{2}|} - \frac{k_C Q}{|\mathbf{x} + \tfrac{\mathbf{d}}{2}|} \approx -\mathbf{E}_0\cdot\mathbf{x} + k_C Q\,\frac{\mathbf{d}\cdot\mathbf{x}}{r^3}. \qquad (6.2.12)$$

In the limit

$$\mathbf{p} = \lim_{\substack{d\to 0 \\ Qd<\infty}} Q\,\mathbf{d}, \qquad\qquad \mathbf{P} = \frac{3}{4\pi R^3}\mathbf{p} = \rho\mathbf{d}$$

not only are the dipole moment and dipole density exact, but also (6.2.12)

$$\phi_i(\mathbf{x}) = -\mathbf{E}_0\cdot\mathbf{x} + \frac{4\pi k_C}{3}\mathbf{x}\cdot\mathbf{P} \qquad\qquad \text{for} \quad r \le R\,,$$

$$\phi_a(\mathbf{x}) = -\mathbf{E}_0\cdot\mathbf{x} + \frac{4\pi k_C}{3}\frac{R^3}{r^3}\mathbf{x}\cdot\mathbf{P} \qquad\qquad \text{for} \quad r > R\,. \qquad (6.2.13)$$

The validity of the solution

$$\mathbf{E}_i = \mathbf{E}_0 - \frac{4\pi k_C}{3}\mathbf{P}, \qquad \mathbf{E}_a = \mathbf{E}_0 - \frac{4\pi k_C}{3}\frac{R^3}{r^3}\Big[\mathbf{P} - 3\frac{(\mathbf{P}\cdot\mathbf{x})\mathbf{x}}{r^2}\Big] \qquad (6.2.14)$$

is shown by verifying that the continuity conditions at the sphere surface are met. For the Laplace equation, we do not show this separately, as we started from the exact solution (2.4.7).

Since on the sphere surface

$$\mathbf{E}_i(R) = \mathbf{E}_0 - \frac{4\pi k_C}{3}\mathbf{P} \quad \text{and} \quad \mathbf{E}_a(R) = \mathbf{E}_i(R) + 4\pi k_C(\mathbf{P}\cdot\mathbf{e}_r)\mathbf{e}_r \qquad (6.2.15)$$

only differ in the normal component, is the continuity of the tangential components obvious. It remains to be shown that the normal component of $\mathbf{D}$ does not jump at the sphere surface, which directly follows from (6.2.15):

$$\big(\mathbf{D}_a(R) - \mathbf{D}_i(R)\big)\cdot\mathbf{e}_r = \big(\mathbf{E}_a(R) - \mathbf{E}_i(R) - 4\pi k_C\mathbf{P}\big)\cdot\mathbf{e}_r = 0\,.$$

The field inside the sphere can be brought to the form

$$\mathbf{E}_i = \mathbf{E}_0 + \mathbf{E}' \qquad \text{with} \qquad \mathbf{E}' = -\frac{4\pi k_C}{3}\mathbf{P} = -\frac{1}{\epsilon_0}\sum_{j=1}^{3} N_j\,\mathbf{P}_j \qquad (6.2.16)$$

bring. $\mathbf{E}'$ is the induced field, which partially shields the external field and N_j are the so-called *depolarization factors* [Kittel, 2005, (16.8)].

Note on more complicated bodies

The potential ϕ_i (6.2.12) of the homogeneously charged sphere is quadratic in $\mathbf{x}$. The small displacement $\mathbf{d}$ of two oppositely charged spheres has led to a linear potential and thus to a homogeneous field $\mathbf{E}'$.

We conclude from this that bodies with $\phi_i \sim x_j^2$ have a homogeneous induced field $\mathbf{E}'$, or in other words, bodies that are more complex than ellipsoids will not be homogeneously polarized.

From the potential of an ellipsoid (6.3.9) (see problems 6.4 and 6.5) a homogeneous field $\mathbf{E}'$ can be derived, where the depolarization factors fulfill the "sum rule"

$$N_1 + N_2 + N_3 = 4\pi k_r \qquad \Rightarrow \qquad \boxed{\text{SI: } N_1 + N_2 + N_3 = 1}\,.$$

By suitable limit formations, the N_i for sphere, wire (stretched rotational ellipsoid) etc. can be derived:

Sphere $\qquad\qquad\qquad\qquad\qquad\qquad N_1 = \frac{4\pi k_r}{3},\ N_2 = \frac{4\pi k_r}{3},\ N_3 = \frac{4\pi k_r}{3}.$
Thin plate in xy-plane $\qquad\qquad\ N_1 = 0,\qquad N_2 = 0,\qquad N_3 = 4\pi k_r.$
Long cylinder (wire on z-axis) $N_1 = 2\pi k_r,\ N_2 = 2\pi k_r,\ N_3 = 0.$

6.2.5 The Clausius-Mossotti Formula

We want to try to relate the atomic polarizability α in a crystal with cubic symmetry to the dielectric constant ϵ_r. For this, we take a plate of the homogeneous solid, which, when viewed microscopically, should consist of atomic dipoles. We now pick out one of these dipoles and imagine a small hollow sphere cut out around it. Within this, the dipoles should be located on the lattice sites and move in a local field $\mathbf{E}_l$, as indicated in Fig. 6.11. To be calculated is the local field $\mathbf{E}_l$ that the dipole sitting in the center of the hollow sphere feels, when the external field $\mathbf{E}_0$ is applied to the solid. Between the

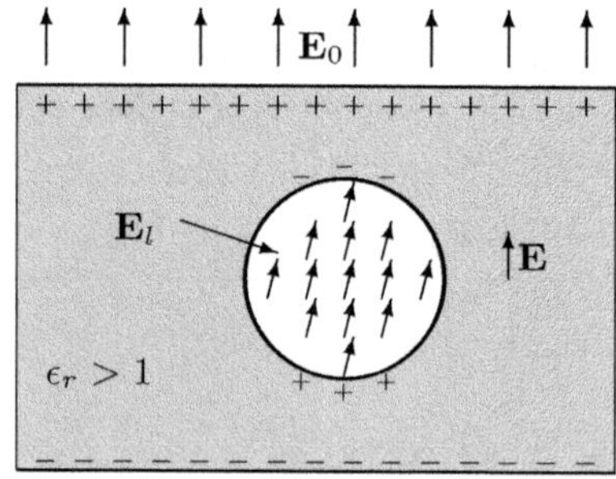

Fig. 6.11. Dielectric plate with external field and small hollow sphere with dipoles; $\mathbf{E}_l = \mathbf{E} + (4\pi/3)\mathbf{P} + \mathbf{E}_n$, where it is to be shown that the field of the dipoles $\mathbf{E}_n = 0$

plates is $\mathbf{E} = \mathbf{E}_0/\epsilon_r$ homogeneous. If there were no dipoles within the hollow sphere, we would have the also homogeneous "far field"

$$\mathbf{E}_f = \mathbf{E} + \frac{4\pi k_C}{3}\mathbf{P},$$

since the surface charge density of the hollow sphere is the negative of a solid sphere (6.2.5)–(6.2.11). The field $(4\pi k_C/3)\,\mathbf{P}$ is called Lorentz field.

The dipole located at the center of the hollow sphere feels besides $\mathbf{E}_f$ also the near field $\mathbf{E}_n$ from the surrounding dipoles (inside the hollow sphere). The field of these dipoles is calculated using (2.2.24), where the dipoles are arranged on a cubic lattice:

$$\mathbf{E}_n = k_C \sum_{k\neq 0} \frac{1}{r_k^3}\left(-\mathbf{p}_0 + 3\mathbf{x}_k \frac{\mathbf{p}_0 \cdot \mathbf{x}_k}{r_k^2}\right).$$

The near field $\mathbf{E}_n$, which the dipole $k = 0$ senses, is composed of the dipole fields of all other dipoles and disappears for reasons of symmetry[1]. We set the z-axis parallel to the dipole $\mathbf{p}_0 = p\,\mathbf{e}_z$ and thus obtain

$$\mathbf{E}_{nx} = k_C \sum_{k\neq 0} \frac{1}{r_k^5}\,3x_k z_k p = 0, \qquad\qquad \mathbf{E}_{ny} = k_C \sum_{k\neq 0} \frac{1}{r_k^5}\,3y_k z_k p = 0,$$

$$\mathbf{E}_{nz} = k_C \sum_{k\neq 0} \frac{1}{r_k^3}\left(-p + 3\frac{p\,z_k^2}{r_k^2}\right) = 0,$$

from which it follows that the local field is given as

$$\mathbf{E}_l = \mathbf{E}_f + \mathbf{E}_n = \mathbf{E} + \frac{4\pi k_C}{3}\mathbf{P}, \qquad\qquad \text{SI: } \mathbf{E}_l = \mathbf{E} + \frac{1}{3\epsilon_0}\mathbf{P}.$$

The isotropy of the cubic lattice is a prerequisite for the disappearance of the near field, which is why this formula also applies to liquids and gases.

The dielectric constant is calculated from the polarizability α in a medium with n particles per unit volume:

$$\mathbf{P} = n\alpha\mathbf{E}_l = n\alpha\left(\mathbf{E} + \frac{4\pi k_C}{3}\mathbf{P}\right) = n\mathbf{p}.$$

After resolving for $\mathbf{P}$, it follows that

$$\mathbf{P} = \frac{n\alpha\mathbf{E}}{1 - (4\pi k_C/3)n\alpha} = \chi_e \epsilon_0 \mathbf{E}. \tag{6.2.17}$$

The susceptibility is thus

$$\chi_e = \frac{n\alpha/\epsilon_0}{1 - (4\pi k_C/3)n\alpha}. \tag{6.2.18}$$

---

[1] E_{nz} disappears because in the sum for each summand with $\mathbf{k}_i = \left(k_x\ k_y\ k_z\right)$ also the summands $\mathbf{n}_j = \left(k_z\ k_x\ k_y\right)$ and $\mathbf{n}_l = \left(k_y\ k_z\ k_x\right)$ exist and the sum of these three terms equals zero.

We can determine the dielectric constant with (5.2.10) from the susceptibility as

$$\epsilon_r = 1 + 4\pi k_r \chi_e = \frac{1 + (8\pi k_C/3)n\alpha}{1 - (4\pi k_C/3)n\alpha}. \tag{6.2.19}$$

The relation between the susceptibility and the atomic polarizability

$$\frac{\epsilon_r - 1}{\epsilon_r + 2} = \frac{4\pi k_C}{3}n\alpha, \qquad\qquad \text{SI:}\ \frac{\epsilon_r - 1}{\epsilon_r + 2} = \frac{1}{3\epsilon_0}n\alpha \tag{6.2.20}$$

is the *Clausius-Mossotti* formula. In optics, where in (6.2.20) for $\sqrt{\epsilon_r}$ the refractive index n is used, this relationship is called *Lorenz-Lorentz* formula [Sommerfeld, 1952, §11.C, p. 75]

For gases, $k_C n\alpha \ll 1$, so it applies

$$\epsilon_r = 1 + 4\pi k_r \chi_e = \frac{1 + (8\pi k_C/3)n\alpha}{1 - (4\pi k_C/3)n\alpha} \approx 1 + 4\pi k_C n\alpha \quad\Longrightarrow\quad \chi_e = \frac{n\alpha}{\epsilon_0}.$$

Notes: In materials with more complex symmetry, the Clausius-Mossotti formula (6.2.20) does not apply in this form; it is then the near field is no longer zero.

Another assumption, which is not always justified, is the polarizability α of the atoms, which corresponds to the gaseous state. In crystals, the atoms can be squeezed, which influences their polarizability.

The choice of a spherical cavity ensures that the depolarization factor $N = 4\pi k_r/3$ is the same for all directions, which then also applies to $\mathbf{E}_f$.

6.3 Energy in the Dielectric

6.3.1 Derivation of the Field Energy in the Dielectric

In deriving the energy in the microscopic case, we assumed that we brought the charges one by one from infinity to their position. The energy required for this could be put into the form

$$U = \frac{1}{2} \int \mathrm{d}^3x\, \rho(\mathbf{x})\, \phi(\mathbf{x}). \tag{2.4.2'}$$

If you replace the microscopic charge density ρ with ρ_f, you get with the help of Gauss's law (5.2.16a) the electrostatic part of the field energy in the already known form of (5.6.7):

$$U = \frac{1}{24\pi k_r} \int \mathrm{d}^3x\, \phi(\mathbf{x})\nabla\cdot\mathbf{D} \stackrel{\text{part. int.}}{=} \frac{1}{8\pi k_r} \int \mathrm{d}^3x\, \mathbf{E}\cdot\mathbf{D}. \tag{6.3.1}$$

This ad hoc approach for linear media will be explained in more detail below.

Supplement to the derivation of field energy

In analogy to the derivation of the internal energy in vacuum, the energy expenditure δU to bring additional charge into a given configuration within a dielectric is to be calculated. To the existing charge density $\rho_f(\mathbf{x})$ and the potential $\phi(\mathbf{x})$ now the additional free charge $\delta\rho_f$ is to be brought in. Unlike in section 2.4, where point charges were assumed, here $\delta\rho_f$ is continuous, so that

$$\delta U = \int \mathrm{d}^3x\, \phi(\mathbf{x})\, \delta\rho_f(\mathbf{x})\,.$$

If we set $\delta\rho_f = \dfrac{\boldsymbol{\nabla}\!\cdot\!\delta\mathbf{D}}{4\pi k_r}$ and $\mathbf{E} = -\boldsymbol{\nabla}\phi$, we obtain after integration by parts

$$\delta U = \frac{1}{4\pi k_r}\int \mathrm{d}^3x\, \phi(\mathbf{x})\,\boldsymbol{\nabla}\cdot(\delta\mathbf{D}) = \frac{1}{4\pi k_r}\int \mathrm{d}^3x\, \mathbf{E}\cdot\delta\mathbf{D}. \qquad (6.3.2)$$

This implies

$$U = \int_0^U \delta U = \frac{1}{4\pi k_r\epsilon_0}\int \mathrm{d}^3x\, \frac{1}{\epsilon_r}\int_0^{\mathbf{D}} \delta\mathbf{D}\cdot\mathbf{D} = \frac{1}{8\pi k_r}\int \mathrm{d}^3x\, \mathbf{E}(\mathbf{x})\cdot\mathbf{D}(\mathbf{x})\,.$$

Then it also applies

$$U = \frac{1}{8\pi k_r}\int \mathrm{d}^3x\, (-\boldsymbol{\nabla}\phi)\cdot\mathbf{D} = \frac{1}{2}\int \mathrm{d}^3x\, \phi\,\rho_f\,.$$

If there are conductors of volumes V_i in the dielectric, (6.3.1) needs to be modified:

$$U = \frac{1}{8\pi k_r}\int \mathrm{d}^3x\, \mathbf{E}\cdot\mathbf{D} = \frac{-1}{8\pi k_r}\int \mathrm{d}^3x\, \phi(\mathbf{x})\boldsymbol{\nabla}\cdot\mathbf{D} \overset{(A.4.3)}{=} \frac{-1}{8\pi k_r}\left\{\oiint_{\partial V} \phi\,\mathrm{d}\mathbf{S}\cdot\mathbf{D}\right.$$

$$\left.-\sum_i \phi_i \oiint_{\partial V_i} \mathrm{d}\mathbf{S}_i\cdot\mathbf{D} - \int \mathrm{d}^3x\, \phi(\mathbf{x})\,\boldsymbol{\nabla}\cdot\mathbf{D}\right\} = \frac{1}{2}\sum_i \phi_i q_i + \frac{1}{2}\int \mathrm{d}^3x\, \phi\,\rho_f. \qquad (6.3.1')$$

Force on charge distribution

The force exerted by the electric field on a volume element in the dielectric can be determined by shifting the medium by $\delta s(\mathbf{x})$ and calculating the work $\delta A = -\delta U$ performed in the process, and withdrawn from the field energy:

$$\delta A = \int \mathrm{d}^3x\, \delta\mathbf{s}\cdot\mathbf{f}_c = -\delta U,$$

$$\delta U = \frac{1}{8\pi k_r\epsilon_0}\delta\!\int \mathrm{d}^3x\, \frac{D^2}{\epsilon_r} = \frac{1}{4\pi k_r}\int \mathrm{d}^3x\, \mathbf{E}\cdot\delta\mathbf{D} - \frac{1}{8\pi k_c}\int \mathrm{d}^3x\, E^2\,\delta\epsilon_r,$$

$$\mathbf{E}\cdot\delta\mathbf{D} = -\boldsymbol{\nabla}\cdot(\phi\delta\mathbf{D}) + \phi\boldsymbol{\nabla}\cdot\delta\mathbf{D} = -\boldsymbol{\nabla}\cdot(\phi\delta\mathbf{D}) + 4\pi k_r\phi\delta\rho.$$

Substituted into δU, after applying Gauss's theorem for a sufficiently large volume V:

$$\delta U = \int_V \mathrm{d}^3x\, \phi\delta\rho - \frac{1}{8\pi k_c}\int_V \mathrm{d}^3x\, E^2\,\delta\epsilon_r. \qquad (6.3.3a)$$

Now if $\rho_f(\mathbf{x})$ was shifted by $\delta\mathbf{s}$, but the volume remains unchanged, then integrate over $\rho'_f(\mathbf{x}) = \rho_f(\mathbf{x} - \delta\mathbf{s})$:

$$\delta\rho = \rho_f(\mathbf{x} - \delta\mathbf{s}) - \rho_f(\mathbf{x}) = -\delta\mathbf{s}\cdot\boldsymbol{\nabla}\rho_f(\mathbf{x}) = -\boldsymbol{\nabla}\cdot(\delta\mathbf{s}\rho_f). \tag{6.3.3b}$$

It is assumed that $\delta\mathbf{s}$ varies only slightly. The same procedure can also be applied to ϵ_r:

$$\delta\epsilon_r = -\delta\mathbf{s}\cdot\boldsymbol{\nabla}\epsilon_r = -\boldsymbol{\nabla}\cdot\epsilon_r\delta\mathbf{s}.$$

Thus, after integration by parts, we obtain

$$\delta U = -\int_V \mathrm{d}^3x\,\delta\mathbf{s}\cdot\rho_f\mathbf{E} + \frac{1}{8\pi k_C}\int \mathrm{d}^3x\,E^2\delta\mathbf{s}\cdot\boldsymbol{\nabla}\epsilon_r. \tag{6.3.3c}$$

From this follows the force density [Greiner, 1991, (7.17)]

$$\mathbf{f}_C(\mathbf{x}) = \rho_f\,\mathbf{E} - \frac{1}{8\pi k_C}E^2\,\boldsymbol{\nabla}\epsilon_r. \tag{6.3.3d}$$

The first term is the force of the field on the free charges. The second term provides a force at the boundary of the dielectric to the vacuum, which wants to pull this into the vacuum. Thus, a dielectric is pulled into a plate capacitor, as will be shown later. If you have two point charges q and q', which are embedded in a constant dielectric, you get the force law for point charges

$$\mathbf{F}_C = q\mathbf{E} = k_C\frac{qq'}{\epsilon_r r^2}\mathbf{e}_r.$$

One can generalize (6.3.3d) by requiring that ϵ_r is a unique function of the mass density ρ_m of matter [Becker, Sauter, 1973, (3.4.9)]:

$$\delta\epsilon_r = \frac{\mathrm{d}\epsilon_r}{\rho_m}\delta\rho_m = -\frac{\mathrm{d}\epsilon_r}{\mathrm{d}\rho_m}\boldsymbol{\nabla}\cdot\rho_m\delta\mathbf{s}.$$

Following the same steps as before (integration by parts), the result now has an additional term

$$\mathbf{f}_C(\mathbf{x}) = \rho_f\,\mathbf{E} - \frac{1}{8\pi k_C}E^2\,\boldsymbol{\nabla}\epsilon_r + \frac{1}{8\pi k_C}\boldsymbol{\nabla}E^2\frac{\mathrm{d}\epsilon_r}{\mathrm{d}\rho_m}\rho_m.$$

The third term does not provide a total force, as shown by the application of Gauss's theorem (A.4.4) but internal forces that can be balanced by volume change, pressure, etc. Dielectric liquids can be described with this model [Panofsky & Phillips, 1962, §6-7].

6.3.2 Energy and Force when Changing the Permittivity

Given is a medium with a predetermined charge density ρ_f and the dielectric constant ϵ_a. The electrostatic energy of the system is then

$$U_a = \frac{1}{8\pi k_r}\int \mathrm{d}^3x\,\mathbf{E}_a\cdot\mathbf{D}_a.$$

Change of the dielectric constant at constant charge

Now the dielectric constant is to be changed to ϵ_r without changing ρ_f. One can imagine that a dielectric is inserted into a capacitor, as shown in Fig. 6.12. The energy difference to be calculated results from the change in the polarizability of the medium, starting from a medium with the dielectric constant ϵ_a and the energy U_a and finally finding ϵ_r and U:

$$\Delta U_q = U - U_a = \frac{1}{8\pi k_r} \int d^3x \left(\mathbf{D}\cdot\mathbf{E} - \mathbf{E}_a\cdot\mathbf{D}_a\right) \quad \text{with} \quad \begin{cases} \mathbf{D}_a = \epsilon_a\epsilon_0\mathbf{E}_a \\ \mathbf{D} = \epsilon_r\epsilon_0\mathbf{E}. \end{cases}$$

Since ρ_f is the same for both media, $\boldsymbol{\nabla}\cdot\mathbf{D} = \boldsymbol{\nabla}\cdot\mathbf{D}_a$ applies. By adding $\mathbf{E}_a\cdot\mathbf{D}$ one obtains a term of the form $\mathbf{E}_a \cdot (\mathbf{D}-\mathbf{D}_a)$. Now $\mathbf{E}_a = -\boldsymbol{\nabla}\phi_a$, and after integration by parts one sees that this contribution disappears. Following the same scheme, the term $-\mathbf{E}\cdot\mathbf{D}_a$ is added:

$$\Delta U_q = \frac{1}{8\pi k_r} \int d^3x \left[\underbrace{(\mathbf{E} + \mathbf{E}_a)}_{-\boldsymbol{\nabla}(\phi+\phi_a)} \cdot (\mathbf{D} - \mathbf{D}_a) + \mathbf{E} \cdot \mathbf{D}_a - \mathbf{E}_a \cdot \mathbf{D}\right]$$

$$= \frac{1}{8\pi k_r} \int d^3x \left[(\phi + \phi_a)\boldsymbol{\nabla} \cdot (\mathbf{D} - \mathbf{D}_a) + \mathbf{E} \cdot \mathbf{D}_a - \mathbf{E}_a \cdot \mathbf{D}\right].$$

If one further sets $\mathbf{D} - \epsilon_0\mathbf{E} = 4\pi k_r\mathbf{P}$, the result in vacuum ($\epsilon_0\mathbf{E}_a = \mathbf{D}_a$) is the energy difference

$$\Delta U_q = \frac{1}{8\pi k_r} \int d^3x \left(\mathbf{E}\cdot\mathbf{D}_a - \mathbf{E}_a \cdot \mathbf{D}\right) \overset{\epsilon_a=1}{=} -\frac{1}{2} \int d^3x\, \mathbf{P} \cdot \mathbf{E}_a. \tag{6.3.4}$$

If the charge is kept constant, the system is isolated and the work δA_q, performed by the system is at the expense of the internal energy

$$\delta\Delta U_q + \delta A_q = 0 \qquad\qquad \text{with} \qquad\qquad \delta A_q = \mathbf{F}_q \cdot \delta\mathbf{x}.$$

$\delta\mathbf{x}$ is a rigid displacement. The variation with constant charge yields

$$\delta\Delta U_q = \frac{1}{8\pi k_r} \int d^3x\, \mathbf{D} \cdot \delta\mathbf{E} \overset{\text{part. int.}}{=} \frac{1}{2} \int d^3x\, \rho_f\delta\phi = \frac{1}{2} \int d^3x\, \rho_f(\delta\mathbf{x}\cdot\boldsymbol{\nabla})\phi.$$

From this follows the force

$$\mathbf{F}_q = \frac{1}{2} \int d^3x\, \rho_f\mathbf{E}, \tag{6.3.5}$$

which could also be determined by $\mathbf{F}_q = -\boldsymbol{\nabla}\Delta U_q$.

Change of the dielectric constant at constant potential

Practically more important than the above-discussed change of the dielectric constants at fixed charges is the one at fixed potential. Here, when introducing the dielectric, charges are supplied to the metals by the battery or drawn from them. Thus, ρ_f changes, while the potential remains unchanged $\mathbf{E} = \mathbf{E}_a = -\boldsymbol{\nabla}\phi$. We start from the vacuum, i.e. that $\epsilon_a = 1$ and $\mathbf{D}_a = \epsilon_0 \mathbf{E}_a$ applies:

$$\Delta U_\phi = \frac{1}{8\pi k_r} \int \mathrm{d}^3 x \left(\mathbf{E}\cdot\mathbf{D} - \mathbf{E}_a\cdot\mathbf{D}_a\right) = \frac{1}{8\pi k_r} \int \mathrm{d}^3 x \left(\mathbf{E}_a\cdot\mathbf{D} - \epsilon_0\mathbf{E}_a\cdot\mathbf{E}\right)$$

$$= \frac{1}{8\pi k_r} \int \mathrm{d}^3 x\, \mathbf{E}_a\cdot(\mathbf{D} - \epsilon_0\mathbf{E}) = \frac{1}{2} \int \mathrm{d}^3 x\, \mathbf{E}_a\cdot\mathbf{P}. \tag{6.3.6}$$

The energy change is inversely equal to the change when the charge density is kept constant. If we now calculate the force on a body with the coordinate $\mathbf{x}$ which acts on a body, we also have to consider the energy change of the voltage source at constant potential. The work done by the system δA is therefore no longer determined solely by the decrease in internal energy δU but also the energy supplied by the voltage source must be included, so that

$$\delta A_\phi + \delta\Delta U_\phi = \delta U_b = \int \mathrm{d}^3 x\, \phi\delta\rho_f.$$

With $U_b > 0$ the energy supplied by the voltage source to the system is denoted. It remains to specify

$$\delta\Delta U_\phi = \frac{1}{8\pi k_r} \int \mathrm{d}^3 x\, \mathbf{E}_a \cdot \delta(\mathbf{D} - \epsilon_0\mathbf{E}) = \frac{-1}{8\pi k_r} \int \mathrm{d}^3 x\, (\boldsymbol{\nabla}\phi) \cdot \delta\mathbf{D}$$

$$\stackrel{\text{part. int.}}{=} \frac{1}{2} \int \mathrm{d}^3 x\, \phi\,\delta\rho_f.$$

If you set $\delta\rho_f = (\delta\mathbf{x} \cdot \boldsymbol{\nabla})\rho_f$, you get

$$\delta A_\phi = \delta U_b - \delta U_\phi = \frac{1}{2} \int \mathrm{d}^3 x\, \phi(\delta\mathbf{x} \cdot \boldsymbol{\nabla})\rho_f \stackrel{\text{part. int.}}{=} \frac{1}{2} \int \mathrm{d}^3 x\, \mathbf{E} \cdot \delta\mathbf{x}\,\rho_f.$$

So the force shows, regardless of whether the charge or the potential is held constant, in the direction with increasing ϵ_r, as sketched in Fig. 6.12

$$\mathbf{F}_\phi = \frac{1}{2} \int \mathrm{d}^3 x\, \mathbf{E}\rho_f. \tag{6.3.7}$$

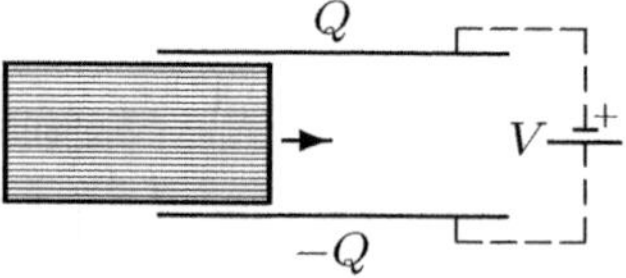

Fig. 6.12. Dielectric in the plate capacitor: The dielectric is drawn into the capacitor both at constant charge on the capacitor plates and at constant voltage V

Note: The empty plate capacitor, as shown in Fig. 6.12, has a homogeneous field, if one disregards the edge corrections. If the dielectric is entirely in the capacitor, it becomes polarized and then has a dipole moment, on which no force acts in the homogeneous field. However, if the dielectric is only partially inserted, there is in the area of the dielectric a surface charge larger by the factor ϵ_r. This creates a stronger and thus inhomogeneous field in the empty part of the edge area, which pulls the dielectric into the capacitor.

Problems for Chapter 6

6.1. *Plate capacitor*: A capacitor (charge Q) is partially filled with a dielectric, as outlined in Fig. 6.6. Verify Tab. 6.2, p. 203, i.e. calculate σ, E, C along with the polarization charge σ_P.

6.2. *Dielectric sphere in a homogeneous field*: A dielectric sphere (dielectric constant ϵ_r, radius R) is located in a homogeneous field **E**. Potential and field for $r \geq R$ are the same as those of a concentric, conducting sphere with radius a, which is to be determined.

6.3. *Charge in front of a dielectric sphere*: The charge q is located, as outlined in Fig. 6.13, in front of a dielectric sphere with permittivity ϵ_r.

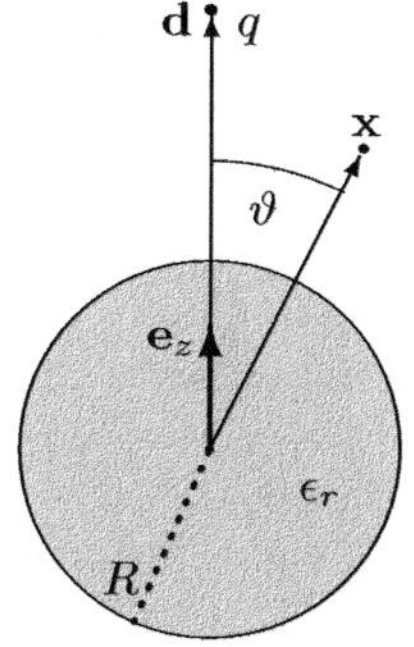

Fig. 6.13. Charge q in front of dielectric sphere with radius R at a distance $\mathbf{d} = (0, 0, d)$ from the center

1. Expand the potentials ϕ^i and ϕ^a inside and outside the sphere using Legendre polynomials and determine the coefficients from the boundary conditions.
2. Form suitable limits to describe the fields for a point charge at a distance $d_0 = R - d$ in front of a dielectric half-plane and
3. a dielectric sphere in a homogeneous field.

Hint: You can compare your results with those of sections 6.2.3, page 202 and 6.2.4, page 204.

6.4. *Homogeneously charged ellipsoid*: Given is an ellipsoid with homogeneous charge density ρ; the corresponding potential in the interior and exterior (ϕ_i and ϕ_a) is sought, which according to Dirichlet [Becker, Sauter, 1973, p. 53] is determined by the approach

$$\phi_i(\mathbf{x}) = c_0 \int_0^\infty d\lambda \, \frac{1}{\sqrt{g(\lambda)}} \left[1 - \sum_{j=1}^3 f_j(x_j, \lambda) \right] \qquad c_0 = k_c \rho \pi a_1 a_2 a_3 , \qquad (6.3.8)$$

$$\phi_a(\mathbf{x}) = c_0 \int_{\lambda_0}^\infty d\lambda \, \frac{1}{\sqrt{g(\lambda)}} \left[1 - \sum_{j=1}^3 f_j(x_j, \lambda) \right] , \qquad (6.3.9)$$

$$f_j(x_j, \lambda) = \frac{x_j^2}{a_j^2 + \lambda} = \frac{x_j^2}{g_j(\lambda)} , \qquad g(\lambda) = \prod_{j=1}^3 g_j(\lambda) , \qquad g_j(\lambda) = a_j^2 + \lambda . \qquad (6.3.10)$$

a_j are the semi-axes of the ellipsoid and λ_0 is defined by

$$\sum_j f_j(x_j, \lambda_0) = 1. \tag{6.3.11}$$

The primary aim is to show that ϕ_i satisfies the Poisson and ϕ_a the Laplace equation in the interior and exterior space and that $\phi_i(\partial V) = \phi_a(\partial V)$ are continuous on the surface.

6.5. *Ellipsoid in a homogeneous field*: If an ellipsoid with the permittivity ϵ_r is placed in a homogeneous external field $\mathbf{E}_0$, the field

$$\mathbf{E} = \mathbf{E}_0 - \frac{1}{\epsilon_0} \sum_{j=1}^{3} N_j \mathbf{P}_j$$

will be established inside. The depolarization factors are to be specified, whereby in particular the relation $\sum_j N_j = 4\pi k_r$ should be verified.
Hint: You get $\mathbf{P}$ when you infinitesimally shift two ellipsoids with the charge densities $\pm\rho$ against each other; the corresponding potentials are given in problem 6.4.

6.6. *Spontaneous polarization: Field lines of a cuboid*
Fig. 6.14 shows the xy-plane of a cuboid extended infinitely in the z-direction. This consists of a medium with the homogeneous polarization $\mathbf{P} = P\mathbf{e}_y$. The cross-section

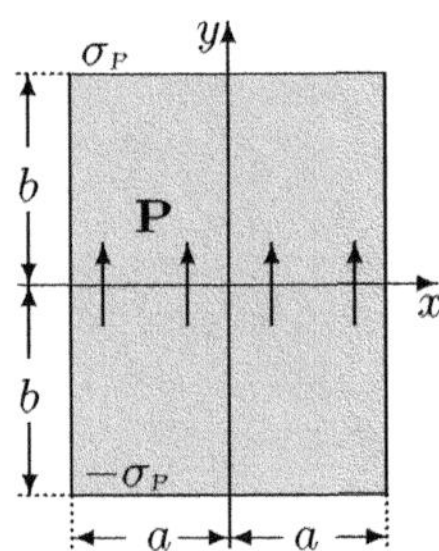

Fig. 6.14. Cross-section $z = 0$ through an infinitely long cuboid of homogeneous polarization $\mathbf{P}$ and the surface charge
$\sigma_P = P > 0$
$\mathbf{P} = P\mathbf{e}_y \big[\theta(y + b) - \theta(y - b)\big]\big[\theta(x + a) - \theta(x - a)\big]$
$\rho_P = -\boldsymbol{\nabla}\cdot\mathbf{P} = -\sigma_P\big[\delta(y + b) - \delta(y - b)\big]\big[\theta(x + a) - \theta(x - a)\big]$

of Fig. 6.14 is similar to that of the bar magnet in Fig. 7.5. This also applies to the field lines $\mathbf{E} \leftrightharpoons \mathbf{H}$ and $\mathbf{D} \leftrightharpoons \mathbf{B}$, which can be calculated analytically for the cuboid.

1. Calculate the fields $\mathbf{E}$ and $\mathbf{D}$.
2. Determine the field lines near the surface charges $y = b$ for $|x| < a$ and sketch the course of the field lines ($\mathbf{E}$ and $\mathbf{D}$).

6.7. *Method of complex analysis: Field lines of a cuboid*
Fig. 6.14 of problem 6.6 shows the xy-plane of a cuboid extended infinitely in the z-direction. Calculate the potential $\phi(x, y)$ and the electric field lines $\psi(x, y) = \text{const.}$ for this cuboid using the method of complex analysis, section 3.5.2.

6.8. *Vector potential of the dielectric displacement*: Fig. 6.15 shows the xy-plane of a z-direction infinitely extended, homogeneously polarized cylinder. $\mathbf{D}$ is a source-free field whose vector potential $\mathbf{A}$ you should calculate; from this determine then $\mathbf{D} = \boldsymbol{\nabla}\times\mathbf{A}$. Sketch the field lines $\mathbf{D}$ and $\mathbf{E}$.
Hint: Start from (4.1.4) and calculate $\boldsymbol{\nabla}\times\mathbf{D}$.

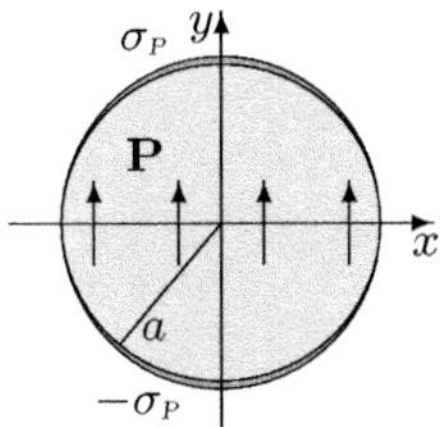

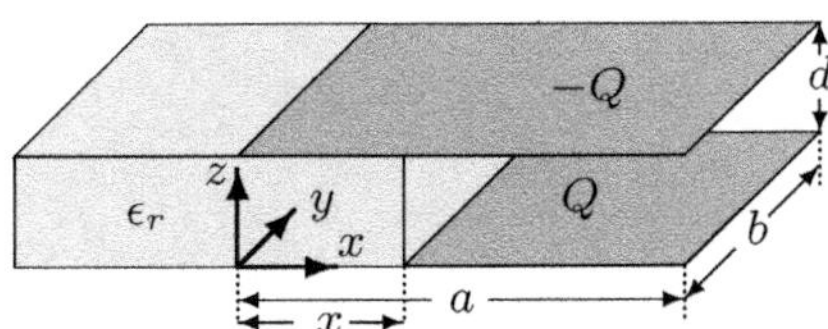

Fig. 6.15. Cross-section $z = 0$ through an infinitely long circular cylinder of homogeneous polarization **P** and the surface charge $\sigma_P > 0$

Fig. 6.16. A dielectric is drawn into (pushed into) the plate capacitor with the charge $Q > 0$

6.9. *Dielectric in the plate capacitor*

1. The charge Q is located on the isolated plate capacitor. Calculate the energy difference $\Delta U_q(x) = U(x) - U_0$ and force $\mathbf{F}_q(x)$ on the dielectric when it is pushed into the plate capacitor by the distance x (see Fig. 6.16). Neglect stray fields.
2. Now the plate capacitor is held at the voltage $V = Q/d$ while the dielectric is inserted. Again, calculate the energy difference $\Delta U_v(x)$ and force $\mathbf{F}_v(x)$.

References

Becker R., Sauter F. *Theorie der Elektrizität 1*, 21th ed. Teubner, Stuttgart (1973)

Greiner W. *Classical Electrodynamics*, 2nd ed. Springer (1991)

Kittel Ch. *Introduction to Solid State Physics* 8th ed., J. Wiley Sons (2005)

Panofsky W., Phillips M. *Classical Electricity and Magnetism*, 2. ed. Addison-Wesley (1962)

Schwabl F *Statistical Mechanics*, 2nd ed. Springer Berlin (2006)

Sommerfeld A. *Electrodynamics*, Academic Press Inc., New York (1952)

7

Magnetostatics in Matter

7.1 Basic Equations of Magnetostatics

We have already seen that for $\dot{\mathbf{D}} = \dot{\mathbf{B}} = 0$ the Maxwell equations (5.2.16) in each two differential equations for electrostatics and magnetostatics decouple. For the latter, the basic equations are

$$\boldsymbol{\nabla} \times \mathbf{H} = \frac{4\pi k_r k_L}{c}\,\mathbf{j}_f \quad \Rightarrow \quad \text{SI: } \boldsymbol{\nabla} \times \mathbf{H} = \mathbf{j}_f, \qquad \boldsymbol{\nabla} \cdot \mathbf{B} = 0, \qquad (7.1.1)$$

which together with the material equation (5.2.17)

$$\mathbf{B} = \mu_r \mu_0 \mathbf{H} = \mu_0(\mathbf{H} + 4\pi k_r \mathbf{M}) = \mu_0 \mathbf{H} + 4\pi k_r \mathbf{J} \qquad (7.1.2)$$

form the basis of magnetostatics. Here we have also given the rarely used magnetic polarization $\mathbf{J} = \mu_0 \mathbf{M}$. If you write the Ampère's law in the form

$$\boldsymbol{\nabla} \times \mathbf{B} = \frac{4\pi k_C}{c k_L}\left(\mathbf{j}_f + \mathbf{j}_M\right) \qquad \text{with} \quad \mathbf{j}_M = \frac{c}{k_L}\boldsymbol{\nabla} \times \mathbf{M}, \qquad (7.1.3)$$

the vector potential follows from this $\qquad\qquad\qquad \text{SI: } \frac{k_C}{c k_L} = \frac{\mu_0}{4\pi}$

$$\mathbf{A}(\mathbf{x}) = \frac{k_C}{c k_L}\int \mathrm{d}^3 x'\, \frac{\mathbf{j}_f(\mathbf{x}') + \mathbf{j}_M(\mathbf{x}')}{|\mathbf{x} - \mathbf{x}'|} \qquad \text{with} \qquad \boldsymbol{\nabla} \cdot \mathbf{A} = 0. \qquad (7.1.4)$$

The magnetic induction is given by $\mathbf{B} = \boldsymbol{\nabla} \times \mathbf{A}$. In linear media (see page 165) one obtains

$$\mathbf{A}(\mathbf{x}) = \frac{k_C}{c k_L}\int \mathrm{d}^3 x'\, \mu_r\, \frac{\mathbf{j}_f(\mathbf{x}')}{|\mathbf{x} - \mathbf{x}'|}. \qquad (7.1.5)$$

Note: From (7.1.1) and (7.1.2) follow vortex and source densities of $\mathbf{H}$ and $\mathbf{B}$:

© The Editor(s) (if applicable) and The Author(s), under exclusive license
to Springer-Verlag GmbH, DE, part of Springer Nature 2025
D. Petrascheck, F. Schwabl, *Electrodynamics*, https://doi.org/10.1007/978-3-662-71502-4_7

$$\omega_H = \frac{1}{4\pi k_r}\boldsymbol{\nabla}\times\mathbf{H} = \frac{k_L}{c}\mathbf{j}_f, \qquad\qquad \rho_H = \frac{1}{4\pi k_r}\boldsymbol{\nabla}\cdot\mathbf{H} \equiv \rho_M = -\boldsymbol{\nabla}\cdot\mathbf{M},$$

$$\omega_B = \frac{1}{4\pi k_r}\boldsymbol{\nabla}\times\mathbf{B} = \frac{k_L}{c}\mu_0\,(\mathbf{j}_f+\mathbf{j}_M), \qquad \rho_B = 0. \tag{7.1.6}$$

For the magnetic charge density, the designation ρ_M is always used instead of ρ_H in the following. If $\mathbf{j}_f = 0$, then $\mathbf{H}$ is given by:

$$\phi_M(\mathbf{x}) = k_r \int \mathrm{d}^3 x' \,\frac{\rho_M(\mathbf{x}')}{|\mathbf{x}-\mathbf{x}'|}, \qquad\qquad \mathbf{H} = -\boldsymbol{\nabla}\phi_M. \tag{7.1.6'}$$

7.1.1 Boundary Conditions at Material Surfaces

The boundary conditions for the two basic equations (7.1.1) were already derived in section 5.2.5. From $\boldsymbol{\nabla}\cdot\mathbf{B} = 0$ follows the continuity of the normal component B_n and from Ampère's law one obtains from (5.2.25) in the given absence of surface currents at the interface the continuity of the tangential components of $\mathbf{H}$.

To briefly repeat it using Fig. 7.1, the continuity of the normal component is shown by the Gauss's theorem through integration over an infinitesimal cylinder (pillbox). If $\boldsymbol{\nabla}\times\mathbf{H} = 0$, the continuity of the tangential components is obtained using the Stokes's theorem by integration over an infinitesimal rectangle:

$$B_\perp^{(1)} = B_\perp^{(2)}, \qquad\qquad \mathbf{H}_\parallel^{(1)} = \mathbf{H}_\parallel^{(2)}. \tag{7.1.7}$$

The use of $\mathbf{H}$ is often convenient, as it is determined by the free currents, which are experimentally well controllable.

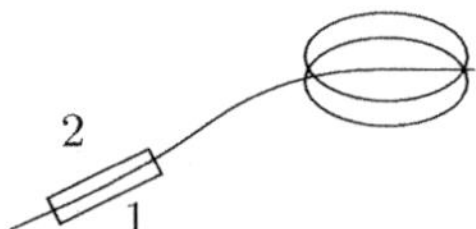

Fig. 7.1. Interface between medium 1 and 2. The rectangle of infinitesimal height is used to determine the transition condition for the tangential components of $\mathbf{H}$, the cylindrical disc for the normal component of $\mathbf{B}$

From $\boldsymbol{\nabla}\cdot\mathbf{B} = \mu_0(\boldsymbol{\nabla}\cdot\mathbf{H}+4\pi k_r\boldsymbol{\nabla}\cdot\mathbf{M})$ it follows

$$\boldsymbol{\nabla}\cdot\mathbf{H} = 4\pi k_r\rho_M \qquad\qquad \text{with} \qquad\qquad \rho_M = -\boldsymbol{\nabla}\cdot\mathbf{M}, \tag{7.1.8}$$

which states that $\mathbf{H}$ has sources and sinks. Since $\boldsymbol{\nabla}\times\mathbf{H}=0$, $\mathbf{H}$ can be represented by a scalar potential ϕ_M. The charge density ρ_M defined in (7.1.8) has its counterpart in electrostatics in ρ_P (5.2.6).

In the case of a permanent magnet, it can be assumed that $\mathbf{M}(\mathbf{x})$ has a jump at the surface. In Fig. 7.2 an infinitesimal cylinder around $z = 0$ is considered. Then, according to (7.1.8), $\mathbf{M}=\mathbf{M}_0\,\theta(-z)$ and $\mathbf{n}=\mathbf{e}_z$:

$$\rho_M = -\boldsymbol{\nabla}\cdot\mathbf{M} = \sigma_M\,\delta(z) \qquad\qquad \text{with} \qquad\qquad \sigma_M = \mathbf{M}_0\cdot\mathbf{n}. \tag{7.1.9}$$

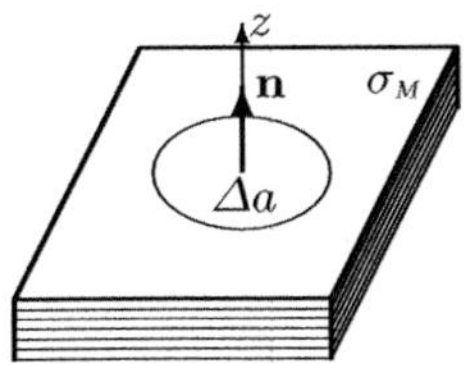

Fig. 7.2. The xy-plane is a surface of the permanent magnet with the surface charge σ_M and the magnetic moment $\mathbf{m} = \sigma_M\,\Delta a\,\mathbf{n}$. Δa is the base area of an infinitesimal cylinder around the origin of the coordinate system

7.1.2 Helmholtz's Decomposition Theorem

A vector field can be decomposed into a curl-free and a source-free vector field, if certain conditions, concerning continuity and asymptotic behavior, are given. With the help of the identities (A.2.38) and (2.1.7)

$$\Delta\mathbf{a} = \boldsymbol{\nabla}(\boldsymbol{\nabla}\cdot\mathbf{a}) - \boldsymbol{\nabla}\times(\boldsymbol{\nabla}\times\mathbf{a}), \qquad \mathbf{a} = \mathbf{v}(\mathbf{x}')/|\mathbf{x}-\mathbf{x}'| \tag{7.1.10}$$

a vector field $\mathbf{v}$, which asymptotically falls off with a power $1/r^{\epsilon}$ with $\epsilon > 0$, can be represented as

$$\mathbf{v}(\mathbf{x}) = \frac{-1}{4\pi}\int \mathrm{d}^3x'\,\Delta\frac{\mathbf{v}(\mathbf{x}')}{|\mathbf{x}-\mathbf{x}'|} \tag{7.1.11}$$

$$= \underbrace{\frac{-1}{4\pi}\int \mathrm{d}^3x'\,\boldsymbol{\nabla}\Big[\boldsymbol{\nabla}\cdot\frac{\mathbf{v}(\mathbf{x}')}{|\mathbf{x}-\mathbf{x}'|}\Big]}_{\mathbf{v}_l(\mathbf{x}) = -\boldsymbol{\nabla}\phi(\mathbf{x})} + \underbrace{\frac{1}{4\pi}\int \mathrm{d}^3x'\,\boldsymbol{\nabla}\times\Big[\boldsymbol{\nabla}\times\frac{\mathbf{v}(\mathbf{x}')}{|\mathbf{x}-\mathbf{x}'|}\Big]}_{\mathbf{v}_t(\mathbf{x}) = \boldsymbol{\nabla}\times\mathbf{A}(\mathbf{x})}.$$

This is a decomposition of $\mathbf{v}$ into a vortex-free field $\mathbf{v}_l$, since $\boldsymbol{\nabla}\times\mathbf{v}_l = 0$, and a source-free field $\mathbf{v}_t$, since $\boldsymbol{\nabla}\cdot\mathbf{v}_t = 0$. This is referred to as the weak form of the decomposition theorem when the associated potentials $\phi(\mathbf{x})$ and $\mathbf{A}(\mathbf{x})$ are not included in the decomposition theorem.

If v decreases stronger than $1/r$, these potentials can be obtained by moving $\boldsymbol{\nabla}$ or $\boldsymbol{\nabla}\times$ in front of the integral. It needs to be shown that the integral

$$\phi(\mathbf{x}) = \frac{1}{4\pi}\int \mathrm{d}^3x'\,\mathbf{v}(\mathbf{x}')\cdot\boldsymbol{\nabla}\frac{1}{|\mathbf{x}-\mathbf{x}'|}, \tag{7.1.12}$$

which extends over a simply connected, unbounded domain $\mathcal{G}\subseteq\mathbb{R}^3$ with a (piecewise) smooth boundary surface, is finite. If one integrates over a sphere S_R with radius R, it is sufficient to show that the integrand asymptotically with $R\to\infty$ vanishes to achieve convergence. Let $\phi_R(\mathbf{x})$ be the portion of the potential of the sphere S_R around $\mathbf{x}$, then the portion ϕ_a outside of it, where $\mathbf{v}(r''\to\infty) = v_\infty\mathbf{x}''/r''^{2+\epsilon}$ forms an upper limit:

$$\phi_a(\mathbf{x}) = \frac{1}{4\pi}\int_{r''>R}\mathrm{d}^3x''\,\mathbf{v}(\mathbf{x}''+\mathbf{x})\cdot\frac{\mathbf{x}''}{r''^3} \leq v_\infty\int_R^\infty \frac{\mathrm{d}r''}{r''^{1+\epsilon}} = \frac{v_\infty}{\epsilon R^\epsilon} \to 0.$$

For weaker decay, the potential ϕ_V is defined for a (simply connected) finite volume V with a smooth boundary

$$\phi_V(\mathbf{x}) = \frac{1}{4\pi} \int_V \mathrm{d}^3 x' \, \mathbf{v}(\mathbf{x}') \cdot \nabla \frac{1}{|\mathbf{x}-\mathbf{x}'|} \tag{7.1.13}$$

and the limit $V \to \infty$ is formed by subtracting the (also diverging) potential of an arbitrarily chosen point $\mathbf{x}_0$, at which $|\mathbf{v}(\mathbf{x}_0)| < \infty$:

$$\phi(\mathbf{x}) \to \phi(\mathbf{x}, \mathbf{x}_0) = \phi(\mathbf{x}) - \phi(\mathbf{x}_0) := \lim_{V \to \infty} \left[\phi_V(\mathbf{x}) - \phi_V(\mathbf{x}_0) \right]. \tag{7.1.14}$$

The potential here is given as the difference to the potential of the point $\mathbf{x}_0$. If $\mathbf{v}$ decreases stronger than $1/r$, one chooses $\mathbf{x}_0$ in infinity, where $\phi(\mathbf{x}_0 \to \infty)$ vanishes. $\phi(\mathbf{x})$ is then the 'absolute value' of the potential. The approach to the vector potential $\mathbf{A}(\mathbf{x})$ is the same as for the scalar potential and the underlying systematics are outlined in appendix A.4.7.

Finally, it should be noted that (7.1.11) also applies to piecewise continuously differentiable vector fields with finite discontinuities, so that the results can be summarized in the form of the *Helmholtz's decomposition theorem:*

Theorem: *Let $\mathbf{v}(\mathbf{x})$ be a piecewise continuously differentiable vector field with the asymptotic behavior*

$$\lim_{r \to \infty} v(r) r^\epsilon < \infty \qquad for \qquad \epsilon > 0,$$

then the decomposition applies

$$\mathbf{v}(\mathbf{x}) = \mathbf{v}_l(\mathbf{x}) + \mathbf{v}_t(\mathbf{x}) = -\nabla \phi(\mathbf{x}) + \nabla \times \mathbf{A}(\mathbf{x}), \tag{7.1.15}$$

where the potentials

$$\phi(\mathbf{x}) = \frac{-1}{4\pi} \int \mathrm{d}^3 x' \, \mathbf{v}(\mathbf{x}') \cdot \nabla' \frac{1}{|\mathbf{x}-\mathbf{x}'|},$$
$$\mathbf{A}(\mathbf{x}) = \frac{1}{4\pi} \int \mathrm{d}^3 x' \, \mathbf{v}(\mathbf{x}') \times \nabla' \frac{1}{|\mathbf{x}-\mathbf{x}'|} \qquad with \qquad \nabla \cdot \mathbf{A}(\mathbf{x}) = 0 \tag{7.1.16}$$

in the range $1 \geq \epsilon > 0$ to be replaced by the difference to the potentials at a freely chosen convergence point $\mathbf{x}_0$:

$$\phi(\mathbf{x}, \mathbf{x}_0) = \phi(\mathbf{x}) - \phi(\mathbf{x}_0), \qquad \mathbf{A}(\mathbf{x}, \mathbf{x}_0) = \mathbf{A}(\mathbf{x}) - \mathbf{A}(\mathbf{x}_0). \tag{7.1.17}$$

This decomposition is unique if $v_l(\mathbf{x})$ vanishes at infinity.

Remarks

1. Without the assumption that $\mathbf{v}_l$ asymptotically vanishes, curl- and source-free vector fields $\mathbf{v}_h$ can be added to the vector field $\mathbf{v}_l$ if they are subtracted from $\mathbf{v}_t$ again, without changing $\mathbf{v}$.

2. It is sufficient to only require the vanishing of $\mathbf{v}$ at infinity, without reference to a power [Blumenthal, 1905]:

$$\phi(\mathbf{x}, \mathbf{x}_0) = \lim_{V \to \infty} \left[\phi_V(\mathbf{x}) - \phi_V(\mathbf{x}_0) - (\mathbf{x} - \mathbf{x}_0) \cdot \nabla_0 \phi_V(\mathbf{x}_0) \right],$$
$$\mathbf{A}(\mathbf{x}, \mathbf{x}_0) = \lim_{V \to \infty} \left[\mathbf{A}_V(\mathbf{x}) - \mathbf{A}_V(\mathbf{x}_0) - (\mathbf{x} - \mathbf{x}_0) \cdot \nabla_0 \mathbf{A}_V(\mathbf{x}_0) \right]. \tag{7.1.18}$$

The price for this is the abandonment of strict uniqueness; the theorem can then also be applied to sublinearly increasing vector fields, as shown in appendix A.4.7. The condition that $\mathbf{v}_l$ must asymptotically vanish cannot be maintained. The subfields $\mathbf{v}_{l,t}$ of a vector field $\mathbf{v}$, which asymptotically vanishes logarithmically, can each (logarithmically) diverge [Blumenthal, 1905; Petrascheck, Folk, 2017]. A consequence of 7.1.18 is the modification of (7.1.15):

$$\mathbf{v}(\mathbf{x}) - \mathbf{v}(\mathbf{x}_0) = -\boldsymbol{\nabla}\phi(\mathbf{x}, \mathbf{x}_0) + \boldsymbol{\nabla}\times\mathbf{A}(\mathbf{x}, \mathbf{x}_0). \tag{7.1.19}$$

3. The theorem also applies when singularities occur, as long as the integrals (7.1.16)–(7.1.18) converge, such as for the field of a point charge (problem 7.1).

Proof of the decomposition theorem

In (7.1.11) the decomposition of a vector field $\mathbf{v}$ into a source field $\mathbf{v}_l$ and a vortex field $\mathbf{v}_t$ was shown, if $\mathbf{v}$ asymptotically at least with a small power $\sim 1/r^\epsilon$ vanishes. The associated scalar potential (7.1.12) only exists if $v \sim 1/r^{1+\epsilon}$ falls off more than linearly. For weaker decay, we only provided the construction of the potential (7.1.14) and refer to the appendix A.4.7, where asymptotically sublinearly increasing vector fields are also discussed. The proofs of existence for the vector potentials are completely analogous to those for the scalar potentials, so a repetition for those is not necessary. So only the proof of the uniqueness of the decomposition remains.

Uniqueness of the decomposition of the vector field into a vortex-free and a source-free part

Let $\mathbf{v} = \mathbf{v}_l + \mathbf{v}_t$ and $\mathbf{v} = \mathbf{v}'_l + \mathbf{v}'_t$ be two different decompositions of $\mathbf{v}$ into source and vortex fields for the same sources and vortices, then $\mathbf{v}_d = \mathbf{v}_l - \mathbf{v}'_l$ is a source- and vortex-free vector field:

$$\boldsymbol{\nabla}\cdot\mathbf{v}_d = \boldsymbol{\nabla}\times\mathbf{v}_d = 0, \qquad \mathbf{v}_d = -\boldsymbol{\nabla}\phi_d, \qquad \Delta\phi_d = 0.$$

The general solution of the Laplace equation in spherical coordinates (3.2.38) is

$$\phi_d(\mathbf{x}) = \sum_{l=0}^{\infty} \sum_{m=-l}^{l} (\alpha_{lm} r^l + \beta_{lm} r^{-l-1}) Y_{lm}(\vartheta, \varphi).$$

Since according to the maximum and minimum principle (see appendix A.4.5) the maximum and minimum values of a harmonic function on the lying on the edge, all $\beta_{lm} = 0$ must be. The radial vector field

$$\mathbf{v}_d\cdot\mathbf{e}_r = -\frac{\partial\phi_d}{\partial r} = -\sum_{l=0}^{\infty}\sum_{m=-l}^{l} \alpha_{lm} l r^{l-1} Y_{lm}(\vartheta, \varphi) \overset{r\to\infty}{=} 0$$

must also asymptotically vanish, since $v_l(r \to \infty) = 0$. Thus, all coefficients $\alpha_{lm} = 0$ vanish except α_{00}, and the solution of the scalar potential is therefore a constant

$$\phi_d(\mathbf{x}) = \alpha_{00} Y_{00} = \alpha_{00}/\sqrt{4\pi}.$$

Thus, $\mathbf{v}_d = 0$, and the solution is unique. We have not shown that $\mathbf{v}_l$, calculated using (7.1.16), vanishes for $r \to \infty$.

Alternative representation of the potentials

The form of the potentials ϕ and $\mathbf{A}$ (7.1.16) follows directly from the decomposition (7.1.11), but is usually replaced by representations resulting from integrations by parts of (7.1.16). The integration is done over an unbounded domain $\mathcal{G} \subseteq \mathbb{R}^3$. To simplify the calculations, a single discontinuity surface S is assumed. This divides the domain $\mathcal{G}$ into the two subvolumes $V^>$ and $V^<$. Since $\mathbf{v}$ is continuous in both subvolumes, Gauss's theorem (A.4.3) can be applied

$$\phi(\mathbf{x}) = \frac{1}{4\pi} \int_{\mathcal{G}\backslash S} d^3 x' \, \frac{\boldsymbol{\nabla}' \cdot \mathbf{v}(\mathbf{x}')}{|\mathbf{x}-\mathbf{x}'|} - \frac{1}{4\pi} \left\{ \oiint_{\partial V_<} d\mathbf{S}' \cdot \frac{\mathbf{v}^<(\mathbf{x}')}{|\mathbf{x}-\mathbf{x}'|} + \oiint_{\partial V_>} d\mathbf{S}' \cdot \frac{\mathbf{v}^>(\mathbf{x}')}{|\mathbf{x}-\mathbf{x}'|} \right\}.$$

The remaining parts of the surface do not contribute as long as $v(r \to \infty) \sim 1/r^{1+\epsilon}$. For a weaker decrease, one resorts to (7.1.17)–(7.1.18). The same transformations can also be made with the vector potential and one obtains, whereby the boundary term is again evaluated using the (extended) Gauss's theorem (A.4.5).

The surface normals of $V^>$ and $V^<$ have different signs, so that we get at the discontinuity surfaces the difference $\mathbf{v}^< - \mathbf{v}^>$. In summary, we get:

$$\begin{aligned}
\phi(\mathbf{x}) &= \frac{1}{4\pi} \int_{\mathcal{G}\backslash S} d^3 x' \, \frac{\boldsymbol{\nabla}' \cdot \mathbf{v}(\mathbf{x}')}{|\mathbf{x}-\mathbf{x}'|} - \frac{1}{4\pi} \oiint_S d\mathbf{S}' \cdot \frac{\mathbf{v}^<(\mathbf{x}') - \mathbf{v}^>(\mathbf{x}')}{|\mathbf{x}-\mathbf{x}'|}, \\
\mathbf{A}(\mathbf{x}) &= \frac{1}{4\pi} \int_{\mathcal{G}\backslash S} d^3 x' \, \frac{\boldsymbol{\nabla}' \times \mathbf{v}(\mathbf{x}')}{|\mathbf{x}-\mathbf{x}'|} - \frac{1}{4\pi} \oiint_S d\mathbf{S}' \times \frac{\mathbf{v}^<(\mathbf{x}') - \mathbf{v}^>(\mathbf{x}')}{|\mathbf{x}-\mathbf{x}'|}.
\end{aligned} \tag{7.1.20}$$

As already mentioned, for a weaker decrease, one uses (7.1.17)–(7.1.18) and expands, if necessary, (7.1.20) to multiple discontinuities. For some applications, such as the calculation of the fields of homogeneously magnetized bodies, (7.1.20) will prove to be advantageous since the volume integral vanishes there.

Helmholtz theorem

The content of the Helmholtz theorem is the unique determination of a vector field $\mathbf{v}$ from its sources and vortices (*div-curl-problem*).

If $\mathbf{v}$ is now known, so are its sources $\rho = \boldsymbol{\nabla}\cdot\mathbf{v}/4\pi$ and vortices $\mathbf{j} = \boldsymbol{\nabla}\times\mathbf{v}/4\pi$, whereby $\mathbf{x} \notin F$. In addition, there are surface charges σ_v and surface vortices (surface currents) $\mathbf{K}_v$, where $\mathrm{d}\mathbf{a}' = \mathrm{d}a'\,\mathbf{n}$ and $\mathbf{n}$ points from $V^<$ outward.

$$\sigma_v = \mathbf{n}\cdot(\mathbf{v}^< - \mathbf{v}^>)/4\pi, \qquad\qquad \mathbf{K}_v = \mathbf{n}\times(\mathbf{v}^< - \mathbf{v}^>)/4\pi. \tag{7.1.21}$$

Inserted into (7.1.20), the potentials from the sources and vortices are obtained, as long as these fall off asymptotically quickly enough.

The contributions of the surface charges/surface vortices are a result of discontinuities in the vector field or singularities of the sources/vortices, which were excluded from the volume integrals in (7.1.20) by means of $\mathcal{G}\backslash S$. However, these can also remain integrated in the volume contributions. Thus, we formulate the Helmholtz theorem in reference to Griffiths [2017, appendix B]:

If the divergence $\rho(\mathbf{x})$ and the curl $\mathbf{j}(\mathbf{x})$ of a vector function $\mathbf{v}(\mathbf{x})$ are specified, and if they both go to zero faster than $1/r^2$ as $r\to\infty$, and if $\mathbf{v}(\mathbf{x})$ goes to zero as $r\to\infty$, then $\mathbf{v}$ is given uniquely by

$$\mathbf{v}(\mathbf{x}) = -\boldsymbol{\nabla}\phi(\mathbf{x}) + \boldsymbol{\nabla}\times\mathbf{A}(\mathbf{x}), \tag{7.1.22}$$

$$\phi(\mathbf{x}) = \int \mathrm{d}^3 x'\, \frac{\rho(\mathbf{x}')}{|\mathbf{x}-\mathbf{x}'|}, \qquad\qquad \mathbf{A}(\mathbf{x}) = \int \mathrm{d}^3 x'\, \frac{\mathbf{j}(\mathbf{x}')}{|\mathbf{x}-\mathbf{x}'|}. \tag{7.1.23}$$

If the source or vortex density decays by $1/r^\epsilon$ in the range $1 < \epsilon \leq 2$, the potentials can be calculated using (7.1.17).

Remark: For the even weaker decay $1/r^\epsilon$ with $0 < \epsilon \leq 1$, the potentials can be calculated using (7.1.18): $\mathbf{v}$ may then asymptotically diverge sublinearly, but is only determined up to a vector constant.

7.1.3 Potentials and Fields in Ferromagnets

The magnetization $\mathbf{M}$ is given. There are no free currents in the considered volume ($\mathbf{j}_f = 0$), so that $\boldsymbol{\nabla}\times\mathbf{H} = 0$. The material equation (7.1.2) is then a decomposition of $\mathbf{M}$ into a vortex-free and a source-free field:

$$4\pi k_r \mathbf{M} = -\mathbf{H} + \mathbf{B}/\mu_0 \qquad\Leftrightarrow\qquad \mathbf{v} = \mathbf{v}_l + \mathbf{v}_t. \tag{7.1.24}$$

So either $\mathbf{H}$ is calculated as a gradient of ϕ_M or $\mathbf{B}$ as the curl of $\mathbf{A}$. At the boundary surfaces of the magnet, $\mathbf{M}$ decreases so strongly that, as sketched in Fig. 7.2, a discontinuity describes the situation better than a continuous function.

The magnet has the volume V. The magnetization disappears in the outer space ($\mathbf{v}^> = 0$). Then one obtains from (7.1.16) taking into account $\mathbf{v}_l = -\mathbf{H} = \boldsymbol{\nabla}\phi_M$:

$$\phi_M(\mathbf{x}) = k_r \int_V \mathrm{d}^3 x'\, \mathbf{M}\cdot\boldsymbol{\nabla}' \frac{1}{|\mathbf{x}-\mathbf{x}'|},$$

$$\mathbf{A}(\mathbf{x}) = k_r \mu_0 \int_V \mathrm{d}^3 x'\, \mathbf{M}\times\boldsymbol{\nabla}' \frac{1}{|\mathbf{x}-\mathbf{x}'|}. \tag{7.1.25}$$

The discontinuity, however, is better represented by the formulation (7.1.20) resulting from integration by parts:

$$\phi_M(\mathbf{x}) = -k_r \int_V \mathrm{d}^3 x' \, \frac{\boldsymbol{\nabla}' \cdot \mathbf{M}}{|\mathbf{x}-\mathbf{x}'|} + k_r \oiint_{\partial V} \frac{\mathrm{d}\mathbf{a}' \cdot \mathbf{M}}{|\mathbf{x}-\mathbf{x}'|},$$

$$\mathbf{A}(\mathbf{x}) = k_r \mu_0 \left\{ \int_V \mathrm{d}^3 x' \, \frac{\boldsymbol{\nabla}' \times \mathbf{M}}{|\mathbf{x}-\mathbf{x}'|} - \oiint_{\partial V} \frac{\mathrm{d}\mathbf{a}' \times \mathbf{M}}{|\mathbf{x}-\mathbf{x}'|} \right\}. \tag{7.1.26}$$

These are the basic expressions for the potentials in magnetostatics. In many cases, the magnet has a constant magnetization; it then only contributes to the potential and consequently to the fields on the surface.

In the permanent magnet, we have the magnetostatic charge/current densities (7.1.8), (7.1.3) and the surface charge/surface current densities (7.1.9)

$$\rho_M(\mathbf{x}) = -\boldsymbol{\nabla} \cdot \mathbf{M}, \qquad\qquad \mathbf{j}_M(\mathbf{x}) = (c/k_L)\boldsymbol{\nabla} \times \mathbf{M},$$

$$\sigma_M(\mathbf{x}) = \mathbf{M} \cdot \mathbf{n}, \qquad\qquad \mathbf{K}_M(\mathbf{x}) = (c/k_L)\mathbf{M} \times \mathbf{n}. \tag{7.1.27}$$

If you insert these into (7.1.26), you get the potentials corresponding to the Helmholtz theorem ($k_r \mu_0 k_L / c = k_C / c k_L \overset{\mathrm{SI}}{=} \mu_0/4\pi$)

$$\phi_M(\mathbf{x}) = k_r \left\{ \int_V \mathrm{d}^3 x' \, \frac{\rho_M(\mathbf{x}')}{|\mathbf{x}-\mathbf{x}'|} + \oiint_{\partial V} \mathrm{d}S' \, \frac{\sigma_M(\mathbf{x}')}{|\mathbf{x}-\mathbf{x}'|} \right\},$$

$$\mathbf{A}(\mathbf{x}) = \frac{k_C}{c k_L} \left\{ \int_V \mathrm{d}^3 x' \, \frac{\mathbf{j}_M(\mathbf{x}')}{|\mathbf{x}-\mathbf{x}'|} + \frac{1}{c} \oiint_{\partial V} \mathrm{d}S' \, \frac{\mathbf{K}_M(\mathbf{x}')}{|\mathbf{x}-\mathbf{x}'|} \right\}. \tag{7.1.28}$$

7.1.4 Applications

Homogeneously magnetized sphere

There are several ways to calculate the magnetic field of a homogeneously magnetized sphere with the radius R ($\mathbf{M} = \mathbf{M}_0 \, \theta(R-r)$). We determine ϕ_M using (7.1.25) and $\boldsymbol{\nabla}' |\mathbf{x}-\mathbf{x}'|^{-1} = -\boldsymbol{\nabla} |\mathbf{x}-\mathbf{x}'|^{-1}$ in spherical coordinates ($\xi' = \cos\vartheta'$):

$$\phi_M(\mathbf{x}) = -k_r \mathbf{M}_0 \cdot \boldsymbol{\nabla} \phi_R(\mathbf{x}), \tag{7.1.29}$$

$$\phi_R(\mathbf{x}) = \int \mathrm{d}^3 x' \, \frac{\theta(R-r')}{|\mathbf{x}-\mathbf{x}'|} = \int_0^R \mathrm{d}r' \int_{-1}^1 \frac{2\pi r'^2 \, \mathrm{d}\xi'}{\sqrt{r^2 + r'^2 - 2rr'\xi'}}$$

$$= -\frac{2\pi}{r} \int_0^R \mathrm{d}r' \, r' \sqrt{r^2 + r'^2 - 2rr'\xi'} \, \Big|_{-1}^1 = \frac{2\pi}{r} \int_0^R \mathrm{d}r' \, r' \left[(r+r') - |r-r'|\right]$$

$$= \frac{4\pi R^3}{3} \begin{cases} 1/r & \text{for } r \geq R \\ 3/2R - r^2/2R^3 & \text{for } r < R. \end{cases} \tag{7.1.30}$$

From this follows with $\mathbf{m} = (4\pi R^3/3)\mathbf{M}_0$

$$\phi_M(\mathbf{x}) = k_r \begin{cases} \mathbf{m} \cdot \dfrac{\mathbf{x}}{r^3} & \text{for } r \geq R \\[2ex] \dfrac{1}{R^3}\,\mathbf{m} \cdot \mathbf{x} & \text{for } r < R. \end{cases} \tag{7.1.31}$$

For the fields then applies

$$\mathbf{H}_a = -\boldsymbol{\nabla}\phi_M = k_r\Big[3\frac{(\mathbf{m}\cdot\mathbf{x})\mathbf{x}}{r^5} - \frac{\mathbf{m}}{r^3}\Big] \qquad\qquad r \geq R,$$

$$\mathbf{H}_i = -k_r\frac{1}{R^3}\mathbf{m} = -\frac{4\pi k_r}{3}\mathbf{M}_0 \qquad\qquad r < R. \tag{7.1.32}$$

This directly implies

$$\mathbf{B}_a = \mu_0\mathbf{H}_a = k_r\mu_0\Big[3\frac{(\mathbf{m}\cdot\mathbf{x})\mathbf{x}}{r^5} - \frac{\mathbf{m}}{r^3}\Big] \qquad\qquad r \geq R,$$

$$\mathbf{B}_i = \mu_0(\mathbf{H}_i + 4\pi k_r\mathbf{M}_0) = \frac{8\pi k_r\mu_0}{3}\mathbf{M}_0 \qquad\qquad r < R.$$

Instead of calculating $\mathbf{H}$ using ϕ_M, $\mathbf{B}$ can be calculated using

$$\mathbf{A}(\mathbf{x}) = \frac{k_C}{k_L^2} \begin{cases} \mathbf{m} \times \dfrac{\mathbf{x}}{r^3} & \text{for } r \geq R \\[2ex] \dfrac{1}{R^3}\,\mathbf{m} \times \mathbf{x} & \text{for } r < R \end{cases} \tag{7.1.31'}$$

can be calculated (problem 7.4), or one can exploit the analogy between electrostatics and magnetostatics

$$\mathbf{P} \leftrightharpoons \mathbf{M}, \qquad\qquad \epsilon_0\mathbf{E} \leftrightharpoons \mathbf{H}, \qquad\qquad \mathbf{D} \leftrightharpoons \mu_0^{-1}\mathbf{B},$$

by using the results of the dielectric sphere, section 6.2.4, for $\mathbf{E}_0 = 0$, but with $\mathbf{P} \neq 0$:

$$\mathbf{E}_i = -\frac{4\pi k_C}{3}\mathbf{P} = -\frac{4\pi k_r}{3\epsilon_0}\mathbf{P} \qquad\Rightarrow\qquad \mathbf{H}_i = -\frac{4\pi k_r}{3}\mathbf{M}. \tag{6.2.7'}$$

Fig. 7.3 shows the field lines of $\mathbf{H}$ and $\mathbf{B}$ of a sphere. The $\mathbf{H}$-field has sources on the surface of the sphere, the $\mathbf{B}$-field has vortices.

Homogeneously magnetized bar magnet

The starting point is a cylinder of length $2l$ with radius a, as outlined in Fig. 7.4. Cylinder coordinates are taken according to the symmetry of the system. The homogeneous magnetization is thus determined by

$$\mathbf{M}(\mathbf{x}) = \theta(l^2 - z^2)\,\theta(a - \varrho)M\mathbf{e}_z.$$

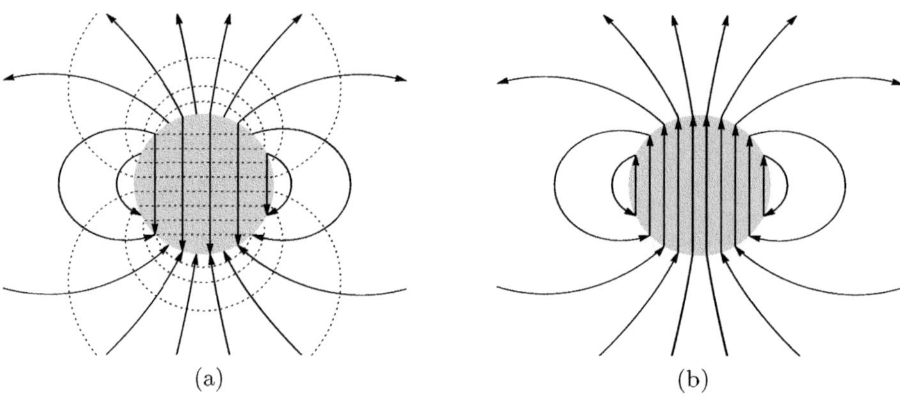

(a)　　　　　　　(b)

Fig. 7.3. Sphere with (a) **H**-field and (b) **B**-field

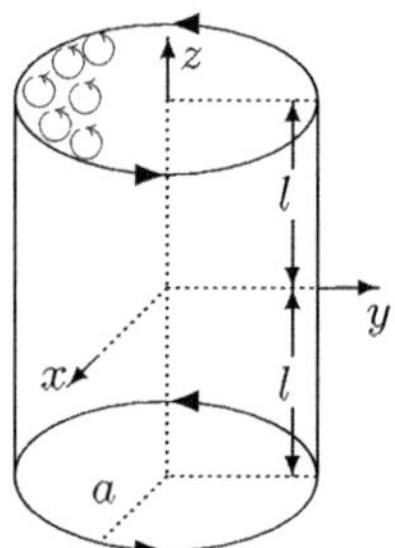

Fig. 7.4. Cylinder of radius a and length $2l$. The magnetic moments are sketched as small circular currents, all with the same direction of rotation

First, we determine the divergence and curl of $\mathbf{M}$, or charge and magnetization current:

$$\rho_M = -\boldsymbol{\nabla}\cdot\mathbf{M}(\mathbf{x}) = -\frac{\partial M_z}{\partial z} = 2z\delta(l^2 - z^2)\,\theta(a-\varrho)M$$

$$= \big[\delta(l-z) - \delta(l+z)\big]\theta(a-\varrho)M,$$

$$\mathbf{j}_M = \frac{c}{k_L}\boldsymbol{\nabla}\times\mathbf{M} = -\frac{c}{k_L}\mathbf{e}_\varphi\frac{\partial M_z}{\partial\varrho} = \frac{c}{k_L}\mathbf{e}_\varphi\,\theta(l^2 - z^2)\,\delta(a-\varrho)M.$$

So there are magnetic charges on both base surfaces and magnetization currents on the entire lateral surface. Since $\mathbf{j}_f = 0$, according to (7.1.4)

$$\mathbf{A} = \frac{k_C}{k_L c}\int \mathrm{d}^3 x'\,\frac{\mathbf{j}_M(\mathbf{x}')}{|\mathbf{x}-\mathbf{x}'|},\quad \mathbf{B} = \frac{k_C}{k_L c}\int \mathrm{d}^3 x'\,\mathbf{j}_M(\mathbf{x}')\times\frac{\mathbf{x}-\mathbf{x}'}{|\mathbf{x}-\mathbf{x}'|^3},\quad \text{SI:}\ \frac{k_C}{k_L c} = \frac{\mu_0}{4\pi}.$$

$\mathbf{j}_M$ corresponds to the current density (4.2.36) in the finite coil. We only have to replace the current per unit length I_n by $(c/k_L)M$. Then the field of the coil (4.2.37) can be adopted.

The surface current density is $j_\varphi = (c/k_L)\theta(l^2 - z^2)\,\delta(a-\varrho)M$. The currents inside compensate each other, as sketched in Fig. 7.4. Let $\mathbf{B}_i$ and $\mathbf{H}_i$ be the fields inside the coil and $\mathbf{B}_a$ and $\mathbf{H}_a$ the fields in the outer space, then it applies

$$\mathbf{B}_i = \mu_0(\mathbf{H}_i + 4\pi k_r \mathbf{M}), \qquad\qquad \mathbf{B}_a = \mu_0 \mathbf{H}_a .$$

The boundary conditions state that the normal component B_n and the tangential components H_{tg} are continuous.

On the base surfaces, the B-lines are continuous, since $\mathbf{M}$ is perpendicular to the surface and thus also the tangential component is continuous. The situation on the lateral surface is more complicated. The two components B_ϱ and B_φ are indeed continuous, but in the z-direction, due to $B_{az} = B_{iz} - 4\pi k_r \mu_0 M$, a change of direction occurs, which turns the field into the negative z-direction as can be seen in Fig. 7.5. The lines have been drawn using (4.2.38) and (4.2.40), since for $M = I_n k_L / c$ the field $\mathbf{B}$ is equal to that of the coil. Now B-lines are always closed, i.e., lines near the lateral surface form a vortex-like structure, as they only have a small normal component there.

Scalar potential

First, we know $\mathbf{H} = \mu_0(\mathbf{B} - 4\pi k_r \mathbf{M})$, since $\mathbf{B}$ is known. Since $\mathbf{M} = M\,\mathbf{e}_z$ is constant, the volume integral with $\boldsymbol{\nabla}\cdot\mathbf{M} = 0$ vanishes and only the surface contribution remains and from this only the integrals over the base and top surface

$$\phi_M = k_r \oiint \frac{\mathrm{d}\mathbf{S}'\cdot\mathbf{M}}{|\mathbf{x}-\mathbf{x}'|} \qquad \Rightarrow \qquad \mathbf{H} = -\boldsymbol{\nabla}\phi_M = k_r \oiint \mathrm{d}\mathbf{S}'\cdot\mathbf{M}\,\frac{\mathbf{x}-\mathbf{x}'}{|\mathbf{x}-\mathbf{x}'|^3} .$$

It turns out that the direct calculation of $\mathbf{H}$ is more difficult than that of $\mathbf{B}$, so here $\mathbf{H}$ is only determined by $\mathbf{H} = \mathbf{B}/\mu_0 - 4\pi k_r \mathbf{M}$.

In Fig. 7.5, the $\mathbf{H}$-field of a cylinder is sketched. The sources are on the base surfaces, and the analogy to the $\mathbf{E}$-field of two homogeneously charged plates is obvious; since $a \ll l$, as in the edge area of a capacitor, there is a tendency for the field lines to deflect outward. On the lateral surface, the H-lines continuously pass through, as there $\mathbf{M}$ is parallel to the surface, and on the base surfaces they begin (end) in the sources (sinks). A homogeneous field is present in the magnet, analogous to electrostatics, in the ellipsoid, i.e. also in the sphere Fig. 7.3.

Note: In older literature [Sommerfeld, 1952, p. 10] one finds for the magnetic auxiliary field $\mathbf{H}$ the term *magnetic excitation* in analogy to the *electrical excitation* for the dielectric displacement. The $\mathbf{H}$-lines in Fig. 7.5 are thus excitation lines.

In the z-axis, $\varrho = 0$, the fields can be given in a simple form [Sommerfeld, 1952, §12, (7b)]:

$$\mathbf{B} = \frac{4\pi k_C}{k_L^2}\frac{M}{2}\Big[\frac{z+l}{\sqrt{a^2+(z+l)^2}} - \frac{z-l}{\sqrt{a^2+(z-l)^2}}\Big]\mathbf{e}_z, \quad \text{SI:}\ \frac{4\pi k_C}{k_L^2} = \mu_0. \qquad (7.1.33)$$

Fig. 7.6 shows the field strengths of $\mathbf{B}$ and $\mathbf{H}$ along the z-axis. In a longer rod magnet, H almost disappears inside. It is said that $\mathbf{H}$ "has a demagnetizing effect". Note in particular that the positive and negative areas of the H-curve

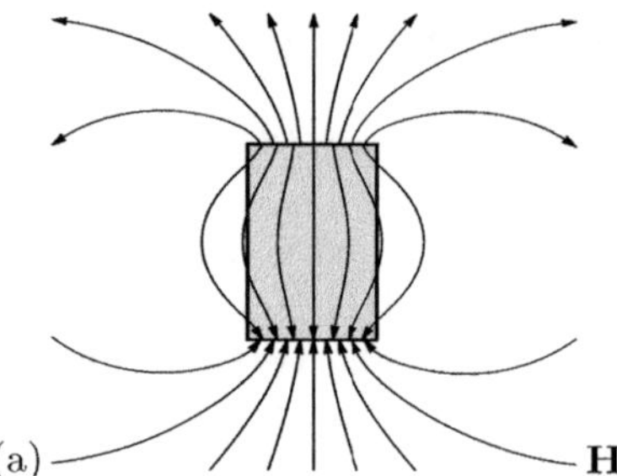
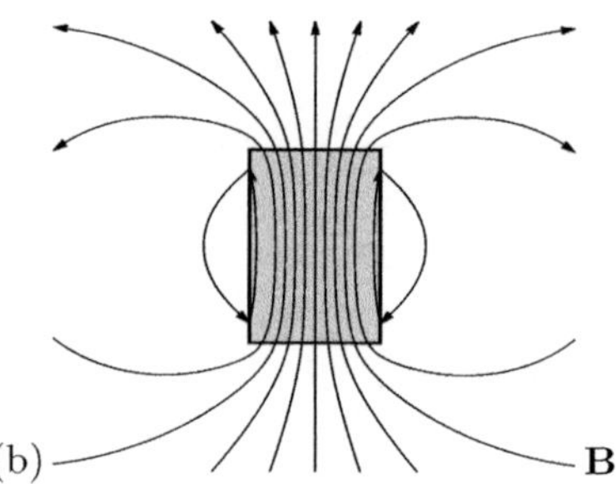

Fig. 7.5. (a) Cylinder with sources (+ or N) at the upper and sinks (– or S) at the lower base surface and corresponding H-field lines; these are, starting from the sources on the top surface, within the cylinder pushed towards the edge.
(b) The B-field lines are continuous at the base surfaces; vortices are on the lateral surface; the field lines are pushed into the interior of the cylinder. In the exterior space, B- and H-field lines are the same

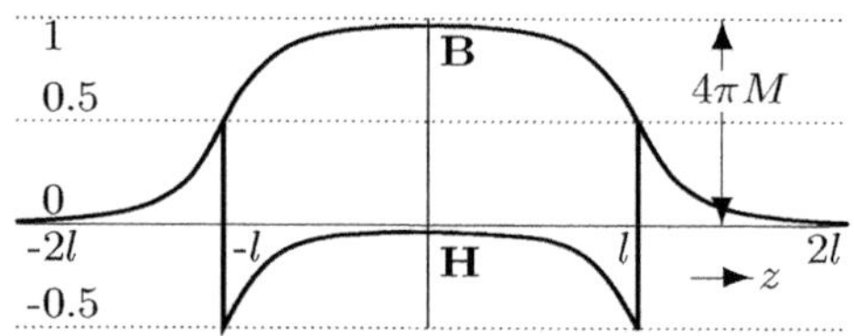

Fig. 7.6. Cylindrical permanent magnet with radius $a = l/4$ and $\mathbf{M} = M\mathbf{e}_z$. The field strengths B_z and H_z on the cylinder axis $\varrho = 0$ are shown.

compensate each other, since $\mathbf{H}$ is vortex-free. The magnetic ring tension Z_B (vortex strength of the magnetic field $\mathbf{B}$) is

$$Z_B = \oint d\mathbf{x} \cdot \mathbf{B} \qquad \overset{\varrho<a}{\Longrightarrow} \qquad Z_B = \int_{-\infty}^{\infty} dz\, B_z = \frac{4\pi k_C}{k_L^2} M\, 2l,$$

$$Z_H = \oint d\mathbf{x} \cdot \mathbf{H} = 0 \qquad \overset{\varrho<a}{\Longrightarrow} \qquad Z_H = \int_{-\infty}^{\infty} dz\, H_z = 0.$$

From the difference of the path integrals, one obtains $Z_B = \dfrac{4\pi k_C}{k_L^2} \oint d\mathbf{x} \cdot \mathbf{M}$. The explicit calculation is required in problem 7.5.

Magnetic field of a ring coil

In Fig. 7.7, a ring coil with radius R is sketched. As long as the gap d is very small, the stray field can also be neglected.

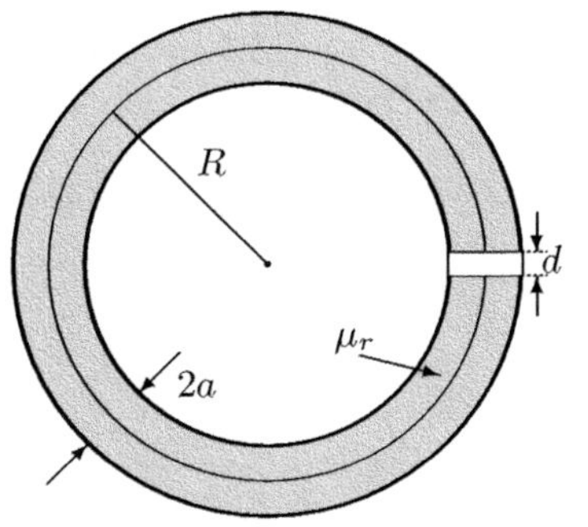

Fig. 7.7. Ring coil with narrow gap d
R Radius of the torus (coil)
      Extension of the coil: $[R - a, R + a]$
N Number of windings.
$n = \dfrac{N}{2\pi R}$ Number of windings per unit length
d Gap width
I Current in the wire
μ_r Permeability of the coil's iron core

Within the coil, the magnetic field $H(\varrho)\,\mathbf{e}_\varphi$ will be largely constant. $\mathbf{H}$ is calculated with Ampère's law (4.1.5), which represents the integral form of the inhomogeneous Maxwell equation (7.1.1):

$$\iint_A \mathbf{da}\cdot(\boldsymbol{\nabla}\times\mathbf{H}) = 4\pi k_r \frac{k_L}{c}\iint_A \mathbf{da}\cdot\mathbf{j}_f \quad\overset{\text{Stokes}}{\Longrightarrow}\quad \oint_{\partial A}\mathbf{ds}\cdot\mathbf{H} = 4\pi k_r\frac{k_L I_A}{c}.$$

$$(7.1.34)$$

I_A is the current passing through the area A. First, the closed coil is considered; the torus lies in the plane $z = 0$ with the center at the origin of the coordinates. The area A is the circular area K_ϱ in the xy-plane, which intersects the coil for circles with $R - a < \varrho < R + a$. With N windings, you get NI inside the coil. If $\varrho > R + a$, the contribution of the outer windings compensates for that of the inner ones and you have $\mathbf{H}=0$ again:

$$\oint_{\partial S_\varrho}\mathbf{ds}\cdot\mathbf{H} = 2\pi\varrho H = 4\pi k_r\frac{k_L}{c}\iint_{S_\varrho}\mathbf{dS}\cdot\mathbf{j}_f = \frac{4\pi k_r k_L NI}{c}\left[\theta(R - a - \varrho) - \theta(R + a - \varrho)\right].$$

The closed ring coil therefore has no stray field.

With a gap d, you replace $2\pi\varrho H$ with $(2\pi\varrho - d)H + dH_s$. Because $\boldsymbol{\nabla}\cdot\mathbf{B} = 0$, the normal component of $\mathbf{B}$ is continuous and you get in the center of the coil for the fields B, H in the magnetic core and B_s, H_s in the gap: $B = B_s$, $\mu_0 H_s = B = \mu_r\mu_0 H$, i.e. $H_s = \mu_r H$:

$$\text{G: } H_s = \mu_r\frac{4\pi}{c}\frac{IN}{(2\pi R - d) + d\mu_r}, \qquad \text{SI: } H_s = \mu_r\frac{IN}{(2\pi R - d) + d\mu_r}. \qquad (7.1.35)$$

Infinitely long coil

We can consider the infinitely long coil as a limit case $R \to \infty$ of the ring coil and then obtain from (7.1.35) that the field in the coil is given by

$$H = \frac{k_L}{c}4\pi k_r I n.$$

The closed ring coil has no field in the outer space, which should also apply to the straight coil of length $l \to \infty$. This is consistent with the result derived in section 4.2.5, page 138 There, $\mathbf{B}$ for an infinitely long, straight coil as limit $l \to \infty$ of the finite coil of length $2l$ was derived using Biot-Savart's law. It is simpler to derive the field from Ampère's law.

In the direct calculation, due to the axial symmetry and the infinite length, we expect that $\mathbf{H} = \mathbf{H}(\varrho)$. The Maxwell equation

$$\boldsymbol{\nabla} \times \mathbf{H}(\varrho) = \frac{k_L}{c} 4\pi k_r\, j\, \mathbf{e}_\varphi$$

further restricts $\mathbf{H} = H(\varrho)\,\mathbf{e}_z$. Now, as sketched in Fig. 7.8, we take out a rectangle of area A and move along the boundary ∂A:

$$\oint_{\partial A} \mathrm{d}\mathbf{x} \cdot \mathbf{H} = \big[H_z(\varrho_1) - H_z(\varrho_2) \big] l_z = \frac{4\pi k_r k_L n I}{c}\, l_z \big[\theta(a - \varrho_1) - \theta(a - \varrho_2) \big].$$

Since $H_z(\varrho_2 \to \infty) = 0$, we obtain inside the coil the homogeneous field

$$\mathbf{B} = \frac{4\pi k_C}{c k_L}\, I n \mu_r\, \theta(a - \varrho)\, \mathbf{e}_z\,, \qquad \text{SI:}\ \mathbf{B} = \mu_0 \mu_r I n\, \theta(a - \varrho)\, \mathbf{e}_z. \qquad (7.1.36)$$

In direct calculation of the vector potential with (7.1.4), $\mathbf{A}$ can diverge, if

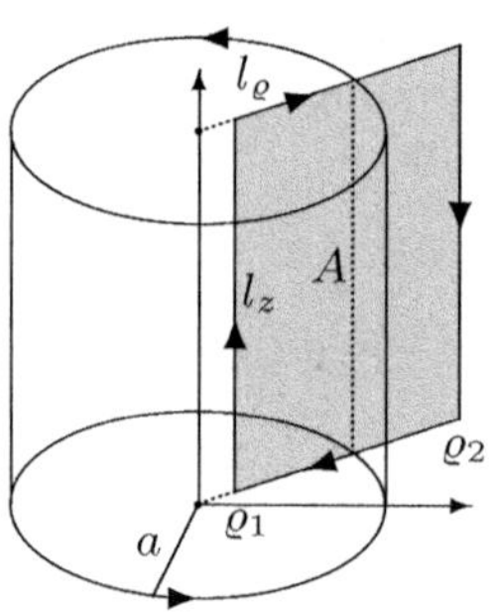

Fig. 7.8. Infinitely long coil with radius a. To calculate H with Ampère's law: If the shaded rectangle A encloses the wires of one side then it is $H\, l_z = 4\pi n I l_z / c$

the currents are not limited to a finite area, which is not the case for the field

$$\mathbf{H} = k_r \frac{k_L}{c} \int \mathrm{d}^3 x'\, \frac{\mathbf{j}_f(\mathbf{x}') \times (\mathbf{x} - \mathbf{x}')}{|\mathbf{x} - \mathbf{x}'|^3} \qquad \text{with} \qquad \mathbf{j}_f = n I \delta(a - \varrho')\, \mathbf{e}_{\varphi'}$$

(Biot-Savart, see section 4.2.5). The vector field associated with the coil is most easily calculated in the current-free space from the definition of the magnetic flux using Stokes' theorem:

$$\Phi_B = \iint_A \mathrm{d}\mathbf{a} \cdot \mathbf{B} = \iint_A \mathrm{d}\mathbf{a} \cdot (\boldsymbol{\nabla} \times \mathbf{A}) = \oint_{\partial A} \mathrm{d}\mathbf{s} \cdot \mathbf{A}.$$

For symmetry reasons, $\mathbf{A} = A(\varrho)\,\mathbf{e}_\varphi$, which we determine from Φ_B

$$2\pi \varrho A_\varphi(\varrho) = \begin{cases} \varrho^2 B \pi & \text{for } \varrho < a \\ a^2 \pi B & \text{for } \varrho > a. \end{cases}$$

From this it follows, since we know B,

$$\mathbf{A} = \mathbf{e}_\varphi \begin{cases} \dfrac{\varrho B}{2} = \dfrac{4\pi k_r k_L}{c}\dfrac{nI\mu_r}{2}\varrho \overset{\text{SI}}{=} \dfrac{nI\mu_r}{2}\varrho, & \varrho < a \\[3mm] \dfrac{a^2 B}{2\varrho} = \dfrac{4\pi k_r k_L}{c}\dfrac{nI\mu_r}{2}\dfrac{a^2}{\varrho} \overset{\text{SI}}{=} \dfrac{nI\mu_r}{2}\dfrac{a^2}{\varrho}, & \varrho > a. \end{cases} \tag{7.1.37}$$

Magnetic field of a semi-infinite cuboid

Here we consider a rather exotic configuration. An infinite cuboid: $-\infty < x, z < 0$ and $-\infty < y < \infty$ is homogeneously magnetized, as sketched in Fig. 7.9:

$$\mathbf{M}(\mathbf{x}) = M\mathbf{e}_z\,\theta(-x)\,\theta(-z). \tag{7.1.38}$$

The magnetic charge density ρ_M and the surface current density are:

$$\rho_M(\mathbf{x}) = -\boldsymbol{\nabla}\cdot\mathbf{M} = M\delta(z)\theta(-x), \quad \mathbf{j}_M(\mathbf{x}) = \frac{c}{k_L}\boldsymbol{\nabla}\times\mathbf{M} = \frac{c}{k_L}M\delta(x)\theta(-z)\mathbf{e}_y.$$

The configuration corresponding to electrostatics is the semi-infinite charged plane. Two oppositely charged half-planes were compared with the plate capacitor in Fig. 2.18, and in this context, the configuration looks somewhat less exotic. Here, the main aim is to show that a vector field ($\mathbf{M}$), which is finite everywhere, has a source field ($\mathbf{H}$) and a vortex field ($\mathbf{B}$) that diverge logarithmically.

We carry out the following calculation using the Helmholtz theorem, according to which the fields apply:

$$4\pi k_r \mathbf{M} = -\mathbf{H} + \mathbf{B}/\mu_0 \qquad\qquad \Leftrightarrow \qquad\qquad \mathbf{v} = \mathbf{v}_l + \mathbf{v}_t.$$

Following the notation of the Helmholtz theorem, it is

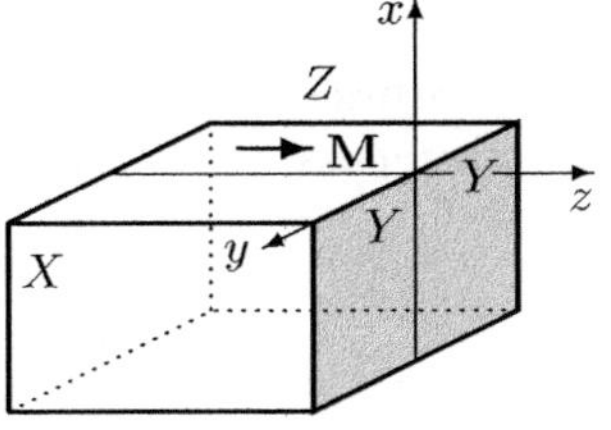

Fig. 7.9. Cuboid with the sides $(X, Y, Z \to \infty)$:
$-X < x < 0$, $-Y < y < Y$ and $-Z < z < 0$
is homogeneously magnetized: $\mathbf{M} = M\mathbf{e}_z$.
For the potential ϕ_M, only the shaded area contributes
for $Z \to \infty$

$$\rho = \frac{\boldsymbol{\nabla}\cdot\mathbf{v}}{4\pi} = -k_r\rho_M, \qquad\qquad \phi = \int \mathrm{d}^3x' \frac{\rho(\mathbf{x}')}{|\mathbf{x}-\mathbf{x}'|} \overset{(7.1.26)}{=} -\phi_M.$$

First, $\phi \to \phi_V$ must be calculated for a finite cuboid. Since the magnetization is constant, only the surface contributes, especially the shaded front surface in Fig. 7.9. The calculation itself is tedious and part of the problem 7.6. Due to the divergence of $\lim_{V\to\infty} \phi_V$ we have to fall back on (7.1.18):

$$\phi_2(\mathbf{x}, \mathbf{x}_0) = \lim_{V\to\infty} \left\{ \phi_V(\mathbf{x}) - \phi_V(\mathbf{x}_0) - (\mathbf{x} - \mathbf{x}_0)\cdot\boldsymbol{\nabla}_0\phi_V(\mathbf{x}_0) \right\}$$

$$= k_r M \left\{ \pi z(\operatorname{sgn} z - \operatorname{sgn} z_0) - 2z\left(\arctan\frac{x}{z} - \arctan\frac{x_0}{z_0} \right) \right. \tag{7.1.39}$$

$$\left. - x\left(\ln(x^2 + z^2) - \ln(x_0^2 + z_0^2) \right) + 2(x - x_0) \right\}.$$

$$\mathbf{v}_l(\mathbf{x}, \mathbf{x}_0) = -\boldsymbol{\nabla}\phi_2(\mathbf{x}, \mathbf{x}_0) = \mathbf{v}_l(\mathbf{x}) - \mathbf{v}(\mathbf{x}_0),$$

$$\mathbf{v}_l(\mathbf{x}) = k_r M \left\{ \mathbf{e}_x \ln(x^2 + z^2) + \mathbf{e}_z \left[2\arctan\frac{x}{z} - \pi\operatorname{sgn} z \right] \right\}. \tag{7.1.40}$$

Returning to magnetostatics, it holds $\mathbf{H}(\mathbf{x}, \mathbf{x}_0) = -\mathbf{v}_l(\mathbf{x}, \mathbf{x}_0)$. The vortex field follows from (7.1.19):

$$\mathbf{B}(\mathbf{x}, \mathbf{x}_0) = 4\pi k_r \mu_0 \left[\mathbf{M}(\mathbf{x}) - \mathbf{M}(\mathbf{x}_0) \right] + \mu_0 \mathbf{H}(\mathbf{x}, \mathbf{x}_0).$$

$\mathbf{B} = \mu_0\mathbf{H}$ therefore only applies if, in addition to $\mathbf{x}$, $\mathbf{x}_0$ is also outside the magnetized space; however, the fields are still dependent on $\mathbf{x}_0$. It is noteworthy that $\mathbf{B}$ and $\mathbf{H}$ diverge logarithmically, even though the distance to the cuboid becomes large for $z \to \infty$.

7.1.5 Magnetostatic Force Law

Already in Newton's time, there were attempts to determine the dependence of the attraction or repulsion of magnetic poles. In 1750, Michell [Michell, 1750, p. 19] stated the force as inversely proportional to the square of the distance. In addition to attempts to determine the repulsion of electric charges using the torsion balance, Coulomb [Coulomb, 1780–1789] also applied the corresponding force law to magnetic poles. The prerequisite is that the magnets are long and thin, so that the opposite poles are not effective [Maxwell, 1892, Art. 373]:

The repulsion between two like magnetic poles is in the straight line joining them, and is numerically equal to the product of the strengths of the poles divided by the square of the distance between them.

$$\mathbf{F}_M = k_M\, p_m\, p_m' \frac{\mathbf{x} - \mathbf{x}'}{|\mathbf{x} - \mathbf{x}'|^3} \overset{(4.2.46')}{=} p_m\, \mathbf{B} = p_B\, \mathbf{H}. \tag{7.1.41}$$

Following Sommerfeld [1952, (2.6)], the pole strength p_m here refers to the magnetic charge of a pole, while Jackson [1998, (6.157)] gives the pole strength as $g = p_B = p_m\mu_0$ [Petrascheck, 2021]. For a thin solenoid, the pole strength in (4.2.46) was given as

$$p_m = (k_L/c)\, I_n\, a^2 \pi$$

where $I_n = I/l$ is the current per unit length. Starting from Tab. C.2, p. 634, the pole strength (7.1.6) can be given as $p_m = (k_L/\sqrt{k_C})\,p_m^*$ if p_m^* is given in Gauss units. Since $k_M p_m p_m'$ is independent of the units, one obtains the relationship

$$k_M = k_C/k_L^2. \tag{7.1.42}$$

The magnetostatic force law was used to define the electromagnetic units (emu) and the Coulomb's law to define the electrostatic units (esu), both of which are discussed in section C.2.1. Two unit poles repel each other at a distance $r = 1\,\mathrm{cm}$ with $F_M = 1\,\mathrm{dyn}$. The current I necessary for the pole strength is in electromagnetic units $1\,\mathrm{Biot}$, where $a^2\pi = 1\,\mathrm{cm}^2$, $l = 1\,\mathrm{cm}$ and $c = c_0\,\mathrm{cm\,s}^{-1}$:

$$p_B = \mu_0 k_L r \sqrt{\frac{F_M}{k_C}}, \quad I = \frac{rlc}{a^2\pi}\sqrt{\frac{F_M}{k_C}} = \begin{cases} \sqrt{4\pi/\mu_0}\,\sqrt{\mathrm{dyn}} = 10\,\mathrm{A}, & \mathrm{SI}, \\ \sqrt{\mathrm{dyn}} = 1\,\mathrm{Biot}, & \mathrm{emu}, \\ c_0\sqrt{\mathrm{cm}^3\,\mathrm{g\,s}^{-2}} = c_0\,\mathrm{statA}, & \mathrm{G,\ esu}. \end{cases}$$

It should be emphasized again that (7.1.41) is very limited in its validity and is only of interest from a historical perspective.

7.2 Induction

From the law of induction we know that with the temporal change of the magnetic flux $(k_L/c)\frac{\mathrm{d}}{\mathrm{d}t}\Phi_B = -\oint_{\partial A} \mathrm{d}s\cdot\mathbf{E}$ an electric vortex field $\mathbf{E}$ is built up, which in turn causes an electric current in a metal wire. This process is referred to as electromagnetic induction. If Φ_B is generated by a stationary current I, then according to the Biot-Savart law $\mathbf{B}$, i.e. also Φ_B, is proportional to I: $\Phi_B = (k_L/c)IL$. L is a coefficient of induction determined by the geometry of the arrangement. This coefficient is accessible to us both through the magnetic energy and through the magnetic flux.

7.2.1 Energy of the Magnetic Field

We obtained the energy of the magnetic field

$$U = \frac{1}{8\pi k_r} \int \mathrm{d}^3x\, \mathbf{B}\cdot\mathbf{H} \tag{7.2.1}$$

as part of the field energy (5.6.7) when creating the energy balance of the electromagnetic field. In the renewed determination, the focus should primarily be on the difference between magnetic and electrostatic field energy.

In electrostatics, the energy of a charge distribution is determined by bringing the charges from infinity into the desired configuration. For a selected particle n, the energy

$$U_n = -e \int_{-\infty}^{\mathbf{x}} \mathrm{d}\mathbf{x}' \cdot \mathbf{E}^{(n)}(\mathbf{x}',t') = -e \int_{t_0}^{t} \mathrm{d}t'\, \mathbf{v}(t') \cdot \mathbf{E}^{(n)}(\mathbf{x}',t')$$

must be applied to move it to the prescribed position. $\mathbf{E}^{(n)}$ is the field of the already existing configuration, with the particles being held in their places The charge is brought to the location $\mathbf{x}$ in the time interval $[t_0, t]$, where $\mathrm{d}\mathbf{x} = \mathbf{v}\,\mathrm{d}t$.

The magnetic energy cannot be understood in this way, since the force $\sim \mathbf{v} \times \mathbf{B}$ is perpendicular to $\mathbf{v}$ and therefore does not do work. However, building up the magnetic field requires energy, which is provided by the currents, or by the source-free part of the electric field. From the induction law (1.3.10)

$$\boldsymbol{\nabla} \times \mathbf{E} = -\frac{k_L}{c}\dot{\mathbf{B}} = -\frac{k_L}{c}\boldsymbol{\nabla} \times \dot{\mathbf{A}}$$

it follows that the source-free part of $\mathbf{E}$ is given by $-(k_L/c)\dot{\mathbf{A}}$.

When a circuit $I(t)$ is switched on, it generates the magnetic field $\mathbf{B}(\mathbf{x},t)$. The power provided by $\mathbf{E}$ is then

$$\frac{\mathrm{d}U_m}{\mathrm{d}t} = -I(t) \oint_{\partial A} \mathbf{E} \cdot \mathrm{d}\mathbf{x} = \frac{k_L}{c}I(t) \oint_{\partial A} \mathrm{d}\mathbf{x} \cdot \dot{\mathbf{A}}.$$

$\oint \mathrm{d}\mathbf{x} \cdot \mathbf{E}$ is the electromotive force (emf) Z_E, which acts on an electron during a revolution on the circular current.

Now we move from the wire to a continuous current distribution and set $\mathbf{j}_f$ from the Ampère-Maxwell law (5.2.16c):

$$\frac{\mathrm{d}U_m}{\mathrm{d}t} = -\int_V \mathrm{d}^3x\, \mathbf{j}_f \cdot \mathbf{E} = -\frac{c}{4\pi k_r k_L} \int_V \mathrm{d}^3x \left((\boldsymbol{\nabla} \times \mathbf{H}) - \frac{k_L}{c}\dot{\mathbf{D}} \right) \cdot \mathbf{E}. \qquad (7.2.2)$$

The last term determines the electric field energy

$$\dot{U}_e = \frac{1}{8\pi k_r} \frac{\mathrm{d}}{\mathrm{d}t} \int_V \mathrm{d}^3x\, \mathbf{D} \cdot \mathbf{E}(\mathbf{x},t) \quad \Rightarrow \quad \int_{t_0}^{t} \mathrm{d}t'\, \dot{U}_e(t') = U_e(t) - U_e(t_0) = 0,$$

which must be applied and which vanishes, since $\mathbf{B}$ is constant in both the initial and final state and thus the source-free field $\mathbf{E} = 0$. The further calculation proceeds completely analogous to that for the energy balance, page 191. If you insert (5.6.2)

$$\mathbf{E} \cdot (\boldsymbol{\nabla} \times \mathbf{H}) = \mathbf{H} \cdot (\boldsymbol{\nabla} \times \mathbf{E}) - \boldsymbol{\nabla} \cdot (\mathbf{E} \times \mathbf{H}) = -\frac{k_L}{c}\mathbf{H} \cdot \dot{\mathbf{B}} - \frac{4\pi k_r k_L}{c} \boldsymbol{\nabla} \cdot \mathbf{S}$$

into (7.2.2), the result after applying Gauss's theorem is

$$\frac{\mathrm{d}U_m}{\mathrm{d}t} = \oiint_{\partial V} \mathrm{d}\mathbf{a} \cdot \mathbf{S} + \frac{1}{4\pi k_r} \int_V \mathrm{d}^3x\, \dot{\mathbf{B}} \cdot \mathbf{H} \stackrel{V \to \infty}{=} \frac{1}{8\pi k_r} \frac{\mathrm{d}}{\mathrm{d}t} \int_V \mathrm{d}^3x\, \mathbf{B} \cdot \mathbf{H}. \qquad (7.2.3)$$

The flux through ∂V disappears for a sufficiently large V, and additionally let $\mu_r(\mathbf{x},t) = \mu_r(\mathbf{x})$. If you integrate (7.2.3) with $U_m(t_0) = 0$, you get for $U_m(t)$ (7.2.1).

The magnetic field energy as a function of the current density

The representations of field energy analogous to electrostatics

$$U_e = \frac{1}{8\pi k_r} \int \mathrm{d}^3x\, \mathbf{E}\cdot\mathbf{D} = \frac{1}{2}\int \mathrm{d}^3x\, \rho_f \phi = \frac{k_C}{2}\int \mathrm{d}^3x \int \mathrm{d}^3x'\, \frac{\rho_f(\mathbf{x})\rho_f(\mathbf{x}')}{\epsilon(\mathbf{x}')|\mathbf{x}-\mathbf{x}'|}$$

can also be found in magnetostatics. Starting from (7.2.1) for a linear medium, U_m using $\mathbf{B} = \boldsymbol{\nabla}\times\mathbf{A}$ can be transformed as follows:

$$U_m = \frac{1}{8\pi k_r}\int \mathrm{d}^3x\, \mathbf{H}\cdot(\boldsymbol{\nabla}\times\mathbf{A}) = \frac{1}{8\pi k_r}\int \mathrm{d}^3x\left[\boldsymbol{\nabla}\cdot(\mathbf{A}\times\mathbf{H})+\mathbf{A}\cdot(\boldsymbol{\nabla}\times\mathbf{H})\right]. \quad (7.2.4)$$

The 1st term on the right side is the divergence of $\mathbf{A}\times\mathbf{H}$ and can be transformed into a surface integral using the Gauss's theorem, which for $r \to \infty$ under the assumption of local currents vanishes:

$$\frac{1}{4\pi}\oiint \mathrm{d}\mathbf{a}\cdot(\mathbf{A}\times\mathbf{H}) \overset{r\to\infty}{\propto} r^2\,\frac{1}{r}\,\frac{1}{r^2} \to 0\,.$$

According to Ampère's law (7.1.1), $\boldsymbol{\nabla}\times\mathbf{H} = (4\pi k_r k_L/c)\,\mathbf{j}_f$. For $\mathbf{A}$ we insert (7.1.5)

$$U_m = \frac{k_L}{2c}\int \mathrm{d}^3x\, \mathbf{A}\cdot\mathbf{j}_f = \frac{k_C}{2c^2}\int \mathrm{d}^3x \int \mathrm{d}^3x'\, \mu_r(\mathbf{x}')\frac{\mathbf{j}_f(\mathbf{x})\cdot\mathbf{j}_f(\mathbf{x}')}{|\mathbf{x}-\mathbf{x}'|}\,. \quad (7.2.5)$$

Magnetic self and interaction energy

Given are $n \geq 2$ spatially separated circuits, through which stationary (time-independent) currents flow. They are embedded in a medium of homogeneous permeability μ_r, as sketched in Fig. 7.10, page 243 for $n = 2$ circuits. Current densities and vector potentials can be specified separately for the circuits:

$$\mathbf{j}_f(\mathbf{x}) = \sum_{i=1}^{n}\mathbf{j}_i\,, \qquad\qquad \mathbf{A}(\mathbf{x}) = \sum_{i=1}^{n}\mathbf{A}_i(\mathbf{x})\,.$$

The field energy generated by these currents (7.2.5)

$$U_m = \sum_{i,k=1}^{n}U_{ik}\,, \qquad\qquad U_{ik} = \frac{k_L}{2c}\int \mathrm{d}^3x\, \mathbf{A}_i(\mathbf{x})\cdot\mathbf{j}_k(\mathbf{x}) \qquad (7.2.6)$$

decomposes into self and interaction energy U_s and U_w in a similar way as in the electrostatics the energy of the charge distributions (see section 2.4):

$$U_m = U_s + U_w\,, \qquad U_s = \sum_{i=1}^{n}U_{ii}\,, \qquad U_w = \sum_{i=1}^{n-1}\sum_{k>i}^{n}(U_{ik}+U_{ki})\,.$$

For the vector potential, one uses (7.1.5) and obtains so

$$
U_s = \sum_{i=1}^{n} \frac{k_C}{2c^2}\mu_r \int \mathrm{d}^3x\,\mathrm{d}^3x'\,\frac{\mathbf{j}_i(\mathbf{x})\cdot\mathbf{j}_i(\mathbf{x}')}{|\mathbf{x}-\mathbf{x}'|}, \qquad \text{SI:}\ \frac{k_C}{c^2}=\frac{\mu_0}{4\pi},
$$

$$
U_w = \sum_{i=1}^{n}\sum_{k\neq i}^{n}\frac{k_L}{2c}\int \mathrm{d}^3x\,\mathbf{A}_i\cdot\mathbf{j}_k = \sum_{i=1}^{n-1}\sum_{k>i}^{n}\frac{k_C}{c^2}\mu_r \int \mathrm{d}^3x\,\mathrm{d}^3x'\,\frac{\mathbf{j}_i(\mathbf{x})\cdot\mathbf{j}_k(\mathbf{x}')}{|\mathbf{x}-\mathbf{x}'|}. \tag{7.2.7}
$$

For line currents, the self-energy is singular, which reminds of the self-energy of point charges and requires the calculation with finite wire cross sections.

To show the equation $U_{ik} = U_{ki}$, only the exchange of $\mathbf{x}$ with $\mathbf{x}'$ is necessary; this relation also has its counterpart in electrostatics and is also referred to here as the *reciprocity theorem*. It is helpful when the interaction energy of a current loop $\mathbf{j}_f$ in an external field should be specified:

$$
U_w = \frac{k_L}{c}\int \mathrm{d}^3x\,\mathbf{A}^{\mathrm{e}}\cdot\mathbf{j}_f. \tag{7.2.8}
$$

In a homogeneous external field ($\mu_r = 1$) one obtains using $\mathbf{A}^{\mathrm{e}} = \mathbf{B}^{\mathrm{e}}\times\mathbf{x}/2$

$$
U_w = \frac{k_L}{2c}\int \mathrm{d}^3x\,(\mathbf{B}^{\mathrm{e}}\times\mathbf{x})\cdot\mathbf{j}_f = \mathbf{B}^{\mathrm{e}}\cdot\frac{k_L}{2c}\int \mathrm{d}^3x\,(\mathbf{x}\times\mathbf{j}_f) \overset{(4.2.1)}{=} \mathbf{B}^{\mathrm{e}}\cdot\mathbf{m}. \tag{7.2.9}
$$

$\mathbf{m}$ is the dipole moment of the current distribution. We notice the difference of the total field energy to potential energy (4.3.5).

The energy when changing the permeability

The magnetic energy changes when a magnetizable body of permeability μ_r is introduced into an area where the fields are $\mathbf{B}_0$ and $\mathbf{H}_0$. One can think of a solenoid into which a core is introduced whose permeability $\mu_r(\mathbf{x})$ is independent of $\mathbf{H}$ (linear medium). The corresponding problem was already treated in electrostatics, page 212, and the methods applied there are also leading here.

Change of permeability with constant potential

The system is isolated, so that the change $\delta\mu_r$ caused by inserting of a core into the coil at constant potential – however this could be achieved[1] – results in a change in the current distribution (see (7.1.5)). We bring $\Delta U_{\mathbf{A}}$ into the desired form by inserting $\mathbf{B} = \mathbf{B}_0$ and then $\mathbf{H} = \mathbf{B}/\mu_0 - 4\pi k_r\mathbf{M}$, where we start from $\mathbf{H}_0 = \mathbf{B}_0/\mu_0$:

$$
\Delta U_{\mathbf{A}} = \frac{1}{8\pi k_r}\int \mathrm{d}^3x\,(\mathbf{H}\cdot\mathbf{B} - \mathbf{H}_0\cdot\mathbf{B}_0) = \frac{-1}{2}\int \mathrm{d}^3x\,\mathbf{B}_0\cdot\mathbf{M}.
$$

---

[1] $\mathbf{A}$ is in classical electrodynamics a non-observable (auxiliary) entity.

The energy is lower when the body is inside the coil, so the (isolated) system does the work $\delta A = -\delta \Delta U_{\mathbf{A}}$ by pulling the body in.

Force on magnetizable body with constant potential

We now calculate the force on a magnetizable body with fixed potential, where we do not use the energy difference $\Delta U_{\mathbf{A}}$, but (7.2.5):

$$\delta A = \mathbf{F_A} \cdot \delta\mathbf{x} = \frac{-k_L}{2c} \int d^3x\, \mathbf{A} \cdot \underbrace{(\delta\mathbf{x}\cdot\boldsymbol{\nabla})\mathbf{j}}_{\delta\mathbf{j}} = \frac{-k_L}{2c} \int d^3x \left\{ (\delta\mathbf{x}\cdot\boldsymbol{\nabla})\mathbf{A}\cdot\mathbf{j} - \mathbf{j}\cdot(\delta\mathbf{x}\cdot\boldsymbol{\nabla})\mathbf{A} \right\}.$$

According to (A.2.33) is

$$(\delta\mathbf{x}\cdot\boldsymbol{\nabla})\,\mathbf{A} = \boldsymbol{\nabla}(\mathbf{A}\cdot\delta\mathbf{x}) - \delta\mathbf{x}\times(\boldsymbol{\nabla}\times\mathbf{A}),$$
$$\mathbf{j}\cdot(\delta\mathbf{x}\cdot\boldsymbol{\nabla})\,\mathbf{A} = \mathbf{j}\cdot\boldsymbol{\nabla}(\mathbf{A}\cdot\delta\mathbf{x}) - \mathbf{j}\cdot(\delta\mathbf{x}\times\mathbf{B}).$$

If we use $\boldsymbol{\nabla}\cdot\mathbf{j} = 0$ and the Gauss's theorem (A.4.4) for scalar functions, we get

$$\delta A = \frac{-k_L}{2c} \int d^3x \left\{ \boldsymbol{\nabla}\cdot\left[\delta\mathbf{x}(\mathbf{A}\cdot\mathbf{j})\right] - \boldsymbol{\nabla}\cdot\left[\mathbf{j}(\mathbf{A}\cdot\delta\mathbf{x})\right] + \mathbf{j}\cdot(\delta\mathbf{x}\times\mathbf{B}) \right\}$$
$$= \frac{-k_L}{2c} \oiint_{\partial V} d\mathbf{S}\cdot\left[\delta\mathbf{x}(\mathbf{A}\cdot\mathbf{j}) - \mathbf{j}(\mathbf{A}\cdot\delta\mathbf{x})\right] + \frac{k_L}{2c} \int d^3x\, \delta\mathbf{x}\cdot(\mathbf{j}\times\mathbf{B}).$$

With $V \to \infty$ the surface integrals vanish and one obtains, not unexpectedly, the Lorentz force:

$$\mathbf{F_A} = -\frac{\partial \Delta U_{\mathbf{A}}}{\partial \mathbf{x}} = \frac{k_L}{2c} \int d^3x\, \mathbf{j}\times\mathbf{B}.$$

Change of permeability with constant currents

We now turn to the physically relevant case that the magnetizable body is intorduced in an area with the fields $\mathbf{H_0} = \mathbf{B_0}/\mu_0$, where the current sources remain unchanged:

$$\Delta U_{\mathbf{j}} = \frac{1}{8\pi k_r} \int d^3x\, (\mathbf{H}\cdot\mathbf{B} - \mathbf{H_0}\cdot\mathbf{B_0}) = \frac{1}{8\pi k_r} \int d^3x\, (\mathbf{B}\cdot\mathbf{H_0} - \mathbf{B_0}\cdot\mathbf{H}) + \bar{U}.$$

If we specify the energy difference in this form, then it disappears

$$\bar{U} = \frac{1}{8\pi k_r} \int_V d^3x\, (\mathbf{B}+\mathbf{B_0})\cdot(\mathbf{H}-\mathbf{H_0}) = \frac{1}{8\pi k_r} \int_V d^3x\, \left[\boldsymbol{\nabla}\times(\mathbf{A}+\mathbf{A_0})\right]\cdot(\mathbf{H}-\mathbf{H_0})$$
$$= \frac{1}{8\pi k_r} \int_V d^3x\, \left\{ \boldsymbol{\nabla}\cdot\left[(\mathbf{A}+\mathbf{A_0})\times(\mathbf{H}-\mathbf{H_0})\right] + (\mathbf{A}+\mathbf{A_0})\cdot\left[\boldsymbol{\nabla}\times(\mathbf{H}-\mathbf{H_0})\right] \right\}$$
$$= \frac{1}{8\pi k_r} \oiint_{\partial V} d\mathbf{S}\cdot\left[(\mathbf{A}+\mathbf{A_0})\times(\mathbf{H}-\mathbf{H_0})\right] + \frac{k_L}{2c} \int_V d^3x\, (\mathbf{A}+\mathbf{A_0})\cdot(\mathbf{j}-\mathbf{j_0}) \right\} = 0.$$

The surface term vanishes, because due to the continuity of the tangential components $(\mathbf{H}-\mathbf{H}_0)\times d\mathbf{S}=0$ and because $\mathbf{j}=\mathbf{j}_0$ the volume integral vanishes. Thus one obtains

$$\Delta U_{\mathbf{j}} = \frac{1}{8\pi k_r}\int d^3x\,(\mathbf{B}\cdot\mathbf{H}_0 - \mathbf{B}_0\cdot\mathbf{H}) \overset{\mathbf{B}=\mu_0(\mathbf{H}+4\pi k_r\mathbf{M})}{=} \frac{1}{2}\int d^3x\,\mathbf{B}_0\cdot\mathbf{M}.$$

This corresponds exactly to the result (6.3.6) when introducing a dielectric into a capacitor at constant voltage, so that we again have the correspondence between current source and voltage in mind.

Force on magnetizable bodies with constant currents

In electrostatics, page 212, the battery provided the necessary energy U_b, to keep the voltage constant when introducing the dielectric into the capacitor. We will now apply the same procedure to magnetic energy. Here, the battery is needed to keep the current constant:

$$\delta\Delta U_{\mathbf{j}} + \delta A = \delta U_b = \frac{k_L}{c}\int d^3x\,\mathbf{j}\cdot\delta\mathbf{A}.$$

The energy difference at constant current is

$$\delta\Delta U_{\mathbf{j}} = \frac{1}{8\pi k_r}\int d^3x\,\mathbf{H}\cdot\delta\mathbf{B} = \frac{1}{8\pi k_r}\int d^3x\,\mathbf{H}\cdot(\boldsymbol{\nabla}\times\delta\mathbf{A}) = \frac{k_L}{2c}\int d^3x\,\mathbf{j}\cdot\delta\mathbf{A}.$$

Here we have used (A.2.34), where the surface contribution for $V \to \infty$ vanishes. The work on the magnetizable body is thus

$$\delta A = \delta\mathbf{x}\cdot\mathbf{F}_{\mathbf{j}} = \frac{k_L}{2c}\int d^3x\,\mathbf{j}\cdot(\delta\mathbf{x}\cdot\boldsymbol{\nabla})\mathbf{A}.$$

We transform the second term using (A.2.33)

$$\boldsymbol{\nabla}(\delta\mathbf{x}\cdot\mathbf{A}) = (\delta\mathbf{x}\cdot\boldsymbol{\nabla})\mathbf{A} + \delta\mathbf{x}\times(\boldsymbol{\nabla}\times\mathbf{A}),$$

$$\delta A = \frac{k_L}{2c}\int d^3x\,\left\{\mathbf{j}\cdot\boldsymbol{\nabla}(\delta\mathbf{x}\cdot\mathbf{A}) - \mathbf{j}\cdot(\delta\mathbf{x}\times\mathbf{B})\right\}$$

$$\overset{\boldsymbol{\nabla}\cdot\mathbf{j}=0}{=} \frac{k_L}{2c}\int d^3x\,\left\{\boldsymbol{\nabla}\cdot[\mathbf{j}(\delta\mathbf{x}\cdot\mathbf{A})] + \delta\mathbf{x}\cdot(\mathbf{j}\times\mathbf{B})\right\}.$$

The first term can be converted into a surface integral using Gauss's theorem, which vanishes for $V \to \infty$. This gives the force

$$\mathbf{F}_{\mathbf{j}} = \frac{k_L}{2c}\int d^3x\,\mathbf{j}\times\mathbf{B} = \left.\frac{\partial\Delta U_{\mathbf{j}}}{\partial\mathbf{x}}\right|_{\mathbf{j}}.$$

Note: For the infinitesimal rotation $\delta\mathbf{x} = \delta\boldsymbol{\varphi}\times\mathbf{x}$ of an external potential around the origin, one obtains from (7.2.9) an energy change of the form

$$\delta U_w = \delta\boldsymbol{\varphi}\cdot\frac{k_L}{2c}\int d^3x\,\mathbf{x}\times(\mathbf{j}\times\mathbf{B}^e) = \delta\boldsymbol{\varphi}\cdot\mathbf{N},$$

where $\mathbf{N}$ is the torque (4.3.7) of the current distribution.

The magnetic energy in the hard ferromagnet

In a permanent magnet, there are no free currents present ($\mathbf{j}_f = 0$). A consequence is that the magnetic field energy (5.6.7), taken over the whole space, vanishes [Jackson, 1998, problem 5.21]:

$$U_m = \frac{1}{8\pi k_r} \int \mathrm{d}^3x\, \mathbf{H}\cdot\mathbf{B} = \frac{-1}{8\pi k_r} \int \mathrm{d}^3x\, \mathbf{B}\cdot\boldsymbol{\nabla}\phi_M \overset{\text{int. by parts}}{=} \frac{1}{8\pi k_r} \int \mathrm{d}^3x\, \phi_M\, \boldsymbol{\nabla}\cdot\mathbf{B} = 0.$$

To determine the magnetic energy in the ferromagnet, we proceed in the same way as for the electrostatic energy and place a dipole at $\mathbf{x}_1$. The field of the dipole at location $\mathbf{x}$ is

$$\mathbf{B}_1(\mathbf{x}) = \frac{k_C}{k_L^2}\left[3\frac{\mathbf{m}_1\cdot(\mathbf{x}-\mathbf{x}_1)(\mathbf{x}-\mathbf{x}_1)}{|\mathbf{x}-\mathbf{x}_1|^5} - \frac{\mathbf{m}_1}{|\mathbf{x}-\mathbf{x}_1|^3} + 4\pi\mathbf{m}_1\,\delta^{(3)}(\mathbf{x}-\mathbf{x}_1)\right]$$

and the potential energy of a magnetic moment (permanent dipole) in the external field $\mathbf{B}$ is according to (4.3.5): $U = -\mathbf{m}\cdot\mathbf{B}$. We obtain

$$\mathbf{m}_1 \to \mathbf{x}_1: \quad U_1 = 0$$
$$\mathbf{m}_2 \to \mathbf{x}_2: \quad U_2 = -\mathbf{m}_2\cdot\mathbf{B}_1(\mathbf{x}_2)$$
$$\vdots \qquad\qquad \vdots$$
$$U' = -\sum_{i<j}\mathbf{m}_j\cdot\mathbf{B}_i = -\frac{1}{2}{\sum_{i,j}}'\,\mathbf{m}_j\cdot\mathbf{B}_i\,,$$

where $'$ means that only over $i \neq j$ may be summed. When transitioning to the continuum, however, the self-energy is included ($i = j$)

$$U = -\frac{1}{2}\int \mathrm{d}^3x\, \mathbf{M}(\mathbf{x})\cdot\mathbf{B}(\mathbf{x}).$$

Now we substitute for $\mathbf{B}$:

$$U = -\frac{\mu_0}{2}\int \mathrm{d}^3x\, \mathbf{M}\cdot(\mathbf{H} + 4\pi k_r\mathbf{M}) = W_0 - \frac{\mu_0}{2}\int \mathrm{d}^3x\, \mathbf{M}\cdot\mathbf{H},$$
$$W_0 = -\frac{4\pi k_r\mu_0}{2}\int \mathrm{d}^3x\, M^2.$$

The self-energy is a quantity independent of the configuration, which applies to the term W_0. Now we substitute for $4\pi k_r\mathbf{M} = \mathbf{B}/\mu_0 - \mathbf{H}$ and use $\int \mathrm{d}^3x\, \mathbf{B}\cdot\mathbf{H} = 0$. The desired expression, which corresponds to the electrostatic energy (2.4.3), is thus

$$U = W_0 + \frac{\mu_0}{8\pi k_r}\int \mathrm{d}^3x\, H^2 = -\frac{k_L^2}{8\pi k_C}\int \mathrm{d}^3x\, B^2.$$

7.2.2 Induction Coefficients

The magnetic field energy of a circuit embedded in a homogeneous medium is according to (7.2.5)

$$U_m = \frac{k_L}{2c} \int d^3x \, \mathbf{A} \cdot \mathbf{j}_f = \frac{k_C \mu_r}{2c^2} \int d^3x \, d^3x' \, \frac{\mathbf{j}_f(\mathbf{x}) \cdot \mathbf{j}_f(\mathbf{x}')}{|\mathbf{x} - \mathbf{x}'|} . \tag{7.2.10}$$

If a current I can be specified for the arrangement of the circuits, then U can be represented by   SI: $k_C/c^2 = \mu_0/4\pi$

$$U_m = \frac{LI^2}{2} \qquad \text{with} \quad L = \frac{k_C}{c^2} \frac{\mu_r}{2I^2} \int d^3x \, d^3x' \, \frac{\mathbf{j}_f(\mathbf{x}) \cdot \mathbf{j}_f(\mathbf{x}')}{|\mathbf{x} - \mathbf{x}'|} , \tag{7.2.11}$$

where L is referred to as self-inductance. However, if the configuration consists of several current loops

$$\mathbf{j}_f(\mathbf{x}) = \sum_{i=1}^{n} \mathbf{j}_i(\mathbf{x}) ,$$

it is often useful to consider the currents of individual current loops separately:

$$U_m = \frac{1}{2} \sum_{i,k=1}^{n} L_{ik} \, I_i I_k , \tag{7.2.12}$$

$$L_{ik} = \frac{k_C \mu_r}{c^2} \int d^3x \, d^3x' \, \frac{1}{|\mathbf{x} - \mathbf{x}'|} \frac{\mathbf{j}_i(\mathbf{x})}{I_i} \cdot \frac{\mathbf{j}_k(\mathbf{x}')}{I_k} . \tag{7.2.13}$$

L_{ii} is called for $1 \le i \le n$ *self-inductance.*
L_{ik} is called for $1 \le i, k \le n$ and $i \ne k$ *mutual inductance.*

The L_{ik} are purely geometric. They only depend on the shape of the conductors and the geometric shape of the streamlines within them. This becomes particularly clear for thin wires, where $\int_V d^3x \, \mathbf{j}/I = A \int_C d s \, \mathbf{j}/jA = \int_C d\mathbf{s}$. The mutual inductances can then be represented by

$$L_{ik} = \frac{k_C \mu_r}{c^2} \int_{C_i} \int_{C_k} \frac{d\mathbf{s}_i \cdot d\mathbf{s}_k}{|\mathbf{x}(s_i) - \mathbf{x}(s_k)|} \qquad \text{for} \quad i \ne k . \tag{7.2.14}$$

C_i indicates the path of the i-th wire. For L_{ii} the corresponding representation would diverge, and one has to calculate with extended conductors.

The determination of the self-inductance is particularly simple if one knows the energy and it can be brought into the following form:

$$U = \sum_i U^{(i)} = \frac{I^2}{2} \sum_i L^{(i)} \quad \Rightarrow \quad L = \sum_i L^{(i)} = \sum_i \frac{2 \, U^{(i)}}{I^2} . \tag{7.2.15}$$

Partial inductances: As can be seen from (7.2.15), one can define self-inductances and mutual inductances on individual parts of a circuit.

We will apply this in the calculation of the self-inductance of a straight wire. The self-inductances of two coils can be added if they are far apart, perpendicular to each other, or closed with iron.

The restriction to a homogeneous permeability μ_r is not severe. In long coils, the field disappears outside and only the area inside with the iron core contributes to the energy.

7.2.3 Magnetic Flux and Inductance

Another approach to the induction coefficients (7.2.14) is obtained via the magnetic flux

$$\Phi_B = \iint_A \mathrm{da} \cdot \mathbf{B} = \iint_A \mathrm{da} \cdot (\boldsymbol{\nabla} \times \mathbf{A}) = \oint_{\partial A} \mathrm{dx} \cdot \mathbf{A} \tag{7.2.16}$$

through an area A. If we can specify the entire current I for the configuration, we obtain in analogy to (7.2.11)

$$\Phi_B = \frac{c}{k_L} IL, \qquad L = \frac{k_C}{c^2} \oint_{\partial A} \mathrm{dx} \cdot \int \mathrm{d}^3 x' \, \frac{\mathbf{j}(\mathbf{x}')/I}{|\mathbf{x} - \mathbf{x}'|}, \qquad \text{SI: } \frac{k_C}{c^2} = \frac{\mu_0}{4\pi}. \tag{7.2.17}$$

The calculation of the self-inductance L requires the consideration of the finite cross-section of the wire. A line wire has a logarithmically singular self-inductance L.

Neumann formula

Two current loops are sketched in Fig. 7.10. The field $\mathbf{B}_2$ originating from the current I_2 of current loop 2 partially permeates loop 1 and generates the flux $\Phi_{B_{12}}$ in it. We assume that we have a closed circuit with the boundary

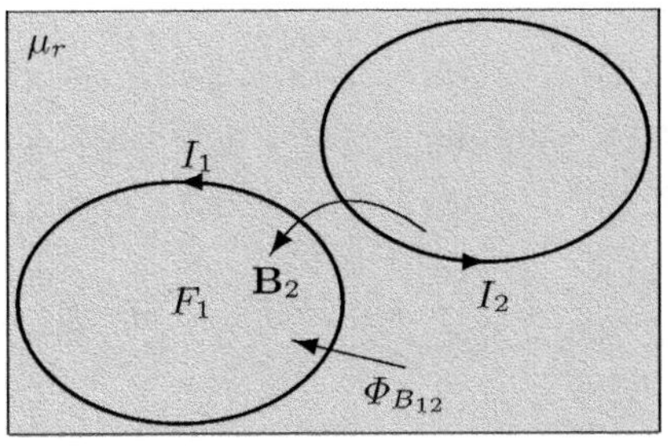

Fig. 7.10. The current I_2 generates the field $\mathbf{B}_2$, a part of which penetrates loop 1 resulting in the flux $\Phi_{B_{12}}$

∂A, consisting of a thin wire for which $\int \mathrm{d}^3 x' \, \mathbf{j}(\mathbf{x}'(s'))... = I \int \mathrm{ds}'...$ applies. Loop 2 has the vector potential

$$\mathbf{A}_2 = \frac{k_C \mu_r}{c k_L} \int d^3 x' \frac{\mathbf{j}_2(\mathbf{x}')}{|\mathbf{x} - \mathbf{x}'|} = \frac{k_C I \mu_r}{c k_L} \oint_{\partial A_2} d\mathbf{s}_2 \frac{1}{|\mathbf{x} - \mathbf{x}'(s_2)|} \, . \qquad (7.2.18)$$

The flux generated by $\mathbf{B}_2$ in loop 1 is according to (7.2.18)

$$\Phi_{B_{12}} = \oint_{\partial A_1} d\mathbf{x}_1 \cdot \mathbf{A}_2(\mathbf{x}_1) = \frac{c}{k_L} L_{12} I_2, \, L_{12} = \frac{k_C \mu_r}{c^2} \oint_{\partial A_1} \oint_{\partial A_2} \frac{d\mathbf{x}_1 \cdot d\mathbf{x}_2}{|\mathbf{x}_1 - \mathbf{x}_2|} \, . \qquad (7.2.19)$$

The equation on the right is the *Neumann formula*, which was derived by *F. Neumann* in 1845 and named after him [Neumann, 1889, p. 10]. This term is not consistently used in the literature; the expression $L_{12} c^2 / \mu_r k_C$ is referred to by Sommerfeld as *Neumann potential* [Sommerfeld, 1952, §15 eq. (20)]. The inductances obey the reciprocity law

$$L_{12} = L_{21} \, , \qquad (7.2.20)$$

as is immediately apparent from the definition of L_{12}.

Several current loops: We can immediately extend (7.2.19) to several circuits $\mathbf{j} = \sum_k \mathbf{j}_k$. The flux Φ_{B_i} through a single current loop i then consists of the contributions of all loops:

$$\Phi_{B\,i} = \frac{k_C \mu_r}{c k_L} \oint_{\partial A_i} \sum_{k=1}^n I_k \oint_{\partial A_k} \frac{d\mathbf{x}_i \cdot d\mathbf{x}'_k}{|\mathbf{x}_i - \mathbf{x}'_k|} = \frac{c}{k_L} \sum_{k=1}^n L_{ik} I_k \, . \qquad (7.2.21)$$

Here too, the self-induction coefficient must be calculated with finite wire cross-sections. A temporal change of the currents entails a change of $\mathbf{B}$ and thus according to the law of induction in the i-th loop the field $\mathbf{E} = -\dot{\mathbf{A}} k_L / c$ is induced:

$$\oint_{\partial A_i} d\mathbf{x} \cdot \mathbf{E} = \iint_{A_i} d\mathbf{a} \cdot \nabla \times \mathbf{E} = -\frac{k_L}{c} \dot{\Phi}_{B\,i} = -\sum_k L_{ik} \dot{I}_k \, . \qquad (7.2.22)$$

7.2.4 The Self-inductances of Selected Configurations

Self-inductance of a coil

Given is a coil of length h, radius a and with n windings per cm, as in Fig. 7.11 outlined. If the coil is long enough, the field is

$$H = \frac{4\pi k_r k_L I n}{c} \qquad \Rightarrow \qquad \text{G: } H = \frac{4\pi I n}{c}, \qquad \text{SI: } H = I n. \qquad (7.2.23)$$

With this, we can calculate the magnetic energy as

$$U_m = \frac{VBH}{8\pi k_r} = \frac{a^2 \pi h}{8\pi k_r} \mu_r \mu_0 \Big(\frac{4\pi k_r k_L I n}{c} \Big)^2 = \Big(\frac{4\pi^2 k_r k_L^2}{c^2} n^2 a^2 h \mu_r \mu_0 \Big) \frac{I^2}{2} = \frac{L I^2}{2} \, .$$

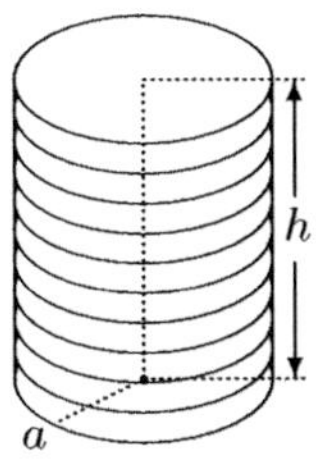

Fig. 7.11. Coil: height h, radius a, n windings per cm, where, as usual, the coil pitch per turn is neglected. $N = nh$ is the number of windings and $l = 2\pi a\,nh$ the length of the wire

Now we replace $k_r k_L^2 \mu_0$ by k_C and obtain for the self-inductance

$$
L = \frac{k_C \mu_r l^2}{c^2 h}, \qquad \text{G: } L = \frac{\mu_r l^2}{c^2 h}, \qquad \text{SI: } L = \frac{\mu_0}{4\pi}\frac{\mu_r l^2}{h} = \mu_0 \frac{\mu_r N^2 A}{h},
$$

where $l = 2\pi a N$ is the wire length and $A = a^2 \pi$ is the area of the coil. As expected, the energy and consequently the inductance may only depend on k_C, but not on k_L.

Straight wire and double line

Given is a wire of length l with the (small) radius a and the permeability $\mu_r^{(i)}$, as shown in Fig. 7.12. The magnetic energy is divided according to (7.2.15) into an inner part $U_m^{(i)}$ and an outer $U_m^{(e)}$ and calculates the two parts separately. The magnetic field of the wire is obtained from Ampère's law (4.1.5) and (4.1.6). For the external space, we can directly use (4.1.10):

$$
2\pi\varrho H_\varphi^{(i)} = \frac{4\pi k_r k_L I}{c}\frac{\varrho^2}{a^2} \stackrel{\text{SI}}{=} I\frac{\varrho^2}{a^2}, \qquad 2\pi\varrho H_\varphi^{(e)} = \frac{4\pi k_r k_L I}{c} \stackrel{\text{SI}}{=} I, \qquad (7.2.24)
$$

$$
L^{(i)} = \frac{2U_m^{(i)}}{I^2} = \frac{\mu_r^{(i)}\mu_0}{4\pi k_r I^2}\int \mathrm{d}^3 x\,|H^{(i)}|^2 = \frac{\mu_r^{(i)}\mu_0 k_r k_L^2}{\pi c^2 a^4}2\pi l \int_0^a \mathrm{d}\varrho\,\varrho^3
$$

$$
= k_r k_L^2 \frac{\mu_r^{(i)}\mu_0 l}{2c^2} = k_C \frac{\mu_r^{(i)} l}{2c^2} \stackrel{\text{SI}}{=} \frac{\mu_r^{(i)} l}{8\pi}. \qquad (7.2.25)
$$

It should be noted that $L^{(i)}$ does not depend on a. The dominant part, however, comes from the external space with $\mu_r^{(e)}$

$$
L^{(e)} = \frac{\mu_r^{(e)}\mu_0}{4\pi k_r I^2}\int \mathrm{d}^3 x\,|H^{(e)}|^2 \approx \frac{\mu_r^{(e)}\mu_0 k_r k_L^2}{\pi c^2}2\pi l \int_a^R \frac{\mathrm{d}\varrho}{\varrho} = \frac{2k_C \mu_r^{(e)} l}{c^2}\ln\frac{R}{a}
$$

and here the finite cross-section of the wire is essential. The divergence comes from very large distances R from the wire. If the circuit is closed, this divergence does not occur, as distances that are larger than the current loop no

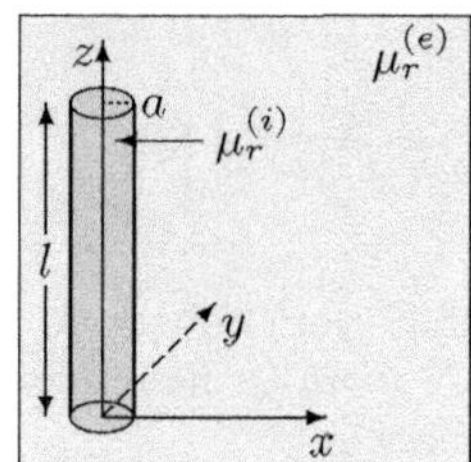

Fig. 7.12. Straight wire of length l, radius a and (internal) permeability $\mu_r^{(i)}$, it is aligned along the z-axis and embedded in a medium of (external) permeability $\mu_r^{(e)}$

longer contribute – and a prerequisite in the derivation of the magnetic energy (7.2.5) were localized currents. One can use $R \sim l$ as a typical distance:

$$\text{G: } L^{(e)} = \frac{2\mu_r^{(e)}\, l}{c^2} \left(\ln \frac{l}{a} + O(1) \right), \quad \text{SI: } L^{(e)} = \frac{2\mu_r^{(e)} \mu_0\, l}{4\pi} \left(\ln \frac{l}{a} + O(1) \right)$$

and denotes $L^{(e)}$ to be of logarithmic accuracy [Landau, Lifshitz 8, 1984, § 34], when the relative error is of the order $1/\ln(l/a)$.

On the self-inductance of a straight wire: Within the wire is

$$L^{(i)} = k_C \frac{\mu_r^{(i)}\, l}{2c^2} \stackrel{\text{SI}}{=\!=} \frac{\mu_0}{8\pi} \mu_r^{(i)}\, l$$

with homogeneous current distribution, as previously shown. For the external part, we start from (7.2.5):

$$U_m = \frac{k_L}{2c} \int \mathrm{d}^3 x\, \mathbf{A}\cdot\mathbf{j}$$

and use that the field in the exterior does not depend on the current distribution within the wire. $U_m^{(e)}$ does not change if one assumes that the current I only flows on the surface of the wire (the field then vanishes inside)

$$\mathbf{j}(\mathbf{x}) = \frac{I}{2\pi a}\, \delta(\varrho - a)\, \mathbf{e}_z \,.$$

The field value on the surface does not change due to this assumption. We thus obtain a line integral along the z-axis:

$$L^{(e)} = \frac{2U_m}{I^2} = \frac{k_L}{cI} \int_0^l \mathrm{d}z\, \mathbf{e}_z \cdot \mathbf{A}(a, z) \,.$$

We calculate the vector potential from (7.1.5)

$$\mathbf{A}(\mathbf{x}) = \frac{k_C \mu_r^{(e)}}{c k_L} \int \mathrm{d}^3 x'\, \frac{\mathbf{j}(\mathbf{x}')}{|\mathbf{x} - \mathbf{x}'|}$$

on the surface of the wire. In a new assumption about the current distribution, we place this in the wire axis:

$$\mathbf{j}(\mathbf{x}') = I\, \delta(x')\, \delta(y')\, \mathbf{e}_z \,.$$

The value of the field does not change as a result. We are looking for $\mathbf{A}$ on the wire surface

$$\mathbf{A}(a, z) = \frac{I\,k_C\,\mu_r^{(e)}}{c k_L} \int_0^l dz' \, \frac{1}{\sqrt{a^2 + (z-z')^2}} \, \mathbf{e}_z \,.$$

In summary, we obtain

$$L^{(e)} = \frac{k_C\,\mu_r^{(e)}}{c^2} \int_0^l dz\,dz' \, \frac{1}{\sqrt{a^2 + (z-z')^2}} \overset{u=z'-z}{=} \frac{k_C\,\mu_r^{(e)}}{c^2} \int_0^l dz \int_{-z}^{l-z} du \, \frac{1}{\sqrt{a^2+u^2}}$$

$$\overset{\text{(B.5.15)}}{=} \frac{k_C\,\mu_r^{(e)}}{c^2} \int_0^l dz \left[\ln\left((l-z) + \sqrt{a^2+(l-z)^2}\right) - \ln\left(-z + \sqrt{a^2+z^2}\right) \right].$$

Now we evaluate the two terms separately

$$\int_0^l dz \, \ln\left((l-z) + \sqrt{a^2+(l-z)^2}\right) \overset{v=l-z}{=} \int_0^l dv \, \ln\left(v + \sqrt{a^2+v^2}\right)$$

$$\overset{\text{(B.5.12)}}{=} v \, \ln\left(v + \sqrt{a^2+v^2}\right) - \sqrt{a^2+v^2} \, \Big|_0^l \overset{l \gg a}{\approx} l \, \ln(2l) - l \,.$$

The 2nd term contributes to $L^{(e)}$ with

$$-\int_0^l dz \, \ln\left(-z + \sqrt{a^2+z^2}\right) \overset{v=-z}{=} \int_0^{-l} dv \, \ln\left(v + \sqrt{a^2+v^2}\right)$$

$$= v \, \ln\left(v + \sqrt{a^2+v^2}\right) - \sqrt{a^2+v^2} \, \Big|_0^{-l} = -l \, \ln\left(-l + \sqrt{a^2+l^2}\right) - \sqrt{a^2+l^2} + a$$

$$\overset{l \gg a}{\approx} -l \, \ln \frac{a^2}{2l} - l \,.$$

It follows that

$$L^{(e)} = k_C\,\mu_r^{(e)} \frac{2l}{c^2} \left(\ln \frac{2l}{a} - 1\right).$$

Now, if we set $\mu_r^{(i)} = \mu_r^{(e)} = 1$, the self-inductance is given by

$$L = L^{(i)} + L^{(e)} = k_C \frac{2l}{c^2}\left(\ln \frac{2l}{a} - \frac{3}{4}\right), \quad \text{SI: } L = \frac{\mu_0}{4\pi} 2l\left(\ln \frac{2l}{a} - \frac{3}{4}\right). \tag{7.2.26}$$

The equation gives an impression of how L increases with the length l of the wire. However, quantitative statements about the self-inductance of the wire can only be made, when the configuration of the entire closed circuit is included.

Self-inductance of the double line

A physically meaningful model is the self-inductance of the double line, as it is sketched in Fig. 7.13. Although the circuit is not closed again, but the

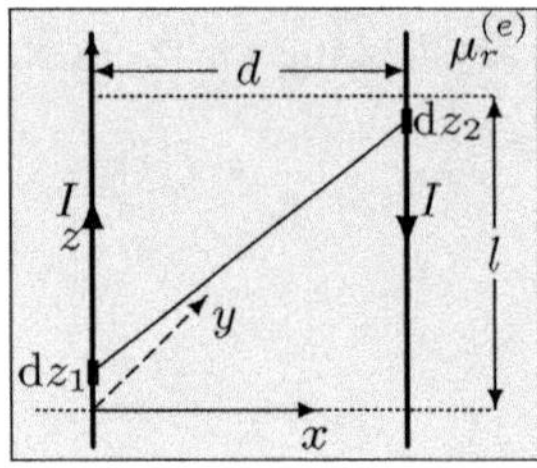

Fig. 7.13. Two straight, parallel wires carrying current I in opposite directions are located at a distance d from each other. They are surrounded by a medium with permeability $\mu_r^{(e)}$. A piece of length l is selected

divergent contributions of the self- and mutual inductances with $\sim 2l \ln(2l)$ cancel each other out. Therefore, the double line as a whole corresponds to a closed circuit at infinity, when the currents of equal magnitude are directed oppositely. According to (7.2.14), the mutual inductance for thin wires ($a \ll d$)

$$
L_{12} = \frac{k_C \mu_r^{(e)}}{c^2} \int_0^l \mathrm{d}z_1 \int_0^l \mathrm{d}z_2 \, \frac{1}{\sqrt{d^2 + (z_2 - z_1)^2}}
$$

$$
\stackrel{\text{(B.5.15)}}{=} \frac{k_C \mu_r^{(e)}}{c^2} \int_0^l \mathrm{d}z_1 \left\{ \ln \left[(l - z_1) + \sqrt{d^2 + (l - z_1)^2} \right] - \ln \left[-z_1 + \sqrt{d^2 + z_1^2} \right] \right\}.
$$

Both terms can be evaluated using (B.5.12); they are the same integrals that we already had for the single wire and finally one obtains in the limit $l \gg d \gg a$ the mutual-inductance

$$
L_{12} = \frac{k_C \mu_r^{(e)}}{c^2} \left\{ \left[d + l \ln \left[l + \sqrt{d^2 + l^2} \right] - \sqrt{d^2 + l^2} \right] \right.
$$

$$
\left. - \left[l \ln \left[-l + \sqrt{d^2 + l^2} \right] + \sqrt{d^2 + l^2} - d \right] \right\} \stackrel{l \gg d}{=} \frac{k_C \mu_r^{(e)}}{c^2} 2l \left(\ln \frac{2l}{d} - 1 \right). \tag{7.2.27}
$$

The total self-inductance of the double line can be obtained using (7.2.26)

$$
L = 2(L_{11} - L_{12}) = k_C 4l \left[\frac{\mu_r^{(e)}}{c^2} \ln \frac{d}{a} + \frac{\mu_r^{(i)}}{4c^2} \right] \tag{7.2.28}
$$

$$
\text{G: } 4l \left[\frac{\mu_r^{(e)}}{c^2} \ln \frac{d}{a} + \frac{\mu_r^{(i)}}{4c^2} \right], \qquad\qquad \text{SI: } \frac{4l\mu_0}{4\pi} \left[\mu_r^{(e)} \ln \frac{d}{a} + \frac{\mu_r^{(i)}}{4} \right],
$$

where $\mu_r^{(i)}$ denotes the self-inductance within the wire (7.2.25) separately.

Self-inductance of a circular wire loop

Given is a wire of permeability $\mu_r^{(i)}$ with radius a, which forms a circular loop with radius b, as shown in Fig. 7.14. The wire is surrounded by a medium of permeability $\mu_r^{(e)}$.

Again, the self-inductance is divided into an inner and an outer part:

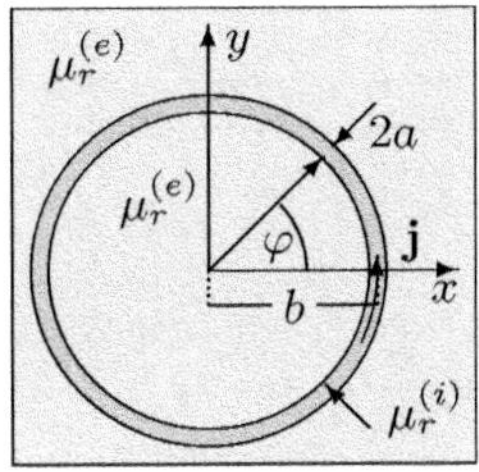

Fig. 7.14. A circular wire of radius a forms a circular loop with radius b

$$L = L^{(i)} + L^{(e)}.$$

We calculate $L^{(i)}$ from the inner energy of the wire, assuming $\mathbf{j}$ to be homogeneous. We can directly refer to the result (7.2.25) of the straight wire, if we replace the length l of the wire with the circumference of the loop $2\pi b$:

$$L^{(i)} = k_{\mathrm{C}} \frac{\mu_r^{(i)} \pi b}{c^2}. \tag{7.2.29}$$

For the outer space we refer to (7.2.14), but we have to consider the finite extent of the wire. This is done by moving the current to the center of the wire for the calculation of $\mathbf{A}$. We have already calculated the vector potential $\mathbf{A}$ for a wire loop with radius b (4.2.18):

$$\mathbf{A} = \mathbf{e}_\varphi \frac{k_{\mathrm{C}}}{c k_L} I b \mu_r^{(e)} \frac{4}{k\sqrt{b\varrho}}\left[(1-\frac{k^2}{2})K(k) - E(k)\right], \quad k^2 = \frac{4b\varrho}{(b+\varrho)^2 + z^2}. \tag{7.2.30}$$

Now we turn to the magnetic flux passing through the circular ring C with radius $b-a$:

$$\Phi_B = \oint_C \mathrm{d}\mathbf{s}\cdot\mathbf{A} = 2\pi \frac{k_{\mathrm{C}}\mu_r^{(e)} I}{c k_L} \frac{4b(b-a)}{k\sqrt{b(b-a)}}\left[(1-\frac{k^2}{2})K(k) - E(k)\right]. \tag{7.2.31}$$

We see from the definition ($\varrho = b - a$ and $z = 0$)

$$k^2 = \frac{4b(b-a)}{(2b-a)^2} \approx 1 - \frac{a^2}{4b^2},$$

that $k \approx 1$ applies. So we replace in the elliptic integral of the 2nd kind $E(k) \approx E(1) = 1$.

To get an approximation for the elliptic integral of the first kind, we introduce the angle δ that

$$\frac{a}{2b} \ll \delta \ll 1$$

fulfills and divide the integral into two parts and set, where it is possible, $k = 1$:

$$K(k) = \int_0^{\pi/2} d\varphi \, \frac{1}{\sqrt{1-k^2\cos^2\varphi}} \approx \int_0^{\delta} d\varphi \, \frac{1}{\sqrt{1-k^2+k^2\varphi^2}} + \int_\delta^{\pi/2} \frac{d\varphi}{\sqrt{1-\cos^2\varphi}}$$

$$= \frac{1}{k}\ln\left(k\varphi + \sqrt{1-k^2+k^2\varphi^2}\right)\Big|_0^{\delta} + \ln\tan\frac{\varphi}{2}\Big|_\delta^{\pi/2}$$

$$\approx \ln\left(\delta+\sqrt{1-k^2+\delta^2}\right) - \ln\sqrt{1-k^2} - \ln\frac{\delta}{2} \approx \ln(2\delta) - \ln\frac{a}{2b} - \ln\frac{\delta}{2} = \ln\frac{8b}{a}.$$

Finally, we used here that $1 - k^2 \ll \delta^2$.

Thus we have

$$\Phi_B = \frac{k_C \mu_r^{(e)} 4\pi b I}{c k_L}\left(\ln\frac{8b}{a} - 2\right). \tag{7.2.32}$$

The self-inductance we get from (7.2.19)

$$L^{(e)} = \frac{\Phi_B k_L}{cI} = \frac{k_C \mu_r^{(e)}}{c^2}\, 4\pi b\left(\ln\frac{8b}{a} - 2\right). \tag{7.2.33}$$

Note: The approximation for $\mathbf{A}$ (4.2.21) is too inaccurate near the wire surface, so we had to start from the exact vector potential.

7.3 Forms of Magnetism

In section 5.2 we introduced the magnetization density $\mathbf{M}$, which is caused by *Ampère's molecular currents*, which in turn can be attributed to the movement of electrons and atomic nuclei. Since the proton is much heavier than the electron with $m_p/m_e \approx 1836$, the contribution of the nuclear moments can be neglected.

Magnetic moment of the electron

In the Bohr atomic model, it is assumed that the electrons move in circular orbits around the nucleus. If a is the radius of such an orbit and $\mathbf{v} = a\omega\,\mathbf{e}_\varphi$ is the speed with which the electron moves around the nucleus, then averaging over the period $T = 2\pi/\omega$ gives the stationary current density

$$\mathbf{j}(\mathbf{x}) = \frac{1}{T}\int_0^T dt\,\rho\,\mathbf{v} = -\frac{e_0}{T}\int_0^T dt\,\delta(\varrho-a)\,\delta(\varphi-\omega t-\varphi_0)\,\delta(z)\,\omega\,\mathbf{e}_\varphi$$

$$= -\left(e_0\omega/2\pi\right)\delta(\varrho-a)\,\delta(z)\,\mathbf{e}_\varphi.$$

Here, $\rho(\mathbf{x}) = -e_0\delta(\varrho-a)\,\delta(z)/2\pi a$. If a slowly varying, homogeneous magnetic field $\mathbf{B}(t) = B(t)\mathbf{e}_z$ is switched on, it follows from the induction law $\boldsymbol{\nabla}\times\mathbf{E} = -\dot{\mathbf{B}}k_L/c$

$$\mathbf{E} = \left(\boldsymbol{\varrho}\times\dot{\mathbf{B}}\right)k_L/2c = -\varrho\,\dot{B}\,\mathbf{e}_\varphi\,k_L/2c. \tag{7.3.1}$$

This induced electric field, which runs everywhere (anti-)parallel to the movement of the electrons, causes the ring current ($m_e\dot{\mathbf{v}} = -e_0\mathbf{E}$)

$$\mathbf{j}^{(\mathrm{ind})}(\mathbf{x},t) = \frac{-e_0}{m_e}\,\rho(\mathbf{x})\int_0^t \mathrm{d}t'\,\mathbf{E}(\mathbf{x},t') = \rho(\mathbf{x})\,\omega_L\,\varrho\,\mathbf{e}_\varphi \quad \text{with} \quad \omega_L = \frac{k_L e_0 B}{2m_e c}.$$

ω_L is called *Larmor frequency*; it is the frequency with which the electron rotates around $\mathbf{B}$. The magnetic moment, which is determined by the ring current, is

$$\boldsymbol{\mu} = \frac{k_L}{2c}\int \mathrm{d}^3x\,\mathbf{x}\times\left(\mathbf{j}(\mathbf{x}) + \mathbf{j}^{(\mathrm{ind})}(\mathbf{x})\right). \tag{7.3.2}$$

If one replaces the charge density $\rho(\mathbf{x}) = -e_0\,\rho_m(\mathbf{x})/m_e$ by the mass density ρ_m, one immediately gets

$$\boldsymbol{\mu} = \boldsymbol{\mu}^L + \boldsymbol{\mu}^{(\mathrm{ind})} = -\frac{k_L e_0}{2m_e c}\left(\mathbf{L} + \mathbf{L}^{(\mathrm{ind})}\right) \tag{7.3.3}$$

with the orbital angular momentum ($\mathbf{v} = a\omega\mathbf{e}_\varphi$)

$$\mathbf{L} = \int \mathrm{d}^3x\,\rho_m\,\mathbf{x}\times\mathbf{v} = \frac{m_e}{a}\int_0^\infty \mathrm{d}\varrho\,\varrho\,\delta(a-\varrho)\boldsymbol{\varrho}\times a\omega\mathbf{e}_\varphi = m_e a^2\,\boldsymbol{\omega} \tag{7.3.4}$$

with $\boldsymbol{\omega} = \omega\mathbf{e}_z$. The induced angular momentum is

$$\mathbf{L}^{(\mathrm{ind})} = \int \mathrm{d}^3x\,\rho_m\,\mathbf{x}\times\mathbf{v} = m_e a^2\,\boldsymbol{\omega}_L \qquad \text{with} \qquad \boldsymbol{\omega}_L = \omega_L\,\mathbf{e}_z. \tag{7.3.5}$$

The orbital angular momentum thus consists of two components, the "impressed" moment $\boldsymbol{\mu}^L$ and the induced moment $\boldsymbol{\mu}^{(\mathrm{ind})}$. $\boldsymbol{\mu}^{(\mathrm{ind})}$ is, regardless of $\boldsymbol{\mu}^L$, according to Lenz's rule, directed against the field $\mathbf{B}$.

The electron also has an intrinsic angular momentum [Postulation of the electron spin by Uhlenbeck & Goudsmit, 1925] the spin $\mathbf{S}$, which contributes to the magnetic moment of the electron

$$\boldsymbol{\mu}^S = -(g k_L e_0/2m_e c)\mathbf{S}$$

with the gyromagnetic factor $g \approx 2$. This factor follows from quantum theory [Schwabl, 2008, §5.3], more precisely from the non-relativistic approximation of the Dirac equation (Pauli equation) and cannot be justified classically. The total magnetic moment of the electron is accordingly

$$\boldsymbol{\mu}^{(\mathrm{el})} = -\mu_B(\hat{\mathbf{L}} + g\hat{\mathbf{S}}) - (k_L e_0 a^2/2c)\,\boldsymbol{\omega}_L. \tag{7.3.6}$$

Here we have replaced $\mathbf{L}$ and $\mathbf{S}$ with the dimensionless vectors $\hat{\mathbf{L}} = \mathbf{L}/\hbar$ and $\hat{\mathbf{S}} = \mathbf{S}/\hbar^2$ and introduced the Bohr magneton

---

[2] Reduced Planck's quantum of action $\hbar = 1.0546 \times 10^{-27}$ erg s.

$$\mu_B = k_L e_0 \hbar / (2 m_e c) \approx 9.27 \times 10^{-21} \mathrm{erg/Gauss}. \qquad (7.3.7)$$

Note: Electrons, which move at the frequencies $\pm\omega$ depending on the direction of rotation, have (assuming $\omega_L \ll \omega$) after the field is switched on the frequencies $\pm\omega + \omega_L$. This corresponds to a splitting of the energies of the electrons, which is known as the *Zeeman* effect.

Larmor's theorem[3]: The influence of a slowly switched on magnetic field $\mathbf{B}$ on the motion of an electron is that after switching on $\mathbf{B}$, the electron performs the same motion with respect to a coordinate system rotating with ω_L as it did before switching on with respect to a resting system.

As already mentioned, it must be noted that $\omega_L \ll \omega$ must be satisfied, which is always fulfilled for atoms. An electron thus orbits the nucleus with $\pm\omega$. If $\mathbf{B} = B\mathbf{e}_z$ is now switched on, the electron after the switch-on time obeys the equation of motion

$$m(\ddot{\mathbf{x}} + \omega^2 \mathbf{x}) = \mathbf{F}_L = (eBk_L/c)\dot{\mathbf{x}} \times \mathbf{e}_z \,.$$

The z-component parallel to $\mathbf{B}$ remains unchanged. Perpendicular to $\mathbf{B}$ we form $x + iy$ and get

$$(\ddot{x} + i\ddot{y}) + \omega^2(x + iy) = 2\omega_L(\dot{y} - i\dot{x}) = -2i\omega_L(\dot{x} + i\dot{y})$$

with the solution

$$x + iy = (Ae^{i\omega t} + Be^{-i\omega t})e^{i\omega_L t} \qquad \text{for} \qquad \omega_L \ll \omega \,.$$

The initial movement is overlaid by $\mathbf{B}$ with a rotation ω_L and the switch-on time is necessary, as only during this an electric field $\mathbf{E}$ is present with which work on the system can be performed. $\mathbf{F}_L$ is always perpendicular to $\mathbf{v}$ and thus does no work.

Magnetic moment of an atom

In an atom, the contributions of the Z electrons add up to a total orbital angular momentum $\mathbf{L} = \sum_i \mathbf{L}_i$ and a Spin $\mathbf{S} = \sum_i \mathbf{S}_i$, where $\mathbf{L}_i$ and $\mathbf{S}_i$ denote the contributions of the individual electrons. The total angular momentum of an atom is thus $\mathbf{J} = \mathbf{L} + \mathbf{S}$. $\mathbf{L}$ and $\mathbf{S}$ are generally not parallel. Since $\mathbf{J}$ is a constant of motion, $\mathbf{S}$ precesses around $\mathbf{J}$ and it contributes to μ only the part parallel to $\mathbf{J}$

$$\mu = -\frac{e_0 k_L}{2 m_e c}\langle \mathbf{J} + \mathbf{S}\rangle = -\frac{e_0 k_L}{2 m_e c}(\mathbf{J} + \mathbf{S})\cdot \mathbf{J}\,\frac{1}{\mathbf{J}^2}\,\mathbf{J} = -g_J \frac{e_0 k_L}{2 m_e c}\mathbf{J} \qquad (7.3.8)$$

with the Landé factor[4]

$$g_J = \frac{(\mathbf{J} + \mathbf{S})\cdot \mathbf{J}}{\mathbf{J}^2} = 1 + \frac{\mathbf{J}^2 + \mathbf{S}^2 - \mathbf{L}^2}{2\mathbf{J}^2}\,. \qquad (7.3.9)$$

Unlike for the individual electrons, μ here is an "average"moment.

---

[3] Formulation according to Becker, Sauter [1959, p. 155]
[4] In quantum mechanics, $\mathbf{J}^2 = \hbar^2 j(j + 1)$ etc.

In calculating the induced magnetic moment, we have placed the electron's orbit in the xy-plane. The removal of this restriction results in

$$\langle \varrho^2 \rangle = \langle r^2 \sin^2 \vartheta \rangle = \langle r^2 \rangle \frac{1}{2} \int_0^\pi \mathrm{d}\vartheta \, \sin^3 \vartheta = \frac{2}{3} \langle r^2 \rangle.$$

For the i-th electron, this gives

$$\boldsymbol{\mu}_i^{(\mathrm{ind})} = -\frac{e_0^2 k_L^2 \langle r_i^2 \rangle}{6 m_e c^2} \mathbf{B} = \frac{k_L^2}{k_C} \frac{r_e}{6} \langle r_i^2 \rangle \mathbf{B}, \qquad r_e = k_C \frac{e_0^2}{m_e c^2},$$

where we have inserted the classical electron radius r_e. If we now sum over all electrons, we obtain for the atom the induced moment

$$\langle \boldsymbol{\mu}^{(\mathrm{ind})} \rangle = -Z \frac{k_L^2}{k_C} \frac{r_e}{6} \mathbf{B} \langle r^2 \rangle, \qquad \langle r^2 \rangle = \frac{1}{Z} \sum_i \langle r_i^2 \rangle. \tag{7.3.10}$$

Magnetism in matter

In section 7.1.1 we rather abruptly introduced materials with strong magnetization as ferromagnets, an assignment that we want to catch up on here with the help of a classification of magnetic materials. However, some basic remarks should precede this classification.

- Spin is a phenomenon that follows from relativistic quantum mechanics and has no classical explanation [Schwabl, 2008, (5.3.29')].
- Even in the derivation of the induced moments (7.3.10), the background is quantum mechanics in the form of Bohr's atomic model with its stable circular orbits.
- The *Bohr-van-Leeuwen theorem* states that within the framework of classical statistics no magnetism occurs.

Proof: The Hamilton function for N electrons in the electromagnetic field is [Schwabl, 2006, (6.1.1)] according to (5.4.11)

$$\mathcal{H}_{kl} = \sum_{i=1}^{N} \frac{1}{2 m_e} \left[\boldsymbol{p}_i - \frac{k_L e}{c} \mathbf{A}(\mathbf{x}_i, t) \right]^2 + V_c(\{\mathbf{x}_i\}, t).$$

V_c is the Coulomb interaction of the electrons. The free energy F is defined by the canonical state integral [Schwabl, 2006, (6.1.32)]:

$$Z_{kl} = \mathrm{e}^{-F/k_B T} = \frac{1}{(2\pi\hbar)^{3N} N!} \int \mathrm{d}^{3N} \mathrm{p} \int \mathrm{d}^{3N} x \, \mathrm{e}^{-\mathcal{H}_{kl}/k_B T}.$$

After the coordinate transformation $\boldsymbol{p}_i' = \mathbf{p}_i - (k_L/c) e \mathbf{A}(\mathbf{x}_i)$ the state integral no longer depends on $\mathbf{A}$, i.e. on the magnetic field $\mathbf{B}$. Thus, the magnetization, which is defined by $\mathbf{M} = -\partial A / \partial \mathbf{B}$ also disappears. Therefore, magnetism cannot be explained within the framework of classical physics.

7.3.1 Diamagnetism

Langevin or Larmor diamagnetism

When a field is applied to an atom, a current is induced according to Lenz's rule, such that $\boldsymbol{\mu}^{(\mathrm{ind})}$ is antiparallel to the external field $\mathbf{B}$. From (7.3.10) one obtains for N atoms per cm^3 with

$$\chi_B = N\mu_0 \frac{\langle \mu^{(\mathrm{ind})}\rangle}{B} = -NZ\mu_0 \frac{k_L^2 e_0^2}{6m_e c^2}\langle r^2\rangle = -NZ\frac{1}{k_r}\frac{r_e}{6}\langle r^2\rangle, \tag{7.3.11}$$

the classical result originating from *Langevin*[5] [Kittel, 2005, (11.5)]. It assumes electrons bound to atoms or molecules and is therefore mainly valid for insulators.

A substance has about 1 neutron and 1 proton per electron; thus one can specify the number of electrons with $n \approx 3\times 10^{23}/\mathrm{g}$. If one again takes $\langle r^2\rangle = a_B^2$ for the average contribution of the individual electron and considers $\chi_m \approx \chi_B$, then the susceptibility per gram is

$$\chi^{(\mathrm{mass})} = \frac{\chi_m}{\rho_m} = -n\frac{r_e a_B^2}{6k_r} \overset{\mathrm{G}}{\approx} -3\times 10^{23}\mathrm{g}^{-1}\frac{2.8\times 10^{-13}\,\mathrm{cm}}{6}5.3^2\times 10^{-18}\,\mathrm{cm}^2$$

$$\overset{\mathrm{G}}{\approx} -0.4\times 10^{-6}\,\mathrm{g}^{-1}\,\mathrm{cm}^3. \tag{7.3.12}$$

$\chi^{(\mathrm{mass})}$ is the so-called *specific susceptibility* or *mass susceptibility*. From Tab. 7.1 one can see that the susceptibilities $\chi^{(\mathrm{mass})}$ do not differ by orders of magnitude and are surprisingly well approximated by (7.3.12). An exception is graphite, in which π-electrons can move within the hexagons of the layer structure. This results in a strongly anisotropic susceptibility, as can be seen from Tab. 7.1.

Landau diamagnetism

Quantum mechanics describes a (spinless) electron, which is in a constant magnetic field, given by $\mathbf{A} = \mathbf{B}\times\mathbf{x}/2$, with the Hamilton operator $\mathcal{H}$ [Schwabl, 2007, (7.2)]

$$\mathcal{H} = \frac{1}{2m_e}\left(\mathbf{p} - \frac{k_L}{c}e\mathbf{A}\right)^2 = \frac{\mathbf{p}^2}{2m_e} - \frac{k_L e}{2m_e c}\mathbf{B}\cdot\mathbf{L} + \frac{k_L^2 e^2 B^2}{8m_e c^2}\varrho^2.$$

The second term is paramagnetic and the third is diamagnetic, where a comparison shows that for $\langle\varrho^2\rangle \sim a_B^2$ the paramagnetic term dominates. For spherically symmetric atoms, i.e. $\langle L_z\rangle = 0$, $\langle\varrho^2\rangle = 2\langle r^2\rangle/3$, we obtain the classical result for the magnetic moment and the susceptibility of the orbital motion of a particle [Kittel, 2005, (11.10)]:

---

[5] Paul Langevin, 1872–1946.

Tab. 7.1. Diamagnetic susceptibilities at room temperature $\chi_m^{\mathrm{SI}} = 4\pi\chi_m$

Substance	$\chi^{(\mathrm{mol})}$ [mol^{-1} cm^3]	Molar mass [g mol^{-1}]	$\chi^{(\mathrm{mass})}$ [g^{-1} cm^3]	ρ_{m} [g cm^{-3}]	χ_m
H$_2$O^a	-13.0×10^{-6}	18.0	-0.72×10^{-6}	1.00	-0.7×10^{-6}
Bi^a	-280.1×10^{-6}	209.0	-1.34×10^{-6}	9.78	-13.1×10^{-6}
Diamond^a	-5.9×10^{-6}	12.0	-0.49×10^{-6}	3.51	-1.7×10^{-6}
He^a	-2.0×10^{-6}	4.0	-0.50×10^{-6}	0.00017	-8.5×10^{-11}
Xe^a	-45.5×10^{-6}	131.3	-0.35×10^{-6}	0.0055	-1.9×10^{-9}
pyrolytic graphite^b $\chi_\perp$	-189.3×10^{-6}	12.0	-15.78×10^{-6}	2.27	-35.8×10^{-6}
pyrolytic graphite^b $\chi_\parallel$	-35.8×10^{-6}	12.0	-2.98×10^{-6}	2.27	-12.6×10^{-6}

^{a} χ^{mol}: *CRC Handbook of Chemistry and Physics*, 79th ed., D.R. Lide, Editor; CRC Press: Boca Raton 1998; chapter 4, pp. 130–135.
^{b} χ_m: M.D. Simon and A.K. Geim J. Appl. Phys. **87**, 6200 (2000).
The anisotropy of pyrolytic graphite can be much more pronounced in samples that have been exposed to higher temperatures, (D.B. Fischbach, Phys.Rev. **123**, 1613 (1961)).

$$\mu = -\frac{\partial E}{\partial \mathbf{B}} = -\frac{k_L^2 e^2 \mathbf{B}}{6m_e c^2}, \qquad \chi_{\mathrm{B}} = \mu_0 \frac{\partial \mu}{\partial B}.$$

In the areas of statistical mechanics and magnetism, one always starts from (external) magnetic fields $\mathbf{H}$, i.e., the above calculation of the magnetic moment starts from the internal energy $E = \langle \mathcal{H} \rangle$, where $\langle ... \rangle$ denotes the thermal average [Schwabl, 2006, (6.1.12)] and

$$\langle \mu \rangle = -\mu_0 \left\langle \frac{\partial \mathcal{H}}{\partial \mathbf{H}} \right\rangle.$$

So one only considers magnetic susceptibilities χ_m. In the previous calculations, this does not matter, as χ_B is almost equal to χ_m.

In metals, the conduction electrons can largely be considered as interaction-free gas, where the spin should be considered separately later. Charged particles moving solely under the influence of a homogeneous magnetic field $\mathbf{B}$, perform a circular motion in the plane perpendicular to the field with the cyclotron frequency ω_c, as shown in (5.4.14). Here quantum mechanics and quantum statistics come into play. On the one hand, the electrons in the material only have a finite number of energy levels available, which are all occupied up to the Fermi energy E_F at sufficiently low temperatures. Occupied here means that each state, as prescribed by the *Fermi statistics*, can only be $2s + 1 = 2$-fold degenerate. However, the spectrum of energy eigenvalues is changed by switching on the magnetic field, since the circular motion of the electrons around the field is quantized (*Landau levels*).

The calculation of the susceptibility goes beyond the scope of this book, which is why we only give the result here [Schwabl, 2006, (6.4.13)]; k_F is the wave number to the Fermi energy.

$$\chi_{\text{Landau}} = -\mu_0 \frac{e_0^2 k_L^2}{12\pi^2 m_e c^2} k_F . \qquad (7.3.13)$$

The spin of the electron provides the paramagnetic contribution (Pauli paramagnetism) $\chi_{\text{P}} = -3\chi_{\text{Landau}}$ to the susceptibility, so that from the electron gas in metals overall a paramagnetic contribution comes. We anticipate here that the susceptibility of the free electrons results from the parts

$$\chi = \chi_P + \chi_{\text{Landau}} + \chi_{\text{osz}},$$

where χ_{osz} is an oscillating part, which describes the de-Haas-van-Alphen oscillations [Schwabl, 2006, section 6.4].

Superconductors

In many metals, any electrical resistance vanishes at very low temperatures $T \leq T_c \sim 10\,\text{K}$ for a direct current. This phenomenon, discovered in 1911 by *Kamerlingh-Onnes*, is explained on a microscopic level by quantum mechanics (BCS theory). But also on a macroscopic level some aspects can be understood, or it can be shown that these do not contradict electrodynamics, as outlined in section 5.4.3.

The current $\mathbf{j}_s$ flows in a superconductor near the surface, as from (5.4.35), page 184 and thus prevents $\mathbf{B}$ from penetrating into the body. $\mathbf{B}$ and $\mathbf{j}_s$ according to the London theory when penetrating into the superconductor exponentially approach zero (see (5.4.34) and (5.4.35), page 184), where the penetration depth $\Lambda \sim 10^2 - 10^3\,\text{Å}$ is.

Inside the superconductor is $\mathbf{B} = 0$ (Meissner effect or Meißner-Ochsenfeld effect). We have found that from (5.4.30) $\boldsymbol{\nabla} \cdot \mathbf{j}_s = 0$ follows. In analogy to $\mathbf{j}_M$, the current can be brought into the form

$$\mathbf{j}_s = (c/k_L)\boldsymbol{\nabla} \times \mathbf{M}_s \qquad (7.3.14)$$

with which (Ampère's law)

$$\boldsymbol{\nabla} \times (\mathbf{B}/\mu_0 - 4\pi k_r \mathbf{M}_s) = 0 .$$

Thus, the tangential components of $\mathbf{H} = \mathbf{B}/\mu_0 + 4\pi k_r \mathbf{M}_s$ are continuous, and on the surface $\mathbf{H}$ is equal to the external $\mathbf{B}_e/\mu_0$. Inside is $\mathbf{B} = \mu_0(1 + 4\pi k_r \chi_m)\mathbf{H} = 0$, i.e.

$$\chi_m = -1/4\pi k_r . \qquad (7.3.15)$$

In Fig. 7.15 the internal field

$$\mathbf{B} = \mathbf{B}_e + 4\pi k_r \mu_0 \mathbf{M}_s$$

is plotted as a function of the external field $\mathbf{B}_e$. If $B_e \geq B_c$, then the superconductivity collapses and the system becomes normally conductive.

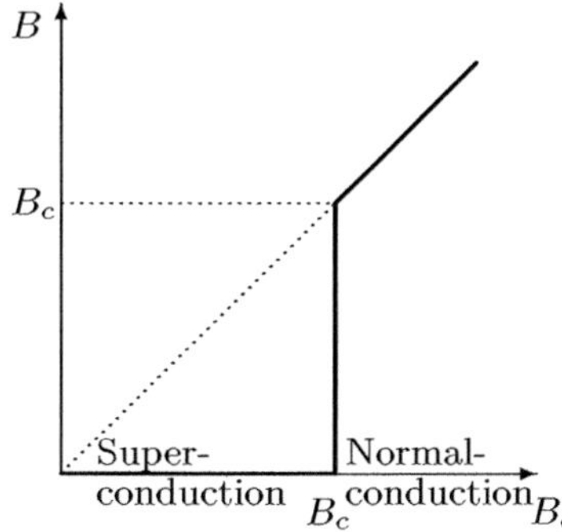

Fig. 7.15. Internal field B as a function of the external field B_e in a type 1 superconductor. If $B_e \leq B_c$ then B_e is shielded so that $B = 0$ (Meissner effect), which means that $4\pi k_r \mu_0 M_s = -B_e$ increases linearly

Superconductors, whose behavior is described by Fig. 7.15, are referred to as type 1 superconductors. These are almost exclusively elements, hardly any alloys, which have a very low T_c in common, so that they are hardly suitable for technical applications.

Notes: Another intuitive explanation can be found for the Meissner effect. According to the BCS theory, the electrons involved in the superconductivity are coupled over the crystal lattice to so-called *Cooper pairs*. These quasi-particles have spin $S = 0$, thus provide no paramagnetic contribution and completely shield **B**, as they due to their integer spins obey the *Bose statistics*, whose states with respect to their occupancy numbers are not subject to any restriction.

The majority of superconducting substances, the so-called type 2 superconductors (*mixed state*), however, behave differently in a magnetic field. Two critical values $\mathbf{B}_{c1}$ and $\mathbf{B}_{c2}$ occur (lower and upper critical field).

$B_e < B_{c1}$: We have a Meissner state as in type 1 superconductor ($\mathbf{B}=0$).

$B_{c1} < B_e < B_{c2}$: There is a coexistence of normal conducting and superconducting areas, resulting in a mixed state known as Shubnikov phase. The curve drops vertically at B_{c1}.

$B_e > B_{c1}$: Flux threads form, which can be imagined as normal conducting lines, around which a circular current flows. The flux threads form a triangular lattice.

Almost all superconducting alloys and all high-temperature superconductors are of type 2; of the elements only niobium is.

7.3.2 Paramagnetism

If the atoms or molecules of a substance have a magnetic moment and the interaction between the moments is so weak that it can be neglected, then the substance is paramagnetic. In an external field, the moments try to align themselves parallel to the field. This is opposed by the temperature movement, which does not prefer any specific direction. Magnetic moment (7.3.8) and magnetic energy of an atom are

$$\boldsymbol{\mu} = -g_J \mu_B \hat{\mathbf{J}}, \qquad\qquad \mathcal{H} = -\boldsymbol{\mu} \cdot \mathbf{B}.$$

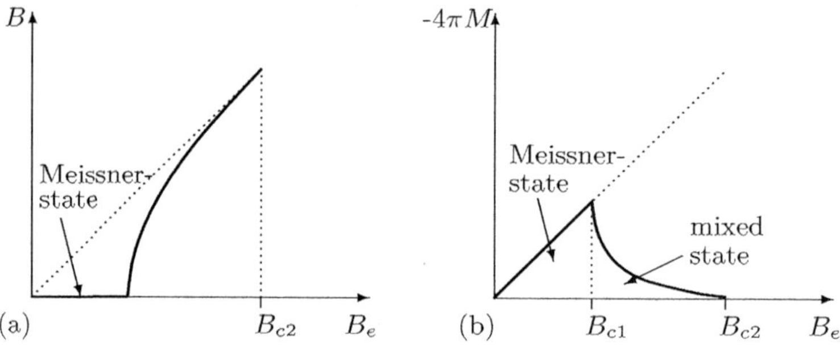

Fig. 7.16. Type 2 superconductor: (a) B as a function of the external field $\mathbf{B}_e$ (b) The magnetization as a function of B_e

$\mu_B = e_0 k_L \hbar / 2 m_e c$ (see (7.3.7)) and $\hat{\mathbf{J}} = \mathbf{J}/\hbar$. From quantum mechanics we take that $\hat{\mathbf{J}}^2 = j(j+1)$ and put that into the Landé factor (7.3.9)

$$g_J = 1 + \frac{j(j+1) + s(s+1) - l(l+1)}{2j(j+1)} . \qquad (7.3.16)$$

Using Boltzmann statistics, the classical state integral

$$Z = \int d\Omega \, e^{\boldsymbol{\mu} \cdot \mathbf{B}/k_B T} = \frac{4\pi k_B T}{\mu B} \sinh\left(\frac{\mu B}{k_B T}\right)$$

can be calculated analogous to the dipole moment on page 187, where $\mathbf{B} = B \mathbf{e}_z$. k_B is the Boltzmann constant and T is the temperature. For the average moment one obtains

$$\langle \boldsymbol{\mu} \rangle = \frac{1}{Z} \int d\Omega \, \boldsymbol{\mu} \, e^{\boldsymbol{\mu} \cdot \mathbf{B}/k_B T} = L\left(\frac{\mu B}{k_B T}\right) \mu \, \mathbf{e}_z , \qquad (7.3.17)$$

where

$$L(u) = \left(\coth u - \frac{1}{u}\right) \overset{u \to 0}{=} \frac{1}{u}\left(1 + \frac{u^2}{3} - \frac{u^4}{45} + ...\right) - \frac{1}{u} = \frac{u}{3}\left(1 - \frac{u^2}{15}\right) + ... \quad (7.3.18)$$

is the Langevin function. Note that $\langle \boldsymbol{\mu} \rangle$ always tries to align parallel to $\mathbf{B}$ since the weight factor is maximal for $\boldsymbol{\mu} \| \mathbf{e}_z$. Let N be the density of atoms/molecules, so is

$$\mathbf{M} = N\langle \boldsymbol{\mu} \rangle = N \begin{cases} (g_J \mu_B J)^2 \, \mathbf{B}/3k_B T & \text{for} \quad g_J J \mu_B B \ll k_B T \\ g_J \mu_B J \mathbf{B}/B & \text{for} \quad g_J J \mu_B B \gg k_B T \end{cases} \qquad (7.3.19)$$

the magnetization. This results in the *Curie's law* for weak fields/high temperatures

$$\chi_m = \mu_0 N (g_J \mu_B)^2 \frac{j(j+1)}{3k_B T} , \qquad (7.3.20)$$

where again $\chi_m \approx \chi_B$ was used.

The calculation of $\langle \mu \rangle$ can only be satisfactorily explained by using quantum theory and statistical mechanics [Schwabl, 2006, section 6.3]

The values of the paramagnetic susceptibility are of the order of magnitude $\chi_m \sim 10^{-5}$ and are therefore particularly at low temperatures about an order of magnitude above those of the diamagnetic susceptibility. Characteristic values at room temperature are in the Tab. 7.2 listed. A saturation of **M** is achieved when all magnetic moments are aligned parallel to the field, which requires field strengths of $10^6\, G$.

Tab. 7.2. Paramagnetic susceptibilities at room temperature. $\chi_m^{\text{SI}} = 4\pi\chi_m$.

Substance	$\chi^{(\text{mol})\,a}$ [mol^{-1} cm^3]	Molarmass [g mol^{-1}]	$\chi^{(\text{mass})}$ [g^{-1} cm^3]	ρ_m [g cm^{-3}]	χ_m
Al	16.5×10^{-6}	27.0	0.61×10^{-6}	2.70	1.65×10^{-6}
Cs	29×10^{-6}	132.9	0.22×10^{-6}	1.87	0.41×10^{-6}
K	21×10^{-6}	39.1	0.53×10^{-6}	0.86	0.46×10^{-6}
O$_2$	3450×10^{-6}	32.0	$107.8\ \times10^{-6}$	0.00143	0.15×10^{-6}
Pd	540×10^{-6}	106.4	5.07×10^{-6}	12.0	$61.0\ \times10^{-6}$
Pt	193×10^{-6}	195.1	0.99×10^{-6}	21.5	$21.0\ \times10^{-6}$
Rb	17×10^{-6}	85.5	0.20×10^{-6}	1.63	0.32×10^{-6}
Sr	92×10^{-6}	87.6	1.05×10^{-6}	2.54	2.67×10^{-6}

[a] Molar susceptibilities from: *CRC Handbook of Chemistry and Physics*, 79th ed., D.R. Lide, Editor; CRC Press: Boca Raton 1998; chapter 4, p. 130–135.

The outlined here *Langevin paramagnetism* is not the only form of paramagnetism. There is also

Pauli paramagnetism: Free (conduction) electrons in metals are considered as free electron gas in the magnetic field, where the electron spins couple to **B**. In addition to Pauli paramagnetism, there is also the *Landau diamagnetism* from the orbital motion of the electrons.

Van Vleck paramagnetism: Substances, whose atoms/molecules in the ground state are characterized by $\mathbf{J} = 0$, can have excited states with $\mathbf{J} \neq 0$ have.
These two forms of paramagnetism are usually weaker than Langevin paramagnetism.

7.3.3 Ferromagnetism

The interaction of atomic magnetic moments on the basis of dipole-dipole interaction (4.3.6) is too weak, to cause a parallel alignment of neighboring magnetic moments. Ferromagnetism is based on the exchange interaction, which can be traced back to the Coulomb interaction and the Pauli exclusion principle. There are several forms of exchange interaction, all of which have in common that they cannot be explained using classical physics and thus go beyond the scope of this text. The exchange interaction is short-ranged, i.e., it is noticeable only between nearest and at best next-nearest neighbors. Depending on the sign, it can prefer a parallel or antiparallel position of the magnetic moments of neighboring atoms. If magnetic moments align parallel, as in Fe, Ni, EuO, Gd, then you have a ferromagnet. Spontaneous magnetization occurs. Some properties:

- The saturation magnetization is reached, compared to the paramagnet, at weak fields (kGauss).
- The linear relation $\mathbf{M} = \chi_m \mathbf{H}$ must be replaced by a functional relationship, as this is outlined in Fig. 7.21, page 265.
- The "initial susceptibility"of the ferromagnet is in the range of $10-10^4$ (paramagnet 10^{-5}).
- After switching off the magnetic field, a residual magnetization remains, depending on the material.
- Above the *Curie temperature* T_c, the strength of the interaction is no longer sufficient for the parallel arrangement of neighboring magnetic moments and the substance becomes paramagnetic.
- Ferromagnets are solid bodies with a crystal structure. In the ferromagnet there exist regions, the so-called *Weiss domains*, in which the orientation of the magnetic moments is the same. The regions are separated by *Bloch walls* in which the orientation of the magnetization changes.

Molecular field approximation

In the Heisenberg model, one assumes spins that are fixed at lattice points and are coupled via an exchange interaction J_{ij}, which here has the dimension of an energy:

$$\mathcal{H} = -\frac{1}{2} \sum_{i,j} J_{ij} \hat{\mathbf{S}}_i \cdot \hat{\mathbf{S}}_j - g\mu_B \sum_i \hat{\mathbf{S}}_i \cdot \mathbf{B} \quad \text{with} \quad J_{ii} = 0 \quad \text{and} \quad \mathbf{S}_i = \hbar \hat{\mathbf{S}}_i. \tag{7.3.21}$$

$\mathbf{B} = B\mathbf{e}_z$ is an external magnetic field that determines the direction of $\mathbf{M} \| \mathbf{B}$. We now change (7.3.21) from $\mathbf{S}$ to

$$\boldsymbol{\mu} = g\mu_B \hat{\mathbf{S}}. \tag{7.3.22}$$

The directions of $\boldsymbol{\mu}$ and $\mathbf{S}$ are the same here and have the tendency to align parallel to $\mathbf{B}$.

$$\mathcal{H} = -\frac{1}{2(g\mu_B)^2} \sum_{i,j} J_{ij} \boldsymbol{\mu}_i \cdot \boldsymbol{\mu}_j - \sum_i \boldsymbol{\mu}_i \cdot \mathbf{B}. \tag{7.3.23}$$

In the molecular field approximation, it is assumed that the magnetic moments

$$\boldsymbol{\mu}_i = \langle \boldsymbol{\mu}_i \rangle + \delta \boldsymbol{\mu}_i$$

deviate only slightly from their mean value. Inserted into (7.3.23), one neglects the fluctuations $\delta\boldsymbol{\mu}_i \cdot \delta\boldsymbol{\mu}_j$ and obtains the Hamilton function in molecular field approximation

$$\mathcal{H}_{MFT} = -\sum_i \boldsymbol{\mu}_i \cdot \mathbf{h}_i + \frac{1}{2(g\mu_B)^2} \sum_{ij} J_{ij}\langle\boldsymbol{\mu}_i\rangle\langle\boldsymbol{\mu}_j\rangle \tag{7.3.24}$$

with the molecular field

$$\mathbf{h}_i = \frac{1}{(g\mu_B)^2} \sum_j J_{ij}\langle\boldsymbol{\mu}_j\rangle + \mathbf{B}\,. \tag{7.3.25}$$

We now also assume that the spin (the magnetic moment) is the same at all lattice sites and also has the same environment[6] ("Bravais lattice"). This makes averages and molecular field independent of the lattice site i:

$$\mathbf{h} = \Big(\frac{\tilde{J}}{(g\mu_B)^2}\langle\mu\rangle + B\Big)\mathbf{e}_z \qquad \text{with} \qquad \langle\boldsymbol{\mu}_i\rangle = \langle\mu\rangle\mathbf{e}_z \qquad \text{and} \qquad \tilde{J} = \sum_j J_{ij}\,.$$

As with the paramagnet, $\boldsymbol{\mu}_i$ moves independently in an average field, only with the difference that this field here is not external, but is formed by the $\langle\boldsymbol{\mu}_j\rangle$ of the surrounding moments. This follows analogously to (7.3.17)

$$\langle\boldsymbol{\mu}\rangle = \frac{1}{Z}\int d\Omega\, \boldsymbol{\mu}\, e^{\boldsymbol{\mu}\cdot\mathbf{h}/k_BT} = \mu_s L\Big(\frac{\mu_s h}{k_BT}\Big)\mathbf{e}_z \qquad \text{with} \qquad \mu_s = g\mu_B S, \tag{7.3.26}$$

where L is again the Langevin function (7.3.18).

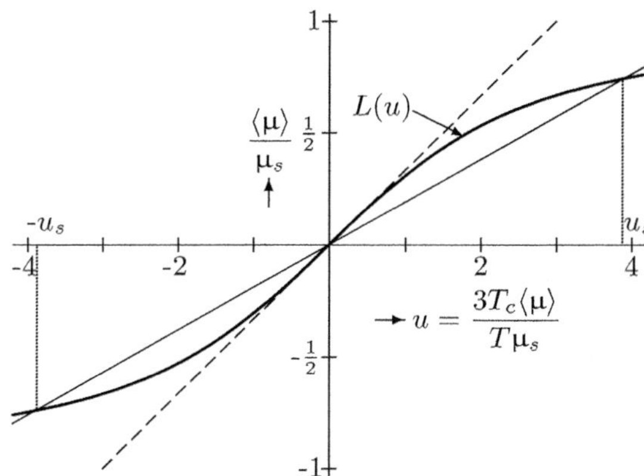

Fig. 7.17. Langevin function $L(u)$ with $L(\infty) = 1$. Every line with a smaller slope than $L'(0) = 1/3$ (dotted line) intersects $L(u)$ at $\pm x_0$; the values $L(\pm u_s)$ are the magnetization $\langle\mu\rangle/\mu_s$ at a given temperature

For $B = 0$, one determines $\langle\mu\rangle$ self-consistently from

---

[6] In the NaCl crystal, the Na^+ ions have neighbors at the same positions as the Cl^- ions, but they are always the other ions; in diamond, a C atom has the same neighboring atoms at the same distances as the adjacent C atom, but in different directions.

$$\frac{\langle\mu\rangle}{\mu_s} = u\,\frac{k_B T}{S^2\tilde{J}} \overset{u\leq 1}{\underset{B=0}{=}} L(u) \frac{u}{3}\Big(1 - \frac{u^2}{15}\Big) \qquad \text{with} \qquad u = \frac{S^2\tilde{J}}{k_B T}\frac{\langle\mu\rangle}{\mu_s}. \tag{7.3.27}$$

$\langle\mu\rangle = 0$ is always a solution and Fig. 7.17 shows that for low temperatures two solutions with finite $\langle\mu\rangle$ are present. The dotted tangent is given by $\langle\mu\rangle/\mu_s = u/3$, which is equal to

$$k_B T_c = S^2\tilde{J}/3. \tag{7.3.28}$$

If we replace $S^2\tilde{J}$ with $3k_B T_c$, then $u = \dfrac{3T_c}{T}\dfrac{\langle\mu\rangle}{\mu_s}$. Substituted into (7.3.27) we obtain

$u = \sqrt{15\dfrac{T_c - T}{T_c}}$. This results in the magnetic moment for $T \lesssim T_c$ and $B = 0$

$$\langle\mu\rangle \approx \pm g\mu_B S\sqrt{\frac{5}{3}\frac{T_c - T}{T_c}}. \tag{7.3.29}$$

So at T_c we have the transition from a phase with finite $\langle\mu\rangle$ to one with $\langle\mu\rangle = 0$. The behavior of the order parameter $\langle\mu\rangle$ as it approaches T_c with $|T_c - T|^{1/2}$ is typical for molecular field theory.

With finite B there is no phase transition, as can be seen from Fig. 7.18, the magnetization $\mathbf{M} = N\langle\mu\rangle$ never disappears (N denotes the spin density). The approximation $L(u) \approx u/3$ is then sufficient, from which the magnetic moment for $T \geq T_c$ follows

$$\frac{\langle\mu\rangle}{\mu_s} = L\Big(\Big(3\frac{\langle\mu\rangle}{\mu_s} + \frac{\mu_s B}{k_B T_c}\Big)\frac{T_c}{T}\Big) \approx \Big(\frac{\langle\mu\rangle}{\mu_s} + \frac{\mu_s B}{3k_B T_c}\Big)\frac{T_c}{T}.$$

The substance is thus in the paramagnetic phase

$$\langle\mu\rangle \approx \frac{\mu_s^2}{3k_B(T - T_c)}\,B\,,$$

from which follows the susceptibility according to the *Curie-Weiss law*

$$\chi_m = \mu_0 N\frac{\mu_s^2}{3k_B(T - T_c)}. \tag{7.3.30}$$

This singular behavior of the susceptibility near the phase transition is characteristic for the molecular field theory.

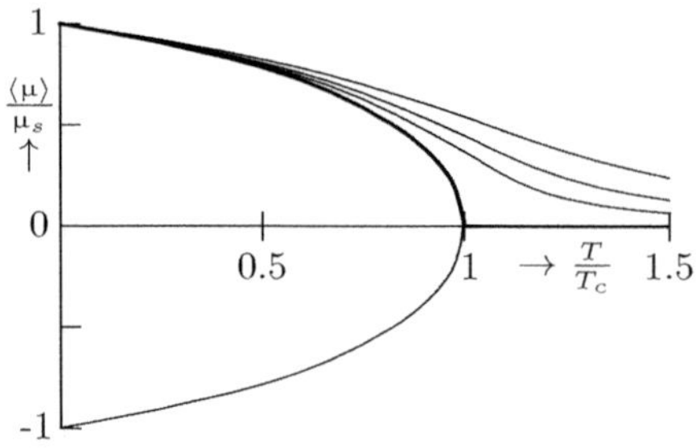

Fig. 7.18. Magnetization in molecular field approximation. For $B = 0$ both settings $\pm\langle\mu\rangle$ are possible. The finite field ($B\mu_s/k_B T_c = 0.1, 0.2, 0.4$) designates a magnetization direction

Weiss domains

The ferromagnet based on the exchange interaction has two adjustment options for $\mathbf{B} \to 0$ with the magnetizations $\pm\mathbf{M}$; both lead to the same minimum (free) energy; without a field all directions are initially equivalent. However, if the exchange interaction is anisotropic, then not all directions are equivalent. In iron, for example, with a body-centered cubic lattice (bcc), it is easier for the spins to align in the direction of the axes $\pm\mathbf{e}_x$, $\pm\mathbf{e}_y$ or $\pm\mathbf{e}_z$ to orient.

One can imagine that the spins in a larger region align parallel, but in another region are oriented in a different direction. These regions are called *Weiss domains*. The boundaries between domains of differently oriented magnetization are the already mentioned Bloch walls, which in polycrystalline structure sometimes can lie at grain boundaries. It is energetically more favorable if the spins between two domains change their orientation only slowly, i.e., the Bloch walls are many atomic layers thick ($\sim 300\,\text{Å}$).

On the energy change at a Bloch wall: The spins are rotated along the z-axis by π. In the yz-plane then for fixed x all spins are parallel. Here we assume that $J \neq 0$ only applies between nearest neighbors and that along the z-axis, i.e., perpendicular to the Bloch wall, only two nearest neighbors are present, as it is sketched in Fig. 7.19. It is therefore sufficient to estimate the energy change of the spin configuration of a linear chain. If the spins for $1, ..., i$ are oriented in the z-direction and $i+1, ...$ in the opposite direction $(-z)$, then the energy change of the chain is $\Delta E = 2J\,S^2$.

If the rotation is distributed over n spins, $\varphi = \pi/n$, then

$$\mathbf{S}_i \cdot \mathbf{S}_{i+1} = \cos\left(\frac{\pi}{n}\right) S^2 \overset{n \gg 1}{\approx} \left(1 - \frac{\pi^2}{2n^2}\right) S^2 \qquad \Rightarrow \qquad \Delta E = nJS^2 \frac{\pi^2}{n^2} \sim \frac{1}{n}.$$

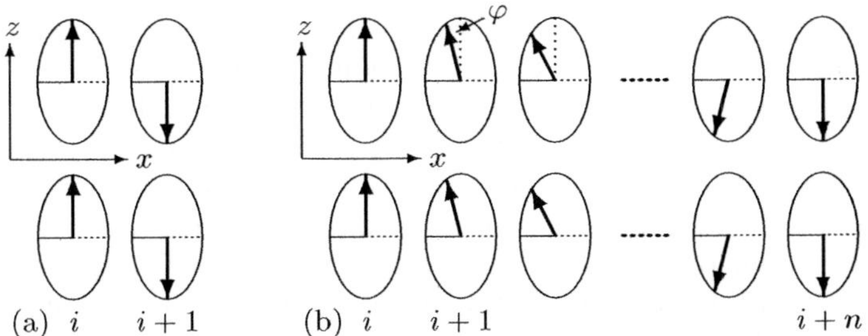

Fig. 7.19. The domain wall is perpendicular to the z-axis   (a) The spin is rotated by π in one atomic layer   (b) The rotation from one atomic layer to the next is $\varphi = \pi/n$ and thus extends over n atomic layers

Within an ideal lattice structure, the Bloch walls are movable in a reversible form when a field is applied, i.e., when a weak field is applied to the ferromagnet, the areas with favorable magnetization initially expand at the expense of the others. Only at high field strengths does the magnetization of entire areas flip in the direction of the field (*Barkhausen jumps*).

If the cuboid outlined in Fig. 7.20 is homogeneously magnetized, $\mathbf{M} = M\,\mathbf{e}_x$, then on the yz-base surfaces magnetic surface charges

$$\rho_M = -\boldsymbol{\nabla}\cdot\mathbf{M} = M\left[\delta(a-x) - \delta(a+x)\right]\theta(b-|y|)\theta(c-|z|)\,,$$

determine the course of the field lines in the external space. If the cuboid is divided into domains in the outlined way, then surface charges can only occur at the diagonally drawn boundary lines. With the help of the normal vector $\mathbf{n}$ on the boundary line between $\mathbf{M}_2$ and $\mathbf{M}_3$ one can show that the normal component of the magnetization continuously goes through this when $\alpha = \pi/4$:

$$\mathbf{n} = -\mathbf{e}_x + \tan\alpha\,\mathbf{e}_y \qquad \Rightarrow \qquad \mathbf{n}\cdot\mathbf{M}_2 = M\tan\alpha, \qquad \mathbf{n}\cdot\mathbf{M}_3 = M\,.$$

So there are no magnetic charges $\rho_M = -\boldsymbol{\nabla}\cdot\mathbf{M} = 0$ and consequently $\boldsymbol{\nabla}\cdot\mathbf{H} = 0$.

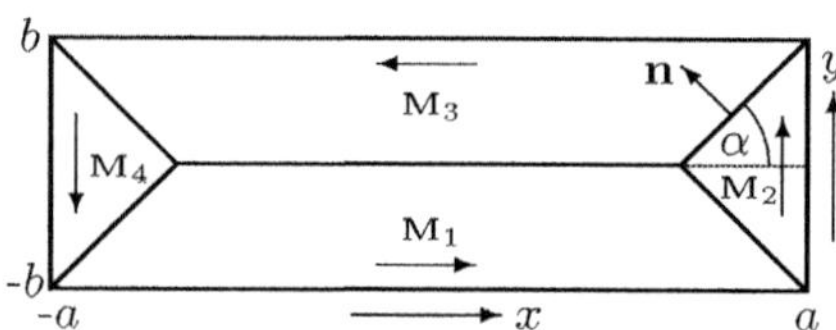

Fig. 7.20. Division of a cuboid into four domains ($|\mathbf{M}_i| = M$, $i = 1,...,4$). For $\alpha = \pi/4$ no field penetrates into the outer space

Soft and hard ferromagnets

If an unmagnetized ferromagnet is brought into a field $\mathbf{H}$, its magnetization will grow, until at a value $\mathbf{H}_s$ the saturation magnetization $\mathbf{M}_s$ is reached, at which all angular momentum moments are aligned parallel to the field. In Fig. 7.21 this is the *virgin magnetization curve* originating from the origin. It is characterized in the lower part by the displacement of the domains; a process that is reversible. This process becomes irreversible with increasing $\mathbf{H}$ until finally the magnetization of the remaining domains is rotated (tilted) in the direction of $\mathbf{H}$. In this part of the virgin curve, the increase of M against H becomes significantly weaker.

If you now remove the field again, a residual magnetization (M_r) remains at $H = 0$. Only at the field strength $-H_c$ (*coercive field strength*) does the magnetization go to zero, to reach again the saturation value at $-H_s$. Instead of in the HM diagram of Fig. 7.21, one can represent magnetization curves in a HB diagram. M_s is determined by the value (H_s, B_s) from which $M_s = (B_s/\mu_0 - H_s)/4\pi k_r$ no longer increases. It is $B_r = 4\pi k_r \mu_0 M_r$. However, the coercive field H_{cb} for $B = 0$ differs from H_c for $M = 0$.
One distinguishes

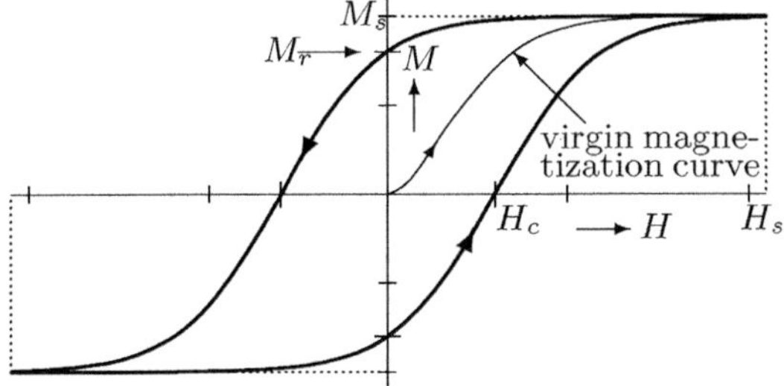

Fig. 7.21. Magnetization of a ferromagnet in the external field **H**; Hysteresis loop: Saturation magnetization $\mathbf{M}_s$; remanent magnetization M_r and coercive field strength H_c

- *soft magnetic substances*: $0.002\,\text{Oe} \leq H_c \leq 1\,\text{Oe}$,
- *hard magnets*: $500\,\text{Oe} \leq H_c \leq 10\,000\,\text{Oe}$.

In Fig. 7.21 the area enclosed by the magnetization curve is proportional to the energy density u, which must be expended in one cycle,

$$u = \frac{1}{8\pi k_r} \oint \mathbf{B}\cdot d\mathbf{H} = \frac{\mu_0}{2} \oint \mathbf{M}\cdot d\mathbf{H}, \tag{7.3.31}$$

since $\oint \mathbf{H}\cdot d\mathbf{H} = 0$. The smaller the area enclosed by a pass through the hysteresis curve, the lower are the losses. The energy density $(BH)_{\text{max}}$ is given in $10^6\,\text{G Oe}$, which is equivalent to $10^6\,[\text{erg/cm}^3]$.

- *soft magnetic substances*: $(BH)_{\text{max}}$: $20-5\,000\,\text{erg/cm}^3$,
- *hard magnetic substances*: $(BH)_{\text{max}}$: $10^5-10^6\,\text{erg/cm}^3$.

χ_m and consequently μ_r presuppose a linear relationship between M and H, which is not given in the ferromagnet . There are various permeabilities.

Tab. 7.3. Saturation magnetization and permeability of ferromagnetic substances at room temperature. J_s is the saturation polarization/magnetization. W_H is the hysteresis loss per cycle.

Substance	Formula (Product)	J_s [kG]	B_r [kG]	H_c [Oe]	μ_{max} 10^5	T_c [°C]	$(BH)_{\text{max}}$ 10^6 G Oe	W_H [erg/cm³]
Supermalloy[a]	79Ni 16Fe 5Mo	.63		0.002	10	673		20
78 Permalloy[a]	78Ni 22Fe	.84		0.05	1	650		500
pure iron[a]		171		0.01	3.5	1040		600
AlNiCo[b]	AlNiCo44/5		13.5	590		850	5.6	
SmCo[b]	Sm32/15–17		11.5	9 500		825	31	
Neodymium (NdFeB)[b]	N45SH		13.5	12 600		340	44	

[a] *CRC Handbook of Chemistry and Physics*, 92th ed., D.R. Lide, Editor; CRC Press: Boca Raton 2012; Chapter 12, p. 112–114.
[b] http://maurermagnetic.ch/PDF/51_E.pdf (Maurer Magnetic AG) downloaded on 9.8. 2014; no information on composition.

Starting from the new curve in Fig. 7.21 defines the *initial permeability*

$$\mu_r^{(i)} = \frac{1}{\mu_0} \frac{\partial B}{\partial H}\bigg|_{H=0}. \tag{7.3.32}$$

The *amplitude permeability* refers to the linear relationship $\mu_r^{(a)} = B/H\mu_0$. $\mu_r^{(a)}$ starts with the value of $\mu_r^{(i)}$, then rises to a maximum, to fall off at $B \lesssim B_s$. If you take the derivative in (7.3.32) at finite values of H, i.e., if you superimpose DC values of H with a weak alternating field, you get the *reversible permeability* $\mu_r^{(\text{rev})} = (1/\mu_0)\partial B/\partial H\big|_H$.

In Tab. 7.3, data for some ferromagnets are listed; especially for Fe the values depend very sensitively on processing and the Fe concentration.

Problems for Chapter 7

7.1. *Helmholtz's Decomposition Theorem*: Given is the field $\mathbf{v}(\mathbf{x}) = q(\mathbf{x}-\mathbf{x}_0)/|\mathbf{x}-\mathbf{x}_0|^3$. Show that the Helmholtz's Decomposition Theorem (7.1.16) remains valid, even though $\mathbf{v}$ is singular at the point $\mathbf{x}_0$.

7.2. *Particle in a homogeneous field*: Take the Hamilton function $\mathcal{H}(\mathbf{x}, \boldsymbol{P})$ of a particle in a homogeneous magnetic field $\mathbf{B}$ and show that the magnetic moment is given by

$$\mathbf{m} = -\frac{\partial \mathcal{H}}{\partial \mathbf{B}} = \mathbf{x} \times \frac{e\mathbf{v}}{2c}.$$

7.3. *Magnetic Torque*: Given is a circuit $\mathbf{j}(\mathbf{x})$ in an external field, given by $\mathbf{A}^e(\mathbf{x})$. Determine the torque of the circuit (4.3.7) from the field energy by imposing an infinitesimal rigid rotation $\delta\mathbf{x} = \delta\boldsymbol{\varphi} \times \mathbf{x}$ on the circuit.

Hint: $\mathbf{j}'(\mathbf{x}') = \mathbf{j}(\mathbf{x}') - \delta\boldsymbol{\varphi} \times \mathbf{j}(\mathbf{x}')$.

7.4. *Vector potential of the homogeneously magnetized sphere*: Verify (7.1.31').

7.5. *Fields in the axis of the bar magnet*: The fields $\mathbf{H}(\varrho, z)$ and $\mathbf{B}(\varrho, z)$ of a homogeneously magnetized bar magnet of radius a and length $2l$ (see Fig. 7.4) are simple when restricted to the z-axis ($\varrho = 0$).

1. Show that the field for $|z| \gg l$ is that of a dipole.
2. Calculate on the z-axis $\mathbf{H}(0, z)$ from the scalar Potential ϕ_M and $\mathbf{B}(0, z)$ from the vector potential $\mathbf{A}(0, z)$. Determine also the circulations Z_H and Z_B (see (1.3.1)).

7.6. *Semi-infinite, magnetized cuboid*: Given is the cuboid sketched in Fig. 7.9, p. 233 with homogeneous magnetization: $\mathbf{M} = M\mathbf{e}_z$. Calculate the scalar potential ϕ_M and the fields $\mathbf{H}(\mathbf{x}, \mathbf{x}_0)$ and $\mathbf{B}(\mathbf{x}, \mathbf{x}_0)$ for the reference point $\mathbf{x}_0$, if $X, Y, Z \to \infty$.

7.7. *Mutual inductance of two circular conductors*:

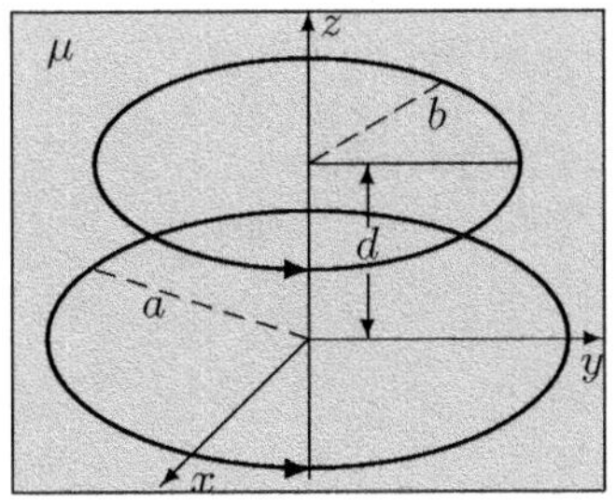

Fig. 7.22. Two concentric circular conductors, embedded in a medium of permeability μ, are shifted by d against each other

1. Show that

$$L_{12} = \frac{4\pi\mu}{c^2}\,\sqrt{ab}\left[\left(\frac{2}{k} - k\right)K(k) - \frac{2}{k}E(k)\right]$$

is the mutual inductance of the two circular conductors outlined in Fig. 7.22. $K(k)$ and $E(k)$ are the complete elliptical integrals of the first and second kind (see (B.5.4) and (B.5.5)) with $k^2 = \dfrac{4ab}{(a+b)^2 + d^2}$.

2. State the result in the limit $d \ll a$ for $a = b$ and compare it with the self-inductance of a wire loop.

References

Becker R., Sauter F. *Theorie der Elektrizität 2*, 8th ed. Teubner Stuttgart (1959)

Blumenthal O. *Ueber die Zerlegung unendlicher Vektorfelder*, Math. Ann. **61**, 235–250 (1905)

Coulomb Charles Augustin de *Septième mémoire sur l'éctricité et le magnétisme*, Mémoires del l'Académie royale des sciences **92**, 455–505 (1789), printed 1793; see *Collections de Mémoires relatifs à la Physique, publié par la société française de physique, Tome 1, Mémoires de Coulomb*, Paris, Gauthier-Villars (1884)

Griffiths D. J. *Electrodynamics* 4th. ed. Cambridge, University Press (2017)

Jackson J. D. *Classical Electrodynamics*, 3rd ed., John Wiley & Sons Inc. (1998)

Kittel Ch. *Introduction to Solid State Physics*, 8th ed., John Wiley & Sons, Inc. (2005)

Landau L. D., Lifshitz E. M. and Pitaevskii L. P. *Elektrodynamics of continuous media*, vol. 8, 2nd ed., Pergamon Press (1984)

Maxwell J. C. *A treatise on electricity and magnetism*, Vol II, 3rd ed. Clarendon Press Oxford (1892)

Michell John *A treatise of artificial magnets*, Cambridge (1750)

Neumann F. *Die mathematischen Gesetze der inducierten elektrischen Ströme*, Ostwald's Klassiker der exakten Wissenschaften **10**, editor C. Neumann, Wilhelm Engelmann in Leipzig (1889)

Petrascheck D., Folk R. *Helmholtz' decomposition theorem and Blumenthal's extension by regularization*, Condens. Matter Phys. **20**, 13002 (2017).

Petrascheck D. *Unit system independent formulation fo electrodynamics*, Eur. J. Phys. **42** 045201 (12pp) (2021)

Schwabl F. *Statistical Mechanics*, 2nd ed. Springer Berlin (2006)

Schwabl F. *Quantum Mechanics*, 4th ed. Springer Berlin (2007)

Schwabl F. *Advanced Quantum Mechanics*, 4th ed. Springer Berlin (2008)
Sommerfeld A. *Electrodynamics*, Academic Press Inc., New York (1952)
Uhlenbeck G. and Goudsmit S. *Naturwissenschaften* **47**, 953 (1925)

Fields of Moving Charges

In this chapter, we will attempt to capture solutions of the Maxwell equations in vacuum in their full time dependence. The main interest is in the periodic movement of a charge distribution and the electromagnetic radiation caused by this.

8.1 Vector and Scalar Potential in Lorenz Gauge

The fields $\mathbf{E}$ and $\mathbf{B}$ are to be determined from the Maxwell equations in vacuum (1.3.21) for given charge and current densities:

$$
\text{(a)} \qquad \boldsymbol{\nabla}\cdot\mathbf{E} = 4\pi k_c\rho, \qquad\qquad \text{(b)} \quad \boldsymbol{\nabla}\times\mathbf{E} + \frac{k_L}{c}\,\dot{\mathbf{B}} = 0,
$$
$$
\text{(c)} \quad k_L\boldsymbol{\nabla}\times\mathbf{B} - \frac{1}{c}\dot{\mathbf{E}} = \frac{4\pi k_c}{c}\,\mathbf{j}, \qquad \text{(d)} \qquad\qquad \boldsymbol{\nabla}\cdot\mathbf{B} = 0. \tag{8.1.1}
$$

$$
\text{SI:} \quad
\begin{aligned}
&\text{(a)} \qquad\qquad \boldsymbol{\nabla}\cdot\mathbf{E} = \rho/\epsilon_0, \qquad\qquad &&\text{(b)} \quad \boldsymbol{\nabla}\times\mathbf{E} + \dot{\mathbf{B}} = 0,\\
&\text{(c)} \quad \boldsymbol{\nabla}\times\mathbf{B} - \mu_0\epsilon_0\dot{\mathbf{E}} = \mu_0\,\mathbf{j}, \qquad &&\text{(d)} \qquad\qquad \boldsymbol{\nabla}\cdot\mathbf{B} = 0.
\end{aligned}
\tag{8.1.1'}
$$

These are differential equations coupled over time derivatives, which were decoupled in statics because of $\dot{\mathbf{E}} = \dot{\mathbf{B}} = 0$. Here we will try to find decoupled equations for suitable defined potentials $\phi(\mathbf{x},t)$ and $\mathbf{A}(\mathbf{x},t)$ to specify, with the help of which the fields can be determined. Because the magnetic field $\mathbf{B}$ is source-free, i.e. $\boldsymbol{\nabla}\cdot\mathbf{B} = 0$, the following applies

$$
\mathbf{B} = \boldsymbol{\nabla}\times\mathbf{A}. \tag{8.1.2}
$$

Thus, the induction equation (8.1.1b) results

© The Editor(s) (if applicable) and The Author(s), under exclusive license
to Springer-Verlag GmbH, DE, part of Springer Nature 2025
D. Petrascheck, F. Schwabl, *Electrodynamics*, https://doi.org/10.1007/978-3-662-71502-4_8

$$\boldsymbol{\nabla} \times \left(\mathbf{E} + \frac{k_L}{c} \dot{\mathbf{A}} \right) = 0 \,.$$

If the curl of a vector vanishes, it can be derived from a scalar potential:

$$\left(\mathbf{E} + \frac{k_L}{c} \dot{\mathbf{A}} \right) = -\boldsymbol{\nabla}\phi \,.$$

The fields are thus determined by the potentials ϕ and $\mathbf{A}$:

$$\mathbf{E} = -\boldsymbol{\nabla}\phi - \frac{k_L}{c}\dot{\mathbf{A}}, \qquad\qquad \mathbf{B} = \boldsymbol{\nabla}\times\mathbf{A}. \tag{8.1.3}$$

Now equations are to be derived with the help of which ϕ and $\mathbf{A}$ can be calculated for given charge- and current densities. The substitution of $\mathbf{E}$ into Gauss's law (8.1.1a) results in

$$\boldsymbol{\nabla}\cdot\mathbf{E} = -\nabla^2\phi - \frac{k_L}{c}\boldsymbol{\nabla}\cdot\dot{\mathbf{A}} = 4\pi k_C \rho. \tag{8.1.4}$$

We add $\ddot{\phi}/c^2$ on both sides and thus obtain

$$\left(\frac{1}{c^2}\frac{\partial^2}{\partial t^2} - \Delta \right)\phi(\mathbf{x},t) \equiv \square\phi(\mathbf{x},t) = 4\pi k_C \rho + \frac{1}{c}\frac{\partial}{\partial t}\left(k_L\boldsymbol{\nabla}\cdot\mathbf{A} + \frac{1}{c}\frac{\partial\phi}{\partial t} \right). \tag{8.1.5}$$

The operator

$$\square = \frac{1}{c^2}\frac{\partial^2}{\partial t^2} - \Delta \tag{8.1.6}$$

is referred to as the d'Alembert operator or sometimes also as "Quabla".

Note: The definition (8.1.6) is related to the used 4-dimensional notation. A space-time point (12.2.1) is given by $x^\mu = (ct, \mathbf{x})$ and $\square$ is the 4-dimensional counterpart to Δ: $\square = \frac{\partial}{\partial x_\mu}\frac{\partial}{\partial x^\mu}$ with $\mu = 0, 1, 2, 3$. The signs in the scalar product are determined by the diagonal elements of the metric tensor $g^{\mu\mu} = (1, -1, -1, -1)$. These are referred to as *signature* according to Rindler [2006, p. 175]. Sexl, Urbantke [2001, section 7.5] in their book call the difference between positive and negative diagonal elements signature; here it has the value –2. The signature +2 or $(-1, 1, 1, 1)$ is also common.

Now $\mathbf{E}$ and $\mathbf{B}$ (8.1.3) are inserted into the Ampère-Maxwell equation (8.1.1c), whereby the identity $\boldsymbol{\nabla}\times\boldsymbol{\nabla}\times\mathbf{A} = \boldsymbol{\nabla}\boldsymbol{\nabla}\cdot\mathbf{A} - \Delta\mathbf{A}$ is used:

$$\boldsymbol{\nabla}\times\mathbf{B} = \boldsymbol{\nabla}\boldsymbol{\nabla}\cdot\mathbf{A} - \Delta\mathbf{A} = \frac{4\pi k_C}{ck_L}\mathbf{j} - \frac{1}{ck_L}\frac{\partial}{\partial t}\left(\boldsymbol{\nabla}\phi + \frac{k_L}{c}\dot{\mathbf{A}} \right).$$

The rearrangement results in

$$\square\mathbf{A} = \frac{4\pi k_C}{ck_L}\mathbf{j} - \boldsymbol{\nabla}\left(\boldsymbol{\nabla}\cdot\mathbf{A} + \frac{1}{ck_L}\frac{\partial\phi}{\partial t} \right). \tag{8.1.7}$$

These are coupled differential equations for ϕ and $\mathbf{A}$. If the potentials fulfill the *Lorenz condition*[1] also called *Lorenz gauge* :

$$\boldsymbol{\nabla}\cdot\mathbf{A}+\frac{1}{ck_L}\frac{\partial}{\partial t}\phi = 0,\tag{8.1.8}$$

then the differential equations for the potentials $(\phi,\mathbf{A})$ decouple:

$$\Box\phi = 4\pi k_C\rho \overset{\text{SI}}{=} \rho/\epsilon_0\,,\qquad\qquad \Box\mathbf{A} = \frac{4\pi k_C}{ck_L}\mathbf{j}\overset{\text{SI}}{=}\mu_0\mathbf{j}.\tag{8.1.9}$$

The Lorenz condition (8.1.8) is not the only possibility to determine the potentials $(\phi,\mathbf{A})$. The fields $\mathbf{E}$ and $\mathbf{B}$ remain unchanged under gauge transformations (gauge invariance)

$$\mathbf{A}' = \mathbf{A}+\boldsymbol{\nabla}\chi,\qquad\qquad \phi' = \phi-\frac{k_L}{c}\dot\chi,\tag{8.1.10}$$

which is evident by inserting into (8.1.3)

$$\begin{aligned}\mathbf{E}' &= -\boldsymbol{\nabla}\Big(\phi-\frac{k_L}{c}\dot\chi\Big) - \frac{k_L}{c}\big(\dot{\mathbf{A}}+\boldsymbol{\nabla}\dot\chi\big) = \mathbf{E},\\ \mathbf{B}' &= \boldsymbol{\nabla}\times\big(\mathbf{A}+\boldsymbol{\nabla}\chi\big) = \mathbf{B}.\end{aligned}\tag{8.1.11}$$

The gauge function χ can therefore be chosen arbitrarily, as long as it is sufficiently smooth (i.e., continuously differentiable enough times). Then the mixed derivatives commute.

We assume that $(\phi,\mathbf{A})$ fulfill the Lorenz condition (8.1.8) and use the freedom in the choice of χ so that $(\phi',\mathbf{A}')$ satisfy a special gauge transformation:

$$\boldsymbol{\nabla}\cdot\mathbf{A}+\frac{1}{ck_L}\dot\phi = \boldsymbol{\nabla}\cdot(\mathbf{A}'-\boldsymbol{\nabla}\chi)+\frac{1}{ck_L}\frac{\partial}{\partial t}\Big(\phi'+\frac{k_L}{c}\frac{\partial}{\partial t}\chi\Big) = \boldsymbol{\nabla}\cdot\mathbf{A}'+\frac{1}{ck_L}\dot\phi'+\Box\chi = 0.$$

The gauge function thus satisfies the inhomogeneous wave equation

$$\Box\chi = -\boldsymbol{\nabla}\cdot\mathbf{A}' - \frac{1}{ck_L}\frac{\partial}{\partial t}\phi'.\tag{8.1.12}$$

If $(\phi',\mathbf{A}')$ fulfill the Lorenz gauge, they are determined up to the solution of the homogeneous wave equation $\Box\chi = 0$. For each specific choice of a gauge, such as the *Coulomb gauge* $\boldsymbol{\nabla}\cdot\mathbf{A}' = 0$ the homogeneous equation can be added to the solution of the inhomogeneous equation (8.1.12).

8.2 Retarded Potentials

8.2.1 The Inhomogeneous Wave Equation

$\rho(\mathbf{x},t)$, $\mathbf{j}(\mathbf{x},t)$ are given and ϕ and $\mathbf{A}$ are sought. Inhomogeneous wave equations of the form must be solved for both potentials

---

[1] Ludvig Lorenz, 1829–1891

$$\Box\psi(\mathbf{x},t) = q(\mathbf{x},t)\,, \tag{8.2.1}$$

as given by (8.1.9). ϕ and $\mathbf{A}$ then fulfill the Lorenz condition (8.1.8). However, the Green's function is first determined, which is a solution of the inhomogeneous wave equation with the inhomogeneity $q(\mathbf{x},t) = \delta^{(3)}(\mathbf{x})\,\delta(t)$:

$$\Box D(\mathbf{x},t) = \delta^{(3)}(\mathbf{x})\,\delta(t)\,. \tag{8.2.2}$$

The solution is then

$$\psi(\mathbf{x},t) = \int \mathrm{d}^3x'\,\mathrm{d}t'\,D(\mathbf{x}-\mathbf{x}',t-t')\,q(\mathbf{x}',t')\,, \tag{8.2.3}$$

as can be verified by applying $\Box$ to ψ using (8.2.2). Solutions of $\Box\psi' = 0$ can still be added to ψ.

The retarded Green function

It turns out to be easier to calculate the Fourier transform first

$$D(\mathbf{k},\omega) = \int \mathrm{d}^3x\,\mathrm{d}t\,\mathrm{e}^{-\mathrm{i}(\mathbf{k}\cdot\mathbf{x}-\omega t)}D(\mathbf{x},t) \tag{8.2.4}$$

and then by back transformation

$$D(\mathbf{x},t) = \int \frac{\mathrm{d}^3k}{(2\pi)^3}\frac{\mathrm{d}\omega}{2\pi}\,\mathrm{e}^{\mathrm{i}(\mathbf{k}\cdot\mathbf{x}-\omega t)}D(\mathbf{k},\omega)\,. \tag{8.2.5}$$

This results in

$$\Box D(\mathbf{x},t) = \int \frac{\mathrm{d}^3k}{(2\pi)^3}\frac{\mathrm{d}\omega}{2\pi}\Big(-\frac{\omega^2}{c^2}+k^2\Big)\mathrm{e}^{\mathrm{i}\mathbf{k}\cdot\mathbf{x}-\mathrm{i}\omega t}D(\mathbf{k},\omega) = \delta^{(3)}(\mathbf{x})\,\delta(t)\,.$$

Multiplying from the left with $\int \mathrm{d}^3x\,\mathrm{d}t\,\mathrm{e}^{-\mathrm{i}\mathbf{k}'\cdot\mathbf{x}+\mathrm{i}\omega't}$ yields

$$\Big(-\frac{\omega'^2}{c^2}+k'^2\Big)D(\mathbf{k}',\omega') = 1 \qquad \Rightarrow \qquad D(\mathbf{k},\omega) = \frac{-c^2}{\omega^2-c^2k^2}\,. \tag{8.2.6}$$

First, we calculate

$$D(\mathbf{k},t) = -c^2\int \frac{\mathrm{d}\omega}{2\pi}\,\mathrm{e}^{-\mathrm{i}\omega t}\frac{1}{\omega^2-c^2k^2} \tag{8.2.7}$$

In (8.2.3), the potential $\psi(\mathbf{x},t)$ could also include contributions from $q(\mathbf{x}',t')$ from times $t' > t$, which would be acausal. For reasons of causality, the integral for $t < t'$ should therefore vanish:

$$D(\mathbf{k},t) = 0 \quad \text{for} \quad t < 0\,.$$

The integration of (8.2.7) is carried out in the complex plane with the *Cauchy's residue theorem* (B.1.10), where causality determines the integration paths C in Fig. 8.1.

If $f(z)$ is analytic in the area enclosed by C, except for m isolated, simple poles, then according to (B.1.10)

$$\oint_C \mathrm{d}z\, f(z) = 2\pi\mathrm{i} \sum_{k=1}^{m} R(z_k) \quad \text{with} \quad R(z_k) = \lim_{z \to z_k} (z - z_k) f(z). \tag{8.2.8}$$

The poles of $f(z) = D(\mathbf{k}, z)\,\mathrm{e}^{-\mathrm{i}zt}/(2\pi)$ are at the positions $z_k = \pm ck$, so that

$$R(\pm ck) = -\frac{c^2}{2\pi} \lim_{z \to \pm ck} (z \mp ck) \frac{\mathrm{e}^{-\mathrm{i}zt}}{z^2 - c^2 k^2} = \mp \frac{c}{4\pi k}\,\mathrm{e}^{\mp \mathrm{i}ckt}. \tag{8.2.9}$$

Now the poles $\omega = \pm ck$ of (8.2.7) are on the real axis and as integration path C the real axis is taken with an ∞-semicircle, which is to be closed in such a way that the integral over the semicircle does not contribute anything. For $t < 0$ the integrand with $\mathrm{e}^{-\mathrm{i}\omega t}$ vanishes exponential in the upper half-plane and for $t > 0$ in the lower one.

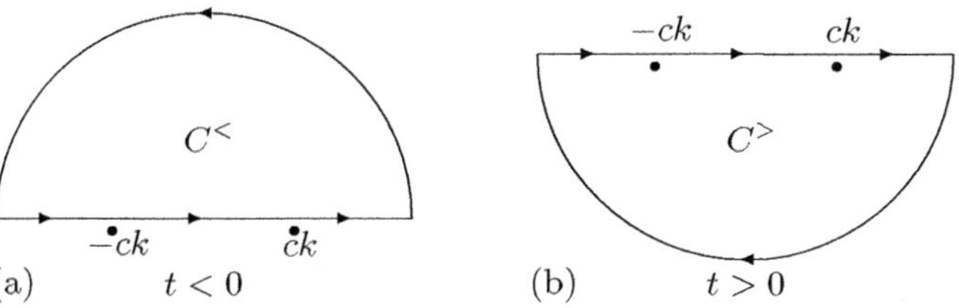

Fig. 8.1. Integration paths for the calculation of the retarded Green function $G(\mathbf{k}, t)$ in the complex ω-plane.   (a) $C^<$ for $t < 0$ (b) $C^>$ for $t > 0$

$t < 0$: The integration path $C^<$ leads along the real ω-axis from $-\infty$ to ∞ and is closed with a semicircle in the upper complex half-plane. $C^<$ must not enclose a pole, as shown in Fig. 8.1 so that $D(\mathbf{k}, t) = 0$

$$\oint_{C^<} \frac{\mathrm{d}\omega}{2\pi}\,\mathrm{e}^{-\mathrm{i}\omega t} D(\mathbf{k}, \omega) = \int_{-\infty}^{\infty} \frac{\mathrm{d}\omega}{2\pi}\,\mathrm{e}^{-\mathrm{i}\omega t} D(\mathbf{k}, \omega) + \underbrace{\int_{\cap} \frac{\mathrm{d}\omega}{2\pi}\,\mathrm{e}^{-\mathrm{i}\omega t} D(\mathbf{k}, \omega)}_{\to\, 0} = 0.$$

$t > 0$: The closed integration path $C^>$ leads on the real axis from $-\infty$ to ∞ and is closed with the infinite semicircle in the lower half-plane so that both poles at $\omega = \pm ck$ are enclosed by $C^<$. (8.2.8) must be preceded by a negative sign because the path is traversed clockwise. The residues are taken from (8.2.9):

$$\oint_{C^>} \frac{\mathrm{d}\omega}{2\pi}\,\mathrm{e}^{-\mathrm{i}\omega t} D(\mathbf{k}, \omega) = \int_{-\infty}^{\infty} \frac{\mathrm{d}\omega}{2\pi}\,\mathrm{e}^{-\mathrm{i}\omega t} D(\mathbf{k}, \omega) + \underbrace{\int_{\cup} \frac{\mathrm{d}\omega}{2\pi}\,\mathrm{e}^{-\mathrm{i}\omega t} D(\mathbf{k}, \omega)}_{\to\, 0}$$

$$= \mathrm{i}\frac{c}{2k}\left(-\mathrm{e}^{\mathrm{i}ckt} + \mathrm{e}^{-\mathrm{i}ckt}\right)$$

$$D(\mathbf{k}, t) = \int_{-\infty}^{\infty} \frac{d\omega}{2\pi}\, e^{-i\omega t}\, D(\mathbf{k}, \omega) = \frac{c}{k}\theta(t)\,\sin(ckt)\,. \tag{8.2.10}$$

Spatial Fourier transformation: Since $D(\mathbf{k}, t)$ only depends on k, we switch to spherical coordinates. The φ-integration yields 2π:

$$D(\mathbf{x}, t) = \frac{c\theta(t)}{(2\pi)^2}\int_0^{\infty} dk\, k\,\sin(ckt)\int_{-1}^{1} d\xi\, e^{ikr\xi} = \frac{c\theta(t)}{2r\pi^2}\int_0^{\infty} dk\,\sin(ckt)\,\sin(kr)$$

$$= \frac{c\theta(t)}{8r\pi^2}\int_{-\infty}^{\infty} dk\,\Big(\cos(kr - ckt) - \cos(kr + ckt)\Big)$$

$$= \frac{c\theta(t)}{8r\pi^2}\int_{-\infty}^{\infty} dk\,\Big(e^{ik(r-ct)} - e^{ik(r+ct)}\Big)\,.$$

Both terms yield δ-functions (see (B.6.14)). Since $r + ct$ never vanishes for $t > 0$, the 2nd term does not contribute to $D(\mathbf{x}, t)$ and we obtain

$$D(\mathbf{x}, t) = \theta(t)\,\frac{c}{4\pi r}\delta(r - ct)\,. \tag{8.2.11}$$

This is the retarded Green's function, which is automatically causal[2]. More generally formulated, we get

$$\Box D(\mathbf{x}-\mathbf{x}', t-t') = \delta^{(3)}(\mathbf{x}-\mathbf{x}')\delta(t-t') \tag{8.2.12}$$

with the retarded (causal) Green's function

$$D(\mathbf{x}-\mathbf{x}', t-t') = \theta(t-t')\,\frac{1}{4\pi|\mathbf{x}-\mathbf{x}'|}\delta\Big(t-t' - \frac{|\mathbf{x}-\mathbf{x}'|}{c}\Big)\,. \tag{8.2.13}$$

Liénard-Wiechert Potentials

Now if we insert (8.2.13) into (8.2.3), we get

$$\psi(\mathbf{x}, t) = \frac{1}{4\pi}\int d^3x' \int_{-\infty}^{t} dt'\, \frac{1}{|\mathbf{x}-\mathbf{x}'|}\,\delta\Big(t-t' - \frac{|\mathbf{x}-\mathbf{x}'|}{c}\Big)q(\mathbf{x}', t')$$

$$= \frac{1}{4\pi}\int d^3x'\, \frac{q(\mathbf{x}', t_r)}{|\mathbf{x}-\mathbf{x}'|} \qquad \text{with}\quad t_r = t - \frac{|\mathbf{x}-\mathbf{x}'|}{c}\,. \tag{8.2.14}$$

The potential at a point $\mathbf{x}, t$ is contributed by charges and currents that were at a point $\mathbf{x}'$ at an earlier time t' at a distance $|\mathbf{x}-\mathbf{x}'| = c(t-t')$ as sketched in Fig. 8.2. $\delta(t-t'-|\mathbf{x}-\mathbf{x}'|/c)$ selects the time t' from the distance $|\mathbf{x}-\mathbf{x}'|$ that light travels in time $t-t'$.

Now if we substitute for the inhomogeneity $q(\mathbf{x}, t) \to 4\pi k_C\rho(\mathbf{x}, t)$ or $q(\mathbf{x}, t) \to \frac{4\pi k_C}{ck_L}\mathbf{j}(\mathbf{x}, t)$, we obtain the retarded potentials

---

[2] When calculating the (acausal) advanced Green's function, both poles are included in the integration over the upper half-circle for $t < 0$: $D_a(\mathbf{x}, t) = -\theta(-t)\frac{c}{4\pi r}\delta(r+ct)$.

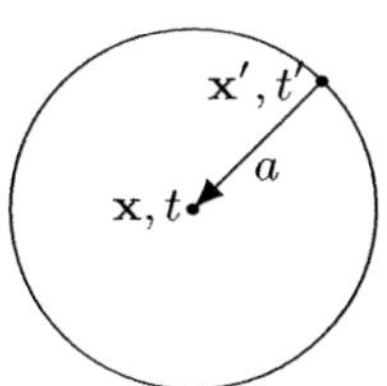

Fig. 8.2. At the point (x,t), only charges/currents $q(\mathbf{x}', t')$ contribute from a time point t', which are on a spherical surface around $(\mathbf{x}, t)$ with the radius $a = c(t - t')$

$$\phi(\mathbf{x}, t) = 4\pi k_C \int d^3 x' \int dt' \, D(\mathbf{x} - \mathbf{x}', t - t') \, \rho(\mathbf{x}', t'),$$

$$\mathbf{A}(\mathbf{x}, t) = \frac{4\pi k_C}{ck_L} \int d^3 x' \, dt' D(\mathbf{x} - \mathbf{x}', t - t') \, \mathbf{j}(\mathbf{x}', t'). \tag{8.2.15}$$

By evaluating the integral over time, one obtains according to (8.2.14)

$$\phi(\mathbf{x}, t) = k_C \int d^3 x' \, \frac{\rho(\mathbf{x}', t_r)}{|\mathbf{x} - \mathbf{x}'|}, \tag{8.2.16}$$

$$\mathbf{A}(\mathbf{x}, t) = \frac{k_C}{ck_L} \int d^3 x' \, \frac{\mathbf{j}(\mathbf{x}', t_r)}{|\mathbf{x} - \mathbf{x}'|}, \qquad t_r = t - \frac{|\mathbf{x} - \mathbf{x}'|}{c}, \qquad \text{SI:} \; \frac{k_C}{ck_L} = \frac{\mu_0}{4\pi}.$$

These are the Liénard-Wiechert potentials. They fulfill the Lorenz gauge (8.1.8), since we started from the decoupled equations (8.1.9).

Remarks: The Liénard-Wiechert potentials can be formulated more compactly using the convolution (B.5.31):

$$\phi(\mathbf{x}, t) = 4\pi k_C (D * \rho)(\mathbf{x}, t), \qquad \mathbf{A}(\mathbf{x}, t) = \frac{4\pi k_C}{ck_L}(D * \mathbf{j})(\mathbf{x}, t). \tag{8.2.17}$$

If we introduce the following four-dimensional vectors according to Tab. 13.2, p. 493

$$(j^\mu) = (c\rho, \mathbf{j}), \qquad (A^\mu) = (\phi/k_L, \mathbf{A}), \qquad \mu = 0, 1, 2, 3, \tag{8.2.18}$$

then we can express the Liénard-Wiechert potentials (8.2.15) in the form

$$A^\mu(\mathbf{x}, t) = A^\mu_h(x) + \frac{4\pi k_C}{ck_L} \int d^4 x' \, \frac{1}{c} D(x - x') j^\mu(x'), \qquad \mu = 0, 1, 2, 3 \tag{8.2.19}$$

Here we have replaced the integration over t' with one over $x'^0 = ct'$, the zeroth component of the event vector, and added a solution of the homogeneous wave equation $A^\mu_h(x)$. This notation is called *covariant* and is used in special relativity.

The retarded potentials, both the scalar and the vector potential were derived by Lorenz [1867, p. 247 and p. 249] and the Lorenz gauge is given on p. 253 for the potentials. However, the work has received little attention [Kragh, 2016]. Soon after the detection of the electron in 1897, Liénard [1898, (12)–(13)] and Wiechert [1900, (25)–(26)] calculated independently of each other the retarded potentials for a point charge.

Fourier decomposition

The Liénard-Wiechert potentials are solutions of the inhomogeneous wave equation (8.2.1). In many cases, a decomposition of the solution $\psi(\mathbf{x},t)$ into its Fourier components is advantageous. If current and charge distributions, represented by $q(\mathbf{x},t)$, are limited to a finite volume then $q(\mathbf{x},t) = 0$ for $r \to \infty$, which is sufficient for the Fourier transformation with respect to $\mathbf{x}$. More interest is in the frequencies ω of a system. The Fourier transformation is again possible if $q(\mathbf{x}, t=0)$ for $t \to \pm\infty$ vanishes[3]. The Fourier component of ψ is then given by the convolution according to (8.2.3)

$$\psi_{\mathbf{k}\omega} = \int \mathrm{d}^3x\,\mathrm{d}t\, e^{-i\mathbf{k}\cdot\mathbf{x}+i\omega t}\psi(\mathbf{x},t) \tag{8.2.20}$$

$$= \int \mathrm{d}^3x\,\mathrm{d}t\, e^{-i\mathbf{k}\cdot\mathbf{x}+i\omega t} \int \mathrm{d}^3x'\,\mathrm{d}t'\, D(\mathbf{x}-\mathbf{x}',t-t')q(\mathbf{x}',t') \quad \begin{cases} \mathbf{x}'' = \mathbf{x}-\mathbf{x}' \\ t'' = t-t' \end{cases}$$

$$= \int \mathrm{d}^3x''\,\mathrm{d}t''\, e^{-i\mathbf{k}\cdot\mathbf{x}''+i\omega t''} \int \mathrm{d}^3x'\,\mathrm{d}t'\, e^{i\mathbf{k}\cdot\mathbf{x}'-i\omega t'}\, D(\mathbf{x}'',t)q(\mathbf{x}',t')$$

$$= D(\mathbf{k},\omega)\, q_{\mathbf{k}\omega}\,.$$

Starting from (8.2.14) we obtain the Fourier components $(t' = t - \frac{k}{\omega}|\mathbf{x}-\mathbf{x}'|)$

$$\psi_\omega(\mathbf{x}) = \int \mathrm{d}t\, e^{i\omega t}\psi(\mathbf{x},t) \overset{(8.2.14)}{=} \frac{1}{4\pi}\int \mathrm{d}t'\, e^{i\omega t'}\int \mathrm{d}^3x'\, \frac{e^{ik|\mathbf{x}-\mathbf{x}'|}}{|\mathbf{x}-\mathbf{x}'|} q(\mathbf{x}',t')$$

$$= \frac{1}{4\pi}\int \mathrm{d}^3x'\, \frac{e^{ik|\mathbf{x}-\mathbf{x}'|}}{|\mathbf{x}-\mathbf{x}'|}\, q_\omega(\mathbf{x}'), \tag{8.2.21}$$

which are determined by the frequency ω alone with $q_\omega(\mathbf{x}')$, where q is either the charge density, $q_\omega(\mathbf{x}) = 4\pi k_C \rho_\omega(\mathbf{x})$, or a component of the current density $q_\omega(\mathbf{x}) = (4\pi k_C/k_L c)j_{\omega\,i}(\mathbf{x})$.

Potentials of a moving point charge

A point charge q moves along a path $\mathbf{s}(t')$ with the velocity $\mathbf{v}(t')$. Charge distribution and current density are thus determined by

$$\rho(\mathbf{x}',t') = q\,\delta^{(3)}\big(\mathbf{x}'-\mathbf{s}(t')\big),$$

$$\mathbf{j}(\mathbf{x}',t') = \rho(\mathbf{x}',t')\,\mathbf{v}(t').$$

Inserted into (8.2.16), it follows first that

$$\mathbf{A}_q(\mathbf{x},t) = \frac{1}{k_L}\boldsymbol{\beta}(t_r)\,\phi_q(\mathbf{x},t) \qquad \text{with} \qquad \boldsymbol{\beta}(t_r) = \frac{\mathbf{v}(t_r)}{c}\,. \tag{8.2.22}$$

---

[3] In no case should $q(\mathbf{x},t)$ for $t \to \pm\infty$ increase.

With the introduction of $\boldsymbol{\beta}$ we have taken into account the fact that the speed only appears in the ratio $\frac{\mathbf{v}}{c}$, which is ignored in the SI system ($k_L = c$). The evaluation of the δ function at the zero point $\mathbf{x}_0 = \mathbf{s}(t_r)$ results in

$$\delta^{(3)}\big(\mathbf{x}' - \mathbf{s}(t_r)\big) = \frac{\delta^{(3)}(\mathbf{x}' - \mathbf{x}_0)}{|\det \mathsf{J}|} \qquad \text{with} \qquad J_{ij} = \frac{\partial\big(x_i' - s_i(t_r)\big)}{\partial x_j'},$$

where for $t_r = t - \frac{1}{c}|\mathbf{x} - \mathbf{x}'|$ and due to $\delta^{(3)}(\mathbf{x}' - \mathbf{s})$ also $\mathbf{x}' = \mathbf{s}$ are to be inserted:

$$J_{ij} = \delta_{ij} - \frac{\mathrm{d}s_i(t_r)}{\mathrm{d}t_r}\frac{\partial t_r}{\partial x_j'} = \delta_{ij} - v_i\frac{x_j - s_j}{c|\mathbf{x} - \mathbf{s}|}.$$

Let $\mathbf{X}$ be the vector from the point charge to the observer, sketched in Fig. 8.3,

$$\mathbf{X}(\mathbf{x}, t_r) = \mathbf{x} - \mathbf{s}(t_r), \qquad R(t_r) = |\mathbf{X}(t_r)|, \qquad \mathbf{e}_R(t_r) = \frac{\mathbf{X}(t_r)}{R(t_r)}, \tag{8.2.23}$$

so the Jacobian matrix can be represented as

$$\mathsf{J} = \mathsf{E} - \boldsymbol{\beta}(t_r) \circ \mathbf{e}_R(t_r),$$

where $\circ$ denotes the tensorial (dyadic) product (A.1.15) of the two vectors. The functional determinant is then

$$J = \det \mathsf{J} = 1 - \boldsymbol{\beta}(t_r) \cdot \mathbf{e}_R(t_r). \tag{8.2.24}$$

Remark: The validity of $\det\big(\mathsf{E} + \mathbf{a} \circ \mathbf{b}\big) = 1 + \mathbf{a} \cdot \mathbf{b}$ can be easily verified in three dimensions. However, the formula holds in $n > 1$ dimensions (see problem A.2).

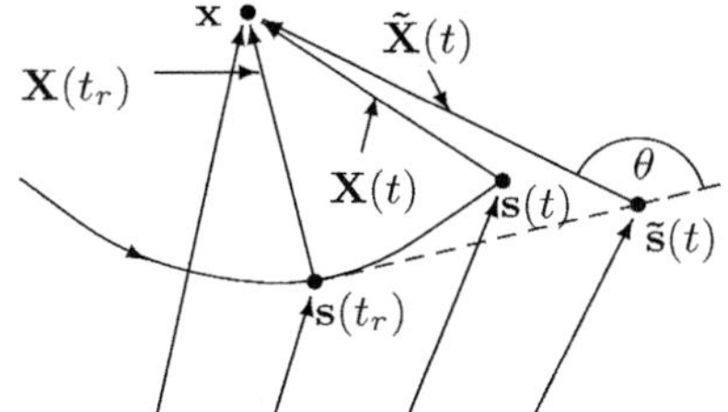

Fig. 8.3. Trajectory of the point charge $\mathbf{s}$, whose distance from the point $\mathbf{x}$ at the times t_r and t is given by $\mathbf{X}(t_r)$ and $\mathbf{X}(t)$; $\tilde{\mathbf{s}}(\mathbf{t})$ is the location where the point charge would be, had it continued to move with $\boldsymbol{\beta}(t_r)$ until time t

Inserted into the δ-function one obtains

$$\delta^{(3)}\Big(\mathbf{x}' - \mathbf{s}\big(t - \frac{|\mathbf{x} - \mathbf{x}'|}{c}\big)\Big) = \frac{\delta^{(3)}\big(\mathbf{x}' - \mathbf{s}(t_r)\big)}{\big|1 - \boldsymbol{\beta}(t_r) \cdot \mathbf{e}_R(t_r)\big|}.$$

Here, the retarded time is determined from

$$t_r = t - \frac{R(t_r)}{c} = t - \frac{|\mathbf{x} - \mathbf{s}(t_r)|}{c}. \tag{8.2.25}$$

The Liénard-Wiechert potentials for a moving point charge are then

$$\phi_q(\mathbf{x},t) = k_C \frac{q}{R(t_r)}\frac{1}{1-\boldsymbol{\beta}(t_r)\cdot\mathbf{e}_R(t_r)} = k_C\frac{q}{R(t_r)\,J(t_r)},$$

$$\mathbf{A}_q(\mathbf{x},t) = \frac{1}{k_L}\boldsymbol{\beta}(t_r)\,\phi_q(\mathbf{x},t). \tag{8.2.26}$$

$\det \mathsf{J}$ is the functional determinant of the coordinate transformation from $\mathbf{x}' \to \mathbf{x}' - \mathbf{s}(t_r)$, which ($|\det \mathsf{J}| < 1$) results in a contraction of the length parallel to $\mathbf{v}$. Fig. 8.7 on page 284 shows that ϕ_q is strongest at a given distance from the point charge in the direction of $\mathbf{v}$.

The scalar potential ϕ_q of a moving point charge at location $\mathbf{x}$ at time t is the potential of the point charge, which was located at time t_r at location $\mathbf{s}(t_r)$, amplified by the factor $1/\det \mathsf{J}$, where the distance to the point charge is given by $R(t_r) = |\mathbf{x} - \mathbf{s}(t_r)|$.

Fields of a moving point charge

With the potentials of a point charge (8.2.26) we have the necessary prerequisites for determining the electromagnetic fields. We note that

$$\phi_q(\mathbf{X},\boldsymbol{\beta}) = \frac{k_C q}{R-\mathbf{X}\cdot\boldsymbol{\beta}} \quad \text{with} \quad \begin{cases}\mathbf{X}(\mathbf{x},t_r)=\mathbf{x}-\mathbf{s}(t_r)\\ \boldsymbol{\beta}(t_r) \;\;=\dot{\mathbf{s}}(t_r)/c\end{cases} \quad \text{and} \quad t_r(\mathbf{x},t)=t-\frac{R(\mathbf{x},t_r)}{c}.$$

The potentials are of the order $1/R$. The derivative with respect to $\mathbf{X}$ leads to terms of the order $1/R^2$ and those with respect to $\boldsymbol{\beta}$ to those of the order $1/R$. The former form the so-called *near fields* ($\mathbf{E}_n$, $\mathbf{B}_n$) and the latter the *far-* or *radiation fields* ($\mathbf{E}_f$, $\mathbf{B}_f$) with $\dot{\boldsymbol{\beta}} \neq 0$, from which we conclude that only the accelerated charge radiates.

The calculation of the electric fields: First, the derivatives of $t_r = t - \frac{R}{c}$ are to be determined consistently:

$$\frac{\partial t_r}{\partial t} = 1 - \frac{1}{c}\frac{\partial R}{\partial \mathbf{X}}\cdot\frac{\partial \mathbf{X}}{\partial t_r}\frac{\partial t_r}{\partial t} = 1 + \mathbf{e}_R\cdot\boldsymbol{\beta}\,\frac{\partial t_r}{\partial t}, \qquad\qquad \frac{\partial t_r}{\partial t} = \frac{1}{J}, \tag{8.2.27}$$

$$\frac{\partial t_r}{\partial x_i} = -\frac{1}{c}\frac{\partial R}{\partial X_j}\left(\delta_{ij} - c\beta_j\frac{\partial t_r}{\partial x_i}\right) = -\frac{e_{Ri}}{c} + \mathbf{e}_R\cdot\boldsymbol{\beta}\,\frac{\partial t_r}{\partial x_i}, \qquad \frac{\partial t_r}{\partial \mathbf{x}} = -\frac{\mathbf{e}_R}{Jc}.$$

Used was $\frac{\partial R}{\partial \mathbf{X}}=\mathbf{e}_R$ and $\frac{\partial \mathbf{X}}{\partial t_r}=-\boldsymbol{\beta}c$. The derivatives with respect to $\mathbf{X}$ are associated with the near field, which decreases with $1/R^2$, while the derivatives with respect to $\boldsymbol{\beta}$ leave the order $1/R$ unchanged and thus form the far field. The individual parts of the fields are

$$\mathbf{E}_n = -\frac{\partial\phi}{\partial\mathbf{X}} - \left(\frac{\partial\phi}{\partial\mathbf{X}}\cdot\frac{\partial\mathbf{X}}{\partial t_r}\right)\left(\frac{\partial t_r}{\partial\mathbf{x}} + \frac{\boldsymbol{\beta}}{c}\frac{\partial t_r}{\partial t}\right), \qquad \mathbf{E}_f = -\left(\frac{\partial\phi}{\partial\boldsymbol{\beta}}\cdot\dot{\boldsymbol{\beta}}\right)\left(\frac{\partial t_r}{\partial\mathbf{x}} + \frac{\boldsymbol{\beta}}{c}\frac{\partial t_r}{\partial t}\right) - \frac{\phi}{c}\dot{\boldsymbol{\beta}}\frac{\partial t_r}{\partial t},$$

$$\mathbf{B}_n = \frac{1}{k_L}\left[\frac{\partial\phi}{\partial\mathbf{X}}\times\boldsymbol{\beta} + \left(\frac{\partial\phi}{\partial\mathbf{X}}\cdot\frac{\partial\mathbf{X}}{\partial t_r}\right)\frac{\partial t_r}{\partial\mathbf{x}}\times\boldsymbol{\beta}\right], \qquad \mathbf{B}_f = \frac{1}{k_L}\left[\left(\frac{\partial\phi}{\partial\boldsymbol{\beta}}\cdot\dot{\boldsymbol{\beta}}\right)\frac{\partial t_r}{\partial\mathbf{x}}\times\boldsymbol{\beta} + \phi\frac{\partial t_r}{\partial\mathbf{x}}\times\dot{\boldsymbol{\beta}}\right].$$

Now we calculate the individual contributions ($\mathbf{X} = \mathbf{x} - \mathbf{s}(t_r)$)

$$\frac{\partial \phi}{\partial \mathbf{X}} = -\frac{k_C q (\mathbf{e}_R - \boldsymbol{\beta})}{(R - \mathbf{X} \cdot \boldsymbol{\beta})^2}, \qquad \frac{\partial \phi}{\partial \mathbf{X}} \cdot \frac{\partial \mathbf{X}}{\partial t_r} = k_C c q \frac{(\mathbf{e}_R - \boldsymbol{\beta}) \cdot \boldsymbol{\beta}}{J^2 R^2}, \qquad \frac{\partial t_r}{\partial \mathbf{x}} + \frac{\boldsymbol{\beta}}{c} \frac{\partial t_r}{\partial t} = -\frac{\mathbf{e}_R - \boldsymbol{\beta}}{Jc},$$

$$\frac{\partial \phi}{\partial \boldsymbol{\beta}} = \frac{k_C q \, \mathbf{X}}{(R - \mathbf{X} \cdot \boldsymbol{\beta})^2}, \qquad \frac{\partial \phi}{\partial \boldsymbol{\beta}} \cdot \dot{\boldsymbol{\beta}} = k_C q \frac{\mathbf{e}_R \cdot \dot{\boldsymbol{\beta}}}{J^2 R}.$$

The summary of the individual terms results in

$$\mathbf{E}_n = \frac{k_C q}{R^2 J^3} \Big[(\mathbf{e}_R - \boldsymbol{\beta})(1 - \boldsymbol{\beta} \cdot \mathbf{e}_R) + \big((\mathbf{e}_R - \boldsymbol{\beta}) \cdot \boldsymbol{\beta} \big) \big(\mathbf{e}_R - \boldsymbol{\beta} \big) \Big] = \frac{k_C q}{R^2 J^3} (\mathbf{e}_R - \boldsymbol{\beta})(1 - \beta^2),$$

$$\mathbf{E}_f = \frac{k_C q}{c R J^3} \Big[(\mathbf{e}_R \cdot \dot{\boldsymbol{\beta}})(\mathbf{e}_R - \boldsymbol{\beta}) - \dot{\boldsymbol{\beta}}(1 - \mathbf{e}_R \cdot \boldsymbol{\beta}) \Big] = \frac{-k_C q}{c R J^3} \Big[\mathbf{e}_R \times (\dot{\boldsymbol{\beta}} \times \mathbf{e}_R) + \mathbf{e}_R \times (\boldsymbol{\beta} \times \dot{\boldsymbol{\beta}}) \Big],$$

$$\mathbf{B}_n = \frac{-k_C q}{k_L R^2 J^3} \Big[\mathbf{e}_R \times \boldsymbol{\beta}(1 - \boldsymbol{\beta} \cdot \mathbf{e}_R) + \big((\mathbf{e}_r - \boldsymbol{\beta}) \cdot \boldsymbol{\beta} \big) \mathbf{e}_R \times \boldsymbol{\beta} \Big] = \frac{-k_C q}{k_L R^2 J^3} (1 - \beta^2) \mathbf{e}_R \times \boldsymbol{\beta},$$

$$\mathbf{B}_f = \frac{-k_C q}{k_L c R J^3} \Big[(\mathbf{e}_R \cdot \dot{\boldsymbol{\beta}}) \mathbf{e}_R \times \boldsymbol{\beta} + (1 - \mathbf{e}_R \cdot \boldsymbol{\beta}) \mathbf{e}_R \times \dot{\boldsymbol{\beta}} \Big]$$

$$= \frac{-k_C q}{k_L c R J^3} \Big[\mathbf{e}_R \times \dot{\boldsymbol{\beta}} + \mathbf{e}_R \times \big(\underbrace{(\mathbf{e}_R \cdot \dot{\boldsymbol{\beta}}) \boldsymbol{\beta} - (\mathbf{e}_R \cdot \boldsymbol{\beta}) \dot{\boldsymbol{\beta}}}_{\mathbf{e}_R \times (\boldsymbol{\beta} \times \dot{\boldsymbol{\beta}})} \big) \Big]. \tag{8.2.28}$$

As a result, we have obtained:

$$\mathbf{E}(\mathbf{x}, t) = \frac{k_C q}{R^2 J^3} \left\{ (1 - \beta^2)(\mathbf{e}_R - \boldsymbol{\beta}) + \frac{R}{c} \mathbf{e}_R \times \big((\mathbf{e}_R - \boldsymbol{\beta}) \times \dot{\boldsymbol{\beta}} \big) \right\} \bigg|_{t_r}, \tag{8.2.29}$$

$$\mathbf{B}(\mathbf{x}, t) = \frac{-k_C q}{k_L R^2 J^3} \left\{ (1 - \beta^2)(\mathbf{e}_R \times \boldsymbol{\beta}) + \frac{R}{c} \big(\mathbf{e}_R \times (\dot{\boldsymbol{\beta}} + \mathbf{e}_R \times (\boldsymbol{\beta} \times \dot{\boldsymbol{\beta}})) \big) \right\} \bigg|_{t_r}.$$

The fields are split into parts without ($\mathbf{E}_n$) and with acceleration ($\mathbf{E}_f$). In other words, they decay into near fields, which fall off with $1/R^2$ and can be characterized by uniform motion, and into radiation fields, which fall off with $1/R$ and are associated with accelerated motion.

Notes: The direction of $\mathbf{E}_n$ can be indicated by the vector

$$\tilde{\mathbf{X}}(t) = \mathbf{X}(t_r) - \boldsymbol{\beta}(t_r) c(t - t_r) \overset{(8.2.25)}{=} R(t_r) \big(\mathbf{e}_R(t_r) - \boldsymbol{\beta}(t_r) \big). \tag{8.2.30}$$

Assuming that the particle would continue to move with $\boldsymbol{\beta}(t_r)$ until time t as sketched in Fig. 8.3, it would reach the location $\tilde{\mathbf{s}}(t)$. The vector $\tilde{\mathbf{X}}(t)$ is parallel to $\mathbf{E}_n(\mathbf{x}, t)$. The near field $\mathbf{E}_n(\mathbf{x}, t)$ is thus the field of a point charge moving uniformly with $\boldsymbol{\beta}(t_r)$.

$\mathbf{B}$ can be determined from $\mathbf{E}$:

$$\mathbf{B}(\mathbf{x}, t) = \frac{1}{k_L} \mathbf{e}_R(t_r) \times \mathbf{E}(\mathbf{x}, t). \tag{8.2.31}$$

Conversely, however, $\mathbf{E}_n \neq k_L \mathbf{B}_n \times \mathbf{e}_R$.

For a charge at rest in time average, which only reaches very small velocities during a periodic movement, $\mathbf{E}_n$ is the electrostatic field of a point charge and $\mathbf{E}_f$ is the radiation field of a dipole:

$$\mathbf{E} = \frac{k_C q\,\mathbf{e}_R}{R^2} - \frac{k_C q}{cR}\,\mathbf{e}_R \times (\dot{\boldsymbol{\beta}} \times \mathbf{e}_R)\Big|_{t_r}, \qquad \mathbf{B} = -\frac{k_C q}{k_L cR}\,\mathbf{e}_R \times \dot{\boldsymbol{\beta}}\Big|_{t_r}. \tag{8.2.32}$$

For the radiation fields decreasing with $1/R$

$$\mathbf{B}_f = \frac{1}{k_L}\mathbf{e}_R \times \mathbf{E}_f \qquad \mathbf{E}_f = k_L \mathbf{B}_f \times \mathbf{e}_R, = -\frac{k_C q}{cR}\big[\dot{\boldsymbol{\beta}} - (\mathbf{e}_R \cdot \dot{\boldsymbol{\beta}})\mathbf{e}_R\big]. \tag{8.2.33}$$

Radiation power

If one calculates the energy flow (5.6.3)

$$\mathbf{S}_f(\mathbf{x},t) = \frac{c}{k_L}\frac{1}{4\pi k_r \mu_0}\,\mathbf{E}_f \times \mathbf{B}_f = \frac{c}{4\pi k_C}\,\mathbf{E}_f \times \big[\mathbf{e}_R(t_r) \times \mathbf{E}_f\big] \tag{8.2.34}$$

through a sphere of radius $R \to \infty$, which is placed around the particle at time t_r, one obtains the energy radiated into the angular element $\mathrm{d}\Omega$ in time $\mathrm{d}t$

$$\frac{\mathrm{d}P_t}{\mathrm{d}\Omega}\mathrm{d}t = R^2(t_r)\,\mathbf{S}_f(\mathbf{x},t) \cdot \mathbf{e}_R(t_r)\mathrm{d}t\,.$$

However, it is obvious to refer the radiated power to t_r, so that

$$\frac{\mathrm{d}P}{\mathrm{d}\Omega} = \frac{\mathrm{d}P_t}{\mathrm{d}\Omega}\frac{\partial t}{\partial t_r} \overset{(8.2.27)}{=} \frac{c}{4\pi k_C}R^2\,E_f^2\,J, \qquad \text{SI:}\ \frac{\mathrm{d}P}{\mathrm{d}\Omega} = \epsilon_0 c R^2\,E_f^2\,J. \tag{8.2.35}$$

It is taken into account that $\mathbf{E}_f \cdot \mathbf{e}_R = 0$. If we substitute for $\mathbf{E}_f$ (8.2.28) in (8.2.35), then

$$\frac{\mathrm{d}P}{\mathrm{d}\Omega} = \frac{k_C q^2}{4\pi c}\frac{1}{J^5}\big[\mathbf{e}_R \times ((\mathbf{e}_R - \boldsymbol{\beta}) \times \dot{\boldsymbol{\beta}})\big]^2\Big|_{t_r}. \tag{8.2.36}$$

The Larmor formula

We start here from velocities in the limit $\beta \to 0$, i.e. $J = 1$ and $\frac{\partial t}{\partial t_r} = 1$. (8.2.34) is thus reduced to

$$\mathbf{S}_f(\mathbf{x},t) = \frac{c}{4\pi k_C}E_f^2\,\mathbf{e}_R\Big|_{t_r} = \frac{k_C}{4\pi}\frac{q^2}{cR^2}\big[\dot{\boldsymbol{\beta}}^2 - (\dot{\boldsymbol{\beta}} \cdot \mathbf{e}_R)^2\big]\mathbf{e}_R\Big|_{t_r}. \tag{8.2.37}$$

We determine the energy penetrating per unit of time through the surface element of a sphere for $R \to \infty$ $\mathrm{d}P = R^2\mathbf{S}_f \cdot \mathbf{e}_R\mathrm{d}\Omega$. The power radiated per unit of angle is

$$\frac{\mathrm{d}P}{\mathrm{d}\Omega} = R^2\,\mathbf{S}_f \cdot \mathbf{e}_R = P_0(t_r)\sin^2\theta \qquad \text{with} \qquad P_0 = \frac{k_C q^2}{4\pi c}\dot{\boldsymbol{\beta}}^2(t_r)\,. \tag{8.2.38}$$

θ is the angle enclosed by $\mathbf{e}_R$ and $\dot{\boldsymbol{\beta}}$. The angle integration over the spherical surface yields the *Larmor formula*

$$P = P_0(t_r)\int_0^{2\pi}\mathrm{d}\varphi\int_0^{\pi}\mathrm{d}\vartheta\sin^3\vartheta = P_0\frac{8\pi}{3} = \frac{2k_C q^2}{3c}\dot{\boldsymbol{\beta}}^2, \tag{8.2.39}$$

$$\text{G:}\ P = \frac{2q^2}{3c^3}\dot{v}^2, \qquad\qquad\qquad \text{SI:}\ P = \frac{\mu_0 q^2}{6\pi c}\dot{v}^2. \tag{8.2.39'}$$

Radiation power of the moving charge

The radiation emitted by a charge that is accelerated to high speeds is relevant in particle accelerators, whether linear or circular. While the total emitted radiation power P depends only on the angle between $\boldsymbol{\beta}$ and $\dot{\boldsymbol{\beta}}$, as the "Liénard formula" (8.2.42) to be derived shows, the directions of $\boldsymbol{\beta}$ and $\dot{\boldsymbol{\beta}}$ with respect to the z-axis are also relevant for the radiation power emitted into the solid angle Ω. We only specify the dependence on Ω for two simple configurations:

$$\frac{\mathrm{d}P}{\mathrm{d}\Omega} = P_0 \frac{\sin^2\vartheta}{(1-\beta\cos\vartheta)^5}, \qquad P_0 = \frac{k_C q^2}{4\pi c}\dot{\beta}^2, \qquad \boldsymbol{\beta}\|\dot{\boldsymbol{\beta}}\|\mathbf{e}_z. \tag{8.2.40}$$

For $\beta = 0$ the radiation intensity is equal to that of the Larmor formula (8.2.38). In this case, the maximum is perpendicular to $\dot{\boldsymbol{\beta}}$. For $\beta > 0$ the maximum becomes more pronounced with increasing speed and moves in the direction of the z-axis.

If $\boldsymbol{\beta} = \beta\mathbf{e}_x$ is perpendicular to $\dot{\boldsymbol{\beta}} = \dot{\beta}\mathbf{e}_z$, one obtains

$$\frac{\mathrm{d}P}{\mathrm{d}\Omega} = P_0 \frac{J^2-(1-\beta^2)\cos^2\vartheta}{J^5}, \qquad J=1-\beta\sin\vartheta\cos\varphi \qquad \begin{cases} \dot{\boldsymbol{\beta}}=\dot{\beta}\mathbf{e}_z \\ \boldsymbol{\beta}=\beta\mathbf{e}_x. \end{cases} \tag{8.2.41}$$

The intensity increases strongly for $\beta \to 1$ around $\boldsymbol{\beta}$, which is the x-axis.

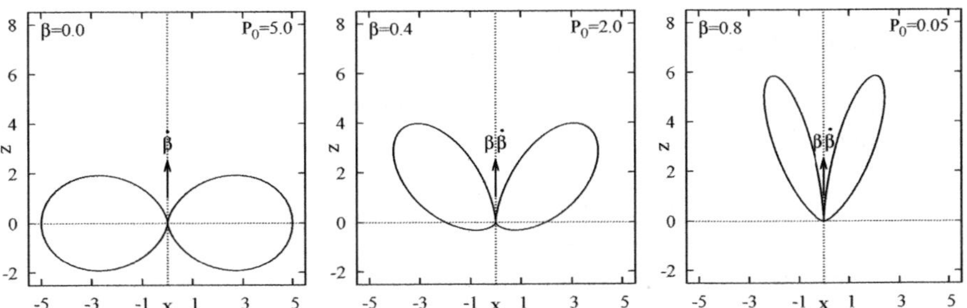

Fig. 8.4. Radiation diagram for $\beta = 0$, 0.4 and $\beta = 0.8$, where $\boldsymbol{\beta}\|\dot{\boldsymbol{\beta}}$. P increases strongly with $1/(1-\beta^2)^3$. In the diagram, this is taken into account by the scaling of $\mathrm{d}P/\mathrm{d}\Omega$ with $P_0 = 5$, 2 and 0.05.

To calculate the total radiation power, one has (8.2.36) over the entire solid angle to integrate, a calculation that is given in the problems 8.10 and 8.11. The calculation is lengthy and we only present the result here:

$$P = \frac{2k_C q^2}{3c}\frac{1}{(1-\beta^2)^3}\left[\dot{\boldsymbol{\beta}}^2 - (\boldsymbol{\beta}\times\dot{\boldsymbol{\beta}})^2\right], \tag{8.2.42}$$

which goes back to Liénard [1898, p. 13] (see also Heaviside [1902]). As outlined in Fig. 8.4, for $\boldsymbol{\beta}\|\dot{\boldsymbol{\beta}}$ the radiation for $v\to c$ increases sharply around $\boldsymbol{\beta}$, i.e. in

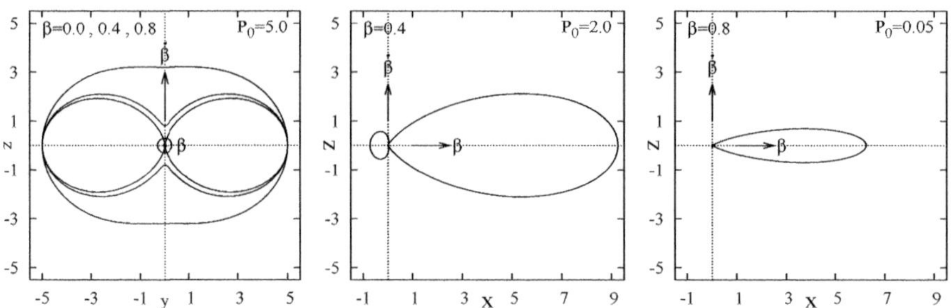

Fig. 8.5. Radiation pattern for the intensity distribution (8.2.41), where $\beta \perp \dot{\beta}$; the increasing intensity is accounted for by the decreasing scaling with P_0. The scales of the axes are arbitrary, but the same in each diagram

the forward direction. Fig. 8.5 shows the radiation power in the plane spanned by $\dot{\beta}$ and β together with the yz-plane perpendicular to it. The intensity increases in the direction of β. In the yz-plane perpendicular to β the intensity changes comparatively little, but is independent of the orientation $\dot{\beta}$.

Note: The laws of electrodynamics, when suitably represented, are form-invariant under the Lorentz transformation. In this, c plays the role of a limit speed that must not be exceeded. This is evident in the Liénard formula (8.2.42), where $\beta \leq 1$ must be. Later we will see that the Liénard formula follows from the Larmor formula in a simple way, if one requires that P is a Lorentz scalar.

The potentials in the linear, homogeneous medium

The Maxwell equations (8.1.1) for a homogeneous medium with constant ϵ_r and μ_r according to (5.2.16):

$$\text{(a)} \qquad \boldsymbol{\nabla} \cdot \mathbf{E} = \frac{4\pi k_C \rho}{\epsilon_r} \qquad\qquad \text{(b)} \quad \boldsymbol{\nabla} \times \mathbf{E} + \frac{k_L}{c}\dot{\mathbf{B}} = 0$$

$$\text{(c)} \quad \boldsymbol{\nabla} \times \mathbf{B} - \frac{\epsilon_r \mu_r}{ck_L}\dot{\mathbf{E}} = \frac{4\pi k_C \mu_r}{ck_L}\mathbf{j} \qquad \text{(d)} \qquad\qquad \boldsymbol{\nabla} \cdot \mathbf{B} = 0.$$

Note: If you replace in Maxwell's equations for the vacuum (8.1.1):

$$c \to \bar{c} = c/n, \quad k_L \to k_L/n, \quad \rho \to \rho/\epsilon_r \quad \text{and} \quad \mathbf{j} \to \mathbf{j}/\epsilon_r, \quad \text{where} \quad n = \sqrt{\epsilon_r \mu_r},$$

the above Maxwell equations are obtained for the medium with the refractive index n. With this method, the Lorenz condition, potentials and fields in the medium can be determined from the corresponding formulas for the vacuum.

Following section 8.1, we obtain the Lorenz condition (8.1.8) in the medium:

$$\boldsymbol{\nabla} \cdot \mathbf{A} + \frac{\epsilon_r \mu_r}{ck_L}\frac{\partial \phi}{\partial t} = 0.$$

$n = \sqrt{\epsilon_r \mu_r}$ is the index of refraction and $\bar{c} = c/n$ is the velocity of light in the medium. The potentials satisfy the wave equations

$$\bar{\Box}\phi = \frac{4\pi k_C \rho}{\epsilon_r}, \qquad\qquad \bar{\Box}\mathbf{A} = \frac{4\pi k_C \mu_r}{ck_L}\mathbf{j}, \qquad\qquad \bar{\Box} = \frac{1}{\bar{c}^2}\frac{\partial^2}{\partial t^2} - \Delta.$$

The retarded potentials are thus determined by (8.2.14) if c is replaced by $\bar{c}$ and t_r by $\bar{t}_r = t - |\mathbf{x} - \mathbf{x}'|/\bar{c}$

$$\phi(\mathbf{x}, t) = \frac{k_C}{\epsilon_r}\int d^3 x' \frac{\rho(\mathbf{x}', \bar{t}_r)}{|\mathbf{x} - \mathbf{x}'|}, \qquad\qquad \mathbf{A}(\mathbf{x}, t) = \frac{k_C \mu_r}{ck_L}\int d^3 x' \frac{\mathbf{j}(\mathbf{x}', \bar{t}_r)}{|\mathbf{x} - \mathbf{x}'|}.$$

Of particular interest are the potentials of point particles, which are now moving in a medium. Here, we can refer back to (8.2.26)

$$\phi_q(\mathbf{x}, t) = k_C \frac{q}{\epsilon_r}\frac{1}{R(\bar{t}_r)|J(\bar{t}_r)|}, \quad J(\bar{t}_r) = 1 - \mathbf{e}_R\cdot\boldsymbol{\beta}n, \quad \mathbf{X}(\bar{t}_r) \equiv \mathbf{X}_r = \mathbf{x} - \mathbf{s}(\bar{t}_r)$$

$$\mathbf{A}_q(\mathbf{x}, t) = \frac{\epsilon_r \mu_r}{k_L}\phi_q(\mathbf{x}, t)\boldsymbol{\beta}(\bar{t}_r), \qquad \bar{t}_r = t - \frac{R_r}{\bar{c}}, \qquad\qquad R_r = |\mathbf{X}_r|. \quad (8.2.44')$$

In a medium with $n > 1$ particles can be faster than light. The radiation emitted by these particles is called *Cherenkov radiation*. The propagation has similarities with the supersonic speed and is addressed in problem 8.3.

8.2.2 Uniformly Moving Charges

Potentials of a uniformly moving point charge

Both the potentials (8.2.26) as well as the fields $\mathbf{E}_n$ and $\mathbf{B}_n$ (8.2.28) of the uniformly moving point charge are already given. Only the time t_r has to be replaced by t. From the sketch Fig. 8.6 one can see, as already noted in (8.2.30), that

$$\mathbf{X}(t_r) - \boldsymbol{\beta}R(t_r) = \mathbf{X}(t).$$

From this, $R_r \equiv R(t_r)$ can be determined by $\mathbf{X} \equiv \mathbf{X}(t)$ if one squares the above relationship and multiplies it with $\boldsymbol{\beta}$:

$$R_r^2 - 2\mathbf{X}_r\cdot\boldsymbol{\beta}R_r + \beta^2 R_r^2 = R^2, \qquad\qquad \mathbf{X}_r\cdot\boldsymbol{\beta} = \mathbf{X}\cdot\boldsymbol{\beta} + \beta^2 R_r.$$

This results in $R_r^2(1 - \beta^2) - 2R_r\mathbf{X}\cdot\boldsymbol{\beta} - R^2 = 0$ with the solution

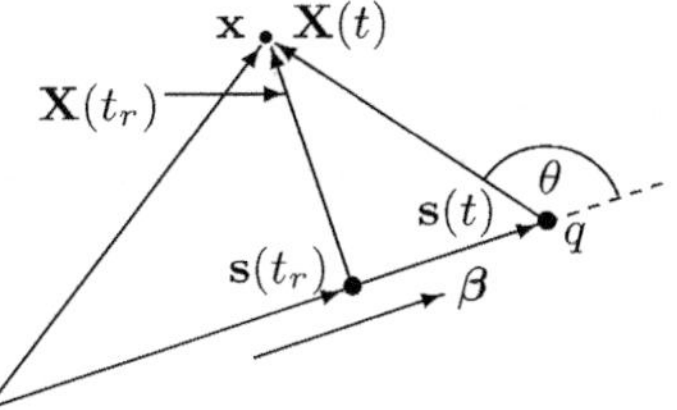

Fig. 8.6. Uniformly moving point charge $\mathbf{s}(t)$; $R = |\mathbf{X}(t)|$ is the distance of the point charge from the observer and $\theta = \sphericalangle\boldsymbol{\beta}, \mathbf{X}(t)$ and $R(t_r) = c(t - t_r)$

$$R_r = \frac{1}{1-\beta^2}\left[(\mathbf{X}\cdot\boldsymbol{\beta}) \pm \sqrt{(\mathbf{X}\cdot\boldsymbol{\beta})^2 + (1-\beta^2)R^2}\right].$$
(8.2.43)

Now one has with

$$(\mathbf{X}\cdot\boldsymbol{\beta})^2 + (1-\beta^2)R^2 = R^2 - \beta^2 R^2 \sin^2\theta = R^2 - (\mathbf{X}\times\boldsymbol{\beta})^2$$

an alternative representation. For the denominator of (8.2.26) it follows

$$JR_r = R_r - \boldsymbol{\beta}\cdot\mathbf{X}_r = (1-\beta^2)R_r - \boldsymbol{\beta}\cdot\mathbf{X} = \sqrt{R^2 - (\mathbf{X}\times\boldsymbol{\beta})^2}.$$

After these somewhat tedious transformations, we have the potentials

$$\phi_q(\mathbf{x},t) = \frac{k_C q}{\sqrt{R^2(t) - (\boldsymbol{\beta}\times\mathbf{X}(t))^2}} = \frac{k_C q}{\sqrt{R^2(t)(1-\beta^2\sin^2\theta)}},$$
(8.2.44)

$$\mathbf{A}_q(\mathbf{x},t) = (\boldsymbol{\beta}/k_L)\phi_q(\mathbf{x},t),$$
(8.2.45)

with $\mathbf{X}(t) = \mathbf{x} - \mathbf{v}t$ obtained. $R(t)$ is the distance of the observer at location $\mathbf{x}$ from the charge, which is located at $\mathbf{s} = \mathbf{v}t$ and θ is, as seen in Fig. 8.6, the angle between $\mathbf{v}$ and $\mathbf{X}$. Fig. 8.7 shows the compression of the equipotential lines in the direction of motion, an effect of 2nd order in β.

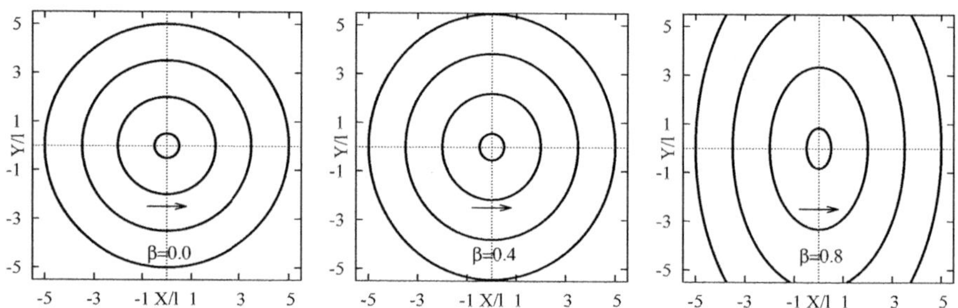

Fig. 8.7. Equipotential lines of a uniformly moving point charge for $v = 0$, $v = 0.4\,c$ and $v = 0.8\,c$; the distance to the point charge $R = \sqrt{X^2 + Y^2}$ is scaled with an arbitrary length l

Fields of a uniformly moving point charge

To determine the fields, it is not necessary to go the way over the derivatives of the potentials. It is easier to insert the expressions for JR_r and $\mathbf{e}_R - \boldsymbol{\beta}$ (8.2.30) directly into $\mathbf{E}_n$ (8.2.28). We thus obtain

$$\mathbf{E}(\mathbf{x},t) = k_C q\frac{(1-\beta^2)\mathbf{X}(t)}{\sqrt{(1-\beta^2)R^2(t) + (\mathbf{X}(t)\cdot\boldsymbol{\beta})^2}^{\,3}}$$
(8.2.46)

and because $\dot{\mathbf{A}} \times \mathbf{A} = 0$ is

$$\mathbf{B}(\mathbf{x}, t) = \boldsymbol{\nabla} \times \mathbf{A}_q = k_L(\boldsymbol{\nabla}\phi_q) \times \boldsymbol{\beta} = k_L\boldsymbol{\beta} \times \mathbf{E}. \qquad (8.2.47)$$

In Fig. 8.8 the lines of constant field strength are drawn for three speeds. Since

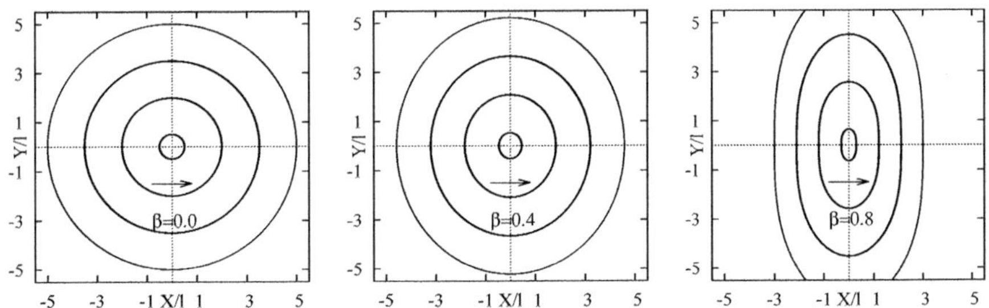

Fig. 8.8. Lines of constant field strength for different speeds $\boldsymbol{\beta} = \beta\mathbf{e}_X$: Field strength $El^2/qk_C = 1$, where X and Y are scaled with the arbitrary length l. For $\beta > 0.8$ the line (area) of constant field strength gets a constriction around the X-axis

the effect is of the order β^2, very high speeds must be assumed for noticeable differences. Unlike with ϕ_q, due to the factor $1 - \beta^2$ the field in the direction of $\boldsymbol{\beta}$ decreases faster with greater speed. The Poynting vector (5.6.3) is then

$$\mathbf{S} = \frac{ck_L}{4\pi k_C}\mathbf{E} \times \mathbf{B} = c\frac{k_L^2}{4\pi k_C}\,\mathbf{E} \times (\boldsymbol{\beta} \times \mathbf{E}), \qquad \text{SI: } \mathbf{S} = \frac{c}{\mu_0}\,\mathbf{E} \times (\boldsymbol{\beta} \times \mathbf{E})\,.$$

$\mathbf{S} \propto 1/R^4$ for $R \to \infty$. For sufficiently large V there is no energy flow through the surface, as one would expect if the charge radiates. This implies that a uniformly moving charge distribution does not radiate.

Electric field lines

The field lines themselves are straight lines that originate from $\mathbf{s}(t)$ and whose density increases perpendicular to the direction of $\mathbf{v}$ ($\theta = \pi/2$, see Fig. 8.6):

$$\left(1 - \beta^2\right)R^2(t) + (\mathbf{X}(t) \cdot \boldsymbol{\beta})^2 = R^2(t)\left(1 - \beta^2 + \beta^2 \cos^2\theta\right).$$

We now place a sphere with radius R_0 around the charge and assume that $\boldsymbol{\beta} = \beta\mathbf{e}_z$. The electric flux

$$\Delta\Phi_E = R_0^2 \int_{\Delta F} d\Omega\, \mathbf{e}_R(t) \cdot \mathbf{E} = 2\pi k_C q(1 - \beta^2) \int_{\vartheta_a}^{\vartheta_b} \frac{d\vartheta\, \sin\vartheta}{\sqrt{1 - \beta^2 + \beta^2 \cos^2\vartheta}^{\,3}}$$

$$= 2\pi k_C q \frac{\cos\vartheta}{\sqrt{1 - \beta^2 + \beta^2 \cos^2\vartheta}}\bigg|_{\vartheta_b}^{\vartheta_a}$$

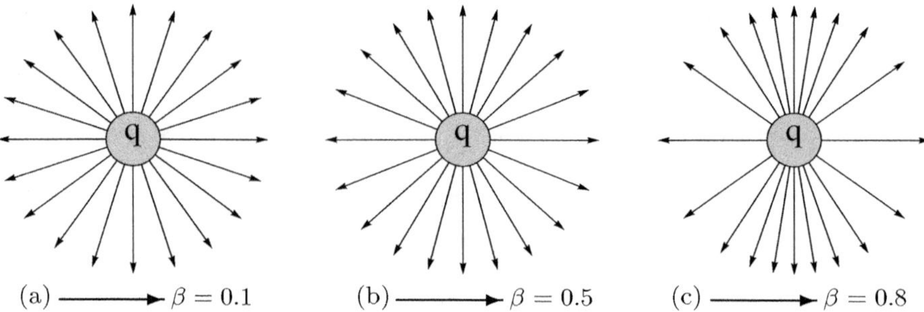

Fig. 8.9. Field lines of a moving positive charge

through the surface $\Delta A = 2\pi(\cos\vartheta_a - \cos\vartheta_b)$ on the unit sphere determines the field line density. It is immediately apparent that the entire flux through the surface

$$\Phi_E = \oiint \mathrm{da}\cdot\mathbf{E} = 4\pi k_C q$$

is independent of β. If we sketch the field lines in the xz plane such that a total of $2n$ lines emanate from the stationary charge, then the electric flux per field line is given by $q\pi/n$, which corresponds to an angular range $\vartheta_b - \vartheta_a = \pi/n$. For the moving charge, starting from ϑ_a, the value of ϑ_b is to be determined such that $\Delta\Phi_E = \Delta A\,(\pi k_C \pi q/n)$. Fig. 8.9 shows the field lines with increasing speed. Since the effect is quadratic in $\mathbf{v}/c$, it only becomes noticeable for speeds $v/c > 0.2$.

8.2.3 Coulomb Gauge

In addition to the Lorenz gauge (8.1.8), the Coulomb gauge also has a greater significance in physics, e.g. in quantum electrodynamics. The Coulomb gauge is also referred to as *radiation gauge* or *transverse gauge*, since $\boldsymbol{\nabla}\cdot\mathbf{A}^{\mathrm{C}} = 0$. We will see that in this case the scalar potential (8.2.49) does not contribute to the radiation and thus only $\mathbf{A}^{\mathrm{C}}$ is used to quantize the electromagnetic field.

If one replaces the Lorenz gauge (8.1.8) with the Coulomb gauge

$$\boldsymbol{\nabla}\cdot\mathbf{A}^{\mathrm{C}} = 0, \tag{8.2.48}$$

then yields from (8.1.4)

$$\boldsymbol{\nabla}\cdot\mathbf{E} = \boldsymbol{\nabla}\cdot\left(-\boldsymbol{\nabla}\phi^{\mathrm{C}} - \frac{k_L}{c}\dot{\mathbf{A}}^{\mathrm{C}}\right) = 4\pi k_C\rho,$$

the Poisson equation $\boldsymbol{\nabla}^2\phi^{\mathrm{C}} = -4\pi k_C\rho$ known from electrostatics with

$$\phi^{\mathrm{C}}(\mathbf{x},t) \equiv \phi^{\mathrm{qs}}(\mathbf{x},t) = k_C\int\mathrm{d}^3x'\,\frac{\rho(\mathbf{x}',t)}{|\mathbf{x}-\mathbf{x}'|}\,. \tag{8.2.49}$$

Here, ϕ^{C} is the instantaneous (quasi-static) Coulomb potential of a charge distribution; hence the term Coulomb gauge.

In the Coulomb gauge, $\mathbf{E}^{\mathrm{qs}} = -\boldsymbol{\nabla}\phi^{\mathrm{C}}$ is the source part and $\mathbf{E}^{(\mathrm{w})} = -\frac{k_L}{c}\dot{\mathbf{A}}^{\mathrm{C}}$ is the vortex part of $\mathbf{E}$. Here, the Coulomb gauge differs from the Lorenz gauge.

The quasi-static (source) field $\mathbf{E}^{\mathrm{qs}}$ vanishes at a great distance, the so-called *far zone* (see section 8.3.6) and thus only contributes in the *near zone* (see section 8.3.5) to the field. The radiation field of the far zone is determined solely by the source-free component. Similarly, the displacement current density

$$\mathbf{j}_d = \frac{\dot{\mathbf{E}}}{4\pi k_C} = -\mathbf{j}_l + \mathbf{j}_w, \qquad \mathbf{j}_l = \frac{\boldsymbol{\nabla}\dot{\phi}^{\mathrm{C}}}{4\pi k_C}, \qquad \mathbf{j}_w = -\frac{k_L\ddot{\mathbf{A}}^{\mathrm{C}}}{4\pi k_C c} \qquad (8.2.50)$$

is composed of the vortex-free $-\mathbf{j}_l$ and the source-free $\mathbf{j}_w$.

From the Ampère-Maxwell equation (8.1.7), by substituting $\boldsymbol{\nabla}\times\mathbf{B} = -\Delta\mathbf{A}^{\mathrm{C}}$ the wave equation follows

$$\Box\mathbf{A}^{\mathrm{C}} = \frac{4\pi k_C}{ck_L}\left(\mathbf{j} - \mathbf{j}_l\right) = \frac{4\pi k_C}{ck_L}\mathbf{j}_t \overset{\text{SI}}{=} \mu_0\mathbf{j}_t\,. \qquad (8.2.51)$$

Its retarded solution is according to (8.2.16)

$$\mathbf{A}^{\mathrm{C}}(\mathbf{x}, t) = \frac{k_C}{ck_L}\int \mathrm{d}^3x'\, \frac{\mathbf{j}_t(\mathbf{x}', t_r)}{|\mathbf{x}-\mathbf{x}'|}. \qquad (8.2.52)$$

Very similar to the properties of $\mathbf{A}^{\mathrm{C}}$ is the Helmholtz potential of the magnetic field $\mathbf{A}_h^b$ with $\boldsymbol{\nabla}\times\mathbf{A}_h^b = \mathbf{B}$ and $\boldsymbol{\nabla}\cdot\mathbf{A}_h^b = 0$. Now, if $\mathbf{B} = \boldsymbol{\nabla}\times\mathbf{A}^{\mathrm{C}}$ is substituted into $\mathbf{A}_h^b$, this results in the equality of both potentials (problem 8.4)

$$\mathbf{A}_h^b(\mathbf{x}, t) = \frac{1}{4\pi}\int \mathrm{d}^3x'\, \mathbf{B}(\mathbf{x}', t)\times\boldsymbol{\nabla}'\frac{1}{|\mathbf{x}-\mathbf{x}'|} = \mathbf{A}^{\mathrm{C}}(\mathbf{x}', t). \qquad (8.2.53)$$

In section 8.3.2, it is shown that for periodic current densities $\mathbf{A}_h^b$ can be represented by the source-free electric field (see (8.3.7b)).

From $\mathbf{j}$, we have subtracted the portion of the displacement current $\mathbf{j}_l$ originating from the charges. Now $\Delta\dot{\phi}^{\mathrm{C}} = -4\pi k_C\dot{\rho}$, from which follows:

$$\boldsymbol{\nabla}\cdot\mathbf{j}_t = \boldsymbol{\nabla}\cdot\mathbf{j}+\dot{\rho} = 0, \qquad\qquad \boldsymbol{\nabla}\times\mathbf{j}_l = \boldsymbol{\nabla}\times\frac{1}{4\pi k_C}\boldsymbol{\nabla}\dot{\phi}^{\mathrm{C}} = 0. \qquad (8.2.54)$$

Thus, the decomposition of $\mathbf{j} = \mathbf{j}_t + \mathbf{j}_l$ into a vortex and a source current density is shown. This designation is more accurate than that of a division into a transverse and a longitudinal current density, since $\mathbf{j}_l$ is generally not perpendicular to $\mathbf{j}_t$.

$\mathbf{A}^{\mathrm{C}}$ thus satisfies the inhomogeneous wave equation (8.2.51), where the vortex-free, longitudinal part of the current vector $(\boldsymbol{\nabla}\times\mathbf{j}_l = 0)$ does not contribute to the vector field, but only $\mathbf{j}_t$. In the far zone, the radiation field is determined solely by $\mathbf{A}^{\mathrm{C}}$ (*radiation gauge*).

Quasi-static potentials

If a potential at time t is determined by the charge or current distribution at the same time, this is referred to as quasi-static, like ϕ^{C} in (8.2.49). The corresponding quasi-static vector potential

$$\mathbf{A}^{\mathrm{qs}}(\mathbf{x}, t) = \frac{k_C}{ck_L} \int \mathrm{d}^3 x' \, \frac{\mathbf{j}(\mathbf{x}', t)}{|\mathbf{x} - \mathbf{x}'|} = \mathbf{A}_t^{\mathrm{qs}}(\mathbf{x}, t) + \mathbf{A}_l^{\mathrm{qs}}(\mathbf{x}, t) \tag{8.2.55}$$

can be separated into the two components $\mathbf{A}_t^{\mathrm{qs}}$ and $\mathbf{A}_l^{\mathrm{qs}}$, which are formed with $\mathbf{j}_t$ and $\mathbf{j}_l$ respectively. For the divergence, a condition similar to the Lorenz gauge (8.1.8) applies:

$$\boldsymbol{\nabla} \cdot \mathbf{A}^{\mathrm{qs}}(\mathbf{x}, t) \overset{\overset{\boldsymbol{\nabla} \to -\boldsymbol{\nabla}'}{\text{int. by parts}}}{=} \frac{k_C}{ck_L} \int \mathrm{d}^3 x' \, \frac{\boldsymbol{\nabla}' \cdot \mathbf{j}(\mathbf{x}', t)}{|\mathbf{x} - \mathbf{x}'|} \overset{(8.2.54)}{=} -\frac{1}{ck_L} \dot{\phi}^{\mathrm{qs}}(\mathbf{x}, t). \tag{8.2.56}$$

As a quasi-static potential, $\mathbf{A}^{\mathrm{qs}}$ satisfies the vector Poisson equation, as follows from (8.2.55)

$$\Delta \mathbf{A}^{\mathrm{qs}}(\mathbf{x}, t) = -\frac{4\pi k_C}{ck_L} \mathbf{j}(\mathbf{x}, t), \qquad\qquad \mathbf{B}^{\mathrm{qs}} = \boldsymbol{\nabla} \times \mathbf{A}^{\mathrm{qs}}. \tag{8.2.57}$$

Using the identity (A.2.38): $\Delta = \boldsymbol{\nabla}\boldsymbol{\nabla} - \boldsymbol{\nabla}\times\boldsymbol{\nabla}\times$ and (8.2.50), $\mathbf{j}(\mathbf{x}, t)$ is decomposed into source and vortex current densities:

$$\begin{aligned}
\mathbf{j}_l(\mathbf{x}, t) &= \frac{-ck_L}{4\pi k_C} \boldsymbol{\nabla}\,\boldsymbol{\nabla} \cdot \mathbf{A}^{\mathrm{qs}}(\mathbf{x}, t) = \frac{-1}{4\pi k_C} \dot{\mathbf{E}}^{\mathrm{qs}}(\mathbf{x}, t), \\
\mathbf{j}_t(\mathbf{x}, t) &= \frac{ck_L}{4\pi k_C} \boldsymbol{\nabla} \times \left(\boldsymbol{\nabla} \times \mathbf{A}^{\mathrm{qs}}(\mathbf{x}, t) \right) = \frac{ck_L}{4\pi k_C} \boldsymbol{\nabla} \times \mathbf{B}^{\mathrm{qs}}(\mathbf{x}, t).
\end{aligned} \tag{8.2.58}$$

Initially, $\mathbf{j}$ is obtained by the sum of the two parts and then inserted into the Ampère-Maxwell equation (8.1.1c):

$$\begin{aligned}
\boldsymbol{\nabla} \times \mathbf{B}^{\mathrm{qs}}(\mathbf{x}, t) &= \frac{4\pi k_C}{ck_L} \mathbf{j}(\mathbf{x}, t) + \frac{1}{ck_L} \dot{\mathbf{E}}^{\mathrm{qs}}(\mathbf{x}, t), \\
\boldsymbol{\nabla} \times \left[\mathbf{B}(\mathbf{x}, t) - \mathbf{B}^{\mathrm{qs}}(\mathbf{x}, t) \right] &= \frac{1}{ck_L} \left[\dot{\mathbf{E}}(\mathbf{x}, t) - \dot{\mathbf{E}}^{\mathrm{qs}}(\mathbf{x}, t) \right] = \frac{1}{ck_L} \dot{\mathbf{E}}_t(\mathbf{x}, t).
\end{aligned} \tag{8.2.59}$$

We will return to this form of the Ampère-Maxwell equation when dealing with Helmholtz potentials in section 8.3.2.

Note: If one assumes a spatially limited current and charge density outside of this, $\mathbf{j}(\mathbf{x}, t) = 0$, i.e. $\mathbf{j}_t = -\mathbf{j}_l$.

For large r one obtains $\mathbf{A}^{\mathrm{qs}} \sim \frac{k_C}{rck_L} \dot{\mathbf{p}}(t)$ and $\mathbf{j}_l \sim \dot{\mathbf{E}}^{(p)}$ (see problem 8.5). $\mathbf{p}$ is the dipole moment of the charge distribution and $\mathbf{E}^{(p)}$ is the field of the dipole (2.2.5).

Only the contributions of individual Fourier components $\mathrm{e}^{\mathrm{i}\mathbf{k}\cdot\mathbf{x}}\, \mathbf{A}^{\mathrm{qs}}(\mathbf{k}, t)$ belonging to current components $\mathbf{j}_l$ and $\mathbf{j}_t$ are orthogonal to each other.

Gauge function and causality in the Coulomb gauge

In section 8.2, the Liénard-Wiechert potentials (8.2.16) were determined in
Lorenz gauge. (8.1.11) shows that $\mathbf{E}$ and $\mathbf{B}$ are not changed by a gauge transformation. Now the gauge function for the Coulomb gauge is to be determined,
which is why $\boldsymbol{\nabla}\cdot\mathbf{A}^{C} = 0$ is inserted in (8.1.12):

$$\Box\chi = -\frac{1}{ck_L}\dot{\phi}^C, \qquad \text{respectively} \qquad \Delta\chi \overset{(8.1.10)}{=} -\boldsymbol{\nabla}\cdot\mathbf{A}, \qquad (8.2.60)$$

where the retarded solution of the d'Alembert equation from the Liénard-
Wiechert potentials (8.2.16) is well known.

The scalar potential given in Coulomb gauge (8.2.49) is acausal, since the
contributions of the electric charge from different locations have no time delay.
Also $\mathbf{A}^C$ is acausal due to the source current (8.2.50) $\mathbf{j}_l = \frac{1}{4\pi k_C}\boldsymbol{\nabla}\dot{\phi}^C$:

$$\mathbf{A}^{C} = \mathbf{A} - \mathbf{A}_l \qquad \text{with} \qquad \mathbf{A}_l(\mathbf{x}, t) = \frac{k_C}{ck_L}\int d^3x' \frac{\mathbf{j}_l(\mathbf{x}', t_r)}{|\mathbf{x} - \mathbf{x}'|} \qquad (8.2.61)$$

$\mathbf{B}$ is however causal, since $\mathbf{A}_l$ does not contribute to the field due to $\boldsymbol{\nabla}\times\mathbf{j}_l = 0$:

$$\mathbf{B} = \boldsymbol{\nabla}\times\mathbf{A}^{C} = \boldsymbol{\nabla}\times\mathbf{A}.$$

For the electric field

$$\mathbf{E} = -\boldsymbol{\nabla}\phi^{C} - \frac{k_L}{c}(\dot{\mathbf{A}} - \dot{\mathbf{A}}_l) \qquad (8.2.62)$$

it can be shown using (8.1.10) that – as it must be – the acausal contributions
cancel out (see problem 8.6).

Plane electromagnetic waves

If no sources are present, one gets from (8.2.49) $\phi^{C} = 0$. A distinction between
$\mathbf{A}^C$ and $\mathbf{A}$ is superfluous. The homogeneous wave equation yields

$$\Box\mathbf{A} = 0 \quad \text{with} \quad \mathbf{A} = \mathbf{e}_1\frac{E_0}{ik_Lk}\,e^{i(\mathbf{k}\cdot\mathbf{x}-\omega t)} \quad \text{and} \quad -\frac{\omega^2}{c^2} + k^2 = 0. \quad (8.2.63)$$

For $\mathbf{A}$ follows from $\boldsymbol{\nabla}\cdot\mathbf{A} = 0$ and the dispersion $\omega = ck$

$$\boldsymbol{\nabla}\cdot\mathbf{A} = \mathbf{e}_3\cdot\mathbf{e}_1\frac{1}{k_L}E_0\,e^{i\mathbf{k}\cdot\mathbf{x}-ikct} = 0 \qquad \text{with} \quad \mathbf{e}_3 = \frac{\mathbf{k}}{k}.$$

The wave vector $\mathbf{k}$ is perpendicular to the polarization vector $\mathbf{e}_1$ (transversality). It is clear that in the plane perpendicular to $\mathbf{k}$ there can only be two
polarization directions.

$$\mathbf{E} = -\frac{k_L}{c}\dot{\mathbf{A}} = \mathbf{e}_1\, E_0\, e^{i(\mathbf{k}\cdot\mathbf{x}-ckt)},$$

$$\mathbf{B} = \boldsymbol{\nabla}\times\mathbf{A} = \mathbf{e}_3\times\mathbf{e}_1\frac{1}{k_L}E_0\, e^{i(\mathbf{k}\cdot\mathbf{x}-ckt)} = \mathbf{e}_2\frac{1}{k_L}E_0\, e^{i(\mathbf{k}\cdot\mathbf{x}-ckt)}.$$

The unit vectors $\mathbf{e}_1$, $\mathbf{e}_2 = \mathbf{e}_3\times\mathbf{e}_1$ and $\mathbf{e}_3 = \mathbf{k}/k$ form an orthonormal coordinate system, which is right-handed because of $\mathbf{e}_2 = \mathbf{e}_3\times\mathbf{e}_1$ ($\mathbf{E}$, $\mathbf{B}$, $\mathbf{k}$).

In light waves, the amplitudes of the electric and magnetic wave are orthogonal to each other $\mathbf{E}\perp\mathbf{B}$ ($\mathbf{e}_1\perp\mathbf{e}_2$) and to the propagation direction $\mathbf{k}$. The average energy transport is determined by the Poynting vector (5.6.3) and this by the average energy density (5.6.4) of the electromagnetic field $\langle u\rangle = |E_0|^2/8\pi k_C$

$$\langle\mathbf{S}\rangle = \frac{ck_L}{8\pi k_C}\mathbf{E}\times\mathbf{B}^* = c\frac{|E_0|^2}{8\pi k_C}\mathbf{e}_3 = c\langle u\rangle\mathbf{e}_3. \tag{8.2.64}$$

Note: The Poynting vector can also be written in the form

$$\text{G: } \mathbf{S} = \frac{c}{4\pi}\mathbf{E}\times\mathbf{H} = \frac{E^2}{Z_0}\mathbf{e}_3, \qquad\qquad \text{SI: } \mathbf{S} = \mathbf{E}\times\mathbf{H} = \frac{E^2}{Z_0}\mathbf{e}_3 \tag{8.2.65}$$

$Z_0 = 4\pi k_C/c$ is the impedance of free space (vacuum impedance):

$$\text{G: } Z_0 = \frac{4\pi}{c} = 4.19\times10^{-10}\,\text{stat}\Omega, \qquad \text{SI: } Z_0 = \frac{1}{c\epsilon_0} = \sqrt{\mu_0/\epsilon_0} = 377\,\Omega. \tag{8.2.65'}$$

The energy flux density $\mathbf{S}$ is a vector quantity of mechanics and thus independent of the system.

8.3 Radiation of a Moving Charge Distribution

We start from a charge and current distribution that is limited to a volume of the size $V \sim d^3$.

A charge and/or current distribution emits radiation during accelerated movement, whereby this movement is usually periodic. So initially it is of interest Maxwell equations and potentials to derive for the temporal Fourier components.

8.3.1 Maxwell Equations for the Fourier Components

The Fourier transforms of the charge and current density are obtained from (8.2.20) and those of the Liénard-Wiechert potentials from (8.2.21)

$$\rho_\omega(\mathbf{x}) = \int dt\, e^{i\omega t}\rho(\mathbf{x},t), \qquad\qquad \mathbf{j}_\omega(\mathbf{x}) = \int dt\, e^{i\omega t}\mathbf{j}(\mathbf{x},t), \tag{8.3.1}$$

$$\phi_\omega(\mathbf{x}) = k_C\int d^3x'\frac{e^{ik|\mathbf{x}-\mathbf{x}'|}}{|\mathbf{x}-\mathbf{x}'|}\rho_\omega(\mathbf{x}'), \quad \mathbf{A}_\omega(\mathbf{x}) = \frac{k_C}{ck_L}\int d^3x'\frac{e^{ik|\mathbf{x}-\mathbf{x}'|}}{|\mathbf{x}-\mathbf{x}'|}\mathbf{j}_\omega(\mathbf{x}'). \tag{8.3.2}$$

From the Lorenz condition (8.1.8) and the continuity equation, the Fourier transforms follow after partial integration of the 1st term

$$\int dt\, e^{i\omega t}\Big[\frac{1}{ck_L}\dot{\phi}(\mathbf{x},t)+\boldsymbol{\nabla}\cdot\mathbf{A}(\mathbf{x},t)\Big]=0 \;\Rightarrow\; -\frac{i\omega}{ck_L}\phi_\omega(\mathbf{x})+\boldsymbol{\nabla}\cdot\mathbf{A}_\omega(\mathbf{x})=0, \quad (8.3.3)$$

$$\int dt\, e^{i\omega t}\big[\dot{\rho}(\mathbf{x},t)+\boldsymbol{\nabla}\cdot\mathbf{j}(\mathbf{x},t)\big]=0 \;\Rightarrow\; -i\omega\,\rho_\omega(\mathbf{x})+\boldsymbol{\nabla}\cdot\mathbf{j}_\omega(\mathbf{x})=0. \quad (8.3.4)$$

Remark: If the density becomes constant for $t\to\pm\infty$, i.e. $\rho(\mathbf{x},t\to\pm\infty)=\rho(\mathbf{x})$, then the boundary term of the partial integration also vanishes in this case:

$$\lim_{T\to\infty} e^{i\omega t}\,\rho(\mathbf{x},t)\Big|_{-T}^{T} = \lim_{T\to\infty}\rho(\mathbf{x})2i\omega\,\frac{\sin(\omega T)}{\omega} \stackrel{(B.6.13)}{=} \rho(\mathbf{x})2\pi i\omega\,\delta(\omega).$$

The temporal Fourier components of the fields are defined as

$$\mathbf{E}_\omega(\mathbf{x}) = \int dt\, e^{i\omega t}\,\mathbf{E}(\mathbf{x},t), \qquad \mathbf{B}_\omega(\mathbf{x}) = \int dt\, e^{i\omega t}\,\mathbf{B}(\mathbf{x},t). \qquad (8.3.5)$$

If you multiply (8.1.1) with $e^{i\omega t}$ and integrate over t, you get the Maxwell equations for the Fourier components

$$\begin{aligned}
&\text{(a)} & \boldsymbol{\nabla}\cdot\mathbf{E}_\omega &= 4\pi k_C\rho_\omega, & \text{(b)}\;\; &\boldsymbol{\nabla}\times\mathbf{E}_\omega - \frac{i\omega k_L}{c}\mathbf{B}_\omega = 0, \\
&\text{(c)}\;\; \boldsymbol{\nabla}\times\mathbf{B}_\omega + \frac{i\omega}{ck_L}\mathbf{E}_\omega &= \frac{4\pi k_C}{ck_L}\mathbf{j}_\omega, & \text{(d)} & &\boldsymbol{\nabla}\cdot\mathbf{B}_\omega = 0.
\end{aligned}$$
$$(8.3.6)$$

8.3.2 Helmholtz Potentials for Source and Vortex Fields

Of particular interest are the radiation fields that occur during the periodic movement of charge and current distributions, whereby the decomposition theorem enables us to decompose these into longitudinal source and transverse vortex fields. It can be shown that in the far zone only the transverse fields survive.

It is assumed that charge and current distributions are local. Then $\mathbf{A}$ falls off asymptotically no weaker than $1/r$, which also applies to the fields. In the following, the frequency ω is omitted in all electromagnetic quantities, i.e. $\rho(\mathbf{x})\equiv\rho_\omega(\mathbf{x})$, and ω is replaced by ck.

Helmholtz potentials of the magnetic field

The scalar Helmholtz potential ϕ_h^b of the magnetic field must vanish because $\mathbf{B}$ is source-free ($\boldsymbol{\nabla}\cdot\mathbf{B}=0$):

$$\phi_h^b(\mathbf{x}) = 0, \qquad\qquad \mathbf{B}_l(\mathbf{x}) = -\boldsymbol{\nabla}\phi_h^b(\mathbf{x}) = 0. \qquad (8.3.7a)$$

The Helmholtz vector potential $\mathbf{A}_h^b$ of the magnetic field is obtained directly from the induction equation (8.3.6b), where the electric field $\mathbf{E} = \mathbf{E}_l + \mathbf{E}_t$ due to $\boldsymbol{\nabla}\times\mathbf{E}_l = 0$ is replaced by its transverse component $\mathbf{E}_t$. Since $\boldsymbol{\nabla}\cdot\mathbf{E}_t = 0$, $\mathbf{E}_t$ is the Helmholtz potential of the magnetic field:

$$\mathbf{A}_h^b(\mathbf{x}) = -(\mathrm{i}/k_L k)\mathbf{E}_t(\mathbf{x}), \quad \boldsymbol{\nabla}\times\mathbf{A}_h^b(\mathbf{x}) = \mathbf{B}(\mathbf{x}), \quad \boldsymbol{\nabla}\cdot\mathbf{A}_h^b(\mathbf{x}) = 0. \quad (8.3.7\mathrm{b})$$

Note: The calculation of $\mathbf{A}_h^b(\mathbf{x})$ using the decomposition theorem (7.1.16) is cumbersome due to the convergence of the integrals, since for $r \to \infty$ not only $B \sim 1/r$, but due to the retardation also the derivatives of $\mathbf{B}$ and $\mathbf{E}$ decrease with $1/r$ (problem 8.7); one substitutes $\mathbf{B}$ from the induction equation.

Helmholtz potentials of the electric field

It is very easy with the Helmholtz theorem to determine the scalar potential $\phi_h^e = \phi^{\mathrm{qs}}$ from the sources, since these are known by the Gauss's law (8.3.6a): $\boldsymbol{\nabla}\cdot\mathbf{E} = 4\pi k_C\rho$

$$\phi_h^e(\mathbf{x}) = \frac{1}{4\pi}\int \mathrm{d}^3x' \, \frac{\boldsymbol{\nabla}\cdot\mathbf{E}(\mathbf{x}')}{|\mathbf{x}-\mathbf{x}'|}, \quad \mathbf{E}^{\mathrm{qs}}(\mathbf{x}) = -\boldsymbol{\nabla}\phi_h^e(\mathbf{x}), \quad \boldsymbol{\nabla}\times\mathbf{E}^{\mathrm{qs}}(\mathbf{x}) = 0. \quad (8.3.8\mathrm{a})$$

Note: The problem becomes more challenging when trying to obtain the potential from the decomposition theorem

$$\phi_h^e(\mathbf{x}) = \frac{-1}{4\pi}\int \mathrm{d}^3x' \, \mathbf{E}(\mathbf{x}')\cdot\boldsymbol{\nabla}'\frac{1}{|\mathbf{x}-\mathbf{x}'|} \stackrel{!}{=} \frac{1}{4\pi}\int \mathrm{d}^3x' \, \frac{\boldsymbol{\nabla}'\cdot\mathbf{E}(\mathbf{x}')}{|\mathbf{x}-\mathbf{x}'|},$$

since the disappearance of the boundary term is not obvious (problem 8.7).

The vector potential of the electric field can be derived directly from the Ampère-Maxwell equation in the form of (8.2.59)

$$\boldsymbol{\nabla}\times\left[\mathbf{B}(\mathbf{x}) - \mathbf{B}^{\mathrm{qs}}(\mathbf{x})\right] = -\mathrm{i}(k/k_L)\mathbf{E}_t(\mathbf{x}), \tag{8.3.8b}$$

$$\mathbf{A}_h^e(\mathbf{x}) = \mathrm{i}(k_L/k)\left[\mathbf{B}(\mathbf{x}) - \mathbf{B}^{\mathrm{qs}}(\mathbf{x})\right], \quad \boldsymbol{\nabla}\times\mathbf{A}_h^e(\mathbf{x}) = \mathbf{E}_t(\mathbf{x}), \quad \boldsymbol{\nabla}\cdot\mathbf{A}_h^e(\mathbf{x}) = 0.$$

$\mathbf{B}^{\mathrm{qs}} = \boldsymbol{\nabla}\times\mathbf{A}^{\mathrm{qs}}$ satisfies as a quasi-static field the vector-Poisson equation (8.2.57) and is source-free, but not vortex-free and disappears at a great distance (far zone) from the current sources ($\sim 1/r^2$).

Note: Even when calculating $\mathbf{A}_h^e$ with the decomposition theorem, due to the boundary term a detour must be taken by inserting $\mathbf{E}$ from the Ampère-Maxwell equation (problem 8.7).

Summarizing our results, the fields $\mathbf{E}_l = \mathbf{E}^{\mathrm{qs}}$ and $\mathbf{B}^{\mathrm{qs}}$ only have a limited range (*near zone*), while $\mathbf{E}_t$ and $\mathbf{B} - \mathbf{B}^{\mathrm{qs}}$ contain the long-range radiation components, with $\mathbf{E}_t$ perpendicular to $\mathbf{B} - \mathbf{B}^{\mathrm{qs}}$. The quasi-static potentials ϕ_h^e and $\mathbf{A}^{\mathrm{qs}}$ are only definitely convergent when ρ or $\mathbf{j}$ decrease stronger than $1/r^2$. The convergence behavior can however be improved by the calculation with G_1 or G_2, (A.4.28).

All Helmholtz potentials could be determined directly from the Maxwell equations and one obtains from the Helmholtz potentials again the Maxwell equations, where $\mathbf{\nabla}\times\mathbf{B}^{\mathrm{qs}} = (4\pi k_C/ck_L)\mathbf{j}-\mathrm{i}(k/k_L)\mathbf{E}_l$:

$$\text{(a)} \quad -\Delta\phi_h^e = \mathbf{\nabla}\cdot\mathbf{E} = 4\pi k_C\rho, \qquad\qquad \text{(b)} \ \mathbf{\nabla}\times\mathbf{A}_h^b = \mathbf{B} = \frac{k_L}{\mathrm{i}k}\mathbf{\nabla}\times\mathbf{E},$$

$$\text{(c)} \ \ \mathbf{\nabla}\times\mathbf{A}_h^e = \mathbf{E}_t = \frac{\mathrm{i}k_L}{k}\left[\mathbf{\nabla}\times\mathbf{B}-\frac{4\pi k_C}{ck_L}\mathbf{j}\right]-\mathbf{E}_l, \quad \text{(d)} \ \ -\Delta\phi_h^b = \mathbf{\nabla}\cdot\mathbf{B}=0.$$

$$(8.3.9)$$

8.3.3 Periodic Movement

We have shown that due to the linearity of the Maxwell equations, the fields for each frequency ω can be determined separately. It is therefore reasonable to assume a time dependence of the form

$$\rho(\mathbf{x},t) = \rho(\mathbf{x})\,\mathrm{e}^{-\mathrm{i}\omega t} \qquad \text{and} \qquad \mathbf{j}(\mathbf{x},t) = \mathbf{j}(\mathbf{x})\,\mathrm{e}^{-\mathrm{i}\omega t} \qquad (8.3.10)$$

without restricting the generality of the solution. The retarded time goes into the potentials (8.2.16)

$$t_r = t - \frac{|\mathbf{x}-\mathbf{x}'|}{c}, \qquad (8.3.11)$$

from which, given the time dependence of (8.3.10) for the density or the current, it follows that $(\omega/c = k)$

$$\rho(\mathbf{x}',t_r) = \rho(\mathbf{x}')\,\mathrm{e}^{-\mathrm{i}\omega\left(t-|\mathbf{x}-\mathbf{x}'|/c\right)} = \rho(\mathbf{x}',t)\,\mathrm{e}^{\mathrm{i}k|\mathbf{x}-\mathbf{x}'|}. \qquad (8.3.12)$$

The continuity equation reads

$$\dot\rho(\mathbf{x},t) + \mathbf{\nabla}\cdot\mathbf{j}(\mathbf{x},t) = 0 \qquad\quad \Rightarrow \qquad\quad \rho(\mathbf{x}) = -\frac{\mathrm{i}}{\omega}\mathbf{\nabla}\cdot\mathbf{j}(\mathbf{x}). \qquad (8.3.13)$$

Potentials

One obtains for the Liénard-Wiechert potentials (8.2.16)

$$\phi(\mathbf{x},t) = \phi(\mathbf{x})\,\mathrm{e}^{-\mathrm{i}\omega t} \quad \text{with} \quad \phi(\mathbf{x}) = k_C\int\mathrm{d}^3x'\,\frac{\mathrm{e}^{\mathrm{i}k|\mathbf{x}-\mathbf{x}'|}}{|\mathbf{x}-\mathbf{x}'|}\rho(\mathbf{x}'),$$

$$(8.3.14)$$

$$\mathbf{A}(\mathbf{x},t) = \mathbf{A}(\mathbf{x})\,\mathrm{e}^{-\mathrm{i}\omega t} \quad \text{with} \quad \mathbf{A}(\mathbf{x}) = \frac{k_C}{ck_L}\int\mathrm{d}^3x'\,\frac{\mathrm{e}^{\mathrm{i}k|\mathbf{x}-\mathbf{x}'|}}{|\mathbf{x}-\mathbf{x}'|}\mathbf{j}(\mathbf{x}').$$

The retardation thus goes with the factor $\mathrm{e}^{\mathrm{i}k|\mathbf{x}-\mathbf{x}'|}$ into $\phi(\mathbf{x})$ and $\mathbf{A}(\mathbf{x})$ and considering that the various parts of the source do not contribute to the potential (field) with the same phase. The starting point were the decoupled wave equations for ϕ and $\mathbf{A}$, which assumes the Lorenz gauge (8.1.8). With $\dot{\phi} = -i\omega\phi$ one obtains for these according to (8.3.3)

$$\frac{1}{ck_L}\dot{\phi}(\mathbf{x}, t) + \nabla \cdot \mathbf{A}(\mathbf{x}, t) = 0 \qquad \Rightarrow \qquad \phi(\mathbf{x}) = -i\frac{k_L}{k}\nabla \cdot \mathbf{A}(\mathbf{x}). \qquad (8.3.15)$$

In the following, approximations for $\mathbf{A}$ are made with a multipole expansion. The same approximations cannot be applied to ϕ without further ado, as the Lorenz condition may be violated. It is therefore sensible to determine ϕ from (8.3.15) or to bypass it altogether. Thus, $\mathbf{E}$ can be calculated from the Ampère-Maxwell equation from $\mathbf{B}$ as will be shown below. This ensures the consistency of the approximation of $\mathbf{E}$ and $\mathbf{B}$.

Fields

According to the Ampère-Maxwell equation (8.3.9), one obtains for $\mathbf{E}$ outside the charge and current distribution, where $\mathbf{j} = 0$:

$$\mathbf{E}(\mathbf{x}, t) = (ik_L/k)\,\nabla \times \mathbf{B}. \qquad (8.3.16)$$

Conversely, one obtains from the Faraday's law of induction

$$\mathbf{B}(\mathbf{x}, t) = -(i/kk_L)\,\nabla \times \mathbf{E}. \qquad (8.3.17)$$

As already mentioned, the scalar potential ϕ is avoided in the calculation of the fields. This is achieved by means of (8.3.16), since $\mathbf{B} = \nabla \times \mathbf{A}$.

The Liénard-Wiechert potentials decrease for $r \to \infty$ with $1/r$. If one forms the derivatives (gradient, rotation) of the potentials for $r \gg r'$, then due to the retardation

$$\nabla\, e^{ik|\mathbf{x}-\mathbf{x}'|} = ik\frac{\mathbf{x}-\mathbf{x}'}{|\mathbf{x}-\mathbf{x}'|}\, e^{ik|\mathbf{x}-\mathbf{x}'|} \approx ik\mathbf{e}_r\, e^{ik|\mathbf{x}-\mathbf{x}'|}$$

the leading term is also of the order $1/r$ and not – as in electrostatics – of $1/r^2$. We conclude from this

$$\begin{aligned}\mathbf{B}(\mathbf{x}, t) &= ik\mathbf{e}_r \times \mathbf{A} + \mathrm{O}(d/r) \\ \mathbf{E}(\mathbf{x}, t) &= -k_L\mathbf{e}_r \times \mathbf{B} + \mathrm{O}(d/r)\end{aligned} \qquad \text{Farzone with} \quad r \gg d\,. \quad (8.3.18)$$

$k = \omega/c = 2\pi/\lambda$ is the wave number and $r' \sim d$ the extension of the source.

8.3.4 Expansion According to Multipoles and Zones

To calculate the potentials (8.3.14), approximations are made. Not only because the potentials are usually not exactly integrable, but because not all special cases are equally interesting and often approximations allow a better insight into the physical processes.

(8.3.16) shows that in periodic motion $\mathbf{E}$ in a region where $\mathbf{j} = 0$, can be calculated from $\boldsymbol{\nabla} \times \mathbf{B}$, i.e. from $\mathbf{A}$. Our interest lies in such regions, so that knowledge of the scalar potential ϕ is not necessary and we therefore limit ourselves to approximations for $\mathbf{A}$.

Note: As in magnetostatics, we start from the expansion of $\mathbf{A}$ according to (4.2.2), but assume that ρ and $\mathbf{j}$ are time-dependent

$$\mathbf{A} = \frac{k_C}{ck_L r} \int \mathrm{d}^3 x' \, \mathbf{j} \Big(1 + \frac{\mathbf{x} \cdot \mathbf{x}'}{r^2} + \ldots \Big) = \frac{k_C}{ck_L} \Big\{ \frac{\dot{\mathbf{p}}}{r} + \frac{c}{k_L} \frac{\mathbf{m} \times \mathbf{x}}{r^3} + \frac{1}{2r^3} \int \mathrm{d}^3 x' \, \dot{\rho} \mathbf{x}'(\mathbf{x} \cdot \mathbf{x}') + \ldots \Big\}.$$

In addition to the magnetic dipole potential (4.2.10), there are now contributions from the electric dipole and the electric quadrupole according to (4.2.8) and (4.2.9), i.e. two contributions that are not present in magnetostatics. $\mathbf{A}$ is a quasi-static potential in this approximation. We have in the multipole expansion still to consider the retardation by the factor $\mathrm{e}^{ik|\mathbf{x}-\mathbf{x}'|}$, which leads to long-range parts in the fields and thus to a division into near, intermediate and far zone.

The multipole expansion is given by (4.2.2)

$$\frac{1}{|\mathbf{x} - \mathbf{x}'|} = \frac{1}{r} \Big[1 + \frac{\mathbf{x} \cdot \mathbf{x}'}{r^2} + \mathrm{O}\big(\frac{d^2}{r^2}\big) \Big].$$

Then we make the approximation

$$k|\mathbf{x}-\mathbf{x}'| = kr - k\frac{\mathbf{x}}{r} \cdot \mathbf{x}' + \underbrace{\frac{k}{2r} \Big(r'^2 - (\frac{\mathbf{x}}{r} \cdot \mathbf{x}')^2 \Big)}_{\mathrm{O}(kd^2/r)} + \cdots$$

$$\mathrm{e}^{ik|\mathbf{x}-\mathbf{x}'|} \approx \mathrm{e}^{ikr} \, \mathrm{e}^{-i\mathbf{k} \cdot \mathbf{x}'} \qquad\qquad kd \ll r/d$$

$$\approx \mathrm{e}^{ikr} \Big[1 - i\mathbf{k} \cdot \mathbf{x}' + \mathrm{O}(k^2 d^2) + \mathrm{O}\big(\frac{kd^2}{r}\big) \Big] \qquad \mathbf{k} = k\mathbf{e}_r \, .$$

The term of order $\mathrm{O}(\frac{kd^2}{r})$ disappears in the far zone. In the near zone, the retardation can be completely neglected. One obtains

$$\frac{\mathrm{e}^{ik|\mathbf{x}-\mathbf{x}'|}}{|\mathbf{x}-\mathbf{x}'|} \approx \frac{\mathrm{e}^{ikr}}{r} \mathrm{e}^{-i\mathbf{k} \cdot \mathbf{x}'} \Big\{ 1 + \mathrm{O}\big(\frac{d}{r}\big) + \mathrm{O}\big(\frac{kd^2}{r}\big) \Big\} \tag{8.3.19a}$$

$$\approx \frac{\mathrm{e}^{ikr}}{r} \Big\{ 1 + \frac{\mathbf{x} \cdot \mathbf{x}'}{r^2}(1 - ikr) + \mathrm{O}\big(\frac{d^2}{r^2}\big) + \mathrm{O}(k^2 d^2) + \mathrm{O}\big(\frac{kd^2}{r}\big) \Big\}. \tag{8.3.19b}$$

In both expansions, the 1st term is the contribution of the electric dipole moment to $\mathbf{A}$, where $\mathrm{e}^{-i\mathbf{k} \cdot \mathbf{x}'}$ still accounts for the finite extent d of the radiation source through different phase factors at the spatial points $\mathbf{x}'$. Thus, $\mathrm{e}^{-i\mathbf{k} \cdot \mathbf{x}}$ contains the far-reaching components of all electric and magnetic multipoles.

The 2nd term of (8.3.19b) determines the contribution of the magnetic dipole and the electric quadrupole terms, where the far zone is taken into account by the factor $(1 - ikr)$. In the intermediate zone (transition zone), all terms of (8.3.19b) must be considered. One distinguishes $(\mathbf{k} \equiv k\mathbf{e}_r)$

(a) $kd < kr \ll 1 \quad \Leftrightarrow \quad d < r \ll \lambda:$ $\qquad \dfrac{1}{|\mathbf{x}-\mathbf{x}'|}$ $\quad$ Near zone,

(b) $kd \ll kr \sim 1 \quad \Leftrightarrow \quad d \ll r \sim \lambda$ $\qquad\qquad\qquad$ Transition zone, $\qquad\qquad$ (8.3.20)

(c) $kd \ll 1 \ll kr \quad \Leftrightarrow \quad d \ll \lambda \ll r:$ $\quad \dfrac{\mathrm{e}^{ikr}}{r}\,\mathrm{e}^{-i\mathbf{k}\cdot\mathbf{x}'}$ $\quad$ Far zone, also radiation or wave zone.

The condition $kd \ll 1$ of the far zone may be too restrictive for some antennas and can be replaced there by $kd \ll r/d$.

8.3.5 Near Zone

If $r \ll \lambda$, then $\mathrm{e}^{ik|\mathbf{x}-\mathbf{x}'|}$ can be approximated by 1 $(r > d)$. The potentials $\phi(\mathbf{x})$ and $\mathbf{A}(\mathbf{x})$ (8.3.14) then become the potentials of electro- and magnetostatics, which is also referred to as the static zone:

$$\phi(\mathbf{x}) = k_C \int \mathrm{d}^3 x' \, \frac{\rho(\mathbf{x}')}{|\mathbf{x}-\mathbf{x}'|}, \qquad\qquad \mathbf{A}(\mathbf{x}) = \frac{k_C}{c k_L} \int \mathrm{d}^3 x' \, \frac{\mathbf{j}(\mathbf{x}')}{|\mathbf{x}-\mathbf{x}'|}.$$

Quasi-static potentials, like the approximations ϕ and $\mathbf{A}$ of the near zone, satisfy the Lorenz gauge, as shown in (8.2.56). For the harmonic time evolution $\sim \mathrm{e}^{-i\omega t}$ assumed here, the Lorenz gauge (8.3.15) can be verified using the continuity equation (8.3.13) for the near field.

8.3.6 Far Zone

Potentials and fields

In the far zone, $d \ll r$ always applies. Another restriction concerns the wavelength. If this becomes small compared to d, the distance from which the far zone applies shifts further outwards. We will apply the approximation for the far zone (8.3.19a) to ϕ and $\mathbf{A}$ and show that the Lorenz gauge in (8.3.21) is only asymptotically fulfilled ($\mathbf{k} = k \mathbf{e}_r$):

$$\phi(\mathbf{x}) = k_C \frac{\mathrm{e}^{ikr}}{r} \int \mathrm{d}^3 x' \, \rho(\mathbf{x}') \, \mathrm{e}^{-i\mathbf{k}\cdot\mathbf{x}'} = k_C \frac{\mathrm{e}^{ikr}}{r} \rho(\mathbf{k}) = k_C \frac{\mathrm{e}^{ikr}}{cr} \mathbf{e}_r \cdot \mathbf{j}(\mathbf{k}). \quad (8.3.21)$$

The relationship between ρ and $\mathbf{j}$ is provided by the continuity equation:

$$\int \mathrm{d}^3 x' \, \mathrm{e}^{-i\mathbf{k}\cdot\mathbf{x}'} \Big[-i\omega \rho(\mathbf{x}') + \boldsymbol{\nabla}' \cdot \mathbf{j}(\mathbf{x}') \Big] = -i\omega \rho(\mathbf{k}) + i\mathbf{k}\cdot\mathbf{j}(\mathbf{k}) = 0,$$

$$\rho(\mathbf{k}) = (1/c)\, \mathbf{e}_r \cdot \mathbf{j}(\mathbf{k}). \qquad\qquad (8.3.21')$$

For the vector potential, one obtains in the far zone:

$$\mathbf{A}(\mathbf{x}) = \frac{k_C}{c k_L} \frac{\mathrm{e}^{ikr}}{r} \int \mathrm{d}^3 x' \, \mathbf{j}(\mathbf{x}') \, \mathrm{e}^{-i\mathbf{k}\cdot\mathbf{x}'} = \frac{k_C}{c k_L} \frac{\mathrm{e}^{ikr}}{r} \mathbf{j}(\mathbf{k}). \qquad (8.3.22)$$

The scalar potential corresponding to (8.3.22) is obtained by inserting

$$\nabla \cdot \mathbf{A} = \frac{k_C}{ck_L} \frac{\mathrm{e}^{\mathrm{i}kr}}{r} \left[\left(\mathrm{i}k - \frac{1}{r}\right) \mathbf{e}_r \cdot \mathbf{j}(\mathbf{k}) + \nabla \cdot \mathbf{j}(\mathbf{k}) \right]$$

into the Lorenz condition (8.3.15). Assuming that the extension of the current source $d \ll \lambda$ is fulfilled, the last term can be neglected and one obtains

$$\phi(\mathbf{x}) = -\mathrm{i}\frac{k_L}{k} \nabla \cdot \mathbf{A} \approx k_C\left(1 + \mathrm{i}\frac{1}{kr}\right) \frac{\mathrm{e}^{\mathrm{i}kr}}{cr} \mathbf{e}_r \cdot \mathbf{j}(\mathbf{k}). \tag{8.3.23}$$

The scalar potential (8.3.21) fulfills the Lorenz condition only asymptotically for $r \to \infty$. $\rho(\mathbf{k})$ is the form factor known from solid state physics of a charge distribution. The finite $\mathbf{k}$ takes into account the phase difference of waves originating from different source points of the charge/current distribution; for point charges (point dipoles) the Fourier transform is independent of $\mathbf{k}$ $(\rho(\mathbf{k}) = \rho_0)$.

The fields corresponding to the potentials are according to (8.3.18) neglecting contributions of the order $1/r^2$

$$\mathbf{B}(\mathbf{x}, t) = \mathrm{i}k\frac{k_C}{ck_L} \frac{\mathrm{e}^{\mathrm{i}kr-\mathrm{i}\omega t}}{r} \mathbf{e}_r \times \mathbf{j}(\mathbf{k}), \tag{8.3.24}$$

$$\mathbf{E}(\mathbf{x}, t) = \mathrm{i}k\frac{k_C}{c} \frac{\mathrm{e}^{\mathrm{i}kr-\mathrm{i}\omega t}}{r} \mathbf{e}_r \times \left(\mathbf{j}(\mathbf{k}) \times \mathbf{e}_r\right). \tag{8.3.25}$$

Here, $\mathbf{E}$, $\mathbf{B}$ and $\mathbf{e}_r$ form a right-handed coordinate system:

$$\mathbf{E} = k_L \mathbf{B} \times \mathbf{e}_r, \qquad \mathbf{B} = \frac{1}{k_L} \mathbf{e}_r \times \mathbf{E}, \qquad \mathbf{e}_r = \frac{k_L}{\mathbf{B}^2} \mathbf{E} \times \mathbf{B}. \tag{8.3.26}$$

Average radiation power

Of interest is the radiation power emitted by the source. It is determined by $\mathbf{S}$. We limit ourselves to the average energy flux density $\langle \mathbf{S} \rangle$ at a great distance from the source. From the fields (8.3.24) and (8.3.25) of the far zone we then obtain the average energy flux $(\mathbf{e}_r \cdot \mathbf{E} = 0)$

$$\langle \mathbf{S} \rangle = \frac{ck_L}{8\pi k_C} \mathbf{E} \times \mathbf{B}^* = \frac{c}{8\pi k_C} \mathbf{E} \times (\mathbf{e}_r \times \mathbf{E}^*) = \frac{c}{8\pi k_C} |\mathbf{E}|^2 \mathbf{e}_r \tag{8.3.27}$$

$$= \frac{k_C k^2}{8\pi c\, r^2} |\mathbf{e}_r \times (\mathbf{j}(\mathbf{k}) \times \mathbf{e}_r)|^2 \mathbf{e}_r = \frac{k_C k^2}{8\pi c\, r^2} \left[|\mathbf{j}(\mathbf{k})|^2 - |\mathbf{e}_r \cdot \mathbf{j}(\mathbf{k})|^2\right] \mathbf{e}_r.$$

The average energy passing through the surface element $\mathbf{da}$ per unit of time is for a sphere with radius r

$$\mathrm{d}\langle P \rangle = \mathbf{da} \cdot \langle \mathbf{S} \rangle, \qquad \mathbf{da} = r^2 \mathrm{d}\Omega\, \mathbf{e}_r = r^2 \sin\vartheta\, \mathrm{d}\vartheta \mathrm{d}\varphi\, \mathbf{e}_r.$$

The average power radiated from the solid angle $\mathrm{d}\Omega$ is thus

$$\frac{\mathrm{d}\langle P\rangle}{\mathrm{d}\Omega} = r^2 \mathbf{e}_r \cdot \langle \mathbf{S}\rangle = \frac{cr^2}{8\pi k_C}\,|\mathbf{E}|^2 = \frac{k_C k^2}{8\pi c}\left(|\mathbf{j}(\mathbf{k})|^2 - |\mathbf{e}_r\cdot\mathbf{j}(\mathbf{k})|^2\right). \tag{8.3.28}$$

The component of $\mathbf{j}$ perpendicular to the surface normal $\mathbf{e}_r$ determines the radiated power. If here for the point dipole $\mathbf{j}(\mathbf{k}) = -\mathrm{i}\omega\mathbf{p}$ is inserted, one directly obtains the average power (8.4.15) of the dipole radiation.

8.4 The Radiation Components of the Multipoles

We now turn to the calculation of the parts of the radiation that can be attributed to the multipole moments. The radiation of the electric dipole makes by far the most important contribution here.

8.4.1 Electric Dipole Radiation

Potentials of the point dipole

Given is the point dipole oscillating at the origin $\mathbf{p}(t) = \mathbf{p}\,\mathrm{e}^{-\mathrm{i}\omega t}$, whose charge density is (2.2.2). The current density is obtained from the continuity equation. Together with their Fourier transforms, this results in

$$\rho_p(\mathbf{x}) = -\mathbf{p}\cdot\boldsymbol{\nabla}\delta^{(3)}(\mathbf{x}), \qquad\qquad \mathbf{j}_p(\mathbf{x}) = -\mathrm{i}\omega\mathbf{p}\,\delta^{(3)}(\mathbf{x}) \tag{8.4.1}$$

$$\rho_p(\mathbf{k}) = \int \mathrm{d}^3 x\, \mathrm{e}^{-\mathrm{i}\mathbf{k}\cdot\mathbf{x}}\rho_p(\mathbf{x}) = -\mathrm{i}\mathbf{k}\cdot\mathbf{p}, \quad \mathbf{j}_p(\mathbf{k}) = \int \mathrm{d}^3 x\, \mathrm{e}^{-\mathrm{i}\mathbf{k}\cdot\mathbf{x}}\mathbf{j}_p(\mathbf{x}) = -\mathrm{i}\omega\mathbf{p}.$$

Now we calculate the Liénard-Wiechert potentials. First, we insert the dipole density into (8.3.14), integrate by parts and shift the differentiation from $\boldsymbol{\nabla}'$ to $-\boldsymbol{\nabla}$:

$$\phi(\mathbf{x}) = -k_C\int \mathrm{d}^3 x'\,\frac{\mathrm{e}^{\mathrm{i}k|\mathbf{x}-\mathbf{x}'|}}{|\mathbf{x}-\mathbf{x}'|}\mathbf{p}\cdot\boldsymbol{\nabla}'\delta^{(3)}(\mathbf{x}') = k_C\int \mathrm{d}^3 x'\,\delta^{(3)}(\mathbf{x}')\mathbf{p}\cdot\boldsymbol{\nabla}'\frac{\mathrm{e}^{\mathrm{i}k|\mathbf{x}-\mathbf{x}'|}}{|\mathbf{x}-\mathbf{x}'|}$$

$$= -k_C\mathbf{p}\cdot\boldsymbol{\nabla}\frac{\mathrm{e}^{\mathrm{i}kr}}{r} = k_C\mathbf{p}\cdot\mathbf{e}_r\frac{\mathrm{e}^{\mathrm{i}kr}}{r^2}\left(1-\mathrm{i}kr\right) \tag{8.4.2a}$$

$$\mathbf{A}(\mathbf{x}) = -\mathrm{i}k\frac{k_C\mathbf{p}}{k_L}\int \mathrm{d}^3 x'\, \mathrm{e}^{\mathrm{i}k|\mathbf{x}-\mathbf{x}'|}\frac{\delta^{(3)}(\mathbf{x}')}{|\mathbf{x}-\mathbf{x}'|} = -\mathrm{i}k\frac{k_C\mathbf{p}}{k_L}\frac{\mathrm{e}^{\mathrm{i}kr}}{r}. \tag{8.4.2b}$$

The dipole part in the multipole expansion

From the leading term of the expansion (8.3.19) we obtain the radiation of the electric dipole moment of the oscillating charge distribution:

$$\mathbf{A}(\mathbf{x}) = \frac{k_C}{ck_L}\int \mathrm{d}^3 x'\,\mathbf{j}(\mathbf{x}')\frac{\mathrm{e}^{\mathrm{i}k|\mathbf{x}-\mathbf{x}'|}}{|\mathbf{x}-\mathbf{x}'|} \approx \frac{k_C}{ck_L}\frac{\mathrm{e}^{\mathrm{i}kr}}{r}\int \mathrm{d}^3 x'\,\mathbf{j} \overset{(4.2.8)}{=} \frac{k_C}{ck_L}\frac{\mathrm{e}^{\mathrm{i}kr}}{r}\int \mathrm{d}^3 x'\,\dot{\rho}\,\mathbf{x}'$$

$$= -\mathrm{i}k\frac{k_C}{k_L}\frac{\mathrm{e}^{\mathrm{i}kr}}{r}\mathbf{p} \qquad\qquad \text{with} \qquad\qquad \dot{\mathbf{p}} = -\mathrm{i}\omega\mathbf{p}. \tag{8.4.3a}$$

All further terms of $\mathbf{A}$ contain at least one additional x'_k in the integrand, leading to higher moments. The above approximation for $\mathbf{A}$ is therefore the exact potential for the electric dipole.

If we compare the approximation for the far-field (8.3.22) to the electric dipole radiation we see

$$\frac{e^{ikr}}{r} \quad\longleftrightarrow\quad \frac{e^{ikr}}{r}\,e^{-i\mathbf{k}\cdot\mathbf{x}'} \qquad\Longrightarrow\qquad \mathbf{j}(0) \quad\longleftrightarrow\quad \mathbf{j}(\mathbf{k}).$$

The potentials of the electric dipole radiation follow from those of the far-field (8.3.22)–(8.3.23), if one replaces the Fourier transform $\mathbf{j}(\mathbf{k})$ by $\mathbf{j}(0)$.

The two cases merge into each other, when $kd \ll 1$, i.e. $d \ll \lambda$, as for example with visible light emitted by atoms.

Atom $d \approx 1$ Å, visible light $\lambda \approx 3.6\times 10^3 \;\; -7.8\times 10^3$Å
$\qquad\qquad$ X-rays[4] $\qquad \lambda \sim 1.0\times 10^{-1} - 3.0\times 10^1$Å.

In the electric dipole radiation, where $kd \ll 1$, it is common to replace the Fourier transforms of ρ and $\mathbf{j}$ by the dipole moment $\mathbf{p}$

$$\dot{\mathbf{p}} = \int_V d^3x\,\mathbf{x}\dot{\rho} = -\int_V d^3x\,\mathbf{x}\,\boldsymbol{\nabla}\cdot\mathbf{j} = \int_V d^3x\,\mathbf{j} \;\Rightarrow\; -i\omega\mathbf{p} = \mathbf{j}(0). \qquad (8.4.3b)$$

We now insert the dipole part of the current into (8.3.21')

$$\rho_p(\mathbf{k}) = \mathbf{k}\cdot\mathbf{j}(0)/\omega = -i\mathbf{k}\cdot\mathbf{p}$$

and thus obtain the charge density ρ_p of the dipole component. The total charge attributable to the dipole $\rho_p(0) = 0$ vanishes as expected.

The charge distribution ρ oscillates with the same phase, i.e. the field of the moving charge is perceived in the far zone as the radiation field of a point dipole (in the general case, $\rho(\mathbf{k})$ leads to phase differences due to the finite extent of the source which are equivalent to a mixture of other multipole moments).

Now we can, starting from (8.3.22) and (8.3.23) specify the potentials of the electric dipole radiation in a suitable form and obtain the already known potentials of the point dipole (8.4.2):

$$\phi(\mathbf{x},t) = -ikk_C\Big(1 - \frac{1}{ikr}\Big)\frac{e^{ikr-i\omega t}}{r}\,\mathbf{e}_r\cdot\mathbf{p}, \qquad (8.4.4)$$

$$\mathbf{A}(\mathbf{x},t) = \frac{k_C}{ck_L}\frac{e^{ikr-i\omega t}}{r}\,\mathbf{j}(0) = -ik\frac{k_C}{k_L}\frac{e^{ikr-i\omega t}}{r}\,\mathbf{p}. \qquad (8.4.5)$$

In the Coulomb gauge, ϕ^C is the quasi-static potential of the dipole, and $\mathbf{A}^C = \mathbf{A}^b_h$ is determined by (8.3.7b):

$$\phi^C(\mathbf{x}) = k_C\frac{\mathbf{p}\cdot\mathbf{x}}{r^3}, \qquad\qquad \mathbf{A}^C(\mathbf{x}) = \frac{1}{ik_Lk}\big[\mathbf{E}(\mathbf{x}) + \boldsymbol{\nabla}\phi^C(\mathbf{x})\big]. \qquad (8.4.5')$$

---

[4] The boundaries, both to gamma and to UV radiation are not precisely defined.

Dipole fields

We are interested in the generally valid expressions for the dipole fields, which follow from the potentials (8.4.4) and (8.4.5), but without the restriction to the far zone. We obtain the magnetic field from the curl of the vector potential (8.4.5):

$$\mathbf{B}(\mathbf{x}, t) = \boldsymbol{\nabla} \times \mathbf{A} = \frac{k_C}{k_L} k^2 \frac{e^{ikr - i\omega t}}{r} \Big(1 - \frac{1}{ikr}\Big) \, \mathbf{e}_r \times \mathbf{p}. \tag{8.4.6}$$

The calculation of the electric field is somewhat more tedious. We start from the Ampère-Maxwell equation. $\boldsymbol{\nabla} \times \mathbf{B} = (k_L/c)\dot{\mathbf{E}} = -ik_L k \mathbf{E}$ yields, since at the locations $\mathbf{x}$ under consideration no current flows ($\mathbf{j}(\mathbf{x}, t) = 0$):

$$\mathbf{E}(\mathbf{x}) \overset{(8.3.16)}{=} (ik_L/k)\boldsymbol{\nabla} \times \mathbf{B}$$

$$= ik_C k \Big\{ \Big[\boldsymbol{\nabla} \frac{e^{ikr}}{r}\Big(1 - \frac{1}{ikr}\Big)\Big] \times (\mathbf{e}_r \times \mathbf{p}) + \Big[\frac{e^{ikr}}{r}\Big(1 - \frac{1}{ikr}\Big)\Big] \boldsymbol{\nabla} \times (\mathbf{e}_r \times \mathbf{p}) \Big\}.$$

$$\boldsymbol{\nabla} \frac{e^{ikr}}{r^n} = \frac{e^{ikr}}{r^{n+1}} (-n + ikr) \, \mathbf{e}_r \,,$$

$$\boldsymbol{\nabla} \frac{e^{ikr}}{r}\Big(1 - \frac{1}{ikr}\Big) = \Big[\frac{e^{ikr}}{r^2}(-1 + ikr) + \frac{e^{ikr}}{ikr^3}(2 - ikr)\Big]\mathbf{e}_r = \frac{e^{ikr}}{r^2}\Big[ikr - 2\Big(1 - \frac{1}{ikr}\Big)\Big]\mathbf{e}_r \,,$$

$$\boldsymbol{\nabla} \times (\mathbf{p} \times \mathbf{e}_r) = \frac{1}{r}\Big[2\mathbf{p} - \mathbf{e}_r \times (\mathbf{p} \times \mathbf{e}_r)\Big].$$

The field can be divided by means of $\mathbf{e}_r \times (\mathbf{p} \times \mathbf{e}_r) = \mathbf{p} - (\mathbf{e}_r \cdot \mathbf{p})\mathbf{e}_r$ into the far-reaching radiation component and the near field of the electric dipole:

$$\mathbf{E}(\mathbf{x}, t) = k_C \frac{e^{ikr - i\omega t}}{r} \Big\{ k^2 \mathbf{e}_r \times (\mathbf{p} \times \mathbf{e}_r) + \frac{1}{r^2}(1 - ikr)\big[3(\mathbf{p} \cdot \mathbf{e}_r)\mathbf{e}_r - \mathbf{p}\big] \Big\}. \tag{8.4.7}$$

Alternatively, $\mathbf{E} = -\boldsymbol{\nabla}\phi - \frac{k_L}{c}\dot{\mathbf{A}}$ can be calculated with the potentials (8.4.4) and (8.4.5). If one neglects the term $\sim 1/r^2$ in (8.4.4), i.e., calculates with ϕ in the same order as with $\mathbf{A}$, then the Lorenz gauge is not fulfilled and the consistency with (8.4.7) is only given for the far zone.

Near zone

In the near zone, $kr \ll 1$ (and $d < r$) applies. If one approximates the fields (8.4.6) and (8.4.7) with this specification, one obtains

$$\mathbf{E}(\mathbf{x}, t) = k_C \frac{3\big(\mathbf{e}_r \cdot \mathbf{p}(t)\big)\mathbf{e}_r - \mathbf{p}(t)}{r^3} \qquad \text{with} \qquad \mathbf{p}(t) = \mathbf{p}\,e^{-i\omega t}, \tag{8.4.8}$$

$$\mathbf{B}(\mathbf{x}, t) = i\frac{k_C}{k_L} kr \, \frac{\mathbf{e}_r \times \mathbf{p}(t)}{r^3}. \tag{8.4.9}$$

The electric field is that of the dipole $\mathbf{p}(t)$ and the magnetic field $k_L \mathbf{B}$ is smaller by the factor kr than the electric field $\mathbf{E}$ and is perpendicular to this ($\mathbf{B} \cdot \mathbf{E} = 0$).

Far zone

We obtain the fields from (8.3.24)–(8.3.27) by replacing $\mathbf{j}(\mathbf{k})$ with $-i\omega\mathbf{p}$ or simpler, from the exact fields (8.4.6) and (8.4.7) by taking out the far fields falling off with $1/r$:

$$\mathbf{B}(\mathbf{x},t) = \frac{k_C}{k_L}k^2\,\frac{e^{ikr-i\omega t}}{r}\,\mathbf{e}_r\times\mathbf{p} \tag{8.4.10}$$

$$\mathbf{E}(\mathbf{x},t) = k_L\mathbf{B}(\mathbf{x},t)\times\mathbf{e}_r = k_C k^2\,\frac{e^{ikr-i\omega t}}{r}\,\mathbf{e}_r\times(\mathbf{p}\times\mathbf{e}_r). \tag{8.4.11}$$

$\mathbf{E}$ and $k_L\mathbf{B}$ are spherical waves of equal strength, where $\mathbf{E}$ lies in the plane spanned by $\mathbf{e}_r$ and $\mathbf{p}$. $\mathbf{B}$ is perpendicular to this plane, as sketched in Fig. 8.10.

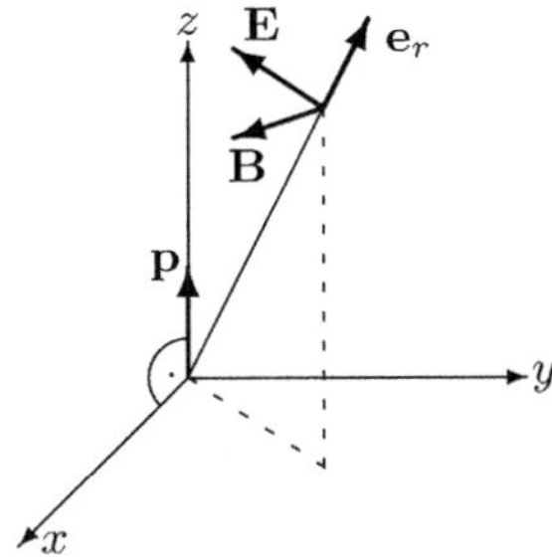

Fig. 8.10. Coordinate system for electric dipole radiation; $\mathbf{E}$ lies in the plane spanned by $\mathbf{e}_r$ and $\mathbf{p}$ and $\mathbf{B}$ is perpendicular to it

Note: The equations for the radiation of a charge distribution moving with the frequency ω can be transformed if in (8.4.10) and (8.4.11)

$$k^2\mathbf{p} = -\frac{1}{c^2}\ddot{\mathbf{p}} \qquad \text{and} \qquad \ddot{\mathbf{p}}(t_r) = \frac{\partial^2}{\partial t^2}\left[\mathbf{p}\,e^{-i\omega t_r}\right] = e^{ikr-i\omega t}\,\ddot{\mathbf{p}}(t)$$

are inserted $(t_r = t - r/c)$:

$$\mathbf{B}(\mathbf{x},t) = \frac{k_C}{k_L c^2 r}\ddot{\mathbf{p}}(t_r)\times\mathbf{e}_r, \quad \mathbf{E}(\mathbf{x},t) = \frac{-k_C}{c^2 r}\mathbf{e}_r\times\left(\ddot{\mathbf{p}}(t_r)\times\mathbf{e}_r\right) = k_L\mathbf{B}\times\mathbf{e}_r. \tag{8.4.12}$$

In Fig. 8.11 one can see a dipole field in the inner circle with $r \lesssim \lambda/2$ at time $t = 0$. Subsequently, the dipole becomes weaker and disappears at $t = \pi/2\omega$, where the field lines detach from the origin. Then a dipole field builds up in the opposite direction, which disappears at $t = 3\pi/2$. Its field lines, now again detached from the origin, are oriented opposite to the previous ones, as can be seen from Fig. 8.11.

The magnetic field is circular in the xy-plane, as also sketched in Fig. 8.11 $(\mathbf{e}_r\times\mathbf{p} = -\mathbf{e}_\varphi)$. If we consider a snapshot of $\mathbf{B}$ at a time $t_0 = 2\pi n/\omega = n/\nu$ (n integer), then for $r = 2\pi m/k = m\lambda$ the field is clockwise and for $r = \pi(2m+1)/k = (m+1/2)\lambda$ it is counterclockwise directed (m integer). For intermediate times t_0 one obtains in the far zone from $\cos(kr - \omega t_0)$ the same picture as with the electric dipole.

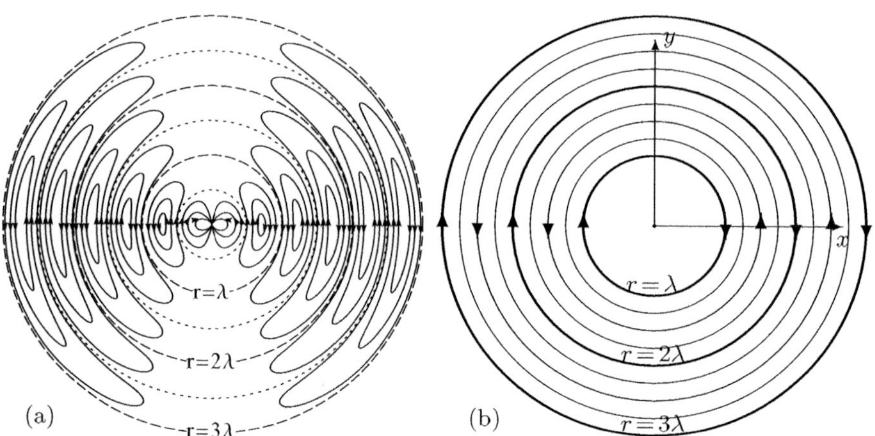

Fig. 8.11. (a) Electric field lines of a dipole in the xz plane, where $\mathbf{p}(t) = p(t)\mathbf{e}_z$. The near field is that of an oscillating dipole. Already for $r > \lambda/2$ radiation fields (vortex fields) are formed, whose intensity is maximal perpendicular to $\mathbf{p}$. (b) Equatorial cross-section of the magnetic field lines of a dipole

Radiation power

With the fields of the far zone, (8.3.24) and (8.3.25) we have determined the average radiation power for a periodically moving charge distribution. Here we limit ourselves to the electric dipole radiation. We only need to replace $\mathbf{j}(\mathbf{k})$ with $\mathbf{j}(0) = -\mathrm{i}\omega\mathbf{p}$ to get the fields and the (average) radiation power for the dipole, (8.4.10) and (8.4.11). From the energy flux density (8.3.27)

$$\mathbf{S}(\mathbf{x},t) = \frac{ck_L}{4\pi k_C}\,\mathrm{Re}\,\mathbf{E}(\mathbf{x},t) \times \mathrm{Re}\,\mathbf{B}(\mathbf{x},t) = \frac{c}{4\pi k_C}(\mathrm{Re}\,\mathbf{E})^2 \mathbf{e}_r$$

one obtains using $\mathrm{d}P = \mathbf{S}\mathrm{d}\mathbf{a}$ the radiation passing through the surface element of a sphere $\mathrm{d}\mathbf{a} = r^2 \mathrm{d}\Omega\,\mathbf{e}_r$ with the radius $r \to \infty$

$$\frac{\mathrm{d}P}{\mathrm{d}\Omega} = \frac{ck_C}{4\pi}k^4 p^2 \sin^2\theta\,\cos^2(kr-\omega t)\,,$$

where $p^2 - (\mathbf{e}_r \cdot \mathbf{p}) = p^2 \sin^2\theta$. The dipole points in a fixed direction (z-axis: $\theta \to \vartheta$). The formula cannot be used without further ado, especially if the direction of the dipole moment varies over time, as is the case with a circulating electron. The integrated intensity is

$$P = \int \mathrm{d}\Omega\,\frac{\mathrm{d}P}{\mathrm{d}\Omega} = k_C\frac{2ck^4}{3}p^2 \cos^2(kr - \omega t).$$

This formula is also obtained by inserting $q\dot{\boldsymbol{\beta}}(t_r) = \frac{1}{c}\ddot{\mathbf{p}}(t_r)$ into the Larmor formula (8.2.39). Expressed through the oscillating point charge $q\mathbf{x}(t) = \mathbf{p}\cos(\omega t)$ with $\mathbf{p} = q\mathbf{x}_0$ results in

$$P = k_\mathrm{C}\frac{2q^2}{3c^3}\,\ddot{\mathbf{x}}^2(t_r) \qquad \text{and} \qquad \langle P\rangle = k_\mathrm{C}\frac{q^2}{3c^3}\,\omega^4\,\mathbf{x}_0^2\,, \tag{8.4.13}$$

which once again shows that only the accelerated charge radiates. The energy that is radiated per period ($\lambdabar = \lambda/2\pi = 1/k$):

$$\mathrm{E} = \int_0^{2\pi/\omega} \mathrm{d}t\, P(t) = k_\mathrm{C}\frac{\pi}{3}\frac{q^2}{\lambdabar}\left(\frac{\mathbf{x}_0}{\lambdabar}\right)^2 = \frac{\pi}{3} \times \underset{\text{at distance } \lambdabar}{\text{Coulomb energy}} \times \underbrace{\left(\frac{\text{amplitude}}{\lambdabar}\right)^2}_{\ll 1}.$$

Average radiated energy

If one averages the just derived quantities like $\mathbf{S}$ and P over a period, then only $\cos^2(kr - \omega t)$ is to be replaced by its average value $\langle\cos^2(kr - \omega t)\rangle = 1/2$. However, since the complex fields are better adapted to the problems, we define the time-averaged energy flux density (8.3.27) with these, limiting ourselves to the far zone (8.4.10) and (8.4.11) or (8.4.12); it would be even simpler in the expressions for the far zone (section 8.3.6) to replace $\mathbf{j}(\mathbf{k})$ with $\mathbf{j}(0) = -\mathrm{i}\omega\mathbf{p}$:

$$\langle\mathbf{S}\rangle = \frac{ck_L}{8\pi k_\mathrm{C}}\,\mathbf{E}\times\mathbf{B}^* = \frac{c}{8\pi k_\mathrm{C}}\,|\mathbf{E}|^2\mathbf{e}_r \overset{(8.4.11)}{=} k_\mathrm{C}\frac{ck^4}{8\pi\, r^2}\big|\mathbf{e}_r\times(\mathbf{p}\times\mathbf{e}_r)\big|^2\,\mathbf{e}_r$$

$$= \frac{k_\mathrm{C}}{8\pi\, c^3\, r^2}\big|\mathbf{e}_r\times(\ddot{\mathbf{p}}(t_r)\times\mathbf{e}_r)\big|^2\,\mathbf{e}_r\,. \tag{8.4.14}$$

In $\langle\mathbf{S}\rangle$, time, and thus also the retarded time, does not enter. With $\langle\mathbf{S}\rangle$ we can again specify the average radiated power that is radiated into the solid angle element $\mathrm{d}\Omega$, where we again start from a sphere of radius r:

$$\frac{\mathrm{d}\langle P\rangle}{\mathrm{d}\Omega} = r^2\langle\mathbf{S}\rangle\cdot\mathbf{e}_r = \frac{cr^2}{8\pi k_\mathrm{C}}\,|\mathbf{E}|^2 = \frac{ck_\mathrm{C}k^4}{8\pi}\big|\mathbf{e}_r\times(\mathbf{p}\times\mathbf{e}_r)\big|^2. \tag{8.4.15}$$

Now according to (A.1.60): $\mathbf{e}_r\times(\mathbf{p}\times\mathbf{e}_r) = \mathbf{p} - (\mathbf{p}\cdot\mathbf{e}_r)\mathbf{e}_r$, so that

$$\big|\mathbf{e}_r\times(\mathbf{p}\times\mathbf{e}_r)\big|^2 = |\mathbf{p} - (\mathbf{e}_r\cdot\mathbf{p})\mathbf{e}_r|^2 = |\mathbf{p}|^2 - |\mathbf{e}_r\cdot\mathbf{p}|^2 = |\mathbf{p}|^2\sin^2\theta,$$

where θ is the angle enclosed by $\mathbf{e}_r$ and $\mathbf{p}$. If the z-axis is fixed by $\mathbf{p}$, then $\theta \equiv \vartheta$ and thus

$$\frac{\mathrm{d}\langle P\rangle}{\mathrm{d}\Omega} = k_\mathrm{C}\frac{ck^4}{8\pi}|p|^2\sin^2\vartheta = \frac{Z_0 k^4}{32\pi^2}c^2|p|^2\sin^2\vartheta, \qquad Z_0 = \frac{4\pi k_\mathrm{C}}{c}. \tag{8.4.16}$$

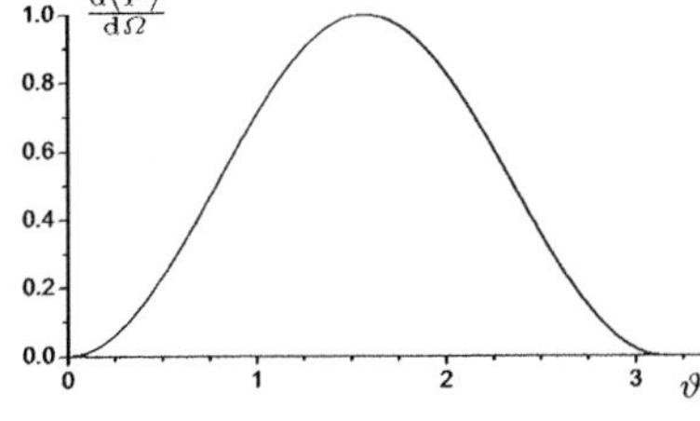

Fig. 8.12. The radiated energy $\mathrm{d}\langle P\rangle/\mathrm{d}\Omega$ is maximum for $\vartheta = \pi/2$ (arbitrary unit for $\mathrm{d}\langle P\rangle/\mathrm{d}\Omega$)

The radiation is maximum at the equator, as outlined in Fig. 8.12. For the total radiation one obtains

$$\langle P \rangle = \int \mathrm{d}\Omega \, \frac{\mathrm{d}\langle P \rangle}{\mathrm{d}\Omega} = k_C \frac{ck^4}{8\pi} |p|^2 \, 2\pi \int_0^\pi \mathrm{d}\vartheta \sin^3 \vartheta = k_C \frac{ck^4}{3} |p|^2 . \tag{8.4.17}$$

$\mathbf{p}$ is the dipole moment of the charge distribution, which oscillates with ω $\mathbf{p}(t) = \mathbf{p}\,\mathrm{e}^{-\mathrm{i}\omega t}$, i.e. $\ddot{\mathbf{p}}(t) = -\omega^2 \mathbf{p}(t)$. Accordingly, (see (8.2.39))

$$\langle P \rangle = \frac{Z_0 k^4}{12\pi} c^2 |\mathbf{p}|^2 = \frac{Z_0}{12\pi c^2} |\ddot{\mathbf{p}}|^2 . \tag{8.4.18}$$

If the electron moves on a circular path, one has a rotating dipole moment (see problem 8.13) with the radiation power

$$\frac{\mathrm{d}\langle P \rangle}{\mathrm{d}\Omega} = k_C \frac{ck^4 |p|^2}{8\pi} (1 + \cos^2 \vartheta), \qquad \langle P \rangle = k_C \frac{2ck^4 |p|^2}{3} . \tag{8.4.19}$$

Note: $\langle P \rangle \sim k^4$ explains the blue color of the sky (Lord Rayleigh): Sun rays excite the air molecules. The radiation is stronger at the blue end (see Rayleigh scattering).

Red coloration of the sun and moon at sunrise and sunset: The blue light they emit is more strongly scattered off course than the red.

8.4.2 Dipole Radiation of an Antenna

Let there be a linear antenna, whose length d is small compared to the wavelength, $kd \ll 1$. The antenna is fed in the middle (coaxial feed), as sketched in Fig. 8.13. Thus, the current spreads symmetrically in the antenna.

This configuration is to be solved under the assumption that the current is known and only flows along the thin wire. Sensible assumptions for the current density are the sinusoidal

$$\mathbf{j}(\mathbf{x}, t) = I \sin\left(\frac{kd}{2} - k|z|\right) \delta(x)\delta(y)\,\theta\left(\frac{d}{2} - |z|\right) \mathrm{e}^{-\mathrm{i}\omega t} \, \mathbf{e}_z \tag{8.4.20}$$

or the linear propagation

$$\mathbf{j} = I\left(1 - \frac{2|z|}{d}\right)\delta(x)\delta(y)\theta\left(\frac{d}{2} - |z|\right)\mathrm{e}^{-\mathrm{i}\omega t} \, \mathbf{e}_z \tag{8.4.21}$$

in the antenna arms. For the vector potential, one obtains after (8.4.3) the dipole part for $kd \ll 1$ in the far zone

$$\mathbf{A}(\mathbf{x}, t) = \frac{k_C}{ck_L} \frac{\mathrm{e}^{\mathrm{i}(kr - \omega t)}}{r} \, \mathbf{j}(0) \qquad \text{with} \qquad \mathbf{j}(0) = \int \mathrm{d}^3 x' \, \mathbf{j}(\mathbf{x}') = -\mathrm{i}kc\mathbf{p} .$$

In the case of linear propagation (8.4.21), which is only meaningful for short antennas, the Fourier transform of the current

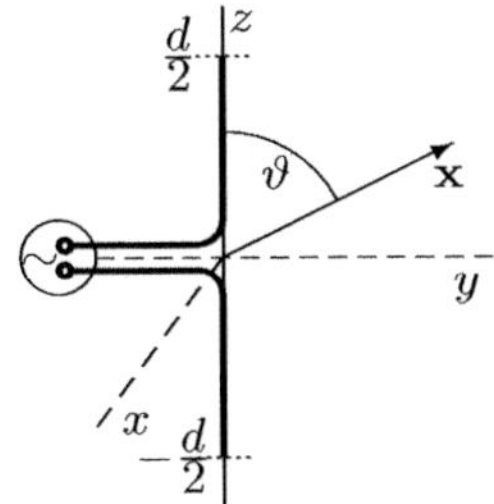

Fig. 8.13. Linear antenna of length d; as an ideal antenna, it is very thin, so that $\mathbf{j}(\mathbf{x},t) = j(z,t)\,\delta(x)\delta(y)\,\mathbf{e}_z$ and the configuration is thus axially symmetric

$$\mathbf{j}(0) = I\mathbf{e}_z \int_{-d/2}^{d/2} \mathrm{d}z\,(1 - \frac{2|z|}{d}) = 2I\mathbf{e}_z\left(z - \frac{z^2}{d}\right)\bigg|_0^{d/2} = \mathbf{e}_z\,\frac{Id}{2},$$

$$\mathbf{p} = \mathrm{i}\frac{Id}{2kc}\,\mathbf{e}_z\,. \tag{8.4.22}$$

For $\mathbf{B}$ and $\mathbf{E}$, one obtains in the far zone $(\boldsymbol{\nabla}r = \mathbf{e}_r)$

$$\mathbf{A}(\mathbf{x}) = \frac{k_C}{ck_L}\frac{\mathrm{e}^{\mathrm{i}kr}}{r}\mathbf{j}(0) = \frac{k_C}{ck_L}\frac{Id}{2}\frac{\mathrm{e}^{\mathrm{i}kr}}{r}\mathbf{e}_z \quad \Rightarrow \quad \begin{aligned} \mathbf{B} &= \boldsymbol{\nabla}\times\mathbf{A} \approx \mathrm{i}k\mathbf{e}_r\times\mathbf{A},\\ \mathbf{E} &= k_L\mathbf{B}\times\mathbf{e}_r\,,\\ \mathbf{E}\times\mathbf{B}^* &= k_L|\mathbf{B}|^2\mathbf{e}_r\,. \end{aligned}$$

This results in the Poynting vector

$$\langle\mathbf{S}\rangle = \frac{ck_L^2}{8\pi k_C}|B|^2\,\mathbf{e}_r = \frac{k_C}{8\pi c}\frac{k^2\,I^2\,d^2}{4r^2}\sin^2\vartheta\,\mathbf{e}_r\,. \tag{8.4.23}$$

The integration over the solid angle results in $4\pi\,(2/3)$, so that the radiated energy, calculated according to (8.4.18), yields:

$$\langle P\rangle = k_C\frac{I^2(kd)^2}{12c} = Z_0\frac{I^2(kd)^2}{48\pi}\,, \tag{8.4.24}$$

where for the impedance of the free space either $\;$ SI: $Z_0 = 1/c\epsilon_0 = 377\,\Omega\;$ or $\;$ G: $Z_0 = 4\pi/c\;$ is to be inserted:

$$\begin{aligned} \langle P\rangle &= 100(d/\lambda)^2\,I^2[\Omega\,\mathrm{A}^2] = 0.15\,\mathrm{W} \;\;\text{at}\;\; \lambda = 50\,\mathrm{m},\;\; d = 2\,\mathrm{m},\;\; I = 1\,\mathrm{A},\\ \langle P\rangle &= \phantom{100(d/\lambda)^2\,I^2[\Omega\,\mathrm{A}^2]} = 1\,\mathrm{W} \;\;\;\;\text{at}\;\; \lambda = 20\,\mathrm{m},\;\; d = 2\,\mathrm{m},\;\; I = 1\,\mathrm{A}. \end{aligned}$$

Using the continuity equation $\mathrm{i}\omega\rho = \boldsymbol{\nabla}\cdot\mathbf{j}$ we obtain the constant line charge density on the antenna arms

$$\rho(\mathbf{x}) = \delta(x)\,\delta(y)\,\mathrm{sgn}(z)\,\frac{2\mathrm{i}I}{\omega d}\,\theta(\frac{d}{2} - |z|)\,.$$

In lossless antennas, the radiation resistance R_s can be defined as the resistance that would generate the radiated power,

$$P = \frac{I^2}{2}\,R_s \qquad\qquad \Rightarrow \qquad\qquad R_s = Z_0\frac{(kd)^2}{24\pi}\,.$$

Contributions of higher multipoles

With the restriction of $kd \ll 1$, only the contributions of electric dipole radiation are considered. For these, it applies that $P \propto k^2$. If $kd \sim 1$, contributions of higher multipoles are added, and one reverts to (8.3.22), which means that one replaces $\mathbf{j}(0)$ with $\mathbf{j}(\mathbf{k})$. For the linear current course (8.4.21) one obtains

$$\mathbf{j}(\mathbf{k}) = \int d^3x'\, e^{-i\mathbf{k}\cdot\mathbf{x}'}\, \mathbf{j}(\mathbf{x}') = I \int_{-d/2}^{d/2} dz'\, e^{-ik_z z'}\left(1 - \frac{2|z'|}{d}\right)\mathbf{e}_z$$

$$= \frac{4I}{dk_z^2}\left[1 - \cos\frac{k_z d}{2}\right]\mathbf{e}_z = \mathbf{j}(0)\,\frac{\sin^2(k_z\frac{d}{4})}{(k_z\frac{d}{4})^2}.$$

In particular, a dependence of the intensity distribution (8.3.28) on the polar angle ϑ is added by $\mathbf{j}(\mathbf{k})$:

$$\frac{d\langle P\rangle}{d\Omega} = k_c \frac{k^2}{8\pi c}|\mathbf{j}(\mathbf{k})|^2 \sin^2\vartheta = \frac{d\langle P_{\text{dipol}}\rangle}{d\Omega}\left|\frac{\sin(k_z\frac{d}{4})}{k_z\frac{d}{4}}\right|^4.$$

The radiation intensity becomes weaker the larger the value of $k_z = k\cos\vartheta$ is. Thus, there always remains a small angular range around $\vartheta = \pi/2$ with only slight weakening. The radiation power is obtained as ($a = kd/4$; $x = \cos\vartheta$):

$$\langle P\rangle = 2\pi \int_{-1}^{1} d\cos\vartheta\, \frac{d\langle P\rangle}{d\Omega} = \langle P_{\text{dipol}}\rangle \frac{3}{2}\int_0^1 dx \left|\frac{\sin(ax)}{ax}\right|^4 (1 - x^2) \qquad (8.4.25)$$

If one calculates the radiation power for $kd = 2\pi$, it has decreased to $\sim 0.75\langle P_{\text{dipol}}\rangle$ accordingly. At the same time, however, the linear current distribution (8.4.21) becomes more unrealistic. For the rod antenna is

$$\mathbf{m} = \frac{k_{\text{L}}}{2c}\int d^3x'\, \mathbf{x}' \times \mathbf{j} = 0\,,$$

so that only electric multipoles contribute to $\langle P\rangle$.

Note: Assumptions about the course of the current in an antenna, given a certain excitation, can only be made in simple cases for very thin and good conductors. This is a complicated boundary value problem.

On the arrangement of antennas

With several appropriately arranged, synchronously operated antennas the angular dependence of the radiation can be influenced. If there are several antennas at the points $\mathbf{a}_j$, the only influence of the different locations $|\mathbf{a}_j| \ll r$ in the far zone can be found in the phase of the spherical wave

$$e^{ik|\mathbf{x}-\mathbf{a}_j|} \approx e^{ikr - i\mathbf{k}\cdot\mathbf{a}_j}\,.$$

Thus, for n identical dipoles (8.4.11)

$$\mathbf{E}_t(\mathbf{x}, t) = \sum_{j=1}^{n} \mathbf{E}_j(\mathbf{x}, t) \approx k_c\, k^2\, \frac{e^{ikr - i\omega t}}{r}\, \mathbf{e}_r \times (\mathbf{p} \times \mathbf{e}_r) \sum_{j=1}^{n} e^{-i\mathbf{k}\cdot\mathbf{a}_j} \tag{8.4.26}$$

$$= \mathbf{E}(\mathbf{x}, t)\, F(\mathbf{k}).$$

The field $\mathbf{E}_t$ of the entire arrangement is equal to the field $\mathbf{E}$ of the individual dipole, multiplied by a structure factor F, which modifies the directionality of the radiation. This behavior is equivalent to the scattering of X-rays in matter, where the structure factor of a crystal lattice allows scattering only in very specific directions.

If n antennas are arranged along a chain, as sketched in Fig. 8.14, this is referred to as a dipole line. With regular distances of the dipoles, the origin

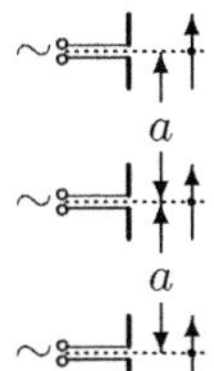

Fig. 8.14. Dipole array: Stacked dipoles (linear antennas). The distance between the dipoles is usually set to $a = \lambda/2$. In a more general arrangement of the dipoles, usually side by side, one speaks of a dipole group or a dipole field

is placed in the middle of the chain to avoid unnecessary phase factors. For n dipoles one obtains for

$$F(\mathbf{k}) = e^{i\mathbf{k}\cdot\mathbf{a}(n+1)/2} \sum_{j=1}^{n} e^{-i\mathbf{k}\cdot\mathbf{a} j} \overset{\substack{\text{geometric} \\ \text{series}}}{=} \frac{\sin\left(\frac{n\mathbf{k}\cdot\mathbf{a}}{2}\right)}{\sin\left(\frac{\mathbf{k}\cdot\mathbf{a}}{2}\right)} \tag{8.4.27}$$

the form of the diffraction function (cardinal sine for $\mathbf{k}\!\cdot\!\mathbf{a} \ll 1$) with a maximum becoming sharper with n. This applies even more strongly to the intensity

$$\frac{\mathrm{d}\langle P_t\rangle}{\mathrm{d}\Omega} = \frac{\mathrm{d}\langle P\rangle}{\mathrm{d}\Omega}\, |F(\mathbf{k})|^2\,, \tag{8.4.28}$$

where the first factor is the intensity of the individual dipole. To describe the directionality of the radiation, some definitions are necessary. So one speaks of an (isotropic) spherical radiator, when $\mathrm{d}\langle P\rangle/\mathrm{d}\Omega = \langle P\rangle/4\pi$. If P is the power of the antenna under consideration, then D is called the directivity, G the gain and η the efficiency:

$$D = \frac{4\pi}{\langle P\rangle} \frac{\mathrm{d}\langle P(\Omega_{\max})\rangle}{\mathrm{d}\Omega}\,, \qquad G = \eta D, \qquad \eta = \frac{R_s}{R_s + R_v}.$$

$\Omega_{\max}$ is the angle of maximum radiation (in the far zone). If the antenna is lossless, i.e. $R_v = 0$, then $\eta = 1$ and $G = D$. Directivity and gain are generally given as logarithms (decibels). The amplitude and phase information of the far field $(r \to \infty)$ is provided by the vector radiation pattern:

$$\mathbf{C}(\Omega) = \frac{\mathbf{E}(r,\Omega)\,\mathrm{e}^{-ikr}}{|\mathbf{E}(r,\Omega_{\max})|} = C_\vartheta(\Omega)\,\mathbf{e}_\vartheta + C_\varphi(\Omega)\,\mathbf{e}_\varphi, \quad C(\Omega) = \frac{|\mathbf{E}(r,\Omega)|}{|\mathbf{E}(r,\Omega_{\max})|}.$$

Instead of the spherical radiator, the Hertzian dipole is also used as a reference object.

Loop antennas

In loop antennas, the wire spans an area (rectangle, circle etc.). We will only discuss the circular current loop here, as outlined in Fig. 8.15. The current

$$\mathbf{j}(\mathbf{x}',t) = I\,\delta(\varrho'-a)\,\delta(z')\,\mathrm{e}^{-i\omega t}\,\mathbf{e}_{\varphi'}$$

flows in a circle of radius a. The magnetic moment of the circular current is

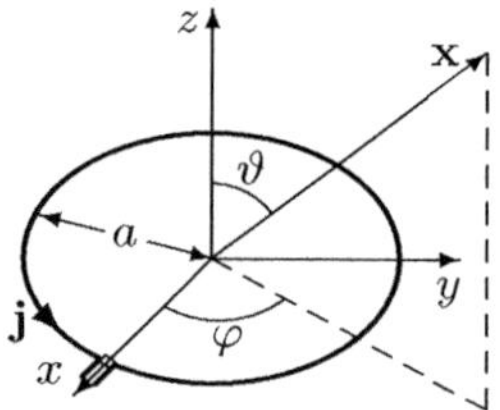

Fig. 8.15. Circular loops or coils are so-called frame antennas; if the entire current $I(t) = I_0\,\mathrm{e}^{-i\omega t}$ within the loop is the same, then it has no electrical dipole moment, and the antenna acts as a magnetic dipole radiator

$\mathbf{m} = (I/c)a^2\pi\mathbf{e}_z$, while the electric dipole moment $\mathbf{p}$ vanishes. In the wave zone, the vector potential is

$$\mathbf{A}(\mathbf{x},t) = \frac{k_C}{ck_L}\frac{\mathrm{e}^{ikr-i\omega t}}{r}\,\mathbf{j}(\mathbf{k}) = \frac{k_C}{ck_L}\frac{\mathrm{e}^{ikr-i\omega t}}{r}\,(-2\pi i a)J_1(ka\sin\vartheta)\,\mathbf{e}_\varphi.$$

J_1 is the Bessel function of the first order (see section B.4.2); the exact calculation is required in problem 8.16. The directional dependence of the radiation power

$$\frac{\mathrm{d}\langle P\rangle}{\mathrm{d}\Omega} = k_C\frac{k^4 c}{8\pi k_L^2}m^2\sin^2\vartheta = \frac{Z_0 k^4}{32\pi^2}\frac{c^2 m^2}{k_L^2}\sin^2\vartheta, \quad \mathbf{m} = \frac{k_L}{c}a^2\pi I\mathbf{e}_z \quad \text{for } ka\ll 1$$

is equal to that of the electric dipole (8.4.18) if m/k_L is replaced by p. For $ka > 1$ additional lobes arise. This spreading of the radiation can be attributed to the increasing proportion of higher multipoles with increasing k.

8.4.3 Magnetic Dipole and Electric Quadrupole Radiation

In the vector potential of a moving charge distribution (8.3.14) is now the 2nd term in the expansion of $\mathrm{e}^{ikr|\mathbf{x}-\mathbf{x}'|}/|\mathbf{x}-\mathbf{x}'|$ according to (8.3.19b)

$$\frac{\mathrm{e}^{ik|\mathbf{x}-\mathbf{x}'|}}{|\mathbf{x}-\mathbf{x}'|} = \frac{\mathrm{e}^{ikr}}{r}\left(1 + \frac{\mathbf{x}'\cdot\mathbf{x}}{r^2}(1-ikr)\right)$$

to consider (the 1st term has already brought the electric dipole radiation)

$$A_i(\mathbf{x}) = \frac{k_C}{ck_L}\frac{\mathrm{e}^{ikr}}{r^3}(1-ikr)\,x_k \int \mathrm{d}^3x'\,x'_k j_i(\mathbf{x}'). \tag{8.4.29}$$

The integral can be evaluated analogously to (4.2.10), whereby once more the auxiliary formula (4.2.9) can be applied ($\boldsymbol{\nabla}\cdot\mathbf{j} = -\dot\rho = i\omega\rho$)[5]:

$$\int \mathrm{d}^3x\,[x_i\,j_k(\mathbf{x}) + x_k\,j_i(\mathbf{x})] = -\int \mathrm{d}^3x\,(\boldsymbol{\nabla}\cdot\mathbf{j})x_i x_k = -i\omega \int \mathrm{d}^3x\,\rho(\mathbf{x})\,x_i x_k\,.$$

This results in

$$A_i(\mathbf{x}) = \frac{k_C \mathrm{e}^{ikr}}{ck_L r^3}(1-ikr)\frac{x_k}{2}\int\mathrm{d}^3x'\left\{x'_k j_i(\mathbf{x}') - x'_i j_k(\mathbf{x}') - i\omega\rho(\mathbf{x}')x'_i x'_k\right\} \tag{8.4.30}$$

two contributions, the first of which is the magnetic dipole radiation and the second is the electric quadrupole radiation. In vector notation, (8.4.30) reads

$$\mathbf{A}(\mathbf{x}) = \frac{k_C \mathrm{e}^{ikr}}{2ck_L r^3}(1-ikr)\int\mathrm{d}^3x'\left\{[\mathbf{j}\circ\mathbf{x}' - \mathbf{x}'\circ\mathbf{j}] - i\omega\rho(\mathbf{x}')\mathbf{x}'\circ\mathbf{x}'\right\}\mathbf{x}. \tag{8.4.31}$$

Now the first term of (8.4.31) is transformed into

$$(\mathbf{j}\circ\mathbf{x}' - \mathbf{x}'\circ\mathbf{j})\mathbf{x} \stackrel{\text{(A.1.16)}}{=} (\mathbf{x}'\times\mathbf{j})\times\mathbf{x}$$

and inserted into the magnetic dipole moment (4.2.1)

$$\mathbf{m} = \frac{k_L}{2c}\int \mathrm{d}^3x'\,\mathbf{x}'\times\mathbf{j}(\mathbf{x}')$$

inserted. We express the last term through the 2nd electric moment M (2.5.8):

$$\mathsf{M} = \int \mathrm{d}^3x'\,\rho(\mathbf{x}')(\mathbf{x}'\circ\mathbf{x}') \qquad \Leftrightarrow \qquad M_{ij} = \int \mathrm{d}^3x'\,\rho(\mathbf{x}')x'_i x'_j.$$

Thus, we obtain the vector potential (8.4.31) separated into a magnetic dipole and an electric quadrupole component

$$\mathbf{A}(\mathbf{x},t) = \frac{k_C}{k_L^2}\frac{\mathrm{e}^{ikr-i\omega t}}{r^2}(1-ikr)\left(\mathbf{m}\times\mathbf{e}_r - \frac{ik_L\omega}{2c}\mathsf{M}\mathbf{e}_r\right). \tag{8.4.32}$$

---

[5] $\int\mathrm{d}^3x\,x_k j_i = \int\mathrm{d}^3x\,x_k\frac{\partial x_i}{\partial x_l}j_l = -\int\mathrm{d}^3x\left[\frac{\partial x_k}{\partial x_l}x_i j_l + x_k x_i\frac{\partial j_l}{\partial x_l}\right] = \frac{1}{2}\int\mathrm{d}^3x\left[x_k j_i - x_i j_k + x_k x_i\dot\rho\right].$

Magnetic dipole radiation

The magnetic dipole component is according to (8.4.32)

$$\mathbf{A}^{(m)}(\mathbf{x},t) = \frac{k_C}{k_L^2}\,\frac{\mathrm{e}^{\mathrm{i}(kr-\omega t)}}{r^2}\,(1-\mathrm{i}kr)\,\mathbf{m}\times\mathbf{e}_r, \qquad \text{SI:}\ \frac{k_C}{k_L^2} = \frac{\mu_0}{4\pi}. \tag{8.4.33}$$

This implies for the electric field

$$\mathbf{E} = -\frac{k_L}{c}\,\dot{\mathbf{A}} = \frac{k_C}{k_L}\,k^2\,\frac{\mathrm{e}^{\mathrm{i}kr-\mathrm{i}\omega t}}{r}\,\Big(1-\frac{1}{\mathrm{i}kr}\Big)\,\mathbf{m}\times\mathbf{e}_r, \tag{8.4.34}$$

where in the wave zone ($kr \gg 1$) only the 1st term contributes. The electric field of the magnetic dipole radiation thus corresponds to the magnetic field of the electric dipole radiation (8.4.6).

The solution for the magnetic dipole is obtained from the electric dipole by replacing there $\mathbf{E} \to k_L\mathbf{B}$, $\mathbf{B} \to -\mathbf{E}/k_L$ and $\mathbf{p} \to \mathbf{m}/k_L$, which is evident from the comparison of (8.4.6) with (8.4.34). In this sense, the magnetic field from (8.4.7) is adopted here

$$\mathbf{B} = \boldsymbol{\nabla}\times\mathbf{A} = \frac{k_C}{k_L^2}\,\frac{\mathrm{e}^{\mathrm{i}kr-\mathrm{i}\omega t}}{r}\left[k^2\mathbf{e}_r\times(\mathbf{m}\times\mathbf{e}_r) + \big(3\mathbf{e}_r(\mathbf{e}_r\cdot\mathbf{m})-\mathbf{m}\big)\frac{1}{r^2}(1-\mathrm{i}kr)\right].$$

The radiation is like that of the electric dipole, i.e. here the near field has the shape of a (magnetic) dipole field. The radiation power has the same form for both types:

$$\langle\mathbf{S}\rangle = \frac{ck_L}{8\pi k_C}\,\mathbf{E}\times\mathbf{B}^* = \frac{k_C}{k_L^2}\,\frac{c\,k^4}{8\pi\,r^2}\,\big(|\mathbf{m}|^2 - |\mathbf{e}_r\cdot\mathbf{m}|^2\big)\,\mathbf{e}_r. \tag{8.4.35}$$

With $\langle\mathbf{S}\rangle$ we can again specify the average radiated power in the solid angle element $\mathrm{d}\Omega$:

$$\frac{\mathrm{d}\langle P\rangle}{\mathrm{d}\Omega} = r^2\langle\mathbf{S}\rangle\cdot\mathbf{e}_r = \frac{k_C}{k_L^2}\,\frac{c\,k^4}{8\pi}\,|\mathbf{m}|^2\,\sin^2\theta, \tag{8.4.36}$$

where θ is the angle enclosed by $\mathbf{m}$ and $\mathbf{e}_r$. The total (average) radiation power is according to (8.4.24)

$$\langle P\rangle = \frac{k_C}{k_L^2}\,\frac{ck^4}{3}|\mathbf{m}|^2 \overset{Z_0=4\pi k_C/c}{=} \frac{Z_0}{12\pi}\,\frac{c^2}{k_L^2}\,k^4|\mathbf{m}|^2. \tag{8.4.37}$$

The electric field $\mathbf{E}$ has different polarization in the wave zone, depending on the observation point $\mathbf{x}$, (see section 10.1). If we start from the electric dipole radiation, the polarization vector of the electric field $\boldsymbol{\epsilon} \sim \mathbf{e}_r \times (\mathbf{p} \times \mathbf{e}_r)$ lies in the plane spanned by $\mathbf{p}$ and $\mathbf{e}_r$. However, in the case of magnetic dipole radiation, the polarization vector $\boldsymbol{\epsilon} \sim \mathbf{m} \times \mathbf{e}_r$ is perpendicular to the plane spanned by $\mathbf{m}$ and $\mathbf{e}_r$. The radiation power of the magnetic dipole radiation

(8.4.37) is equal to that of the electric dipole radiation, if in (8.4.18) the electric dipole moment $\mathbf{p}$ is replaced by the magnetic $\mathbf{m}$. However, in general, the radiation power of the magnetic dipole radiation is significantly weaker than that of the electric dipole radiation.

If you take a circular current $I(t)=Ie^{-i\omega t}$ as the basis for the magnetic moment with the diameter d, then according to (4.2.14) $m = k_L I(t) d^2\pi/4c$. A linear antenna of length d with the current I has the dipole moment $p = Id/2kc$ according to (8.4.22). It is therefore $m/k_L p = k_L(\pi/2)\, dk$. As long as $d\omega \ll c$, the electric dipole radiation is stronger.

Electric quadrupole radiation

The last term in (8.4.32) is the contribution of the electric quadrupole of a charge distribution to the radiation:

$$\mathbf{A}^{(Q)}(\mathbf{x},t) = -\mathrm{i}\frac{k_C}{k_L}\frac{k}{2}\frac{e^{\mathrm{i}kr-\mathrm{i}\omega t}}{r^2}(1-\mathrm{i}kr)(\mathsf{M}\mathbf{e}_r). \qquad (8.4.38)$$

Now we move on to the traceless quadrupole moments defined in (2.5.9) $3\mathsf{M} = \mathsf{Q} + M\mathsf{E}$ with $M = \mathrm{tr}\,\mathsf{M}$ and the unit tensor E:

$$\mathbf{A}^{(Q)}(\mathbf{x},t) = -\mathrm{i}\frac{k_C}{k_L}\frac{k}{6}\frac{e^{(\mathrm{i}kr-\omega t)}}{r^2}(1-\mathrm{i}kr)\left(\mathsf{Q}\mathbf{e}_r + M\,\mathbf{e}_r\right).$$

M contributes nothing to the fields $\mathbf{B}$ and $\mathbf{E}$, so we omit this term

$$\mathbf{A}^{(Q)}(\mathbf{x}) = -\mathrm{i}\frac{k_C}{k_L}\frac{k}{6}\,f(r)\,\mathsf{Q}\,\mathbf{x} \qquad \text{with} \qquad f(r) = \frac{e^{\mathrm{i}kr}}{r^3}(1-\mathrm{i}kr). \qquad (8.4.39)$$

For $\mathbf{B}$, taking into account $\boldsymbol{\nabla}\times\mathsf{Q}\mathbf{x} = 0$

$$\mathbf{B}(\mathbf{x}) = -\mathrm{i}\frac{k_C}{k_L}\frac{k}{6}\frac{f'(r)}{r}(\mathbf{x}\times\mathsf{Q}\,\mathbf{x}) \qquad \text{with} \qquad \frac{f'(r)}{r} = \frac{e^{\mathrm{i}kr}}{r^5}\left(k^2r^2+3\mathrm{i}kr-3\right).$$

For $\mathbf{E}$, we get $(\boldsymbol{\nabla}\cdot\mathsf{Q}\mathbf{x} = 0)$

$$\mathbf{E}(\mathbf{x}) = \frac{k_C}{k_L}\frac{\mathrm{i}k_L}{k}\boldsymbol{\nabla}\times\mathbf{B} = k_C\frac{1}{6}\,\tilde{f}(r)\,\mathbf{x}\times(\mathbf{x}\times\mathsf{Q}\,\mathbf{x}) - \frac{1}{2}\frac{f'(r)}{r}\mathsf{Q}\,\mathbf{x},$$

where we have defined

$$\tilde{f}(r) = \frac{1}{r}\frac{\mathrm{d}}{\mathrm{d}r}\frac{f'}{r} = \frac{e^{\mathrm{i}kr}}{r^7}\left(\mathrm{i}k^3r^3 - 6k^2r^2 - 15\mathrm{i}kr + 15\right)$$

The spatial part of the near field $(kr = 0)$

$$\mathbf{E}_n(\mathbf{x}) = k_C\frac{1}{2}\frac{1}{r^7}\left\{5\left[(\mathbf{x}\cdot\mathsf{Q}\,\mathbf{x})\,\mathbf{x} - r^2\,\mathsf{Q}\,\mathbf{x}\right] + r^2\,\mathsf{Q}\,\mathbf{x}\right\}$$

is, unsurprisingly, equal to the electrostatic quadrupole field (2.5.9).

Quadrupole radiation in the far field

It is helpful to write down the vector potential (8.4.38) for the far field as this reveals a similarity to the electric dipole radiation (8.4.5):

$$\mathbf{A}_f^{(Q)}(\mathbf{x},t) = -\frac{k_C}{k_L}\frac{k^2}{6}\frac{e^{ikr-i\omega t}}{r}\,\mathbf{Q}\,\mathbf{e}_r. \tag{8.4.40}$$

If you replace the vector $\mathbf{p}$ in the dipole potential with the vector $-ik(\mathbf{Q}\mathbf{e}_r)$, you get $\mathbf{A}_f^{(Q)}$. The far field of $\mathbf{B}$ is given by the 1st term of f'/r and that of $\mathbf{E}$ by the 1st term of $\tilde{f}(r)$:

$$\mathbf{B}_f(\mathbf{x}) = -i\frac{k_C}{k_L}\frac{k^3}{6}\frac{e^{ikr}}{r}\,(\mathbf{e}_r\times\mathbf{Q}\mathbf{e}_r), \tag{8.4.41}$$

$$\mathbf{E}_f(\mathbf{x}) = ik_C\,\frac{k^3}{6}\frac{e^{ikr}}{r}\,\mathbf{e}_r\times(\mathbf{e}_r\times\mathbf{Q}\mathbf{e}_r). \tag{8.4.42}$$

The energy flow density $\langle\mathbf{S}\rangle$ multiplied with r^2 results in the radiation

$$\begin{aligned}
\frac{\langle dP\rangle}{d\Omega} &= \frac{ck_L}{8\pi k_C}\frac{1}{k_L}k_C^2\frac{k^6}{36}\big[|\mathbf{Q}\mathbf{e}_r|^2 - |\mathbf{e}_r\cdot(\mathbf{Q}\mathbf{e}_r)|^2\big]\\
&= \frac{ck_C}{8\pi}\frac{k^6}{36}(\mathbf{Q}\mathbf{e}_r)^2\,\sin^2\theta.
\end{aligned} \tag{8.4.43}$$

In the 2nd line, a real tensor $\mathbf{Q}$ is assumed. θ is then the angle between the vectors $\mathbf{e}_r$ and $\mathbf{Q}\mathbf{e}_r$. The formulas become simpler when the quadrupole tensor is in principal axis form and is axially symmetric:

$$Q_{ij} = \frac{3}{2}\,Q_0\big(\delta_{iz}\,\delta_{jz} - \frac{1}{3}\delta_{ij}\big). $$

In this definition, $Q_0 = Q_{zz}$, from which

$$\mathbf{E}_f(\mathbf{x}) = ik_C k^3\,\frac{Q_0}{4}\frac{e^{ikr}}{r}\,\mathbf{e}_r\times(\mathbf{e}_r\times\mathbf{e}_z)\cos\vartheta, \tag{8.4.44}$$

$$\mathbf{B}_f(\mathbf{x}) = -i\frac{k_C}{k_L}k^3\,\frac{Q_0}{4}\frac{e^{ikr}}{r}\,(\mathbf{e}_r\times\mathbf{e}_z)\cos\vartheta \tag{8.4.45}$$

follows. In the wave zone, the time-averaged energy flux density

$$\langle\mathbf{S}\rangle = \frac{ck_L^2}{8\pi k_C}|\mathbf{B}|^2\,\mathbf{e}_r = \frac{ck_C}{8\pi}\frac{k^6 Q_0^2}{r^2 16}\,\sin^2\vartheta\,\cos^2\vartheta\,\mathbf{e}_r \tag{8.4.46}$$

and the resulting radiation power

$$\frac{d\langle P\rangle}{d\Omega} = r^2\langle\mathbf{S}\rangle\cdot\mathbf{e}_r = \frac{ck_C}{8\pi}\frac{k^6 Q_0^2}{16}\,\sin^2\vartheta\,\cos^2\vartheta \tag{8.4.47}$$

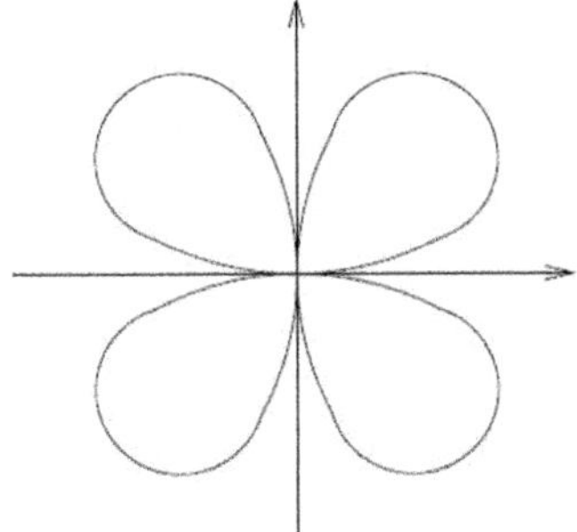

Fig. 8.16. Angular distribution of the quadrupole radiation for an axially symmetric quadrupole in the xz-plane

have the angular distribution typical for a quadrupole, sketched in Fig. 8.16. The total radiation power is obtained by integrating over the solid angle ($\xi = \cos\vartheta$):

$$\langle P\rangle = \frac{ck_C}{8\pi}\frac{k^6Q_0^2}{16}\int_0^{2\pi}\mathrm{d}\varphi\int_{-1}^1\mathrm{d}\xi(1-\xi^2)\xi^2 = \frac{ck_C}{4}\frac{k^6Q_0^2}{60} = \frac{c^2Z_0k^6Q_0^2}{960\pi}. \quad (8.4.48)$$

In atoms, the electric quadrupole radiation is much weaker than the dipole radiation and can generally be neglected. As a simple model, we take a charge e (electron), which moves on a circular path of radius a with the frequency ω, as sketched in Fig. 8.17a. The charge $-e$ rests in the center, so that the atom is electrically neutral. We therefore have a rotating dipole moment $\mathbf{p}(t) = ea(\mathbf{e}_x + \mathrm{i}\mathbf{e}_y)\mathrm{e}^{-\mathrm{i}\omega t}$ and can specify the corresponding average radiated energy (8.4.19) $P = 2k_Cce^2a^2k^4/3$.

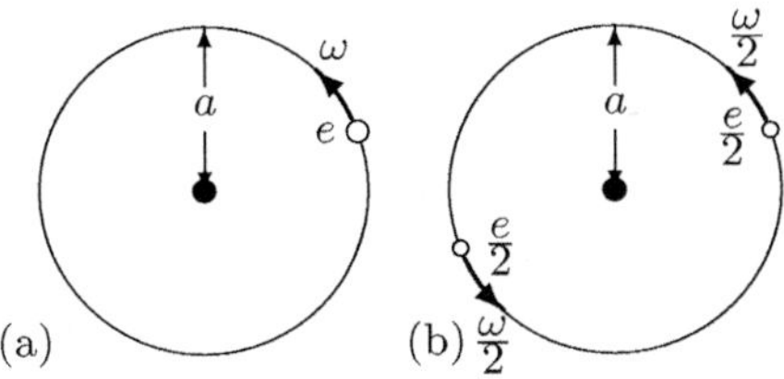

Fig. 8.17. (a) When a charge e moves on a circular path, it has a rotating dipole moment.
(b) If the charge is divided into two points always opposite each other on the circular path, it has a rotating Quadrupole moment

If 2 charges $e/2$ always move opposite each other on a circle, as sketched in Fig. 8.17b, they form a quadrupole of the form Fig. 2.5b, which rotates in the xy-plane. The calculation of the radiation is given in the problem 8.20. One obtains

$$\frac{\mathrm{d}\langle P\rangle}{\mathrm{d}\Omega} = \frac{ck_C}{8\pi}\frac{e^2a^4k^6}{16}\sin^2\vartheta(1+\cos^2\vartheta), \qquad \langle P\rangle = ck_C\frac{e^2a^4k^6}{40}. \quad (8.4.49)$$

The ratio

$$\langle P\rangle/\langle P^{\mathrm{Dipol}}\rangle = 3a^2k^2/80$$

tells us that in atoms the quadrupole radiation can be neglected, unless the wavelengths $\lambda < 2\pi a$ of the radiation are of the size of the circumference - and these are at most a few Å. For atomic transitions with wavelengths $\sim 10^2$ to 10^3 Å, the quadrupole radiation is only noticeable when the corresponding dipole transition is forbidden.

8.4.4 Polarization Potentials

Finally, a brief introduction to an alternative method for calculating radiation fields will be given. H. Hertz introduced another vector potential in 1889, the so-called *Hertz vector* (see for example Stratton [1941, §1.11], Simonyi [1973, §4.6] or Born, Wolf [1986, §2.2.2]). In the description of the charges $\rho(\mathbf{x},t)$ and $\mathbf{j}(\mathbf{x},t)$ we refer to the material equations (5.2.17). Charge and current density (5.2.2) and (5.2.3) we represent in formal analogy to the (bound) currents and charges in the medium by their polarization densities:

$$\rho(\mathbf{x},t) = -\boldsymbol{\nabla}\cdot\mathbf{P}_e(\mathbf{x},t) \qquad \mathbf{j}(\mathbf{x},t) = \dot{\mathbf{P}}_e(\mathbf{x},t) + \frac{c}{k_L}\boldsymbol{\nabla}\times\mathbf{M}_e(\mathbf{x},t). \qquad (8.4.50)$$

The index e denotes the external polarizations, where $\mathbf{P}_e$ can be assumed to be curl-less. It is easy to convince oneself that (8.4.50) automatically satisfies the continuity equation:

$$\dot{\rho}(\mathbf{x},t) + \boldsymbol{\nabla}\cdot\mathbf{j}(\mathbf{x},t) = -\boldsymbol{\nabla}\cdot\dot{\mathbf{P}}_e(\mathbf{x},t) + \boldsymbol{\nabla}\cdot\dot{\mathbf{P}}_e(\mathbf{x},t) = 0\,.$$

We now define *polarization potentials*, the so-called *Hertz vectors*[6] according to (8.2.14)

$$\mathbf{Z}_q(\mathbf{x},t) = k_C \int \mathrm{d}^3x' \frac{\mathbf{P}_e(\mathbf{x}',t_r)}{|\mathbf{x}-\mathbf{x}'|}, \qquad\qquad (8.4.51)$$

$$\mathbf{Z}_m(\mathbf{x},t) = \frac{k_C}{k_L^2} \int \mathrm{d}^3x' \frac{\mathbf{M}_e(\mathbf{x}',t_r)}{|\mathbf{x}-\mathbf{x}'|}, \qquad\qquad t_r = t - \frac{|\mathbf{x}-\mathbf{x}'|}{c},$$

where t_r is the retarded time (8.2.14). These vector potentials are causal solutions of the wave equations

$$\Box\mathbf{Z}_q(\mathbf{x},t) = 4\pi k_C\mathbf{P}_e(\mathbf{x},t), \qquad \Box\mathbf{Z}_m(\mathbf{x},t) = \frac{4\pi k_C}{k_L^2}\mathbf{M}_e(\mathbf{x},t). \qquad (8.4.52)$$

We verify that ϕ and $\mathbf{A}$ are given by the $\mathbf{Z}_q$ and $\mathbf{Z}_m$:

$$\phi(\mathbf{x},t) = -\boldsymbol{\nabla}\cdot\mathbf{Z}_q(\mathbf{x},t) \qquad\Rightarrow\qquad \Box\phi = -4\pi k_C\boldsymbol{\nabla}\cdot\mathbf{P}_e = 4\pi k_C\rho, \quad (8.4.53)$$

$$\mathbf{A}(\mathbf{x},t) = \frac{\dot{\mathbf{Z}}_q(\mathbf{x},t)}{ck_L} + \boldsymbol{\nabla}\times\mathbf{Z}_m(\mathbf{x},t) \;\Rightarrow\; \Box\mathbf{A} = \frac{4\pi k_C}{ck_L}\left(\dot{\mathbf{P}}_e + \frac{c}{k_L}\boldsymbol{\nabla}\times\mathbf{M}_e\right) = \frac{4\pi k_C}{ck_L}\mathbf{j}$$

---

[6] Jackson [1998, §6.13] defines: $\boldsymbol{\Pi}_e = \mathbf{Z}_q\epsilon_0$ and $\boldsymbol{\Pi}_m = \mathbf{Z}_m/\mu_0$.

and with these definitions automatically fulfill the Lorenz gauge:

$$\frac{1}{ck_L}\dot\phi + \boldsymbol{\nabla}\cdot\mathbf{A} = -\frac{1}{ck_L}\,\boldsymbol{\nabla}\cdot\dot{\mathbf{Z}}_q + \frac{1}{ck_L}\,\boldsymbol{\nabla}\cdot\dot{\mathbf{Z}}_q = 0.$$

The electric and magnetic fields are then

$$\mathbf{B}(\mathbf{x},t) = \boldsymbol{\nabla}\times\mathbf{A} = \frac{1}{ck_L}\boldsymbol{\nabla}\times\dot{\mathbf{Z}}_q(\mathbf{x},t) + \boldsymbol{\nabla}\times\big[\boldsymbol{\nabla}\times\mathbf{Z}_m(\mathbf{x},t)\big],$$

$$\mathbf{E}(\mathbf{x},t) = -\boldsymbol{\nabla}\phi - \frac{k_L}{c}\dot{\mathbf{A}} = \boldsymbol{\nabla}(\boldsymbol{\nabla}\cdot\mathbf{Z}_q) - \frac{1}{c^2}\ddot{\mathbf{Z}}_q - \frac{k_L}{c}\boldsymbol{\nabla}\times\dot{\mathbf{Z}}_m, \qquad (8.4.54)$$

$$\mathbf{D}(\mathbf{x},t) = \epsilon_0\mathbf{E} + 4\pi k_r\mathbf{P}_e \overset{(8.4.52)}{=} \epsilon_0\boldsymbol{\nabla}\times(\boldsymbol{\nabla}\times\mathbf{Z}_q) - \epsilon_0\frac{k_L}{c}\boldsymbol{\nabla}\times\dot{\mathbf{Z}}_m(\mathbf{x},t).$$

Gauge transformation

We have already learned at the beginning of this chapter that there are different gauges for ϕ and $\mathbf{A}$, which lead to the same electric and magnetic fields. A comparable freedom in the Hertz vector potentials is given by

$$\mathbf{Z}'_q = \mathbf{Z}_q + \frac{k_L}{c}\boldsymbol{\nabla}\chi + \boldsymbol{\nabla}\times\mathbf{V}, \qquad\qquad \mathbf{Z}'_m = \mathbf{Z}_m - \frac{1}{ck_L}\dot{\mathbf{V}}$$

given, where χ and $\mathbf{V}$ are solutions of the homogeneous wave equation and

$$\phi' = -\boldsymbol{\nabla}\cdot\mathbf{Z}'_q = \phi - \frac{k_L}{c}\Delta\chi, \qquad\qquad \mathbf{A}' = \frac{\dot{\mathbf{Z}}'_q}{ck_L} + \boldsymbol{\nabla}\times\mathbf{Z}'_m = \mathbf{A} + \frac{1}{c^2}\boldsymbol{\nabla}\dot\chi$$

fulfill the Lorenz gauge (8.1.8).

Electric dipole

The rather complex equations for $\mathbf{B}$ and $\mathbf{E}$ become simpler when one calculates the radiation fields for the electric point dipole:

$$\rho(\mathbf{x},t) = -\mathbf{p}(t)\cdot\boldsymbol{\nabla}\delta^{(3)}\big[\mathbf{x}-\mathbf{x}_0(t)\big] \overset{(5.2.6)}{\Longrightarrow} \mathbf{P}_e(\mathbf{x},t) = \mathbf{p}(t)\,\delta^{(3)}\big[\mathbf{x}-\mathbf{x}_0(t)\big],$$

$$\mathbf{Z}_q(\mathbf{x},t) = k_C\int \mathrm{d}^3x\,\frac{\mathbf{p}(t_r)\,\delta^{(3)}\big[\mathbf{x}-\mathbf{x}_0(t_r)\big]}{|\mathbf{x}-\mathbf{x}_0(t_r)|} = k_C\frac{\mathbf{p}(t_r)}{|\mathbf{x}-\mathbf{x}_0(t_r)|}.$$

$\mathbf{Z}_m$ vanishes in this case, so that

$$\mathbf{B}(\mathbf{x},t) = \frac{1}{ck_L}\boldsymbol{\nabla}\times\dot{\mathbf{Z}}_q(\mathbf{x},t) \qquad\text{and}\qquad \mathbf{E}(\mathbf{x},t) = \boldsymbol{\nabla}(\boldsymbol{\nabla}\cdot\mathbf{Z}_q) - \frac{1}{c^2}\ddot{\mathbf{Z}}_q.$$

Since the Hertz vectors satisfy the Lorenz gauge, the potentials and radiation fields calculated with them are consistent in every order.

Magnetic dipole

$$\mathbf{M}_e(\mathbf{x}, t) = \mathbf{m}(t)\, \delta^{(3)}\big[\mathbf{x} - \mathbf{x}_0(t)\big]$$

$$\mathbf{Z}_m(\mathbf{x}, t) = \frac{k_C}{k_L^2} \int d^3x'\, \frac{\mathbf{m}(t_r)\, \delta^{(3)}\big[\mathbf{x}' - \mathbf{x}_0(t_r)\big]}{|\mathbf{x} - \mathbf{x}'|} = \frac{k_C}{k_L^2}\, \frac{\mathbf{m}(t_r)}{|\mathbf{x} - \mathbf{x}_0(t_r)|}.$$

Taking into account $\mathbf{Z}_q(\mathbf{x}, t) = 0$, the radiation fields are

$$\mathbf{B}(\mathbf{x}, t) = \boldsymbol{\nabla} \times \big[\boldsymbol{\nabla} \times \mathbf{Z}_m(\mathbf{x}, t)\big], \qquad \mathbf{E}(\mathbf{x}, t) = -\frac{k_L}{c}\, \boldsymbol{\nabla} \times \dot{\mathbf{Z}}_m(\mathbf{x}, t).$$

8.5 Radiation Reaction

An accelerated charge radiates energy via the electromagnetic fields, which can only occur at the expense of its mechanical energy. For a charged particle of mass m, on which an external force $\mathbf{F}_{\text{ext}}$ acts, the energy radiation by a dissipative force $\mathbf{F}_{\text{rad}}$ must be taken into account in the Newtonian equation of motion:

$$m\dot{\mathbf{v}} = \mathbf{F}_{\text{ext}} + \mathbf{F}_{\text{rad}}, \tag{8.5.1}$$

where $\mathbf{F}_{\text{rad}}$ is determined from the energy balance. Multiplication with $\mathbf{v}$ yields

$$\frac{1}{2}\frac{d}{dt}mv^2 = \mathbf{v} \cdot \left(\mathbf{F}_{\text{ext}} + \mathbf{F}_{\text{rad}}\right).$$

The Larmor formula (8.2.39)

$$P = k_C \frac{2e^2}{3c^3}\, \dot{v}^2 = -\mathbf{v} \cdot \mathbf{F}_{\text{rad}}$$

indicates the radiation energy for a moving point charge in the limit $v \to 0$. (8.4.13) is the corresponding formula for an oscillating charge. The integration yields

$$\int_{t_1}^{t_2} dt\, \mathbf{v} \cdot \mathbf{F}_{\text{rad}} = -k_C \frac{2e^2}{3c^3} \int_{t_1}^{t_2} \dot{v}^2 = -k_C \frac{2e^2}{3c^3} \left(\mathbf{v} \cdot \dot{\mathbf{v}}\Big|_{t_1}^{t_2} - \int_{t_1}^{t_2} dt\, \ddot{\mathbf{v}} \cdot \mathbf{v}\right).$$

If the boundary term disappears, as is the case when $\mathbf{v} \perp \dot{\mathbf{v}}$, then the self-force due to radiation reaction (*Abraham-Lorentz-force*)

$$\mathbf{F}_{\text{rad}} = k_C \frac{2e^2}{3c^3}\, \ddot{\mathbf{v}} = m\tau_0\, \ddot{\mathbf{v}}, \tag{8.5.2}$$

$$\tau_0 = k_C \frac{2}{3c}\frac{e^2}{mc^2} \overset{m=m_e}{=} \frac{2r_e}{3c} \approx 6 \times 10^{-24}\,\text{sec}. \tag{8.5.3}$$

(8.5.2) is approximately valid when, as with limited or even periodic motion, the boundary term only vanishes on average over time. τ_0 is the (very short) time it takes for light to travel 2/3 of the classical electron radius r_e. For particles other than electrons, τ_0 is correspondingly smaller according to the mass ratio. If we now substitute (8.5.2) into (8.5.1), then we obtain the *Abraham-Lorentz equation of motion*

$$m\dot{\mathbf{v}} = \mathbf{F}_{\text{ext}} + \tau_0 m \ddot{\mathbf{v}}, \tag{8.5.4}$$

which is referred to by Rohrlich [1997, (2.3)], where $\mathbf{F}_{\text{ext}}$ is the Lorentz force (1.2.5), as the *Lorentz equation*. We use this name for the undamped case (5.4.6). Without external force, the (homogeneous) equation of motion has, in addition to the force-free motion with constant speed $\mathbf{v} = \mathbf{v}_0$, even when $\mathbf{v}(0) = 0$, exponentially increasing solutions:

$$\dot{\mathbf{v}} = \dot{\mathbf{v}}_0\, e^{t/\tau_0} \qquad\qquad \Rightarrow \qquad\qquad \mathbf{v} = \tau_0 \dot{\mathbf{v}}_0\, e^{t/\tau_0}\,,$$

so-called "runaway solutions"(see problem 8.22). These solutions are incompatible with the assumption that

$$\mathbf{v} \cdot \dot{\mathbf{v}} = \tau_0\, \dot{v}_0^2\, e^{2t/\tau_0}$$

vanishes on average over time and are therefore omitted. Let's take another look at the energy balance of the Abraham-Lorentz equation of motion (8.5.4), by multiplying it with $\mathbf{v}$:

$$\frac{\mathrm{d}}{\mathrm{d}t}\frac{mv^2}{2} = \mathbf{F}_{\text{ext}} \cdot \mathbf{v} + m\tau_0\big[\mathbf{v}\cdot\ddot{\mathbf{v}} + \dot{v}^2 - \dot{v}^2\big] \tag{8.5.5}$$

$$= \mathbf{F}_{\text{ext}} \cdot \mathbf{v} + \tau_0\frac{\mathrm{d}^2}{\mathrm{d}t^2}\frac{mv^2}{2} - P, \qquad P = k_c\frac{2e^2}{3c^3}\dot{v}^2 = \tau_0 m\dot{v}^2\,.$$

The second term on the right side is the so-called *Schott term*. This takes into account internal energy rates, which can be both negative and positive. $P(t)$ is the radiation power (8.2.39) according to the Larmor formula.

Lorentz model for an electron bound in an atom

An electron in an atom is harmonically bound and therefore fulfills the conditions for (8.5.2), which lead to (8.5.4)

$$m_e\big(\ddot{\mathbf{x}} - \tau_0\dddot{\mathbf{x}} + \omega_0^2\,\mathbf{x}\big) = \mathbf{F}_{\text{ext}}(t)\,. \tag{8.5.6}$$

We examine the solutions of the homogeneous equation. With the Eulerian approach

$$\mathbf{x} = \mathbf{x}_0\, e^{-\alpha t}$$

we obtain

$$\tau_0\alpha^3 = -(\alpha^2 + \omega_0^2)\,.$$

Every cubic equation with real coefficients has a real solution, which here must be negative. $\alpha < 0$ but means an exponentially growing solution (*runaway solution*), which in turn is not consistent with the assumptions. The other two (here conjugate complex) solutions can be determined in good approximation by expanding the coefficients α with respect to τ_0 for $\omega_0 \tau_0 \ll 1$ up to the 1st order

$$\alpha = \alpha_0 + \tau_0 \alpha_1 \qquad \Rightarrow \qquad \tau_0 \alpha_0^3 = -(\alpha_0^2 + \omega^2) - 2\tau_0 \alpha_0 \alpha_1 + O(\tau_0^2).$$

One obtains by comparing coefficients $\alpha_0 = \pm i\omega_0$ and $\alpha_1 = -\alpha_0^2/2$ and thus the solution

$$\mathbf{x}(t) = \mathbf{x}_0\, e^{\pm i\omega_0 t - (\tau_0 \omega_0^2/2)t} \qquad \Rightarrow \qquad \dddot{\mathbf{x}} = -\omega_0^2 \dot{\mathbf{x}} + O(\tau_0). \tag{8.5.7}$$

$\mathbf{x}(t)$ is the solution of the damped harmonic oscillator, whose equation of motion one obtains, when one substitutes for $\dddot{\mathbf{x}}$ in (8.5.6)

$$m_e\big(\ddot{\mathbf{x}} + \tau_0 \omega_0^2\, \dot{\mathbf{x}} + \omega_0^2 \mathbf{x}\big) = \mathbf{F}_{\text{ext}}(t)\,. \tag{8.5.8}$$

The radiation reaction in a general motion remains unanswered here.

Note: We have de facto treated the radiation reaction in a dissipative, not closed system. The produced radiation disappears and never comes back into the system; the fields are not dynamic variables of this system, but we have only considered their energy.

8.5.1 More general approach to radiation reaction in the non-relativistic case

The failure to calculate the radiation reaction for a general motion is primarily due to the point-like charge with the associated diverging electrostatic self-energy. If one takes into account the finite charge distribution, differential-difference equations occur.

We now follow Rohrlich [2008] to obtain a differential equation satisfying the 2nd Newton's law. If the charge distribution for the external force is to act like a point charge, the force may only vary slightly over the extent of the charge distribution:

$$|\tau_0 \dot{\mathbf{F}}_{\text{ext}}(t)| \ll |\mathbf{F}_{\text{ext}}(t)|. \tag{8.5.9}$$

If the external force satisfies this condition, the radiation loss $m\dot{\mathbf{v}} \approx \dot{\mathbf{F}}_{\text{ext}}$ is a small correction to the external force. If one now inserts this expression into the Abraham-Lorentz's equation of motion (8.5.4), one obtains in 1st order in τ_0

$$m\dot{\mathbf{v}} = \mathbf{F}_{\text{ext}}(t) + \tau_0 \dot{\mathbf{F}}_{\text{ext}}(t)\,. \tag{8.5.10}$$

This equation has no runaway solutions if, with the forces, as far as these satisfy (8.5.9), the acceleration of the particle also asymptotically vanishes. We still determine the energy balance by multiplying (8.5.10) with $\mathbf{v}$

$$\frac{\mathrm{d}}{\mathrm{d}t}\frac{mv^2}{2} = \mathbf{F}_{\text{ext}} \cdot \mathbf{v} + \tau_0 \dot{\mathbf{F}}_{\text{ext}} \cdot \mathbf{v} = \mathbf{F}_{\text{ext}} \cdot \mathbf{v} + \tau_0 \frac{\mathrm{d}}{\mathrm{d}t}\dot{\mathbf{F}}_{\text{ext}} \cdot \mathbf{v} - P',$$

$$P' = \tau_0 \mathbf{F}_{\text{ext}} \cdot \dot{\mathbf{v}} \approx m\tau_0 \mathbf{v} \cdot \dot{\mathbf{v}}. \tag{8.5.11}$$

Within the linear approximation in τ_0, the radiation reaction is again given by the Larmor formula.

8.5.2 Finite Charge Distribution

We now assume that the particle, an electron, has a mass m_0 independent of the charge. In addition, the electron has a rigid charge distribution ρ of finite extent with the fields $\mathbf{E}_s$ and $\mathbf{B}_s$. $\mathbf{F}_s$ is the Lorentz force exerted by the fields $\mathbf{E}_s$ and $\mathbf{B}_s$ on ρ. In the Newtonian equation of motion for the electron, the external force $\mathbf{F}_{\text{ext}}$ must now be added to the self-force $\mathbf{F}_s$:

$$m_0 \frac{\mathrm{d}\mathbf{v}}{\mathrm{d}t} = \mathbf{F}_{\text{ext}} + \int \mathrm{d}^3x\, \rho\big(\mathbf{x} - \mathbf{s}(t)\big) \Big[\mathbf{E}_s(\mathbf{x},t) + \frac{k_L \mathbf{v}(t)}{c} \times \mathbf{B}_s(\mathbf{x},t)\Big]. \tag{8.5.12}$$

$\mathbf{s}(t)$ is the location of the electron and $\mathbf{v}(t) = \dot{\mathbf{s}}(t)$ its velocity. If one takes a spherical shell for ρ and neglects all non-linear terms, then one obtains (problem 8.23)

$$m_0 \dot{\mathbf{v}}(t) = \ \mathbf{F}_{\text{ext}}(t) + k_C \frac{e^2}{3a^2 c}\Big[\mathbf{v}\big(t - \frac{2a}{c}\big) - \mathbf{v}(t)\Big] \tag{8.5.13}$$

$$\overset{(8.5.3)}{=} \ \mathbf{F}_{\text{ext}}(t) + \frac{m_e c^2 \tau_0}{2a^2}\Big[\mathbf{v}\big(t - \frac{2a}{c}\big) - \mathbf{v}(t)\Big].$$

One now makes a Taylor expansion of $\mathbf{v}(t - \frac{2a}{c})$ and neglects the terms $O(a^3)$, since these vanish for $a \to 0$:

$$\Big[m_0 + \frac{m_e c \tau_0}{a}\Big]\dot{\mathbf{v}}(t) = \mathbf{F}_{\text{ext}}(t) + m_e \tau_0 \ddot{\mathbf{v}}(t). \tag{8.5.14}$$

The 2nd term on the left side is referred to as electromagnetic mass

$$m_{\text{em}} = m_e \frac{c\tau_0}{a} \overset{(8.5.3)}{=} k_C \frac{2e^2}{3ac^2}. \tag{8.5.15}$$

The sum $m_0 + m_{\text{em}}$ is interpreted as the physical mass (rest mass) m_e; thus (8.5.14) is the Abraham-Lorentz equation of motion (8.5.4).

If one replaces in (8.5.13) m_0 by $m_0 = m_e - m_{\text{em}}$, one obtains without external forces

$$\big(1 - \frac{c\tau_0}{a}\big)\dot{\mathbf{v}}(t) = \frac{\tau_0 c^2}{2a^2}\Big[\mathbf{v}\big(t - \frac{2a}{c}\big) - \mathbf{v}(t)\Big]. \tag{8.5.16}$$

This equation has runaway solutions only as long as a is smaller than $c\tau_0 = (2/3)r_e$. Then $m_0 < 0$ (and the Hamilton function is not positively definite; Moniz, Sharp [1977]). This makes it clear that the increasing electrostatic self-energy in point-like charge distributions is responsible for the acausal behavior. With increasing concentration of the charge, one also expects stronger repulsion forces within the electron. Already Poincaré tried to achieve stability with the help of so-called *Poincaré tensions*. In the context of electrostatics, we have already assigned the mass $m_{es} = e^2/2ac^2$ to the electrostatic self-energy of a spherical shell (2.4.10) using the Einstein formula (14.1.2), which differs from the electromagnetic mass $m_{em} = (4/3)m_{es}$. Neither the question of how to solve this inequality of electromagnetic to electrostatic mass, the so-called (4/3)-problem, nor how the electrostatic repulsion can be balanced by internal binding forces, will be addressed here [Yaghjian, 2006]. The radiation reaction, which was only treated here in the limit of small velocities, will be revisited on p. 523 for finite velocities in covariant form.

Problems for Chapter 8

8.1. *Lorenz gauge of the Liénard-Wiechert potentials*: Show explicitly that the Liénard-Wiechert potentials (8.2.16) fulfill the Lorenz condition (8.1.8).

8.2. *Fields of the moving point charge*: Starting from the potentials of a moving point charge $\mathbf{A}$ and ϕ, specify the radiation fields in the far zone ($r \gg s$).

8.3. *Cherenkov radiation*: A point particle with charge q moves at a uniform speed $\mathbf{v}$ in a homogeneous isotropic linear medium (ϵ_r, μ frequency-independent).

1. Determine ϕ_q and $\mathbf{E}$ as functions of t especially in the range $1 \le \bar{\beta} = n\beta < n$.
2. Determine, i.e. also sketch, again for $1 \le \bar{\beta} < n$, the area where $\mathbf{E}$ does not vanish and indicate the direction of maximum intensity on.

8.4. *Coulomb–Helmholtz*: Show that the Helmholtz potential (8.2.53) resulting from the decomposition theorem is equal to $\mathbf{A}^C$ assuming that sources/vortices are local i.e., that $\mathbf{A}$ for $r \to \infty$ does not decrease less than $1/r$.

8.5. *On the quasi-static potential*: Show that $\mathbf{A}^{qs}(\mathbf{x}, t) = \dfrac{k_C}{rck_L}\dot{\mathbf{p}}$ for $r \gg d$, if d characterizes the area to which charges and currents are limited. $\mathbf{A}^{qs}$ is the potential of a point charge for which you should calculate $\mathbf{j}(\mathbf{x})$. Then show that $\mathbf{j}_t(\mathbf{x}) = -\mathbf{j}_l(\mathbf{x})$ for $r > 0$ and $\mathbf{j}_l(\mathbf{k}) \cdot \mathbf{j}_t(\mathbf{k}) = 0$.

8.6. *Causality in Coulomb gauge*: Show directly that the acausal terms of $\mathbf{E}$ in (8.2.62) cancel out.

8.7. *Calculation of the Fourier components $\mathbf{A}_h^b(\mathbf{x})$, $\phi_h^e(\mathbf{x})$ and $\mathbf{A}_h^e(\mathbf{x})$ using the decomposition theorem:* The starting point is fields that do not decrease less than $1/r$; but due to retardation the derivatives can also decrease with $1/r$.

8.8. *Electric dipole field*: Calculate the electric dipole field (8.4.7) using $\mathbf{E} = -\boldsymbol{\nabla}\phi - k_L\dot{\mathbf{A}}/c$.

8.9. *Radiation of a rotating electron*

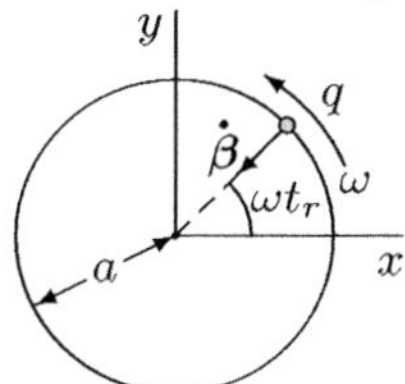

An electron moves on a circular path in the xy plane with the radius a and the frequency ω. Calculate the radiation powers $\langle\frac{\mathrm{d}P}{\mathrm{d}\Omega}\rangle$ and P, where $\langle...\rangle$ denotes time averaging. Neglect $\dot{\beta}$: $\dot{\beta}=0$.

8.10. *Liénard formula, part 1.* The calculation of the radiation power of a moving point charge (8.2.42) we do in two steps. First, we calculate with the fields $\mathbf{E}_f$ and $\mathbf{B}_f$ the radiation power $\partial P/\partial\cos\vartheta$ by integration of (8.2.36) over the azimuth angle φ.

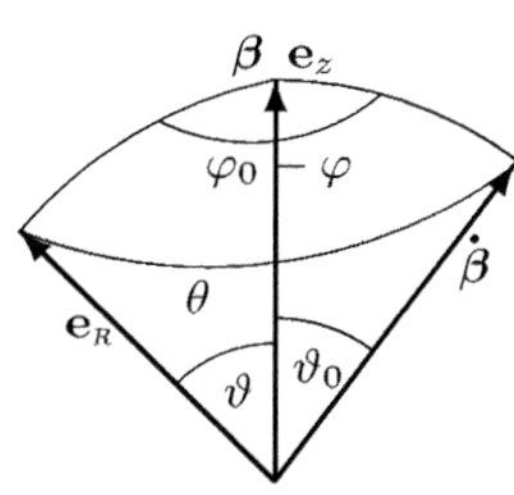

Positions of the vectors $\mathbf{e}_R=\mathbf{X}/R$, $\boldsymbol{\beta}$ and $\dot{\boldsymbol{\beta}}$. As sketched in the adjacent figure, $\boldsymbol{\beta}$ lies in the z-axis. The angle ϑ_0 between $\boldsymbol{\beta}$ and $\dot{\boldsymbol{\beta}}$ is given.

Hint: After performing the φ-integration, you should obtain:

$$\frac{\partial P}{\partial\cos\vartheta}=\frac{q^2\dot{\beta}^2}{2cJ^5}\left\{-(1-\beta^2)\left[\cos^2\vartheta_0\cos^2\vartheta+\frac{1}{2}\sin^2\vartheta_0\sin^2\vartheta\right]\right.$$
$$\left.+2J\,\cos^2\vartheta_0\,\beta\cos\vartheta+J^2\right\}. \tag{8.5.17}$$

8.11. *Liénard's formula, part 2.* Now verify, starting from $\partial P/\partial\cos\vartheta$ (8.5.17) the Liénard's formula (8.2.42).

Hints: $J=1-\mathbf{e}_R\cdot\boldsymbol{\beta}=1-\beta\xi$. Integrals of the form appear:

$$I_{n+1}=\int_{-1}^{1}\frac{\mathrm{d}\xi}{(1-\beta\xi)^{n+1}}=\frac{2}{n(1-\beta^2)^n}\sum_{k=0}^{\lfloor\frac{n-1}{2}\rfloor}\binom{n}{2k+1}\beta^{2k}.$$

You obtain $P=\dfrac{2k_cq^2\dot{\beta}^2}{3c}\gamma^6\left(1-\beta^2\sin^2\vartheta_0\right)$.

8.12. *Frequency spectrum of a point charge on a circular path:* A particle with charge q moves on a circular path in the xy-plane with radius a and frequency ω. $\mathbf{A}(t)=\mathbf{A}(t+T)$ is a periodic function with $T=2\pi/\omega$ and therefore can be developed into a Fourier series [see Panofsky & Phillips, 1975, chap. 20-4]

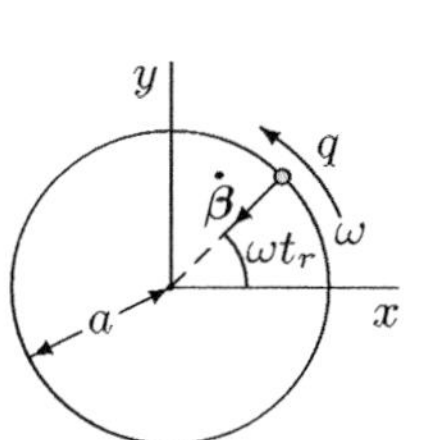

$$\mathbf{A}=\sum_{n=-\infty}^{\infty}(\psi_n,\mathbf{A})\,\psi_n \qquad\text{with}\qquad \psi_n(t)=\mathrm{e}^{-\mathrm{i}\omega nt},$$

$$\mathbf{A}^n=(\psi_n,\mathbf{A})=\frac{1}{T}\int_0^T\mathrm{d}t\,\psi_n^*(t)\,\mathbf{A}.$$

You get a line spectrum with the frequencies $n\omega$, where n is the order of the harmonic frequency.

1. Calculate the Fourier coefficients $\mathbf{A}^n$ for $r \to \infty$.

   Hints: You only know $\mathbf{A}$ as a function $\mathbf{A}(\mathbf{X}(t_r), \boldsymbol{\beta}(t_r))$, i.e. you have to make the corresponding transformation from t to t_r. Integral representation of the integer Bessel function of the first kind and recursion relations:

$$J_n(n\zeta) = \frac{1}{2\pi} \int_{-\pi}^{\pi} \mathrm{d}t\, e^{-int + in\zeta \sin t}, \qquad\qquad \zeta = ka\sin\vartheta \quad \text{with } k = \frac{\omega}{c},$$

$$\frac{\mathrm{d}J_n(n\zeta)}{\mathrm{d}n\zeta} = \frac{1}{2}\big[J_{n-1}(n\zeta) - J_{n+1}(n\zeta)\big], \quad J_n(n\zeta) = \frac{\zeta}{2}\big[J_{n-1}(n\zeta) + J_{n+1}(n\zeta)\big],$$

$$\mathbf{A}^n = \frac{k_C}{k_L} q\beta e^{in(\varphi - \frac{\pi}{2})} \frac{e^{inkr}}{r}\Big[\mathbf{e}_\varrho \frac{1}{\zeta} + i\mathbf{e}_\varphi \frac{\mathrm{d}}{\mathrm{d}n\zeta}\Big] J_n(n\zeta) \qquad \text{(result)}.$$

2. Calculate the asymptotic fields $\mathbf{B}^n$ and (time-averaged) $\langle \frac{\mathrm{d}P^n}{\mathrm{d}\Omega}\rangle$.

8.13. *Point dipole*: Given is a (Hertzian) point dipole $\mathbf{p}(t) = \mathbf{p}_0\, e^{-i\omega t}$.

1. Specify $\rho(\mathbf{x}, t)$ and $\mathbf{j}(\mathbf{x}, t)$ for the point dipole.
   Hint: You can determine $\mathbf{j}$ using the continuity equation.
2. Calculate the retarded potentials $\phi(\mathbf{x}, t)$ and $\mathbf{A}(\mathbf{x}, t)$ in Lorenz gauge.
3. Calculate $\mathbf{E}$ and $\mathbf{B}$ for the near and far zone.
4. Determine the average radiated power $\dfrac{\mathrm{d}\langle P\rangle}{\mathrm{d}\Omega}$ and $\langle P\rangle$ for the cases
   a) $\mathbf{p}$ lies in the z-axis
   b) $\mathbf{p}$ rotates in the xy plane.
      Note: You can define $\mathbf{p}$ complex (analogous to circular polarization); orient yourself to (10.2.1)

8.14. *Vector potential in Coulomb gauge*: Calculate the vector potential $\mathbf{A}^{(C)}(\mathbf{x}, t)$ of an oscillating point dipole.

8.15. *Decomposition of the vector field of electric dipole radiation*: Given is the vector field (8.4.7)

$$\mathbf{v}(\mathbf{x}) = \mathbf{E}(\mathbf{x}) = k_C \frac{e^{ikr}}{r}\left\{k^2 \mathbf{e}_r \times (\mathbf{p} \times \mathbf{e}_r) + \frac{1}{r^2}(1 - ikr)\big[3(\mathbf{p}\cdot\mathbf{e}_r)\mathbf{e}_r - \mathbf{p}\big]\right\}. \tag{8.4.7}$$

Determine the following quantities using the Helmholtz decomposition theorem:

1. Calculate the sources ρ_H and vortices $\mathbf{j}_H$ of the vector field $\mathbf{v}$ and establish the connection to the density ρ and the magnetic field $\mathbf{B}$.
2. Calculate the potentials $\phi_H(\mathbf{x})$ and $\mathbf{A}_H(\mathbf{x})$. Show that $\phi_H(\mathbf{x})$ is the quasi-static potential in Coulomb gauge and express $\mathbf{A}_H$ through $\mathbf{B}$.
3. Now calculate $\mathbf{v}_l$ and $\mathbf{v}_t$ using the potentials.

Hint: For the calculation of $\mathbf{A}_H$, you can find the integral (B.5.24) in the appendix:

8.16. *Magnetic dipole radiation*:

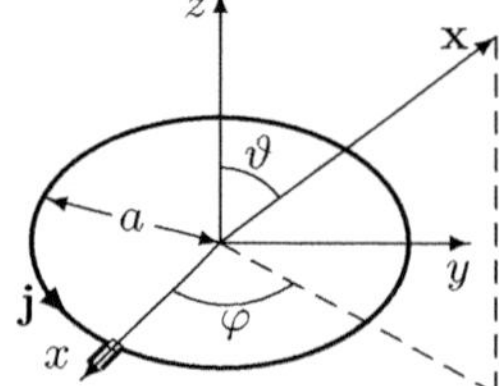

The circular loop or the coil are so-called *frame antennas*. In a circle of radius a the current flows

$$\mathbf{j}(\mathbf{x}', t) = I\,\delta(\varrho' - a)\,\delta(z')\,e^{-i\omega t}\,\mathbf{e}_{\varphi'}\,.$$

1. Calculate the far fields with the help of

$$J_n(n\zeta) = \frac{1}{2\pi} \int_{-\pi}^{\pi} dt\, e^{-int+in\zeta \sin t}, \qquad\qquad \zeta = ka\sin\vartheta \quad \text{with } k = \frac{\omega}{c},$$

$$J_1(\zeta) \approx \zeta/2, \qquad\qquad\qquad \zeta \ll 1,$$

$$J_1(\zeta) \approx \sqrt{2\pi/\zeta}\, \cos(\zeta - \frac{3\pi}{4}), \qquad \zeta \gg \pi.$$

2. Calculate $\langle \frac{dP}{d\Omega} \rangle$ and $\langle P \rangle$.

   Hint: The following integral is given by Schwinger [1945]

$$\int_0^\pi d\vartheta\, \sin\vartheta\, J_1^2(ka\sin\vartheta) = \frac{1}{ka} \int_0^{2ka} dx\, J_2(x),$$

$$\int_0^\infty dx\, J_n(x) = 1, \qquad\qquad J_2(x) \approx \frac{x^2}{4} \qquad\qquad \text{for } x \ll 1.$$

3. Show that there is no electric multipole radiation and give the contributions $\langle \frac{dP}{d\Omega} \rangle$ and $\langle P \rangle$ of the magnetic dipole radiation.

8.17. *Loop antenna:* Given again is the wire loop with the radius a, as in problem 8.16 outlined. The electric and magnetic Dipole radiation are to be determined in the wave zone for the following currents

1. $\mathbf{j}(\mathbf{x},t) = I \sin\varphi\, \delta(\varrho - a)\, \delta(z)\, e^{-i\omega t}\, \mathbf{e}_\varphi$,
2. $\mathbf{j}(\mathbf{x},t) = I \sin\frac{\varphi}{2}\, \delta(\varrho - a)\, \delta(z)\, e^{-i\omega t}\, \mathbf{e}_\varphi$.

Hint: In the 1st case, only the electric dipole radiation occurs, in the 2nd case both electric as well as magnetic, with the radiation powers add up.

8.18. *Dipole line/Dipole group:* n equal and synchronous antennas (Dipoles) form a linear chain with the lattice constant a. The dipoles point in the z-direction. Try to justify, why the distance is usually given with $a = \lambda/2$ and sketch the radiation characteristic, taking $n = 5$ dipoles.

1. The linear chain is oriented in the z-direction (Dipole line).
2. Now the $n = 5$ dipoles (also oriented in z-direction) are lined up along the x-axis (dipole group).

8.19. *Radiation of a rotationally symmetric charge:* The charge and current distribution of a sphere of radius R is rotationally symmetric. Show that the sphere does not radiate.

8.20. *Quadrupole radiation of two rotating charges:* Two negative electric charges $-e_0/2$ move on a circular path with the radius a and the frequency $\frac{\omega}{2}$ in the mathematically positive sense. The point charges are always opposite, i.e. they form with the positive core charge e a straight line. Calculate

1. the retarded vector potential in the far zone,
2. the radiation power $dP/d\Omega$ and the total radiated power.

8.21. *Hertz vectors:* Show that by inserting the Hertz vectors $\mathbf{Z}_q$ and $\mathbf{Z}_m$ (8.4.51) into

$$\phi(\mathbf{x},t) = -k_C \int d^3x' \left. \frac{\boldsymbol{\nabla}'\cdot\mathbf{P}_e(\mathbf{x}',t')}{|\mathbf{x}-\mathbf{x}'|}\right|_{t'=t_r}, \qquad t_r = t - \frac{|\mathbf{x}-\mathbf{x}'|}{c},$$

$$\mathbf{A}(\mathbf{x},t) = \frac{k_C}{ck_L} \int d^3x' \left. \frac{\dot{\mathbf{P}}_e(\mathbf{x}',t') + (c/k_L)\boldsymbol{\nabla}'\times\mathbf{M}_e(\mathbf{x}',t')}{|\mathbf{x}-\mathbf{x}'|}\right|_{t'=t_r}$$

directly the equations $\phi = -\boldsymbol{\nabla}\cdot\mathbf{Z}_q$ and $\mathbf{A} = \dot{\mathbf{Z}}_q/ck_L + \boldsymbol{\nabla}\times\mathbf{Z}_m$ are obtained, while these in (8.4.53) were verified by applying the d'Alembert operator.

8.22. *Runaway solution of the Abraham-Lorentz equation of motion*: We restrict ourselves to the solution of (8.5.4) in one dimension.

1. For uncharged particles, the acceleration at a discontinuity of the force $F = F_0\theta(t - t_0)$ is also discontinuous (but v is continuous). Show that for charged particles also $\dot{v}$ is continuous, as long as the force does not contain a δ-function .
2. Now assume that F_0 is switched off at time $t_1 > t_0$. Determine the general solution for $\dot{v}$.
3. If you choose the initial condition so that $\dot{v}(t_1) = 0$, the acceleration does not increase after the force is switched off. Also specify $v(-\infty) = 0$ and determine $\dot{v}(t)$ and $v(t)$.

8.23. *Radiation Reaction of the Spherical Shell*: Calculate the radiation reaction for an electron, when $v \ll c$ and the charge distribution is a spherically symmetric spherical shell, i.e. start from (8.5.12) and verify (8.5.13).
Hint: Neglect terms of higher order in $\mathbf{s}, \dot{\mathbf{s}}$ etc., to obtain a linear differential equation.

8.24. *Momentum Change due to Radiation*: The electron oscillates around a fixed point in space, i.e. $\mathbf{v}$ can be set to zero. Show that the momentum (sum of mechanical and field momentum) does not change due to "radiation".

References

Born M., Wolf E. *Principles of Optics*, 6th ed. Pergamon Press, Oxford (1986)

Heaviside O. *The Waste of Energy from a Moving Electron*, Nature **6**, 6 (1902)

Jackson J. D. *Classical Electrodynamics*, 3rd ed., John Wiley & Sons Inc. (1998)

Kragh H. *Ludvig Lorenz, Electromagnetism, and the Theory of Telephone Currents* arXiv:1606.00205v1 [physics.hist-ph], 1–16 (2016)

Liénard J. A. *Champ Électrique et Magnétique* in *L'Éclairage Électrique* **16**, 5–14 (1898)

Lorenz L. V. *Ueber die Identität der Schwingungen des Lichts mit den elektrischen Strömen*, Ann. Phys. Chem. **131**, 243–263 (1867)

Moniz E. J. and Sharp D. H., Phys. Rev. D **15**,2850 (1977)

Panofsky W. & Phillips M. *Classical Electricity and Magnetism*, Addison-Wesley Publishing Company, Reading, Massachusetts (1975)

Rindler W. *Relativity: special, general, and cosmological*, 2nd ed., Oxford University Press (2006)

Rohrlich F. *The dynamics of a charged sphere and the electron*, Am. J. Phys. **65**, 1051-1056 (1997)

Rohrlich F. *Dynamics of a charged particle*, Phys. Rev. E **77**, 046609 (2008)

Schwinger J. *On Radiation by Electrons in a Betatron*, transcribed by M. A. Furman, Accelerator and Fusion Research Division, Berkeley (1996), doi.org/10.2172/1195620

Sexl U. R., Urbantke H. K. *Relativity, Groups, Particles* Springer Wien (2001)

Stratton J. A. *Electromagnetic Theory*, McGraw-Hill Inc., New York (1941)

Simonyi K. *Theoretische Elektrotechnik*, 8th ed., VEB Deutscher Verlag der Wissenschaften, Berlin (1980)

Sommerfeld A. *Electrodynamics*, Academic Press Inc., New York (1952)

Wiechert E. *Elektrodynamische Elementargesetze* Archives Néerlandaises, Série II, **5**, 549–573 (1900). Lecture on January 7, 1897

Yaghjian A. D. *Relativistic Dynamics of a Charged Sphere*, Lecture Notes in Physics m11, 2nd ed. Springer Berlin (2006)

9

Quasi-stationary Currents

9.1 The Quasi-stationary Approximation

In the systems considered here, the electrical conduction is limited to wires, which should be thin, as then $\mathbf{E}$ and $\mathbf{j} = \sigma\mathbf{E}$ are largely homogeneous. Within a conductor stretch without branching, resistances and inductances can be summarized. The conductive stretches may also be interrupted by capacitors, as will be further explained.

Our system thus consists of resistances, inductances, capacitors and voltage sources, the components of an electrical network and its branching points. Instead of exact knowledge of the geometry of the system, it is sufficient to know the topology of the network.

The fields in such networks are slowly variable in many applications, which in this context means that within a time τ characteristic for the system

$$c\tau \gg l,$$

where l represents the size of the network. For periodic processes, τ is the oscillation period.

If l no longer meets this condition, as is the case with long lines, by subdividing the system into smaller units, the quasi-stationary approximation can still be applied. This is used in the derivation of the telegraph equation in section 9.2.5.

Under the above condition, the retardation

$$t_r = t - \frac{|\mathbf{x} - \mathbf{x}'|}{c} \approx t \qquad \text{with} \qquad \frac{|\mathbf{x} - \mathbf{x}'|}{c} \sim \frac{l}{c} \ll \tau$$

can be neglected. This part of the quasi-static approximation is achieved by neglecting the derivative of the vortex field $\dot{\mathbf{E}}_{\mathbf{w}} = -\frac{1}{c}\ddot{\mathbf{A}}^{\mathrm{c}} = 0$. The Coulomb

D. Petrascheck, F. Schwabl, *Electrodynamics*, https://doi.org/10.1007/978-3-662-71502-4_9

gauge, section 8.2.3, is used, since in this source and vortex parts are separated and the scalar potential ϕ^{C} is already quasi-static (see (8.2.49)). The retardation and the associated radiation is limited to $\mathbf{A}^{\mathrm{C}}$. With the neglect of the vortex component of the displacement current (8.2.50)

$$\mathbf{A}^{\mathrm{C}}(\mathbf{x},t) \xrightarrow{\ddot{\mathbf{A}}^{\mathrm{c}}=0} \mathbf{A}_t^{\mathrm{qs}}(\mathbf{x},t) = \frac{k_C}{ck_L} \int \mathrm{d}^3 x' \frac{\mathbf{j}_t(\mathbf{x}',t)}{|\mathbf{x}-\mathbf{x}'|}, \qquad \mathbf{\nabla}\cdot\mathbf{A}_t^{\mathrm{qs}} \overset{(8.2.54)}{=} 0 \qquad (9.1.1)$$

the retardation disappears. Thus, there are no longer any long-range fields and no radiation. From here, the quasi-static approximation for the conductor system (resistances and inductances) is to be considered separately from that for capacitors. In the quasi-stationary (quasi-static) approximation when dealing with the capacity in the induction equation $\dot{\mathbf{B}}$ and with the inductance in the Ampère-Maxwell equation $\dot{\mathbf{D}}$ is neglected.

9.1.1 Approximation for the Inductive Part of a Network

In a conductor, within time intervals $\Delta t \ll \tau$ the displacement of charges, i.e. the source component of the displacement current $\mathbf{j}_l = \frac{1}{4\pi}\mathbf{\nabla}\dot{\phi}^{\mathrm{qs}}$ (8.2.50), is negligibly small:

$$\mathbf{\nabla}\cdot\mathbf{j}(\mathbf{x},t) = \mathbf{\nabla}\cdot\mathbf{j}_l \overset{(8.2.50)}{=} -\dot{\rho} = 0.$$

$\mathbf{j}$ is then source-free and the current strength at each point of a unbranched conductor is equal. $\mathbf{j}$ is referred to as *quasi-stationary current*. In (9.1.1) so $\mathbf{j}_t = \mathbf{j} - \mathbf{j}_l$ is to be replaced by $\mathbf{j}$, which gives the quasi-static vector potential (8.2.55). The neglect of the displacement current turns the Ampère-Maxwell law (1.3.15) into Ampère's law of magnetostatics (4.1.1a–ba), where the time is a parameter in the system.

Maxwell's equations

In general, the quasi-static approximation is understood to mean solely the neglect of the displacement current:

$$
\begin{array}{lll}
\text{(a)} & \mathbf{\nabla}\cdot\mathbf{D} = 4\pi k_r \rho_f, & \qquad \text{(b)} \quad \mathbf{\nabla}\times\mathbf{E}+\dfrac{k_L}{c}\dot{\mathbf{B}} = 0, \\[2ex]
\text{(c)} & \mathbf{\nabla}\times\mathbf{H} = \dfrac{k_L}{c}4\pi k_r\mathbf{j} \overset{\mathrm{SI}}{=\!=} \mathbf{j}, & \qquad \text{(d)} \qquad\qquad \mathbf{\nabla}\cdot\mathbf{B} = 0.
\end{array}
\qquad (9.1.2)
$$

So we have replaced the Ampère-Maxwell equation with Ampère's law of magnetostatics. This allows the conductive parts, especially the inductances, to be specified. For homogeneous ϵ_r and μ_r, (8.2.49) and (8.2.55)

$$\phi^{\mathrm{qs}}(\mathbf{x},t) = \frac{k_C}{\epsilon_r} \int \mathrm{d}^3 x' \frac{\rho_f(\mathbf{x}',t)}{|\mathbf{x}-\mathbf{x}'|}, \qquad\qquad (9.1.3)$$

$$\mathbf{A}^{\mathrm{qs}}(\mathbf{x},t) = \frac{k_C\mu_r}{ck_L} \int \mathrm{d}^3 x' \frac{\mathbf{j}(\mathbf{x}',t)}{|\mathbf{x}-\mathbf{x}'|} \qquad \text{with} \qquad \mathbf{\nabla}\cdot\mathbf{A}^{\mathrm{qs}} = 0. \qquad (9.1.4)$$

The change of the Ampère-Maxwell equation implies $\boldsymbol{\nabla}\cdot\mathbf{j}=0$. Consequently, the current I is constant along the entire conductor. Substituted into (9.1.3) one verifies that the Coulomb gauge $\boldsymbol{\nabla}\cdot\mathbf{A}^{\mathrm{qs}}=0$ is automatically fulfilled. The fields have the usual form

$$\mathbf{E}=-\boldsymbol{\nabla}\phi^{\mathrm{qs}}-\frac{k_L}{c}\dot{\mathbf{A}}^{\mathrm{qs}}, \qquad\qquad \mathbf{B}=\boldsymbol{\nabla}\times\mathbf{A}^{\mathrm{qs}}. \tag{9.1.5}$$

9.1.2 Approximation for the Capacitive Part of the Network

When charging and discharging the capacitor, a magnetic field is created according to the law of induction. The quasi-static approximation for the capacitor consists in neglecting the vortex component $\mathbf{E}_w$ or equivalently from $\dot{\mathbf{B}}$ in the law of induction:

$$\boldsymbol{\nabla}\times\mathbf{E}=0 \qquad\qquad \Rightarrow \qquad\qquad \mathbf{E}(\mathbf{x},t)=-\boldsymbol{\nabla}\phi^{\mathrm{qs}}(\mathbf{x},t).$$

The current state of the capacitor is according to (2.2.30)

$$Q(t)=CV(t) \tag{9.1.6}$$

determined by electrostatics. From the continuity equation, the discharge current $I(t)=-\dot{Q}(t)$ is obtained, where here $I(t)$ is the current flowing away from the charge $Q>0$ (see Fig. 9.3 on page 332).

Maxwell's equations

By neglecting of induction, i.e. the source-free part of the current, one obtains

$$\begin{array}{llll}\text{(a)} & \boldsymbol{\nabla}\cdot\mathbf{D}=4\pi k_r\rho_f, & \text{(b)} & \boldsymbol{\nabla}\times\mathbf{E}=0, \\[2mm] \text{(c)} & \boldsymbol{\nabla}\times\mathbf{H}=\dfrac{k_L}{c}\big(4\pi k_r\mathbf{j}+\dot{\mathbf{D}}\big), & \text{(d)} & \boldsymbol{\nabla}\cdot\mathbf{B}=0.\end{array} \tag{9.1.7}$$

We can transform the Ampère-Maxwell equation by substituting for the displacement current:

$$\mathbf{j}_d=\frac{1}{4\pi k_r}\dot{\mathbf{D}}=-\mathbf{j}_l \overset{(8.2.50)}{=}-\frac{\epsilon_r}{4\pi k_C}\boldsymbol{\nabla}\dot{\phi}^{\mathrm{C}},$$

where $\phi^{\mathrm{C}}=\phi^{\mathrm{qs}}$. Thus, the vector potential satisfies the vector Poisson equation (8.2.57):

$$\Delta\mathbf{A}_t^{\mathrm{qs}}=-\frac{k_L}{c}4\pi k_r\mu_0\mu_r\big(\mathbf{j}-\mathbf{j}_l\big)=-\frac{k_L}{c}4\pi k_r\mu_0\mu_r\mathbf{j}_t \overset{\mathrm{SI}}{=}-\mu_r\mu_0\mathbf{j}_t \;\text{with}\; \boldsymbol{\nabla}\cdot\mathbf{j}_t=0.$$

Thus, we obtain the potentials

$$\phi^{\mathrm{qs}}(\mathbf{x},t)=\frac{k_C}{\epsilon_r}\int\mathrm{d}^3x'\,\frac{\rho(\mathbf{x}',t)}{|\mathbf{x}-\mathbf{x}'|},$$

$$\mathbf{A}_t^{\mathrm{qs}}(\mathbf{x},t)=\frac{\mu_r k_C}{ck_L}\int\mathrm{d}^3x'\,\frac{\mathbf{j}_t(\mathbf{x}',t)}{|\mathbf{x}-\mathbf{x}'|} \quad\text{with}\quad \boldsymbol{\nabla}\cdot\mathbf{A}_t=0. \tag{9.1.8}$$

For the fields, it follows that we only have a source field

$$\mathbf{E}=-\boldsymbol{\nabla}\phi^{\mathrm{qs}}, \qquad\qquad \mathbf{B}=\boldsymbol{\nabla}\times\mathbf{A}^{\mathrm{qs}}. \tag{9.1.9}$$

9.1.3 Kirchhoff's Rules

In the 2nd Kirchhoff's rule, the loop voltage of a closed circuit is considered, to which naturally also a voltage source belongs, one of the components from which a network consists.

Voltage source

In a galvanic cell (battery), one assumes electrochemical processes that cause the transport of charge carriers q and thus generate an electrostatic field $\mathbf{E}$ that counteracts the transport, as outlined in Fig. 9.1. The force on the charge carriers is described by the *impressed field* $\mathbf{E}^e$. This is typically only different from zero in a small area of the battery. If no current flows, then in the battery $\mathbf{E}^e + \mathbf{E} = 0$. As a result of the impressed voltage V^e, there is an equal large, but oppositely directed, electric voltage [Becker, Sauter, 1973, section 4.3]. The electromotive force $\mathfrak{E}$ is defined as loop voltage

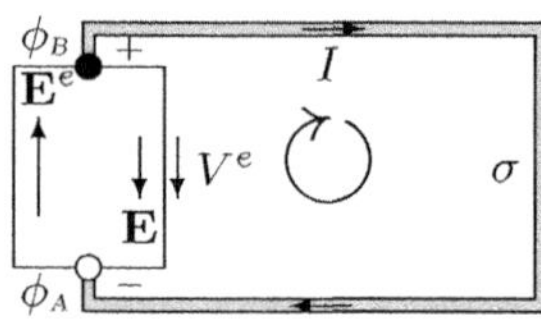

Fig. 9.1. Voltage source (battery) and resistance form a circuit with the terminal voltage $\phi_B - \phi_A$

$$V^e = \oint_C \mathbf{ds} \cdot (\mathbf{E}^e + \mathbf{E}) = \oint_C \mathbf{ds} \cdot \mathbf{E}^e = \mathfrak{E} = \int_A^B \mathbf{ds} \cdot \mathbf{E}^e. \tag{9.1.10}$$

The path C is closed over the connection poles A and B of the battery. As long as no current flows,

$$V^e = \int_A^B \mathbf{ds} \cdot \mathbf{E}^e = -\int_A^B \mathbf{ds} \cdot \mathbf{E} = \phi_B - \phi_A. \tag{9.1.11}$$

If the current $I = jA$ flows in the wire with the conductance σ and the cross-section A, where $\mathbf{j}$ is the current density with $\nabla \cdot \mathbf{j} = 0$, then according to Ohm's law $\mathbf{j} = \sigma \mathbf{E}$ applies:

$$\oint_C \mathbf{ds} \cdot (\mathbf{E}^e + \mathbf{E}) = \int_A^B \mathbf{ds} \cdot \mathbf{E}^e = \oint_C \mathbf{ds} \cdot \frac{jA}{\sigma A} = RI.$$

Here R is the resistance of the circuit. If the voltage source still has the internal resistance R_i, then

$$\int_A^B \mathbf{ds} \cdot (\mathbf{E}^e + \mathbf{E}) = R_i I = V^e - \phi_B + \phi_A.$$

Thus, the voltage at the poles is $V^e - R_i I = \phi_B - \phi_A$.

The source-free impressed field $\mathbf{E}^e$ can certainly be of a nature other than electrochemical. Temperature gradients can cause an electric field, the pressure in piezoelectric crystals or the photoelectric effect in photovoltaics etc. By far the most important example, however, is the induced field in the alternating current generator.

1. Kirchhoff's rule

In quasi-stationary approximation, $\boldsymbol{\nabla} \cdot \mathbf{j} = 0$, from which it follows that the current, even if it changes over time, in an unbranched network is the same at every point in time. We limit ourselves to networks in which the current only flows within wires, as sketched in Fig. 9.2, and in which around the considered areas only ohmic resistances play a role.

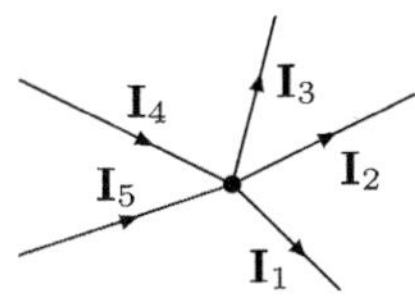

Fig. 9.2. Branching point (node) in a network; according to the 1st Kirchhoff's rule, also known as Kirchhoff's junction rule or Kirchhoff's current law, is $I_1 + I_2 + I_3 = I_4 + I_5$

A direct consequence of $\boldsymbol{\nabla} \cdot \mathbf{j} = 0$ is the *1st Kirchhoff's rule*, which states that at a node of an electrical network, the sum of the incoming currents is equal to the outgoing ones:

$$\sum_{j,\text{incoming}} I_j = \sum_{j,\text{outgoing}} I_j \, . \tag{9.1.12}$$

In a time $t \ll \tau$, the charge $Q_j = I_j t$ has passed a measuring point in conductor j. It must then be

$$\sum_{j,\text{inflowing}} Q_j = \sum_{j,\text{outflowing}} Q_j$$

which is equivalent to the conservation of charge.

Note: The (network) nodes do not have to be simple branching points as sketched in Fig. 9.2, but can include individual components, circuits or parts of a network. In any case, the sum of the currents flowing into the network nodes must be equal to that of the outgoing ones.

2. Kirchhoff's rule

The basis of the *Kirchhoff's 2nd rule* is the law of induction in its integral form

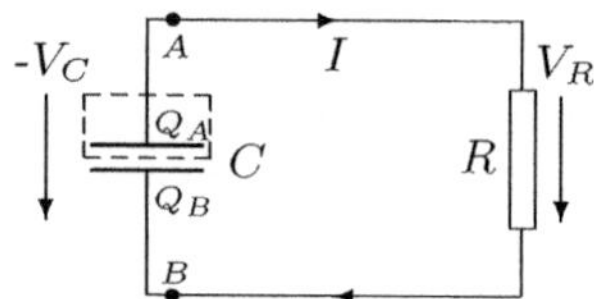

Fig. 9.3. Circuit with capacitor and resistor. The direction of circulation results from the assumption $Q_A = Q > 0$, from which $Q_B = -Q$ follows. Then $V_R > 0$ and $V_C = -V_R < 0$.

$$\oint_C \mathrm{ds} \cdot \mathbf{E} = -\frac{k_L}{c}\,\dot{\Phi}_B\,. \qquad (9.1.13)$$

C is a closed path in the network, as sketched in Fig. 9.3. Now integrate (9.1.13) starting at point A and consider that $\dot{\Phi}_B = 0$

$$V_R + V_C = IR + \frac{Q_B}{C} = 0\,. \qquad (9.1.14)$$

The voltage at the capacitor is, as the arrow indicates, opposite to that at the effective resistance. It is assumed that $Q_A = -Q_B > 0$, so that $I(t)$ has the direction shown in Fig. 9.3. $I(t)$ is obtained from the continuity equation, integrating over the volume V_A dashed in Fig. 9.3:

$$\int_{V_A} \mathrm{d}^3x\,\dot{\rho}_A(\mathbf{x},t) = \dot{Q}_A = -\oiint_{\partial V_A} \mathrm{da}\cdot\mathbf{j} = -I(t) \quad \Rightarrow \quad \dot{Q}_B = I\,. \qquad (9.1.15)$$

The discharge current is determined by differentiating (9.1.14)

$$\dot{I}R + IC = 0 \qquad\qquad \text{with} \qquad\qquad I(t) \propto \mathrm{e}^{-RCt}\,. \qquad (9.1.16)$$

With a longer wire in Fig. 9.3, an inductance L is also connected and thus in (9.1.13) for $\Phi_B = cLI$ (7.2.17) is to be inserted. For the circuit with capacitor, effective resistance and inductance, the result is

$$L\dot{I} + RI + \frac{Q_B}{C} = 0\,. \qquad (9.1.17)$$

Note: The charge of the plate to which the current is supplied (here Q_B) is used to determine the voltage V_C. The current is then given by $I = \dot{Q}_B$, instead of $I = -\dot{Q}$, as one is used to from the continuity equation. This is the usual notation in electrical engineering.

For a circuit with voltage source, resistance, inductance and capacitor (see Fig. 9.5, page 333) applies

$$V_L(t) + V_R(t) + V_C(t) = V^e(t)\,. \qquad (9.1.18)$$

An equation of this kind can be set up in a network for each closed path, i.e. for each loop. It states that the voltages on a closed path add up to zero. Fig. 9.4 shows so-called loops with resistances and a direct current source. For each mesh i, the 2nd Kirchhoff's rule (loop rule) applies

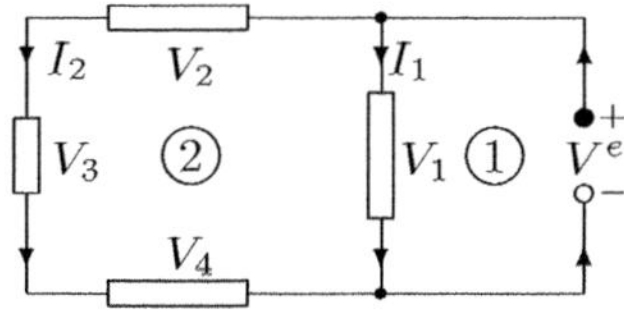

Fig. 9.4. Circuit with two sub-circuits (thus a total of three meshes) and two nodes

$$\sum_j V_j^{(i)} = 0\,,$$

where the signs of $V_j^{(i)}$ are to be chosen according to the direction of the current.

9.2 Oscillation Equation

We consider a circuit, consisting of a resistor R, an inductance L, a capacitance C and an electromotive force (EMF) V^e (see Fig. 9.5). All components of this RLC oscillator are connected in series; they are therefore traversed by the same current $I(t)$. Based on Ohm's law, we will determine the voltages that drop across the individual elements.

In the presence of a battery, a generator, etc., the Ohm's law also includes the current density $\sigma \mathbf{E}^e$ generated by these:

$$\mathbf{j} = \sigma(\mathbf{E} + \mathbf{E}^e). \tag{9.2.1}$$

We integrate over the entire conductor, i.e. from A to B (see Fig. 9.5)

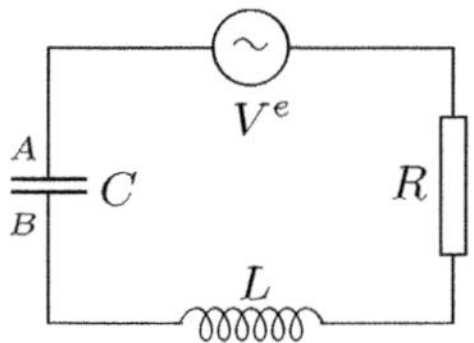

Fig. 9.5. Circuit with EMF, resistance, coil and capacitor

$$\int_A^B \mathrm{ds} \cdot \mathbf{E} + \int_A^B \mathrm{ds} \cdot \mathbf{E}^e = \int_A^B \mathrm{ds} \cdot \frac{\mathbf{j}F}{\sigma F} = RI\,.$$

Resistance: $R = \int_A^B \mathrm{ds}\, \dfrac{1}{\sigma F}$,

impressed electric field: $\mathbf{E}^e$,

electromotive force[1]: $V^e = \int_A^B \mathrm{ds} \cdot \mathbf{E}^e.$

---

[1] also *impressed electric voltage* or *terminal voltage*

From $\quad \mathbf{E} = -\boldsymbol{\nabla}\phi - \dfrac{k_L}{c}\dot{\mathbf{A}}\quad$ follows

$$\int_A^B \mathrm{d}\mathbf{s}\cdot\mathbf{E} = -(\phi_B - \phi_A) - \frac{k_L}{c}\int_A^B \mathrm{d}\mathbf{s}\cdot\dot{\mathbf{A}}$$

$$= -\phi_B + \phi_A - \frac{k_C\mu_r}{c^2}\frac{\partial}{\partial t}\int_A^B \mathrm{d}\mathbf{s}\cdot\int \mathrm{d}^3x'\,\frac{\mathbf{j}(\mathbf{x}',t)}{|\mathbf{x}-\mathbf{x}'|} = -\frac{Q}{C} - L\dot{I}.$$

Here $Q = (\phi_B - \phi_A)C$, where Q is the charge of B and C is the capacitance. As before, we set

$$\mathrm{d}^3x'\,\mathbf{j} = A\,\mathrm{d}\mathbf{s}'\,\mathbf{j} = I\,A\mathrm{d}\mathbf{s}'\,\frac{\mathbf{j}}{A\,j} = I\,\mathrm{d}\mathbf{s}'\,.$$

The self-inductance is thus given by

$$L = \frac{k_C\mu_r}{c^2}\int \mathrm{d}^3x'\int_A^B \mathrm{d}\mathbf{s}\cdot\frac{\mathbf{j}(\mathbf{x}',t)}{I|\mathbf{x}-\mathbf{x}'|} = \frac{k_C\mu_r}{c^2}\iint_A^B \frac{\mathrm{d}\mathbf{s}\cdot\mathrm{d}\mathbf{s}'}{|\mathbf{x}(s) - \mathbf{x}'(s')|}\,.$$

We then obtain using

$$L\dot{I} + RI + \frac{Q}{C} = V^e \tag{9.2.2}$$

the voltages occurring at the coil, resistor and capacitor (terminal voltages). If we differentiate with respect to t and set $I = \dot{Q}$, the following results

$$L\ddot{I} + R\dot{I} + \frac{I}{C} = \dot{V}^e. \tag{9.2.3}$$

We thus find the equation of motion of a damped harmonic oscillator[2], where $\dot{V}^e$ corresponds to the force, L to the mass, R to the damping, C^{-1} to the spring constant and I to the deflection.

9.2.1 Free Oscillations

We are looking for solutions of (9.2.3). In the case of free oscillations, the force $V^e = 0$. One makes the assumption

$$I = I_0\,\mathrm{e}^{-\mathrm{i}\omega_0 t} \qquad \text{and obtains} \qquad -L\omega_0^2 - \mathrm{i}\omega_0 R + \frac{1}{C} = 0\,.$$

From this follows

$$\omega_0 = -\mathrm{i}\frac{R}{2L} \pm \sqrt{\frac{1}{CL} - \frac{R^2}{4L^2}}\,. \tag{9.2.4}$$

---

[2] $\;m\ddot{x} + \gamma\dot{x} + kx = F\quad$ with $\quad\omega_0^2 = k/m$
m is the mass, γ the damping, k the spring constant of the oscillator, on which the external force F acts.

First, we examine the case without damping, in which $R=0$. It is then

$$\omega_0 = \frac{1}{\sqrt{CL}} \qquad \text{and thus} \qquad \tau = 2\pi\sqrt{CL}. \qquad (9.2.5)$$

(9.2.5) is referred to as the *Kirchhoff-Thomson formula*, a special case of (9.2.4). For

$$\frac{R}{2L} \geq \frac{1}{\sqrt{CL}}$$

the discriminant of (9.2.4) is less than zero and thus ω_0 is imaginary and the oscillation is aperiodic, as sketched in Fig. 9.6a. Otherwise, we are dealing with a damped periodic motion (see Fig. 9.6b).

For a small damping, $\omega_0 = -\mathrm{i}\dfrac{R}{2L} \pm \dfrac{1}{\sqrt{CL}}$.

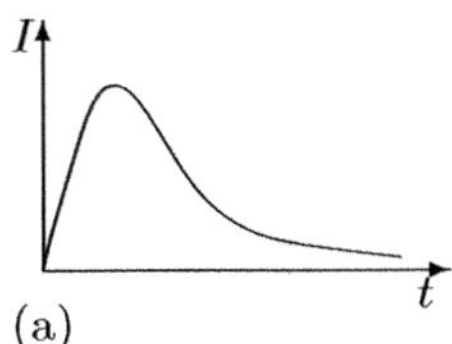
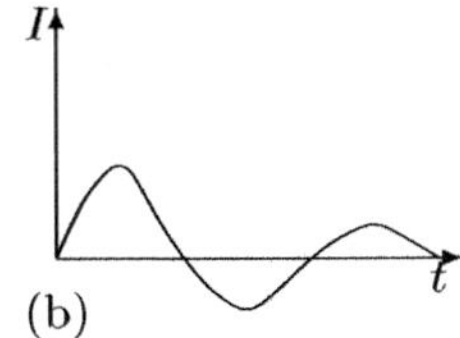

Fig. 9.6. Current I versus t:
(a) aperiodic case
(b) damped oscillation

9.2.2 Forced Oscillations

We now insert the alternating current source into (9.2.3).

$$V^e(t) = V_0^e \mathrm{e}^{-\mathrm{i}\omega t}$$

with the angular frequency ω. With the approach

$$I(t) = I_0\,\mathrm{e}^{-\mathrm{i}\omega t}$$

one obtains from (9.2.3)

$$\left(-L\omega^2 - \mathrm{i}\omega R + \frac{1}{C}\right)I = -\mathrm{i}\omega V^e.$$

We define with

$$Z = R - \mathrm{i}\left(\omega L - \frac{1}{\omega C}\right) = |Z|\,\mathrm{e}^{-\mathrm{i}\varphi} \qquad (9.2.6)$$

the impedance, i.e. the alternating current resistance and thus obtain Ohm's law for alternating current

$$V^e = Z\,I.\tag{9.2.7}$$

The impedance has two components, the *resistance* R and the *reactance*, $\omega L - 1/\omega C$. The relationship between current and voltage has an additional phase

$$\tan\varphi = -\frac{\operatorname{Im}Z}{\operatorname{Re}Z} = \frac{1}{R}\Big(\omega L - \frac{1}{\omega C}\Big),\tag{9.2.8}$$

when capacitors and/or coils are present in the circuit. The magnitude of the impedance

$$|Z| = \sqrt{R^2 + \Big(\omega L - \frac{1}{\omega C}\Big)^2}$$

is the so-called *apparent resistance*. We now apply the real voltage

$$V_r^e(t) = \operatorname{Re}V^e(t) = V_0^e\cos(\omega t)$$

to the circuit and obtain the real current

$$I_r(t) = \operatorname{Re}\left(|Z|^{-1}e^{i\varphi}V_0^e e^{-i\omega t}\right) = \frac{V_0^e\,\cos(\omega t - \varphi)}{\sqrt{R^2 + \Big(\omega L - \frac{1}{\omega C}\Big)^2}}.\tag{9.2.9}$$

The phase shift is $\varphi = 0$, when $\omega = \omega_0 = 1/\sqrt{LC}$. At the same time, $|Z| = R$ at the Thomson frequency ω_0 has its minimum value and according to (9.2.9) the current has its maximum, as sketched in Fig. 9.7.

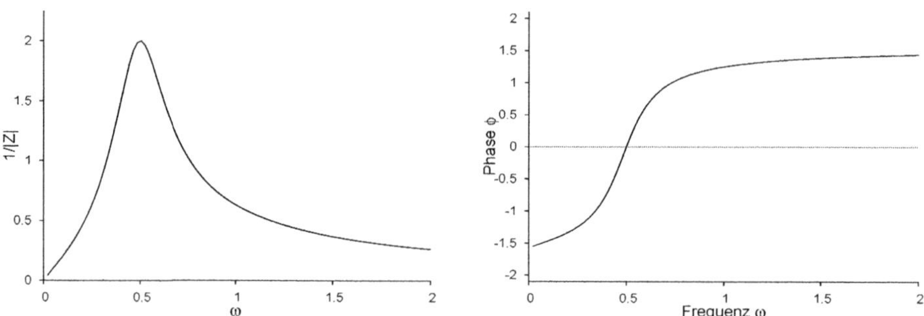

Fig. 9.7. (a) $\frac{1}{|Z|}$ versus ω: Maximum at $\omega = \frac{1}{\sqrt{LC}}$ (b) φ (in sketch ϕ) against ω with zero at $\omega = \frac{1}{\sqrt{LC}}$; used were $R = 1/2$ and $L = C = 2$

$$
\begin{aligned}
\text{For } \omega < \omega_0 \ \text{ and } \varphi < 0 &\qquad \text{current rushes ahead,}\\
\omega > \omega_0 \ \text{ and } \varphi > 0 &\qquad \text{current lags behind,}\\
\omega \to 0: \qquad \varphi = -\pi/2,&\\
\omega \to \infty: \qquad \varphi = \pi/2.&
\end{aligned}
$$

To understand the phase shift, one must consider $I = \dot{Q}$:

$$I = Z^{-1}V^e = -i\omega Q \quad \Rightarrow \quad V^e = -i\omega QZ = -i\omega Q\Big[R - i(\omega L - 1/\omega C)\Big]$$

$$= (Q/C)\Big[-i\omega RC - i\omega^2 LC + 1\Big] = Z\,I.$$

(a) $\omega \ll \omega_0$: At low frequencies, the term with ωL is neglected

$$V^e = \frac{Q}{C}(1 - \mathrm{i}\omega RC) \approx \frac{Q}{C}\,\mathrm{e}^{-\mathrm{i}\omega RC} \quad \text{for } \omega \ll \frac{1}{RC}.$$

As shown in Fig. 9.8, we first have the oscillation set by the current source

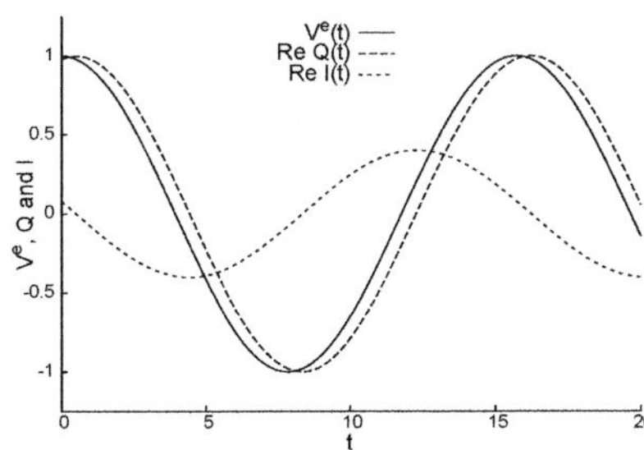

Fig. 9.8. V^e, Q and I versus t for small frequencies ($\omega \ll 1/(RC)$; $V_0^e = C = 1$; $\omega = 0.4$ and $R = 0.5$)

$\cos(\omega t)$. The voltage at the current source is slightly higher than at the capacitor during charging. The current therefore flows to the capacitor. With a delay of ωRC, the charge of the capacitor follows this oscillation. The current I in turn becomes minimal when Q has reached its maximum value, changes its direction and becomes maximum with the discharge of the capacitor when the sign of the charge changes at Q.

$$Q \approx V^e C \mathrm{e}^{\mathrm{i}\omega RC} = V_0^e\, C\, \mathrm{e}^{\mathrm{i}\omega(RC - t)},$$

$$\operatorname{Re} Q = V_0^e C \cos(\omega t - \omega RC),$$

$$\operatorname{Re} I = -\omega V_0^e C \sin(\omega t - \omega RC).$$

(b) At high frequencies $\omega \gg \omega_0$ or $\omega L \gg 1/\omega C$ the ωL-term dominates and the phase shift becomes positive; this is a consequence of Lenz's rule, as the current is changed (amplified) by the applied voltage. This affects the magnetic field, whereby a voltage is induced by the change in the magnetic field, which is opposite to the applied one.

Tab. 9.1. Resistance, capacitance and inductance

Symbol(s)	Designation
$Z =$	Impedance $\big(R + \mathrm{i}(\omega L - 1/\omega C)\big)$
$|Z| =$	Apparent resistance
$\operatorname{Re} Z = R =$	Active resistance (ohmic resistance)
$\operatorname{Im} Z = \omega L - \frac{1}{\omega C} =$	Reactance
$\omega L =$	Inductance (positive reactance)
$1/(\omega C) =$	Capacitance (negative reactance)

9.2.3 Energy Conditions

Multiplying (9.2.2) by I, we get

$$\frac{L}{2}\frac{\mathrm{d}}{\mathrm{d}t}I^2 + RI^2 + \frac{1}{2C}\frac{\mathrm{d}}{\mathrm{d}t}Q^2 = IV^e.$$

The integration over a period $\tau = 2\pi/\omega$ yields

$$\frac{1}{\tau}\int_0^\tau \mathrm{d}t\, RI^2 = \frac{1}{\tau}\int_0^\tau \mathrm{d}t\, IV^e \qquad\Longleftrightarrow\qquad \overline{RI^2} = \overline{IV^e}.$$

Here, the effective current is introduced[3]

$$I_{\text{eff}} = \sqrt{\overline{I^2}} = \sqrt{\frac{1}{\tau}\int_0^\tau \mathrm{d}t\, I^2(t)}\,.$$

The average power

$$\overline{IV^e} = R\,I_{\text{eff}}^2,$$

which must be provided by the EMF, depends only on the active resistance R. Power is defined as

$$\int \mathrm{d}^3x\, \mathbf{j}(\mathbf{x})\cdot \mathbf{E}^e(\mathbf{x}) = \int \mathrm{d}\mathbf{s}\cdot \mathbf{E}^e I = IV^e.$$

It differs from (7.2.2) in the sign and compensates for the losses occurring in the resistors. Inserting Ohm's law (9.2.1) into Joule's heat, we get

$$W_{\text{Joule}} = \int \mathrm{d}^3x\, \mathbf{j}(\mathbf{x})\cdot \left(\mathbf{E}(\mathbf{x}) + \mathbf{E}^e(\mathbf{x})\right) = \int \mathrm{d}^3x\, \frac{\mathbf{j}^2(\mathbf{x})}{\sigma}$$

$$= \int \mathrm{d}s\, A\, \frac{A}{\sigma A}\, \mathbf{j}^2 = I^2 R.$$

The reactive resistance does not enter into Joule's heat. The current is shifted by φ compared to the voltage, which has no influence on

$$\overline{I^2} = \frac{1}{\tau}\int_0^\tau \mathrm{d}t\, I_0^2\, \cos^2(\omega t - \varphi) = \frac{1}{2}I_0^2,$$

so that the effective current and the effective voltage are given by

$$I_{\text{eff}} = \frac{1}{\sqrt{2}}\, I_0 \qquad\text{and}\qquad V_{\text{eff}} = \frac{1}{\sqrt{2}}\, V^e.$$

To determine the angle φ, as sketched in the vector diagram Fig. 9.9, the average power must be determined:

---

[3] also *RMS*-value according to *Root Mean Square*

$$\overline{V^e I} = \frac{1}{\tau} \int_0^\tau dt\, V_0^e I_0\, \cos(\omega t)\, \cos(\omega t - \varphi)$$

$$= \frac{1}{\tau} V^e I_0 \left\{ \int_0^\tau dt\, \cos^2(\omega t)\, \cos(\varphi) + \int_0^\tau dt\, \cos(\omega t)\, \sin(\omega t)\, \sin(\varphi) \right\}$$

$$= \frac{1}{2} V^e I_0 \cos\varphi = I_{\text{eff}}\, V_{\text{eff}}\, \cos\varphi = R I_{\text{eff}}^2 .$$

From this follows

$$R I_{\text{eff}} = V_{\text{eff}} \cos\varphi = \frac{R V^e}{\sqrt{2}\sqrt{R^2 + \left(\omega L - \frac{1}{\omega C}\right)^2}},$$

$$\cos\varphi = \frac{1}{\sqrt{1 + \tan^2\varphi}} = \frac{R}{\sqrt{R^2 + \left(\omega L - \frac{1}{\omega C}\right)^2}}. \tag{9.2.10}$$

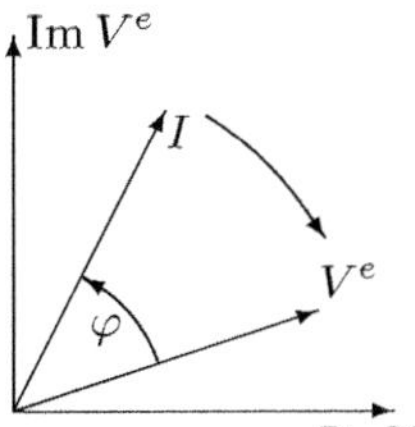

Fig. 9.9. I follows the EMF

Vector diagram: For positive φ, the current I lags behind the EMF by the angle φ. Both "pointers"rotate at the frequency ω in the indicated direction $(e^{-i\omega t})$. The actual current and EMF can be found by projecting onto the real axis. This representation is mainly used in electrical engineering.

9.2.4 Coupled Circuits

We can apply Kirchhoff's 2nd rule (9.1.18) for an arbitrarily composed circuit (loop) in the form

$$\sum_k (\pm R_k I_k) + \sum_j (\pm \frac{Q_j}{C_j}) - \sum_l (\pm V_l^e) = -\frac{k_L}{c} \dot{\Phi}_B .$$

The signs take into account that currents and charges are positive. Fig. 9.10 shows two inductively coupled circuits, assuming that the coupling is weak, i.e. no iron core connects the two coils. The flux through the loop i is according to (7.2.21): $\Phi_{Bi} = c(I_1 L_{i1} + I_2 L_{i2})$. It follows

$$L_{11}\dot{I}_1 + L_{12}\dot{I}_2 + R_1 I_1 + \frac{Q_1}{C_1} = V_1^e \tag{9.2.11}$$

$$L_{12}\dot{I}_1 + L_{22}\dot{I}_2 + R_2 I_2 + \frac{Q_2}{C_2} = V_2^e .$$

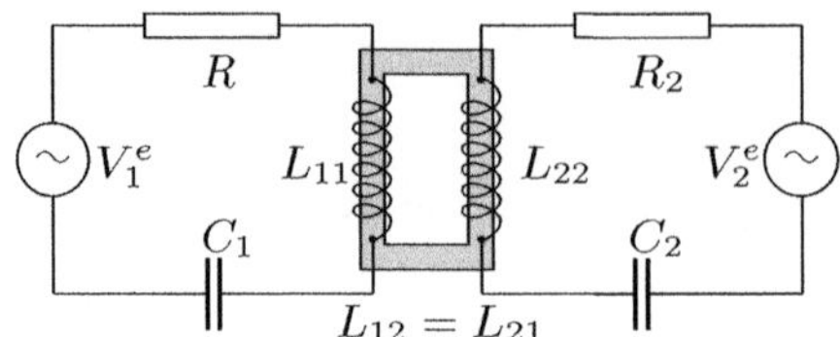

Fig. 9.10. Inductive coupling of two circuits with EMF, resistance, coil and capacitor; the mutual inductances $L_{12} = L_{21}$ are with an iron core (shaded area) correspondingly stronger

In a transformer – and this case is considered – are $V_2^e = 0$ and $V_1^e = V_0 \mathrm{e}^{-\mathrm{i}\omega t}$. Furthermore, the capacitances in both circuits are neglected. One makes the approach $I_k = I_{0k}\, \mathrm{e}^{-\mathrm{i}\omega t + \mathrm{i}\varphi_k}$ for $k = 1, 2$ and gets

$$(R_1 - \mathrm{i}\omega L_{11})I_{01}\, \mathrm{e}^{\mathrm{i}\varphi_1} - \mathrm{i}\omega L_{12}\, I_{02}\, \mathrm{e}^{\mathrm{i}\varphi_2} = V_0$$
$$-\mathrm{i}\omega L_{12}\, I_{01}\, \mathrm{e}^{\mathrm{i}\varphi_1} + (R_2 - \mathrm{i}\omega L_{22})I_{02}\, \mathrm{e}^{\mathrm{i}\varphi_2} = 0 \,.$$

For the currents results

$$I_{01}\, \mathrm{e}^{\mathrm{i}\varphi_1} = V_0\, \frac{R_2 - \mathrm{i}\omega L_{22}}{(R_1 - \mathrm{i}\omega L_{11})(R_2 - \mathrm{i}\omega L_{22}) + \omega^2 L_{12}^2}$$
$$I_{02}\, \mathrm{e}^{\mathrm{i}\varphi_2} = V_0\, \frac{\mathrm{i}\omega L_{12}}{(R_1 - \mathrm{i}\omega L_{11})(R_2 - \mathrm{i}\omega L_{22}) + \omega^2 L_{12}^2} \,.$$

One obtains for the ratio of the two currents

$$\frac{I_{01}}{I_{02}}\, \mathrm{e}^{\mathrm{i}(\varphi_1 - \varphi_2)} = \frac{R_2 - \mathrm{i}\omega L_{22}}{\mathrm{i}\omega L_{12}} \,. \tag{9.2.12}$$

We mainly note here that I_{02} increases linearly with the strength L_{12} of the coupling.

The Wheatstone bridge

The measurement bridge of *Wheatstone* can be used to measure resistances, capacitances, and inductances. The *Maxwell-Wien bridge* sketched in Fig. 9.11 is a variant of a Wheatstone bridge specially designed for the measurement of inductances. In a coil, in addition to the inductance L, there is always the ohmic resistance R of the wire, which is indicated by the equivalent circuit diagram of the series connection of R and L. The resistance R_d and the capacitance C_d, both of which are precisely measurable, are adjusted so that no current flows through the galvanometer. Then

$$I_a = I_b, \qquad I_c = I_d, \qquad I_a Z_a = I_c Z_c \qquad \text{and} \qquad I_b Z_b = I_d Z_d,$$

from which it follows that

$$\frac{I_c}{I_a} = \frac{Z_a}{Z_c} = \frac{Z_b}{Z_d} \,. \tag{9.2.13}$$

Z_d is the impedance of a parallel circuit. If I_d' and I_d'' are the partial currents for which $I_d = I_d' + I_d''$ applies, then according to (9.2.7)

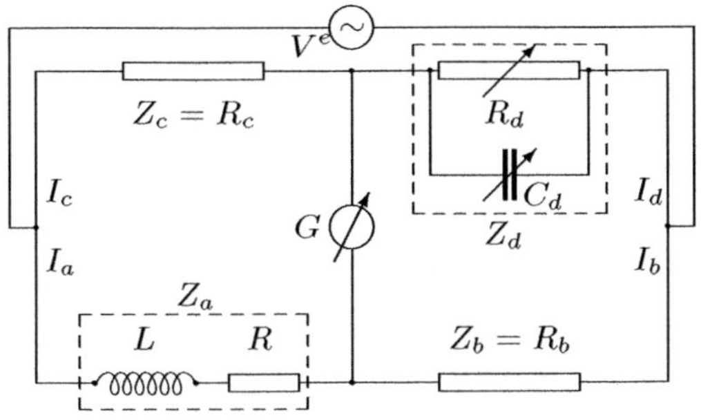

Fig. **9.11.** Maxwell-Wien bridge for the measurement of R and L. R_d and C_d are adjusted until the galvanometer no longer shows any current

$$Z_d I_d = Z_d' I_d' = Z_d'' I_d'' .$$

From this we derive:

$$\frac{1}{Z_d} = \frac{1}{Z_d'} + \frac{1}{Z_d'} = \frac{1}{R_d} - i\omega C_d .$$
(9.2.14)

L and R have impedance in series according to (9.2.6) $Z_a = R - i\omega L$. Thus, one obtains from (9.2.13)

$$Z_a = \frac{R_b R_c}{R_d}\left(1 - i\omega R_d C_d\right) \qquad \Rightarrow \qquad \begin{cases} R &= R_b R_c / R_d \\ L &= R_b R_c C_d . \end{cases}$$
(9.2.15)

Note: For purely ohmic resistances, direct current is sufficient. If capacitance and inductance are arranged diagonally, as in Fig. 9.11, the measurement does not depend on the frequency. The same applies to two inductances when they are next to each other. In a general configuration, however, the frequency enters the measurement.

9.2.5 Telegraph Equation

Very long double lines and/or coaxial lines certainly do not meet the condition necessary for the validity of the quasi-stationary approximation, that their length l is less than $c\tau$, where τ is a characteristic oscillation period. This is particularly true for submarine cables for signal transmission (telegraphy), the laying of which began around 1850. The first calculations, even before Maxwell, go back to *W. Thomson*[4] [Thomson, 1855].

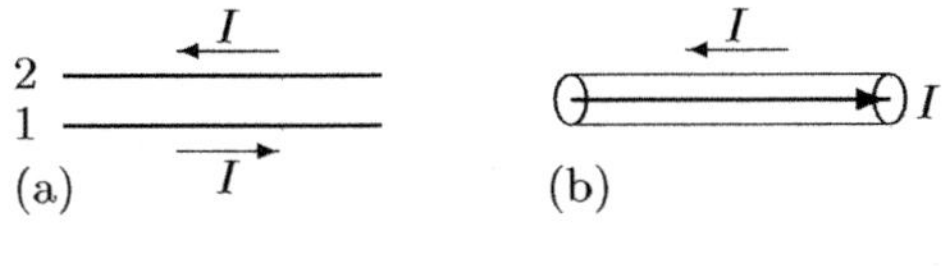

Fig. **9.12.** (a) Double line and (b) Coaxial cable

In Fig. 9.12 a section of a double line is sketched, which as a whole does not fulfill the condition $c\tau > l$ at all. Nevertheless, we can use the considerations

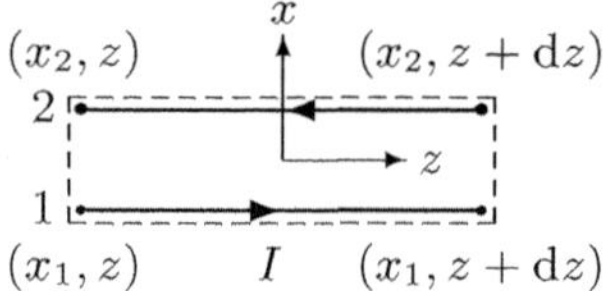

Fig. 9.13. Section $[z\,,\,z+\mathrm{d}z]$ of a double line; integrated along the dashed path

made in the previous section, albeit in differential form. Therefore, as outlined in Fig. 9.13, we consider a small section of a double line and integrate along the dashed path, starting at (x_1, z) using the induction law and the Neumann formula (7.2.19):

$$\oint_{\square} \mathrm{dx}' \cdot \mathbf{E} = \frac{j}{\sigma}\,\mathrm{d}z + \left[\phi_1(x_1, z+\mathrm{d}z) - \phi_2(x_2, z+\mathrm{d}z)\right] + \frac{j}{\sigma}\,\mathrm{d}z$$

$$+ \left[\phi_2(x_2, z) - \phi_1(x_1, z)\right] = -\frac{k_L}{c}\dot{\Phi}_B = -L\dot{I}\,\mathrm{d}z.$$

In the integration over the conducting stretches, we have substituted $\mathbf{E} = \mathbf{j}/\sigma$. There are

$$
\begin{array}{ll}
L & \text{Self-inductance per unit length,}\\
R = 2/(\sigma A) & \text{Resistance of both wires per unit length and}\\
V(z) = \phi_1(x_1, z) - \phi_2(x_2, z) & \text{Potential difference between the wires.}
\end{array}
$$

Thus, after the Taylor expansion of $\phi_{1,2}$, we get

$$RI + \frac{\partial V}{\partial z} + L\dot{I} = 0\,. \tag{9.2.16}$$

The second equation results from the source freedom of the total current:

$$\boldsymbol{\nabla}\cdot\boldsymbol{\nabla}\times\mathbf{H} = 0 = \boldsymbol{\nabla}\cdot\frac{k_L}{c}\frac{1}{k_L^2}\left(4\pi k_r\mathbf{j} + \epsilon_0\dot{\mathbf{E}}\right).$$

Within the conductor 1 we get

$$\frac{\partial I}{\partial z} = -\frac{A}{4\pi k_C}\frac{\partial}{\partial t}\boldsymbol{\nabla}\cdot\mathbf{E} = -A\dot{\rho}(\mathbf{x}) = -C\frac{\partial}{\partial t}V(\mathbf{x}).$$

Here are

$$
\begin{array}{ll}
\lambda = A\,\rho(z,t) & \text{Charge per unit length (line charge) and}\\
C = \lambda/V & \text{Capacitance per unit length.}
\end{array}
$$

$$\frac{\partial I}{\partial z} + C\frac{\partial}{\partial t}V + GV = 0\,. \tag{9.2.17}$$

A loss term G was added here, which comes from the currents through the insulating material and is proportional to the potential difference V.

---

[4] William Thomson, ennobled Lord Kelvin, 1824–1907

The derivation of (9.2.17) with respect to z and the insertion of $\frac{\partial V}{\partial z}$ from (9.2.16) yields the telegraph equation

$$\left\{ LC\frac{\partial^2}{\partial t^2} + (RC + LG)\frac{\partial}{\partial t} + RG - \frac{\partial^2}{\partial z^2} \right\} I = 0 \,. \tag{9.2.18}$$

For $R = G = 0$ we get with

$$\left\{ LC\frac{\partial^2}{\partial t^2} - \frac{\partial^2}{\partial z^2} \right\} I = 0 \tag{9.2.19}$$

the wave equation with the speed

$$v = \frac{1}{\sqrt{LC}} \,. \tag{9.2.20}$$

Note: At first glance, it may seem surprising that (9.2.20) defines a velocity, namely the propagation velocity v of the signal, while the (formally identical) Kirchhoff-Thomson formula (9.2.5) determines the natural frequency ω_0 of the undamped oscillating circuit.

We only remind here that in the case of the double line, unlike with the oscillating circuit, the specified quantities (R, L, C, G) are all defined per unit of length so that v in (9.2.20) has the dimension of a velocity.

(9.2.16) corresponds to the equation derived for the oscillating circuit (9.2.2), but, solely for dimensional reasons, in (9.2.16) the voltage (V^e) is replaced by the voltage change along the cable.

Solutions of the wave equation are of the form

$$I = f(z - vt),$$

where the corresponding value of V in this case is

$$V = Lvf(z - vt) = \sqrt{\frac{L}{C}}\, I. \tag{9.2.21}$$

Voltage and current are therefore in the ratio

$$Z_W = \sqrt{L/C}\,, \tag{9.2.22}$$

which is referred to as *wave resistance*, which is independent of z and t. For finite double lines, it follows from (9.2.21) that if these are terminated at the end by an ohmic resistance the of size of the wave resistance, no discontinuities in current and voltage and thus no reflections occur.

Current and voltage propagate undistorted in the line. With R and the loss term G (9.2.18), additional damping occurs. The question is whether current and voltage can also move undistorted, but damped through the cable. I must therefore satisfy the ansatz

$$I = e^{-\alpha z} f(z - vt). \tag{9.2.23}$$

If you insert into (9.2.18), you get

$$\alpha = R\sqrt{\frac{C}{L}} = \frac{R}{Z_W}\,,$$

but the condition $RC = LG$ must be fulfilled.

(9.2.23) inserted into (9.2.18), results in

$$\left\{\left(LC\,v^2 - 1\right)f'' + \left(2\alpha - v(RC + LG)\right)f' + \left(RG - \alpha^2\right)f\right\}e^{-\alpha z} = 0.$$

The prefactor of f'' vanishes identically, while the prefactor of f for $\alpha = \sqrt{RG}$ is equal to zero. One convinces oneself that then the prefactor of f'

$$\frac{1}{\sqrt{LC}}\left(2\sqrt{RGLC} - RC - LG\right) = \frac{-1}{\sqrt{LC}}\left(\sqrt{RC} - \sqrt{LG}\right)^2$$

only vanishes for $RC = LG$.

9.3 Magnetohydrodynamics

In magnetohydrodynamics (MHD) the hydrodynamics of plasmas, which are ionized fluids (gases, liquids), are described.

The question may arise why MHD is located here, near electrical networks. The similarities with the networks of the previous sections are not too obvious: In networks, the currents are confined to wires, here they exist throughout the plasma, which does not make a big difference. But above all, the temporal changes in both systems are slow, compared to their expansion. This allows the application of the quasi-stationary approximation in both, and that is the reason for the placement of MHD in this chapter.

9.3.1 The Basic Equations

Hydrodynamics: A fluid, whose velocity field $\mathbf{v}(\mathbf{x}, t)$ is, has the mass density $\rho_m(\mathbf{x}, t)$, the shear viscosity ζ and the extensional viscosity η. The continuity equation is used to describe the dynamics

$$\frac{\partial \rho_m}{\partial t} + \boldsymbol{\nabla} \cdot (\mathbf{v}\,\rho_m) = 0 \qquad \overset{\text{incompressible fluid}}{\Longrightarrow} \qquad \boldsymbol{\nabla} \cdot \mathbf{v} = 0 \qquad (9.3.1)$$

and the *Navier-Stokes equation* [Landau, Lifshitz 6, 1987, §15]

$$\rho_m\frac{\mathrm{d}\mathbf{v}}{\mathrm{d}t} = \rho_m\left[\frac{\partial \mathbf{v}}{\partial t} + (\mathbf{v}\cdot\boldsymbol{\nabla})\,\mathbf{v}\right] = -\boldsymbol{\nabla}p + \eta\Delta\mathbf{v} + \left(\zeta + \frac{\eta}{3}\right)\boldsymbol{\nabla}\left(\boldsymbol{\nabla}\cdot\mathbf{v}\right) + \mathbf{f}. \quad (9.3.2)$$

p is the pressure of the fluid for which there is a thermal equation of state not further defined here[5] $p = p(\rho_m, T)$, and $\mathbf{f}$ is a volume force such as gravity and/or the Lorentz force, with only the latter being considered here. If one neglects the internal friction in the fluid, i.e. the viscosities $\eta = \zeta = 0$, this results in the *Euler's equation*

$$\rho_m \left[\frac{\partial \mathbf{v}}{\partial t} + (\mathbf{v} \cdot \boldsymbol{\nabla}) \mathbf{v}\right] = -\boldsymbol{\nabla} p + \rho \mathbf{E} + \frac{k_L}{c} \mathbf{j} \times \mathbf{B}, \tag{9.3.3}$$

where $\mathbf{E}$ and $\mathbf{B}$ are the fields of the moving charges.

Electrodynamics: The magnetic field is determined using the induction equation (1.3.10), for which some assumptions are made:

1. Electrical neutrality: There are an equal number of positive and negative charges. Then the charge density ρ' and the electric field $\mathbf{E}'$ in the rest system of the fluid disappear; simplifying $\epsilon_r = \mu_r = 1$ is assumed.
2. Non-relativistic approximation: It is always $\beta = v/c \ll 1$, from which it follows that $\gamma = 1/\sqrt{1 - \beta^2} \approx 1$.
3. Quasi-stationary approximation: Let τ be a time characteristic for the development of the system and l the extension of the system, then $l \ll c\tau$. Then there is no retardation and the displacement current $\mathbf{j}_d = \dot{\mathbf{E}}/4\pi k_c$ disappears (see section 9.1, page 327).
4. The validity of Ohm's law $\mathbf{j}' = \sigma \mathbf{E}'$ in the rest system of the fluid is assumed, or $\mathbf{j} = \sigma(\mathbf{E} + k_L \boldsymbol{\beta} \times \mathbf{B})$.

In the rest system S' of the plasma, which moves relative to the laboratory system S with $\mathbf{v}$, due to charge neutrality, the averaged charge density $\rho' = 0$. Thus, there is only a magnetic field $\mathbf{B}'$, but no electric field $\mathbf{E}'$.

To calculate the fields in S, we draw the corresponding results of the STR (13.1.29) and (13.1.30) for $\beta \ll 1$:

$$\mathbf{E} = \mathbf{E}' - k_L \boldsymbol{\beta} \times \mathbf{B}', \qquad\qquad \mathbf{B} = \mathbf{B}' + k_L \boldsymbol{\beta} \times \mathbf{E}' \approx \mathbf{B}'. \tag{9.3.4}$$

Furthermore, Ohm's law should apply in the plasma. If you go from the rest system of the fluid S' to the laboratory system S, it follows

$$\mathbf{j}' = \sigma \mathbf{E}' \qquad \overset{(9.3.4)}{\Longrightarrow} \qquad \mathbf{j} = \sigma\big(\mathbf{E} + k_L \boldsymbol{\beta} \times \mathbf{B}\big). \tag{9.3.5}$$

Given the assumed high conductivity of the plasma ($\sigma \to \infty$: ideal MHD) it follows

$$\mathbf{E} = -k_L \boldsymbol{\beta} \times \mathbf{B}.$$

Inserted into the induction equation, by eliminating $\mathbf{E}$, you get

$$\frac{\partial \mathbf{B}}{\partial t} = -\frac{c}{k_L} \boldsymbol{\nabla} \times \mathbf{E} = -\frac{c}{k_L} \boldsymbol{\nabla} \times \left[\frac{\mathbf{j}}{\sigma} - k_L \boldsymbol{\beta} \times \mathbf{B}\right].$$

---

[5] For the ideal gas, $p = n k_B T = \rho_m k_B T / m$.

$\mathbf{j}$ is taken from the Ampére's law (4.1.1a–b)

$$\boldsymbol{\nabla}\times\mathbf{j} = \frac{ck_L}{4\pi k_C}\boldsymbol{\nabla}\times(\boldsymbol{\nabla}\times\mathbf{B}) = -\frac{ck_L}{4\pi k_C}\Delta\mathbf{B},$$

from which the magnetic field of the plasma

$$\frac{\partial\mathbf{B}}{\partial t} = \frac{c^2}{4\pi k_C\sigma}\Delta\mathbf{B} + \boldsymbol{\nabla}\times(\mathbf{v}\times\mathbf{B}), \qquad \mathrm{SI:}\ \frac{c^2}{4\pi k_C} = \frac{1}{\mu_0} \tag{9.3.6}$$

results. If the plasma is incompressible ($\boldsymbol{\nabla}\cdot\mathbf{v} = 0$),

$$\boldsymbol{\nabla}\times(\boldsymbol{\beta}\times\mathbf{B}) \overset{(A.2.35)}{=} (\mathbf{B}\cdot\boldsymbol{\nabla})\boldsymbol{\beta} - (\boldsymbol{\beta}\cdot\boldsymbol{\nabla})\mathbf{B}\,,$$

then the magnetic field is determined by the induction equation in the form

$$\frac{\partial\mathbf{B}}{\partial t} = \frac{c^2}{4\pi k_C\sigma}\Delta\mathbf{B} + (\mathbf{B}\cdot\boldsymbol{\nabla})\mathbf{v} - (\mathbf{v}\cdot\boldsymbol{\nabla})\mathbf{B}. \tag{9.3.7}$$

Still to be evaluated is

$$\frac{1}{c}\big[\mathbf{j}\times\mathbf{B}\big]_i = \frac{-ck_L}{4\pi k_C}\big[\mathbf{B}\times(\boldsymbol{\nabla}\times\mathbf{B})\big]_i = \frac{-ck_L}{4\pi k_C}\epsilon_{ijk}B_j\epsilon_{klm}\nabla_l B_m$$
$$= \frac{-ck_L}{4\pi k_C}(\delta_{il}\delta_{jm}-\delta_{im}\delta_{jl})B_j\nabla_l B_m = \frac{-ck_L}{4\pi k_C}\Big[\frac{1}{2}\nabla_i B^2 - (\mathbf{B}\cdot\boldsymbol{\nabla})B_i\Big].$$

Since $|\mathbf{E}| \ll k_L|\mathbf{B}|$ the term $\rho\mathbf{E}$ in the Euler equation (9.3.3) is neglected and one obtains for $\mathbf{v}$ the Euler equation

$$\rho_m\Big[\frac{\partial\mathbf{v}}{\partial t} + (\mathbf{v}\cdot\boldsymbol{\nabla})\mathbf{v}\Big] = -\boldsymbol{\nabla}\Big(p + \frac{k_L^2}{8\pi k_C}B^2\Big) + \frac{k_L^2}{4\pi k_C}(\mathbf{B}\cdot\boldsymbol{\nabla})\mathbf{B}. \tag{9.3.8}$$

The contribution $k_L^2 B^2/8\pi k_C$, which indicates the magnetic energy density, is referred to here as *magnetic pressure*.

9.3.2 Magnetic Diffusion

If one takes a stationary plasma, $\mathbf{v} = 0$, then the hydrodynamic equations reduce to those of hydrostatics, which couples to the induction equation (9.3.6). Thus, a diffusion equation is created for the magnetic field with D_m as the diffusion coefficient:

$$\frac{d\mathbf{B}}{dt} = D_m\Delta\mathbf{B}, \qquad D_m = \frac{c^2}{4\pi k_C\sigma}. \tag{9.3.9}$$

An existing magnetic field decays in a time $\tau = l/D_m$, where l is a characteristic length for the system. This defines a timescale for $\mathbf{B}$, in which for the earth usually a time of $\tau \sim 10^4$ years and for the sun $\tau \sim 10^{10}$ years is given.

For times $t \ll \tau$ or for $\sigma \to \infty$ in moving plasma (9.3.7) the diffusion term can be neglected and one obtains

$$\frac{\partial \mathbf{B}}{\partial t} \overset{(1.3.6)}{=} (\mathbf{B} \cdot \boldsymbol{\nabla}) \mathbf{v} - (\mathbf{v} \cdot \boldsymbol{\nabla}) \mathbf{B} \qquad \Rightarrow \qquad \frac{\partial \mathbf{B}}{\partial t} = (\mathbf{B} \cdot \boldsymbol{\nabla}) \mathbf{v}. \tag{9.3.10}$$

In this case, the magnetic field moves with the fluid, it sticks to it, so to speak, which in hydrodynamics is referred to as *advection*.

9.3.3 Magnetohydrodynamic Waves

We again neglect the diffusion term as in (9.3.10), taking into account but now that the plasma is flowing, i.e., the Euler equation (9.3.8) for $\mathbf{v}$. If one develops $\mathbf{B}$ around an equilibrium value $\mathbf{B} = \mathbf{B}_0 + \delta \mathbf{B}$ and linearizes for $\mathbf{B}$ and $\mathbf{v}$, one obtains

$$\frac{\partial \delta \mathbf{B}}{\partial t} = (\mathbf{B}_0 \cdot \boldsymbol{\nabla}) \mathbf{v} - (\mathbf{v} \cdot \boldsymbol{\nabla}) \delta \mathbf{B}, \tag{9.3.11}$$

$$\rho_m \frac{d\mathbf{v}}{dt} = \rho_m \frac{\partial \mathbf{v}}{\partial t} = -\frac{k_L^2}{4\pi k_C} \boldsymbol{\nabla} (\mathbf{B}_0 \cdot \delta \mathbf{B}) + \frac{k_L^2}{4\pi k_C} (\mathbf{B}_0 \cdot \boldsymbol{\nabla}) \delta \mathbf{B}. \tag{9.3.12}$$

It is assumed that the pressure p vanishes. Now differentiate the Euler equation with respect to t:

$$\rho_m \frac{\partial^2 \mathbf{v}}{\partial t^2} = -\boldsymbol{\nabla} \Big(\frac{k_L^2}{4\pi k_C} \mathbf{B}_0 \cdot \frac{\partial \delta \mathbf{B}}{\partial t} \Big) + \frac{1}{4\pi} (\mathbf{B}_0 \cdot \boldsymbol{\nabla}) \frac{\partial \delta \mathbf{B}}{\partial t} \, .$$

If one places a coordinate axis parallel to $\mathbf{B}_0$, divides $\mathbf{v}$ into its components parallel and perpendicular to $\mathbf{B}_0$ and substitutes for $\partial \delta \mathbf{B} / \partial t$ from the first equation, one obtains:

$$\frac{\partial \delta \mathbf{B}}{\partial t} = B_0 \nabla_{\parallel} (\mathbf{v}_{\parallel} + \mathbf{v}_{\perp}),$$

$$\rho_m \frac{\partial^2 \mathbf{v}}{\partial t^2} = -\frac{1}{4\pi} \Big[\boldsymbol{\nabla} (B_0^2 \nabla_{\parallel} v_{\parallel}) - (B_0 \nabla_{\parallel}) B_0 \nabla_{\parallel} (\mathbf{v}_{\parallel} + \mathbf{v}_{\perp}) \Big].$$

Thus, for the components perpendicular to $\mathbf{B}_0$ one obtains the wave equation

$$\frac{1}{c_A^2} \frac{\partial^2 \mathbf{v}_{\perp}}{\partial t^2} = \nabla_{\parallel}^2 \mathbf{v}_{\perp}, \qquad\qquad c_A = \frac{k_L B_0}{\sqrt{4\pi k_C \varrho_m}}.$$

Their solutions are the Alfvén waves. These are transverse waves that propagate in the plasma at the speed c_A, which is usually smaller than the speed of sound of the fluid, unless one has systems with very low densities ρ_m, as can occur in astrophysical contexts.

Problems for chapter 9

9.1. *Resistance calculation*: Show that the network outlined in Fig. 9.14 at the frequency

$$\omega^2 = \frac{C_b - \frac{L}{R^2}}{LC_b(C_a + C_b)} > 0$$

only has one active resistance and calculate this.

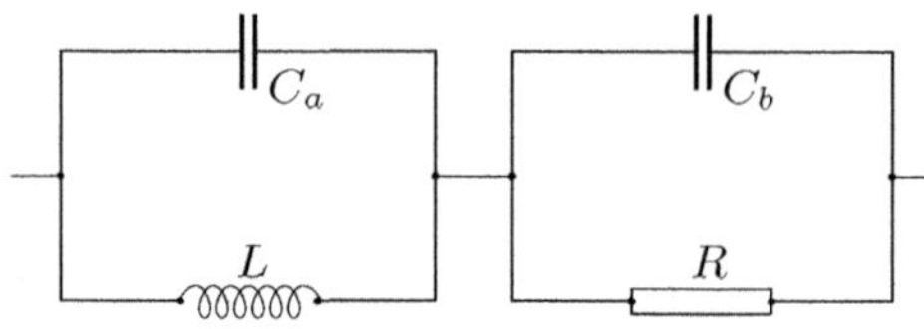

Fig. 9.14. *LC*-circuit and *RC*-circuit in series

9.2. *Wheatstone Bridge*: Sketch a Wheatstone bridge for measuring an inductance along with its Ohmic resistance, using a coil of known inductance instead of the capacitor in Fig. 9.11; R and L should be independent of the frequency of the alternating voltage.

9.3. *Telegraph equation*: Show that the telegraph equation (9.2.18) has solutions with a "loss term" $G = RC/L$ that represent a distortion-free, but damped wave propagation

$$I(t) = e^{-\alpha z} f(z - vt)$$

References

Becker R., Sauter F. *Theorie der Elektrizität 1*, 21th ed. Teubner, Stuttgart (1973)
Landau L. D., Lifshitz E. M. *Fluid Mechanics*, vol 6, 2nd ed., Pergamon Press (1987)
Thomson W. *On the theory of the electric telegraph*, Proc. Roy. Soc. **7**, 382–399 (1855) and Thomson W. *On peristaltic induction of electric currents in submarine telegraph wires*, BA Report, 21–22 (1855)

10

Electromagnetic Waves

10.1 Plane Waves in a Homogeneous Medium

We assume that neither charges nor currents are present and ϵ_r and μ_r are constant. Then we replace in the Maxwell equations $\mathbf{D}$ by $\epsilon_r \mathbf{E}$ and $\mathbf{H}$ by $\mu_r^{-1}\mathbf{B}$:

$$
\begin{array}{llll}
\text{(a)} & \boldsymbol{\nabla}\cdot\mathbf{E} = 0 & \text{(b)} & \boldsymbol{\nabla}\times\mathbf{E} = -\dfrac{k_L}{c}\,\dot{\mathbf{B}} \\[2mm]
\text{(c)} & \boldsymbol{\nabla}\times\mathbf{B} = \dfrac{\epsilon_r \mu_r}{ck_L}\,\dot{\mathbf{E}} & \text{(d)} & \boldsymbol{\nabla}\cdot\mathbf{B} = 0\,.
\end{array}
\tag{10.1.1}
$$

First, we note that the media in question are charge-neutral ($\rho = 0$) and that we have limited ourselves to insulators ($\mathbf{j} = \sigma\mathbf{E} = 0$).

From the induction equation it follows that $\boldsymbol{\nabla}\times(\boldsymbol{\nabla}\times\mathbf{E}) = -k_L \boldsymbol{\nabla}\times\dot{\mathbf{B}}/c = -\epsilon_r\mu_r\ddot{\mathbf{E}}/c^2$, where we have inserted the Ampère-Maxwell equation on the right. Now $\boldsymbol{\nabla}\times(\boldsymbol{\nabla}\times\mathbf{E}) = \boldsymbol{\nabla}\boldsymbol{\nabla}\cdot\mathbf{E} - \Delta\mathbf{E}$, and one obtains

$$
\left(\frac{1}{\bar{c}^2}\frac{\partial^2}{\partial t^2} - \Delta\right)\left\{\begin{matrix}\mathbf{E}\\\mathbf{B}\end{matrix}\right\} \equiv \bar{\Box}\left\{\begin{matrix}\mathbf{E}\\\mathbf{B}\end{matrix}\right\} = 0 \qquad \text{with} \qquad \bar{c} = \frac{c}{\sqrt{\epsilon_r\mu_r}}\,.
\tag{10.1.2}
$$

$\bar{\Box}$ is the d'Alembert operator with $\bar{c} = c/n$, the speed of light in a medium with $n = \sqrt{\mu_r\epsilon_r} \geq 1$. n is the dimensionless refractive index (also index of refraction) of the medium relative to vacuum. The influence of the homogeneous medium is only noticeable in the phase velocity $\bar{c}$. The wave equation (10.1.2) has as (particular) solution plane waves with the dispersion

$$
\omega = \bar{c}k = ck/n, \qquad\qquad n = \sqrt{\epsilon_r\mu_r}
\tag{10.1.3}
$$

and the refractive index n. There is no scalar potential, so the electric field consists only of the source-free component:

© The Editor(s) (if applicable) and The Author(s), under exclusive license to Springer-Verlag GmbH, DE, part of Springer Nature 2025
D. Petrascheck, F. Schwabl, *Electrodynamics*, https://doi.org/10.1007/978-3-662-71502-4_10

$$\mathbf{E} = -\frac{k_L}{c}\dot{\mathbf{A}} \quad \text{and} \quad \mathbf{B} = \boldsymbol{\nabla} \times \mathbf{A} \quad \text{with} \quad \boldsymbol{\nabla} \cdot \mathbf{A} = 0 \, . \tag{10.1.4}$$

If one substitutes (10.1.4) into the Ampère-Maxwell equation, one obtains again the wave equation

$$\left(\frac{1}{\bar{c}^2}\frac{\partial^2}{\partial t^2} - \Delta\right)\mathbf{A}(\mathbf{x}, t) = \bar{\Box}\mathbf{A}(\mathbf{x}, t) = 0, \tag{10.1.5}$$

whose particular solutions are the plane waves

$$\mathbf{A} = -\mathrm{i}\frac{c}{k_L\omega}\boldsymbol{\epsilon}\, E_0\, \mathrm{e}^{\mathrm{i}(\mathbf{k}\cdot\mathbf{x}-\omega t)} \quad \text{with} \quad \boldsymbol{\epsilon}\cdot\mathbf{k} = 0. \tag{10.1.6}$$

Due to the transverse gauge ($\boldsymbol{\nabla}\cdot\mathbf{A}=0$) is $\boldsymbol{\epsilon}\perp\mathbf{k}$. The polarization is determined by the unit vector $\boldsymbol{\epsilon}$. One obtains from (10.1.4) and (10.1.3)

$$\begin{aligned}
\mathbf{E} &= \boldsymbol{\epsilon}\, E_0\, \mathrm{e}^{\mathrm{i}(\mathbf{k}\cdot\mathbf{x}-\omega t)}, \\
\mathbf{B} &= \hat{\mathbf{k}} \times \boldsymbol{\epsilon}(n/k_L)E_0\, \mathrm{e}^{\mathrm{i}(\mathbf{k}\cdot\mathbf{x}-\omega t)} \quad \text{with} \quad \hat{\mathbf{k}} = \mathbf{k}/k \, .
\end{aligned} \tag{10.1.7}$$

$\mathbf{E}$ and $\mathbf{B}$ are both transverse waves and $\mathbf{E}$, $\mathbf{B}$ and $\mathbf{k}$ form a right-handed coordinate system. The amplitude E_0 is complex, however $\mathbf{E}$ and $\mathbf{B}$ are in phase and $k_L B_0 = \sqrt{\epsilon_r \mu_r}\, E_0$. The propagation direction of the plane waves is $\mathbf{k}$. The time-averaged Poynting vector (5.6.3) is

$$\langle\mathbf{S}\rangle = \frac{c}{8\pi k_r k_L}\mathbf{E}\times\mathbf{H}^* = \frac{c\epsilon_0}{8\pi k_r}\sqrt{\frac{\epsilon_r}{\mu_r}}E_0^2\,\hat{\mathbf{k}} = \langle u\rangle\bar{c}\hat{\mathbf{k}}. \tag{10.1.8}$$

Here, $\mathbf{H}$ was calculated from (10.1.7) and $\epsilon_0 = 1/k_L^2\mu_0$ was inserted. The averaged energy density obtained is

$$\langle u\rangle = \frac{1}{16\pi k_r}\left(\epsilon_r\epsilon_0\mathbf{E}\cdot\mathbf{E}^* + \frac{1}{\mu_r\mu_0}\mathbf{B}\cdot\mathbf{B}^*\right) = \frac{\epsilon_r\epsilon_0}{8\pi k_r}E_0^2. \tag{10.1.9}$$

From (10.1.7) it is clear that in the medium, magnetic and electric energy density of the plane waves are equal.

10.2 Polarization of Electromagnetic Waves

10.2.1 Linear and Circular Polarization

The plane waves discussed so far are linearly polarized, i.e. $\mathbf{E}$ points in a fixed direction. A generally polarized wave is obtained by superposition of two linearly polarized waves of the same frequency:

$$\begin{aligned}
\mathbf{E}_1 &= \boldsymbol{\epsilon}_1 E_1\, \mathrm{e}^{\mathrm{i}(\mathbf{k}\cdot\mathbf{x}-\omega t)}, \\
\mathbf{E}_2 &= \boldsymbol{\epsilon}_2 E_2\, \mathrm{e}^{\mathrm{i}(\mathbf{k}\cdot\mathbf{x}-\omega t)}, && \boldsymbol{\epsilon}_2 = \hat{\mathbf{k}}\times\boldsymbol{\epsilon}_1, && \hat{\mathbf{k}} = \mathbf{k}/k, \\
k_L\mathbf{B}_i &= \sqrt{\mu_r\epsilon_r}\,\hat{\mathbf{k}}\times\mathbf{E}_i, && i = 1, 2, \\
\mathbf{E} &= \left(\boldsymbol{\epsilon}_1 E_1 + \boldsymbol{\epsilon}_2 E_2\right)\mathrm{e}^{\mathrm{i}(\mathbf{k}\cdot\mathbf{x}-\omega t)}.
\end{aligned}$$

Linear polarization

If E_1 and E_2 have the same phase, then the resulting wave $\mathbf{E}$ is also linearly polarized and oscillates in the plane spanned by $\mathbf{E}$ and $\mathbf{k}$. This is, as sketched in Fig. 10.1, rotated by $\varphi = \arctan(E_2/E_1)$ against $\boldsymbol{\epsilon}_1$.

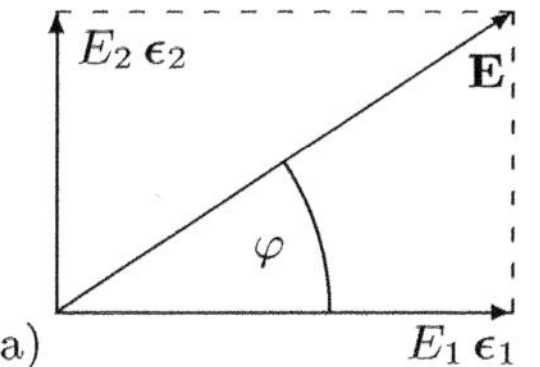

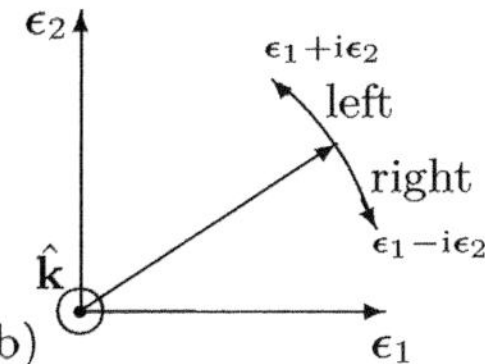

Fig. 10.1. (a) Linear polarization   (b) Circularly polarized light

Note: The term *plane of polarization* is not always used the same way. According to Born, Wolf [1986, p. 28] or Born [1933, p. 24] is – for historical reasons – the direction of the vector of polarization equal $\mathbf{B} = \hat{\mathbf{k}} \times \mathbf{E}$. The plane spanned by $\mathbf{B}$ and $\mathbf{k}$ is then referred to as the plane of polarization.

Circular polarization

If the amplitudes satisfy $E_2 = E_1\,\mathrm{e}^{\pm\mathrm{i}\frac{\pi}{2}}$, then

$$\mathbf{E} = (\boldsymbol{\epsilon}_1 \pm \mathrm{i}\boldsymbol{\epsilon}_2)E_1\,\mathrm{e}^{\mathrm{i}(\mathbf{k}\cdot\mathbf{x}-\omega t)}, \tag{10.2.1}$$
$$\mathrm{Re}\,\mathbf{E} = E_1\big[\boldsymbol{\epsilon}_1\,\cos(\mathbf{k}\cdot\mathbf{x}-\omega t) \mp \boldsymbol{\epsilon}_2\,\sin(\mathbf{k}\cdot\mathbf{x}-\omega t)\big].$$

If the wave moves in the z-direction towards the observer, and rotates counterclockwise we speak of left-polarized light or light of positive helicity (see Fig. 10.1)

$$\mathrm{Re}\,\mathbf{E} = E_1 \begin{pmatrix} \cos\omega t & \mp\sin\omega t \\ \pm\sin\omega t & \cos\omega t \end{pmatrix} \begin{pmatrix} \cos\mathbf{k}\cdot\mathbf{x} \\ \mp\sin\mathbf{k}\cdot\mathbf{x} \end{pmatrix}.$$

$\mathbf{E}$ rotates on a circle: $\begin{array}{l}+\ \text{left} \\ -\ \text{right}\end{array}$ polarized $=\begin{array}{l}\text{positive} \\ \text{negative}\end{array}$ helicity.

If we specify the base vector of the left-circularly polarized wave $\boldsymbol{\epsilon}_+ = (\boldsymbol{\epsilon}_1 + \mathrm{i}\boldsymbol{\epsilon}_2)/\sqrt{2}$, the unitary transformation for $\boldsymbol{\epsilon}_-$ follows:

$$\begin{pmatrix} \boldsymbol{\epsilon}_+ \\ \boldsymbol{\epsilon}_- \end{pmatrix} = \frac{1}{\sqrt{|a|^2+|b|^2}} \begin{pmatrix} a & b \\ -b^* & a^* \end{pmatrix} \begin{pmatrix} \boldsymbol{\epsilon}_1 \\ \boldsymbol{\epsilon}_2 \end{pmatrix} \stackrel{\substack{a=1\\b=\mathrm{i}}}{=\!=} \frac{1}{\sqrt{2}} \begin{pmatrix} \boldsymbol{\epsilon}_1 + \mathrm{i}\boldsymbol{\epsilon}_2 \\ \mathrm{i}(\boldsymbol{\epsilon}_1 - \mathrm{i}\boldsymbol{\epsilon}_2) \end{pmatrix}. \tag{10.2.2'}$$

Then $\boldsymbol{\epsilon}_+ \times \boldsymbol{\epsilon}_- = \hat{\mathbf{k}}$ and $\hat{\mathbf{k}} \times \boldsymbol{\epsilon}_\pm = \pm\boldsymbol{\epsilon}_\mp^*$, as well as $\boldsymbol{\epsilon}_+^* \cdot \boldsymbol{\epsilon}_- = \hat{\mathbf{k}} \cdot \boldsymbol{\epsilon}_\pm = 0$.

General (elliptical) polarization

We can also use the two circularly polarized plane waves as basis solutions:

$$\boldsymbol{\epsilon}_\pm = \frac{1}{\sqrt{2}}(\boldsymbol{\epsilon}_1 \pm \mathrm{i}\boldsymbol{\epsilon}_2) \qquad \begin{cases} \boldsymbol{\epsilon}_1 & = (\boldsymbol{\epsilon}_+ + \boldsymbol{\epsilon}_-)/\sqrt{2} \\[4pt] \mathrm{i}\boldsymbol{\epsilon}_2 & = (\boldsymbol{\epsilon}_+ - \boldsymbol{\epsilon}_-)/\sqrt{2}. \end{cases} \tag{10.2.2}$$

These complex unit vectors have the properties $(\hat{\mathbf{k}} = \mathbf{e}_z)$

$$\boldsymbol{\epsilon}_\pm^* \cdot \boldsymbol{\epsilon}_\mp = 0, \quad \boldsymbol{\epsilon}_\pm \cdot \hat{\mathbf{k}} = 0, \quad \boldsymbol{\epsilon}_\pm^* \cdot \boldsymbol{\epsilon}_\pm = 1, \quad \boldsymbol{\epsilon}_- \times \boldsymbol{\epsilon}_+ = \mathrm{i}\hat{\mathbf{k}}, \quad \boldsymbol{\epsilon}_\pm \times \hat{\mathbf{k}} = \pm \mathrm{i}\boldsymbol{\epsilon}_\pm.$$

A plane wave then has the form

$$\mathbf{E} = \left(E_+ \boldsymbol{\epsilon}_+ + E_- \boldsymbol{\epsilon}_-\right) \mathrm{e}^{\mathrm{i}(\mathbf{k}\cdot\mathbf{x}-\omega t)}. \tag{10.2.3}$$

In general, $E_\pm$ are complex amplitudes. However, if they have the same phase, $E_- = r\,E_+$, then (10.2.3) is an elliptically polarized wave

$$\mathbf{E} = \frac{1}{\sqrt{2}}E_+\left\{(1+r)\boldsymbol{\epsilon}_1 + \mathrm{i}(1-r)\boldsymbol{\epsilon}_2\right\}\mathrm{e}^{\mathrm{i}(\mathbf{k}\cdot\mathbf{x}-\omega t)},$$

$$\mathrm{Re}\,\mathbf{E} = \frac{1}{\sqrt{2}}E_+\left\{(1+r)\boldsymbol{\epsilon}_1\cos(\mathbf{k}\cdot\mathbf{x}-\omega t) - (1-r)\boldsymbol{\epsilon}_2\sin(\mathbf{k}\cdot\mathbf{x}-\omega t)\right\}.$$

The main axis ratio is $(1+r)/(1-r)$. In the limit $r = \pm 1$ one has linear polarization. If $E_- = E_+ r\mathrm{e}^{\mathrm{i}\alpha}$, the plane wave is also elliptically polarized, but the ellipse is rotated by the angle $\alpha/2$, as sketched in Fig. 10.2.

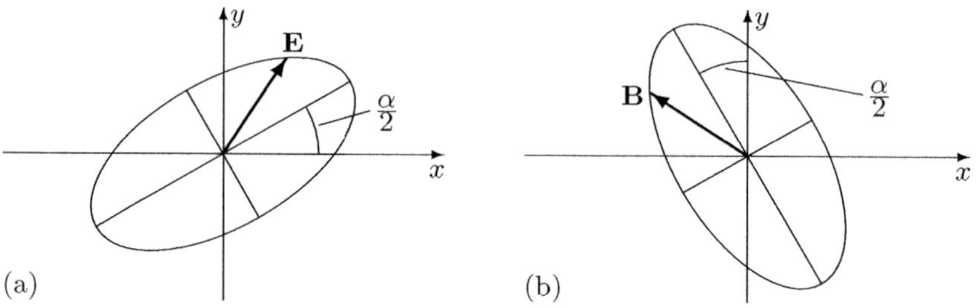

Fig. 10.2. Orientation of the ellipse of $\mathbf{E}$ and $\mathbf{B}$ for α

10.2.2 Stokes Parameters

In most cases, the light wave and its polarization state are not specified, but the polarization of some wave has to be determined.

We start from the basis vectors $\boldsymbol{\epsilon}_1$ and $\boldsymbol{\epsilon}_2$ of linear polarization, where $\mathbf{k}$ is perpendicular to the plane spanned by $\boldsymbol{\epsilon}_1$ and $\boldsymbol{\epsilon}_2$:

$$\mathbf{E} = (a_1 e^{i\alpha_1} \, \boldsymbol{\epsilon}_1 + a_2 e^{i\alpha_2} \, \boldsymbol{\epsilon}_2) e^{i\mathbf{k}\cdot\mathbf{x} - i\omega t} \, . \tag{10.2.4}$$

$a_{1,2}$ are the real amplitudes of the wave and $\alpha_{1,2}$ their phases in the direction of the orthogonal axes $\boldsymbol{\epsilon}_{1,2}$. The Stokes parameters are according to Born, Wolf [1986, p. 30]

$$
\begin{aligned}
s_0 &= |\boldsymbol{\epsilon}_1 \cdot \mathbf{E}|^2 + |\boldsymbol{\epsilon}_2 \cdot \mathbf{E}|^2 = a_1^2 + a_2^2, \\
s_1 &= |\boldsymbol{\epsilon}_1 \cdot \mathbf{E}|^2 - |\boldsymbol{\epsilon}_2 \cdot \mathbf{E}|^2 = a_1^2 - a_2^2, \\
s_2 &= 2\,\mathrm{Re}\left\{(\boldsymbol{\epsilon}_1 \cdot \mathbf{E})^* (\boldsymbol{\epsilon}_2 \cdot \mathbf{E})\right\} = 2 a_1 a_2 \cos(\alpha_2 - \alpha_1), \\
s_3 &= 2\,\mathrm{Im}\left\{(\boldsymbol{\epsilon}_1 \cdot \mathbf{E})^* (\boldsymbol{\epsilon}_2 \cdot \mathbf{E})\right\} = 2 a_1 a_2 \sin(\alpha_2 - \alpha_1).
\end{aligned}
\tag{10.2.5}
$$

For the plane wave, there are only three independent parameters:

$$s_0^2 = s_1^2 + s_2^2 + s_3^2 \, . \tag{10.2.6}$$

In the basis of the linear polarization vectors, the polarization is

$$P = E_2/E_1 = (a_2/a_1)\, e^{i(\alpha_2 - \alpha_1)}. \tag{10.2.7}$$

It can be determined by measuring the Stokes parameters, where the linear and circular polarization are characterized by

$$
\begin{aligned}
&\text{linear polarization}: \quad \alpha_1 = \alpha_2, \qquad\qquad\qquad\qquad s_3 = 0, \\
&\text{circular polarization}: \; a_1 = a_2 \text{ and } \alpha_2 - \alpha_1 = \pm\pi/2 \; s_1 = 0 \text{ and } s_2 = 0.
\end{aligned}
$$

With the Stokes parameters, we can also specify a *coherence matrix* (sometimes called polarization matrix), whose matrix elements $J_{ik} = E_i E_k^*$ are:

$$
\begin{aligned}
\mathsf{J} &= \begin{pmatrix} a_1^2 & a_1 a_2\, e^{-i(\alpha_2 - \alpha_1)} \\ a_1 a_2 e^{i(\alpha_2 - \alpha_1)} & a_2^2 \end{pmatrix} = \frac{1}{2}\begin{pmatrix} s_0 + s_1 & s_2 - i s_3 \\ s_2 + i s_3 & s_0 - s_1 \end{pmatrix} \\
&= \frac{1}{2}\left(s_0 \mathsf{E} + s_1 \sigma_z + s_2 \sigma_x + s_3 \sigma_y\right).
\end{aligned}
\tag{10.2.8}
$$

Here, E is the unit matrix, σ_i are the Pauli matrices[1] and $\det \mathsf{J} = 0$. Thus, an eigenvalue vanishes and

$$\mathbf{E} = e^{i(\alpha_2 + \alpha_1)/2} \, I \, \boldsymbol{\epsilon}_1' \, e^{i\mathbf{k}\cdot\mathbf{x} - i\omega t} \tag{10.2.9}$$

can be specified by a single eigenvector of a complex orthonormal basis $(\boldsymbol{\epsilon}_{1,2}')$ which is an expression of complete polarization. The eigenvalue, the intensity

$$I = \mathrm{tr}\,\mathsf{J} = a_1^2 + a_2^2 \tag{10.2.10}$$

is determined by the invariance of the trace $\mathrm{tr}\,\mathsf{J}$. The calculation of the basis vectors is problem 10.2.

---

[1] $\sigma_x = \begin{pmatrix} 0 & 1 \\ 1 & 0 \end{pmatrix}, \qquad \sigma_y = \begin{pmatrix} 0 & -i \\ i & 0 \end{pmatrix}, \qquad \sigma_z = \begin{pmatrix} 1 & 0 \\ 0 & -1 \end{pmatrix}$

Quasi-monochromatic light

Quasi-monochromatic light of frequency $\bar{\omega}$ refers to a light wave

$$E_1(\mathbf{x}, t) = a_1(t)\, e^{i\alpha_1(t) + i\mathbf{k}\cdot\mathbf{x} - i\bar{\omega}t} \qquad E_2(\mathbf{x}, t) = a_2(t)\, e^{i\alpha_2(t) + i\mathbf{k}\cdot\mathbf{x} - i\bar{\omega}t},$$

$$(10.2.11)$$

where the amplitudes and phases are slowly changing, but their differences remain small at all times. The coherence matrix (10.2.8) is in this case determined by time averages $E_k E_l^* \to \langle E_k E_l^* \rangle$, but otherwise remains the same

$$\mathsf{J} = \begin{pmatrix} \langle a_1^2 \rangle & \langle a_1 a_2 e^{-i(\alpha_2 - \alpha_1)} \rangle \\ \langle a_1 a_2 e^{i(\alpha_2 - \alpha_1)} \rangle & \langle a_2^2 \rangle \end{pmatrix}. \qquad (10.2.12)$$

For the averages, the Schwarz inequality now applies instead of the equal sign

$$\sqrt{J_{11}}\sqrt{J_{22}} \geq |J_{12}| \qquad \Rightarrow \qquad \det \mathsf{J} \geq 0, \qquad (10.2.13)$$

which states that $\mathbf{E}$ is no longer fully polarized. Expressed through the Stokes parameters, one obtains

$$s_0^2 \geq s_1^2 + s_2^2 + s_3^2.$$

For unpolarized (incoherent) light, there is no fixed phase relationship between E_1 and E_2, i.e., $J_{12} = J_{21} = 0$ and the intensity is direction-independent $J_{11} = J_{22}$. One can decompose J in a unique way into an unpolarized and a polarized part:

$$\mathsf{J} = \mathsf{J}_1 + \mathsf{J}_2 = \frac{1}{2}\begin{pmatrix} I_1 & 0 \\ 0 & I_1 \end{pmatrix} + \begin{pmatrix} J_{11} - I_1/2 & J_{12} \\ J_{12}^* & J_{22} - I_1/2 \end{pmatrix} \quad \text{with} \quad \det \mathsf{J}_2 = 0.$$

Thus, taking into account $I = \operatorname{tr}\mathsf{J} = J_{11} + J_{22}$ from $\det \mathsf{J}_2 = 0$, we obtain:

$$\det \mathsf{J} - \frac{I\,I_1}{2} + \frac{I_1^2}{4} = 0 \qquad \Rightarrow \qquad I_1 = I - \sqrt{I^2 - 4\det\mathsf{J}}.$$

The other solution of the quadratic equation is not considered, as then $I_1 > I$ would be. The polarized part of the wave is $I_2 = I - I_1$ and the ratio

$$\mathcal{P} = \frac{I_2}{I} = \sqrt{1 - \frac{4\det\mathsf{J}}{(J_{11} + J_{22})^2}} \qquad (10.2.14)$$

indicates the *degree of polarization* of the wave $\mathbf{E}$.

10.3 Reflection and Refraction Law for Insulators

10.3.1 Snell's Law of Refraction

A scalar plane wave of the form $\psi = \psi_0\, e^{i\mathbf{k}\cdot\mathbf{x} - i\omega t}$ hits the interface of two media with the refractive indices n and n'.

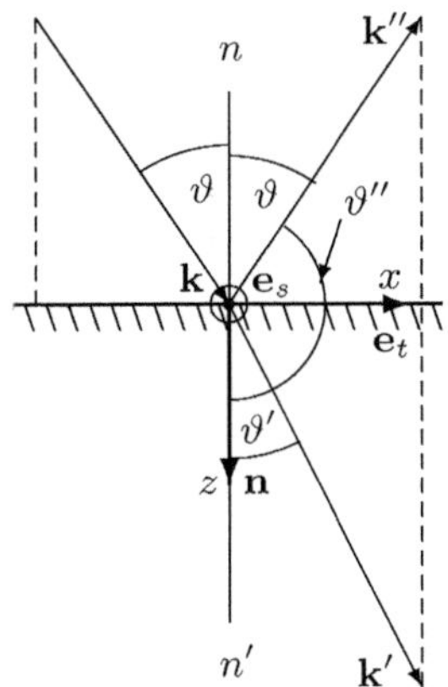

Fig. 10.3. Refraction and reflection of a scalar wave $\psi(\mathbf{x}, t) = \psi_0\, e^{i\mathbf{k}\cdot\mathbf{x} - i\omega t}$ in the xz-plane. It is $k/k' = n/n'$, where for electromagnetic waves $n = \sqrt{\epsilon_r \mu_r}$ and $n' = \sqrt{\epsilon'_r \mu'_r}$ applies. The vectors $\mathbf{e}_t$, $\mathbf{e}_s$, $\mathbf{n}$ span a right-handed coordinate system

The *plane of incidence* is according to Fig. 10.3 the plane spanned by $\mathbf{k}$ and $\mathbf{n}$. Thus, the tangential component $k_s = \mathbf{k}\cdot\mathbf{e}_s = 0$ perpendicular to the plane of incidence vanishes.

Requirement: Incident and scattered wave are continuous at the interface.

With the approach of a refracted ψ' and/or a reflected wave ψ'' can continuity be achieved at a flat interface:

$$\psi_0\, e^{i\mathbf{k}\cdot\mathbf{x} - i\omega t} = \psi'_0\, e^{i\mathbf{k}'\cdot\mathbf{x} - i\omega t} + \psi''_0\, e^{i\mathbf{k}''\cdot\mathbf{x} - i\omega t}\bigg|_{\mathbf{n}\cdot\mathbf{x}=0}, \tag{10.3.1}$$

however, the scalar wave is not yet uniquely determined by this:

1. The continuity is independent of t, therefore ω is the same for all waves.
2. The continuity of the wave over the entire interface is only possible if the tangential components $k_t = \mathbf{e}_t\cdot\mathbf{k}$ of all waves are the same (see Fig. 10.3)

   $$k \sin\vartheta = k' \sin\vartheta' = k \sin\vartheta''.$$

From this follows the *Snell's law of refraction*

$$\frac{\sin\vartheta}{\sin\vartheta'} = \frac{k'}{k} = \frac{n'}{n} \qquad \text{and} \qquad \vartheta'' = \pi - \vartheta. \tag{10.3.2}$$

We have introduced the refractive index n of a wave relative to the vacuum by $k = n k_0$, if k_0 is the wave number in the vacuum.

For the amplitudes at the surface one obtains

$$\psi_0 = \psi'_0 + \psi''_0 ,$$

which states that the refraction of a scalar wave is not yet determined by the continuity of ψ.

(10.3.2) applies to all waves, be they vectorial electromagnetic or elastic waves or wave functions in quantum theory [2].

---

[2] From the continuity of the derivative of the wave it follows that in addition to the refracted wave also a reflected one occurs.

10.3.2 Transition Conditions for Electromagnetic Waves

The previous considerations were of a kinematic nature, which is why they are also valid for different types of waves.

In the following we limit ourselves to electromagnetic fields and also consider dynamic properties, i.e. conditions that result from the Maxwell equations:

1. Dispersion relation $\omega = ck$ and $n = \sqrt{\epsilon_r \mu_r}$,
2. continuity of the tangential components $\mathbf{E}_\parallel$ and $\mathbf{H}_\parallel$,
   continuity of the normal component $\mathbf{D}_\perp$ and $\mathbf{B}_\perp$,
3. homogeneous medium: $\epsilon_r = \text{const.}$ and $\mu_r = \text{const.}$,
4. plane electromagnetic waves satisfy (10.1.7) $k_L \mathbf{B} = n\hat{\mathbf{k}} \times \mathbf{E}$.

These conditions are complete insofar as they also include the kinematic conditions, i.e. the Snell's law of refraction.

Continuity conditions

We use the continuity conditions for

$$\mathbf{E}(\mathbf{x}, t) = -\frac{k_L}{c}\,\dot{\mathbf{A}} = \mathbf{E}_0(\mathbf{x})\,e^{i\mathbf{k}\cdot\mathbf{x} - i\omega t} \qquad \text{with} \qquad \mathbf{k}\cdot\mathbf{A} = 0\,.$$

The vectors $\mathbf{k}$ and $\mathbf{n}$ span the plane of incidence. The two tangential components are determined by the unit vector $\mathbf{e}_t$ in the plane of incidence and $\mathbf{e}_s$ perpendicular to it. $(\mathbf{e}_t,\,\mathbf{e}_s,\,\mathbf{n})$ form a right-handed coordinate system. For $\hat{\mathbf{k}} = \mathbf{k}/k$ applies:

$$\hat{\mathbf{k}} \cdot \mathbf{e}_t = \sin\vartheta, \qquad\qquad \hat{\mathbf{k}} \cdot \mathbf{e}_s = 0, \qquad\qquad \hat{\mathbf{k}} \cdot \mathbf{n} = \cos\vartheta.$$

The vector $\mathbf{a} \times \mathbf{n}$ lies on the interface and has the tangential components

$$(\mathbf{a} \times \mathbf{n}) \cdot \mathbf{e}_t \stackrel{\text{cyclic}}{=} (\mathbf{n} \times \mathbf{e}_t) \cdot \mathbf{a} = \mathbf{e}_s \cdot \mathbf{a} = a_s\,,$$
$$(\mathbf{a} \times \mathbf{n}) \cdot \mathbf{e}_s \stackrel{\text{cyclic}}{=} (\mathbf{n} \times \mathbf{e}_s) \cdot \mathbf{a} = -\mathbf{e}_t \cdot \mathbf{a} = -a_t\,.$$

Since all fields have the same frequency ω and the same tangential components $\mathbf{k}_\parallel$, the conditions apply for the amplitudes $\mathbf{E}_0(\mathbf{x})$ etc. It is $\hat{\mathbf{k}} = \mathbf{k}/k$

$$
\begin{aligned}
&1.\ \mathbf{E}_\parallel: && \left(\mathbf{E}_0 + \mathbf{E}_0'' - \mathbf{E}_0'\right) \times \mathbf{n} = 0, \\[4pt]
&2.\ \mathbf{D}_\perp = \epsilon_r \epsilon_0 \mathbf{E}_\perp: && \left(\epsilon_r(\mathbf{E}_0 + \mathbf{E}_0'') - \epsilon_r' \mathbf{E}_0'\right) \cdot \mathbf{n} = 0, \\[4pt]
&3.\ \mathbf{H}_\parallel = \frac{n}{\mu_r k_L \mu_0}(\hat{\mathbf{k}} \times \mathbf{E})_\parallel: && \left(\frac{n}{\mu_r}(\hat{\mathbf{k}} \times \mathbf{E}_0 + \hat{\mathbf{k}}'' \times \mathbf{E}_0'') - \frac{n'}{\mu_r'}\hat{\mathbf{k}}' \times \mathbf{E}_0'\right) \times \mathbf{n} = 0, \\[4pt]
&4.\ \mathbf{B}_\perp = (n/k_L)(\hat{\mathbf{k}} \times \mathbf{E})_\perp: && \left(n(\hat{\mathbf{k}} \times \mathbf{E}_0 + \hat{\mathbf{k}}'' \times \mathbf{E}_0'') - n'\,\hat{\mathbf{k}}' \times \mathbf{E}_0'\right) \cdot \mathbf{n} = 0.
\end{aligned}
\tag{10.3.3}
$$

10.3.3 Fresnel's Formulas

When an electromagnetic wave refracts at the interface of two homogeneous media, in the plane of incidence, also called *plane of reflection R*, there are still $\mathbf{k}'$ and and $\mathbf{k}''$. The continuity conditions of the fields are now applied to the two cases, $\mathbf{E}^\pi$ in the plane of incidence ($\mathbf{E}^\pi \in R$) and perpendicular to it ($\mathbf{E}^\sigma \perp R$). By superposition of these two cases, the general position of $\mathbf{E}$ can be constructed. The laws for the refraction of the electric field at the interface of homogeneous media are the Fresnel's formulas. Since the refraction depends on a number of parameters, $\vartheta, \epsilon_r, \mu_r, n$ for the incident wave and $\vartheta', \epsilon_r', \mu_r', n'$ for the refracted wave, and the parameters are not (completely) independent of each other, so the Fresnel formulas can be represented in different ways.

Refraction with electric field perpendicular to the plane of incidence

We will now not use the general representation of the continuity conditions for the tangential components that we have just derived, but will determine these directly with $\mathbf{e}_t$ and $\mathbf{e}_s$ according to Fig. 10.4:

$$\mathbf{E} = \mathbf{E}^\sigma \perp R \qquad \Longleftrightarrow \qquad \mathbf{E}_0 = E_0\,\mathbf{e}_s \qquad \Longrightarrow \qquad \mathbf{B} \in R.$$

The determination of E_0' and E_0'' from the continuity conditions is simple and

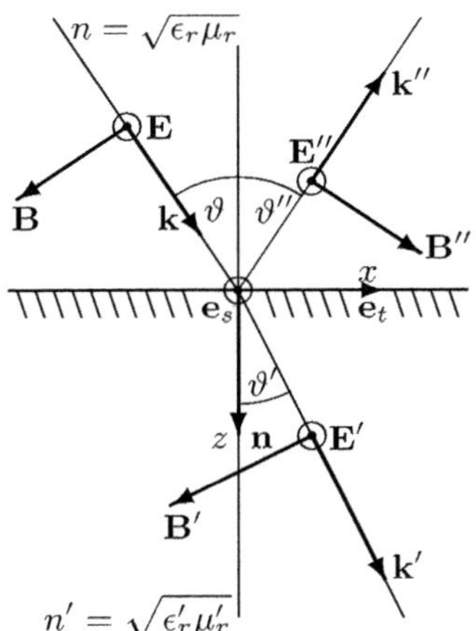

Fig. 10.4. Transmission and reflection, when $\mathbf{E} \perp$ plane of incidence. At the interface, there is the unit vector $\mathbf{e}_t$ in the plane of incidence and $\mathbf{e}_s$ perpendicular to this (σ-polarization): $\mathbf{e}_t$, $\mathbf{e}_s$, $\mathbf{n}$ form a right-handed coordinate system. In the sketch, $\mathbf{E}_0 = E_0\,\mathbf{e}_s$, i.e., $\mathbf{E}$ is perpendicular to the paper plane and directed forward (upwards)

does not need to be followed in all details. Some of these conditions contain no (new) information, so that ultimately only 3 equations remain, from which E_0', E_0'' and Snell's law of refraction emerge.

Given are $\hat{\mathbf{k}} = \mathbf{e}_t \sin\vartheta + \mathbf{n}\cos\vartheta$ with $0 < \vartheta < \pi/2$ and $\mathbf{E} = E_0\mathbf{e}_s$. This implies

$$\mathbf{B}_0 = \frac{n}{k_L}E_0\hat{\mathbf{k}}\times\mathbf{e}_s = \frac{n}{k_L}E_0\big(\mathbf{e}_t\times\mathbf{e}_s\,\sin\vartheta+\mathbf{n}\times\mathbf{e}_s\,\cos\vartheta\big) = \frac{n}{k_L}E_0\big(\mathbf{n}\,\sin\vartheta-\mathbf{e}_t\,\cos\vartheta\big).$$

$$\hat{\mathbf{k}} = \mathbf{e}_t\sin\vartheta+\mathbf{n}\cos\vartheta \quad\Rightarrow\quad \hat{\mathbf{k}}' = \mathbf{e}_t\sin\vartheta'+\mathbf{n}\cos\vartheta', \quad \hat{\mathbf{k}}'' = \mathbf{e}_t\sin\vartheta-\mathbf{n}\cos\vartheta,$$
$$\hat{\mathbf{E}}_0 = E_0\,\mathbf{e}_s \quad\Rightarrow\quad \hat{\mathbf{E}}_0' = \mathbf{e}_s, \quad\qquad\qquad \hat{\mathbf{E}}_0'' = \mathbf{e}_s,$$
$$\hat{\mathbf{B}}_0 = -\mathbf{e}_t\cos\vartheta+\mathbf{n}\sin\vartheta \Rightarrow \hat{\mathbf{B}}_0' = -\mathbf{e}_t\cos\vartheta'+\mathbf{n}\sin\vartheta', \hat{\mathbf{B}}_0'' = \mathbf{e}_t\cos\vartheta+\mathbf{n}\sin\vartheta.$$

From this, the continuity conditions result

1. $\quad (\mathbf{E}_0 + \mathbf{E}_0'')\cdot\mathbf{e}_t = \mathbf{E}_0'\cdot\mathbf{e}_t \quad\Rightarrow$ due to $\mathbf{E}\cdot\mathbf{e}_t = 0$ automatically fulfilled,

   $\quad (\mathbf{E}_0 + \mathbf{E}_0'')\cdot\mathbf{e}_s = \mathbf{E}_0'\cdot\mathbf{e}_s \quad\Rightarrow \qquad\qquad (E_0 + E_0'') = E_0'.$

2. $\quad \epsilon_r(\mathbf{E}_0 + \mathbf{E}_0'')\cdot\mathbf{n} = \epsilon_r'\mathbf{E}_0'\cdot\mathbf{n} \quad\Rightarrow$ due to $\mathbf{E}\cdot\mathbf{n} = 0$ automatically fulfilled.

3. $\quad \frac{1}{\mu_r}(\mathbf{B}_0 + \mathbf{B}_0'')\cdot\mathbf{e}_t = \frac{1}{\mu_r'}\mathbf{B}_0'\cdot\mathbf{e}_t \Rightarrow \frac{n}{\mu_r}(-E_0 + E_0'')\cos\vartheta = -\frac{n'}{\mu_r'}E_0'\cos\vartheta',$

   $\quad \frac{1}{\mu_r}(\mathbf{B}_0 + \mathbf{B}_0'')\cdot\mathbf{e}_s = \frac{1}{\mu_r'}\mathbf{B}_0'\cdot\mathbf{e}_s \Rightarrow$ due to $\mathbf{B}\cdot\mathbf{e}_s = 0$ automatically fulfilled.

4. $\quad (\mathbf{B}_0 + \mathbf{B}_0'')\cdot\mathbf{n} = \mathbf{B}_0'\cdot\mathbf{n} \quad\Rightarrow\quad n(E_0 + E_0'')\sin\vartheta = n'E_0'\sin\vartheta'.$

So, the continuity conditions are

(a) $\qquad\qquad E_0 + E_0'' = E_0'$

(b) $\mu_r'n(E_0 - E_0'')\cos\vartheta = \mu_r n'\,E_0'\cos\vartheta'$

(c) $\qquad n\sin\vartheta(E_0+E_0'') = n'\sin\vartheta'\,E_0'$

to be evaluated. From (a) and (c) follows the law of refraction (10.3.2)

$$n\sin\vartheta = n'\sin\vartheta' \qquad\Rightarrow\qquad \frac{n'\cos\vartheta'}{n\cos\vartheta} = \frac{\cot\vartheta'}{\cot\vartheta} = \frac{\tan\vartheta}{\tan\vartheta'}\,.$$

Substituting (a) into (b) yields

$$\mu_r'n(E_0 - E_0'')\cos\vartheta = \mu_r n'\,(E_0 + E_0'')\cos\vartheta'\,.$$

The rest are simple transformations, by eliminating the refractive index, to achieve the desired representation:

$$\frac{E_0''}{E_0} = \frac{\mu_r'n\cos\vartheta - \mu_r n'\cos\vartheta'}{\mu_r n'\cos\vartheta' + \mu_r'n\cos\vartheta} = \frac{\mu_r'\tan\vartheta' - \mu_r\tan\vartheta}{\mu_r'\tan\vartheta' + \mu_r\tan\vartheta}\,. \qquad (10.3.4)$$

Multiplying (a) by $\mu_r'n\cos\vartheta$ and adding (b) gives

$$2\mu_r'nE_0\cos\vartheta = E_0'(\mu_r'n\cos\vartheta + \mu_r n'\cos\vartheta')\,.$$

Again, eliminate n to get

$$\frac{E_0'}{E_0} = \frac{2\mu_r'n\cos\vartheta}{\mu_r'n\cos\vartheta + \mu_r n'\cos\vartheta'} = \frac{2\mu_r'\tan\vartheta'}{\mu_r'\tan\vartheta' + \mu_r\tan\vartheta}\,. \qquad (10.3.5)$$

(10.3.4) and (10.3.5) are generally held expressions for the Fresnel formulas. In (light) optics, almost always $\mu_r = \mu_r'[= 1]$. If you substitute $\mu_r = \mu_r'$ into (10.3.4) and (10.3.5), you get the Fresnel formulas for $\mathbf{E} \perp R$:

$$\frac{E_0'}{E_0} = \frac{2\cos\vartheta\,\sin\vartheta'}{\sin(\vartheta+\vartheta')}, \qquad\qquad \frac{E_0''}{E_0} = \frac{\sin(\vartheta'-\vartheta)}{\sin(\vartheta'+\vartheta)}. \tag{10.3.6}$$

Perpendicular incidence

Instead of n, you can eliminate ϑ in the Fresnel formulas. This is especially done with perpendicular incidence ($\vartheta = 0$), where you refer back to (10.3.4) and (10.3.5), since $\cos\vartheta = \cos\vartheta' = 1$. With $\mu_r = \mu_r'$ you get

$$E_0'' = \frac{n-n'}{n+n'}\,E_0, \qquad\qquad E_0' = \frac{2n}{n+n'}\,E_0. \tag{10.3.7}$$

Refraction with electric field in the plane of incidence

If $\mathbf{E}$ is in the plane of incidence, then the electric field is π-polarized:

$$\mathbf{B}\perp R \qquad\Longleftrightarrow\qquad \mathbf{B}_0 = \mathbf{e}_s\,B_0 \qquad\Longrightarrow\qquad \mathbf{E} = \mathbf{E}^\pi \in R.$$

According to the sketch Fig. 10.5, $\mathbf{B}$ is parallel to $\mathbf{e}_s$. Then it is

$$\mathbf{E}_0 = \frac{k_L}{n}\,\mathbf{B}_0\times\hat{\mathbf{k}} = E_0\,\mathbf{e}_s\times\hat{\mathbf{k}} = E_0\big(\mathbf{e}_s\times\mathbf{n}\cos\vartheta + \mathbf{e}_s\times\mathbf{e}_t\sin\vartheta\big)$$

$$= E_0\big(\mathbf{e}_t\cos\vartheta - \mathbf{n}\sin\vartheta\big) \qquad \text{with } k_L B_0 = nE_0.$$

Again, the continuity conditions are to be evaluated, which is rather tedious and therefore is carried out as an auxiliary calculation.

Given are the vectors $\mathbf{k}$ with $0 < \vartheta < \pi/2$ and $k_L\mathbf{B} = nE_0\,\mathbf{e}_s$. For their unit vectors applies

$$\begin{aligned}
\hat{\mathbf{k}} &= \mathbf{e}_t\sin\vartheta + \mathbf{n}\cos\vartheta \Rightarrow \hat{\mathbf{k}}' = \mathbf{e}_t\sin\vartheta' + \mathbf{n}\cos\vartheta', \quad \hat{\mathbf{k}}'' = \mathbf{e}_t\sin\vartheta - \mathbf{n}\cos\vartheta,\\
\hat{\mathbf{E}}_0 &= \mathbf{e}_t\cos\vartheta - \mathbf{n}\sin\vartheta \Rightarrow \hat{\mathbf{E}}_0' = \mathbf{e}_t\cos\vartheta' - \mathbf{n}\sin\vartheta', \quad \hat{\mathbf{E}}_0'' = -\mathbf{e}_t\cos\vartheta - \mathbf{n}\sin\vartheta,\\
\hat{\mathbf{B}}_0 &= \mathbf{e}_s \qquad\qquad\qquad\;\; \Rightarrow \hat{\mathbf{B}}_0' = \mathbf{e}_s, \qquad\qquad\quad\; \hat{\mathbf{B}}_0'' = \mathbf{e}_s.
\end{aligned}$$

The continuity conditions can be deduced from this:

1. $\quad(\mathbf{E}_0 + \mathbf{E}_0'')\cdot\mathbf{e}_t = \mathbf{E}_0'\cdot\mathbf{e}_t \quad\Rightarrow\quad (E_0 - E_0'')\cos\vartheta = E_0'\cos\vartheta',$

   $\quad(\mathbf{E}_0 + \mathbf{E}_0'')\cdot\mathbf{e}_s = \mathbf{E}_0'\cdot\mathbf{e}_s \quad\Rightarrow$ because $\mathbf{E}\cdot\mathbf{e}_s = 0$ automatically fulfilled.

2. $\quad\epsilon_r(\mathbf{E}_0 + \mathbf{E}_0'')\cdot\mathbf{n} = \epsilon_r'\mathbf{E}_0'\cdot\mathbf{n} \quad\Rightarrow \epsilon_r(E_0 + E_0'')\sin\vartheta = \epsilon_r'E_0'\sin\vartheta'.$

3. $\quad\dfrac{1}{\mu_r}(\mathbf{B}_0 + \mathbf{B}_0'')\cdot\mathbf{e}_t = \dfrac{1}{\mu_r'}\mathbf{B}_0'\cdot\mathbf{e}_t \Rightarrow$ due to $\mathbf{B}\cdot\mathbf{e}_t = 0$ automatically fulfilled,

   $\quad\dfrac{1}{\mu_r}(\mathbf{B}_0 + \mathbf{B}_0'')\cdot\mathbf{e}_s = \dfrac{1}{\mu_r'}\mathbf{B}_0'\cdot\mathbf{e}_s \Rightarrow \quad \dfrac{n}{\mu_r}(E_0 + E_0'') = \dfrac{n'}{\mu_r'}\,E_0'.$

4. $\quad n(\mathbf{B}_0 + \mathbf{B}_0'')\cdot\mathbf{n} = \epsilon_r'\mathbf{B}_0'\cdot\mathbf{n} \quad\Rightarrow$ due to $\mathbf{B}\cdot\mathbf{n} = 0$ automatically fulfilled.

The three continuity conditions are

(a) $\quad(E_0 - E_0'')\cos\vartheta = E_0'\cos\vartheta',$

(b) $\quad\epsilon_r(E_0 + E_0'')\sin\vartheta = \epsilon_r'E_0'\sin\vartheta',$

(c) $\quad n\mu_r'(E_0 + E_0'') = n'\mu_r E_0'.$

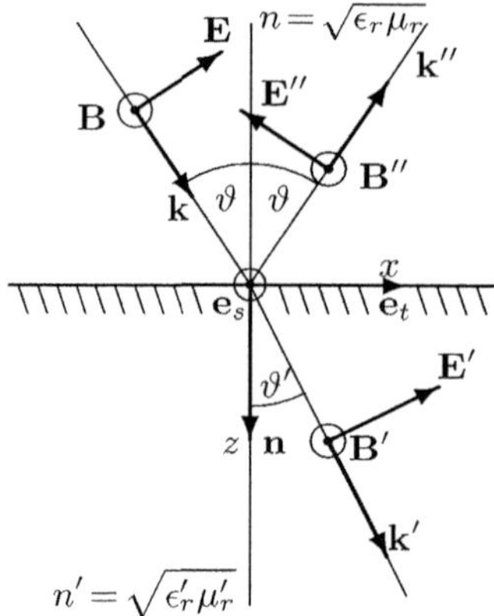

Fig. 10.5. Refraction and reflection, when $\mathbf{E} \in$ incident plane (π-polarization)

Of these three equations, only two are needed to express E_0' and E_0'' through E_0 in terms of. If you substitute $(E_0 + E_0'')$ from (c) into (b), then you reproduce – as in the previous case – the law of refraction.

First, multiply (a) by $n'\mu_r$ and (c) by $-\cos\vartheta'$ and add

$$n'\mu_r(E_0 - E_0'')\cos\vartheta - n\mu_r'(E_0 + E_0'')\cos\vartheta' = 0 \,.$$

If you replace the refractive index with the angles, you get

$$\frac{E_0''}{E_0} = \frac{n'\mu_r\,\cos\vartheta - n\mu_r'\,\cos\vartheta'}{n'\mu_r\,\cos\vartheta + n\mu_r'\,\cos\vartheta'} = \frac{\mu_r\,\sin(2\vartheta) - \mu_r'\,\sin(2\vartheta')}{\mu_r\,\sin(2\vartheta) + \mu_r'\,\sin(2\vartheta')}\,. \tag{10.3.8}$$

To calculate E_0', multiply (a) by $n\mu_r'$ and (c) by $\cos\vartheta$ and add

$$2n\mu_r' E_0 \cos\vartheta = E_0'(n\mu_r'\,\cos\vartheta' + n'\mu_r\,\cos\vartheta)\,.$$

Again, the refractive index is replaced by the angle:

$$\frac{E_0'}{E_0} = \frac{2n\mu_r'\,\cos\vartheta}{n'\mu_r\,\cos\vartheta + n\mu_r'\,\cos\vartheta'} = \frac{4\mu_r'\,\cos\vartheta\,\sin\vartheta'}{\mu_r\,\sin(2\vartheta) + \mu_r'\,\sin(2\vartheta')}\,. \tag{10.3.9}$$

The Fresnel formulas for $\mathbf{E} \in R$ are the limit case $\mu_r' = \mu_r$ of (10.3.9) and (10.3.8):

$$\frac{E_0'}{E_0} = \frac{2\cos\vartheta\,\sin\vartheta'}{\sin(\vartheta+\vartheta')\,\cos(\vartheta-\vartheta')}\,, \qquad \frac{E_0''}{E_0} = \frac{\tan(\vartheta-\vartheta')}{\tan(\vartheta+\vartheta')}\,. \tag{10.3.10}$$

Perpendicular incidence

Then it applies $\vartheta = 0 \Rightarrow \vartheta' = 0$. From (10.3.9) and (10.3.8) it follows for $\mu_r = \mu_r'$ directly

$$\frac{E_0'}{E_0} = \frac{2n}{n+n'}\,, \qquad \frac{E_0''}{E_0} = \frac{n'-n}{n+n'}\,. \tag{10.3.11}$$

Remarks: In the case of perpendicular incidence, the plane of reflection is no longer well defined and we expect that E_0' and E_0'' in the two cases $\mathbf{E} \perp R$ and $\mathbf{E} \in R$ are the same, but note that E_0'' in (10.3.7) and (10.3.11) have different signs. This is due to the definition of $\mathbf{E}_0''$, since for $\mathbf{E} \perp R$ it holds that $\mathbf{E}_0 \parallel \mathbf{E}_0''$, while in the other case $\mathbf{E}_0 \parallel -\mathbf{E}_0''$.

In the case of reflection on an optically denser medium $n' > n$, as can be seen from (10.3.7), $\mathbf{E}_0$ and $\mathbf{E}_0''$ are oppositely directed, i.e. a phase shift of $\pi/2$ occurs.

$$H_2O : \quad n = \frac{4}{3} \quad n' = 1 , \quad \frac{E_0''}{E_0} = \frac{1}{7} \quad \Rightarrow \quad \text{Reflectivity} : \quad \left| \frac{E_0''}{E_0} \right|^2 = 2\% .$$

For waves of lower frequency, one might expect that from the Fresnel's formulas the refraction of $\mathbf{E}$ known from electrostatics at interfaces can be read out, which is de facto not possible.

A fundamental difference to electrostatics is that waves of a given frequency, whether scalar or vectorial, propagate in a medium with a characteristic wavelength for this medium and a continuous transition at the interface is only possible together with a reflected wave (Snell's law of refraction) and this does not disappear for finite frequencies. The continuity conditions for $\mathbf{E}$ and $\mathbf{D}$ are indeed met for all frequencies, but only for all three waves together (see p. 356).

However, in special cases (Brewster angle, (10.3.12)) the reflected wave disappears, then the continuity conditions known from statics apply for all frequencies; one should only switch from the angles $\hat{\mathbf{k}} \cdot \mathbf{n}$ to $\hat{\mathbf{E}} \cdot \mathbf{n}$.

10.3.4 Brewster Angle

There is an angle of incidence for a polarization vector lying parallel to the plane of incidence, at which no reflection occurs. For simplicity, we limit our-selves to $\mu_r = \mu_r'$. The reflected, parallel polarized wave disappears ($E_0'' = 0$) for $\tan(\vartheta + \vartheta') \to \infty$, as can be seen from (10.3.10):

$$\vartheta + \vartheta' = \frac{\pi}{2} .$$

From the law of refraction, it follows

$$n \sin \vartheta = n' \sin(\frac{\pi}{2} - \vartheta) = n' \cos \vartheta,$$

$$\vartheta_B = \arctan \frac{n'}{n} , \qquad \text{Brewster's angle.} \qquad (10.3.12)$$

For this angle of incidence, the reflected wave is completely polarized with the polarization vector ϵ'' perpendicular to the plane of incidence. Also for other angles, the "normal component" is stronger than the parallel one in the reflected light.

Air/glass with $n'/n = 1.5$ has a Brewster's angle of 56 degrees. The reflected beam is completely linearly polarized with ϵ perpendicular to the plane of incidence.

10.3.5 Total Reflection

When transitioning to an optically less dense medium ($n \to n'$ with $n > n'$) the beam is refracted from the normal ($\vartheta < \vartheta'$), as outlined in Fig. 10.6. If you increase ϑ, then at ϑ_0 the angle of refraction $\vartheta' = \pi/2$ is reached. The beam is totally reflected.

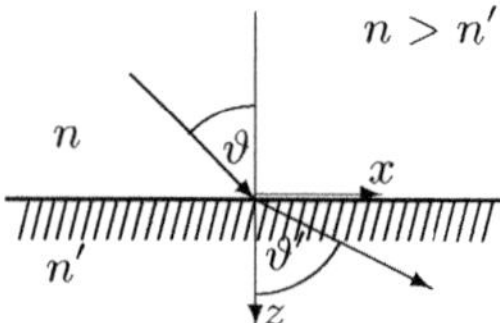

Fig. 10.6. Refraction from perpendicular

The continuity of the tangential components is

$$k \sin \vartheta = k' \sin \vartheta' \quad \overset{\vartheta' = \pi/2}{\Longrightarrow} \quad k \sin \vartheta_0 = k',$$

$$\sin \vartheta_0 = \frac{k'}{k} = \frac{n'}{n} \qquad \text{or} \qquad \vartheta_0 = \arcsin \frac{n'}{n}. \qquad (10.3.13)$$

Let $\vartheta > \vartheta_0$: Then $\sin \vartheta' = \dfrac{n}{n'} \sin \vartheta > 1$ and from this follows

$$\cos \vartheta' = \sqrt{1 - \sin^2 \vartheta'} = i \sqrt{\left(\frac{n}{n'}\right)^2 \sin^2 \vartheta - 1} = i \sqrt{\frac{\sin^2 \vartheta}{\sin^2 \vartheta_0} - 1}.$$

For the wave penetrating into the medium, the following applies

$$e^{i \mathbf{k'} \cdot \mathbf{x}} = e^{i k' (x \sin \vartheta' + z \cos \vartheta')} = e^{i k x \sin \vartheta - k z \sqrt{\sin^2 \vartheta - \sin^2 \vartheta_0}/\sin \vartheta_0}.$$

The refracted wave decays exponentially into the interior of the optically thinner medium. We define the penetration depth δ as the value at which the amplitude has fallen to the value e^{-1}:

$$\delta = \frac{1}{k} \frac{\sin \vartheta_0}{\sqrt{\sin^2 \vartheta - \sin^2 \vartheta_0}}.$$

The penetration of the wave into the medium causes a lateral shift of the reflected beam by $d_{\parallel} = 2\delta \tan \vartheta$. This lateral shift of the totally reflected beam is referred to as the *Goos-Hänchen effect*. Finally, it should be noted that the energy flow $\langle \mathbf{S} \cdot \mathbf{n} \rangle$ into the medium vanishes.

10.3.6 Geometric Optics and Wave Optics

Geometric optics becomes applicable when $\lambda \ll l$, where l is a characteristic length for the system. Monochromatic light can be represented by the scalar wave field

$$\psi(\mathbf{x}, t) = A(\mathbf{x})\, e^{i[S(\mathbf{x}) - \omega t]}. \qquad (10.3.14a)$$

The amplitude $A(\mathbf{x})$ is a spatially slowly varying function, and the phase $S(\mathbf{x})$ should be a function not deviating too much from linearity. S is then referred to as *eikonal*. Substituting ψ into the wave equation

$$\left(\frac{1}{\bar{c}^2(\mathbf{x})} \frac{\partial^2}{\partial t^2} - \nabla^2 \right) \psi(\mathbf{x}, t) = 0 \qquad \text{with} \quad \bar{c}(\mathbf{x}) = \frac{c}{n(\mathbf{x})}$$

results in using $\nabla\psi = e^{iS(\mathbf{x}) - i\omega t}\left(\nabla A + iA\nabla S \right)$

$$-\frac{\omega^2}{\bar{c}^2(\mathbf{x})} + \left(\nabla S \right)^2 - i\Delta S + 2i\frac{(\nabla S) \cdot (\nabla A)}{A} + \frac{\nabla^2 A}{A} = 0.$$

This equation can be separated into real and imaginary parts:

$$-\frac{\omega^2}{\bar{c}^2(\mathbf{x})} + \left(\nabla S \right)^2 + \frac{\nabla^2 A}{A} = 0, \qquad -\Delta S + \frac{(\nabla S) \cdot (\nabla A)}{A} = 0.$$

Under the condition that

$$\left| \frac{\nabla^2 A}{A} \right| \ll \frac{\omega^2}{\bar{c}^2(\mathbf{x})} = \frac{\omega^2}{c^2} n^2(\mathbf{x})$$

the variation of the amplitude compared to the phase can be neglected:

$$-\frac{\omega^2}{c^2} n^2(\mathbf{x}) + \left(\nabla S \right)^2 = 0.$$

If one defines ϕ by

$$S(\mathbf{x}) = \frac{\omega}{c}\, \phi(\mathbf{x}),$$

one obtains the eikonal equation

$$\left(\nabla\phi(\mathbf{x}) \right)^2 = n^2(\mathbf{x}). \qquad (10.3.14b)$$

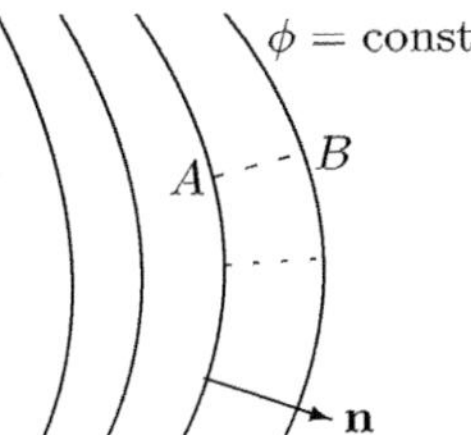

Fig. 10.7. Wavefront or wave surface is the surface of constant phase $\phi(\mathbf{x}) =$ const. and the ray is the line normal to the wave surface (gradient)

The gradient has the magnitude of the refractive index n and is perpendicular to the wavefront

$$\nabla\phi = n(\mathbf{x})\,\mathbf{n},$$

as sketched in Fig. 10.7 ϕ=const. is the surface of constant phase, and $-(\mathrm{i}\,\omega/c)\,\nabla\phi$ the local wave vector ($\sim$momentum). The integral

$$s = \int_A^B d\mathbf{l}\cdot\nabla\phi(\mathbf{x}) = \int_A^B dl\,n(\mathbf{x}) = \int_A^B d\phi = \phi(B) - \phi(A)$$

gives the optical path for the ray. For the ray calculated from the eikonal equation, $n(\mathbf{x})\,\mathbf{n}$ can be derived from the scalar potential ϕ and is therefore independent of the path. The wave surface is a surface of constant phase and the ray is the line that is normal to the wave surface, i.e., the gradient.

Fermat's principle

The propagation of rays in the stationary case can also be derived from Fermat's principle. According to this, the line integral

$$s = \int_A^B dl\,n(\mathbf{x}) = \text{Extremum}$$

for the path of the ray between the points in space A and B is a minimum. For the ray perpendicular to the surface of constant phase, it applies

$$dl\,n = d\mathbf{l}\cdot\nabla\phi = d\phi.$$

If dl is not perpendicular to the wave surface, then

$$dl\,n > d\phi \qquad \Rightarrow \qquad \int_A^B dl\,n > \phi(B) - \phi(A)$$

for any other path.

Note: The Fermat's principle is the analogue to the principle of least action known from mechanics

$$\delta\int_A^B d\mathbf{l}\cdot\mathbf{p} = 0\,,$$

where $\mathbf{p}$ is the momentum of the particle. The variation of the optical path with fixed endpoints leads to

$$\delta s = \delta\int_A^B dl\,n(\mathbf{x})\delta s = \int_A^B \left(dl\,\delta n + n\,\delta dl\right).$$

If $\delta\mathbf{x}$ is the displacement of the path, then

$$\delta dl = \hat{\mathbf{l}}\cdot d\delta\mathbf{x} \qquad\qquad \text{and} \qquad\qquad \delta n = \delta\mathbf{x}\cdot\nabla n\,.$$

$\hat{\mathbf{l}}$ is the unit vector parallel (tangential) to the ray. The second term is transformed (integrated by parts) into

$$n\,\delta \mathrm{d}l = n\,\hat{\mathbf{l}} \cdot \mathrm{d}\delta\mathbf{x} = \mathrm{d}\big(n\,\hat{\mathbf{l}} \cdot \delta\mathbf{x}\big) - \mathrm{d}\big(n\,\hat{\mathbf{l}}\big) \cdot \delta\mathbf{x}\,,$$

where the boundary term disappears, as $\delta\mathbf{x} = 0$ at the boundary points. Thus,

$$\delta s = \int_A^B \Big(\mathrm{d}l\,\boldsymbol{\nabla} n - \mathrm{d}(n\,\hat{\mathbf{l}})\Big)\delta\mathbf{x}\,\delta s = \int_A^B \mathrm{d}l\,\Big(\boldsymbol{\nabla} n - \frac{\mathrm{d}(n\,\hat{\mathbf{l}})}{\mathrm{d}l}\Big)\delta\mathbf{x} = 0\,.$$

For the integral to vanish for all $\delta\mathbf{x}$, it must be

$$\frac{\mathrm{d}(n\,\hat{\mathbf{l}})}{\mathrm{d}l} = \boldsymbol{\nabla} n \qquad \text{or} \qquad \frac{\mathrm{d}\hat{\mathbf{l}}}{\mathrm{d}l} = \frac{1}{n}\,\boldsymbol{\nabla} n - \frac{1}{n}\,\hat{\mathbf{l}}\,\frac{\mathrm{d}n}{\mathrm{d}l}\,.$$

The evaluation of this equation [Landau, Lifshitz 8, 1984, §85] shows that the beam is curved in the direction of the increasing refractive index.

On refraction at lenses

First, geometric optics is applied to the refraction at a cylindrical surface (circle), outlined in Fig. 10.8. The starting point is Snell's law (10.3.2)

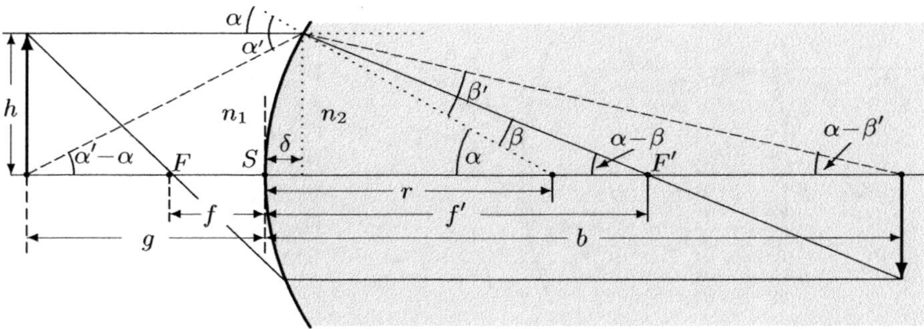

Fig. 10.8. Refraction at the circular sector: $f > 0$ is the object-side and $f' > 0$ the image-side focal length; $n = n_2/n_1 \approx 4$

$n_1 \sin\alpha' = n_2 \sin\beta'$ for small angles: $n_1\alpha' = n_2\beta'$

$$n_1[\alpha + (\alpha' - \alpha)] = n_2[\alpha - (\alpha - \beta')] \quad \Rightarrow \quad n_1\Big[\frac{h}{r-\delta} + \frac{h}{g+\delta}\Big] = n_2\Big[\frac{h}{r-\delta} - \frac{h}{b-\delta}\Big].$$

In this approximation, $\delta = 0$ can be set. The consistently positive distances start from the vertex S, as shown in Fig. 10.8:

$$\frac{1}{g} + \frac{n}{b} = \frac{n-1}{r} = \frac{n}{f'} = nf, \qquad f' = \frac{nr}{n-1}, \qquad n = \frac{n_2}{n_1}. \qquad (10.3.15)$$

The substitution of the sine and tangent functions by their arguments has the disadvantage that the accuracy is difficult to estimate. The refraction at a lens of thickness $d=0$ can be composed of two successive simple refractions (10.3.15) can be composed (see Fig. 10.8 and 10.9, [Pedrotti, 1997, §3]):

$$\frac{1}{g_1}+\frac{n}{b_1}=\frac{n-1}{r_1}, \quad \frac{n}{g_2}+\frac{1}{b_2}=\frac{n-1}{r_2} \quad \overset{b_1=d-g_2}{\Longrightarrow} \quad \frac{1}{g_1}+\frac{1}{b_2}=\frac{n-1}{r_1}+\frac{n-1}{r_2}.$$

If you set the object and image distance $g=g_1$ and $b=b_2$, you get:

$$\frac{1}{g}+\frac{1}{b}=\frac{1}{f'}, \qquad\qquad \frac{1}{f}=\frac{1}{f'}=\frac{n-1}{r_1}+\frac{n-1}{r_2}. \qquad (10.3.16)$$

Supplement: Consideration of the thickness of the lens

In Fig. 10.9 a biconvex lens of thickness d is sketched. The refractive indices to the left and right of the lens are n_1, within the lens n_2. Relevant again is only $n=n_2/n_1$. A ray incident on the lens parallel to the optical axis is refracted at both the front and back and hits, regardless of the distance h to the optical axis, at the focal point F' on this.

The refractions on the front and back of the lens can formally be replaced by a single one at the main plane H_2, as sketched in Fig. 10.9. For the corresponding calculations, the following will be used again the angle functions are replaced by their angles:

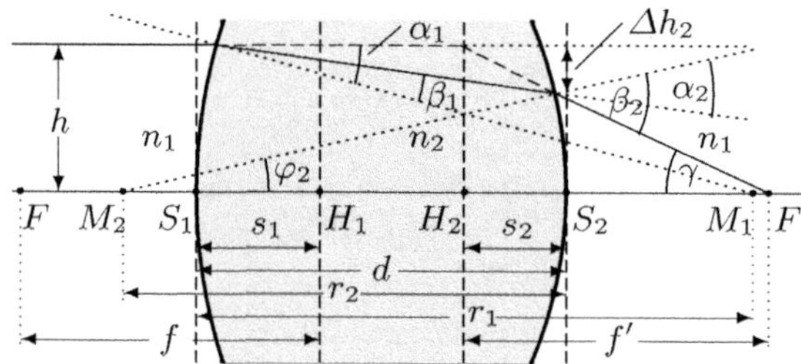

Fig. 10.9. Biconvex lens: $n=n_2/n_1\approx 2$. Circle centers $M_{1,2}$: $r_{1,2}$, Principal planes $H_{1,2}$, Vertex planes $S_{1,2}$, Focal points F, F'. Angle sum in the triangle: $\varphi_2+(\pi-\beta_2)+\gamma=\pi$

$$\alpha_1 = n\beta_1, \quad \alpha_2 = \alpha_1-\beta_1+\varphi_2, \quad \beta_2 = n\alpha_2, \quad \gamma = \beta_2-\varphi_2=(\alpha_1+\varphi_2)(n-1).$$

Now the angles are replaced by the distances r_1, r_2, d and h:

$$\frac{\Delta h_2}{h}=\frac{d}{h}(\alpha_1-\beta_1)=\frac{d(n-1)}{nr_1}, \quad \alpha_1=\frac{h}{r_1}, \quad \varphi_2=\frac{h-\Delta h_2}{r_2}=h\frac{nr_1-d(n-1)}{nr_1r_2}.$$

It is shown that all rays incident parallel to the optical axis are refracted so that they intersect at the focal point F':

$$h = f'\gamma, \qquad \gamma = (\alpha_1+\varphi_2)(n-1) = h[\frac{1}{r_1}+\frac{nr_1-d(n-1)}{nr_1\,r_2}](n-1)$$

From this one obtains the Gullstrand formula (left) and for thin lenses $(d\to 0)$ the lensmaker's formula (right). $D = 1/f'$ is called refractive power:

$$D = \frac{1}{f'} = \frac{n-1}{r_1} + \frac{n-1}{r_2} - \frac{d(n-1)^2}{nr_1r_2} \overset{d\to 0}{\Longrightarrow} \frac{n-1}{r_1} + \frac{n-1}{r_2}. \tag{10.3.17}$$

The positions of the principal planes $H_{1,2}$ still need to be determined:

$$s_2 = \frac{\Delta h_2}{\gamma} = \frac{\Delta h_2}{h}f' = \frac{d(n-1)}{nr_1}f', \qquad f = f', \qquad s_1 = \frac{d(n-1)}{nr_2}f.$$

The object-side distance g and the focal length f end in H_1, while b and f' are defined from H_2. The calculation of the lens shown here can be replaced for more complex systems by the matrix method in which refractions are successively calculated using 2×2 matrices [Pedrotti, 1997, §4].

Wave optics

For quasi-monochromatic light, one makes, starting from the wave equation (10.1.2), a separation approach of the form $\mathbf{E}(\mathbf{x},t) = \mathbf{E}(\mathbf{x})\,e^{-i\omega t}$, where the fields should have a time- and location-independent orientation (scalar wave optics): $\mathbf{E}(\mathbf{x}) = \mathcal{E}(\mathbf{x})\mathbf{e}_z$. The Helmholtz equation is thus obtained

$$(\Delta + k^2)\mathcal{E} = 0, \qquad\qquad k^2 = \omega^2/c^2. \tag{10.3.18}$$

Solutions are the plane waves $e^{i\mathbf{k}\cdot\mathbf{x}}$ or the spherical wave $\mathcal{E} = \mathcal{E}_0\,\frac{1}{r}e^{ikr}$, a solution of (10.3.18), which has no preferred propagation direction.

Bessel bundle

If the amplitude of the wave is independent of the propagation direction, one obtains from (10.3.18) the two-dimensional Poisson equation

$$(\Delta_\perp + k_\perp^2)u(\varrho) = 0; \qquad \mathcal{E}(\mathbf{x}) = e^{ik_z z}\,u(\varrho), \qquad k^2 = k_\perp^2 + k_z^2.$$

As in section 3.4.1, the separation approach $u(\varrho) = u(\varrho)e^{il\varphi}$ is made, where l must be an integer $(u(\varrho,\varphi) = u(\varrho,\varphi+2\pi))$:

$$[\frac{d^2\varrho}{d\varrho^2} + \frac{1}{\varrho}\frac{d}{d\varrho} - \frac{l^2}{\varrho^2} + k_\perp^2)]u_l(\varrho) = 0. \tag{10.3.19}$$

This is the Bessel's differential equation (3.4.6) or (B.4.1) whose regular at the origin solutions are the Bessel functions $u_l(\varrho) \sim J_l(k_\perp\varrho)$. The intensity is ring-shaped, caused by the zeros of J_l and decreases only slowly: $\lim\limits_{\varrho\to\infty}|u_l|^2 \sim 1/\varrho$. A structure results from the superposition:

$$u(\mathbf{x}) = \sum_l a_l e^{il\varphi}J_l(k_\perp\varrho), \qquad J_{-l}(k_\perp\varrho) = (-1)^l J_l(k_\perp\varrho) \tag{10.3.20}$$

The paraxial approximation

We are interested in waves that propagate in a small angle range around $\pm\mathbf{e}_z$. The variation of the amplitude in the z-direction should be small, suggesting an approach of the form $\mathcal{E} = \mathrm{e}^{ikz}\,u(\mathbf{x})$. Inserted into the Helmholtz equation, it follows:

$$(\Delta+k^2)\mathcal{E} = \mathrm{e}^{ikz}\Big[\Delta_\perp - k^2 + 2ik\frac{\partial}{\partial z} + \frac{\partial^2}{\partial z^2} + k^2\Big]u = 0, \qquad \Big|\frac{\partial u}{\partial z}\Big| \ll |2iku|.$$

One neglects $\partial^2 u/\partial z^2$ and thus obtains the paraxial equation:

$$\Big(\Delta_\perp + 2ik\frac{\partial}{\partial z}\Big)u = 0, \qquad \Delta_\perp = \frac{\partial^2}{\partial\varrho^2} + \frac{1}{\varrho}\frac{\partial}{\partial\varrho} + \frac{1}{\varrho^2}\frac{\partial^2}{\partial\varphi^2}. \tag{10.3.21}$$

Using the separation approach $u(\mathbf{x})=u_l(\varrho,z)\,\mathrm{e}^{il\varphi}$ one obtains

$$\Big(\frac{\partial^2}{\partial\varrho^2} + \frac{1}{\varrho}\frac{\partial}{\partial\varrho} - \frac{l^2}{\varrho^2} + 2ik\frac{\partial}{\partial z}\Big)u_l(\varrho,z) = 0. \tag{10.3.22}$$

Gauss bundle

For $l = 0$ a solution of the paraxial equation (10.3.22) is sought. If one takes a narrow conical section of the spherical wave around the z-axis $r = \sqrt{z^2+\varrho^2} \approx z+\varrho^2/2z$:

$$\mathcal{E} = \frac{1}{r}\mathrm{e}^{ikr} \sim \frac{1}{z}\,\mathrm{e}^{ik(z+\varrho^2/2z)} \qquad\Longrightarrow\qquad u_0 \sim \frac{1}{z}\,\mathrm{e}^{ik\varrho^2/2z},$$

this is a solution of the paraxial equation, as can be shown by substituting into (10.3.21):

$$\Delta_\perp u_0 = \frac{iku_0}{z} + \Big(\frac{ik}{z} - \frac{k^2\varrho^2}{z^2}\Big)u_0, \qquad 2ik\frac{\partial u_0}{\partial z} = 2ik\Big(\frac{-1}{z} - \frac{ik\varrho^2}{2z^2}\Big)u_0.$$

If one now replaces z with $q(z) = z-iz_R$, then $u(\mathbf{x})$ is still a solution of the paraxial equation, but has a finite width (*waist*) for $z = 0$:

$$u_0(\mathbf{x}) = c_0\frac{q_0}{q(z)}\,\mathrm{e}^{ik\varrho^2/2q(z)}, \qquad \begin{aligned}q_0 &= q(0)\\ q(z) &= z-iz_R\end{aligned}, \qquad \mathcal{E} = \mathrm{e}^{ikz}\,u(\mathbf{x}). \tag{10.3.23}$$

In addition to the wave number k, another parameter has been added, the *Rayleigh length* z_R, which is responsible for the finite width $w(z)$ of the wave bundle and is the distance at which the beam cross-section, starting from its waist $w_0=w(0)$, doubles – see Fig. 10.10. If one decomposes the exponential factor of (10.3.23) into real and imaginary part

$$\frac{ik\varrho^2}{2(z-iz_R)} = \frac{ik\varrho^2(z+iz_R)}{2(z^2+z_R^2)} = \frac{ik\varrho^2}{2(z+z_R/z)} - \frac{z_R k\varrho^2}{2(z^2+z_R^2)} = \frac{ik\varrho^2}{2R} - \frac{\varrho^2}{w^2},$$

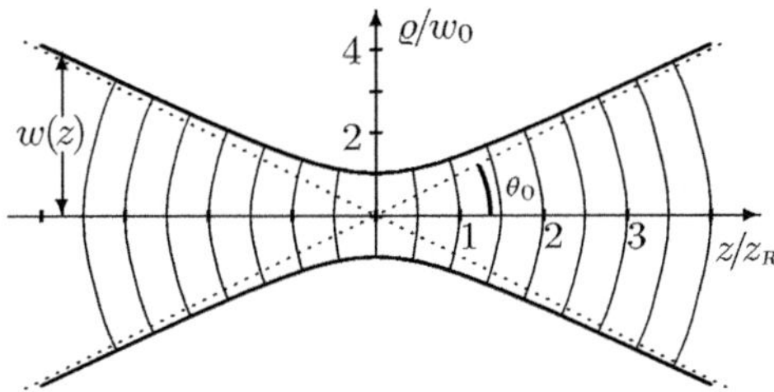

Fig. 10.10. Gaussian beam: Opening angle θ_0 with $\tan\theta_0 = \sqrt{2/kz_R} = 1/2$, i.e. $kz_R = 8$. Phase of the beam:
$$\phi(\varrho,z) = kz - \arctan\frac{z}{z_R} + \frac{k\varrho^2}{2R(z)}.$$
The phases of the drawn wavefronts are due to the Gouy phase $\arctan\frac{z}{z_R}$ not equidistant

this results in

$$u_0 = c_0\frac{q_0}{q}\mathrm{e}^{\mathrm{i}k\varrho^2/2R - \varrho^2/w^2}$$

with the definitions of the parameters of the Gaussian beam (see Fig. 10.10):

$$w^2(z) = \frac{2(z^2+z_R^2)}{kz_R} = \frac{2z_R}{k}\Big(1+\frac{z^2}{z_R^2}\Big) = w_0^2\Big(1+\frac{z^2}{z_R^2}\Big) \quad \text{Beam radius,} \quad (10.3.24)$$

$$w_0 = w(0) = \sqrt{\frac{2z_R}{k}} \qquad\qquad \text{Waist radius,}$$

$$R(z) = \frac{z^2+z_R^2}{z} = z_R\Big(\frac{z}{z_R}+\frac{z_R}{z}\Big) = \frac{z_R^2}{z}\frac{w^2}{w_0^2} \qquad \begin{array}{l}\text{Radius of curvature}\\\text{of the wavefront,}\end{array}$$

$$\phi_G(z) = \arctan\frac{z}{z_R} \qquad\qquad \text{Gouy phase.}$$

It should be noted that w is the double standard deviation of the intensity ($w = 2\sigma$). The Gaussian beam, compared to a plane wave of the same wavenumber, lags slightly behind, so that a phase difference ϕ_G develops compared to this.

Normalization of the beam: For finite z_R, u_0 is normalizable in the xy plane:

$$\int dx\,dy\,|u_0|^2 = \frac{2\pi c_0^2 z_R^2}{z^2+z_R^2}\int_0^\infty d\varrho\,\varrho\,\mathrm{e}^{-z_R k\varrho^2/(z^2+z_R^2)} = \frac{2\pi c_0^2 z_R^2}{z^2+z_R^2}\frac{z^2+z_R^2}{2kz_R}$$

$$= \frac{2\pi c_0^2 z_R}{2k} = 1 \qquad \Rightarrow \qquad c_0 = \sqrt{\frac{k}{\pi z_R}} = \frac{1}{w_0}\sqrt{\frac{2}{\pi}}. \qquad (10.3.25)$$

The intensity $|u_0|^2$ is a Gaussian distribution, centered around the optical axis.

Transformation of the prefactor q_0/q: Let $\phi_G = \arctan\frac{z}{z_R}$ (*Gouy phase*), then

$$\mathrm{e}^{-\mathrm{i}\phi_G} = \cos\phi_G\,(1-\mathrm{i}\tan\phi_G) = \frac{1-\mathrm{i}\tan\phi_G}{\sqrt{1+\tan^2\phi_G}} = \frac{1-\mathrm{i}z/z_R}{\sqrt{1+z^2/z_R^2}}.$$

$$\frac{q_0}{q} = \frac{-\mathrm{i}z_R}{z-\mathrm{i}z_R} = \frac{1}{1+\mathrm{i}\frac{z}{z_R}} = \frac{1-\mathrm{i}z/z_R}{1+z^2/z_R^2} = \frac{w_0}{w}\mathrm{e}^{-\mathrm{i}\phi_G}. \qquad (10.3.26)$$

$$u_0 = c_0\frac{q_0}{q}\mathrm{e}^{\mathrm{i}k\varrho^2/2q} = c_0\frac{q_0}{q}\mathrm{e}^{\mathrm{i}k\varrho^2/2R - \varrho^2/w^2}, \qquad c_0\frac{q_0}{q} = \frac{1}{w}\sqrt{\frac{2}{\pi}}\,\mathrm{e}^{-\mathrm{i}\phi_G}. \qquad (10.3.27)$$

The biconvex lens

We limit ourselves here to thin lenses, cylindrical symmetry, and transmission. Reflection at the interfaces is neglected. When the wave with wave number k passes through a medium of thickness d with the refractive index n, the phase experiences the shift $\delta = k(n-1)d$.

Fig. 10.11 shows a thin lens of thickness $d(0)$ with the curvature radii r_1 and r_2. A plane wavefront hits this. At a distance ϱ from the optical axis, the wave in the medium covers the distance

$$d(\varrho) = d(0) - (\varrho_1 - \sqrt{\varrho_1^2 - \varrho^2}) - (\varrho_1 - \sqrt{\varrho_1^2 - \varrho^2}) \approx d(0) - \varrho^2 \left(\frac{1}{2r_1} + \frac{1}{2r_2}\right)$$

back. The phase shift is then

$$\delta(\varrho) = k(n-1)d(\varrho) = \delta(0) - k\frac{\varrho^2}{2f'}, \qquad \frac{1}{f'} = (n-1)\left(\frac{1}{r_1} + \frac{1}{r_2}\right). \qquad (10.3.28)$$

The wavefront has the radius of curvature behind the lens, i.e. the focal length f'. Now, the wavefront hitting the lens is not flat, but a Gaussian bundle to

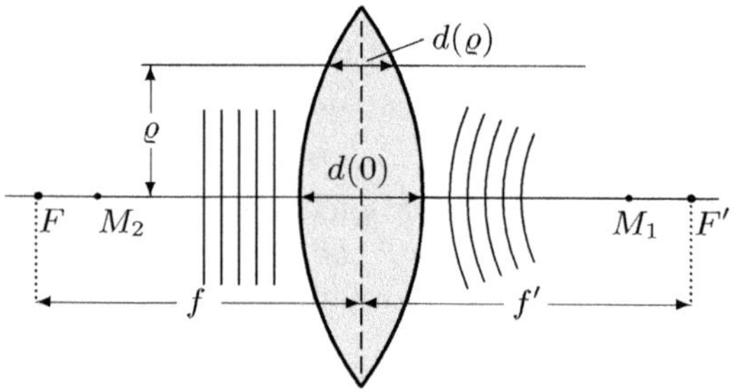

Fig. **10.11.** Plane wave falls on thin biconvex lens. $M_{1,2}$ are the centers of the radii r_1 and r_2; ($n \approx 1.5$)

which the effect of the lens must be added:

$$k\varrho^2\left(\frac{1}{q} - \frac{1}{f'}\right) = \frac{k\varrho^2}{2q'}, \qquad\qquad \frac{1}{q'} = \frac{1}{q} - \frac{1}{f'}. \qquad (10.3.29)$$

A Gaussian bundle therefore remains a Gaussian bundle with changed parameters.

Hermite-Gaussian bundle

If the Gaussian bundle u_0, (10.3.23), is represented in Cartesian coordinates

$$u_0 = u_{0x}(x, z)\, u_{0y}(y, z), \quad u_{0x} = \sqrt{c_0 \frac{q_0}{q}}\, e^{ikx^2/2q}, \quad u_{0y} = \sqrt{c_0 \frac{q_0}{q}}\, e^{iky^2/2q},$$

then u_{0x} and u_{0y} satisfy the paraxial equation separately. We are now looking for solutions with an inner structure $u_x = v_x\, u_{0x}$ with a polynomial v_x and determine the differential equation that v_x satisfies so that $v_x u_x$ also satisfies the paraxial equation:

$$[\Delta_\perp + 2\mathrm{i}k\frac{\partial}{\partial z}]u_{0x}v_x = v_x[\Delta_\perp + 2\mathrm{i}k\frac{\partial}{\partial z}]u_{0x} + u_{0x}[\Delta_\perp + 2\mathrm{i}k\frac{\partial}{\partial z}]v_x + 2\frac{\partial u_{0x}}{\partial x}\frac{\partial v_x}{\partial x}$$

$$= u_{0x}\Big[\frac{\partial^2}{\partial x^2} + 2\frac{\mathrm{i}kx}{q}\frac{\partial}{\partial x} + 2\mathrm{i}k\frac{\partial}{\partial z}\Big]v_x \overset{!}{=} 0. \qquad (10.3.30)$$

Determination of the polynomials v_x: A clue is given by $u_{0x} \sim \mathrm{e}^{-x^2/w^2}$, which can be understood as a weight function of the Hermite polynomials:

$$\int_{-\infty}^{\infty} \mathrm{d}\xi\, \mathrm{e}^{-\xi^2}\, H_l(\xi)\, H_m(\xi) = \delta_{lm}\, l!\, 2^l \sqrt{\pi} \qquad (10.3.31)$$

and therefore suggests that $\mathrm{e}^{-\xi^2/2} = \mathrm{e}^{-x^2/w^2} \Rightarrow \xi = \frac{x\sqrt{2}}{w}$. It turns out that in v_x to H_l a phase factor comes: $v_x = \mathrm{e}^{-\mathrm{i}l\phi_G} H_l(\xi)$

$$\Big\{\frac{\partial^2}{\partial x^2} + 2\mathrm{i}k[\frac{x}{q}\frac{\partial}{\partial x} + \frac{\partial}{\partial z}]\Big\}v_x = \mathrm{e}^{-\mathrm{i}l\phi_G}\Big\{\frac{2}{w^2}\frac{\mathrm{d}^2}{\mathrm{d}\xi^2} + 2\mathrm{i}k[\frac{\xi}{q} - \frac{\mathrm{i}z\,\xi}{z^2+z_R^2}]\frac{\mathrm{d}}{\mathrm{d}\xi} + \frac{4l}{w^2}\Big\}H_l(\xi).$$

The evaluation results in the differential equation for Hermite polynomials

$$\Big[\frac{\mathrm{d}^2}{\mathrm{d}\xi^2} - 2\xi\frac{\mathrm{d}}{\mathrm{d}\xi} + 2l\Big] H_l(\xi) = 0. \qquad (10.3.32)$$

As a solution of the paraxial equation one obtains: $u_H(\mathbf{x}) = u_l(x, z)\, u_m(y, z)$

$$u_l(x, z) = c_l \frac{\sqrt{q_0}}{\sqrt{q}} H_l\Big(\frac{x\sqrt{2}}{w}\Big)\mathrm{e}^{-\mathrm{i}l\phi_G}\, \mathrm{e}^{\mathrm{i}kx^2/q}, \qquad c_l = \frac{1}{\sqrt{l!\, 2^l}}, \qquad (10.3.33)$$

$$\int \mathrm{d}x\, |u_l(x)|^2 = |c_l|^2 c_0 \Big|\frac{q_0}{q}\Big|\frac{w}{\sqrt{2}} \int \mathrm{d}\xi\, H_l^2(\xi)\, \mathrm{e}^{-\xi^2} = |c_l|^2 \frac{1}{\sqrt{\pi}}\, l!\, 2^l \sqrt{\pi} = 1.$$

At the zeros of the polynomials, the intensity disappears and the beam becomes wider at the same time with higher order polynomials.

Laguerre-Gaussian beams

Rotationally symmetric solutions of the paraxial equation (10.3.22) for $l \neq 0$ are obtained using associated Laguerre polynomials $L_p^l(\eta)$, which must have a weight function of the form $\mathrm{e}^{-\eta/2}$ to ensure orthogonality. Accordingly, $\eta = 2\varrho^2/w^2$. One makes the approach:

$$u_L = c_L\,\eta^{|l|/2}\, L_p^{|l|}(\eta)\mathrm{e}^{-\mathrm{i}(|l|+2p)\phi_G}\mathrm{e}^{\mathrm{i}\varphi l}u_0, \quad u_0 = c_0\frac{q_0}{q}\mathrm{e}^{\mathrm{i}k\varrho^2/2R - \varrho^2/2w^2}. \quad (10.3.34)$$

Inserted into the paraxial equation, this approach results in the differential equation for the Laguerre polynomials, which shows that u_L is correct. As already with the Hermite-Gaussian polynomials, the retardation increases the

Gouy phase with increasing order. This goes hand in hand with a broadening of the beam. The zeros of the polynomials form dark rings here, from which the angular momentum can be read. The wave packets are eigenfunctions of the angular momentum operator:

$$L_z u_L = -i\hbar \frac{\partial}{\partial\varphi} u_L = \hbar l \, u_L.$$

In the following auxiliary calculation, it is verified that u_L is a solution of the paraxial equation.

Auxiliary calculation: Now u_L is defined as a product of $v_l e^{-i(|l|+2p)\phi_G}$ with u_0:

$$u_L = c_L \, v_l \, e^{-i(|l|+2p)\phi_G} \, e^{i\varphi l} \, u_0.$$

When inserting the product approach into the paraxial equation, we can directly refer to (10.3.22):

$$\Big[\frac{\partial^2}{\partial\varrho^2} + \Big(\frac{1}{\varrho} + \frac{2ik\varrho}{q}\Big)\frac{\partial}{\partial\varrho} + 2ik\frac{\partial}{\partial z} - \frac{l^2}{\varrho^2}\Big] v_l \, e^{-i(|l|+2p)\phi_G} = 0. \qquad (10.3.35)$$

The Gouy phase contributes a constant to the differential equation:

$$e^{-i(|l|+2p)\phi_G}\Big[\frac{\partial^2}{\partial\varrho^2} + \frac{1}{\varrho}\frac{\partial}{\partial\varrho} + 2ik\Big(\frac{\varrho}{q}\frac{\partial}{\partial\varrho} + \frac{\partial}{\partial z}\Big) - \frac{l^2}{\varrho^2} + \frac{4|l|+8p}{w^2}\Big] v_l(\eta) = 0.$$

We expect that the two terms proportional to $2ik$ are real:

$$2ik\Big[\frac{\varrho}{q}\frac{\partial\varrho}{\partial\eta}\frac{\partial}{\partial\eta} + \frac{\partial z}{\partial\eta}\frac{\partial}{\partial\eta}\Big] v_l(\eta) = 2ik\Big[\frac{z+iz_R}{z^2+z_R^2}2\eta - \frac{2z\eta}{z^2+z_R^2}\Big]\frac{\partial v_l}{\partial\eta} = \frac{-4k\eta z_R}{z^2+z_R^2}\frac{\partial v_l}{\partial\eta}.$$

Now we generally switch to $\eta = 2\varrho^2/w^2$ and multiply from the left with ϱ^2:

$$\Big[4\eta^2\frac{\partial^2}{\partial\eta^2} + (2\eta - 4\eta^2)\frac{\partial}{\partial\eta} - l^2 + (2|l|+4p)\eta\Big] v_l(\eta) = 0.$$

In the last step, we set $v_l = \eta^{|l|/2} v(\eta)$:

$$\frac{\partial v_l}{\partial\eta} = \eta^{|l|/2}\Big[\frac{\partial}{\partial\eta} + \frac{|l|}{2\eta}\Big]v, \qquad \frac{\partial^2 v_l}{\partial\eta^2} = \eta^{|l|/2}\Big[\frac{\partial^2}{\partial\eta^2} + \frac{|l|}{\eta}\frac{\partial}{\partial\eta} + \frac{|l|(|l|-2)}{4\eta^2}\Big]v.$$

$$\eta^{|l|/2}\Big[4\eta^2\frac{\partial^2 v}{\partial\eta^2} + \big(2|l|+2-2\eta\big)2\eta\frac{\partial}{\partial\eta} - l^2 + (2|l|+4p)\eta + l^2 - 2|l| + |l|(2-2\eta)\Big]v = 0.$$

Dividing from the left by 4η results in the differential equation for associated Laguerre polynomials $v = L_p^l(\eta)$ with the following orthogonality conditions:

$$\Big[\eta\frac{\partial^2 v}{\partial\eta^2} + (|l|+1-\eta)\frac{\partial}{\partial\eta} + p\Big]L_p^{|l|} = 0, \qquad \int_0^\infty d\eta\, e^{-\eta}\eta^l\, L_p^l L_s^l = \frac{(p+l)!}{p!}\delta_{ps}.$$

We set: $(c_0 w_0)^2 = 2/\pi$

$$2\pi\int_0^\infty d\varrho\,\varrho\,|u_L|^2 = 2\pi\frac{w^2}{4}c_L^2\frac{c_0^2 w_0^2}{w^2}\int_0^\infty d\eta\, e^{-\eta}\eta^{|l|}\big(L_p^{|l|}\big)^2 = c_L^2\frac{(p+|l|)!}{p!} = 1.$$

Now u_L – (10.3.34) – can be suitably represented:

$$u_L(\mathbf{x}) = c_L c_0 \frac{q_0}{q} \left(\frac{\varrho\sqrt{2}}{w}\right)^{|l|} L_p^{|l|}\left(\frac{2\varrho^2}{w^2}\right) e^{-\varrho^2/w^2} e^{il\varphi - i(|l|+2p)\phi_G + ik\varrho^2/2R},$$

$$c_L c_0 \frac{q_0}{q} \overset{(10.3.27)}{=} \sqrt{\frac{2\,p!}{\pi\,(p+|l|)!}} \frac{1}{w} e^{-i\phi_G}. \tag{10.3.36}$$

In Fig. 10.12, intensity profiles with $p = 0$ are depicted ($L_0^l = 1$). With increasing l, the beam becomes wider, while at the same time the intensity in the center disappears, which is attributed to the polynomial $(\sqrt{2}\varrho/w)^l$.

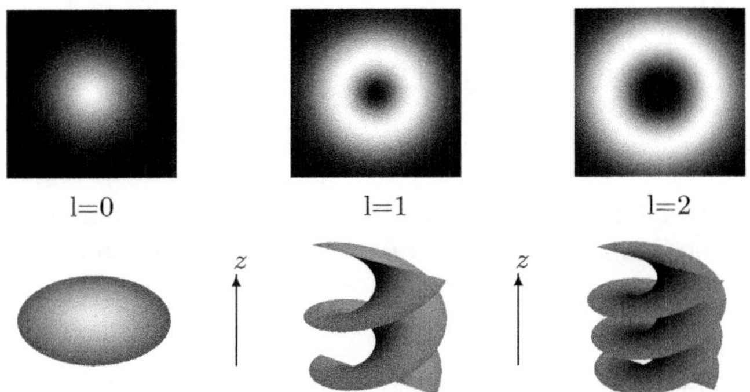

Fig. 10.12. Intensity profile and phase progression of a Laguerre-Gaussian beam according to (10.3.36) with $L_0^l = 1$; in the sketches for the phase, the curvature of the wavefront $R(z)$ is neglected

However, our interest lies in the phase of the beam, which reflects the angular momentum. The first associated Laguerre polynomials are ($l \geq 0$):

$$L_0^l(\eta) = 1, \qquad L_1^l(\eta) = 1 + l - \eta, \qquad L_2^l(\eta) = \frac{(l+1)(l+2)}{2} - (2+l)\eta + \frac{\eta^2}{2},$$

where only values with $\varrho \leq w$, i.e. $\eta \leq 2$ are considered. In the representation

$$\mathcal{E} = \operatorname{sgn} L_p^{|l|} |u_L| e^{i\phi}, \quad \phi = kz + l\varphi - (|l|+1+2p)\arctan\frac{z}{z_R} - \frac{z}{z_R}\frac{\varrho^2}{w^2} \tag{10.3.37}$$

it should be noted that at the zeros of the phase ϕ of the polynomials, a jump of π is added. This is not the case in the phase courses shown in Fig. 10.12. Rather, it is shown how from a Gaussian beam, in which the phase is the same in discrete (slightly curved) planes, with the help of angular momentum a contiguous spiral is created.

10.4 Waves in Conductors

The dielectric constant of metals diverges with $\omega \to 0$. Thus, the wavelength in the metal with $\delta \sim 2\pi c/(\omega\sqrt{|\epsilon_r|})$ becomes small compared to the vacuum wavelength of $2\pi c/\omega$. When δ is also small compared to the curvature of the metal surface this greatly simplifies the reflection at a metal surface.

10.4.1 The Equations for Wave Propagation in Metals

To describe waves in (homogeneous) conductors, in the Maxwell equations (10.1.1) $\mathbf{j}$ must first be included in the Ampère-Maxwell equation (5.2.16)

$$\mathbf{\nabla}\times\mathbf{H} = (k_L/c)\big(4\pi k_r\mathbf{j} + \dot{\mathbf{D}}\big),$$

where according to Ohm's law $\mathbf{j}=\sigma\mathbf{E}$

$$\mathbf{\nabla}\times\mathbf{B} = \mu_r\mu_0(k_L/c)\big(4\pi k_r\sigma\mathbf{E} + \epsilon_r\epsilon_0\,\dot{\mathbf{E}}\big)$$

is. Here, the last term indicates the so-called displacement current. If you differentiate the Ampère-Maxwell equation with respect to t, you get

$$\mathbf{\nabla}\times\dot{\mathbf{B}} = (1/ck_L)\big(4\pi k_C\mu_r\sigma\dot{\mathbf{E}} + \epsilon_r\mu_r\ddot{\mathbf{E}}\big).$$

Now you set for $\dot{\mathbf{B}}$ from the induction law $\dot{\mathbf{B}} = -(c/k_L)\mathbf{\nabla}\times\mathbf{E}$ and take into account that $\mathbf{\nabla}\cdot\mathbf{E} = 0$:

$$\Delta\mathbf{E} = \frac{1}{c^2}\big(4\pi k_C\mu_r\sigma\dot{\mathbf{E}} + \epsilon_r\mu_r\ddot{\mathbf{E}}\big). \tag{10.4.0}$$

We consider a wave incident perpendicular to the metal, as sketched in Fig. 10.13:

Incident wave :	$\mathbf{E}$	$= E_0\,\mathbf{e}_t\,e^{ikz-i\omega t},$
reflected wave :	$\mathbf{E}_r$	$= E_{0r}\,\mathbf{e}_t\,e^{-ikz-i\omega t},$
penetrating wave :	$\mathbf{E}_d$	$= E_{0d}\,\mathbf{e}_t\,e^{iqz-i\omega t}.$

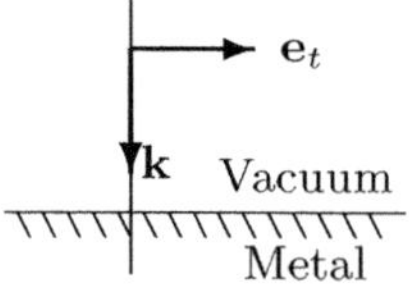

Fig. 10.13. An electric field hits the interface vacuum-metal perpendicularly

With the above approach for $\mathbf{E}_d$ in metals, one obtains

$$-q^2 = -i\omega\frac{4\pi k_C\mu_r\sigma}{c^2} - \frac{\epsilon_r\mu_r}{c^2}\,\omega^2 = -\frac{\mu_r\omega}{c^2}\big(\epsilon_r\omega + i\,4\pi k_C\sigma\big).$$

Of interest is the size ratio of the two terms on the right side.

Low conductivity: $\epsilon_r \omega \gg 4\pi k_C \sigma$

The contribution of $\sigma \mathbf{E}$ can be neglected and one obtains the usual wave equation

$$\Big(\Delta - \frac{\epsilon_r \mu_r}{c^2}\frac{\partial^2}{\partial t^2}\Big)\mathbf{E} = 0\,.$$

The medium behaves like a dielectric.

High conductivity: $\epsilon_r \omega \ll 4\pi k_C \sigma$

If the conductivity is large and the frequency not too high, then the contribution of the displacement current can be neglected (quasi-stationary approximation)

$$\omega \ll 4\pi k_C\,\frac{\sigma}{\epsilon_r} \qquad \Rightarrow \qquad \Big(\Delta - \frac{4\pi k_C \mu_r \sigma}{c^2}\frac{\partial}{\partial t}\Big)\mathbf{E} = 0\,. \qquad (10.4.1)$$

In this range, the medium behaves like a good conductor. From (10.4.1) it follows

$$q^2 = +\mathrm{i}\omega\,\frac{4\pi k_C \mu_r \sigma}{c^2} \qquad \Rightarrow \qquad q = \frac{1+\mathrm{i}}{\sqrt{2}}\,\frac{\sqrt{4\pi k_C \mu_r \sigma \omega}}{c}\,.$$

Inserted into

$$\mathbf{E}_d = E_{0d}\,\mathrm{e}^{\mathrm{i}qz} = E_{0d}\,\mathrm{e}^{(\mathrm{i}-1)\frac{\sqrt{2\pi k_C \mu_r \sigma \omega}}{c}\,z}$$

one obtains the penetration depth

$$\delta = \frac{c}{\sqrt{2\pi k_C \mu_r \sigma \omega}}\,, \qquad\qquad \text{SI: } \delta = \sqrt{\frac{2}{\mu_r \mu_0 \sigma \omega}} \qquad (10.4.2)$$

as the value at which the amplitude of $\mathbf{E}_d$ has decreased to the value $1/e$. For the magnetic field $\mathbf{B}$, the same value for the penetration depth is obtained.

In copper, the penetration depth δ at a frequency of $50\,\mathrm{Hz}$ is approximately $\delta = 9.4\,\mathrm{mm}$, so that at "normal" cross sections ($4\,\mathrm{mm}^2 \leftrightarrow 1.28\,\mathrm{mm}$ diameter) the current is homogeneously distributed in the wire; at $500\,\mathrm{kHz}$, the lower end of the medium wave, is $\delta = 0.1\,\mathrm{mm}$.

(10.4.1) can be brought into the form

$$\Big(\Delta - \frac{2}{\delta^2 \omega}\frac{\partial}{\partial t}\Big)\mathbf{E} = 0 \qquad\qquad \text{with} \qquad \frac{4\pi k_C \mu_r \sigma \omega}{c^2} = \frac{2}{\delta^2} \qquad (10.4.3)$$

and the inequality $\omega\epsilon_r \ll 4\pi k_C \sigma$ for the validity range of the metallic behavior is then

$$\omega\epsilon_r\,\frac{\mu_r\omega}{c^2} = k^2\epsilon_r\mu_r \ll 4\pi k_C\sigma\,\frac{\mu_r\omega}{c^2} = \frac{2}{\delta^2} \qquad \Rightarrow \qquad n = \sqrt{\epsilon_r\mu_r} \ll \frac{\sqrt{2}}{k\,\delta}.$$

We have extended the inequality with $(\omega\mu_r/c^2)$ and $k = \omega/c$, the wave number in vacuum, inserted. $\sqrt{\epsilon_r\mu_r}$ is the refractive index at small σ and δ the penetration depth.

Average conductivity: $\omega\epsilon_r \lesssim 4\pi k_C\sigma$

In media with poorer conductivity, the displacement current cannot be neglected (at low penetration depths) and the term with $\ddot{\mathbf{E}}$ must also be taken into account due to the high frequencies:

$$\left[\Delta - \left(\frac{\epsilon_r\mu_r}{c^2} + \frac{2i}{\delta^2\,\omega^2}\right)\frac{\partial^2}{\partial t^2}\right]\mathbf{E} = \left[\Delta - \frac{1}{c^2}\left(\epsilon_r\mu_r + \frac{2i}{\delta^2\,k^2}\right)\frac{\partial^2}{\partial t^2}\right]\mathbf{E} = 0. \quad (10.4.4)$$

Assuming that $\mathbf{E} \sim e^{-i\omega t}$ $\dot{\mathbf{E}}$ can be replaced by $(i/\omega)\,\ddot{\mathbf{E}}$. This results in a wave equation with a complex, frequency-dependent refractive index.

The root of a complex number[3] we approximate according to

$$\sqrt{a + ib} = \begin{cases} \sqrt{a}\left(1 + i\dfrac{b}{2a}\right) & a \gg b \geq 0 \\[2mm] \dfrac{1 + i}{\sqrt{2}}\,\sqrt{b}\left(1 - i\dfrac{a}{2b}\right) & b \gg a \geq 0. \end{cases}$$

For the complex refractive index we get

$$n + i\kappa = \sqrt{\epsilon_r\mu_r + \frac{2i}{k^2\delta^2}}$$

$$\approx \sqrt{\epsilon_r\mu_r}\left(1 + \frac{i}{k^2\delta^2\,\epsilon_r\mu_r}\right) \qquad\qquad \text{for} \quad \epsilon_r\mu_r \gg \frac{2}{k^2\,\delta^2}$$

$$\approx \frac{1}{k\delta}\left(1 + \frac{\epsilon_r\mu_r k^2\delta^2}{4}\right) + \frac{i}{k\delta}\left(1 - \frac{\epsilon_r\mu_r k^2\delta^2}{4}\right) \quad \text{for} \quad \epsilon_r\mu_r \ll \frac{2}{k^2\,\delta^2}.$$

In the second case, the good conductor, applies

$$n \approx \kappa \approx \frac{1}{k\delta} \gg 1 \qquad\qquad \text{for } 1 \leq \sqrt{\epsilon_r\mu_r} \ll \frac{\sqrt{2}}{\delta k}. \qquad\qquad (10.4.5)$$

Perpendicular incidence

We take the Fresnel's formulas (10.3.11) for the perpendicular incidence with $n = 1$ and $n' = n + i\kappa$. With this we calculate the intensities

$$E_r = \frac{n + i\kappa - 1}{1 + n + i\kappa}E_0 \qquad\qquad \text{and} \qquad\qquad E_d = \frac{2}{n + i\kappa + 1}E_0\,,$$

---

[3] exact: $\sqrt{a + ib} = \sqrt{\dfrac{1}{2}\left(\sqrt{a^2 + b^2} + a\right)} + i\sqrt{\dfrac{1}{2}\left(\sqrt{a^2 + b^2} - a\right)}$

$$\left|\frac{E_r}{E_0}\right|^2 = \frac{(n-1)^2 + \kappa^2}{(n+1)^2 + \kappa^2} = 1 - \frac{4n}{(n+1)^2 + \kappa^2},$$

For good conductors, therefore, according to (10.4.5) $\left|\frac{E_d}{E_0}\right| \to 0$. Almost the entire wave is reflected and we have a metal mirror in front of us:

$$e^{i\mathbf{q}\cdot\mathbf{x}} = e^{ik(n+i\kappa)z} \approx e^{(i-1)z/\delta}.$$

10.4.2 Cylindrical Wire (Skin Effect)

We consider the distribution of current density in a conductor (wire) through which a current flows. The previous results suggest that the current will not be evenly distributed over the cross-section, but more or less on the surface, because due to the penetration depth δ of the electric field, this will be weaker inside. This effect is referred to as the *skin effect*.

For the electric field in a conductor, we can use (10.4.1)

$$\left(\Delta - \frac{4\pi k_C \mu_r \sigma}{c^2}\frac{\partial}{\partial t}\right)\mathbf{E} \overset{(10.4.3)}{=} \left(\Delta - \frac{2}{\delta^2 \omega}\frac{\partial}{\partial t}\right)\mathbf{E} = 0.$$

The cross-section of the wire (see Fig. 10.14) is circular and for symmetry reasons $\mathbf{E}$=const. on the surface. In addition, in the outer space $\boldsymbol{\nabla}\cdot\mathbf{E} = 0$ and $\boldsymbol{\nabla}\times\mathbf{E} = 0$. We take cylindrical coordinates, according to the symmetry of the system $\mathbf{E} = E(\varrho)e^{-i\omega t}\mathbf{e}_z$ and differentiate with respect to t

$$\left(\frac{\partial^2}{\partial\varrho^2} + \frac{1}{\varrho}\frac{\partial}{\partial\varrho} + \frac{2i}{\delta^2}\right)E(\varrho) = 0. \tag{10.4.6}$$

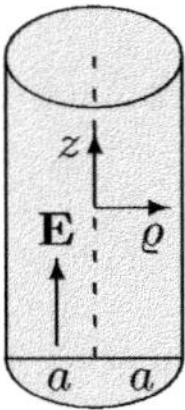

Fig. 10.14. Piece of a straight wire with radius a; wire axis and $\mathbf{E}$ point in z-direction.

$$\mathbf{E} = E(\varrho)\,e^{-i\omega t}\,\mathbf{e}_z$$

We define $\kappa^2 = \dfrac{2i}{\delta^2}$, i.e. $\kappa = \dfrac{1+i}{\delta}$, and substitute in (10.4.6) $\varrho = \frac{z}{\kappa}$. This gives us

$$\left(\frac{d^2}{dz^2} + \frac{1}{z}\frac{d}{dz} + 1\right)E(\frac{z}{\kappa}) = 0 .$$

This is the Bessel's differential equation (B.4.1) with the Bessel functions $Z_\nu(z)$ for $\nu = 0$. So $E(\frac{z}{\kappa}) = Z_0(z)$ or $E(\varrho) = Z_0(\kappa\varrho)$, and for Z_0 the function J_0, which is regular at the origin J_0, can be used:

$$E(\varrho) = C\, J_0(\kappa\varrho)\,.$$

C is determined from the total current I:

$$I = \sigma \iint \mathrm{d}f\, E(\varrho) = \sigma 2\pi C \int_0^a \mathrm{d}\varrho\, \varrho\, J_0(\kappa\varrho) = 2\pi\sigma C \frac{1}{\kappa^2} \int_0^{a\kappa} \mathrm{d}u\, u\, J_0(u)$$

$$= 2\pi\sigma C \frac{1}{\kappa^2}\, u\, J_1(u)\Big|_0^{a\kappa} = 2\pi a\sigma C \frac{1}{\kappa}\, J_1(a\kappa)\,.$$

From (B.4.8) the formula used here $\dfrac{\mathrm{d}}{\mathrm{d}z}\left(z^n J_n(z)\right) = z^n J_{n-1}(z)$ can be derived. Thus, C and $\mathbf{E}$ are determined (and the current density):

$$C = \frac{I\kappa}{2\pi a\sigma}\,\frac{1}{J_1(\kappa a)}\,, \qquad\qquad E(\varrho) = \frac{I\kappa}{2\pi a\sigma}\,\frac{J_0(\kappa\varrho)}{J_1(\kappa a)}\,. \qquad (10.4.7)$$

In the largest area of the wire, $\kappa a \gg 1$, so we use the asymptotic expansions for the Bessel functions:

$$J_\nu(\kappa\varrho) = \sqrt{\frac{2}{\pi\kappa\varrho}}\,\cos\left(\kappa\varrho - \frac{\nu\pi}{2} - \frac{\pi}{4}\right) \sim \frac{1}{2}\sqrt{\frac{2}{\pi\kappa\varrho}}\,\mathrm{e}^{-\mathrm{i}\left(\kappa\varrho - \frac{\nu\pi}{2} - \frac{\pi}{4}\right)}\,,$$

taking into account that $\mathrm{e}^{\mathrm{i}\kappa\varrho} = \mathrm{e}^{\mathrm{i}(1+\mathrm{i})\varrho/\delta} \to 0$. This results in

$$\frac{J_0(\kappa\varrho)}{J_1(a\kappa)} = \sqrt{\frac{a}{\varrho}}\,\mathrm{e}^{-\mathrm{i}\frac{\pi}{2}}\,\mathrm{e}^{\mathrm{i}\kappa(a-\varrho)} = (-\mathrm{i})\sqrt{\frac{a}{\varrho}}\,\mathrm{e}^{\frac{-1+\mathrm{i}}{\delta}(a-\varrho)}\,.$$

In this asymptotic form for the electric field

$$E(\varrho) = -\mathrm{i}\,\frac{I\kappa}{2\pi a\sigma}\sqrt{\frac{a}{\varrho}}\,\mathrm{e}^{-\frac{a-\varrho}{\delta} + \mathrm{i}\frac{a-\varrho}{\delta}} \qquad (10.4.8)$$

the already announced effect that the field is pushed outwards, can be seen. Now we give the resistance for $\varrho = a$:

$$\mathfrak{R} = \frac{E(a)}{I} = \frac{\kappa}{2\pi a\sigma}\,\frac{J_0(\kappa a)}{J_1(\kappa a)} = (-\mathrm{i})\frac{\kappa}{2\pi a\sigma}$$
$$= (-\mathrm{i})R_0\frac{\kappa a}{2} = R_0\frac{a}{2\delta}(1-\mathrm{i})\,. \qquad (10.4.9)$$

Compared to the direct current resistance

$$R_0 = \frac{E(a)}{I} = \frac{1}{\pi a^2\sigma}$$

we have in (10.4.9) a cross-section that is about δ/a smaller. The effect becomes significant especially at higher frequencies; in households, at a frequency of $50\,\mathrm{Hz}$ and wires with a cross-section of $\leq 10\,\mathrm{mm}^2$ the current density is still homogeneous.

Simplified calculation: We take the asymptotic form of (10.4.6):

$$\Big(\frac{\mathrm{d}^2}{\mathrm{d}\varrho^2} - \kappa^{*2}\Big)E(\varrho) = 0, \qquad \kappa^* = \frac{1-\mathrm{i}}{\delta}, \qquad E(\varrho) = E_0\,\mathrm{e}^{\kappa^*\varrho} = E_0\,\mathrm{e}^{(1-\mathrm{i})\varrho/\delta},$$

$$I = \sigma \iint \mathrm{d}f\, E(\varrho) = 2\pi\sigma \int_0^a \mathrm{d}\varrho\varrho\,\mathrm{e}^{\kappa^*\varrho} = 2\pi\sigma\Big(\frac{a}{\kappa^*} - \frac{1}{\kappa^{*2}}\Big)\mathrm{e}^{\kappa^*a} \approx \frac{2\pi a\sigma}{\kappa^*}\,\mathrm{e}^{\kappa^*a}$$

This results in

$$\mathfrak{R} = \frac{E(a)}{I} = \frac{\kappa^*}{2\pi a\sigma} = \frac{1-\mathrm{i}}{2\pi a\sigma\delta} = \frac{1-\mathrm{i}}{2}\frac{a}{\delta}R_0 \qquad \text{with} \qquad R_0 = \frac{1}{\pi a^2\sigma}.$$

10.5 Waves in Cavity Resonators and Waveguides

Within a cavity or a tube with good conductive walls, let $\mathbf{j}_f = 0$ and $\varrho_f = 0$. We keep open the possibility that the space is filled with a homogeneous dielectric. So the Maxwell equations in the form of (10.1.1) apply and the wave equations (10.1.2)-(10.1.5).

The penetration depth of electromagnetic waves $\delta = c/\sqrt{2\pi k_C\sigma\mu_r\omega} \overset{\sigma\to\infty}{\longrightarrow} 0$ in a metal decreases with increasing conductivity and frequency. So we assume an ideal conductor with $\delta = 0$, whose walls are reflective.

From the general boundary conditions it follows that at the metal surface the tangential components of $\mathbf{E}$ and the normal component of $\mathbf{B}$ vanish:

$$\mathbf{E}\times\mathbf{n} = 0 \qquad\qquad \text{and} \qquad\qquad \mathbf{B}\cdot\mathbf{n} = 0, \qquad\qquad (10.5.1)$$

where $\mathbf{n}$ should be the normal vector to the metal surface.

10.5.1 Standing Waves in a Cavity Resonator

The cavity should be a parallelepiped (cube) made of conductive, i.e. reflective material. For simplicity, we assume that there is no dielectric in the cavity ($\bar{c} = c$). Inside, the fields $\mathbf{E}$ and $\mathbf{B}$ must satisfy the homogeneous wave equation:

$$\Box\mathbf{E} = 0 \qquad\qquad\qquad \Box\mathbf{B} = 0,$$

as explained in section 10.1. The boundary conditions (10.5.1) require that the tangential components of $\mathbf{E}$ and the normal component of $\mathbf{B}$ at the wall vanish. This will lead to standing (transversal) waves with discrete frequencies in the cavity.

Rectangle with reflective walls

Fig. 10.15 shows a rectangle with ideally conductive sides. The fields $\mathbf{E}$ and $\mathbf{B}$ are to be determined in such a way that they meet the boundary conditions. We limit ourselves here to the calculation of the electric field. The magnetic field $\mathbf{B}$ can then be determined using the induction equation (10.1.1):

$$\mathbf{E}(\mathbf{x}, t) = \mathbf{E}(\mathbf{x}) \cos(\omega t) \quad E_x(x, 0) = E_x(x, L_y) = 0 \quad E_y(0, y) = E_y(L_x, y) = 0.$$

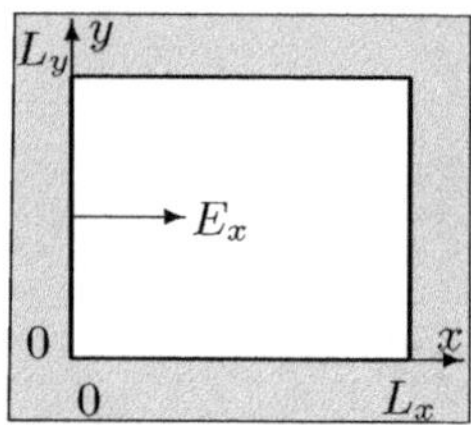

Fig. 10.15. Cavity radiation: Rectangle with reflective sides of length L_x and L_y

These continuity conditions are obviously met by the approach

$$\begin{aligned}
E_x(x, y) &= A_x(\mathbf{k}) \cos(k_x\, x) \sin(k_y\, y) & k_x L_x &= n_x \pi \\
E_y(x, y) &= A_y(\mathbf{k}) \sin(k_x\, x) \cos(k_y\, y) & k_y L_y &= n_y \pi
\end{aligned} \tag{10.5.2}$$

fulfills – as does the wave equation itself. These are standing waves that arise

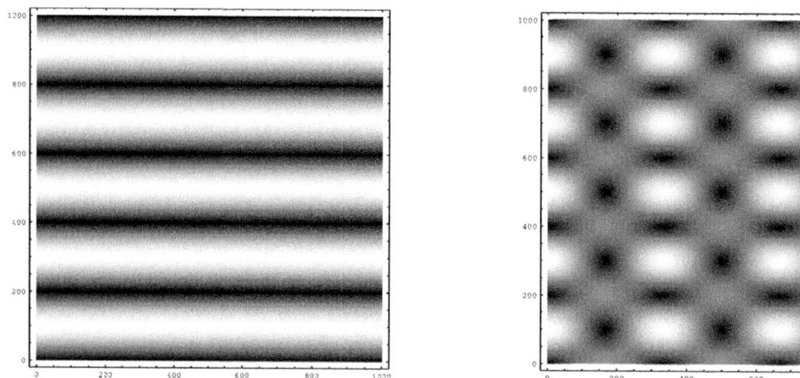

Fig. 10.16. Intensity of the electric field (10.5.2), which forms standing waves in a square. Calculated was $|\mathbf{E}|$ with the amplitudes $A_{x,y} = \pm k_{x,y}/k$ for $n_x = 0$ and 3 at $n_y = 5$; with increasing intensity of $|\mathbf{E}|$ the image becomes brighter.

from the superposition of back and forth running waves and form a regular interference pattern, as shown in Fig. 10.16 depicted. This pattern oscillates with $\cos(\omega t)$.

We explicitly show here that the pattern oscillating with $\cos(\omega t)$ of the Fig. 10.16 consists of back and forth running transverse electric waves. First, we decompose using the addition theorem for trigonometric functions

$$
\begin{aligned}
E_x(x,y,t) &= A_x(\mathbf{k},t)\,\frac{1}{2}\Big(\sin(\mathbf{k}\cdot\mathbf{x}) - \sin(\mathbf{k}'\cdot\mathbf{x})\Big) \qquad \begin{cases} \mathbf{k} = (k_x\,,\ k_y) \\ \mathbf{k}' = (k_x\,,\ -k_y) \end{cases} \\
&= A_x(\mathbf{k},t)\,\frac{1}{4}\Big(\big[\sin(\mathbf{k}\cdot\mathbf{x}+\omega t) + \sin(\mathbf{k}\cdot\mathbf{x}-\omega t)\big] \\
&\qquad\quad -\big[\sin(\mathbf{k}'\cdot\mathbf{x}+\omega t) + \sin(\mathbf{k}'\cdot\mathbf{x}-\omega t)\big]\Big),
\end{aligned}
$$

$$
\begin{aligned}
E_y(x,y,t) &= A_y(\mathbf{k},t)\,\frac{1}{2}\Big(\sin(\mathbf{k}\cdot\mathbf{x}) + \sin(\mathbf{k}'\cdot\mathbf{x})\Big) \\
&= A_y(\mathbf{k},t)\,\frac{1}{4}\Big(\big[\sin(\mathbf{k}\cdot\mathbf{x}+\omega t) + \sin(\mathbf{k}\cdot\mathbf{x}-\omega t)\big] \\
&\qquad\quad +\big[\sin(\mathbf{k}'\cdot\mathbf{x}+\omega t) + \sin(\mathbf{k}'\cdot\mathbf{x}-\omega t)\big]\Big).
\end{aligned}
$$

We now represent the electric field as the sum of two forward (in x-direction: $\mathbf{E}_{1,2}$) and two backward running waves

$$
\mathbf{E} = \sum_{i=1}^{4} \mathbf{E}_i \qquad \text{with} \qquad
\begin{aligned}
\mathbf{E}_{1,3}(\mathbf{x},t) &= \frac{1}{4}\begin{pmatrix} A_x \\ A_y \end{pmatrix}\sin(\mathbf{k}\cdot\mathbf{x}\mp\omega t), \\[1em]
\mathbf{E}_{2,4}(\mathbf{x},t) &= \frac{1}{4}\begin{pmatrix} -A_x \\ A_y \end{pmatrix}\sin(\mathbf{k}'\cdot\mathbf{x}\mp\omega t).
\end{aligned}
$$

The amplitudes are now to be determined so that the waves are transverse:

$$
\mathbf{k}\cdot\mathbf{E}_1 = 0 \quad\Rightarrow\quad k_x A_x + k_y A_y = 0 \quad\Rightarrow\quad A_x = \frac{A\,k_y}{k} \quad\text{and}\quad A_y = -\frac{A\,k_x}{k}\,.
$$

We see that $\mathbf{k}'\cdot\mathbf{E}_2 = 0$ is fulfilled. The two waves $\mathbf{E}_{1,2}$ are therefore transverse and thus also $\mathbf{E}_{3,4}$. For each mode $\mathbf{k}$, $\mathbf{E}$ (except for A) is uniquely determined, since there is only one transverse polarization direction in two dimensions.

The particular solution of the wave equation with (n_x, n_y) nodes is therefore a standing wave, which can be decomposed into partial waves that run back and forth in the cavity and are reflected at the walls.

Standing waves in a cuboid

The boundary conditions for the tangential component $\mathbf{E}_x$ at the boundary surfaces $y = 0, L_y$ and $z = 0, L_z$ are:

$$
E_x(x,0,z) = E_x(x,L_y,z) = 0 \quad\text{and}\quad E_x(x,y,0) = E_x(x,y,L_z) = 0\,.
$$

For (y,z) the corresponding equations apply. The (real) electric field

$$
\mathbf{E}(\mathbf{x},t) = \mathbf{E}(\mathbf{x})\,\cos(\omega t),
$$

$$
E_i(\mathbf{x}) = A_i\,\cos(k_i x_i)\prod_{j\neq i}\sin(k_j x_j) \qquad \begin{cases} \mathbf{k}\cdot\mathbf{e}_i\,L_i = n_i\,\pi \quad n_i \geq 0 \\ \mathbf{k}\cdot\mathbf{A} = 0 \end{cases} \tag{10.5.3}
$$

now has two (transverse) polarization directions. The magnetic field is calculated with the induction equation:

$$\frac{k_L}{c}\,\dot{B}_i = -\epsilon_{ijl}\nabla_j E_l = -\cos(\omega t)\,\epsilon_{ijl}k_j A_l\,\sin(k_i x_i)\,\cos(k_j x_j)\,\cos(k_l x_l)\,.$$

From $\mathbf{B}(\mathbf{x},t) = \mathbf{B}(\mathbf{x})\,\sin(\omega t)$ it follows

$$B_i(\mathbf{x},t) = -\frac{1}{k_L}\sin(\omega t)\,\epsilon_{ijl}\frac{k_j A_l}{k}\,\sin(k_i x_i)\,\cos(k_j x_j)\,\cos(k_l x_l). \qquad (10.5.4)$$

The normal components of $\mathbf{B}$ disappear at the edge and $\nabla\cdot\mathbf{B}=0$.

10.5.2 Electromagnetic Waves in Waveguides

The Maxwell equations for the waveguide

Within axisymmetric, good conductive tubes, one expects in the direction of the axis parallel to the tube, which should always be the z-axis, a continuous wave, which suggests the approach

$$\mathbf{E}(\mathbf{x},t) = \mathbf{E}_\perp + \mathbf{E}_z = \Big[\mathbf{E}_{0\perp}(x,y) + \mathbf{E}_{0z}(x,y)\Big]\psi(z,t),$$

$$\psi(z,t) = e^{ik_z z - i\omega t}, \qquad\qquad (10.5.5)$$

$$\mathbf{B}(\mathbf{x},t) = \mathbf{B}_\perp + \mathbf{B}_z = \Big[\mathbf{B}_{0\perp}(x,y) + \mathbf{B}_{0z}(x,y)\Big]\psi(z,t)$$

suggests. This approach must satisfy the Maxwell equations (10.1.1) for the propagation of waves in a homogeneous medium:

$$\text{(a)}\quad \nabla\cdot\mathbf{E} = 0 \qquad\qquad \text{(b)}\quad \nabla\times\mathbf{E} = \frac{i\omega}{\sqrt{\epsilon_r\mu_r}}\frac{k_L}{\bar{c}}\mathbf{B}$$

$$\text{(c)}\quad \nabla\times\mathbf{B} = -\frac{i\omega}{\bar{c}k_L}\sqrt{\epsilon_r\mu_r}\,\mathbf{E} \quad \text{(d)}\quad \nabla\cdot\mathbf{B} = 0. \qquad (10.5.6)$$

(10.5.6) are invariant under the exchange of $k_L\mathbf{B} \leftrightharpoons -n\mathbf{E}$. Above all, one can switch from vacuum to dielectric in (10.5.6) if one replaces in the formulas derived for the vacuum with

$$c \to \bar{c} = \frac{c}{n}, \qquad \mathbf{E} \to n\mathbf{E} \qquad \text{with} \qquad n = \sqrt{\epsilon_r\mu_r}. \qquad (10.5.6')$$

It is therefore sufficient to specify the wave propagation in vacuum. This is determined by the homogeneous wave equations (10.1.2) $\Box\mathbf{E}=0$ and $\Box\mathbf{B}=0$. For the fields in waveguides, one is primarily interested in the longitudinal components E_z or B_z interested, for which applies

$$\Box E_z = \Big[\frac{1}{c^2}\frac{\partial^2}{\partial t^2} - \Delta\Big]E_z(x,y)\psi(z,t) = 0.$$

We now use the product rule for

$$\Delta E_{0z}\psi = \psi\Delta E_{0z} + 2(ik_z\mathbf{e}_z\psi)\cdot(\boldsymbol{\nabla}E_{0z}) - E_{0z}k_z^2 = (\Delta_\perp - k_z^2)E_z$$

and get $(\Delta_\perp = \nabla_x^2 + \nabla_y^2)$

$$\left(\Delta_\perp + \frac{\omega^2}{c^2} - k_z^2\right)\mathbf{E}_z(\mathbf{x},t) = 0. \tag{10.5.7}$$

The same equation also applies to $\mathbf{B}_z$. The transverse components are obtained using (10.5.9) from the longitudinal ones. The following side calculation represents $\mathbf{E}_\perp$ and $\mathbf{B}_\perp$ as a function of E_z and B_z.

The fields are already divided into a longitudinal and a transverse part in (10.5.5), which should also apply to the operators $\boldsymbol{\nabla} = \boldsymbol{\nabla}_\perp + \boldsymbol{\nabla}_z$.

(a) $\qquad\qquad \boldsymbol{\nabla}_\perp\cdot\mathbf{E}_\perp = -\boldsymbol{\nabla}_z\cdot\mathbf{E}_z \qquad$ (b) $\qquad\qquad \boldsymbol{\nabla}_\perp\times\mathbf{E}_\perp = (ik_L\omega/c)\mathbf{B}_z$

$$\boldsymbol{\nabla}_\perp\times\mathbf{E}_z + \boldsymbol{\nabla}_z\times\mathbf{E}_\perp = (ik_L\omega/c)\mathbf{B}_\perp$$

(c) $\qquad\quad \boldsymbol{\nabla}_\perp\times\mathbf{B}_\perp = -(i\omega/ck_L)\mathbf{E}_z \quad$ (d) $\qquad\qquad \boldsymbol{\nabla}_\perp\cdot\mathbf{B}_\perp = -\boldsymbol{\nabla}_z\cdot\mathbf{B}_z$

$$\boldsymbol{\nabla}_\perp\times\mathbf{B}_z + \boldsymbol{\nabla}_z\times\mathbf{B}_\perp = -(i\omega/ck_L)\mathbf{E}_\perp.$$

$$\tag{10.5.8}$$

We multiply the transversal part of the induction equation (b) vectorially from the left with $\boldsymbol{\nabla}_z\times$

$$\boldsymbol{\nabla}_z\times(\boldsymbol{\nabla}_\perp\times\mathbf{E}_z) = \boldsymbol{\nabla}_\perp(\boldsymbol{\nabla}_z\cdot\mathbf{E}_z) - (\boldsymbol{\nabla}_z\cdot\boldsymbol{\nabla}_\perp)\mathbf{E}_z = ik_z\boldsymbol{\nabla}_\perp E_z,$$

$$\boldsymbol{\nabla}_z\times(\boldsymbol{\nabla}_z\times\mathbf{E}_\perp) = \boldsymbol{\nabla}_z(\boldsymbol{\nabla}_z\cdot\mathbf{E}_\perp) - (\boldsymbol{\nabla}_z\cdot\boldsymbol{\nabla}_z)\mathbf{E}_\perp = k_z^2\,\mathbf{E}_\perp,$$

$$\boldsymbol{\nabla}_z\times\mathbf{B}_\perp = -(i\omega/ck_L)\,\mathbf{E}_\perp - \boldsymbol{\nabla}_\perp\times\mathbf{B}_z,$$

where we have substituted from the Ampère-Maxwell equation (c) in the last line. This results, when we bring the terms with $\mathbf{E}_\perp$ to the left side,

$$\left(\frac{\omega^2}{c^2} - k_z^2\right)\mathbf{E}_\perp = -ik_z\boldsymbol{\nabla}_\perp E_z + \frac{ik_L\omega}{c}\left(\boldsymbol{\nabla}_\perp\times\mathbf{B}_z\right),$$

$$\left(\frac{\omega^2}{c^2} - k_z^2\right)\mathbf{B}_\perp = -ik_z\boldsymbol{\nabla}_\perp B_z - \frac{i\omega}{ck_L}\left(\boldsymbol{\nabla}_\perp\times\mathbf{E}_z\right). \tag{10.5.9}$$

With the help of these equations, the transversal components can be determined from the longitudinal ones.

Boundary conditions for TE and TM waves

If you set $E_z = 0$, the electric field is purely transversal. These waves are the so-called TE waves, for which the boundary conditions are to be determined, while of course (10.5.1) must be adhered to. If you multiply the 2nd line of (10.5.9) scalarly with the normal vector of the surface $\mathbf{n}$ and take into account that $\mathbf{n}\cdot\mathbf{B} = 0$, you get $\left.\dfrac{\partial B_z}{\partial n}\right|_S = 0$; S denotes the surface of the waveguide.

In the other case, you set $B_z = 0$ and thus obtain a purely transversal magnetic field. The associated waves are the TM waves. From (10.5.1) it follows that $E_z\big|_S = 0$, since E_z is a tangential component of **E**.

Rectangular waveguides

For a waveguide with a rectangular cross-section (see Fig. 10.15) with the sides L_x and L_y, the conditions at the edge are

$$
\begin{aligned}
B_x &= 0, & E_y &= 0, & E_z &= 0, & \quad\text{for}\quad x &= 0 & &\text{and}\quad x = L_x, \\
B_y &= 0, & E_z &= 0, & E_x &= 0, & \quad\text{for}\quad y &= 0 & &\text{and}\quad y = L_y.
\end{aligned}
$$

We refer back to the solution approach for **E** in the rectangle (10.5.2) and take for the z-direction a continuous, complex wave:

$$
\begin{aligned}
E_x(\mathbf{x}, t) &= A_x\, e^{ik_z z - i\omega t}\, \cos(k_x x)\, \sin(k_y y), & k_x &= n_x \pi / L_x, \\
E_y(\mathbf{x}, t) &= A_y\, e^{ik_z z - i\omega t}\, \sin(k_x x)\, \cos(k_y y), & k_y &= n_y \pi / L_y, \qquad (10.5.10) \\
E_z(\mathbf{x}, t) &= A_z\, e^{ik_z z - i\omega t}\, \sin(k_x x)\, \sin(k_y y).
\end{aligned}
$$

The Gauss's law

$$
\nabla \cdot \mathbf{E} = \Big[ik_z A_z - (k_x A_x + k_y A_y) \Big] e^{ik_z z - i\omega t}\, \sin(k_x x)\, \sin(k_y y) = 0 \quad (10.5.11)
$$

reduces the number of independent amplitudes (polarization directions) to two. From the induction equation $k_L \mathbf{B} = -\frac{i}{k}\, \nabla \times \mathbf{E}$ it follows

$$
\begin{aligned}
B_x(\mathbf{x}, t) &= -\frac{i}{k_L}(\hat{k}_y A_z - i\hat{k}_z A_y)\, e^{ik_z z - i\omega t}\, \sin(k_x x)\, \cos(k_y y), & \hat{k}_i &= k_i / k, \\
B_y(\mathbf{x}, t) &= -\frac{i}{k_L}(i\hat{k}_z A_x - \hat{k}_x A_z)\, e^{ik_z z - i\omega t}\, \cos(k_x x)\, \sin(k_y y), & & (10.5.12) \\
B_z(\mathbf{x}, t) &= -\frac{i}{k_L}(\hat{k}_x A_y - \hat{k}_y A_x)\, e^{ik_z z - i\omega t}\, \cos(k_x x)\, \cos(k_y y),
\end{aligned}
$$

with the dispersion relation

$$
\frac{n_x^2 \pi^2}{L_x^2} + \frac{n_y^2 \pi^2}{L_y^2} + k_z^2 = \frac{\omega^2}{c^2}. \tag{10.5.13}
$$

Classification of the solution

We distinguish the solutions

1. TE waves: The transverse electric wave has $E_z = 0$ and $B_z \neq 0$; it satisfies the two-dimensional Laplace equation $\Delta_\perp \mathbf{E} = 0$. At the boundary S is $\dfrac{\partial B_z}{\partial n} = 0$.

2. TM waves: The transverse magnetic wave has $B_z = 0$ and $E_z \neq 0$ and fulfills $\Delta_\perp \mathbf{B} = 0$. At the boundary is $E_z = 0$.

3. TEM waves: There are no waves with $E_z = B_z = 0$ in the waveguide (but there are in the coaxial conductor).

The general solution is a superposition of TE and TM waves.

First, we note that for the transverse wave numbers $k_x = k_y = 0$ the fields $\mathbf{E} = \mathbf{B} = 0$ disappear. So there is a lowest critical frequency

$$\omega_c^{\mathrm{TE}} = c\frac{\pi}{L_{\max}} \qquad\qquad L_{\max} = \max(L_x, L_y)\,, \tag{10.5.14}$$

where it is evident from (10.5.10) and (10.5.12) that the solution is a TE wave. If now $n_x = n_y = 1$, we can achieve with $A_z = 0$ that $E_z = 0$ and the solution is the TM wave with the lowest, i.e. the critical frequency

$$\omega_c^{\mathrm{TM}} = c\sqrt{\frac{\pi^2}{L_x^2} + \frac{\pi^2}{L_y^2}}\,. \tag{10.5.15}$$

Below this, no waves of the corresponding type exist in the waveguide.

Cutting speed

For each mode (n_x, n_y) there is a minimum frequency as a function of n_x and n_y:

$$\omega_\perp = \sqrt{\frac{n_x^2\pi^2}{L_x^2} + \frac{n_y^2\pi^2}{L_y^2}} \qquad\Rightarrow\qquad k_z = \frac{1}{c}\sqrt{\omega^2 - \omega_\perp^2}\,,$$

$$v_{\mathrm{P}} = \frac{\omega}{k_z} = \frac{c}{\sqrt{1 - \frac{\omega_\perp^2}{\omega^2}}} = \frac{c}{\sin\epsilon} \geq c. \tag{10.5.16}$$

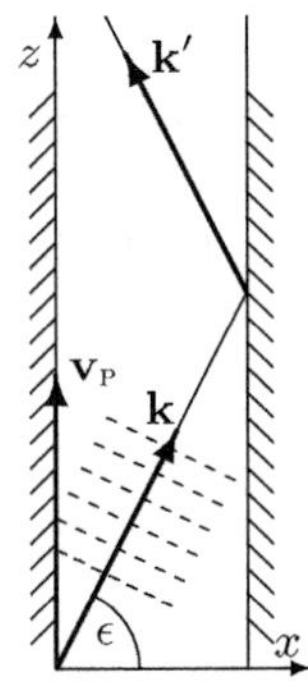

Fig. 10.17. In the waveguide, the wavefront moves with c along $\mathbf{k} = (k_x, 0, k_z)$ or $\mathbf{k}' = (-k_x, 0, k_z)$ and the phase velocity in the z-direction is then $v_{\mathrm{P}} = c/\sin\epsilon$

v_{P} is the propagation speed of the wave (phase velocity) in the z-direction. For a better understanding, we consider the TE wave with $n_x = 1, n_y = 0$. The only non-vanishing component of $\mathbf{E}$,

$$E_y = \frac{A_y}{2i}\left(\mathrm{e}^{\mathrm{i}\mathbf{k}\cdot\mathbf{x}} - \mathrm{e}^{\mathrm{i}\mathbf{k'}\cdot\mathbf{x}}\right)\mathrm{e}^{-\mathrm{i}\omega t} \quad \text{with} \quad \mathbf{k} = \left(\frac{\pi}{L_x}, 0, k_z\right) \quad \text{and} \quad \mathbf{k'} = \left(\frac{-\pi}{L_x}, 0, k_z\right),$$

consists of the superposition of two waves in the xz-plane with the angle ϵ to the x-axis, as sketched in Fig. 10.17 ($\cos\epsilon = \pm\omega_\perp/\omega$). We can imagine E_y as being created from the continuous reflection of the waves at the interfaces, where the propagation speed v_P is equal to the *intersection speed* of the wave plane with the surfaces $x = 0$ and $x = L_x$.

Wave packet and group velocity

If you superimpose plane waves, this results in a wave packet

$$\Psi(\mathbf{x}, t) = \int \frac{\mathrm{d}^3 k}{(2\pi)^3}\, F(\mathbf{k})\, \mathrm{e}^{\mathrm{i}\mathbf{k}\cdot\mathbf{x} - \mathrm{i}\omega t},$$

which is localized according to the weight of the amplitudes $F(\mathbf{k})$ of the plane partial wave $\mathrm{e}^{\mathrm{i}\mathbf{k}\cdot\mathbf{x} - \mathrm{i}\omega t}$ and the dispersion $\omega = \omega(k)$ accordingly. A wide $\mathbf{k}$ distribution leads to a wave packet limited to a narrow space. Conversely, a wave packet with a narrow $\mathbf{k}$ distribution has a large extension in real space. Assuming that $f(\mathbf{k})$ only vanishes in a small range around $\mathbf{k}_0$, one obtains from the Taylor expansion

$$\omega(k) \approx \omega(k_0) + \mathbf{v}_g \cdot (\mathbf{k} - \mathbf{k}_0), \qquad\qquad \mathbf{v}_g = \frac{\partial\omega(k_0)}{\partial\mathbf{k}}, \qquad\qquad (10.5.17)$$

where $\mathbf{v}_g$ is the *group velocity* of the wave packet. Substituting for $\mathbf{k} = \mathbf{k}_0 + \mathbf{k'}$ and introducing a distribution centered around $\mathbf{k'} = 0$ $f(\mathbf{k'}) = F(\mathbf{k}_0 + \mathbf{k'})$, we get

$$\Psi(\mathbf{x}, t) = \mathrm{e}^{\mathrm{i}\mathbf{k}_0\cdot\mathbf{x} - \mathrm{i}\omega_0 t}\, \psi(\mathbf{x}, t), \qquad \psi(\mathbf{x}, t) = \int \frac{\mathrm{d}^3 k'}{(2\pi)^3}\, f(\mathbf{k'})\, \mathrm{e}^{\mathrm{i}\mathbf{k'}\cdot(\mathbf{x} - \mathbf{v}_g t)}.$$

$\psi(\mathbf{x}, t)$, the envelope of the wave packet, moves with the group velocity $\mathbf{v}_g$. This movement is superimposed by a plane wave, the so-called pilot wave, with the phase velocity $v_p = \omega_0/k_0$.

Returning to the waveguide, the wave packet is one-dimensional (k_z) with the group velocity

$$v_g = \left.\frac{\mathrm{d}\omega}{\mathrm{d}k_z}\right|_{\omega_0} = c\sqrt{1 - \frac{\omega_\perp^2}{\omega_0^2}},$$

which, as it must be, is always less than c.

Waveguide with circular cross-section

We now distinguish from the outset between TM waves and TE waves and examine these separately.

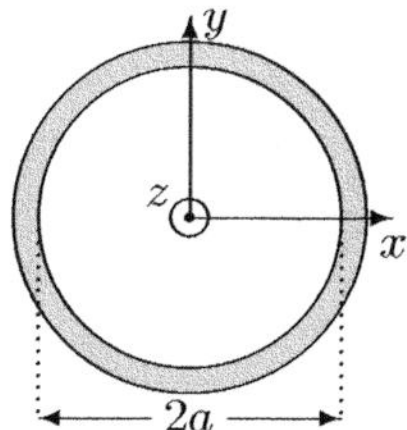

Fig. 10.18. Waveguide with circular cross-section of radius a

TM waves

For TM waves, $B_z = 0$, so that the component E_z is of primary interest, for which a continuous plane wave is assumed in the z-direction

$$E_z(\mathbf{x}, t) = E_z(\varrho, \varphi)\, \mathrm{e}^{\mathrm{i}k_z z - \mathrm{i}\omega t}\,. \tag{10.5.18}$$

Adjusted to the symmetry of the configuration, one assumes for the wave equation $\square \mathbf{E} = 0$ the Laplace operator in cylindrical coordinates (3.4.1) and separates the z-component:

$$\left(-\frac{\omega^2}{c^2} - \frac{1}{\varrho}\frac{\partial}{\partial\varrho}\varrho\frac{\partial}{\partial\varrho} - \frac{1}{\varrho^2}\frac{\partial^2}{\partial\varphi^2} + k_z^2 \right) E_z(\varrho, \varphi) = 0\,. \tag{10.5.19}$$

As demonstrated in section 3.4.1, we make for

$$E_z(\varrho, \varphi) = R(\varrho)\,\Phi(\varphi) \qquad \text{with} \qquad \Phi(\varphi) = A_n \cos(n\varphi) + B_n \sin(n\varphi)$$

a product approach and obtain

$$\left(\frac{1}{\varrho}\frac{\partial}{\partial\varrho}\varrho\frac{\partial}{\partial\varrho} + k_\perp^2 - \frac{n^2}{\varrho^2} \right) R(\varrho) = 0 \qquad \text{with} \qquad k_\perp^2 = k^2 - k_z^2\,. \tag{10.5.20}$$

With $k_\perp > 0$ (10.5.20) is the Bessel's differential equation with the solutions regular at the origin $J_n(k_\perp\varrho)$. If we also set the coordinate system so that $B_n = 0$, then

$$E_z(\varrho, \varphi) = A_n\, J_n(k_\perp\varrho)\, \cos(n\varphi) \qquad \text{with} \qquad J_n(k_\perp a) = 0\,. \tag{10.5.21}$$

Let x_{nl} be the l-th zero $J_n(x_{nl}) = 0$, then $k_{nl} = x_{nl}/a$ is the associated wave number with $n \geq 0$ and $l \geq 1$. We again have discrete, transverse wave numbers. The critical (lowest) frequency of the TM wave is determined by $n = 0$ and $l = 1$

$$\omega_c^{\mathrm{TM}} = c\, k_{01}\,. \tag{10.5.22}$$

Some values for x_{nl} are listed in Tab. B.3, p. 622. For the phase velocity in the z-direction applies

$$v_{\mathrm{P}} = \frac{\omega}{k_z} = \frac{\omega}{\sqrt{k^2 - k_{nl}^2}}\,, \qquad \omega = c\sqrt{k_z^2 + k_{nl}^2}\,, \qquad v_g = \frac{\mathrm{d}\omega}{\mathrm{d}k_z} = c\frac{k_z}{k}\,,$$

since $\omega = c\sqrt{k_z^2 + k_{nl}^2}$.

Note: Solutions with $k_\perp^2 < 0$ are the modified Bessel functions, of which only I_n at the origin are regular. However, these have no zeros, so that $E_z(a) = 0$ cannot be fulfilled on the whole circle. For TM/TE waves, $\mathbf{E}_\perp$ and $\mathbf{B}_\perp$ are orthogonal to each other.

TE waves

In TE waves, $E_z = 0$. When calculating B_z, one can directly refer to the results of the TM waves. With the approach (10.5.18)

$$B_z(\mathbf{x}, t) = B_z(\varrho, \varphi)\, e^{ik_z z - i\omega t} \tag{10.5.23}$$

one obtains

$$B_z(\varrho, \varphi) = A_n\, J_n(k_\perp \varrho)\, \cos(n\varphi) \qquad \text{with} \qquad \frac{dJ_n(k_\perp a)}{d\varrho} = 0 \,. \tag{10.5.24}$$

From (10.5.9) it follows that in TE waves

$$\left(\frac{\omega^2}{c^2} - k_z^2\right)\mathbf{B}_\perp = -ik_z\, \boldsymbol{\nabla}_\perp B_z \,.$$

This results in the boundary condition

$$\mathbf{n} \cdot \mathbf{B} = 0 \qquad\qquad \Rightarrow \qquad\qquad \frac{\partial B_z}{\partial \varrho} = 0 \,.$$

The transverse frequencies are now determined by the maxima and minima ξ_{nl} and $k_\perp = \xi_{nl}/a$. The critical frequency is $\omega = c\xi_{11}/a$.

Coaxial conductor

In Fig. 10.19 the cross section of a coaxial cable is sketched. Unlike in waveguides here besides TE and TM waves also possible TEM waves. The coaxial

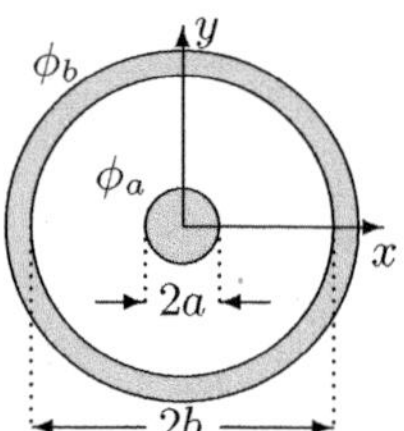

Fig. 10.19. Cross-section through a coaxial cable

conductor can transmit TEM waves "of any"frequency with c or $\bar{c}$, if there is a medium in the coaxial cable, transmit, i.e., it is suitable for transmitting wide frequency bands.

TEM waves

We start from the Maxwell equations (10.5.8) decomposed into transverse and longitudinal components and set, according to the definition of TEM waves $E_z = B_z = 0$

$$
\begin{array}{ll}
\text{(a)} \ \ \boldsymbol{\nabla}_\perp \cdot \mathbf{E}_\perp = 0 & \text{(b)} \ \ \boldsymbol{\nabla}_\perp \times \mathbf{E}_\perp = 0 \\
 & \ \ \ \ \ \boldsymbol{\nabla}_z \times \mathbf{E}_\perp = (i\omega k_L/c)\,\mathbf{B}_\perp \\
\text{(c)} \ \ \boldsymbol{\nabla}_\perp \times \mathbf{B}_\perp = 0 & \text{(d)} \ \ \boldsymbol{\nabla}_\perp \cdot \mathbf{B}_\perp = 0 \\
\ \ \ \ \ \boldsymbol{\nabla}_z \times \mathbf{B}_\perp = -(i\omega/ck_L)\,\mathbf{E}_\perp .
\end{array}
\tag{10.5.25}
$$

If we set in (10.5.9) $E_z = B_z = 0$, we obtain for TEM waves the dispersion

$$
\left(\frac{\omega^2}{c^2} - k_z^2\right)\mathbf{E}_\perp = 0 \qquad \Rightarrow \qquad \left(\frac{\omega^2}{c^2} - k_z^2\right) = 0 .
\tag{10.5.26}
$$

Furthermore, it follows from $\Box\mathbf{E} = 0$ with (10.5.26) that $\Delta_\perp \mathbf{E}_\perp = 0$. It is sufficient accordingly $\mathbf{E}_\perp$ and $\mathbf{B}_\perp$ of the two-dimensional Laplace equation. According to (10.5.25), $\boldsymbol{\nabla}\times\mathbf{E}_\perp^{(0)} = 0$, which is why

$$
\mathbf{E}_\perp^{(0)} = -\boldsymbol{\nabla}\phi(x, y).
$$

From $\boldsymbol{\nabla}_\perp \cdot \mathbf{E}_\perp^{(0)} = 0$ it follows $\Delta_\perp \phi = 0$. The potential is constant on an ideal conductor, i.e. $\phi_a = \phi(a)$ and $\phi_b = \phi(b)$, which is why ϕ only depends on ϱ:

$$
\Delta_\perp \phi(\varrho) = \frac{1}{\varrho}\frac{\partial}{\partial\varrho}\frac{1}{\varrho}\frac{\partial}{\partial\varrho}\phi(\varrho) = 0 \qquad \Rightarrow \qquad \phi = A\ln\varrho + C
\tag{10.5.27}
$$

$$
\phi_a = A\ln a + C \quad \phi_b = A\ln b + C \quad \Rightarrow \quad A = \frac{\phi_b - \phi_a}{\ln b - \ln a} \quad C = \frac{\phi_a \ln b - \phi_b \ln a}{\ln b - \ln a} .
$$

For the TEM fields one obtains using (10.5.25)

$$
\mathbf{E}_\perp(\mathbf{x}, t) = \frac{\phi_a - \phi_b}{\ln b - \ln a}\frac{1}{\varrho}\,\mathrm{e}^{\mathrm{i}k_z z - \mathrm{i}\omega t}\,\mathbf{e}_\varrho = \frac{a}{\varrho}\,E_0\,\mathrm{e}^{\mathrm{i}k_z z - \mathrm{i}\omega t}\,\mathbf{e}_\varrho ,
$$

$$
\mathbf{B}_\perp(\mathbf{x}, t) = \frac{c}{\mathrm{i}k_L\omega}\boldsymbol{\nabla}_z \times \mathbf{E}_\perp = \frac{ck_z}{k_L\omega}\frac{a}{\varrho}\,E_0\,\mathrm{e}^{\mathrm{i}k_z z - \mathrm{i}\omega t}\,\mathbf{e}_\varphi ,
\tag{10.5.28}
$$

$$
E_0 = E_\perp^{(0)}(a) = \frac{\phi_a - \phi_b}{\ln b - \ln a}\frac{1}{a} .
\tag{10.5.29}
$$

E_0 is the strength of the electric field on the surface of the inner conductor. Now $ck_z/\omega = 1$. In a medium, the right side would still have to be multiplied by n. Without a metal wire, ϕ would be singular for $\varrho = 0$ and there would be, as in the waveguide, no TEM solution.

Problems for chapter 10

10.1. *Linear, circular and elliptical polarization:* Two linearly polarized monochromatic waves $\mathbf{E}_{a,b}(\mathbf{x}, t) = \mathbf{E}_{0a,0b}\, \mathrm{e}^{\mathrm{i}\mathbf{k}\cdot\mathbf{x}-\mathrm{i}\omega t}$ with $\mathbf{k} = k\,\mathbf{e}_z$, the first of which is polarized in the x direction, the second in the y direction, have different amplitudes and phases. Determine the different possibilities of polarization of $\mathbf{E} = \mathbf{E}_a + \mathbf{E}_b$, which result from the phase difference and the amplitudes, especially in the case with $E_a = E_b$.

10.2. *Basis of a monochromatic wave:* For the monochromatic wave

$$\mathbf{E} = \left(a_1 \mathrm{e}^{\mathrm{i}\alpha_1}\, \epsilon_1 + a_2 \mathrm{e}^{\mathrm{i}\alpha_2}\, \epsilon_2\right) \mathrm{e}^{\mathrm{i}\mathbf{k}\cdot\mathbf{x}-\mathrm{i}\omega t} = I\, \epsilon_1'\, \mathrm{e}^{\mathrm{i}\mathbf{k}\cdot\mathbf{x}-\mathrm{i}\omega t}$$

specify the orthogonal basis ϵ_1' and ϵ_2' .

10.3. *Coherence matrix:* Show that the coherence matrix of independent light waves $E^{(n)}$ is the sum of the coherence matrices of the individual light waves.

10.4. *Natural vibrations inside a conductive hollow sphere*
Using the Hertz vector, determine the electric and magnetic fields inside a hollow sphere. For the natural vibrations, you get a transcendental equation.

Hint: $\mathbf{Z}_q$ should also be regular at the center of the sphere, i.e., one overlays the outgoing spherical wave with an incoming one ($\propto \sin(kr)/r$).

References

Born M. *Optik*, Springer Berlin (1933)
Born M. & Wolf E. *Principles of Optics*, 6. ed. Pergamon Press, Oxford (1986)
Landau L. D. & Lifshitz E. M. *Elektrodynamics of continuous media*, vol 8, 2nd ed. Pergamon Press (1984)
Pedrotti F. & Pedrotti L. *Introduction to Optics* 2. ed. Prentice Hall (1997)

X-Ray Scattering

On the Scattering of Electromagnetic Waves

Thomson scattering

A free electron oscillates with the electric field of the light/X-ray wave and thus generates dipole radiation of the same frequency. The scattered wave is a spherical wave with the angular distribution typical for dipole fields. The strength of the interaction is determined by the classical electron radius r_e.

Scattering on a charge distribution

In the scattering on an atom (molecule), the scattered waves originate from different points in space and therefore have different phases. The scattered waves interfere, with the phase differences depending on the direction from which the atom is observed. The incident and scattered wave have the same frequency.

Scattering on a crystal lattice

When the electromagnetic wave hits a crystal lattice, the asymptotic scattered waves of the individual atoms have no phase differences in the forward direction and their intensity is maximal there. Even with small deviations from the forward direction the waves interfere destructively, so we observe no radiation there. However, one can choose the wave vector $\mathbf{k}$ of the incident wave so that in another direction the interference becomes constructive again. Then, in addition to the transmitted beam, a diffracted beam also appears. This is the case when the difference between the incident beam $\mathbf{k}$ and the diffracted beam $\mathbf{k}'$ equals a vector $\mathbf{g}$ from the reciprocal lattice, however, since the scattering is elastic, $|\mathbf{k}'| = |\mathbf{k}|$. Then the *Bragg condition* is fulfilled. Graphically, this can be represented using a sphere of radius k, the so-called *Ewald sphere*.

D. Petrascheck, F. Schwabl, *Electrodynamics*, https://doi.org/10.1007/978-3-662-71502-4_11

k, whose direction and length are given, points from the center of the sphere to a point of the reciprocal lattice. If another lattice point lies on the surface of the *Ewald sphere*, the Bragg condition is fulfilled, since both vectors are of the same length and the difference is a vector from the reciprocal lattice.

In this description of scattering, it is assumed that neither the the incident wave is weakened by scattering, nor are the scattered waves scattered by further lattice atoms. The scattering is kinematic.

Dynamical diffraction

Now we assume that both the incident wave and the scattered wave are in a medium with a periodic dielectric function ϵ_r, where for X-ray radiation the average dielectric constant is almost 1 ($|1 - \epsilon_r| \lesssim 10^{-5}$).

We are now looking for solutions for the Maxwell equations in the medium, a single crystal, when a incident plane wave hits it. If one is far from a Bragg condition, the wave (almost) passes through the crystal undisturbed. Near the Bragg condition, the (approximate) solution consists of two wave fields with slightly different wave vectors, which interfere and thus form intensity oscillations (*Pendellösungen*). Characteristic is also an anomalous absorption, as the two wave fields in the crystal are damped differently: One wave field has the maxima between the atoms, one at the atoms; the latter is weakened more. Each of these wave fields consists of a transmitted and a reflected wave, which individually are not solutions of the Maxwell equations. At the back surface of the crystal, the outgoing waves separate into a transmitted and a reflected beam.

In very thin crystals, significantly below $100\,\mu$m, the intensity of the scattered wave increases just like in kinematic scattering with thickness. Beyond that, the conservation of energy flow becomes noticeable after which in the absence of absorption the total intensity of transmitted and reflected beam must be equal to that of the incident beam. However, since the transmitted and the reflected waves in the crystal are coherent, there are oscillations between transmitted and reflected intensity.

The dynamical theory of electron and especially of neutron diffraction is very similar to that of X-ray diffraction. In particle beams, however, instead of the Maxwell equations, the Schrödinger equation is responsible for the dynamics The solution method itself has analogies to weakly bound electrons in the solid[1].

The particles in the crystal near a Bragg plane have a very low effective mass, i.e., a very small momentum is sufficient to drastically change the direction of the particles.

---

[1] The energy eigenvalues of weakly bound particles are near the zone boundary (Bragg plane) given by a two-level system (2×2 matrix) with small splitting. In scattering, one has the same system of equations, only one is looking for the wave vectors for a given energy.

11.1 Scattering of Light by Electrons

11.1.1 Scattering on Free Electrons

The basic mechanism for the scattering of light on stationary electrons goes back to J. J. Thomson [1893]. The electrons follow the electric field vector of the light and are thus excited to oscillate. You then have oscillating dipoles that emit Hertzian dipole radiation. As long as the frequency of the incident radiation is not too high, the dipole radiation has the same frequency as the incident wave. If the frequency becomes too high, the probability increases that the electron from the light wave, i.e. from the photon, gets a hit and flies away; the scattered radiation then has a correspondingly lower frequency. This is the *Compton effect*, which is discussed in more detail in section 14.3.2.

For our purposes, it is sufficient to keep in mind that the following considerations only apply to wavelengths that are significantly above the Compton wavelength (see Tab. C.7, p. 645) $\lambda_c = 2.43 \times 10^{-10}$ cm. Light with $\lambda < \lambda_c$ is likely to knock the electron away. It is then better to speak of particles (photons) than of light waves.

The incident electromagnetic waves are usually X-rays, whose wavelengths are in the angstrom range. The electrons of the electron shell of atoms with binding energies on the order of electron volts can, compared to the energy of 10 keV of the X-ray, be considered free. At this energy, as explained in the introduction, one has a wavelength of 1.24 Å and a frequency of $\nu = c/\lambda = 2.41 \cdot 10^{18}$ Hz.

The potentials of an incident plane wave are

$$\phi(\mathbf{x}, t) = 0 \qquad \text{and} \qquad \mathbf{A}(\mathbf{x}, t) = \mathbf{A}_0 \, e^{i(\mathbf{k}\cdot\mathbf{x} - \omega t)} . \qquad (11.1.1)$$

This results in the fields

$$\mathbf{E}(\mathbf{x}, t) = -\frac{k_L}{c}\frac{\partial \mathbf{A}}{\partial t} - \boldsymbol{\nabla}\phi = -\frac{k_L}{c}\frac{\partial \mathbf{A}}{\partial t} = i k_L \frac{\omega}{c}\mathbf{A} \qquad (11.1.2)$$

$$\mathbf{B}(\mathbf{x}, t) = \boldsymbol{\nabla}\times\mathbf{A} = i\mathbf{k}\times\mathbf{A} = (1/k_L)\hat{\mathbf{k}}\times\mathbf{E}. \qquad (11.1.3)$$

Here $\omega = kc$ and $\hat{\mathbf{k}} = \mathbf{k}/k$. The average energy flow of the incident beam is given by the Poynting vector (8.2.64):

$$\langle \mathbf{S} \rangle = \frac{ck_L}{8\pi k_C}\mathbf{E}\times\mathbf{B}^* \stackrel{(11.1.3)}{=} \frac{c}{8\pi k_C}|\mathbf{E}_0|^2\hat{\mathbf{k}} = \frac{1}{2Z_0}|\mathbf{E}_0|^2\hat{\mathbf{k}}. \qquad (11.1.4)$$

Here, $Z_0 = \frac{4\pi k_C}{c}$ is the vacuum impedance (8.2.65') and

$$\mathbf{E}(\mathbf{x}, t) = \mathbf{E}_0 \, e^{i(\mathbf{k}\cdot\mathbf{x} - \omega t)}. \qquad (11.1.5)$$

The initially stationary electron in the origin is subject to the Coulomb force

$$m_e \ddot{\mathbf{s}} = e\mathbf{E}(\mathbf{s}, t) \approx e\mathbf{E}_0 \, e^{-i\omega t} \qquad (11.1.6)$$

and forces the electron into oscillatory movements. The situation can be seen in Fig. 11.1.

We assume that the electron only reaches velocities $v_0 \ll c$. The Lorentz component $e(k_L/c)\,\mathbf{v} \times \mathbf{B}$ can then be neglected. For (11.1.6) the approach $\mathbf{s} = \mathbf{s}_0 e^{-i\omega t}$ leads to

$$|\dot{\mathbf{s}}_0| = v_0 = \frac{e_0 E_0}{m_e \omega} \qquad \text{and} \qquad s_0 = \frac{e_0 E_0}{m_e \omega^2} = \frac{v_0}{ck}. \tag{11.1.7}$$

This justifies the assumption $\mathbf{E}(\mathbf{s}, t) = \mathbf{E}_0(0, t)$, since with $v_0/c \ll 1$ also $ks_0 = v_0/c \ll 1$.

For X-ray radiation of $\hbar\omega = 10\,\mathrm{keV}$ ($\lambda = 1.24\,\text{Å}$) field strengths of $E_0 \approx 8.7 \times 10^9\,\mathrm{statV\,cm^{-1}} \widehat{=} 2.6 \times 10^{14}\,\mathrm{V\,m^{-1}}$ necessary for $v_0/c = 0.01$. With this field strength, $s_0 = 2 \times 10^{-11}\,\mathrm{cm}$. It is not only visible that all prerequisites for the classical treatment are met, but also that for an atom/molecule the phase differences between the scattered waves of the individual electrons are relevant. The oscillating dipole moment is created by the movement of the

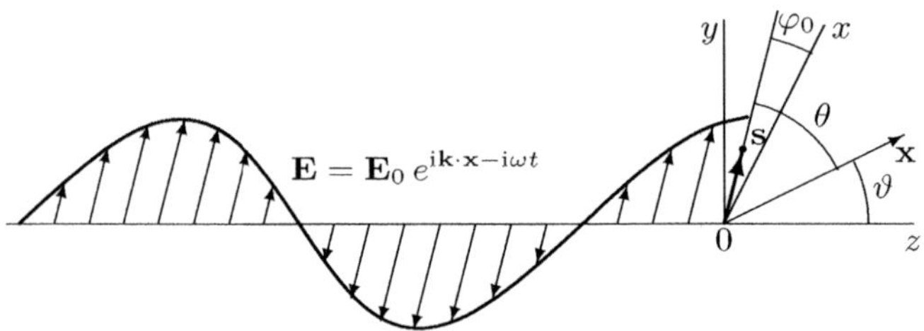

Fig. 11.1. The electric field $\mathbf{E}$ of the incident radiation excites the electron to Hertzian dipole oscillations with the deflection $\mathbf{s}(t)$, where the maximum deflection $s_0 \ll \lambda$

charge

$$\mathbf{p}(t) = e\,\mathbf{s}(t) = \frac{e_0^2}{m_e \omega^2}\,\mathbf{E}_0 e^{-i\omega t}. \tag{11.1.8}$$

The radiation fields of this dipole moment are according to (8.4.12)

$$\mathbf{B}_s(\mathbf{x}, t) = \frac{k_C}{k_L c^2 r}\,\ddot{\mathbf{p}}(t_r) \times \mathbf{e}_r = \frac{r_e}{k_L r}\,e^{ikr - i\omega t}\,\mathbf{E}_0 \times \mathbf{e}_r\,,$$

$$\mathbf{E}_s(\mathbf{x}, t) = -\frac{k_C}{c^2 r}\,\mathbf{e}_r \times \left(\ddot{\mathbf{p}}(t_r) \times \mathbf{e}_r\right) = -\frac{r_e}{r}\,\mathbf{e}_r \times \left(e^{ikr - i\omega t}\,\mathbf{E}_0 \times \mathbf{e}_r\right). \tag{11.1.9}$$

$t_r = t - r/c$ is the retarded time and $r_e = k_C e^2/m_e c^2$ the classical electron radius. The scattered waves are spherical waves with r_e as pre-factor.

Intensity and polarization

In scattering, one considers the energy radiated into the solid angle element $d\Omega$. For the electric field of moving electrons, we have an electric dipole radiation in the far zone, for which we have derived the intensities in (8.4.14) and (8.4.15):

$$\langle \mathbf{S}_s \rangle = \frac{k_L}{2Z_0} \mathbf{E}_s \times \mathbf{B}_s^* = \frac{1}{2Z_0} \frac{r_e^2}{r^2} |\mathbf{E}_s|^2 \mathbf{e}_r . \tag{11.1.10}$$

From this, we get for the radiation scattered into the solid angle element $d\Omega$, divided by the flux of the incoming radiation (11.1.4)

$$\frac{d\sigma}{d\Omega} = \frac{|\langle \mathbf{S}_s \rangle|}{|\langle \mathbf{S} \rangle|} r^2 = \frac{|\mathbf{E}_s|^2}{|\mathbf{E}_0|^2} = r_e^2 \sin^2 \theta . \tag{11.1.11}$$

According to Fig. 11.1, the propagation direction of the plane wave is $\mathbf{k} = k\mathbf{e}_z$. The transverse oscillation plane of the electric field is defined in the xy-plane by φ_0:

$$\mathbf{E}_0 = E_0 \left(\cos \varphi_0 \, \mathbf{e}_x + \sin \varphi_0 \, \mathbf{e}_y \right) .$$

Parallel to $\mathbf{E}$ oscillates $\mathbf{p}$. θ is determined by the scalar product

$$\mathbf{E}_0 \cdot \mathbf{e}_r = E_0 \cos \theta = \cos \varphi_0 \mathbf{e}_x \cdot \mathbf{e}_r + \sin \varphi_0 \mathbf{e}_x \cdot \mathbf{e}_r$$
$$= E_0 \sin \vartheta \left(\cos \varphi_0 \cos \varphi + \sin \varphi_0 \sin \varphi \right) = E_0 \sin \vartheta \cos(\varphi - \varphi_0) .$$

For unpolarized X-rays the φ_0 are statistically distributed and average out:

$$\langle \cos^2 \theta \rangle = \langle \sin^2 \vartheta \, \cos^2 (\varphi_0 - \varphi) \rangle = \frac{\sin^2 \vartheta}{2} .$$

For the differential scattering cross section of unpolarized X-ray radiation on a free electron, we obtain the Thomson scattering formula shown in Fig. 11.2

$$\langle \frac{d\sigma}{d\Omega} \rangle = r_e^2 \frac{1}{2}(1 + \cos^2 \vartheta) . \tag{11.1.12}$$

$r_e^2 \approx 8 \times 10^{-26} \mathrm{cm}^2 = 0.08$ barn is the area characteristic for the differential cross section. The same angular dependence is obtained for an electron orbiting in the xy plane. The intensity itself is axially symmetric, but this does not apply to the individual incident linearly polarized waves. A wave incident at φ_0 is with

$$\sin^2 \theta = 1 - \sin^2 \vartheta \, \cos^2(\varphi - \varphi_0) \tag{11.1.13}$$

preferably scattered in the direction perpendicular to φ_0, $\varphi = \varphi_0 \pm \pi/2$. The scattered radiation of the unpolarized incident X-rays is thus partially polarized.

Total cross section

The angular integration gives the total Thomson cross section

$$\sigma_t = \frac{r_e^2}{2} \int_0^{2\pi} \mathrm{d}\varphi \int_0^{\pi} \sin\vartheta \, \mathrm{d}\vartheta \left(1 + \cos^2\vartheta\right) = \frac{8\pi r_e^2}{3} = 0.665 \times 10^{-24}\,\mathrm{cm}^2. \quad (11.1.14)$$

Thomson scattering is elastic, i.e., the electron oscillates exactly with the frequency of the incident radiation. This only applies to "large" wavelengths. For sufficiently small wavelength $\lambda \sim \lambda_c \approx 2.4 \times 10^{-2}\,\text{Å}$, the Compton wavelength of the electron (Tab. C.7, p. 645), the photon imparts an (elastic) impact to the electron. Using energy-momentum conservation, the frequency of the scattered radiation is calculated in section 14.3.2.

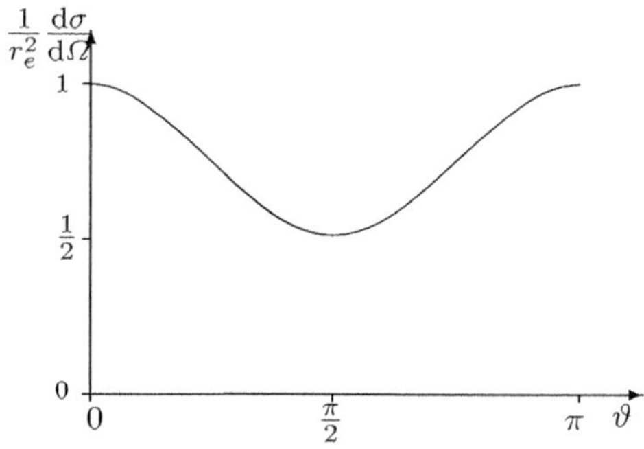

Fig. 11.2. The (average) differential scattering cross section of Thomson scattering depending on the scattering angle ϑ

The Thomson formula is therefore only valid for wavelengths $\lambda \gg \lambda_c$. For X-rays with $\lambda > 0.5\,\text{Å}$ this condition is fulfilled. The characteristic radiations used in X-ray scattering lie between $0.55\,\text{Å}$ (Ag) and $2.3\,\text{Å}$ (Cr).

11.1.2 Scattering on Weakly Bound Electrons

A phenomenological approach to the polarizability of atoms is given by the Lorentz-Drude or oscillator model (5.5.5). The electrons j are bound to the nucleus with springs $k_j = m_e \omega_j^2$ depending on the strength of the binding forces and are subject to a damping γ_j, in which all processes are included, through which the electron loses energy

$$m_e\left(\ddot{\mathbf{x}}_j + \gamma_j \dot{\mathbf{x}}_j + \omega_j^2 \mathbf{x}_j\right) = -e_0\,\mathbf{E}_0 e^{-\mathrm{i}\omega t}.$$

Again, the electric field $\mathbf{E}$ is periodic in time, which then also applies to the dipole moment of the electron, which oscillates with $\mathbf{E}$:

$$\mathbf{p}_j = -e_0\,\mathbf{x}_j = \frac{e_0^2}{m_e}\,\frac{1}{\omega_j^2 - \omega^2 - \mathrm{i}\gamma_j\omega}\,\mathbf{E} = \alpha_j(\omega)\,\mathbf{E}. \quad (11.1.15)$$

The contribution of the j-th electron to the polarizability (5.5.6) is thus

$$\alpha_j(\omega) = \frac{e_0^2}{m_e} \frac{1}{\omega_j^2 - \omega^2 - i\gamma_j\omega}.$$ (11.1.16)

We now substitute $\mathbf{p}$ (11.1.15) into the scattering field (11.1.9):

$$\mathbf{E}_s = r_e\,\alpha_j(\omega)\,\mathbf{e}_r \times (\mathbf{E}_0 \times \mathbf{e}_r)\,\frac{\mathrm{e}^{ikr-i\omega t}}{r}.$$ (11.1.17)

With regard to the scattering field (11.1.9), only the polarizability of the electron has changed, by replacing $\alpha = -e^2/m_e\,\omega^2$ with (11.1.16). Accordingly, the scattered waves of the free and the weakly bound electrons differ only by the factor $-\alpha_j\,\omega^2$, which is why the differential scattering cross section (11.1.12) for the weakly bound electrons only needs to be multiplied by $|\omega^2\,\alpha_j|^2$:

$$\frac{\mathrm{d}\sigma}{\mathrm{d}\Omega} = \left(\frac{\mathrm{d}\sigma}{\mathrm{d}\Omega}\right)_t |\omega^2\alpha_j(\omega)|^2 = r_e^2\,\frac{1}{2}\,(1+\cos^2\vartheta)\,\frac{\omega^4}{(\omega^2-\omega_j^2)^2 + \gamma_j^2\,\omega^2}.$$ (11.1.18)

The right side again assumes incoming unpolarized light. However, the scattered light is not axially symmetric with respect to its polarization according to (11.1.13).

Rayleigh scattering

For frequencies that are smaller than the excitation frequencies ω_j, $\omega \ll \omega_j$, (11.1.18) simplifies to

$$\frac{\mathrm{d}\sigma}{\mathrm{d}\Omega} = \left(\frac{\mathrm{d}\sigma}{\mathrm{d}\Omega}\right)_t \frac{\omega^4}{\omega_j^4}.$$ (11.1.19)

The frequencies in question here are in the visible range and below. Sunrays that hit air particles generate oscillating electric moments in them, which in turn emit light. Shorter wavelengths are scattered more strongly, so that visible light experiences a frequency shift towards the blue color.

Rayleigh scattering is used to explain both the blue color of the sky and the red color of the setting sun. In the latter case, the short-wave component is scattered more on the longer path through the atmosphere, so that the long-wave component predominates.

Resonance scattering

For frequencies $\omega \approx \omega_j$, the differential scattering cross section (11.1.18) has a resonance, the width of which depends on the strength of the damping γ_j. If a bound electron emits scattering radiation, the dissipative force $\mathbf{F}_{\mathrm{rad}}$ (8.5.2) must be taken into account in the equation of motion, which is given by the Abraham-Lorentz equation:

$$\mathbf{F}_{\mathrm{rad}} = k_C \frac{2e^2}{3c^3} \,\dddot{\mathbf{v}} \overset{(8.5.7)}{=} -k_C \frac{2e^2\omega_j^2}{3c^3}\,\mathbf{v} = -m_e\gamma_j\,\mathbf{v}$$

(we have omitted the index j of the considered electron in force and velocity). It was used that for bound motion $\ddot{\mathbf{v}} = -\omega_j^2\mathbf{v}$ (see (8.5.7)). Now we insert into the Abraham-Lorentz equation of motion (8.5.4) and obtain the equation of motion (8.5.8) of the damped harmonic oscillator

$$m_e\,\dot{\mathbf{v}} = \mathbf{F}_{\mathrm{ext}} - m_e\gamma_j\mathbf{v}\,. \tag{11.1.20}$$

The damping resulting solely from radiation reaction is very small, so that the resonance in (11.1.18) becomes very sharp. In Fig. 11.3 the course of the

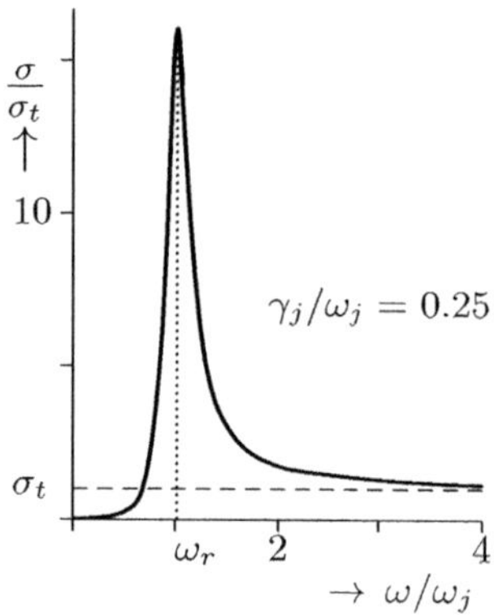

Fig. 11.3. Resonance scattering on an electron; σ_t is the Thomson scattering cross section, ω_j is the frequency of the oscillator and γ_j its damping

total scattering cross section

$$\sigma(\omega) = \sigma_t \frac{\omega^4}{(\omega^2-\omega_j^2)^2 + \gamma_j^2\omega^2} \qquad \Rightarrow \qquad \omega_r^2 \approx \frac{\omega_j^2}{1-\gamma_j^2/2\omega_j^2},$$

$$\sigma_r = \sigma(\omega_r) = \sigma_t \frac{\omega_j^2}{\gamma_j^2}\frac{1}{1-\gamma_j^2/4\omega_j^2} \tag{11.1.21}$$

sketched as a function of ω, where the frequency of the resonance ω_r for stronger damping γ_j is slightly shifted to higher frequencies, but the sharpness of the resonance decreases drastically.

The linewidth of the resonance is determined by γ_j. We try an estimation, where we use the H-atom as a reference. ω_j is determined by the transition from the ground state with $n = 1$ to the state $n = 2$ (Lyman series), whose energy according to (C.2.14) is given by

$$k_C^2\frac{m_e e^4}{2\hbar^2}\Big(1 - \frac{1}{4}\Big) = \hbar\omega_0 \quad \Rightarrow \quad \hbar\omega_0 = \frac{3}{8}m_e c^2\alpha_f^2 \quad \text{and} \quad \alpha_f = k_C\frac{e^2}{\hbar c} \approx \frac{1}{137}.$$

α_f is the fine structure constant. $\omega_j = \omega_0$ is now the oscillator frequency relevant for the model. If there is no further damping than the radiation reaction, then

$$\gamma_0 = k_C \frac{2e^2}{3m_e c^3}\omega_0^2 = \frac{2}{3}\frac{\hbar\omega_0}{m_e c^2}\,\alpha_f\,\omega_0 = \frac{1}{4}\,\alpha_f^3\,.$$

The dimensionless factor for the damping and the height of the resonance are

$$\frac{\gamma_0}{\omega_0} = \frac{1}{4}\alpha_f^3 \qquad \text{and} \qquad \sigma_r = \frac{8\pi}{3}\,r_e^2\,\frac{16}{\alpha_f^6}\sigma_r = \frac{2\pi}{3}\left(\frac{8\,a_B}{\alpha_f}\right)^2. \qquad (11.1.22)$$

In this case, γ_0 is referred to as the *natural linewidth*. If the scattering cross section is given with $\sigma_r = \pi R^2$, then $R \sim 900\,a_B$, where $a_B = 0.529\,\text{Å}$ is the Bohr radius.

A broadening of the lines, associated with a frequency shift, is due, among other things, to the movement of the atoms (Doppler broadening) or to the recoil that the emitted (absorbed) photon (γ-quantum) transfers to the atom. In some crystals, however, the recoil can be absorbed by the crystal lattice as by ^{57}Fe at $\hbar\omega_0 = 14.4\,\text{keV}$. This is the Mössbauer effect. Lines are achieved that are so sharp ($\gamma_0/\omega_0 \sim 10^{-13}$), that the energy loss of the photon in the gravitational field of the earth can be detected on them.

11.1.3 Scattering on a Charge Distribution

After the contribution of a single electron to the scattering, the contribution of an atom, i.e. a charge distribution, is now calculated by summing the scattering waves of the individual electrons. For this, one needs the expression for the scattered wave (11.1.9), when the electron is not at the origin, but at location $\mathbf{x}'$, as sketched in Fig. 11.4.

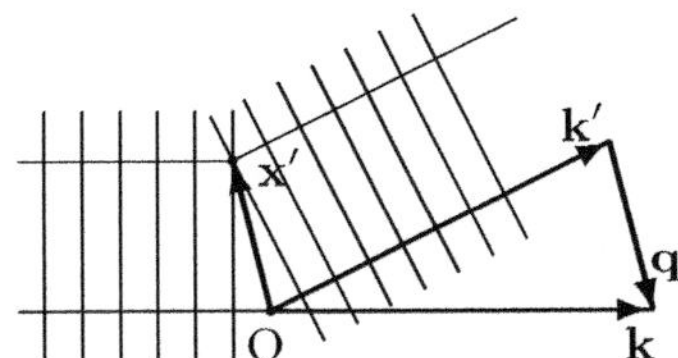

Fig. 11.4. The scattering on a charge distribution; around O one sees the phase shift $\mathbf{q} \cdot \mathbf{x}'$ between O and $\mathbf{x}'$, it is here small; $\mathbf{q}$ is the scattering vector

The incident wave, which had the value $\mathbf{E}_0$ at the origin, has at location $\mathbf{x}'$ with $\mathbf{E}_0\,e^{i\mathbf{k}\cdot\mathbf{x}'}$ an additional phase factor received. In addition, one replaces

$$\mathbf{e}_r = \frac{\mathbf{x}}{r} \quad \text{by} \quad \frac{\mathbf{x}-\mathbf{x}'}{|\mathbf{x}-\mathbf{x}'|}:$$

$$\mathbf{E}_s(\mathbf{x},\mathbf{x}',t) = -r_e\left\{\frac{\mathbf{x}-\mathbf{x}'}{|\mathbf{x}-\mathbf{x}'|}\times\left(\mathbf{E}_0\,e^{i\mathbf{k}\cdot\mathbf{x}'}\times\frac{\mathbf{x}-\mathbf{x}'}{|\mathbf{x}-\mathbf{x}'|}\right)\right\}\frac{e^{-i\omega t+ik|\mathbf{x}-\mathbf{x}'|}}{|\mathbf{x}-\mathbf{x}'|}.$$

The difference $|\mathbf{x}-\mathbf{x}'|$ of r only needs to be considered in the exponent, which corresponds to the far zone of the multipole radiation of a charge distribution:

$$k|\mathbf{x} - \mathbf{x}'| \approx kr - k\,\frac{\mathbf{x} \cdot \mathbf{x}'}{r} = kr - \mathbf{k}' \cdot \mathbf{x}' \qquad \text{with} \qquad \mathbf{k}' = k\frac{\mathbf{x}}{r} = k\mathbf{e}_r\,.$$

The summation of the scattered waves results in

$$\mathbf{E}_s(\mathbf{x}, t) = -r_e \int \mathrm{d}^3 x' \left(\mathbf{e}_r \times (\mathbf{E}_0\, \mathrm{e}^{\mathrm{i}\mathbf{k}\cdot\mathbf{x}'} \times \mathbf{e}_r) \right) \frac{n(\mathbf{x}')}{r}\, \mathrm{e}^{\mathrm{i}kr - \mathrm{i}\mathbf{k}'\cdot\mathbf{x}' - \mathrm{i}\omega t}\,.$$

$\rho(\mathbf{x}) = e\,n(\mathbf{x})$ is the charge density of the electrons of the atom (molecule). The atomic form factor

$$f_0(\mathbf{q}) = \int \mathrm{d}^3 x'\, n(\mathbf{x}')\, \mathrm{e}^{-\mathrm{i}\mathbf{q}\cdot\mathbf{x}'} \tag{11.1.23}$$

with the scattering vector $\mathbf{q} = \mathbf{k}' - \mathbf{k}$ is defined as the Fourier transform of the electron density. It takes into account the finite extent of the system through the different phases with which the individual areas contribute to the scattering wave.

Note: Calculations of the atomic form factors can be found in the International Tables for Crystallography. The following sum of exponential functions is an approximation for the form factor of silicon (see Tab. 10.5.9, p. 383):

$$f\!\left(\frac{\sin\theta}{\lambda}\right) = \sum_{i=1}^{4} a_i\, \mathrm{e}^{-b_i(\sin\theta/\lambda)^2} + c\,.$$

Fig. 11.5 shows the form factor for silicon calculated with the above approximation.

Tab. 11.1. Atomic form factor; coefficients for silicon

$a_1 = 6.2915$	$a_2 = 3.0353$	$a_3 = 1.9891$	$a_4 = 1.5410$
$b_1 = 2.4386$	$b_2 = 32.3337$	$b_3 = 0.6785$	$b_4 = 81.6937$ $c = 1.1407$

The atomic scattering wave is now determined as

$$\mathbf{E}_s(\mathbf{x}, t) = -r_e\, f_0(\mathbf{q})\, (\mathbf{e}_r \times (\mathbf{E}_0 \times \mathbf{e}_r)) \frac{1}{r}\, \mathrm{e}^{\mathrm{i}kr - \mathrm{i}\omega t}\,. \tag{11.1.24}$$

The differential cross section for the atom is given as

$$\left(\frac{\mathrm{d}\sigma}{\mathrm{d}\Omega}\right)_a = \left(\frac{\mathrm{d}\sigma}{\mathrm{d}\Omega}\right)_e |f_0(\mathbf{q})|^2\,. \tag{11.1.25}$$

Dispersion corrections

In (11.1.23)–(11.1.25) the electrons are free particles, which is justified when comparing the $\sim 10\,\mathrm{keV}$ of X-ray radiation with $\sim$eV binding energies of the electrons.

In a more precise study, however, one would also take into account the influence of the binding forces of the electrons to the nucleus which leads to so-called dispersion corrections. Especially from the inner shells, where the electrons are more localized, we expect contributions to $f_0(\mathbf{q})$, which decrease less with q than (11.1.23):

$$f(\mathbf{q}) = f_0(\mathbf{q}) + \Delta f' - \mathrm{i}\Delta f'' . \tag{11.1.26}$$

$\Delta f'(\mathbf{q})$ and $\Delta f''(\mathbf{q})$ are dispersion or *Hönl* corrections. They naturally depend on ω, but are usually given for a specific X-radiation, like Cu-K$_{\alpha 1}$, i.e. for a specific ω, so that the ω-dependence is not listed separately.

In the Lorentz model (5.5.7), the binding of the electrons is given by harmonic forces between electron shells and the nucleus. The shells are represented as point-like particles by their center of charge. If we now combine the continuous density distribution $n_k(\mathbf{x})$ of the k-th electron with the oscillator model (5.5.6), then we obtain a generalized electron density

$$\tilde{n}(\mathbf{x}) = \sum_k g_k\, n_k(\mathbf{x}) \quad \text{with} \quad g_k = \frac{f_k\omega^2}{\omega^2 - \omega_k^2 + \mathrm{i}\gamma_k\omega} \approx f_k\left(1 + \frac{\omega_k^2}{\omega^2} - \mathrm{i}\frac{\gamma_k}{\omega}\right),$$

in which the g_k contain the resonances of the electron-photon interaction. From the g_k it is evident that the absorption via $\Delta f''$ decreases with $1/\omega$

$$f(\mathbf{q}) = \sum_k f_k \int \mathrm{d}^3 x\, \mathrm{e}^{-\mathrm{i}\mathbf{x}\cdot\mathbf{q}}\, n_k(\mathbf{x})\left(1 + \frac{\omega_k^2}{\omega^2} - \mathrm{i}\frac{\gamma_k}{\omega}\right) = f_0(\mathbf{q}) + \frac{\overline{\omega^2}}{\omega^2} - \mathrm{i}\frac{\gamma}{\omega}.$$

Such phenomenological approaches are not sufficient for quantitative statements.

Thus, X-ray scattering provides information about the charge distribution in solids. The scattering considered here is coherent because the amplitudes of the scattered waves of the individual charges are superimposed. However, if the intensities are overlaid, the scattering is referred to as incoherent.

Interference effects only occur in coherent radiation. The extension of the

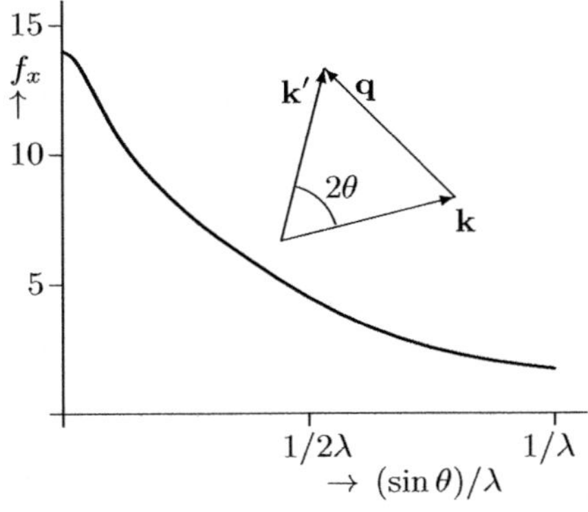

Fig. 11.5. Atomic form factor of Si (Z=14). θ is the half scattering angle; the curve was drawn for Mo-K$_\alpha$ radiation with $\lambda = 0.7107$ Å

scattering cross-section to the (coherent) scattering on several atoms (liquid, crystal) is evident. It will be discussed later. The corresponding theory is known from electrodynamics and solid state physics. The scattering at the atom can be understood as the sum of the scattering amplitudes of the individual electrons. Far from the atom, the scattering wave of the atom can be considered as a spherical wave. The superposition of the spherical waves from regularly arranged atoms leads to interferences, which find their expression in the Bragg condition. This is a purely geometric relationship between the wavelength of the incident beam and the rigid crystal lattice of the solid.

The next step, which goes beyond the geometry of the scattering process, the kinematics, was made independently by Ewald [1916] and by Laue (1917). This dynamical theory still forms the essential part of X-ray diffraction. It is the basis for crystal optics with X-rays. So much for elastic scattering.

Recently, however, inelastic, i.e. diffuse scattering has come more to the fore. Although the excitation energies in the solid (phonons, etc.) with 0.01–0.1 eV are five orders of magnitude smaller than the energy of the incident beam, the ever more advanced measurement methods allow an ever better evaluation of inelastic scattering.

To justify the classical non-relativistic approach, one should also estimate the speed with which electrons move around the nucleus. The starting point is a hydrogen-like atom with the atomic number Z and the potential $V = -Ze^2/r$. According to the virial theorem, the expectation values are

$$2\langle T \rangle = \langle \mathbf{x} \cdot \boldsymbol{\nabla} V \rangle = -\langle V \rangle \qquad \Rightarrow \qquad \mathrm{E} = \langle T + V \rangle = -\langle T \rangle.$$

The energy eigenvalues are known to be [Schwabl, 2007, §6.3]

$$\mathrm{E}_n = -k_\mathrm{C}^2 \frac{m_e Z^2 e^4}{2\hbar^2 n^2} \qquad n = 1, 2, \ldots \tag{C.2.14}$$

On circular orbits, the electron has a constant velocity v, which is given by $\langle T \rangle = m_e v^2/2$, from which it follows

$$v = k_\mathrm{C} \frac{Z\,e^2}{n\,\hbar} = \frac{Z}{n} c\alpha_f \qquad \text{with} \qquad \alpha_f = k_\mathrm{C} \frac{e^2}{\hbar c} \approx \frac{1}{137}.$$

From this, it becomes clear that the velocities of the electrons in heavier nuclei need to be treated relativistically.

11.1.4 Scattering on the Lattice

After we were able to specify the scattering on the atom by adding the contributions of the individual electrons, we now determine the scattering of a light wave on a crystal by summing up the contributions of the individual, regularly arranged atoms to the scattered wave.

Analogous to scattering on the atom, where we had to take into account the phases of the incident wave at the locations of the electrons, we now have to add the phases of the incident wave at the lattice points.

Linear chain

We now start from the atomic scattering wave (11.1.24), where the atoms are supposed to form a linear chain of $N = 2M + 1$ atoms. The chain with the lattice constant a lies on the x-axis.

$$\mathbf{E}_s(\mathbf{x}, t) = -r_e\, f_0(\mathbf{q}) \sum_{n=-M}^{M} \mathbf{e}_r \times \left(\mathbf{E}_0\, \mathrm{e}^{-i\mathbf{q}\cdot a n \mathbf{e}_x} \times \mathbf{e}_r\right) \frac{1}{r}\, \mathrm{e}^{ikr - i\omega t}. \qquad (11.1.27)$$

Now is $(\mathbf{q} = \mathbf{k}' - \mathbf{k})$

$$S(q_x) = \sum_{n=-M}^{M} \mathrm{e}^{-iq_x a n} = \mathrm{e}^{iq_x a M} \sum_{n=0}^{2M} \mathrm{e}^{-iq_x a n} = \mathrm{e}^{iq_x a M}\, \frac{\mathrm{e}^{-iq_x a(2M+1)} - 1}{\mathrm{e}^{-iq_x a} - 1}$$

$$= \frac{\sin\left(\frac{q_x a}{2} N\right)}{\sin\frac{q_x a}{2}} \underset{m\text{ whole}}{=} \frac{\sin\left[\left(\frac{q_x a}{2} - m\pi\right)N\right]}{\sin\left(\frac{q_x a}{2} - \pi m\right)} = \frac{\sin\left[\frac{a}{2}\left(q_x - \frac{2\pi m}{a}\right)N\right]}{\sin\left[\frac{a}{2}\left(q_x - \frac{2\pi m}{a}\right)\right]}.$$

As long as (in the limit $N \to \infty$) the denominator is finite, disappears due to the rapid variation of the numerator every integral in this area. For values $q_x \approx 2\pi m/a$ the sine in the denominator can be approximated by its argument and we have the usual representation for a δ-function (see Tab. B.4, p. 627). For the structure factor we get so

$$S(q_x) = \sum_{m=-\infty}^{\infty} \lim_{N \to \infty} \frac{\sin\left[\frac{a}{2}\left(q_x - \frac{2\pi m}{a}\right)N\right]}{\frac{a}{2}\left(q_x - \frac{2\pi m}{a}\right)} = \sum_{m=-\infty}^{\infty} \frac{2\pi}{a}\, \delta\left(q_x - \frac{2\pi m}{a}\right). \qquad (11.1.28)$$

$2\pi/a$ is the lattice constant of the reciprocal chain to the linear chain and $g = 2\pi m/a$ is a lattice point of this chain.

The above structure factor is also referred to as *lattice delta function* because their contributions in the form of δ-functions only come from points of the reciprocal lattice.

Three-dimensional Bravais lattice

A Bravais lattice is a point lattice in which all lattice points have the same environment. There are three base vectors $\mathbf{a}_i$ with which you can reach every lattice point:

$$\mathbf{a_n} = \sum_{i=1}^{3} n_i\, \mathbf{a}_i \qquad \text{with} \qquad n_i = 0, \pm 1, \pm 2, \ldots$$

A primitive unit cell contains exactly one lattice point, like the parallelepiped that is spanned by the three base vectors. If you extend the structure factor (11.1.28) to three dimensions, one obtains

$$S(\mathbf{q}) = \frac{(2\pi)^3}{v_c} \sum_{\mathbf{g}} \delta^{(3)}(\mathbf{q} - \mathbf{g}).$$

(11.1.29)

v_c is the volume of the unit cell, which is formed by the basis vectors $\mathbf{a}_i$ with $i = 1, 2, 3$: $v_c = \mathbf{a}_1 \cdot (\mathbf{a}_2 \times \mathbf{a}_3)$. The basis vectors of the reciprocal lattice are

$$\mathbf{g}_1 = \frac{2\pi}{v_c} \mathbf{a}_2 \times \mathbf{a}_3, \qquad \mathbf{g}_2 = \frac{2\pi}{v_c} \mathbf{a}_3 \times \mathbf{a}_1, \qquad \mathbf{g}_3 = \frac{2\pi}{v_c} \mathbf{a}_1 \times \mathbf{a}_2, \text{ d.h.}$$

$$\mathbf{g}_i = \frac{\pi}{v_c} \epsilon_{ijk} \mathbf{a}_j \times \mathbf{a}_k.$$

(11.1.30)

The reciprocity is shown in

$$\mathbf{g}_i \cdot \mathbf{a}_j = 2\pi \delta_{ij}.$$

(11.1.31)

Coming back to the scattered wave, we have obtained for this

$$\mathbf{E}_s(\mathbf{x}, t) = -r_e \, f_0(\mathbf{q}) \, S(\mathbf{q}) \, \mathbf{e}_r \times (\mathbf{E}_0 \times \mathbf{e}_r) \frac{1}{r} \, e^{ikr - i\omega t}.$$

(11.1.32)

The differential scattering cross section results from this as

$$\left(\frac{d\sigma}{d\Omega}\right)_N = \frac{r^2 |E_s|^2}{|E_0|^2} = \left(\frac{d\sigma}{d\Omega}\right)_a |S(\mathbf{q})|^2.$$

(11.1.33)

The first term is the differential scattering cross section of the atom (11.1.25). If you multiply (11.1.33) by $1/N$, you get with (11.1.29)   ($|S(\mathbf{q})|^2 = N\,S(\mathbf{q})$) for the differential scattering cross section of a single atom:

$$\frac{d\sigma}{d\Omega} = \left(\frac{d\sigma}{d\Omega}\right)_a \frac{(2\pi)^3}{v_c} \sum_{\mathbf{g}} \delta^{(3)}(\mathbf{q} - \mathbf{g}).$$

(11.1.34)

This is the well-known Bragg scattering on crystals, where one only receives contributions when the momentum transfer between the incident and scattered wave meets the Bragg condition

$$\mathbf{q} = \mathbf{k}' - \mathbf{k} = \mathbf{g}.$$

(11.1.35)

The intensity increases linearly with the volume of the crystal detected by the X-ray beam. However, the increase in intensity is weaker than linear in larger single crystals, which is referred to as *primary extinction*. Since the total intensity of the transmitted and diffracted wave is preserved, the intensity of the diffracted wave cannot increase arbitrarily. However, it must be added that for a given wavelength the range in which diffraction occurs only has the width of arcseconds.

A real crystal (mosaic crystal) consists of many small crystallites, whose orientation differs by $\lesssim 1°$ differs. The primary extinction then mainly depends on the size of these crystallites. Since the individual crystallites scatter incoherently with each other, the total intensity is again proportional to the volume.

If several crystallites in the crystal are exactly parallel, the ones further back crystallites receive a weaker beam, which also leads to a deviation from linearity. This is called *secondary extinction*. In a good mosaic crystal, the secondary extinction also only plays a subordinate role.

On the other hand, X-ray scattering is always affected by more or less strong absorption. Graphically, the Bragg condition (11.1.35) is represented using the *Ewald sphere*, Fig. 11.6.

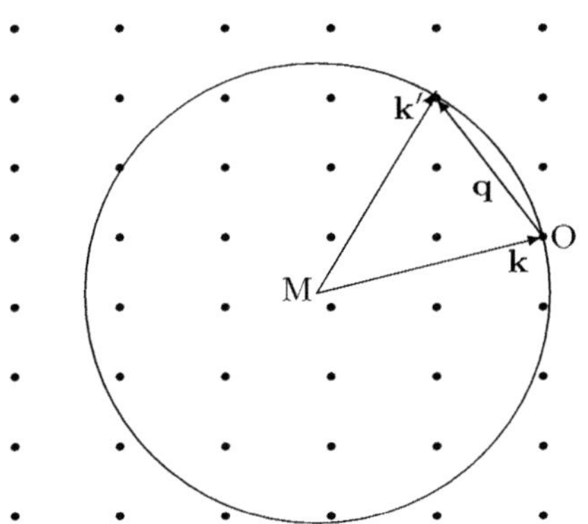

Fig. 11.6. Elastic scattering. The scattering surface is the Ewald sphere; if, in addition to the origin, another reciprocal lattice point is on the Ewald sphere then $\mathbf{q}=\mathbf{k}'-\mathbf{k}$ is a vector from the reciprocal lattice and Bragg scattering occurs

---

Laue conditions: We start from the (Miller) indices (h,k,l), which determine the reciprocal lattice point responsible for the scattering

$$\mathbf{g} = h\mathbf{g}_1 + k\mathbf{g}_2 + l\mathbf{g}_3 \, .$$

The Laue conditions are then $\mathbf{a}_1\cdot\mathbf{q}=2\pi h$, $\quad \mathbf{a}_2\cdot\mathbf{q}=2\pi k \quad$ and $\quad \mathbf{a}_3\cdot\mathbf{q}=2\pi l$.

Non-primitive lattice

If you have two different types of atoms, as is the case with NaCl, it is evident that the Na^+ ions have a different environment than the Cl^- ions. But even with only one type of atom, it may be that the crystal lattice is not a Bravais lattice, as is the case with the diamond lattice of a Si crystal, where neighboring atoms see their respective neighbors under different orientations.

In the crystal lattice, we start from a unit cell, where the total r atoms are located at positions $\mathbf{b}_s$ relative to the origin of the cell. In the scattering wave (11.1.27), only the atoms within the unit cell are to be inserted:

$$\mathbf{E}_s(\mathbf{x},t) = -r_e \sum_{n=-M}^{M} e^{-i\mathbf{q}\cdot\mathbf{a}_n} \sum_{s=1}^{r} e^{-i\mathbf{q}\cdot\mathbf{b}_s} f_s(\mathbf{q})\mathbf{e}_r \times \left(\mathbf{E}_0 e^{-i\mathbf{q}\cdot\mathbf{a}_n} \times \mathbf{e}_r\right) \frac{1}{r} e^{ikr-i\omega t} \, .$$

We have replaced the atomic form factor f_0 with the structure factor of the unit cell

$$f_0(\mathbf{q}) \to F(\mathbf{q}) = \sum_{s=1}^{r} e^{-i\mathbf{q}\cdot\mathbf{b}_s}\, f_s(\mathbf{q}) \quad f_s(\mathbf{q}) = \int d^3x\, n(\mathbf{x})\, e^{-i\mathbf{q}\cdot\mathbf{x}}. \quad (11.1.36)$$

This results in

$$\left(\frac{d\sigma}{d\Omega}\right)_N = r_e^2\, |F(\mathbf{q})|^2\, |S(\mathbf{q})|^2 . \quad (11.1.37)$$

With the dispersion corrections (11.1.26) one also obtains here

$$F(\mathbf{q}) = \sum_{s=1}^{r} e^{-i\mathbf{q}\cdot\mathbf{b}_s} \left[f_{0s}(\mathbf{q}) + \Delta f_s'(\mathbf{q}) - i\Delta f_s''(\mathbf{q}) \right]$$
$$= F_0(\mathbf{q}) + \Delta F'(\mathbf{q}) - i\Delta F''(\mathbf{q}). \quad (11.1.38)$$

Again, the dependence on ω in $\Delta F'$ and $\Delta F''$ is not indicated.

Note: We have only dealt with scattering on rigid lattices. If one takes into account the interaction of the X-rays with the phonons of the sample, this must be considered in thermal equilibrium

$$\langle e^{-i\mathbf{q}\cdot\mathbf{x}_\mathbf{n}}(t)\rangle = e^{-i\mathbf{q}\cdot(\mathbf{a}_\mathbf{n}+\mathbf{b}_s)} \langle e^{-i\mathbf{q}\cdot\mathbf{u}_{\mathbf{n}s}(t)}\rangle ,$$

where $\mathbf{u}_{ns(t)}$ is the displacement of the atom $\mathbf{n}s$ from its equilibrium position. The Debye-Waller factor $W_s(\mathbf{q})$ can be calculated exactly for harmonic crystals and has for cubic symmetry the form

$$e^{-W_s(\mathbf{q})} = \langle e^{-i\mathbf{q}\cdot\mathbf{u}_{\mathbf{n}s}}\rangle \approx e^{-\frac{q^2}{6}\langle u_s^2\rangle} \qquad \text{with} \qquad \langle u_s^2\rangle \propto T .$$

$e^{-2W_s(\mathbf{g})}$ gives the attenuation of the intensity of the scattered radiation due to its interaction with the phonons of the sample.

11.2 Dynamical Theory of X-Ray Diffraction

In the last section, we assumed that incident plane waves penetrate a rather small sample and thereby excite the electrons of the atoms to oscillations and thus to the emission of scattered waves. We assumed that the scattered waves themselves propagate unhindered, without to excite electrons to radiation or to interfere with the incoming wave. With the assumed regular arrangement of the atoms, the superposition of the scattered waves only results in finite intensity in the directions characterized by the Bragg condition (11.1.35).

However, in large ideal crystals, the X-rays, whether incident or diffracted, must be solutions of the Maxwell equations in a periodic charge distribution. So, it is the equations of motion for electromagnetic radiation, the Maxwell equations, in the crystal to solve.

As already mentioned in the introduction to this chapter, the dynamical theory plays a central role in crystal optics, with the differences between X-ray, neutron, and electron beams being small. What they have in common is the weak interaction with the atoms of a (dislocation-free) crystal lattice. Also mentioned was the proximity of the fundamental equations of dynamical theory to the theory of weakly bound electrons in a periodic potential.

11.2.1 Electromagnetic Waves in the Crystal

In the propagation of X-rays, the Maxwell equations (5.2.16) are used, where no free charges ρ_f and no free currents $\mathbf{j}_f$ are present:

$$\text{(a)} \quad \boldsymbol{\nabla}\cdot\mathbf{D} = 0 \qquad\qquad \text{(b)} \quad \boldsymbol{\nabla}\times\mathbf{E} = \frac{-k_L}{c}\,\dot{\mathbf{B}}$$
$$\text{(c)} \quad \boldsymbol{\nabla}\times\mathbf{H} = \frac{k_L}{c}\,\dot{\mathbf{D}} \qquad \text{(d)} \quad \boldsymbol{\nabla}\cdot\mathbf{B} = 0. \tag{11.2.1}$$

The interaction can be described by the polarization of the medium (5.2.3) through the electric field, where $\mu_r = 1$, i.e. $\mu_0\mathbf{H} = \mathbf{B}$ is assumed. The relevant material equation (5.2.17) is

$$\mathbf{D} = \epsilon_0\mathbf{E} + 4\pi k_r\mathbf{P} = (1+4\pi k_r\chi_e)\epsilon_0\mathbf{E}. \tag{11.2.2}$$

$\chi_e(\mathbf{x})$ is here a scalar, but lattice-periodic function, over which is not averaged, since the wavelengths of the X-rays are comparable to the atomic distances (Cu-K$_{\alpha 1}$ = 1.54056 Å).

The propagation of X-rays in matter is usually described by $\mathbf{D}$, which is source-free, where it is used that the interaction is very weak ($\sim 10^{-5}$), so that $\epsilon_0\mathbf{E} \approx (1-4\pi k_r\chi_e)\mathbf{D}$.

We now substitute $\mathbf{D}$ into the induction equation, form the curl and use subsequently the Ampère-Maxwell equation:

$$\boldsymbol{\nabla}\times\left[\boldsymbol{\nabla}\times\frac{1-4\pi k_r\chi_e}{\epsilon_0}\mathbf{D}\right] = -\frac{k_L}{c}\frac{\partial}{\partial t}\,\boldsymbol{\nabla}\times\mu_0\mathbf{H} = -\frac{k_L^2\mu_0}{c^2}\frac{\partial^2\mathbf{D}}{\partial t^2}. \tag{11.2.3}$$

From this follows with the help of (A.2.38) $\boldsymbol{\nabla}\times(\boldsymbol{\nabla}\times\mathbf{D}) = -\Delta\mathbf{D}$:

$$-\Delta\mathbf{D} + \frac{1}{c^2}\frac{\partial^2\mathbf{D}}{\partial t^2} = 4\pi k_r\boldsymbol{\nabla}\times(\boldsymbol{\nabla}\times\chi_e\mathbf{D}). \tag{11.2.4}$$

The only difference in the various systems here is in the prefactor of the susceptibility χ_e, which can be eliminated with the help of the definition

$$\chi(\mathbf{x}) = 4\pi k_r\chi_e(\mathbf{x}) \qquad \Rightarrow \qquad \text{G: } \chi = 4\pi\chi_e, \qquad \text{SI: } \chi = \chi_e. \tag{11.2.2'}$$

Now we make the assumption

$$\mathbf{D}(\mathbf{x},t) = \mathbf{D}(\mathbf{x})\,\mathrm{e}^{-i\omega t}$$

and get ($k = \omega/c$)

$$(\Delta + k^2)\mathbf{D} = -\boldsymbol{\nabla} \times (\boldsymbol{\nabla} \times \chi\,\mathbf{D}). \tag{11.2.5}$$

We expand χ into a Fourier series

$$\chi(\mathbf{x}) = \sum_{\mathbf{g}} \chi_{\mathbf{g}}\, e^{i\mathbf{g}\cdot\mathbf{x}} \tag{11.2.6}$$

and make a Bloch ansatz for $\mathbf{D}$:

$$\mathbf{D}(\mathbf{x}) = e^{i\mathbf{K}\cdot\mathbf{x}}\mathbf{d}(\mathbf{x}) = e^{i\mathbf{K}\cdot\mathbf{x}}\sum_{\mathbf{g}} \mathbf{d}_{\mathbf{g}}\, e^{i\mathbf{g}\cdot\mathbf{x}}, \tag{11.2.7}$$

where $\mathbf{d}(\mathbf{x})$ is lattice-periodic. For the left side of (11.2.5) one obtains

$$(\Delta + k^2)\mathbf{D}(\mathbf{x}) = -\sum_{\mathbf{g}}(K_{\mathbf{g}}^2 - k^2)\,e^{i\mathbf{K}_{\mathbf{g}}\cdot\mathbf{x}}\mathbf{d}_{\mathbf{g}}, \qquad \mathbf{K}_{\mathbf{g}} = \mathbf{K} + \mathbf{g}. \tag{11.2.8}$$

The evaluation of the right side of (11.2.5) is somewhat more complex. First, the following expressions are evaluated ($\mathbf{g}'' = \mathbf{g} - \mathbf{g}'$):

$$\chi\mathbf{D}(\mathbf{x}) = \sum_{\mathbf{g}''} \chi_{\mathbf{g}''}\, e^{i\mathbf{g}''\cdot\mathbf{x}} \sum_{\mathbf{g}'} \mathbf{d}_{\mathbf{g}'}\, e^{i\mathbf{K}_{\mathbf{g}'}\cdot\mathbf{x}} = \sum_{\mathbf{g}} e^{i\mathbf{K}_{\mathbf{g}}\cdot\mathbf{x}} \sum_{\mathbf{g}'} \chi_{\mathbf{g}-\mathbf{g}'}\,\mathbf{d}_{\mathbf{g}'}, \tag{11.2.9}$$

$$\boldsymbol{\nabla} \times \left[\boldsymbol{\nabla} \times \chi\mathbf{D}(\mathbf{x})\right] = \sum_{\mathbf{g}} e^{i\mathbf{K}_{\mathbf{g}}\cdot\mathbf{x}} \sum_{\mathbf{g}'} \chi_{\mathbf{g}-\mathbf{g}'}\, i\mathbf{K}_{\mathbf{g}} \times \left[i\mathbf{K}_{\mathbf{g}} \times \mathbf{d}_{\mathbf{g}'}\right].$$

Brought to the right side of (11.2.5), one obtains [Kato, 1974, (4-9a)]

$$\sum_{\mathbf{g}} e^{i\mathbf{K}_{\mathbf{g}}\cdot\mathbf{x}} \left\{ (K_{\mathbf{g}}^2 - k^2)\mathbf{d}_{\mathbf{g}} - \sum_{\mathbf{g}'} \chi_{\mathbf{g}-\mathbf{g}'}\, \mathbf{K}_{\mathbf{g}} \times (\mathbf{d}_{\mathbf{g}'} \times \mathbf{K}_{\mathbf{g}}) \right\} = 0.$$

If this equation is to hold for all $\mathbf{x}$, each individual summand must vanish separately. Thus, one has a homogeneous linear system of equations for the Fourier coefficients. With the definition[2]

$$\mathbf{d}_{\mathbf{g}'[\mathbf{g}]} = \hat{\mathbf{K}}_{\mathbf{g}} \times (\mathbf{d}_{\mathbf{g}'} \times \hat{\mathbf{K}}_{\mathbf{g}}), \qquad \hat{\mathbf{K}}_{\mathbf{g}} = \mathbf{K}_{\mathbf{g}}/K_{\mathbf{g}}, \qquad \mathbf{d}_{\mathbf{g}} = \mathbf{d}_{\mathbf{g}[\mathbf{g}]} \tag{11.2.10}$$

it takes the following form:

$$\left[K_{\mathbf{g}}^2 - k^2\right]\mathbf{d}_{\mathbf{g}} - K_{\mathbf{g}}^2 \sum_{\mathbf{g}'} \chi_{\mathbf{g}-\mathbf{g}'}\, \mathbf{d}_{\mathbf{g}'[\mathbf{g}]} = 0. \tag{11.2.11}$$

These are the *fundamental equations of dynamical theory*. The condition on the right in (11.2.10) is obtained from the Gauss's law:

$$\boldsymbol{\nabla}\cdot\mathbf{D} = i\sum_{\mathbf{g}} e^{i\mathbf{K}_{\mathbf{g}}\cdot\mathbf{x}}\, \mathbf{K}_{\mathbf{g}}\cdot\mathbf{d}_{\mathbf{g}} = 0 \qquad\qquad \Rightarrow \qquad\qquad \mathbf{K}_{\mathbf{g}}\cdot\mathbf{d}_{\mathbf{g}} = 0.$$

---

[2] For $\mathbf{d}_{\mathbf{g}'[\mathbf{g}]}$ see: Authier [2002, §5.1–5.3]; also common is $[\mathbf{d}_{\mathbf{g}'}]_{\mathbf{g}}$.

Thus, $\mathbf{d_{g[g]}} = \hat{\mathbf{K}}_{\mathbf{g}} \times (\mathbf{d_g} \times \hat{\mathbf{K}}_{\mathbf{g}}) = \mathbf{d_g}$. The partial waves of the displacement field (11.2.7) are therefore all transversal, and they have (each) two polarization directions.

Simple statements about polarization are possible when in (11.2.12) only one term with $\mathbf{g} \neq 0$ contributes (*two-beam approximation*). Then there exists only one scattering plane R spanned by $\mathbf{K}$ and $\mathbf{K_g}$ and one can proceed as with the refraction of $\mathbf{E}$ in the homogeneous field (Fresnel's formulas, section 10.3.3, p. 357), where the reflection at the homogeneous medium for a linearly polarized electric field perpendicular to the scattering plane ($\mathbf{E}^\sigma$) and in the scattering plane ($\mathbf{E}^\pi$) was treated separately. The general field is a superposition of the two fields.

One thus sets the polarization directions for $\mathbf{d_g}$ once perpendicular (σ) and once parallel (π) to the scattering plane and inserts these into the fundamental equations (11.2.11). It is then sufficient to calculate the strength of the amplitude $|\mathbf{d_g}|$.

$$\mathbf{d_g} \cdot \mathbf{d_{g[g]}} = d_g^2, \ \ \mathbf{d_{g'[g]}} \cdot \mathbf{d_{g[g]}} = \left[\mathbf{d_{g'}} - (\hat{\mathbf{K}}_{\mathbf{g}} \cdot \mathbf{d_{g'}})\hat{\mathbf{K}}_{\mathbf{g}}\right] \cdot \mathbf{d_g} = \mathbf{d_{g'}} \cdot \mathbf{d_g} = C_{g'g}\, d_g\, d_{g'},$$

where $|C_{g'g}| \leq 1$ are 'polarization factors'. The fundamental equations for d_g are obtained by the scalar multiplication of (11.2.11) with $\mathbf{d_{g'[g]}}$:

$$(K_{\mathbf{g}}^2 - k^2)\, d_g - K_{\mathbf{g}}^2 \sum_{\mathbf{g'}} \chi_{\mathbf{g} - \mathbf{g'}}\, C_{g'g}\, d_{g'} = 0. \tag{11.2.12}$$

This is a linear homogeneous system of equations for calculating the Fourier coefficients d_g. Non-trivial solutions are only obtained when the determinant of the coefficient matrix vanishes. This determines $\mathbf{K}$, though not entirely. Additionally, it should be noted that the incident wave must meet the boundary conditions at the crystal surface must fulfill.

If one neglects all summands in (11.2.12) except $\mathbf{g'} = \mathbf{0}, \mathbf{g}$, the system of equations is reduced to a 2×2-matrix (two-beam approximation) and there are only two partial waves propagating in the directions $\mathbf{K}_{1,2}$ and $\mathbf{K}_{1,2} + \mathbf{g}$, where $\mathbf{K}_1$ and $\mathbf{K}_2$ differ only slightly from the wave vector $\mathbf{k}$ of the incident wave.

Then there is also only one scattering plane R present and the polarization factors C are

$$
\begin{aligned}
\mathbf{d_g} \text{ and } \mathbf{d}_0 \perp R \quad &\Longrightarrow C = 1, & \sigma - polarization, \\
\mathbf{d_g} \text{ and } \mathbf{d}_0 \parallel R \quad &\Longrightarrow C = |\cos 2\theta_B|, & \pi - polarization.
\end{aligned}
$$

$2\theta_B$ is the angle enclosed by $\mathbf{K}$ and $\mathbf{K_g}$. In the dynamical theory, it is common to define $C \geq 0$; this is possible because in the two-beam approximation (11.2.19) only C^2 appears.

Polarizability

The predominant part of the interaction of the electric field with matter is due to the Thomson scattering, where the frequencies ω are far above the excitation energies ω_j of the (outer) electrons (11.1.16). In a medium with the charge density $\rho(\mathbf{x}) = e\,n(\mathbf{x})$ one obtains for the polarization in a time-varying field $\mathbf{E} = \mathbf{E}_0\,e^{-i\omega t}$

$$\mathbf{P} = \rho(\mathbf{x})\,\mathbf{x} = -\frac{e_0^2\,n(\mathbf{x})}{m_e\omega^2}\,\mathbf{E} = \chi_e\epsilon_0\,\mathbf{E} \quad\Rightarrow\quad \chi = 4\pi k_r\chi_e = -4\pi k_C\frac{e_0^2\,n(\mathbf{x})}{m_e\omega^2}.$$

For free electrons, $\mathbf{x} = (e_0/m_e\omega^2)\,\mathbf{E}$. We have defined the susceptibility χ_e for visible light as a quantity averaged over many atoms, which is isotropic for most substances.

X-rays have wavelengths comparable to atomic distances, so that in χ the periodicity of the crystal lattice must be taken into account as follows:

$$\rho(\mathbf{x}) = \sum_{\mathbf{n},s}{}' \rho_s(\mathbf{x}-\mathbf{a_n}-\mathbf{b}_s).$$

The development (11.2.6) of χ into a Fourier series has the coefficients

$$\chi(\mathbf{g}) = \frac{1}{V}\int_V \mathrm{d}^3x\,e^{-i\mathbf{g}\cdot\mathbf{x}}\,\chi(\mathbf{x}) = -4\pi k_C\frac{e_0^2}{v_c m_e\omega^2}\sum_s\int \mathrm{d}^3x\,e^{-i\mathbf{g}\cdot\mathbf{x}}n_s(\mathbf{x})$$

$$= -4\pi\frac{r_e}{v_c k^2}\,F(\mathbf{g}). \tag{11.2.13}$$

Here, $\rho_s(\mathbf{x}) = -e_0\,n_s(\mathbf{x})$ is the charge density of the atom type s, $r_e = k_C e_0^2/m_e c^2$ (classical electron radius), $k = \omega/c$, v_c is the volume of the unit cell and $F(\mathbf{g})$ is the structure factor of the unit cell (11.1.38). In this case, the dispersion correction $\Delta F''$ due to absorption in large crystals should not be neglected:

$$\chi(\mathbf{x}) = 4\pi k_r\chi_e(\mathbf{x}) = \chi_r(\mathbf{x}) + i\chi_i(\mathbf{x}). \tag{11.2.14}$$

With the separation of $\chi(\mathbf{x})$ into a real and an (absorptive) imaginary part the Fourier transforms are

$$\chi_{\mathbf{g}} = \chi(\mathbf{g}) = \chi_r(\mathbf{g}) + i\chi_i(\mathbf{g}) \tag{11.2.15}$$

for reflections without inversion symmetry $\chi_r(\mathbf{g})$ and $\chi_i(\mathbf{g})$ are also complex. In the Tab. 11.2 values for $\chi_{\mathbf{g}}$ are given.

Note: Sometimes the conductivity σ_P is included in the interaction [Vartanyants, Kovalchuk, 2001]: Ohm's law (5.3.2") and isotropic local coupling ($\sigma(\mathbf{x},\mathbf{x}') = \delta^{(3)}(\mathbf{x}-\mathbf{x}')\sigma(\mathbf{x})$) result in

$$\mathbf{j}_P(\mathbf{x},t) = \int \mathrm{d}t'\,\sigma_P(\mathbf{x},t-t')\mathbf{E}(\mathbf{x},t') \quad\Rightarrow\quad \mathbf{j}_P(\mathbf{x},\omega) = \sigma_P(\mathbf{x},\omega)\mathbf{E}(\mathbf{x},\omega).$$

The electric field polarizes the medium according to (5.2.3) with the current density

$$\mathbf{j}_P(\mathbf{x},t) = \dot{\mathbf{P}}(\mathbf{x},t) \quad\Rightarrow\quad \mathbf{j}_P(\mathbf{x},\omega) = -i\omega\mathbf{P}(\mathbf{x},\omega) = -i\omega\chi_e(\mathbf{x},\omega)\epsilon_0\mathbf{E}(\mathbf{x},\omega).$$

So we obtain $\sigma_P(\mathbf{x},\omega) = -i\epsilon_0\omega\chi_e(\mathbf{x},\omega)$.

Tab. 11.2. Experimental values for the scattering of Cu-K$_{\alpha 1}$ radiation in silicon and germanium[a]. The 'Pendellösung' lengths Δ_0 (11.2.34) are measured at symmetrical reflection ($\gamma_i = \gamma_g$); the given values refer to Cu-K$_{\alpha 1}$-radiation ($\lambda = 1.54$ Å) and Mo-K$_{\alpha 1}$-radiation ($\lambda = 0.71$ Å)

	λ [Å]	$\chi_{r0} \times 10^6$	$\chi_{i0} \times 10^6$	$|\chi_{r\,220}| \times 10^6$	$\chi_{i\,220} \times 10^6$	$\Delta_0(220)$ [μm]	μ_0 [mm^{-1}]
Si	1.54	-15.1	0.35	9.13	0.34	15.4	14.4
Ge	1.54	-28.7	0.86	20.3	0.83	7.0	35.3
Si	0.71	-3.16	0.0165	1.90	0.159	36.6	1.46
Ge	0.71	-6.40	0.36	4.60	0.35	15.2	31.9

[a] The values are from Pinsker [1978, tables 4.2–4.4]

11.2.2 Procedure for Solving the Fundamental Equations

The fundamental equations (11.2.12) form a homogeneous linear system of equations

$$\left[K_{\mathbf{g}}^2(1 - \chi_0) - k^2\right] d_g - K_{\mathbf{g}}^2 \sum_{\mathbf{g}' \neq \mathbf{g}} \chi_{\mathbf{g}-\mathbf{g}'}\, C_{g'g}\, d_{g'} = 0. \tag{11.2.16}$$

For non-trivial solutions, the secular determinant (determinant of the coefficient matrix) must vanish. The interaction is of the order of magnitude $|\chi_{\mathbf{g}}| \sim 10^{-5} \ll 1$, as can be seen from Tab. 11.2 for silicon and germanium, i.e., the non-diagonal matrix elements are smaller by this order of magnitude than the diagonal ones.

The (average) refractive index of X-rays from vacuum to crystal is according to (11.2.2) $n = \sqrt{\epsilon_r} \approx 1 + \chi_0/2$ almost 1, which is why the continuous wave with $\mathbf{K}$ should deviate only minimally from the incident wave with $\mathbf{k}$, i.e. $K \approx k$. For partial waves $\mathbf{K_g}$, which are excited in the crystal, i.e. have a finite amplitude d_g, applies

$$|K_{\mathbf{g}}^2 - k^2| \sim k^2 |\chi_0|.$$

All points $\mathbf{g}$ of the reciprocal lattice, which fulfill this condition, lie on (near) the Ewald sphere Fig. 11.6.

1. K can always be chosen so that

$$K^2(1 - \chi_0) - k^2 = 0.$$

   If no other diagonal element is small, the secular determinant of (11.2.16) reduces to

$$K^2 - K_0^2 = 0 \qquad \text{with} \qquad K_0^2 = k^2/(1 - \chi_0). \tag{11.2.17}$$

2. If there is another point ($\mathbf{g}$) near the Ewald sphere in addition to the origin, as is the case with Bragg scattering, then (11.2.16) reduces to the 2×2 matrix [Kato, 1974, (4-18)]

$$\left(K^2 - K_0^2\right) d_0 - \frac{K^2}{1-\chi_0} C\chi_{-\mathbf{g}}\, d_g = 0,$$
$$\frac{-K_{\mathbf{g}}^2}{1-\chi_0} C\chi_{\mathbf{g}}\, d_0 + \left(K_{\mathbf{g}}^2 - K_0^2\right) d_g = 0. \tag{11.2.18}$$

We replace in the off-diagonal elements $K^2/(1-\chi_0)$ or $K_{\mathbf{g}}^2/(1-\chi_0)$ by k^2 and thus obtain the secular determinant

$$\left(K^2 - K_0^2\right)\left(K_{\mathbf{g}}^2 - K_0^2\right) - k^4 C^2 \chi_{\mathbf{g}}\chi_{-\mathbf{g}} = 0. \tag{11.2.19}$$

Note: For (photonic) crystals, one would have the same secular equation for a given $\mathbf{K}$, which then however satisfies periodic boundary conditions. From this, the dispersion (band structure) in the crystal ($\omega = kc$) would have to be calculated.

3. At special points of high symmetry, more than two points of the reciprocal lattice can lie close to the Ewald sphere Fig. 11.6. In this case, one speaks of multiple-beam cases and has a correspondingly more complicated secular determinant to solve. We do not go into such solutions.

11.2.3 Refraction in the Single-Beam Case

If only the origin is near the Ewald sphere, as described by (11.2.17):

$$K^2 = K_0^2 = k^2(1 + \chi_0) = k^2\epsilon_r = k^2 n^2, \tag{11.2.20}$$

then the single-beam case is present. In the medium, the X-ray senses the average "potential"χ_0. The refractive index of the medium is $n = \sqrt{\epsilon_r} \approx 1 + \chi_0/2 < 1$. The beam is therefore refracted from the perpendicular, but n differs only imperceptibly from 1. From the continuity of the tangential component, it follows first

$$K_{0\perp}^2 = k_\perp^2 + (K_0^2 - k^2)$$

and from this, as can be seen from Fig. 11.7,

$$\mathbf{K}_0 \approx \mathbf{k} + \frac{k\chi_0}{2\gamma}\mathbf{n} = \mathbf{k} + \frac{k\chi_{0r}}{2\gamma}\mathbf{n} + \mathrm{i}\frac{\mu_0}{2\gamma}\mathbf{n} \quad \text{with} \quad \mu_0 = k\chi_{0i}. \tag{11.2.21}$$

Here, $\gamma = \cos\vartheta = \mathbf{k}\cdot\mathbf{n}/k$ is the cosine of the angle of incidence and μ_0 is the linear absorption coefficient, which we obtain from the complex susceptibility

$$\chi_0 = \chi_{0r} + \mathrm{i}\chi_{0i}.$$

(11.2.21) follows from Snell's law of refraction, (10.3.2) $\sin\vartheta = \sqrt{1+\chi_{0r}}\,\sin\vartheta'$ in media with absorption. So far, we have only calculated $\mathbf{K}_0$ in the medium. The amplitudes are to be calculated from the continuity conditions (10.3.3), where we refer to the Fresnel formulas (10.3.9) and (10.3.5) for the σ- or π-polarization. As outlined in Fig. 11.7, in addition to the continuous wave, a reflected wave with $\mathbf{k}'' = \mathbf{k}_\parallel - \mathbf{k}_\perp$ occurs, whose amplitude, however, in both cases (σ- and π-polarization) is of the size $\sim \chi_0$ and can therefore be neglected, as long as no grazing incidence is present.

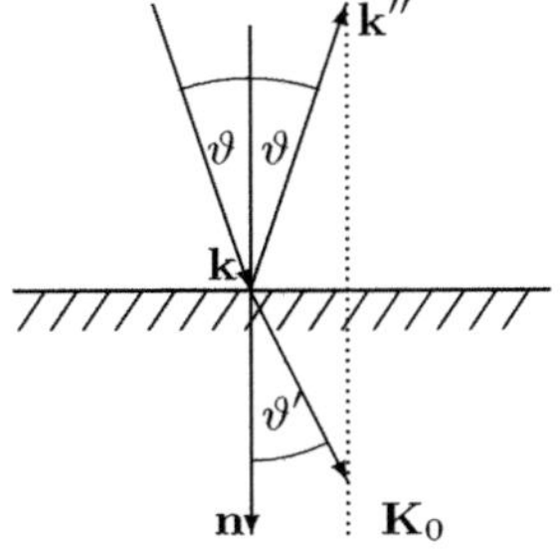

Fig. 11.7. Scattering on a crystal, far from any Bragg condition ($\vartheta \approx \vartheta'$).
$\mathbf{k}'' = (\mathbf{k}_\parallel, -\mathbf{k}_\perp)$ is the wave vector of the reflected wave, which, however, due to its small amplitude compared to the incident wave $D_0''/D_0 \sim \chi_0$ can be neglected

11.2.4 The Two-Beam Case

In addition to the origin, there is now another point of the reciprocal lattice on the Ewald sphere. Thus, we have the 2×2 matrix (11.2.18) in front of us, whose secular determinant (11.2.19) is to be calculated. Again, we exclude grazing incidence. In the two-beam approximation, the polarization C only occurs together with $\chi_{\pm\mathbf{g}}$:

$$\tilde{\chi}_{\mathbf{g}} = C \chi_{\mathbf{g}}, \tag{11.2.22}$$

so that the secular equation (11.2.19) is now

$$\left(K^2 - K_0^2\right)\left(K_{\mathbf{g}}^2 - K_0^2\right) - k^4 \tilde{\chi}_{\mathbf{g}} \tilde{\chi}_{-\mathbf{g}} = 0. \tag{11.2.23}$$

A prerequisite for the validity of the secular equation is that $\mathbf{k}$ almost fulfills a Bragg condition, i.e. $|K^2 - K_0^2| \ll k^2$ and $|K_{\mathbf{g}}^2 - K_0^2| \ll k^2$. This reduces the number of solutions to two. We therefore expect in the crystal a splitting of the incident wave $\mathbf{k} \to \mathbf{K}_{1,2}$, which is determined by $C\chi_{\mathbf{g}}$. So we have two wave fields whose superposition leads to the interference phenomena of the dynamical theory.

Internal angles: We first try to solve (11.2.23) without considering the surface of the crystal. To do this, we transform:

$$K_{\mathbf{g}}^2 - K_0^2 = K^2 - K_0^2 + 2(\mathbf{K}-\mathbf{k})\cdot\mathbf{g} + (2\mathbf{k}+\mathbf{g})\cdot\mathbf{g}. \tag{11.2.24}$$

If $\mathbf{k}$ is the wave vector of the incident wave, the last term parameterizes the angular deviation of the incident wave from the Bragg angle. We orient ourselves according to Fig. 11.8, where $\mathbf{k}_B$ meets the Bragg condition exactly for $|\mathbf{k}_B| = k$,

$$(\mathbf{k}-\mathbf{k}_B)\cdot\mathbf{g} = kg\left[\cos\left(\theta+\frac{\pi}{2}\right)-\cos(\theta_B+\frac{\pi}{2})\right] = kg\left[\sin\theta_B - \sin\left(\theta-\theta_B+\theta_B\right)\right]$$

$$\approx kg\cos\theta_B\,(\theta_B-\theta) = k^2\sin(2\theta_B)\,(\theta_B-\theta). \tag{11.2.25}$$

In reference to Max von Laue [1960, (28.8)] we define the deviation parameter

$$\alpha_L = \frac{1}{k^2}\left(\mathbf{k}+\frac{\mathbf{g}}{2}\right)\cdot\mathbf{g} = \frac{(\mathbf{k}-\mathbf{k}_B)\cdot\mathbf{g}}{k^2} \approx \sin(2\theta_B)\,(\theta_B-\theta). \tag{11.2.26}$$

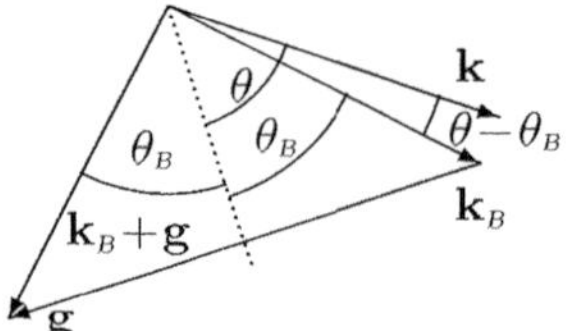

Fig. 11.8. The internal angles θ and θ_B are determined by $\mathbf{g}$, $\mathbf{k}$ and $\mathbf{k}_B$. $\mathbf{k}$ and $\mathbf{k}_B$ with $|\mathbf{k}_B| = k$ are assigned to the incident wave

For backscattering ($\theta_B = \pi/2$) is the angular range in which the waves are reflected (acceptance range), significantly larger: $\alpha_L = (\theta_B - \theta)^2$.

The inclusion of the surface

The secular equation (11.2.23) cannot be solved without reference to the surface(s) of the crystal. $\mathbf{n}$ is the normal vector of the flat front surface of the crystal, which points into the crystal, as shown in Fig. 11.9. The incident wave has the wave vector $\mathbf{k}$ in the outer space, and the tangential components of the wave vectors are continuous at the entrance surface:

$$\mathbf{k}_\| = \mathbf{K}_{0\|} = \mathbf{K}_\| \,.$$

$\mathbf{K}_0$ is the wave vector resulting from refraction in the medium (11.2.21). The vectors $\mathbf{K}$ or $\mathbf{K}_0$ can therefore only differ from $\mathbf{k}$ in the normal component.

For computational reasons, it is convenient to replace the vector of the incident wave $\mathbf{k}$ in (11.2.24) by $\mathbf{K}_0$. We then define by means of

$$\mathbf{K} - \mathbf{K}_0 = k\epsilon\mathbf{n} \tag{11.2.27}$$

the dimensionless parameter ϵ, which determines the splitting of $\mathbf{K}_{1,2}$ with respect to $\mathbf{K}_0$ parallel to the normal vector $\mathbf{n}$ of the entrance surface. $\epsilon_L = (K-k)/k$ is referred to by Max von Laue as *excitation error*. If the parameter of the angular deviation α_L is now replaced by

$$k^2\alpha = \left(\mathbf{K}_0 + \frac{\mathbf{g}}{2}\right)\cdot\mathbf{g} \overset{(11.2.21)}{=} k^2\alpha_L + \frac{k^2\chi_0}{4\mathbf{n}\cdot\mathbf{k}}\,\mathbf{n}\cdot\mathbf{g}, \tag{11.2.28}$$

one obtains for (11.2.24)

$$(\mathbf{K}+\mathbf{g})^2 = K^2 - K_0^2 + 2k\epsilon\,\mathbf{n}\cdot\mathbf{g} + 2k^2\alpha.$$

External angles: Experimentally directly accessible are the angle ϑ_i of the incident and the angle ϑ_g of the diffracted beam in relation to the surface normal. There are different cases, depending on whether the diffracted beam enters the crystal (*Laue case*), see Fig. 11.9, or is reflected by it (*Bragg case*). The incidence angle ϑ_i is limited to $0 \le \vartheta_i < \pi/2$. While θ and θ_B are always positive, ϑ_i and ϑ_g are given in the counterclockwise direction, starting from the surface normal. Thus, ϑ_g has the range $-\pi/2 < \vartheta_g < 3\pi/2$. In Fig. 11.9, four different configurations are sketched for the diffracted beam.

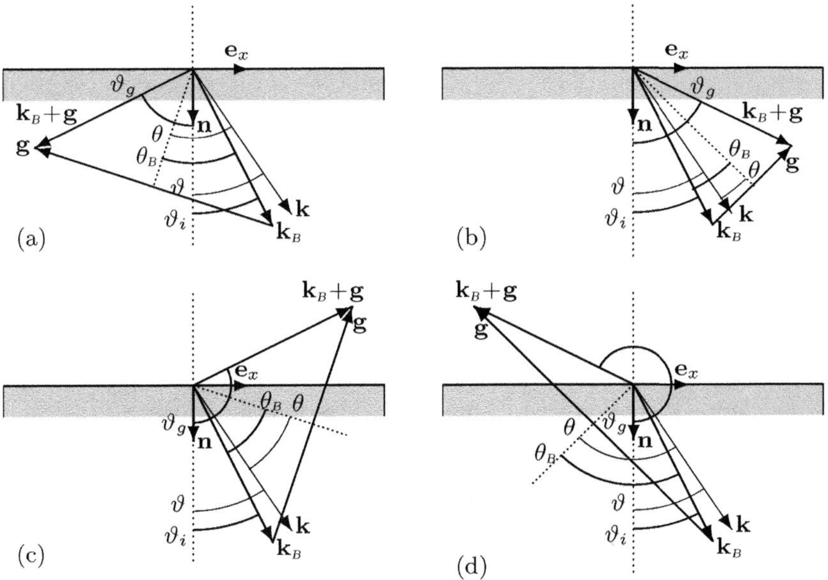

Fig. 11.9. ϑ_i, ϑ_g and θ_B are determined by the vector $\mathbf{k}_B$ which falls exactly in the Bragg direction ($k_B = k$).
Laue case $\gamma_g > 0$: (a) $-\pi/2 < \vartheta_g < \vartheta_i$ and (b) $\vartheta_i < \vartheta_g < \pi/2$
Bragg case $\gamma_g < 0$: (c) $\pi/2 < \vartheta_g < 3\pi/2 - \vartheta_i$ and (d) $3\pi/2 - \vartheta_i < \vartheta_g < 3\pi/2$

Cases (b) and (d) differ from (a) and (c) in a change of sign when transitioning from the inner angles $\theta_B - \theta$ to the outer angles $\vartheta - \vartheta_i$, which is of subordinate importance but, to avoid ambiguities, in the following, when the Laue case is mentioned, only case (a) is meant, while the Bragg case only refers to case (c). In these two cases, $(\gamma_g = \cos \vartheta_g)$

$$\vartheta_i - \vartheta = \operatorname{sgn} \gamma_g \left(\theta_B - \theta \right) \stackrel{(11.2.26)}{=} \frac{\alpha \operatorname{sgn} \gamma_g}{\sin(2\theta_B)}, \quad 2\theta_B = \operatorname{sgn} \gamma_g (\vartheta_i - \vartheta_g). \quad (11.2.29)$$

Now some geometric relationships need to be noted, using Fig. 11.9 for orientation:

$$\gamma_i = \cos \vartheta_i \qquad \text{and} \qquad \gamma_g = \cos \vartheta_g,$$
$$\mathbf{n} \cdot \mathbf{k}_B = k \gamma_i, \qquad \mathbf{n} \cdot (\mathbf{k}_B + \mathbf{g}) = k \gamma_g, \qquad \mathbf{n} \cdot \mathbf{g} = k(\gamma_g - \gamma_i). \quad (11.2.30)$$

These relations can usually be used within the required accuracy also for $\mathbf{k}$, $\mathbf{K}_0$ or $\mathbf{K}$; so is $\mathbf{k} \cdot \mathbf{n} = k \gamma_i$.

Wave vectors in the two-beam case

Now ϵ can be determined from (11.2.23), whereby the secular equation (11.2.23) only applies if the diagonal elements are small, i.e., if $\epsilon \ll 1$ is:

$$K^2 - K_0^2 = (\mathbf{K} - \mathbf{K}_0)\cdot(\mathbf{K} + \mathbf{K}_0) \approx 2k^2\,\gamma_i\epsilon,$$
$$K_{\mathbf{g}}^2 - K_0^2 \approx 2k^2\,\gamma_i\epsilon + 2k^2\,\alpha + 2k^2(\gamma_g - \gamma_i)\epsilon. \tag{11.2.31}$$

From this, the characteristic equation (11.2.23) results

$$4\gamma_i\epsilon(\epsilon\gamma_g + \alpha) - \tilde{\chi}_{\mathbf{g}}\tilde{\chi}_{-\mathbf{g}} = 0 \quad \Rightarrow \quad \epsilon^2 + \frac{\epsilon\alpha}{\gamma_g} - \nu^2\frac{|\tilde{\chi}_{\mathbf{g}}\tilde{\chi}_{-\mathbf{g}}|}{4\gamma_i\gamma_g} = 0. \tag{11.2.32}$$

The absorptive part of $\chi_{\mathbf{g}}$ is taken into account in (11.2.32) by

$$\nu^2 = \frac{\chi_{\mathbf{g}}\chi_{-\mathbf{g}}}{|\chi_{\mathbf{g}}\chi_{-\mathbf{g}}|} \quad \Rightarrow \quad \begin{cases} \nu = \sqrt{1 - \kappa^2} + i\kappa \\ \kappa = \mathrm{Im}\,\nu \overset{\chi_{\mathbf{g}} = \chi_{-\mathbf{g}}}{\approx} \chi_{\mathbf{g}i}/\chi_{\mathbf{g}r}. \end{cases} \tag{11.2.33}$$

We also define a characteristic length, the Pendellösung length

$$\Delta_0 = \frac{2\pi}{k}\,\frac{\sqrt{|\gamma_i\gamma_g|}}{\sqrt{|\tilde{\chi}_{\mathbf{g}}\tilde{\chi}_{-\mathbf{g}}|}} \tag{11.2.34}$$

and thus obtain the solution

$$\begin{aligned}
\epsilon_{1,2} &= -\frac{\alpha}{2\gamma_g} \pm \sqrt{\frac{\alpha^2}{4\gamma_g^2} + \mathrm{sgn}\,\gamma_g\left(\frac{\nu\pi}{k\Delta_0}\right)^2} \\
&= \frac{\pi}{k\Delta_0}\left(-\zeta \pm \sqrt{\zeta^2 + \nu^2\,\mathrm{sgn}\,\gamma_g}\right).
\end{aligned} \tag{11.2.35}$$

The scattering intensity is measured by rotating the sample through the Bragg reflex (rocking curve), where one can start from the deviation to the exact Bragg position to then parameterize the angle appropriately. We take the definition from (11.2.35) [Zachariasen, 1945, (3.141)]

$$\zeta = \frac{k\Delta_0}{\pi}\,\frac{\alpha}{2\gamma_g} \overset{(11.2.28)}{=} \frac{k\Delta_0}{\pi}\left(\frac{\alpha_L}{2\gamma_g} + \frac{\chi_0}{4k\gamma_i\gamma_g}\mathbf{n}\cdot\mathbf{g}\right) = y + i\eta. \tag{11.2.36}$$

The decomposition of ζ into real part and imaginary part results in

$$y = \frac{k\Delta_0}{\pi}\left(\frac{\sin(2\theta_B)(\theta_B - \theta)}{2\gamma_g} + \chi_{0r}\frac{\gamma_g - \gamma_i}{4\gamma_i\gamma_g}\right), \tag{11.2.37}$$

$$\eta = \frac{k\Delta_0}{\pi}\chi_{0i}\frac{\gamma_g - \gamma_i}{4\gamma_i\gamma_g} = \mathrm{sgn}\,\gamma_g\,\frac{\chi_{0i}}{\sqrt{|\tilde{\chi}_{\mathbf{g}}\tilde{\chi}_{-\mathbf{g}}|}}\frac{\gamma_g - \gamma_i}{2\sqrt{\gamma_i|\gamma_g|}}. \tag{11.2.38}$$

y is the relevant angle and η only has an influence on the absorption. Δ_0 is the already mentioned Pendellösung length, which is the characteristic length of the theory. The solution gives for the wave numbers in the crystal

$$\begin{aligned}
\mathbf{K}_{1,2} &= \mathbf{K}_0 + k\epsilon_{1,2}\,\mathbf{n} = \mathbf{K}_0 + \frac{\pi}{\Delta_0}\left(-\zeta \pm \sqrt{\zeta^2 + \nu^2\,\mathrm{sgn}\,\gamma_g}\right)\mathbf{n}, \\
\mathbf{K}_0 &= \mathbf{k} + k\chi_0/(2\gamma_i)\,\mathbf{n}.
\end{aligned} \tag{11.2.39}$$

Notes:

1. The difference $\mathbf{K}_1 - \mathbf{K}_2 = (2\pi/\Lambda_0)\sqrt{\zeta^2 + \text{sgn}\,\gamma_g\,\nu^2}\,\mathbf{n}$ leads to interferences, i.e. to intensity oscillations of the characteristic length Λ_0. However, this not only depends on the reflex $\mathbf{g}$, but also on $\mathbf{k}$ and the orientation of the surface.

$$\Lambda = \frac{g}{k^2\,\sqrt{|\chi_{\mathbf{g}}\,\chi_{-\mathbf{g}}|}} \overset{(11.2.13)}{=} \frac{v_c}{4\pi r_e}\,\frac{1}{\sqrt{|F(\mathbf{g})\,F(-\mathbf{g})|}} \tag{11.2.40}$$

is a characteristic length that depends only on $\mathbf{g}$ and the strength of the interaction (r_e). For the deviation $\mathbf{k} - \mathbf{k}_B$, only the component parallel to $\mathbf{g}$ is relevant – and its order of magnitude is determined by Λ:

$$\zeta = \text{sgn}\,\gamma_g\,\sqrt{\frac{\gamma_i}{|\gamma_g|}}\left[\Lambda(\mathbf{k}-\mathbf{k}_B)\cdot\frac{\mathbf{g}}{g} + \frac{\gamma_g-\gamma_i}{2|\gamma_i\gamma_g|}\,\frac{\chi_0}{\sqrt{|\tilde{\chi}_{\mathbf{g}}\tilde{\chi}_{-\mathbf{g}}|}}\right]. \tag{11.2.41}$$

The relationship to the Pendellösung length is

$$\Lambda = \frac{g}{2\pi k}\,\frac{C}{\sqrt{|\gamma_i\gamma_g|}}\,\Lambda_0\,. \tag{11.2.42}$$

2. For $\theta_B = \pi/4$ is $C = 0$, and there is no refracted π-wave. If a wave hits a surface at an angle ϑ_i and ϑ_g is the angle of the refracted wave (see Fig. 11.9(a)), then in the Laue case for $\vartheta_i + |\vartheta_g| = 2\theta_B = \pi/2$ there is a certain similarity to the Brewster angle, where also at π-polarization for $\vartheta + \vartheta' = \pi/2$ the reflected wave disappears.

3. The parameterization of the angle (11.2.37) can be rewritten in

$$y = \frac{\theta_B-\theta}{\delta_0} + \frac{\Delta\vartheta_0}{\delta_0}, \qquad \delta_0 = \frac{\lambda\,\gamma_g}{\Lambda_0\,\sin(2\theta_B)}, \qquad \Delta\theta_0 = \frac{\chi_{0r}(\gamma_g-\gamma_i)}{2\gamma_g\sin(2\theta_B)}. \tag{11.2.43}$$

The half-width of the averaged reflection curve (11.3.13) for the Laue case, given by $y = \pm 1$, can be expressed by δ_0:

$$|\theta(y{=}1) - \theta(y{=}{-}1)| = |-2\delta_0|.$$

This is also the width of the plateau in the Bragg case (*Darwin width*) (11.3.8), which is also determined by $y = \pm 1$. The shift of the maximum of the reflection from the Bragg angle is $\theta-\theta_B = \Delta\theta_0$.

4. Unfortunately, the dynamical theory does not have a uniform notation. Susceptibilities χ_0 and χ_g or the Bragg angles θ_B are denoted the same everywhere, but the differences begin with the reciprocal lattice vectors ($\mathbf{g} \Leftrightarrow 2\pi\mathbf{h}$). The angle deviation y and the Pendellösung length Λ_0 are used by "all"authors, whether Zachariasen [1945]; Kato [1974] or Authier [2002] parameterized differently. Here we mainly follow the notation of Rauch, Petrascheck [1978], who in turn orient themselves on Zachariasen and establish a reference to Authier's definitions η and Λ_0: $\eta = \zeta/\nu$ and $\Lambda_0 = \Delta_0/\nu$.

Amplitude ratios

For the calculation of the fields, one uses the amplitude ratios, which are obtained from the fundamental equations (11.2.18):

$$X = \frac{d_{\mathbf{g}}}{d_0} = \frac{K^2 - K_0^2}{K^2 C \chi_{-\mathbf{g}}}(1-\chi_0) \approx \frac{2\gamma_i \epsilon}{C \chi_{-\mathbf{g}}}.$$

Here we have replaced $K^2 - K_0^2$ using (11.2.31) by the excitation error $\epsilon \to \epsilon_{1,2}$ (11.2.35). We transform the factor $\pi/(k\Delta_0)$ of $\epsilon_{1,2}$ using (11.2.34), so that

$$X_{1,2} = \frac{d_{1,2}(\mathbf{g})}{d_{1,2}(0)} = \frac{\sqrt{|\chi_{\mathbf{g}}\chi_{-\mathbf{g}}|}}{\chi_{-\mathbf{g}}}\sqrt{\frac{\gamma_i}{|\gamma_g|}}\left(-\zeta \pm \sqrt{\zeta^2 + \nu^2\,\mathrm{sgn}\,\gamma_g}\right), \qquad (11.2.44)$$

$$X_1 X_2 = -\frac{\chi_{\mathbf{g}}}{\chi_{-\mathbf{g}}}\frac{\gamma_i}{\gamma_g}. \qquad (11.2.45)$$

We have now calculated the wave vectors $\mathbf{K}_{1,2}$ from the secular equation (11.2.19) (11.2.39) for the two-beam case, when the wave hits the crystal surface at the angle ϑ (see Fig. 11.9, p. 415). Depending on the position of the Bragg planes, the wave can penetrate into the crystal ($\gamma_g > 0$) or be reflected ($\gamma_g < 0$). Due to the small difference between $\mathbf{K}_1$ and $\mathbf{K}_2$, one will have to deal with interferences on the scale of the Pendellösung length $\Delta_0 \lesssim 100\,\mu\mathrm{m}$.

We now know the amplitude ratios $X_{1,2}$, (11.2.44), but not the amplitudes $d_{1,2}(0)$ and $d_{1,2}(\mathbf{g})$, which are only determined with the continuity conditions at the exit surface, which is the task of the next section.

11.3 Laue and Bragg Case

A classic application of the dynamical theory is the diffraction of an incident beam on a plane-parallel plate, as sketched in Fig. 11.10. The incident wave should be linearly polarized (σ or π) and normalized (amplitude $= 1$)

$$\mathbf{D}_i(\mathbf{x}) = \boldsymbol{\epsilon}_{\sigma,\pi}\,e^{i\mathbf{k}\cdot\mathbf{x}} = \boldsymbol{\epsilon}_{\sigma,\pi}\,\psi_i(\mathbf{x}).$$

The general dielectric displacement is then a superposition of these two polarizations, perpendicular to the reflection plane (σ) and in the reflection plane (π). For our further considerations, the vector character is not of relevance; the two polarizations are treated separately. We define crystal waves for the directions $\mathbf{k}_B$ and $\mathbf{k}_B + \mathbf{g}$

$$\Psi_0(\mathbf{x}) = d_1(0)e^{i\mathbf{K}_1\cdot\mathbf{x}} + d_2(0)e^{i\mathbf{K}_2\cdot\mathbf{x}}, \qquad (11.3.1)$$

$$\Psi_g(\mathbf{x}) = d_1(\mathbf{g})\,e^{i(\mathbf{K}_1+\mathbf{g})\cdot\mathbf{x}} + d_2(\mathbf{g})\,e^{i(\mathbf{K}_2+\mathbf{g})\cdot\mathbf{x}}.$$

At the entrance surface, continuity implies $\Psi_0 = \psi_i$. For Ψ_g, it is either vanishing at the entrance surface (Fig. 11.10a) or at the rear surface (Fig. 11.10b) With these two boundary conditions, the problem is determined. Although in the following $\Psi_{0,g}$ are determined, the solutions of the fundamental equations are the wave fields $\Psi_{1,2}$.

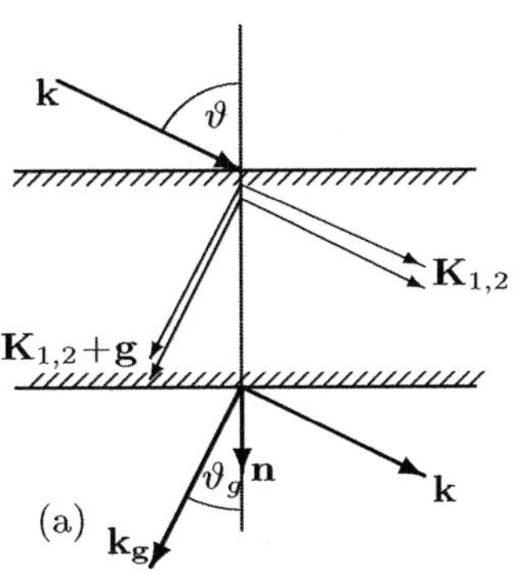
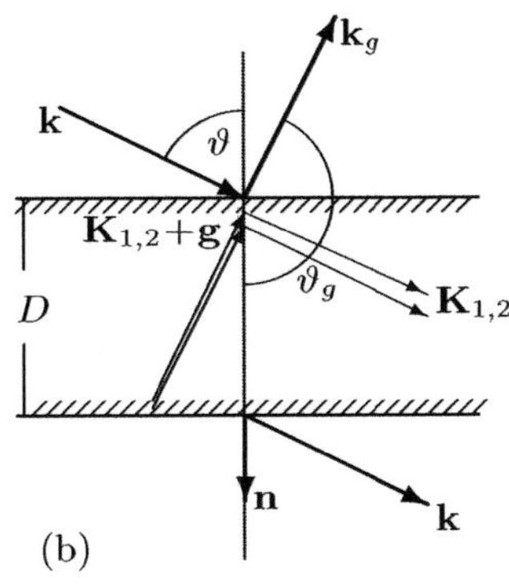

Fig. 11.10. Diffraction on a crystal plate:
(a) Laue case: $\gamma_g = \cos\vartheta_g > 0$. The diffracted beam exits at the rear surface
(b) Bragg case: $\gamma_g < 0$. The diffracted beam exits at the entry surface

11.3.1 Diffraction in one Dimension

It makes sense to study the scattering on a one-dimensional model, as here some mechanisms are particularly simple.

In one dimension, it is obvious that the diffracted beam can only come out on the front side and thus according to Fig. 11.10 is to be assigned to the Bragg geometry. We will also only describe the absorption-free case, but would like to point out that the intensities obtained for the one-dimensional case remain unchanged for three dimensions, only that the parameter y in three dimensions characterizes the angular deviation from the Bragg position (instead of the energy).

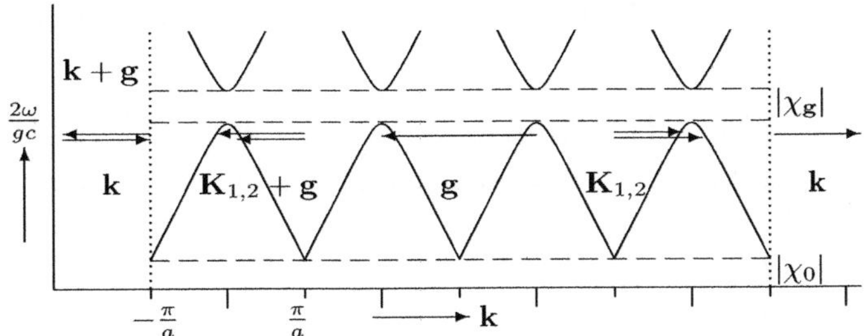

Fig. 11.11. Extended zone scheme of a one-dimensional structure with linear dispersion. Within the band gap, no waves can propagate and there is total reflection; just above and below interferences of $K_{1,2}$ occur in the reflected and transmitted beam; at a further distance from the band gap, (almost) nothing is reflected

If the energy of the incident beam is far from any Bragg condition, the beam will penetrate into the medium with slightly changed wave vector $|\mathbf{K_0}|$.

Fig. 11.11 shows the periodic zone scheme for a particle (photon) with linear dispersion in the one-dimensional reciprocal lattice. Due to the weak interaction with the lattice, the linear dispersion is only changed near the zone boundary. The curvature becomes very strong there, and a forbidden zone is created. The band gap states that in this energy range no radiation can linger, which means total reflection for incoming radiation.

Now an attempt is made to mathematically grasp these qualitative statements. We start with an incident wave $\psi(x) = e^{i\mathbf{k}\cdot\mathbf{x}}$ with linear dispersion $c|k| = \omega$. In the crystal, the rays then propagate

$$\Psi_0 = d_1(0)\, e^{i\mathbf{K}_1\cdot\mathbf{x}} + d_2(0)\, e^{i\mathbf{K}_2\cdot\mathbf{x}},$$

$$\Psi_g = d_1(\mathbf{g})\, e^{i(\mathbf{K}_1+\mathbf{g})\cdot\mathbf{x}} + d_2(\mathbf{g})\, e^{i(\mathbf{K}_2+\mathbf{g})\cdot\mathbf{x}}.$$

If we set $\gamma_i = 1$ and $\gamma_g = -1$ in (11.2.42), (11.2.36) and (11.2.39), we obtain

$$\mathbf{K}_{1,2} = \left(K_0 + \frac{\pi}{\Delta_0}\left(-\zeta \pm \sqrt{\zeta^2 - \nu^2}\right)\right)\mathbf{e}_x .$$

$$\zeta = y + i\eta = -\Lambda\left(k - \frac{|g|}{2}\right) - \frac{\chi_0}{|\chi_\mathbf{g}|} .$$

According to (11.2.21), $K_0 = k + \chi_0/2 < k$, due to the average potential of the medium. We note that waves with $\mathbf{k} = \mathbf{k}_B$ in the crystal have wave vectors that with $y \geq 1$ lie outside the range of total reflection (band gap). At the front surface $x = 0$, the transmitted wave is continuous:

$$d_1(0) + d_2(0) = 1$$

and at the back surface with $x = D$, Ψ_g vanishes:

$$d_1(g)\, e^{i(K_1+g)D} + d_2(g)\, e^{i(K_2+g)D} = 0 .$$

However, we limit ourselves here to the case without absorption and calculate with the $X_{1,2}$ from (11.2.44) and $X_1 X_2 = 1$ the wave functions explicitly.

$$X_1\, d_1(0)\, e^{iK_1 D} + X_2\left(1 - d_1(0)\right) e^{iK_2 D} = 0 . \tag{11.3.2}$$

It is useful to introduce the parameterized crystal thickness [Zachariasen, 1945, (3.140)]

$$A = \frac{D}{\Lambda} = \frac{\pi D}{\Delta_0} . \tag{11.3.3}$$

The amplitudes are then

$$d_{1,2}(0) = \frac{\mp X_{2,1} e^{iK_{2,1}D}}{X_1 e^{iK_1 D} - X_2 e^{iK_2 D}} = \frac{1}{2}\,\frac{(\pm y + \sqrt{y^2-1})e^{\mp iA\sqrt{y^2-1}}}{\sqrt{y^2-1}\,\cos\left(A\sqrt{y^2-1}\right) - iy\sin\left(A\sqrt{y^2-1}\right)}, \tag{11.3.4}$$

$$d_{1,2}(g) = \frac{\mp X_1 X_2 e^{iK_{2,1}D}}{X_1 e^{iK_1 D} - X_2 e^{iK_2 D}} = \frac{1}{2}\,\frac{\mp e^{\mp iA\sqrt{y^2-1}}}{\sqrt{y^2-1}\,\cos\left(A\sqrt{y^2-1}\right) - iy\sin\left(A\sqrt{y^2-1}\right)} .$$

The wave function propagating in the direct direction is obtained by adding the corresponding amplitudes $(d_{1,2}(0))$ at the back:

$$\Psi_0(D) = \frac{\sqrt{y^2-1}}{\sqrt{y^2-1}\,\cos\left(A\sqrt{y^2-1}\right) - iy\,\sin\left(A\sqrt{y^2-1}\right)}\,e^{iK_0 D - iAy}. \tag{11.3.5}$$

The intensity is the absolute square of the wave function:

$$P_0 = |\Psi_0(D)|^2 = \frac{y^2-1}{y^2 - \cos^2\left(A\sqrt{y^2-1}\right)}. \tag{11.3.6}$$

The calculation of the deflected beam is completely analogous, only that here the wave function on the front surface must be taken:

$$\Psi_g(0) = \frac{i\,\sin A\sqrt{y^2-1}}{\sqrt{y^2-1}\,\cos\left(A\sqrt{y^2-1}\right) - iy\,\sin\left(A\sqrt{y^2-1}\right)}. \tag{11.3.7}$$

Here, the beam intensity is given, where the intensity of the incident and deflected (refracted) wave is based on the same cross-section ($P_g = |\Phi_g|^2\,\gamma_g/\gamma_i$):

$$P_g = |\Psi_g(0)|^2 = \frac{\sin^2\left(A\sqrt{y^2-1}\right)}{y^2 - \cos^2\left(A\sqrt{y^2-1}\right)}. \tag{11.3.8}$$

The conservation of energy flow is expressed in

$$P_0(y) + P_g(y) = 1. \tag{11.3.9}$$

The intensity distribution $P_g(y)$ is shown in Fig. 11.12. A is a parameter for the thickness of the crystal. To resolve the intensity oscillations, which are due to the interference of the two wave fields, the crystal should not be too thick. y is a parameter for the distance of K_0 from the Bragg condition. If $-1 \le y \le 1$, the beam is totally reflected. The $K_{1,2}$ are imaginary, and the beam only penetrates to a depth of the order of magnitude Δ_0. The relative deviations y from the Bragg condition are small, i.e. of the size χ_0. In one dimension, y probes the energy in the range around the band gap at $k \approx |g|/2$.

In three dimensions, the same intensity profile is obtained when the diffracted beam leaves the crystal again at the front surface. This is called the *Bragg case*.

However, the frequency of the incident beam is then fixed, and one gets Fig. 11.12 by rotating through the Bragg position. The width of the reflex is again determined by y, which is now a measure for the deviation in arcseconds. In the crystal, standing waves are the solutions for $\mathbf{D}$ (11.2.7), the Bloch waves $\mathbf{D}_j(\mathbf{x})$ for each wave field $j = 1, 2$. If a wave field has its nodes at the locations $\mathbf{a}_n$ of the atoms, it can be expected that the interaction of $\mathbf{D}_j$ with the crystal is low, i.e. the absorption is rather weak. This thought was picked up by Borrmann [1950] who showed that $|\mathbf{D}_1(\mathbf{a}_n)|^2$ at the lattice points $(\cos \mathbf{a}_n \cdot \mathbf{g} = 1)$ is minimal and $|\mathbf{D}_2(\mathbf{a}_n)|^2$ is maximal. This only applies in this form for the Laue case, which is to be shown in problem 11.4.

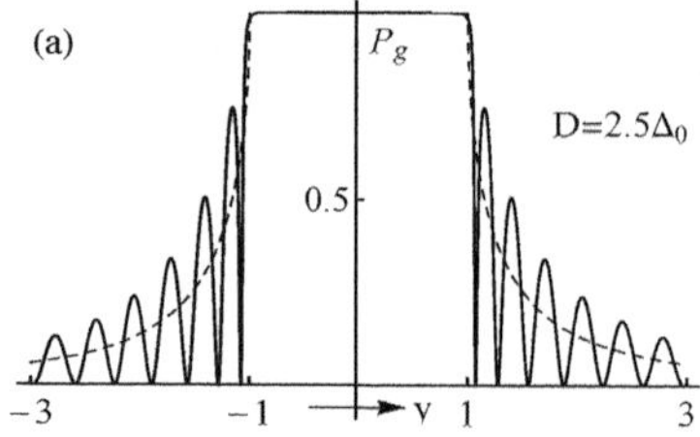 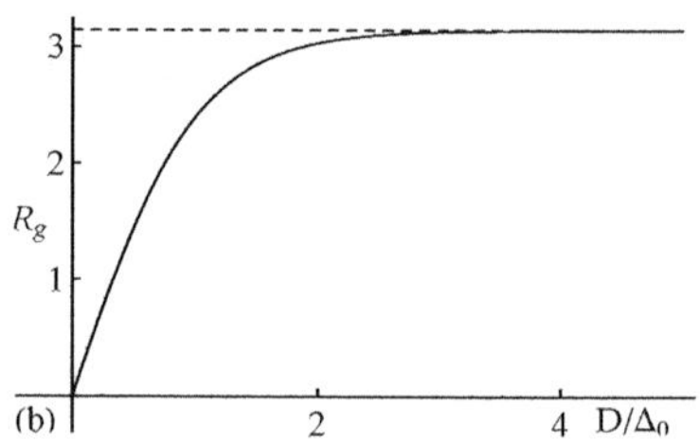

Fig. 11.12. (a) Intensity oscillations in the case of backscattering (Bragg case) for a crystal of thickness $D = 2.5\Delta_0$ compared to the rocking curve averaged over the oscillations (dashed). (b) Integrated intensity depending on the thickness

Integral reflectivities

Even in the Bragg case, the integral reflectivity is exactly calculable. In the case without absorption, one obtains

$$R_g^y = \int_{-\infty}^{\infty} dy \, \frac{\sin^2\left(A\sqrt{y^2-1}\right)}{y^2 - \cos^2\left(A\sqrt{y^2-1}\right)} = \pi \tanh A \,. \tag{11.3.10}$$

The calculation of the integral was carried out by *by Laue*[3] (see problem 11.2). Again, one has the linearity valid for the kinematic theory in small (thin) crystals (see Fig. 11.12). However, saturation is reached very quickly after which the intensity remains unchanged.

11.3.2 Laue Geometry

This case (see Fig. 11.10a) has no equivalent in one dimension, since there cannot be achieved by lateral deflection of the beam, that the Bragg condition is fulfilled. The field vector $\mathbf{D}_0$ of the incident beam is either perpendicular to the scattering plane (σ-) or lies in the scattering plane (π-polarization). The incident wave and the transmitted crystal wave (11.3.1) are continuous at the front surface ($z=0$), while the diffracted crystal wave disappears there:

$$d_1(0) + d_2(0) = 1 \,,$$
$$d_1(\mathbf{g}) + d_2(\mathbf{g}) = X_1 \, d_1(0) + X_2 \, d_2(0) = 0 \,.$$

For $\gamma_g > 0$ (Laue case, see Fig. 11.10) follows from (11.2.44) and (11.2.45)

$$d_{1,2}(0) = \frac{\mp X_{2,1}}{X_1 - X_2} = \frac{\pm\zeta + \sqrt{\zeta^2 + \nu^2}}{2\sqrt{\zeta^2 + \nu^2}} \,,$$

$$d_{1,2}(\mathbf{g}) = \frac{\mp X_1 X_2}{X_1 - X_2} = \frac{\chi_{\mathbf{g}}}{\sqrt{|\chi_{\mathbf{g}}\chi_{-\mathbf{g}}|}} \sqrt{\frac{\gamma_i}{\gamma_g}} \frac{\pm 1}{2\sqrt{\zeta^2 + \nu^2}} \,.$$

---

[3] Max von Laue, 1879–1960, Nobel prize 1914, [Max von Laue, 1960]

The wave functions (11.3.1) are obtained using (11.2.39) and (11.2.36):

$$\Psi_0(\zeta, D) = \left(\cos\left(A\sqrt{\zeta^2 + \nu^2}\right) + \frac{iz}{\sqrt{\zeta^2 + \nu^2}} \sin\left(A\sqrt{\zeta^2 + \nu^2}\right) \right) e^{i\mathbf{K}_0 \cdot \mathbf{x} - iA\zeta},$$

$$\Psi_g(\zeta, D) = \frac{\chi_{\mathbf{g}}}{\sqrt{|\chi_{\mathbf{g}}\chi_{-\mathbf{g}}|}} \sqrt{\frac{\gamma_i}{\gamma_g}} \frac{i\sin(A\sqrt{\zeta^2 + \nu^2})}{\sqrt{\zeta^2 + \nu^2}} \, e^{i(\mathbf{K}_0 + \mathbf{g}) \cdot \mathbf{x} - iA\zeta}. \tag{11.3.11}$$

The absolute square of the wave function gives the intensity, whereby the absorption ($\zeta \to y$) should be neglected here. However, we have always given the radiant intensity $P_g = |\Psi_g|^2 \gamma_g / \gamma_i$ instead of the intensity, because this refers the intensity to the cross section of the incident beam, which is only relevant for the diffracted beam ($\gamma_i \neq \gamma_g$):

$$P_0(y, D) = 1 - \frac{\sin^2(A\sqrt{1 + y^2})}{1 + y^2}. \tag{11.3.12}$$

From the conservation of the energy flow $P_0(y, D) + P_g(y, D) = 1$ it follows immediately that

$$P_g(y) = \frac{\sin(A\sqrt{1 + y^2})}{1 + y^2} \qquad \Longrightarrow \qquad \overline{P}_g = \frac{1}{2(1 + y^2)}. \tag{11.3.13}$$

In particular, the rocking curve $P_g(y)$ for the deflected direction is experimentally accessible. It is shown in Fig. 11.13 for a thin crystal plate.

In "thick" crystals, the oscillations become very narrow and cannot be resolved experimentally. The averaged distribution is a Lorentz curve.

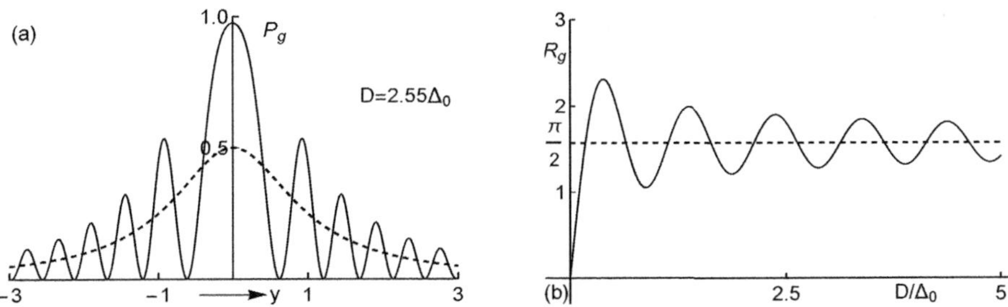

Fig. 11.13. (a) Rocking curve in the Laue case with a crystal thickness of $D/\Delta_0 = 2.55$. The averaged curve is dashed. (b) Integrated intensity

Reflection curves with weak absorption

The significance of the Pendellösung oscillations decreases rapidly with noticeable absorption, since one of the two wave fields is very much more attenuated so that the interferences (oscillations) disappear. This is then referred to as anomalous absorption.

The imaginary part of $\mathbf{K}_0 \cdot \mathbf{x} - A\zeta$ in (11.3.11) is calculated at the back surface. This results in

$$\mathrm{Im}(\mathbf{K}_0 \cdot \mathbf{x} - A\zeta) = \mu_0 \frac{D_e}{2} \quad \text{with} \quad D_e = \left(\frac{1}{\gamma_i} + \frac{1}{\gamma_g}\right) \frac{D}{2}. \tag{11.3.14}$$

The effective thickness D_e indicates the distance that the beam travels in the crystal to reach the back surface, if it chooses a zigzag path in the two directions $\mathbf{k}_B$ and $\mathbf{k}_B + \mathbf{g}$. The normal attenuation of the beam thus results from the effective distance traveled in the crystal.

In Fig. 11.14, the rocking curves for different absorption are shown. With normal attenuation, in the case of $\mu_0 D_e = 5$ there would be no noticeable intensity in either the diffracted or the transmitted beam. The fact that noticeable intensity still comes through, as can be seen from Fig. 11.14, is due to the anomalous absorption, the different damping of the two wave fields

$$\mathbf{K}_{1,2} \cdot \mathbf{x}\big|_{\mathbf{n} \cdot \mathbf{x} = D} = \mathbf{K}_{0r} \cdot \mathbf{x} + i\frac{\mu_0 D}{2\gamma_i} + \frac{\pi D}{\Delta_0}\left(-\zeta \pm \sqrt{\zeta^2 + \nu^2}\right).$$

From this it follows with reference to (11.2.36)

$$\mathrm{Im}\left(\mathbf{K}_{1,2} \cdot \mathbf{x}\right)\bigg|_{\mathbf{n} \cdot \mathbf{x} = D} = \frac{\mu_0 D}{2\gamma_i} \mp \frac{\kappa A}{\sqrt{y^2 + 1}}. \tag{11.3.15}$$

From this it is immediately apparent that, when $2\kappa A = \mu_0 D_e$ (i.e. $\chi_0 = \chi_g$), the damping for the wave field 1 disappears (almost) for small y. At the same time, it is noted that the Pendellösung oscillations rapidly lose significance with increasing absorption. The averaged beam strengths are

$$P_g(y) = \frac{e^{-\mu_0 D_e}}{2(1 + y^2)} \cosh \frac{2\kappa A}{\sqrt{1 + y^2}}, \tag{11.3.16}$$

$$P_0(y) = \frac{e^{-\mu_0 D_e}}{2(1 + y^2)} \left((1 + 2y^2) \cosh \frac{2\kappa A}{\sqrt{1 + y^2}} + 2y\sqrt{1 + y^2} \sinh \frac{2\kappa A}{\sqrt{1 + y^2}}\right).$$

Integral reflectivities

If you integrate the rocking curves, you get the integral reflectivities

$$R_g^y = \int_{-\infty}^{\infty} dy\, P_g(y) = \frac{\pi}{2} \int_0^{2A} dx\, J_0(x) \approx \frac{\pi}{2} \begin{cases} 1 & \text{for } A \to \infty \\ 2A & \text{for } A \to 0. \end{cases} \tag{11.3.17}$$

$J_0(x)$ is a Bessel function, *Waller's formula*, which is treated in more detail in appendix B.4; for small arguments, we have the expansion

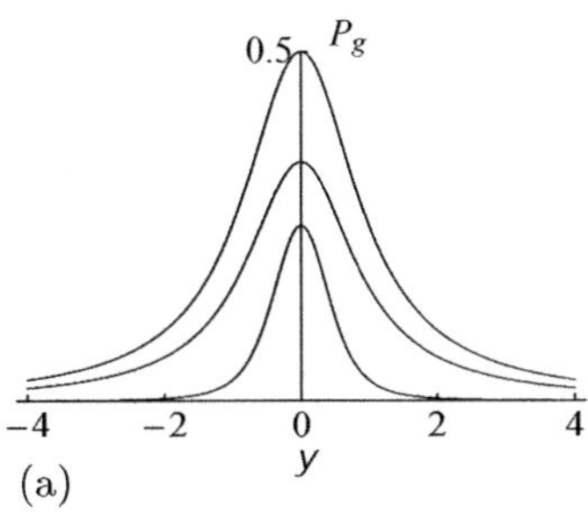
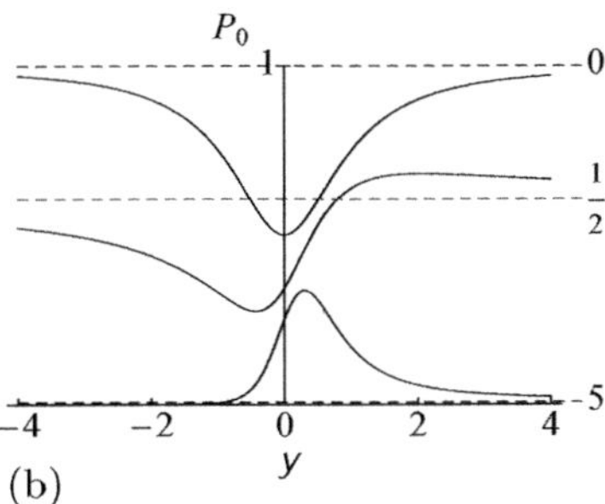

Fig. 11.14. Average intensities in the Laue case with increasing absorption $\mu_0\,D_e$. The scales are based on the (11.3.16) specified reduced beam strengths. The indicated curves have the normal attenuation $N_a = \mathrm{e}^{-\mu D_e}$ of $0, 0.5$ and 5.
(a) Reflection curves.   (b) Transmission curves: The dashed line indicates the expected for N_a (normal) attenuation

$$J_0(z) = \sum_{k=0}^{\infty} (-1)^k \frac{z^{2k}}{2^{2k}(k!)^2} \approx 1 - \frac{z^2}{2}, \qquad |\arg z| < \pi.$$

These results are very well supported experimentally. The course calculated according to (11.3.17) is recorded in Fig. 11.13b. Up to thicknesses $D < \Delta_0/2$ the intensity increases linearly. In this area, the kinematic theory is valid. The deviation from linearity is referred to as *primary extinction*.

In the experiment, the crystal is rotated through the Bragg angle and the intensity as a function of the specular angle θ or $\theta - \theta_B$ is measured:

$$R_g^\theta = \int \mathrm{d}(\theta - \theta_B) P_g = \int \mathrm{d}y\, P_g(y) \left| \frac{\mathrm{d}\theta}{\mathrm{d}y} \right| = R_g^y \left| \frac{\mathrm{d}\theta}{\mathrm{d}y} \right|. \tag{11.3.18}$$

The integral reflectivities R_g^y we have given for both the Laue case (11.3.17) as well as for the Bragg case (11.3.10).

11.3.3 Bragg Geometry

The results derived in the one-dimensional case largely also apply in three dimensions. An extension is only necessary by considering absorption. The boundary conditions are analogous to the one-dimensional case:

1. Continuity of the wave function at the front surface ($z=0$):

$$1 = d_1(0) + d_2(0).$$

2. Disappearance of the diffracted wave at the back surface ($z=D$):

$$0 = d_1(\mathbf{g})\, \mathrm{e}^{ik\epsilon_1 D} + d_2(\mathbf{g})\, \mathrm{e}^{ik\epsilon_2 D}.$$

The calculations are completely analogous to the one-dimensional case, and one obtains (see problem 11.3) for the amplitudes:

$$d_{1,2}(0) = \frac{\left(\pm\zeta + \sqrt{\zeta^2 - \nu^2}\right) e^{\mp iA\sqrt{\zeta^2 - \nu^2}}}{\left(\zeta + \sqrt{\zeta^2 - \nu^2}\right) e^{-iA\sqrt{\zeta^2 - \nu^2}} - \left(-\zeta + \sqrt{\zeta^2 - \nu^2}\right) e^{iA\sqrt{\zeta^2 - \nu^2}}},$$

$$d_{1,2}(\mathbf{g}) = \frac{\chi_{\mathbf{g}}}{\sqrt{|\chi_{\mathbf{g}}\chi_{-\mathbf{g}}|}} \sqrt{\frac{\gamma_i}{|\gamma_g|}} \tag{11.3.19}$$

$$\times \frac{\pm e^{\mp iA\sqrt{\zeta^2 - \nu^2}}}{\left(\zeta + \sqrt{\zeta^2 - \nu^2}\right) e^{-iA\sqrt{\zeta^2 - \nu^2}} - \left(-\zeta + \sqrt{\zeta^2 - \nu^2}\right) e^{iA\sqrt{\zeta^2 - \nu^2}}}.$$

The waves at the front ($\mathbf{n}\cdot\mathbf{x} = 0$) and back surface ($\mathbf{n}\cdot\mathbf{x} = D$) are

$$\Psi_0(\zeta, D) = d_1(0)\, e^{i\mathbf{K}_1\cdot\mathbf{x}} + d_2(0)\, e^{i\mathbf{K}_2\cdot\mathbf{x}},$$

$$\Psi_g(\zeta, 0) = [d_1(\mathbf{g}) + d_2(\mathbf{g})]e^{i(\mathbf{k}+\mathbf{g})\cdot\mathbf{x}}. \tag{11.3.20}$$

Even more than in the Laue case, in the Bragg case only the reflected wave is of interest. As one would expect with reflection from the surface, in Ψ_g the term with the normal attenuation is missing. It is somewhat tedious to break down the complex angle functions into real and imaginary parts, which is why further evaluation is omitted here

$$P_g(y) = \left|\frac{\chi_{\mathbf{g}}}{\chi_{-\mathbf{g}}}\right| \left|\frac{\sin A\sqrt{\zeta^2 - \nu^2}}{\sqrt{\zeta^2 - \nu^2}\,\cos A\sqrt{\zeta^2 - \nu^2} - i\zeta\,\sin A\sqrt{\zeta^2 - \nu^2}}\right|^2. \tag{11.3.21}$$

With vanishing absorption, one obtains the results of reflection on the one-dimensional lattice (11.3.8) shown in Fig. 11.12, page 422. The Pendellösung generally plays a lesser role in the Bragg case than in the Laue case, as only about 15% of the intensity is affected, which penetrate into the crystal and is not immediately reflected near the surface. Also, the calculation of the averaged distribution (see problem 11.1) is a bit more tedious.

For the intensities averaged over the Pendellösung oscillations, we obtain

$$P_g(y) = \begin{cases} 1 & \text{for } |y| \leq 1 \\ 1 - \sqrt{1 - y^{-2}} & \text{for } |y| > 1. \end{cases} \tag{11.3.22}$$

Fig. 11.12 shows the average intensity.

The reflection curves in the case of weak absorption

Absorption has a slightly different meaning in the Bragg reflection than in the Laue case. In the former, the majority is reflected at the surface, with the penetration depth proportional to Δ_0. With low absorption, this contribution is hardly weakened. From the back surface comes another contribution – about 15% – which is absorbed more strongly. But this only applies to contributions with $|y| > 1$, i.e., the half-width of the rocking curve becomes a little narrower.

To make more precise statements, we consider a simpler geometric arrangement, the so-called *symmetric Bragg case*. In this case, the lattice planes are parallel to the surface, so that for the associated lattice vector $\hat{\mathbf{g}} = \mathbf{g}/g = -\mathbf{n}$ applies, where we refer to Fig. 11.10. Then $\gamma_g = -\gamma_i$. Furthermore, the reflection under consideration should have inversion symmetry, from which $\chi_{-\mathbf{g}} = \chi_{\mathbf{g}}$ follows.

For the susceptibility, $|\chi_{\mathbf{g}}| \leq |\chi_0|$ always applies. In the situation considered here, we take $\chi_{\mathbf{g}} = \chi_0$ with the polarization $C = 1$. We then get, referring to (11.2.37),

$$\zeta = y + i\eta \qquad \begin{cases} y &= -\Lambda\,(\mathbf{k} - \mathbf{k}_B) \cdot \hat{\mathbf{g}} + \chi_{0r}/|\chi_{\mathbf{g}}| \\ \eta &= \chi_{0i}/|\chi_{\mathbf{g}}|. \end{cases} \tag{11.3.23}$$

With these assumptions, the absorption is determined by $\mu_0 D_e = 2\eta\Lambda$ and $\kappa = \eta$ (see (11.2.33)). First, a fact that we have not mentioned so far: The waves falling exactly in the Bragg direction $\mathbf{k} \equiv \mathbf{k}_B$ are, except in the symmetric Laue case, all asymmetric with respect to the reflection curves. Here $\mathbf{k}_B$ has the value $y = 1$, so it is on the edge of the plateau. We now return

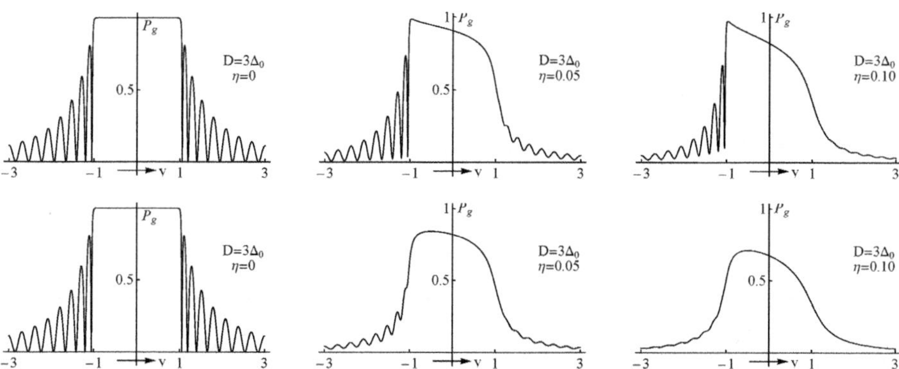

Fig. 11.15. Bragg reflection on an absorbing crystal with inversion symmetry. It is assumed that $\chi_{\mathbf{g}} = \chi_0$. The absorption is described by $\eta = \chi_{0i}/|\chi_{\mathbf{g}}|$. The top row describes the symmetric Bragg case, while the second row indicates an asymmetric situation with $(\gamma_i - \gamma_g)/(2\sqrt{\gamma_i|\gamma_g|}) = 2$

to the field strengths $|\mathbf{D}_j(\mathbf{a}_n)|^2$ at the lattice points $\mathbf{a}_n$ (see problem 11.4). In the range $|y| < 1$ is $|\mathbf{D}_1(\mathbf{a}_n)|^2 = |\mathbf{D}_2(\mathbf{a}_n)|^2$, i.e., both intensities are equal, but one observes in the range $-1 < y < 1$, starting at $y = -1$, a decrease in intensity and thus an increase in absorption, as can also be seen in Fig. 11.15.

11.4 Dynamical Diffraction of Spherical Waves

The starting point for the spherical theory should be the spherical wave. The exact form is not essential here, because the dynamic equations select from the offered wide beam a very narrow *Bragg window* in which diffraction occurs. The rest goes through unhindered. It is assumed that within this narrow window the amplitudes of the partial waves are constant. The "spherical"waves in the diffracted or transmitted direction are formed by superposition of the plane partial waves from the area of the Bragg window:

$$\Phi_{0,g}(\mathbf{x}) = \int_{-\infty}^{\infty} dy\, \Psi_{0,g}(y), \tag{11.4.1}$$

where for $\Psi_{0,g}(y)$ the wave functions of the Laue or Bragg case are to be used.

The incident wave

The crystal waves to be calculated with (11.4.1) can be caused by an incident wave of the form

$$\Phi_i = \int_{-\infty}^{\infty} dy\, e^{i\mathbf{k}\cdot\mathbf{x}} = \int_{-\infty+i\eta}^{\infty+i\eta} d\zeta\, e^{i\mathbf{k}\cdot\mathbf{x}}. \tag{11.4.2}$$

For both, (11.4.1) and (11.4.2), is $\mathbf{k}$ to be determined as a function of y. We orient ourselves to the sketch Fig. 11.16 and develop up to the 1st order in $\vartheta - \vartheta_i$:

$$\mathbf{k} = k\{\cos\vartheta\,\mathbf{n} + \sin\vartheta\,\mathbf{e}_x\} \approx \mathbf{k}_B + k(\vartheta-\vartheta_i)\cos\vartheta_i\,(\mathbf{e}_x - \tan\vartheta_i\,\mathbf{n}). \tag{11.4.3}$$

We now replace using (11.2.29) $\alpha \to y$:

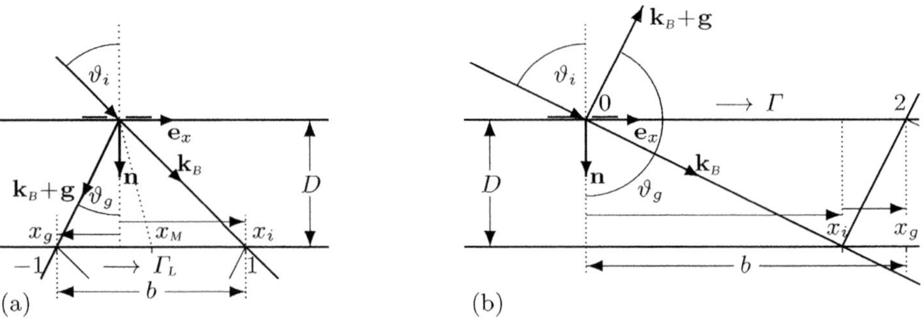

Fig. 11.16. Γ_L or Γ parameterize the width b of the Borrmann fan.
(a) Laue case: $b = x_i - x_g$ and $\Gamma_L = 2(x - x_M)/b$. (b) Bragg case: $b = x_i + x_g$ and $\Gamma = 2x/b$

$$\vartheta - \vartheta_i = -\operatorname{sgn}\gamma_g\,\frac{\alpha}{\sin(2\theta_B)} \overset{(11.2.36)}{=} -\operatorname{sgn}\gamma_g\,\frac{2\gamma_g}{\sin(2\theta_B)}\left(\frac{\pi y}{k\Delta_0} - \chi_{0r}\,\frac{\gamma_g - \gamma_i}{4\gamma_i\gamma_g}\right)$$

and obtain

$$\mathbf{k} = \mathbf{k}_B - \left[\frac{2|\gamma_g|\gamma_i}{\sin(2\theta_B)}\frac{y\pi}{\Delta_0} - \operatorname{sgn}\gamma_g\, k\chi_{0r}\frac{\gamma_g - \gamma_i}{2\sin(2\theta_B)}\right](\mathbf{e}_x - \tan\vartheta_i\,\mathbf{n})\,. \quad (11.4.4)$$

Based on Fig. 11.16 we can transform the expression

$$D\frac{\sin(2\theta_B)}{|\gamma_i\gamma_g|} = D\frac{\sin(\vartheta_i - \vartheta_g)}{\gamma_i\gamma_g} = D\tan\vartheta_i - D\tan\vartheta_g = x_i - \operatorname{sgn}\gamma_g\, x_g = b.$$

Thus we define the parameter

$$\Gamma = \frac{2\gamma_i|\gamma_g|}{\sin(2\theta_B)}\frac{x}{D} = \frac{2x}{b} \quad (11.4.5)$$

and obtain

$$\mathbf{k}\cdot\mathbf{x} = \mathbf{k}_B\cdot\mathbf{x} - Ay\left(\Gamma - \frac{2z}{b}\tan\vartheta_i\right) + \frac{k\chi_{0r}}{2b}\frac{\gamma_g - \gamma_i}{\gamma_i\gamma_g}(x - z\tan\vartheta_i)\,. \quad (11.4.6)$$

Thus, for the incident wave, one obtains

$$\phi_i(\mathbf{x}) = \lim_{y_0\to\infty}\int_{-y_0}^{y_0} dy\, e^{i\mathbf{k}\cdot\mathbf{x}} = e^{i\mathbf{k}_B\cdot\mathbf{x}}\, 2\pi\,\frac{b}{2A}\,\delta(x - z\tan\vartheta_i)\,. \quad (11.4.7)$$

This is a wave packet that is localized along the Bragg direction and hits the crystal exactly at one point (line). There, the bundle spreads out fan-like within the *Borrmann fan*, the angular range of $2\theta_B$, which is bounded by the vectors $\mathbf{k}_B$ and $\mathbf{k}_B + \mathbf{g}$. It thus treats the propagation from a point-like (line-like) source on the crystal front surface. We normalize the intensity of the incident wave to

$$I_i = \frac{1}{2y_0}\int_{-\infty}^{\infty} dx\,|\phi_i(\mathbf{x})|^2 = \frac{b\Delta_0}{D}\,. \quad (11.4.8)$$

Phase factor

The phase of the incident wave is given in (11.4.6), but it can be brought into a more illustrative form. It is assumed that the beam in the crystal takes a zigzag path in the directions of $\mathbf{k}_B$ and $\mathbf{k}_B + \mathbf{g}$ to reach a point x on the surface. In Fig. 11.17 this path covered in the crystal is summarized to

$$s = s_0 + s_g = \frac{D - d_g}{\gamma_i} + \frac{d_g}{\gamma_g} = \frac{D}{\gamma_i} - \operatorname{sgn}\gamma_g\,\frac{D\tan\vartheta_i - x}{\sin(2\theta_B)}(\gamma_g - \gamma_i)\,.$$

The analogous consideration can also be made for the Bragg case and one obtains for the path in the crystal

$$s = \frac{z}{\gamma_i} + \operatorname{sgn}\gamma_g\,\frac{x - z\tan\vartheta_i}{\sin(2\theta_B)}(\gamma_g - \gamma_i), \qquad z = 0 \quad \text{or} \quad z = D. \quad (11.4.9)$$

This results in

$$\mathbf{k}\cdot\mathbf{x} = \mathbf{k}_B\cdot\mathbf{x} - Ay\left(\Gamma - \frac{2z}{b}\tan\vartheta_i\right) + \frac{k\chi_{0r}}{2}\left(s - \frac{z}{\gamma_i}\right), \tag{11.4.10}$$

$$\mathbf{K}_0\cdot\mathbf{x} = \mathbf{k}_B\cdot\mathbf{x} - A\zeta\left(\Gamma - \frac{2z}{b}\tan\vartheta_i\right) + \frac{k\chi_0}{2}s.$$

11.4.1 Laue Case

If you have a divergent incident beam, which as shown in Fig. 11.17 enters through a narrow slit, it spreads within the two directions $\mathbf{k}_B$ and $\mathbf{k}_g$. This *Borrmann fan* has an opening angle of $2\theta_B$ and a width b at the back surface. Thus, in this area, we obtain a spatial intensity profile. We assume that

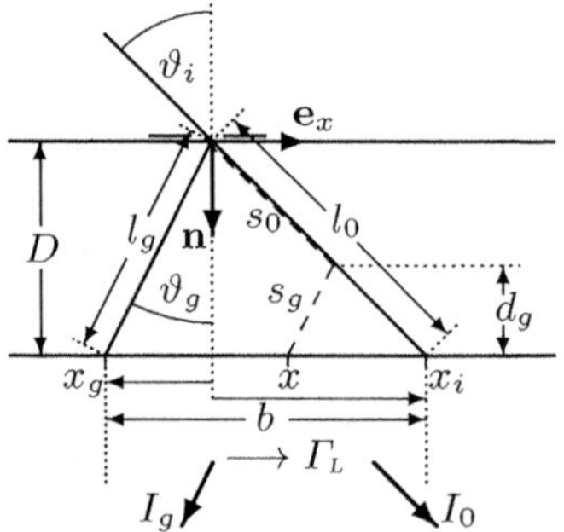

Fig. 11.17. In the Laue case, the beam spreads out within the Borrmann fan for point or line entry. To get to x, the beam covers the distance s_0 in the direction of $\mathbf{k}_B$ and s_g in that of $\mathbf{k}_B+\mathbf{g}$.
Laue case: $b = x_i - x_g$, $\Gamma_L = 2(x - x_M)/b$, $l_{0,g} = D/\gamma_{0,g}$

our beam is monochromatic and all plane partial waves are coherent. If we disregard the normalization, the crystal wave (11.4.1) can be formed by simple superposition of the solutions of the plane partial waves of the dynamical theory (11.3.11):

$$\Phi_g(\mathbf{x}) = c_g \int_{-\infty+i\eta}^{\infty+i\eta} d\zeta\, \frac{\sin\left(A\sqrt{\zeta^2+\nu^2}\right)}{\sqrt{\zeta^2+\nu^2}}\, e^{-iA\Gamma_L\zeta} \tag{11.4.11}$$

$$\Phi_0(\mathbf{x}) = c_0 \int_{-\infty+i\eta}^{\infty+i\eta} d\zeta\left[\cos\left(A\sqrt{\zeta^2+\nu^2}\right) + \frac{i\zeta}{\sqrt{\zeta^2+\nu^2}}\sin\left(A\sqrt{\zeta^2+\nu^2}\right)\right]e^{-iA\Gamma_L\zeta}$$

$$= \frac{c_0}{c_g}\left(\frac{\partial}{\partial A} - \frac{1+\Gamma_L}{A}\frac{\partial}{\partial\Gamma_L}\right)\Phi_g(\mathbf{x}). \tag{11.4.12}$$

The prefactors and Γ_L are given by

$$c_0 = e^{i\mathbf{k}_B\cdot\mathbf{x}+ik\chi_0 s/2}, \qquad c_g = \frac{i\chi_\mathbf{g}}{\sqrt{|\chi_\mathbf{g}\chi_{-\mathbf{g}}|}}\, e^{i(\mathbf{k}_B+\mathbf{g})\cdot\mathbf{x}+ik\chi_0 s/2},$$

$$\Gamma_L = \frac{2(x-x_M)}{b} = \frac{2(x-x_i)+b}{b} = \Gamma-\Gamma_i+1.$$

The integral (11.4.11) has no cut in the complex ζ-plane, since in the power series expansions of the integrand $\sqrt{\zeta^2 + \nu^2}^{\,n}$ only even powers occur and the path C in Fig. 11.18 does not enclose a pole, so that according to the Cauchy's integral theorem

$$\oint_C \mathrm{d}\zeta\,\Psi_g = 0 \qquad \Rightarrow \qquad \int_{-\infty+i\eta}^{\infty+i\eta} \mathrm{d}\zeta\,\Psi_g = \int_{-\infty}^{\infty} \mathrm{d}\zeta\,\Psi_g .$$

This Fourier integral is exactly solvable and can be looked up in integral tables

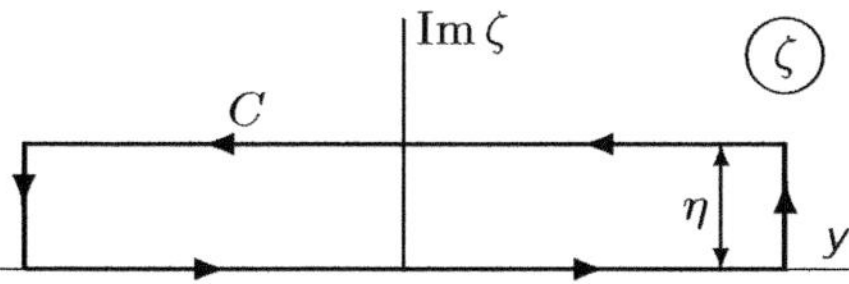

Fig. 11.18. Integration path C in the complex ζ-plane

[Gradshteyn, Ryzhik, 1965, Ziff. 3.876-1]:

$$\Phi_g(\mathbf{x}) = c_g \pi\, J_0(\nu A\sqrt{1-\Gamma_L^2})\,\theta(1-|\Gamma_L|) . \tag{11.4.13}$$

J_0 and J_1 are Bessel functions. For the transmitted wave function one obtains

$$\Phi_0(\mathbf{x}) = c_0\left[\frac{2\pi}{A}\delta(1-\Gamma_L) - \pi\nu^2(1+\Gamma_L)\frac{J_1(\nu A\sqrt{1-\Gamma_L^2})}{\nu A\sqrt{1-\Gamma_L^2}}\,\theta(1-|\Gamma_L|)\right]. \tag{11.4.14}$$

Since only a very small part is diffracted from the range $-\infty < y < \infty$, almost everything goes through undisturbed, if one disregards the phase shift through the average crystal potential. The δ-function is an expression for the fact that almost everything passes through undisturbed.

Intensity profiles

In the limit $A \to \infty$ the asymptotic expansions of the Bessel functions can be used:

$$\Phi_g(\mathbf{k}_0, \mathbf{x}) \approx c_g\left[\frac{2\pi}{\nu A\sqrt{1-\Gamma^2}}\right]^{\frac{1}{2}} \sin\left(\nu A\sqrt{1-\Gamma^2} + \frac{\pi}{4}\right)\theta(1-|\Gamma|). \tag{11.4.15}$$

Ignoring the absorption we get

$$P_0(\Gamma) = \frac{(1+\Gamma_L)\cos^2\left(A\sqrt{1-\Gamma_L^2} + \frac{\pi}{4}\right)}{(1-\Gamma_L)\sqrt{1-\Gamma_L^2}} , \tag{11.4.16}$$

$$P_g(\Gamma) = \frac{\sin^2\left(A\sqrt{1-\Gamma_L^2} + \frac{\pi}{4}\right)}{\sqrt{1-\Gamma_L^2}} . \tag{11.4.17}$$

For the case without absorption, Fig. 11.19 shows the intensity profile for the diffracted beam formed from (11.4.13) and (11.4.8):

$$P_g(\Gamma_L) = \frac{A\pi}{2}\, J_0^2(A\sqrt{1-\Gamma_L^2}).$$

The crystal plate is thin with $D/\Delta_0 = 5.6$. The asymptotic expansion of J_0 for thick crystals leads to the result of the ray considerations of the previous section. The significance of the Pendellösung oscillations decreases with increasing absorption quickly, i.e. the averaged intensity profiles are the physically more relevant quantities. An exception are neutrons, which have very low losses due to absorption in many crystals.

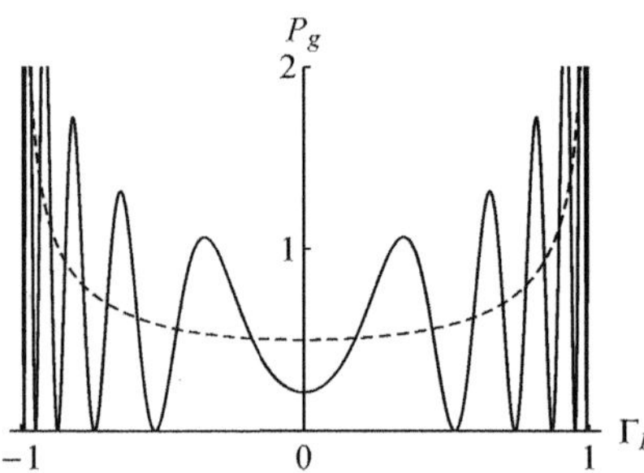

Fig. 11.19. Spatial intensity profile for $D/\Delta_0 = 5.6$ and averaged intensity (dashed line)

Approximate calculation of the intensity profiles

Method of stationary phase

A slightly better insight is obtained through the approximate solution of the equations with the method of stationary phase. This is used for the evaluation of integrals whose integrands oscillate strongly. It is therefore used primarily in optics. Given is the integral

$$I = \int_a^b \mathrm{d}u\, A(u)\mathrm{e}^{\mathrm{i}f(u)}.$$

In this notation, A(u) is a slowly varying function compared to the rapid oscillations of the factor $\mathrm{e}^{\mathrm{i}f(u)}$. The integration limits are chosen so that the minimum of $f(u)$ lies within the integration range and no cutoff effects occur due to the limits. If one expands around the minimum ($f''(u_0) > 0$) at u_0, it follows

$$I \sim A(u_0)\,\mathrm{e}^{\mathrm{i}f(u_0)} \int_{-\infty}^{\infty} \mathrm{d}u\, \mathrm{e}^{(\mathrm{i}/2)f''(u_0)(u-u_0)^2}$$
$$= A(u_0)\,\sqrt{2\pi/f''(u_0)}\,\mathrm{e}^{\mathrm{i}f(u_0)+\mathrm{i}\pi/4}. \tag{11.4.18}$$

The integral appearing here is a complete Fresnel integral

$$\int_{-\infty}^{\infty} du\, e^{i\alpha u^2} = \sqrt{\frac{\pi}{|\alpha|}}\, e^{(i\pi/4)\,\text{sgn}\,\alpha}.$$

The contribution to the integral comes from a narrow range around the stationary phase u_0.

The integrand of (11.4.11) becomes a very rapidly varying function with increasing thickness A. If one decomposes the sine into its two exponential components

$$\Phi_g(\mathbf{x}) = c_g \int_{-\infty}^{\infty} \frac{dy}{2i\sqrt{y^2 + \nu^2}} \left[e^{iA\left(\sqrt{y^2+\nu^2}-\Gamma y\right)} - e^{iA\left(-\sqrt{y^2+\nu^2}-\Gamma y\right)} \right],$$

thus, Φ_g offers itself for calculation with the method of stationary phase:

$$\frac{\partial}{\partial y} A\left(\pm\sqrt{\nu^2 + y^2} - \Gamma y\right) = 0 \qquad \Rightarrow \qquad y_{1,2} = \frac{\pm\nu\Gamma}{\sqrt{1-\Gamma^2}}. \qquad (11.4.19)$$

If one now evaluates the integral with the stationary phases $y_{1,2}$, one obtains the asymptotic wave function (11.4.15).

Calculation of the intensity profile with plane waves

$y_{1,2}$ not only gives the value of the stationary phase, but also the direction Γ of energy transport (Poynting vector) of the respective wave field. If one sums the partial waves that belong to a fixed value Γ, one again obtains

$$P_{0,g}(\Gamma) = \left| \Psi_{0,g}^1(y_1)\sqrt{\frac{dy_1}{d\Gamma}}\, e^{i\pi/4} + \Psi_{0,g}^2(y_2)\sqrt{\left|\frac{dy_2}{d\Gamma}\right|}\, e^{-i\pi/4} \right|^2$$

for the wave function the already known result (11.4.15) and for the intensity (11.4.17). Thus, there is no discrepancy between the spherical theory and this theory of plane waves [Shull, Oberteuffer, 1972]

11.4.2 The Intensity Profiles in the Bragg Case

The intensity profiles in the Bragg case are of lesser importance than those in the Laue case, as the main contribution is almost independent of the thickness of the crystal plate and is immediately reflected at the point of entry and does not show a clear interference structure.

The following calculation of the spherical wave [Kato, 1974] is relatively long and contains technical details that are only of interest to some readers. We recommend continuing with section 11.4.4, or to focus on the final results (11.4.47) and (11.4.48).

The amplitudes in the Bragg case

In the Bragg case, the integrals cannot be calculated directly as in the Laue case by integrating the wave function $\Psi_g(\zeta)$ (see (11.4.11)). With the expansion of the amplitudes (11.3.19) into a power series, as will become clear later, the fact is taken into account that waves are reflected at the back of the crystal plate, which are spatially separated from the portion reflected at the front surface (see Fig. 11.22). First, it is noted that

$$\frac{1}{\zeta + \sqrt{\zeta^2 - \nu^2}} = \frac{\zeta - \sqrt{\zeta^2 - \nu^2}}{\nu^2}.$$

This relation is useful for the expansion of (11.3.19) into a geometric series:

$$d_1(\mathbf{g}) = \hat{d}_g \frac{\zeta - \sqrt{\zeta^2 - \nu^2}}{\nu^2} \frac{1}{1 - \left(\frac{\zeta - \sqrt{\zeta^2 - \nu^2}}{\nu}\right)^2 e^{2iA\sqrt{\zeta^2 - \nu^2}}}, \tag{11.4.20}$$

$$d_2(\mathbf{g}) = -\hat{d}_g \frac{\zeta - \sqrt{\zeta^2 - \nu^2}}{\nu^2} \frac{e^{2iA\sqrt{\zeta^2 - \nu^2}}}{1 - \left(\frac{\zeta - \sqrt{\zeta^2 - \nu^2}}{\nu}\right)^2 e^{2iA\sqrt{\zeta^2 - \nu^2}}},$$

where we have introduced the abbreviation

$$\hat{d}_g = \frac{\chi_{\mathbf{g}}}{\sqrt{|\chi_{\mathbf{g}}\chi_{-\mathbf{g}}|}} \sqrt{\frac{\gamma_i}{|\gamma_g|}} \tag{11.4.21}$$

:

$$d_1(\mathbf{g}) + d_2(\mathbf{g}) = \frac{\hat{d}_g}{\nu}\left(1 - e^{2iA\sqrt{\zeta^2 - \nu^2}}\right) \sum_{n=0}^{\infty} \left(\frac{\zeta - \sqrt{\zeta^2 - \nu^2}}{\nu}\right)^{2n+1} e^{2iAn\sqrt{\zeta^2 - \nu^2}}. \tag{11.4.22}$$

The wave function

The plane wave in the crystal is

$$\Psi(\mathbf{x}) = \left(d_1(\mathbf{g})\, e^{i\mathbf{K}_1 \cdot \mathbf{x}} + d_2(\mathbf{g})\, e^{i\mathbf{K}_2 \cdot \mathbf{x}}\right) e^{i\mathbf{g} \cdot \mathbf{x}}. \tag{11.4.23}$$

We need the wave vectors (11.4.10) at the front surface $\mathbf{x} = (x, 0)$ for which applies

$$\mathbf{K}_{1,2} \cdot \mathbf{x} = \mathbf{k} \cdot \mathbf{x} = \mathbf{k}_B \cdot \mathbf{x} - A y \Gamma + \frac{k\chi_{0r}}{2} s. \tag{11.4.24}$$

The spherical wave at the front surface is then

$$\Phi_g(x) = e^{i(\mathbf{k}_B + \mathbf{g}) \cdot \mathbf{x} + \frac{k\chi_{0r}}{2} s} \int_{-\infty}^{\infty} dy\, e^{-iA\Gamma y} \left(d_1(\mathbf{g}) + d_2(\mathbf{g})\right). \tag{11.4.25}$$

s is again the zigzag path to the point x. We now go from y to ζ. We have already determined Γ in (11.4.5), where D is the thickness of the crystal plate. We now define

$$\tilde{d}_g = \hat{d}_g\, e^{i(\mathbf{k}_B + \mathbf{g})\cdot\mathbf{x} + isk\chi_0/2} \tag{11.4.26}$$

and obtain for the wave function

$$\Phi_g(x) = \frac{\tilde{d}_g}{\nu} \int_{-\infty+i\eta}^{\infty+i\eta} d\zeta\, e^{-iA\Gamma\zeta}\left(1 - e^{2iA\sqrt{\zeta^2-\nu^2}}\right)$$
$$\times \sum_{n=0}^{\infty} \left(\frac{\zeta - \sqrt{\zeta^2-\nu^2}}{\nu}\right)^{2n+1} e^{2iAn\sqrt{\zeta^2-\nu^2}}. \tag{11.4.27}$$

To calculate Φ_g, one has to compute integrals of the form

$$R_m(s_1, s_2) = \frac{1}{\nu} \int_{-\infty+i\eta}^{\infty+i\eta} d\zeta \left(\frac{\zeta - \sqrt{\zeta^2-\nu^2}}{\nu}\right)^m e^{-is_2\zeta}\, e^{is_1\sqrt{\zeta^2-\nu^2}} \tag{11.4.28}$$

because

$$\Phi_g(x) = \tilde{d}_g \sum_{n=0}^{\infty}\left[R_{2n+1}(2nA, A\Gamma) - R_{2n+1}(2(n+1)A, A\Gamma)\right] \tag{11.4.29}$$
$$= \tilde{d}_g \left\{ R_1(0, A\Gamma) + \sum_{n=1}^{\infty}\left[R_{2n+1}(2nA, A\Gamma) - R_{2n-1}(2nA, A\Gamma)\right]\right\}.$$

11.4.3 Evaluation of the Integral R_m

The following calculation of the integral R_m will only be of interest to a few readers. Nevertheless, we show the derivation in detail, as it is rarely found and due to the inconsistent notations the tracing of the calculation in other books is even more tedious:

$$R_m(s_1, s_2) = \frac{1}{\nu} \int_{-\infty+i\eta}^{\infty+i\eta} d\zeta \left(\frac{\zeta - \sqrt{\zeta^2-\nu^2}}{\nu}\right)^m e^{-is_2\zeta + is_1\sqrt{\zeta^2-\nu^2}}$$
$$= \frac{1}{\nu} \oint_{C'} d\zeta \left(\frac{\zeta - \sqrt{\zeta^2-\nu^2}}{\nu}\right)^m e^{-is_2\zeta + is_1\sqrt{\zeta^2-\nu^2}}. \tag{11.4.30}$$

Now both $s_1 = 2nA > 0$ and $s_2 = A\Gamma > 0$. According to Fig. 11.20 is the integral over the lower semicircle C_- to be closed, when

$$s_2 - s_1 = A(\Gamma - 2n) > 0, \tag{11.4.31}$$

since then C_- contributes nothing. The integrand has no poles in the enclosed area, so that according to the Cauchy's integral theorem[4] the path can be contracted to C'. In the next step, we perform the transformation $w = \zeta/\nu$:

---

[4] $\oint_{\partial F} dz\, f(z) = 0$, if the $f(z)$ is analytic (holomorphic) in F.

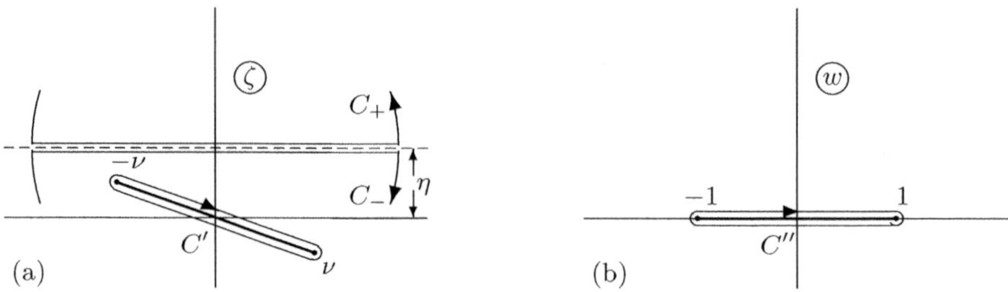

Fig. 11.20. (a) $\eta > |\kappa| > 0$; in the sketch $\mathrm{Im}\,\nu < 0$; for $s_2 - s_1$ C_- does not contribute and the path can be contracted to C'.   (b) C'' is the path C' in the w-plane

$$R_m(s_1, s_2) = \oint_{C''} \mathrm{d}w \left(w - \sqrt{w^2-1}\right)^m \mathrm{e}^{-is_2\nu w + is_1\nu\sqrt{w^2-1}}. \tag{11.4.32}$$

The coordinate transformation

$$\varrho\,\mathrm{e}^{i\varphi} = w + \sqrt{w^2-1} \qquad \text{with} \qquad \varrho = \sqrt{\frac{s_2+s_1}{s_2-s_1}} \tag{11.4.33}$$

turns the clockwise path C'' into a circle with the radius ϱ, which is traversed counterclockwise. It is

$$s_2 w - s_1\sqrt{w^2-1} = \frac{s_2+s_1}{2}\left(w - \sqrt{w^2-1}\right) + \frac{s_2-s_1}{2}\left(w + \sqrt{w^2-1}\right)$$

$$= \frac{s_2+s_1}{2}\,\frac{1}{\varrho}\mathrm{e}^{-i\varphi} + \frac{s_2-s_1}{2}\,\varrho\,\mathrm{e}^{i\varphi} = \sqrt{s_2^2 - s_1^2}\,\cos\varphi,$$

$$R_m(s_1, s_2) = \frac{i}{2}\oint \mathrm{d}\varphi\left(\varrho\,\mathrm{e}^{i\varphi} - \frac{1}{\varrho\,\mathrm{e}^{i\varphi}}\right)\varrho^{-m}\,\mathrm{e}^{-i\varphi m}\,\mathrm{e}^{-i\nu\sqrt{s_2^2-s_1^2}\,\cos\varphi}. \tag{11.4.34}$$

Here one resorts to an integral representation of the Bessel functions [Gradshteyn, Ryzhik, 1965, Ziff. 8.411]:

$$J_n(z) = \frac{1}{2\pi}\int_{-\pi}^{\pi}\mathrm{d}\phi\,\mathrm{e}^{-in\phi + iz\sin\phi} \overset{\phi=\phi'-\pi/2}{=} \frac{\mathrm{e}^{in\pi/2}}{2\pi}\int_{-\pi}^{\pi}\mathrm{d}\phi'\,\mathrm{e}^{-in\phi' - iz\cos\phi'}. \tag{11.4.35}$$

After that is

$$R_m(s_1, s_2) = i\pi\,(-i)^{m-1}\left(\sqrt{\frac{s_2-s_1}{s_2+s_1}}\right)^{m-1} \tag{11.4.36}$$

$$\times\left\{J_{m-1}\left(\nu\sqrt{s_2^2-s_1^2}\right) + \frac{s_2-s_1}{s_2+s_1}J_{m+1}\left(\nu\sqrt{s_2^2-s_1^2}\right)\right\}\theta(s_2-s_1).$$

The θ function takes into account that for $s_2 < s_1$ the integral (11.4.28) is evaluated over the semicircle C_+, where it vanishes.

11.4.4 The Entire Wave Function

We combine the individual summands:

$$\Phi_g(x) = \tilde{d}_g \left\{ R_1(0, A\Gamma) + \underbrace{\sum_{n=1}^{\infty} \Big(R_{2n+1}(2nA, A\Gamma) - R_{2n-1}(2nA, A\Gamma) \Big)}_{\text{i}\pi\,\Phi_{gn}} \right\}.$$

$$(11.4.37)$$

We pull out the factor iπ and get

$$\Phi_g(x) = \text{i}\pi\, \tilde{d}_g \left\{ \Phi_{g0}(\Gamma) + \sum_{n=1}^{\infty} \Phi_{gn}(\Gamma) \right\}.$$

$$(11.4.38)$$

The prefactor

$$\tilde{d}_g = \frac{\chi_{\mathbf{g}}}{\sqrt{|\chi_{\mathbf{g}}\chi_{-\mathbf{g}}|}} \sqrt{\frac{\gamma_i}{|\gamma_g|}}\, \text{e}^{\text{i}\mathbf{k}_B\cdot\mathbf{x} + \text{i}sk\chi_0/2}$$

$$(11.4.39)$$

was already defined in (11.4.26). The distance s, that a beam has to travel to get to point x is sketched in Fig. 11.21.

The Φ_{gn} functions are the contributions of the waves reflected n times at the back, as can be seen from the sketch Fig. 11.21 or Fig. 11.22.

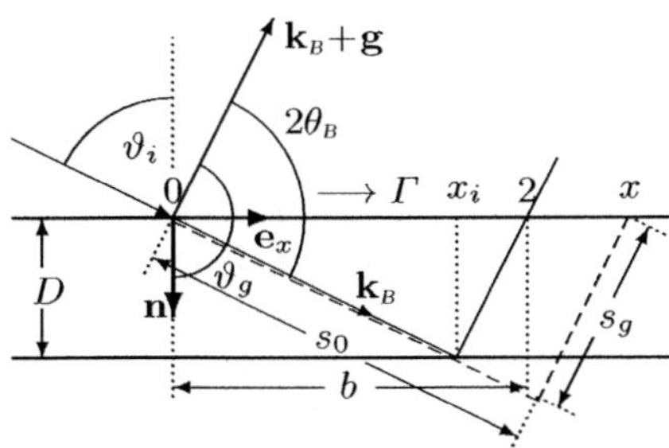

Fig. 11.21. Length $s = s_0 + s_g$ of the zigzag path (dashed) to the point x; unlike in the sketch it is assumed that the beam only travels short distances in the directions $\mathbf{k}_B$ and $\mathbf{k}_B + \mathbf{g}$ within the crystal; it is immediately apparent that the normal damping plays only a minor role in the wave Φ_{g0} diffracted at the front surface; $\Gamma = 2x/b$

Reflection at the front surface

The by far predominant contribution comes from $n = 0$:

$$\Phi_{g0}(x) = \tilde{d}_g\, R_1(0, A\Gamma) = \tilde{d}_g\, \text{i}\pi \Big[J_0(\nu A\Gamma) + J_2(\nu A\Gamma) \Big] \theta(\Gamma)$$

$$(11.4.40)$$

$$J_0(x) + J_2(x) = 2J_1(x)/x.$$

$$\Phi_{g0}(x) = \tilde{d}_g\, \frac{2\pi\text{i}}{\nu A\Gamma}\, J_1(\nu A\Gamma)\, \theta(\Gamma),$$

$$(11.4.41)$$

$$\tilde{d}_g = \frac{\chi_{\mathbf{g}}}{\sqrt{|\chi_{\mathbf{g}}\chi_{-\mathbf{g}}|}} \sqrt{\frac{\gamma_i}{|\gamma_g|}}\, \text{e}^{\text{i}\mathbf{k}_B\cdot\mathbf{x} + \text{i}sk\chi_0/2} \, .$$

This is the portion directly reflected at the front surface, which according to (11.4.20) is the 1st term in the expansion of $d_1(\mathbf{g})$. However, if only one wave field contributes to the intensity, one cannot expect an interference structure corresponding to the Pendellösung lengths.

The half-value width of the amplitude of Φ_{0g} is independent of the thickness D of the plate and is approximately $A\Gamma \approx 2.2$, which corresponds to a width $x = (2.2\,b/2\pi D)\Delta_0 \sim \Delta_0$. $\tilde{d}_g$ contains the normal attenuation given by $\mu_0 = k\chi_{0i}$, which is proportional to the length s of the zigzag path to the point x. In Fig. 11.21, s is shown as a dashed line.

Contributions from the back surface

We start here with the idea that these parts of the spherical wave enter the crystal and are (partially) reflected at the back surface. If one follows the path analogous to ray optics, the contributions start with $\Gamma = 2$ or $x = b$:

$$
\begin{aligned}
\Phi_{gn}(x) = \tilde{d}_g\,(-i\pi)\,(-1)^{n-1}\Bigg[& \left(\frac{\Gamma-2n}{\Gamma+2n}\right)^{n-1} J_{2n-2}\big(\nu A\sqrt{\Gamma^2-4n^2}\big) \\
& + 2\left(\frac{\Gamma-2n}{\Gamma+2n}\right)^{n} J_{2n}\big(\nu A\sqrt{\Gamma^2-4n^2}\big) \\
& + \left(\frac{\Gamma-2n}{\Gamma+2n}\right)^{n+1} J_{2n+2}\big(\nu A\sqrt{\Gamma^2-4n^2}\big) \Bigg]\theta(\Gamma-2n).
\end{aligned}
\tag{11.4.42}
$$

The intensity profile

The procedure for calculating the intensity is simple, as one only has to specify the absolute square of the wave function Φ_g. Φ_g is obtained by inserting the individual summands (11.4.42) into (11.4.38). Our interest is less in the intensity related to the x-component than in the radiant intensity normalized with the intensity I_i (11.4.8) of the incident wave and parameterized with Γ (11.4.5):

$$
P_g(\Gamma) = \frac{|\gamma_g|}{\gamma_i}\frac{|\Phi_g|^2}{I_i}\frac{dx}{d\Gamma} = \frac{|\gamma_g|}{\gamma_i}\frac{D}{b\Delta_0}\frac{b}{2}|\tilde{d}_g|^2\,\pi^2\left|\sum_{n=0}^{\infty}\Phi_{gn}(\Gamma)\right|^2 .
\tag{11.4.43}
$$

In thicker crystals, $D/\Delta_0 \gg 1$, the wave function Φ_{g0} (11.4.41) as a function of Γ drops sharply, and there is practically no overlap with Φ_{gn} for $n \geq 1$. It also holds for not too large n, that Φ_{gn} with $n \geq 1$ has no overlap with $\Phi_{gn'}$ for $n' \geq n$. In addition, contributions with increasing n to the total intensity decrease rapidly. The total intensity can therefore be approximated with the sum of the intensities $|\Phi_{gn}|^2$:

$$
P_g(\Gamma) = \left|\frac{\chi_{\mathbf{g}}}{\chi_{-\mathbf{g}}}\right| e^{-\mu_0 s}\frac{\pi A}{2}\left|\sum_{n=0}^{\infty}\Phi_g^n(\Gamma)\right|^2 \approx e^{-\mu_0 s}\frac{\pi A}{2}\sum_{n=0}^{\infty}\left|\Phi_g^n(\Gamma_g)\right|^2 .
\tag{11.4.44}
$$

$\mu_0 = k\chi_{0i}$ is the linear attenuation coefficient and s is the path length shown in dashed lines in Fig. 11.21, which the beam travels to point x on the front surface:

$$s = \frac{x}{\sin(2\theta_B)}(\gamma_i - \gamma_g).$$

(11.4.45)

$\Gamma = 2x/b$ is also shown in Fig. 11.22. The main contribution to the intensity comes from

$$P_{g0}(\Gamma) = \frac{\pi A}{2}\left|\Phi_{g0}(\Gamma)\right|^2\theta(\Gamma) = \left|\frac{2\,J_1(A\Gamma)}{A\Gamma}\right|^2\theta(\Gamma).$$

(11.4.46)

The half-value width is at $A\Gamma \approx 1.6$, where $A\Gamma = 2\pi\,(x/\Delta_0)\,(D/b)$ with $D/b = |\gamma_i\gamma_g|/\sin(2\theta_B)$. $P_{g0}(x)$ is thus independent of D. The width of the reflection curve in Fig. 11.22 is $\sim \Delta_0$. The amounts with $n \geq 1$ have a Pendellösung structure over which can be averaged. The dashed line in Fig. 11.22 indicates this averaging. With it, it is also possible to specify the size of the individual contributions to the total intensity:

$$\overline{P}_g(\Gamma) = \frac{\pi A}{2}\left|\frac{2J_1(A\Gamma)}{A\Gamma}\right|^2\theta(\Gamma) + 8\sum_{n=1}^{\infty}\left(\frac{2n}{\Gamma+2n}\right)^4\left(\frac{\Gamma-2n}{\Gamma+2n}\right)^{2n-2}\frac{\theta(\Gamma-2n)}{\sqrt{\Gamma^2-4n^2}}.$$

(11.4.47)

If you integrate the intensities, you get (as it must be)

$$R_g = \frac{8}{3} + \sum_{n=1}^{\infty}\frac{3\cdot 16}{(16\,n^2 - 9)(16\,n^2 - 1)} = \pi.$$

(11.4.48)

The total reflectivity must of course be equal to that of the plane waves, but one sees that – assuming no absorption – about 15% of the intensity comes from the range $\Gamma > 2$, but only 4‰ from the $\Gamma > 4$. The equations (11.4.47) and (11.4.48) suggest, that the transmitted wave in the crystal is reflected back and forth multiple times and exits at locations that are spatially far apart (see Fig. 11.22). The individual components hardly interfere, so that the intensity can be interpreted as a sum of n-times reflected waves.

This is not to say that the partial beams Φ_{gn} are incoherent with each other; it is only certain that they do not interfere due to the spatial separation. Therefore, in spectrometers that use multiple Bragg reflections, such as the Bonse-Hart camera [Villa et al., 2003] care must be taken that the contributions from the back surface do not interfere disturbingly.

11.5 Takagi-Taupin Equations

In section 11.2 the wave equation (11.2.5) was transformed using the Bloch approach (11.2.7) into a linear, homogeneous system of equations, which near

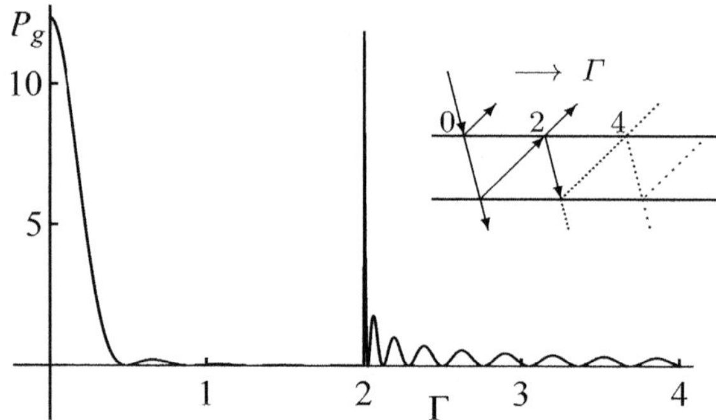

Fig. 11.22. Intensity profile in the Bragg case. Noteworthy is the portion of radiation from the rear surface, which, being spatially separated from the main part, is not always attributed to the rocking curves and integral intensities

a Bragg condition is reduced to the 2×2 matrix (11.2.18) and could be solved. While the tangential component of the wave vector remains unchanged upon entering the crystal, a split occurs in the normal component into two very closely spaced vectors $\mathbf{K}_i$, which lead to interferences (Pendellösung oscillations). The X-rays propagate in the crystal with the wavelengths typical for them of $\sim 1\,\text{Å} = 10^{-4}\,\mu m$, to which Pendellösung oscillations of $\sim 10\,\mu m$ are superimposed.

If slightly altered conditions are now present, as they are, for example, due to a deformed crystal lattice, the Bloch approach must be modified, in order to achieve a separation of the two oscillations. Takagi [1969] has in his approach for the closely spaced values of $\mathbf{K}_i$ specified a value of $\mathbf{K}_0$ that is close to the Bragg condition, but 'free' to choose. The Bloch amplitudes are then indeed location-dependent, but vary only slightly within a few unit cells, which is equivalent to an eikonal approximation (10.3.14a) for the individual beam directions.

11.5.1 Ideal Crystals

The diffraction of a plane wave on a crystal plate is derived here using Takagi's eikonal approximation:

$$\mathbf{D}(\mathbf{x}) = \sum_{\mathbf{g}} e^{iS_g(\mathbf{x})}\mathbf{D}_g(\mathbf{x}), \qquad S_g(\mathbf{x}) = \mathbf{K}_g\cdot\mathbf{x}, \qquad \mathbf{K}_g = \mathbf{K}_0 + \mathbf{g}. \qquad (11.5.1)$$

Here, $\mathbf{K}_0$ is the wave vector with which the X-rays propagate in a homogeneous medium or, equivalently, in a crystal far from any Bragg condition:

$$K_0 = nk = (1+\chi_0/2)k, \qquad\qquad \mathbf{K}_{0\|} = \mathbf{k}_\|. \qquad (11.5.2)$$

(11.5.1) corresponds to the approach (10.3.14a) for each beam of direction $\mathbf{K}_g$. $\mathbf{D}_{\mathbf{g}}(\mathbf{x})$ is the slowly varying amplitude, whose 2nd derivative is neglected.

Now the approach (11.5.1) is inserted into (11.2.5). Using (11.2.9), the Takagi-Taupin equations are obtained in the (11.2.11) corresponding form, where in the sum K_g was replaced by k:

$$2\mathrm{i}\mathbf{K}_0\cdot\boldsymbol{\nabla}\,\mathbf{D}_{\mathbf{g}}(\mathbf{x}) = \left[K_{\mathbf{g}}^2 - k^2 - k^2\chi_0\right]\mathbf{D}_{\mathbf{g}}(\mathbf{x}) - k^2 \sum_{\mathbf{g}'\neq\mathbf{g}} \chi_{\mathbf{g}-\mathbf{g}'}\,\mathbf{D}_{\mathbf{g}'[\mathbf{g}]}(\mathbf{x}). \quad (11.5.3)$$

For the vortex field $\mathbf{D}$, since $|\boldsymbol{\nabla}\cdot\mathbf{D}_{\mathbf{g}}|\ll|\mathbf{K}_{\mathbf{g}}\cdot\mathbf{D}_{\mathbf{g}}|$:

$$\boldsymbol{\nabla}\cdot\mathbf{D}_{\mathbf{g}} = \sum_{\mathbf{g}} e^{\mathrm{i}\mathbf{K}_{\mathbf{g}}\cdot\mathbf{x}}(\mathrm{i}\mathbf{K}_{\mathbf{g}}+\boldsymbol{\nabla})\cdot\mathbf{D}_{\mathbf{g}} = 0 \qquad \Rightarrow \qquad \mathbf{K}_{\mathbf{g}}\cdot\mathbf{D}_{\mathbf{g}} = 0. \qquad (11.5.4)$$

Now, analogous to the procedure for (11.2.12), we multiply scalarly with $\mathbf{D}_{\mathbf{g}}$ ($\mathbf{D}_{\mathbf{g}'[\mathbf{g}]}\cdot\mathbf{D}_{\mathbf{g}} = D_{\mathbf{g}}D_{\mathbf{g}'}C_{g'g}$):

$$\hat{\mathbf{K}}_{\mathbf{g}}\cdot\boldsymbol{\nabla}\,D_{\mathbf{g}}(\mathbf{x}) = -\mathrm{i}k\beta_g D_{\mathbf{g}}(\mathbf{x}) + \frac{\mathrm{i}k}{2}\sum_{\mathbf{g}'\neq\mathbf{g}} \chi_{\mathbf{g}-\mathbf{g}'}\,C_{g'g}D_{\mathbf{g}'}(\mathbf{x}),$$

$$\beta_g = \frac{1}{2k^2}\left[K_{\mathbf{g}}^2 - k^2(1+\chi_0)\right]. \qquad (11.5.5)$$

The designation β_g follows Authier [2002, (11.6)] and is equal to α (11.2.28).

The two-beam case

If the incident wave fulfills a Bragg condition only for $\mathbf{g}$, the sum in (11.5.5) reduces to a single term, and it is $C_{g0} = C_{0g} = C$, i.e. $\tilde{\chi}_{\pm\mathbf{g}} = C\chi_{\pm\mathbf{g}}$, and $\beta_0 = 0$. At the same time, we switch to the oblique coordinates $\hat{\mathbf{s}}_0 = \hat{\mathbf{K}}_0$, and $\hat{\mathbf{s}}_g = \hat{\mathbf{K}}_{\mathbf{g}}$, which are oriented along the beam directions, as can be seen in Fig. 11.23. is to be taken. The Takagi-Taupin (TT) equations are now

$$\mathrm{i}\frac{\partial}{\partial s_0}D_0(\mathbf{x}) = -\frac{k}{2}\tilde{\chi}_{-\mathbf{g}}D_{\mathbf{g}}(\mathbf{x}), \qquad (11.5.6)$$

$$\mathrm{i}\frac{\partial}{\partial s_g}D_{\mathbf{g}}(\mathbf{x}) = k\beta_g D_{\mathbf{g}}(\mathbf{x}) - \frac{k}{2}\tilde{\chi}_{\mathbf{g}}D_0(\mathbf{x}).$$

Following the sketch Fig. 11.23 is ($\gamma_g = \cos\vartheta_g$)

$$\hat{\mathbf{s}}_0 \equiv \hat{\mathbf{K}}_0 = \mathbf{e}_x\sin\vartheta_i + \mathbf{e}_z\gamma_i, \qquad\qquad \hat{\mathbf{s}}_g \equiv \hat{\mathbf{K}}_{\mathbf{g}} = \mathbf{e}_x\sin\vartheta_g + \mathbf{e}_z\gamma_g.$$

In the Laue case, $\gamma_g > 0$, but usually $\sin\vartheta_g < 0$. $P(x,z)$ has the coordinates:

$$x = s_0\sin\vartheta_i + s_g\sin\vartheta_g, \qquad\qquad z = s_0\cos\vartheta_i + s_g\cos\vartheta_g. \qquad (11.5.7)$$

It is $s_g < 0$, i.e. $s_g\sin\vartheta_g > 0$. Continuity at the entry surface $\mathbf{x}_e$ is ensured, if $D_0(\mathbf{x}_e)$ does not depend on x. This suggests the approach:

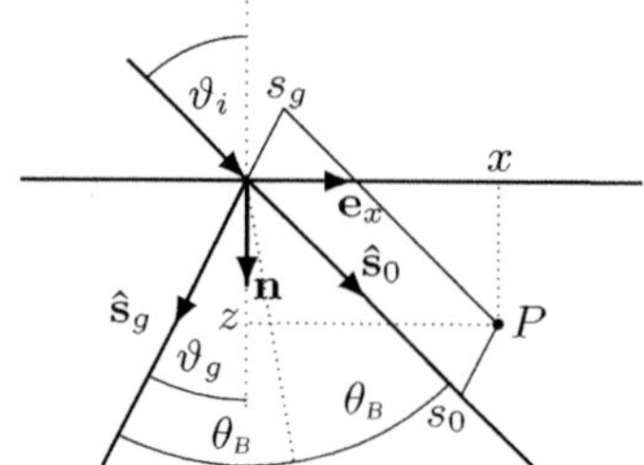

Fig. 11.23. Laue case: $P(x, z)$ is a point in the crystal. Here $\vartheta_i = \vartheta_0 > 0$ and $\vartheta_g < 0$. s_0 and s_g are paths along the directions $\hat{\mathbf{s}}_0$ and $\hat{\mathbf{s}}_g$

$$D_0(\mathbf{x}) = D_0(z) = e^{ik\epsilon(\gamma_i s_0 + \gamma_g s_g)} \quad \Rightarrow \quad D_{\mathbf{g}}(z) = \frac{2\epsilon\gamma_i}{\tilde{\chi}_{-\mathbf{g}}} D_0(z). \tag{11.5.8}$$

Substituted into (11.5.6) follows

$$-k\epsilon\gamma_i D_0 + \frac{k}{2}\tilde{\chi}_{-\mathbf{g}} D_{\mathbf{g}} = 0, \tag{11.5.9}$$

$$\frac{k}{2}\chi_{\mathbf{g}} D_0 - k(\epsilon\gamma_g + \beta_g) D_{\mathbf{g}} = 0.$$

For the homogeneous equation to have non-trivial solutions, the secular determinant must vanish:

$$\epsilon\gamma_i(\epsilon\gamma_g + \beta_g) - |\tilde{\chi}_g|^2/4 = 0. \tag{11.5.10}$$

This is the secular equation (11.2.32), since $\beta_g = \alpha$. Therefore, the result (11.2.35) can be adopted, noting that the Bragg case is still included here, since only the continuity of the tangential component at the entrance surface has been considered as a boundary condition. This results in two solutions (wave fields) D_{01} and D_{02}, where ϵ is given in (11.2.39):

$$k\epsilon_{1,2}z = \frac{\pi z}{\Delta_0}\left[-y \pm \sqrt{y^2 + \operatorname{sgn}\gamma_g}\,\right].$$

In the next step, the boundary conditions (continuity of the transmitted wave; no diffracted wave at the entrance surface) must be considered, with the total solution being a superposition of the two wave fields:

$$
\begin{aligned}
a_1 D_{01}(0) + a_2 D_{02}(0) &= a_1 + a_2 = 1, \\
a_1 D_{\mathbf{g}1}(0) + a_2 D_{\mathbf{g}2}(0) &= (a_1\epsilon_1 + a_2\epsilon_2)/\chi_{-\mathbf{g}} = 0.
\end{aligned}
\tag{11.5.11}$$

If you insert the solutions

$$a_1 = \epsilon_2/(\epsilon_2 - \epsilon_1), \qquad\qquad a_2 = \epsilon_1/(\epsilon_1 - \epsilon_2)$$

into the amplitudes $D_0(z)$ and $D_{\mathbf{g}}(z)$, you get the already known wave functions (11.3.11). The same procedure can of course also be applied to the Bragg case. The advantages of the eikonal approximation only become evident with more complex boundary conditions.

11.5.2 Slightly Distorted Crystal Lattice

Starting from the space point $\mathbf{x}_0$, a distortion $\mathbf{u}(\mathbf{x}_0)$ of the crystal lattice is assumed, so that $\mathbf{x} = \mathbf{x}_0 + \mathbf{u}(\mathbf{x}_0)$. Inserted into the lattice-periodic interaction (11.2.6) one gets

$$\chi'(\mathbf{x}) = \chi(\mathbf{x} - \mathbf{u}(\mathbf{x}_0)) = \sum_{\mathbf{g}} \chi_{\mathbf{g}}\, e^{i\mathbf{g}\cdot(\mathbf{x}-\mathbf{u})}, \tag{11.5.12}$$

where the displacement vector $\mathbf{u}$ changes only slightly over microscopic distances, so that in (11.5.12) $\mathbf{u}(\mathbf{x}_0)$ can be replaced by $\mathbf{u}(\mathbf{x})$. The distorted lattice is considered in the displacement field by the approach

$$\mathbf{D}(\mathbf{x}) = \sum_{\mathbf{g}} e^{i\mathbf{K}_g\cdot(\mathbf{x}-\mathbf{u})}\, \mathbf{D}'_g(\mathbf{x}) \tag{11.5.13}$$

is taken into account. For the base vector of the distorted lattice, in first order (problem 11.5)

$$\mathbf{a}'_i(\mathbf{x}) = \mathbf{a}_i + \mathbf{a}_i\cdot\boldsymbol{\nabla}\mathbf{u}(\mathbf{x}), \qquad\qquad i = 1, 2, 3,$$

where $\mathbf{a}_i$ is the base vector of the undistorted lattice. The base vectors $\mathbf{b}'_i$ of the distorted reciprocal lattice fulfill in first order in $\mathbf{u}$

$$\mathbf{a}'_i\cdot\mathbf{b}'_j = 2\pi\delta_{ij}, \qquad \mathbf{b}'_i = \mathbf{b}_i - \mathbf{b}_i\cdot\boldsymbol{\nabla}\mathbf{u}, \qquad v'_c = v_c(1 + \boldsymbol{\nabla}\cdot\mathbf{u}). \tag{11.5.14}$$

$v_c = \mathbf{a}_1\cdot(\mathbf{a}_2\times\mathbf{a}_3)$ or v'_c is the volume of the unit cell of the ideal or the distorted lattice. The (general) reciprocal lattice vector of the distorted lattice is then according to (11.5.14)

$$\mathbf{g}'(\mathbf{x}) = \mathbf{g} - \boldsymbol{\nabla}\mathbf{g}\cdot\mathbf{u}(\mathbf{x}). \tag{11.5.15}$$

$\mathbf{D}$ has no sources (11.5.4), which means $(\mathbf{K}'_{\mathbf{g}} = \mathbf{K}_0 + \mathbf{g}'(\mathbf{x}))$:

$$i(\mathbf{K}'_{\mathbf{g}} + \boldsymbol{\nabla})\cdot\mathbf{D}'_g = 0, \qquad\qquad |\boldsymbol{\nabla}\cdot\mathbf{D}'_g| \ll |\mathbf{K}'_{\mathbf{g}}\cdot\mathbf{D}'_g|. \tag{11.5.16}$$

Since the direction of $\mathbf{K}'_g$ differs only slightly from $\mathbf{K}_g$ and thus from $\mathbf{s}_g$, the TT equations (11.5.5) can be adopted almost unchanged:

$$\frac{\partial}{\partial s_g}D'_{\mathbf{g}}(\mathbf{x}) = -ik\beta'_g D'_{\mathbf{g}}(\mathbf{x}) + \frac{ik}{2}\sum_{\mathbf{g}'\neq\mathbf{g}}\chi_{\mathbf{g}-\mathbf{g}'}\, C_{g'g}D'_{\mathbf{g}'}(\mathbf{x}), \tag{11.5.17}$$

$$\beta'_g = \left[K'^{\,2}_{\mathbf{g}} - k^2(1+\chi_0)\right]/2k^2.$$

The relevant difference to (11.5.5) is the spatial dependence of the coefficients β'_g. Restricted to the two-beam case, one obtains

$$\frac{\partial}{\partial s_0}D'_0 = -ik\beta'_0 D'_0 + i\frac{k}{2}\chi_{-\mathbf{g}}\, C\, D'_g,$$

$$\frac{\partial}{\partial s_g}D'_g = -ik\beta'_g D'_g + i\frac{k}{2}\chi_0\, C\, D'_0. \tag{11.5.18}$$

Boundary conditions for the two-beam case

The entrance surface $\mathbf{x}_e$, which was previously infinitely extended and flat, is now 'quasi-flat', i.e., in the area where the incident wave does not vanish, $\mathbf{x}_e$ can be considered flat

$$D_i(\mathbf{x}) = \psi_i(\mathbf{x})\, e^{i\mathbf{k}\cdot\mathbf{x}}. \tag{11.5.19}$$

This is generally a spherical wave or a wave bundle of finite width. The continuity on the input surface can be written as

$$\psi_i(\mathbf{x}_e)\, e^{i\mathbf{k}\cdot\mathbf{x}_e} = D_0'(\mathbf{x}_e)e^{i\mathbf{K}_0\cdot\mathbf{x}_e} + D_g'(\mathbf{x}_e)\, e^{i\mathbf{K}_g\cdot\mathbf{x}_e - i\mathbf{g}\cdot\mathbf{u}}. \tag{11.5.20}$$

Using the definition

$$\phi_i(\mathbf{x}) = \psi_i(\mathbf{x})\, e^{i(\mathbf{k}-\mathbf{K}_0)\cdot\mathbf{x}}$$

the boundary condition (11.5.20) can be rewritten:

$$\left[D_0'(\mathbf{x}_e) - \phi_i(\mathbf{x}_e)\right] + D_g'(\mathbf{x}_e)e^{i\mathbf{g}\cdot(\mathbf{x}_e-\mathbf{u})} = 0 \quad\Rightarrow\quad \begin{cases} D_0'(\mathbf{x}_e) = \phi_i(\mathbf{x}_e) \\ D_g'(\mathbf{x}_e) = 0. \end{cases} \tag{11.5.21}$$

Substituted into the TT equations, one obtains the boundary conditions for the derivatives, if $\beta_0 = 0$

$$\frac{\partial D_0'(\mathbf{x}_e)}{\partial s_0} = 0, \qquad\qquad \frac{\partial D_g'(\mathbf{x}_e)}{\partial s_g} = \frac{ik}{2}\tilde{\chi}_g'\phi_i(\mathbf{x}_e). \tag{11.5.22}$$

Problems for Chapter 11

11.1. *Averaging the rocking curve in the Bragg case*: Calculate the averaged intensity $\bar{P}_g$ (11.3.22) starting from (11.3.8).

Hint: Form the average for $y > 1$ by integrating over a thickness interval ΔA.

11.2. *Integral reflectivity in the Bragg case*: Verify the integral intensity (11.3.10) by integrating P_g.

Instructions: First, transform the integral:

$$R = -i \int_{-\infty}^{\infty} dy\, \frac{1}{y}\, \frac{1}{\sqrt{y^2 - 1}\, \cot\left(A\sqrt{y^2 - 1}\right) - iy}.$$

Then show that the integrand on the real axis only has a pole at $y = 0$ and on the upper half of the complex plane ($\mathrm{Im}\, y > 0$) is regular. This can be done with the transformation

$$u = i\sinh y = \xi + i\eta \qquad\qquad \Rightarrow \qquad\qquad \mathrm{Im}\, y = \sinh\xi\,\cos\eta > 0.$$

Then you can evaluate the integral using the residue theorem.

11.3. *Calculation of the wave function for the Bragg case:*

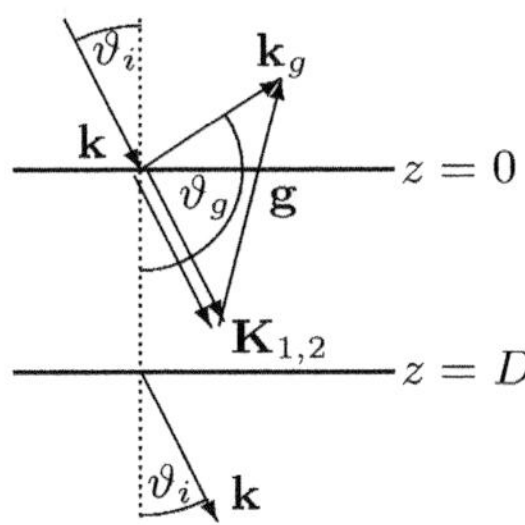

The adjacent sketch shows a crystal plate on which the incident beam $\psi_i(\mathbf{x}) = e^{i\mathbf{k}\cdot\mathbf{x}}$ is reflected according to the Bragg case.

Verify with the continuity conditions given in section 11.3.3 the amplitudes of the reflected and the transmitted wave (11.3.19) and provide the corresponding wave functions.

11.4. *Anomalous Absorption:* Show that in the Laue case, the wave field 1 has its nodes near the atomic positions, while the nodes of the wave field 2 are located between the atoms. Analyze the situation also for the Bragg case, where such a simple distinction is no longer useful.

Hint: Calculate the intensity $|\mathbf{D}(\mathbf{x})|^2$ for the Bloch waves (11.2.7) of inversion-symmetric reflections.

11.5. *Slightly distorted reciprocal lattice:* A Bravais lattice with the basis vectors $\mathbf{a}_i$ and the corresponding reciprocal basis vectors $\mathbf{b}_i$, $i = 1, 2, 3$ will be distorted: $\mathbf{x}' = \mathbf{x} + \mathbf{u}(\mathbf{x})$. Determine the basis vectors of the distorted lattice $\mathbf{a}'_i$, $\mathbf{b}'_i$ and the volume v'_c of the distorted unit cell to first order in $\mathbf{u}$.

References

Authier A. *Dynamical Theory of X-Ray Diffraction*, Oxford University Press (2002)

Borrmann G. *Die Absorption von Röntgenstrahlen im Fall der Interferenz*, Z. Physik **127**, 297-323 (1950)

Ewald P.P. *Zur Begründung der Kristalloptik; I: Theorie der Dispersion*, Ann. Physik **49**,1–38 (1916)

Gradshteyn I.S. and Ryzhik I.M. *Table of Integrals, Series, and Products*, Academic Press N.Y.(1965)

Kato N. *Dynamical Theory for perfect crystals* in L. Azaroff et al. *X-Ray Diffraction*, McGraw-Hill (1974)

Laue M. von *Röntgenstrahlinterferenzen*, 3rd ed., Akadem. Verlagsges. Frankfurt (1960)

Pinsker Z.G. *Dynamical Scattering of X-Rays in Crystals*, Solid State Sciences **3**, Springer Berlin (1978)

Rauch H. and Petrascheck D. *Dynamical Neutron Diffraction and its Application* in Topics of Current Physics (*Neutron Diffraction*) **6**, 303–351, Springer (1978)

Schwabl F. *Quantum Mechanics*, 4th ed. Springer Berlin (2007)

Shull C.G. and Oberteuffer J. *Spherical-Wave Neutron Propagation and Pendellösung Fringe Structure in Silicon*, Phys.Rev.Lett. **29**, 871–874 (1972)

Takagi S. *A Dynamical Theory of Diffraction for a Distorted Crystal* J. Phys. Soc. Japan **26**, 1239–1253 (1969)

Thomson J. J. *Notes on recent researches in electricity and magnetism*, Clarendon Press, Oxford (1893)

Vartanyants I. A. and Kovalchuk M. V., *Theory and applications of x-ray standing waves in real crystals*, Rep. Prog. Phys. **64**, 1009–1084 (2001)

Villa M. et al., J. Appl. Cryst. **36**, 769 (2003)

Zachariasen W. H. *Theory of X-Ray Diffraction in Crystals*, John Wiley & Sons, London (1945)

12

Special Theory of Relativity

The concept of the propagation of light or electromagnetic waves was associated with a medium, the so-called *(light-)ether*. Therefore, it was initially not very disturbing that the Maxwell equations were not invariant under different speed moving inertial systems. There was a coordinate system (CS), which was distinguished, since in this the ether rested.

After all attempts had failed to determine the movement against the ether, the discovery of the Lorentz transformation (LT), under which the Maxwell equations remain invariant, gained importance. It took some attempts until the LT was established.

The equivalence of all uniformly moving systems required a revision of the concepts of space and time, whose acceptance took a longer time even among physicists. This is one reason why the LT was given a lot, perhaps too much, space.

Electrodynamics and the Galilei transformation

The laws of classical mechanics are invariant under the translation, rotation (rotation matrix R) and relative speed of two CS. This general transformation is called *Galilei transformation*, a term often used for the sole transformation of speed. For the latter, we use the term *boost* and thus also have the designation Lorentz transformation not solely occupied with the speed transformation. The general (homogeneous) Galilei transformation is

$$
\begin{array}{lcl}
t' = t & \xrightarrow{\text{boost}} & t' = t \\
\mathbf{x}' = \mathsf{R}\mathbf{x} - \mathbf{v}t & & \mathbf{x}' = \mathbf{x} - \mathbf{v}t.
\end{array}
\qquad (12.0.1)
$$

It will now be examined based on the wave equation whether also the laws of electrodynamics under the Galilei transformation are invariant

$$
\Box\phi = \left(\frac{1}{c^2}\frac{\partial^2}{\partial t^2} - \nabla^2\right)\phi = 0\,.
\qquad (12.0.2)
$$

For the boost (12.0.1) we get

$$\frac{\partial}{\partial t} = \frac{\partial t'}{\partial t}\frac{\partial}{\partial t'} + \frac{\partial x'_i}{\partial t}\frac{\partial}{\partial x'_i} = \frac{\partial}{\partial t'} - v_i\frac{\partial}{\partial x'_i}, \qquad \frac{\partial}{\partial x_i} = \frac{\partial t'}{\partial x_i}\frac{\partial}{\partial t'} + \frac{\partial x'_k}{\partial x_i}\frac{\partial}{\partial x'_k} = \frac{\partial}{\partial x'_i},$$

from which it follows that the d'Alembert operator does not remain invariant:

$$\Box = \Box' + \frac{1}{c^2}\Big[-2\mathbf{v}\cdot\nabla'\frac{\partial}{\partial t'} + (\mathbf{v}\cdot\nabla')^2\Big].$$

If the Galilean transformation were the correct transformation between two reference systems that move relative to each other with $\mathbf{v}$, then the laws of electrodynamics would be different in different inertial systems. The Galilean transformation is already the most general linear transformation in three spatial dimensions, from which it follows that a transformation that leaves the laws of electrodynamics invariant, must include time.

12.1 Invariance Properties and the Principle of Relativity

To describe natural processes, we need a reference system in which we determine the positions of the particles at given times. Among the various reference systems, the inertial systems occupy a distinguished position:

A reference system is called an inertial system if in it force-free particles move uniformly.

This requires that

1. space and time are homogeneous and
2. space is isotropic.

Therefore, no space-time point (event) and no direction should be distinguished. From the first point it follows that a transformation between two inertial systems S and S' must be linear. Any other power, apart from constants, distinguishes a space-time point. Isotropy ensures that the orientation of the coordinate system (CS) can be arbitrary.

We have no way of determining the absolute speed of a CS, but can only determine relative speeds between different CS. Inertial systems are completely equivalent, which is expressed in the *principle of relativity* as follows:

1. *Special principle of relativity*[1]: It is based solely on the equivalence of two inertial systems. Two inertial systems differ only in the relative speed ($\mathbf{v} \leftrightharpoons -\mathbf{v}$).

   Note: Einstein [1916, p. 770] gave this principle of relativity the epithet 'special' and thus distinguished it from the *general principle of relativity*, which also includes gravity, [Einstein, 1916, p. 776].

---

[1] Sometimes also called *universal principle of relativity* [Schröder, 2014, p. 17].

2. *Galilean principle of relativity* or *principle of relativity of classical mechanics*: The laws of classical mechanics apply. In these, the time differences in all inertial systems are the same and consequently also the spatial distances of simultaneous events (space-time points). The special principle of relativity is thus restricted.
3. *Einstein's principle of relativity*: The special principle of relativity is here restricted by the laws of electrodynamics, e.g. the wave equation.

Note: The Galilean and Einsteinian principles of relativity are only compatible at small speeds. For larger speeds, the laws of mechanics need to be adapted to the Einsteinian principle of relativity.

The *postulate of the constancy of the speed of light* has not been addressed so far. It is not absolutely necessary for the derivation of the Lorentz transformation.

12.1.1 Construction of the Lorentz Transformation

The transformation, the so-called *Lorentz transformation*, is to be determined, under which the laws of electrodynamics, as represented by the wave equation, remain form-invariant. As already explained, the transformation must be linear and four-dimensional.

We limit ourselves to a boost, in which the inertial system S' moves against S with $\mathbf{v} = v\mathbf{e}_x$ and the coordinate axes are parallel to each other. No space-time point is distinguished, which is why the transformation is linear. At time $t = t' = 0$, the origin of S' should coincide with S, and we take into account only homogeneity and isotropy with the following approach:

$$t' = \alpha_0 t + \alpha_1 x + \alpha_2 y + \alpha_3 z, \qquad y' = \beta_2 t + d_{21} x + d_{22} y + d_{23} z,$$
$$x' = \gamma(-vt + x) + d_{12} y + d_{13} z, \qquad z' = \beta_3 t + d_{31} x + d_{32} y + d_{33} z.$$

The coefficients $\alpha_2 = \alpha_3 = 0$ vanish, as these would lead to direction-dependent times t' at a given distance from the x-axis. $\beta_2 = \beta_3 = 0$, otherwise the directions $\perp \mathbf{v}$ would be unequal. The axes are parallel, so for $i \neq j$: $d_{ij} = 0$. Furthermore, $d_{22} = d_{33} = d_\perp$. α_1 changes its sign with v, as t' must remain unchanged for $v \to -v$ and $x \to -x$. Therefore, we set $\alpha_1 = -v\alpha$. The coefficients γ, α_0, α and $d_\perp$ depend only on $|v|$:

$$t' = \alpha_0 t - v\alpha x, \qquad y' = d_\perp y, \qquad (12.1.1)$$
$$x' = \gamma(-vt + x), \qquad z' = d_\perp z.$$

We will derive the Lorentz Transformation (LT) in several ways, gradually restricting the assumptions.

Principle of relativity and constancy of light speed

First, an elegant and simple derivation is outlined, which is based on Einstein [1905] and is essentially followed here. The inertial system S' moves with

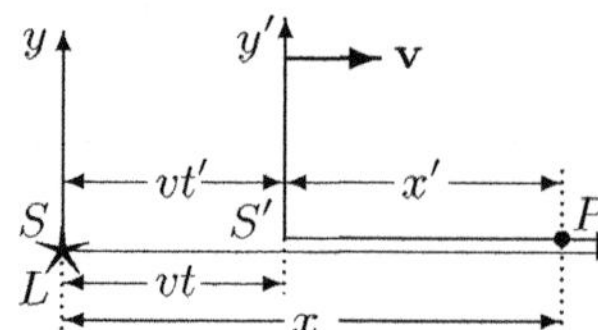

Fig. 12.1. S' moves with $\mathbf{v} = v\mathbf{e}_x$ relative to S. At time $t = t' = 0$ a light flash L is emitted from $\mathbf{x} = \mathbf{x}' = 0$, which is observed in S' at point $P(\mathbf{x}', t')$. The axes of S and S' are parallel

$\mathbf{v} = v\mathbf{e}_x$ against S, as indicated in Fig. 12.1. The coordinate axes of the two systems are parallel, so at best it can be $y' = d_\perp\, y$. However, according to the principle of relativity, $y = d_\perp\, y'$ must also be true, from which $d_\perp = 1$ follows.

At time $t = t' = 0$ a light flash L is emitted from $\mathbf{x} = \mathbf{x}' = 0$. The light flash is observed from a point $P(\mathbf{x}', t')$ in S' at time t'. During this time, S has moved away by $-vt'$. For the observer in S', the distance in S is therefore given by

$$x = \gamma(x' + vt') \qquad \overset{\text{principle of relativity}}{\Longrightarrow} \qquad x' = \gamma(x - vt). \qquad (12.1.2)$$

In the special case of a light signal, after applying the postulate of the constancy of the speed of light $x = ct$ and $x' = ct'$, which is inserted into the previous formulas:

$$ct = \gamma(c + v)t', \qquad\qquad ct' = \gamma(c - v)t.$$

Multiplying the two equations together yields

$$\gamma = \frac{1}{\sqrt{1 - \beta^2}}, \qquad\qquad \beta = \frac{v}{c}. \qquad (12.1.3)$$

γ is the so-called *Lorentz factor*. The transformation behavior of time is still missing. The starting point is the reverse transformation

$$x = \gamma(x' + vt') \qquad \Rightarrow \qquad t' = \frac{x - \gamma x'}{\gamma v} = \frac{x - \gamma^2(x - vt)}{\gamma v} = -x\frac{\gamma v}{c^2} + \gamma t.$$

We have used that $\frac{\gamma^2 - 1}{\gamma v} = \frac{\gamma v}{c^2}$. The LT (boost) is therefore:

$$ct' = \gamma(ct - \beta x), \qquad x' = \gamma(x - \beta ct), \qquad y' = y, \qquad z' = z. \qquad (12.1.4)$$

The inverse transformation only differs by $v \to -v$. Before Einstein, Larmor [1900, section XI], Lorentz [1904, p. 812] and Poincaré [1905, p. 1505], who named the transformation after Lorentz, derived the LT.

Einstein's principle of relativity

Now the Lorentz transformation is derived using the form invariance of the wave equation in inertial systems S and S', which move against each other with $\mathbf{v} = v\mathbf{e}_x$. The demand for the constancy of the speed of light is not necessary. This follows automatically when the principle of relativity is applied to the wave equation. We can therefore start from (12.1.2) and restrict ourselves to x and t:

$$x' = \gamma(x - vt), \qquad \Longrightarrow \qquad x = \gamma(x' + vt'). \tag{12.1.5}$$

If you solve the second equation for t' and eliminate x', you get

$$t' = \frac{1}{\gamma v}\left(x - \gamma x'\right) = \frac{1 - \gamma^2}{\gamma v}\, x + \gamma t. \tag{12.1.6}$$

This is a linear transformation, in which the parameter γ is still free. $\gamma = 1$ represents the Galilei transformation. By means of the demand for the invariance of the wave equation under the transformation, γ is determined. For this purpose

$$\frac{\partial}{\partial t} = \frac{\partial t'}{\partial t}\frac{\partial}{\partial t'} + \frac{\partial x'}{\partial t}\frac{\partial}{\partial x'} = \gamma\frac{\partial}{\partial t'} - \gamma v\frac{\partial}{\partial x'}\,,$$

$$\frac{\partial}{\partial x} = \frac{\partial t'}{\partial x}\frac{\partial}{\partial t'} + \frac{\partial x'}{\partial x}\frac{\partial}{\partial x'} = \frac{1 - \gamma^2}{\gamma v}\frac{\partial}{\partial t'} + \gamma\frac{\partial}{\partial x'}$$

are calculated and inserted into the d'Alembert operator. The term with the mixed derivatives is set to zero, from which the rest follows:

$$\Box = \frac{1}{c^2}\frac{\partial^2}{\partial t^2} - \frac{\partial^2}{\partial x^2} = \frac{1}{c^2}\left(\gamma\frac{\partial}{\partial t'} - \gamma v\frac{\partial}{\partial x'}\right)^2 - \left(\frac{1 - \gamma^2}{\gamma v}\frac{\partial}{\partial t'} + \gamma\frac{\partial}{\partial x'}\right)^2$$

$$= \underbrace{\left(\frac{\gamma^2}{c^2} - \frac{(1 - \gamma^2)^2}{\gamma^2 v^2}\right)}_{\Rightarrow 1/c^2}\frac{\partial^2}{\partial t'^2} - 2\underbrace{\left(\frac{\gamma\gamma v}{c^2} + \frac{\gamma(1 - \gamma^2)}{\gamma v}\right)}_{0}\frac{\partial^2}{\partial t'\partial x'} + \underbrace{\left(\frac{\gamma^2 v^2}{c^2} - \gamma^2\right)}_{\Rightarrow -1}\frac{\partial^2}{\partial x'^2}\,.$$

The evaluation results in

$$\gamma = \frac{1}{\sqrt{1 - v^2/c^2}} \qquad \Rightarrow \qquad \frac{1}{c^2}\frac{\partial^2}{\partial t^2} - \frac{\partial^2}{\partial x^2} = \frac{1}{c^2}\frac{\partial^2}{\partial t'^2} - \frac{\partial^2}{\partial x'^2}\,. \tag{12.1.7}$$

If you insert γ into (12.1.6), you get the LT in the form of (12.1.4):

$$t' = \gamma(t - vx/c^2), \qquad\qquad x' = \gamma(-vt + x). \tag{12.1.8}$$

The light has the speed c in S' again, which was not explicitly required. Furthermore, the transformation assumes speeds $|v| \leq c$. The deviations ($\gamma > 1$) from the Galilean transformation ($\gamma = 1$) are of the order v^2/c^2, which is why classical mechanics remains valid for not too high speeds.

Equivalence of inertial systems

The question arises as to what the transformation between inertial systems looks like on the sole basis of the principle of relativity without reference to physical laws. This transformation can be determined up to a universal speed, which was noted early on by Ignatowsky [1910], Frank and Rothe [1911]. The calculation related to this is more complex than the derivation of LT based on the postulates of Einstein's principle of relativity and the constancy of the speed of light [Einstein, 1905], but of fundamental interest. Most textbooks on electrodynamics do not cover this, rather those on the theory of relativity, such as Sexl, Urbantke [2001, §1.3] or Schröder [2014, §3.4.1].

S and S' differ only in the relative speed. Therefore, from the reverse transformation $d_\perp = 1$ we get: The components perpendicular to $\mathbf{v}$ remain unchanged. We now set $\alpha = \gamma\eta$ in (12.1.1):

$$\begin{pmatrix} t' \\ x' \end{pmatrix} = \mathsf{L}(v) \begin{pmatrix} t \\ x \end{pmatrix}, \qquad \mathsf{L} = \begin{pmatrix} \alpha_0 & -v\gamma\eta \\ -\gamma v & \gamma \end{pmatrix}. \tag{12.1.9}$$

Since the inverse transformation $\mathsf{L}^{-1}(v)$ must be equal to $\mathsf{L}(-v)$, we get

$$\mathsf{L}^{-1}(v) = \frac{1}{\det \mathsf{L}} \begin{pmatrix} \gamma & v\gamma\eta \\ v\gamma & \alpha_0 \end{pmatrix} = \mathsf{L}(-v). \tag{12.1.10}$$

From the comparison it follows that $\alpha_0 = \gamma$ and $\det \mathsf{L} = \gamma^2(1-v^2\eta) = 1$:

$$\mathsf{L}(v) = \gamma \begin{pmatrix} 1 & -v\eta \\ -v & 1 \end{pmatrix}, \qquad \gamma = \frac{1}{\sqrt{1-v^2\eta}}. \tag{12.1.11}$$

For $\eta = 0$ this is the Galilei transformation.

Addition of parallel velocities

If two transformations are performed consecutively, one must again obtain a transformation of type (12.1.9):

$$\mathsf{L}_2\,\mathsf{L}_1 = \gamma_1\gamma_2 \begin{pmatrix} 1+v_1v_2\eta_2 & -v_1\eta_1-v_2\eta_2 \\ -v_1-v_2 & 1+v_1v_2\eta_1 \end{pmatrix} = \mathsf{L}_3 = \gamma_3 \begin{pmatrix} 1 & -v_3\eta_3 \\ -v_3 & 1 \end{pmatrix}. \tag{12.1.12}$$

The right side only represents a transformation (12.1.9) if $\eta_1 = \eta_2 = \eta_3$, i.e. η is a universal constant. Then it applies

$$v_3 = \frac{v_1+v_2}{1+\eta v_1 v_2}, \qquad \gamma_3 = \frac{1+\eta v_1 v_2}{\sqrt{(1-\eta v_1^2)(1-\eta v_2^2)}} = \frac{1}{\sqrt{1-\eta v_3^2}}. \tag{12.1.13}$$

If you insert $\eta = 1/c^2$ here, you have the well-known addition theorem for parallel velocities (12.4.16) in front of you. The transformations L form a (commutative) group with multiplication as a link. Unit element $\mathsf{L}(0)$ and inverse element $\mathsf{L}(-v)$, (12.1.10) exist, and it applies the associative law[2].

---

[2] $(\mathsf{L}_1\,\mathsf{L}_2)\mathsf{L}_3 = \mathsf{L}_1(\mathsf{L}_2\,\mathsf{L}_3)$.

If you define with $V = 1/\sqrt{|\eta|}$ a universal velocity, the transformation takes the form

$$t' = \gamma\Big(t - \operatorname{sgn}\eta\,\frac{vx}{V^2}\Big), \qquad\qquad \gamma = \frac{1}{\sqrt{1 - \operatorname{sgn}\eta\,v^2/V^2}}. \qquad (12.1.14)$$
$$x' = \gamma(-vt + x),$$

The scalar product of the four-vector $(Vt, \mathbf{x})$:

$$s^2 = V^2 t^2 - \operatorname{sgn}\eta\,r^2 \qquad\qquad \Rightarrow \qquad\qquad s'^2 = s^2 \qquad (12.1.15)$$

is an invariant of the transformation. For $\operatorname{sgn}\eta = -1$ the geometry of space-time is Euclidean, but the addition for $v_1 v_2 = V^2$ has an (infinite) discontinuity; it is therefore not unique. There v_3 becomes singular and changes its sign. This case is excluded. $\eta = 0$ has already been assigned to the Galilean transformation, so only $\operatorname{sgn}\eta = 1$ remains. The associated geometry is called pseudo-Euclidean, and its properties will be discussed in detail in section 12.2, where the LT is derived from the invariance of the scalar product (12.1.15). Here we have obtained the Lorentz transformation from the equivalence of the inertial systems.

On the general Lorentz transformation

To derive the general linear and homogeneous transformation between two inertial systems, it seems reasonable to first construct the transformation for a general direction of $\boldsymbol{\beta} = \mathbf{v}/c$, i.e. $\Lambda(\boldsymbol{\beta}, 0)$. S' moves relative to S with $\mathbf{v}$ as shown in the following sketch.

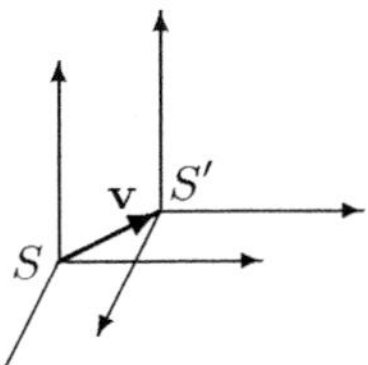

Fig. 12.1': S' moves with $\mathbf{v}$ relative to S. The coordinates of $\mathbf{x}$ are transformed into the component parallel to $\mathbf{v}$: $\mathbf{x}_\parallel = \mathbf{x}\cdot\hat{\mathbf{v}}$ with the unit vector $\hat{\mathbf{v}} = \mathbf{v}/v$ and divided into the components perpendicular to it $\mathbf{x}_\perp = \mathbf{x} - \mathbf{x}_\parallel$. According to (12.1.8) we get

$$t' = \gamma\Big(t - \frac{\mathbf{v}\cdot\mathbf{x}}{c^2}\Big), \quad \mathbf{x}'_\parallel = \gamma(\mathbf{x}_\parallel - \mathbf{v}t) = \gamma(\mathbf{x}\cdot\hat{\mathbf{v}} - \mathbf{v}t), \quad \mathbf{x}'_\perp = \mathbf{x}_\perp.$$

We switch to $\boldsymbol{\beta} = \mathbf{v}/c$ with the unit vector $\hat{\boldsymbol{\beta}} = \boldsymbol{\beta}/\beta$ and obtain the LT for arbitrary orientation of $\mathbf{v}$:

$$\begin{aligned} ct' &= \gamma(ct - \mathbf{x}\cdot\boldsymbol{\beta}) \\ \mathbf{x}' &= -\gamma\boldsymbol{\beta}ct + \mathbf{x} + (\gamma - 1)(\mathbf{x}\cdot\hat{\boldsymbol{\beta}})\,\hat{\boldsymbol{\beta}} \end{aligned} \qquad \Rightarrow \qquad \begin{pmatrix} ct' \\ \mathbf{x}' \end{pmatrix} = \Lambda(\boldsymbol{\beta}, 0)\begin{pmatrix} ct \\ \mathbf{x} \end{pmatrix}. \qquad (12.1.16)$$

The spatial rotation $\Lambda(0, \boldsymbol{\alpha})$ will only be discussed much later in section 12.4 and the combination of both transformations results in the general Lorentz transformation.

Remark: The determination of the LT becomes more complicated when starting from the general transformation. Then you have 16 parameters. Six parameters ($\boldsymbol{\beta}$ and rotation angle $\boldsymbol{\alpha}$) are to be specified, the rest can be determined using the invariance of the d'Alembert operator, which is the subject of problems 12.1 and 12.2.

Voigt [1887] has examined the conditions under which the wave equation remains invariant and thus found the Lorentz transformation long before Larmor and Lorentz and that in a general form, but with the beauty flaw that he set $\gamma = 1$. Therefore, the variables must be multiplied by a scale factor.

12.1.2 On the Ether Theory

It was initially Huygens [1690][3], who presented a wave theory of light. A wave requires – analogous to a sound wave – a medium in which it can propagate, which according to the conception of the time should be a fine, invisible inponderable fluid. However, in the 18th century the wave theory could not prevail against the *Newton*[4] attributed corpuscular theory.

The corpuscular theory was supported, among other things, by the simple explanation of the aberration observed by Bradley around 1728: When observing light coming from a star, one must, since one is moving with the earth around the sun at speed v, tilt the telescope by the angle $\alpha \approx v/c$ to be able to observe the star. The discovery of aberration was thus also an important indication for the movement of the earth around the sun. In Fig. 12.2a, the beam paths of a light source L, which moves with v relative to an observation point F are sketched. The last piece of the path goes through a telescope, whereby in the time t, which a light particle needs for the distance within the telescope, F shifts by vt.

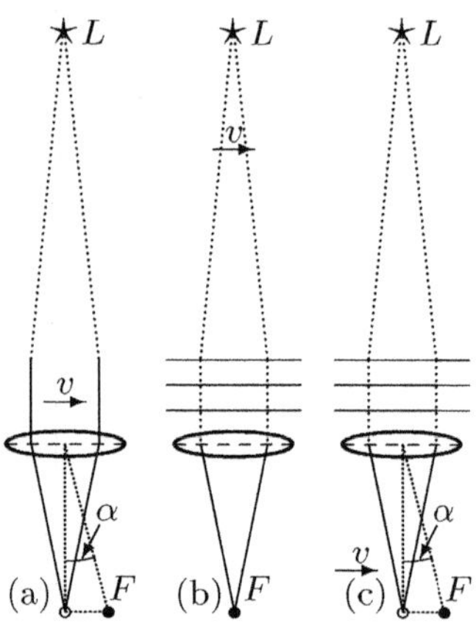

Fig. 12.2. On aberration: L is the light source (star) and F is the observation point; since one can apply geometric optics in the telescope, regardless of whether the light is wave or particle, only rays are sketched under (b) and (c).
(a) Light consists of corpuscles and F moves towards L with v: Inclined position to compensate for $\alpha \approx v/c$
(b) F is at rest in the ether: The movement of L has no influence
(c) F moves with v relative to the ether: Geometric optics show the same aberration as under (a)

Through the ether, in a wave theory, the equivalence of the inertial systems of light source and observer is initially abolished, since the system in which the ether is at rest has a preferred position: If the observer is at rest relative to the ether, he will always measure the speed c, regardless of whether the light source is moving or not. But if the observer moves against the ether,

---

[3] Christiaan Huygens, 1629–1695
[4] Sir Isaac Newton, 1643–1727

he should measure a changed speed. As explained in Fig. 12.2b, one will not notice any aberration at all in the rest system of the ether, as the waves in the ether spread concentrically from L and so the telescope must be directed perpendicular to the wave front. A movement of L plays no role (except that one must follow the movement of L with the telescope).

Only when moving with v against the ether will one have to tilt the telescope by the same angle α so that the waves meet at the focal point. In Fig. 12.2c the wavefront emanating from L is sketched by horizontal lines in front of the telescope. However, the telescope continues to move in the time it takes for the waves to pass through the telescope. For the sketching of the movement of the light waves in the telescope, however, geometric optics was used.

At the beginning of the 19th century, *Young*[5] helped the view that light is a wave phenomenon to break through with his diffraction experiments on the double slit. Fresnel found through experiments with polarized light that light waves are purely transversal, without a longitudinal component. Because of the transversality, the ether now had to be assigned properties of a solid body, such as elasticity. This is not easily compatible with a movement of the earth relative to the ether and the penetration of the ether into matter.

Maxwell recognized that light waves are electromagnetic waves, and so the light ether became an electromagnetic ether. Maxwell himself constructed complicated mechanical models[6]. It should also be mentioned that according to Kragh [2016] Lorenz was one of the few exceptions who did not believe in the existence of the ether.

However, the fundamental view was that electrodynamics only applies in the rest system of the ether. Within this classical conception, one would expect that the speed of light depends on the speed of the observer.

In the absolutely resting system S a light wave is emitted, which propagates with the speed c. The coordinates of the wavefront on the x-axis are $\pm ct$.

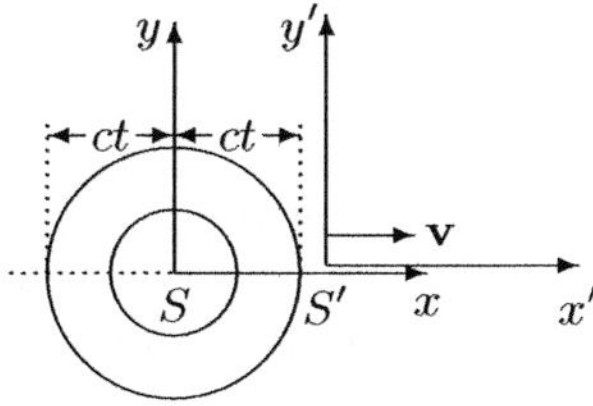

Fig. **12.3.** Moving system S' in the (stationary) ether: $x' = x - vt$.
$$x' = ct - vt = (c-v)t, \quad \text{right wavefront,}$$
$$x' = -ct - vt = -(c+v)t, \quad \text{left wavefront}$$

The system S' moves to the right at speed v, as sketched in Fig. 12.3. It was believed based on these considerations that one could determine the move-

---

[5] Thomas Young, 1773–1829

[6] Rotating vortex elements were intended for **B**, between which small polarizable spheres ("molecules") were located for **E** [Schöpf, 1982]

ments against the absolute ether by measuring the speed of light in different directions.

12.1.3 Michelson-Morley Experiment

The rest system of the light ether is distinguished according to the previous considerations, since only in this the laws of electrodynamics would remain unchanged. Michelson[7] tried to determine the velocity $\mathbf{v}$ of the earth with the arrangement sketched in Fig. 12.4. The following described experiment was in 1887 repeated by Morley[8] with higher accuracy.

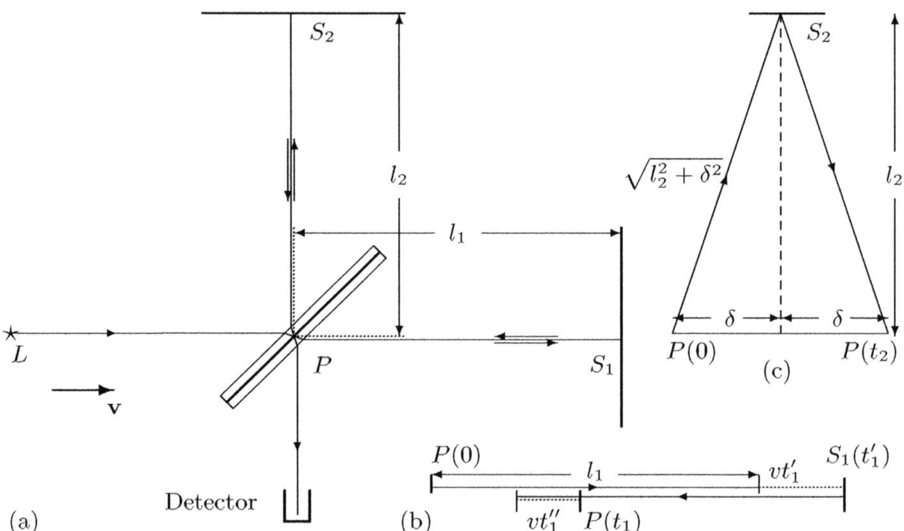

Fig. 12.4. (a) Experimental setup for the Michelson-Morley experiment; L is the (coherent) light source, P the semi-transparent mirror
(b) Path $P \to S_1 \to P$ seen from the stationary ether: $t_1 = t_1' + t_1''$
(c) Path $P \to S_2 \to P$ seen from the stationary ether

Assumption: As outlined in Fig. 12.4, the laboratory is moving to the right with v. The time t_1, which the light beam needs from $P \to S_1 \to P$ and the time t_2 for the path from $P \to S_2 \to P$ are considered from the system of the stationary ether. In this, the light has the speed c, but in the time that the light needs to pass the distance l_1, the mirror S_1 has moved away. For this, P comes towards the beam on the return path. The way there and back to S_2 takes the same time, whereby but P has moved on by the distance $2\delta = vt_2$:

[7] Albert Abraham Michelson, 1852–1932, Nobel Prize 1907
[8] Edward Morley, 1838–1923

$$P \to S_1 : \quad ct_1' = l_1 + vt_1' \qquad \Rightarrow \quad t_1' = l_1/(c-v),$$
$$S_1 \to P : \quad ct_1'' = l_1 - vt_1'' \qquad \Rightarrow \quad t_1'' = l_1/(c+v),$$
$$P \to S_2 : \quad ct_2' = \sqrt{l_2^2 + (vt_2')^2} \quad \Rightarrow \quad t_2'' = l_2/\sqrt{c^2 - v^2}.$$

$$P \to S_1 \to P : \quad t_1 = t_1' + t_2' = 2l_1 c/(c^2 - v^2),$$
$$P \to S_2 \to P : \quad t_2 = 2t_2' = 2l_2/\sqrt{c^2 - v^2}.$$

This results in a runtime difference

$$\Delta t = t_1 - t_2 = \frac{2l_1}{c\left(1 - v^2/c^2\right)} - \frac{2l_2}{c\sqrt{1 - v^2/c^2}}. \tag{12.1.17}$$

If you rotate the interferometer by $90°$ clockwise, then the time difference between $\bar{t}_1$ for the beam going to mirror S_1 and $\bar{t}_2$ for the beam going to S_2 is given by

$$\overline{\Delta t} = \bar{t}_1 - \bar{t}_2 = \frac{2l_1}{c\sqrt{1 - v^2/c^2}} - \frac{2l_2}{c\left(1 - v^2/c^2\right)}.$$

When rotating the apparatus by $90°$, the total time difference is

$$\Delta t - \overline{\Delta t} = \frac{2(l_1 + l_2)}{c}\left(\frac{1}{\left(1 - v^2/c^2\right)} - \frac{1}{\sqrt{1 - v^2/c^2}}\right).$$

For $l_1 = l_2 = l$ one obtains the time difference

$$\Delta t = \frac{2l}{c}\left[\frac{1}{1 - v^2/c^2} - \frac{1}{\sqrt{1 - v^2/c^2}}\right] \stackrel{v \ll c}{=} l\,\frac{v^2}{c^3}.$$

Between the two rays, there is thus a time difference that depends on the speed, but is only of 2nd order. Since the rays, which come from different points of the light source, are not exactly parallel, an interference pattern always results in this interferometer; maximum amplification for $\omega(t_1 - t_2) = 2n\pi$ and extinction for $\omega(t_1 - t_2) = (2n+1)\pi$.

When the interferometer is rotated by $90°$, the sign of Δt changes, and the total displacement becomes twice as large for $l = l_1 = l_2$. The interference pattern would have to shift if v changes.

Assuming that the stationary ether is determined by the solar system, the earth moves at about $30\,\mathrm{km/s}$ relative to the ether. This corresponds to a path difference $\Delta s = c(t_1 - t_2) = l \times 10^{-8}$. The Michelson interferometer had $l = 11\,\mathrm{m}$, which, taking into account a rotation of the apparatus by $90°$ results in a path difference of $\Delta s = 220\,\mathrm{nm}$, which would be perceptible in visible light ($\lambda \sim 500\,\mathrm{nm}$).

No path difference could be detected, which implies that the speed of light is independent of the direction.

Actually, the requirement of the validity of electrodynamics (principle of relativity) in all inertial systems should be sufficient to drop the ether hypothesis;

with the experimental determination of the independence of the speed of light from the orientation and thus also from the motion of the earth, the Lorentz transformation (LT) can be derived as a transformation between inertial systems.

1. *Principle of relativity*: The laws of nature are the same in all inertial systems.
2. *Constancy of the speed of light*: The speed of light is independent of the motion of the light source and the observer.

12.1.4 Fizeau's Experiment

The influence of the motion of a transparent fluid (water) on the speed of light is investigated by bringing rays to interference in an interferometer, as outlined in Fig. 12.5, that have traveled the same path with and against the flow of the fluid. This experiment was conducted by Fizeau[9] in 1851.

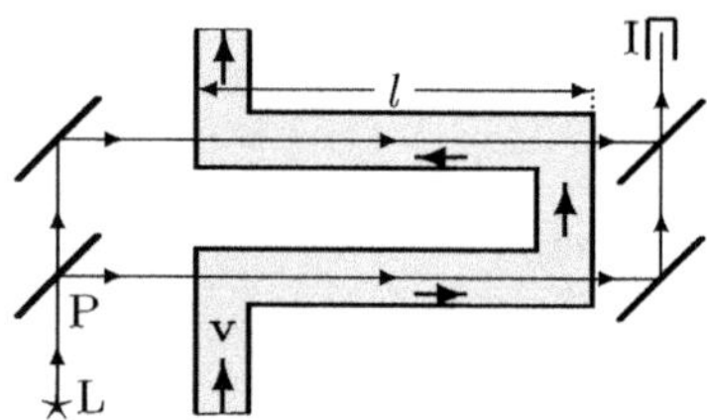

Fig. 12.5. L is the light source, P is a semi-transparent mirror and I is an interferometer

The experimental result yielded a speed w for the lower path, where $\bar{c} = c/n$ is the speed of light in the stationary liquid, with

$$w = \bar{c} + vf \qquad \text{and} \qquad f = 1 - 1/n^2. \qquad (12.1.18)$$

f is the Fresnel's drag coefficient. The explanation of a partial drag of the ether in the moving fluid by f does not seem particularly elegant, but the formula itself is confirmed by the velocity addition (12.4.16) of the STR when one assumes $\bar{c} \gg v$:

$$w = \frac{\bar{c} + v}{1 + \frac{\bar{c}v}{c^2}} = \frac{\bar{c} + v}{1 + \frac{v}{n^2\bar{c}}} \approx (\bar{c} + v)(1 - \frac{v}{n^2\bar{c}}) \approx \bar{c} + v(1 - \frac{1}{n^2}).$$

Within the framework of the STR, there is therefore the (full) drag of light through moving bodies, which naturally must comply with the addition theorem for velocities.

---

[9] Hippolyte Fizeau, 1819–1896

12.2 The Lorentz Transformation

We have already derived the Lorentz transformation (LT) from several perspectives, by determining the conditions under which the wave equation remains invariant or with the help of a flash of light, which is observed in the systems S and S'. Above all, however, the LT could be determined solely from the equivalence of inertial systems up to a speed $V \to c$, where according to (12.1.15) the quadratic form $s^2 = c^2t^2 - \mathbf{x}^2$ is invariant under LT and leads to a pseudo-Euclidean geometry.

Here we proceed directly from considerations of invariance. As before, at the time $t = t' = 0$ the systems S and S' coincide. Because of the Einstein's principle of relativity, the transformation must leave the wave equation, i.e. the d'Alembert operator, invariant (12.1.7):

$$\frac{\partial^2}{c^2 \partial t^2} - \boldsymbol{\nabla}^2 = \frac{\partial^2}{c^2 \partial t'^2} - \boldsymbol{\nabla}'^2 .$$

This is a quadratic form of the four-dimensional gradient vector. The quadratic form invariant under linear transformations is the scalar product. The d'Alembert operator is therefore the scalar product of the gradient vector in the pseudo-Euclidean geometry.

In vector calculus, we distinguish between co- and contravariant vectors in oblique coordinates in Euclidean space. In non-Euclidean geometries, we always do this. In the appendix A.1 you can find a representation of vector calculus. The four-vector of an event is defined by the coordinates

$$x^0 = ct, \quad x^1 = x, \quad x^2 = y, \quad x^3 = z, \quad \Leftrightarrow \quad (x^\mu) = (x^0, \mathbf{x}) \quad \text{contravariant,}$$
$$x_0 = ct, \quad x_1 = -x, \quad x_2 = -y, \quad x_3 = -z, \quad \Leftrightarrow \quad (x_\mu) = (x_0, -\mathbf{x}) \quad \text{covariant,} \quad (12.2.1)$$

where greek indices μ always run from 0 to 3. Indices in latin script, like k or l go as before from 1 to 3. The different sign of time and space components becomes obvious from the metric tensor g:

$$\mathbf{g} = (g_{\mu\nu}) = (g^{\mu\nu}) = \begin{pmatrix} 1 & 0 & 0 & 0 \\ 0 & -1 & 0 & 0 \\ 0 & 0 & -1 & 0 \\ 0 & 0 & 0 & -1 \end{pmatrix}, \tag{12.2.2}$$

$$x_\mu = g_{\mu\nu}\, x^\nu, \qquad\qquad x^\mu = g^{\mu\nu}\, x_\nu . \tag{12.2.3}$$

Note: The diagonal elements of g have the signs $(+, -, -, -)$. The signature of a matrix is the set of the signs of the diagonal elements (principal values) [Landau, Lifshitz 2 , 1987, § 84], which can be indicated by the number of positive and negative elements. In the STR, these are the signatures $(1,3)$ [Jackson, 1998, (11.81)] and $(3,1)$ [Griffiths, 2017, (12.33)]:

$$(g_{\mu\nu}) = k_S \begin{pmatrix} 1 & \mathbf{0}^T \\ \mathbf{0} & -(\delta_{mn}) \end{pmatrix}, \qquad \begin{cases} k_S = 1 & \text{Signature (1,3): "timelike convention"} \\[4pt] k_S = -1 & \text{Signature (3,1): "spacelike convention".} \end{cases} \tag{12.2.2'}$$

The scalar product of the gradient vector (12.2.6) and the line element are thus:

$$\Box = \partial^\mu \partial_\mu = k_S \left(\frac{1}{c^2} \frac{\partial^2}{\partial t^2} - \boldsymbol{\nabla}^2 \right), \qquad\qquad \mathrm{d}s^2 = \mathrm{d}x^\mu \mathrm{d}x_\mu = k_S(c^2 \mathrm{d}t^2 - \mathrm{d}\mathbf{x}^2).$$

In the "timelike" convention, which will be used in the following, the line element with $\mathrm{d}s^2 > 0$ is timelike, see Fig. 12.13. It has the signature (1,3) for the $k_S = 1$.

The coordinates of two uniformly moving reference systems must be connected by a linear transformation:

$$x'^\mu = \Lambda^\mu{}_\nu \, x^\nu + a^\mu. \tag{12.2.4}$$

For the covariant components, this results in the transformation

$$x'_\mu = \Lambda_\mu{}^\nu \, x_\nu + a_\mu, \qquad\qquad \Lambda_\mu{}^\nu = g_{\mu\mu'} \, \Lambda^{\mu'}{}_{\nu'} \, g^{\nu'\nu}. \tag{12.2.5}$$

With the four-vector of the gradient

$$\partial_\mu = \frac{\partial}{\partial x^\mu}, \qquad\qquad \partial^\mu = \frac{\partial}{\partial x_\mu} \tag{12.2.6}$$

the d'Alembert operator can be represented as

$$\Box = g^{\mu\nu} \, \partial_\mu \partial_\nu = \partial_\mu \, \partial^\mu. \tag{12.2.7}$$

From (12.2.4) and (12.2.5) one obtains the inverse transformations

$$\partial_\lambda = \frac{\partial x'^\mu}{\partial x^\lambda} \frac{\partial}{\partial x'^\mu} = \Lambda^\mu{}_\lambda \partial'_\mu, \qquad\qquad \partial^\lambda = \frac{\partial x'_\mu}{\partial x_\lambda} \frac{\partial}{\partial x'_\mu} = \Lambda_\mu{}^\lambda \partial'^\mu, \tag{12.2.8}$$

where the special position of the signs in the gradient vector should be pointed out: $(\partial^\mu) = (\partial^0, -\boldsymbol{\nabla})$ and $(\partial_\mu) = (\partial_0, \boldsymbol{\nabla})$. According to the principle of relativity, the wave equation is the same in all inertial systems, i.e.

$$\partial_\lambda \, \partial^\lambda = \Lambda^\mu{}_\lambda \partial'_\mu \, \Lambda_\nu{}^\lambda \partial'^\nu = \partial'_\mu \, \partial'^\mu, \tag{12.2.9}$$
$$= \partial'_\mu \, \Lambda^\mu{}_\lambda \, g^{\lambda\rho} \Lambda^\sigma{}_\rho \, \partial'_\sigma = \partial'_\mu \, g^{\mu\sigma} \, \partial'_\sigma$$

is invariant under the Lorentz transformation. From the 1st line it follows that

$$\Lambda^\mu{}_\lambda \, \Lambda_\nu{}^\lambda = \delta^\mu{}_\nu = \begin{cases} 1 & \text{for } \mu = \nu \\ 0 & \text{otherwise} \end{cases} \qquad\Rightarrow\qquad \Lambda_\nu{}^\lambda = \left(\Lambda^{-1}\right)^\lambda{}_\nu. \tag{12.2.10}$$

From the 2nd line of (12.2.9) we deduce that for the invariance of $\Box$

$$\Lambda^\mu{}_\lambda g^{\lambda\rho} \Lambda^\sigma{}_\rho = g^{\mu\sigma}, \qquad \Lambda^\sigma{}_\rho = (\Lambda^T)_\rho{}^\sigma \qquad\Rightarrow\qquad \mathbf{\Lambda} \mathbf{g}^{-1} \mathbf{\Lambda}^T = \mathbf{g}^{-1} \tag{12.2.11}$$

must be fulfilled. This condition defines the LT, where a matrix notation was used, which will be discussed in more detail.

Supplement: The conditions (12.2.10)–(12.2.11) are obtained in a slightly modified form also by means of $x'^{\mu} x'_{\mu} = x^{\lambda} x_{\lambda}$:

$$\Lambda^{\mu}{}_{\lambda}\, \Lambda_{\mu}{}^{\nu} = (\Lambda^{T})_{\lambda}{}^{\mu}\, \Lambda_{\mu}{}^{\nu} = \delta_{\lambda}{}^{\nu}, \qquad \Lambda_{\mu}{}^{\nu} = (\Lambda^{-1})^{\nu}{}_{\mu} = ((\Lambda^{-1})^{T})_{\mu}{}^{\nu} := (\Lambda^{C})_{\mu}{}^{\nu}.$$

$\Lambda^{C} = \Lambda^{-1\,T}$ is the contragredient matrix to Λ (inverse and transposed). Furthermore, the LT must meet the requirement

$$\Lambda^{\mu}{}_{\lambda}\, g_{\mu\nu}\, \Lambda^{\nu}{}_{\rho} = g_{\lambda\rho} \qquad\qquad \Rightarrow \qquad\qquad \Lambda^{T} g \Lambda = g. \tag{12.2.12}$$

(12.2.11) and (12.2.12) are equivalent, only in (12.2.11) we started from the inverse transformation.

Matrix notation

Although an orthonormal basis $(\{\vec{e}_{\lambda}\})$ is applied to Minkowski space, but the vector notation in the form $\vec{x} = x^{\mu} \vec{e}_{\mu} = x_{\nu} \vec{e}^{\nu}$ is rarely used, instead the notation is reduced to components:

$$\vec{a} \cdot \vec{b} \qquad\qquad \Longleftrightarrow \qquad\qquad a^{\mu} b_{\mu}.$$

Let's start with the vectors. We have already anticipated in (8.2.18) that the Liénard-Wiechert potentials $\phi(\mathbf{x}, t)/k_{\mathrm{L}}$, $\mathbf{A}(\mathbf{x}, t)$ are components of a four-dimensional vector field $(A^{\mu}(\mathbf{x}, t))$.

For tensors of 2nd degree, in addition to the notation in components, one in matrix symbolism [Sexl, Urbantke, 2001, §3] is used. In this, some equations are clearer, although we immediately state that the raising and lowering of the indices from the mixed matrix $\Lambda = (\Lambda^{\mu}{}_{\nu})$ and the Einstein summation convention should be apparent:

$$\Lambda \equiv (\Lambda^{\alpha}{}_{\beta}), \qquad\qquad \mathsf{E} = (g^{\alpha}{}_{\beta}) = (\delta^{\alpha}{}_{\beta}) = (\delta_{\beta}{}^{\alpha}). \tag{12.2.13}$$

The term g (12.2.2) stands for both the contravariant $(\mathsf{g}^{-1} = (g^{\mu\nu}))$ and the covariant $(\mathsf{g} = (g_{\mu\nu}))$ metric tensor, which are both equal. The index position should be given by Λ. In matrix notation, (12.2.5) reads

$$\Lambda_{\mu}{}^{\nu} = g_{\mu\mu'}\, \Lambda^{\mu'}{}_{\nu'}\, g^{\nu'\nu} = (\mathsf{g}\Lambda\mathsf{g}^{-1})_{\mu}{}^{\nu} \quad \Leftrightarrow \quad \Lambda^{C} := (\Lambda^{-1})^{T} = \mathsf{g}\,\Lambda\,\mathsf{g}^{-1}. \tag{12.2.14}$$

The elements of the transposed and the contragredient matrix Λ^{T} and Λ^{C} are of the form $(\Lambda^{T})_{\mu}{}^{\nu}$ and $(\Lambda^{C})_{\mu}{}^{\nu}$, i.e. $(\Lambda^{\mu}{}_{\nu})^{T} = \Lambda_{\nu}{}^{\mu} = (\Lambda^{C})_{\nu}{}^{\mu} = (\Lambda^{-1})^{\mu}{}_{\nu}$. In addition to (12.2.14) we note

$$\Lambda^{\lambda}{}_{\nu}\, \Lambda_{\mu}{}^{\nu} = \Lambda^{\lambda}{}_{\nu}\, (g^{\nu\nu'}\, \Lambda^{\mu'}{}_{\nu'}\, g_{\mu'\mu}) = \delta^{\lambda}{}_{\mu} \quad \Leftrightarrow \quad \Lambda \Lambda^{-1} = \Lambda(\mathsf{g}^{-1}\Lambda^{T}\mathsf{g}) = \mathsf{E}. \tag{12.2.15}$$

With the definitions $\Lambda = (\Lambda^{\mu}{}_{\nu})$ and $\mathsf{g} = (g_{\mu\nu})$ the ad hoc introduced tensorial form of the conditions (12.2.11)–(12.2.12) can be verified

$$\Lambda\,\mathsf{g}\,\Lambda^{T} = \mathsf{g} \qquad\qquad \text{or} \qquad\qquad \Lambda^{T} \mathsf{g}\, \Lambda = \mathsf{g}. \tag{12.2.16}$$

Now we introduce a shortened notation for the inhomogeneous LT

$$x'^{\mu} = \Lambda^{\mu}{}_{\nu} x^{\nu} + a^{\mu} \quad \Rightarrow \quad x' = \Lambda x + a \quad \Leftrightarrow \quad x' = (\Lambda, a)x. \qquad (12.2.17)$$

The inhomogeneous Lorentz transformations, also called Poincaré transformations, form a group, i.e., they fulfill the following four properties necessary for the formation of a group.

1. *Binary operation*: The multiplication of two LTs $(\bar{\Lambda}, \bar{a})$ and (Λ, a) results in

$$x'' = \bar{\Lambda} x' + \bar{a} = \bar{\Lambda}\big(\Lambda x + a\big) + \bar{a} = \bar{\Lambda}\Lambda x + (\bar{\Lambda} a + \bar{a}),$$

$$(\bar{\bar{\Lambda}}, \bar{\bar{a}}) = (\bar{\Lambda}, \bar{a})\,(\Lambda, a) = (\bar{\Lambda}\Lambda,\ \bar{\Lambda} a + \bar{a}), \qquad (12.2.18)$$

where $\bar{\bar{\Lambda}} = \bar{\Lambda}\Lambda$ has the property (12.2.16) of an LT: $\bar{\bar{\Lambda}}\,\mathrm{g}\,\bar{\bar{\Lambda}}^{T} = \mathrm{g}$.

2. *Associativity*: $(\bar{\bar{\Lambda}}, \bar{\bar{a}})\,\big((\bar{\Lambda}, \bar{a})(\Lambda, a)\big) = \big((\bar{\bar{\Lambda}}, \bar{\bar{a}})(\bar{\Lambda}, \bar{a})\big)\,(\Lambda, a)$.
3. *Identity element*: $(\mathrm{E}, 0)$.
4. *Inverse element*: $(\Lambda, a)^{-1} = (\Lambda^{-1}, -\Lambda^{-1} a)$ with $\Lambda^{-1} = \mathrm{g}^{-1}\Lambda^{T}\mathrm{g}$.

Naturally, the homogeneous LTs also form a subgroup of the Poincaré group. It is evident that the translations (E, a) form a subgroup, the translation group. Another subgroup of the inhomogeneous Lorentz group is the rotation group $(\Lambda, \mathbf{0})$. For rotations $(\mathrm{R}^{-1} = \mathrm{R}^{T})$ the condition applies

$$\Lambda = \begin{pmatrix} 1 & \mathbf{0}^{T} \\ \mathbf{0} & \mathrm{R} \end{pmatrix} : \qquad \Lambda \mathrm{g}\Lambda^{T} = \begin{pmatrix} 1 & \mathbf{0}^{T} \\ \mathbf{0} & -\mathrm{R} \end{pmatrix}\begin{pmatrix} 1 & \mathbf{0}^{T} \\ \mathbf{0} & \mathrm{R}^{T} \end{pmatrix} = \begin{pmatrix} 1 & \mathbf{0}^{T} \\ \mathbf{0} & -\mathrm{E} \end{pmatrix} = \mathrm{g}.$$

12.2.1 Classification of the Lorentz Group

The homogeneous Lorentz transformation is defined according to (12.2.16) by $\Lambda \mathrm{g}\Lambda^{T} = \mathrm{g}$. From this it follows for the determinants that

$$\det \Lambda \ \det \mathrm{g}\ \det \Lambda^{T} = \det \mathrm{g}$$

and from this that $\det \Lambda = \pm 1$. For the matrix element $\Lambda^{0}{}_{0}$ it holds

$$\Lambda^{0}{}_{\mu}\,g^{\mu\nu}\,\Lambda^{0}{}_{\nu} = \left(\Lambda^{0}{}_{0}\right)^{2} - \sum_{k}\left(\Lambda^{0}{}_{k}\right)^{2} = 1,$$

which is why $|\Lambda^{0}{}_{0}| \geq 1$ and $\operatorname{sgn}\Lambda^{0}{}_{0} = \pm 1$. This distinction is essential because a transformation with $\Lambda^{0}{}_{0} \leq -1$ includes a time reflection T. If in addition $\det \Lambda = 1$, then a space reflection P is also present[10] [Sexl, Urbantke, 2001, §6.3] or [Schwabl, 2008, §6.1].

---

[10] The representations of time and space reflection are

$$\mathrm{T} = \begin{pmatrix} -1 & & & \\ & 1 & & \\ & & 1 & \\ & & & 1 \end{pmatrix}, \qquad \mathrm{P} = \begin{pmatrix} 1 & & & \\ & -1 & & \\ & & -1 & \\ & & & -1 \end{pmatrix} \qquad \rightarrow \qquad \mathrm{P}\cdot\mathrm{T} = -1.$$

Tab. 12.1. Classification of Lorentz transformations into classes and groups

Lorentz transformation	sgn $\Lambda^0{}_0$	det Λ	Representation by $\mathsf{L}^\uparrow_+$
proper orthochronous LT	1	1	$\mathsf{L}^\uparrow_+$
improper orthochronous LT (Space inversion)	1	-1	$\mathsf{L}^\uparrow_- = \mathsf{P} \cdot \mathsf{L}^\uparrow_+$
Space-time inversions	-1	1	$\mathsf{L}^\downarrow_+ = \mathsf{P}{\cdot}\mathsf{T} \cdot \mathsf{L}^\uparrow_+$
Time inversions	-1	-1	$\mathsf{L}^\downarrow_- = \mathsf{T} \cdot \mathsf{L}^\uparrow_+$

Lorentz Group	Representation by Lorentz classes
Proper orthochronous Lorentz group	$\mathsf{L}^\uparrow_+$
Proper Lorentz group	$\mathsf{L}_+ = \mathsf{L}^\uparrow_+ \cup \mathsf{L}^\downarrow_+$
Orthochronous Lorentz group	$\mathsf{L}^\uparrow = \mathsf{L}^\uparrow_+ \cup \mathsf{L}^\uparrow_-$
Orthochoric Lorentz group	$\mathsf{L}_0 = \mathsf{L}^\uparrow_+ \cup \mathsf{L}^\downarrow_-$

LTs that contain a spatial reflection, but no time reflection $(L^\uparrow_-)$, do not form a group, but belong to a (sub-)class of the Lorentz group L. All elements of the classes $\mathsf{L}^\uparrow_\pm$, $\mathsf{L}^\downarrow_\pm$ products of $\mathsf{L}^\uparrow_+$ with P and T can be represented, as shown in Tab. 12.1. In this notation, the arrow stands for the time direction and $\mp$ for the (im)proper rotation. Not all of these classes form groups; these are formed by the union of classes that are also listed in Tab. 12.1.

The full Lorentz group L contains all transformations including spatial and time reflections. If no spatial reflections are allowed, the subgroup obtained is the proper Lorentz group L_+. If time reflections are also excluded, then the remaining Lorentz transformations form the proper orthochronous (also restricted) Lorentz group $\mathsf{L}^\uparrow_+$.

12.2.2 The Proper Orthochronous Lorentz Group

This group consists of the elements det $\Lambda = 1$ and $\Lambda^0{}_0 \geq 1$ and includes

a) Rotations,
b) Lorentz transformations in the narrower sense.

We consider once again a proper orthochronous LT (det $\Lambda = 1$, $\Lambda^0{}_0 \geq 1$) in which the y- and the z-axis remain unchanged. The general form is

$$\Lambda = \begin{pmatrix} L^0{}_0 & L^0{}_1 & 0 & 0 \\ L^1{}_0 & L^1{}_1 & 0 & 0 \\ 0 & 0 & 1 & 0 \\ 0 & 0 & 0 & 1 \end{pmatrix} : \quad \begin{matrix} \Lambda^T \mathsf{g} \Lambda = \mathsf{g} \\ \det \Lambda = 1 \end{matrix} \quad \Rightarrow \quad L^0{}_0 = L^1{}_1, \quad L^1{}_0 = L^0{}_1. \quad (12.2.20)$$

The real 2×2 matrix L with $L^0{}_0 = \cosh \eta \geq 1$ and det L $= 1$ is then

$$\mathsf{L} = \begin{pmatrix} \cosh\eta & -\sinh\eta \\ -\sinh\eta & \cosh\eta \end{pmatrix} = \begin{pmatrix} \gamma & -\beta\gamma \\ -\beta\gamma & \gamma \end{pmatrix} \quad \text{with} \begin{cases} \gamma = \cosh\eta \\ \beta = \tanh\eta \end{cases} \qquad (12.2.21)$$

and

$$\gamma = \cosh\eta = \frac{1}{\sqrt{1-\tanh^2\eta}} = \frac{1}{\sqrt{1-\beta^2}} \quad \text{with} \begin{cases} -\infty < \eta < \infty \\ -1 < \beta < 1 \,. \end{cases}$$

A boost can thus be formally represented as a "rotation", where as a result of the pseudo-Euclidean metric the angle of rotation is imaginary ($\cos(\mathrm{i}\eta) = \cosh\eta$). The transformation is now

$$x'^0 = \gamma(x^0 - \beta x^1), \qquad\qquad x'^1 = \gamma(-\beta x^0 + x^1).$$

We now have to relate β and γ to v and c. In the limit $c \to \infty$ the LT must transition into the Galilean transformation, which means that

$$t' = t\gamma - x\frac{\beta\gamma}{c} \to t \qquad\qquad \Longrightarrow \gamma = 1 \quad \text{and} \quad \frac{\beta}{c} = 0\,,$$
$$x' = x\gamma - \gamma\beta ct \to x - vt \quad \Longrightarrow \beta c = v.$$

Thus, we have, albeit somewhat cumbersome, derived the LT (12.1.4).

12.3 Concept of Space-Time

Moving train and concept of time

The principle of relativity together with the constancy of the speed of light implies that we must change our concept of time. We can clarify this with a simple example [Einstein, 2009, p. 16]. Given is a train of length $2l_0$, which moves with the speed v (see Fig. 12.6).

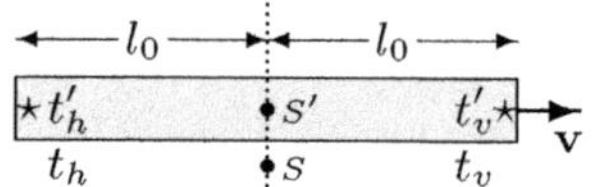

Fig. 12.6. The observers in the train (S') and on the platform (S) both observe the flashes of light at time $t' = t = 0$.

A signal from the rear end of the train and one from the front end arrive exactly at time $t = 0$ in the middle of the train.

Observer in the train: The speed of light in my system is c, so the signal was sent at the same time $t'_v = t'_h = -l_0/c$ from both ends.

Observer on the platform: The speed of light in my system is c; the signal from the front end had a shorter path than the one from the rear end, so it was sent later: ($\beta = v/c$ and $l = l_0\sqrt{1-\beta^2}$)

$$t_h = \frac{vt_h}{c} + \frac{l}{c} = \frac{l}{c}\frac{1}{1-\beta} = \frac{l_0}{c}\sqrt{\frac{1+\beta}{1-\beta}}, \quad t_v = \frac{-vt_v}{c} + \frac{l}{c} = \frac{l}{c}\frac{1}{1+\beta} = \frac{l_0}{c}\sqrt{\frac{1-\beta}{1+\beta}}\,.$$

12.3.1 Synchronization of Clocks

After the example given in the preceding section, the concept of simultaneity loses its character independent of the state of motion, which is why we cannot simply synchronize clocks within a system by transporting them from one place to another. Einstein [1905] therefore, in the derivation of the LT, he preceded the definition of simultaneity along with the synchronization of clocks.

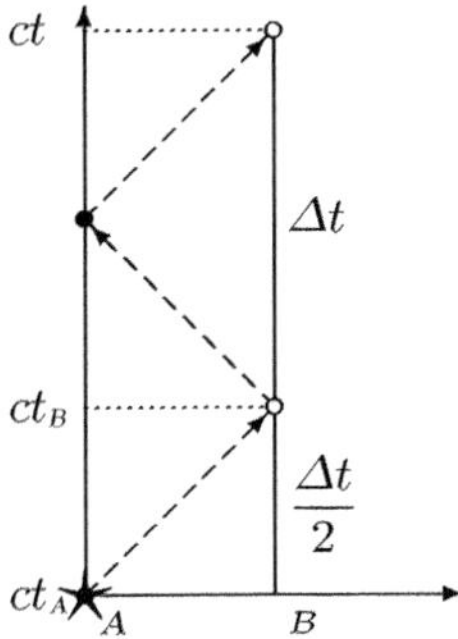

Fig. 12.7. Space-time diagram for the synchronization of clock B with clock A. The light flash (dashed line) moves in the time span $\Delta t/2 = t_B - t_A$ (under 45°) from A to B, is reflected there for the first time and arrives at B again after another reflection at A. The observer at B can measure the time $\Delta t = t - t_B$, that the light needed to return to B; if the observer at B also knows the time t_A at which the light flash was emitted he can set the clock to the time $t = t_A + 3\Delta t/2$.

To synchronize a clock at location B with a clock at location A, as outlined in Fig. 12.7, a light signal is sent from A at time t_A; this arrives after the time $\Delta t/2 = (x_B - x_A)/c$ at B. The clock in B is then set to $t_B = (t_A + t_B)/2$ and you have synchronous clocks. This not very elegant method requires both knowledge of the distance and the speed of light. If you want to synchronize a clock located at B with the one in A without precise knowledge of the distance and the speed of light, it is sufficient to inform the observer in B that a light flash will be sent at time t_A. In B, after reflections of the beam in B and subsequently in A, the time to be set can be calculated according to Fig. 12.7.

Simultaneity: Two events taking place at locations A and B are simultaneous if the synchronized clocks located at these places display the same time.

The requirement that two signals arrive at the same time in the middle $(x_B - x_A)/2$ can be used both for synchronization and for determining simultaneity. Therefore, the definition of simultaneity is based on the constancy of the speed of light.

12.3.2 Space-Time Diagram

For the graphical representation of events (space-time point) in the STR, Minkowski[11] developed diagrams that represent an event in two inertial systems S and S' moving against each other in a unique way. The focus is on the spatial component x, which is chosen parallel to $\mathbf{v}$. The system S' moving

---

[11] Hermann Minkowski, 1864–1909

against the "resting"system S is not only oblique, but its unit scales, which lie on hyperbolas due to the pseudo-Euclidean geometry, are longer than in the system S, which has orthonormal axes.

Galilean transformation

If one describes space only by the component x, then an event that takes place in the stationary inertial system S at space-time (x,t) will have the coordinates $P(x - vt, t)$ in the inertial system S' moving with v. Lines of equal times $t = t'$ are horizontal in Fig. 12.8, i.e. the length of the time unit on the t'-axis is greater than that on the t-axis.

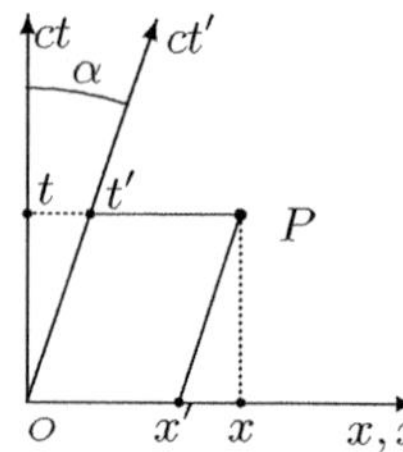

Fig. 12.8. Space-time diagram for the Galilean transformation; the t'-axis indicates the location of the origin $(x'=0)$ of the system S' at time t in S; for dimensional reasons the t- and t'-axes are scaled with any speed, here c, $:\tan\alpha = v/c$, where $v > c$ is quite possible.
$$S' :\ t' = t,$$
$$x' = x - vt$$

Lorentz transformation

The inclination angle α of the t'-axis is the same in Fig. 12.8 and Fig. 12.9. However, with the LT, the x'^1-axis, given by $x'^0 = 0$, is also inclined with α relative to the x^1-axis. Also different in the two sketches are the lengths in the moving system S', which will be discussed further below.

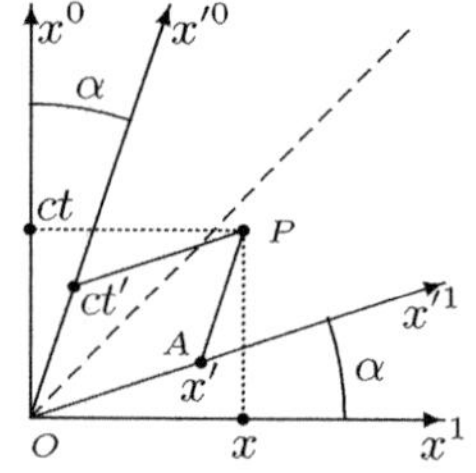

Fig. 12.9. Space-time diagram for the Lorentz transformation.
$$x'^0 = \gamma(x^0 - \beta x^1), \qquad x'^1 = \gamma(-\beta x^0 + x^1),$$
$$x'^0\text{-axis} : x'^1 = 0 \quad \rightarrow \quad x^1 = \beta x^0,$$
$$x'^1\text{-axis} : x'^0 = 0 \quad \rightarrow \quad x^0 = \beta x^1.$$
Slope $= \tan\alpha = \beta = v/c$

In S', A has the coordinates $(x', 0)$, from which it is apparent that there O and A are simultaneous. In S, A has the time $x^0 = \beta x$ $(t = xv/c^2)$, which is also the time difference between O and A.

Unit scales

In Euclidean systems, the unit scale is the unit circle (the unit sphere), which follows from the invariance of the scalar product. In the non-Euclidean geometry of STR this results in hyperbolic branches (hyperboloids), as sketched in Fig. 12.10.

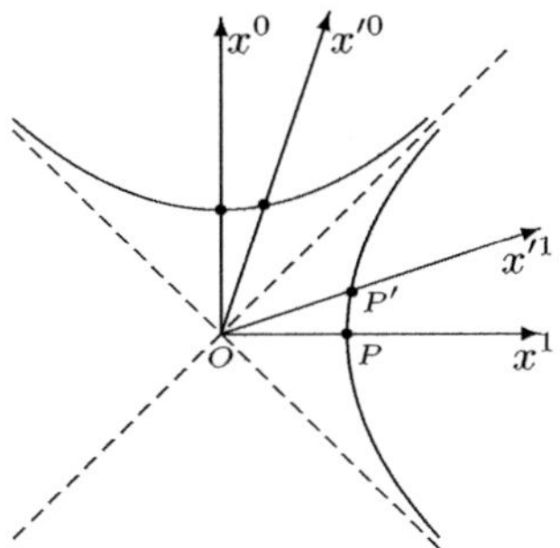

Fig. 12.10. Space-time diagram for unit lengths. Hyperbola around time axis (x^0), where $y = z = 0$:

$$(x'^0)^2 - (x'^1)^2 = (x^0)^2 - (x^1)^2 = 1 \,.$$

Hyperbola around space-axis (x^1), where $y = z = 0$:

$$(x'^0)^2 - (x'^1)^2 = (x^0)^2 - (x^1)^2 = -1$$

From the invariance of the scalar product, it follows for the unit length

$$x'^\mu x'_\mu = x^\nu x_\nu = \pm 1 \,.$$

The unit lengths in the systems S and S' are different. In S, the hyperbola intersects the x'^1-axis at $P' = (\gamma, \beta\gamma)$. The length of the unit $\overline{OP'}$ in S is $\gamma\sqrt{1 + \beta^2}$.

Lorentz contraction

Given are the systems S and S'. A ruler moves with S', its initial and final coordinates are x'_a and x'_b and its length is thus $l_0 = x'_b - x'_a$.

Measurement of the length of the ruler in S: The positions of the initial and final point x_a and x_b are determined at the same time t, where $l = x_b - x_a$, as shown in Fig. 12.11. If one expresses $x_b - x_a$ through the coordinates in S',

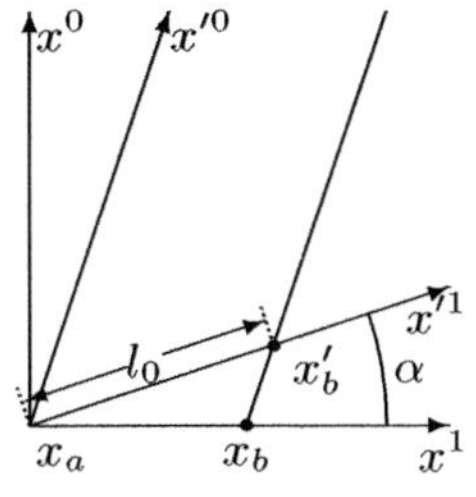

Fig. 12.11. Space-time diagram for a ruler of length l_0. x_a and x_b are measured at time t. The coordinates of the ruler in S' are:

$$x'_a = \gamma(x_a - vt), \qquad\qquad x'_b = \gamma(x_b - vt).$$

The difference yields $l = x_b - x_a = (x'_b - x'_a)/\gamma = l_0/\gamma$

one obtains the Lorentz contraction

$$l = l_0/\gamma = l_0\sqrt{1 - \beta^2} \,, \tag{12.3.1}$$

as explicitly calculated in Fig. 12.11. The contraction is measured by simultaneously determining the positions x_a and x_b of the moving ruler in system S.

From the observer's point of view in S', the marking of the starting point x'_a occurs at time $t' = 0$ and the marking of the end point x'_b at an earlier time $t' < 0$; hence the shortening. The Lorentz contraction is reciprocal. A ruler at rest in S appears contracted in S'.

Measurement times and their difference as seen from S':

$$ct'_a = \gamma(ct - \beta x_a), \quad ct'_b = \gamma(ct - \beta x_b) \quad \text{and} \quad ct'_b - ct'_a = -\gamma\beta(x_b - x_a).$$

Comparing the length $x_b - x_a$ with l_0 in Fig. 12.11, we obtain with the sine theorem[12]

$$\frac{x_b - x_a}{l_0} = \frac{\sin(\frac{\pi}{2} - 2\alpha)}{\sin(\frac{\pi}{2} + \alpha)} = \frac{\cos 2\alpha}{\cos \alpha} = \frac{\cos \alpha}{\gamma^2} = \frac{1}{\gamma^2 \sqrt{1 + \beta^2}}, \qquad \tan \alpha = \beta.$$

With the unit scales, we have seen that in Fig. 12.10 the unit $\overline{OP'}$ in S appears longer by the factor $\gamma\sqrt{1 + \beta^2}$ than the unit $\overline{OP}$. Therefore, the correct Lorentz contraction follows only through multiplication with this factor $x_b - x_a = l_0/\gamma$. These space-time diagrams thus distort the length ratios somewhat.

Time dilation

Two events t'_a and t'_b occur at the same location x' in S'. For the times and their difference in S we get

$$t_a = \gamma\left(t'_a + \frac{v}{c^2} x'\right), \quad t_b = \gamma\left(t'_b + \frac{v}{c^2} x'\right) \quad \text{and} \quad \Delta t = \gamma\left(t'_b - t'_a\right) = \gamma\tau.$$

While time τ passes in S', the longer time passes in S $\Delta t = \gamma\tau$ (see Fig. 12.12): *Moving clocks run slower.* This effect is also reciprocal.

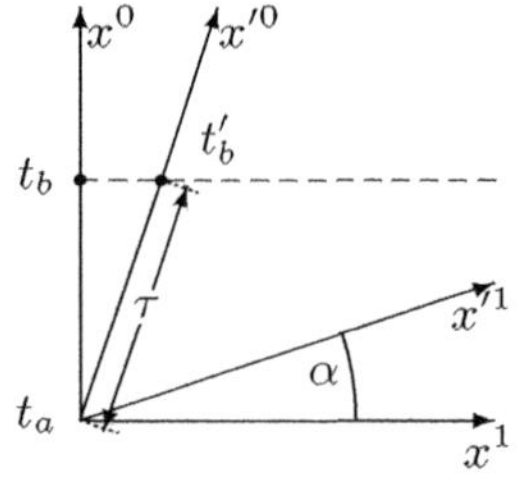

Fig. 12.12. Space-time diagram for the observation of two events in S at times t_a and t_b, which in S' took place at the same location $\mathbf{x}' = 0$ at times $t'_a = 0$ and t'_b. In S for $\Delta t = t_b - t_a$ one obtains: $\Delta t = \tau\gamma\sqrt{1 + \beta^2} \cos \alpha = \gamma\tau$. $\tau\gamma\sqrt{1 + \beta^2}$ is, due to the distortion of the scales, the length of τ in units of S and $\cos \alpha = 1/\sqrt{1 + \beta^2}$

One can see the time dilation in the lifespan of elementary particles:

$$\pi^+ - \text{Mesons } \tau_0 = 2.56 \times 10^{-8} \text{ s (at rest)}$$
$$\tau = 1.5 \ \tau_0 \approx 3.8 \ \times 10^{-8} \text{ s } (for \ v = 0.75\,c)$$

$$\mu - \quad \text{Mesons } \tau_0 = 2.2 \times 10^{-6} \text{ s}$$
$$\text{Range } \lambda_0 = c\tau_0 = 6.6 \times 10^4 \text{ cm} = 0.66\,\text{km}$$

Since $v \lesssim c$ the range is actually greater: $\lambda \approx 10 - 20\,\text{km}$[13].

---

[12] In a triangle with sides a, b, c and the angles α, β, γ opposite the sides: $\sin \alpha/a = \sin \beta/b = \sin \gamma/c$.

[13] μ-Mesons are created at a height of 10–20 km and reach the surface of the earth.

Present-future

Lengths and time differences change during a Lorentz transformation; there is no absolute simultaneity. But what is invariant is

$$s^2 = \Delta x_\mu \Delta x^\mu \qquad \text{with} \quad \Delta x^\mu = x_b^\mu - x_a^\mu \,.$$

The areas with $s^2 > 0$ (timelike) and $s^2 < 0$ (spacelike) are marked in Fig. 12.13. An LT therefore only moves within the range specified by s^2:

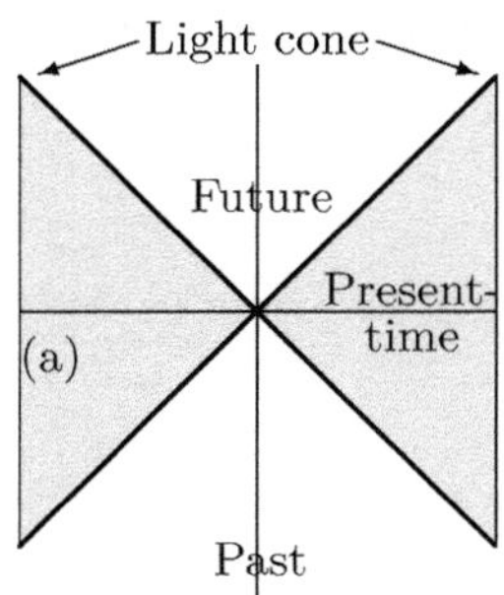

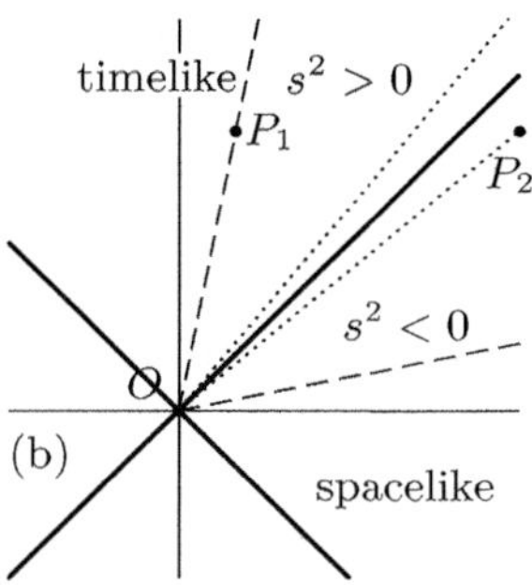

Fig. 12.13. (a) Space-time diagram with light cones.
(b) The events in O and P_1 occur in the dashed CS at the same place at different times while the events in O and P_2 in the dotted CS occur simultaneously, but at different places occur

$s^2 > 0$: timelike vector; there is a system in which the two events occur at the same place at different times

$s^2 = 0$: lightlike vector

$s^2 < 0$: spacelike vector; here there is a system in which the two events occur at the same time, but remain spatially separated

Events in the future cone can be influenced by O.
Events in the past light cone can influence the events in O.
Events in the present area are independent of O.

Proper time

The time difference and the spatial distance depend on the frame of reference. Only invariant is

$$s^2 = (x_b^\mu - x_a^\mu) g_{\mu\nu} (x_b^\nu - x_a^\nu) \,.$$

If the events are infinitesimally adjacent, $\mathrm{d}x^\mu = x_b^\mu - x_a^\mu$, the scalar product results in the invariant line element $\mathrm{d}s^2 = \mathrm{d}x^\mu \mathrm{d}x_\mu$. We assume that $\mathrm{d}x^\mu$ is timelike ($\mathrm{d}s^2 > 0$). In the frame of reference, in which the components of both events are the same, thus in the instantaneous rest system of the particle, the time difference is

$$\mathrm{d}s = c\mathrm{d}\tau \equiv \text{time difference on co-moving clock.}$$

From the definition of the line element

$$(\mathrm{d}s)^2 = (\mathrm{d}x^0)^2\left(1 - \left(\frac{\mathrm{d}\mathbf{x}}{\mathrm{d}x^0}\right)^2\right)$$

follows for timelike events

$$\mathrm{d}s = \mathrm{d}x^0\sqrt{1-\beta^2} = c\mathrm{d}t/\gamma \qquad \Longleftrightarrow \qquad \mathrm{d}\tau = \mathrm{d}t/\gamma\,. \tag{12.3.2}$$

Definition: The proper time τ is the time shown by a co-moving clock:

$$\tau = \int_0^t \mathrm{d}t'\sqrt{1-\beta^2} \le t\,. \tag{12.3.3}$$

It is better suited to Minkowski diagrams and many other cases to specify the proper time τ by the path s:

$$s = \int_0^s \mathrm{d}s' = c\int_0^t \mathrm{d}t'\sqrt{1-\beta^2} = c\tau \le ct\,. \tag{12.3.4}$$

Moving clocks run slower; this is not limited to uniform motion. We have related the time difference $\mathrm{d}t$ in a fixed system with the difference in the respective instantaneous rest system of the clock.

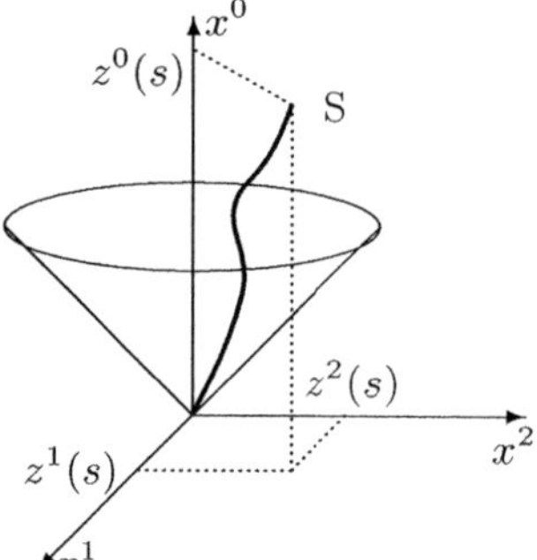

Fig. 12.14. Space-time diagram of a world line and the light cone

Moving clocks are found in satellites. The differences in rate compared to the clocks positioned on earth have to be taken into account in navigation systems like GPS [Campbell, 2006]. The effects of the satellite's relative speed of approx. $4\,\mathrm{km/s}$ are of the size $\sim 10^{-10}$; in addition, there are various corrections due to the gravitational potential, so that the onboard clocks lag by approx. $40\,\mu\mathrm{s}$ per day.

The path $\mathbf{x}(t)$ of a particle is referred to as a world line (see Fig. 12.14). For it, $\mathrm{d}x_\mu \mathrm{d}x^\mu > 0$ applies. If we know $\mathbf{x}(t)$, we can calculate the proper time from (12.3.3).

It is convenient to specify the world line of a particle not by $\mathbf{x} = \mathbf{x}(t)$, but as a function of the proper time or the path $s = c\tau$: $t = t(s)$ and $\mathbf{x} = \mathbf{x}(t(s))$.

$$(z^{\mu}(s)) = (x^{\mu}(t(s))) = (c\,t(s),\ \mathbf{x}(t(s)))$$

Four-vector of the world line.

$$(12.3.5)$$

Clock paradox

In the Clock paradox, the time of a clock that moves away from the earth at a speed β and then returns to earth is compared with the clock that remained on earth, as sketched in Fig. 12.15a. From (12.3.3)

$$t_B - t_A = \frac{1}{c} \int_{s(A)}^{s(B)} ds\, \sqrt{1 - \beta^2(s)}$$

we conclude that moving clocks run slower, but the faster movement also manifests itself in a (in the plane of the drawing) longer world line.

In other words – as the so-called *twin paradox* – the twin, who starts the journey at point A, will find at point B, that the twin remaining stationary at S has aged more than he has. If we don't consider a twin (since he cannot withstand arbitrarily large accelerations), but equalize the clocks in the starting point A in the inertial system S' and do the same from S'' at the turning point P, then when comparing the times at the end point B, the time difference $t_1 - t_2 = t_1\left(1 - \sqrt{1 - \beta^2}\right)/c$ will be noticed, which has come about without acceleration phases. t_1 is to be assigned to the clock remaining stationary in S.

The apparent paradox now is that, viewed from the inertial system S', time in system S passes more slowly and the traveling twin knows his remaining brother ages slower; to meet him, he must switch to S'' and thus cannot describe his journey as being stationary in an inertial system.

If we now, as sketched in Fig. 12.15b, consider the system S' as the rest system of the traveling twin, then this one has to set off at P according to the addition theorem for speeds with $-2\beta/(1 + \beta^2)$, to reach the stationary twin. Again, the world line of the traveling twin is "longer "and thus the elapsed time is shorter.

The strong accelerations mentioned in the twin paradox can be avoided by relying on the comfortable form of a journey with uniform acceleration/deceleration at the strength of earth's gravity ($g = 981\,\mathrm{cm/s^2}$) (see problem 12.4).

Four-velocity

Velocity is defined by

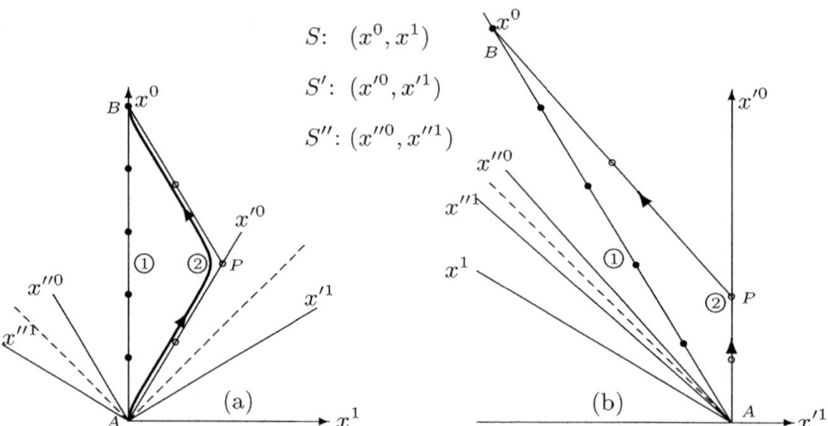

Fig. 12.15. Clock paradox; the dashed lines are light cones, and the dots are units of time of the respective resting systems: in (a) S; in (b) S'

(a) Clock 1 is at rest in S, while clock 2 moves away with $\beta = 0.6$ to P and then returns to B with $\beta' = -0.6$; the thick line indicates the same path with acceleration phases at A, P and B

(b) Initially, Clock 2 is at rest in S', while Clock 1 is moving away with $\beta = -0.6$. From the turning point P Clock 2 approaches Clock 1 with $\beta \approx -0.882$

$$\mathrm{v}^\mu(\tau) = \frac{\mathrm{d}x^\mu}{\mathrm{d}\tau} = \frac{\mathrm{d}x^\mu}{\mathrm{d}t}\frac{\mathrm{d}t}{\mathrm{d}\tau} = v^\mu \gamma \qquad \Leftrightarrow \qquad (\mathrm{v}^\mu) = \gamma \begin{pmatrix} c \\ \mathbf{v} \end{pmatrix}, \qquad (12.3.6)$$

where we have substituted $\mathrm{d}t/\mathrm{d}\tau = \gamma$ according to (12.3.2). Both for diagrams and for notation, the dimensionless four-velocity is often (analogous to β) more suitable

$$u^\mu(s) = \dot{x}^\mu(s) = \frac{\mathrm{d}x^\mu}{\mathrm{d}t}\frac{\mathrm{d}t}{\mathrm{d}s} = \frac{\mathrm{v}^\mu}{c} \qquad \Leftrightarrow \qquad (u^\mu) = \gamma \begin{pmatrix} 1 \\ \boldsymbol{\beta} \end{pmatrix}. \qquad (12.3.7)$$

As shown in Fig. 12.16, the instantaneous velocity of a particle relative to an inertial system S can be read from the slope of the world line: (u^μ) is a unit vector that is tangent to the world line. From the scalar product follows

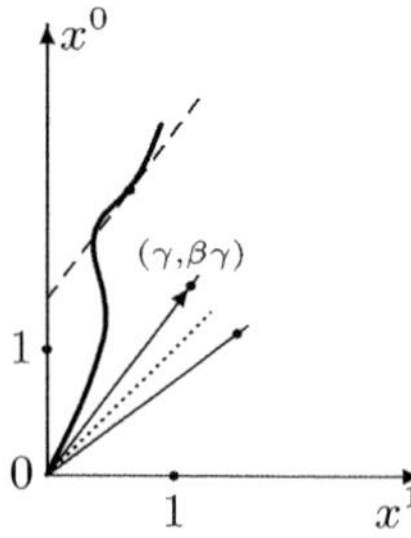

Fig. 12.16. Four-velocity at a point on a world line: The tangent at the point is moved parallel to the origin and there it results in the instantaneous moving system S'. The unit length of the time axis of S', i.e., the point $(\gamma, \beta\gamma)$ determines the four-velocity u (here $\beta = 0.75$)

$u^\mu u_\mu = 1$, which can be taken from Fig. 12.16, where u is the length of the unit scale in the (local) moving system.

Note: In the following, the derivatives $\dot{x}^\mu(t)$, $\dot{x}^\mu(\tau)$ or $\dot{x}^\mu(s)$ indicate by the argument whether it is derived after t, the proper time τ or the path s. Thus, it applies

$$\dot{x}^\mu(t) = \frac{\mathrm{d}x^\mu}{\mathrm{d}t}, \quad \dot{x}^\mu(\tau) = \frac{\mathrm{d}x^\mu}{\mathrm{d}\tau} = v^\mu = \gamma\dot{x}^\mu(t), \quad \dot{x}^\mu(s) = \frac{\mathrm{d}x^\mu}{\mathrm{d}s} = u^\mu = \frac{\gamma}{c}\dot{x}^\mu(t). \quad (12.3.8)$$

This rule is naturally not limited to x^μ.

12.3.3 Observation of Fast Moving Bodies

The Lorentz contraction of bodies along their direction of motion, already postulated before the establishment of the theory of relativity by FitzGerald [1889], was long assumed to be directly observable without further ado.

The finite speed of light means that light that arrives at the observer (camera) at a certain time, has originated from different places of a fast moving body, at different times, i.e. from different positions of the body. This has, with or without Lorentz contraction, a changed appearance as a result, and the effects of the time differences are usually larger than those of Lorentz contraction, as they are of first order in β.

In the article *How does a moving rod appear to a stationary observer according to the theory of relativity* this was discussed by Lampa [1924]. Surprisingly, the work was not noticed, so that Gamow [1940] could make an incorrect representation of moving bodies.

The picture was only corrected with articles by Terrell [1959] and Penrose [1959], who showed that a passing ball always appears as a ball to the observer or the camera, albeit twisted. Subsequently, numerous articles on this topic were written and it was included in textbooks [Ruder, 1993] A distinction is made between effects that affect the shape and those that affect brightness and color, with only the former being discussed here.

Since the appearance of rapidly moving bodies does not directly relate to the theory of relativity and especially to electrodynamics we will limit ourselves to outlining the simplest effects.

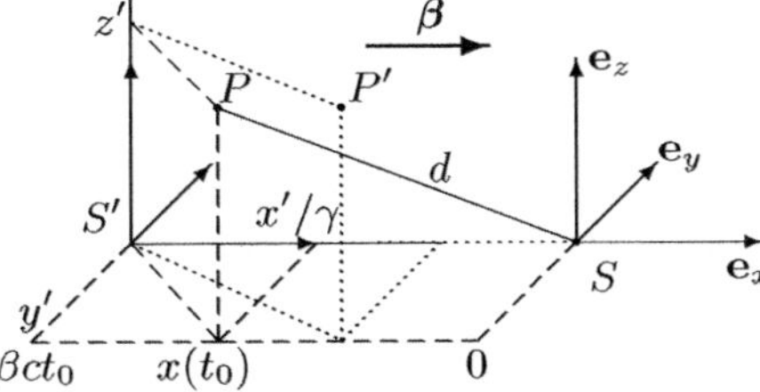

Fig. 12.17. A light source is moving with $\beta = \beta\mathbf{e}_x$ towards S; From the system S' it is always at the point P'; from S it is at the location P at time t_0 at the location P. d is the distance that the light has to travel to the point S, where the camera is

In Fig. 12.17 a point-like light source $P'(\mathbf{x}')$ is shown, which is moving with β towards S and whose light beam reaches the camera at S at time t_1. The location $x(t_0)$ of the light source is to be determined when at time t_1 is photographed:

$$x(t_0) = \frac{x'}{\gamma} + \beta\, ct_0\,, \qquad\qquad y(t_0) = y'\,, \qquad\qquad z(t_0) = z'\,.$$

t_0 is negative here, as S' is flying towards S. At time t_1 the light emitted at t_0 reaches the point S:

$$ct_1 = ct_0 + d \quad \Rightarrow \quad x(t_1) = \beta ct_1 + \frac{x'}{\gamma} = x(t_0) + \beta\sqrt{x^2(t_0) + y^2 + z^2}\,.$$

One determines $x(t_0)$ as a function of t_1 or $x(t_1)$ and can thus specify the location of the light source at time t_1:

$$x(t_0) = \gamma^2\, x(t_1) - \beta\gamma\sqrt{\gamma^2 x^2(t_1) + y^2 + z^2} \quad \text{with} \quad x(t_1) = \beta ct_1 + \frac{x'}{\gamma}\,.$$

The sign of the root results from $x(t_0) < x(t_1)$. Thus, for any body given in S', its appearance in S can be determined at any time t_1 [Kern et. al., 1997].

Approaching rod

An approaching, glowing rod, as shown in Fig. 12.18a, is observed (filmed) from a point P. The light from the points further out has a longer path to P and therefore comes from earlier times, when the rod was further away from P. Thus, one sees in P a hyperbola of the form

$$\frac{(x-x_0)^2}{a^2} + \frac{y^2}{b^2} = 1, \qquad x_0 = \frac{-\varrho_0}{1+\beta}, \qquad a = \frac{\varrho_0\,\beta}{1+\beta}, \qquad b = \varrho_0\sqrt{\frac{1-\beta}{1+\beta}}\,.$$

If the rod does not move centrally towards P, but flies past it ($y > 0$), it appears rotated at a greater distance. In the longitudinal direction, as sketched in Fig. 12.18b, the rod flying towards P appears elongated/shortened, according to

$$l = l_0\sqrt{1 \pm \beta}/\sqrt{1 \mp \beta}\,.$$

This only applies if P lies in the rod axis and not, as in Fig. 12.18b, slightly next to it, which is necessary if you want to observe the elongation (shortening).

Note: The rod flying towards the observer seems to move faster than the one moving away. We consider a grid of length $s = c\Delta t$. So if the light travels the distance s, the rod has covered $s\beta$. For the observer in P, this means that the approaching rod after $(1-\beta)c\Delta t$ has passed the next mark of the grid, or that the rod apparently approaches point P at the speed $s/\Delta t = 1/(1-\beta)$. This can lead to apparent superluminal speeds, as observed in quasars [Kraus, 2005, p. 40]. Conversely, the receding rod seems to move slower.

It should also be noted that a fast-moving camera can also see "backwards". Light rays, whose speed parallel to the camera is less than that of the camera ($\mathbf{c}\cdot\hat{\boldsymbol{\beta}} < \beta$), are overtaken by the camera and thus brought onto the film.

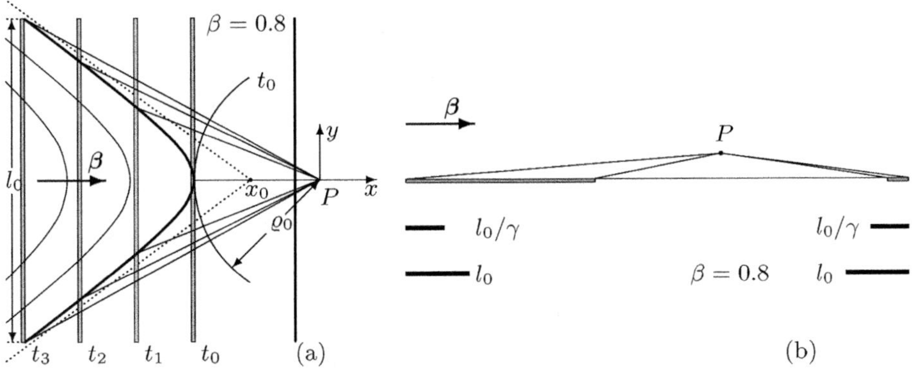

Fig. 12.18. (a) A rod of length l_0 flying transversely towards the observer at $\beta = 0.8$ is seen as a hyperbola at P. The light rays emanating from the two points of the rod at times t_i all hit the circle at time t_0 and thus arrive simultaneously at P; the dotted line is the asymptote, outside of which the light beam directed towards P moves slower parallel to β than the rod. At the time of observing the hyperbola, the rod is already very close to P (black line)

(b) The longitudinally oriented rod appears elongated when it flies towards P and shortened when it moves away; l_0 is the rest length and l_0/γ the measured, Lorentz-contracted length

Terrell rotation

We have already seen in the example of the rod that approaching/receding bodies appear larger/smaller in snapshots, also twisted when they fly by sideways and distorted when they are close. The rod flying by sideways, when the camera is perpendicular to the direction of motion, will show the Lorentz contraction.

How the image changes when a three-dimensional body, a cube, flies by, is sketched in Fig. 12.19. It is assumed that all light rays emanating from the cube and coming to observer P at a certain point in time, can be considered parallel.

The bodies are then rather small and far enough away. At time t_1 the corner point D is in position D_1. Light that is emitted from there reaches the front side of the cube after $c(t_2 - t_1) = l_0$.

Thus, all rays that originate from $\overline{D_1 A_2}\,\overline{A_2 B_2}$ arrive at the same time at the (infinitely distant) observer P.

In Fig. 12.19c it is shown that the observed image corresponds to a cube rotated by the angle α: $\sin\alpha = \beta$ and $\cos\alpha = \gamma^{-1}$.

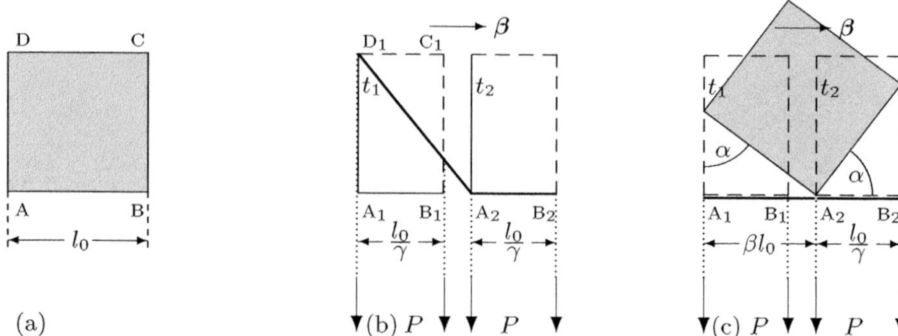

Fig. 12.19. Cube of length l_0; The camera P is far enough away so that all rays are parallel. The cube moves with $\beta = 0.8$; visible are the back and bottom surface; the front surface is hidden.

(a) Side surface of a cube of length l_0 in the rest system

(b) The light rays that originate from the solid line in the interval $[t_1, t_2]$ from the back surface, all arrive simultaneously with the rays from the bottom surface at P

(c) The image on the film can be interpreted as a rotated cube: $\tan \alpha = \beta\gamma$

12.4 Composition of Lorentz Transformations

12.4.1 Lorentz transformation for arbitrary orientation of the relative velocity

The restricted LT ($\Lambda^0{}_0 = 1$ and $\det \Lambda = 1$) consists of a rotation R and a velocity transformation, the boost, together. It is characterized by the velocity $\boldsymbol{\beta} = \mathbf{v}/c$ and the rotation $\boldsymbol{\alpha}$:

$$\Lambda = \Lambda(\boldsymbol{\beta}, \boldsymbol{\alpha}) \equiv \Lambda(\mathbf{0}, \boldsymbol{\alpha})\,\Lambda(\boldsymbol{\beta}, \mathbf{0}) = \Lambda(\boldsymbol{\beta}', \mathbf{0})\,\Lambda(\mathbf{0}, \boldsymbol{\alpha}). \tag{12.4.1}$$

If the rotation $R(\boldsymbol{\alpha})$ is performed first, then $\boldsymbol{\beta}$ is also rotated: $\boldsymbol{\beta}' = R\boldsymbol{\beta}$.

Boost

The movement from S' to S is not limited to the x-direction as sketched in Fig. 12.20 and calculated on page 453.

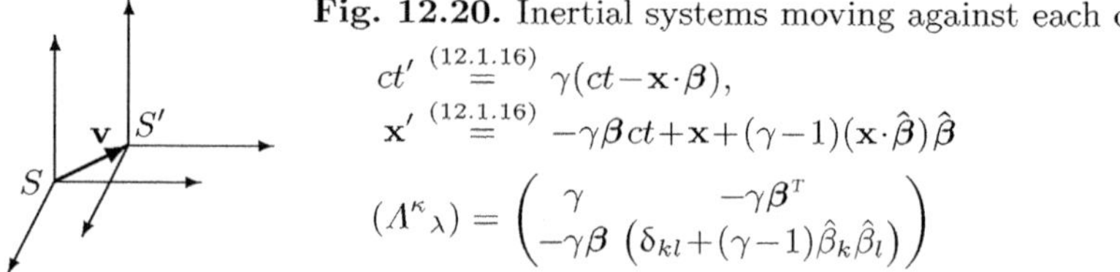

Fig. 12.20. Inertial systems moving against each other with $\mathbf{v}$

$$ct' \overset{(12.1.16)}{=} \gamma(ct - \mathbf{x}\cdot\boldsymbol{\beta}),$$
$$\mathbf{x}' \overset{(12.1.16)}{=} -\gamma\boldsymbol{\beta}ct + \mathbf{x} + (\gamma - 1)(\mathbf{x}\cdot\hat{\boldsymbol{\beta}})\hat{\boldsymbol{\beta}} \tag{12.4.2}$$

$$(\Lambda^\kappa{}_\lambda) = \begin{pmatrix} \gamma & -\gamma\boldsymbol{\beta}^T \\ -\gamma\boldsymbol{\beta} & \left(\delta_{kl} + (\gamma-1)\hat{\beta}_k\hat{\beta}_l\right) \end{pmatrix}$$

Matrix representation: Using (12.4.2) we obtain from $(x'^\mu) = \Lambda\,(x^\mu)$:

$$\Lambda(\boldsymbol{\beta}, \mathbf{0}) = \begin{pmatrix} \gamma & -\gamma \boldsymbol{\beta}^T \\ -\gamma \boldsymbol{\beta} & \mathsf{E} + (\gamma - 1)\hat{\boldsymbol{\beta}} \circ \hat{\boldsymbol{\beta}} \end{pmatrix}. \tag{12.4.3}$$

The symbol $\circ$ denotes the tensorial (dyadic) product (A.1.15). For small values of β, $\gamma - 1 \approx \beta^2/2$.

Rotation

The rotation here is an operation that only acts on the spatial dimensions of the LT: $\mathbf{x}' = \mathsf{R}\mathbf{x}$. It can be understood as

1. an abstract operation that leaves the scalar product $\mathbf{x} \cdot \mathbf{x} = \mathbf{x}' \cdot \mathbf{x}'$ invariant, as
2. a matrix operation in which a point P is rotated relative to a fixed basis (active rotation), or as
3. a matrix operation in which the basis is twisted relative to a space-fixed point (passive rotation).

In the notation used here, no distinction is made between active and passive rotation[14]. If $\boldsymbol{\alpha}$ is the arbitrarily oriented axis of rotation and α is the angle of rotation given in the mathematically positive sense (right-hand rule), then [§1.3, Sexl, Urbantke, 2001; Iro, 2015, (8.13)]

$$\mathbf{x}' = \mathbf{x} \cos\alpha + (\hat{\boldsymbol{\alpha}} \cdot \mathbf{x})\hat{\boldsymbol{\alpha}}(1 - \cos\alpha) + \hat{\boldsymbol{\alpha}} \times \mathbf{x} \sin\alpha, \qquad \hat{\boldsymbol{\alpha}} = \boldsymbol{\alpha}/\alpha. \tag{12.4.4}$$

The matrix of a rotation is

$$\begin{pmatrix} 1 & \mathbf{0}^T \\ \mathbf{0} & \mathsf{R} \end{pmatrix} \quad \text{with} \quad R^k{}_l = \delta_{kl}\cos\alpha + \hat{\alpha}_k\hat{\alpha}_l(1 - \cos\alpha) + \epsilon_{kjl}\,\hat{\alpha}_j \sin\alpha. \tag{12.4.5}$$

Since $\boldsymbol{\beta}$ and $\boldsymbol{\alpha}$ are vectors from the three-dimensional Euclidean space, we do not distinguish between co- and contravariant components in these. Rotation matrices are orthogonal, i.e. $\mathsf{R}^T = \mathsf{R}^{-1}$. To determine the angle of rotation, the invariance of the trace can be used:

$$\cos\alpha = \frac{1}{2}\Big(\sum_k R^k{}_k - 1\Big), \tag{12.4.6}$$

where the angle of rotation varies between $0 \le \alpha \le \pi$. Rotations with $\alpha > \pi$ are described with $\alpha' = 2\pi - \alpha$.

The axis of rotation then changes direction: $\boldsymbol{\alpha}' = -\boldsymbol{\alpha}$. This uniquely describes the rotation, except for $\alpha = \pi$, where $\Lambda(\boldsymbol{\beta}, \boldsymbol{\alpha}) = \Lambda(\boldsymbol{\beta}, -\boldsymbol{\alpha})$. The axis of rotation is determined from the difference $R^k{}_l - R^l{}_k$:

$$\hat{\alpha}_i \sin\alpha = -\frac{1}{4}\epsilon_{ikl}\big(R^k{}_l - R^l{}_k\big). \tag{12.4.7}$$

---

[14] One can distinguish between the rotation of the vector $x_k \to x_k'$ and the rotation of the CS: $\mathbf{e}_k \to \mathbf{e}_k' \Leftrightarrow x_k \to x_{k'}$.

Boost and rotation

The general LT (12.4.1) consists of boost and rotation [Sexl, Urbantke, 2001, §6.1]

$$\begin{pmatrix} \bar{x}^0 \\ \bar{\mathbf{x}} \end{pmatrix} = \begin{pmatrix} 1 & \mathbf{0}^T \\ \mathbf{0} & \mathsf{R} \end{pmatrix} \begin{pmatrix} \gamma & -\gamma\boldsymbol{\beta}^T \\ -\gamma\boldsymbol{\beta} & \mathsf{E}+(\gamma-1)\hat{\boldsymbol{\beta}}\circ\hat{\boldsymbol{\beta}} \end{pmatrix} \begin{pmatrix} x^0 \\ \mathbf{x} \end{pmatrix}$$

$$= \begin{pmatrix} \gamma & -\gamma\boldsymbol{\beta}^T \\ -\gamma\mathsf{R}\boldsymbol{\beta} & \mathsf{R}+(\gamma-1)(\mathsf{R}\hat{\boldsymbol{\beta}})\circ\hat{\boldsymbol{\beta}} \end{pmatrix} \begin{pmatrix} x^0 \\ \mathbf{x} \end{pmatrix}. \tag{12.4.8}$$

If rotation is performed first, the result remains unchanged because $\boldsymbol{\beta}'=\mathsf{R}\boldsymbol{\beta}$. Of particular interest is the reverse path, i.e., deducing velocity and rotation from the elements of a given LT. We assume that Λ does not contain any time or space reflections[15]:

$$\gamma = \Lambda^0{}_0, \qquad \beta_k = -\Lambda^0{}_k/\gamma, \qquad \sum_k \left(\Lambda^0{}_k\right)^2 = \beta^2\gamma^2. \tag{12.4.9}$$

$$\mathsf{R} = \Lambda - \frac{\gamma-1}{\beta^2}(\mathsf{R}\boldsymbol{\beta})\circ\boldsymbol{\beta} \qquad \Leftrightarrow \qquad R^k{}_l = \Lambda^k{}_l - \frac{1}{\gamma+1}\Lambda^k{}_0\Lambda^0{}_l. \tag{12.4.10}$$

On the right side we have inserted

$$(\gamma - 1)/\beta^2 = \gamma^2/(\gamma + 1). \tag{12.4.11}$$

The detailed calculation is problem 12.5. Rotation angle and rotation axis are to be determined by (12.4.6) or (12.4.7).

12.4.2 Addition of Velocities

General velocity addition theorem

In the inertial system S', a particle moves with $\boldsymbol{\beta}' = \mathbf{v}'/c$, as sketched in Fig. 12.21. S' in turn moves with $\boldsymbol{\beta} = \mathbf{v}/c$ relative to S. The velocity $\boldsymbol{\beta}'' = \mathrm{d}\mathbf{x}/\mathrm{d}x^0$, with which the particle moves in S, is sought.

First, the reverse transformation of (12.4.2) is needed, which is obtained by replacing $\mathbf{v} \to -\mathbf{v}$ in (12.4.2):

$$x^0 = \gamma(x'^0 + \boldsymbol{\beta}\cdot\mathbf{x}'), \tag{12.4.12}$$

$$\mathbf{x} = \gamma\boldsymbol{\beta}x'^0 + \mathbf{x}' + (\gamma-1)(\hat{\boldsymbol{\beta}}\cdot\mathbf{x}')\hat{\boldsymbol{\beta}}.$$

At time $t'=0$ the particle was at the origin of S', so that $\mathbf{x}'=\boldsymbol{\beta}'x'^0$:

$$x^0 = \gamma(x'^0 + \boldsymbol{\beta}\cdot\boldsymbol{\beta}'\,x'^0), \tag{12.4.13}$$

$$\mathbf{x} = \gamma\boldsymbol{\beta}x'^0 + \boldsymbol{\beta}'x'^0 + (\gamma-1)(\hat{\boldsymbol{\beta}}\cdot\boldsymbol{\beta}')\hat{\boldsymbol{\beta}}\,x'^0.$$

---

[15] For $L^0{}_0 < 0$ a time reflection T must be performed first; the same applies to space reflection.

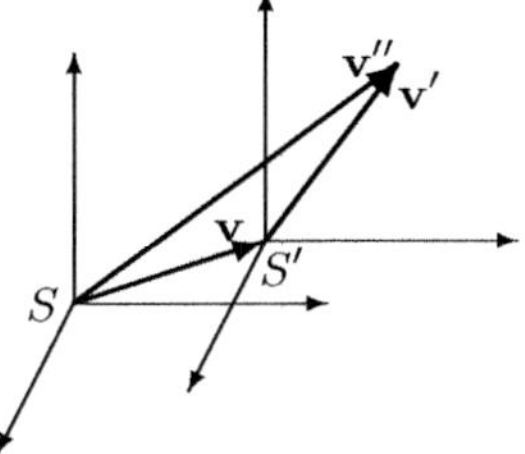

Fig. 12.21. In S' a particle moves with $\mathbf{v}'$, while S' opposite S the speed $\mathbf{v}$ has. The speed $\mathbf{v}'' = \mathrm{d}\mathbf{x}/\mathrm{d}t$ of the particle in S is to be calculated.
Not sketched is the shortening of the axes seen from S of S' (and vice versa); so the particle would also be in a different place if swapped from $\mathbf{v} \leftrightarrows \mathbf{v}'$ at a different place

From the first equation, as a result of time dilation, one obtains

$$\frac{\mathrm{d}x'^0}{\mathrm{d}x^0} = \frac{\mathrm{d}t'}{\mathrm{d}t} = \frac{1}{\gamma(1 + \boldsymbol{\beta}\cdot\boldsymbol{\beta}')} \, . \tag{12.4.14}$$

With the chain rule $\beta'' = \dfrac{\mathrm{d}\mathbf{x}}{\mathrm{d}x^0} = \dfrac{\mathrm{d}\mathbf{x}}{\mathrm{d}x'^0} \dfrac{\mathrm{d}x'^0}{\mathrm{d}x^0}$ results from (12.4.13)

$$\beta'' = \frac{\gamma\boldsymbol{\beta} + \boldsymbol{\beta}' + (\gamma - 1)(\hat{\boldsymbol{\beta}} \cdot \boldsymbol{\beta}')\hat{\boldsymbol{\beta}}}{\gamma(1 + \boldsymbol{\beta} \cdot \boldsymbol{\beta}')} = \frac{\boldsymbol{\beta} + \boldsymbol{\beta}'_\parallel + \boldsymbol{\beta}'_\perp/\gamma}{1 + \boldsymbol{\beta} \cdot \boldsymbol{\beta}'} \, . \tag{12.4.15}$$

(12.4.15) is the *general velocity addition theorem* for the addition of arbitrarily directed velocities. The right-hand expression is easier to remember, as first $\boldsymbol{\beta}'_\parallel = (\boldsymbol{\beta}'\cdot\hat{\boldsymbol{\beta}})\hat{\boldsymbol{\beta}}$ is added to $\boldsymbol{\beta}$ according to the "usual" addition theorem and then $\boldsymbol{\beta}'_\perp = \hat{\boldsymbol{\beta}}\times(\boldsymbol{\beta}' \times \hat{\boldsymbol{\beta}})$ is added, whereby the factor γ^{-1} is due to time dilation, as mentioned below.

If one swaps in Fig. 12.21 $\mathbf{v} \leftrightarrows \mathbf{v}'$, then $|\boldsymbol{\beta}''|$ remains unchanged, but the direction of $\boldsymbol{\beta}''$ is different. We will address this when multiplying two Lorentz transformations, page 482.

Addition of parallel velocities

If now $\boldsymbol{\beta}'$ is parallel to $\boldsymbol{\beta}$, then (12.4.15) simplifies to the *velocity addition theorem*

$$\beta'' = \frac{\beta + \beta'}{1 + \beta\beta'}\,\hat{\boldsymbol{\beta}} \qquad \Leftrightarrow \qquad v'' = \frac{v + v'}{1 + vv'/c^2} \, , \tag{12.4.16}$$

where v and v' have opposite signs for antiparallel motion. If $v/c = \beta = \tanh\eta$, then the velocity addition reads

$$\beta'' = \tanh\eta'' = \tanh(\eta + \eta').$$

Since $|\tanh\eta| \leq 1$, the statement of this equation is that the addition of two velocities, each of which is less than or equal to the speed of light, results in a speed $v'' \leq c$. The speed of light is therefore a limiting speed. It holds

$$\begin{aligned}
&\text{for} \quad v, v' \ll c &&\Rightarrow \quad v'' = v + v', \\
&\text{for} \quad v \ \text{and/or} \ v' = c &&\Rightarrow \quad v'' = \frac{v + v'}{1 + vv'/c^2} = c.
\end{aligned}$$

Addition of orthogonal velocities

If $\beta' \perp \beta$, then (12.4.15) simplifies to

$$\beta'' = \beta + \beta'/\gamma. \tag{12.4.17}$$

Since in S' the components perpendicular to β are unshortened, it is understandable, that the slowing down of the component of the speed perpendicular to β is due to time dilation: $\mathrm{d}x'/\mathrm{d}t' = \mathrm{d}x'/(\gamma \mathrm{d}t)$.

Unattainability of the speed of light

A consequence of the velocity addition (12.4.16) was that c is a limit speed, which cannot be reached by adding two speeds $\beta = v/c = 1 - \epsilon$ and $\beta' = v'/c = 1 - \epsilon'$:

$$\beta'' = \frac{\beta + \beta'}{1 + \beta\beta'} = \frac{2 - \epsilon - \epsilon'}{2 - \epsilon - \epsilon' + \epsilon\epsilon'} < 1.$$

In a rather cumbersome calculation, one obtains from (12.4.15)

$$\gamma'' = \frac{1}{\sqrt{1 - \beta''^2}} = \gamma\gamma'(1 + \boldsymbol{\beta} \cdot \boldsymbol{\beta}').$$

The simpler way would be the multiplication of two boosts (12.4.21). It is immediately apparent that β'' has its maximum value when $\boldsymbol{\beta}\|\boldsymbol{\beta}'$ and its minimum when $\boldsymbol{\beta}$ and $\boldsymbol{\beta}'$ are antiparallel. For parallel velocities, we have already shown that c is a limit speed. We can also verify that $\gamma'' \geq 1$, regardless of the directions of the velocities:

$$\gamma''^2_{\min} = \frac{(1 - \beta\beta')^2}{(1 - \beta^2)(1 - \beta'^2)} \geq 1.$$

Therefore, a superluminal speed is not reached. This would result in acausal behavior, as a signal, which is sent out at time $t' = t = 0$, would arrive at time $t'(P) < 0$ in $x'(P)$.

Velocity of a particle in different inertial systems

A particle moves in S with the velocity $\mathbf{w} = \mathrm{d}\mathbf{x}/\mathrm{d}t$. In the system S' moving against S with $\mathbf{v}$, the particle has the velocity $\mathbf{w}' = \mathrm{d}\mathbf{x}'/\mathrm{d}t$.

This is exactly the situation we encountered in the velocity addition (see Fig. 12.21, page 479) had: $\mathbf{w}\,\widehat{=}\,\mathbf{v}''$ and $\mathbf{w}'\,\widehat{=}\,\mathbf{v}'$. Using the general addition theorem for velocities (12.4.15) we can specify $\mathbf{w} = \mathbf{v} + \mathbf{w}'$, remembering that this addition is generally not commutative:

$$w_\| = \frac{w'_\| + v}{1 + \mathbf{v} \cdot \mathbf{w}'/c^2}, \qquad \mathbf{w}_\perp = \frac{\mathbf{w}'_\perp}{\gamma(1 + \mathbf{v} \cdot \mathbf{w}'/c^2)}, \tag{12.4.18}$$

$$w'_\| = \frac{w_\| - v}{1 - \mathbf{v} \cdot \mathbf{w}/c^2}, \qquad \mathbf{w}'_\perp = \frac{\mathbf{w}_\perp}{\gamma(1 - \mathbf{v} \cdot \mathbf{w}/c^2)}. \tag{12.4.19}$$

When swapping $\mathbf{w} \rightleftharpoons \mathbf{w}'$, only $\mathbf{v}$ needs to be replaced by $-\mathbf{v}$.

Aberration of light

Light is emitted from a body in the direction $\mathbf{w}'$ with $w' = c$. This body moves with $\mathbf{v} = v\mathbf{e}_x$ towards an observer, for whom the light seems to come from the direction $\mathbf{w}$ ($w = c$). As long as $\mathbf{v}$ is not parallel (or antiparallel) to $\mathbf{w}'$, the light is deflected, which means that $\mathbf{w}$ and $\mathbf{w}'$ are not parallel. This deflection from $\mathbf{w}'$ to $\mathbf{w}$ is referred to as aberration.

So given are two reference systems, S and S', with S' moving against S with $\mathbf{v}$. A particle moves with $\mathbf{w}'$ relative to the system S'; in S it then has the speed $\mathbf{w} = \mathbf{v} + \mathbf{w}'$.

The relationship of $\mathbf{w}$ with $\mathbf{w}'$ is given by (12.4.18). Of interest here are the different angles that $\mathbf{w}$ and $\mathbf{w}'$ make with the x-axis. Fig. 12.22 sketches the special case with $\alpha = 90°$. It applies

$$\tan\alpha = \frac{w_z}{w_x} \qquad \begin{cases} w_x = w\cos\alpha \\ w_y = 0 \\ w_z = w\sin\alpha \end{cases}, \qquad \tan\alpha' = \frac{w'_z}{w'_x} \qquad \begin{cases} w'_x = w'\cos\alpha' \\ w'_y = 0 \\ w'_z = w'\sin\alpha' \end{cases},$$

$$\tan\alpha = \frac{w_z}{w_x} = \frac{w'_z}{\gamma(w'_x + v)} = \frac{w'\sin\alpha'}{w'\cos\alpha' + v}\sqrt{1 - \beta^2},$$

$$\tan\alpha' = \frac{w'_z}{w'_x} = \frac{w_z}{\gamma(w_x - v)} = \frac{w\sin\alpha}{w\cos\alpha - v}\sqrt{1 - \beta^2}.$$

A light beam falling perpendicularly in S is considered

$$w = c, \qquad\qquad \alpha = \frac{\pi}{2}, \qquad\qquad \tan\alpha' = -\frac{c}{v}\sqrt{1 - \frac{v^2}{c^2}}.$$

In the system S' the beam no longer falls in vertically, but is around the

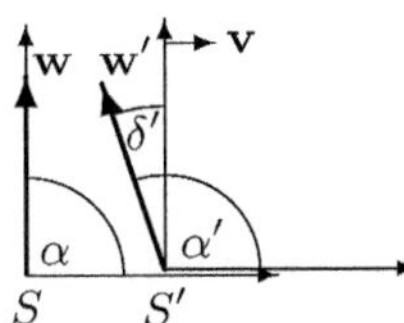

Fig. 12.22. S' moves against S with $\mathbf{v} = v\mathbf{e}_x$. A particle, which moves with $\mathbf{w}'$ in the xz-plane in S' has in S the speed $\mathbf{w}$. It is assumed here that the particle is a beam of light, which in S is observed under $\alpha = 90°$. It appears rotated by δ'

angle $\delta' = \alpha' - \frac{\pi}{2}$ rotated, which is referred to as aberration.

$$\tan\delta' = \tan(\alpha' - \frac{\pi}{2}) = \frac{-\cos\alpha'}{\sin\alpha'} = \frac{-1}{\tan\alpha'} = \frac{v/c}{\sqrt{1 - v^2/c^2}} \approx \frac{v}{c} \approx 10^{-4}.$$

Headlight effect

If an observer moves very fast, he will see the environment in front of him in a small angular range around the direction of $\mathbf{v}$ appears brighter than the lateral areas.

The parts behind him, on the other hand, appear darker. It's like illuminating the surroundings in the direction of movement with a headlight (headlight effect).

We are considering here a very fast object passing close by. Its speed is $\mathbf{v} = v\,\mathbf{e}_x$ and the light $\mathbf{w}'$ emitted by it at the angle α' has in S the direction $\mathbf{w}$ and is there, as in Fig. 12.22 sketched, observed at the angle α ($\beta = v/c$)

$$\tan\alpha = \frac{w_z}{w_x} = \frac{c\sin\alpha'}{c\cos\alpha' + v}\sqrt{1-\beta^2} \overset{\alpha'\ll 1}{\approx} \tan\alpha'\,\frac{1}{1+\beta}\sqrt{1-\beta^2} = \tan\alpha'\sqrt{\frac{1-\beta}{1+\beta}}.$$

The angular range $[-\alpha', \alpha']$ from S' is thus mapped into a $\sqrt{(1-\beta)/(1+\beta)}$ smaller range and therefore correspondingly brighter. Conversely, the fleeing object ($v \to -v$) is seen darker by the same factor. It should be mentioned that the additionally occurring Doppler effect shifts the light of approaching objects to higher frequencies and that of fleeing objects to lower ones.

12.4.3 Multiplication of two Boosts

Here again, the general velocity addition theorem (12.4.15) from the multiplication of two boosts is derived (see Fig. 12.23) and in particular the rotation associated with the addition is calculated:

$$\begin{pmatrix} \gamma'' & -\mathbf{a}^T \\ -\mathbf{b} & \mathsf{D} \end{pmatrix} = \begin{pmatrix} \gamma' & -\gamma'\boldsymbol{\beta}'^T \\ -\gamma'\boldsymbol{\beta}' & \mathsf{E}+(\gamma'-1)\hat{\boldsymbol{\beta}}'\circ\hat{\boldsymbol{\beta}}' \end{pmatrix}\begin{pmatrix} \gamma & -\gamma\boldsymbol{\beta}^T \\ -\gamma\boldsymbol{\beta} & \mathsf{E}+(\gamma-1)\hat{\boldsymbol{\beta}}\circ\hat{\boldsymbol{\beta}} \end{pmatrix}$$

$$= \begin{pmatrix} \gamma'' & -\gamma''\boldsymbol{\beta}''^T \\ -\gamma''\mathsf{R}\boldsymbol{\beta}'' & \mathsf{R}+(\gamma''-1)(\mathsf{R}\hat{\boldsymbol{\beta}}'')\circ\hat{\boldsymbol{\beta}}'' \end{pmatrix}. \tag{12.4.20}$$

The multiplication of the two boosts results in the first line

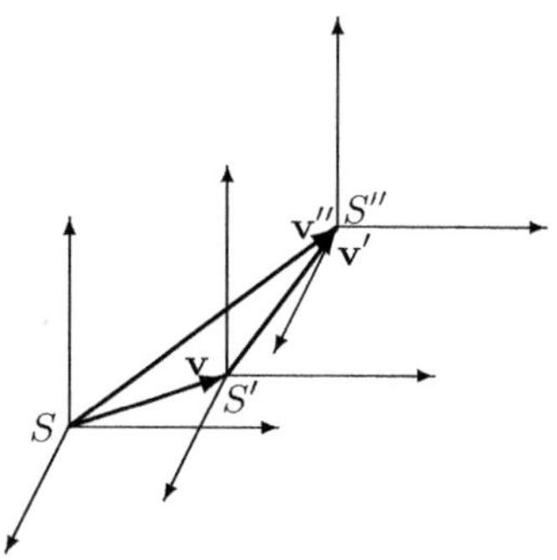

Fig. 12.23. S' is moving relative to S with $\mathbf{v}$ and S'' relative to S' with $\mathbf{v}'$; as seen from S, the axes of S'' are not only shortened, but also rotated, which is not shown

$$\gamma'' = \gamma'\gamma(1 + \boldsymbol{\beta}'\cdot\boldsymbol{\beta}), \tag{12.4.21}$$

$$\boldsymbol{\beta}'' = \frac{\mathbf{a}}{\gamma''} = \frac{1}{\gamma\gamma'(1 + \boldsymbol{\beta}\cdot\boldsymbol{\beta}')}\,\gamma'\left[\gamma\boldsymbol{\beta} + \boldsymbol{\beta}' + (\gamma-1)(\hat{\boldsymbol{\beta}}\cdot\boldsymbol{\beta}')\hat{\boldsymbol{\beta}}\right].$$

With this, we have simply derived the general velocity addition theorem and at the same time shown that β'' does not depend on the order of the boosts. The calculation of the remaining lines, especially of D is somewhat tedious[16]:

$$\mathbf{b} = \gamma\left[\boldsymbol{\beta} + \gamma'\boldsymbol{\beta}' + (\gamma'-1)(\boldsymbol{\beta}\cdot\hat{\boldsymbol{\beta}}')\hat{\boldsymbol{\beta}}'\right], \tag{12.4.22}$$

$$\mathsf{D} = \mathsf{E} + (\gamma-1)(\hat{\boldsymbol{\beta}}\circ\hat{\boldsymbol{\beta}}) + (\gamma'-1)(\hat{\boldsymbol{\beta}}'\circ\hat{\boldsymbol{\beta}}') + \gamma\gamma'\left[1 + \frac{\gamma}{\gamma+1}\,\frac{\gamma'}{\gamma'+1}\,\boldsymbol{\beta}\cdot\boldsymbol{\beta}'\right]\boldsymbol{\beta}'\circ\boldsymbol{\beta}.$$

The rotation is given by (12.4.10)

$$\mathsf{R} = \mathsf{D} - (\gamma''-1)(\mathsf{R}\hat{\boldsymbol{\beta}}'')\circ\hat{\boldsymbol{\beta}}'' = \mathsf{D} - \frac{1}{\gamma''+1}\,\mathbf{b}\circ\mathbf{a} \tag{12.4.23}$$

given. The axis of rotation can be determined from the antisymmetric part of the rotation matrix (12.4.5). The terms must be of the form

$$\left[\boldsymbol{\beta}\circ\boldsymbol{\beta}' - \boldsymbol{\beta}'\circ\boldsymbol{\beta}\right]_{ij} = \epsilon_{ijk}\left[\boldsymbol{\beta}\times\boldsymbol{\beta}'\right]_k.$$

$\boldsymbol{\alpha} \propto \pm\boldsymbol{\beta}\times\boldsymbol{\beta}'$ indicates the axis of rotation, where the sign follows from the rotation angle $0 \le \alpha \le \pi$.

Of interest is the case where $\beta' \ll 1$, so that in $\Lambda(\boldsymbol{\beta}'',\boldsymbol{\alpha})$ only terms linear in $\boldsymbol{\beta}'$ need to be considered. The corresponding development of (12.4.23) – problem 12.7 – results in the infinitesimal rotation $\boldsymbol{\alpha}$:

$$\mathsf{R}^i{}_j = \delta^i{}_j + \frac{\gamma}{\gamma+1}\,\epsilon_{ijk}\left[\boldsymbol{\beta}\times\boldsymbol{\beta}'\right]_k = \delta^i{}_j - \epsilon_{ijk}\alpha_k. \tag{12.4.24}$$

The (infinitesimal) rotation is according to (12.4.4)

$$\mathbf{x}' = \mathbf{x} + \boldsymbol{\alpha}\times\mathbf{x} \;\Rightarrow\; \boldsymbol{\alpha} = \frac{\gamma}{\gamma+1}\boldsymbol{\beta}'\times\boldsymbol{\beta}, \quad \alpha = \frac{\gamma-1}{\gamma}\frac{\beta'}{\beta}\sin\left(\sphericalangle\boldsymbol{\beta}',\boldsymbol{\beta}\right). \tag{12.4.25}$$

Thomas precession

We now proceed from the idea that S' is not an inertial system, but performs a constant circular rotation with respect to S. In S, S' then has after a short time Δt the speed in 1st order of Δt:

$$\boldsymbol{\beta}'' = \boldsymbol{\beta}(t) + \Delta\boldsymbol{\beta} \approx \boldsymbol{\beta} + \frac{\mathrm{d}\boldsymbol{\beta}}{\mathrm{d}t}\Delta t \overset{(12.4.17)}{=} \boldsymbol{\beta} + \frac{\boldsymbol{\beta}'}{\gamma}.$$

Thus, the rotation (12.4.25) that occurred in time Δt along with the associated angular velocity $\lim\limits_{\Delta\to 0}\Delta\boldsymbol{\alpha}/\Delta t$ can be specified:

$$\Delta\boldsymbol{\alpha} = \frac{\gamma^2}{\gamma+1}\Delta\boldsymbol{\beta}\times\boldsymbol{\beta} \qquad\Rightarrow\qquad \boldsymbol{\omega}_T = -\frac{\gamma^2}{\gamma+1}\boldsymbol{\beta}\times\frac{\mathrm{d}\boldsymbol{\beta}}{\mathrm{d}t}. \tag{12.4.26}$$

---

[16] $(\mathbf{v}\circ\mathbf{v})(\mathbf{w}\circ\mathbf{w}) = (\mathbf{v}\cdot\mathbf{w})(\mathbf{v}\circ\mathbf{w})$

ω_T is the so-called *Thomas precession*; it is a consequence of the rotation occurring in the multiplication due to the space-time coupling of the LT of two boosts. The Thomas precession is thus a kinematic, relativistic effect; that it is a relativistic effect is also clear from $\lim\limits_{c\to\infty} \omega_T = 0$. We will return to the Thomas precession in section 14.3.3 to calculate the spin-orbit coupling of an electron.

Note: Without the restriction to a circular motion with constant speed β, one obtains using the addition theorem (12.4.15) the contributions linear to β'

$$\frac{\beta'}{\gamma} = \Delta\beta + \frac{\beta\cdot\beta'}{1+\gamma}\beta,$$

where the additional term is proportional to β and thus contributes to ω_T.

12.4.4 Doppler Effect

If a moving light source emits radiation of frequency ω_0, the stationary observer sees the radiation at the frequency ω. This is the analogous effect to sound waves, where we hear the whistle of an approaching train at a higher pitch than that of the departing one.

The light source moves in the x-direction, as sketched in Fig. 12.24, with the speed $\mathbf{v} = v\,\mathbf{e}_x$ and emits light of frequency ω_0. The four-dimensional

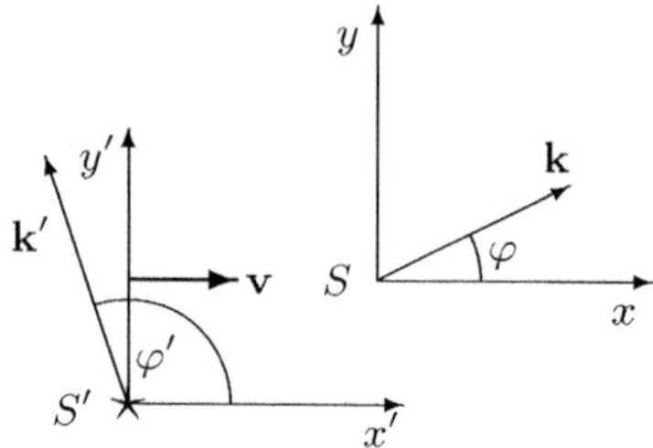

Fig. 12.24. A light source S' moves with $\mathbf{v}$ against S and emits light of frequency $\omega' = \omega_0$. Light that goes from S' in direction φ', is observed in S under φ

wave vector has the form

$$(k^\mu) = (\omega/c, \mathbf{k}) \qquad \text{with} \qquad k_\mu k^\mu = 0 \,. \qquad (12.4.27)$$

(k^μ) is a light-like vector with the dispersion relation $\omega = kc$ as an invariant. To describe the frequency shift in the system S in general, it is sufficient to restrict oneself in the system S' to the $x'y'$ plane.

We start from a frequency ω and an angle φ under which we observe the source S', and can thus conclude on $\omega' = k'c$ and $\mathbf{k}' = k'\,\cos\varphi'\,\mathbf{e}_{x'} + \sin\varphi'\,\mathbf{e}_{y'}$:

$$(k'^\mu) = \begin{pmatrix} \gamma & -\beta\gamma & 0 & 0 \\ -\beta\gamma & \gamma & 0 & 0 \\ 0 & 0 & 1 & 0 \\ 0 & 0 & 0 & 1 \end{pmatrix} k \begin{pmatrix} 1 \\ \cos\varphi \\ \sin\varphi \\ 0 \end{pmatrix} = k \begin{pmatrix} \gamma(1 - \beta\cos\varphi) \\ \gamma(-\beta + \cos\varphi) \\ \sin\varphi \\ 0 \end{pmatrix} . \qquad (12.4.28)$$

In detail, this results in

$$\omega' \equiv \omega_0 = \omega\gamma(1 - \beta\cos\varphi), \tag{12.4.29}$$

$$\cos\varphi' = \frac{k\gamma}{k'}\big(-\beta + \cos\varphi\big) = \frac{-\beta + \cos\varphi}{1 - \beta\cos\varphi}, \tag{12.4.30}$$

$$\sin\varphi' = \frac{k}{k'}\sin\varphi = \frac{\sin\varphi}{\gamma(1 - \beta\cos\varphi)}. \tag{12.4.31}$$

In the resting system S, the frequency of the light source is according to (12.4.29)

$$\omega = \frac{\omega_0}{\gamma(1 - \beta\cos\varphi)}. \tag{12.4.32}$$

The deflection experienced by the signal emanating from S' under φ', the aberration, is determined by (12.4.30) and (12.4.31) We distinguish:

1. *Longitudinal Doppler effect*

   The light source moves in the x-direction with the speed $\mathbf{v} = v\,\mathbf{e}_x$ towards an observer in S. Observation is under $\varphi = 0$, i.e. $(k^\mu) = (k\ k\ 0\ 0)$. In S one thus observes

$$\omega = \omega_0\sqrt{\frac{1+\beta}{1-\beta}} \qquad \text{and} \qquad \cos\varphi' = 1, \qquad \text{i.e.} \qquad \varphi' = 0.$$

   If the light source moves away under $\varphi = \pi$, then $\varphi' = \pi$ and $\omega = \omega_0\gamma(1 - \beta)$.

2. *Transverse Doppler effect*

   The light source continues to move with $\mathbf{v} = v\mathbf{e}_x$, only the light emitted from S' is observed at an angle $\varphi = \pi/2$. Thus, $(k^\mu) = (k\ 0\ k\ 0)$ and it is $\mathbf{k} \perp \mathbf{v}$. From (12.4.32) it follows

$$\omega = \omega_0\sqrt{1 - \beta^2}. \tag{12.4.33}$$

   From $\cos\varphi' = -\beta$ it follows that in S' the waves "are emitted backwards"and only through the aberration are seen by the stationary observer under $\pi/2$. The frequency $\omega \approx \omega_0(1 - v^2/(2c^2))$ is therefore lower, regardless of the sign of $\mathbf{v}$ and we have a second order effect in $\beta = v/c$ before us. In non-relativistic approximation, there is no frequency shift in the transverse case. Only signals that are emitted in S' in the range $\pi/2 < \varphi' < \pi$ contribute to the transverse Doppler effect, i.e. are directed "backwards"and are observed through the aberration under $\varphi = \pi/2$.

   The decrease in frequency in the transverse Doppler effect is due to time dilation and was therefore of fundamental interest. The experiment by Ives-Stilwell [1938] was the first evidence of the transverse Doppler effect, using canal rays (H-ions) with speeds of $\sim 10^6\,\mathrm{m\,s^{-1}}$.

The factor of $\sqrt{1-\beta^2}$ in (12.4.29) attributable to time dilation is isotropic and therefore also measurable in the longitudinal direction. The first direct evidence in the transverse direction was provided by Hasselkamp, Mondry, Scharmann [1979].

3. *General Doppler effect*

For the receding light source, $\cos\varphi < 0$, which according to (12.4.32) results in a decrease in frequency. If the light source is moving towards the observer, then, when

$$\gamma(1 - \beta\cos\varphi) = 1\,,$$

no frequency shift ($\omega = \omega_0$) is to be expected. In the classical non-relativistic Doppler effect, this is always at $\varphi_0 = \pi/2$. The shift $\varphi_0 < \pi/2$ is a consequence of STR, and for a given β_0 it is

$$\varphi_0 = \arccos\Big(\frac{1 - \sqrt{1-\beta_0^2}}{\beta_0}\Big)\,.$$

Fig. 12.25 shows this curve, which separates the case $\omega > \omega_0$ from $\omega < \omega_0$. In S', the waves falling under φ_0 without frequency shift are radiated at an angle $\varphi_0' = \pi - \varphi_0$ ($k_x' = -k_x$ and $k_y' = k_y$). We now proceed from the

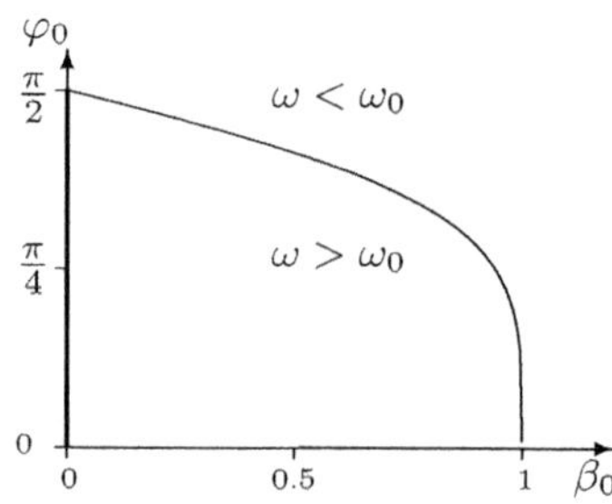

Fig. 12.25. Angle φ_0 under which an object of speed $\beta_0 < 1$ can be observed without frequency shift - including the trivial case $\beta_0 = 0$. The signal is emitted under $\varphi_0' = \pi - \varphi_0$

light source S' and determine (k^μ) from the inverse LT to (12.4.28)

$$\omega = \omega_0\gamma(1 + \beta\cos\varphi') \qquad \text{and} \qquad \cos\varphi = \frac{\beta + \cos\varphi'}{1 + \beta\cos\varphi'}\,.$$

The relationship between $\omega \leftrightharpoons \omega_0$ and $\varphi \leftrightharpoons \varphi'$ is given by the replacement of $\mathbf{v}$ by $-\mathbf{v}$, as it must be according to the principle of relativity. This is not the case with the non-relativistic Doppler effect, where one must distinguish whether the observer or the source is moving relative to the medium (gas).

For sources with very high speeds, almost all signals originating from S' are observed in a small range $\varphi \approx 0$. The received frequencies ω depend on φ', so that a shift from blue to red sets in around the forward direction.

For a fast passing object, in addition to the distortion and rotation caused by the Terrell rotation, to the change in brightness due to the headlight effect, color changes due to the Doppler effect can also be observed.

In the non-relativistic approximation, $\mathbf{x}' = \mathbf{x} - \mathbf{v}t$. The frequency shift is obtained from

$$\psi' = \mathrm{e}^{\mathrm{i}(\mathbf{k}'\cdot\mathbf{x}'-\omega_0 t)} = \mathrm{e}^{\mathrm{i}(\mathbf{k}'\cdot(\mathbf{x}-t\mathbf{v})-\omega_0 t)} = \mathrm{e}^{\mathrm{i}(\mathbf{k}'\cdot\mathbf{x}-\omega t)} \,,$$

which gives the frequency shift of

$$\omega = \omega_0 + \mathbf{k}'\cdot\mathbf{v}.$$

The Doppler effect occurring in non-relativistic physics is also described by (12.4.32) but with $\gamma = 1$ and is thus the 1st order approximation in β.

Problems for Chapter 12

12.1. *Invariance of the wave equation: Conditions for the transformation.* In reference to a work by Voigt [1887] we try the transformation Λ (12.1.16) to narrow down so far that the d'Alembert operator ($x_0 = ct$) remains invariant:

$$\Box = \frac{\partial^2}{\partial x_0^2} - \frac{\partial^2}{\partial \mathbf{x}^2} = \Box' = \frac{\partial^2}{\partial x_0'^2} - \frac{\partial^2}{\partial \mathbf{x}'^2} \qquad \begin{cases} x_0' &= \gamma x_0 - \mathbf{a}\cdot\mathbf{x} \\ \mathbf{x}' &= -\mathsf{D}\boldsymbol{\beta}\, x_0 + \mathsf{D}\,\mathbf{x} \,. \end{cases}$$

Show that $\mathbf{a} = \gamma\boldsymbol{\beta}$ and the coefficients of D additionally have to fulfill the six conditions

$$\gamma^2(1-\beta^2) = \mathbf{q}_i^2 - (\mathbf{q}_i\cdot\boldsymbol{\beta})^2 \qquad\qquad \text{for} \quad i = 1,2,3\,,$$
$$(\mathbf{q}_i\cdot\boldsymbol{\beta})(\mathbf{q}_j\cdot\boldsymbol{\beta}) = \mathbf{q}_i\cdot\mathbf{q}_j \qquad\qquad \text{for} \quad i < j \qquad\qquad (12.4.34)$$

must be fulfilled. Here, $\mathbf{q}_i^T = \begin{pmatrix} d_{i1} & d_{i2} & d_{i3} \end{pmatrix}$ are the Row vectors of D.

12.2. *Invariance of the wave equation: Determination of the transformation.*
1. Show that from (12.4.34) it follows that $\det\mathsf{D} = \gamma$ and $\det\Lambda = \gamma^2(1-\beta^2)$.
2. Using (12.4.34), determine the condition equations for R, if you perform the decomposition $\mathsf{D} = \mathsf{R}\mathsf{Q}$ with $\mathsf{Q} = \mathsf{E} + (\gamma-1)\hat{\boldsymbol{\beta}}\circ\hat{\boldsymbol{\beta}}$ (see (12.4.3)). Q replaces, applied to $\mathbf{x}$, the component parallel to $\boldsymbol{\beta}$ Component $\mathbf{x}_\parallel$ by $\gamma\mathbf{x}_\parallel$.

$$\det\mathsf{R}=1 \quad \text{u.} \quad \mathbf{r}_i\cdot\mathbf{r}_j = \mathbf{q}_i\cdot\mathbf{q}_j - \frac{\gamma^2-1}{\gamma^2\beta^2}(\mathbf{q}_i\cdot\boldsymbol{\beta})(\mathbf{q}_j\cdot\boldsymbol{\beta}) \quad \text{with} \quad \mathbf{r}_i^T = \text{row vector of } \mathsf{R}$$

3. To be determined is γ as a function of β, which can be done using $\det\Lambda = 1$. Show that then the general transformation is a boost, followed by a rotation.

12.3. *Universal speed in LT.* In problem 12.2 we have the transformation

$$\begin{pmatrix} x_0' \\ \mathbf{x}' \end{pmatrix} = \begin{pmatrix} 1 & \mathbf{0}^T \\ \mathbf{0} & \mathsf{R} \end{pmatrix} \begin{pmatrix} \gamma & -\gamma\boldsymbol{\beta}^T \\ -\gamma\boldsymbol{\beta} & \mathsf{E} + (\gamma-1)\hat{\boldsymbol{\beta}}\circ\hat{\boldsymbol{\beta}} \end{pmatrix} \begin{pmatrix} x_0 \\ \mathbf{x} \end{pmatrix} , \qquad \gamma = \frac{1}{\sqrt{1-\beta^2}}$$

derived under which the wave equation remains invariant. We have used that $x_0 = ct$, but made no statement about x_0'. A different choice than $x_0' = ct'$ see (12.4.3) contradicts the principle of relativity. Show explicitly that for inertial systems also in S' the speed of light must be c, i.e., that there is only one (universal) speed in LT.

12.4. *Twin paradox*: Castor embarks on a journey to Sirius and immediately returns upon arrival. His spaceship accelerates for the first half of the journey to Sirius uniformly with $b_c = 981\,\mathrm{cm\,s^{-2}}$ and then decelerates at the same rate, so that it comes to a standstill at Sirius. The return journey proceeds in the same way as the outward journey. The distance to Sirius $l = 8.6$ light years (or 2.64 parsec). 1 parsec$= 3.0857 \times 10^{18}$ cm. The year, in turn, has approx. 3.1536×10^7 s. What is the maximum speed the spaceship reaches, how long does the journey take for Castor and how long has the left-behind twin brother Pollux waited for Castor? Provide the world line of the journey (including sketch). *Hint*: The uniform acceleration is discussed on page 520, see (14.1.12).

12.5. *Calculation of the rotation* R *from a restricted LT*: $\Lambda(\beta, \alpha)$ is a restricted LT $(\Lambda^0{}_0 \geq 1 , \det \Lambda = 1)$. Represent this transformation as a product of a rotation and a boost and determine the axis of rotation, i.e., verify (12.4.10).

12.6. *On general velocity addition*: S' moves with $\mathbf{v} = v\mathbf{e}_x$ relative to S and S'' with $\mathbf{v}'$ relative to S', where the direction of $\mathbf{v}'$ is to be kept general. Calculate by multiplication of $\Lambda\Lambda'$ the velocity $\mathbf{v}'' = \mathbf{v} + \mathbf{v}'$ and γ'' and show that $|\mathbf{v}''| \leq c$.

1. Show that for $\mathbf{v}' = v'\mathbf{e}_x$ you get the formula for the velocity addition (12.4.16).
2. Calculate $|\mathbf{v}''| = |\mathbf{v}+\mathbf{v}'|$ for a generally held direction of $\mathbf{v}'$ and show that $v'' \leq c$. *Hint*: It is sufficient to calculate $\Lambda''^0{}_0$ (why?).
3. Calculate $\tilde{\mathbf{v}} = \mathbf{v}' + \mathbf{v}$. How does $\tilde{\mathbf{v}}$ differ from $\mathbf{v}''$?

12.7. *Rotation*: Again, two pure boosts $\Lambda(\beta, \mathbf{0})$ with $\beta = \mathbf{v}/c$ and $\Lambda(\beta', \mathbf{0})$ with $\beta' = \mathbf{v}'/c$ are given. The resulting LT contains, in addition to the speed β'', also a rotation α: $\Lambda(\beta'', \alpha)$. Let

$$\begin{pmatrix} \gamma'' & -\mathbf{a}^T \\ -\mathbf{b} & \mathsf{D} \end{pmatrix} = \begin{pmatrix} \gamma' & -\gamma'\beta'^T \\ -\gamma'\beta' & 1 + (\gamma'-1)\hat{\beta}'\circ\hat{\beta}' \end{pmatrix} \begin{pmatrix} \gamma & -\gamma\beta^T \\ -\gamma\beta & 1 + (\gamma-1)\hat{\beta}\circ\hat{\beta} \end{pmatrix} .$$

Let $\beta' \ll 1$, so that only the 1st order in β' needs to be considered. Calculate the rotation matrix R for this case and specify the rotation axis α and rotation angle α.

12.8. *Charge conservation*: Show by integrating the continuity equation that the total charge Q of a closed system is conserved.

12.9. *Rod and hole*: Given is a hole of length d_0. A rod, which in its rest system has the length l_0, moves with $\beta = \sqrt{3}/2$ relative to the hole. Does the rod fit into the hole if $d = l$? Describe the situation from S, the inertial system of the hole, and from S', the inertial system of the rod, as seen.

References

Campbell J. *GPS im Schatten des Uhrenparadoxon – Betrachtungen zu den Auswirkungen der Relativitätstheorie bei den Satellitennavigationssystemen* in "Festschrift 125 Jahre Geodäsie und Geoinformatik". Wiss. Arbeiten der Fachrichtung Geodäsie und Geoinformatik der Universität Hannover, Heft **263**, 129–146 (2006)

Einstein A. *Zur Elektrodynamik bewegter Körper*, Ann. Physik **17**, 891-921 (1905)

Einstein A. *Die Grundlage der allgemeinen Relativitätstheorie*, Ann. Physik **49**, 769–822 (1916)

Einstein A. *Über die spezielle und die allgemeine Relativitätstheorie* 24th ed. Springer (2009); 1st ed. (1917)

FitzGerald G. F. *The Ether and the Earth's Atmosphere*, Science **30**, 390 (1889)

Frank Ph., Rothe H. *Über die Transformation von Raumzeitkoordinaten von ruhenden und bewegten Systemen*, Ann. Physik **34**, 825–855 (1911)

Gamow G. *Mr. Tompkins in Wonderland* Cambridge University Press (1940)

Griffiths D. *Introduction to Electrodynamics*, 4th ed. Cambridge University Press (2017)

Hasselkamp D., Mondry E, Scharmann E. *Direct observation of the transversal Doppler-shift* Z. Physik A, **289**, 151–155 (1979)

Huygens C. *Traité de la lumière*, Marchand Libraire, Leide (1690)

Ignatowsky W. von *Das Relativitätsprinzip*, Archiv für Mathmatik und Physik **17**, 1–24 (1910) und **18**, 17–40 (1911)

Iro H. *A Modern Approach to Classical Mechanics*, 2nd ed., World Scientific (2015)

Ives H. E. and Stilwell G. R. *An Experimental Study of the Rate of a Moving Atomic Clock*, J. Opt. Soc. Am. **28**, 215-226 (1938)

Jackson J. D. *Classical Electrodynamics*, 3rd ed., John Wiley & Sons Inc. (1998)

Kern J., Kraus U., Lehle B., Rau R. and Ruder H. *Aussehen relativistisch bewegter Objekte*, Praxis der Naturwissenschaften Physik **46**, Heft 2, 2-6 (1997)

Kragh H. *Ludvig Lorenz, Electromagnetism, and the Theory of Telephone Currents* arXiv:1606.00205v1 [physics.hist-ph], 1–16 (2016)

Kraus U. *Bewegung am kosmischen Tempolimit* in *Sterne und Weltraum*, August 2005, 40–46, Spektrum.de (2005)

Lampa A. *Wie erscheint nach der Relativitätstheorie ein bewegter Stab einem ruhenden Beobachter*, Z. Physik **27**, 138–148 (1924)

Landau L. D. and Lifshitz E. M. *The Classical Theory of Fields*, vol. 2, 4th ed., Butterworth-Heinemann, reprinted (1994)

Landau L. D., Lifshitz E. M. *Elektrodynamics of Continuous Media*, Vol 8, 2nd ed. Pergamon Press, Oxford (1984)

Larmor J. in *Aether and Matter*, Cambridge University Press (1900)

Lorentz H. *Die relative Bewegung der Erde und des Äthers* in Abhandlungen über Theoretische Physik, Teubner Leipzig, 443–447 (1892/1907)

Lorentz H. *Electromagnetic phenomena in a system moving with any velocity smaller than that of light*, KNAW, Proceedings **6**, 809–831 (1904)

Penrose R. *The apparent shape of a relativistically moving sphere*, Proc. Cambridge Phil. Soc. **55**, 137–139 (1959)

Poincaré H. *Sur la dynamique de l'électron*, Comptes rendus de l'Academie de schiences **140**, 1504-1508 (1905)

Ruder H. & M. *Die spezielle Relativitätstheorie*, Vieweg (1993)

Schwabl F. *Advanced Quantum Mechanics*, 4th ed. Springer Berlin (2008)

Schöpf H. G. *Maxwell Äthertheorien*, Astron. Nachr. **303**, 29–37 (1982)

Schröder U. *Spezielle Relativitätstheorie*, 5. Aufl. Harri Deutsch, Frankfurt (2014)

Sexl R. U. and Urbantke H. K. *Relativity, Groups, Particles*, Springer Wien (2001)

Terrell J. *Invisibility of the Lorentz Contraction*, Phys. Rev. **116**, 1041–1045 (1959)

Voigt W. *Über das Dopplersche Prinzip*, Nachr. Ges. Wiss. Göttingen **8**, 41–51 (1887)

Covariant Electrodynamics

13.1 Maxwell's Equations in Covariant Form

We have already seen that a point charge, which only has an electrostatic field at rest, is surrounded by electric and magnetic fields when in motion ((8.2.46) and (8.2.47)) whose strength depends on the speed, and conclude from this that $\mathbf{E}$ and $\mathbf{B}$ cannot simply be supplemented to four-vectors separately. Before however, it is shown that $\mathbf{E}$ and $\mathbf{B}$ can be summarized in the form of an antisymmetric field strength tensor $F^{\mu\nu}$, which has six independent elements, the transformation behavior of tensors and tensor fields will be examined.

13.1.1 Tensor Properties

Tensors are objects that are defined in each reference system by the same number of components (indices), whereby between the reference systems the individual components are linked by a linear transformation, the LT, according to the rules that apply to each vector. Tab. 13.1 shows tensors and their transformation properties for different levels.

During the transformation of fields, as illustrated in Fig. 13.1 using a rotation in three-dimensional space the point P of the event remains fixed, but has the coordinates $x' = \Lambda x$ in the system S': $x'^{\mu} = \Lambda^{\mu}{}_{\nu}x^{\nu}$. The analogy to rotation can also be extended to other transformations such as the boost.

Transformation properties of fields

1. Scalar fields: $\phi'(P) = \phi(P)$,

D. Petrascheck, F. Schwabl, *Electrodynamics*, https://doi.org/10.1007/978-3-662-71502-4_13

2. Vector fields: $A'^\mu(P) = \Lambda^\mu{}_\nu A^\nu(P)$,
$A'^\mu(x')$ results from $A^\mu(x)$ (with $x'^\rho = \Lambda^\rho{}_\sigma x^\sigma$), by transforming the vector components $A^\mu(x)$ along with the coordinate system:

$$A'^\mu(x') = \Lambda^\mu{}_\nu A^\nu(x), \qquad\qquad A'_\mu(x') = \Lambda_\mu{}^\nu A_\nu(x).$$

3. Tensor fields of the nth level:

$$T'^{\mu_1\cdots}{}_{\mu_l\dots}{}^{\cdots\mu_n}(x') = \Lambda^{\mu_1}{}_{\bar\mu_1}\dots\Lambda_{\mu_l}{}^{\bar\mu_l}\dots\Lambda^{\mu_n}{}_{\bar\mu_n}\, T^{\bar\mu_1\cdots}{}_{\bar\mu_l\dots}{}^{\cdots\bar\mu_n}(x)\,.$$

4. By contraction of a tensor transforms the following third-order tensor like a vector; used was $\Lambda_\nu{}^{\bar\lambda}\Lambda^\nu{}_{\bar\nu} = \delta^{\bar\lambda}{}_{\bar\nu}$:

$$T'^{\mu\nu}{}_{,\nu} = \Lambda_\nu{}^{\bar\lambda}\,\Lambda^\mu{}_{\bar\mu}\,\Lambda^\nu{}_{\bar\nu}\,T^{\bar\mu\bar\nu}{}_{,\bar\lambda} = \Lambda^\mu{}_{\bar\mu}\,T^{\bar\mu\bar\nu}{}_{,\bar\nu}\,. \tag{13.1.1}$$

In Tab. 13.1 the transformation properties are summarized.

Tab. 13.1. Transformation behavior of tensors and tensor fields

contra-/covariant vector	$v'^\mu = \Lambda^\mu{}_\nu v^\nu$	$v'_\mu = \Lambda_\mu{}^\nu v_\nu$
2nd order tensor	$T'^{\mu\nu} = \Lambda^\mu{}_{\bar\mu}\Lambda^\nu{}_{\bar\nu}T^{\bar\mu\bar\nu}$	
3rd order tensor	$T'^{\mu\nu}{}_{,\lambda} = \Lambda_\lambda{}^{\bar\lambda}\Lambda^\mu{}_{\bar\mu}\Lambda^\nu{}_{\bar\nu}\partial_{\bar\lambda}T^{\bar\mu\bar\nu}$	
	$\cdots\qquad\cdots$	
scalar field	$\phi'(x') = \phi(x)$	$\phi(x) = \phi(\Lambda^{-1}x')$
vector field	$A'^\mu(x') = \Lambda^\mu{}_\nu A^\nu(x)$	
2nd order tensor field	$T'^{\mu\nu}(x') = \Lambda^\mu{}_{\bar\mu}\Lambda^\nu{}_{\bar\nu}T^{\bar\mu\bar\nu}(x)$	
	$\cdots\qquad\cdots$	

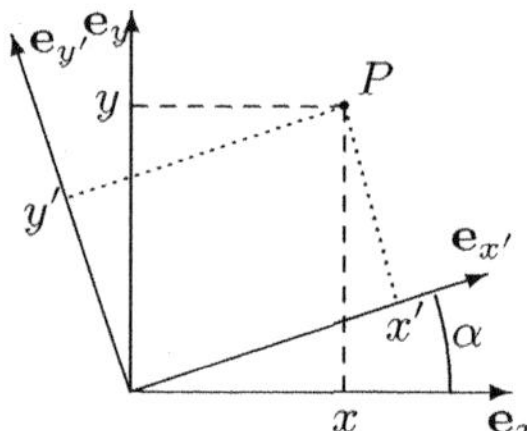

$$\begin{aligned}
\phi'(P) &= \phi(P) & \Leftrightarrow \quad & \phi'(\mathbf{x}') = \phi(\mathsf{R}^{-1}\mathbf{x}') = \phi(\mathbf{x})\\
A'_i(P) &= R_{ij}\,A_j(P) & \Leftrightarrow \quad & A'_i(\mathbf{x}') = R_{ij}\,A_j(\mathsf{R}^{-1}\mathbf{x})\\
& & & \quad = R_{ij}\,A_j(\mathbf{x})\,.
\end{aligned}$$

Fig. 13.1. (Passive) rotation $\mathsf{R}(\alpha)$ around the z-axis by the angle α; ϕ is a scalar field and $\mathbf{A}$ a vector field

Transformation of four-vectors

In the previous formulation of the STR, we have not yet needed tensors of higher rank, but four-vectors were sufficient. For these, the transformation properties (12.2.21) apply

$$\left(u'^\mu\right) = \left(\gamma(u^0 - \beta u^1),\ \gamma(u^1 - \beta u^0),\ u^2,\ u^3\right), \tag{13.1.2}$$

when $\beta = v_x/c$. For the general orientation, one obtains from (12.4.2)

$$u'^0 = \gamma(u^0 - \mathbf{u}\cdot\boldsymbol{\beta}), \tag{13.1.3}$$

$$\mathbf{u}' = \gamma(\mathbf{u}_\| - u^0\boldsymbol{\beta}) + \mathbf{u}_\perp \quad \text{with} \quad \mathbf{u}_\| = \frac{\mathbf{u}\cdot\boldsymbol{\beta}}{\beta} \quad \text{and} \quad \mathbf{u}_\perp = \frac{\boldsymbol{\beta}\times(\mathbf{u}\times\boldsymbol{\beta})}{\beta^2}.$$

In the reverse transformation, simply replace β with $-\beta$. The following Tab. 13.2 provides an overview of four-vectors. Here, the invariant scalar product provides information about whether a vector is spacelike, lightlike or timelike. Timelike means that the spatial component can be made to disappear through an LT. In the case of (x^μ), this is the same location at different times. For spacelike vectors at most the zeroth component can disappear. In the case of (x^μ) then an event can indeed occur at the same time, but only at different places, and for acceleration, it means that it cannot be made to disappear by any LT. The gradient vector does not fit into the above scheme.

Tab. 13.2. Four-vectors

Four-vector	Reference	contravariant vector		scalarproduct
Event	(12.2.1)	x^μ	$(ct, \mathbf{x})$	$s^2 = c^2 t^2 - r^2$
Worldline	(12.3.5)	$z^\mu = x^\mu(s)$	$(ct(s), \mathbf{x}(s))$	$s^2 \geq 0$
Gradient	(12.2.6)	$\partial^\mu = \dfrac{\partial}{\partial x_\mu}$	$\left(\dfrac{1}{c}\dfrac{\partial}{\partial t}, -\boldsymbol{\nabla}\right)$	$\Box = \dfrac{1}{c^2}\dfrac{\partial^2}{\partial t^2} - \nabla^2$
Wavenumber	(12.4.27)	k^μ	$(\omega/c, \mathbf{k})$	$\omega^2/c^2 - \mathbf{k}^2 = 0$
Velocity	(12.3.7)	$u^\mu = \dot{x}^\mu(s)$	$\gamma(1, \boldsymbol{\beta})$	1
	(12.3.6)	$\mathrm{v}^\mu = \dot{x}^\mu(\tau)$	$\gamma(c, \mathbf{v})$	c^2
Momentum	(14.1.1)	$p^\mu = mcu^\mu$	$(\mathrm{E}/c, \boldsymbol{p})$	$m^2 c^2$
Acceleration	(14.1.6)	$a^\mu = \dot{u}^\mu(s)$	$\left(\dot{\gamma}(s), \dot{\gamma}(s)\boldsymbol{\beta} + \gamma\dot{\boldsymbol{\beta}}(s)\right)$	$-\gamma^{-2}\dot{\gamma}^2 - \gamma^2\dot{\boldsymbol{\beta}}^2$
	(14.1.5)	$\mathrm{b}^\mu = \dot{\mathrm{v}}^\mu(\tau)$		
Current density	(13.1.4)	j^μ	$(c\rho, \rho\mathbf{v})$	$c^2 \rho^2 \gamma^{-2}$
Potential	(13.1.10)	A^μ	$(\phi/k_{\mathrm{L}}, \mathbf{A})$	$\phi_q^2/\gamma^2 k_{\mathrm{L}}^2$
Force density	(14.1.25)	f^μ	$\rho\left(\boldsymbol{\beta}\cdot\mathbf{E}, \mathbf{E} + k_{\mathrm{L}}\boldsymbol{\beta}\times\mathbf{B}\right)$	$-\rho^2 E'^2 \gamma^{-2}$

The event vector is space-, time-, or light-like, the world line is time- or light-like.
The velocity vectors, momentum and current density, are timelike.
Acceleration and force density are spacelike; ρ is the density of a narrowly limited charge distribution (point) and $\mathbf{E}'$ is the field (13.1.29) in the rest system of the particle.
(A_q^μ) is the four-potential of a point charge (8.2.22), which is timelike.

13.1.2 Covariant Tensors of Electrodynamics

Current density

To achieve a covariant formulation of electrodynamics, we first combine the charge density ρ and the current density $\mathbf{j}$, as we have already done in (8.2.18):

$$(j^\mu) = \begin{pmatrix} c\rho(\mathbf{x},t) \\ \mathbf{j}(\mathbf{x},t) \end{pmatrix} = \sum_n e_n \begin{pmatrix} c \\ \mathbf{v}_n \end{pmatrix} \delta^{(3)}(\mathbf{x} - \mathbf{x}_n(t)). \tag{13.1.4}$$

However, it must be shown that the current density j^μ behaves like a four-vector (transforms). For this purpose, the representation

$$j^\mu(x) = c\sum_n e_n \int_{-\infty}^{\infty} ds\, \dot{z}_n^\mu(s)\, \delta^{(4)}(x - z_n(s)) \tag{13.1.5}$$

is verified. Here is[1] after (12.3.7)

$$\left| \frac{\mathrm{d}}{\mathrm{d}s}(x^0 - z^0(s)) \right| = \dot{z}^0(s) = \gamma, \qquad \delta\left(x^0 - z^0(s)\right) = \delta\left(s - s(t)\right)\frac{1}{\gamma},$$

$$j^\mu(x) = c\sum_n e_n \dot{z}_n^\mu(s(t))\delta^{(3)}\left(\mathbf{x} - \mathbf{z}_n(s(t))\right)\frac{1}{\gamma} = \sum_n e_n \begin{pmatrix} c \\ \mathbf{v}_n \end{pmatrix} \delta^{(3)}\left(\mathbf{x} - \mathbf{z}_n(t)\right).$$

j^μ is a four-vector, since $\dot{z}^\mu$ is a four-vector and according to (B.6.11)

$$\delta^{(4)}(x') = \delta^{(4)}(\Lambda x) = \delta^{(4)}(x') \cdot |\det \Lambda|^{-1} = \delta^{(4)}(x). \tag{13.1.6}$$

$\delta^{(4)}(x)$ transforms under LT according to the rules for scalar fields.

Continuity equation

The continuity equation (1.1.15) is the four-divergence of the current density and thus invariant under the LT:

$$\dot{\rho} + \nabla \cdot \mathbf{j} = 0 \qquad\qquad \Leftrightarrow \qquad\qquad \partial_\mu j^\mu \equiv j^\mu{}_{,\mu} = 0. \tag{13.1.7}$$

Scalar product

The current density is $\mathbf{j} = \rho\,\mathbf{v}$, from which

$$j^\mu j_\mu = c^2 \rho^2(1 - \beta^2) > 0$$

follows that (j^μ) is timelike. In the co-moving system, $\mathbf{j}=0$.

Charge density of a point charge

A point particle $\mathbf{x}_q'$ has the density in the rest system S'

$$n'(\mathbf{x}',t') = \delta^{(3)}(\mathbf{x}' - \mathbf{x}_q') = \int ds\, \delta^{(4)}(\Lambda(x - x_q)) \tag{13.1.8}$$

$$= \int ds\, \frac{\delta^{(3)}(\mathbf{x} - \mathbf{x}_q)\delta(x^0 - x_q^0(s))}{|\det \Lambda|} = \frac{\delta^{(3)}(\mathbf{x} - \mathbf{x}_q)}{\gamma} = \frac{n(\mathbf{x},t)}{\gamma}.$$

---

[1] $\delta(f(x)) = \sum_i \delta(x - x_i)\left| \dfrac{\mathrm{d}f(x)}{\mathrm{d}x} \right|^{-1}$, see (B.6.7).

We have oriented ourselves on the proof of the covariance of (j^μ) (13.1.5). For the charge density, it follows from $\rho(\mathbf{x}, t) = e\, n(\mathbf{x}, t)$: $\rho' = \rho/\gamma$; the charge density appears increased by the Lorentz contraction compared to the rest system.

Charge invariance

If one calculates from the charge density ρ' present in the co-moving system S' the one in the laboratory system S, one has to consider the Lorentz contraction for the volume element: $\mathrm{d}^3 x' = \gamma \mathrm{d}^3 x$. This implies

$$\rho'\, \mathrm{d}^3 x' = \rho\, \mathrm{d}^3 x\,, \tag{13.1.9}$$

which is an expression of charge invariance. We will come back to this a little later in section 13.2.3.

Vector potential

We have already mentioned, see (8.2.18), that not only charge and current density can be combined into a four-vector, but also the scalar and the vector potential, where we refer back to (8.2.19):

$$(A^\mu) = \begin{pmatrix} \phi/k_L \\ \mathbf{A} \end{pmatrix}, \qquad A^\mu(\mathbf{x}, t) = \frac{4\pi k_C}{ck_L} \int \mathrm{d}^4 x' \,\frac{1}{c} D(x - x') j^\mu(x'). \tag{13.1.10}$$

$D(x - x')$ is the solution of the inhomogeneous wave equation (8.2.12)

$$\Box D(x) = c\, \delta^{(4)}(x)\,.$$

Under LT, $\Box$ is invariant and $\delta^{(4)}(x)$ is a scalar, from which it can be concluded that also $D(x)$ is a scalar field. Since we have shown that j^μ is a four-vector, A^μ must also be one. This can be read directly from the wave equation

$$\partial^\nu \partial_\nu A^\mu = \Box A^\mu = \frac{4\pi k_C}{ck_L}\, j^\mu\,. \tag{13.1.11}$$

Lorenz gauge

A gauge transformation leaves the wave equation unchanged:

$$A^\mu \to \bar{A}^\mu = A^\mu + \partial^\mu \chi \quad \overset{\text{Lorenz gauge}}{\longrightarrow} \quad \partial_\mu \partial^\mu \chi = \Box \chi = 0.$$

(8.1.8) is the four-divergence of (A^μ) and thus invariant under LT:

$$\frac{1}{ck_L}\frac{\partial}{\partial t}\phi + \boldsymbol{\nabla}\cdot\mathbf{A} = 0 \qquad \Leftrightarrow \qquad \partial_\mu A^\mu \equiv A^\mu{}_{,\mu} = 0. \tag{13.1.12}$$

Scalar product and point particles

The Liénard-Wiechert potential for a moving point charge (8.2.26) has the form $(\phi_q, \boldsymbol{\beta}\phi_q)$, from which it follows that (A_q^μ) is a timelike vector:

$$A_q^\mu A_{q\,\mu} = \phi_q^2/\gamma^2 k_L^2 > 0. \tag{13.1.13}$$

13.1.3 Field Strength Tensor

As mentioned at the beginning of this section, $\mathbf{E}$ and $\mathbf{B}$ are certainly not to be represented as four-vectors, since for example, the movement of a charge creates a magnetic field and so under an LT $\mathbf{E}$ and $\mathbf{B}$ must be transformed together This draws our attention to second order tensors. The elements of the field strength tensor[2]

$$F^{\mu\nu} = \partial^\mu A^\nu - \partial^\nu A^\mu = A^{\nu,\mu} - A^{\mu,\nu} \tag{13.1.14}$$

are components of the fields $\mathbf{E}$ and $\mathbf{B}$

$$F^{k0} = \partial^k A^0 - \partial^0 A^k = -\frac{\partial\phi}{\partial x^k} - \frac{\partial A^k}{c\partial t} = E_k/k_L, \tag{13.1.15}$$

$$F^{kl} = \partial^k A^l - \partial^l A^k = (\delta_{ka}\delta_{lb} - \delta_{kb}\delta_{la})\partial^a A^b = \epsilon_{klj}\epsilon_{abj}\partial^a A^b = -\epsilon_{klj}B_j.$$

The (latin) indices like k, l are assigned the values 1 to 3. A distinction between co- and contravariant components is obsolete for E_k and B_k, but also for ϵ_{ijk}, as these are not four-vectors (tensors) in pseudo-Euclidean space. From (13.1.15) it follows

$$\left(F^{\mu\nu}\right) = \begin{pmatrix} 0 & -\mathbf{E}^T/k_L \\ \mathbf{E}/k_L & -(\epsilon_{mnk}B_k) \end{pmatrix}. \tag{13.1.16}$$

$F^{\mu\nu} = -F^{\nu\mu}$ is a contravariant antisymmetric 4×4-tensor and thus has six independent elements, the fields $\mathbf{E}$ and $\mathbf{B}$. For the covariant field strength tensor applies

$$F_{\mu\nu} = g_{\mu\rho}g_{\nu\sigma}F^{\rho\sigma} \quad\Longrightarrow\quad F_{k0} = -F^{k0} \qquad F_{kl} = F^{kl}. \tag{13.1.17}$$

Now we can specify these two field strength tensors directly:

$$\left(F^{\mu\nu}\right) = \begin{pmatrix} 0 & \frac{-E_x}{k_L} & \frac{-E_y}{k_L} & \frac{-E_z}{k_L} \\ E_x/k_L & 0 & -B_z & B_y \\ E_y/k_L & B_z & 0 & -B_x \\ E_z/k_L & -B_y & B_x & 0 \end{pmatrix}, \quad \left(F_{\mu\nu}\right) = \begin{pmatrix} 0 & \frac{E_x}{k_L} & \frac{E_y}{k_L} & \frac{E_z}{k_L} \\ -E_x/k_L & 0 & -B_z & B_y \\ -E_y/k_L & B_z & 0 & -B_x \\ -E_z/k_L & -B_y & B_x & 0 \end{pmatrix}. \tag{13.1.18}$$

For further calculations, the fields $\mathbf{E}$ and $\mathbf{B}$ are required as functions of $F^{\mu\nu}$:

$$E_k = k_L F^{k0}, \qquad B_k = -\epsilon_{klm}\partial^l A^m = -\frac{1}{2}\epsilon_{klm}F^{lm}. \tag{13.1.19}$$

---

[2] Einstein [1916, (1)], Becker, Sauter [1973, (11.1.11)] or Schwabl [2008, (14.1.7)] have defined $F^{\mu\nu}$ with opposite sign: $F^{\mu\nu} = A^{\mu,\nu} - A^{\nu,\mu}$.

The dual field strength tensor

The tensor $(\tilde{F}^{\mu\nu})$ which is dual to $(F^{\mu\nu})$ is defined by

$$\tilde{F}^{\mu\nu} = \frac{1}{2}\,\epsilon^{\mu\nu\rho\sigma}\,F_{\rho\sigma}\,, \qquad\qquad F^{\mu\nu} = -\frac{1}{2}\,\epsilon^{\mu\nu\rho\sigma}\,\tilde{F}_{\rho\sigma}\,. \tag{13.1.20}$$

In the appendix A.1.3, p. 561 the properties of the totally antisymmetric tensor (Levi-Civita symbol) are listed:

$$\epsilon_{\mu\nu\rho\sigma} = -\epsilon^{\mu\nu\rho\sigma} = \begin{cases} 1 & \text{for all even permutations (0123)} \\ -1 & \text{for odd permutations} \\ 0 & \text{otherwise.} \end{cases} \tag{13.1.21}$$

The sign difference between co- and contravariant tensor comes from $\det g = -1$. The following detailed calculation shows that $\tilde{F}^{\mu\nu}$ arises from $F^{\mu\nu}$ by replacing $\mathbf{E}/k_L \to -\mathbf{B}$ and $\mathbf{B} \to \mathbf{E}/k_L$

$$\left(\tilde{F}^{\mu\nu}\right) = \begin{pmatrix} 0 & \mathbf{B}^T \\ -\mathbf{B} & -(\epsilon_{mnk}E_k/k_L) \end{pmatrix}. \tag{13.1.22}$$

The non-diagonal elements are $\epsilon^{mn0l} = \epsilon^{0mnl} = -\epsilon_{mnl})$

$$\tilde{F}^{0n} = \frac{1}{2}\epsilon^{0nkl}F_{kl} = \frac{-1}{2}\epsilon_{nkl}F^{kl} = B_n\,, \quad \tilde{F}^{mn} = \frac{1}{2}\epsilon^{mn\kappa\lambda}F_{\kappa\lambda} = \epsilon^{mn0l}F_{0l} = -\epsilon_{mnl}\frac{E_l}{k_L}.$$

For $\tilde{F}^{mn}$, either $\kappa = 0$ or $\lambda = 0$, otherwise the index 0 does not appear in the ϵ-tensor. The two terms are equal. Used were (13.1.15), (13.1.17) and (13.1.21).

Analogous to the relations (13.1.19) for (F^{lk}) apply to the dual tensor

$$B_k = -\tilde{F}^{k0} = \tilde{F}_{k0}\,, \qquad \tilde{F}^{k0} = -B_k\,, \qquad \tilde{F}_{k0} = -\tilde{F}^{k0}\,,$$

$$E_k = -\frac{1}{2}\epsilon_{klm}\tilde{F}^{lm}\,, \qquad \tilde{F}^{ij} = -\epsilon_{ijk}E_k\,, \qquad \tilde{F}_{kl} = \tilde{F}^{kl}. \tag{13.1.23}$$

Note: While $(F^{\mu\nu})$ is largely uniformly defined, there are differences in the dual tensor, which are due to a different definition of the antisymmetric tensor. Here, according to Sexl, Urbantke [2001, (5.50)] or Scheck [2016, (2.50)] the definition

$$\epsilon_{0123} = 1 \qquad \Rightarrow \qquad \epsilon^{0123} = -1$$

is used, while in the books by Landau, Lifshitz 2 [1987, (6,8)] or Jackson [1998, (11.139)] $\epsilon^{0123} = 1$. This has no effect on the (homogeneous) Maxwell equations (13.1.25). $\tilde{F}^{\mu\nu}$ changes the sign when x_4 is the time axis: $\epsilon_{0123} = -\epsilon_{1230} \hat{=} -\epsilon_{1234}$.

13.1.4 Maxwell's Equations

We will now, analogous to the previous procedure, insert four-vectors and field tensors into the Maxwell equations to arrive at a suitable covariant notation. The four inhomogeneous Maxwell equations are

$$\boldsymbol{\nabla}\cdot\mathbf{E} = 4\pi k_C\rho \qquad \Rightarrow \qquad \partial_k F^{k0} = \frac{4\pi k_C}{ck_L}j^0,$$

$$\boldsymbol{\nabla}\times\mathbf{B} = \frac{4\pi k_C\mathbf{j}+\dot{\mathbf{E}}}{ck_L} \quad \Rightarrow \quad \frac{-1}{2}\epsilon_{ijk}\partial_j\epsilon_{klm}F^{lm} = \partial_j F^{ji} = \frac{4\pi k_C}{ck_L}j^i - \partial_0 F^{0i}.$$

This results in the covariant equations[3]

$$F^{\nu\mu}{}_{,\nu} = \frac{4\pi k_C}{ck_L}\,j^\mu. \tag{13.1.24}$$

The four homogeneous Maxwell equations can be written as

$$\boldsymbol{\nabla}\cdot\mathbf{B}=0 \implies \qquad -\partial_k\tilde{F}^{k0}=0,$$

$$\boldsymbol{\nabla}\times\mathbf{E}+\frac{k_L}{c}\dot{\mathbf{B}}=0 \xrightarrow{\times 1/k_L} -\frac{1}{2}\epsilon_{ijk}\partial_j\epsilon_{klm}\tilde{F}^{lm}-\partial_0\tilde{F}^{i0} = -\partial_j\tilde{F}^{ij}-\partial_0\tilde{F}^{i0}=0.$$

From this follow the homogeneous Maxwell equations in covariant form

$$\tilde{F}^{\nu\mu}{}_{,\nu} = 0. \tag{13.1.25}$$

These can also be represented by

$$F_{\lambda\mu,\nu} + F_{\nu\lambda,\mu} + F_{\mu\nu,\lambda} = 0. \tag{13.1.26}$$

The above equation is antisymmetric with respect to the exchange of two indices. The left side disappears identically when two indices are the same. Non-trivial conditions are obtained only when all three indices are different. These are the four homogeneous Maxwell equations – problem 13.2.

The covariance of the Maxwell equations

We have shown that the current density is a four-vector. According to the principle of relativity the inhomogeneous wave equation $\Box A^\mu = \frac{4\pi k_C}{ck_L}j^\mu$ must have the same form in all inertial systems which means that A^μ must be a four-vector, since $\Box$ is invariant. Thus, $F^{\mu\nu} = \partial^\mu A^\nu - \partial^\nu A^\mu$ must be a tensor of 2nd rank. It therefore follows

$$F'^{\nu\mu}{}_{,\nu} = \Lambda_\nu{}^\lambda\partial_\lambda\,\Lambda^\nu{}_{\bar{\nu}}\Lambda^\mu{}_{\bar{\mu}}F^{\bar{\nu}\bar{\mu}} = \partial_{\bar{\nu}}\,\Lambda^\mu{}_{\bar{\mu}}F^{\bar{\nu}\bar{\mu}} = \frac{4\pi k_C}{ck_L}\Lambda^\mu{}_{\bar{\mu}}j^{\bar{\mu}} = \frac{4\pi k_C}{ck_L}j'^\mu, \quad (13.1.27)$$

which is proven by the covariance of the inhomogeneous Maxwell equations.

---

[3] If you add the factor $k_s = \pm 1$ on the right side of (13.1.24), so $k_s = -1$ takes into account the signature (3,1): see (12.2.2').

13.1.5 Transformation of the Electromagnetic Field

To determine the transformation behavior of the electromagnetic fields, one starts from the field tensor:

$$F'^{\mu\nu} = \Lambda^\mu{}_{\bar\mu}\Lambda^\nu{}_{\bar\nu}F^{\bar\mu\bar\nu}.$$

The explicit calculation of the $F'^{\mu\nu}$ is not very stimulating, but for the sake of completeness it is given below.

LT of the field strength tensor: The $\Lambda^\mu{}_\nu$ of the LT (12.4.3) are separated into parts

$$\Lambda^0{}_0 = \gamma, \qquad\qquad \Lambda^0{}_l = \Lambda^l{}_0 = -\gamma\beta_l, \qquad\qquad \Lambda^k{}_l = \delta_{kl} + (\gamma-1)\hat\beta_k\hat\beta_l,$$

to calculate the $F'^{\mu\nu}$ step by step using (13.1.15):

$$\begin{aligned}
F'^{i0} &= \Lambda^i{}_\kappa\Lambda^0{}_\lambda F^{\kappa\lambda} = \Lambda^i{}_0\Lambda^0{}_l F^{0l} + \Lambda^i{}_k\Lambda^0{}_0 F^{k0} + \Lambda^i{}_k\Lambda^0{}_l F^{kl} && (13.1.28') \\
&= \gamma^2\beta_i\beta_l F^{0l} + \gamma[F^{i0} + (\gamma-1)\hat\beta_i\hat\beta_k F^{k0}] - \gamma\beta_l[F^{il} + (\gamma-1)\hat\beta_i\hat\beta_k F^{kl}] \\
&= -\gamma^2\beta_i\boldsymbol{\beta}{\cdot}\mathbf{E}/k_{\mathrm{L}} + \gamma E_i/k_{\mathrm{L}} + \gamma(\gamma-1)\hat\beta_i\hat{\boldsymbol{\beta}}{\cdot}\mathbf{E}/k_{\mathrm{L}} + \gamma(\boldsymbol{\beta}\times\mathbf{B})_i.
\end{aligned}$$

$$\begin{aligned}
F'^{ij} &= \Lambda^i{}_\kappa\Lambda^j{}_\lambda F^{\kappa\lambda} = \Lambda^i{}_0\Lambda^j{}_l F^{0l} + \Lambda^i{}_k\Lambda^j{}_0 F^{k0} + \Lambda^i{}_k\Lambda^j{}_l F^{kl} && (13.1.28'') \\
&= -\gamma\beta_i F^{0j} - \gamma\beta_j F^{i0} + [F^{ij} + (\gamma-1)\hat\beta_i\hat\beta_k F^{kj} + (\gamma-1)\hat\beta_j\hat\beta_l F^{il}] \\
&= \gamma\beta_i E_j/k_{\mathrm{L}} - \gamma\beta_j E_i/k_{\mathrm{L}} - \epsilon_{ijm}B_m - (\gamma-1)[\epsilon_{kjm}\hat\beta_i\hat\beta_k + \epsilon_{ilm}\hat\beta_j\hat\beta_l]B_m.
\end{aligned}$$

The interest is less in the tensor F' than in the fields, which are divided into components parallel and perpendicular to $\boldsymbol{\beta}$:

$$\begin{aligned}
E'_n &= k_{\mathrm{L}}F'^{n0} = \hat\beta_n\hat{\boldsymbol{\beta}}{\cdot}\mathbf{E} + \gamma(E_n - \hat\beta_n\hat{\boldsymbol{\beta}}{\cdot}\mathbf{E}) + \gamma k_{\mathrm{L}}(\boldsymbol{\beta}\times\mathbf{B})_n, \\
B'_n &= -\frac{1}{2}\epsilon_{nij}F'^{ij} = -\gamma(\boldsymbol{\beta}\times\mathbf{E}/k_{\mathrm{L}})_n + B_n + (\gamma-1)[B_n - \hat\beta_n(\hat{\boldsymbol{\beta}}{\cdot}\mathbf{B})]
\end{aligned}$$

For arbitrary directions of $\mathbf{v}$ one obtains

$$\mathbf{E}' = \mathbf{E}_\parallel + \gamma(\mathbf{E}_\perp + k_{\mathrm{L}}\boldsymbol{\beta}\times\mathbf{B}), \tag{13.1.29}$$
$$\mathbf{B}' = \mathbf{B}_\parallel + \gamma(\mathbf{B}_\perp - \boldsymbol{\beta}\times\mathbf{E}/k_{\mathrm{L}}). \tag{13.1.30}$$

The decomposition into components parallel and perpendicular to $\mathbf{v}$ is

$$\mathbf{E} = \mathbf{E}_\parallel + \mathbf{E}_\perp = (\mathbf{E}{\cdot}\hat{\boldsymbol{\beta}})\,\hat{\boldsymbol{\beta}} + \hat{\boldsymbol{\beta}}\times(\mathbf{E}\times\hat{\boldsymbol{\beta}}).$$

Invariants of the field strength tensors

Since the trace of an antisymmetric tensor vanishes, only the 2nd level remains with the two invariants $\mathrm{tr}\{\mathsf{F}\mathsf{F}^T\}$ and $\mathrm{tr}\{\tilde{\mathsf{F}}\mathsf{F}^T\}$:

$$F^{\kappa\lambda}F_{\kappa\lambda} = 2F^{0l}F_{0l} + F^{kl}F_{kl} = -2E^2/k_L^2 + \epsilon_{kli}B_i\epsilon_{klj}B_j$$
$$= -2(E^2/k_L^2 - B^2), \tag{13.1.31}$$

$$\tilde{F}^{\kappa\lambda}F_{\kappa\lambda} = 2\tilde{F}^{0l}F_{0l} + \tilde{F}^{kl}F_{kl} = 2B_lE_l/k_L + \epsilon_{kli}(E_i/k_L)\epsilon_{klj}B_j$$
$$= 4\mathbf{E}\cdot\mathbf{B}/k_L. \tag{13.1.32}$$

Remarks:

If $\mathbf{B} = 0$ and $\mathbf{E} \neq 0$ in S, then $|\mathbf{E}'/k_L| > |\mathbf{B}'|$ in S'.

If $\mathbf{E} \perp \mathbf{B}$ in S, then $\mathbf{E}' \perp \mathbf{B}'$ in S'. $(\epsilon_{\alpha\beta\gamma\delta})$ is a pseudotensor, which implies that (13.1.32) is a Lorentz pseudoscalar.

13.2 Covariant Electrodynamics in Media

13.2.1 Maxwell Equations in Matter

In matter, the homogeneous Maxwell equations remain unchanged, compared to the vacuum and in the inhomogeneous equations $\mathbf{E}$ is replaced by $\mathbf{D}$ and $\mathbf{B}$ by $\mathbf{H}$. This suggests the introduction of a field strength tensor $H^{\mu\nu}$ which differs from $F^{\mu\nu}$ only by replacing $\mathbf{E}/k_L$ with $\mathbf{D}k_L$ and $\mathbf{B}$ with $\mathbf{H}$:

$$\left(H^{\mu\nu}\right) = \begin{pmatrix} 0 & -k_L\mathbf{D}^T \\ k_L\mathbf{D} & -(\epsilon_{ijk}H_k) \end{pmatrix}. \tag{13.2.1}$$

From this directly follow the inhomogeneous equations

$$H^{\nu\mu}{}_{,\nu} = \frac{k_L}{c}4\pi k_C\epsilon_0 j^{\mu}. \tag{13.2.2}$$

The homogeneous equations (13.1.25) or (13.1.26) remain unchanged. However, the fields $\mathbf{D}$ and $\mathbf{H}$ are only identical with the dielectric displacement $\mathbf{D} = \epsilon_r\epsilon_0\mathbf{E}$ and the magnetic field $\mathbf{H} = \mu_r\mu_0\mathbf{B}$ if we are in the rest system of the matter. Thus, the fields $(\hat{\boldsymbol{\beta}} = \boldsymbol{\beta}/\beta)$ transform

$$\mathbf{D}' = \mathbf{D}_\| + \gamma\left(\mathbf{D}_\perp + \boldsymbol{\beta}\times\mathbf{H}/k_L\right),$$
$$\mathbf{H}' = \mathbf{H}_\| + \gamma\left(\mathbf{H}_\perp - k_L\boldsymbol{\beta}\times\mathbf{D}\right) \tag{13.2.3}$$

for $\mathbf{D}$ and $\mathbf{H}$ analogous to $\mathbf{E}$ and $\mathbf{B}$, (13.1.29) and (13.1.30), but the material equations (constitutive equations) (5.2.17) only apply to fields in the rest system of the matter. The dual tensor

$$\tilde{H}^{\mu\nu} = \frac{1}{2}\,\epsilon^{\mu\nu\rho\sigma}\,H_{\rho\sigma}$$

arises from $(H^{\mu\nu})$ by replacing $k_L\mathbf{D} \to -\mathbf{H}$ and $\mathbf{H} \to k_L\mathbf{D}$. Under the restrictions applicable to the fields $\mathbf{D}$ and $\mathbf{H}$ one obtains the equations corresponding to the invariants (13.1.31) and (13.1.32) by substitution of $\mathbf{E}/k_L \to k_L\mathbf{D}$ and $\mathbf{B} \to \mathbf{H}$.

13.2.2 Material Equations

We now assume that the (homogeneous) matter is at rest in the system S', i.e. $\mathbf{D}' = \epsilon_r\epsilon_0\mathbf{E}'$ and $\mathbf{B}' = \mu_r\mu_0\mathbf{H}'$, and thus obtain the constitutive equations from the LT (13.2.3).

Auxiliary calculation for the material equations in moving matter:

$$\mathbf{D}'_\| = \mathbf{D}_\| = \epsilon_r\epsilon_0\mathbf{E}'_\| \overset{(13.1.29)}{=} \epsilon_r\epsilon_0\mathbf{E}_\|, \quad \mathbf{H}'_\| = \mathbf{H}_\| = \mu_r\mu_0\mathbf{B}'_\| \overset{(13.1.30)}{=} \mu_r\mu_0\mathbf{B}_\|.$$

For the normal components, one starts from

1. $\mathbf{D}_\perp + \boldsymbol{\beta}\times\mathbf{H}_\perp/k_L = \epsilon_r\epsilon_0\big(\mathbf{E}_\perp + k_L\boldsymbol{\beta}\times\mathbf{B}_\perp\big),$
2. $\mathbf{H}_\perp - k_L\boldsymbol{\beta}\times\mathbf{D}_\perp = (\mu_r\mu_0)^{-1}\big(\mathbf{B}_\perp - \boldsymbol{\beta}\times\mathbf{E}_\perp/k_L\big)$

and multiplies vectorially from the left with $\boldsymbol{\beta}\times$ or $(1/k_L)\boldsymbol{\beta}\times$

3. $\boldsymbol{\beta}\times\mathbf{D}_\perp - \beta^2\mathbf{H}_\perp/k_L = \epsilon_r\epsilon_0\big[\boldsymbol{\beta}\times\mathbf{E}_\perp + k_L\boldsymbol{\beta}\times(\boldsymbol{\beta}\times\mathbf{B}_\perp)\big],$
4. $\boldsymbol{\beta}\times\mathbf{H}_\perp/k_L + \beta^2\mathbf{D}_\perp = (\mu_r\mu_0)^{-1}\big[\boldsymbol{\beta}\times\mathbf{B}_\perp/k_L - \boldsymbol{\beta}\times(\boldsymbol{\beta}\times\mathbf{E}_\perp)/k_L^2\big].$

We have also used $\boldsymbol{\beta}\times(\boldsymbol{\beta}\times\mathbf{H}) = -\beta^2\mathbf{H}_\perp$. If you subtract the 4th equation from the 1st (or add the 3rd to the 2nd) and expand the right side with $-\epsilon_r\epsilon_0(\beta^2\mathbf{E}_\perp + \boldsymbol{\beta}\times(\boldsymbol{\beta}\times\mathbf{E}_\perp))$, you get

$$\mathbf{D}_\perp(1-\beta^2) = \epsilon_r\,\epsilon_0\big\{\mathbf{E}_\perp(1-\beta^2) + (1-1/\epsilon_r\mu_r)\big[k_L\boldsymbol{\beta}\times\mathbf{B}_\perp - \boldsymbol{\beta}\times(\boldsymbol{\beta}\times\mathbf{E}_\perp)\big]\big\}$$

$$\mathbf{H}_\perp(1-\beta^2) = (\mu_r\mu_0)^{-1}\big\{\mathbf{B}_\perp(1-\beta^2) + (\epsilon_r\mu_r-1)\big[\boldsymbol{\beta}\times\mathbf{E}_\perp/k_L + \boldsymbol{\beta}\times(\boldsymbol{\beta}\times\mathbf{B}_\perp)\big]\big\}.$$

In summary, this results in the material equations for linear media [Becker, Sauter, 1973, (11.1.23)]

$$\mathbf{D} = \epsilon_r\epsilon_0\big[\mathbf{E} + \gamma^2(1-\tfrac{1}{\epsilon_r\mu_r})\,\boldsymbol{\beta}\times(k_L\mathbf{B} - \boldsymbol{\beta}\times\mathbf{E})\big],$$

$$\mathbf{H} = (\mu_r\mu_0)^{-1}\big[\mathbf{B} + \gamma^2(\epsilon_r\mu_r-1)\,\boldsymbol{\beta}\times(\mathbf{E}/k_L + \boldsymbol{\beta}\times\mathbf{B})\big]. \tag{13.2.4}$$

If the matter is at rest in S', then there $\mathbf{D}' = \epsilon_r\epsilon_0\mathbf{E}'$ applies, where for $\mathbf{E}'$ and $\mathbf{D}'$ (13.1.29) and (13.2.3) are to be inserted. The components parallel to $\mathbf{v}$ remain unchanged and cancel out. Thus one obtains, as Minkowski [1908, (C),(D)] has shown, the constitutive equations:

$$\mathbf{D} + \frac{1}{k_L}\boldsymbol{\beta}\times\mathbf{H} = \epsilon_r\epsilon_0[\mathbf{E} + k_L\boldsymbol{\beta}\times\mathbf{B}], \quad \mathbf{H} - k_L\boldsymbol{\beta}\times\mathbf{D} = \frac{1}{\mu_r\mu_0}\Big[\mathbf{B} - \frac{1}{k_L}\boldsymbol{\beta}\times\mathbf{E}\Big].$$

These are the so-called *Minkowski equations* [Sommerfeld, 1952, (34.5)].

Magnetization-polarization tensor

For some applications, the fields in the rest system of matter are better described by the material equations (5.2.17)

$$\mathbf{D} = \epsilon_0(\mathbf{E} + 4\pi k_c \mathbf{P}) \qquad \text{and} \qquad \mathbf{H} = \mathbf{B}/\mu_0 - 4\pi k_r \mathbf{M}.$$

than by permittivity ϵ and permeability μ. Becker, Sauter [1973, (11.3.3)] define a tensor whose components are magnetization and polarization

$$(M^{\mu\nu}) = \begin{pmatrix} 0 & -k_L \mathbf{P}^T \\ k_L \mathbf{P} & (\epsilon_{mnk} M_k) \end{pmatrix}. \tag{13.2.5}$$

Then the material equations are given by

$$\mathsf{H} = k_L^2 \epsilon_0 \mathsf{F} + 4\pi k_r \mathsf{M}.$$

Note: Vanderlinde [2003, (11.22)] and Bartelmann et al. [2018, (8.114)] use the notation P for M, which indicates polarization. Panofsky, Phillips [1962, (18-62)] designate the tensor $\mathsf{M}_{\text{Panofsky}} = -\mathsf{M}/c$ as the *moment tensor*. Chaichian et al. [2016, (8.87)] in turn define their *polarization four tensor* with the opposite sign: $\mathsf{M}_{\text{Chaichian}} = -\mathsf{M}$.

If the matter is at rest in S', which moves with $\boldsymbol{\beta}$ relative to S, then

$$\begin{aligned} \mathbf{P} &= \mathbf{P}'_{\parallel} + \gamma\big(\mathbf{P}'_{\perp} + \boldsymbol{\beta}\times\mathbf{M}'/k_L\big), \\ \mathbf{M} &= \mathbf{M}'_{\parallel} + \gamma\big(\mathbf{M}'_{\perp} - k_L\boldsymbol{\beta}\times\mathbf{P}'\big). \end{aligned} \tag{13.2.6}$$

This connection is remarkable insofar as a polarized, but not magnetized body ($\mathbf{M}' = 0$) gets a finite magnetization through movement. On the other hand, a non-polarized body, such as a permanent magnet moving relative to S, appears polarized in S.

Bound currents

In matter, the bound polarization charge density ρ_P and the bound current densities of polarization $\mathbf{j}_P$ and magnetization $\mathbf{j}_M$ are added to the free charge density ρ_f and the free current density $\mathbf{j}_f$:

$$\rho_P = -\boldsymbol{\nabla}\cdot\mathbf{P}, \qquad\qquad \mathbf{j}_b = \dot{\mathbf{P}} + (c/k_L)\boldsymbol{\nabla}\times\mathbf{M}. \tag{5.2.2–5.2.3}$$

$\mathbf{j}_b$ can be represented as a four-vector. The continuity equation (5.2.5) is then given as

$$(j_b^{\mu}) = (c\rho_P, \mathbf{j}_b): \qquad j_{b\ ,\mu}^{\mu} = 0 \qquad\qquad \Leftrightarrow \qquad\qquad \dot{\rho}_P + \boldsymbol{\nabla}\cdot\mathbf{j}_b = 0.$$

Using $(M^{\mu\nu})$, (5.2.2–5.2.3) can be written in covariant form:

$$M^{\mu\nu}{}_{,\mu} = -(k_L/c)j_b^{\nu}.$$

Boundary conditions

In S', the system in which the matter is at rest, the usual continuity conditions apply: Continuity of the tangential components $\mathbf{E}'_t$ and $\mathbf{H}'_t$ and of the normal components $\mathbf{D}_n$ and $\mathbf{B}_n$. However, since $\mathbf{\nabla}\cdot\mathbf{D} = 0$ and $\mathbf{\nabla}\cdot\mathbf{B} = 0$ also apply in S, the c<ontinuity of the normal components D_n and B_n remains unchanged.

The tangential components of the fields (13.1.29) and (13.1.30), which are linear in β, are then continuous given by

$$\mathbf{E} + k_L\boldsymbol{\beta}\times\mathbf{B} \qquad\text{and}\qquad \mathbf{H} - k_L\boldsymbol{\beta}\times\mathbf{D}.$$

In this approximation, the boundary conditions at the interface of media 1 and 2 are determined by

$$\mathbf{n}\times(\mathbf{E}_1 - \mathbf{E}_2) = k_L(\boldsymbol{\beta}\cdot\mathbf{n})(\mathbf{B}_1 - \mathbf{B}_2), \tag{13.2.7}$$
$$\mathbf{n}\times(\mathbf{H}_1 - \mathbf{H}_2) = -k_L(\boldsymbol{\beta}\cdot\mathbf{n})(\mathbf{D}_1 - \mathbf{D}_2). \tag{13.2.8}$$

For the fields $\mathbf{D}$ and $\mathbf{H}$, (13.2.4) can be used.

13.2.3 Charge Transport in Moving Conductors

Charge and current form a vector $(j^\mu) = (c\rho, \mathbf{j})$, which according to

$$c\rho' = \gamma(c\rho - \boldsymbol{\beta}\cdot\mathbf{j}), \qquad\qquad \mathbf{j}' = \gamma(\mathbf{j}_\parallel - \rho\mathbf{v}) + \mathbf{j}_\perp \tag{13.2.9}$$

transforms when S' moves with $\mathbf{v}$ relative to S. Here, (j^μ) always satisfies the continuity equation $\partial_\mu j^\mu = 0$, and $\mathbf{j}_\parallel$ is parallel to $\mathbf{v}$.

Charge conservation

The conservation and the equality of the magnitude of the charge of proton and electron are well-established experimental facts. The velocities of electrons and atomic nuclei in matter are different, so a charge change caused by the LT would have been observed.

It should be considered that in a system S with $\rho = 0$ but with existing current $\mathbf{j} \neq 0$ in a system S' moving relative to S with $v\mathbf{e}_x$, the charge density $\rho' \neq 0$. On the other hand, a stationary charge ρ in an otherwise matter-free space is perceived by a moving system S' as a current:

$$j^0 = c\rho, \qquad j^1 = 0, \qquad j'^0 = \gamma j^0, \qquad j'^1 = -\gamma\beta j^0. \tag{13.2.10}$$

The charge density j'^0 is indeed larger when viewed from S', but due to the Lorentz contraction, the volume over which the charge is distributed is smaller by the same factor:

$$Q = \int \mathrm{d}^3x\,\rho, \qquad\qquad Q' = \int \mathrm{d}^3x'\,\gamma\rho = \int \mathrm{d}^3x\,\rho.$$

The total charge Q is thus a relativistic invariant. This applies to the total charge of any closed system, as can be demonstrated by integrating the continuity equation over the system volume.

Current of electrons in a conductor

In a wire, sketched in Fig. 13.2, a current $\mathbf{j}$ flows, caused by electrons moving in the wire. The system is charge neutral (positive ions). If we divide the

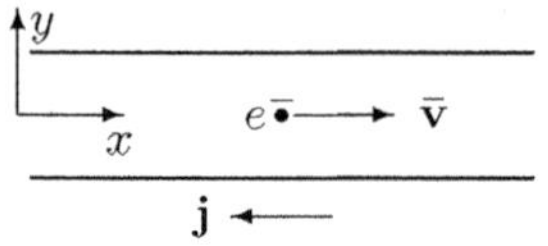

$\bar{\mathbf{v}}$: Velocity of the electrons
$\mathbf{j} = -ne_0\bar{\mathbf{v}}$: Currentdensity.

Fig. 13.2. Current of electrons in a conductor

system into electrons and ions, then in the rest system of the wire

$$(j^0, j^1) = (j^0_{\text{ion}}, j^1_{\text{ion}}) + (j^0_{\text{el}}, j^1_{\text{el}}) = ne_0(c, 0) - ne_0(c, \bar{v}) = -ne_0(0, \bar{v}),$$

where $j^1_{\text{el}} = -ne_0\bar{v}$ is the conduction current of the electrons. From the system S' moving with v we get from $j^0 = c\rho_{\text{ion}}$ the convection current j'^1_{ion} (see (13.2.10)):

$$j'^0_{\text{ion}} = \gamma c ne_0, \qquad\qquad\qquad j'^1_{\text{el}} = -\gamma c\beta ne_0.$$

The electronic conduction current changes according to (13.2.9)

$$j'^0_{\text{el}} = -ne_0\gamma(c - \beta\bar{v}), \qquad\qquad j'^1_{\text{el}} = -ne_0\gamma(\bar{v} - \beta c).$$

The sum gives the total current

$$j'^0 = ne_0\gamma\beta\bar{v} = -\gamma\beta j^1, \qquad\qquad j'^1 = -ne_0\gamma\bar{v} = \gamma j^1.$$

We notice that in S' the charge density no longer disappears and so the system is no longer electrically neutral, but has a positive charge density.

From j^0_{el} also originates a convection current, which compensates for that of the ions. What remains is a conduction current amplified by γ.

From now on, we assume that $\bar{\mathbf{v}}=\mathbf{v}$. In Fig. 13.3, S' is then the rest system of the electrons. The world lines of electrons and ions are shwon. It can be seen that the x'^1-axis (at time $t' = 0$) intersects fewer world lines of electrons than of ions. As a result of the Lorentz contraction, this results in an increased charge density of the ions. From the charge density present in S', one could

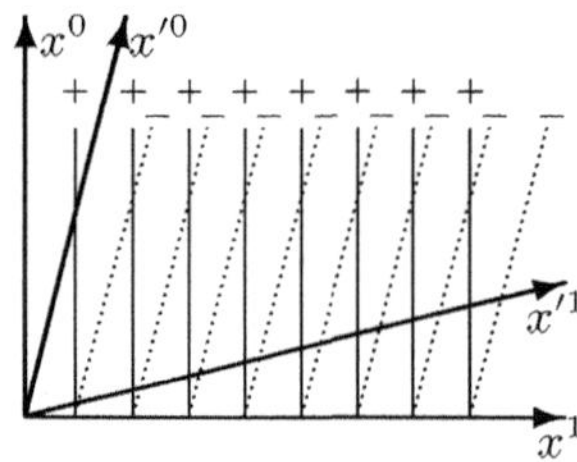

Fig. 13.3. From S', here the rest system of the e^-, since $\bar{\mathbf{v}}=\mathbf{v}$, it is observed that $j'^0 > 0$, i.e. the system is not neutral [Becker, Sauter, 1973, §11.2]

conclude that a force acts on a test charge q introduced into the system, which is at rest in S ($\mathbf{v}_q = 0$).

In Fig. 13.4a it is sketched that the current density $\mathbf{j} = -I\delta(y)\delta(z)\,\mathbf{e}_x$ generates a field $\mathbf{B}$, which according to Biot-Savart's law (4.1.11) is given by

$$\mathbf{B} = \frac{k_C}{ck_L}\frac{2I}{\varrho}\left(0\,,\,\frac{z}{\varrho}\,,\,-\frac{y}{\varrho}\right) \qquad \text{with} \qquad \varrho = \sqrt{y^2 + z^2}\,.$$

Due to charge neutrality, there is no electric field ($\mathbf{E}=0$) and since the test charge is at rest $\mathbf{v}_q = 0$, no force acts on the particle in S.

In Fig. 13.4b, S' is the system in which the electrons are at rest. We now have a current from the moving ions and the test particle also has the velocity $-\mathbf{v}$. The force acting on the test charge is sought.

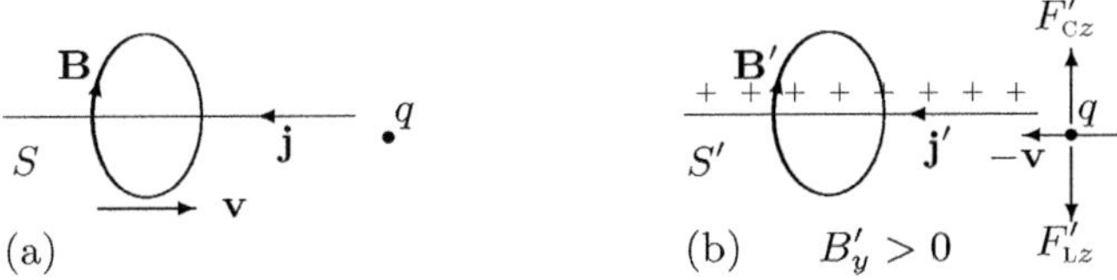

Fig. 13.4. Force on a test charge q
(a) S: Electrons move with $\mathbf{v}$, q is at rest
(b) S': Electrons are at rest, ions and q move with $-\mathbf{v}$

S: Force on test particle disappears, since $\mathbf{E} = 0$ and $\mathbf{v}_q = 0$:
$$\mathbf{F} = q\left(\mathbf{E} + k_L\boldsymbol{\beta}_q \times \mathbf{B}\right) = 0.$$

S': On q act $\mathbf{E}' \neq 0$ and $\mathbf{B}' \neq 0$, but the forces compensate each other

$$\mathbf{E}' \overset{(13.1.29)}{=} \gamma k_L \boldsymbol{\beta} \times \mathbf{B}, \quad \mathbf{B}' \overset{(13.1.30)}{=} \mathbf{B}_\| + \gamma \mathbf{B}_\perp \;\Rightarrow\; \mathbf{F}' = q(\mathbf{E}' - k_L\boldsymbol{\beta} \times \mathbf{B}') = 0.$$

The test particle also experiences no force in S'.

Moving current loop

A current loop, sketched in Fig. 13.5, moves with the speed v in the positive x-direction (constant cross-section A_0).

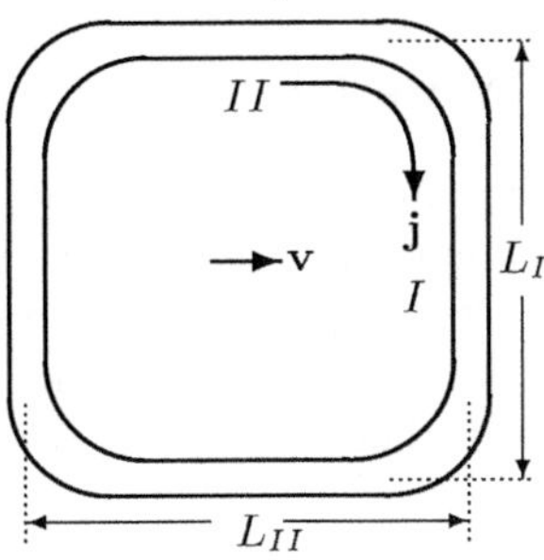

Fig. 13.5. Current loop moving with $\mathbf{v}$.
In the co-moving system S' is $j'^\mu = \left(0\ \mathbf{j}\right)$.

We divide the loop into two sections, in the first $\mathbf{j}$ is perpendicular to $\mathbf{v}$ (I) and in the second $\mathbf{j}$ is parallel to $\mathbf{v}$ (II)

$$I: \ \mathbf{j} \perp \mathbf{v}, \quad (j'^{\mu}) = (0, 0, -j, 0), \qquad (j^{\mu}) = (0, 0, -j, 0),$$
$$II: \ \mathbf{j}\|\mathbf{v}, \quad (j'^{\mu}) = (0, j, 0, 0), \qquad (j^{\mu}) = (\gamma\beta j, \gamma j, 0, 0).$$

The current loop therefore has the electric dipole moment in the system S $p = vj A_0 L_{II} L_I$ and according to (4.2.14) the magnetic dipole moment $m = k_L j A_0 L_{II} L_I/c$. It can be shown that every magnetic moment $\mathbf{m}$ is associated with an electric dipole:

$$\mathbf{p} = \beta \times \mathbf{m}/k_L. \tag{13.2.11}$$

Now the current loop, as indicated in Fig. 13.6, should move with $\mathbf{v}$.

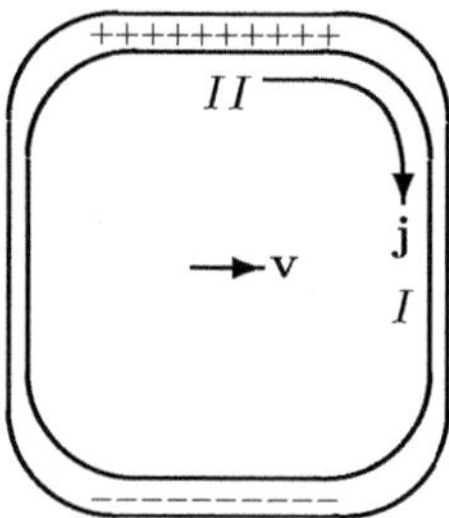

Fig. 13.6. Dipole moment of a current loop moving with $\mathbf{v}$ Due to the Lorentz contraction, the currents ($I = A\,j$) are:

$$I: \ \mathbf{j} \perp \mathbf{v}, \quad A = A_0 \sqrt{1-\beta^2}, \quad I_I = A_0 \sqrt{1-\beta^2}\,j,$$
$$II: \ \mathbf{j}\|\mathbf{v}, \quad A = A_0, \quad I_{II} = \gamma A_0 j$$

This seems at first glance to contradict charge conservation. In fact, but when integrating over a fixed volume V, it follows that

$$\frac{\mathrm{d}}{\mathrm{d}t} \int_V \mathrm{d}^3 x\, \rho(\mathbf{x}, t) = - \oiint \mathrm{d}\mathbf{a}\cdot\mathbf{j}.$$

It is

$$\oiint \mathrm{d}\mathbf{a}\cdot\mathbf{j} = A_0 j \left(\sqrt{1-\beta^2} - \gamma\right) = -A_0 j \gamma \beta^2$$

and ($j^0 = c\rho$)

$$\frac{\mathrm{d}}{\mathrm{d}t} \int_V \mathrm{d}^3 x\, \rho(\mathbf{x}, t) = A_0 \frac{v}{c} \gamma\beta j.$$

So, the left and right sides are equal. The current must ensure that the charge is built up at the places where the loop is moving. The charge density is therefore not constant over time.

13.2.4 Maxwell's Equations for Non-Magnetic Material

It is instructive to derive Maxwell's equations in 'slowly' moving (linear in $\mathbf{v}$), non-magnetic material ($\mu = 1$). In the homogeneous equations, the material does not appear directly and the induction equation (1.3.10) as well as the source freedom of $\mathbf{B}$ (1.3.19) are valid unchanged in moving material. It makes sense to represent current, polarization, magnetization in their rest system S'. According to (13.2.9), one obtains in 1st order in β:

$$c\rho_f = c\rho'_f + \boldsymbol{\beta}\cdot\mathbf{j}'_f, \qquad\qquad \mathbf{j}_f = \mathbf{j}'_f + c\rho'_f\boldsymbol{\beta}.$$

The charge density resulting from the movement relative to the current $\boldsymbol{\beta}\cdot\mathbf{j}'_f/c$, a consequence of simultaneity in the LT, is generally a small effect. $\rho'_f\mathbf{v}$ is the convection current of the free charges. Inserted into (13.2.2) it follows:

$$H^{\kappa 0}{}_{,\kappa} = k_L\boldsymbol{\nabla}\cdot\mathbf{D} = \frac{k_L}{c}4\pi k_r j_f^0 = \frac{k_L}{c}4\pi k_r\,(c\rho'_f + \boldsymbol{\beta}\cdot\mathbf{j}'_f)$$

$$H^{\kappa l}{}_{,\kappa} = \boldsymbol{\nabla}\times\mathbf{H} - \frac{k_L}{c}\dot{\mathbf{D}} = \frac{k_L}{c}4\pi k_r j^l = \frac{k_L}{c}4\pi k_r\,(j_f'^l + c\rho'_f\beta).$$

Now the polarization and magnetization (13.2.6) of the slowly moving, non-magnetized matter are still to be added:

$$\mathbf{P}(\mathbf{x},t) = \mathbf{P}'(\mathbf{x}',t'), \qquad\qquad \mathbf{M}(\mathbf{x},t) = -\boldsymbol{\beta}\times\mathbf{P}'(\mathbf{x}',t').$$

From this follow the Maxwell equations

$$\boldsymbol{\nabla}\cdot\mathbf{D} = 4\pi k_r\,(\rho'_f + \frac{1}{c}\boldsymbol{\beta}\cdot\mathbf{j}'_f)$$

$$\boldsymbol{\nabla}\times\mathbf{B} - \frac{k_L}{c}\dot{\mathbf{D}} = \frac{4\pi k_C}{k_L}\left[\frac{1}{c}\mathbf{j}'_f + \boldsymbol{\beta}\,\rho'_f + \boldsymbol{\nabla}'\times(\mathbf{P}'\times\boldsymbol{\beta})\right].$$

When neglecting the charge density $\boldsymbol{\beta}\cdot\mathbf{j}'_f/c$, these are the Maxwell equations going back to Lorentz and from Panofsky, Phillips [1962, (9-18)] specified in slowly moving, non-magnetized matter. The last term of the current density, which is to be assigned to the polarization, is referred to as *Röntgen current*.

13.2.5 Ohm's Law

In the system S' in which the conductor is at rest, Ohm's law (5.3.2) applies in the form: $\mathbf{j}' = \sigma\mathbf{E}'$, where $\mathbf{j}'$ is the conduction current. Viewed from the laboratory system, the result from (13.2.9) and (13.1.29) is

$$\mathbf{j}'_\| = \sigma\mathbf{E}'_\|: \qquad (\mathbf{j}-c\rho\boldsymbol{\beta})_\| = \sigma\gamma^{-1}\mathbf{E}_\| = \sigma\gamma(1-\beta^2)\,(\mathbf{E}+k_L\boldsymbol{\beta}\times\mathbf{B})_\|,$$

$$\mathbf{j}'_\perp = \sigma\mathbf{E}'_\perp: \qquad (\mathbf{j}-c\rho\boldsymbol{\beta})_\perp = \sigma\gamma(\mathbf{E}+k_L\boldsymbol{\beta}\times\mathbf{B})_\perp.$$

Taking into account that

$$\beta^2(\mathbf{E}+k_L\boldsymbol{\beta}\times\mathbf{B})_\| = \boldsymbol{\beta}\,\boldsymbol{\beta}\cdot(\mathbf{E}+k_L\boldsymbol{\beta}\times\mathbf{B}),$$

one obtains Ohm's law for moving conductors in the form given by Sommerfeld [1952, (34.9)]:

$$\mathbf{j}-c\rho\boldsymbol{\beta} = \sigma\gamma\big[\mathbf{E}+k_L\boldsymbol{\beta}\times\mathbf{B} - \boldsymbol{\beta}\,\boldsymbol{\beta}\cdot(\mathbf{E}+k_L\boldsymbol{\beta}\times\mathbf{B})\big], \qquad (13.2.12)$$

$$j^i - c\rho\beta_i = \sigma k_L\big[F^{i\lambda}u_\lambda - \beta_i F^{0\lambda}u_\lambda\big].$$

In the laboratory system, the convection current is added to the conduction current, which has nothing to do with the conductivity in the metal and therefore must be subtracted from $\mathbf{j}$: $\mathbf{j}_L = \mathbf{j} - \mathbf{j}_K$.

Sommerfeld refers to $\rho\mathbf{v}$ as convection current, which however does not quite agree with the following definition, according to which no convection current should occur for $\rho' = 0$.

Covariant formulation of Ohm's law

First, the four-vector of the convection current (j_K^μ) is defined:

$$j_K^\mu := c\rho'\, u^\mu \stackrel{(13.2.9)}{=} \gamma(c\rho - \mathbf{j}\cdot\boldsymbol{\beta})u^\mu = j^\lambda u_\lambda u^\mu. \tag{13.2.13}$$

Then (13.2.12) suggests the following covariant formulation for Ohm's law [(8.108) Chaichian et al., 2016; Bartelmann et al., 2018, §8.6]:

$$j_L^\mu = j^\mu - j_K^\mu = j^\mu - (j^\lambda u_\lambda)u^\mu = \sigma k_L F^{\mu\lambda} u_\lambda. \tag{13.2.14}$$

For the zeroth component of the conduction current one obtains

$$c\rho_L = \mathbf{j}_L \cdot \boldsymbol{\beta} = \sigma\gamma \mathbf{E}\cdot\boldsymbol{\beta}.$$

If you insert this into (13.2.12), the agreement is with (13.2.14) easy to verify. In addition, $j_L^\mu u_\mu = 0$ applies.

13.3 Unipolar Induction

The Barlow's wheel

Barlow [1822] has shown using the arrangement shown in Fig. 13.7 that a gear wheel, whose tips dip into mercury, begins to rotate when this is in a magnetic field that is parallel to the axis of rotation and a current flows from the wheel hub (W) to the tip. In this part of the circuit, the Lorentz force acts on the conduction electrons and sets the wheel in motion in a tangential direction. This device can be considered the first (unipolar) electric motor. At the same time, Faraday [1822] observed the movement of a current-carrying wire around a magnetic pole.

A little later, Arago [1824] hung a magnetic needle just above a copper disc, separated by glass. The magnetic field of the needle generates eddy currents in the rotating disc whose field causes the needle to follow the disc (see eddy current brake, problem 5.8). The experiment was soon replicated [Babbage, Herschel, 1825] and only confirmed for good conductors[4]. Faraday [1832, §4] explained it using induction. The copper disc is sometimes referred to as the *Arago disc.*

---

[4] Arago experimented with various materials

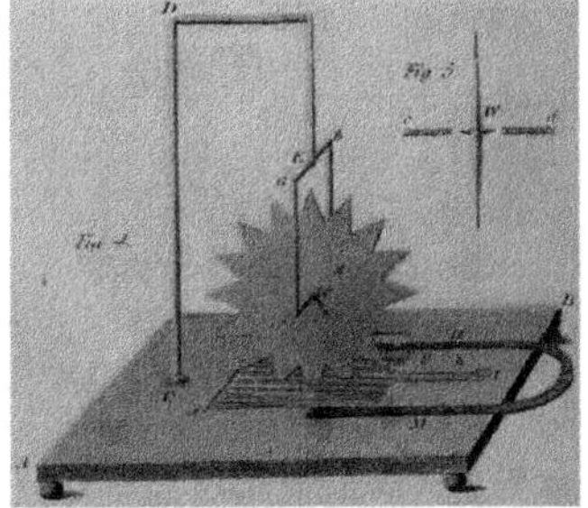

Fig. 13.7. Experimental setup according to Barlow [1822]: Around the mercury basin (f,g,i) there is a horseshoe magnet (H,M). The Cu gear wheel (W) is mounted on a copper frame (a,b,c,d) so that it can rotate freely. If a DC voltage is applied to the wheel axle and the Hg basin (D and i), the wheel starts to rotate

13.3.1 Induction and EMF

Stationary currents cannot be maintained by conservative electrostatic fields, since the energy rate (5.3.6) $\dot{u}_{\mathrm{mech}} = \mathbf{j} \cdot \mathbf{E}$ cannot be provided by these. If one decomposes the electric field into a conservative source component $\mathbf{E}^q$ and a non-conservative (vortex) component $\mathbf{E}^w$, then the electromotive force (1.3.1) is given by

$$\mathfrak{E} = \oint_C \mathrm{d}\mathbf{x} \cdot (\mathbf{E}^q + \mathbf{E}^w) = \oint_C \mathrm{d}\mathbf{x} \cdot \mathbf{E}^w, \tag{13.3.1}$$

where the curve C can partially lie within a medium, a wire or in a vacuum. On the closed path C, only a part of the path should contribute to the EMF. This corresponds to the path from the axis to the edge of the cylinder $(A \rightarrow B)$ in the unipolar generator sketched in Fig. 13.8. Furthermore, the current should disappear:

$$\int_A^B \mathrm{d}\mathbf{x} \cdot (\mathbf{E}^q + \mathbf{E}^w) \overset{(5.3.2)}{=} \int_A^B \mathrm{d}\mathbf{x} \cdot \mathbf{j}/\sigma = 0.$$

From this, one obtains

$$\int_A^B \mathrm{d}\mathbf{x} \cdot \mathbf{E}^w = \oint \mathrm{d}\mathbf{x} \cdot \mathbf{E}^w = \mathfrak{E} = -\int_A^B \mathrm{d}\mathbf{x} \cdot \mathbf{E}^q = \phi_B - \phi_A. \tag{13.3.2}$$

It follows that in an open circuit, the voltage between two points is equal to the EMF. It also follows that in a medium, when no current flows, for non-conservative forces $\mathbf{E}^w = -\mathbf{E}^q$. The non-conservative fields (caused e.g. by chemical potentials) are then equal to the electrostatic fields that are caused by these [Panofsky, Phillips, 1962, Sec. 9-1].

13.3.2 The Unipolar Generator

The unipolar generator goes back to an experiment by Faraday [1832]; see for example Montgomery [1999]. Here, a magnetized cylinder was immersed in a mercury bath and set into rotation. If one replaces the mercury bath with sliding contacts, this corresponds to the configuration sketched in Fig. 13.8[5].

---

[5] The magnetized cylinder can be replaced by a copper disc (Faraday disc), which rotates in an (homogeneous) external magnetic field $\mathbf{B}$

The magnet at rest is not polarized ($\mathbf{P'} = 0$) and magnetized parallel to the

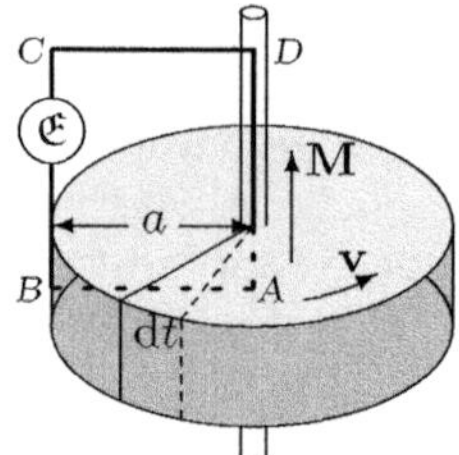

Fig. 13.8. The unipolar generator consists of a cylindrical permanent magnet ($\mathbf{M} = M\mathbf{e}_z$) with a radius a and the thickness (length) d with sliding contacts on the disc and shaft. The magnet rotates with the angular velocity $\boldsymbol{\omega} = \omega\,\mathbf{e}_z$, so that $\mathbf{v} = \boldsymbol{\omega} \times \mathbf{x} = \omega\varrho\,\mathbf{e}_\varphi$. In the time $\mathrm{d}t$, the integration path also rotates with the cylinder, and there is an increase in the magnetic flux according to $\mathrm{d}\varphi = \omega\mathrm{d}t$

axis of rotation: In motion, however, the magnetized disk appears polarized in the laboratory system S according to (13.2.6):

$$\mathbf{M} = \gamma\mathbf{M'} = M\mathbf{e}_z\,\theta(a-\varrho)\big[\theta(z)-\theta(z-d)\big],$$

$$\mathbf{P} = \gamma\boldsymbol{\beta} \times \mathbf{M'}/k_L = \boldsymbol{\beta} \times \mathbf{M}/k_L = (M\omega/ck_L)\varrho\,\theta(a-\varrho)\big[\theta(z)-\theta(z-d)\big].$$

Since $\beta = v/c \ll 1$, $\gamma = 1$ and $M = M'$. Accordingly, for the magnetic field one obtains

$$\mathbf{B} = (1-\gamma)\hat{\boldsymbol{\beta}} \cdot \mathbf{B'}\,\hat{\boldsymbol{\beta}} + \gamma\mathbf{B'} \overset{\gamma=1}{=} \mathbf{B'}, \qquad\qquad \hat{\boldsymbol{\beta}} = \mathbf{v}/v. \qquad (13.3.3)$$

We orient ourselves according to Becker, Sauter [1973, §11.3] and Panofsky, Phillips [1962, §9-5]. The induction law in integral form (1.3.4) reads, when one substitutes for the convective derivative (1.3.6)

$$\mathfrak{E} = \oint_{\partial A} \mathrm{d}\mathbf{x}\cdot\mathbf{E} = -\frac{k_L}{c}\frac{\mathrm{d}}{\mathrm{d}t}\iint_A \mathrm{d}\mathbf{a}\cdot\mathbf{B} = -\frac{k_L}{c}\iint_A \mathrm{d}\mathbf{a}\cdot\big[\dot{\mathbf{B}} - \boldsymbol{\nabla}\times(\mathbf{v}\times\mathbf{B})\big]$$

$$= \frac{k_L}{c}\oint_{\partial F} \mathrm{d}\mathbf{x}\cdot(\mathbf{v}\times\mathbf{B}) = \frac{k_L\omega}{c}\int_0^a \mathrm{d}\varrho\,\varrho\,\mathbf{B}\cdot\mathbf{e}_z = \frac{k_L\omega}{2\pi c}\Phi_B(a), \qquad (13.3.4)$$

where we have neglected the radius of the wave. A calculation of $\mathbf{B}$ from the given $\mathbf{M}$ is not carried out here. However, one can assume that in the middle of a long cylinder the field is almost constant and $\Phi_B = 4\pi k_C\epsilon_0 M\,a^2\pi$. The polarization charge

$$\rho_P = -\boldsymbol{\nabla}\cdot\mathbf{P} = \frac{M\omega}{ck_L}\big[-2\theta(a-\varrho) + \delta(a-\varrho)\big]\big[\theta(z)-\theta(z-d)\big]$$

is negative within the cylinder, accompanied by a positive surface charge on the cylinder jacket. Thus, charge neutrality is also established.

Implication from the Lorentz force

The electrons move with the permanent magnet. Thus, the Lorentz force $\mathbf{F}_L$ acts on these. The displacement of the electrons, here into the center of the disk, creates a restoring electric field $\mathbf{E}$. However, in equilibrium, no force $\mathbf{F}$ should act on the electrons:

$$\mathbf{F} = e\left(\mathbf{E} + \frac{k_L}{c}\mathbf{v}\times\mathbf{B}\right) = 0 \quad \Rightarrow \quad \mathbf{E} = -\frac{k_L}{c}\mathbf{v}\times\mathbf{B} = -\frac{k_L\omega}{c}B\varrho. \tag{13.3.5}$$

The voltage arising at the cylinder (radius a) is therefore

$$\phi_B - \phi_A = -\int_0^a \mathrm{d}\mathbf{s}\cdot\mathbf{E} = \frac{k_L\omega}{2\pi c}\Phi_B. \tag{13.3.6}$$

In common consideration, the Lorentz force is associated with an electric particle that cuts the magnetic field lines, which seems difficult when you imagine the magnetic field as fixed on the cylinder, so that it rotates with the electron. Now $\mathbf{B} = \mathbf{B}'$ is time constant and only the electron moves.

The rotating sphere

In reference to [Landau, Lifshitz 8, 1984, §63], the homogeneously magnetized sphere sketched in Fig. 13.9 is considered. The magnetic field $\mathbf{B}_i = (8\pi k_r\mu_0/3)\mathbf{M}$ is homogeneous within the sphere. Outside the sphere, we have a dipole field. The axis of rotation is arranged parallel to the magnetization $\mathbf{M}$, so that

$$\mathbf{E}_i = k_L\mathbf{B}_i\times\mathbf{v}/c = \frac{8\pi k_r k_L}{3}\mu_0\mathbf{M}\times\boldsymbol{\beta} = -\frac{8\pi k_r k_L\mu_0 M\omega}{3c}\varrho \quad r < a. \tag{13.3.7}$$

Due to the rotation, the sphere appears polarized with the charge density ρ_P:

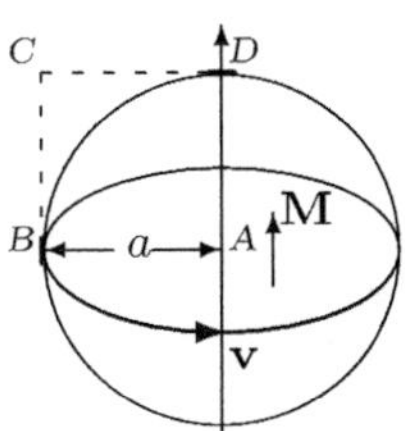

Fig. 13.9. The axis of rotation of the homogeneously magnetized sphere is parallel to the magnetization. $\mathbf{v} = \boldsymbol{\omega}\times\mathbf{x}$. $\mathbf{B}_i = (8\pi k_r\mu_0/3)\,\mathbf{M}$ for $r < a$

$$\mathbf{P} = \boldsymbol{\beta}\times\mathbf{M}/k_L, \quad \rho_P = -\boldsymbol{\nabla}\cdot\mathbf{P} = -\frac{2M\omega}{ck_L}\theta(a-r) + \frac{M\omega a}{ck_L}\sin^2\vartheta\,\delta(a-r).$$

The sphere has a homogeneous negative polarization charge, which is balanced by a positive polarization charge on the surface of the sphere. For the electric field, the situation is more complicated. Although the charge density within the sphere is also negative, and $\mathbf{E}$ is vortex-free

$$\rho = \frac{\boldsymbol{\nabla}\cdot\mathbf{E}}{4\pi k_C} = -\frac{4M\omega}{3ck_L}, \qquad\qquad \boldsymbol{\nabla}\times\mathbf{E} = 0,$$

but the surface charge is not so easy to determine, since the field $\mathbf{E}_a$ outside the sphere is not known. Due to the symmetry, it must be the field of a quadrupole, whose calculation is to be done in problem 13.3. One obtains the scalar potential

$$\phi(\mathbf{x}) = k_C \frac{4\pi M\omega}{9ck_L} \begin{cases} 2(r^2-a^2)+(r^2-3z^2) & r < a \\ \dfrac{a^5}{r^5}(r^2-3z^2) & r > a. \end{cases} \tag{13.3.8}$$

The potential is short-range and negative on the axis of rotation.

13.3.3 Motion of an Infinitely Long Cuboid

We initially described the unipolar induction using (13.3.4) as an induction phenomenon and then identified the Lorentz force as the cause of $\mathbf{E}$ using (13.3.5). In addition, we have given the fields $\mathbf{E}_i$ and $\mathbf{P}$ using the formulas resulting from the Lorentz transformation in (13.3.7). In the configuration to be discussed now, the unipolar induction should be treated only on the basis of the LT, i.e. the STR.

We now consider a uniformly moving and long permanent magnet. The magnetization direction should be, as shown in Fig. 13.10, the z-direction. The reason for this example is that we have two inertial systems for which the Lorentz transformation certainly applies. We refer in this context to Becker, Sauter [1973, §11.3]. The starting point is the permanent magnet sketched in Fig. 13.10. In the rest system of the magnet, the magnetization $\mathbf{M}' = M'\mathbf{e}_z$

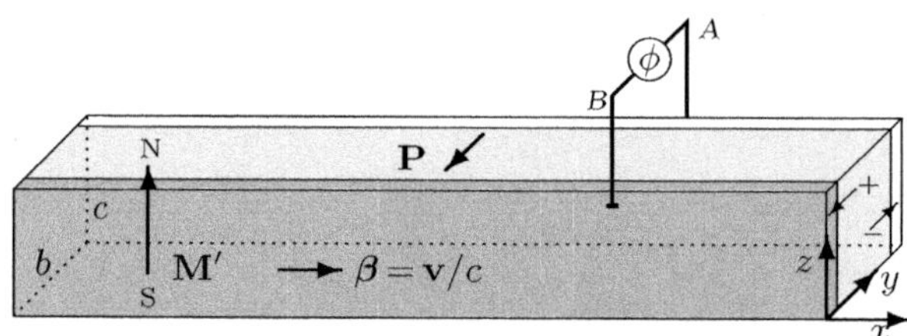

Fig. 13.10. Polarization of a long, magnetized and linearly moving cuboid: $\mathbf{M}' = M'\mathbf{e}_z$, $\boldsymbol{\beta} = \beta\mathbf{e}_x$ and $\mathbf{P} = \gamma\boldsymbol{\beta}\times\mathbf{M}'/k_L = -\gamma\beta M'\mathbf{e}_y/k_L$. The polarization charges are located on the side surfaces. The stationary conductor loop is equipped with sliding contacts to the moving magnet

and the polarization $\mathbf{P}' = 0$. If the magnet moves with $\mathbf{v} = v\mathbf{e}_x$, it appears polarized according to (13.2.6) to the stationary observer:

$$\begin{aligned} \mathbf{M} &= (1-\gamma)\mathbf{M}'\cdot\hat{\boldsymbol{\beta}}\,\hat{\boldsymbol{\beta}} + \gamma\mathbf{M}' = \gamma M'\mathbf{e}_z, \\ \mathbf{P} &= \gamma\boldsymbol{\beta}\times\mathbf{M}'/k_L = \boldsymbol{\beta}\times\mathbf{M}/k_L = -\gamma\beta M'\mathbf{e}_y/k_L. \end{aligned} \tag{13.3.9}$$

For the fields ($\mathbf{E}' = 0$) this means

$$\mathbf{B} = (1-\gamma)\hat{\boldsymbol{\beta}}\cdot\mathbf{B}'\,\hat{\boldsymbol{\beta}} + \gamma\mathbf{B}' \equiv \mathbf{B}'_\parallel + \gamma\mathbf{B}'_\perp,$$

$$\mathbf{E} = -k_L\gamma\boldsymbol{\beta}\times\mathbf{B}' = -k_L\boldsymbol{\beta}\times\mathbf{B} = -\frac{1}{k_L\epsilon_0}\boldsymbol{\beta}\times(\mathbf{H}+4\pi k_r\mathbf{M}). \tag{13.3.10}$$

The component parallel to $\boldsymbol{\beta}$ does not contribute to $\mathbf{E}$. So in the following formulas: $\mathbf{B} \equiv \gamma\mathbf{B}'_\perp$. If one defines

$$\psi(y,z) = \big[\theta(y)-\theta(y-b)\big]\big[\theta(z)-\theta(z-c)\big] \qquad \Rightarrow \qquad \mathbf{M} = \mathbf{M}_0\psi,$$

one obtains for $\mathbf{P}$ and the polarization charges

$$\mathbf{P} = \boldsymbol{\beta}\times\mathbf{M}/k_L = -\beta M\psi\,\mathbf{e}_y/k_L,$$

$$\rho_P = -\boldsymbol{\nabla}\cdot\mathbf{P} = \beta(M/k_L)\big[\delta(y)-\delta(y-b)\big]\big[\theta(z)-\theta(z-c)\big]. \tag{13.3.11}$$

$\mathbf{P}$ is parallel to the cover surfaces (bottom and top), which is why $\boldsymbol{\nabla}\cdot\mathbf{P}$ disappears there. Inside the cuboid there are no polarization charges, but there are surface charges on the side surfaces. The next question concerns the (total) charge density $\rho = \rho_f + \rho_P$:

$$\rho = \frac{\boldsymbol{\nabla}\cdot\mathbf{E}}{4\pi k_C} = k_L\frac{\boldsymbol{\nabla}\cdot(\mathbf{B}\times\boldsymbol{\beta})}{4\pi k_C} = k_L\frac{\boldsymbol{\beta}\cdot(\boldsymbol{\nabla}\times\mathbf{B})}{4\pi k_C}$$

$$\overset{\boldsymbol{\nabla}\times\mathbf{H}=0}{=} \frac{1}{k_L}\boldsymbol{\beta}\cdot(\boldsymbol{\nabla}\times\mathbf{M}) = \rho_P. \tag{13.3.12}$$

This results, not unexpectedly, in $\rho_f = 0$, i.e. $\boldsymbol{\nabla}\cdot\mathbf{E} = -4\pi k_C\boldsymbol{\nabla}\cdot\mathbf{P}$. Now we ask about the vortex densities:

$$\boldsymbol{\nabla}\times\mathbf{P} = -(\beta M/k_L)(\boldsymbol{\nabla}\psi)\times\mathbf{e}_y = (\beta M/k_L)\mathbf{e}_z\cdot\boldsymbol{\nabla}\psi = \beta\boldsymbol{\nabla}\cdot\mathbf{M}/k_L,$$

$$\boldsymbol{\nabla}\times\mathbf{E} = k_L\boldsymbol{\nabla}\times(\mathbf{B}\times\boldsymbol{\beta}) \overset{(A.2.35)}{=} k_L[(\boldsymbol{\beta}\cdot\boldsymbol{\nabla})\mathbf{B} - \boldsymbol{\beta}\boldsymbol{\nabla}\cdot\mathbf{B}] = k_L(\boldsymbol{\beta}\cdot\boldsymbol{\nabla})\mathbf{B}. \tag{13.3.13}$$

The polarization has vortices at the end faces. If we divide $\mathbf{P}$ into a vortex part $\mathbf{P}_w$ and a source part $\mathbf{P}_q$, then within the beam $\mathbf{E} = -4\pi k_C\mathbf{P}_q$ and $\mathbf{D} = 4\pi k_r\mathbf{P}_w$ and outside the beam $\mathbf{D} = \epsilon_0\mathbf{E}$.

The electric field $\mathbf{E}$ is vortex-free due to the infinite length of the beam and the jump of $\mathbf{E}$ in the normal component is determined by the jump of the tangential component of $\mathbf{B}$ (multiplied with β). We calculate the voltage according to

$$\phi = -\int_A^B \mathbf{ds}\cdot\mathbf{E} = \frac{k_L}{c}\int_A^B \mathbf{ds}\cdot(\mathbf{v}\times\mathbf{B}) = -k_L\beta\int_{y_A}^{y_B} dy\,B_z. \tag{13.3.14}$$

Problems for Chapter 13

13.1. *Moving point charge*: A point charge q is at rest in the origin of the inertial system S'. S' moves with $\mathbf{v}$ relative to S, where at the time $t = t' = 0$ also $\mathbf{x} = \mathbf{x}' = 0$

coincide. Calculate the field $\mathbf{E}$ of the point charge in S using Λ and compare it with the field resulting from the Liénard-Wiechert potentials (8.2.46):

$$\mathbf{E}(\mathbf{x},t) = \frac{k_c q\left(1 - \beta^2\right)\mathbf{X}(t)}{\sqrt{\left(1-\beta^2\right)R^2(t) + (\mathbf{X}(t)\cdot\boldsymbol{\beta})^2}^{\,3}}\,, \qquad \mathbf{X}(t) = \mathbf{x} - \boldsymbol{\beta}ct.$$

13.2. *Homogeneous Maxwell equations*: Show that the sum

$$F_{\mu\nu,\lambda} + F_{\lambda\mu,\nu} + F_{\nu\lambda,\mu} = 0 \tag{13.1.26}$$

vanishes and describes the homogeneous Maxwell equations.

13.3. *Potential and field of a rotating magnetized sphere*: A homogeneously magnetized sphere (radius a), as shown in Fig. 13.8, rotates with the angular velocity ω, where the magnetization $\mathbf{M}$ is parallel to the axis of rotation. Calculate the electric fields inside and outside the sphere along with the associated potentials.

References

Arago M., note to a lecture of Nov. 22, 1824 in the Académie royale des Sciences, Ann. chim. phys. **27**, 363 (1824), published as *Sur la découverte d'une novelle action magnétique* in Nouveau Bulletin des Sciences, 5–6 (1825)

Babbage C., Herschel J. *Account of the Repetition of M. Arago's Experiments on the Magnetism Manifested by Various Substances during the Act of Rotation*, Phil. Trans. R. Soc. London **115**, 467–496 (1825)

Barlow P. *A curious electro-magnetic Experiment*, Phil. Mag. **59**, 241–242 (1822)

Bartelmann M., Feuerbacher B., Krüger T., Lüst E., Rebhan A., Wipf A. *Theoretische Physik 2/ Elektrodynamik*, Springer Spektrum (2018)

Becker R., Sauter F. *Theorie der Elektrizität 1*, 21th ed. Teubner, Stuttgart (1973)

Chaichian M., Merches I., Radu D., Tureanu A. *Electrodynamics*, Springer Berlin (2016)

Einstein A. *Eine neue formale Deutung der Maxwellschen Feldgleichungen der Elektrodynamik*, S.B. preuss. Akad. Wiss. 184–188 (1916)

Faraday M. *On some new Electro-Magnetical Motions, and on the Theory of Magnetism*, Quaterly Journal of Science **12**, 74–96 (1822) and other articles herein.

Faraday M. *Experimental Researches in Electricity*, Phil. Trans. R. Soc. Lond. **122**, 125–162 (1832)

Jackson J. D. *Classical Electrodynamics*, 3rd ed., John Wiley & Sons Inc. (1998)

Landau L. D. and Lifshitz E. M. *The Classical Theory of Fields*, vol. 2, 4th ed., Butterworth-Heinemann, reprinted (1994)

Landau L. D., Lifshitz E. M. and Pitaevskii L. P. *Elektrodynamics of continuous media*, vol. 8, 2nd ed., Pergamon Press (1984)

Larmor J. in *Aether and Matter*, Cambridge University Press (1900)

Minkowski H. *Die Grundgleichungen für die elektromagnetischen Vorgänge in bewegten Körpern*, Nachrichten von der Königlichen Ges. d. Wissenschaften zu Göttingen, 53–111 (1908)

Montgomery H. *Unipolar induction: a neglected topic in the teaching of electromagnetism*, Eur. J. Phys. **20**, 271–280 (1999)

Panofsky W., Phillips M. *Classical electricity and magnetism*, 2nd ed. Addison – Wesley, Reading (1962)

Scheck F. *Theoretische Physik 3* 4th ed., Springer Spektrum (2017)

Schwabl F. *Advanced Quantum Mechanics*, 4th ed. Springer Berlin (2008)

Sexl R. U. and Urbantke H. K. *Relativity, Groups, Particles*, Springer Wien (2001)

Sommerfeld A. *Electrodynamics*, Academic Press Inc., New York (1952)

Vanderlinde J. *Classical Electromagnetic Theory* 2nd ed. Springer (2004)

14

Relativistic Mechanics

14.1 Newton's Lex Secunda

According to Mach [1933, p. 240] the 2nd Newton's law remains unchanged in relativistic mechanics:

The change in motion is proportional to the action of the moving force and occurs in the direction of the straight line along which that force acts.

We are looking for the covariant form of this (and other) laws of classical mechanics. First, starting from the velocity (v^μ), the covariant vectors for momentum, acceleration, and (Lorentz) force are defined.

14.1.1 Velocity, Momentum, and Acceleration

Four-velocity and four-momentum

In relativistic mechanics, it is assumed that in the co-moving system, in which a body is at rest, the laws of classical mechanics remain unchanged. The time in this system (12.3.2), the so-called proper time τ or $s = c\tau$, defines the velocity (v^μ) (12.3.6) or $(u^\mu) = (v^\mu)/c$ (12.3.7) as the derivative of the world line (x^μ) with respect to τ or s

$$(v^\mu) = \left(\frac{dx^\mu}{d\tau}\right) = \gamma \begin{pmatrix} c \\ \mathbf{v} \end{pmatrix} \qquad \text{and} \qquad (u^\mu) = \left(\frac{dx^\mu}{ds}\right) = \gamma \begin{pmatrix} 1 \\ \boldsymbol{\beta} \end{pmatrix}. \qquad (14.1.1')$$

We have thus obtained the four-velocity directly in covariant form. The vector is timelike, $u^\mu u_\mu = 1$, i.e. its zeroth component is greater than the spatial part, which can thus be made to disappear by a suitable Lorentz transformation, which is the case in the co-moving system.

© The Editor(s) (if applicable) and The Author(s), under exclusive license
to Springer-Verlag GmbH, DE, part of Springer Nature 2025
D. Petrascheck, F. Schwabl, *Electrodynamics*, https://doi.org/10.1007/978-3-662-71502-4_14

Note: It is often useful to specify the argument that was differentiated: $v^l = \dot{x}^l(\tau)$ or $v^l = \dot{x}^l(t)$. Moreover, the designation of the four-velocity is not uniform[1].

If you multiply (v^μ) by m, you get the four-momentum

$$\mathrm{p}^\mu = m\mathrm{v}^\mu(\tau) = mc u^\mu \qquad \Leftrightarrow \qquad (\mathrm{p}^\mu) = m\gamma \begin{pmatrix} c \\ \mathbf{v} \end{pmatrix} = \begin{pmatrix} \mathrm{p}^0 \\ \boldsymbol{p} \end{pmatrix}. \qquad (14.1.1)$$

m is the rest mass, often also denoted by $m(0)$ or m_0. For the energy E, the relationship known as the *Einstein formula* applies

$$\mathrm{E} = \gamma mc^2 = m(v)\,c^2 \qquad \text{with} \quad m(v) = \gamma m. \qquad (14.1.2)$$

The relativistic mass $m\gamma$, which enters the equations of motion as inertial mass, is speed-dependent. As the kinetic energy of a particle increases, it becomes heavier, i.e. more inert against acceleration. Using (14.1.2) we get

$$\mathrm{p}^0 = \mathrm{p}_0 = \gamma mc = \mathrm{E}/c, \qquad\qquad \boldsymbol{p} = \gamma m\mathbf{v}. \qquad (14.1.3)$$

By contraction, using $u^\mu u_\mu = 1$, we obtain the invariant

$$\mathrm{p}^\mu\,\mathrm{p}_\mu = (\mathrm{p}^0)^2 - \boldsymbol{p}^2 = m^2 c^2 = \mathrm{E}^2/c^2 - \boldsymbol{p}^2. \qquad (14.1.4)$$

As can be seen from (14.1.4), the energy as a function of $\boldsymbol{p}$ or $\mathbf{v}$

$$\mathrm{E} = \sqrt{m^2 c^4 + c^2 \boldsymbol{p}^2} = mc^2/\sqrt{1-\beta^2}$$

is sketched in Fig. 14.1. Near the speed of light, i.e. when $|\boldsymbol{p}| \gg mc$, is $\mathrm{E} \approx |\boldsymbol{p}|c$.

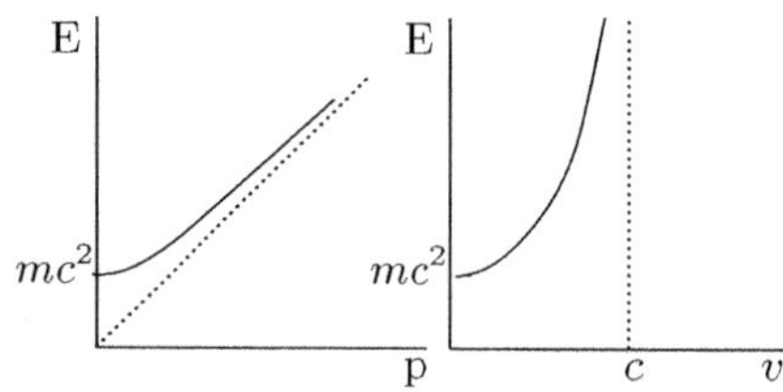

Fig. 14.1. Energy as a function of $\mathrm{p} = |\boldsymbol{p}|$ and v; E(p) is a hyperboloid, the so-called *mass shell*

The four-momentum of the photon

The four-vector (12.4.27) of the wave number has already been encountered in the Doppler effect. Now the photon has the speed $|\mathbf{v}| = c$, which is only possible in connection with the rest mass $m = 0$ is possible. This results in

[1]
$$v^\lambda = \frac{\mathrm{d}x^\lambda}{\mathrm{d}\tau}$$

u^λ Becker, Sauter [1973, (10.5.12)]	u^λ Nolting [2017, (2.39)]
U^λ Jackson [1998, (11.36)]	η^λ Griffiths [2017, (12.41)]

$$u^\lambda = \frac{\mathrm{d}x^\lambda}{\mathrm{d}s}$$

u^λ Panofsky, Phillips [1962, (17-27)]	u^λ Sexl, Urbantke [2001, (4.3)]
u^λ Landau, Lifshitz 2 [1987, (7,1)]	u^λ Chaichian et al. [2016, (7.65)]

$$(\mathrm{p}^\mu) = \big(|\boldsymbol{p}|,\, \boldsymbol{p}\big) \qquad \text{with} \qquad \mathrm{E} = c|\boldsymbol{p}|.$$

$\boldsymbol{p}$ and $\mathrm{E} = c|\boldsymbol{p}|$ are momentum and energy of the photon, where for massless particles $\mathrm{p}^\mu \mathrm{p}_\mu = 0$ applies. The energy of the photon is determined by quantum theory $\mathrm{E} = \hbar\omega$. Thus,

$$\mathrm{p}^0 = \hbar\omega/c = \hbar k \qquad \Rightarrow \qquad \boldsymbol{p} = \hbar\mathbf{k} \qquad \text{and} \qquad (\mathrm{p}^\mu) = \hbar\,(k,\, \mathbf{k})\,.$$

Four-acceleration

The derivative of the velocity with respect to proper time

$$\mathrm{b}^\mu = \frac{\mathrm{d}^2 x^\mu}{\mathrm{d}\tau^2} = \dot{\mathrm{v}}^\mu(\tau) = a^\mu\, c^2 \tag{14.1.5}$$

defines the acceleration, where

$$a^\mu = \frac{\mathrm{d}^2 x^\mu}{\mathrm{d}s^2} \qquad \Rightarrow \qquad (a^\mu) = (\dot{u}^\mu(s)) = (\dot{\gamma}(s),\, \dot{\gamma}(s)\boldsymbol{\beta}+\gamma\dot{\boldsymbol{\beta}}) \tag{14.1.6}$$

is the derivative with respect to s of the dimensionless velocity (u^μ). From

$$\frac{\mathrm{d}}{\mathrm{d}s} u^\mu u_\mu = 0 \qquad \text{follows} \qquad \dot{a}^\mu u_\mu = 0.$$

Four-velocity and four-acceleration are thus orthogonal to each other. For the scalar product applies

$$a^\mu a_\mu = -\gamma^{-2}\,\dot{\gamma}^2(s) - \gamma^2\dot{\boldsymbol{\beta}}^2(s) \le 0\,,$$

i.e. the acceleration is spacelike. This result was already expected due to the orthogonality $u^\mu \dot{u}_\mu = 0$, since (u^μ) is timelike. With an LT, one cannot switch to an inertial system in which the spatial component of the acceleration vanishes. With th help of $\dot{\gamma}(s)=\gamma^3\boldsymbol{\beta}\cdot\dot{\boldsymbol{\beta}}(s)$

$$a^\mu a_\mu = -\gamma^4\big[(1-\beta^2)\dot{\boldsymbol{\beta}}^2(s)+(\boldsymbol{\beta}\cdot\dot{\boldsymbol{\beta}})^2\big] = -\gamma^4\big[\dot{\boldsymbol{\beta}}^2(s)-(\boldsymbol{\beta}\times\dot{\boldsymbol{\beta}})^2\big] \tag{14.1.7}$$

can be brought into a form that we encountered with radiation power (8.2.42).

Equation of motion

In its simplest form, Newton's second law states that force equals mass times acceleration

$$m\mathrm{b}^\mu = \dot{\mathrm{p}}^\mu(\tau) = K^\mu \quad \Leftrightarrow \quad mc^2 a^\mu = c\dot{\mathrm{p}}^\mu(s) = K^\mu \quad \Rightarrow \quad \dot{\boldsymbol{p}}(t) = \mathbf{K}/\gamma. \tag{14.1.8}$$

$\mathbf{K}_N = \mathbf{K}/\gamma$ is the Newtonian force. Depending on the question, it is more convenient to express the equation of motion through the path s, the proper time τ or the time t. The four-force (K^μ) is the so-called *Minkowski force* [Sommerfeld, 1952, §32 (5)], where according to (14.1.10): $K^0 = \mathbf{K} \cdot \boldsymbol{\beta}$.

Free Particle

Before considering the effect of a force on the particle, one assures oneself about the force-free motion:

$$ mc\dot{u}^\mu = 0 \quad \Rightarrow \quad p^\mu = mcu^\mu = \text{const.} \quad \Rightarrow \quad x^\mu(s) = x^\mu(0) + u^\mu s. \qquad (14.1.9) $$

The world line is a straight line. (14.1.9) is an expression of the *1st Newton's law* (*principle of inertia*), here in the formulation by Mach [1933, p. 240]:

A body persists in its state of rest or of uniform straight-line motion, unless it is compelled by forces acting upon it, to change its state.

Motion under the influence of a force

$$ mc^2 \ddot{x}^\mu(s) = mc^2 \, a^\mu(s) = K^\mu. $$

The meaning of the zeroth component of the *Minkowski*-force:

$$ \dot{u}^\mu u_\mu = 0 \qquad \Rightarrow \qquad mc^2 \dot{u}^\mu u_\mu = K^\mu u_\mu = 0 \,. $$

From this it follows that $K_0 c$ is equal to the work done by $\mathbf{K}$ on the particle:

$$ K^0 = \mathbf{K} \cdot \boldsymbol{\beta} = mc^2 \frac{\mathrm{d}\gamma}{\mathrm{d}s} = \frac{1}{c} \frac{\mathrm{d}E}{\mathrm{d}\tau} \,. \qquad (14.1.10) $$

Uniformly accelerated reference system

If in the rest system S' of a particle its acceleration (here in the x^1 direction) is constant, $(b'^\mu) = (0, b, 0, 0)$, then the particle is uniformly accelerated. We ask ourselves how the time in the "stationary laboratory system" compares to the proper time in the uniformly accelerated system, if this moves away at time $t = 0$ with the initial velocity $\mathbf{v} = 0$.

First, we transform according to (13.1.2) the acceleration from the co-moving system into the laboratory system, where it is given by (14.1.5):

$$ b^\mu = \frac{\mathrm{d}^2 x^\mu}{\mathrm{d}\tau^2} = \left(c\frac{\mathrm{d}\gamma}{\mathrm{d}\tau} \,, \frac{\mathrm{d}\gamma v}{\mathrm{d}\tau} \,, 0 \,, 0 \right) = (\gamma\beta b \,, \gamma b \,, 0 \,, 0) \,. $$

From this it follows for the spatial part by integration using $\gamma\mathrm{d}\tau = \mathrm{d}t$ (12.3.2)

$$ \frac{1}{\gamma} \frac{\mathrm{d}\gamma v}{\mathrm{d}\tau} = b \quad \Rightarrow \quad \int_0^t \mathrm{d}t' \frac{\mathrm{d}\gamma v}{\mathrm{d}t'} = \frac{v(t)}{\sqrt{1 - v(t)^2/c^2}} = bt \,. \qquad (14.1.11) $$

As long as $v \ll c$ we have the classical result $v = bt$, a result that for $t \to \infty$ unsurprisingly transitions into $v = c$. The resolution according to v and a further integration results in

$$v = \frac{bt}{\sqrt{1 + b^2 t^2 / c^2}} \qquad \Rightarrow \qquad x = \int_0^t dt'\, v(t') = b\,\frac{c^2}{b^2}\left(\sqrt{1 + \frac{b^2 t^2}{c^2}} - 1\right).$$

Again, we have for short times (or small velocities) the classical result $x = b\,t^2/2$, which for $t \to \infty$ transitions into $x = ct$.

The proper time is obtained by means of[2]: (12.3.3)

$$\tau = \int_0^t dt'\, \sqrt{1 - \beta^2} = \int_0^t \frac{dt'}{\sqrt{1 + (bt'/c)^2}} = \frac{c}{b}\,\ln\left(\frac{bt}{c} + \sqrt{1 + \left(\frac{bt}{c}\right)^2}\right). \tag{14.1.12}$$

For short times or small velocities, $bt \ll c$, the proper time $\tau \lesssim t$ as is classically expected, while for $t \to \infty$ the proper time only increases logarithmically: $\tau \sim (c/b)\,\ln(2bt/c)$. In this limit case, although $v \approx c$, it never reaches c, as then τ should not increase further.

14.1.2 Radiation Power and Radiation Reaction

On the radiation power of the point charge

One can expect that the radiation power P of the moving charge is invariant under the LT. If one starts from the Larmor formula (8.2.39), then according to (14.1.7) $\dot{u}^\mu(\tau)\dot{u}_\mu(\tau) = -\ddot{\beta}^2(\tau)$, when $\beta = 0$. This implies

$$P = -\frac{2k_C e^2}{3c}\frac{du^\mu}{d\tau}\frac{du_\mu}{d\tau} = -\frac{2k_C e^2}{3m^2 c^3}\frac{dp^\mu}{d\tau}\frac{dp_\mu}{d\tau}. \tag{14.1.13}$$

Since P is a scalar, its form must be preserved under Lorentz transformations, i.e. for finite β. $\dot{u}^\mu \dot{u}_\mu$ was calculated for finite β in (14.1.7). If one replaces $d\tau$ with dt/γ, then from (14.1.13) follows Liénard's result (8.2.42). The radiation in the relativistic case is of particular interest in particle accelerators, where one distinguishes between linear and circular accelerators.

Linear acceleration

In the case of linear motion, β and $\dot{\beta}$ are parallel, so that (14.1.7) simplifies accordingly:

$$\frac{d\gamma\beta}{dt} = \gamma^3 \beta\cdot\dot{\beta}\beta + \gamma\dot{\beta} = \gamma^3\dot{\beta} \quad \Rightarrow \quad \frac{du^\mu}{ds}\frac{du_\mu}{ds} = -\left(\gamma^2\frac{d\beta}{ds}\right)^2 = -\left(\frac{1}{c}\frac{d\gamma\beta}{dt}\right)^2.$$

Velocity and momentum are $(u^\mu) = (\gamma, \beta\gamma)$ and $p^\mu = mcu^\mu$, from which follows:

$$P = \frac{2k_C e^2}{3c}\left(\frac{du}{dt}\right)^2 = \frac{2k_C e^2}{3m^2 c^3}\left(\frac{dp}{dt}\right)^2.$$

---

[2] Auxiliary formula (B.5.15): $\int dx/\sqrt{1 + a^2 x^2} = (1/a)\ln\left(ax + \sqrt{1 + a^2 x^2}\right) \qquad a > 0$.

Of interest is the ratio of the radiation power to the power of the external forces [Schwinger, 1949], for which we do some intermediate calculations:

$$\frac{\mathrm{dE}}{\mathrm{dt}} = mc^2 \frac{\mathrm{d}\gamma}{\mathrm{dt}} = mc^2 \gamma^3 \boldsymbol{\beta} \cdot \frac{\mathrm{d}\boldsymbol{\beta}}{\mathrm{dt}} = mc^2 \gamma^3 \beta\dot{\beta},$$

$$\frac{\mathrm{d}\boldsymbol{p}}{\mathrm{dt}} = mc\frac{\mathrm{d}\gamma\boldsymbol{\beta}}{\mathrm{dt}} = mc\gamma^3 \dot{\boldsymbol{\beta}} = \frac{\boldsymbol{\beta}}{\beta^2 c}\frac{\mathrm{dE}}{\mathrm{dt}}.$$

We now assume that m is the electron mass, so that r_e is the classical electron radius and $\mathrm{E}_0 = mc^2 \approx 0.511\,\mathrm{MeV}$ is the rest energy of the electron:

$$\frac{P}{\mathrm{dE}/\mathrm{dt}} = \frac{2k_C e^2}{3m^2 c^3}\frac{1}{\beta^2 c^2}\frac{\mathrm{dE}}{\mathrm{dt}} = \frac{2r_e}{3\mathrm{E}_0}\frac{1}{\beta^2 c}\frac{\mathrm{dE}}{\mathrm{dt}}. \tag{14.1.14}$$

In a further step, we measure the ratio of P to dE/dt with the energy change per unit length $\mathrm{d}x = \beta c\,\mathrm{dt}$:

$$\frac{P}{\mathrm{dE}/\mathrm{dt}} = \frac{2r_e}{3\mathrm{E}_0}\frac{1}{\beta}\frac{\mathrm{dE}}{\mathrm{d}x}. \tag{14.1.15}$$

However, the energy gain over a distance r_e is much smaller than E_0, so that the radiation in the linear accelerator should hardly play a role.

Synchrotron

If the electron is kept on a circular path, it is assumed that despite the energy change in the direction of motion $\dot{\boldsymbol{\beta}} \perp \boldsymbol{\beta}$. For P one then obtains using (14.1.7) and $\dot{\gamma} = 0$

$$P = \frac{2k_C e^2}{3c}\gamma^4\,\dot{\boldsymbol{\beta}}(t)^2 = \frac{2k_C e^2}{3c}\,\gamma^2\,\dot{\mathbf{u}}(t)^2. \tag{14.1.16}$$

If R is the radius of the circular path and ω is the angular velocity with which the electron moves, then

$$|\dot{\mathbf{u}}(t)| = \gamma\,\beta\,\omega = \gamma\,\beta\,(c\beta/R) \qquad \text{and} \qquad \gamma = \mathrm{E}/\mathrm{E}_0.$$

This results in

$$P = k_C \frac{2}{3}\,\omega\,\frac{e^2}{R}\,\beta^3\left(\frac{\mathrm{E}}{\mathrm{E}_0}\right)^4. \tag{14.1.17}$$

In a full revolution with $v \approx c$ the energy

$$\frac{\Delta\mathrm{E}}{\mathrm{E}_0} = \frac{4\pi}{3}\frac{r_e}{R}\left(\frac{\mathrm{E}}{\mathrm{E}_0}\right)^4 \tag{14.1.18}$$

is radiated. This can be seen from the fact that large radii R are necessary to achieve high energies. In a synchrotron, where one wants to achieve higher radiation powers, one will superimpose the circular path with wave lines that are generated by undulators.

Radiation reaction

In section 8.5, the reaction of the radiation on the electron was treated for the limit $|\mathbf{v}| \to 0$. Now the more general case of finite speeds is to be treated, trying to derive from considerations of covariance the generalization of the Abraham-Lorentz motion equation (8.5.4), the relativistic *Lorentz-Abraham-Dirac equation*, the so-called LAD equation, to obtain. We again orient ourselves on the work of Rohrlich [2008].

We remember that the radiated energy for a fast moving electron is not given by the Larmor formula, but by the Liénard formula (8.2.42) or (14.1.13)

$$P = -\frac{2k_C e^2}{3c^3}\,\dot{\mathbf{v}}^\mu(\tau)\,\dot{\mathbf{v}}_\mu(\tau) \overset{(8.5.3)}{=} -m\tau_0\,\dot{\mathbf{v}}^\mu\dot{\mathbf{v}}_\mu = m\tau_0\ddot{\mathbf{v}}^\mu\mathbf{v}_\mu. \tag{14.1.19}$$

The dots here consistently denote derivatives with respect to τ. On the right, we have used the fact that $\mathbf{v}^\mu(\tau)\dot{\mathbf{v}}_\mu(\tau) = 0$, i.e.

$$\frac{\mathrm{d}}{\mathrm{d}\tau}\dot{\mathbf{v}}^\mu\mathbf{v}_\mu = \dot{\mathbf{v}}^\mu\dot{\mathbf{v}}_\mu + \ddot{\mathbf{v}}^\mu\mathbf{v}_\mu.$$

A motion equation of the form

$$m\dot{\mathbf{v}}^\mu(\tau) = m\tau_0\ddot{\mathbf{v}}^\mu(\tau)$$

is not covariant, since only the left side disappears when multiplied by $\mathbf{v}_\mu$. This can be "repaired" by substracting the contribution resulting from multiplication with $\mathbf{v}_\mu$:

$$m\dot{\mathbf{v}}^\mu = m\tau_0\Big[\ddot{\mathbf{v}}^\mu + \frac{1}{c^2}\dot{\mathbf{v}}^\nu\dot{\mathbf{v}}_\nu\,\mathbf{v}^\mu\Big].$$

The zeroth component of this equation multiplied by c must represent the energy balance, at least for $v \to 0$:

$$\frac{\mathrm{d}\mathrm{E}}{\mathrm{d}\tau} = \tau_0\frac{\mathrm{d}^2\mathrm{E}}{\mathrm{d}\tau^2} - P(\tau)\frac{\mathbf{v}^0}{c}.$$

Here we have substituted the relativistic energy $\mathrm{E} = mc^2\gamma$. The comparison with (8.5.5) immediately shows that the 1st term on the right side is the Schott term and the 2nd term indicates the energy loss according to the Liénard formula. Now we add an external force F_{ext}^μ and obtain the LAD equation

$$m\dot{\mathbf{v}}^\mu(\tau) = F_{\mathrm{ext}}^\mu + m\tau_0\Big(\ddot{\mathbf{v}}^\mu + \frac{1}{c^2}\dot{\mathbf{v}}^\nu\dot{\mathbf{v}}_\nu\,\mathbf{v}^\mu\Big). \tag{14.1.20}$$

It should be noted that the LAD equation also includes the momentum changes of the electron due to the radiation. As in the non-relativistic case, there are also solutions with acausal behavior here. Attempts are made in the same way to eliminate these by requirements on the external forces. To simplify the notation, we introduce a projection tensor:

$$P^{\mu\nu} = g^{\mu\nu} - \frac{1}{c^2}\mathsf{v}^{\mu}\mathsf{v}^{\nu} \qquad\Rightarrow\qquad m\dot{\mathsf{v}}^{\mu}(\tau) = F^{\mu}_{\text{ext}} + m\tau_0 P^{\mu\nu}\ddot{\mathsf{v}}_{\nu}.$$

We differentiate the LAD equation to reintroduce $\ddot{\mathsf{v}}_{\mu}$ into it, neglecting the contributions of order $O(\tau_0^2)$:

$$m\dot{\mathsf{v}}^{\mu} = F^{\mu}_{\text{ext}} + \tau_0 P^{\mu\nu}\dot{F}_{\text{ext}\,\nu} \qquad \text{with} \qquad |\tau_0 P^{\mu\nu}\dot{F}_{\text{ext}\,\nu}| \ll |F^{\mu}_{\text{ext}}|. \qquad (14.1.21)$$

The condition on the right is the relativistic formulation of (8.5.9), which is intended to prevent too rapid a variation of the external force.

14.1.3 Lorentz Force

According to Lorentz's electron theory, the force per unit volume on a spatially limited charge distribution (point charge, see sect. 5.1)

$$\mathbf{f}(\mathbf{x}, t) = \rho\big(\mathbf{E} + k_L \boldsymbol{\beta}\times\mathbf{B}\big) = \frac{1}{c}\big(j^0\,\mathbf{E} + k_L\mathbf{j}\times\mathbf{B}\big). \qquad (14.1.22)$$

$\mathbf{E}$ and $\mathbf{B}$ are external fields. If you substitute for E_i and B_k the components of the field strength tensor (13.1.19) and take into account that $j_j = -j^j$ is the covariant component, you get

$$\begin{aligned}
f^i &= \frac{1}{c}\big(E_i j_0 - k_L \epsilon_{ijk} j_j B_k\big) = \frac{k_L}{c}\Big(F^{i0}\,j_0 + \frac{1}{2}\epsilon_{ijk}\epsilon_{klm}j_j F^{lm}\Big)\\
&= \frac{k_L}{c}\big(F^{i0}j_0 + F^{ij}\,j_j\big) = \frac{k_L}{c}\,F^{i\lambda}j_\lambda.
\end{aligned} \qquad (14.1.23)$$

This is obviously the spatial part of a four-vector of force density, whose zeroth component

$$f^0 = \frac{k_L}{c}F^{0l}j_l = \frac{1}{c}\,\mathbf{E}\cdot\mathbf{j} \qquad (14.1.24)$$

is the power density of the current multiplied by $1/c$ ((5.3.6): $\dot{u}_{\text{mech}} = \mathbf{j}\cdot\mathbf{E}$):

$$f^{\mu} = \frac{k_L}{c}F^{\mu\nu}j_\nu. \qquad (14.1.25)$$

For a point charge, (13.1.4) applies

$$(j^{\mu}) = \rho_q(x)\,(c\,\mathbf{v}) = \rho_q c\gamma^{-1}\,(u^{\mu}), \qquad \rho_q(x) = q\delta^{(3)}(\mathbf{x}-\mathbf{x}_q(t)),$$

from which the force density

$$(f^{\mu}) = k_L\rho_q\,\gamma^{-1}(F^{\mu\nu}u_\nu) = \rho_q\,(\mathbf{E}\cdot\boldsymbol{\beta},\ \mathbf{E} + k_L\boldsymbol{\beta}\times\mathbf{B})$$

follows. The total force on the point charge, the Lorentz force (1.2.5)

$$\mathbf{F} = \int \mathrm{d}^3 x\, \mathbf{f} = q(\mathbf{E} + k_L \boldsymbol{\beta} \times \mathbf{B}) \tag{14.1.26}$$

becomes covariant by multiplying with γ. This results in the Minkowski force:

$$(K^\mu) = \gamma(\boldsymbol{\beta}\cdot\mathbf{F},\ \mathbf{F}) = \gamma q(\boldsymbol{\beta}\cdot\mathbf{E},\ \mathbf{E} + k_L \boldsymbol{\beta}\times\mathbf{B}) \stackrel{(14.1.23)}{=} qk_L(F^{\mu\nu}u_\nu). \tag{14.1.27}$$

In the rest system of the charge S' the non-relativistic (Newtonian) force $\mathbf{K}'$ acts on it

$$(K'^\mu) \stackrel{(13.1.3)}{=} (0,\ \mathbf{F}_\parallel + \gamma\mathbf{F}_\perp) \stackrel{(13.1.29)}{=} (0,\ q\mathbf{E}') \ \Rightarrow\ \mathbf{F} = \mathbf{K}'_\parallel + \gamma^{-1}\mathbf{K}'_\perp. \tag{14.1.28}$$

There are differences of opinion as to whether $\mathbf{K}'$ [Weinberg, 1972, (2.3.5)] or $\mathbf{F}$ [Nolting, 2017, (2.47)] is the classical non-relativistic force, which Fließbach [2018, §4] explains in detail. The equation of motion for an electron in the electromagnetic field (5.4.6)

$$\frac{\mathrm{d}\boldsymbol{p}}{\mathrm{d}t} = \mathbf{F} = e(\mathbf{E} + k_L \boldsymbol{\beta}\times\mathbf{B}) \tag{14.1.29}$$

we have referred to the acting force $\mathbf{F}$ as the Lorentz equation. (14.1.29) is the spatial part of a covariant equation of motion, although both the left and the right side are not formulated covariantly. If (14.1.29) is multiplied by γ and expanded according to (14.1.25) with the power provided by the field $\boldsymbol{\beta}\cdot\mathbf{E}$, covariance is obtained

$$\frac{\mathrm{d}p^\mu}{\mathrm{d}\tau} = K^\mu \stackrel{(14.1.27)}{=} ek_L F^{\mu\nu}u_\nu. \tag{14.1.30}$$

$\mathbf{K} = \gamma\mathbf{F}$ is the spatial part of the covariant force; finally, the fields $\mathbf{E}$ and $\mathbf{B}$ have been replaced by the field strength tensor.

If the radiation reaction is also included in the motion equation, we obtain in the non-relativistic case the Abraham-Lorentz equation (8.5.4) and in covariant form the Lorentz-Abraham-Dirac equation (14.1.20).

14.1.4 Energy-Momentum Tensor

The stress tensor and the energy-momentum balance

In section 5.6 on energy and momentum balance, it was shown in (5.6.12) that the total force density, i.e., the "mechanical" Lorentz force density plus the force density of the field is given by the divergence of the Maxwell stress tensor (T_{ij}) (5.6.11) (momentum theorem). An attempt is now made to find the covariant tensor $(T_M^{\mu\nu})$ associated with T_{ij}, whose divergence is the "mechanical" force density f^i:

$$\dot{p}_{\text{mech}\,i} = -\dot{p}_{\text{Feld}\,i} + T_{ni,n} \;\Rightarrow\; f^i = T_M^{\nu i}{}_{,\nu} : \quad T_M^{0i} = -c p_{\text{Feld}\,i}, \quad T_M^{ni} = T_{ni}.$$

f^0 is determined using Poynting's theorem (5.6.5):

$$\dot{u}_{\text{mech}} = -\dot{u}_{\text{Feld}} - \mathbf{\nabla}\cdot\mathbf{S} \;\Rightarrow\; f^0 = T_M^{\nu 0}{}_{,\nu} : \quad T_M^{00} = -u_{\text{Feld}}, \quad T_M^{n0} = -S_n/c.$$

Energy and momentum balance are determined by the divergence of the tensor referred to by Sommerfeld [1952, §35] as the *stress-energy tensor*

$$f^\mu = T_M^{\nu\mu}{}_{,\nu}\,, \qquad\qquad (T_M^{\mu\nu}) = \begin{pmatrix} -u_{\text{Field}} & -c\boldsymbol{p}_{\text{Field}}^{\mathsf{T}} \\ -\mathbf{S}/c & (T_{mn}) \end{pmatrix}. \tag{14.1.31}$$

In the derivation of the tensor, which goes back to Minkowski [1908, (74)] we start from the Lorentz force density (14.1.25) and insert into this the four-current from the inhomogeneous Maxwell equation (13.2.2):

$$j_f^\mu = \frac{c}{k_L}\frac{1}{4\pi k_r} H^{\nu\mu}{}_{,\nu}\,,$$

$$f^\mu = \frac{k_L}{c} F^{\mu\nu} j_{f\nu} = \frac{1}{4\pi k_r} F^{\mu\nu} g_{\nu\rho} H^{\sigma\rho}{}_{,\sigma} = \frac{1}{4\pi k_r}\Big[\partial_\sigma\big(F^{\mu\nu} g_{\nu\rho} H^{\sigma\rho}\big) - F^{\mu\nu}{}_{,\sigma} g_{\nu\rho} H^{\sigma\rho}\Big].$$

Now the 2nd term is reshaped using (13.1.26) $(F_{\lambda\rho,\sigma} + F_{\sigma\lambda,\rho} = F_{\sigma\rho,\lambda})$:

$$F^{\mu\nu}{}_{,\sigma} g_{\nu\rho} H^{\sigma\rho} = g^{\mu\lambda} F_{\lambda\rho,\sigma} H^{\sigma\rho} \overset{\sigma\leftrightharpoons\rho}{=} \frac{g^{\mu\lambda}}{2} H^{\sigma\rho}\big(F_{\lambda\rho,\sigma} - F_{\lambda\sigma,\rho}\big)$$

$$= \frac{g^{\mu\lambda}}{2} H^{\sigma\rho} F_{\sigma\rho,\lambda} \overset{\epsilon,\mu=\text{const.}}{=} \frac{g^{\mu\lambda}}{4}\partial_\lambda H^{\sigma\rho} F_{\sigma\rho}\,.$$

The sought tensor, which fulfills (14.1.31), thus has the form

$$T_M^{\mu\nu} = \frac{1}{4\pi k_r}\big(F^{\mu\lambda} g_{\lambda\rho} H^{\nu\rho} - \frac{g^{\mu\nu}}{4} H^{\lambda\rho} F_{\lambda\rho}\big). \tag{14.1.32}$$

T_M is trace-free $(T_M^{\mu\nu} g_{\mu\nu} = 0)$ and does not depend on the signature (12.2.2').

The representation of the $T_M^{\mu\nu}$ by $\mathbf{E}$, $\mathbf{D}$ and $\mathbf{B}$, $\mathbf{H}$ is somewhat tedious. We start here with the 1st term $(T_1^{\mu\nu})$ of (14.1.32):

$$4\pi k_r T_1^{00} = -F^{0l} H^{0l} = -\mathbf{E}\cdot\mathbf{D}, \qquad 4\pi k_r T_1^{0n} = -F^{0l} H^{nl} = -E_l\,\epsilon_{nlk} H_k/k_L,$$

$$4\pi k_r T_1^{mn} = F^{m0} H^{n0} - F^{ml} H^{nl}, \qquad 4\pi k_r T_1^{m0} = -F^{ml} H^{0l} = -\epsilon_{mlk} B_k D_l k_L,$$

$$= E_m D_n - \epsilon_{mlr}\epsilon_{nls} B_r H_s = E_m D_n - \delta_{mn}\mathbf{B}\cdot\mathbf{H} + H_m B_n.$$

In the last term we substitute for the invariant: $H^{\sigma\rho} F_{\sigma\rho} = 2(\mathbf{B}\cdot\mathbf{H} - \mathbf{E}\cdot\mathbf{D})$. Summarized, this results in the stress tensor (14.1.31)

$$(T_M^{\mu\nu}) = \frac{1}{4\pi k_r}\begin{pmatrix} -(\mathbf{E}\cdot\mathbf{D} + \mathbf{B}\cdot\mathbf{H})/2 & -k_L\mathbf{D}\times\mathbf{B} \\ -\mathbf{E}\times\mathbf{H}/k_L & \big(E_m D_n + H_m B_n - \delta_{mn}(\mathbf{E}\cdot\mathbf{D} + \mathbf{B}\cdot\mathbf{H})/2\big) \end{pmatrix}.$$

It was not addressed that T_M in matter is asymmetric, which (in analogy to elasticity theory) could lead to torques [Sommerfeld, 1952, §35]. The symmetric tensor proposed by Abraham [1909] [Pauli, 1921, (303)] has led to the so-called Abraham-Minkowski debate McDonald [2017] indicates in the literature.

The symmetric energy-momentum tensor

The energy-momentum tensor of a system consists of the sum of the mechanical and the electromagnetic parts: $T_S^{\nu\mu} = T_{\text{mech}}^{\nu\mu} + T^{\nu\mu}$. It fulfills the continuity equation $T_S^{\nu\mu}{}_{,\nu} = 0$ [Schwabl, 2008, (12.4.3)]. The "mechanical" part is determined by $\dot{p}^\mu$. Thus, the sign changes in (14.1.31):

$$f^\mu = (k_L/c)F^{\mu\nu}j_\nu = -T^{\nu\mu}{}_{,\nu}. \tag{14.1.31'}$$

In the microscopic theory of electrodynamics, the energy-momentum tensor derived using (14.1.31) is symmetric. $(T^{\mu\nu})$ is determined up to $T^{\nu\mu}{}_{,\nu} = 0$. One sets in (14.1.32) $\epsilon_r = \mu_r = 1$ and changes the sign:

$$(T^{\kappa\lambda}) = \frac{-k_L^2}{4\pi k_C}\left(F^{\kappa\nu}g_{\nu\rho}F^{\lambda\rho} - \frac{g^{\kappa\lambda}}{4}F^{\nu\rho}F_{\nu\rho}\right) = \begin{pmatrix} u_{\text{Feld}} & \mathbf{S}^{\mathrm{T}}/c \\ \mathbf{S}/c & -(\sigma_{kl}) \end{pmatrix}. \tag{14.1.33}$$

The stress T_{kl} is henceforth called σ_{kl}, so as not to conflict with $(T_{\kappa\lambda})$. The row or column added to the three-dimensional tensor are the field energy density (5.6.4) and the energy flux density (5.6.3):

$$\sigma_{kl} \overset{(5.6.11)}{=} \frac{1}{4\pi k_C}\left[E_k E_l + k_L^2 B_k B_l - \frac{1}{2}(E^2 + k_L^2 B^2)\delta_{kl}\right]. \tag{14.1.34}$$

$$u_{\text{Field}} = \frac{1}{8\pi k_C}(E^2 + k_L^2 B^2), \qquad \mathbf{S} = c^2 \boldsymbol{p}_{\text{Field}} = \frac{k_L c}{4\pi k_C}\mathbf{E}\times\mathbf{B}, \tag{14.1.35}$$

The zeroth component of the four-divergence of the energy-momentum tensor

$$f^0 = -\partial_\lambda T^{\lambda 0} \qquad \Leftrightarrow \qquad \dot{u}_{\text{mech}}(t) = \mathbf{j}\cdot\mathbf{E} = -\dot{u}_{\text{Field}}(t) - \boldsymbol{\nabla}\cdot\mathbf{S} \tag{14.1.36}$$

results in the energy balance (5.6.5). $cf^0 = \mathbf{j}\cdot\mathbf{E}$ is the power density provided by $\mathbf{E}$ (see (14.1.24) or (5.3.6)). The spatial components of the four-divergence of $T^{\kappa\lambda}$

$$f^k = -\partial_\lambda T^{\lambda k} \qquad \Leftrightarrow \qquad f^k = \rho E_k + (\mathbf{j}\times\mathbf{B})_k = -\dot{p}_{\text{Field}k} + \nabla_l\sigma_{lk} \tag{14.1.37}$$

represent the balance equation for the momentum densities (5.6.12).

From all previous presentations on electrodynamics, it was clear that the covariance is built into the laws, and all equations and conservation quantities can be derived in a simple and elegant way from the tensors (j^μ), (A^μ) and $(F^{\mu\nu})$.

14.2 Lagrange Formalism

It is necessary to adapt the dynamics of the theory of relativity, which is done here using the principle of least action already used in section 5.4, with a Lagrange function suitable for the theory of relativity. Following Jackson [1998, §12], in section 14.2.1 the Lagrange function is determined analogously to the non-relativistic theory in section 5.4, while in section 14.2.2 a derivation based only on covariance is applied. This necessitates a certain degree of duplication.

14.2.1 Relativistic Lagrange Function

We proceed unchanged from the action integral (5.4.1). In a relativistic theory, S must be a Lorentz scalar:

$$S = \int_{t_1}^{t_2} dt\, L(\mathbf{x}, \mathbf{v}) = \frac{1}{c} \int_{s_1}^{s_2} ds\, L_r \qquad \text{with} \qquad L_r = \gamma\, L. \tag{14.2.1}$$

The path element $ds = \sqrt{x_\mu x^\mu}$ of the world line is a Lorentz scalar. It follows that $L_r = \gamma\, L$ is also such a scalar.

We do not take the opportunity here to minimize the world line of the particle by means of L_r by varying x^μ and u^μ, but return to the integral over $dt\, L$ and obtain by variation $\delta S{=}0$ of $\mathbf{x}$ and $\mathbf{v}$ the already known Euler-Lagrange equations (5.4.2):

$$\frac{\partial L}{\partial \mathbf{x}} - \frac{d}{dt}\left(\frac{\partial L}{\partial \mathbf{v}}\right) = 0. \tag{14.2.2}$$

Lagrange function for a free particle

L must not depend on the location for free particles, but only on the velocity. The only invariant that can be formed with the velocity (12.3.7) is $u^\mu u_\mu {=} 1$. Accordingly, L_r must be a constant of the dimension of an energy: $L_r = -\alpha m c^2$. L must still be compatible with the non-relativistic form $L_{\mathrm{nr}} = m v^2/2$ for $v \ll c$, which is the case for $\alpha = 1$:

$$L = -\frac{mc^2}{\gamma} \overset{v \ll c}{=} -mc^2 + \frac{mv^2}{2}. \tag{14.2.3}$$

The rest energy mc^2 has no influence on the variation of the action.

Note: In the Lagrange function $L = K - V$, the portion for the free particle is referred to as the *kinetic potential* [Sommerfeld, 1952, §32]. The derivatives $p_i = \partial K/\partial v_i$ are the momenta. In classical, non-relativistic mechanics K is the kinetic energy T. This disappears with $v \to 0$. The definition

$$L_{\mathrm{frei}} = K = mc^2\left(1 - 1/\gamma\right) \tag{14.2.4}$$

would be more adequate in many respects, since $K(v \ll c) \approx T = mv^2/2$. The abbreviation was used again here

$$\gamma = 1/\sqrt{1-\beta^2} \qquad \text{with} \quad \boldsymbol{\beta} = \mathbf{v}/c. \tag{14.2.5}$$

For the momentum, one obtains from the Euler-Lagrange equations (14.2.2)

$$\boldsymbol{p} = \frac{\partial \mathrm{L}}{\partial \mathbf{v}} = m\gamma\mathbf{v}, \tag{14.2.6}$$

a result that we have already used. From (14.2.2):

$$\frac{\mathrm{d}}{\mathrm{d}t}\frac{\partial \mathrm{L}}{\partial \mathbf{v}} = \frac{\mathrm{d}\boldsymbol{p}}{\mathrm{d}t} = \frac{\partial \mathrm{L}}{\partial \mathbf{x}} = 0$$

it follows also, that the free particle does not experience any acceleration, so its speed remains constant. The energy E of the particle follows from

$$\mathrm{E} = \boldsymbol{p} \cdot \mathbf{v} - \mathrm{L} = m\gamma v^2 + \frac{mc^2}{\gamma} = mc^2\gamma \overset{\beta\to0}{\approx} mc^2 + \frac{mv^2}{2}. \tag{14.2.7}$$

The first term is the rest energy of the particle. If you insert (14.2.6) into (14.2.5), you get

$$\gamma = \sqrt{m^2c^2 + \mathrm{p}^2}/mc \tag{14.2.8}$$

and the energy as a function of momentum

$$\mathrm{H} = c\sqrt{m^2c^2 + \mathrm{p}^2}. \tag{14.2.9}$$

Particle in the electromagnetic field

The Lorentz transformation was derived under the assumption that the laws of electrodynamics apply in all inertial systems. Therefore, we could directly take the part of the electromagnetic field from the Lagrange function (5.4.9):

$$\mathrm{L}_{\mathrm{el}} = -e\big(\phi - k_L\boldsymbol{\beta}\cdot\mathbf{A}\big) = -ek_L\gamma^{-1}\,u_\mu\,A^\mu. \tag{14.2.10}$$

As expected, $\gamma\,\mathrm{L}_{el}$ is a Lorentz scalar. We get a more direct access to the action integral via the electromagnetic potential:

$$\mathrm{S}_{\mathrm{el}} = -e\frac{k_L}{c}\int_1^2 \mathrm{d}x^\mu\,A_\mu = -e\frac{k_L}{c}\int_{s_1}^{s_2}\mathrm{d}s\,\frac{\mathrm{d}x^\mu}{\mathrm{d}s}A_\mu \;\Rightarrow\; \mathrm{L}_{\mathrm{el}} = -\frac{ek_L}{\gamma}u^\mu A_\mu. \tag{14.2.11}$$

The sign is determined by $\mathrm{L} = T - V$, where V here is the electromagnetic potential. Thus,

$$\mathrm{L} = -\frac{mc^2}{\gamma} - \frac{ek_L}{\gamma}u^\mu A_\mu = -\frac{mc^2}{\gamma} - e\big(\phi - \frac{k_L}{c}\mathbf{A}\cdot\mathbf{v}\big). \tag{14.2.12}$$

This results in the generalized (canonical) momentum

$$\boldsymbol{P} = \frac{\partial \mathrm{L}}{\partial \mathbf{v}} = \boldsymbol{p} + e\frac{k_L}{c}\mathbf{A} \tag{14.2.13}$$

with $\boldsymbol{p} = m\gamma\mathbf{v}$, the spatial components of the kinetic four-momentum. The Euler-Lagrange equation (14.2.2) is

$$\dot{\boldsymbol{P}}(t) = \dot{\boldsymbol{p}} + e\Big(\frac{k_L}{c}\frac{\partial \mathbf{A}}{\partial t} + k_L(\boldsymbol{\beta}\cdot\boldsymbol{\nabla})\mathbf{A}\Big) = \frac{\partial \mathrm{L}}{\partial \mathbf{x}} = -e\boldsymbol{\nabla}\big(\phi - k_L\boldsymbol{\beta}\cdot\mathbf{A}\big).$$

The individual terms can be brought into the more compact form[3] of the Lorentz equation (14.1.30)

$$\dot{\boldsymbol{p}}(t) = e\big(\mathbf{E} + k_L\boldsymbol{\beta}\times\mathbf{B}\big) = \mathbf{F} = \mathbf{K}/\gamma \tag{14.2.14}$$

where, unlike in (5.4.6), the momentum is relativistic. Multiplying both sides by $\boldsymbol{\beta} = \boldsymbol{p}/\mathrm{p}^0$ results in (problem 14.1)

$$\boldsymbol{\beta}\cdot\dot{\boldsymbol{p}}(t) = \dot{\mathrm{p}}^0(t) = K^0/\gamma = \boldsymbol{\beta}\cdot\mathbf{E} \qquad \Leftrightarrow \qquad \frac{\mathrm{dE}}{\mathrm{d}t} = \mathbf{v}\cdot\mathbf{E}. \tag{14.2.15}$$

Covariant becomes (14.2.14) only when both sides are multiplied by γ:

$$\big(\dot{\mathrm{p}}^\mu(\tau)\big) = \big(K^\mu\big) = e\gamma\big(\boldsymbol{\beta}\cdot\mathbf{E},\ \mathbf{E}+k_L\boldsymbol{\beta}\times\mathbf{B}\big).$$

For the Hamilton function (see section 5.4) one obtains [Jackson, 1998, (12.17)]

$$\mathcal{H} = \boldsymbol{P}\cdot\mathbf{v} - \mathrm{L} = mc^2\gamma + e\phi = c\sqrt{m^2c^2 + (\boldsymbol{P}-e\frac{k_L}{c}\mathbf{A})^2} + e\phi.$$

14.2.2 Covariant Formulation of Hamilton's Principle

So far, we have been guided by the idea that the classical laws apply in the rest system of the particle which is why we assume that in the action integral (14.2.1) t is to be replaced by the proper time τ or $s = c\tau$ to determine the world line by varying $x^\mu(s)$ and $u^\mu(s)$, which is why in (14.2.1) we are now directly asking for the minimum:

$$\mathrm{S} = \frac{1}{c}\int_{s_1}^{s_2} \mathrm{d}s\, \mathrm{L}_r(x, u). \tag{14.2.16}$$

In this notation, S, $\mathrm{d}s = \sqrt{\mathrm{d}x_\mu \mathrm{d}x^\mu}$ and $\mathrm{L}_r(x, u)$ are all scalars. A variation after x^μ and s^μ leads directly to covariant equations, but it should be noted, that the variation of the speed is subject to the condition $u_\mu u^\mu = 1$ ([Jackson, 1998, §12.1.B]).

---

[3] Auxiliary formula: $\mathbf{b}\cdot\boldsymbol{\nabla}_i\mathbf{a} - \mathbf{b}\cdot\boldsymbol{\nabla}\,a_i = [\mathbf{b}\times(\boldsymbol{\nabla}\times\mathbf{a})]_i$

Principle of the shortest path

The variation of the path of the free particle

$$\delta \int_{t_1}^{t_2} \frac{\mathrm{d}t}{\gamma} = \delta \int_{\tau_1}^{\tau_2} \mathrm{d}\tau$$

has been referred to by Sommerfeld [1952, §32.C] as the *principle of shortest proper time*. However, we will switch from $\mathrm{d}\tau$ to $\mathrm{d}s = c\mathrm{d}\tau$, so that

$$\delta \int_{s_1}^{s_2} \mathrm{d}s = \int_{s_1}^{s_2} \delta\mathrm{d}s \tag{14.2.17}$$

according to Sommerfeld, represents the *principle of the shortest path* or the *principle of the geodesic path*. In the special theory of relativity with constant metric coefficients g_{ik}, the geodesic is a straight line and also describes the path of the free particle, which we will return to later.

Principle of least action

We now turn to the variation of the action integral (14.2.16):

$$\delta \mathrm{S} = \frac{1}{c} \int_{s_1}^{s_2} \delta\big(\mathrm{d}s\, \mathrm{L}_r\big). \tag{14.2.18}$$

Here, as already in (14.2.17), the path element is to be varied. The identity

$$\mathrm{d}s = \sqrt{\mathrm{d}x_\mu \mathrm{d}x^\mu} = \sqrt{u_\mu u^\mu}\, \mathrm{d}s \tag{14.2.19}$$

shows that without the restriction $u_\mu u^\mu = 1$ the variation could be carried out at constant path element. For this purpose, we switch from s to a variable λ, which like s should be monotonically increasing

$$\mathrm{d}s = \sqrt{w_\mu w^\mu}\, \mathrm{d}\lambda, \qquad w^\mu(\lambda) = \frac{\mathrm{d}x^\mu}{\mathrm{d}\lambda} = u^\mu \frac{\mathrm{d}s}{\mathrm{d}\lambda} = u^\mu \sqrt{w_\nu w^\nu}. \tag{14.2.20}$$

The velocity $w^\mu(\lambda)$ is now no longer subject to any constraint, and S is a functional of the world line $x^\mu(\lambda)$ and the velocity $w^\mu(\lambda)$:

$$\mathrm{S} = \frac{1}{c} \int_{\lambda_1}^{\lambda_2} \mathrm{d}\lambda \tilde{\mathrm{L}}_r(x, w) \quad \text{with} \quad \tilde{\mathrm{L}}_r(x, w) = \sqrt{w^\mu w_\mu}\, \mathrm{L}_r(x, \frac{w}{\sqrt{w^\mu w_\mu}}). \tag{14.2.21}$$

The variation is now carried out in the usual way

$$\delta \mathrm{S} = \frac{1}{c} \int_{\lambda_1}^{\lambda_2} \mathrm{d}\lambda \delta \tilde{\mathrm{L}}_r(x, w) = \frac{1}{c} \int_{\lambda_1}^{\lambda_2} \mathrm{d}\lambda \Big[\frac{\partial \tilde{\mathrm{L}}_r}{\partial x^\mu} \delta x^\mu + \frac{\partial \tilde{\mathrm{L}}_r}{\partial w^\mu} \delta w^\mu \Big] = 0. \tag{14.2.22}$$

One integrates the term with δw^μ partially:

$$\delta \mathrm{S} = \frac{1}{c} \frac{\partial \tilde{\mathrm{L}}_r}{\partial w^\mu} \delta x^\mu \Big|_{\lambda_1}^{\lambda_2} + \frac{1}{c} \int_{\lambda_1}^{\lambda_2} \mathrm{d}\lambda \Big(\frac{\partial \tilde{\mathrm{L}}_r}{\partial x^\mu} - \frac{\mathrm{d}}{\mathrm{d}\lambda} \frac{\partial \tilde{\mathrm{L}}_r}{\partial w^\mu} \Big) \delta x^\mu = 0 \tag{14.2.23}$$

and takes into account that the variation at the boundary $\delta x^\mu(\lambda_1) = \delta x^\mu(\lambda_2) = 0$, which is why the boundary term vanishes. One obtains the covariant Euler-Lagrange equations

$$\frac{\partial \tilde{L}_r}{\partial x^\mu} = \frac{\mathrm{d}}{\mathrm{d}\lambda} \frac{\partial \tilde{L}_r}{\partial w^\mu} \,. \tag{14.2.24}$$

Lagrange function for a free particle

For a free particle, L_r is a scalar, since the only invariant $u_\mu u^\mu = 1$ is a scalar and L_r must not depend on x^μ. L_r must be negative so that the true path becomes a minimum; the straight line between s_1 and s_2 is the maximum path. In addition, L_r must have the dimension of an energy, so that $L_r \propto -mc^2$. We get the exact relationship for the free particle using (14.2.1) and (14.2.3):

$$L_r = \gamma\, L = -mc^2 \,. \tag{14.2.25}$$

With this, we determine using the Euler-Lagrange equations (14.2.24)

$$\tilde{L}_r = -mc^2 \sqrt{w_\mu w^\mu} \tag{14.2.26}$$

the equation of motion for the free particle:

$$-mc^2 \frac{\mathrm{d}}{\mathrm{d}\lambda} \frac{\partial}{\partial w_\mu} \sqrt{w_\nu w^\nu} = -mc^2 \frac{\mathrm{d}}{\mathrm{d}\lambda} \frac{w^\mu}{\sqrt{w_\nu w^\nu}} \overset{(14.2.20)}{=} -mc^2 \sqrt{w_\lambda w^\lambda} \frac{\mathrm{d}}{\mathrm{d}s} u^\mu(s) = 0$$

and with $u^\mu(s) = $const we get a straight-line motion

$$L_r = -mc^2 \qquad \text{and} \qquad m\,c\dot{u}^\mu(s) = \mathrm{p}^\mu(s) = 0. \tag{14.2.27}$$

Lagrange function for a particle in the electromagnetic field

The Lagrange function, whose Euler-Lagrange equation is the Lorentz equation has, is according to (5.4.9) or (14.2.12)

$$L = \frac{m\,v^2}{2} + e\frac{k_L}{c}\mathbf{v}\cdot\mathbf{A} - e\phi \overset{(14.2.3)}{\longrightarrow} -\frac{mc^2}{\gamma} - \frac{ek_L}{\gamma} u_\mu A^\mu. \tag{14.2.28}$$

Now $L_r = \gamma L$, from which for the electron in the electromagnetic field

$$L_r = -mc^2 - ek_L u_\mu A^\mu, \qquad \tilde{L}_r = -mc^2 \sqrt{w_\mu w^\mu} - ek_L w_\mu A^\mu \tag{14.2.29}$$

follows. Inserted into (14.2.24) we get

$$-ek_L w_\lambda \partial^\mu A^\lambda = -\frac{\mathrm{d}}{\mathrm{d}\lambda}\left(mc^2 \frac{w^\mu}{\sqrt{w^\lambda w_\lambda}} + ek_L A^\mu \right).$$

We go back to the variable s here, where we replace w^λ by u^λ:

$$-ek_L u_\lambda \partial^\mu A^\lambda = -\frac{\mathrm{d}}{\mathrm{d}s}\left(mc^2\, u^\mu + ek_L A^\mu\right).$$

Now we use that $\dfrac{\mathrm{d}A^\mu}{\mathrm{d}s} = \dfrac{\partial A^\mu}{\partial x_\nu}\dfrac{\mathrm{d}x_\nu}{\mathrm{d}s}$ and get

$$mc\frac{\mathrm{d}u^\mu}{\mathrm{d}s} = e\frac{k_L}{c}\left(\frac{\partial A^\nu}{\partial x_\mu} - \frac{\partial A^\mu}{\partial x_\nu}\right)u_\nu \qquad \Leftrightarrow \qquad \frac{\mathrm{d}p^\mu}{\mathrm{d}s} = e\frac{k_L}{c}F^{\mu\nu}\,u_\nu. \qquad (14.2.30)$$

This is the Lorentz equation (14.1.30) in covariant form for a particle of charge e in the electromagnetic field.

Note: If one varies (see problem 14.3)

$$\delta\left(\mathrm{d}s\,\mathrm{L}_r\right) = u_\mu\,\delta\mathrm{d}x^\mu\,\mathrm{L}_r + \left[\mathrm{d}s\frac{\partial\mathrm{L}_r}{\partial x^\mu}\delta x^\mu + \frac{\partial\mathrm{L}_r}{\partial u^\mu}\left(\delta^\mu{}_\nu - u^\mu\,u_\nu\right)\delta\mathrm{d}x^\nu\right] \qquad (14.2.31)$$

directly, one obtains modified Euler-Lagrange equations

$$\frac{\partial\mathrm{L}_r}{\partial x^\mu} = \frac{\mathrm{d}}{\mathrm{d}s}\left(\frac{\partial\mathrm{L}_r}{\partial u^\nu}\left(\delta^\nu{}_\mu - u^\nu\,u_\mu\right) + \mathrm{L}_r\,u_\mu\right). \qquad (14.2.32)$$

If one substitutes for L_r (14.2.29), one again obtains the covariant Lorentz equations.

14.2.3 Electromagnetic Field Equations

The action integral of a particle in a given field (see (14.2.16) and (14.2.29)) can be extended to a system of particles by superposition:

$$\mathrm{S} = -\frac{1}{c}\int_{s_1}^{s_2}\mathrm{d}s\sum_n\left(m_n c^2 + e_n k_L u_n^\mu A_\mu\right). \qquad (14.2.33)$$

We now want to add the contribution of the electromagnetic field to the action integral. This must be a Lorentz scalar of the dimension of an energy density. Therefore, only the invariant $F^{\mu\nu}F_{\mu\nu}$ (13.1.31) is in question, as the other invariant $\tilde{F}^{\mu\nu}F_{\mu\nu}$ (13.1.32) is a pseudoscalar. For a stationary charge distribution, i.e. $\mathbf{B}=0$, the Lagrange density of the field $\mathcal{L}_{\text{Feld}}$ should be equal to the electrostatic energy density:

$$\alpha F^{\mu\nu}F_{\mu\nu} = -2\alpha E^2/k_L^2 = \frac{1}{8\pi k_C}E^2 \qquad \Rightarrow \qquad \alpha = -\frac{k_L^2}{16\pi k_C}.$$

With the correct factor α one obtains

$$\mathrm{S}_{\text{Feld}} = \frac{1}{c}\int\mathrm{d}^4x\,\mathcal{L}_{\text{Feld}} \qquad \text{with} \qquad \mathcal{L}_{\text{Feld}} = -\frac{k_L^2}{16\pi k_C}F^{\mu\nu}F_{\mu\nu}. \qquad (14.2.34)$$

The question is no longer about the movement of particles in a given field, but concerns the determination of the field given a specific charge and current distribution. The first term of (14.2.33) concerning free particles is therefore no longer relevant and is omitted. In the second term, we will now switch from the point charges e_n to ρ, where we refer back to the current density (13.1.4):

$$\sum_n e_n u_n^\mu = \int d^3x \sum_n e_n \delta^{(3)}(\mathbf{x} - \mathbf{x}_n(s))\, u_n^\mu = \frac{\gamma}{c} \int d^3x\, j^\mu.$$

If we substitute $ds \to dx^0/\gamma$, we obtain together with S_{Feld}

$$S = -\frac{1}{c} \int d^4x \left(\frac{k_L}{c} j^\mu A_\mu + \frac{k_L^2}{16\pi k_C} F^{\mu\nu} F_{\mu\nu} \right). \tag{14.2.35}$$

$\mathcal{L}$ is varied according to the fields A_μ and the field derivatives $A_{\mu,\nu}$:

$$\delta S = \frac{1}{c} \int d^4x\, \delta\mathcal{L} = \frac{1}{c} \int d^4x \left(\frac{\partial\mathcal{L}}{\partial A_{\mu,\nu}} \delta A_{\mu,\nu} + \frac{\partial\mathcal{L}}{\partial A_\mu} \delta A_\mu \right)$$

$$\stackrel{\substack{\text{Gauss's}\\\text{theorem}}}{=} \frac{1}{c} \oiint dO_\nu \frac{\partial\mathcal{L}}{\partial A_{\mu,\nu}} \delta A_\mu - \frac{1}{c} \int d^4x \left(\partial_\nu \frac{\partial\mathcal{L}}{\partial A_{\mu,\nu}} - \frac{\partial\mathcal{L}}{\partial A_\mu} \right) \delta A_\mu = 0.$$

The spatial integration limits are at infinity, where no currents and fields are and thus do not contribute to the surface term. According to the principle of least action, the variation of the potentials vanishes at the boundaries of time integration, so that overall the surface term contributes nothing. The Euler-Lagrange equations of the variation problem, the field equations [Schwabl, 2008, (12.2.15)]

$$\partial_\nu \frac{\partial\mathcal{L}}{\partial A_{\mu,\nu}} = \frac{\partial\mathcal{L}}{\partial A_\mu} \tag{14.2.36}$$

are for the Lagrange density

$$\mathcal{L} = -\frac{k_L}{c} j_\mu A^\mu - \frac{k_L^2}{16\pi k_C} F^{\mu\nu} F_{\mu\nu} \tag{14.2.37}$$

the inhomogeneous Maxwell equations (13.1.24)

$$F^{\nu\mu}{}_{,\nu} = \frac{4\pi k_C}{c k_L} j^\mu. \tag{14.2.38}$$

For the variation of the field term:

$$\frac{\partial F^{\mu\nu} F_{\mu\nu}}{\partial A_{\alpha,\beta}} = 2 F^{\mu\nu} \frac{\partial F_{\mu\nu}}{\partial A_{\alpha,\beta}} = 2 F^{\mu\nu} \frac{\partial A_{\nu,\mu} - \partial A_{\mu,\nu}}{\partial A_{\alpha,\beta}} = 4\, F^{\beta\alpha}.$$

The canonical energy-momentum tensor

If no charges and currents are present, then the Lagrange density of the free electromagnetic field according to (14.2.34) and (13.1.31) is:

$$\mathcal{L} = \mathcal{L}_{\text{Feld}} = \frac{-k_L^2}{16\pi k_C} F^{\mu\nu} F_{\mu\nu} = \frac{1}{8\pi k_C}(E^2 - k_L^2 B^2), \tag{14.2.39}$$

$$\text{SI: } \mathcal{L} = -\frac{1}{4\mu_0} F^{\mu\nu} F_{\mu\nu} = \frac{1}{2}(\epsilon_0 E^2 - \frac{1}{\mu_0} B^2).$$

If one defines the momentum conjugate to the A^μ by means of

$$\Pi^\mu = \frac{\partial \mathcal{L}}{\partial A_{\mu,0}} = -\frac{k_L^2}{4\pi k_C} F^{0\mu} = \frac{k_L}{4\pi k_C} \begin{cases} 0 & \mu = 0 \\ E_m & \mu = m > 0, \end{cases} \tag{14.2.40}$$

where $F^{\mu\nu} = A^{\nu,\mu} - A^{\mu,\nu}$, then the Hamilton density of the radiation field is:

$$\mathcal{H} = \Pi^\mu A_{\mu,0} - \mathcal{L} = \frac{\partial \mathcal{L}}{\partial A_{\lambda,0}} A_{\lambda,0} - \mathcal{L} = \frac{E^2 + k_L^2 B^2}{8\pi k_C} = u_{\text{Feld}}(\mathbf{x}, t). \tag{14.2.41}$$

Note: The conjugate momentum is often referred to as $\Pi^\mu = \partial \mathcal{L}/\partial \dot{A}_\mu$ defined. For $\mathcal{H} = \Pi^\mu \dot{A}_\mu - \mathcal{L}$ nothing changes.

The extension of the Hamilton density to a covariant tensor [Jackson, 1998, (12.103)] suggests the following form:

$$T^\mu_{C\ \nu} = \frac{\partial \mathcal{L}}{\partial A_{\lambda,\mu}} A_{\lambda,\nu} - \delta^\mu_{\ \nu} \mathcal{L}.$$

This is the canonical energy-momentum tensor; to evaluate the differentiation we refer to (14.2.36) and obtain:

$$T^{\mu\nu}_C = g^{\nu\nu'} T^\mu_{C\ \nu'} = \frac{-k_L^2}{16\pi k_C}\left[g^{\nu\nu'} 4 F^{\mu\lambda} A_{\lambda,\nu'} + g^{\mu\nu} F^{\rho\sigma} F_{\rho\sigma}\right] \tag{14.2.42}$$

If you set in the first term $A_{\lambda,\nu'} = F_{\nu'\lambda} + A_{\nu',\lambda}$, you get the symmetric energy-momentum tensor $(T^{\mu\nu})$: (14.1.33) and an asymmetric part $(T^{\mu\nu}_{as})$, whose divergence vanishes $T^{\mu\nu}_{as\ ,\mu} = 0$

$$T^{\mu\nu}_C = T^{\mu\nu} + T^{\mu\nu}_{as}, \qquad T^{\mu\nu}_{as} = -\frac{k_L^2}{4\pi k_C} g^{\nu\nu'} F^{\mu\lambda} A_{\nu',\lambda}. \tag{14.2.43}$$

14.2.4 The Free Electromagnetic Field and its Quantization

The radiation field

According to the theorem of Helmholtz, section 7.1.2, 'any' vector field can be decomposed into a source-free (transverse) and a curl-free (longitudinal) field. Free radiation fields $(\mathbf{E}, \mathbf{B}, \mathbf{A})$ are divergence-free and therefore transverse. In the following, the Coulomb gauge $\nabla \cdot \mathbf{A} = 0$ is used. The free radiation field thus satisfies the homogeneous wave equation $\Box \mathbf{A} = 0$. The real field $\mathbf{A}$ can be expanded for periodic boundary conditions in a volume $V = L^3$ according to partial waves. The occurring factors are chosen so that the Fourier coefficients $\mathbf{A_k}$ are dimensionless:

$$\mathbf{A}(\mathbf{x},t) = \frac{c\sqrt{4\pi k_C}}{k_L\sqrt{2V}} \sum_\mathbf{k} \frac{\sqrt{\hbar}}{\sqrt{\omega_k}} \left(e^{i\mathbf{k}\cdot\mathbf{x}}\mathbf{A_k}(t) + e^{-i\mathbf{k}\cdot\mathbf{x}}A_\mathbf{k}^*(t)\right),$$

$$\mathbf{A_k}(t) = \sum_\lambda \boldsymbol{\epsilon}_{\mathbf{k}\lambda} a_{\mathbf{k}\lambda}(t), \qquad a_{\mathbf{k}\lambda}(t) = e^{-i\omega_k t} a_{\mathbf{k}\lambda}, \qquad \omega_k = ck, \qquad (14.2.44)$$

where $a_{\mathbf{k}\lambda}$ are the dimensionless amplitudes of the partial waves. From the transversality $\boldsymbol{\nabla}\cdot\mathbf{A} = 0$ it follows, that all factors $\mathbf{k}\cdot\mathbf{A_k}(t) = 0$. The partial waves $\mathbf{A_k}$ are therefore polarized perpendicular to their propagation direction $\mathbf{k}$ (transverse gauge) and the two polarization vectors $\boldsymbol{\epsilon}_{\mathbf{k}\lambda}$ form with $\hat{\mathbf{k}} = \mathbf{k}/k$ an orthogonal coordinate system:

$$\boldsymbol{\epsilon}_{\mathbf{k}\lambda}^* \cdot \boldsymbol{\epsilon}_{\lambda'} = \delta_{\lambda,\lambda'}, \qquad\qquad \hat{\mathbf{k}}\cdot\boldsymbol{\epsilon}_{\mathbf{k}\lambda} = 0, \qquad\qquad \lambda = 1, 2. \qquad (14.2.45)$$

The polarization vectors are complex, especially in the case of circular polarization, see section 10.2. In the Coulomb gauge, the covariant notation is not necessary, as can be seen in (14.2.44). Until the end of the section, the designation A_i is used instead of A^i. From the Fourier expansion for $\mathbf{A}$, the expansions for $\mathbf{E}$ and $\mathbf{B}$ follow in the following form:

$$\mathbf{E}(\mathbf{x},t) = \frac{-k_L}{c}\dot{\mathbf{A}} = \frac{i\sqrt{4\pi k_C}}{\sqrt{2V}} \sum_\mathbf{k} \sqrt{\hbar\omega_k}\left[e^{i\mathbf{k}\cdot\mathbf{x}}\mathbf{A_k}(t) - e^{-i\mathbf{k}\cdot\mathbf{x}}\mathbf{A}_\mathbf{k}^*(t)\right],$$

$$(14.2.46)$$

$$\mathbf{B}(\mathbf{x},t) = \boldsymbol{\nabla}\times\mathbf{A} = \frac{i\sqrt{4\pi k_C}}{k_L\sqrt{2V}} \sum_\mathbf{k} \sqrt{\hbar\omega_k}\left[e^{i\mathbf{k}\cdot\mathbf{x}}(\hat{\mathbf{k}}\times\mathbf{A_k}(t)) - e^{-i\mathbf{k}\cdot\mathbf{x}}(\hat{\mathbf{k}}\times\mathbf{A}_\mathbf{k}^*(t))\right].$$

The field energy

The Hamilton function is obtained by integrating over the energy density $\mathcal{H}$ (problem 14.4):

$$\mathrm{H} = \int \mathrm{d}^3x\, \frac{E^2 + k_L^2 B^2}{8\pi k_C} = \frac{1}{2} \sum_{\mathbf{k},\lambda} \hbar\omega_\mathbf{k}(a_{\mathbf{k}\lambda}a_{\mathbf{k}\lambda}^* + a_{\mathbf{k}\lambda}^* a_{\mathbf{k}\lambda}). \qquad (14.2.47)$$

The energies of the electric and magnetic fields vary with t, but their sum is time-independent. Formally considered, terms like $\sim \mathbf{A_k}\cdot\mathbf{A}_{-\mathbf{k}}$ only cancel out in $E^2 + k_L B^2$. (14.2.47) is a sum of harmonic oscillators. The energy of the entire system is thus as a sum of the energies $\hbar\omega_k$ of the partial waves without any reference to a spatial distribution, and the dependence on electromagnetic units has also disappeared.

Note: Represented by the conjugate variables P and Q, one obtains the Hamilton function

$$Q_{\mathbf{k}\lambda} = \frac{\sqrt{\hbar}}{\sqrt{2\omega_k}}[a_{\mathbf{k}\lambda}(t)+a_{\mathbf{k}\lambda}^*(t)], \qquad P_{\mathbf{k}\lambda} = -\mathrm{i}\frac{\sqrt{\hbar\omega_k}}{\sqrt{2}}[a_{\mathbf{k}\lambda}(t)-a_{\mathbf{k}\lambda}^*(t)],$$

$$\mathrm{H} = \frac{1}{2}\sum_{\mathbf{k},\lambda}(P_{\mathbf{k}\lambda}^2 + \omega_k^2 Q_{\mathbf{k}\lambda}^2). \tag{14.2.48}$$

For the conjugate variables, the Hamilton's equations of motion apply

$$\dot{Q}_{\mathbf{k}\lambda} = \{Q_{\mathbf{k}\lambda},\mathrm{H}\} = \frac{\partial \mathrm{H}}{\partial P_{\mathbf{k}\lambda}}, \qquad\qquad \dot{P}_{\mathbf{k}\lambda} = \{P_{\mathbf{k}\lambda},\mathrm{H}\} = -\frac{\partial \mathrm{H}}{\partial Q_{\mathbf{k}\lambda}}.$$

The Poisson bracket, the classical counterpart to the commutator is

$$\{Q_{\mathbf{k}\lambda},P_{\mathbf{k}'\lambda'}\} = \sum_{\mathbf{q}\nu}\left(\frac{\partial Q_{\mathbf{k}\lambda}}{\partial Q_{\mathbf{q}\nu}}\frac{\partial P_{\mathbf{k}'\lambda'}}{\partial P_{\mathbf{q}\nu}} - \frac{\partial Q_{\mathbf{k}\lambda}}{\partial P_{\mathbf{q}\nu}}\frac{\partial P_{\mathbf{k}'\lambda'}}{\partial Q_{\mathbf{q}\nu}}\right) = \delta_{\mathbf{k},\mathbf{k}'}\delta_{\lambda,\lambda'}.$$

This representation of the field energy as a sum of harmonic oscillators already points to the scheme of the quantization known from mechanics.

The field momentum

In the same sense, the momentum, calculated with the help of the momentum density (5.6.15), is represented as the sum of the momenta of the partial waves, with reference to the following explicit calculation:

$$\boldsymbol{P}_{\mathrm{Field}} = \frac{k_L}{4\pi c k_C}\int \mathrm{d}^3x\,\mathbf{E}\times\mathbf{B} = \sum_{\mathbf{k}\lambda}\hbar\mathbf{k}\,a_{\mathbf{k}\lambda}^* a_{\mathbf{k}\lambda}, \quad \mathrm{SI:}\ \frac{k_L}{4\pi c k_C} = \epsilon_0. \tag{14.2.49}$$

Auxiliary calculation: The evaluation carried out with the help of (14.2.46) yields:

$$\boldsymbol{P}_{\mathrm{Field}} = \frac{-\hbar}{2cV}\sum_{\mathbf{k},\mathbf{q}}\sqrt{\omega_k\omega_q}\int \mathrm{d}^3x\,\left\{\mathbf{A}_{\mathbf{k}}\times(\hat{\mathbf{q}}\times\mathbf{A}_{\mathbf{q}})\,\mathrm{e}^{\mathrm{i}(\mathbf{k}+\mathbf{q})\cdot\mathbf{x}} + ...\right\}$$

$$= \frac{\hbar}{2}\sum_{\mathbf{k}}\mathbf{k}\left\{\mathbf{A}_{\mathbf{k}}\cdot\mathbf{A}_{-\mathbf{k}}\,\mathrm{e}^{-2\mathrm{i}\omega_k t} + \mathbf{A}_{\mathbf{k}}^*\cdot\mathbf{A}_{\mathbf{k}} + \mathbf{A}_{\mathbf{k}}\cdot\mathbf{A}_{\mathbf{k}}^* + \mathbf{A}_{\mathbf{k}}^*\cdot\mathbf{A}_{-\mathbf{k}}^*\mathrm{e}^{2\mathrm{i}\omega_k t}\right\}$$

It was exploited that $\mathbf{A}$ is transversal, i.e. $\nabla\cdot\mathbf{A} = 0$, from which it follows that $\mathbf{k}\cdot\mathbf{A}_{\mathbf{k}} = 0$ and $\mathbf{A}_{\mathbf{k}}\times(\hat{\mathbf{k}}\times\mathbf{A}_{\mathbf{k}}^*) = \hat{\mathbf{k}}(\mathbf{A}_{\mathbf{k}}\cdot\mathbf{A}_{\mathbf{k}}^*)$. The contributions from $\hat{\mathbf{k}}(\mathbf{A}_{\mathbf{k}}\cdot\mathbf{A}_{-\mathbf{k}})$ and $\hat{\mathbf{k}}(\mathbf{A}_{\mathbf{k}}^*\cdot\mathbf{A}_{-\mathbf{k}}^*)$ also vanish, since which are antisymmetric in $\mathbf{k}$. Thus, one obtains (14.2.49).

The quantization of the free electromagnetic field

If one replaces the amplitudes $a_{\mathbf{k}\lambda}, a_{\mathbf{k}\lambda}^*$ with operators $\hat{a}_{\mathbf{k}\lambda}, \hat{a}_{\mathbf{k}\lambda}^\dagger$ with the commutation rules

$$[a_{\mathbf{k}\lambda}, a_{\mathbf{k}'\lambda'}^\dagger] = \delta_{\mathbf{k},\mathbf{k}'}\delta_{\lambda,\lambda'}, \qquad [a_{\mathbf{k}\lambda}, a_{\mathbf{k}'\lambda'}] = [a_{\mathbf{k}\lambda}^\dagger, a_{\mathbf{k}'\lambda'}^\dagger] = 0, \tag{14.2.50}$$

then the field energy (14.2.47) can be represented by

$$H = \sum_{\mathbf{k},\lambda} \hbar\omega_{\mathbf{k}}(\hat{a}^{\dagger}_{\mathbf{k}\lambda}\hat{a}_{\mathbf{k}\lambda} + \frac{1}{2}).$$

$$(14.2.51)$$

The excitations are photons. The operator $\hat{a}^{\dagger}_{\mathbf{k}\lambda}$ generates a photon of frequency ω_k; their number is not limited, i.e., photons obey Bose statistics.

Note: For the conjugate operators P and Q (14.2.48) the commutators are: $[\hat{Q}_{\mathbf{k}\lambda}, \hat{P}_{\mathbf{k}'\lambda'}] = i\hbar\delta_{\mathbf{k},\mathbf{k}'}\delta_{\lambda,\lambda'}$ and $[\hat{Q}_{\mathbf{k}\lambda}, \hat{Q}_{\mathbf{k}'\lambda'}] = [\hat{P}_{\mathbf{k}\lambda}, \hat{P}_{\mathbf{k}'\lambda'}] = 0$.

One can expect that in the continuous space of locations the commutators will contain delta functions, instead of the Kronecker delta of the discrete k-space. However, the transversality of the fields brings an additional restriction, so that one obtains a *transverse delta function*

$$[\hat{A}_i(\mathbf{x},t), \hat{E}_j(\mathbf{x}',t)] = -\,i\hbar\frac{4\pi k_C}{2V}\sum_{\mathbf{k},\lambda,\mathbf{q},\lambda'}\frac{\sqrt{\omega_q}}{\sqrt{\omega_k}}\{e^{i\mathbf{k}\cdot\mathbf{x}-i\omega_k t}\,\epsilon_{\mathbf{k}\lambda,i}\,e^{-i\mathbf{q}\cdot\mathbf{x}'+i\omega_q t}\,\epsilon^{*}_{\mathbf{q}\lambda',j}$$

$$[\hat{a}_{\mathbf{k}\lambda}, \hat{a}^{\dagger}_{\mathbf{q}\lambda'}] - e^{-i\mathbf{k}\cdot\mathbf{x}+i\omega_k t}\,\epsilon^{*}_{\mathbf{k}\lambda,i}\,e^{i\mathbf{q}\cdot\mathbf{x}'-i\omega_q t}\,\epsilon_{\mathbf{q}\lambda',j}\,[\hat{a}^{\dagger}_{\mathbf{k}\lambda}, \hat{a}_{\mathbf{q}\lambda'}]\}$$

$$= -\,i\hbar\frac{4\pi k_C}{2V}\sum_{\mathbf{k},\lambda}\{e^{i\mathbf{k}\cdot(\mathbf{x}-\mathbf{x}')}\,\epsilon_{\mathbf{k}\lambda,i}\,\epsilon^{*}_{\mathbf{k}\lambda,j} + e^{-i\mathbf{k}\cdot(\mathbf{x}-\mathbf{x}')}\,\epsilon^{*}_{\mathbf{k}\lambda,i}\,\epsilon_{\mathbf{k}\lambda,j}\} \quad (14.2.52)$$

The polarization vectors form an orthonormal coordinate system with $\hat{\mathbf{k}}$, so that

$$\sum_{\lambda}\epsilon^{*}_{\mathbf{k}\lambda,i}\epsilon_{\mathbf{k}\lambda,j} + \hat{k}_i\hat{k}_j = \delta_{ij}.$$

Inserted into (14.2.52), one obtains ($\hat{\mathbf{k}} = \mathbf{k}/k$)

$$[\hat{A}_i(\mathbf{x},t), \hat{E}_j(\mathbf{x}',t)] = -i\hbar\frac{4\pi k_C}{V}\sum_{\mathbf{k}}(\delta_{ij} - \hat{k}_i\hat{k}_j)e^{i\mathbf{k}\cdot(\mathbf{x}-\mathbf{x}')}$$

$$= (-i\hbar)4\pi k_C\delta^{T}_{ij}(\mathbf{x}-\mathbf{x}').$$

$$(14.2.53)$$

Here the transverse delta function is defined, both for discrete $\mathbf{k}$-values in a finite volume and also for a continuous k-spectrum

$$\delta^{T}_{ij}(\mathbf{x}) = \frac{1}{V}\sum_{\mathbf{k}}(\delta_{ij} - \hat{k}_i\hat{k}_j)e^{i\mathbf{k}\cdot\mathbf{x}} = \int\frac{d^3k}{(2\pi)^3}\,(\delta_{ij} - \hat{k}_i\hat{k}_j)e^{i\mathbf{k}\cdot\mathbf{x}}.$$

$$(14.2.54)$$

Further commutators are [Schwabl, 2008, (14.4.12)]:

$$[\hat{E}_i(\mathbf{x},t), \hat{B}_j(\mathbf{x}',t)] = \epsilon_{jln}\nabla'_l[\hat{E}_i(\mathbf{x},t), \hat{A}_n(\mathbf{x}',t)] = i\hbar 4\pi k_C\,\epsilon_{ijl}\nabla'_l\delta^{(3)}(\mathbf{x}-\mathbf{x}'),$$

$$[\hat{A}_i(\mathbf{x},t), \hat{A}_j(\mathbf{x}',t)] = [\hat{A}_i(\mathbf{x},t), \hat{B}_j(\mathbf{x}',t)] = 0.$$

$$(14.2.55)$$

14.2.5 The Angular Momentum

If $\mathbf{x}_0$ sets the reference point for the angular momentum, then the total angular momentum is determined by

$$\boldsymbol{J} = \boldsymbol{J}_{\text{Field}} - \boldsymbol{J}_0 = \int \mathrm{d}^3 x\,(\mathbf{x} - \mathbf{x}_0) \times \boldsymbol{P}_{\text{Field}}, \qquad \boldsymbol{J}_0 = \mathbf{x}_0 \times \boldsymbol{P}_{\text{Field}},$$

$$\boldsymbol{J}_{\text{Field}} = \boldsymbol{L} + \boldsymbol{S} = \int \mathrm{d}^3 x\,\mathbf{x} \times (\mathbf{E} \times \mathbf{B}). \tag{14.2.56}$$

$\boldsymbol{J}_{\text{Field}}$ is the so-called transverse momentum, since $\mathbf{A}$ and $\mathbf{E}$ are transverse. The integrand can be decomposed as follows:

$$\bigl(\mathbf{x} \times (\mathbf{E} \times (\boldsymbol{\nabla} \times \mathbf{A}))\bigr)_r = \epsilon_{rsi} x_s (E_j \nabla_i A_j - E_j \nabla_j A_i) \tag{14.2.57}$$

$$= E_j (\mathbf{x} \times \boldsymbol{\nabla})_r A_j - \mathbf{E} \cdot \boldsymbol{\nabla}(\mathbf{x} \times \mathbf{A})_r + (\mathbf{E} \times \mathbf{A})_r.$$

Orbital angular momentum and intrinsic angular momentum are determined by:

$$\boldsymbol{L} = \frac{k_L}{4\pi c k_C} \int \mathrm{d}^3 x\, E_j (\mathbf{x} \times \boldsymbol{\nabla}) A_j, \qquad \boldsymbol{L}_2 = \frac{-k_L}{4\pi c k_C} \int \mathrm{d}^3 x\, \mathbf{E} \cdot \boldsymbol{\nabla}(\mathbf{x} \times \mathbf{A}),$$

$$\boldsymbol{S} = \frac{k_L}{4\pi c k_C} \int \mathrm{d}^3 x\,(\mathbf{E} \times \mathbf{A}). \tag{14.2.58}$$

It remains to be shown that for transverse fields $\boldsymbol{L}_2 = 0$.

The spin

The intrinsic angular momentum is defined according to (14.2.58) as

$$\boldsymbol{S} = \frac{k_L}{4\pi c k_C} \int \mathrm{d}^3 x\,(\mathbf{E} \times \mathbf{A}). \tag{14.2.59}$$

The vector product can only point in the direction of $\hat{\mathbf{k}}$ since $\mathbf{E}$ and $\mathbf{A}$ are transverse:

$$\boldsymbol{S} = \frac{\mathrm{i}\hbar}{2V} \sum_{\mathbf{k},\mathbf{q}} \frac{\sqrt{\omega_k}}{\sqrt{\omega_q}} \int \mathrm{d}^3 x\, \bigl(\mathrm{e}^{\mathrm{i}\mathbf{k}\cdot\mathbf{x}} \mathbf{A}_{\mathbf{k}} - \mathrm{e}^{-\mathrm{i}\mathbf{k}\cdot\mathbf{x}} \mathbf{A}_{\mathbf{k}}^*\bigr) \times \bigl(\mathrm{e}^{\mathrm{i}\mathbf{q}\cdot\mathbf{x}} \mathbf{A}_{\mathbf{q}} + \mathrm{e}^{-\mathrm{i}\mathbf{q}\cdot\mathbf{x}} \mathbf{A}_{\mathbf{q}}^*\bigr)$$

$$= \frac{\mathrm{i}}{2} \sum_{\mathbf{k}} \hbar\bigl(\mathbf{A}_{\mathbf{k}} \times \mathbf{A}_{-\mathbf{k}} - \mathbf{A}_{\mathbf{k}}^* \times \mathbf{A}_{\mathbf{k}} + \mathbf{A}_{\mathbf{k}} \times \mathbf{A}_{\mathbf{k}}^* - \mathbf{A}_{\mathbf{k}}^* \times \mathbf{A}_{-\mathbf{k}}^*\bigr).$$

$\mathbf{A}_{\mathbf{k}} \times \mathbf{A}_{-\mathbf{k}}$ and $\mathbf{A}_{\mathbf{k}}^* \times \mathbf{A}_{-\mathbf{k}}^*$ vanish, as they are antisymmetric in $\mathbf{k}$:

$$\boldsymbol{S} = -\mathrm{i} \sum_{\mathbf{k}} \hbar \mathbf{A}_{\mathbf{k}}^* \times \mathbf{A}_{\mathbf{k}} = -\mathrm{i} \sum_{\mathbf{k},\lambda,\lambda'} \hbar a_{\mathbf{k}\lambda}^* a_{\mathbf{k}\lambda'} (\boldsymbol{\epsilon}_{\mathbf{k}\lambda}^* \times \boldsymbol{\epsilon}_{\mathbf{k}\lambda'}). \tag{14.2.60}$$

Adapted to the symmetry are circularly polarized modes (10.2.2') with $\lambda = \pm$

$$\epsilon_{\mathbf{k},\pm} = \frac{\mp 1}{\sqrt{2}}(\epsilon_{\mathbf{k},1} \pm i\epsilon_{\mathbf{k},2}), \qquad \epsilon_{\mathbf{k}\lambda}^* \cdot \epsilon_{\mathbf{k}\lambda'} = \delta_{\lambda,\lambda'}, \qquad \epsilon_{\mathbf{k}\lambda}^* \times \epsilon_{\mathbf{k}\lambda'} = \lambda i\hat{\mathbf{k}}\delta_{\lambda,\lambda'}$$

with their help, the angular momentum takes the form

$$\boldsymbol{S} = \sum_{\mathbf{k}} \hbar\hat{\mathbf{k}}\left(a_{\mathbf{k}+}^* a_{\mathbf{k}+} - a_{\mathbf{k}-}^* a_{\mathbf{k}-}\right). \tag{14.2.61}$$

The intrinsic angular momentum of a partial wave is therefore parallel or antiparallel to the direction of propagation.

The orbital angular momentum

The calculation of $\boldsymbol{L}$ in a finite volume V with periodic boundary conditions implies restrictions, as then $\boldsymbol{L}$ is not an exact constant of motion is [Lenstra, Mandel, 1982]. With the Fourier transformation, one can avoid the difficulties [Simmons, Guttmann, 1970, Appendix VI.2]:

$$\mathbf{A}(\mathbf{x},t) = \frac{c\sqrt{4\pi k_c \hbar}}{k_L \sqrt{2}} \int \frac{\mathrm{d}^3 k}{\sqrt{2\pi}^3} \frac{1}{\sqrt{\omega_k}}\left[\mathrm{e}^{i\mathbf{k}\cdot\mathbf{x}}\mathbf{A}(\mathbf{k},t) + \mathrm{e}^{-i\mathbf{k}\cdot\mathbf{x}}\mathbf{A}^*(\mathbf{k},t)\right],$$

$$\mathbf{A}(\mathbf{k},t) = \sum_{\lambda} a_\lambda(\mathbf{k},t)\,\epsilon_\lambda(\mathbf{k}), \qquad a_\lambda(\mathbf{k},t) = \mathrm{e}^{-i\omega_k t}\,a_\lambda(\mathbf{k}). \tag{14.2.62}$$

The Fourier transformation is symmetrically defined here; when switching from the sum to the integral, one only has to replace V with $(2\pi)^3$.

Auxiliary calculation: It has to be shown that $\boldsymbol{L}_2$ (14.2.58) vanishes:

$$\boldsymbol{L}_2 = \frac{-k_L}{4\pi c k_c}\int \mathrm{d}^3 x\, E_j(\mathbf{x}\times\nabla_j\mathbf{A}) - \boldsymbol{S} = \frac{-i\hbar}{2}\int \frac{\mathrm{d}^3 k\,\mathrm{d}^3 q}{(2\pi)^3}\sqrt{\omega_k}\int \mathrm{d}^3 x$$

$$[\mathrm{e}^{i\mathbf{k}\cdot\mathbf{x}}A_j(\mathbf{k}) - \mathrm{e}^{-i\mathbf{k}\cdot\mathbf{x}}A_j^*(\mathbf{k})][\mathrm{e}^{i\mathbf{q}\cdot\mathbf{x}}(\mathbf{x}\times\mathbf{A}(\mathbf{q})) - (\mathbf{x}\times\mathbf{A}^*(\mathbf{q}))\mathrm{e}^{-i\mathbf{q}\cdot\mathbf{x}}]\frac{iq_j}{\sqrt{\omega_q}} - \boldsymbol{S}.$$

Now the factors $\mathrm{e}^{\pm i\mathbf{k}\cdot\mathbf{x}}$ are shifted to the right and replaced by $\mathbf{x}\,\mathrm{e}^{\pm i\mathbf{k}\cdot\mathbf{x}} = \mp i\frac{\partial}{\partial\mathbf{k}}\mathrm{e}^{\pm i\mathbf{k}\cdot\mathbf{x}}$. Then the integration over $\mathrm{d}^3 x$ can be carried out. After eliminating $\mathbf{q}$ using $\delta^{(3)}(\mathbf{k}\pm\mathbf{q})$, one obtains

$$\boldsymbol{L}_2 = \frac{i}{2}\int \mathrm{d}^3 k\,\sqrt{\omega_k}\Big\{A_j(\mathbf{k})[(\frac{\partial}{\partial\mathbf{k}}\times\mathbf{A}(-\mathbf{k})) + (\frac{\partial}{\partial\mathbf{k}}\times\mathbf{A}^*(\mathbf{k}))]$$

$$-A_j^*(\mathbf{k})[(\frac{\partial}{\partial\mathbf{k}}\times\mathbf{A}(\mathbf{k})) + (\frac{\partial}{\partial\mathbf{k}}\times\mathbf{A}^*(-\mathbf{k}))]\Big\}\frac{k_j}{\sqrt{\omega_k}} - \boldsymbol{S} = 0.$$

The contributions of the form $A_j(\mathbf{k})\,k_j\,(\frac{\partial}{\partial\mathbf{k}}\times\mathbf{A}(-\mathbf{k}))\frac{1}{\sqrt{\omega_k}}$ vanish due to transversality. Thus, only the vector products of the form $\mathbf{A}(\mathbf{k})\times\mathbf{A}(-\mathbf{k})$ remain, which according to (14.2.60) yield $\boldsymbol{S}$. This shows that $\boldsymbol{L}_2 = 0$.

The orbital angular momentum $\boldsymbol{L}$, as it results from the division (14.2.58), reads

$$L = \frac{k_L}{4\pi c k_C} \int \mathrm{d}^3 x \, E_j (\mathbf{x} \times \boldsymbol{\nabla}) A_j = \frac{\mathrm{i}\hbar}{2} \int \frac{\mathrm{d}^3 k \, \mathrm{d}^3 q}{(2\pi)^3} \frac{\sqrt{\omega_k}}{\sqrt{\omega_q}} \tag{14.2.63}$$

$$\int \mathrm{d}^3 x \, [\mathrm{e}^{\mathrm{i}\mathbf{k}\cdot\mathbf{x}} A_j(\mathbf{k}) - \mathrm{e}^{-\mathrm{i}\mathbf{k}\cdot\mathbf{x}} A_j^*(\mathbf{k})] (\mathbf{x} \times \boldsymbol{\nabla}) [\mathrm{e}^{\mathrm{i}\mathbf{q}\cdot\mathbf{x}} A_j(\mathbf{q}) + \mathrm{e}^{-\mathrm{i}\mathbf{q}\cdot\mathbf{x}} A_j^*(\mathbf{q})].$$

The simplification of this expression is similar to that of $\boldsymbol{L}_2$, as shown in the auxiliary calculation.

Auxiliary calculation: It is shown that $\boldsymbol{L}_1$ does not contribute

$$\boldsymbol{L}_1 = \frac{\mathrm{i}\hbar}{2} \int \frac{\mathrm{d}^3 k \, \mathrm{d}^3 q}{(2\pi)^3} \frac{\sqrt{\omega_k}}{\sqrt{\omega_q}} \int \mathrm{d}^3 x \, \mathbf{A}(\mathbf{k}) \cdot \mathbf{A}(\mathbf{q}) \, (\mathbf{x} \times \mathrm{i}\mathbf{q}) \, \mathrm{e}^{\mathrm{i}(\mathbf{k}+\mathbf{q})\cdot\mathbf{x}}.$$

$\boldsymbol{L}_1$ is symmetrized using $\mathbf{q} = \frac{\hat{\mathbf{q}}\omega_q}{c}$, by taking half the contribution, then swapping $\mathbf{k}$ with $\mathbf{q}$ and adding it to the original half:

$$\boldsymbol{L}_1 = \frac{\mathrm{i}\hbar}{4c} \int \frac{\mathrm{d}^3 k \, \mathrm{d}^3 q}{(2\pi)^3} \sqrt{\omega_k \omega_q} \int \mathrm{d}^3 x \, \mathbf{A}(\mathbf{k}) \cdot \mathbf{A}(\mathbf{q}) \, (\mathbf{x} \times \mathrm{i}(\hat{\mathbf{k}}+\hat{\mathbf{q}})) \mathrm{e}^{\mathrm{i}(\mathbf{k}+\mathbf{q})\cdot\mathbf{x}}.$$

As already in the calculation of $\boldsymbol{L}_2$, $\mathbf{x}\,\mathrm{e}^{\mathrm{i}\mathbf{k}\cdot\mathbf{k}}$ is replaced by $-\mathrm{i}\frac{\partial}{\partial\mathbf{k}}\,\mathrm{e}^{\mathrm{i}\mathbf{k}\cdot\mathbf{k}}$ and the integration over $\mathrm{d}^3 x$ is carried out taking into account that $\frac{\partial}{\partial\mathbf{k}} \times \hat{\mathbf{k}} = 0$:

$$\boldsymbol{L}_1 = \frac{\mathrm{i}\hbar}{4c} \int \mathrm{d}^3 k \, \sqrt{\omega_k} A_j(\mathbf{k}) \frac{\partial}{\partial\mathbf{k}} \times \int \mathrm{d}^3 q \, (\hat{\mathbf{k}}+\hat{\mathbf{q}}) \, A_j(\mathbf{q}) \sqrt{\omega_q} \, \delta^{(3)}(\mathbf{k}+\mathbf{q}) = 0. \tag{14.2.64}$$

Since $\boldsymbol{L}_1$ vanishes, the following terms remain from (14.2.63):

$$L = \frac{\mathrm{i}\hbar}{2} \int \mathrm{d}^3 k \, \big[A_j^*(\mathbf{k}) (\frac{\partial}{\partial\mathbf{k}} \times \mathbf{k}) A_j(\mathbf{k}) - A_j(\mathbf{k}) (\frac{\partial}{\partial\mathbf{k}} \times \mathbf{k}) A_j^*(\mathbf{k}) \big]. \tag{14.2.65}$$

We have omitted the time dependence as it does not contribute. Plane waves have no orbital angular momentum. For further discussion, please refer to Mandel, Wolf [1995, (10.6-21)] or Cohen-Tannoudji et al [1989, (C.18)].

Note: In the field quantization in the continuous (Fock-) space [Mandel, Wolf, 1995, §10.10] one obtains the Hamiltonian and the commutators:

$$H = \sum_\lambda \int \mathrm{d}^3 k \, \hbar\omega_k \big[\hat{a}_\lambda^\dagger(\mathbf{k}) \, \hat{a}_\lambda(\mathbf{k}) + \frac{1}{2} \big] \tag{14.2.66}$$

$$[\hat{a}_\lambda(\mathbf{k}), \hat{a}_{\lambda'}^\dagger(\mathbf{q})] = \delta_{\lambda,\lambda'} \, \delta^{(3)}(\mathbf{k}-\mathbf{q}), \qquad [\hat{a}_\lambda(\mathbf{k}), \hat{a}_{\lambda'}(\mathbf{q})] = [\hat{a}_\lambda^\dagger(\mathbf{k}), \hat{a}_{\lambda'}^\dagger(\mathbf{q})] = 0.$$

14.3 Kinematic Effects

14.3.1 Energy-Momentum Conservation Law

Two particles interact with each other, as sketched in Fig. 14.2:

$$c^2 m_1 \ddot{x}_1^\mu(s_1) = f_1^\mu(s_1) \qquad \text{and} \qquad c^2 m_2 \ddot{x}_2^\mu(s_2) = f_2^\mu(s_2)\,.$$

Particle 1 experiences the recoil from the "exchange particle" (photon,..) be-

Fig. 14.2. Interaction (collision) process, in which a particle (photon, gluon etc.) is exchanged

fore particle 2 feels the collision. The principle *action=reaction*, the third Newtonian law, is formulated by Mach [1933, §2.7] as:

The action is always equal to the reaction, or the effects of two bodies on each other are always equal and in opposite directions,
which can only apply in this simple form for instantaneous interaction. The generalized principle is

$$\int_{-\infty}^{\infty} \mathrm{d}s_1\, f_1^\mu(s_1) + \int_{-\infty}^{\infty} \mathrm{d}s_2\, f_2^\mu(s_2) = 0\,. \tag{14.3.1}$$

Energy and momentum conservation

From (14.3.1) follows for the conservation of momentum

$$\mathrm{p}_1^\mu(\infty) + \mathrm{p}_2^\mu(\infty) = \mathrm{p}_1^\mu(-\infty) + \mathrm{p}_2^\mu(-\infty). \tag{14.3.2}$$

The zeroth component of the four-momentum indicates the energy E/c of a particle, so that with (p^μ) energy and momentum conservation are given.

Collision process

$$m_i\, \boldsymbol{p}_i(t) = \mathbf{K}_i(t) \qquad i = 1, 2\,. \tag{14.3.3}$$

The forces are non-zero only at the time of the collision t_0, and then it holds

$$\mathbf{K}_1(t_0) + \mathbf{K}_2(t_0) = 0\,.$$

The conservation of momentum is ensured by (14.3.2), where the times immediately before and after the collision at t_0 can be used.

14.3.2 Compton Scattering

The most well-known and certainly one of the simplest examples of relativistic kinematics is the *Compton effect*, which explains the scattering of light on electrons solely by means of energy-momentum conservation. In section 11.1.1, the scattering of electric waves on free electrons, the so-called Thomson scattering, was derived. In this case, the electron is excited by the electric field to oscillate, so that it emits scattered radiation of the same frequency. In experiments, it was found that in addition x-ray radiation of lower frequency occurred, for which a different mechanism had to be responsible.

The *Compton scattering* refers to the collision process of an electron with a photon, whose kinematics are examined here. Unlike in Thomson scattering, where the electron was excited by $\mathbf{E}$ to oscillate, it is repelled here. As sketched in Fig. 14.3, a photon with the momentum $(\vec{q} \equiv (q^\mu))$ is scattered on a electron $\vec{p} \equiv (p^\mu)$. After the collision, the photon has momentum $\vec{q}\,'$ and the electron has momentum $\vec{p}\,'$.

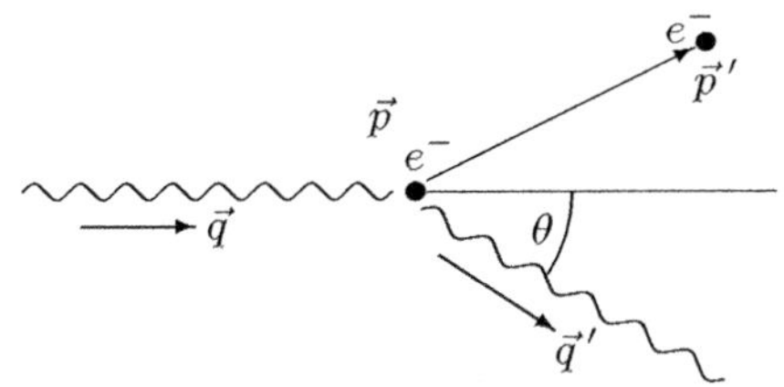

Fig. 14.3. Scattering process of a photon $\vec{q} = \hbar(\omega/c, \mathbf{k})$ with a stationary electron $\vec{p} = (m_e c, \mathbf{0})$

Energy and momentum conservation yield

$$\vec{p} + \vec{q} = \vec{p}\,' + \vec{q}\,' + \vec{q}\,'' \qquad \Longrightarrow \qquad \vec{p} + \vec{q} - \vec{q}\,' = \vec{p}\,'.$$

From this, by squaring $(\vec{p} \cdot \vec{q} \equiv p^\mu q_\mu = q_\mu p^\mu$ and $\vec{p}^{\,2} \equiv p^\mu p_\mu)$

$$(\vec{p} + \vec{q} - \vec{q}\,')^2 = \vec{p}^{\,2} + \vec{q}^{\,2} + 2\vec{p} \cdot (\vec{q} - \vec{q}\,') - 2\vec{q} \cdot \vec{q}\,' + \vec{q}^{\,\prime 2} = \vec{p}^{\,\prime 2}.$$

If you insert the invariants $p^2 = p'^2 = m^2 c^2$ and $q^2 = q'^2 = 0$:

$$\vec{p} \cdot (\vec{q} - \vec{q}\,') = \vec{q} \cdot \vec{q}\,',$$

then we have eliminated the coordinates of the scattered electron, which is usually not measured. Now we go into the rest system of the electron with $\vec{p} = (m_e c, \mathbf{0})$, $\vec{q} = (\hbar\omega/c, \hbar\mathbf{k})$, $|\mathbf{k}| = \omega/c$ and obtain

$$m_e\,\hbar(\omega - \omega') = \frac{\hbar^2}{c^2}\big(\omega\omega' - c^2 \mathbf{k} \cdot \mathbf{k}'\big) = \frac{\omega\omega'}{c^2}(1 - \cos\theta).$$

Now we insert the Compton frequency (see Tab. C.7, p. 645) $\omega_c = m_e\,c^2/\hbar$ into the equation, so it is

$$\omega_c(\omega - \omega') = \omega\omega'(1 - \cos\theta)\,,$$

or expressed differently:

$$\omega' = \frac{\omega}{1 + 2\frac{\omega}{\omega_c}\sin^2\frac{\theta}{2}}\,. \tag{14.3.4}$$

If we characterize the photons by their wavelengths, then the Compton scattering formula reads:

$$\lambda' - \lambda = 4\pi\lambda_c \sin^2(\theta/2) \tag{14.3.5}$$

with the Compton wavelength (see Tab. C.7, p. 645) $\lambda_c = h/m_e c$.

The intensity cannot be calculated from the conservation laws. However, one can expect that for low energies and small frequency difference the Thomson scattering cross section (11.1.14) must approximately apply, which can be seen from Fig. 14.4. The calculation of Compton scattering in 2nd order perturbation theory goes back to Klein, Nishina [1929]:

$$\frac{\mathrm{d}\sigma}{\mathrm{d}\Omega} = \frac{r_e^2}{2}\left(\frac{\omega'}{\omega}\right)^2\left(\frac{\omega'}{\omega} + \frac{\omega}{\omega'} - \sin^2\theta\right)\,, \tag{14.3.6}$$

where for ω'/ω the Compton formula (14.3.4) is to be inserted. The derivation of the Klein-Nishina formula can be found in textbooks on relativistic quantum mechanics [Bjørken, Drell, 1964, (7.74)]. As already mentioned in the Thomson scattering formula, forward scattering is more pronounced at higher energy than in Thomson scattering. For $\omega = \omega'$ one obtains the differential scattering cross section (11.1.12).

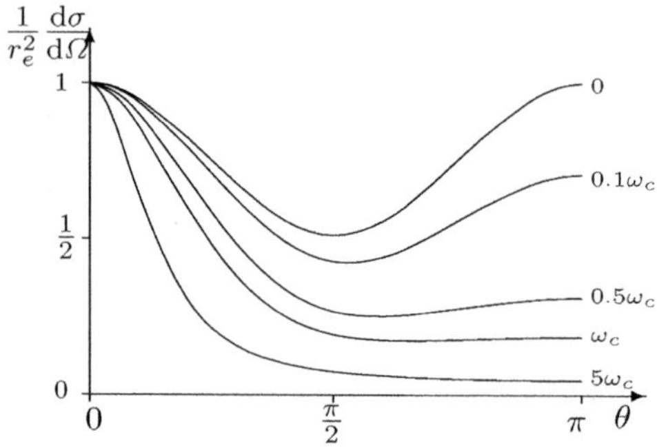

Fig. **14.4.** Compton Scattering (14.3.6): For $\omega = 0$ one obtains the Thomson formula; the deviations from this are already at $\omega = 0.1\omega_c$ considerable

The experiments were conducted by Compton in 1923. Their significance lay in the confirmation of $\mathbf{q} = \hbar\mathbf{k}$.

The term *inverse Compton effect* refers to the scattering of a high-energy electron on a low-energy photon, which gains energy as a result. The effect occurs when high-energy electrons scatter on the cosmic background radiation.

14.3.3 The Motion of the Electron around the Nucleus

Uhlenbeck and *Goudsmit* were able to explain the anomalous Zeeman effect
in 1925 by attributing an intrinsic angular momentum to the electron

$$\boldsymbol{\mu} = \frac{k_L g e}{2 m_e c}\, \mathbf{S}, \tag{14.3.7}$$

giving, where the gyromagnetic factor (also Landé factor) had the value
$g = 2$[4]. However, this could not explain the fine structure splitting, whose
theoretical value was twice as large as measured in the experiment. *Thomas*
showed in 1927 that with relativistic motion of an angular momentum on
a circular path, a precession ω_T occurs, which reduces the influence of the
splitting[5].

The Thomas precession (12.4.26) is a kinematic effect. The electron experiences constant acceleration perpendicular to its motion as it revolves around
the nucleus. In order for the electron to remain in a momentary rest system,
Lorentz transformations around perpendicular axes must be constantly performed.

Two velocity transformations (boost) with non-collinear $\mathbf{v}$ and $\mathbf{w}$ result in
an LT that is not a pure boost, but also includes a rotation around the axis
$\mathbf{v} \times \mathbf{w}$.

We consider the motion of an electron moving in an orbit around the atomic
nucleus, as sketched in Fig. 14.5. The energy (4.3.5) of a magnetic moment $\boldsymbol{\mu}$
in a magnetic field is

$$U = -\boldsymbol{\mu}\cdot\mathbf{B} = -\frac{k_L g e}{2 m_e c}\, \mathbf{S}\cdot\mathbf{B}.$$

According to (13.1.30), the spin of the moving electron senses the electrostatic field of the nucleus as a weak magnetic field, which causes a spin-orbit
interaction.

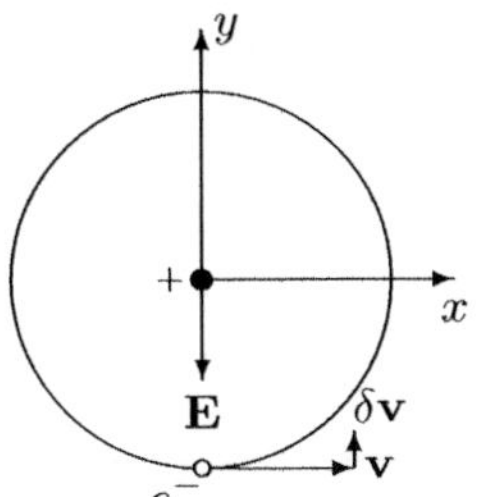

Fig. 14.5. According to the sketch, the electron moves at
time t with $\mathbf{v} = v\mathbf{e}_x$. After the time span δt the electron has
changed its speed to $\mathbf{v} + \delta\mathbf{v}$, where $\delta\mathbf{v} = \delta v\,\mathbf{e}_y$.
From U follows the equation of motion for a magnetic mo-
ment (Spin):
$$d\mathbf{S}/dt = \mathbf{N} = \boldsymbol{\mu}\times\mathbf{B}$$

---

[4] $g = 2.00232$ with corrections from quantum electrodynamics.
[5] The Dirac equation found by *Dirac* in 1928, or its non-relativistic approximation,
the Pauli equation, however, automatically lead to the correct results, both in terms
of the factor $g = 2$ and the spin-orbit interaction.

Note: The equation of motion for the spin can be seen in analogy to the orbital angular momentum $\boldsymbol{L}$ of classical mechanics:

$$\mathrm{d}\boldsymbol{L}/\mathrm{d}t = \mathbf{N}$$

or can be understood from the Heisenberg equation [Schwabl, 2007, (8.56)]

$$\frac{\mathrm{d}\mathbf{S}}{\mathrm{d}t} = \frac{\mathrm{i}}{\hbar}\,[\mathrm{H},\mathbf{S}] \qquad \text{with} \qquad \mathrm{H} = -\boldsymbol{\mu}\cdot\mathbf{B} = -\frac{egk_L}{2m_e c}\,\mathbf{S}\cdot\mathbf{B}$$

and the commutation rules [Schwabl, 2007, (9.9)]

$$[S_i, S_j] = \mathrm{i}\hbar\epsilon_{ijk}\,S_k\,,$$

$$\frac{\mathrm{d}S_i}{\mathrm{d}t} = -\frac{\mathrm{i}k_L ge}{2m_e c\hbar}[S_j, S_i]B_j = -\frac{\mathrm{i}k_L ge}{2m_e c\hbar}\,\epsilon_{jik}\mathrm{i}\hbar S_k B_j = \frac{k_L ge}{2m_e c}\,\epsilon_{ikj}S_k B_j = \left(\boldsymbol{\mu}\times\mathbf{B}\right)_i = N_i\,.$$

It is assumed that the electron is moving at time t with the velocity $\mathbf{v} = v\,\mathbf{e}_x$. In the rest system of the electron, i.e. in the body-fixed system, we thus have the equation of motion [Jackson, 1998, §11.8]

$$\frac{\mathrm{d}\mathbf{S}}{\mathrm{d}t} = \mathbf{N}' = \boldsymbol{\mu}\times\mathbf{B}'. \tag{14.3.8}$$

The fields $\mathbf{E}'$ and $\mathbf{B}'$ can be determined with an LT from the fields in the laboratory system, where for our purposes $\mathbf{B}'$ (13.1.30) in the approximation with $\gamma = 1$ is sufficient:

$$\mathbf{B}' \approx \mathbf{B} - \boldsymbol{\beta}\times\mathbf{E}/k_L.$$

However, the electron is not in an inertial system, but experiences on its path around the atomic nucleus a acceleration perpendicular to its direction of motion, which corresponds to rotation around the nucleus. This is incorporated into the equation of motion in the same way as in classical mechanics.

This is the analogous effect to a vector $\mathbf{a}$, which performs a rotational motion:

$$\left(\frac{\mathrm{d}\mathbf{a}}{\mathrm{d}t}\right)_{\text{Labor}} = \left(\frac{\mathrm{d}\mathbf{a}}{\mathrm{d}t}\right)'_{\text{body-fixed}} + \boldsymbol{\omega}\times\mathbf{a}\,.$$

The last term comes from the rotation of the body-fixed coordinate system [Iro, 2015, §8.5]. If one differentiates $\mathbf{a}(t) = a_i(t)\mathbf{e}_i = a_i'(t)\mathbf{e}_i'(t)$ with respect to t, one obtains $\dot{\mathbf{a}} = \dot{a}_i'\mathbf{e}_i' + a_i'\dot{\mathbf{e}}_i'$, where $\dot{\mathbf{e}}_i' = \boldsymbol{\omega}\times\mathbf{e}_i'$.

For an equation of motion of the form

$$\left(\frac{\mathrm{d}\mathbf{S}}{\mathrm{d}t}\right)_{\text{Laboratory}} = \left(\frac{\mathrm{d}\mathbf{S}}{\mathrm{d}t}\right)'_{\text{body-fixed}} + \boldsymbol{\omega}\times\mathbf{S} = \frac{gek_L}{2m_e c}\,\mathbf{S}\times\mathbf{B}' - \mathbf{S}\times\boldsymbol{\omega}_T$$

we can directly specify the internal energy:

$$U = U' + \mathbf{S}\cdot\boldsymbol{\omega}_{\mathrm{T}} = -\boldsymbol{\mu}\cdot\left(\mathbf{B} - \boldsymbol{\beta}\times\mathbf{E}/k_L\right) + \mathbf{S}\cdot\boldsymbol{\omega}_{\mathrm{T}}. \tag{14.3.9}$$

In an atomic environment, $\mathbf{E}$ can be derived from the potentials of the nuclei, where we assume $\phi = \phi(r)$:

$$\mathbf{E} = -\boldsymbol{\nabla}\phi = -\left(\frac{\mathrm{d}\phi}{\mathrm{d}r}\right)\mathbf{e}_r. \tag{14.3.10}$$

Thus, we have

$$U = -\frac{k_L g e}{2m_e c}\,\mathbf{S}\cdot\mathbf{B} - \frac{1}{r}\frac{\mathrm{d}\phi}{\mathrm{d}r}\,\frac{ge}{2m_e c}\,\mathbf{S}\cdot(\boldsymbol{\beta}\times\mathbf{x}) + \boldsymbol{\omega}_{\mathrm{T}}\cdot\mathbf{S}. \tag{14.3.11}$$

Now we introduce the orbital angular momentum of the electron

$$\boldsymbol{L} = \mathbf{x}\times m_e\boldsymbol{\beta}\,c$$

and obtain

$$U = -\frac{k_L g e}{2m_e c}\,\mathbf{S}\cdot\mathbf{B} + \frac{1}{r}\frac{\mathrm{d}\phi}{\mathrm{d}r}\,\frac{ge}{2m_e^2 c^2}\,\mathbf{S}\cdot\boldsymbol{L} + \boldsymbol{\omega}_{\mathrm{T}}\cdot\mathbf{S}. \tag{14.3.12}$$

The second and third terms determine the interaction of the spin with the orbit. The frequency ω_{T}, which indicates the rotation of the spin (electron) around the nucleus, we have already determined in (12.4.26):

$$\boldsymbol{\omega}_{\mathrm{T}} = -\frac{\gamma^2}{\gamma+1}\,\boldsymbol{\beta}\times\frac{\mathrm{d}\boldsymbol{\beta}}{\mathrm{d}t}.$$

The acceleration $\dot{\boldsymbol{\beta}}\perp\mathbf{v}$ is given by the electric field $\mathbf{E}$ from the atomic nucleus (14.3.10), around which the electron moves $\dot{\boldsymbol{\beta}} = e\mathbf{E}/m_e c$. This implies

$$\boldsymbol{\omega}_{\mathrm{T}} = -\frac{\gamma^2}{\gamma+1}\,\frac{e}{m_e c^2}\,\mathbf{v}\times\mathbf{E} = -\frac{1}{r}\frac{\mathrm{d}\phi(r)}{\mathrm{d}r}\,\frac{e}{2m_e^2 c^2}\,\boldsymbol{L}. \tag{14.3.13}$$

Substituted into (14.3.12) one obtains

$$U = -\boldsymbol{\mu}\cdot\mathbf{B} + \frac{e}{r}\frac{\mathrm{d}\phi}{\mathrm{d}r}\,\frac{g-1}{2m_e^2 c^2}\,\mathbf{S}\cdot\boldsymbol{L}. \tag{14.3.14}$$

The strength of the spin-orbit coupling is halved by the factor $\gamma^2/(\gamma+1) = 1/2$ "halved" $(g \to g-1)$.

Problems for Chapter 14

14.1. *Lorentz equation*: Calculate the zeroth component $\mathrm{d}p^0/\mathrm{d}t$ from the spatial part $\mathrm{d}\boldsymbol{p}/\mathrm{d}t$ to verify (14.2.15).

14.2. *Lagrange density*: Determine the Euler-Lagrange equations for

$$\mathcal{L}_L = -\frac{k_L^2}{4\pi k_C}\frac{1}{2}A_{\mu,\nu}\,A^{\mu,\nu} - \frac{k_L}{c}A_\mu j^\mu.$$

14.3. *Euler-Lagrange equation*: Solve the variation problem

$$\delta S = \frac{1}{c}\int_{s_1}^{s_2}\delta\big(\mathrm{d}s\,\mathrm{L}_r\big)$$

directly by calculating $\delta\mathrm{d}s$ and $\delta\mathrm{L}_r$ as functions of δu^μ (or $\delta\mathrm{d}x^\mu$) and δx^μ; i.e., verify (14.2.32) and calculate the equation of motion with it.

Hint: First show: $\qquad\qquad\qquad \delta\mathrm{d}s = u_\mu\mathrm{d}\delta x^\mu, \qquad\qquad \delta u^\mu = \big(\delta^\mu{}_\nu - u^\mu u_\nu\big)\dfrac{\mathrm{d}\delta x^\nu}{\mathrm{d}s}.$

14.4. *Hamilton function*: Calculate the electric and magnetic field energy separately and verify (14.2.47).

14.5. *Inverse Compton scattering*: A high-energy electron and a long-wavelength photon move in the x-direction, where the photon is reflected in a "frontal"collision (reversing its direction of motion). Show that the scattered photon approximately has the following frequency:

$$\omega' = \frac{4\omega\gamma^2}{1 + 4\hbar\omega\gamma/m_e c^2}.$$

References

Abraham M. *Zur Elektrodynamik bewegter Körper*, Rend. Circ. Matem. Palermo **28**,1–28 (1909)

Becker R., Sauter F. *Theorie der Elektrizität 1*, 21th ed. Teubner, Stuttgart (1973)

Bjørken J. and Drell S. *Relativistische Quantenmechanik*, Bibliographisches Institut, Mannheim (1964)

Chaichian M., Merches I., Radu D., Tureanu A. *Electrodynamics*, Springer Berlin (2016)

Cohen-Tannoudji C., Dupont-Roc J., Grynberg G. *Photons and Atoms: Introduction to Electrodynamics*, Wiley-VCH (1989)

Fließbach T. *Die relativistische Masse*, Springer Spektrum, Berlin (2018)

Griffiths D. *Introduction to Electrodynamics*, 4th ed., Cambridge University Press (2017)

Iro H. *A Modern Approach to Classical Mechanics*, 2nd ed., World Scientific (2015)

Jackson J. D. *Classical Electrodynamics*, 3rd ed., John Wiley & Sons Inc. (1998)

Klein O. and Nishina Y. *Über die Streuung von Strahlung durch freie Elektronen nach der neuen relativistischen Quantendynamik von Dirac*, Z. Physik **52**, 853–868 (1929)

Landau L. D. and Lifshitz E. M. *The Classical Theory of Fields*, vol. 2, 4th ed. Butterworth-Heinemann, reprinted (1994)

Lenstra D., Mandel L. *Angular momentum of the quantized electromagnetic field with periodic boundary conditions*, Phys. Rev. A **26**, 3428–3437 (1982)

Mach E. *Die Mechanik in ihrer Entwicklung*, 9th ed., Brockhaus Leipzig (1933)

Mandel L., Wolf E. *Optical coherence and quantum optics*, Cambridge University Press, N.Y. (1995)

McDonald K. T. *Bibliography on the Abraham-Minkowski Debate* (2017) url: https://physics.princeton.edu/~mcdonald/examples/ambib.pdf, last accessed on January 2, 2022

Minkowski H. *Die Grundgleichungen für die elektromagnetischen Vorgänge in bewegten Körpern*, Nachr. Königl. Ges. Wiss. Göttingen, 53–111 (1908)

Nolting W. *Special Theory of Relativity*, 1st ed. Springer (2017)

Panofsky W., Phillips M. *Classical electricity and magnetism*, 2nd ed. Addison – Wesley, Reading (1962)

Pauli W. *Theory of Relativity*, Pergamon Press (1958) translated from *Relativitätstheorie*, Encyklopädie der Mathematischen Wissenschaften **V19**, Teubner (1921)

Rohrlich F. *Dynamics of a charged particle* Phys. Rev. E **77**, 046609 (2008)

Schröder U. *Spezielle Relativitätstheorie*, 5th ed. Harri Deutsch, Frankfurt (2014)

Schwabl F. *Quantum Mechanics*, 4th ed. Springer Berlin (2007)

Schwabl F. *Advanced Quantum Mechanics*, 4th ed. Springer Berlin (2008)

Schwinger J. *On the Classical Radiation of accelerated Electrons*, Phys. Rev. **75**, 1912–1925 (1949)

Sheppard C. J. and Kemp B. A. *Relativistic analysis of field-kinetic and canonical electromagnetic systems*, Phys. Rev. A **93**, 053832 (2016)

Simmons J. and Guttmann M. *States, Waves and Photons: Modern Introduction to Light*, Addison-Wesley (1970)

Sommerfeld A. *Electrodynamics*, Academic Press Inc., New York (1952)

Sexl R. U. and Urbantke H. K. *Relativity, Groups, Particles*, Springer Wien (2001)

Weinberg S. *Gravitation and Cosmology*, John Wiley & Sons, N.Y. (1972)

A

Vectors, Vector Analysis and Integral Theorems

The mathematical description of physical processes must be independent of the reference system. It must therefore be representable by concepts that are invariant with respect to linear coordinate transformations in Euclidean space $\mathbb{R}^n$.

Examples are the length of a line, the area of a planar figure or the volume. All mentioned objects are scalars (tensors of 0th order). If the direction is added to length, we speak of a vector, a tensor of 1st order. If one changes from one reference system to another using a linear coordinate transformation, the coefficients with which the coordinates of the new reference system are determined by the old one form a tensor of 2nd order.

We will not proceed completely systematically here and define all quantities and operations, but will only touch on matrices and matrix multiplication.

A.1 Vector Calculus in Euclidean Space

A.1.1 Vectors

(Euclidean) vectors are defined by their length, their direction (see Fig. A.1) and the following arithmetic operations:

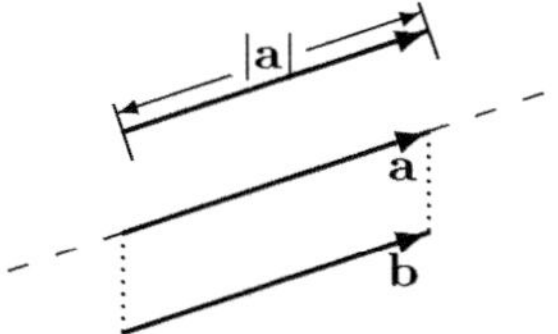

Fig. A.1. Length and direction define a vector; thus $\mathbf{a} = \mathbf{b}$. The arrow indicates the direction and the scalar $a = |\mathbf{a}|$ the length

D. Petrascheck, F. Schwabl, *Electrodynamics,* https://doi.org/10.1007/978-3-662-71502-4_15

1. Vector addition: $\mathbf{c} = \mathbf{a} + \mathbf{b}$.

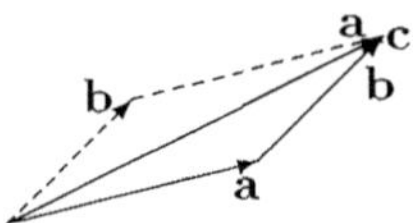

Fig. A.2. Vector addition: $\mathbf{c} = \mathbf{a} + \mathbf{b}$, simultaneously showing the commutative law: $\mathbf{c} = \mathbf{b} + \mathbf{a}$

2. Multiplication with a scalar: $\mathbf{b} = \alpha\,\mathbf{a}$.

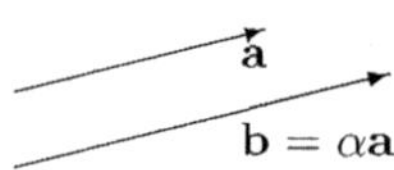

Fig. A.3. Multiplication of $\mathbf{a}$ with a real number ($\alpha = 1.5$)

3. Scalar product: $\alpha = \mathbf{a}\cdot\mathbf{b} = ab\cos\theta$.

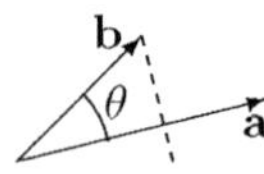

Fig. A.4. Dot product: $\mathbf{a}\cdot\mathbf{b} = ab\cos\theta$

The inner product of two vectors, the dot product, is a scalar that results from the multiplication of the length of $\mathbf{a}$ with the length of the projection of $\mathbf{b}$ onto $\mathbf{a}$ ($b\cos\theta$) is composed of: $\alpha = ab\cos\theta$.

These calculations fulfill the following rules:

1. Addition
   a) $\mathbf{a} + \mathbf{b} = \mathbf{b} + \mathbf{a}$ commutative law
   b) $\mathbf{a} + (\mathbf{b} + \mathbf{c}) = (\mathbf{a} + \mathbf{b}) + \mathbf{c}$ associative law
   c) $\mathbf{a} + \mathbf{0} = \mathbf{a}$ zero vector (neutral element)
   d) $\mathbf{a} + (-\mathbf{a}) = \mathbf{0}$ inverse element of addition
2. Multiplication with a scalar
   a) $1\,\mathbf{a} = \mathbf{a}$ unit element
   b) $\alpha(\beta\mathbf{a}) = (\alpha\beta)\mathbf{a}$ associative law
   c) $(\alpha + \beta)\mathbf{a} = \alpha\mathbf{a} + \beta\mathbf{a}$ distributive law for scalar addition
   d) $\alpha(\mathbf{a} + \mathbf{b}) = \alpha\mathbf{a} + \alpha\mathbf{b}$ distributive law for vector addition
3. Dot product
   a) $\mathbf{a} \cdot \mathbf{b} = \mathbf{b} \cdot \mathbf{a}$ commutative law
   b) $(\alpha\mathbf{a}) \cdot \mathbf{b} = \mathbf{a} \cdot (\alpha\mathbf{b})$ associative law
   c) $\mathbf{a} \cdot (\mathbf{b} + \mathbf{c}) = \mathbf{a} \cdot \mathbf{b} + \mathbf{a} \cdot \mathbf{c}$ distributive law
   d) $\mathbf{a} \cdot \mathbf{b} = 0 \;\; \forall \mathbf{b} \rightsquigarrow \mathbf{a} = \mathbf{0}$ zero vector

Vector space   V

1. If all vectors fulfill the above conditions, they are Euclidean vectors, and V is a Euclidean vector space.

2. If the vectors only fulfill the first 8 arithmetic operations, i.e., no scalar product is defined, then these are affine vectors, and V is an affine vector space.

Example: Single-line (single-column) matrices are affine vectors.

The vector space is associated with a basis, with which each vector can be uniquely described by coordinates. The number of basis vectors is the dimension of the vector space.

Point space

The n-dimensional point space $\mathbb{R}^n$ is the set of all n-tuples of real numbers $P(x_1, ..., x_n) = P(\mathbf{x})$.

Each pair of points (P_1, P_2) from $\mathbb{R}^n$ should be assigned a vector from the vector space V with the following properties:

1. $\overrightarrow{P_1 P_2} = -\overrightarrow{P_2 P_1}$,
2. $\overrightarrow{P_1 P_2} = \overrightarrow{P_1 P_3} + \overrightarrow{P_3 P_2}$,
3. if $O \in \mathbb{R}^n$, then for every vector $\mathbf{x} \in V$ there is exactly one point $X \in \mathbb{R}^n$, such that $\mathbf{x} = \overrightarrow{O X}$.

The point space $\mathbb{R}^n$ is an

1. Euclidean point space, if it is associated with the Euclidean vector space,
2. or an affine point space, if it is associated with the affine vector space.

Linear dependence

$\mathbf{a}$ and $\mathbf{b}$ are linearly dependent if they are (anti-)parallel to each other: $\mathbf{b} = \alpha \mathbf{a}$. More generally, $\mathbf{a}_1, ..., \mathbf{a}_n$ are linearly dependent if there are real numbers $\alpha^1, ..., \alpha^n$ such that

$$\sum_{k=1}^{n} \alpha^k \mathbf{a}_k = \mathbf{0},$$

where not all $\alpha^k = 0$ may be. The superscript of the indices of the real numbers α^k has no deeper meaning here.

Dimension and basis

A vector space V is n-dimensional if there are n linearly independent vectors $\mathbf{a}_n$ and all $n+1$ vectors are linearly dependent

$$\sum_{k=1}^{n} \alpha^k \mathbf{a}_k = \mathbf{x}.$$

The $\mathbf{a}_k$ with $k = 1, ..., n$ form a basis of the vector space, and α^k are the coordinates. We use $\mathbf{h}_k$ for (non-orthogonal) basis vectors and x^k for the coordinates

$$\mathbf{x} = \sum_{k=1}^{n} x^k \, \mathbf{h}_k = x^k \mathbf{h}_k \, . \tag{A.1.1}$$

We sum over doubly occurring indices (Einstein's summation convention). The (holonomic) basis vectors $\mathbf{h}_k$ form an oblique coordinate system.

A.1.2 Vectors in the N-Dimensional Space

Scalar product and metric

The scalar product of two vectors is given by

$$\mathbf{x} \cdot \mathbf{y} = x^i \, \mathbf{h}_i \cdot y^j \mathbf{h}_j = x^i y^j \, g_{ij} \qquad \text{with} \qquad g_{ij} = \mathbf{h}_i \cdot \mathbf{h}_j \, . \tag{A.1.2}$$

Note: With $\mathbf{a}$ a column vector is denoted. The scalar product is then

$$\mathbf{a} \cdot \mathbf{b} \equiv \begin{pmatrix} a_1 & \ldots & a_n \end{pmatrix} \begin{pmatrix} b_1 \\ \vdots \\ b_n \end{pmatrix} = \mathbf{a}^{\mathrm{T}} \mathbf{b} \, . \tag{A.1.3}$$

Distance

The distance between two points is represented by

$$d = |\overrightarrow{AB}| \qquad\qquad \rightsquigarrow \qquad\qquad d^2 = \overrightarrow{AB} \cdot \overrightarrow{AB} \, .$$

With $\mathbf{a} = \overrightarrow{OA}$ and $\mathbf{b} = \overrightarrow{OB}$ we get

$$\overrightarrow{AB} = \overrightarrow{AO} + \overrightarrow{OB} = \mathbf{b} - \mathbf{a} := \mathbf{c} \, .$$

This results in

$$d^2 = \mathbf{c} \cdot \mathbf{c} = c^i c^j \, g_{ij} \, .$$

The distance $d \geq 0$ between two points is positive and only disappears when the points coincide ($\mathbf{c}=0$). Now, $\mathbf{a} \cdot \mathbf{b} = ab \cos\theta$, where, as can be seen from Fig. A.4, $\theta = \sphericalangle \mathbf{a}, \mathbf{b}$:

$$\cos\theta = \frac{\mathbf{a} \cdot \mathbf{b}}{ab} = \frac{g_{ij} a^i b^j}{\sqrt{g_{ij} a^i a^j} \, \sqrt{g_{ij} b^i b^j}} \, .$$

This is the Cauchy-Schwarz inequality, from which the triangle inequality follows:

$$|\mathbf{a} \cdot \mathbf{b}| \leq |\mathbf{a}| \, |\mathbf{b}| \qquad \Rightarrow \qquad |\mathbf{a} + \mathbf{b}| \leq |\mathbf{a}| + |\mathbf{b}| \, . \tag{A.1.4}$$

Covariant and contravariant basis

We now define contravariant basis vectors (dual) $\mathbf{h}^i$ by

$$\mathbf{h}_i \cdot \mathbf{h}^j = g_i{}^j = \delta_i{}^j = \begin{cases} 1 & i = j \\ 0 & i \neq j \end{cases}. \tag{A.1.5}$$

First, we note that from $\mathbf{x} = x^k \, \mathbf{h}_k$ follows

$$\mathbf{h}^i \cdot \mathbf{x} = x^k \, \mathbf{h}^i \cdot \mathbf{h}_k = x^k \, \delta^i{}_k = x^i \,.$$

So we can represent $\mathbf{x}$ by

$$\mathbf{x} = (\mathbf{x} \cdot \mathbf{h}^i) \, \mathbf{h}_i = (\mathbf{x} \cdot \mathbf{h}_i) \, \mathbf{h}^i \,. \tag{A.1.6}$$

The contravariant basis vectors $\mathbf{h}^i$ are therefore in the covariant basis

$$\mathbf{h}^i = (\mathbf{h}^i \cdot \mathbf{h}^j) \, \mathbf{h}_j = g^{ij} \, \mathbf{h}_j \qquad \text{with} \quad g^{ij} = \mathbf{h}^i \cdot \mathbf{h}^j \,. \tag{A.1.7}$$

The contravariant basis vectors are thus determined by the (contravariant) metric tensor g. From the multiplication of (A.1.7) from the right with $\mathbf{h}_k$ follows

$$\mathbf{h}^i \cdot \mathbf{h}_k = \delta^i{}_k = g^{ij} \, \mathbf{h}_j \cdot \mathbf{h}_k = g^{ij} \, g_{jk} \,. \tag{A.1.8}$$

(g^{ij}) is thus the inverse of (g_{ij}). We now multiply (A.1.7) with g_{ki}:

$$g_{ki}\mathbf{h}^i = g_{ki} \, g^{ij} \, \mathbf{h}_j = \delta_k{}^j \mathbf{h}_j = \mathbf{h}_k \,.$$

Similarly, we obtain with (A.1.6)

$$g_{ik} \, x^k = g_{ik} \, \mathbf{h}^k \cdot \mathbf{x} = \mathbf{h}_i \cdot \mathbf{x} = x_i \,.$$

The metric tensor (g_{ij}) can thus be used to "lower"and (g^{ij}) to "raise"the indices of the basis vectors, whereby the method is not only for vector quantities, but also applies to tensors.

A crystal lattice is described with a covariant basis. The basis vectors thus have the dimension of a length [l]. The associated contravariant basis is the reciprocal lattice, whose basis vectors have the dimension $[l^{-1}]$ and describe wave numbers. The dual vector space $\mathbf{V}^*$ is associated with the contravariant basis.

The (g^{ij}) are inverse to the (g_{ij}). Therefore, to calculate the (g^{ij}) the determinant g of the covariant metric tensor $(g_{ij})$is required:

$$g = \|\mathbf{g}\| = \frac{1}{\|(g^{ij})\|} = \det \mathbf{g} = \begin{vmatrix} g_{11} & \cdots\cdots & g_{1n} \\ \cdots & \cdots\cdots & \cdots \\ \cdots & \cdots\cdots & \cdots \\ g_{n1} & \cdots\cdots & g_{nn} \end{vmatrix}. \tag{A.1.9}$$

Two-dimensional oblique basis in Euclidean space.

The previous considerations can be clearly demonstrated in the simplest system, the two-dimensional Euclidean space.

Vectors from the two-dimensional Euclidean vector space, represented by a covariant basis or from the dual (reciprocal) vector space, represented by a contravariant basis:

covariant basis vectors: $\mathbf{h}_1$ and $\mathbf{h}_2$ with $|\mathbf{h}_1| = |\mathbf{h}_2| = 1$,

contravariant basis vectors: $\mathbf{h}^1$ and $\mathbf{h}^2$ with $|\mathbf{h}^1| = |\mathbf{h}^2| = 1/\sin\theta$,

Vector in covariant basis: $\mathbf{x} = (\mathbf{x} \cdot \mathbf{h}^1)\,\mathbf{h}_1 + (\mathbf{x} \cdot \mathbf{h}^2)\,\mathbf{h}_2$,

Vector in contravariant basis: $\mathbf{x} = (\mathbf{x} \cdot \mathbf{h}_1)\,\mathbf{h}^1 + (\mathbf{x} \cdot \mathbf{h}_1)\,\mathbf{h}^2 = x^1\,\mathbf{h}_1 + x^2\,\mathbf{h}_2$.

For the length of the basis vectors, according to Fig. A.5: $\mathbf{h}^1 \mathbf{h}_1 = h^1 h_1 \cos(\frac{\pi}{2} - \theta) = 1$, from which $h^1 = 1/(h_1 \sin\theta)$ and $h^2 = 1/(h_2 \sin\theta)$ follow.

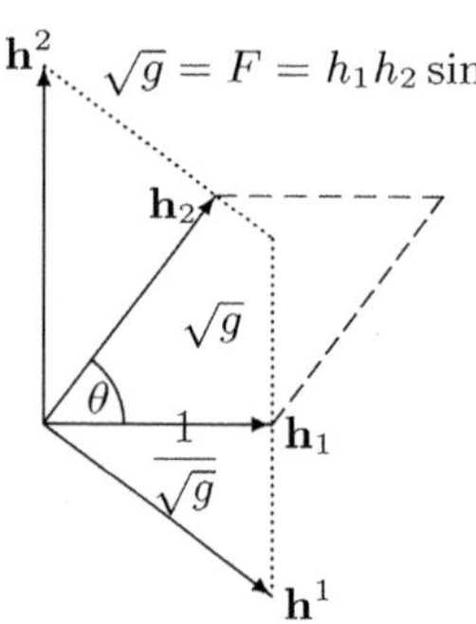

Fig. A.5. Two-dimensional, oblique grid. Contra- and covariant basis along with unit cells (parallelograms); $h^i = 1/(h_i \sin\theta)$

The metric tensor is given by

$$(g_{ik}) = \begin{pmatrix} \mathbf{h}_1 \cdot \mathbf{h}_1 & \mathbf{h}_1 \cdot \mathbf{h}_2 \\ \mathbf{h}_2 \cdot \mathbf{h}_1 & \mathbf{h}_2 \cdot \mathbf{h}_2 \end{pmatrix} = \begin{pmatrix} h_1^2 & h_1 h_2 \cos\theta \\ h_1 h_2 \cos\theta & h_2^2 \end{pmatrix} \quad \text{with} \quad g = h_1^2 h_2^2 \sin^2\theta.$$

Generalized cross product

Another operation is the generalized cross product. We define it by

$$\mathbf{c} = \mathbf{a}^{(2)} \times \mathbf{a}^{(3)} \times \ldots \times \mathbf{a}^{(n)} = \sqrt{g}\,\epsilon^{j_1 \cdots j_n}\,\mathbf{h}_{j_1}\,a_{j_2}^{(2)} \cdots a_{j_n}^{(n)}$$

$$= \frac{1}{\sqrt{g}} \begin{vmatrix} \mathbf{h}_1 & \cdots \cdots & \mathbf{h}_n \\ a_1^{(2)} & \cdots \cdots & a_n^{(2)} \\ \vdots & \cdots \cdots & \vdots \\ a_1^{(n)} & \cdots \cdots & a_n^{(n)} \end{vmatrix}. \tag{A.1.10}$$

Here, $\epsilon(j_1, \ldots, j_n) = g\,\epsilon^{j_1, \cdots j_n}$ is the totally antisymmetric tensor (the Levi-Civita symbol (A.1.36)). (A.1.10) is a generalization of the vector product $\mathbf{c} = \mathbf{a}^{(2)} \times \mathbf{a}^{(3)}$ from three dimensions, however, this notation is not suitable for two dimensions.

Generalized scalar triple product

In addition to the generalized cross product, a generalized scalar product can be given by scalar multiplying the cross product with another vector:

$$[\mathbf{a}^{(1)} \ldots \mathbf{a}^{(n)}] = \mathbf{a}^{(1)} \cdot [\mathbf{a}^{(2)} \times \ldots \times \mathbf{a}^{(n)}] = \sqrt{g}\, e^{j_1 \cdots j_n} (\mathbf{a}^{(1)} \cdot \mathbf{h}_{j_1})\, a_{j_2}^{(2)} \ldots a_{j_n}^{(n)}$$

$$= \frac{1}{\sqrt{g}} \begin{vmatrix} a_1^{(1)} & \cdots\cdots & a_n^{(1)} \\ \vdots & \cdots\cdots & \vdots \\ a_1^{(n)} & \cdots\cdots & a_n^{(n)} \end{vmatrix}, \tag{A.1.11}$$

where we have substituted $a_{j_1}^{(1)} = \mathbf{h}_{j_1} \cdot \mathbf{a}^{(1)}$. w only vanishes if all $\mathbf{a}^{(i)}$ are linearly independent and there is no further linearly independent vector. If $\mathbf{a}^{(i)}$ are the basis vectors $\mathbf{h}^i$, then the $a_j^{(i)} = (\mathbf{h}_j \cdot \mathbf{h}^i) = \delta_j{}^i$, and one obtains:

$$w = [\mathbf{h}^1 \ldots \mathbf{h}^n] = \mathbf{h}^1 \cdot [\mathbf{h}^2 \times \ldots \times \mathbf{h}^n] = \frac{1}{\sqrt{g}} \begin{vmatrix} 1 & 0 & \cdots\cdots & 0 \\ 0 & 1 & \cdots\cdots & 0 \\ \vdots & \vdots & \cdots\cdots & \vdots \\ 0 & 0 & \cdots\cdots & 1 \end{vmatrix} = \frac{1}{\sqrt{g}}. \tag{A.1.12}$$

For the covariant basis vectors $\mathbf{a}^{(i)} = \mathbf{h}_i$ the matrix elements are $a_j^{(i)} = (\mathbf{h}_j \cdot \mathbf{h}_i) = g_{ji}$ and thus the cross product is

$$v = [\mathbf{h}_1 \ldots \mathbf{h}_n] = \frac{1}{\sqrt{g}} \begin{vmatrix} g_{11} & \cdots & g_{1n} \\ \vdots & \cdots & \vdots \\ g_{n1} & \cdots & g_{nn} \end{vmatrix} = \sqrt{g}. \tag{A.1.13}$$

Cross product in two dimensions

The factor $\sqrt{g} = F = h_1 h_2 |\sin\theta|$ in two dimensions is the area of the parallelogram spanned by the base vectors $\mathbf{h}_1$ and $\mathbf{h}_2$, as shown in Fig. A.5. The cross product of $\mathbf{a}$ is given by

$$\mathbf{b} = \frac{1}{\sqrt{g}} \begin{vmatrix} \mathbf{h}_1 & \mathbf{h}_2 \\ a_1 & a_2 \end{vmatrix} = \frac{1}{\sqrt{g}}(\mathbf{h}_1 a_2 - \mathbf{h}_2 a_1),$$

$$|\mathbf{b}|^2 = (\mathbf{h}^1 a^2 - \mathbf{h}^2 a^1) \cdot (\mathbf{h}_1 a_2 - \mathbf{h}_2 a_1) = |\mathbf{a}|^2.$$

The scalar triple product of the $\mathbf{h}_i$ yields the area spanned by these

$$[\mathbf{h}_1\, \mathbf{h}_2] = \mathbf{h}_1 \cdot \frac{1}{\sqrt{g}}(\mathbf{h}_1 g_{22} - \mathbf{h}_2 g_{21}) = \frac{1}{\sqrt{g}}(g_{11}g_{22} - g_{12}g_{21}) = \sqrt{g}.$$

Dyadic product

The *dyadic* or *tensorial product* of two vectors

$$\mathbf{a} \circ \mathbf{b} = \mathbf{a}\,\mathbf{b}^T \qquad \Leftrightarrow \qquad \begin{pmatrix} a_1 \\ \vdots \\ a_n \end{pmatrix} (b_1 \ldots b_n) = \begin{pmatrix} a_1 b_1 & \ldots & a_1 b_n \\ \vdots & \ldots & \vdots \\ a_n b_1 & \ldots & a_n b_n \end{pmatrix} \tag{A.1.14}$$

is defined by the linear transformation in the form[1]

$$(\mathbf{a} \circ \mathbf{b})\mathbf{x} = \mathbf{a}(\mathbf{b} \cdot \mathbf{x}). \tag{A.1.15}$$

From the definition it follows directly that the dyadic product is not commutative. Rather, as is to be shown in problem A.1, that

$$\big[(\mathbf{a} \circ \mathbf{b}) - (\mathbf{b} \circ \mathbf{a})\big]\mathbf{x} = (\mathbf{b} \times \mathbf{a}) \times \mathbf{x}. \tag{A.1.16}$$

In (A.1.14) the dyadic product is represented as a matrix. Its elements depend on the basis. This can be a cartesian coordintae system $\mathbf{e}_i$, as well as an oblique one with $\mathbf{a} = a^k \mathbf{h}_k$ and $\mathbf{b}^T = b_l \mathbf{h}^{T\,l}$:

$$(\mathbf{a} \circ \mathbf{b})_{ij} = \mathbf{e}_i^T (\mathbf{a} \circ \mathbf{b}) \mathbf{e}_j = (\mathbf{e}_i \cdot \mathbf{a})(\mathbf{b} \cdot \mathbf{e}_j) = a_i b_j\,,$$

$$(\mathbf{a} \circ \mathbf{b})^i{}_j = \mathbf{h}^{i\,T} (\mathbf{a} \circ \mathbf{b}) \mathbf{h}_j = (\mathbf{h}^i \cdot \mathbf{a})(\mathbf{b} \cdot \mathbf{h}_j) = a^i b_j\,. \tag{A.1.17}$$

The significance of the dyadic product also lies in its use as a projection operator. Let $\boldsymbol{\epsilon}$ be a vector of length $|\boldsymbol{\epsilon}| = 1$, then

$$(\boldsymbol{\epsilon} \circ \boldsymbol{\epsilon})\mathbf{a} = \boldsymbol{\epsilon}\,(\boldsymbol{\epsilon} \cdot \mathbf{a}) \tag{A.1.18}$$

is a vector in the direction of $\boldsymbol{\epsilon}$ with the length of the projection of $\mathbf{a}$ onto the direction of $\boldsymbol{\epsilon}$. The operator $\mathsf{P} = (\mathsf{E} - \boldsymbol{\epsilon} \circ \boldsymbol{\epsilon})$ projects $\mathbf{a}$ onto the plane perpendicular to $\boldsymbol{\epsilon}$.

With the unit tensor $\sum_k (\mathbf{e}_k \circ \mathbf{e}_k)$ $\mathbf{a}' = \mathsf{T}\mathbf{a}$ can be decomposed into its components:

$$a_i' = \mathbf{e}_i^T \mathbf{a}' = \mathbf{e}_i^T \mathsf{T} \sum_k (\mathbf{e}_k \circ \mathbf{e}_k)\mathbf{a} = (\mathbf{e}_i^T \mathsf{T} \mathbf{e}_k)(\mathbf{e}_k \cdot \mathbf{a}) = T_{ik} a_k\,. \tag{A.1.19}$$

The determinant of the dyadic product of two vectors always vanishes since their rows or columns are proportional to each other:

$$\det\,(\mathbf{a} \circ \mathbf{b}) = 0 \qquad \text{and} \qquad \det\,(\mathsf{E} + \mathbf{a} \circ \mathbf{b}) = 1 + \mathbf{a} \cdot \mathbf{b}. \tag{A.1.20}$$

The proof of the second relation, which is needed for the calculation of the potential of a moving point charge (8.2.24), is the problem A.2.

---

[1] Analogous to the scalar product $\mathbf{a} \cdot \mathbf{b} = \mathbf{a}^T \mathbf{b}$ is the tensor product $\mathbf{a} \circ \mathbf{b} = \mathbf{a}\,\mathbf{b}^T$; frequently used is also the notation $\mathbf{a} \otimes \mathbf{b}$.

Change of basis

In a linear, homogeneous coordinate transformation, the vector remains unchanged, but one moves to a new basis. This is referred to as a passive transformation, which is defined here by

$$\bar{\mathbf{h}}^i = a^i{}_j \, \mathbf{h}^j \qquad\qquad \Rightarrow \qquad\qquad \mathbf{h}^k = (\mathsf{a}^{-1})^k{}_i \, \bar{\mathbf{h}}^i \tag{A.1.21}$$

and establishes a new contravariant basis. On the right is the inverse transformation, which arises from the left equation by multiplication with $(\mathsf{a}^{-1})^k{}_i$. The transformation (A.1.21) must, like the original basis, fulfill the condition

$$\bar{\mathbf{h}}^i \cdot \bar{\mathbf{h}}_j = \delta^i{}_j \,, \tag{A.1.22}$$

where we use for the covariant basis

$$\bar{\mathbf{h}}_i = b_i{}^j \, \mathbf{h}_j \qquad\qquad \Rightarrow \qquad\qquad \mathbf{h}_k = (\mathsf{b}^{-1})_k{}^i \, \bar{\mathbf{h}}_i \tag{A.1.23}$$

and determine the $b_i{}^j$ by (A.1.22). Since the contravariant basis is dual to the covariant one, we can expect that this applies in a similar form for the matrices a and b. For (A.1.22) we get

$$\bar{\mathbf{h}}_j \cdot \bar{\mathbf{h}}^i = \mathbf{h}_l \, (\mathsf{b}^T)^l{}_j \cdot a^i{}_k \, \mathbf{h}^k = (\mathsf{b}^T)^k{}_j \, a^i{}_k = \delta^i{}_j \,,$$

from which follows

$$(\mathsf{b}^T)^k{}_j = (\mathsf{a}^{-1})^k{}_j \qquad\qquad \Leftrightarrow \qquad\qquad \mathsf{b}^T = \mathsf{a}^{-1} \,. \tag{A.1.24}$$

b is contragredient to a. Thus, the transformations (A.1.23) are

$$\bar{\mathbf{h}}_i = (\mathsf{a}^{-1})^j{}_i \, \mathbf{h}_j \qquad\qquad \Rightarrow \qquad\qquad \mathbf{h}_k = a^i{}_k \, \bar{\mathbf{h}}_i \,. \tag{A.1.25}$$

The metric coefficients for the basis $\bar{\mathbf{h}}^i$ are (see (12.2.11)

$$\bar{g}^{ij} = \bar{\mathbf{h}}^i \cdot \bar{\mathbf{h}}^j = a^i{}_k \, g^{kl} \, a^j{}_l \qquad\qquad \Leftrightarrow \qquad\qquad \bar{\mathsf{g}}^{-1} = \mathsf{a}\,\mathsf{g}^{-1}\,\mathsf{a}^T, \tag{A.1.26}$$

where $\mathsf{a} = (a^j{}_i)$ and $\mathsf{g}^{-1} = (g^{kl})$. The same applies analogously for the contravariant basis

$$\bar{g}_{ij} = \bar{\mathbf{h}}_i \cdot \bar{\mathbf{h}}_j = (\mathsf{a}^{-1})^k{}_i \, g_{kl} \, (\mathsf{a}^{-1})^l{}_j \qquad\qquad \Leftrightarrow \qquad\qquad \bar{\mathsf{g}} = \mathsf{a}^{-1^T} \mathsf{g}\,\mathsf{a}^{-1} \tag{A.1.27}$$

with $\mathsf{g} = (g_{kl})$. This relation was to be expected, as $\bar{g}^{ij}$ must be inverse to $\bar{g}_{ij}$.

Transformation behavior of the components

In the passive transformation, the vector remains unchanged, so that after (A.1.21)

$$\mathbf{x} = \bar{\mathbf{h}}_i \, \bar{x}^i = \mathbf{h}_j \, x^j = \bar{\mathbf{h}}_i \, a^i{}_j \, x^j$$

applies. The contravariant components transform

$$\bar{x}^i = a^i{}_j\, x^j \tag{A.1.28}$$

just like the contravariant basis vectors, which applies accordingly for the covariant components:

$$\mathbf{x} = \bar{\mathbf{h}}^i\, \bar{x}_i = \mathbf{h}^j x_j = \bar{\mathbf{h}}^i\, (\mathsf{a}^{-1})^j{}_i\, x_j\,,$$
$$\bar{x}_i = (\mathsf{a}^{-1\,T})_i{}^j\, x_j\,. \tag{A.1.29}$$

x^i transforms by multiplication with a and x_i with the *contragredient* matrix $\mathsf{a}^{-1\,T}$. The transformation of the covariant components is *contragredient* to the transformation of the contravariant components.

We explicitly show the invariance of the scalar product:

$$\bar{x}^i\, \bar{x}_i = a^i{}_j\, x^j\, (a^{-1\,T})_i{}^k\, x_k = (a^{-1})^k{}_i\, a^i{}_j\, x^j\, x_k = x^k\, x_k\,.$$

Rotations

In rotations (rotary reflections), the length of the base vectors remains unchanged $|\bar{\mathbf{h}}_i| = |\mathbf{h}_i|$. Therefore, according to (A.1.27) for the determinants

$$\det\bar{\mathsf{g}} = (\det\mathsf{a})^2\,\det\mathsf{g} \qquad\Rightarrow\qquad \det\mathsf{a} = \pm 1\,.$$

Supplement to matrix symbolism

We have the transformation matrix a as a matrix in the form

$$\mathsf{a} = (a^i{}_k) \tag{A.1.30}$$

defined; the matrix b is defined as a "mixed" matrix (co- and contravariant indices) the same as a, but it is not further used as a construction aid, i.e., a is the only "mixed" matrix. However, it is not a tensor of 2nd rank in the vector space.

Tensor objects of n-th rank are defined by the transformation properties

$$\bar{T}^{\bar{i}_1,\cdots}{}_{\bar{i}_r\ldots}{}^{\cdots\bar{i}_n} = a^{\bar{i}_1}{}_{i_1}\ldots a_{\bar{i}_r}{}^{i_r}\ldots a^{\bar{i}_n}{}_{i_n}\, T^{i_1,\cdots}{}_{i_r\ldots}{}^{\cdots i_n} \tag{A.1.31}$$

defined. For tensors $n > 2$ the matrix symbolism is not applicable. The index position (co- or contravariant) ultimately results from a and the summation convention. Here we have defined $\mathsf{g} = (g_{kl})$ as a covariant tensor.

Calculation rules: The transposed matrix elements, even of mixed matrices, are determined according to the usual scheme of swapping the indices:

$$c^T_{ij} = (a_{ik}\, b_{kj})^T \equiv (\mathsf{ab})^T_{ij} = (\mathsf{b}^T\mathsf{a}^T)_{ij} = b^T_{ik}\, a^T_{kj} = a_{jk}\, b_{ki} = c_{ji}\,,$$
$$(c^i{}_j)^T = (a^i{}_k\, b^k{}_j)^T \equiv (\mathsf{ab})^{T i}{}_j = (\mathsf{b}^T\mathsf{a}^T)^i{}_j = b^{T i}{}_k\, a^{T k}{}_j = a_j{}^k\, b_k{}^i = c_j{}^i\,,$$
$$(d^i)^T = (a^i{}_k\, b^k)^T = (\mathsf{a}\,\mathsf{b})^{T\,i} = (\mathsf{b}^T\,\mathsf{a}^T)^i = b^{T\,k}\, a^T{}_k{}^i = a^i{}_k\, b^{T\,k} = d^i\,,$$
$$(a_k{}^l)^T = (a_{km}\, g^{ml})^T = g^{lm}\, a_{mk} = a^l{}_k\,.$$

Orthonormal basis and Cartesian coordinates

In the Cartesian coordinate system (CS) the basis vectors are perpendicular to each other and have a length of 1. We denote such vectors with $\mathbf{e}_i$ and know that they are based on an orthonormal coordinate system [2]:

$$\mathbf{h}_i \cdot \mathbf{h}_j \rightarrow g_{ij} = \mathbf{e}_i \cdot \mathbf{e}_j = \delta_{ij} = \begin{cases} 1 & \text{for } i = j \\ 0 & \text{otherwise}. \end{cases} \tag{A.1.32}$$

Thus, g^{ij} is also a unit tensor, and for all basis vectors $\mathbf{e}_i = \mathbf{e}^i$. For the coordinates, it also follows $x_i = g_{ij}x^j = x^i$.

The distinction between co- and contravariant indices is thus not necessary in orthogonal coordinate systems and is not made.

The vector is then given by

$$\mathbf{x} = x^i\,\mathbf{e}_i = x_i\,\mathbf{e}_i = \sum_{i=1}^{n} x_i\,\mathbf{e}_i\,. \tag{A.1.33}$$

This results in the length of a vector

$$r = \sqrt{g_{ij}x^i x^j} = \sqrt{x^j x_j} = \sqrt{(x_j)^2},$$

the path element

$$\mathrm{d}\mathbf{x} = \mathbf{e}_i\,\mathrm{d}x^i = \sum_{i=1}^{n}\mathrm{d}x_i\,\mathbf{e}_i, \tag{A.1.34}$$

and the square of the distance

$$\mathrm{d}\mathbf{x}^2 = \mathrm{d}x^i\,\mathrm{d}x_i = \sum_{i=1}^{n}\mathrm{d}x_i^2\,. \tag{A.1.35}$$

A.1.3 Levi-Civita Symbol

The Levi-Civita symbol, permutation symbol, ϵ-tensor or totally antisymmetric tensor, is defined by

$$\epsilon(i_1, ..., i_n) \equiv \epsilon_{i_1...i_n} = \begin{cases} -1 & P(i_1, ..., i_n)\text{ odd permutation} \\ 1 & P(i_1, ..., i_n)\text{ even permutation} \\ 0 & \text{otherwise}. \end{cases} \tag{A.1.36}$$

An equivalent definition is the specification of the following properties

---

[2] δ_{ij} is referred to as the Kronecker symbol.

1. $\epsilon(1, 2, ..., n) = 1$,

2. under exchange of two indices, the sign changes.   (A.1.37)

From the 2nd property, it follows that $\epsilon_{i_1...i_n} = 0$, when two indices are equal.

Now the determinant of the unit matrix $\det \mathsf{E} = 1$ (matrix elements δ_{ij}). If you swap two rows, the sign changes just like in the 2nd property of (A.1.37). If two rows are equal, the determinant vanishes. So it applies, when $\mathbf{e}_j$ is the j-th unit vector:

$$\epsilon(i_1, ..., i_n) = \det(\mathbf{e}_{i_1}, ..., \mathbf{e}_{i_j}) = \begin{vmatrix} \delta_{i_1 1} & ..... & \delta_{i_1 n} \\ ... & ..... & ... \\ ... & ..... & ... \\ \delta_{i_1 n} & ..... & \delta_{i_n n} \end{vmatrix} = \begin{vmatrix} \delta_{1 i_1} & ..... & \delta_{1 i_n} \\ ... & ..... & ... \\ ... & ..... & ... \\ \delta_{n i_1} & ..... & \delta_{n i_n} \end{vmatrix}. \quad (A.1.38)$$

The determinant of a matrix a is given by

$$\det \mathsf{a} = \begin{vmatrix} a_{11} & ..... & a_{1n} \\ ... & ..... & ... \\ ... & ..... & ... \\ a_{n1} & ..... & a_{nn} \end{vmatrix} = \epsilon(j_1 ..... j_n)\, a_{1 j_1}\, a_{2 j_2} ... a_{n j_n}. \quad (A.1.39)$$

The calculation of the determinant of a 3×3 matrix with the Levi-Civita symbol:

$$\begin{vmatrix} a_1 & a_2 & a_3 \\ b_1 & b_2 & b_3 \\ c_1 & c_2 & c_3 \end{vmatrix} = \epsilon(i, j, k)\, a_i b_j c_k = a_1(b_2 c_3 - b_3 c_2) + a_2(-b_1 c_3 + b_3 c_1) + a_3(b_1 c_2 - b_2 c_1)$$

$$= a_1 \begin{vmatrix} b_2 & b_3 \\ c_2 & c_3 \end{vmatrix} + a_2 \begin{vmatrix} b_1 & b_3 \\ c_1 & c_3 \end{vmatrix} + a_3 \begin{vmatrix} b_1 & b_2 \\ c_1 & c_2 \end{vmatrix}.$$

The scheme for calculating the determinant using subdeterminants (Laplace's expansion theorem) can be extended to higher dimensions.

If you take a different arrangement of the row vectors in (A.1.39): $a_{k j_k} \rightarrow a_{i_k j_k}$, then only the sign of the determinant can change according to the permutation of the i_k, or the determinant vanishes if two indices ($i_k = i_l$) are equal. Thus, the relation holds

$$\epsilon(j_1, ..., j_n)\, a_{i_1 j_1} ... a_{i_n j_n} = \epsilon(i_1, ..., i_n)\, \det \mathsf{a}. \quad (A.1.40)$$

In (A.1.36) the covariant ϵ-tensor was equated with the Levi-Civita symbol The contravariant ϵ-tensor still needs to be determined, which is done using (A.1.40):

$$\epsilon^{i_1 \cdots i_n} = g^{i_1 j_1} ... g^{i_n j_n}\, \epsilon_{j_1 \cdots j_n} = \begin{vmatrix} g^{i_1 1} & ..... & g^{i_1 n} \\ \vdots & \vdots\vdots & \vdots \\ g^{i_n 1} & ..... & g^{i_n n} \end{vmatrix} = \det(g^{ij})\, \epsilon(i_1, ..., i_n). \quad (A.1.41)$$

We thus obtain

$$
\epsilon^{i_1 \cdots i_n}\, \epsilon_{k_1 \ldots k_n} = \frac{1}{g}
\begin{vmatrix}
\delta^{i_1}{}_{k_1} & \cdots & \delta^{i_1}{}_{k_n} \\
\vdots & \vdots & \vdots \quad \vdots \\
\delta^{i_n}{}_{k_1} & \cdots & \delta^{i_n}{}_{k_n}
\end{vmatrix}
\qquad \text{with} \quad g = \det(g_{ij}) . \qquad (A.1.42)
$$

Calculation rules

We need the transvections[3] of the ϵ tensor especially for the three-dimensional Euclidean space and the four-dimensional (pseudo-Euclidean) Minkowski space of the STR. In the 1st case, the following applies:

$$
\epsilon^{ijk}\, \epsilon_{ilm} = \frac{1}{g}
\begin{vmatrix}
\delta^{j}{}_{l} & \delta^{j}{}_{m} \\
\delta^{k}{}_{l} & \delta^{k}{}_{m}
\end{vmatrix}
= \frac{1}{g}\left(\delta^{j}{}_{l}\, \delta^{k}{}_{m} - \delta^{j}{}_{m}\, \delta^{k}{}_{l} \right) ,
$$

$$
\epsilon^{ijk}\, \epsilon_{ijm} = \frac{1}{g}\, 2\, \delta^{k}{}_{m} , \qquad\qquad\qquad (A.1.43)
$$

$$
\epsilon^{ijk}\, \epsilon_{ijk} = \frac{1}{g}\, 3! .
$$

In four dimensions, it should be noted that in Minkowski space $g = -1$.

$$
\epsilon^{\alpha\beta\gamma\delta}\, \epsilon_{\alpha\nu\varrho\sigma} = \frac{1}{g}
\begin{vmatrix}
\delta^{\beta}{}_{\nu} & \delta^{\beta}{}_{\varrho} & \delta^{\beta}{}_{\sigma} \\
\delta^{\gamma}{}_{\nu} & \delta^{\gamma}{}_{\varrho} & \delta^{\gamma}{}_{\sigma} \\
\delta^{\delta}{}_{\nu} & \delta^{\delta}{}_{\varrho} & \delta^{\delta}{}_{\sigma}
\end{vmatrix} ,
$$

$$
\epsilon^{\alpha\beta\gamma\delta}\, \epsilon_{\alpha\beta\varrho\sigma} = \frac{1}{g}\, 2!
\begin{vmatrix}
\delta^{\gamma}{}_{\varrho} & \delta^{\gamma}{}_{\sigma} \\
\delta^{\delta}{}_{\varrho} & \delta^{\delta}{}_{\sigma}
\end{vmatrix} ,
$$

$$
\epsilon^{\alpha\beta\gamma\delta}\, \epsilon_{\alpha\beta\gamma\sigma} = \frac{1}{g}\, 3!\, \delta^{\delta}{}_{\sigma} , \qquad\qquad\qquad (A.1.44)
$$

$$
\epsilon^{\alpha\beta\gamma\delta}\, \epsilon_{\alpha\beta\gamma\delta} = \frac{1}{g}\, 4! .
$$

A.1.4 Determinants

A *bilinear form* $f(\mathbf{x}, \mathbf{y})$, i.e., a function of two vectors, has the following properties:

$$
\begin{aligned}
f(\mathbf{x}+\mathbf{y}, \mathbf{z}) &= f(\mathbf{x}, \mathbf{z}) + f(\mathbf{y}, \mathbf{z}), & f(\mathbf{x}, \mathbf{y}+\mathbf{z}) &= f(\mathbf{x}, \mathbf{y}) + f(\mathbf{x}, \mathbf{z}), \\
f(a\mathbf{x}, \mathbf{y}) &= f(\mathbf{x}, \mathbf{y})a, & f(\mathbf{x}, b\mathbf{y}) &= f(\mathbf{x}, \mathbf{y})b .
\end{aligned} \qquad (A.1.45)
$$

Now, if we substitute for $\mathbf{x} = x^i \mathbf{h}_i$ and $\mathbf{y} = y^j \mathbf{h}_j$, we get

$$
f(\mathbf{x}, \mathbf{y}) = x^i\, y^j\, f(\mathbf{h}_i, \mathbf{h}_j) = c_{ij}\, x^i\, y^j , \qquad\qquad (A.1.46)
$$

where we sum over duplicate indices.

---

[3] transvection is a contraction of indices belonging to different factors, e.g. $c = a_i b^i$.

An antisymmetric bilinear form has the additional property

$$f(\mathbf{x}+\mathbf{y},\mathbf{x}+\mathbf{y})=0 \qquad \Rightarrow \qquad f(\mathbf{y},\mathbf{x})=-f(\mathbf{x},\mathbf{y})\,. \tag{A.1.47}$$

For *antisymmetric multilinear forms*, it holds that these vanish if two columns are equal $f(..,\mathbf{x},..,\mathbf{x},..)=0$. If we start from $c_{1..i..j..n}=c$ and swap any two indices, only the sign changes $c_{1..j..i..n}=-c$. However, if at least two indices are equal, the corresponding coefficient vanishes. We can choose $c=1$ and then obtain as multilinear form the so-called *Leibniz formula*

$$D(\mathbf{x},\mathbf{y},\mathbf{z},...)=\sum_{i=(i_1,...,i_n)}(-1)^{\sigma(i)}\,x^{i_1}y^{i_2}z^{i_3}...,\tag{A.1.48}$$

where we sum over all permutations i and $\sigma(i)$ is the number of necessary swaps, to bring the permutation i to the order $1,2,..,n$. For the basis vectors, according to (A.1.46), $D(\mathbf{h}_1,...,\mathbf{h}_n)=1$. This can be stated in the following theorem [van der Waerden, 2003, § 25]:

There is a single antisymmetric multilinear form D, which for the basis vectors $\mathbf{h}_1,...,\mathbf{h}_n$ has the value 1. Every antisymmetric bilinear form f arises from D by multiplication with $c=f(\mathbf{h}_1,...,\mathbf{h}_n)$:

$$f(\mathbf{x},\mathbf{y},...)=x^iy^j...\,f(\mathbf{h}_i,\mathbf{h}_j,...)=f(\mathbf{h}_1,\mathbf{h}_2,...)\,D(\mathbf{x},\mathbf{y},...)\,.\tag{A.1.49}$$

$D(\mathbf{x},\mathbf{y},...)$ is called the determinant of the n vectors $\mathbf{x},\mathbf{y},...$ to the basis $\mathbf{h}_i$. Let $\mathbf{x}=\mathbf{b}_1,\ \mathbf{y}=\mathbf{b}_2,...,$ then

$$D(\mathbf{b}_1,...,\mathbf{b}_n)=\epsilon_{ijk...}\,b^i{}_1b^j{}_2b^k{}_3...=\begin{vmatrix}b^1{}_1 & ... & b^1{}_n \\ \vdots & \ddots & \vdots \\ b^n{}_1 & ... & b^n{}_n\end{vmatrix}\,.$$

Of relevance is the multiplication theorem for determinants. Given are two square $n\times n$ matrices A and B. For their determinants applies

$$\det(\mathsf{A}\,\mathsf{B})=\det(\mathsf{A})\,\det(\mathsf{B})\,.\tag{A.1.50}$$

For the proof, we note that after (A.1.49)

$$D(\mathsf{A}\mathbf{b}_1,...,\mathsf{A}\mathbf{b}_n)=D(\mathsf{A}\mathbf{h}_1,...,\mathsf{A}\mathbf{h}_n)D(\mathbf{b}_1,...,\mathbf{b}_n)\ \text{ with }\ \mathsf{A}\mathbf{h}_k=\mathbf{h}_ia^i{}_k=\begin{pmatrix}a^1{}_k\\a^2{}_k\\\vdots\end{pmatrix}\,.$$

A method for calculating the determinant is the Laplace's expansion theorem

$$\det(\mathsf{A})=\sum_{k=1}^{n}(-1)^{i+k}a^i{}_k\,\det(\mathsf{A}_{ik})\,,\tag{A.1.51}$$

where A_{ik} is the submatrix of A that arises, when the $i-$th row and the k-th column are crossed out. (A.1.51) is complex for larger determinants due to the number of calculations and is (numerically) inaccurate.

A.1.5 Three-Dimensional Vectors

Due to the importance of three-dimensional space for physics, it is justified to treat this case separately and thus accept some repetitions, such as cross and scalar product. This is subsequently useful for the curvilinear coordinates, which are only treated in three dimensions.

Vector product

In three-dimensional space, the cross product (vector product) (A.1.10) is given by

$$\mathbf{c} = \mathbf{a} \times \mathbf{b} = \sqrt{g}\epsilon^{ijk}\,\mathbf{h}_i a_j b_k = \frac{1}{\sqrt{g}}\begin{vmatrix} \mathbf{h}_1 & \mathbf{h}_2 & \mathbf{h}_3 \\ a_1 & a_2 & a_3 \\ b_1 & b_2 & b_3 \end{vmatrix} \tag{A.1.52}$$

$$= \sqrt{g}\epsilon_{ijk}\,\mathbf{h}^i a^j b^k = \sqrt{g}\begin{vmatrix} \mathbf{h}^1 & \mathbf{h}^2 & \mathbf{h}^3 \\ a^1 & a^2 & a^3 \\ b^1 & b^2 & b^3 \end{vmatrix},$$

which represents a vector perpendicular to $\mathbf{a}$ and $\mathbf{b}$. This can be seen from (A.1.52), when multiplied by $\mathbf{a}$ or $\mathbf{b}$ scalarly:

$$\mathbf{a} \cdot \mathbf{c} = \mathbf{a} \cdot (\mathbf{a} \times \mathbf{b}) = \sqrt{g}\,\epsilon^{ijk}\,a_i a_j b_k = 0.$$

The magnitude c is obtained using (A.1.43) from

$$c^2 = g\,\epsilon^{ijk}\mathbf{h}_i a_j b_k\,\epsilon_{lmn}\mathbf{h}^l a^m b^n = g\,\epsilon^{ijk}\,\epsilon_{imn} a_j b_k a^m b^n$$
$$= \left(\delta^j{}_m \delta^j{}_n - \delta^j{}_m \delta^j{}_n\right) a_j b_k a^m b^n = a^2 b^2 - (\mathbf{a}\cdot\mathbf{b})^2 = a^2 b^2 \sin^2\theta.$$

$\mathbf{c}$ thus has as magnitude $(c = ab\sin\vartheta)$ the area of the in Fig. A.6 of the drawn parallelogram. At the same time, we have demonstrated the validity of the *Lagrange identity*

$$(\mathbf{a} \times \mathbf{b}) \cdot (\mathbf{c} \times \mathbf{d}) = (\mathbf{a} \cdot \mathbf{c})(\mathbf{b} \cdot \mathbf{d}) - (\mathbf{a} \cdot \mathbf{d})(\mathbf{b} \cdot \mathbf{c}) \tag{A.1.53}$$

for $\mathbf{c}=\mathbf{a}$ and $\mathbf{b}=\mathbf{d}$. The direction of $\mathbf{c}$ is defined such that $\mathbf{a}$, $\mathbf{b}$ and $\mathbf{c}$ form a right-handed coordinate system (see Fig. A.6).

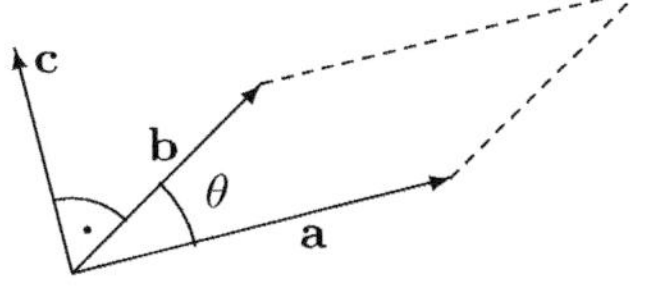

Fig. A.6. Vector product (cross product): $|\mathbf{c}| = ab\sin\theta$ and $\mathbf{c}$ is perpendicular to the surface spanned by $\mathbf{a}$ and $\mathbf{b}$

Calculation rules for the vector product

a) $\mathbf{a} \times \mathbf{b} = -\mathbf{b} \times \mathbf{a}$ anti-commutative law
b) $(\mathbf{a}\alpha) \times \mathbf{b} = \mathbf{a} \times (\mathbf{b}\alpha)$ associative law
c) $\mathbf{a} \times (\mathbf{b} + \mathbf{c}) = \mathbf{a} \times \mathbf{b} + \mathbf{a} \times \mathbf{c}$ distributive law
d) $\mathbf{a} \times \mathbf{b} = 0$ collinearity condition, if $\mathbf{a}, \mathbf{b} \neq 0$.

Polar and axial vectors

Under an inversion, *polar* vectors transform according to $\mathbf{x} \to -\mathbf{x}$, while *axial* vectors retain their sign $\mathbf{c} \to \mathbf{c}$. An example of an axial vector is the vector product of two polar vectors

$$\mathbf{c} = \mathbf{a} \times \mathbf{b} \xrightarrow{\text{Inversion}} (-\mathbf{a}) \times (-\mathbf{b}) = \mathbf{c},$$

as is the magnetic field $\mathbf{B} = \nabla \times \mathbf{A}$.

Scalar triple product

The scalar multiplication is referred to as the triple product

$$[\mathbf{abc}] = \mathbf{a} \cdot (\mathbf{b} \times \mathbf{c}) = \sqrt{g}\, \epsilon_{ijk}\, a^i b^j c^k = \sqrt{g}\, \epsilon^{ijk}\, a_i b_j c_k$$

$$= \frac{1}{\sqrt{g}} \begin{vmatrix} a_1 & a_2 & a_3 \\ b_1 & b_2 & b_3 \\ c_1 & c_2 & c_3 \end{vmatrix} \tag{A.1.54}$$

is referred to. From the definition, the invariance of the triple product with respect to cyclic permutation is immediately apparent:

$$\mathbf{a} \cdot (\mathbf{b} \times \mathbf{c}) = \mathbf{b} \cdot (\mathbf{c} \times \mathbf{a}) = \mathbf{c} \cdot (\mathbf{a} \times \mathbf{b}) = -\mathbf{a} \cdot (\mathbf{c} \times \mathbf{b}). \tag{A.1.55}$$

$\mathbf{a} \times \mathbf{b}$ is a vector that is perpendicular to the parallelogram formed by $\mathbf{a}$ and $\mathbf{b}$ and has the magnitude of the area of this parallelogram. If you multiply it scalarly with $\mathbf{c}$, you get the volume of the parallelepiped spanned by $\mathbf{a}$, $\mathbf{b}$ and $\mathbf{c}$

$$v = \mathbf{h}_1 \times \mathbf{h}_2 \cdot \mathbf{h}_3 = \frac{1}{\sqrt{g}} \begin{vmatrix} h_{11} & h_{12} & h_{13} \\ h_{21} & h_{22} & h_{23} \\ h_{31} & h_{32} & h_{33} \end{vmatrix} = \sqrt{g}, \tag{A.1.56}$$

since $\mathbf{h}_i = h_{ik}\, \mathbf{h}^k$ with $h_{ik} = \mathbf{h}_k \cdot \mathbf{h}_i = g_{ki}$.

Covariant and contravariant basis

The contravariant basis vectors are obtained from the covariant ones by Conditions $\mathbf{h}_i \cdot \mathbf{h}^j = \delta_i{}^j$. From

$$\mathbf{c} = \mathbf{h}_1 \times \mathbf{h}_2 \quad \text{follow} \quad \mathbf{c} \cdot \mathbf{h}_1 = \mathbf{c} \cdot \mathbf{h}_2 = 0 \quad \text{and} \quad \mathbf{c} \cdot \mathbf{h}_3 = \alpha \quad \Rightarrow \quad \mathbf{c} = \frac{1}{\alpha} \mathbf{h}^3.$$

The contravariant basis vectors are thus given by

$$\mathbf{h}^i\,\epsilon_{ijk} = \frac{1}{\sqrt{g}}\,\mathbf{h}_j \times \mathbf{h}_k\,. \tag{A.1.57}$$

From this follows

$$\mathbf{h}^i = \frac{\sqrt{g}}{2}\,\epsilon^{ijk}\,\mathbf{h}_j \times \mathbf{h}_k\,.$$

Conversely, it also applies

$$\mathbf{h}_i\,\epsilon^{ijk} = \frac{1}{\sqrt{g}}\,\mathbf{h}^j \times \mathbf{h}^k\,. \tag{A.1.58}$$

Using the given definitions for the vector product (A.1.52) the following relations can be verified:

$$\mathbf{a} \times (\mathbf{b} \times \mathbf{c}) + \mathbf{b} \times (\mathbf{c} \times \mathbf{a}) + \mathbf{c} \times (\mathbf{a} \times \mathbf{b}) = 0 \qquad \text{Jacobi-Identity,} \quad (A.1.59)$$

$$\mathbf{a} \times (\mathbf{b} \times \mathbf{c}) = (\mathbf{a} \cdot \mathbf{c})\mathbf{b} - (\mathbf{a} \cdot \mathbf{b})\mathbf{c} \qquad \text{Grassmann identity,} \quad (A.1.60)$$

$$(\mathbf{a} \times \mathbf{b}) \cdot (\mathbf{c} \times \mathbf{d}) = (\mathbf{a} \cdot \mathbf{c})(\mathbf{b} \cdot \mathbf{d}) - (\mathbf{a} \cdot \mathbf{d})(\mathbf{b} \cdot \mathbf{c}) \quad \text{Lagrange identity.} \quad (A.1.61)$$

Metric tensor

Two scalar products (mixed products) can be multiplied according to the following rule:

$$[\mathbf{a}\,\mathbf{b}\,\mathbf{c}][\mathbf{d}\,\mathbf{e}\,\mathbf{f}] = \begin{vmatrix} \mathbf{a} \cdot \mathbf{d} & \mathbf{a} \cdot \mathbf{e} & \mathbf{a} \cdot \mathbf{f} \\ \mathbf{b} \cdot \mathbf{d} & \mathbf{b} \cdot \mathbf{e} & \mathbf{b} \cdot \mathbf{f} \\ \mathbf{c} \cdot \mathbf{d} & \mathbf{c} \cdot \mathbf{e} & \mathbf{c} \cdot \mathbf{f} \end{vmatrix}\,.$$

If you square a mixed product, you get the *Gram determinant*

$$D(\mathbf{a}, \mathbf{b}, \mathbf{c}) = [\mathbf{a}\mathbf{b}\mathbf{c}]^2 = \begin{vmatrix} \mathbf{a} \cdot \mathbf{a} & \mathbf{a} \cdot \mathbf{b} & \mathbf{a} \cdot \mathbf{c} \\ \mathbf{b} \cdot \mathbf{a} & \mathbf{b} \cdot \mathbf{b} & \mathbf{b} \cdot \mathbf{c} \\ \mathbf{c} \cdot \mathbf{a} & \mathbf{c} \cdot \mathbf{b} & \mathbf{c} \cdot \mathbf{c} \end{vmatrix} \geq 0\,. \tag{A.1.62}$$

D is the square of the volume of the parallelepiped, which no longer depends on the order of the vectors and is positive semidefinite. If the Gram determinant vanishes, then the vectors are linearly dependent. This also applies to the metric tensor

$$\det \mathbf{g} = [\mathbf{h}_1\mathbf{h}_2\mathbf{h}_3]^2 = \begin{vmatrix} \mathbf{h}_1 \cdot \mathbf{h}_1 & \mathbf{h}_1 \cdot \mathbf{h}_2 & \mathbf{h}_1 \cdot \mathbf{h}_3 \\ \mathbf{h}_2 \cdot \mathbf{h}_1 & \mathbf{h}_2 \cdot \mathbf{h}_2 & \mathbf{h}_2 \cdot \mathbf{h}_3 \\ \mathbf{h}_3 \cdot \mathbf{h}_1 & \mathbf{h}_3 \cdot \mathbf{h}_2 & \mathbf{h}_3 \cdot \mathbf{h}_3 \end{vmatrix} \geq 0\,,$$

which is non-negative for Euclidean geometry.

A.2 Vector Analysis and Local Coordinates

Vectors, like **a**, were previously independent of the spatial point. If this restriction is dropped, so $\mathbf{a} = \mathbf{a}(\mathbf{x})$, then it will often make sense to switch to base vectors $\mathbf{h}_i = \mathbf{h}_i(\mathbf{x})$, which also depend on **x**.

A.2.1 Curvilinear Coordinates

The base vectors that form the coordinate sysytem did not have to be orthogonal but were independent of the spatial point. Fig. A.7 shows for the two-dimensional space the curve families $\xi^i(\mathbf{x})$=const. with $i = 1, 2$. The tangents to the curves at point $P(\mathbf{x})$ indicate the directions of the local covariant base vectors.

In three-dimensional space, the ξ^i=const. represent surfaces. The intersection lines of the surfaces determine the directions of the base vectors. The prerequisite is that the $\xi^i(\mathbf{x})$ are reversibly unique in an region G.

Covariant base vectors: $\mathbf{h}_1$, $\mathbf{h}_2$, $\mathbf{h}_3$.

Contravariant coordinates: ξ^1, ξ^2, ξ^3 with $\xi^i = \xi^i(\mathbf{x})$.

$$\text{Path element:} \quad d\mathbf{x} = \mathbf{h}_1\, d\xi^1 + \mathbf{h}_2\, d\xi^2 + \mathbf{h}_3\, d\xi^3 . \tag{A.2.1}$$

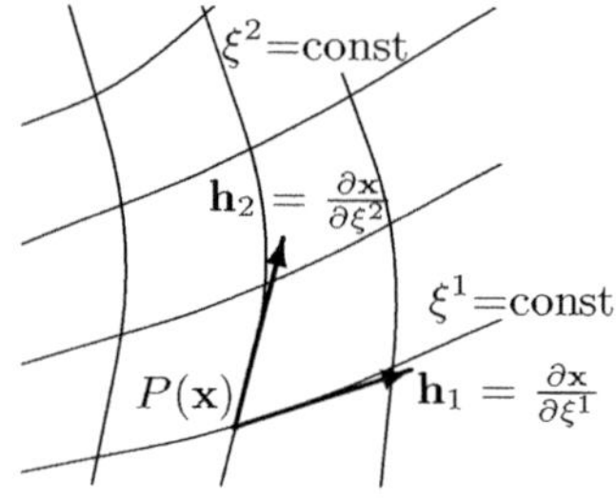

Fig. A.7. Curvilinear, contravariant coordinates $\xi^{1,2}(\mathbf{x}) =$ const. with the covariant basis vectors $\mathbf{h}_{1,2}(\mathbf{x})$

Basis vectors

At the point $P(\mathbf{x})$, the surfaces ξ^i=const intersect. The covariant basis vectors $\mathbf{h}_i(\mathbf{x})$ are defined as tangents to the intersection lines of the surfaces $\xi_{j\neq i}$=const:

$$\mathbf{h}_i = \frac{\partial \mathbf{x}}{\partial \xi^i} = \frac{\partial x_\alpha}{\partial \xi^i}\, \mathbf{e}^\alpha = h_{i\alpha}\, \mathbf{e}^\alpha \qquad \text{with} \quad \mathbf{x} = x_\alpha \mathbf{e}^\alpha . \tag{A.2.2}$$

The $\mathbf{e}^\alpha = \mathbf{e}_\alpha$ form a Cartesian coordinate system in which the contra- and covariant basis vectors coincide. Furthermore, it is $dx_\alpha = \dfrac{\partial x_\alpha}{\partial \xi^i}\, d\xi^i$. The square of the distance is calculated as

$$\mathrm{d}\mathbf{x}^2 = \mathbf{h}_i \cdot \mathbf{h}_j \,\mathrm{d}\xi^i \,\mathrm{d}\xi^j = g_{ij}\,\mathrm{d}\xi^i \,\mathrm{d}\xi^j \qquad \text{with} \quad g_{ij} = \mathbf{h}_i \cdot \mathbf{h}_j \,. \tag{A.2.3}$$

Once you have the local basis vectors $\mathbf{h}_i(\mathbf{x})$ with (A.2.2), then all arithmetic operations and relations that were derived for the oblique systems apply to them.

A curvilinear coordinate system is called orthogonal if the $\mathbf{h}_i$ are orthogonal at every point $P(\mathbf{x})$. The metric tensor $\mathbf{g}$ is then diagonal, and it is convenient to define the $\mathbf{h}_i$ as unit vectors $\mathbf{e}_\alpha = \mathbf{e}^\alpha$. Since the $\mathbf{h}_i$ do not all have to have the same dimension, $\mathbf{g}$ is not the unit matrix.

For the contravariant basis vectors, (A.1.57) applies

$$\mathbf{h}^i = \frac{v}{2}\,\epsilon^{ijk}\mathbf{h}_j \times \mathbf{h}_k \qquad \text{with} \quad v = [\mathbf{h}_1\mathbf{h}_2\mathbf{h}_3]\,, \tag{A.2.4}$$

where v is also the volume spanned by the basis vectors (spatial product $[\mathbf{h}_1\mathbf{h}_2\mathbf{h}_3]$). Using $\mathrm{d}\mathbf{x}$ one obtains

$$\mathrm{d}\mathbf{x} = \mathbf{h}_j\,\mathrm{d}\xi^j \qquad \Rightarrow \qquad \mathbf{h}^i \cdot \mathrm{d}\mathbf{x} = \mathbf{h}^i \cdot \mathbf{h}_j\,\mathrm{d}\xi^j = \delta^i{}_j\,\mathrm{d}\xi^j = \mathrm{d}\xi^i,$$

$$\mathbf{h}_i \cdot \mathrm{d}\mathbf{x} = \mathbf{h}_i \cdot \mathbf{h}_j\,\mathrm{d}\xi^j = g_{ij}\,\mathrm{d}\xi^j = \mathrm{d}\xi_i\,.$$

In general, this differential relation is not integrable, i.e., there is no function $\xi_i = \xi_i(\xi^j)$.

The spatial product $[\mathbf{h}_1\mathbf{h}_2\mathbf{h}_3]$ is calculated with the Cartesian components of $\mathbf{h}_i$, ($h_{i\alpha} = \partial x_\alpha/\partial \xi^i$), where $\bar{g} = 1$:

$$v = \sqrt{g} = [\mathbf{h}_1\mathbf{h}_2\mathbf{h}_3] = \sqrt{\bar{g}}\,\begin{vmatrix} \dfrac{\partial x}{\partial \xi^1} & \dfrac{\partial x}{\partial \xi^2} & \dfrac{\partial x}{\partial \xi^3} \\[2mm] \dfrac{\partial y}{\partial \xi^1} & \dfrac{\partial y}{\partial \xi^2} & \dfrac{\partial y}{\partial \xi^3} \\[2mm] \dfrac{\partial z}{\partial \xi^1} & \dfrac{\partial z}{\partial \xi^2} & \dfrac{\partial z}{\partial \xi^3} \end{vmatrix}. \tag{A.2.5}$$

$v = \sqrt{g}$ is the determinant of the Jacobian matrix J, the so-called functional determinant of the transformation from Cartesian to curvilinear coordinates

$$\det \mathsf{J} = \left| \frac{\partial(x,y,z)}{\partial(\xi^1,\xi^2,\xi^3)} \right|. \tag{A.2.6}$$

For the volume element, one obtains

$$\mathrm{d}^3x = \sqrt{g}\,\mathrm{d}\xi^1\mathrm{d}\xi^2\mathrm{d}\xi^3\,. \tag{A.2.7}$$

A surface element is given by $\mathrm{d}\mathbf{a}^1 = \mathbf{h}_2 \times \mathbf{h}_3\,\mathrm{d}\xi^2\mathrm{d}\xi^3$. From this follows with (A.1.57)

$$\mathrm{d}\mathbf{a} = \mathbf{h}^i\,\mathrm{d}a_i \qquad \text{with} \quad \mathrm{d}a_i = \frac{\sqrt{g}}{2}\,|\epsilon_{ijk}|\mathrm{d}\xi^j\mathrm{d}\xi^k. \tag{A.2.8}$$

Note: The sign of a surface element can be chosen differently and is determined by the (physical) circumstances. Thus, the normal on the surface of a closed volume points into the exterior space.

Since the $\mathbf{h}_i$ depend on the location, expressions of the form $\partial \mathbf{h}_i / \partial \xi_j$ occur in differential operations. From (A.2.2) follows

$$\frac{\partial \mathbf{h}_i}{\partial \xi^j} = \frac{\partial^2 \mathbf{x}}{\partial \xi^j \partial \xi^i} = \frac{\partial \mathbf{h}_j}{\partial \xi^i} \, . \tag{A.2.9}$$

A.2.2 Differential Operations

Del operator

The differential operator

$$\boldsymbol{\nabla} = \mathbf{e}^\alpha \, \nabla_\alpha = \mathbf{e}^\alpha \, \partial_\alpha = \mathbf{e}^\alpha \, \frac{\partial}{\partial x^\alpha} = \mathbf{e}_\alpha \, \frac{\partial}{\partial x_\alpha} \tag{A.2.10}$$

is referred to as the *del operator*. $\mathbf{e}^\alpha = \mathbf{e}_\alpha$ are the unit vectors of a Cartesian coordinate system. In the coordinate transformation (A.1.21) of an oblique basis, the components transform according to (A.1.29)

$$\bar{x}^i = a^i{}_j \, x^j \qquad \text{with} \quad x^j = (\mathbf{a}^{-1})^j{}_k \, \bar{x}^k$$

and one obtains

$$\frac{\partial}{\partial \bar{x}^i} = \frac{\partial x^j}{\partial \bar{x}^i} \frac{\partial}{\partial x^j} = (\mathbf{a}^{-1})^j{}_k \frac{\partial \bar{x}^k}{\partial \bar{x}^i} \frac{\partial}{\partial x^j} = (\mathbf{a}^{-1\,\mathrm{T}})_i{}^j \frac{\partial}{\partial x^j} \, . \tag{A.2.11}$$

The del operator thus transforms like a vector or more precisely: The derivatives with respect to the contravariant coordinates are the covariant components of the del operator, which is described by $\partial_i = \dfrac{\partial}{\partial x^i}$.

In curvilinear coordinates (A.2.2), the del operator is given by

$$\boldsymbol{\nabla} = \mathbf{h}^k \, (\mathbf{h}_k \cdot \mathbf{e}_\alpha) \, \frac{\partial}{\partial x_\alpha} = \mathbf{h}^k \, h_{k\alpha} \, \frac{\partial}{\partial x_\alpha} = \mathbf{h}^k \, \frac{\partial x_\alpha}{\partial \xi^k} \, \frac{\partial}{\partial x_\alpha} = \mathbf{h}^k \, \frac{\partial}{\partial \xi^k} \, . \tag{A.2.12}$$

Gradient

The application of the del operator to a scalar function $\phi(\mathbf{x})$ is the gradient of this function:

$$\operatorname{grad}\phi = \boldsymbol{\nabla}\phi = \mathbf{e}^\alpha \, \frac{\partial \phi}{\partial x^\alpha} = \mathbf{h}^k \, \frac{\partial \phi}{\partial \xi^k} \, . \tag{A.2.13}$$

Meaning of the gradient:

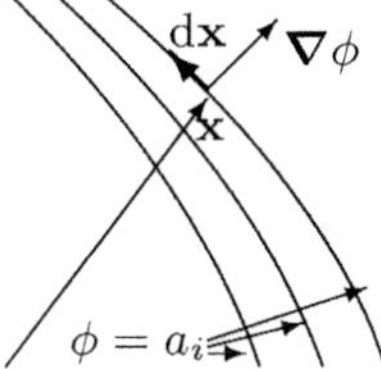

Fig. A.8. Surfaces $\phi(\mathbf{x})$=const and gradient

1. Family of surfaces $\phi(x) = a$

$$\phi(\mathbf{x} + \mathrm{d}\mathbf{x}) = \phi(\mathbf{x}) + \mathrm{d}\mathbf{x} \cdot \nabla\phi.$$

   If $\mathbf{x}+\mathrm{d}\mathbf{x}$ also lies on the surface, then $\mathrm{d}\mathbf{x} \cdot \nabla\phi = 0$, from which it follows that $\nabla\phi$ is perpendicular to the surface.
2. From $\mathrm{d}\phi = \mathrm{grad}\,\phi \cdot \mathrm{d}\mathbf{x}$ it can be seen that the change of ϕ, $\mathrm{d}\phi$, is greatest when $\mathrm{d}\mathbf{x}$ is parallel to $\nabla\phi$. Thus, $\mathrm{grad}\,\phi$ indicates the direction of the strongest change of $\phi(\mathbf{x})$, as sketched in Fig. A.8.
3. Contour line diagram in cartography. The normal lines indicate the fall lines along which $h = \phi(x, y)$ changes the most.

Vector gradient

The location change of a vector-valued function

$$\mathbf{v}(\mathbf{x} + \mathrm{d}\mathbf{x}) - \mathbf{v}(\mathbf{x}) = \mathrm{d}\mathbf{v}(\mathbf{x}) = \left(\mathrm{d}\mathbf{x} \cdot \nabla\right)\mathbf{v}(\mathbf{x})$$

is in Cartesian coordinates

$$\mathrm{d}\mathbf{v} = \mathrm{d}x_\beta \frac{\partial}{\partial x_\beta}\mathbf{e}_\alpha v_\alpha = \mathbf{e}_\alpha \mathrm{d}x_\beta \frac{\partial v_\alpha}{\partial x_\beta} = \mathbf{e}_\alpha T_{\alpha\beta}\mathrm{d}x_\beta. \tag{A.2.14}$$

T is referred to as the vector gradient or derivative tensor and in Cartesian coordinates is equal to the Jacobian matrix:

$$T_{\alpha\beta} = \frac{\partial v_\alpha}{\partial x_\beta} = v_{\alpha,\beta}. \tag{A.2.15}$$

The notation $\partial v_\alpha/\partial x_\beta = v_{\alpha,\beta}$ for the derivative of a vector, scalar or tensor is used on a case-by-case basis. With non-Cartesian coordinates, the subscript and superscript of the indices for co- and contravariant derivatives must be observed.

In curvilinear coordinates, an additional contribution is obtained through the location dependence of the $\mathbf{h}_i$:

$$\mathrm{d}\mathbf{v} = \mathrm{d}\xi^k \frac{\partial}{\partial \xi^k}\mathbf{h}_i v^i = \mathrm{d}\xi^k \left(\frac{\partial v^i}{\partial \xi^k}\mathbf{h}_i + v^i \frac{\partial \mathbf{h}_i}{\partial \xi^k}\right).$$

In the 2nd contribution, i is replaced by j and transformed to

$$\frac{\partial \mathbf{h}_j}{\partial \xi^k} = \mathbf{h}^i(\mathbf{h}_i \cdot \frac{\partial \mathbf{h}_j}{\partial \xi^k}) = \mathbf{h}^i\, \Gamma_{i|jk} = \mathbf{h}_i\, g^{il}\Gamma_{l|jk} = \mathbf{h}_i\, \Gamma^i{}_{jk}\,, \qquad (A.2.16)$$

from which it follows:

$$\mathrm{d}\mathbf{v} = \mathrm{d}\xi^k\left(\frac{\partial v^i}{\partial \xi^k} + v^j\, \Gamma^i{}_{jk}\right)\mathbf{h}_i = \mathbf{h}_i\, T^i{}_k\, \mathrm{d}\xi^k\,. \qquad (A.2.17)$$

The vector gradient is accordingly in curvilinear coordinates

$$T^i{}_k = \frac{\partial v^i}{\partial \xi^k} + v^j\, \Gamma^i{}_{jk}\,. \qquad (A.2.18)$$

The Christoffel symbols of the 1st (3-index symbol) and 2nd kind were introduced

$$\Gamma_{i|jk} = \mathbf{h}_i \cdot \frac{\partial \mathbf{h}_j}{\partial \xi^k} = \Gamma_{i|kj}\,, \qquad\qquad \Gamma^i{}_{jk} = \mathbf{h}^i \cdot \mathbf{h}_{j,k} = g^{il}\Gamma_{l|jk}\,. \qquad (A.2.19)$$

Properties of the Christoffel symbols

If you set in (A.2.19) $\mathbf{h}_j = \partial \mathbf{x}/\partial \xi^j$, the symmetry $j \leftrightarrows k$ directly follows from the exchange of derivatives:

$$\Gamma_{i|jk} = \mathbf{h}_i \cdot \frac{\partial \mathbf{h}_j}{\partial \xi^k} = \mathbf{h}_i \cdot \frac{\partial^2 \mathbf{x}}{\partial \xi^k \partial \xi^j} = \Gamma_{i|kj}\,. \qquad (A.2.20)$$

Note: In the notation used here for the Christoffel symbols the vertical line indicates that Γ is symmetric in the indices j and k:

$$\Gamma_{jk|i} = \mathbf{h}_{j,k} \cdot \mathbf{h}_i = \Gamma_{i|jk} = \mathbf{h}_i \cdot \mathbf{h}_{j,k}\,.$$

If one forms the derivative of the metric tensor and inserts (A.2.19), it follows

$$g_{ij,k} = \mathbf{h}_{i,k} \cdot \mathbf{h_j} + \mathbf{h}_i \cdot \mathbf{h}_{j,k} = \Gamma_{j|ik} + \Gamma_{i|jk}\,.$$

If one adds to $g_{ij,k}$ the derivatives $g_{ik,j}$ and $g_{jk,i}$ and uses the symmetry $\mathbf{h}_{i,j} = \mathbf{h}_{j,i}$, one obtains

$$\Gamma_{i|jk} = \frac{1}{2}\left(g_{ij,k} + g_{ik,j} - g_{jk,i}\right)\,. \qquad (A.2.21)$$

In orthogonal systems, such as cylindrical coordinates or spherical coordinates, $\Gamma_{i|jk}$ always vanishes when all three indices are different, since the g_{ij} are diagonal.

Returning to (A.2.5) and (A.1.57) we calculate in an intermediate step

$$\frac{\partial \sqrt{g}}{\partial \xi^k} = \frac{\partial}{\partial \xi^k}\left(\mathbf{h}_1 \cdot \mathbf{h}_2 \times \mathbf{h}_3\right) = \left(\frac{\partial \mathbf{h}_1}{\partial \xi^k} \cdot \mathbf{h}^1 + \frac{\partial \mathbf{h}_2}{\partial \xi^k} \cdot \mathbf{h}^2 + \frac{\partial \mathbf{h}_3}{\partial \xi^k} \cdot \mathbf{h}^3\right)\sqrt{g}\,.$$

From this follows the sought relationship

$$\mathbf{h}^i \cdot \frac{\partial \mathbf{h}_i}{\partial \xi^k} = \frac{1}{\sqrt{g}} \frac{\partial \sqrt{g}}{\partial \xi^k} = \Gamma^i{}_{ik}\,. \tag{A.2.22}$$

The derivative of the contravariant basis vectors corresponding to (A.2.16) is obtained by

$$\mathbf{h}_i \cdot \mathbf{h}^k = \delta_i{}^k \qquad \Rightarrow \qquad \mathbf{h}_{i,j} \cdot \mathbf{h}^k + \mathbf{h}_i \cdot \mathbf{h}^k{}_{,j} = 0$$

$$\frac{\partial \mathbf{h}^k}{\partial \xi^j} = \mathbf{h}^i \left(\mathbf{h}_i \cdot \frac{\partial \mathbf{h}^k}{\partial \xi^j} \right) = -\mathbf{h}^i \left(\mathbf{h}_{i,j} \cdot \mathbf{h}^k \right) = -\mathbf{h}^i \, \Gamma^k{}_{ij}\,. \tag{A.2.23}$$

Divergence

The significance of divergence – as well as rotation – will be discussed later in the integral theorems. Here, only the definitions in Cartesian and curvilinear coordinates are presented, starting again with Cartesian coordinates:

$$\operatorname{div} \mathbf{v} = \boldsymbol{\nabla} \cdot \mathbf{v} = \frac{\partial v_\alpha}{\partial x_\alpha} = v_{\alpha,\alpha}\,, \tag{A.2.24}$$

$$\boldsymbol{\nabla} \cdot \mathbf{v} = \mathbf{h}^k \cdot \frac{\partial}{\partial \xi^k} \mathbf{h}_j \, v^j = \frac{\partial v^j}{\partial \xi^j} + \mathbf{h}^k \cdot \frac{\partial \mathbf{h}_j}{\partial \xi^k} \, v^j = \frac{\partial v^j}{\partial \xi^j} + \mathbf{h}^k \cdot \frac{\partial \mathbf{h}_k}{\partial \xi^j} \, v^j\,.$$

It was used that $\mathbf{h}_{k,j} = \mathbf{h}_{j,k}$. With (A.2.22) one obtains for the divergence

$$\boldsymbol{\nabla} \cdot \mathbf{v} = \frac{\partial v^j}{\partial \xi^j} + \frac{1}{\sqrt{g}} \frac{\partial \sqrt{g}}{\partial \xi^j} \, v^j = \frac{1}{\sqrt{g}} \frac{\partial \sqrt{g}\, v^j}{\partial \xi^j}\,. \tag{A.2.25}$$

Curl

If the first vector in the vector product (A.1.52) is replaced by the del operator, one obtains for Cartesian coordinates with $\bar{g} = 1$

$$\operatorname{curl} \mathbf{v} = \boldsymbol{\nabla} \times \mathbf{v} = \sqrt{\bar{g}} \epsilon^{\alpha\beta\gamma} \mathbf{e}_\alpha \frac{\partial}{\partial x^\beta} v_\gamma = \frac{1}{\sqrt{\bar{g}}} \begin{vmatrix} \mathbf{e}_1 & \mathbf{e}_2 & \mathbf{e}_3 \\ \frac{\partial}{\partial x^1} & \frac{\partial}{\partial x^2} & \frac{\partial}{\partial x^3} \\ v_1 & v_2 & v_3 \end{vmatrix}\,. \tag{A.2.26}$$

In curvilinear coordinates, we substitute the del operator (A.2.12) and obtain

$$\boldsymbol{\nabla} \times \mathbf{v} = \mathbf{h}^j \frac{\partial}{\partial \xi^j} \times \mathbf{h}^k v_k = \mathbf{h}^j \times \mathbf{h}^k \frac{\partial v_k}{\partial \xi^j} + v_k\, \mathbf{h}^j \times \frac{\partial \mathbf{h}^k}{\partial \xi^j}\,.$$

The contribution from the differentiation of the base vectors, the second term on the right side, vanishes, since according to (A.2.23) when swapping $i \leftrightarrows j$ $\Gamma^k{}_{ij}$ is symmetric and the vector product is antisymmetric:

$$\mathbf{h}^j \times \frac{\partial \mathbf{h}^k}{\partial \xi^j} = -\mathbf{h}^j \times \mathbf{h}^i \, \Gamma^k{}_{ij} = 0\,.$$

With (A.1.58) one obtains

$$\boldsymbol{\nabla} \times \mathbf{v} = \mathbf{h}^j \times \mathbf{h}^k \frac{\partial v_k}{\partial \xi^j} = \sqrt{g}\, \epsilon^{ijk}\, \mathbf{h}_i \frac{\partial v_k}{\partial \xi^j} = \frac{1}{\sqrt{g}} \begin{vmatrix} \mathbf{h}_1 & \mathbf{h}_2 & \mathbf{h}_3 \\ \frac{\partial}{\partial \xi^1} & \frac{\partial}{\partial \xi^2} & \frac{\partial}{\partial \xi^3} \\ v_1 & v_2 & v_3 \end{vmatrix}. \tag{A.2.27}$$

Laplace operator, applied to scalar function

The Laplace operator, applied to a scalar function ϕ, is given by

$$\Delta \phi = \boldsymbol{\nabla} \cdot \boldsymbol{\nabla}\phi = \boldsymbol{\nabla} \cdot \boldsymbol{\nabla} \phi. \tag{A.2.28}$$

In Cartesian coordinates one obtains

$$\Delta = \frac{\partial}{\partial x^\alpha} \frac{\partial}{\partial x_\alpha} = \frac{\partial^2}{\partial x_\alpha^2} \tag{A.2.29}$$

and in curvilinear coordinates using (A.2.22) and (A.2.9)

$$\Delta = \mathbf{h}^i \frac{\partial}{\partial \xi^i} \cdot \mathbf{h}_j \frac{\partial}{\partial \xi_j} = \left(\mathbf{h}^i \cdot \frac{\partial \mathbf{h}_j}{\partial \xi^i} + \frac{\partial}{\partial \xi^j} \right) \frac{\partial}{\partial \xi_j} = \left(\frac{1}{\sqrt{g}} \frac{\partial \sqrt{g}}{\partial \xi^j} + \frac{\partial}{\partial \xi^j} \right) \frac{\partial}{\partial \xi_j}.$$

Thus, the Laplace operator is given by

$$\Delta = \frac{1}{\sqrt{g}} \frac{\partial}{\partial \xi^j} \sqrt{g}\, g^{ji} \frac{\partial}{\partial \xi^i} = \frac{1}{\sqrt{g}} \left(\sqrt{g}\, g^{ji} \right)_{,j} \frac{\partial}{\partial \xi^i} + g^{ij} \frac{\partial^2}{\partial \xi^i \partial \xi^j}. \tag{A.2.30}$$

Applied to a scalar function ϕ, one obtains

$$\Delta \phi = \frac{1}{\sqrt{g}} \frac{\partial}{\partial \xi^j} \sqrt{g}\, g^{ij} \frac{\partial \phi}{\partial \xi^i}. \tag{A.2.31}$$

Laplace operator, applied to vector-valued function

To anticipate: The calculation of $\Delta \mathbf{v}$ is with $\Delta \mathbf{v} = \boldsymbol{\nabla}\boldsymbol{\nabla}\cdot\mathbf{v} - \boldsymbol{\nabla}\times\boldsymbol{\nabla}\times\mathbf{v}$ (A.2.38) more handy than with the formula derived here

$$\Delta \mathbf{v} = \Delta v^k \mathbf{h}_k = \mathbf{h}_k \Delta v^k + v^k \Delta \mathbf{h}_k + 2 g^{ij}\, v^k{}_{,i}\, \mathbf{h}_{k,j}.$$

In the last term, we substitute (A.2.16) $\mathbf{h}_{k,j} = \mathbf{h}_l \Gamma^l{}_{jk}$. To enable a component-wise calculation, we still need to transform the 2nd term:

$$v^k (\Delta \mathbf{h}_k) = v^k \left[\frac{1}{\sqrt{g}} \left(\sqrt{g} g^{ij} \right)_{,j} \mathbf{h}_{k,i} + g^{ij} \frac{\partial}{\partial \xi^j} \mathbf{h}_{k,i} \right]$$

$$= \left[\frac{1}{\sqrt{g}} \frac{\partial \sqrt{g} g^{ij}}{\partial \xi^j} \mathbf{h}_l \Gamma^l{}_{ik} + g^{ij} \left(\mathbf{h}_m \Gamma^m{}_{jl} \Gamma^l{}_{ik} + \mathbf{h}_l \Gamma^l{}_{ik,j} \right) \right] v^k.$$

Summarized, this gives the expression

$$\Delta \mathbf{v} = \mathbf{h}_l \left[\Delta v^l + \frac{1}{\sqrt{g}} \frac{\partial \sqrt{g} g^{ij}}{\partial \xi^j} \Gamma^l{}_{ik} + g^{ij} \left(\Gamma^l{}_{jm} \Gamma^m{}_{ik} + \Gamma^l{}_{ik,j} \right) v^k + 2 g^{ij} \Gamma^l{}_{jk} v^k{}_{,i} \right]. \tag{A.2.32}$$

Vector identities

$$\boldsymbol{\nabla}(\mathbf{a}\cdot\mathbf{b}) = (\mathbf{a}\cdot\boldsymbol{\nabla})\mathbf{b} + \mathbf{a}\times(\boldsymbol{\nabla}\times\mathbf{b}) + (\mathbf{b}\cdot\boldsymbol{\nabla})\mathbf{a} + \mathbf{b}\times(\boldsymbol{\nabla}\times\mathbf{a}), \qquad (A.2.33)$$

$$\boldsymbol{\nabla}\cdot(\mathbf{a}\times\mathbf{b}) = \mathbf{b}\cdot(\boldsymbol{\nabla}\times\mathbf{a}) - \mathbf{a}\cdot(\boldsymbol{\nabla}\times\mathbf{b}), \qquad (A.2.34)$$

$$\boldsymbol{\nabla}\times(\mathbf{a}\times\mathbf{b}) = \mathbf{a}(\boldsymbol{\nabla}\cdot\mathbf{b}) - (\mathbf{a}\cdot\boldsymbol{\nabla})\mathbf{b} - \mathbf{b}(\boldsymbol{\nabla}\cdot\mathbf{a}) + (\mathbf{b}\cdot\boldsymbol{\nabla})\mathbf{a}, \qquad (A.2.35)$$

$$\boldsymbol{\nabla}\times\boldsymbol{\nabla}\phi = \mathbf{0}, \qquad (A.2.36)$$

$$\boldsymbol{\nabla}\cdot(\boldsymbol{\nabla}\times\mathbf{a}) = 0, \qquad (A.2.37)$$

$$\boldsymbol{\nabla}\times(\boldsymbol{\nabla}\times\mathbf{a}) = \boldsymbol{\nabla}(\boldsymbol{\nabla}\cdot\mathbf{a}) - \nabla^2\mathbf{a}. \qquad (A.2.38)$$

Note: The proofs of these identities are generally simple, if one bases them on a Cartesian coordinate system ($g = 1$) (see problem A.3):

$$\left(\boldsymbol{\nabla}\times\boldsymbol{\nabla}\phi\right)_\alpha = \epsilon_{\alpha\beta\gamma}\,\nabla_\beta\nabla_\gamma\phi = 0,$$

since when exchanging $\beta \rightleftharpoons \gamma$ the ϵ-tensor changes the sign but the entire expression remains unchanged. For the same reason, the following also applies

$$\left(\boldsymbol{\nabla}\cdot(\boldsymbol{\nabla}\times\mathbf{a})\right)_\alpha = \epsilon_{\alpha\beta\gamma}\,\nabla_\alpha\nabla_\beta a_\gamma = 0,$$

$$\left(\boldsymbol{\nabla}\times(\boldsymbol{\nabla}\times\mathbf{a})\right)_\alpha = \epsilon_{\alpha\beta\gamma}\,\nabla_\beta\,\epsilon_{\gamma\delta\lambda}\,\nabla_\delta a_\lambda = (\delta_{\alpha\delta}\,\delta_{\beta\lambda} - \delta_{\alpha\lambda}\,\delta_{\beta\delta})\,\nabla_\beta\nabla_\delta a_\lambda$$

$$= \nabla_\alpha(\boldsymbol{\nabla}\cdot\mathbf{a}) - \boldsymbol{\nabla}\cdot\boldsymbol{\nabla}\,a_\alpha.$$

Special cases ($\mathbf{p}$ is constant):

$$\boldsymbol{\nabla}(\mathbf{p}\cdot\mathbf{e}_r) = (1/r)\left[\mathbf{p} - (\mathbf{p}\cdot\mathbf{e}_r)\mathbf{e}_r\right], \qquad (A.2.39)$$

$$\boldsymbol{\nabla}(\mathbf{p}\cdot\mathbf{x}/r^3) = (1/r^3)\left[\mathbf{p} - 3(\mathbf{p}\cdot\mathbf{e}_r)\mathbf{e}_r\right], \qquad (A.2.40)$$

$$\boldsymbol{\nabla}\times(\mathbf{p}\times\mathbf{x}) = 2\mathbf{p}. \qquad (A.2.41)$$

A.3 Orthogonal Curvilinear Coordinate Systems

In important cases, curvilinear coordinate systems have orthogonal basis vectors, as with cylindrical and spherical coordinates. In such cases, some of the vector operations simplify somewhat, which is now referred to. In orthogonal coordinate systems, the metric tensor is diagonal:

$$g_{ij} = \mathbf{h}_i\cdot\mathbf{h}_j = \delta_{ij}\,h_i^2, \qquad\qquad \sqrt{g} = h_1 h_2 h_3, \qquad\qquad h_i = |\mathbf{h}_i|. \qquad (A.3.1)$$

If you determine the Cartesian coordinates a_α from the a^i of the holonomic basis, the transformation is determined by the Jacobian matrix

$$a_\alpha = (\mathbf{e}_\alpha\cdot\mathbf{h}_i)a^i = J_{\alpha i}\,a^i, \qquad\qquad J_{\alpha i} = \mathbf{e}_\alpha\cdot\frac{\partial\mathbf{x}}{\partial\xi^i} = \frac{\partial x_\alpha}{\partial\xi^i}.$$

Not limited to orthogonal coordinate systems is

$$J_{i\alpha}\,J_{\alpha j} = \mathbf{h}_i \cdot \mathbf{h}_j = g_{ij} \qquad\qquad \Leftrightarrow \qquad\qquad \mathsf{J}^T\,\mathsf{J} = \mathsf{g}\,.$$

In orthogonal coordinate systems instead of the holonomic basis $\mathbf{h}_i$ the (local) orthonormal basis $\mathbf{e}_i$ is used. For these, $\mathbf{e}^i = \mathbf{e}_i$ applies equally for the Cartesian $\mathbf{e}^\alpha = \mathbf{e}_\alpha$:

$$\mathbf{h}_i = h_i\,\mathbf{e}_i\,, \qquad\qquad \mathbf{a} = a^i\,\mathbf{h}_i = a_{\xi^i}\,\mathbf{e}_i\,, \qquad\qquad a_{\xi^i} = a^i h_i\,. \qquad (\mathrm{A}.3.2)$$

The transformation to the Cartesian coordinates is now given by the rotation S:

$$a^\alpha = (\mathbf{e}_\alpha \cdot \mathbf{e}_i)\,a_{\xi^i} = S_{\alpha i}\,a_{\xi^i}\,, \qquad\qquad J_{\alpha i} = h_i S_{\alpha i}\,. \qquad (\mathrm{A}.3.3)$$

For cylindrical coordinates are $a_{\xi^i} = a_\varrho,\ a_\varphi$ and a_z .

A.3.1 Cylindrical Coordinates

The starting point is a Cartesian coordinate system with which the cylindrical coordinates (see Fig. A.9) share the z-axis:

$$\mathbf{x} = \varrho\cos\varphi\,\mathbf{e}_x + \varrho\sin\varphi\,\mathbf{e}_y + z\,\mathbf{e}_z \qquad \begin{cases} 0 \le \varrho < \infty \\ 0 \le \varphi < 2\pi\,. \end{cases} \qquad (\mathrm{A}.3.4)$$

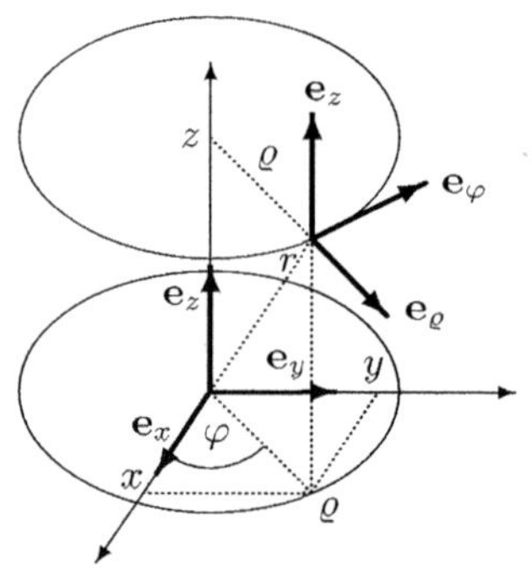

Fig. A.9. Cylindrical coordinates: $\xi^i = \varrho,\ \varphi,\ z$ and
$$\mathbf{h}_i = \frac{\partial\mathbf{x}}{\partial\xi^i}$$
$$\mathbf{h}_1 = \mathbf{e}_\varrho = \cos\varphi\,\mathbf{e}_x + \sin\varphi\,\mathbf{e}_y\,,$$
$$\mathbf{h}_2 = \varrho\mathbf{e}_\varphi = \varrho(-\sin\varphi\,\mathbf{e}_x + \cos\varphi\,\mathbf{e}_y)\,,$$
$$\mathbf{h}_3 = \mathbf{e}_z$$

If you solve (A.3.4) for $\xi^i = \varrho,\ \varphi,\ z$, you get

$$\xi^1 = \varrho = \sqrt{x^2+y^2} \quad \xi^2 = \varphi = \begin{cases} \arctan\frac{y}{x} + \pi\theta(-xy) + \pi\theta(-y) & \\ \arccos\frac{x}{\varrho} & y \ge 0 \\ 2\pi - \arccos\frac{x}{\varrho} & y < 0 \end{cases} \quad \xi^3 = z. \quad (\mathrm{A}.3.5)$$

The Cartesian coordinates $h_{i\alpha}$ of $\mathbf{h}_i$ are obtained from

$$\mathbf{h}_i = \frac{\partial\mathbf{x}}{\partial\xi^i} = \mathbf{e}_\alpha\frac{\partial x_\alpha}{\partial\xi^i}\,.$$

$$\mathbf{h}_1 = \left(\cos\varphi\ \sin\varphi\ 0\right),\quad \mathbf{h}_2 = \varrho\left(-\sin\varphi\ \cos\varphi\ 0\right),\quad \mathbf{h}_3 = \left(0\ 0\ 1\right),$$

$$h_1 = 1,\qquad\qquad\qquad h_2 = \varrho,\qquad\qquad\qquad h_3 = 1,\qquad (A.3.6)$$

$$\mathbf{e}_\varrho = \mathbf{h}_1 = \mathbf{h}^1,\qquad\quad \mathbf{e}_\varphi = \frac{1}{\varrho}\,\mathbf{h}_2 = \varrho\,\mathbf{h}^2,\qquad\quad \mathbf{e}_z = \mathbf{h}_3 = \mathbf{h}^3.\quad (A.3.7)$$

Conversely, it follows from $\mathbf{e}_\alpha = \left(\mathbf{e}^i \cdot \mathbf{e}_\alpha\right)\mathbf{e}_i$

$$\mathbf{e}_x = \cos\varphi\,\mathbf{e}_\varrho - \sin\varphi\,\mathbf{e}_\varphi,\qquad\qquad \mathbf{e}_y = \sin\varphi\,\mathbf{e}_\varrho + \cos\varphi\,\mathbf{e}_\varphi \qquad (A.3.8)$$

The coordinates $(a^i h_i)$ for the basis $\mathbf{e}_i$ and the diagonal metric tensor (A.3.1) are still to be specified:

$$
\begin{aligned}
a_\varrho &= a^1 h_1 & a_\varphi &= a^2 h_2 & a_z &= a^3 h_3\\
&= a_x \sin\varphi - a_y \cos\varphi, & &= a_x \cos\varphi + a_y \sin\varphi, & a_z &= a^3,\quad (A.3.9)\\
g_{\varrho\varrho} &= 1, & g_{\varphi\varphi} &= \varrho^2, & g_{zz} &= 1.
\end{aligned}
$$

Jacobian matrix and metric tensor

$$\mathsf{J} = \frac{\partial(x\ y\ z)}{\partial(\varrho\ \varphi\ z)} = \begin{pmatrix} \cos\varphi & -\varrho\sin\varphi & 0\\ \sin\varphi & \varrho\cos\varphi & 0\\ 0 & 0 & 1 \end{pmatrix},\qquad \mathbf{g} = \begin{pmatrix} 1 & 0 & 0\\ 0 & \varrho^2 & 0\\ 0 & 0 & 1 \end{pmatrix}.\qquad (A.3.10)$$

Functional determinant (A.2.5)

$$v = \sqrt{g} = \varrho. \qquad (A.3.11)$$

Volume and surface element (A.2.7) and (A.2.8)

$$\mathrm{d}^3 x = \varrho\,\mathrm{d}\varrho\mathrm{d}\varphi\mathrm{d}z,$$
$$\mathrm{d}\mathbf{a} = \mathbf{e}_\varrho\,\varrho\,\mathrm{d}\varphi\mathrm{d}z + \mathbf{e}_\varphi\,\mathrm{d}\varrho\mathrm{d}z + \mathbf{e}_z\,\varrho\mathrm{d}\varrho\mathrm{d}\varphi. \qquad (A.3.12)$$

Line element (A.2.1)

$$\mathrm{d}\mathbf{s} = \mathbf{e}_\varrho\,\mathrm{d}\varrho + \mathbf{e}_\varphi\,\varrho\mathrm{d}\varphi + \mathbf{e}_z\,\mathrm{d}z, \qquad (A.3.13)$$
$$\mathrm{d}s^2 = \mathrm{d}\varrho^2 + \varrho^2\mathrm{d}\varphi^2 + \mathrm{d}z^2. \qquad (A.3.14)$$

Delta function (B.6.11)

$$\delta^{(3)}(\mathbf{x} - \mathbf{x}') = \frac{1}{\varrho}\,\delta(\varrho - \varrho')\,\delta(\varphi - \varphi')\,\delta(z - z'). \qquad (A.3.15)$$

Del operator (A.2.12)

$$\boldsymbol{\nabla} = \mathbf{h}^k\,\frac{\partial}{\partial\xi^k} = \mathbf{e}_\varrho\,\frac{\partial}{\partial\varrho} + \frac{1}{\varrho}\,\mathbf{e}_\varphi\,\frac{\partial}{\partial\varphi} + \mathbf{e}_z\,\frac{\partial}{\partial z}. \qquad (A.3.16)$$

Gradient (A.2.13)

$$\boldsymbol{\nabla}\phi = \mathbf{h}^k\,\frac{\partial\phi}{\partial\xi^k} = \Big(\mathbf{e}_\varrho\,\frac{\partial}{\partial\varrho} + \frac{1}{\varrho}\,\mathbf{e}_\varphi\,\frac{\partial}{\partial\varphi} + \mathbf{e}_z\,\frac{\partial}{\partial z}\Big)\phi\,. \tag{A.3.17}$$

Divergence (A.2.25)

$$\boldsymbol{\nabla}\cdot\mathbf{v} = \frac{1}{\sqrt{g}}\,\frac{\partial\sqrt{g}\,v^j}{\partial\xi^j} = \frac{1}{\varrho}\,\frac{\partial\varrho v_\varrho}{\partial\varrho} + \frac{1}{\varrho}\,\frac{\partial v_\varphi}{\partial\varphi} + \frac{\partial v_z}{\partial z}\,. \tag{A.3.18}$$

Curl (A.2.27)

$$\boldsymbol{\nabla}\times\mathbf{v} = \frac{1}{\varrho}\begin{vmatrix} \mathbf{e}_\varrho & \varrho\mathbf{e}_\varphi & \mathbf{e}_z \\ \dfrac{\partial}{\partial\varrho} & \dfrac{\partial}{\partial\varphi} & \dfrac{\partial}{\partial z} \\ v_\varrho & \varrho v_\varphi & v_z \end{vmatrix} \tag{A.3.19}$$

$$= \mathbf{e}_\varrho\Big(\frac{1}{\varrho}\frac{\partial v_z}{\partial\varphi} - \frac{\partial v_\varphi}{\partial z}\Big) + \mathbf{e}_\varphi\Big(\frac{\partial v_\varrho}{\partial z} - \frac{\partial v_z}{\partial\varrho}\Big) + \mathbf{e}_z\frac{1}{\varrho}\Big(\frac{\partial\varrho v_\varphi}{\partial\varrho} - \frac{\partial v_\varrho}{\partial\varphi}\Big)\,.$$

Laplace Operator (A.2.30)

$$\Delta = \frac{1}{\sqrt{g}}\,\frac{\partial}{\partial\xi^j}\Big(\sqrt{g}\,g^{ji}\,\frac{\partial}{\partial\xi^i}\Big) = \Big(\frac{1}{\varrho}\frac{\partial}{\partial\varrho}\varrho\frac{\partial}{\partial\varrho} + \frac{1}{\varrho^2}\frac{\partial^2}{\partial\varphi^2} + \frac{\partial^2}{\partial z^2}\Big)\,. \tag{A.3.20}$$

This determines $\Delta\phi$. $\Delta\mathbf{a}$ is determined with (A.2.38). Let it be

$$\mathbf{b} = \boldsymbol{\nabla}\times\mathbf{a} = \begin{cases} b_\varrho &= \dfrac{1}{\varrho}\dfrac{\partial}{\partial\varphi}\,a_z - \dfrac{\partial}{\partial z}\,a_\varphi \\[2mm] b_\varphi &= \dfrac{\partial}{\partial z}\,a_\varrho - \dfrac{\partial}{\partial\varrho}\,a_z \\[2mm] b_z &= \dfrac{1}{\varrho}\Big(\dfrac{\partial}{\partial\varrho}(\varrho\,a_\varphi) - \dfrac{\partial}{\partial\varphi}\,a_\varrho\Big)\,. \end{cases}$$

One now calculates the components of $\boldsymbol{\nabla}\,\boldsymbol{\nabla}\cdot\mathbf{a}$ and $\boldsymbol{\nabla}\times(\boldsymbol{\nabla}\times\mathbf{a})$

$$\operatorname{grad}_\varrho \boldsymbol{\nabla}\cdot\mathbf{a} = \frac{\partial}{\partial\varrho}\underbrace{\Big(\frac{1}{\varrho}\,a_\varrho + \frac{\partial a_\varrho}{\partial\varrho} + \frac{1}{\varrho}\frac{\partial a_\varphi}{\partial\varphi} + \frac{\partial a_z}{\partial z}\Big)}_{\boldsymbol{\nabla}\mathbf{a}},$$

$$\big(\boldsymbol{\nabla}\times(\boldsymbol{\nabla}\times\mathbf{a})\big)_\varrho = \frac{1}{\varrho}\frac{\partial}{\partial\varphi}\underbrace{\Big(\frac{1}{\varrho}\,a_\varphi + \frac{\partial}{\partial\varrho}a_\varphi - \frac{1}{\varrho}\frac{\partial}{\partial\varphi}a_\varrho\Big)}_{b_z} - \frac{\partial}{\partial z}\underbrace{\Big(\frac{\partial}{\partial z}a_\varrho - \frac{\partial}{\partial\varrho}a_z\Big)}_{b_\varphi}.$$

The difference yields

$$(\Delta\mathbf{a})_\varrho = \underbrace{\Big(\frac{\partial^2}{\partial\varrho^2} + \frac{1}{\varrho}\frac{\partial}{\partial\varrho} + \frac{1}{\varrho^2}\frac{\partial^2}{\partial\varphi^2} + \frac{\partial^2}{\partial z^2}\Big)}_{\Delta}a_\varrho - \frac{1}{\varrho^2}\,a_\varrho - \frac{2}{\varrho}\frac{\partial}{\partial\varphi}\,a_\varphi\,.$$

For the other two terms, one obtains

$$(\Delta\mathbf{a})_\varphi = \Delta\,a_\varphi + \frac{1}{\varrho^2}\,a_\varrho + \frac{2}{\varrho}\frac{\partial}{\partial\varphi}\,a_\varphi\,,$$

$$(\Delta\mathbf{a})_z = \Delta\,a_z\,.$$

A.3.2 Spherical Coordinates

The surfaces $\xi^i=$const. are in spherical coordinates (see Fig. A.10) the spherical surface with r, the conical shell formed by the polar angle ϑ and the plane formed by the azimuth angle φ and the z-axis. The starting point is

$$\mathbf{x} = r\sin\vartheta\,\cos\varphi\,\mathbf{e}_x + r\sin\vartheta\,\sin\varphi\,\mathbf{e}_y + r\cos\vartheta\,\mathbf{e}_z, \quad \begin{array}{l} 0 \le r < \infty, \\ 0 \le \vartheta \le \pi, \\ 0 \le \varphi < 2\pi. \end{array} \quad \text{(A.3.21)}$$

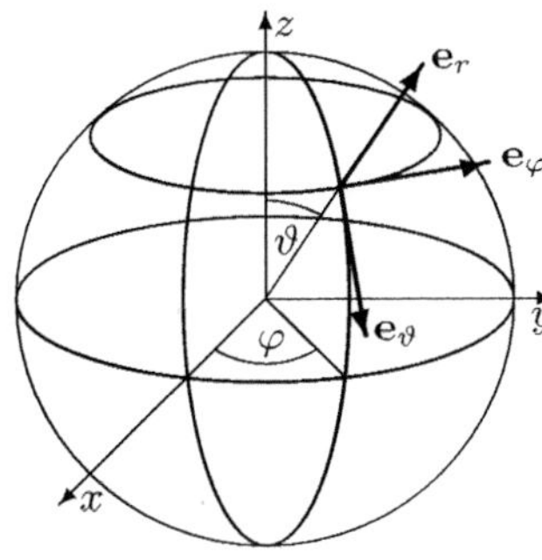

Fig. A.10. Spherical coordinates: $\xi^i = r, \vartheta, \varphi$ and
$$\mathbf{h}_i = \mathbf{e}_\alpha\,\frac{\partial x_\alpha}{\partial \xi^i}\,,$$
$$\mathbf{h}_1 = \mathbf{e}_r = \sin\vartheta\,(\cos\varphi\,\mathbf{e}_x + \sin\varphi\,\mathbf{e}_y) + \cos\vartheta\,\mathbf{e}_z\,,$$
$$\mathbf{h}_2 = r\mathbf{e}_\vartheta = r\cos\vartheta(\cos\varphi\,\mathbf{e}_x + \sin\varphi\,\mathbf{e}_y) - r\sin\vartheta\,\mathbf{e}_z$$
$$\mathbf{h}_3 = r\sin\vartheta\,\mathbf{e}_\varphi = r\sin\vartheta\,(-\sin\varphi\,\mathbf{e}_x + \cos\varphi\,\mathbf{e}_y)$$

Conversely, one obtains ($\xi^1 = r,\ \xi^2 = \vartheta,\ \xi^3 = \varphi$)

$$\varrho = \sqrt{x^2 + y^2}\,, \qquad \vartheta = \arccos\frac{z}{r}\,, \qquad\qquad \text{(A.3.22)}$$

$$r = \sqrt{\varrho^2 + z^2}\,, \qquad \varphi = \arctan\frac{y}{x} + \pi\theta(-x) + 2\pi\theta(x)\theta(-y)\,.$$

θ is the step function (B.6.17). In the legend of Fig. A.10, the base vectors $\mathbf{h}_i$ are given. From this we read:

$$h_1 = 1\,, \qquad\qquad h_2 = r\,, \qquad\qquad h_3 = r\sin\vartheta\,,$$

$$\mathbf{e}_r = \mathbf{h}_1 = \mathbf{h}^1\,, \qquad \mathbf{e}_\vartheta = \frac{1}{r}\mathbf{h}_2 = r\mathbf{h}^2\,, \qquad \mathbf{e}_\varphi = \frac{1}{r\sin\vartheta}\mathbf{h}_3 = r\sin\vartheta\,\mathbf{h}^3,$$

$$\mathbf{e}_r = \begin{pmatrix} \sin\vartheta\,\cos\varphi \\ \sin\vartheta\,\sin\varphi \\ \cos\vartheta \end{pmatrix}, \quad \mathbf{e}_\vartheta = \begin{pmatrix} \cos\vartheta\,\cos\varphi \\ \cos\vartheta\,\sin\varphi \\ -\sin\vartheta \end{pmatrix}, \quad \mathbf{e}_\varphi = \begin{pmatrix} -\sin\varphi \\ \cos\varphi \\ 0 \end{pmatrix}, \quad \text{(A.3.23)}$$

$$g_{rr} = 1\,, \qquad\qquad g_{\vartheta\vartheta} = r^2\,, \qquad\qquad g_{\varphi\varphi} = r^2\sin^2\vartheta\,.$$

Jacobian matrix and metric tensor

$$\mathsf{J} = \frac{\partial(x\,y\,z)}{\partial(r\,\vartheta\,\varphi)} = \begin{pmatrix} \sin\vartheta\cos\varphi & r\cos\vartheta\cos\varphi & -r\sin\vartheta\sin\varphi \\ \sin\vartheta\sin\varphi & r\cos\vartheta\sin\varphi & r\sin\vartheta\cos\varphi \\ \cos\vartheta & -r\sin\vartheta & 0 \end{pmatrix}, \quad \mathsf{g} = \begin{pmatrix} 1 & 0 & 0 \\ 0 & r^2 & 0 \\ 0 & 0 & r^2\sin\vartheta \end{pmatrix}.$$

$$\text{(A.3.24)}$$

Functional determinant (A.2.5)

$$v = \sqrt{g} = r^2 \sin \vartheta \,. \tag{A.3.25}$$

The unit vectors are

$$\begin{aligned}
\mathbf{e}_x &= \sin \vartheta \, \cos \varphi \, \mathbf{e}_r + \cos \vartheta \, \cos \varphi \, \mathbf{e}_\vartheta - \sin \varphi \, \mathbf{e}_\varphi \,, \\
\mathbf{e}_y &= \sin \vartheta \, \sin \varphi \, \mathbf{e}_r + \cos \vartheta \, \sin \varphi \, \mathbf{e}_\vartheta + \cos \varphi \, \mathbf{e}_\varphi \,, \\
\mathbf{e}_z &= \cos \vartheta \, \mathbf{e}_r \qquad - \sin \vartheta \, \mathbf{e}_\vartheta \,.
\end{aligned} \tag{A.3.26}$$

The coordinates of $\mathbf{v}$ are calculated with (A.3.2)

$$\begin{aligned}
a_r &= a^1 h_1 = a_x \sin \vartheta \cos \varphi + a_y \sin \vartheta \sin \varphi + a_z \cos \vartheta \,, \\
a_\vartheta &= a^2 h_2 = a_x \cos \vartheta \cos \varphi + a_y \cos \vartheta \sin \varphi - a_z \sin \vartheta \,, \\
a_\varphi &= a^3 h_3 = -a_x \sin \varphi + a_y \cos \varphi \,.
\end{aligned} \tag{A.3.27}$$

Volume- (A.2.7) and *surface element* (A.2.8)

$$\mathrm{d}^3 x = \sqrt{g} \, \mathrm{d}\xi^1 \mathrm{d}\xi^2 \mathrm{d}\xi^3 = r^2 \sin \vartheta \, \mathrm{d}r \, \mathrm{d}\vartheta \, \mathrm{d}\varphi \,, \tag{A.3.28}$$

$$\mathrm{d}\mathbf{a} = \frac{\sqrt{g}}{2} \, \epsilon_{ijk} \mathbf{h}^i \mathrm{d}\xi^j \mathrm{d}\xi^k = \mathbf{e}_r \, r^2 \sin \vartheta \, \mathrm{d}\vartheta \mathrm{d}\varphi + \mathbf{e}_\vartheta \, r \, \sin \vartheta \mathrm{d}r \mathrm{d}\varphi + \mathbf{e}_\varphi \, r \, \mathrm{d}r \mathrm{d}\varphi \,.$$

Line element (A.2.1)

$$\mathrm{d}\mathbf{s} = \mathbf{e}_r \, \mathrm{d}r + \mathbf{e}_\vartheta \, r \, \mathrm{d}\vartheta + \mathbf{e}_\varphi \, r \, \sin \vartheta \, \mathrm{d}\varphi \,, \tag{A.3.29}$$

$$\mathrm{d}s^2 = \mathrm{d}r^2 + r^2 \mathrm{d}\vartheta^2 + r^2 \sin^2 \vartheta \mathrm{d}\varphi^2 \,. \tag{A.3.30}$$

Delta function (B.6.11)

$$\delta^{(3)}(\mathbf{x} - \mathbf{x}') = \frac{1}{r^2 \sin \vartheta} \, \delta(r - r') \, \delta(\vartheta - \vartheta') \, \delta(\varphi - \varphi') \,. \tag{A.3.31}$$

Del operator (A.2.12):

$$\boldsymbol{\nabla} = \mathbf{h}^k \frac{\partial}{\partial \xi^k} = \mathbf{e}_r \frac{\partial}{\partial r} + \mathbf{e}_\vartheta \frac{1}{r} \frac{\partial}{\partial \vartheta} + \mathbf{e}_\varphi \frac{1}{r \sin \vartheta} \frac{\partial}{\partial \varphi} \,. \tag{A.3.32}$$

Gradient (A.2.13):

$$\boldsymbol{\nabla}\phi = \mathbf{h}^k \frac{\partial \phi}{\partial \xi^k} = \left(\mathbf{e}_r \frac{\partial}{\partial r} + \mathbf{e}_\vartheta \frac{1}{r} \frac{\partial}{\partial \vartheta} + \mathbf{e}_\varphi \frac{1}{r \sin \vartheta} \frac{\partial}{\partial \varphi} \right) \phi \,. \tag{A.3.33}$$

Christoffel symbols: Starting from the metric coefficients (A.3.23)

$$g_{11} = 1, \qquad\qquad g_{22} = r^2, \qquad\qquad g_{33} = r^2 \sin^2 \vartheta$$

one calculates their non-vanishing derivatives

$$g_{22,1} = 2r, \qquad\qquad g_{33,1} = 2r \sin^2 \vartheta, \qquad\qquad g_{33,2} = 2r^2 \sin \vartheta \cos \vartheta.$$

With (A.2.21) we can determine the non-vanishing 3-index symbols:

$$\Gamma_{1|22} = -\frac{g_{22,1}}{2} = -r, \qquad\qquad \Gamma_{1|33} = -\frac{g_{33,1}}{2} = -r\sin^2\vartheta,$$

$$\Gamma_{2|12} = \frac{g_{22,1}}{2} = r, \qquad\qquad \Gamma_{2|33} = -\frac{g_{33,2}}{2} = -r^2\sin\vartheta\,\cos\vartheta,$$

$$\Gamma_{3|13} = \frac{g_{33,1}}{2} = r\sin^2\vartheta, \qquad\qquad \Gamma_{3|23} = \frac{g_{33,2}}{2} = r^2\sin\vartheta\,\cos\vartheta.$$

When multiplied by g^{ii}, one obtains

$$\Gamma^1{}_{22} = -r, \qquad \Gamma^1{}_{33} = -r\sin^2\vartheta, \qquad \Gamma^2{}_{12} = \frac{1}{r}, \qquad \Gamma^2{}_{33} = -\cot\vartheta,$$

$$\Gamma^3{}_{13} = \frac{1}{r}, \qquad \Gamma^3{}_{23} = \cot\vartheta.$$

The non-vanishing derivatives of the Christoffel symbols are

$$\Gamma^1{}_{22,1} = -1, \qquad \Gamma^1{}_{33,1} = -\sin^2\vartheta, \quad \Gamma^1{}_{33,2} = -r\sin 2\vartheta, \quad \Gamma^2{}_{12,1} = -\frac{1}{r^2},$$

$$\Gamma^2{}_{33,2} = -\frac{1}{\sin^2\vartheta}, \quad \Gamma^3{}_{13,1} = -\frac{1}{r^2}, \qquad \Gamma^3{}_{23,2} = \frac{1}{\sin^2\vartheta}.$$

Divergence (A.2.25)

$$\boldsymbol{\nabla}\cdot\mathbf{v} = \frac{1}{\sqrt{g}}\,\frac{\partial\sqrt{g}\,v^j}{\partial\xi^j} = \frac{1}{r^2\sin\vartheta}\left(\frac{\partial r^2\sin\vartheta v_r}{\partial r} + \frac{\partial r\sin\vartheta v_\vartheta}{\partial\vartheta} + \frac{\partial r v_\varphi}{\partial\varphi}\right)$$

$$= \frac{1}{r^2}\frac{\partial r^2 v_r}{\partial r} + \frac{1}{r\sin\vartheta}\frac{\partial\sin\vartheta v_\vartheta}{\partial\vartheta} + \frac{1}{r\sin\vartheta}\frac{\partial v_\varphi}{\partial\varphi}. \qquad (A.3.34)$$

Curl (A.2.27)

$$\boldsymbol{\nabla}\times\mathbf{v} = \frac{1}{r^2\sin\vartheta}\begin{vmatrix} \mathbf{e}_r & r\mathbf{e}_\vartheta & r\sin\vartheta\mathbf{e}_\varphi \\ \dfrac{\partial}{\partial r} & \dfrac{\partial}{\partial\vartheta} & \dfrac{\partial}{\partial\varphi} \\ v_r & r v_\vartheta & r\sin\vartheta v_\varphi \end{vmatrix} = \frac{1}{r^2\sin\vartheta}\Bigg[\mathbf{e}_r\Big(\frac{\partial r\sin\vartheta\,v_\varphi}{\partial\vartheta} - \frac{\partial r v_\vartheta}{\partial\varphi}\Big)$$

$$+\, r\mathbf{e}_\vartheta\Big(\frac{\partial v_r}{\partial\varphi} - \frac{\partial r\sin\vartheta\,v_\varphi}{\partial r}\Big) + r\sin\vartheta\,\mathbf{e}_\varphi\Big(\frac{\partial r v_\vartheta}{\partial r} - \frac{\partial v_r}{\partial\vartheta}\Big)\Bigg]. \qquad (A.3.35)$$

Laplace operator (A.2.30)

$$\Delta = \frac{1}{\sqrt{g}}\frac{\partial}{\partial\xi^i}\Big(\sqrt{g}\,g^{ij}\frac{\partial}{\partial\xi^j}\Big) \qquad\qquad (A.3.36)$$

$$= \frac{1}{r^2}\frac{\partial}{\partial r}r^2\frac{\partial}{\partial r} + \frac{1}{r^2\sin\vartheta}\frac{\partial}{\partial\vartheta}\sin\vartheta\frac{\partial}{\partial\vartheta} + \frac{1}{r^2\sin^2\vartheta}\frac{\partial^2}{\partial\varphi^2}.$$

It is often useful to split the Laplace operator into a radial and an angle-dependent operator: $\Delta = \Delta_r - \hat{\mathbf{L}}^2/r^2$,

$$
\Delta_r = \frac{1}{r^2} \frac{\partial}{\partial r} r^2 \frac{\partial}{\partial r} = \frac{\partial^2}{\partial r^2} + \frac{2}{r} \frac{\partial}{\partial r} = \frac{1}{r} \frac{\partial^2}{\partial r^2} r,
$$

$$
\hat{\mathbf{L}} = -\mathrm{i}\mathbf{x} \times \boldsymbol{\nabla} = \mathrm{i}\mathbf{e}_\vartheta \frac{1}{\sin\vartheta} \frac{\partial}{\partial\varphi} - \mathrm{i}\mathbf{e}_\varphi \frac{\partial}{\partial\vartheta}.
$$

$$(\text{A.3.36'})$$

Application of the Laplace operator to a vector

$\Delta\mathbf{v}$ is determined with (A.2.32), where only $l=1$, i.e. $\mathbf{h}_1 = \mathbf{e}_r$, is calculated.

1. $\Delta v^1 = \Delta v_r$.
2. $v^2 = r^{-1} v_\vartheta$ and $v^3 = (r\sin\vartheta)^{-1} v_\varphi$.

$$
\frac{1}{\sqrt{g}} \frac{\partial \sqrt{g}g^{ij}}{\partial\xi^j} \Gamma^1{}_{ik}v^k = \frac{1}{\sqrt{g}} \frac{\partial \sqrt{g}g^{22}}{\partial\xi^2} \Gamma^1{}_{22}v^2 + \frac{1}{\sqrt{g}} \frac{\partial \sqrt{g}g^{33}}{\partial\xi^3} \Gamma^1{}_{33}v^3
$$

$$
= \frac{1}{r^2\sin\vartheta} \frac{\partial\sin\vartheta}{\partial\vartheta} \left(-\frac{1}{r}\right)v^2 = -\frac{\cot\vartheta}{r^2} v_\vartheta \,.
$$

3. $g^{ij} \Gamma^1{}_{jn}\Gamma^n{}_{ik}v^k = g^{22} \Gamma^1{}_{22}\Gamma^2{}_{21}v^1 + g^{33} \Gamma^1{}_{33}\Gamma^3{}_{31}v^1 = -\frac{1}{r^2} v^1 - \frac{1}{r^2} v^1 = -\frac{2}{r^2} v_r$.
4. $g^{ij}\mathbf{h}_1\Gamma^1{}_{ik,j}v^k = 0$.
5. $2g^{ij} \Gamma^1_{jk} v^k{}_{,i} = 2\left[g^{22} \Gamma^1_{22} \frac{\partial v^2}{\partial\vartheta} + g^{33} \Gamma^1_{33} \frac{\partial v^3}{\partial\varphi}\right] = -2\frac{1}{r^2} \left(\frac{\partial v_\vartheta}{\partial\vartheta} + \frac{1}{\sin\vartheta} \frac{\partial v_\varphi}{\partial\varphi}\right)$.

Thus, we have obtained that

$$
(\Delta\mathbf{v})_r = \mathbf{e}_r\left[(\Delta v_r) - \frac{2}{r^2}v_r - \frac{2}{r^2}\cot\vartheta\, v_\vartheta - 2\frac{1}{r^2}\left(\frac{\partial v_\vartheta}{\partial\vartheta} + \frac{1}{\sin\vartheta}\frac{\partial v_\varphi}{\partial\varphi}\right)\right].
$$

The analogous calculation for the other components leads to

$$
\Delta\mathbf{v} = \mathbf{e}_r\left[\Delta v_r - \frac{2}{r^2}v_r - \frac{2\cot\vartheta}{r^2}v_\vartheta - \frac{2}{r^2}v_{\vartheta,\vartheta} - \frac{2}{r^2\sin\vartheta}v_{\varphi,\varphi}\right]
$$
$$
+ \mathbf{e}_\vartheta\left[\Delta v_\vartheta - \frac{1}{r^2\sin^2\vartheta}v_\vartheta + \frac{2}{r^2}v_{r,\vartheta} - \frac{2\cos\vartheta}{r^2\sin^2\vartheta}v_{\varphi,\varphi}\right]
$$
$$
+ \mathbf{e}_\varphi\left[\Delta v_\varphi - \frac{1}{r^2\sin^2\vartheta}v_\varphi + \frac{2}{r^2\sin\vartheta}v_{r,\varphi} + \frac{2\cos\vartheta}{r^2\sin^2\vartheta}v_{\varphi,\varphi}\right].
$$

$$(\text{A.3.37})$$

A.3.3 Elliptical Coordinates

Elliptical coordinates are rather rarely used in electrodynamics, so we will only briefly touch on them here. This is also because there are several variants and the definitions are not always uniform.

Elliptical coordinates for a stretched rotation ellipsoid

We start from the elliptical coordinates of the stretched rotation ellipsoid, as sketched in Fig. A.11:

$$x = l\sqrt{(\xi^2 - 1)(1 - \eta^2)}\,\cos\varphi, \qquad\qquad 1 \le \xi \le \infty,$$
$$y = l\sqrt{(\xi^2 - 1)(1 - \eta^2)}\,\sin\varphi, \qquad\qquad -1 \le \eta \le 1, \qquad\qquad \text{(A.3.38)}$$
$$z = l\xi\eta, \qquad\qquad 0 \le \varphi < 2\pi.$$

The local variables are $\xi^1 = \eta$, $\xi^2 = \xi$, $\xi^3 = \varphi$. From this follow the base

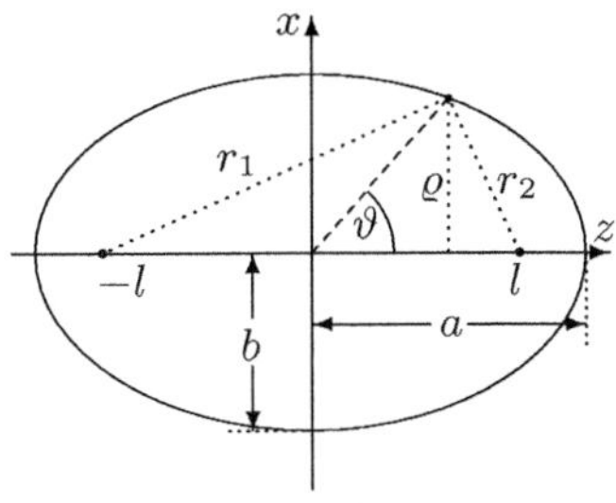

Fig. A.11. Ellipse in the zx-plane. $r_{1,2}$ are the radii with $r_1 + r_2 = 2a$ and l is the linear eccentricity. The variables $\xi = \cosh u$ and $\eta = \cos\vartheta$ with the polar angle ϑ.
The semi-axes are $a = l\xi$ and $b = l\sqrt{\xi^2 - 1}$

vectors

$$\mathbf{h}_1 = \frac{\partial \mathbf{x}}{\partial \eta} = -l\eta\sqrt{\frac{\xi^2 - 1}{1 - \eta^2}}\left[\cos\varphi\mathbf{e}_x + \sin\varphi\mathbf{e}_y\right] + l\xi\mathbf{e}_z,$$

$$\mathbf{h}_2 = \frac{\partial \mathbf{x}}{\partial \xi} = l\xi\sqrt{\frac{1 - \eta^2}{\xi^2 - 1}}\left[\cos\varphi\mathbf{e}_x + \sin\varphi\mathbf{e}_y\right] + l\eta\mathbf{e}_z, \qquad\qquad \text{(A.3.39)}$$

$$\mathbf{h}_3 = \frac{\partial \mathbf{x}}{\partial \varphi} = l\sqrt{(\xi^2 - 1)(1 - \eta^2)}\left[-\sin\varphi\mathbf{e}_x + \cos\varphi\mathbf{e}_y\right].$$

The scalar triple product, calculated from the $\mathbf{h}_i$, results in

$$\mathbf{v} = \mathbf{h}_1\cdot(\mathbf{h}_2 \times \mathbf{h}_3) = l^3\left[\xi^2 - \eta^2\right] = \sqrt{\det \mathbf{g}}.$$

The metric tensor is diagonal, since the $\mathbf{h}_i$ are local, orthogonal coordinates:

$$g_{11} = \mathbf{h}_1\cdot\mathbf{h}_1 = l^2\eta^2\frac{\xi^2 - 1}{1 - \eta^2} + l^2\xi^2 = l^2\frac{\xi^2 - \eta^2}{1 - \eta^2} = \frac{v}{l(1 - \eta^2)},$$

$$g_{22} = \mathbf{h}_2\cdot\mathbf{h}_2 = l^2\xi^2\frac{1 - \eta^2}{\xi^2 - 1} + l^2\eta^2 = l^2\frac{\xi^2 - \eta^2}{\xi^2 - 1} = \frac{v}{l(\xi^2 - 1)}, \qquad\qquad \text{(A.3.40)}$$

$$g_{33} = \mathbf{h}_3\cdot\mathbf{h}_3 = l^2(\xi^2 - 1)(1 - \eta^2).$$

Now it is easy to specify the scalar product:

$$v = \sqrt{g} = l^3(\xi^2 - \eta^2). \tag{A.3.41}$$

The orthonormal unit vectors are given by $\mathbf{h}_i = h_i\,\mathbf{e}_i$

$$h_1 = l\sqrt{\frac{\xi^2 - \eta^2}{\xi^2 - 1}}, \qquad h_2 = l\sqrt{\frac{\xi^2 - \eta^2}{1 - \eta^2}}, \qquad h_3 = l\sqrt{(\xi^2 - 1)(1 - \eta^2)}. \tag{A.3.42}$$

The gradient is given by

$$\boldsymbol{\nabla} = \mathbf{h}^i\,\frac{\partial}{\partial \xi^i} = \mathbf{e}_i\,\frac{1}{h_i}\,\frac{\partial}{\partial \xi^i} \tag{A.3.43}$$

$$= \mathbf{e}_\xi\,\frac{1}{l}\sqrt{\frac{\xi^2 - 1}{\xi^2 - \eta^2}}\,\frac{\partial}{\partial \xi} + \mathbf{e}_\eta\,\frac{1}{l}\sqrt{\frac{1 - \eta^2}{\xi^2 - \eta^2}}\,\frac{\partial}{\partial \eta} + \mathbf{e}_\varphi\,\frac{1}{l}\,\frac{1}{\sqrt{(1 - \eta^2)(\xi^2 - 1)}}\,\frac{\partial}{\partial \varphi}\,.$$

Without going into the calculation, we also give the Laplace operator (A.2.30):

$$\Delta = \frac{1}{l^2(\xi^2 - \eta^2)}\left[\frac{\partial}{\partial \eta}(1 - \eta^2)\frac{\partial}{\partial \eta} + \frac{\partial}{\partial \xi}(\xi^2 - 1)\frac{\partial}{\partial \xi}\right] + \frac{1}{l^2(\xi^2 - 1)(1 - \eta^2)}\,\frac{\partial^2}{\partial \varphi^2}\,. \tag{A.3.44}$$

Elliptical coordinates for a flattened rotation ellipsoid

We limit ourselves here to the definition of the local variables $\xi^1 = \xi$, $\xi^2 = \eta$ and $\xi^3 = \varphi$:

$$\begin{aligned}
x &= l\sqrt{(1 + \xi^2)(1 - \eta^2)}\,\cos\varphi, & 0 &\le \xi < \infty, \\
y &= l\sqrt{(1 + \xi^2)(1 - \eta^2)}\,\sin\varphi, & -1 &\le \eta \le 1, \\
z &= l\xi\eta, & 0 &\le \varphi < 2\pi\,.
\end{aligned} \tag{A.3.45}$$

Again, $\eta = \cos\vartheta$ can be substituted by the polar angle, while $\xi = \sinh u$. For the length of the unit vectors and the g_{ij} one obtains

$$h_1 = l\sqrt{\frac{\xi^2 + \eta^2}{1 + \xi^2}}, \qquad h_2 = l\sqrt{\frac{\xi^2 + \eta^2}{1 - \eta^2}}, \qquad h_3 = l\sqrt{(1 + \xi^2)(1 - \eta^2)},$$

$$g_{11} = l^2\frac{\xi^2 + \eta^2}{1 + \xi^2}, \qquad g_{22} = l^2\frac{\xi^2 + \eta^2}{1 - \eta^2}, \qquad g_{33} = l^2(1 + \xi^2)(1 - \eta^2). \tag{A.3.46}$$

A.4 Vector Fields and Integral Theorems

A vector field $\mathbf{v}(\mathbf{x})$ is a function that assigns a vector to each point in space. Examples are the electromagnetic fields, the gravitational field or the velocity field of a moving fluid.

To understand electrodynamics, knowledge of the properties of vector fields $\mathbf{v}(\mathbf{x}, t)$ is a prerequisite. Historically, vector fields first appeared in physics in the field of hydrodynamics. This is also reflected in the terms used such as flow, source and sink or vortex. As a vector field, one has in mind the velocity $\mathbf{v}$ of the flow or $\rho\,\mathbf{v}$, where ρ is the density.

A.4.1 Gauss's Theorem and Divergence

If $\mathbf{v}$ is the velocity of the flow and $\rho(\mathbf{x},t)$ is the density, then

$$\rho\,\mathbf{v}(\mathbf{x},t)\cdot\mathrm{da}\,\mathrm{d}t$$

is the mass that flows through the surface element da in the time interval $\mathrm{d}t$. Applied to a finite surface A, the flow Φ is defined by

$$\Phi = \iint_A \mathrm{da}\cdot\mathbf{v}(\mathbf{x},t),$$

as shown in Fig. A.12. If the vector field $\mathbf{v}$ is replaced again by $\rho\,\mathbf{v}$ with $\mathbf{v}$ as

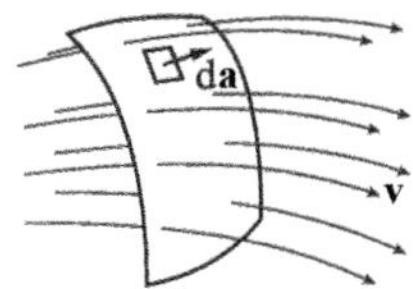

Fig. A.12. Flow through a surface (surface element da)

velocity, then Φ is the mass of the fluid flowing through the surface per unit of time.

Integral representation of the divergence

If a volume V with the surface ∂V is given, the flow through the surface

$$\Phi = \oiint_{\partial V} \mathrm{dS}\cdot\mathbf{v}$$

is also the yield of all sources and sinks contained in V. The normal vector $\mathbf{n}$ on the surface element $\mathrm{dS} = \mathrm{d}S\,\mathbf{n}$ is always directed outward.

If $\Phi = 0$, the flow penetrating (from one side) compensates the flow exiting (on the other side), i.e., there are no sources or sinks inside V, or they compensate each other. If you now divide Φ by V and perform the limit $V \to 0$, you get a *source density*

$$q = \lim_{V\to0}\frac{\Phi}{V} = \lim_{V\to0}\frac{1}{V}\oiint_{\partial V}\mathrm{dS}\cdot\mathbf{v}. \tag{A.4.1}$$

Since $q = \boldsymbol{\nabla}\cdot\mathbf{v}$, (A.4.1) is the integral representation of the divergence. This expression can be easily calculated in Cartesian coordinates, in which the volume $V = \mathrm{d}x\,\mathrm{d}y\,\mathrm{d}z$. In the direction of the x-axis, the flux is the difference

$$\frac{1}{\mathrm{d}x\mathrm{d}y\mathrm{d}z}\big(v_x(x+\mathrm{d}x,y,z)\,\mathrm{d}y\,\mathrm{d}z - v_x(x,y,z)\,\mathrm{d}y\,\mathrm{d}z\big) = \frac{\partial v_x}{\partial x},$$

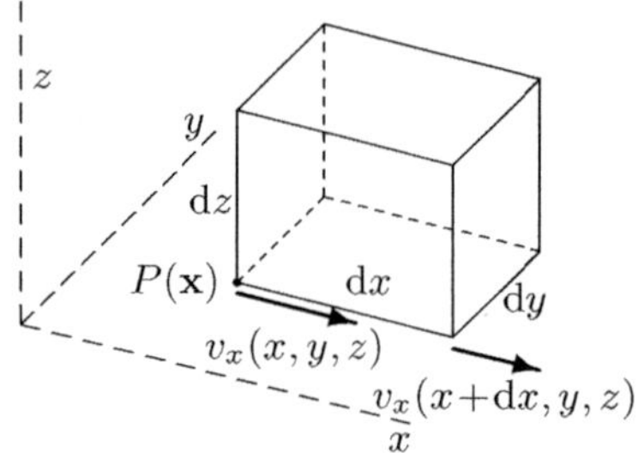

Fig. A.13. Divergence of a cuboid with $V = \mathrm{d}x\,\mathrm{d}y\,\mathrm{d}z$; the surface normal is always directed outwards

as can be seen from Fig. A.13. The analogous result is obtained for the two other directions, so that:

$$q = \frac{\partial v_x}{\partial x} + \frac{\partial v_y}{\partial y} + \frac{\partial v_z}{\partial z} = \boldsymbol{\nabla}\cdot\mathbf{v}. \tag{A.4.2}$$

The source density is referred to as the divergence of $\mathbf{v}$. The divergence is a scalar differential invariant that is independent of the reference system.

Gauss's theorem

If you divide V into two sub-volumes, $V = V_1 + V_2$, with the surfaces $\partial V_{1,2}$, then

$$\Phi = \Phi_1 + \Phi_2 = \sum_{i=1}^{2} \oiint_{\partial V_i} \mathrm{d}\mathbf{S}\cdot\mathbf{v} = \oiint_{\partial V} \mathrm{d}\mathbf{S}\cdot\mathbf{v},$$

as the contributions of the common boundary of V_1 and V_2 compensate. $\mathbf{n}_1 = -\mathbf{n}_2$ at the contact surfaces. If one refines the division, one obtains with $V_i = \Delta V$ according to (A.4.2)

$$\Phi = \lim_{n\to\infty} \sum_{i=1}^{n} \Delta V\, q_i = \int_V \mathrm{d}^3 x\, q(\mathbf{x}) = \lim_{n\to\infty} \sum_{i=1}^{n} \oiint_{\partial V_i} \mathrm{d}\mathbf{S}\cdot\mathbf{v}.$$

Thus, for the flux, one obtains the relationship that the integral over sources and sinks $\boldsymbol{\nabla}\cdot\mathbf{v}$ is equal to the total flux through the surface of V:

$$\Phi = \int_V \mathrm{d}^3 x\, \boldsymbol{\nabla}\cdot\mathbf{v} = \oiint_{\partial V} \mathrm{d}\mathbf{S}\cdot\mathbf{v}, \tag{A.4.3}$$

which represents the well-known Gauss's theorem.

The vector field $\mathbf{v}$ is not yet fully determined by its sources and sinks alone, i.e. by the scalar quantity $q(\mathbf{x}, t) = \boldsymbol{\nabla}\cdot\mathbf{v}(\mathbf{x}, t)$.

General form of Gauss's theorem

First, we replace in (A.4.3) $\mathbf{v}$ by $\mathbf{c}\phi(\mathbf{x})$, where $\mathbf{c}$ is a constant vector

$$\int_V \mathrm{d}^3 x\, \boldsymbol{\nabla}\!\cdot\!\mathbf{c}\,\phi = \oiint_{\partial V} \mathrm{d}\mathbf{S}\!\cdot\!\mathbf{c}\,\phi.$$

From this follows, if we draw $\mathbf{c}$ in front of the integral on both the left and the right side, Gauss's theorem for scalar fields

$$\int_V \mathrm{d}^3 x\, \boldsymbol{\nabla}\phi = \oiint_{\partial V} \mathrm{d}\mathbf{S}\,\phi. \tag{A.4.4}$$

If one replaces in (A.4.3) $\boldsymbol{\nabla}\!\cdot\!\mathbf{v}$ by $\mathbf{c}\!\cdot\!(\boldsymbol{\nabla}\!\times\!\mathbf{v})$, where $\mathbf{c}$ is to be constant, one obtains using

$$\boldsymbol{\nabla}\!\cdot\!(\mathbf{v}\!\times\!\mathbf{c}) \overset{\text{cyclic}}{=} \mathbf{c}\!\cdot\!(\boldsymbol{\nabla}\!\times\!\mathbf{v})$$

$$\mathbf{c}\!\cdot\!\int \mathrm{d}^3 x\, \boldsymbol{\nabla}\!\times\!\mathbf{v} = \oiint \mathrm{d}\mathbf{S}\cdot(\mathbf{v}\!\times\!\mathbf{c}) = \mathbf{c}\!\cdot\!\oiint (\mathrm{d}\mathbf{S}\!\times\!\mathbf{v}).$$

Since this applies for any arbitrary constant vector $\mathbf{c}$, it is

$$\int_V \mathrm{d}^3 x\, \boldsymbol{\nabla}\!\times\!\mathbf{v} = \oiint_{\partial V} (\mathrm{d}\mathbf{S}\!\times\!\mathbf{v}). \tag{A.4.5}$$

In an even more general form, one replaces v_i with $T_{i\ldots}$:

$$\int_V \mathrm{d}^3 x\, \nabla_i\, T_{i\ldots} = \oiint_{\partial V} \mathrm{d}S_i\, T_{i\ldots}\,. \tag{A.4.6}$$

If one substitutes $T_{ij} = \epsilon_{jik} v_k$ into (A.4.6), one obtains (A.4.5).

Mnemonic rule: $\mathrm{d}^3 x\, \nabla_i\, T_{i\ldots} \to \mathrm{d}S_i\, T_{i\ldots}$.

Mathematical prerequisites for the Gauss theorem

$\mathbf{v}$ should be continuously differentiable in V and ∂V should be smooth, i.e., on each sub-surface of ∂V there exists a normal vector that is directed outward (from the enclosed volume). V may also contain holes and consist of several sub-volumes V_i.

Notes: A singular point is excluded by an infinitesimal sphere (hole), whose surface brings an additional contribution. If the singularity is a δ-function (distribution), it can be treated like a continuous function, if it is considered as the limit of a (continuous) test function (see Tab. B.4, p. 627)

According to the formulation by Großmann [Großmann, 2012, p. 281] V should be regular and ∂V should be regular: A region in $\mathbb{R}^3$ is called regular if it is finite and closed and only surrounded by finite many closed regular surfaces.

A surface is called regular if it consists at most of finite many surface pieces with a smooth boundary curve is composed.

A surface piece is called regular if at least one representation $z = f(x, y)$ with continuously differentiable function f exists.

Regular regions with a regular surface are also referred to as smooth. Infinitely extended regions are allowed if the integrals are represented as the limit of smooth, finite regions (improper integrals).

Potential field

A scalar field $\phi(\mathbf{x})$ is referred to as a potential field when $\phi(\mathbf{x})$ is the potential of a vector field $\mathbf{v}(\mathbf{x}) = -\boldsymbol{\nabla}\phi(\mathbf{x})$:

Potential flows are solutions of the vector field $\mathbf{v}$, which can be derived from a scalar potential $\phi(\mathbf{x}, t)$. Let $\mathbf{E}$ be such a vector field

$$\mathbf{E} = -\boldsymbol{\nabla}\,\phi(\mathbf{x}, t).$$

The gradient $-\boldsymbol{\nabla}\phi$ is perpendicular to the equipotential surfaces and thus indicates the direction of the flow. The potential must satisfy the given source density $q(\mathbf{x}, t)$:

$$\boldsymbol{\nabla}\cdot\boldsymbol{\nabla}\phi = \Delta\phi = -\boldsymbol{\nabla}\cdot\mathbf{E} = -q(\mathbf{x}, t), \tag{A.4.7}$$

which is known as the Poisson equation. If one forms the line integral[4]

$$\int_{\mathbf{x}_1}^{\mathbf{x}_2} \mathrm{d}\mathbf{s}\cdot\mathbf{E} = -\int_{\mathbf{x}_1}^{\mathbf{x}_2} \mathrm{d}\phi = \phi(\mathbf{x}_1) - \phi(\mathbf{x}_2) , \tag{A.4.8}$$

then this is independent of the path. Thus, the line integral vanishes on every closed path C:

$$\oint_C \mathrm{d}\mathbf{s}\cdot\mathbf{E} = 0. \tag{A.4.9}$$

One also sees that for every field $\mathbf{E} = -\boldsymbol{\nabla}\phi$ the curl

$$\boldsymbol{\nabla}\times\mathbf{E} = \boldsymbol{\nabla}\times\boldsymbol{\nabla}\phi = 0 \tag{A.4.10}$$

vanishes. A vector field $\mathbf{E}$, which can be derived from a scalar potential ϕ (potential field), thus satisfies the conditions (A.4.9) and (A.4.7) and is therefore vortex-free. There are then no closed field lines, but only those with the starting points in the sources and the end points in the sinks.

It is assumed that $\mathbf{v}$ has continuous partial derivatives in a simply connected region G.

The vector field $\mathbf{v}(\mathbf{x})$ is then irrotational according to (A.2.36), $\boldsymbol{\nabla}\times\mathbf{v} = 0$ and the following statements are equivalent:

1. $\boldsymbol{\nabla}\times\mathbf{v} = 0$ at every point in G,

2. $\oint \mathrm{d}\mathbf{s}\cdot\mathbf{v} = 0$ on every closed path in G,

3. $\displaystyle\int_{\mathbf{x}_0}^{\mathbf{x}} \mathrm{d}\mathbf{s}\cdot\mathbf{v}$ is independent of the path,

4. $\mathbf{v} = -\boldsymbol{\nabla}\phi$ with $\phi(\mathbf{x}) = -\displaystyle\int_{\mathbf{x}_0}^{\mathbf{x}} \mathrm{d}\mathbf{s}\cdot\mathbf{v}$.

---

[4] $\mathrm{d}\phi = \dfrac{\partial\phi}{\partial x}\,\mathrm{d}x + \dfrac{\partial\phi}{\partial y}\,\mathrm{d}y + \dfrac{\partial\phi}{\partial z}\,\mathrm{d}z = \mathrm{d}\mathbf{s}\cdot\boldsymbol{\nabla}\phi$

A.4.2 Curl and Stokes' Theorem

If a field has closed field lines, then the line integral

$$Z = \oint_C \mathbf{v} \cdot d\mathbf{s} \tag{A.4.11}$$

has a finite value for one revolution on the closed curve C. Z is called circulation. If $\mathbf{v}$ is the velocity field of a fluid, then Z is its vortex strength, whereas for a magnetic field $\mathbf{B}$, $\mathbf{v} \to \mathbf{B}$ is Z the magnetic circulation.

Integral representation of the curl

Analogous to the source density, a circulation density, the vorticity z,

$$z(\mathbf{x}, t) = \lim_{A \to 0} \frac{Z}{A} = \frac{dZ}{da} = \lim_{A \to 0} \frac{1}{A} \oint_{\partial A} d\mathbf{s} \cdot \mathbf{v}$$

can be defined by shrinking C around a point. A is the area enclosed by C.

The value dZ of the circulation in the partial area $d\mathbf{a} = da\,\mathbf{n}$ is to be calculated. The value is characterized by a vector $\boldsymbol{\zeta} = \boldsymbol{\nabla} \times \mathbf{v}$, which is referred to as the curl or rotor of $\mathbf{v}$:

$$dZ = z d\mathbf{a} = \boldsymbol{\zeta} \cdot \mathbf{n}\,da = \boldsymbol{\nabla} \times \mathbf{v} \cdot \mathbf{n}\,da.$$

z is thus the component of the curl of $\mathbf{v}$ perpendicular to the surface normal $\mathbf{n}$. For the vortex density, one obtains

$$z(\mathbf{x}, t) = \mathbf{n} \cdot (\boldsymbol{\nabla} \times \mathbf{v}) = \lim_{\Delta A \to 0} \frac{1}{\Delta A} \oint_{\partial A} \mathbf{v} \cdot d\mathbf{s}. \tag{A.4.12}$$

(A.4.12) is an integral representation of the curl of $\mathbf{v}$.

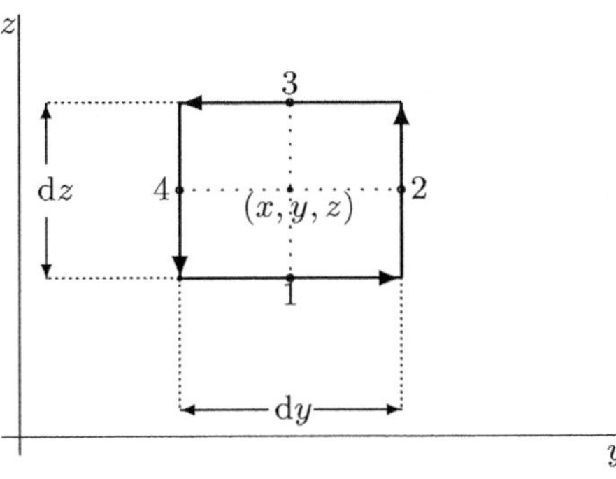

Fig. A.14. The area element is a rectangle in the yz plane. v_y is multiplied by dy at points 1 and 3, and v_z is multiplied by dz at points 2 and 4 to calculate the line integral.

If one calculates the line integral for the rectangle shown in Fig. A.14, one obtains with a Taylor expansion around (x, y, z)

$$\oint_\square \mathbf{ds}\cdot\mathbf{v} = \mathrm{d}y\left[v_y(x,y,z-\frac{\mathrm{d}z}{2}) - v_y(x,y,z+\frac{\mathrm{d}z}{2})\right]$$

$$+ \mathrm{d}z\left[v_z(x,y+\frac{\mathrm{d}y}{2},z) - v_z(x,y-\frac{\mathrm{d}y}{2},z)\right] \approx \mathrm{d}z\mathrm{d}y\left(\frac{\partial v_z}{\partial y} - \frac{\partial v_y}{\partial z}\right).$$

This implies for $\mathbf{n} = \mathbf{e}_x = (1\,0\,0)$

$$\boldsymbol{\nabla}\times\mathbf{v}\cdot\mathbf{n} = (\boldsymbol{\nabla}\times\mathbf{v})_x = \frac{1}{\mathrm{d}a}\oint_\square \mathbf{ds}\cdot\mathbf{v} = \frac{\partial v_z}{\partial y} - \frac{\partial v_y}{\partial z}.$$

The calculation can be done analogously for the other two components, which implies:

$$\boldsymbol{\nabla}\times\mathbf{v} = \left(\frac{\partial v_z}{\partial y} - \frac{\partial v_y}{\partial z}, \frac{\partial v_x}{\partial z} - \frac{\partial v_z}{\partial x}, \frac{\partial v_y}{\partial x} - \frac{\partial v_x}{\partial y}\right).$$

Stokes' theorem

If one divides the area enclosed by C into two subareas, as sketched in Fig. A.15a, one can see that the dividing line of the two curves is traversed in the opposite direction. Therefore, it holds that

$$Z = Z_1 + Z_2 = \oint_{C_1} \mathbf{v}\cdot\mathbf{ds} + \oint_{C_2} \mathbf{v}\cdot\mathbf{ds}.$$

When subdividing into small partial areas $\mathrm{d}a$, as shown in Fig. A.15b, the

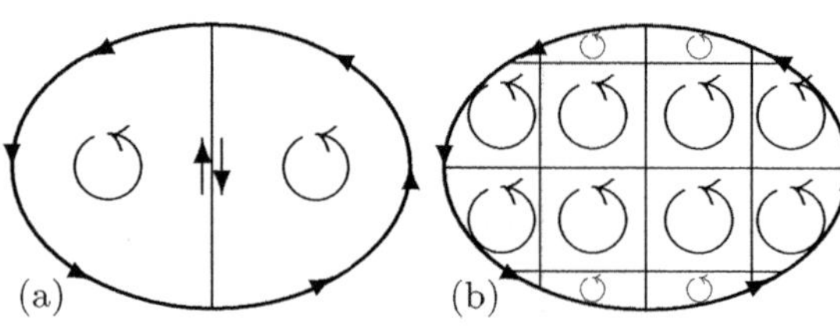

Fig. A.15. (a) If the enclosed area A is divided, then the paths inside are each traversed in the opposite direction
(b) With finer division, only the boundary curve ∂A remains

contributions of the inner lines, which separate the partial areas from each other, are compensated due to the opposite direction in which they are traversed and ultimately only the contribution of the outer edge, the curve C, remains. The integral of the circulation density $z(\mathbf{x})$ (A.4.12) is formed, where $\mathbf{n}\,\mathrm{d}a = \mathbf{d}a$ is substituted. One then obtains the *Stokes's theorem*

$$Z = \iint_A \mathbf{d}a\cdot\boldsymbol{\nabla}\times\mathbf{v} = \oint_{\partial A} \mathbf{ds}\cdot\mathbf{v}. \tag{A.4.13}$$

The magnetic field, which originates from the current in a linear conductor, forms such closed lines. For a circulation along the field line, energy must be expended, which was previously referred to as circulation. The field lines cannot be described with a unique scalar potential. For this, the curl vanishes:

$$\boldsymbol{\nabla} \times \boldsymbol{\nabla}\phi = 0. \tag{A.4.14}$$

Note: Z must not depend on the choice of the surface ∂V.

Suppose A_1 and A_2 are two surfaces with the common boundary curve ∂A. A_1 and A_2 should not overlap, so that they enclose a volume V. The difference is given by

$$\oiint_{\partial V} \mathrm{d}\mathbf{a}\cdot\boldsymbol{\nabla}\times\mathbf{v} = \iint_{A_1} \mathrm{d}\mathbf{a}\cdot\boldsymbol{\nabla}\times\mathbf{v} - \iint_{A_2} \mathrm{d}\mathbf{a}\cdot\boldsymbol{\nabla}\times\mathbf{v}.$$

According to Gauss's theorem, for the volume V enclosed by A_1 and A_2

$$\oiint_{\partial V} \mathrm{d}\mathbf{a}\cdot\boldsymbol{\nabla}\times\mathbf{v} = \int_V \mathrm{d}^3x, \boldsymbol{\nabla}\cdot\boldsymbol{\nabla}\times\mathbf{v} = 0,$$

since, according to (A.2.36) $\boldsymbol{\nabla}\cdot\boldsymbol{\nabla}\times\mathbf{v} = 0$.

From (A.2.37) it follows that the divergence of any vector field $\mathbf{B}$, which is the curl of another vector field $\mathbf{A}$, vanishes:

$$\mathbf{B} = \boldsymbol{\nabla}\times\mathbf{A} \quad \overset{(\mathrm{A.2.37})}{\Longrightarrow} \quad \boldsymbol{\nabla}\cdot\mathbf{B} = \boldsymbol{\nabla}\cdot(\boldsymbol{\nabla}\times\mathbf{A}) = 0.$$

In Fig. A.16 two vector fields $\mathbf{v}_a = \alpha x\,\mathbf{e}_x$ and $\mathbf{v}_b = \beta y\,\mathbf{e}_x$ are sketched, of

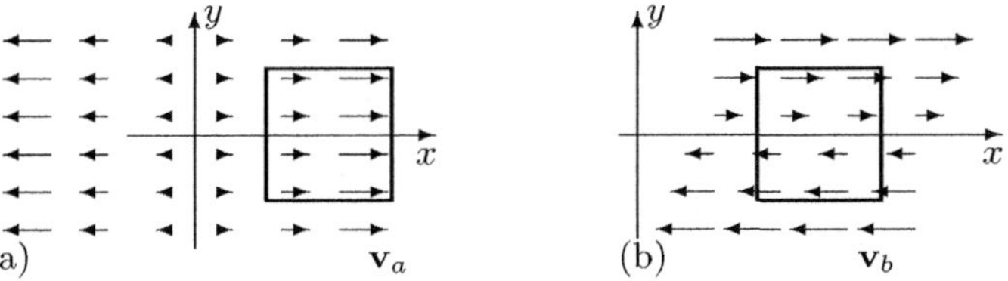

Fig. A.16. Vector fields   (a) $\mathrm{div}\,\mathbf{v}_a \neq 0$ and $\boldsymbol{\nabla}\times\mathbf{v}_a = 0$ (b) $\boldsymbol{\nabla}\cdot\mathbf{v}_b = 0$ and $\boldsymbol{\nabla}\times\mathbf{v}_b \neq 0$

which the first is irrotational and has a constant divergence ($\boldsymbol{\nabla}\times\mathbf{v}_a = 0$ and $\boldsymbol{\nabla}\cdot\mathbf{v}_a = \alpha$) and the second has no sources, but a constant, finite curl ($\boldsymbol{\nabla}\times\mathbf{v}_b = -\beta$ and $\boldsymbol{\nabla}\cdot\mathbf{v}_b = 0$).

The flux Φ through the indicated volume depends for $\mathbf{v}_a$ only on its size and not on location and shape.

Similarly, the vorticity Z of $\mathbf{v}_b$ is only determined by the size of the area in the xy-plane and not by the shape of the boundary curve and its location.

General form of Stokes' theorem

Just as in the case of Gauss's theorem, a more general form can also be given for Stokes' theorem. We first insert $\mathbf{v} = \phi\mathbf{c}$ into (A.4.13), where $\mathbf{c}$ is an arbitrary constant vector:

$$\mathbf{c}\cdot\int_{\partial A}\mathrm{d}\mathbf{s}\,\phi = \iint_A \mathrm{d}\mathbf{a}\cdot(\boldsymbol{\nabla}\phi\times\mathbf{c}) \overset{\mathrm{cyclic}}{=} \mathbf{c}\cdot\iint_A \mathrm{d}\mathbf{a}\times\boldsymbol{\nabla}\phi.$$

From this follows

$$\int_{\partial A}\mathrm{d}\mathbf{s}\,\phi = \iint_A \mathrm{d}\mathbf{a}\times\boldsymbol{\nabla}\phi. \tag{A.4.15}$$

Now we replace $\mathbf{v}$ with $\mathbf{v}\times\mathbf{c}$ and get

$$\mathbf{c}\cdot\int_{\partial A}\mathrm{d}\mathbf{s}\times\mathbf{v} = \iint_A \mathrm{d}\mathbf{a}\cdot(\boldsymbol{\nabla}\times(\mathbf{v}\times\mathbf{c})) \overset{\mathrm{cyclic}}{=} \iint_A (\mathrm{d}\mathbf{a}\times\boldsymbol{\nabla})\cdot(\mathbf{v}\times\mathbf{c})$$
$$\overset{\mathrm{cyclic}}{=} \mathbf{c}\cdot\iint_A ((\mathrm{d}\mathbf{a}\times\boldsymbol{\nabla})\times\mathbf{v}).$$

From this follows

$$\int_{\partial A}\mathrm{d}\mathbf{s}\times\mathbf{v} = \iint_A (\mathrm{d}\mathbf{a}\times\boldsymbol{\nabla})\times\mathbf{v}. \tag{A.4.16}$$

We can generalize again:

$$\int_{\partial A}\mathrm{d}s_i\,T_{i\ldots} = \epsilon_{ijk}\iint_A \mathrm{d}a_j\nabla_k T_{i\ldots}\,. \tag{A.4.17}$$

In this notation, (A.4.15) and (A.4.16) are the special cases $T_{i\ldots} = \phi$ and $T_{i\ldots} = \epsilon_{lim}v_m$. The mnemonic rule here is $\mathrm{d}s_i = \epsilon_{ijk}\,\mathrm{d}a_j\nabla_k$.

A.4.3 The Green's Theorems

Given is a vector field of the form

$$\mathbf{v}(\mathbf{x},t) = \psi(\mathbf{x},t)\,\boldsymbol{\nabla}\phi(\mathbf{x},t)\,. \tag{A.4.18}$$

ψ and ϕ are scalar fields for which then applies:

$$\boldsymbol{\nabla}\cdot\mathbf{v} = \boldsymbol{\nabla}\cdot(\psi\,\boldsymbol{\nabla}\phi) = (\boldsymbol{\nabla}\psi)\cdot(\boldsymbol{\nabla}\phi) + \psi\,\nabla^2\phi\,.$$

Substituting into Gauss's theorem (A.4.3) we get

$$\oiint_{\partial V}\mathrm{d}\mathbf{S}\cdot(\psi\,\boldsymbol{\nabla}\phi) = \int_V \mathrm{d}^3x\left[(\boldsymbol{\nabla}\psi)\cdot(\boldsymbol{\nabla}\phi) + \psi\,\nabla^2\phi\right]. \tag{A.4.19}$$

This equation is known as the 1st Green's theorem. From the interchange of ψ with ϕ and subsequent subtraction we get the 2nd Green's theorem:

$$\oiint_{\partial V}\mathrm{d}\mathbf{S}\cdot(\psi\,\boldsymbol{\nabla}\phi - \phi\,\boldsymbol{\nabla}\psi) = \int_V \mathrm{d}^3x\left(\psi\,\nabla^2\phi - \phi\,\nabla^2\psi\right). \tag{A.4.20}$$

A.4.4 Green Function of the Laplace Operator

The Green function of a linear differential operator L, which in our case is the Laplace operator, is defined by

$$\mathrm{L}G(\mathbf{x}) = a\delta^{(3)}(\mathbf{x}) \quad \overset{\substack{\mathrm{L}=\Delta \\ a=-4\pi}}{\Longrightarrow} \quad \Delta G(\mathbf{x}-\mathbf{x}') = -4\pi\delta^{(3)}(\mathbf{x}-\mathbf{x}') . \qquad (A.4.21)$$

G is therefore the solution of the inhomogeneous differential equation with the δ-function as inhomogeneity. In general, $a=1$, but not always.

The Green's function of the Laplace operator

$$G(\mathbf{x}-\mathbf{x}') = \frac{1}{|\mathbf{x}-\mathbf{x}'|} \qquad (A.4.22)$$

was already derived in section 2.1 (see (2.1.6)). Here, an alternative path is chosen for the derivation again:

If one represents $G(\mathbf{x})$ by its Fourier transform $G(\mathbf{k})$ and substitutes in (A.4.21) one obtains

$$\Delta G(\mathbf{x}) = \Delta \int \frac{\mathrm{d}^3 k}{(2\pi)^3}\, \mathrm{e}^{\mathrm{i}\mathbf{k}\cdot\mathbf{x}}\, G(\mathbf{k}) = -\int \frac{\mathrm{d}^3 k}{(2\pi)^3}\, k^2\, \mathrm{e}^{\mathrm{i}\mathbf{k}\cdot\mathbf{x}}\, G(\mathbf{k}) = -4\pi\delta^{(3)}(\mathbf{x}) .$$

If one multiplies from the left with $\int \mathrm{d}^3 x\, \mathrm{e}^{-\mathrm{i}\mathbf{q}\cdot\mathbf{x}}$, it follows

$$G(\mathbf{q}) = 4\pi/q^2 . \qquad (A.4.23)$$

From this, one calculates $G(\mathbf{x})$ in spherical coordinates ($\xi=\cos\vartheta$) using $\displaystyle\int_0^\infty \mathrm{d}x\,\frac{\sin x}{x} = \frac{\pi}{2}$:

$$G(\mathbf{x}) = \int \frac{\mathrm{d}^3 k}{(2\pi)^3}\, \mathrm{e}^{\mathrm{i}\mathbf{k}\cdot\mathbf{x}}\, \frac{4\pi}{k^2} = \frac{1}{\pi}\int_0^\infty \mathrm{d}k \int_{-1}^1 \mathrm{d}\xi\, \mathrm{e}^{\mathrm{i}kr\xi} = \frac{2}{\pi}\int_0^\infty \mathrm{d}k\, \frac{\sin(kr)}{kr} = \frac{1}{r} .$$

A.4.5 Mean Value Theorem

If one substitutes into the 2nd Green's theorem (A.4.20) for $\phi(\mathbf{x}')$ a solution function of the Laplace equation ($\Delta'\phi(\mathbf{x}')=0$ for $\mathbf{x}'\in V$) and for $\psi=G(\mathbf{x}-\mathbf{x}')$ the Green function (A.4.22), one obtains

$$\int_V \mathrm{d}^3 x\, \big[G\,\Delta'\phi - \phi\,\Delta'G\big] = 4\pi\phi(\mathbf{x}) = \oiint_{\partial V} \mathrm{d}\mathbf{S}'\cdot\big[G\,\boldsymbol{\nabla}'\phi - \phi\,\boldsymbol{\nabla}'G\big] \quad \mathbf{x}\in V .$$

One takes for V a sphere S_R around $\mathbf{x}$ with the radius $R=|\mathbf{x}-\mathbf{x}'|$ and obtains

$$\phi(\mathbf{x}) = \frac{1}{4\pi}\oiint_{\partial S_R} \mathrm{d}\mathbf{S}'\cdot\Big[\frac{1}{R}\boldsymbol{\nabla}'\phi(\mathbf{x}') - \frac{\mathbf{R}}{R^3}\phi(\mathbf{x}')\Big] = \frac{1}{4\pi}\oiint_{\partial S_R} \mathrm{d}\Omega'\, \phi(\mathbf{x}') . \qquad (A.4.24)$$

In the 1st term, we draw the constant R in front of the integral, convert the surface integral with Gauss's theorem into a volume integral and note that this vanishes because $\Delta'\phi = 0$. In the 2nd term, we have substituted $d\mathbf{S}' = -R\mathbf{R}\,d\Omega'$.

Therefore, $\phi(\mathbf{x})$ is the arithmetic mean of the values of $\phi(\partial S_R)$ on the sphere surface of radius R around $\mathbf{x}$, which is referred to as the *mean value theorem of potential theory*.

Principle of maximum and minimum

From the mean value theorem, it follows directly that the largest and smallest values of ϕ are on the edge of V. If we had a local maximum at location $\mathbf{x}$ in V, then according to the mean value theorem, we would have values on the surface of a small sphere, which is still entirely in V, that are greater than or equal to the value of $\mathbf{x}$. There is therefore no maximum (minimum) inside V.

Both theorems have their correspondence in the complex analysis (see section B.1.2).

A.4.6 Solution of the Poisson Equation with Green Functions

Solutions to linear inhomogeneous differential equations can be found using the method of Green functions. Let L be again a linear differential operator, the Laplace operator, whose Green function is determined by (A.4.21). L applied to ϕ results in $a\rho$:

$$\mathrm{L}\phi(\mathbf{x}) = a\,\rho(\mathbf{x}) \quad \overset{\substack{\mathrm{L}=\Delta\\ a=-4\pi}}{\Longrightarrow} \quad \Delta\phi(\mathbf{x}) = -4\pi\rho(\mathbf{x}). \tag{A.4.25}$$

A solution ϕ_0 of this differential equation is

$$\phi_0(\mathbf{x}) = \int d^3x'\, G_0(\mathbf{x},\mathbf{x}')\rho(\mathbf{x}') \quad \text{with} \quad G_0(\mathbf{x},\mathbf{x}') = \frac{1}{|\mathbf{x}-\mathbf{x}'|}. \tag{A.4.26}$$

By calculating $\mathrm{L}\phi_0$ one can verify that (A.4.26) satisfies the Poisson equation (A.4.25); the general solution is obtained by adding an arbitrary harmonic solution ϕ_h of the Laplace equation.

Method of regularization

The solution ϕ_0 of the Poisson equation (A.4.26) is limited by the required finiteness of the integral. The density ρ is assumed to be finite, or the integration over a singularity (point charge) provides a finite contribution, so that a singular contribution is solely due to the asymptotic behavior of the product $G_0\rho$. If ρ falls off for $r \to \infty$ with $1/r^2$ or weaker, the integral will generally diverge.

Blumenthal [1905] has shown that the Green function (A.4.21) can be modified so that it falls off more rapidly at infinity. The potential (A.4.26) can then be given for a correspondingly weaker decreasing density ρ. This is achieved by the approach

$$G_i(\mathbf{x}-\mathbf{x}_0,\mathbf{x}'-\mathbf{x}_0) = \frac{1}{|\mathbf{x}'-\mathbf{x}|} - \sum_{l=0}^{i-1}\frac{1}{l!}\left((\mathbf{x}_0-\mathbf{x})\cdot\nabla'\right)^l\frac{1}{|\mathbf{x}'-\mathbf{x}_0|}. \qquad (A.4.27)$$

Here, the first i terms of the Taylor expansion of G_0 were subtracted from G_0 around a largely freely selectable convergence point (or regularization point):

$$G_1(\mathbf{x}-\mathbf{x}_0,\mathbf{x}'-\mathbf{x}_0) = \frac{1}{|\mathbf{x}'-\mathbf{x}|} - \frac{1}{|\mathbf{x}'-\mathbf{x}_0|},$$

$$G_2(\mathbf{x}-\mathbf{x}_0,\mathbf{x}'-\mathbf{x}_0) = G_1(\mathbf{x}-\mathbf{x}_0,\mathbf{x}'-\mathbf{x}_0)-(\mathbf{x}_0-\mathbf{x})\cdot\nabla'\frac{1}{|\mathbf{x}'-\mathbf{x}_0|}, \qquad (A.4.28)$$

$$G_3(\mathbf{x}-\mathbf{x}_0,\mathbf{x}'-\mathbf{x}_0) = G_2(\mathbf{x}-\mathbf{x}_0,\mathbf{x}'-\mathbf{x}_0) - \frac{1}{2}\left[(\mathbf{x}-\mathbf{x}_0)\cdot\nabla'\right]^2\frac{1}{|\mathbf{x}'-\mathbf{x}_0|}.$$

This achieves an asymptotic decay $G_i \sim 1/r'^{1+i}$. However, the development has the catch that G_i becomes singular around the regularization point according to $1/|\mathbf{x}'-\mathbf{x}_0|^i$ so that the potential generally diverges for $i>3$. Another weak point in the development is that G_i for $i>2$ is not a Green function of the original Poisson equation:

$$\Delta G_i(\mathbf{x}-\mathbf{x}_0,\mathbf{x}'-\mathbf{x}_0) = -4\pi\left[\delta^{(3)}(\mathbf{x}-\mathbf{x}')-\delta_{i,3}\delta^{(3)}(\mathbf{x}'-\mathbf{x}_0)\right], \qquad i\leq 3 \quad (A.4.29)$$

$$\Delta\phi_i(\mathbf{x},\mathbf{x}_0) = -4\pi\left[\rho(\mathbf{x})-\delta_{i,3}\rho(\mathbf{x}_0)\right].$$

For $i=3$ the additional term can be taken into account by a function of 2nd degree in $\mathbf{x}$, whose linear part is freely selectable, since this is a harmonic solution. It is convenient to choose

$$\phi_\epsilon(\mathbf{x}) = \frac{2\pi}{3}\rho(\mathbf{x}_0)|\mathbf{x}-\mathbf{x}_0|^2 \qquad \Rightarrow \qquad \Delta\phi_c(\mathbf{x}) = 4\pi\rho(\mathbf{x}_0),$$

since then $\phi_\epsilon(\mathbf{x})$ is the part of $\phi_3(\mathbf{x})$ that originates from the integral of an infinitesimal sphere S_ϵ around the singular point $\mathbf{x}' = x_0$. Thus it is

$$\bar{\phi}_3(\mathbf{x},\mathbf{x}_0) = \phi_3(\mathbf{x},\mathbf{x}_0)-\phi_\epsilon(\mathbf{x}) \qquad\qquad (A.4.30)$$

the solution of the Poisson equation. From (A.4.27) an auxiliary formula can be derived, which we will apply several times in the following to switch in the integrands from ∇ to ∇':

$$\nabla G_{i+1}(\mathbf{x}-\mathbf{x}_0,\mathbf{x}'-\mathbf{x}_0) = -\nabla' G_i(\mathbf{x}-\mathbf{x}_0,\mathbf{x}'-\mathbf{x}_0). \qquad (A.4.31)$$

A.4.7 Additions to the Helmholtz Decomposition Theorem

The decomposition theorem was by Stokes [1849] demonstrated and completed by Helmholtz [1858]. It was introduced into the textbooks of electrodynamics by Föppl [1894]. He assumed that the sources and vortices (and discontinuity surfaces) lie in the finite; the associated vector field $\mathbf{v}$ then asymptotically decreases at least with $1/r^2$. Already in 1905, Blumenthal [Blumenthal, 1905] showed that it is sufficient to assume that the vector field asymptotically "somehow"vanishes.

In physics, unlike in mathematics, Blumenthal's findings were largely ignored. The Helmholtz decomposition presented here and in section 7.1.2 is based on the work of Petrascheck, Folk [2015]; Petrascheck [2016]. The decomposition is referred to as (Helmholtz's) decomposition theorem, as (Helmholtz's) main theorem of vector analysis ([Blumenthal, 1905]; [Großmann, 2012, p. 368]; [Fließbach, 2008, p. 27]) or as the fundamental theorem of vector calculus ([Sommerfeld II, 1950, § 20])

Related to the decomposition of a vector field is the *Div-curl-problem*. The problem here is to determine the vector field $\mathbf{v}$ given sources and vortices while adhering to certain asymptotic conditions. This is referred to as the Helmholtz theorem ([Griffiths, 2017, §B]; [Arfken & Weber, 2005, §1.16]), where the distinction between the two theorems is often not clearly made, since not only the vector field $\mathbf{v}$, but also its source and vortex components $\mathbf{v}_{l,t}$ can be determined.

Sublinearly diverging vector fields

Vector fields can generally, regardless of their asymptotic behavior, be decomposed into source and vortex components. This is rarely pointed out. We take this as an opportunity to sketch the proof, following a work by Gregory [1996], without going into the prerequisites. The starting point are the vector Poisson equation and the identity (A.2.38):

$$\mathbf{v} = \Delta\mathbf{a} = \boldsymbol{\nabla}(\boldsymbol{\nabla}\cdot\mathbf{a}) - \boldsymbol{\nabla}\times(\boldsymbol{\nabla}\times\mathbf{a}) = -\boldsymbol{\nabla}\phi + \boldsymbol{\nabla}\times\mathbf{A} \qquad (A.4.32)$$

with $\phi = -\boldsymbol{\nabla}\cdot\mathbf{a}$ and $\mathbf{A} = -\boldsymbol{\nabla}\times\mathbf{a}$. Each solution $\mathbf{a}$ thus determines a decomposition of $\mathbf{v}$ into a vortex-free and a source-free vector field. Now $\mathbf{a}$ (and $\mathbf{v}$) is expanded according to spherical functions:

$$\mathbf{a}(\mathbf{x}) = \sum_{l=0}^{\infty}\sum_{m=-l}^{l} \tilde{\mathbf{a}}_{lm}(r)\, Y_{lm}(\vartheta,\varphi), \qquad \tilde{\mathbf{a}}_{lm}(r) = \int d\Omega\, \mathbf{a}(\mathbf{x})\, Y_{lm}^{*}(\vartheta,\varphi)$$

and applies the Laplace operator (A.2.30) to $\mathbf{a}$. This results in ordinary differential equations for the radial part of the form

$$\left[\frac{d}{dr}\,r^2\,\frac{d}{dr} - l(l+1)\right]\tilde{\mathbf{a}}_{lm}(r) = \tilde{\mathbf{v}}_{lm}(r).$$

Thus, a decomposition is in principle possible, although not quite as simple as it seems here [Gregory, 1996]. Therefore, we continue with the application of regularization for the Poisson equation, in which solution methods for certain asymptotic conditions can be specified.

In section 7.1.2 (7.1.11) it was shown that (A.4.32) for

$$\mathbf{a} = \mathbf{v}(\mathbf{x}')\, G_i(\mathbf{x}, \mathbf{x}')$$

with $i = 0$ causes the decomposition of $\mathbf{v}$ into a vortex-free and a source-free part, when integrating over $\mathbf{x}'$. For $i \leq 3$ and $\mathbf{x}_0 = 0$ one obtains for (7.1.11) using (A.4.29):

$$\mathbf{v}(\mathbf{x}) - \delta_{i,3}\mathbf{v}(0) = \frac{-1}{4\pi} \int d^3x'\, \Delta\mathbf{v}(\mathbf{x}')\, G_i(\mathbf{x}, \mathbf{x}') \tag{A.4.33}$$

$$= -\boldsymbol{\nabla} \underbrace{\frac{1}{4\pi} \int d^3x' \left[\boldsymbol{\nabla}\cdot\mathbf{v}(\mathbf{x}')G_i(\mathbf{x}, \mathbf{x}') \right]}_{\phi(\mathbf{x}, 0)} + \boldsymbol{\nabla}\times \underbrace{\frac{1}{4\pi} \int d^3x' \left[\boldsymbol{\nabla}\times\mathbf{v}(\mathbf{x}')G_i(\mathbf{x}, \mathbf{x}') \right]}_{\mathbf{A}(\mathbf{x}, 0)}.$$

The asymptotic behavior of $\mathbf{v}$ is influenced by the Green function G_i. For $i=0$ and $i=1$, v must decrease stronger than $1/r$, while for $i=2$ a decrease with a small power is sufficient, and for $i=3$ even with a sublinear increase of v the integrals in (A.4.33) remain finite. The application of (A.4.31) $\boldsymbol{\nabla}G_{i+1} = -\boldsymbol{\nabla}'G_i$ with subsequent partial integration results in allowing $\mathbf{x}_0 \neq 0$:

$$\phi_i(\mathbf{x}, \mathbf{x}_0) = -\frac{1}{4\pi} \int d^3x'\, \mathbf{v}(\mathbf{x}')\cdot\boldsymbol{\nabla}'G_i(\mathbf{x}-\mathbf{x}_0, \mathbf{x}'-\mathbf{x}_0),$$

$$\mathbf{A}_i(\mathbf{x}, \mathbf{x}_0) = \frac{1}{4\pi} \int d^3x'\, \mathbf{v}(\mathbf{x}')\times\boldsymbol{\nabla}'G_i(\mathbf{x}-\mathbf{x}_0, \mathbf{x}'-\mathbf{x}_0). \tag{A.4.34}$$

The decomposition (A.4.33) of $\mathbf{v}$ is now

$$\mathbf{v}(\mathbf{x}) - \delta_{i,2}\mathbf{v}(\mathbf{x}_0) = -\boldsymbol{\nabla}\phi_i(\mathbf{x}, \mathbf{x}_0) + \boldsymbol{\nabla}\times\mathbf{A}_i(\mathbf{x}, \mathbf{x}_0) \qquad i \leq 2. \tag{A.4.35}$$

We can now formulate the decomposition theorem almost identically to section 7.1.2

Theorem: *Let $\mathbf{v}(\mathbf{x})$ be a piecewise continuously differentiable vector field with the asymptotic behavior*

$$\lim_{r\to\infty} v(r)r^{\epsilon+1-i} < \infty \qquad with \qquad \epsilon > 0 \quad and \quad i \leq 2,$$

then the decomposition applies

$$\mathbf{v}(\mathbf{x}) - \delta_{i,2}\mathbf{v}(\mathbf{x}_0) = -\boldsymbol{\nabla}\phi_i(\mathbf{x}, \mathbf{x}_0) + \boldsymbol{\nabla}\times\mathbf{A}_i(\mathbf{x}, \mathbf{x}_0) \tag{A.4.36}$$

into a vortex-free and a source-free vector field with the potentials (A.4.34), where the Green functions G_i are given by (A.4.27). This decomposition is unique for $i<2$, when v_l asymptotically vanishes, and unique up to a constant for $i=2$, when v_l increases weaker than linearly.

Remark: The asymptotic conditions for $\mathbf{v}_l$ are ensured by the potentials (A.4.34); they only prevent harmonic vector fields $\mathbf{v}_h$ from being attributed to the vortex-free field $\mathbf{v}_l$ and subtracted from $\mathbf{v}_t$.

Proof: The existence of the decomposition has already been shown. The uniqueness of this decomposition was already shown for $i \leq 1$ in section 7.1.2. It is therefore sufficient to prove that this is unique for $i = 2$ up to a vectorial constant. One again starts from two different decompositions and defines the source-free and vortex-free field $\mathbf{v}_d = \mathbf{v}_l - \mathbf{v}'_l$. This can be written as the gradient of a scalar potential ϕ_d (see (3.2.38)). The radial vector field is then $\mathbf{e}_r \cdot \mathbf{v}_d = -\mathbf{e}_r \cdot \boldsymbol{\nabla} \phi_d$. For $r \to \infty$ all coefficients with $l-1 > 1-\epsilon$ and $\epsilon > 0$ must vanish, as the corresponding terms would lead to a stronger divergence. So only the contributions with $l = 0$ and $l = 1$ remain:

$$\phi_d(r, \vartheta, \varphi) = \alpha_{00} Y_{00} + \sum_{m=-1}^{1} \alpha_{1m} Y_{1m}(\vartheta, \varphi) \, r = \frac{\alpha_{00}}{\sqrt{4\pi}} - \mathbf{w} \cdot \mathbf{x}. \qquad (A.4.37)$$

We thus obtain

$$\mathbf{v}_d(\mathbf{x}) = -\boldsymbol{\nabla} \phi_d = \mathbf{w}.$$

Therefore, $\mathbf{v}_d = \mathbf{w}$ is unique, except for a vectorial constant.

On the calculation of the potentials

It is not always clear from the outset which G_i leads to finite integrals; it is only certain that for $\lim_{r \to \infty} v(r) r^{\epsilon + 1 - i} < \infty$ the potentials calculated with G_i converge. The starting point is the potential $\phi_v(\mathbf{x})$ calculated for a finite volume V with smooth boundary (7.1.13):

$$\phi_v(\mathbf{x}) = \frac{1}{4\pi} \int_V \mathrm{d}^3 x' \, \mathbf{v}(\mathbf{x}') \cdot \boldsymbol{\nabla} \frac{1}{|\mathbf{x} - \mathbf{x}'|}, \qquad (A.4.38)$$

$$\phi_0(\mathbf{x}) = \lim_{V \to \infty} \phi_v(\mathbf{x}).$$

This immediately follows using (A.4.28)

$$\phi_{v1}(\mathbf{x}, \mathbf{x}_0) = \frac{1}{4\pi} \int_V \mathrm{d}^3 x' \, \mathbf{v}(\mathbf{x}') \cdot \boldsymbol{\nabla} G_1(\mathbf{x} - \mathbf{x}_0, \mathbf{x}' - \mathbf{x}_0) \qquad (A.4.39)$$

$$= \frac{1}{4\pi} \int_V \mathrm{d}^3 x' \, \mathbf{v}(\mathbf{x}') \cdot \boldsymbol{\nabla} \left[\frac{1}{|\mathbf{x} - \mathbf{x}'|} - \frac{1}{|\mathbf{x}_0 - \mathbf{x}'|} \right] = \phi_v(\mathbf{x}) - \phi_v(\mathbf{x}_0),$$

$$\phi_1(\mathbf{x}, \mathbf{x}_0) = \lim_{V \to \infty} \left[\phi_{v1}(\mathbf{x}) - \phi_{v1}(\mathbf{x}_0) \right]. \qquad (A.4.40)$$

Similarly, one obtains

$$\phi_{v2}(\mathbf{x}, \mathbf{x}_0) = \frac{1}{4\pi} \int_V \mathrm{d}^3 x' \, \mathbf{v}(\mathbf{x}') \cdot \boldsymbol{\nabla} \left[G_1(\mathbf{x} - \mathbf{x}_0, \mathbf{x}' - \mathbf{x}_0) - (\mathbf{x} - \mathbf{x}_0) \cdot \boldsymbol{\nabla}_0 \frac{1}{|\mathbf{x}_0 - \mathbf{x}'|} \right]$$

$$= \phi_v(\mathbf{x}) - \phi_v(\mathbf{x}_0) - (\mathbf{x} - \mathbf{x}_0) \boldsymbol{\nabla}_0 \cdot \phi_v(\mathbf{x}_0), \qquad (A.4.41)$$

$$\phi_2(\mathbf{x}, \mathbf{x}_0) = \lim_{V \to \infty} \left[\phi_{v1}(\mathbf{x}) - (\mathbf{x} - \mathbf{x}_0) \cdot \boldsymbol{\nabla}_0 \phi_v(\mathbf{x}_0) \right].$$

The path to ϕ_1 and ϕ_2 can be simplified somewhat by developing and dividing ϕ_V in the limit $V \to \infty$ into

$$\phi_V(\mathbf{x}) = \phi_r(\mathbf{x}) + \boldsymbol{\alpha}_2(V) \cdot \mathbf{x} + \alpha_1(V). \tag{A.4.42}$$

ϕ_r must be finite for $V \to \infty$, but $\boldsymbol{\alpha}_2$ and α_1 can diverge:

$$\begin{aligned}
\phi_1(\mathbf{x}) &= \phi_r(\mathbf{x}) - \phi_r(\mathbf{x}_0) + (\mathbf{x} - \mathbf{x}_0) \cdot \boldsymbol{\alpha}_2(V), \\
\phi_2(\mathbf{x}) &= \phi_r(\mathbf{x}) - \phi_r(\mathbf{x}_0) - (\mathbf{x} - \mathbf{x}_0) \cdot \boldsymbol{\nabla}_0 \phi_r(\mathbf{x}_0).
\end{aligned} \tag{A.4.43}$$

So it is only necessary to calculate the potential $\phi_V(\mathbf{x})$ with G_0 and develop for $V \to \infty$ which should keep the extra effort for ϕ_1 and ϕ_2 within limits.

Analogous equations can be obtained for the vector potentials:

$$\begin{aligned}
\mathbf{A}_V(\mathbf{x}) &= \frac{1}{4\pi} \int_V \mathrm{d}^3 x' \, \mathbf{v}(\mathbf{x}') \times \boldsymbol{\nabla}' \frac{1}{|\mathbf{x} - \mathbf{x}'|}, \\
\mathbf{A}_{V1}(\mathbf{x}, \mathbf{x}_0)) &= \lim_{V \to \infty} \big[\mathbf{A}_V(\mathbf{x}) - \mathbf{A}_V(\mathbf{x}_0) \big], \\
\mathbf{A}_{V2}(\mathbf{x}, \mathbf{x}_0)) &= \lim_{V \to \infty} \big[\mathbf{A}_V(\mathbf{x}) - \mathbf{A}_V(\mathbf{x}_0) - (\mathbf{x} - \mathbf{x}_0) \cdot \boldsymbol{\nabla}_0 \mathbf{A}_V(\mathbf{x}_0) \big].
\end{aligned} \tag{A.4.44}$$

Problems for Appendix A

A.1. *Dyads are not commutative*: Prove the validity of the relation (A.1.16), which is particularly applied in magnetostatics.

A.2. *Calculation of a determinant*: In the calculation of the potentials of the moving point charge, a functional determinant of the form $|\mathsf{E} + \mathbf{a} \circ \mathbf{b}|$ appears (see (8.2.24)). Show that

$$\det \left(\mathsf{E} + \mathbf{a} \circ \mathbf{b} \right) = 1 + \mathbf{a} \cdot \mathbf{b}.$$

Hint: You can conduct the proof with complete induction.

A.3. *Proof of identities*:

1. $\boldsymbol{\nabla}(\mathbf{a} \cdot \mathbf{b}) = (\mathbf{a} \cdot \boldsymbol{\nabla}) \mathbf{b} + \mathbf{a} \times (\boldsymbol{\nabla} \times \mathbf{b}) + (\mathbf{b} \cdot \boldsymbol{\nabla}) \mathbf{a} + \mathbf{b} \times (\boldsymbol{\nabla} \times \mathbf{a})$.
2. $\boldsymbol{\nabla} \cdot (\mathbf{a} \times \mathbf{b}) = \mathbf{b} \cdot (\boldsymbol{\nabla} \times \mathbf{a}) - \mathbf{a} \cdot (\boldsymbol{\nabla} \times \mathbf{b})$.
3. $\boldsymbol{\nabla} \times (\mathbf{a} \times \mathbf{b}) = \mathbf{a} \times (\boldsymbol{\nabla} \times \mathbf{b}) - \mathbf{b} \times (\boldsymbol{\nabla} \times \mathbf{a}) = \mathbf{a}(\boldsymbol{\nabla} \cdot \mathbf{b}) - \mathbf{b}(\boldsymbol{\nabla} \cdot \mathbf{a}) + (\mathbf{b} \cdot \boldsymbol{\nabla}) \mathbf{a} - (\mathbf{a} \cdot \boldsymbol{\nabla}) \mathbf{b}$.

A.4. *Application of differential operators on unit vectors*: In curved KS, the effect of the differential operators on the base vectors must also be taken into account. In this context, verify the following relations

Cylindricalcoordinates :

$$\begin{aligned}
\boldsymbol{\nabla} \cdot \mathbf{e}_\varrho &= \frac{1}{\varrho}, & \boldsymbol{\nabla}\boldsymbol{\nabla} \cdot \mathbf{e}_\varrho &= -\frac{1}{\varrho^2} \mathbf{e}_\varrho, & \boldsymbol{\nabla} \times \mathbf{e}_\varrho &= 0, & \boldsymbol{\nabla} \times \boldsymbol{\nabla} \times \mathbf{e}_\varrho &= 0, \\
\boldsymbol{\nabla} \cdot \mathbf{e}_\varphi &= 0, & \boldsymbol{\nabla}\boldsymbol{\nabla} \cdot \mathbf{e}_\varphi &= 0, & \boldsymbol{\nabla} \times \mathbf{e}_\varphi &= \frac{1}{\varrho} \mathbf{e}_z, & \boldsymbol{\nabla} \times \boldsymbol{\nabla} \times \mathbf{e}_\varphi &= \frac{1}{\varrho^2} \mathbf{e}_\varphi, \\
\boldsymbol{\nabla} \cdot \mathbf{e}_z &= 0, & \boldsymbol{\nabla}\boldsymbol{\nabla} \cdot \mathbf{e}_z &= 0, & \boldsymbol{\nabla} \times \mathbf{e}_z &= 0, & \boldsymbol{\nabla} \times \boldsymbol{\nabla} \times \mathbf{e}_z &= 0.
\end{aligned} \tag{A.4.45}$$

Spherical coordinates:

$$\nabla \cdot \mathbf{e}_r = \frac{2}{r}, \qquad \nabla \nabla \cdot \mathbf{e}_r = -\frac{2}{r^2} \, \mathbf{e}_r, \qquad\qquad \nabla \times \mathbf{e}_r = 0,$$

$$\nabla \cdot \mathbf{e}_\vartheta = \frac{\cot \vartheta}{r}, \qquad \nabla \nabla \cdot \mathbf{e}_\vartheta = -\frac{\cot \vartheta}{r^2} \, \mathbf{e}_r - \frac{\mathbf{e}_\vartheta}{r^2 \sin^2 \vartheta}, \qquad \nabla \times \mathbf{e}_\vartheta = \frac{\mathbf{e}_\varphi}{r}, \qquad (A.4.46)$$

$$\nabla \cdot \mathbf{e}_\varphi = 0, \qquad \nabla \nabla \cdot \mathbf{e}_\varphi = 0, \qquad\qquad \nabla \times \mathbf{e}_\varphi = \frac{\cot \vartheta}{r} \, \mathbf{e}_r - \frac{\mathbf{e}_\vartheta}{r},$$

$$\nabla \times \nabla \times \mathbf{e}_r = 0, \qquad \nabla \times \nabla \times \mathbf{e}_\vartheta = \frac{\cot \vartheta}{r^2} \, \mathbf{e}_r, \qquad \nabla \times \nabla \times \mathbf{e}_\varphi = \frac{\mathbf{e}_\varphi}{r^2 \sin^2 \vartheta}. \qquad (A.4.47)$$

A.5. *Angular momentum in spherical coordinates*: Show that

$$\hat{\mathbf{L}} = -i\mathbf{x} \times \nabla = i \frac{1}{\sin \vartheta} \mathbf{e}_\vartheta \frac{\partial}{\partial \varphi} - i\mathbf{e}_\varphi \frac{\partial}{\partial \vartheta}$$

and calculate the scalar product $\hat{\mathbf{L}}^2$ using $\hat{\mathbf{L}}$.

A.6. *Differential operators applied to* $\mathbf{a} = f(r)\mathbf{p}$: In most cases, the constant vector $\mathbf{p}$ will point in the z-direction, which is not required here. Verify the following relations, where $f'(r) = \mathrm{d}f/\mathrm{d}r$ and $f'' = \mathrm{d}^2 f/\mathrm{d}r^2$:

$$\nabla \cdot \mathbf{a} = f' \, \mathbf{e}_r \cdot \mathbf{p}, \qquad\qquad \nabla \nabla \cdot \mathbf{a} = \left(f'' - \frac{f'}{r}\right)(\mathbf{e}_r \cdot \mathbf{p}) \, \mathbf{e}_r + \frac{f'}{r} \, \mathbf{p},$$

$$\nabla \times \mathbf{a} = f' \, \mathbf{e}_r \times \mathbf{p}, \qquad\qquad \nabla \times \nabla \times \mathbf{a} = \left(f'' - \frac{f'}{r}\right)(\mathbf{e}_r \cdot \mathbf{p}) \, \mathbf{e}_r - \left(f'' + \frac{f'}{r}\right) \mathbf{p},$$

$$\Delta \mathbf{p} = \left(f'' + \frac{2f'}{r}\right) \mathbf{p}.$$

A.7. *Elliptical coordinates*: Given are the elliptical coordinates of the elongated rotation ellipsoid in the form

$$x = l\sqrt{(\xi^2 - 1)(1 - \eta^2)} \, \cos \varphi, \qquad\qquad 1 \leq \xi \leq \infty,$$

$$y = l\sqrt{(\xi^2 - 1)(1 - \eta^2)} \, \sin \varphi, \qquad\qquad -1 \leq \eta \leq 1,$$

$$z = l\xi\eta, \qquad\qquad 0 \leq \varphi < 2\pi \, .$$

1. Calculate the metric tensors (g^{ij}) and (g_{ij}) and $\sqrt{g}$ with $g = \det(g_{ij})$.
   Hint: Use the sizes $\xi^1 = \eta, \xi^2 = \xi, \xi^3 = \varphi$. With this, you can calculate $\mathbf{h}_i = \partial \mathbf{x}/\partial \xi^i$; then $g_{ij} = \mathbf{h}_i \cdot \mathbf{h}_j$ and g^{ij}.
2. Calculate the Laplace operator in these coordinates

A.8. *On the Levi-Civita symbol*:

1. Calculate, starting from $\epsilon_{\mu\nu\rho\sigma}$, using the metric tensor g the contravariant tensor $\epsilon^{\mu\nu\rho\sigma}$.
2. Calculate

$$\epsilon^{\alpha\beta\gamma\delta} \, \epsilon_{\alpha\beta\rho\sigma}, \qquad\qquad \epsilon^{\alpha\beta\gamma\delta} \, \epsilon_{\alpha\beta\gamma\sigma}, \qquad\qquad \epsilon^{\alpha\beta\gamma\delta} \, \epsilon_{\alpha\beta\gamma\delta} \, .$$

A.9. *On the uniqueness of the decomposition of a vector field*: Show using the 2nd Green's theorem that the decomposition of a vector field $\mathbf{v}$, which diverges asymptotically sublinearly, into a vortex-free field $\mathbf{v}_l$ and a source-free field $\mathbf{v}_t$ is unique up to a vectorial constant.

Hint: Insert into (A.4.20) $\phi(\mathbf{x}') = \phi_d(\mathbf{x}')$ with $\Delta\phi_d = 0$ and $\psi(\mathbf{x}') = G_2(\mathbf{x}, \mathbf{x}')$.

A.10. *Potential and field of a surface charge*: Given is the charge density of the half-plane

$$\rho(\mathbf{x}) = \sigma \delta(z)\,\theta(x)$$

The potential and electric field of the configuration are to be calculated. Also show that $\nabla \cdot \mathbf{E} = 4\pi k_C \rho$. Auxiliary integral:

$$\int dx\, \ln\left(b + \sqrt{a^2 + x^2}\right) = x\ln[b + \sqrt{a^2 + x^2}] + b\ln[x + \sqrt{a^2 + x^2}] - x$$
$$+ \sqrt{a^2 - b^2}\left\{ \arctan[\frac{x}{\sqrt{a^2 - b^2}}] - \arctan\left[\frac{bx}{\sqrt{a^2 - b^2}\sqrt{a^2 + x^2}}\right] \right\}.$$

The potential remains up to a linear function and the field up to a constant undetermined.

References

Arfken G. & Weber H. *Mathematical Methods for Physicists*, 6th ed. Elsevier (2005)

Blumenthal O. *Ueber die Zerlegung unendlicher Vektorfelder*, Math. Ann. **61**, 235–250 (1905)

Fließbach T. *Elektrodynamik*, 5th ed. Springer Spektrum (2008)

Föppl A. *Einführung in die Maxwellsche Theorie der Elektrizität*, Teubner Leipzig (1894)

Griffiths D. *Introduction to Electrodynamics*, 4th ed., Cambridge University Press (2017)

Gregory R. D. *Helmholtz's theorem when the domain is infinite and when the field has singular points*, Q Jl Mech. appl. Math. **49**, 439–450, 1996

Großmann S. *Mathematischer Einführungskurs für die Physik*, 10th ed. Springer-Vieweg (2012)

Helmholtz H. *Über die Integrale der Hydrodynamischen Gleichungen, welche den Wirbelbewegungen entsprechen*, Journal für die reine und angewandte Mathematik, **55**, 25–55 (1858)

Petrascheck D. and Folk R. *The Helmholtz Decomposition of decreasing and increasing vector fields* arXiv:1506.00235 [physics.class-ph] (2015)

Petrascheck D. *The Helmholtz decomposition revisited*, Eur. J. Phys. **37**, 015201 (2016)

Sommerfeld A. *Mechanics of deformable bodies*, Academic Press (1950)

Stokes G. *On the dynamical theory of diffraction*, Trans. Cambridge Phil. Soc., **9**,1, Compl. Works vol. II, see p. 10, item 8; presented 1849 and published 1856

van der Waerden B. L. *Algebra, Volume I*, Springer New York Inc. (2003)

B

Mathematical Tools

B.1 Elements of the Theory of Functions

In electrodynamics, especially in electrostatics (potential theory), is in some areas the complex analysis very helpful. For this purpose, the conformal mapping, the mean value properties from electrostatics, the Kramers-Kronig dispersion relations of the dielectric function and the calculation of integrals with the residue theorem in dynamic diffraction are mentioned.

B.1.1 Analytic Functions

We decompose a complex function $f(z)$ into real and imaginary part:

$$f(z) = \phi(x,y) + i\psi(x,y) \qquad \text{with} \qquad z = x + iy \qquad x, y \in \mathbb{R}. \qquad \text{(B.1.1)}$$

Complex derivative

A unique function $f(z)$ in the area $A \in \mathbb{C}$ is differentiable at the point z if

$$f'(z) \equiv \frac{\mathrm{d}f(z)}{\mathrm{d}z} = \lim_{h \to 0} \frac{f(z+h) - f(z)}{h} \qquad z \in A \qquad \text{(B.1.2)}$$

is unique and independent of the direction.

Cauchy-Riemann differential equations

From the direction independence of the derivative it follows that the real and imaginary part are not independent:

$$\frac{\partial f}{\partial x} = \frac{\partial f}{\partial (iy)} \qquad \Rightarrow \qquad \frac{\partial \phi}{\partial x} = \frac{\partial \psi}{\partial y} \qquad \frac{\partial \phi}{\partial y} = -\frac{\partial \psi}{\partial x}. \qquad \text{(B.1.3)}$$

D. Petrascheck, F. Schwabl, *Electrodynamics*, https://doi.org/10.1007/978-3-662-71502-4_16

These are the Cauchy-Riemann differential equations. From these we obtain the Laplace equations

$$\frac{\partial^2 \phi}{\partial x^2} + \frac{\partial^2 \phi}{\partial y^2} = 0 \qquad \text{and} \qquad \frac{\partial^2 \psi}{\partial x^2} + \frac{\partial^2 \psi}{\partial y^2} = 0, \tag{B.1.4}$$

whose solutions are harmonic functions.

Definition of analytic functions

A function $f(z)$ is analytic (holomorphic) in an open area $A \subseteq \mathbb{C}$, if one of the following properties is fulfilled at every point $z \in A$.

- $f(z)$ is unique and differentiable.
- Real and imaginary parts fulfill the Cauchy-Riemann differential equations.
- $f(z)$ is arbitrarily often differentiable.
- $f(z)$ can be represented by a power series (Taylor series).
- $\oint_C \mathrm{d}z\, f(z) = 0 \qquad \forall C \in A$ (*Morera's theorem*).

B.1.2 Properties of Analytic Functions

Basic integral of function theory

$$J_k = \oint \mathrm{d}z\, (z - z_0)^k = \mathrm{i}\varrho^{k+1} \int_0^{2\pi} \mathrm{d}\varphi\, \mathrm{e}^{\mathrm{i}(k+1)\varphi} = 2\pi\mathrm{i}\delta_{k,-1} \quad k \text{ integer.} \tag{B.1.5}$$

Cauchy's integral theorem

If $f(z)$ is analytic in a simply connected region A, then

$$\oint_C \mathrm{d}z\, f(z) = 0 \tag{B.1.6}$$

along any closed curve C entirely within A. If there is a singular point in A, the Cauchy's tntegral theorem can be applied, if, as sketched in Fig. B.1, no singularity is enclosed by the integration path. The method can be extended

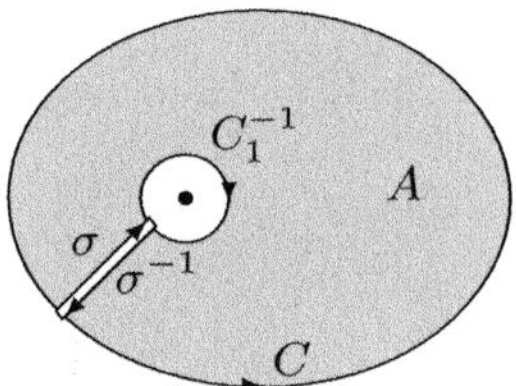

Fig. B.1. The region becomes doubly connected by a singularity. It is bypassed as sketched, so that the entire path is within A

to multiple isolated singularities

$$\oint_C \mathrm{d}z\, f(z) + \oint_{C_1^{-1}} \mathrm{d}z\, f(z) = 0 \quad \Rightarrow \quad \oint_C \mathrm{d}z\, f(z) = \sum_k \oint_{C_k} \mathrm{d}z\, f(z)\,. \quad \text{(B.1.7)}$$

Cauchy's integral formula

The function $f(z)$ is analytic in $z \in A$. Thus, $f(z)/(z - z_0)$ has a pole at the point z_0. According to (B.1.7) and the basic integral (B.1.5) is

$$f(z_0) = \frac{1}{2\pi \mathrm{i}} \oint_{C_0} \mathrm{d}z\, \frac{f(z)}{z - z_0}\,. \qquad\qquad \text{(B.1.8)}$$

One can verify the formula by shrinking the path C_0 to an infinitesimal circle around z_0.

Higher derivatives of analytic functions

$f(z)$ is not only simply differentiable, but all higher order derivatives exist, as shown by (B.1.8):

$$\frac{\mathrm{d}^n f(z)}{\mathrm{d}z^n} = \frac{1}{2\pi \mathrm{i}} \oint_C \mathrm{d}\zeta\, \frac{\mathrm{d}^n}{\mathrm{d}z^n} \frac{f(\zeta)}{(\zeta - z)^n} = \frac{n!}{2\pi \mathrm{i}} \oint_C \mathrm{d}\zeta\, \frac{f(\zeta)}{(\zeta - z)^n}\,. \qquad \text{(B.1.9)}$$

Residue theorem

The C_k in (B.1.7) are infinitesimal circles around the pole at z_k. If the pole is simple, then $g(z) = f(z)(z - z_k)$ is an analytic function around z_k and according to the Cauchy integral formula (B.1.8) is

$$R_k = \frac{1}{2\pi \mathrm{i}} \oint_{C_k} \mathrm{d}z\, \frac{g(z)}{z - z_k} = g(z_k) = \lim_{z \to z_k} f(z_k)(z - z_k)\,.$$

If the pole is of order $n > 1$, then the analytic function $g(z) = f(z)(z - z_k)^n$ is expanded into a Taylor series

$$R_k = \frac{1}{2\pi \mathrm{i}} \oint_{C_k} \mathrm{d}z\, f(z) = (z - z_k)^n \sum_{l=1}^{\infty} (z - z_k)^l \frac{1}{l!} \frac{\mathrm{d}^l g(z_k)}{\mathrm{d}z^l}$$

$$\overset{\text{(B.1.5)}}{=} \frac{1}{(n - 1)!} \frac{\mathrm{d}^{n-1} g(z_k)}{\mathrm{d}z^{n-1}}$$

and thus obtains the *residue theorem*

$$\oint_C \mathrm{d}z\, f(z) = \frac{1}{2\pi \mathrm{i}} \sum_k R_k(z_k)\,,$$

$$R_k(z_k) = \frac{1}{(n-1)!} \frac{\mathrm{d}^{n-1}}{\mathrm{d}z^{n-1}} \left[f(z)(z - z_k)^n \right]\Big|_{z = z_k}\,. \qquad \text{(B.1.10)}$$

The poles must not be branching points, since then one orbit would lead to another Riemann sheet.

Cauchy principal value

$f(x)$ has a singularity at the point $x = x_0$ on the real axis as sketched in Fig. B.2a, and the integral extends over a range with $a < x_0 < b$. The principal value is defined by

$$P \int_a^b \mathrm{d}x\, f(x) = \lim_{\epsilon \to 0} \int_a^{x_0-\epsilon} \mathrm{d}x\, f(x) + \int_{x_0+\epsilon}^b \mathrm{d}x\, f(x). \qquad (\mathrm{B.1.11})$$

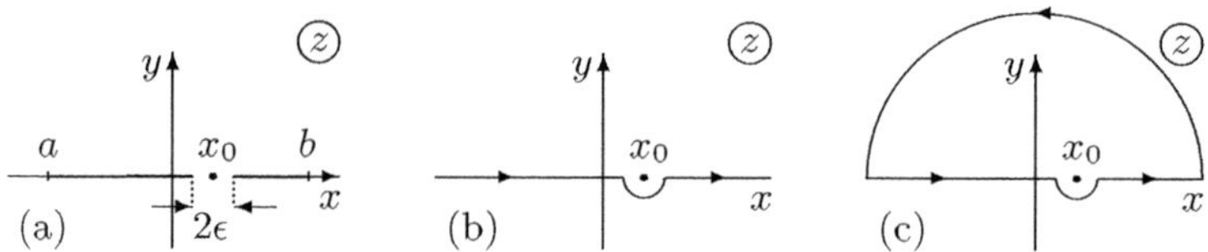

Fig. B.2. (a) Integration path for the principal value integral $a < x_0 < b$
(b) Integration path for the Plemelj relation, if the pole is infinitesimally in the upper half-plane
(c) For the calculation of the Kramers-Kronig relations, the integration path is closed over the upper semicircle, which does not contribute to the integral

Plemelj relation

If there is a simple pole x_0 on the real axis, one has to take the principal value $P\frac{1}{x - x_0}$ in an integration and to bypass the pole in an infinitesimal semicircle below, as shown in Fig. B.2b, or above. One obtains

$$\lim_{\epsilon \to 0} \frac{1}{x - (x_0 \pm \mathrm{i}\epsilon)} = P\Big(\frac{1}{x - x_0}\Big) \pm \mathrm{i}\pi\delta(x - x_0)\,, \qquad (\mathrm{B.1.12})$$

where the second contribution comes from the semicircle. This is the so-called *Plemelj* or *Plemelj-Sokhotsky relation*.

Dispersion relations

Let $f(z)$ be a function analytic in the upper half-plane with $\lim\limits_{|z|\to\infty} f(z) = 0$, which has a simple pole $z = x_0 + \mathrm{i}\epsilon$. We choose the integration path sketched in Fig. B.2(c). According to the Cauchy's integral formula (B.1.8) is

$$f(z) = \frac{1}{2\pi\mathrm{i}} \left\{ \int_{-\infty}^{\infty} \mathrm{d}x'\, \frac{f(x')}{x' - (x_0 + \mathrm{i}\epsilon)} + \oint \mathrm{d}z'\, \frac{f(z')}{z' - (x_0 + \mathrm{i}\epsilon)} \right\}.$$

The second term, the integral over the infinite semicircle, vanishes, and in the first integral we insert the Plemelj formula (B.1.12). This results in

$$\mathrm{Re}\, f(x_0) = \frac{1}{\pi} P \int_{-\infty}^{\infty} \mathrm{d}x\, \frac{\mathrm{Im}\, f(x)}{x - x_0} \qquad \mathrm{Im}\, f(x_0) = \frac{-1}{\pi} P \int_{-\infty}^{\infty} \mathrm{d}x\, \frac{\mathrm{Re}\, f(x)}{x - x_0}. \qquad \text{(B.1.13)}$$

The real and imaginary parts of an analytical function are not independent. These relationships are known in physics as Kramers-Kronig relations or dispersion relations.

Morera's theorem

Morera's theorem is the reversal of Cauchy's integral theorem and states that $f(z)$ in A is an analytical function, if

$$\oint_C \mathrm{d}z\, f(z) = 0$$

for every closed path C in a simply connected area A.

The mean value theorem

If $f(z)$ is analytic and unique in A and C is a circle entirely in A around the point z, then $f(z)$ is equal to the arithmetic mean of the function values on the circle.

The principle of maximum and minimum

The absolute value of a unique analytical function in a closed area A $f(z)$ reaches its greatest and, if $f(z)$ is zero-free in A, also its smallest value on the edge of A, if $f(z)$ in A is not constant.

Liouville's Theorem

If $f(z)$ is analytic, unique and bounded in the entire z-plane, then $f(z)$ is constant.

B.1.3 The Conformal Mapping

Definition: An analytical function $f(z)$ provides at each point, where $f'(z) \neq 0$ a conformal, i.e. angle and length preserving mapping.

Proof: $z(t)$ is the parametric representation of a curve in the complex z-plane by the real parameter t. Given is $w(t) = f(z(t))$.

1. Angle preservation:

$$\dot{w}(t) = \frac{\mathrm{d}w}{\mathrm{d}t} = \frac{\mathrm{d}f}{\mathrm{d}z}\frac{\mathrm{d}z}{\mathrm{d}t} = f'(z)\,\dot{z}(t)\,.$$

For the argument then applies

$$\arg \dot{w} = \arg(f'(z)) + \arg(\dot{z})\,.$$

Two curves, $z(t)$ and $z(s)$, intersecting at a point z_0 are rotated in the w-plane each by the same angle $\arg(f'(z_0))$.

2. Length preservation: Consider the arc length $s(t)$ of a mapping at the location z_0 given by t_0:

$$\left(\frac{\mathrm{d}s}{\mathrm{d}t}\right)^2 = |z(t_0)|^2 \qquad\Rightarrow\qquad \frac{\mathrm{d}s}{\mathrm{d}t} = |\dot z(t_0)|\,.$$

For the ground length $S(t)$ of the mapping $w(t) = f(z(t))$ applies analogously

$$\left(\frac{\mathrm{d}S}{\mathrm{d}t}\right) = |\dot w(t_0)| = |f'(z_0)|\,|\dot z(t_0)| \qquad\overset{\text{Stretch ratio}}{\Longrightarrow}\qquad \frac{\mathrm{d}S}{\mathrm{d}s} = |f'(z_0)|\,.$$

The *stretch ratio* depends only on the point z_0, but not on the curve $z(t)$. $\frac{\mathrm{d}S}{\mathrm{d}s}$ is also called *linear mapping modulus* or *scale* at the location z_0.

Elementary transformations

- *Translation:* $w = z + b \quad b \in \mathbb{C}$.
  Stretch factor: $|f'(z)| = 1$,   rotation angle: $\arg f'(z) = 0$.

- *Rotation:* $w = z\,\mathrm{e}^{\mathrm{i}\alpha} \quad \alpha \in \mathbb{R}$.
  Stretch factor: $s|f'(z)| = 1$,   rotation angle: $\arg f'(z) = \alpha$.

- *Rotation and scaling:* $w = a\,z \quad a \in \mathbb{C}$.
  Scaling factor: $|f'(z)| = |a|$.   rotation angle: $\arg f'(z) = \arg a$.

- *Inversion:* $w = \dfrac{1}{z}$.
  Scaling factor: $|f'(z)| = |f(z)|^2$,   rotation angle: $\arg f'(z) = \pi + 2\arg f(z)$.

More general transformations, such as the fractional linear (Möbius-) transformation

$$w = \frac{az+b}{cz+d}, \qquad\qquad ad - bc \neq 0\,, \qquad\qquad a, b, c, d \in \mathbb{C}$$

are composed of these elementary transformations. Thus

$$w = \frac{z + a}{z - a}\,, \quad a \in \mathbb{R}$$

maps the imaginary axis onto a unit circle, as can be seen from $w\,w* = 1$.

B.2 Legendre Polynomials

The Legendre polynomials are unique and regular solutions of the Legendre differential equation (3.2.18)

$$\left((1 - \xi^2)\frac{\mathrm{d}^2}{\mathrm{d}\xi^2} - 2\xi\frac{\mathrm{d}}{\mathrm{d}\xi} + l(l+1)\right)P_l(\xi) = 0 \tag{B.2.1}$$

in the base domain $[-1, 1]$. Some of their properties, concerning recursion relations, orthogonality, completeness etc. are the subject of this section.

B.2.1 Rodrigues Formula

The direct path of verifying the Rodrigues formula (B.2.2) through the P_l from (3.2.20) is not chosen here, as we are interested in an independent alternative path to determine the P_l.

First, it is shown that the polynomials defined with the Rodrigues formula

$$P_l(\xi) = \frac{1}{2^l\, l!} \frac{\mathrm{d}^l}{\mathrm{d}\xi^l} (\xi^2 - 1)^l \tag{B.2.2}$$

fulfill the Legendre differential equation (B.2.1). Then, with (B.2.2) P_l is explicitly calculated, leading to (3.2.20) results.

Proof of the Rodrigues formula

For abbreviation we introduce:

$$c_l = \frac{1}{2^l\, l!}, \qquad \text{from which follows} \quad P_l = c_l \frac{\mathrm{d}^l}{\mathrm{d}\xi^l}(\xi^2 - 1)^l.$$

The product rule is also strained

$$\frac{\mathrm{d}^l}{\mathrm{d}\xi^l}(f\, g) = \sum_{k=0}^{l} \binom{l}{k} f^{(k)}\, g^{(l-k)},$$

where $f^{(k)}$ means the k-th derivative of f. So it applies

$$
\begin{aligned}
P_{l+1} &= c_{l+1} \frac{\mathrm{d}^{l+1}}{\mathrm{d}\xi^{l+1}}(\xi^2-1)(\xi^2-1)^l \\
&= c_{l+1}\Big[(\xi^2-1)\frac{\mathrm{d}^{l+1}}{\mathrm{d}\xi^{l+1}}(\xi^2-1)^l + (l+1)2\xi\frac{\mathrm{d}^l}{\mathrm{d}\xi^l}(\xi^2-1)^l \\
&\quad + (l+1)l\frac{\mathrm{d}^{l-1}}{\mathrm{d}\xi^{l-1}}(\xi^2-1)^l\Big] \\
&= \frac{1}{2(l+1)}(\xi^2-1)\frac{\mathrm{d}}{\mathrm{d}\xi}P_l + \xi\, P_l + c_l\frac{l}{2}\frac{\mathrm{d}^{l-1}}{\mathrm{d}\xi^{l-1}}(\xi^2-1)^l\Big].
\end{aligned}
\tag{B.2.3}
$$

The last term is not yet simply representable by the P_l, but since we need an expression for $(1 - \xi^2)P_l''$, we differentiate (B.2.3):

$$P_{l+1}' = \frac{1}{2(l+1)}\Big[2\xi\, P_l' + (\xi^2-1)P_l''\Big] + P_l + \xi\, P_l' + \frac{l}{2}\, P_l. \tag{B.2.4}$$

From the Rodrigues formula (B.2.2) we derive by differentiation

$$P_{l+1}' = c_{l+1}\frac{\mathrm{d}^{l+1}}{\mathrm{d}\xi^{l+1}}\Big[2\xi(l+1)(\xi^2-1)^l\Big]$$

and subsequent use of the product rule

$$
\begin{aligned}
P'_{l+1} &= c_{l+1}\, 2(l+1)\left[\xi\,\frac{\mathrm{d}^{l+1}}{\mathrm{d}\xi^{l+1}}(\xi^2-1)^l + (l+1)\frac{\mathrm{d}^l}{\mathrm{d}\xi^l}(\xi^2-1)^l\right]\\
&= \xi\,P'_l + (l+1)P_l
\end{aligned}
\tag{B.2.5}
$$

we derive an expression that we insert into (B.2.4) and solve for $(1-\xi^2)P''_l$:

$$
\begin{aligned}
(1-\xi^2)P''_l &= 2\xi\,P'_l + 2(l+1)\left[P_l + \xi P'_l + \frac{l}{2}P_l - P'_{l+1}\right]\\
&= 2\xi\,P'_l - 2(l+1)\left[P_l + \frac{l}{2}P_l - (l+1)P_l\right] = 2\xi\,P'_l + l(l+1)P_l\,.
\end{aligned}
$$

Thus, the polynomials defined by the Rodrigues formula (B.2.2) satisfy the Legendre differential equation (B.2.1), which was to be shown.

Determination of the P_l from (B.2.2) using the binomial theorem

$$
\begin{aligned}
P_l(\xi) &= \frac{1}{2^l\,l!}\,\frac{\mathrm{d}^l}{\mathrm{d}\xi^l}\sum_{m=0}^{l}\binom{l}{m}(-1)^{l-m}\,\xi^{2m}\\[2mm]
&= \frac{1}{2^l\,l!}\sum_{m=\lfloor\frac{l}{2}\rfloor}^{l}(-1)^{l-m}\frac{l!}{m!\,(l-m)!}\,\underbrace{2m(2m-1)\dots(2m-l+1)}_{(2m)!/(2m-l)!}\,\xi^{2m-l}\\[2mm]
&= \sum_{n}\frac{(-1)^{\frac{1}{2}(l-n)}}{2^l\left(\frac{l+n}{2}\right)!\left(\frac{l-n}{2}\right)!}\,\frac{(n+l)!}{n!}\,\xi^n
\end{aligned}
$$

We have substituted: $n = 2m - l, \quad m = (n+l)/2$.

B.2.2 The Generating Function of the Legendre Polynomials

The generating function of the P_l is

$$
(1-2\xi t + t^2)^{-\frac{1}{2}} = \sum_{l=0}^{\infty}P_l(\xi)\,t^l \qquad\text{for}\qquad |t| < 1\,.
\tag{B.2.6}
$$

If we differentiate (B.2.6) l times and then set $t = 0$, we get

$$
\frac{\partial^l}{\partial t^l}(1-2\xi t + t^2)^{-\frac{1}{2}}\bigg|_{t=0} = l!\,P_l(\xi)\,.
\tag{B.2.7}
$$

If we can prove the validity of this relation, it is also shown that (B.2.6) is the generating function of the P_l. For the proof, we need the Schläfli integral representation of the P_l.

Schläfli integral representation

The Schläfli integral representation is a very special representation of the P_l, which is only treated in detail because it leads to the generating function of the P_l. The starting point is the Cauchy integral formula

$$f(z) = \frac{1}{2\pi i} \oint dt\, \frac{f(t)}{t-z}\,.$$

$f(z)$ is an analytic function and may not have any poles in the area enclosed by the path around z. The n-fold differentiation results in

$$\frac{d^n}{dz^n} f(z) = \frac{1}{2\pi i} \oint dt\, \frac{f(t)\, n!}{(t-z)^{n+1}}\,. \tag{B.2.8}$$

If we substitute the Rodrigues formula (B.2.2) for $f(z)$, we get with

$$P_l(z) = \frac{1}{2^l\, l!} \frac{d^l}{dz^l} (z^2 - 1)^l = \frac{1}{2^l} \frac{1}{2\pi i} \oint dt\, \frac{(t^2-1)^l}{(t-z)^{l+1}} \tag{B.2.9}$$

the Schläfli representation of the $P_l(z)$. Now we return to the proof of the generating function (B.2.6) back and insert (B.2.7)

$$f(z) = \frac{1}{l!} \left. \frac{1}{\sqrt{1 - 2\xi z + z^2}} \right|_{z=0}$$

into (B.2.8):

$$P_l(\xi) = \frac{1}{l!} \frac{\partial^l}{\partial z^l} (1 - 2\xi z + z^2)^{-\frac{1}{2}} \Big|_{z=0} = \frac{1}{2\pi i} \oint_{z=0} dt\, \frac{1}{t^{l+1}} \frac{1}{\sqrt{1 - 2\xi t + t^2}}\,. \tag{B.2.10}$$

The transformation to the variable y results in:

$$\sqrt{1 - 2\xi t + t^2} = 1 - ty\,,$$

$$1 - 2\xi t + t^2 = 1 - 2ty + t^2 y^2 \qquad \Rightarrow \qquad t = 2\frac{y - \xi}{y^2 - 1}\,.$$

Therefore,

$$\frac{dt}{dy} = 2\frac{y^2 - 1 - 2y(y - \xi)}{(y^2 - 1)^2} = -2\frac{y^2 - 2y\xi + 1}{(y^2 - 1)^2}\,,$$

$$\sqrt{1 - 2\xi t + t^2} = \frac{y^2 - 1 - 2y^2 + 2y\xi}{y^2 - 1} = -\frac{y^2 - 2y\xi + 1}{y^2 - 1}\,.$$

Substituted into (B.2.10) this is the Schläfli representation (B.2.9)

$$P_l(\xi) = \frac{1}{2\pi i} \oint_\xi dy\, 2^{-l}\, \frac{(y^2 - 1)^l}{(y - \xi)^{l+1}}\,. \tag{B.2.11}$$

This proves that the representation of the generating function (B.2.6) is valid.

B.2.3 Properties of the Legendre Polynomials

Symmetry

The P_l are even or odd functions, as already shown in the representation (3.2.20) emerges:

$$P_l(-\xi) = (-1)^l P_l(\xi)\,. \tag{B.2.12}$$

Special cases are

$$P_0 = 1\,, \qquad P_1 = \xi\,, \qquad P_2 = \frac{1}{2}(3\xi^2 - 1)\,, \qquad P_3 = \frac{1}{2}(5\xi^3 - 3\xi)\,.$$

Orthogonality

The P_l form a complete and orthogonal system

$$\int_{-1}^{1} \mathrm{d}\xi\, P_l(\xi)\, P_{l'}(\xi) = \frac{2}{2l+1}\, \delta_{ll'}\,. \tag{B.2.13}$$

To prove the orthogonality relation, we start from (see (3.2.17) or (B.2.19))

$$\frac{\mathrm{d}}{\mathrm{d}\xi}(1 - \xi^2)\frac{\mathrm{d}}{\mathrm{d}\xi} P_l(\xi) = -l(l+1)P_l(\xi).$$

It follows

$$-l(l+1)\int_{-1}^{1} \mathrm{d}\xi\, P_{l'}(\xi)\, P_l(\xi) = \int_{-1}^{1} \mathrm{d}\xi\, P_{l'}(\xi) \frac{\mathrm{d}}{\mathrm{d}\xi}(1 - \xi^2)\frac{\mathrm{d}P_l}{\mathrm{d}\xi}\,. \tag{B.2.14}$$

A partial integration of the right side leads to

$$-\int_{-1}^{1} \mathrm{d}\xi\, \frac{\mathrm{d}P_{l'}}{\mathrm{d}\xi}(1 - \xi^2)\frac{\mathrm{d}P_l}{\mathrm{d}\xi}\,,$$

since the boundary term vanishes due to $(1-\xi^2)$. A further partial integration leads to the original integral, where l appears swapped with l'

$$\int_{-1}^{1} \mathrm{d}\xi\, P_l(\xi) \frac{\mathrm{d}}{\mathrm{d}\xi}(1 - \xi^2)\frac{\mathrm{d}P_{l'}}{\mathrm{d}\xi}\,.$$

Thus, (B.2.14) must satisfy the following equation:

$$-l(l+1)\int_{-1}^{1} \mathrm{d}\xi\, P_{l'}(\xi)\, P_l(\xi) = -l'(l'+1)\int_{-1}^{1} \mathrm{d}\xi\, P_{l'}(\xi) P_l(\xi).$$

Therefore, the integral vanishes for $l \neq l'$.

The normalization factor still needs to be determined. Starting from the Rodrigues formula (B.2.2) we calculate the normalization factor, where we initially integrate partially l times :

$$\int_{-1}^{1} d\xi \left[P_l(\xi)\right]^2 = \frac{1}{(2^l\, l!)^2} \int_{-1}^{1} d\xi \left(\frac{d^l}{d\xi^l}(\xi^2-1)^l\right)\left(\frac{d^l}{d\xi^l}(\xi^2-1)^l\right)$$

$$= \frac{(-1)^l}{2^{2l}\,(l!)^2} \int_{-1}^{1} d\xi (\xi^2-1)^l \frac{d^{2l}}{d\xi^{2l}}(\xi^2-1)^l$$

$$= \frac{(2l)!}{2^{2l}(l!)^2} \int_{-1}^{1} d\xi (1-\xi^2)^l\,.$$

Here in the last line we have substituted for $\frac{d^{2l}}{d\xi^{2l}}(\xi^2-1)^l = (2l)!$. The integral remains to be evaluated [Gradshteyn, Rhyzhik, 1965, §3.621]

$$\int_{-1}^{1} d\xi (1-\xi^2)^l = \int_{0}^{\pi} d\vartheta \, \sin^{2l+1}\vartheta = \frac{2^{2l+1}\,(l!)^2}{(2l+1)!}\,.$$

This results in the orthogonality relation given in (B.2.13).

Completeness

$$\sum_{l=0}^{\infty} \frac{2l+1}{2}\, P_l(\xi)\, P_l(\xi') = \delta(\xi-\xi'). \tag{B.2.15}$$

The completeness of the P_l can be verified most easily by expanding a function from the range $[-1,1]$ in terms of Legendre polynomials. If one multiplies (B.2.15) from the left with

$$\int_{-1}^{1} d\xi'\, f(\xi')\,,$$

one obtains

$$f(\xi) = \sum_{l=0}^{\infty} f_l\, P_l(\xi) \qquad \text{with} \qquad f_l = \frac{2l+1}{2} \int_{-1}^{1} d\xi'\, P_l(\xi')\, f(\xi')\,. \tag{B.2.16}$$

Now we multiply (B.2.16) with

$$\frac{2l'+1}{2} \int_{-1}^{1} d\xi\, P_{l'}(\xi)$$

and use the orthogonality on the right side:

$$f_{l'} = \frac{2l'+1}{2} \sum_{l=0}^{\infty} f_l \int_{-1}^{1} d\xi\, P_{l'}(\xi)\, P_l(\xi) = f_{l'}\,. \qquad\qquad \text{q.e.d.}$$

Recursion relations

For all orthogonal polynomials, therefore also for the Legendre polynomials, there are recursion formulas of the form

$$(l+1)P_{l+1}(\xi) - (2l+1)\xi P_l(\xi) + lP_{l-1}(\xi) = 0 \quad \text{with} \quad \begin{cases} P_{-1} = 0 \\ P_0 = 1 \,, \end{cases} \quad \text{(B.2.17)}$$

which is to be shown in the problems B.2 and B.3. Further recursion formulas also exist for the derivatives, whereby the first one was already derived in the proof of the Rodrigues formula (B.2.5):

$$\frac{\mathrm{d}P_{l+1}(\xi)}{\mathrm{d}\xi} - \xi\frac{\mathrm{d}P_l(\xi)}{\mathrm{d}\xi} = (l+1)P_l(\xi),$$

$$(1-\xi^2)\frac{\mathrm{d}P_l(\xi)}{\mathrm{d}\xi} = -l\big[\xi P_l(\xi) - P_{l-1}(\xi)\big]. \qquad \text{(B.2.18)}$$

B.2.4 Associated Legendre Polynomials

The associated Legendre polynomials are solutions of the polar part of the Laplace equation (3.2.17):

$$\left[\frac{\mathrm{d}}{\mathrm{d}\xi}(1-\xi^2)\frac{\mathrm{d}}{\mathrm{d}\xi} - \frac{m^2}{1-\xi^2} + l(l+1)\right]P_l^m(\xi) = 0\,. \qquad \text{(B.2.19)}$$

The P_l^m were determined in section 3.2.4, page 90 so that only some additions are made here.

Generating function

If you differentiate (B.2.6) m times and multiply by $\sqrt{1-\xi^2}\,t^{-m}/(2m-1)!!$, you get the generating function of the associated Legendre polynomials:

$$\frac{1}{\sqrt{1-2\xi t+t^2}}\left(\frac{\sqrt{1-\xi^2}}{1-2\xi t+t^2}\right)^m = \frac{1}{(2m-1)!!}\sum_{l=0}^{\infty} P_l^m(\xi)\,t^{l-m}\,. \qquad \text{(B.2.20)}$$

By integrating the product of two generating functions at different values (s and t) one could verify the orthogonality (3.2.30).

The associated polynomials for $m < 0$

With the identity

$$P_l^{-m}(\xi) = (-1)^m \frac{(l-m)!}{(l+m)!}\,P_l^m(\xi) \qquad \text{(B.2.21)}$$

is the orthogonality relation (3.2.30) also fulfilled for P_l^{-m}. We can show the identity by comparing the leading term of P_l^m and P_l^{-m} with (3.2.28), p. 93, [Schwabl, 2007, Appendix C]:

$$(1-\xi^2)^{m/2} P_l^m = \frac{1}{2^l\, l!}(1-\xi^2)^m \frac{\mathrm{d}^{l+m}}{\mathrm{d}\xi^{l+m}}(\xi^2-1)^l = \frac{(-1)^m}{2^l\, l!}\frac{(2l)!}{(l+m)!}\xi^{l+m}+\cdots,$$

$$(1-\xi^2)^{m/2} P_l^{-m} = \frac{1}{2^l\, l!}\frac{\mathrm{d}^{l-m}}{\mathrm{d}\xi^{l-m}}(\xi^2-1)^l = \frac{1}{2^l\, l!}\frac{(2l)!}{(l-m)!}\xi^{l+m}+\cdots.$$

By comparing the terms, the validity of (B.2.21) can be seen.

Recursion relations

The given relations come from the integral table Gradshteyn, Rhyzhik [1965, §8.733-2. and §8.733-4.]:

$$(l-m+1)P_{l+1}^m(\xi) + (l+m)P_{l-1}^m(\xi) = (2l+1)\xi P_l^m(\xi), \quad |m|\le l-1, \quad \text{(B.2.22)}$$

$$P_{l+1}^m(\xi) - P_{l-1}^m(\xi) = -(2l+1)\sqrt{1-\xi^2}\,P_l^{m-1}(\xi), \qquad |m|\le l-1. \quad \text{(B.2.23)}$$

B.3 Spherical Harmonics

The Y_{lm} are solutions of the angle-dependent part (3.2.9) of the Laplace equation

$$\left(\frac{1}{\sin\vartheta}\frac{\partial}{\partial\vartheta}\sin\vartheta\frac{\partial}{\partial\vartheta} + \frac{1}{\sin^2\vartheta}\frac{\partial^2}{\partial\varphi^2} + l(l+1)\right)Y_{lm}(\vartheta,\varphi) = 0 \qquad \text{(B.3.1)}$$

and as such are harmonic functions. They are given by (3.2.32)

$$Y_{lm}(\vartheta,\varphi) = \Theta_{lm}(\cos\vartheta)\,\Phi_m(\varphi) \qquad\qquad \text{(B.3.2)}$$

with $\Phi_m = \mathrm{e}^{\mathrm{i}m\varphi}/\sqrt{2\pi}$ and the normalized P_l^m

$$\Theta_{lm}(\cos\vartheta) = (-1)^m \sqrt{\frac{2l+1}{2}\frac{(l-m)!}{(l+m)!}}\,P_l^m(\cos\vartheta), \qquad \text{(B.3.3)}$$

where the normalization factor is given by the orthogonality relation (3.2.30):

$$Y_{lm}(\vartheta,\varphi) = \frac{(-1)^{m+l}}{2^l l!}\left[\frac{2l+1}{4\pi}\frac{(l-m)!}{(l+m)!}\right]^{\frac{1}{2}}\sin^m\vartheta\,\frac{\mathrm{d}^{l+m}\sin^{2l}\vartheta}{\mathrm{d}\cos\vartheta^{l+m}}\,\mathrm{e}^{\mathrm{i}m\varphi}. \quad \text{(B.3.4)}$$

Recursion Relations

We have already shown using the P_l that for all orthogonal polynomials recursion relations can be derived. Here we give two relations, the first of which directly follows from (B.2.22) when rewritten for the normalized Θ_{lm}. The second recursion formula ultimately follows from (B.2.23):

$$\cos\vartheta\, Y_{lm}(\vartheta,\varphi) = \frac{\sqrt{(l-m+1)(l+m+1)}}{\sqrt{(2l+3)(2l+1)}}\, Y_{l+1m}(\vartheta,\varphi)$$
$$+ \frac{\sqrt{(l-m)(l+m)}}{\sqrt{(2l+1)(2l-1)}}\, Y_{l-1m}(\vartheta,\varphi), \tag{B.3.5}$$

$$e^{\pm i\varphi}\sin\vartheta\, Y_{lm\mp1}(\vartheta,\varphi) = \mp\frac{\sqrt{(l\pm m)(l\pm m+1)}}{\sqrt{(2l+3)(2l+1)}}\, Y_{l+1m}(\vartheta,\varphi)$$
$$\pm \frac{\sqrt{(l\mp m+1)(l\mp m)}}{\sqrt{(2l+1)(2l-1)}}\, Y_{l-1m}(\vartheta,\varphi). \tag{B.3.6}$$

The relations also apply for $l = m$, since there the prefactor of Y_{l-1l} vanishes.

Addition theorem for spherical harmonics

In section 3.3.2 the addition theorem for spherical harmonics (3.3.4)

$$\sum_{m=-l}^{l} Y_{lm}(\vartheta,\varphi)\, Y_{lm}^{*}(\vartheta',\varphi') = \frac{2l+1}{4\pi} P_l(\cos\theta) \tag{B.3.7}$$

was used to expand the Green function with respect to spherical harmonics. Here, the theorem is to be proven.

The polynomial $P_l(\cos\theta)$ with $\cos\theta = \mathbf{e}_r\cdot\mathbf{e}_{r'}$ – see Fig. B.3 – is a solution of the angle-dependent part (3.2.9) of the Laplace equation

$$\left[\hat{\mathbf{L}}^2 - l(l+1)\right]Y(\Omega) = 0. \tag{B.3.8}$$

We now set $\mathbf{e}_{r'}$ to the z-axis $\mathbf{e}_z$. Then, according to Fig. B.3, the angle $\vartheta = \theta$, and $Y(\Omega) = P_l(\cos\vartheta)$ is a solution of the above equation.

Now a rotation D is performed so that $\mathbf{e}_{r'}$ comes into the position shown in Fig. B.3:

$$\mathsf{D}\left[\hat{\mathbf{L}}^2 - l(l+1)\right]Y(\Omega) = \left[\hat{\mathbf{L}}'^2 - l(l+1)\right]Y'(\Omega) = 0.$$

Due to the invariance of the scalar product under rotations $(\hat{\mathbf{L}}' = \mathsf{D}\hat{\mathbf{L}}\mathsf{D}^{-1})$ Y' belongs to the same l as Y:

$$\mathsf{D}Y(\Omega) = Y'(\Omega) = Y(\bar{\Omega}) \qquad \Rightarrow \qquad \mathsf{D}\, P_l(\cos\vartheta) = P_l(\cos\theta).$$

The Legendre polynomial

$$P_l(\cos\theta) = f(\Omega,\Omega') \qquad \text{with } \Omega = (\vartheta,\varphi) \quad \text{and } \Omega' = (\vartheta',\varphi')$$

can be considered as a function of Ω with the parameter Ω' (and vice versa), since

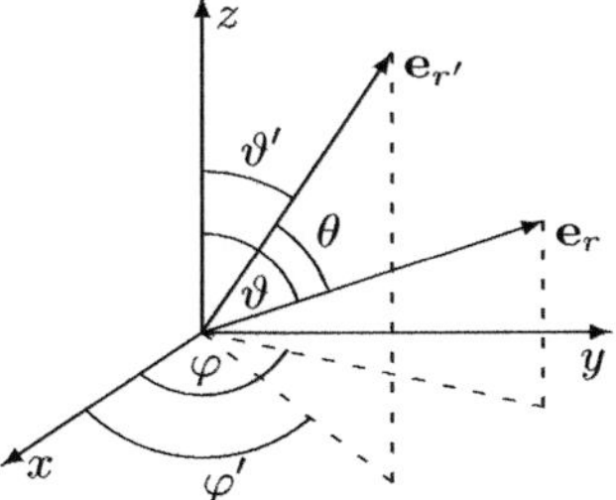

Fig. B.3. Position of the vectors $\mathbf{x}$ and $\mathbf{x}'$ in relation to the $\mathbf{e}_z$-axis: $\cos\theta = \mathbf{e}_r \cdot \mathbf{e}_{r'}$

$$\cos\theta = \cos\vartheta\,\cos\vartheta' + \sin\vartheta\,\sin\vartheta'\,\cos(\varphi-\varphi') \qquad \text{(spherical cosine theorem)}.$$

If $P_l(\cos\theta)$ is expanded as a function of Ω or Ω' according to spherical harmonics,

$$P_l(\cos\theta) = \sum_{m=-l}^{l} A_{lm}(\Omega')\,Y_{lm}(\Omega) = \sum_{m=-l}^{l} A_{lm}(\Omega)\,Y_{lm}(\Omega'), \qquad \text{(B.3.9)}$$

it suffices, since $P_l(\cos\theta)$ is also a solution of (B.3.8), only to take the sum over the subspace of m:

$$A_{lm}(\Omega') = \oiint \mathrm{d}\Omega''\, Y_{lm}^*(\Omega'')\, f(\Omega'',\Omega')\,.$$

The integration is over the surface of the unit sphere. The two developments according to Ω or Ω' are only compatible if A_{lm} is proportional to Y_{lm} or $Y_{l\,-m} = (-1)^m Y_{lm}^*$. Now the individual summands may only depend on the difference $\varphi - \varphi'$, which is why

$$A_{lm}(\Omega') = a_m Y_{lm}^*(\Omega') \quad \Rightarrow \quad P_l(\cos\theta) = \sum_{m=-l}^{l} a_m Y_{lm}^*(\Omega')\,Y_{lm}(\Omega). \qquad \text{(B.3.10)}$$

In the limit case $\theta = 0$ $\mathbf{e}_r$ and $\mathbf{e}_{r'}$ coincide and $\Omega = \Omega'$. The integration over Ω yields, since $P_l(1) = 1$,

$$4\pi = \sum_{m=-l}^{l} a_m \oiint \mathrm{d}\Omega\,|Y_{lm}(\Omega)|^2 \overset{(3.2.36)}{=} \sum_{m=-l}^{l} a_m\,. \qquad \text{(B.3.11)}$$

Now (B.3.10) is squared and integrated over $\mathrm{d}\Omega$. The z-axis of the integration variables Ω is aligned parallel to $\mathbf{e}_{r'}$. Let $\bar{\Omega} = (\theta, \phi)$, then

$$\oiint \mathrm{d}\Omega\,|P_l(\cos\theta)|^2 = \oiint \mathrm{d}\bar{\Omega}\,|P_l(\cos\theta)|^2 \overset{(\text{B.2.13})}{=} \frac{4\pi}{2l+1} = \sum_{m=-l}^{l} a_m^2\,|Y_{lm}(\Omega')|^2\,.$$

After integration over Ω', one obtains another condition:

$$4\pi \frac{4\pi}{2l+1} = \sum_{m=-l}^{l} a_m^2 \,. \tag{B.3.12}$$

(B.3.11) and (B.3.12) are sufficient to determine the a_m. For this, we define the two $2l+1$-dimensional vectors $\mathbf{a}$ with the components a_i and $\mathbf{b}$ with $b_i = 1$. The Cauchy-Schwarz inequality states

$$(\mathbf{a}\cdot\mathbf{a})(\mathbf{b}\cdot\mathbf{b}) \geq (\mathbf{a}\cdot\mathbf{b})^2 \qquad \Rightarrow \qquad (2l+1) \sum_{m=-l}^{l} a_m^2 \geq \left[\sum_{-l}^{l} a_m\right]^2.$$

The equality sign only applies when the two vectors are collinear, i.e. $\mathbf{a} = a_0 \mathbf{b}$. If you substitute for the sums in the right equation (B.3.11) and (B.3.12), both sides are equal, i.e., that $a_m = a_0 = 4\pi/(2l+1)$. Thus, the addition theorem (B.3.7) is proven.

B.4 Bessel Functions

B.4.1 Bessel's Differential Equation

The Bessel functions are solutions to Bessel's differential equation

$$\left(x^2 \frac{\mathrm{d}^2}{\mathrm{d}x^2} + x \frac{\mathrm{d}}{\mathrm{d}x} + x^2 - \nu^2\right) J_\nu(x) = 0 \,. \tag{B.4.1}$$

ν, the order of the Bessel function, is real here, usually even integer and $x = k\varrho \geq 0$; both quantities can be complex.

Solution approach

To get to solutions of (B.4.1), one considers the asymptotic behavior for $x \to 0$ and obtains

$$J_{\pm\nu}(x) \sim x^{\pm\nu} \,.$$

Subsequently, one makes a power series approach, whereby the asymptotic behavior is explicitly shown:

$$J_\nu(x) = x^\nu \sum_{j=0}^{\infty} a_j x^j \,.$$

For (B.4.1) to be fulfilled, the factor of each power of x must vanish. This results in a recursion formula for the coefficients a_j

$$x^{j+\nu}\left\{\left[(\nu+j)^2 - \nu^2\right]a_j + a_{j-2}\right\} = 0 \,.$$

For $j = 1$ it must hold that $a_1 = 0$, which causes all odd a_j to vanish. a_0 is given by $a_0 = 1/[2^\nu\, \Gamma(\nu+1)]$. For the recursion relation one gets then

$$a_{2j} = -\frac{1}{4j(j+\nu)}\, a_{2j-2} = \frac{(-1)^j\, \Gamma(\nu+1)}{2^{2j}\, j!\, \Gamma(j+\nu+1)}\, \frac{1}{2^\nu \Gamma(\nu+1)}\,.$$

Now all steps apply equally for $-\nu$, so that the two linearly independent solutions are

$$J_{\pm\nu}(x) = \left(\frac{x}{2}\right)^{\pm\nu} \sum_{j=0}^{\infty} \frac{(-1)^j}{j!\, \Gamma(j\pm\nu+1)} \left(\frac{x}{2}\right)^{2j}. \tag{B.4.2}$$

$\Gamma(x)$ is the gamma function, which satisfies the functional equation

$$\Gamma(x+1) = x\Gamma(x), \qquad \Gamma(1) = 1, \qquad \Gamma(\tfrac{1}{2}) = \sqrt{\pi}. \tag{B.4.3}$$

For integers $n > 0$ holds $\Gamma(n+1) = n!$.

In general, the Bessel functions $J_\nu(x)$ and the Neumann functions

$$N_\nu(x) = \frac{\cos(\nu\pi)J_\nu(x) - J_{-\nu}(x)}{\sin(\nu\pi)}, \qquad N_n(x) = \lim_{\nu\to n} N_\nu(x), \qquad n \text{ integer}$$

$$\tag{B.4.4}$$

are given as linearly independent solutions (for integer ν the limit has to be formed). The Bessel functions of the 2nd kind are also referred to as Weber functions $Y_\nu(x)$ – see Tab. B.1: $Y_\nu(x) = N_\nu(x)$.

Tab. B.1. Bessel or cylinder functions

$J_\nu(x)$		Bessel function of the 1st kind
$N_\nu(x)$	$= \dfrac{\cos(\nu\pi)J_\nu(x) - J_{-\nu}(x)}{\sin(\nu\pi)}$	Neumann function (Bessel function of the 2nd kind)
$Y_\nu(x)$	$= N_\nu(x)$	Weber function
$H_\nu^{(1)}(x)$	$= J_\nu(x) + iN_\nu(x)$	Hankel function of the 1st kind (Bessel function of the 3rd kind)
$H_\nu^{(2)}(x)$	$= J_\nu(x) - iN_\nu(x)$	Hankel function of the 2nd kind (Bessel function of the 3rd kind)
$I_\nu(x)$	$= (-i)^\nu J_\nu(ix)$	modified Bessel function of the 1st kind (hyperbolic Bessel function)
$K_\nu(x)$	$= \dfrac{\pi i}{2} H_\nu^{(1)}(ix)$	modified Bessel function of the 2nd kind (MacDonald or Basset function)

Modified Bessel's differential equation

If you replace x with $\mathrm{i}x$ in (B.4.1), you get

$$\left(x^2 \frac{\mathrm{d}^2}{\mathrm{d}x^2} + x\frac{\mathrm{d}}{\mathrm{d}x} - x^2 - \nu^2\right) I_\nu(x) = 0. \tag{B.4.5}$$

The solutions, the modified Bessel functions, are defined as

$$I_\nu(x) = \mathrm{i}^{-\nu} J_\nu(\mathrm{i}x), \qquad K_\nu = \frac{\pi}{2}\mathrm{i}^{\nu+1} \underbrace{\left[J_\nu(\mathrm{i}x) + \mathrm{i}N_\nu \mathrm{i}x)\right]}_{H_\nu^{(1)}(\mathrm{i}x)}. \tag{B.4.6}$$

$H_\nu^{(1)}(x)$ is the *Hankel function of the 1st kind.*

B.4.2 Properties of the Bessel Functions

The Bessel functions or modified Bessel functions of the 1st and 2nd kind are the functions $J_\nu(x)$ and $N_\nu(x)$ or $I_\nu(x)$ and $K_\nu(x)$. They are represented for the integer values $\nu = n = 0, 1, 2$ in Fig. B.4 and Fig. B.5.

For $x \to 0$, $J_n(x)$ and $I_n(x)$ are regular and for $x \to \infty$ J_n and N_n oscillate with the amplitude $\sim 1/\sqrt{x}$, while I_n diverges exponentially.

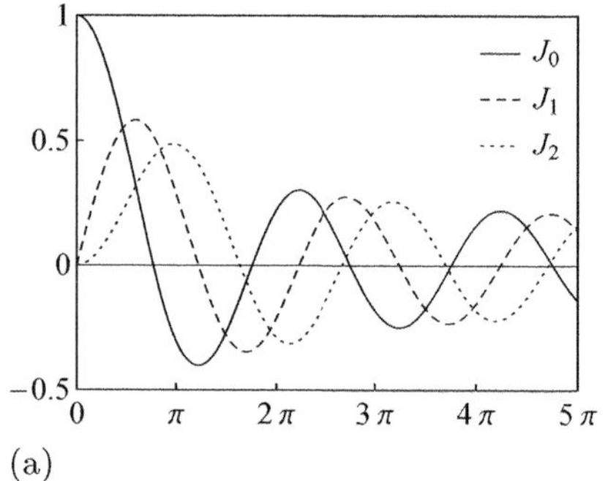
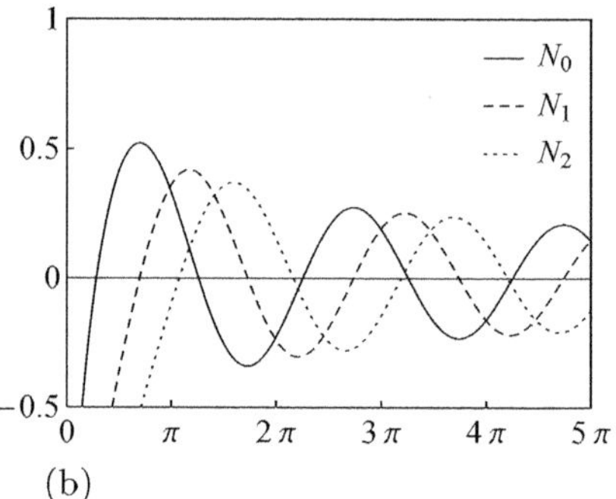

Fig. B.4. (a) Bessel functions of the first kind $J_n(x)$ for $n = 0, 1, 2$ (b) Bessel functions of the second kind (Neumann functions) $N_n(x) \equiv Y_n(x)$

Symmetry

All integer Bessel functions (J_n and N_n) and the modified Bessel functions (I_n and K_n) have the symmetry

$$\begin{aligned}
J_{-n}(x) &= (-1)^n J_n(x), & I_{-n}(x) &= I_n(x), \\
N_{-n}(x) &= (-1)^n N_n(x), & K_{-n}(x) &= K_n(x).
\end{aligned} \tag{B.4.7}$$

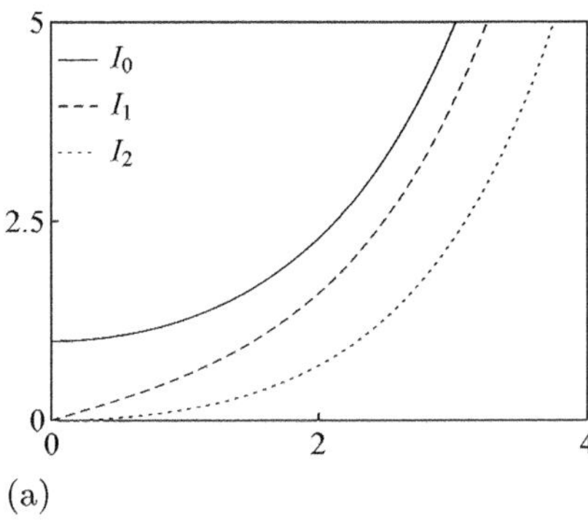

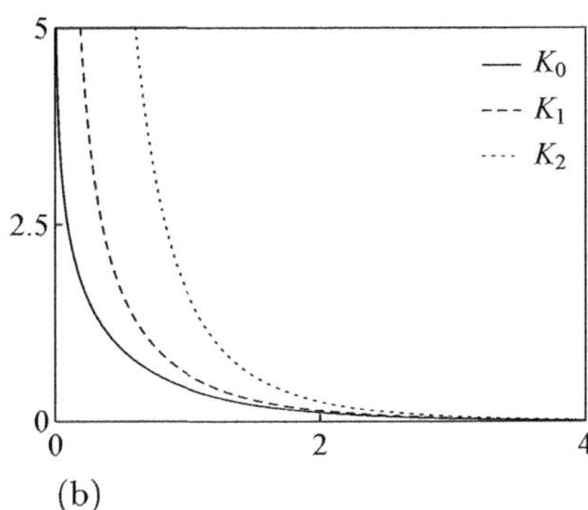

Fig. B.5. (a) Modified Bessel functions of the first kind $I_n(x)$ for $n = 0, 1, 2$ (b) Modified Bessel functions of the second kind $K_n(x)$

This may seem surprising at first glance, when considering the asymptotic behavior for $x \to 0$ from Tab.B.2. From (B.4.2) we see that for integer $n > 0$ the sum for J_{-n} starts with $j = n$.

Recursion relations

The Bessel functions satisfy recursion relations (functional equations), where we use Z_ν for J_ν, $N_\nu(x)$ and $H_\nu^{(1,2)}(x)$ with $Z'_\nu(x) = \frac{\mathrm{d}Z_\nu(x)}{\mathrm{d}x}$:

$$x\big[Z_{\nu-1}(x) + Z_{\nu+1}(x)\big] = 2\nu Z_\nu(x), \qquad Z_{\nu-1}(x) - Z_{\nu+1}(x) = 2Z'_\nu(x),$$

$$x\big[I_{\nu-1}(x) - I_{\nu+1}(x)\big] = 2\nu I_\nu(x), \qquad I_{\nu-1}(x) + I_{\nu+1}(x) = 2I'_\nu(x) \quad \text{(B.4.8)}$$

$$x\big[K_{\nu-1}(x) - K_{\nu+1}(x)\big] = -2\nu K_\nu(x), \quad K_{\nu-1}(x) + K_{\nu+1}(x) = -2K'_\nu(x).$$

Asymptotic behavior

The Bessel functions of the first kind $J_\nu(x)$, depicted in Fig. B.4, are regular at the origin for $\nu \geq 0$, while the second kind, the Neumann functions $N_\nu(x)$, are singular at the origin. Tab. B.2 indicates the asymptotic behavior for $x \to 0$. For the further procedure, we need the positions of the zeros of the Bessel functions, especially of the J_n. The first zeros, listed in Tab. B.3 are to be determined numerically. For large arguments, the ones in Tab. B.2 listed formulas apply, where the relationships

$$J_n(x) = \sqrt{\frac{2}{\pi x}} \cos(x - \frac{n\pi}{2} - \frac{\pi}{4}), \qquad N_n(x) = \sqrt{\frac{2}{\pi x}} \sin(x - \frac{n\pi}{2} - \frac{\pi}{4})$$

arise from Fig. B.4. One obtains for $J_n(x_{nl}) = 0$ for $l \gg n$ approximately

$$x_{nl} = l\pi + \frac{\pi}{2}\left(n - \frac{1}{2}\right) - \frac{4n^2 - 1}{(8l + 4n - 2)\pi} + \dots \tag{B.4.9}$$

Tab. B.2. Asymptotic behavior of the Bessel functions

Behavior for $x \ll 1$ and $n > 0$	

$$J_0(x) = 1 \qquad\qquad N_0(x) = \frac{2}{\pi}\ln x$$

$$J_n(x) = \frac{1}{n!}\left(\frac{x}{2}\right)^n \qquad\qquad N_n(x) = -\frac{(n-1)!}{\pi}\left(\frac{x}{2}\right)^{-n}$$

$$I_0(x) = 1 \qquad\qquad K_0(x) = -\ln x + C \;^{a}$$

$$I_n(x) = \frac{1}{n!}\left(\frac{x}{2}\right)^n \qquad\qquad K_n(x) = \frac{(n-1)!}{2}\left(\frac{x}{2}\right)^{-n}$$

Behavior for $x \gg 1$ and $n \geq 0$

$$J_n(x) = \sqrt{\frac{2}{\pi x}}\cos\!\left(x - \frac{n\pi}{2} - \frac{\pi}{4}\right) \quad N_n(x) = \sqrt{\frac{2}{\pi x}}\sin\!\left(x - \frac{n\pi}{2} - \frac{\pi}{4}\right)$$

$$I_n(x) = \frac{1}{\sqrt{2\pi x}}\,\mathrm{e}^{x} \qquad\qquad K_n(x) = \sqrt{\frac{\pi}{2x}}\,\mathrm{e}^{-x}$$

^{a} $C = 0.577215...$ (Euler-Mascheroni constant)

Tab. B.3. Zeros x_{nk} of the Bessel functions

Zeros of $J_n(x_{nk})$			Zeros of $N_n(x_{nk})$		
J_0: x_{0k} 2.40483	5.52008	8.65373	N_0: x_{0k} 0.89358	3.95768	7.08605
J_1: x_{1k} 3.83171	7.01559	10.17347	N_1: x_{1k} 2.19714	5.42968	8.59601
J_2: x_{2k} 5.13562	8.41724	11.61984	N_2: x_{2k} 3.38424	6.79381	10.02348

B.5 Integrals

B.5.1 Elliptic Integrals

In potential theory, i.e. in electrostatics and magnetostatics, elliptic integrals often occur in configurations with axial symmetry. There are sometimes small differences in the definitions, so it seems reasonable to specify the definitions used.

The elliptic integrals are listed here in the *Legendre's normal form*; here we distinguish between the elliptic integrals of the first, second and third kind (or genus). They are defined in this order by

$$F(\varphi, k) = \int_0^{\varphi} \mathrm{d}\alpha\, \frac{1}{\sqrt{1 - k^2 \sin^2 \alpha}}, \qquad 0 \leq k \leq 1,\; 0 \leq \varphi \leq \frac{\pi}{2}, \tag{B.5.1}$$

$$E(\varphi, k) = \int_0^{\varphi} \mathrm{d}\alpha\, \sqrt{1 - k^2 \sin^2 \alpha}, \tag{B.5.2}$$

$$\Pi(\varphi, n, k) = \int_0^{\varphi} \mathrm{d}\alpha\, \frac{1}{(1 + n\sin^2 \alpha)\sqrt{1 - k^2 \sin^2 \alpha}}. \tag{B.5.3}$$

k is also referred to as the *modulus* of the elliptic integrals and n as the parameter of the integrals of the 3rd kind. φ is an amplitude. If this has the value $\varphi = \pi/2$, then we have a *complete elliptic integral* in front of us:

$$K(k) = F(\frac{\pi}{2}, k) = \int_0^{\pi/2} d\alpha \, \frac{1}{\sqrt{1 - k^2 \sin^2 \alpha}}, \tag{B.5.4}$$

$$E(k) = E(\frac{\pi}{2}, k) = \int_0^{\pi/2} d\alpha \, \sqrt{1 - k^2 \sin^2 \alpha}, \tag{B.5.5}$$

$$\Pi(n, k) = \Pi(\frac{\pi}{2}, n, k) = \int_0^{\pi/2} d\alpha \, \frac{1}{(1 + n \sin^2 \alpha)\sqrt{1 - k^2 \sin^2 \alpha}}. \tag{B.5.6}$$

Functional equations

The following and other relations can be found in Gradshteyn, Rhyzhik [1965, section 8.1]

$$\frac{dK(k)}{dk} = \frac{1}{k}\Big[\frac{E(k)}{1 - k^2} - K(k)\Big]. \tag{B.5.7}$$

On the other hand, by differentiating (B.5.4) one gets

$$\frac{dK(k)}{dk} = \frac{1}{k} \int_0^{\pi/2} d\alpha \, \frac{k^2 \sin^2 \alpha}{\sqrt{1 - k^2 \sin^2 \alpha}^3} = \frac{1}{k}\big[\Pi(-k^2, k) - K(k)\big]. \tag{B.5.8}$$

From the last two equations it follows

$$\Pi(-k^2, k) = \int_0^{\pi/2} d\alpha \, \frac{1}{\sqrt{1 - k^2 \sin^2 \alpha}^3} = \frac{E(k)}{1 - k^2}. \tag{B.5.9}$$

B.5.2 Integrals for Potential Theory

Here are listed some integrals that occur in electrostatics and magnetostatics, without however, their derivation. The integrals are essentially taken from the integral tables Gradshteyn, Rhyzhik [1965]:

$$\int dx \, \frac{1}{a^2 + b^2 x^2} = \frac{1}{ab} \arctan \frac{bx}{a}, \qquad a \geq 0, \tag{B.5.10}$$

$$\int dx \, \ln\left(a^2 + x^2\right) = x \ln(a^2 + x^2) - 2x + 2a \arctan \frac{x}{a}, \tag{B.5.11}$$

$$\int dx \, \ln(x + \sqrt{a^2 + x^2}) = x \ln(x + \sqrt{a^2 + x^2}) - \sqrt{a^2 + x^2} \tag{B.5.12}$$

$$\int_0^\pi d\varphi \, \cos\varphi \ln\left(1 + a^2 \pm 2a \cos\varphi\right) = -\pi\Big[a \, \theta(1 - a) + \frac{1}{a} \theta(a - 1)\Big]. \tag{B.5.13}$$

$$\int \mathrm{d}x \, \sqrt{a^2 + x^2} = \frac{a^2}{2} \ln(x + \sqrt{a^2 + x^2}) + \frac{x}{2} \sqrt{a^2 + x^2}, \tag{B.5.14}$$

$$\int \mathrm{d}x \, \frac{1}{\sqrt{a^2 + x^2}} = \ln(x + \sqrt{a^2 + x^2}) = \operatorname{arsinh} \frac{x}{a} + \ln a, \tag{B.5.15}$$

$$\int \mathrm{d}x \, \frac{1}{\sqrt{a^2 + x^2}^3} = \frac{1}{a^2} \frac{x}{\sqrt{a^2 + x^2}}, \tag{B.5.16}$$

$$\int \mathrm{d}x \, \frac{x^2}{\sqrt{a^2 + x^2}^3} = -\frac{x}{\sqrt{a^2 + x^2}} + \ln(x + \sqrt{a^2 + x^2}), \tag{B.5.17}$$

$$\int_0^\pi \mathrm{d}\varphi \, \frac{1}{\alpha^2 - 2\alpha\beta \cos\varphi + \beta^2} = \frac{\pi}{|\alpha^2 - \beta^2|}, \qquad \alpha^2 \neq \beta^2, \tag{B.5.18}$$

$$\int_0^\pi \mathrm{d}\varphi \, \frac{\cos\varphi}{\alpha^2 - 2\alpha\beta \cos\varphi + \beta^2} = \frac{\pi}{|\alpha^2 - \beta^2|} \begin{cases} \alpha/\beta, & \alpha^2 < \beta^2, \\ \beta/\alpha, & \alpha^2 > \beta^2, \end{cases} \tag{B.5.19}$$

$$\int_0^\pi \mathrm{d}\varphi \, \frac{\cos^2\varphi}{\alpha^2 - 2\alpha\beta \cos\varphi + \beta^2} = \frac{\alpha^2 + \beta^2}{|\alpha^2 - \beta^2|} \begin{cases} \pi/2\beta^2, & \alpha^2 < \beta^2, \\ \pi/2\alpha^2, & \alpha^2 > \beta^2, \end{cases} \tag{B.5.20}$$

$$\int_0^\pi \mathrm{d}\varphi \, \frac{\alpha - \beta \cos\varphi}{\alpha^2 - 2\alpha\beta \cos\varphi + \beta^2} = \frac{\pi}{\alpha} \theta(\alpha - \beta), \qquad 0 < \alpha, \qquad 0 < \beta, \tag{B.5.21}$$

$$\int_0^\pi \mathrm{d}\varphi \, \frac{\alpha \cos\varphi - \beta \cos^2\varphi}{\alpha^2 - 2\alpha\beta \cos\varphi + \beta^2} = \begin{cases} -\pi/2\beta, & 0 < \alpha < \beta, \\ \pi\beta/2\alpha^2, & \alpha > \beta > 0. \end{cases} \tag{B.5.22}$$

Surface integrals

Integration over the surface of the sphere: $\oiint \mathrm{d}\Omega' = \int_0^{2\pi} \mathrm{d}\varphi' \int_{-1}^1 \mathrm{d}\xi'$ with $\xi' = \cos\vartheta'$.

$$\frac{1}{4\pi} \oiint \mathrm{d}\Omega' \, \frac{1}{|\mathbf{x}' - \mathbf{x}|} = \frac{1}{r} \theta(r - r') + \frac{1}{r'} \theta(r' - r), \tag{B.5.23}$$

$$\frac{1}{4\pi} \oiint \mathrm{d}\Omega' \, \frac{\xi'}{|\mathbf{x}' - \mathbf{x}|} = \frac{1}{3} \Big[\frac{r'}{r^2} \theta(r - r') + \frac{r}{r'^2} \theta(r' - r) \Big], \tag{B.5.24}$$

$$\frac{1}{4\pi} \oiint \mathrm{d}\Omega' \, \frac{\xi'^2}{|\mathbf{x}' - \mathbf{x}|} = \frac{1}{3} \Big\{ \Big(\frac{1}{r} + \frac{2r'^2}{5r^3} \Big) \theta(r - r') + \Big(\frac{1}{r'} + \frac{2r^2}{5r'^3} \Big) \theta(r' - r) \Big\}. \tag{B.5.25}$$

$G_i(\mathbf{x}, \mathbf{x}')$ with $i = 1$ and $i = 2$ are the Green functions (A.4.27):

$$G_1(\mathbf{x}, \mathbf{x}') = \frac{1}{|\mathbf{x}' - \mathbf{x}|} - \frac{1}{r'}, \qquad G_2(\mathbf{x}, \mathbf{x}') = G_1(\mathbf{x}, \mathbf{x}') - \frac{\mathbf{x} \cdot \mathbf{x}'}{r'^3}.$$

$$\frac{1}{4\pi}\oiint \mathrm{d}\Omega'\, G_i(\mathbf{x},\mathbf{x}') = \left(\frac{1}{r}-\frac{1}{r'}\right)\theta(r-r'), \qquad i=1,2\,, \quad \text{(B.5.26)}$$

$$\frac{1}{4\pi}\oiint \mathrm{d}\Omega'\, \xi'\, G_1(\mathbf{x},\mathbf{x}') = \frac{1}{3}\left[\frac{r'}{r^2}\theta(r-r') + \frac{r}{r'^2}\theta(r'-r)\right], \qquad \text{(B.5.27)}$$

$$\frac{1}{4\pi}\oiint \mathrm{d}\Omega'\, \xi'\, G_2(\mathbf{x},\mathbf{x}') = \frac{1}{3}\left(\frac{r'}{r^2}-\frac{r}{r'^2}\right)\theta(r-r'), \qquad \text{(B.5.28)}$$

$$\frac{1}{4\pi}\oiint \mathrm{d}\Omega'\, \xi'^2\, G_i(\mathbf{x},\mathbf{x}') = \frac{1}{3}\left\{\left[\left(\frac{1}{r}-\frac{1}{r'}\right) + \frac{2}{5}\frac{r'^2}{r^3}\right]\theta(r-r') + \frac{2}{5}\frac{r^2}{r'^3}\theta(r'-r)\right\},$$

$$\text{(B.5.29)}$$

$$\int \mathrm{d}^3x'\, \frac{f(r')}{|\mathbf{x}-\mathbf{x}'|} = 4\pi\left\{\frac{1}{r}\int_0^r \mathrm{d}r'\, r'^2 f(r') + \int_r^\infty \mathrm{d}r'\, r' f(r')\right\}. \qquad \text{(B.5.30)}$$

B.5.3 Convolution

Depending on the dimension, convolution is understood as an operation of the form

$$(f*g)(t) = \int_{-\infty}^{\infty} \mathrm{d}t'\, f(t-t')\, g(t'),$$

$$(f*g)(\mathbf{x}) = \int_{-\infty}^{\infty} \mathrm{d}^3x'\, f(\mathbf{x}-\mathbf{x}')\, g(\mathbf{x}'),$$

$$(f*g)(\mathbf{x},t) = \int_{-\infty}^{\infty} \mathrm{d}^3x'\mathrm{d}t'\, f(\mathbf{x}-\mathbf{x}',t-t')\, g(\mathbf{x}',t') \qquad \text{(B.5.31)}$$

$$= \int_{-\infty}^{\infty} \frac{\mathrm{d}^3k\,\mathrm{d}\omega}{(2\pi)^4}\, \mathrm{e}^{\mathrm{i}\mathbf{k}\cdot\mathbf{x}-\mathrm{i}\omega t}\, f(\mathbf{k},\omega)\, g(\mathbf{k},\omega).$$

The convolution satisfies the calculation rules:

$$\begin{aligned}
(f*g) &= (g*f) &&\text{Commutativity}\\
(f*g)*h &= f*(g*h) &&\text{Associativity}\\
f*(g+h) &= f*g+f*h &&\text{Distributivity}\\
c(f*g) &= (cf*g) &&\text{Multiplication with scalar.}
\end{aligned}$$

Let $\mathcal{D}$ be a differential operator, then

$$\mathcal{D}(f*g) = (\mathcal{D}f*g) = (f*\mathcal{D}g).$$

Except for the last line of (B.5.31), the definitions in the literature are uniform. However, the Fourier transformation can be defined differently, so caution is required here. If the Fourier transformation is denoted by $\mathcal{F}$

$$\mathcal{F}\{f(\mathbf{x},t)\} = f(\mathbf{k},\omega) = \int_{-\infty}^{\infty} \mathrm{d}^3x\mathrm{d}t\, \mathrm{e}^{-\mathrm{i}\mathbf{k}\cdot\mathbf{x}+\mathrm{i}\omega t}\, f(\mathbf{x},t)\,,$$

then the *convolution theorem* is

$$\mathcal{F}\{(f*g)(\mathbf{x},t)\} = f(\mathbf{k},\omega)\, g(\mathbf{k},\omega)\,. \qquad \text{(B.5.32)}$$

B.6 Distributions

B.6.1 The Dirac Delta Function

The δ-function is not a function in the strict sense, but a distribution, which for our purposes can be sufficiently defined as a sequence of functions δ_n with

$$\lim_{n \to \infty} \delta_n(x - x_0) = \delta(x - x_0).$$

We are dealing with sequences of functions whose area (integral) remains constant, but with $n \to \infty$ they get an increasingly sharp peak around x_0. In the end, only the singular point x_0 contributes to the integral; this may be called "sifting property":

$$\int_a^b \mathrm{d}x \, \delta(x - x_0) \, f(x) = \int_a^b \mathrm{d}x \, \delta(x_0 - x) \, f(x) = f(x_0), \quad a < x_0 < b. \quad \text{(B.6.1)}$$

$f(x)$ should be continuous and of bounded variation around x_0. Therefore, δ-functions are required to: $\delta(x) = \lim_{n \to \infty} \delta_n(x)$:

1. $\displaystyle\int_{-\infty}^{\infty} \mathrm{d}x \, \delta_n(x) = 1.$

   Consequence of this condition: $\delta_n(ax) = \dfrac{1}{a} \delta_n(x).$

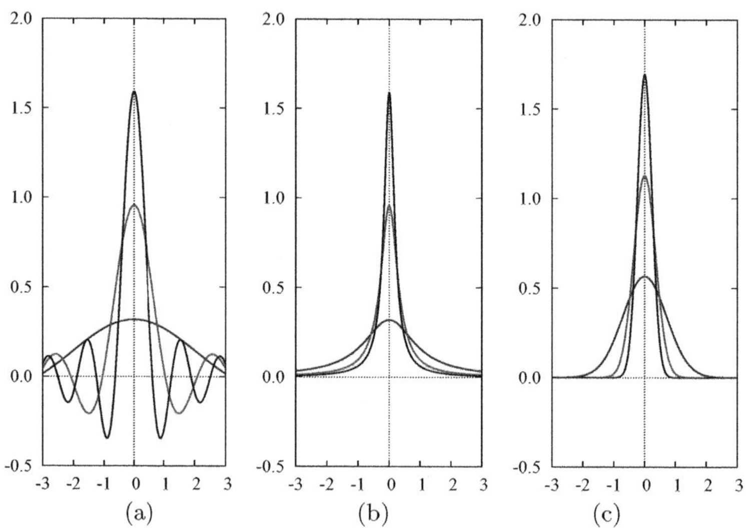

(a)   (b)   (c)

Fig. B.6. $\delta_n(x)$ for $n = 1, 3, 5$.   (a) $\dfrac{n}{\pi} \, \mathrm{si}(\mathrm{n}x)$ (b) $\dfrac{1}{\pi} \dfrac{n}{n^2 x^2 + 1}$ (c) $\dfrac{n}{\sqrt{\pi}} \, \mathrm{e}^{-n^2 x^2}$

2. It seems useful to require the symmetry

$$\delta_n(x) = \delta_n(-x).$$

At $x = 0$ an antisymmetric part contributes nothing, and for $x \neq 0$ δ_n should vanish as much as possible.

Together with the 1st point, it follows: $\delta_n(ax) = \dfrac{1}{|a|}\delta_n(x),$ $a \neq 0$, real.

3. $\lim\limits_{n\to\infty} \delta_n(x) = 0$ for $x \neq 0$.

4. $I = \displaystyle\int_a^b \mathrm{d}x\,\delta(x)\,f(x) = \lim\limits_{n\to\infty} \int_a^b \mathrm{d}x\,\delta_n(x)\,f(x) = f(0),$ $a < 0 < b$.   (B.6.2)

In the Tab. B.4 four sequences of functions $\delta_n(x)$ are given, which are suitable for representing the δ function. Fig. B.6 shows how these functions become narrower with increasing n while maintaining the area under the curves.

Tab. B.4. $\delta_n(x)$ functions for representing the delta function; for all Fourier transforms, $\lim\limits_{n\to\infty} \delta_n(k) = 1$.

	$\delta_n(x)$	Designation	$\delta_n(k) = \displaystyle\int_{-\infty}^{\infty} \mathrm{d}x\, e^{-ikx}\,\delta_n(x)$
(a)	$\dfrac{1}{\pi}\dfrac{\sin(nx)}{x}$	Cardinal sine[1]	$\theta(n-|k|)$
(b)	$\dfrac{1}{\pi}\dfrac{n}{n^2x^2+1}$	Lorentz function	$e^{-|k|/n}$
(c)	$\dfrac{n}{\sqrt{\pi}}\,e^{-x^2n^2}$	Gaussian distribution	$e^{-k^2/(4n^2)}$
(d)	$\dfrac{n}{2\cosh^2(nx)}$	Pöschl-Teller potential	$\dfrac{k\pi/2n}{\sinh(k\pi/2n)}$

[1] Diffraction function at the gap or "sampling function": $\mathrm{si}\,(x) \equiv \mathrm{sinc}\,(x) := \dfrac{\sin x}{x}$

The proof that $\delta_n(x)$ behaves like a δ-function in the limit $n \to \infty$, we show using the sinc function and (B.6.2). Let $a < 0 < b$, so that $x = 0$ lies in the integration interval. Then it holds

$$I = \lim\limits_{n\to\infty}\frac{n}{\pi}\int_a^b \mathrm{d}x\, f(x)\,\mathrm{si}(nx) \overset{t=nx}{=} \lim\limits_{n\to\infty}\frac{1}{\pi}\int_{na}^{nb} \mathrm{d}t\, f(\frac{t}{n})\,\mathrm{si}(t) = f(0)\,\frac{1}{\pi}\underbrace{\int_{-\infty}^{\infty} \mathrm{d}t\,\frac{\sin t}{t}}_{1}.$$

Properties of the delta function

The basic property we have already presented in (B.6.1):

$$\int_a^b dx\, \delta(x-x_0)\, f(x) = \begin{cases} f(x_0) & a < x_0 < b \\ 0 & x_0 < a \quad \text{and } b < x_0\,. \end{cases} \tag{B.6.3}$$

If we replace here $f(x) \to f(x)\,x$ and set $x_0 = 0$, we get

$$\int_a^b dx\, x\, \delta(x)\, f(x) = 0 \qquad \text{or} \quad x\,\delta(x) = 0\,, \tag{B.6.4}$$

as long as $f(0)$ is regular. Furthermore, we note that

$$\int_{-\infty}^a dx\, \delta(x) = \theta(a) = \begin{cases} 0 & \text{for } a < 0 \\ 1 & \text{for } a > 0\,. \end{cases} \tag{B.6.5}$$

From the requirements placed on the δ_n functions (or from (B.6.1)) it follows

$$\begin{aligned} \delta(x) &= \delta(-x)\,, \\ \delta(ax) &= \frac{1}{|a|}\,\delta(x)\,, & a \neq 0 \quad \text{and} \quad a \in \mathbb{R}\,, \\ \delta(x^2 - a^2) &= \delta((x-a)(x+a)) = \frac{1}{2|a|}\left\{\delta(x-a) + \delta(x+a)\right\}\,. \end{aligned} \tag{B.6.6}$$

Now let $f(x)$ be a function with isolated zeros x_i, so that a Taylor expansion can be made around these. Then it applies

$$\begin{aligned} \delta\big(f(x)\big) &= \sum_i \delta\Big(f(x_i) + (x - x_i)\, f'(x_i) + \frac{1}{2}\, f''(x_i)\,(x - x_i)^2 + \dots \Big) \\ &= \sum_i \delta\Big((x-x_i)\left[f'(x_i) + \frac{1}{2}\, f''(x_i)\,(x-x_i) + \dots \right] \Big) = \sum_i \frac{\delta(x-x_i)}{|f'(x_i)|}\,, \end{aligned}$$

$$\delta\big(f(x)\big) = \sum_i \frac{\delta(x-x_i)}{|f'(x_i)|} \qquad \text{with} \quad f(x_i) = 0\,. \tag{B.6.7}$$

For the derivatives it applies

$$\int_a^b dx\, f(x)\, \delta'(x-x_0) \overset{\text{part.int.}}{=} -\int_a^b dx\, f'(x)\, \delta(x-x_0) = -f'(x_0)\,. \tag{B.6.8}$$

The boundary term of the partial integration vanishes, since $\delta(a) = \delta(b) = 0$. From this it follows for the nth derivative

$$\int_a^b dx\, f(x)\, \frac{d^n}{dx^n}\delta(x - x_0) = (-1)^n\, \frac{d^n}{dx^n} f(x)\Big|_{x=x_0}\,. \tag{B.6.9}$$

Multidimensional delta functions

In Cartesian coordinates, the n-dimensional delta function is defined by the product

$$\delta^{(n)}(\mathbf{x}) = \delta(x_1)\delta(x_2)....\delta(x_n)\,. \tag{B.6.10}$$

We most frequently encounter the δ function in three dimensions, sometimes also in the form

$$\delta^{(3)}\big(\mathbf{f}(\mathbf{x})\big) = \delta\big(f_1(\mathbf{x})\big)\,\delta\big(f_2(\mathbf{x})\big)\,\delta\big(f_3(\mathbf{x})\big)$$

with the zeros $\mathbf{f}(\mathbf{x}_i) = 0$. The (B.6.7) corresponding relation is then

$$\delta^{(3)}\big(\mathbf{f}(\mathbf{x})\big) = \sum_i \frac{\delta^{(3)}(\mathbf{x}-\mathbf{x}_i)}{|J|} \quad \text{with} \quad J = \frac{\partial(f_1, f_2, f_3)}{\partial(x,y,z)} = \begin{vmatrix} \frac{\partial f_1}{\partial x} & \cdots & \frac{\partial f_1}{\partial z} \\ \cdots & \cdots & \cdots \\ \frac{\partial f_3}{\partial x} & \cdots & \frac{\partial f_3}{\partial z} \end{vmatrix}. \tag{B.6.11}$$

J is here the Jacobian determinant (functional determinant). This relation can be made understandable if one considers the coordinate transformation

$$\int \mathrm{d}^3x\,\delta^{(3)}\big(\mathbf{f}(\mathbf{x})\big) = \int \mathrm{d}^3f \left|\frac{\partial(x,y,z)}{\partial(f_1,f_2,f_3)}\right| \delta^{(3)}(\mathbf{f}) = \int \mathrm{d}^3f\, \frac{\delta^{(3)}(\mathbf{f})}{|J|} = \frac{1}{|J(\mathbf{x}_0)|}$$

for a single zero $\mathbf{f}(\mathbf{x}_0) = 0$. In three dimensions, one obtains in detail

$$\begin{aligned} \delta^{(3)}(\mathbf{x}-\mathbf{x}') &= \delta(x-x')\delta(y-y')\delta(z-z'), \quad \text{Cartesian coordinates,} \\ &= \frac{\delta(r-r')\delta(\vartheta-\vartheta')\delta(\varphi-\varphi')}{r\sin\vartheta}, \quad \text{spherical coordinates,} \\ &= \frac{\delta(\varrho-\varrho')\delta(\varphi-\varphi')\delta(z-z')}{\varrho}, \quad \text{cylindrical coordinates.} \end{aligned} \tag{B.6.12}$$

Integral representation

The integral representations

$$\begin{aligned} \text{(a)} \quad \delta_n(x) &= \frac{1}{\pi}\frac{\sin(nx)}{x} &&= \int_{-n}^{n} \frac{\mathrm{d}k}{2\pi}\, \mathrm{e}^{\mathrm{i}kx}, \\ \text{(b)} \quad \delta_n(x) &= \frac{1}{\pi}\frac{\frac{1}{n}}{x^2 + \frac{1}{n^2}} &&= \int_{-\infty}^{\infty} \frac{\mathrm{d}k}{2\pi}\, \mathrm{e}^{\mathrm{i}kx-|k|/n}, \\ \text{(c)} \quad \delta_n(x) &= \frac{n}{\sqrt{\pi}}\, \mathrm{e}^{-x^2 n^2} &&= \int_{-\infty}^{\infty} \frac{\mathrm{d}k}{2\pi}\, \mathrm{e}^{\mathrm{i}kx-k^2/4n^2} \end{aligned} \tag{B.6.13}$$

we take from the Tab.B.4, in which the $\delta_n(k)$ are also given. We conclude from this that $\delta(k) = \lim\limits_{n\to\infty} \delta_n(k) = 1$,

$$\delta(x) = \int_{-\infty}^{\infty} \frac{\mathrm{d}k}{2\pi}\, \mathrm{e}^{\mathrm{i}kx}\,. \tag{B.6.14}$$

One-sided delta-function

We decompose the integral representation of $\delta_n(x)$ with the Lorentz function into

$$\delta_n(x) = \int_{-\infty}^{0} \frac{dk}{2\pi}\, e^{ikx-|k|/n} + \int_{0}^{\infty} \frac{dk}{2\pi}\, e^{ikx-|k|/n} = \delta_{n-}(x) + \delta_{n+}(x)\,,$$

where we have defined according to the decomposition

$$\delta_{n\pm}(x) = \int_{0}^{\infty} \frac{dk}{2\pi}\, e^{\pm ikx-k/n} = \frac{1}{2\pi}\, \frac{e^{\pm ikx-k/n}}{\pm ix - \frac{1}{n}}\bigg|_{0}^{\infty}\,.$$

Now we make the limit $n \to \infty$. The convergence factor $\epsilon = 1/n$ in the exponent guarantees that the upper limit disappears:

$$\delta_\pm(x) = \mp \frac{1}{2\pi i}\, \lim_{\epsilon \to 0} \frac{1}{x \pm i\epsilon}\,. \tag{B.6.15}$$

In the denominator, ϵ remains, to make it clear how to perform the integration at the pole $1/x$. For $\delta_-(x)$, the pole at $x = 0$ is circumvented in a semicircle below the real axis, as sketched in Fig. B.7b :

$$\int_{a}^{b} dx\, f(x)\, \delta_-(x) = \frac{1}{2\pi i}\left\{ \int_{a}^{-\epsilon} dx\, \frac{f(x)}{x} + \int_{\epsilon}^{b} dx\, \frac{f(x)}{x} + f(0) \int_{\pi}^{2\pi} d\varphi\, i \right\}$$

$$= \frac{1}{2\pi i}\, P \int_{a}^{b} dx\, \frac{f(x)}{x} + \frac{f(0)}{2}\,.$$

During the integration on the real axis, the pole of $\frac{1}{x}$ is omitted, which is referred to as a principal value integral and is marked with a P. The semicircle around the pole provides half the contribution of a δ-function. In summary, it is

$$\delta_\pm(x) = \frac{1}{2}\, \delta(x) \pm \frac{i}{2\pi}\, P\Big(\frac{1}{x}\Big) = \int_{0}^{\infty} \frac{dk}{2\pi}\, e^{\pm ikx}\,. \tag{B.6.16}$$

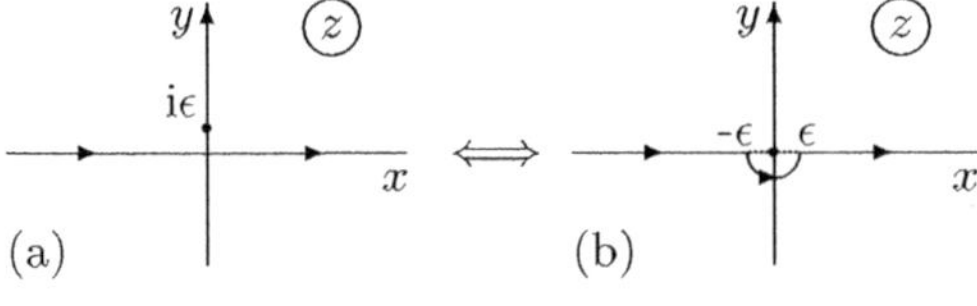

Fig. B.7. Equivalent integration paths for $\delta_-(x)$

B.6.2 Step Function

The step or Heaviside function is defined as

$$\theta(x) = \begin{cases} 0 & x < 0 \\ 1 & x > 0 \,. \end{cases} \qquad (B.6.17)$$

The value $\theta(0)$ does not need to be defined separately, but is usually given as $1/2$. A test function for the distribution $\theta(x)$ is

$$\theta_n(x) = \frac{1}{2}\left[1+\tanh(nx)\right], \quad \theta(x) = \lim_{n\to\infty}\theta_n(x) = \begin{cases} 0 & x < 0 \\ 0.5 & x = 0. \\ 1 & x > 0 \end{cases} \qquad (B.6.18)$$

Signum function

$$\mathrm{sgn}(x) = -1 + 2\theta(x) = \theta(x) - \theta(-x) = \begin{cases} -1 & x < 0 \\ 0 & x = 0 \,. \\ 1 & x > 0 \end{cases} \qquad (B.6.19)$$

Derivative of the step function

From the test function $\theta_n(x)$ follows the test function (d) of the Tab.B.4:

$$\theta'_n(x) = \frac{n}{2}\,\frac{1}{\cosh^2(nx)} = \delta_n(x) \,. \qquad (B.6.20)$$

Therefore, in the limit $n \to \infty$ we have

$$\frac{\mathrm{d}}{\mathrm{d}x}\theta(x) = \delta(x) \qquad \text{and} \qquad \frac{\mathrm{d}}{\mathrm{d}x}\,\mathrm{sgn}(x) = 2\delta(x) \,. \qquad (B.6.21)$$

Integral representation of the step function

If the derivative of the step function is now the δ-function, it is obvious to integrate the integral representation (B.6.14):

$$\theta(x) = \lim_{\epsilon\to 0}\int_{-\infty}^{\infty}\frac{\mathrm{d}k}{2\pi}\,\frac{e^{ikx}}{\mathrm{i}(k-\mathrm{i}\epsilon)}\begin{cases} 0 & x < 0 \\ 0.5 & x = 0 \,, \\ 1 & x > 0 \end{cases} \qquad (B.6.22)$$

where, however, with $\epsilon \to 0_+$ the pole should lie infinitesimally above the real axis. Thus, the pole contributes for $x > 0$, where the integral is closed over the upper semicircle, to the integral, as sketched in Fig. B.8.

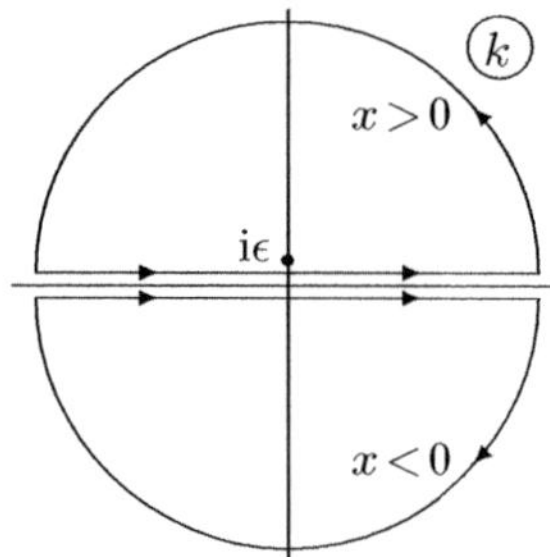

Fig. B.8. Integration paths for the integral representation of the $\theta(x)$-function. With $|k| \to \infty$ the integral over the upper half-circle vanishes for $x > 0$ and for $x < 0$ over the lower half-circle

Problems for Appendix B

B.1. *Determination of the Legendre polynomials*: Show that you can obtain the P_l from the recursion relation (3.2.19).

Hint: You can determine the a_n by complete induction. Assumption of induction:

$$a_n = (-1)^{\frac{l-n}{2}} \frac{(l+n-1)!!}{(l-n)!!} \frac{1}{n!} \, .$$

B.2. *Recursion relation for orthonormal polynomials*: Show that orthonormal polynomials $p_k(x)$ of rank k, which are defined on the base domain $[\alpha, \beta]$ and with a normalized weight function $w(x)$ satisfy the orthonormality relations

$$\int_\alpha^\beta \mathrm{d}x\, w(x)\, p_k(x)p_l(x) = \delta_{kl} \tag{B.6.23}$$

fulfill a recursion relation of the form

$$p_l(x) = \big(a_l\, x + b_l\big)p_{l-1}(x) - c_l\, p_{l-2}(x) \tag{B.6.24}$$

The recursion starts with $l = 1$, where $p_{-1} = 0$ and p_0.

Hint: Make an approach of the form

$$p_l(x) = \big(a_l\, x + b_l\big)p_{l-1}(x) - c_l\, p_{l-2}(x) + \sum_{j=0}^{l-3} \gamma_{lj}\, p_j(x)$$

and use the orthonormality relation (B.6.23) for $k \leq l$.

B.3. *Recursion relation for Legendre polynomials*: In problem B.2 the general form of the recursion relation (B.6.24) for orthonormal polynomials was to be derived. Now verify this recursion relation (B.2.17) for the Legendre polynomials $P_l(x)$.

Note: Use in (B.6.24) the approach $p_l = q_l x^l + s_l x^{l-1} + \dots$, where these coefficients are known to you from (3.2.20).

References

Gradshteyn I. S. & Ryzhik I. M. *Table of Integrals, Series, and Products* Academic Press N.Y. (1965)

Schwabl F. *Quantum Mechanics*, 4th ed. Springer Berlin (2007)

C

Systems of Units in Electrodynamics

C.1 Mechanical Units

In the 19th century, the system of measurement was also a focus of interest. This did not only concern electromagnetic units, but also the mechanical base units (length L, mass M, time T). The definition of electromagnetic quantities

Tab. C.1. Mechanical base units in relation to electrodynamics

	MKS[a] 1889	CGS[b] 1874	practical system[c]	BAAS[d] 1863	W. Thomson[c]	Weber/Gauß[c] 1832
Length	$1\,\mathrm{m}$	$1\,\mathrm{cm}$	$1\times10^{7}\,\mathrm{m}$	$1\,\mathrm{m}$	$1\,\mathrm{cm}$	$1\,\mathrm{mm}$
Mass	$1\,\mathrm{kg}$	$1\,\mathrm{g}$	$1\times10^{-11}\,\mathrm{g}$	$1\,\mathrm{g}$	$1\,\mathrm{g}$	$1\,\mathrm{g}$
Time	$1\,\mathrm{s}$	$1\,\mathrm{s}$	$1\,\mathrm{s}$	$1\,\mathrm{s}$	$1\,\mathrm{s}$	$1\,\mathrm{s}$

Definitions of the base units in the MKS system.

$1\,\mathrm{m} = \dfrac{cs}{299\,792\,458}$	1983	Distance that light travels in a vacuum in $1\,\mathrm{s}$
$1\,\mathrm{kg} = \dfrac{h\times10^{34}\,\mathrm{s\,m^{-2}}}{6.626\,070\,15}$	2019	Multiple of Planck's quantum of action
$1\,\mathrm{s} = \dfrac{9\,192\,631\,770}{\Delta\nu}$	1967	$\Delta\nu = $ Frequency of the hyperfine structure transition in Cs^{133}

Derived units of the CGS system in the MKS system

Force	$1\,\mathrm{dyn}=1\,\mathrm{g\,cm\,s^{-2}}$	$=1\times10^{-5}\,\mathrm{N(ewton)}$
Energy	$1\,\mathrm{erg}=1\,\mathrm{dyn\,cm}=1\,\mathrm{g\,cm^{2}\,s^{-2}}$	$=1\times10^{-7}\,\mathrm{J(oule)}$
Power	$1\,\mathrm{erg/s}=1\,\mathrm{g\,cm^{2}\,s^{-3}}$	$=1\times10^{-7}\,\mathrm{J\,s^{-1}} = 1\times10^{-7}\,\mathrm{W(att)}$

[a] Forster et al. [1890] [b] W. Thomson et al. [1875, p. 255]
[c] Maxwell [1892, Art. 629] [d] Wheatstone et al. [1864, p. 131]

© The Editor(s) (if applicable) and The Author(s), under exclusive license
to Springer-Verlag GmbH, DE, part of Springer Nature 2025
D. Petrascheck, F. Schwabl, *Electrodynamics*, https://doi.org/10.1007/978-3-662-71502-4_17

through the three mechanical base units led to very large or small numerical values. An attempt, which Maxwell also supported, concerned the so-called *practical system* with the earth quadrant as the unit of length and a very small unit of mass. Since the historical development should not be completely omitted here, various mechanical systems are listed in Tab. C.1. It should be mentioned that the names of the mechanical units have remained the same since 1889, but their definitions were subject to several changes.

C.2 Systems of Electrodynamics

Tab. C.2 is a summary of the most important systems. The electrostatic (esu) and the electromagnetic (emu) systems are primarily of historical interest.

Tab. C.2. Conversion of electromagnetic quantities, starting from the Gauss system. Permittivity and permeability of the vacuum are not independent: $\mu_0 = 1/k_L^2 \epsilon_0$

System	k_C	k_L		$\rho' = \dfrac{1}{\sqrt{k_C}}\rho$	$\mathbf{j}' = \dfrac{1}{\sqrt{k_C}}\mathbf{j}$	$\mathbf{E}' = \sqrt{k_C}\,\mathbf{E}$	$\mathbf{B}' = \dfrac{\sqrt{k_C}}{k_L}\mathbf{B}$
Gauss	1	1		ρ	$\mathbf{j}$	$\mathbf{E}$	$\mathbf{B}$
Heaviside-Lorentz	$\dfrac{1}{4\pi}$	1		$\rho^{\mathrm{H}} = \sqrt{4\pi}\rho$	$\mathbf{j}^{\mathrm{H}} = \sqrt{4\pi}\mathbf{j}$	$\mathbf{E}^{\mathrm{H}} = \dfrac{\mathbf{E}}{\sqrt{4\pi}}$	$\mathbf{B}^{\mathrm{H}} = \dfrac{1}{\sqrt{4\pi}}\mathbf{B}$
SI	$\dfrac{1}{4\pi\epsilon_0}$	c		$\rho^{\mathrm{SI}} = \sqrt{4\pi\epsilon_0}\rho$	$\mathbf{j}^{\mathrm{SI}} = \sqrt{4\pi\epsilon_0}\mathbf{j}$	$\mathbf{E}^{\mathrm{SI}} = \dfrac{\mathbf{E}}{\sqrt{4\pi\epsilon_0}}$	$\mathbf{B}^{\mathrm{SI}} = \sqrt{\dfrac{\mu_0}{4\pi}}\mathbf{B}$
elstat. units	1	c		$\rho^{\mathrm{esu}} = \rho$	$\mathbf{j}^{\mathrm{esu}} = \mathbf{j}$	$\mathbf{E}^{\mathrm{esu}} = \mathbf{E}$	$\mathbf{B}^{\mathrm{esu}} = \dfrac{1}{c}\mathbf{B}$
elmagn. units	c^2	c		$\rho^{\mathrm{emu}} = \dfrac{1}{c}\rho$	$\mathbf{j}^{\mathrm{emu}} = \dfrac{1}{c}\mathbf{j}$	$\mathbf{E}^{\mathrm{emu}} = c\mathbf{E}$	$\mathbf{B}^{\mathrm{emu}} = \mathbf{B}$

System	k_C	k_L	ϵ_0	μ_0	$\mathbf{P}' = \dfrac{1}{\sqrt{k_C}}\mathbf{P}$	$\mathbf{M}' = \dfrac{k_L}{\sqrt{k_C}}\mathbf{M}$	$\mathbf{D}' = \epsilon_0\sqrt{k_C}\mathbf{D}$	$\mathbf{H}' = \dfrac{\sqrt{k_C}}{\mu_0 k_L}\mathbf{H}$
Gauss	1	1	1	1	$\mathbf{P}$	$\mathbf{M}$	$\mathbf{D}$	$\mathbf{H}$
Heaviside-Lorentz	$\dfrac{1}{4\pi}$	1	1	1	$\mathbf{P}^{\mathrm{H}} = \sqrt{4\pi}\mathbf{P}$	$\mathbf{M}^{\mathrm{H}} = \sqrt{4\pi}\mathbf{M}$	$\mathbf{D}^{\mathrm{H}} = \dfrac{\mathbf{D}}{\sqrt{4\pi}}$	$\mathbf{H}^{\mathrm{H}} = \dfrac{\mathbf{H}}{\sqrt{4\pi}}$
SI	$\dfrac{1}{4\pi\epsilon_0}$	c	ϵ_0	μ_0	$\mathbf{P}^{\mathrm{SI}} = \sqrt{4\pi\epsilon_0}\mathbf{P}$	$\mathbf{M}^{\mathrm{SI}} = \sqrt{\dfrac{4\pi}{\mu_0}}\mathbf{M}$	$\mathbf{D}^{\mathrm{SI}} = \sqrt{\dfrac{\epsilon_0}{4\pi}}\mathbf{D}$	$\mathbf{H}^{\mathrm{SI}} = \dfrac{\mathbf{H}}{\sqrt{4\pi\mu_0}}$
elstat. units	1	c	1	$\dfrac{1}{c^2}$	$\mathbf{P}^{\mathrm{esu}} = \mathbf{P}$	$\mathbf{M}^{\mathrm{esu}} = c\mathbf{M}$	$\mathbf{D}^{\mathrm{esu}} = \mathbf{D}$	$\mathbf{H}^{\mathrm{esu}} = c\mathbf{H}$
elmagn. units	c^2	c	$\dfrac{1}{c^2}$	1	$\mathbf{P}^{\mathrm{emu}} = \dfrac{1}{c}\mathbf{P}$	$\mathbf{M}^{\mathrm{emu}} = \mathbf{M}$	$\mathbf{D}^{\mathrm{emu}} = \dfrac{1}{c}\mathbf{D}$	$\mathbf{H}^{\mathrm{emu}} = \mathbf{H}$

Note: In electrodynamics, three base units are sufficient and then there are two free parameters (k_L, k_C). With four base vectors, there is only one independent parameter. In the rational MKSA system, $k_C = 10^{-7}c^2\mathrm{N/A}^2$ and k_L is free. With $k_L = c$ the SI system has been chosen and with $k_L = 1$ its symmetric variant. To define the ampere, and consequently k_C, the force between parallel currents is used; but one could also specify F_C. With the maximum number of five base vectors, there is no more free parameter: If tesla is to be the fifth base unit, it should be noted that in a field of $1\,\mathrm{T}$ on a particle with a charge of $1\,\mathrm{C}$ and the speed $v = 1\,\mathrm{m/s}$ the force $F_L = 1\,\mathrm{N}$ works. In the MKSAT system, $k_L \approx 3 \times 10^8\,\mathrm{N/sAT}$ would be.

C.2.1 Systems of Electrodynamics with three Base Units

Electrostatic and electromagnetic system

Maxwell [1892, Art. 620] notes that every electromagnetic quantity can be defined in relation to the three units M,L,T and seeks the connection with these. It should be noted that Maxwell [1892, Art. 599] defines the field (*electromotive intensity*), that a moving particle feels in the magnetic field, with $\mathbf{E}_1 = \mathbf{v} \times \mathbf{B}$. Thus, $k_L = c$ is determined and there is only one parameter to define the dimensions of the electromagnetic quantities.

If one introduces the charge $[Q]$ as the fourth base unit, one obtains the *electric unit system*. However, the unit of charge can be derived from the Coulomb force (1.2.1) using $k_C = 1$: $[Q] = [\sqrt{\mathrm{LML}}\,\mathrm{T}^{-1}]$. Thus, all quantities have been reduced to 3 units (M,L,T), which form the so-called *electrostatic unit system* esu.

With the pole strength $[p_m]$ as the fourth fundamental unit, one obtains the *magnetic unit system*. The unit pole is determined from the magnetostatic force law (7.1.41), where $k_M = k_C/k_L^2 = 1$, i.e. $k_C = c^2$. Now the pole strength has the dimension $[p_m] = [\sqrt{\mathrm{LML}}\,\mathrm{T}^{-1}]$ and the corresponding system with 3 units is the *electromagnetic unit system* emu. In summary, this results in the coefficients listed in Tab. C.2.

Electrostatic and electromagnetic system in matter

For the electrostatic and electromagnetic units, we refer to the textbooks by Maxwell and Föppl. In both books, the dielectric displacement is defined as

$$\mathbf{D} = K\mathbf{E}/4\pi, \quad [\text{Maxwell, 1892, Art. 619] and [Föppl, 1894, (115)]} \quad (\text{C.2.1})$$

The dielectric constant is therefore denoted by K and corresponds in our notation $K \hat{=} \epsilon_r \epsilon_0$. We omit the additional factor $1/4\pi$, as it does not fit into our scheme (see also Jackson [1998, Tab. A2]). Furthermore, we define the polarization by $\mathbf{D} = \epsilon_0 \mathbf{E} + \mathbf{P}$ and thus go beyond the two books that only describe linear dielectric media. For magnetic systems we read at Föppl [1894, (210)]

$$\mathbf{B} = \mu_0(\mathbf{H} + 4\pi\mathbf{J}), \qquad\qquad \mathbf{J} \hat{=} \mathbf{M}. \qquad (\text{C.2.2})$$

The corresponding equation from Maxwell [1892, Art. 619]: $\mathbf{B} = \mathbf{H} + 4\pi\mathbf{J}$ only fits for emu into our scheme, i.e., we stick to Föppl and can thus classify the esu and emu also for matter in Tab. C.2.

Electrostatic and electromagnetic units

Both systems are built on the CGS system like the Gaussian system. Electrostatic units esu have the syllable *stat* before the name of the unit and electromagnetic units emu the word *absolute*, abbreviated with *ab*. For charge and voltage this means according to (C.2.11):

$$1\,\mathrm{statC} = 1\sqrt{\mathrm{cm^3 g}}\,\mathrm{s}^{-1} = c^{-1}\mathrm{abC} = c^{-1}\sqrt{\mathrm{cmg}} \approx \tfrac{1}{3}\times 10^{-10}\,\mathrm{abC} = \tfrac{1}{3}\times 10^{-9}\,\mathrm{C},$$

$$1\,\mathrm{statV} = 1\sqrt{\mathrm{cmg}}\,\mathrm{s}^{-1} = c\,\mathrm{abV} = c\sqrt{\mathrm{cm^{-1}g}} \approx 3\times 10^{10}\,\mathrm{abV} = 300\,\mathrm{V}.$$

esu and emu have different dimensions; more precisely, they differ by powers of c. Characteristic units for the electromagnetic system are $1\,\mathrm{Biot} = 1\,\mathrm{abA} = 0.1\,\mathrm{A}$ and furthermore Gilbert, Oersted, Gauss and Maxwell (see Tab. C.4, p. 638), which are also attributed to the Gauss system.

The speed of light in electrodynamics

A crucial point in Maxwell's equations is the field $\mathbf{E}_1 = \mathbf{v}\times\mathbf{B}$, which acts on moving charged particles. On the one hand, it was only recognized late that this is the Lorentz force (Clausius [1882] did not yet know the force on moving particles), on the other hand, a consequence of this definition of $\mathbf{E}_1$ ($k_L = c$), is that the dimensions of electromagnetic quantities, calculated on the basis of the Coulomb force (esu) and the magnetic force (emu), are different.

Comparing the Coulomb force with the magnetic force (7.1.41), we get

$$K c q^2 = (k_C/k_L^2)p_m^2 \qquad \Rightarrow \qquad p_m = cq \qquad \Rightarrow \qquad q^{\mathrm{esu}} = c q^{\mathrm{emu}}. \tag{C.2.3}$$

Measurements of a quantity in esu and emu differ in powers of a velocity c. Weber & Kohlrausch [1856] first measured c as $c = 310\,740\,\mathrm{km/s}$. Maxwell [1892, Art. 771] writes about the procedure that the same charge was first measured electrostatically and then electromagnetically:

One introduces into a Leyden jar of known capacity C the charge $q = VC$ (see (2.2.30)) and measures the voltage difference with an electrometer in esu. To measure the charge in emu, it is discharged through a galvanometer. The sudden discharge causes a swing of the magnet from which the amount of electricity can be calculated. According to (C.2.3), the value of c can be determined.

It should be added that Weber had his own formulations of the laws of electrodynamics, so that the comparison was not so simple [Stille, 1957]. As a result of the speed of light, the charge unit in emu (*absolute Coulomb*, abC) has a much larger numerical value than in esu (*statcoulomb*, statC): $1\,\mathrm{abC} \approx 3\times 10^{10}\,\mathrm{statC}$, which is not limited to the charge alone, as can be seen from Tab. C.3.

Practical units

The electromagnetic units were considered more important than the electrostatic ones, but with the usual base units (CGS,...) a reasonable order of magnitude for the electromagnetic units could not be achieved. For this reason, "practical units"[1] (Tab. C.1, p. 633) were introduced. These are electromagnetic units with $L = 10^7\,\mathrm{m}$ and $M = 10^{-11}\,\mathrm{g}$. According to these units, also propagated by Maxwell [1892, Art. 629] and adopted in 1881 [Fischer, 1961, p. 78] the voltage in the electromagnetic system is $1\,\mathrm{V} = 10^8\,\mathrm{cm}^{3/2}\,\mathrm{g}^{1/2}\,\mathrm{s}^{-2}$. The practical unit for resistance is $1\,\mathrm{ohm} = 10^9\,\mathrm{cm\,s}$ and consequently, the ampere, here a derived unit, $1\,\mathrm{A} = 1\mathrm{V}/\mathrm{ohm} = 0.1\,\mathrm{cm}^{1/2}\,\mathrm{g}^{1/2}\,\mathrm{s}$, as can be taken from Tab. C.3. Because of the earth quadrant as a unit of length, the "practical

Tab. C.3. Practical electrical units (quadrant units) in electromagnetic (emu) and electrostatic (esu) units; $c \approx 3 \times 10^{10}\,\mathrm{cm}$

(el.) Size	Unit	Quadrant unit	emu	esu
Capacity	farad	$F = (10^9\,\mathrm{cm})^{-1}\mathrm{s}^2$	$= 10^{-9}\ \mathrm{cm}^{-1}\mathrm{s}^2$	$\times c^2$
Resistance	ohm	$\Omega = 10^9\,\mathrm{cm\,s}$	$= 10^9\ \mathrm{cm\,s}$	$\times c$
Charge	coulomb	$C = \sqrt{10^9\,\mathrm{cm}}\,\sqrt{10^{-11}\,\mathrm{g}}$	$= 10^{-1}\ \mathrm{cm}^{1/2}\mathrm{g}^{1/2}$	$\times c$
Electric current	ampere	$A = \sqrt{10^9\,\mathrm{cm}}\,\sqrt{10^{-11}\,\mathrm{g}}\,\mathrm{s}$	$= 10^{-1}\ \mathrm{cm}^{1/2}\mathrm{g}^{1/2}\mathrm{s}$	$\times c$
Energy[a]	joule	$J = (10^9\,\mathrm{cm})^2\,10^{-11}\mathrm{g\,s}^{-2}$	$= 10^7\ \mathrm{erg}$	
Power[a]	watt	$W = (10^9\,\mathrm{cm})^2\,10^{-11}\mathrm{g\,s}^{-3}$	$= 10^7\ \mathrm{erg\,s}^{-1}$	
Voltage	volt	$V = \sqrt{10^9\,\mathrm{cm}}^{\,3}\,\sqrt{10^{-11}\,\mathrm{g}}\,\mathrm{s}^{-2}$	$= 10^8\ \mathrm{cm}^{3/2}\mathrm{g}^{1/2}\mathrm{s}^{-2}$	$/c$
Mag. flux[b]	weber	$Wb = \sqrt{(10^9\,\mathrm{cm})^3\,10^{-11}\,\mathrm{g}}\,\mathrm{s}$	$= 10^8\ \mathrm{cm}^{3/2}\mathrm{g}^{1/2}\mathrm{s}$	$/c$
Inductance[a]	henry	$H = 10^9\,\mathrm{cm}$	$= 10^9\ \mathrm{cm}$	$/c^2$

[a] Added in 1889, see Stille [1961, p. 214]; W=VA.
[b] Proposal by Clausius [1882] to name the unit for Φ_B after Weber.

units"led to "impractical"numerical values in the case of charge or current density and were displaced by CGS units.

To calibrate the electrical quantities, one needed etalons (standard sizes) that should come as close as possible to the "absolute"units. The voltage (Daniell element) and the ohm (using silver wire) were defined by the *British Association for the Advancement of Science* in 1863.

Gaussian unit system

The base units of the Gaussian measurement system are, as already mentioned, those of the CGS system. All electromagnetic quantities are expressed through these three base units. In the symmetric CGS system, also known as the Gaussian system, the factors $k_C = k_M = k_C/k_L^2 = 1$. This implies that for

---

[1] Q.E.: Quadrant of the earth's meridian $\simeq 10^7\,\mathrm{m}$ (definition of the meter from 1799).

all quantities in which only k_C occurs, the electrostatic units are equal to the Gaussian, i.e., symmetric CGS units. In the other quantities, k_L only appears as $k_M = k_C/k_L^2$, which in electromagnetic units is equal to one. A Gaussian unit is therefore either equal to an esu or emu.

The unit of the magnetic dipole moment is determined by the potential energy (4.3.5) $U = -\mathbf{m}\cdot\mathbf{B}$: $[m] = [\mathrm{erg/G}]$. Tab. C.4 lists electromagnetic units that are also used in the Gaussian system. There are only a few specific Gaus-

Tab. C.4. Electromagnetic units that are also Gaussian units

magnetomotive force	ϕ_M	gilbert	$1\,\mathrm{Gb}$	$= 1\,\sqrt{\mathrm{cm\,g}}\,s^{-1}$	$\frac{1}{4\pi}\times 10\,\mathrm{A}$
magn. auxiliary field	$\mathbf{H}$	oersted	$1\,\mathrm{Oe}$	$= 1\,\sqrt{\mathrm{cm^{-1}\,g}}\,s^{-1}$	$\frac{1}{4\pi}\times 10^3\,\mathrm{A/m}$
magn. field	$\mathbf{B}$	gauss	$1\,\mathrm{G}$	$= 1\,\sqrt{\mathrm{cm^{-1}\,g}}\,s^{-1}$	$10^{-4}\,\mathrm{T}$
magn. flux	$\varPhi_B$	maxwell	$1\,\mathrm{Mx}$	$= 1\,\sqrt{\mathrm{cm^3\,g}}\,s^{-1}$	$10^{-8}\,\mathrm{Wb}$
magn. dipole	$\mathbf{m}$	erg/gauss	$1\,\mathrm{erg/G}$	$= 1\,\sqrt{\mathrm{cm^5\,g}}\,s^{-1}$	$1\times 10^{-3}\,\mathrm{A\,m^2}$

The symbols Gb, Oe, Gs (here G) and Mx were established in 1935 by the IEC (International Electrotechnical Commission) [Stille, 1961, p. 212].

sian units, such as franklin (Fr) for the charge, but we will use the statcoulomb unit instead of franklin.

Heaviside-Lorentz System

If you take the prefactor $k_C = 1/4\pi$, the inhomogeneous Maxwell equations no longer have 4π factors. The homogeneous equations remain unchanged. This system is called rational. Of course, then 4π factors appear with other quantities, such as the potentials. The units here are irrelevant, as on the one hand the conversion to SI units is simple, on the other hand, in quantum electrodynamics, natural units ($c=1$) are usually used. The rational system is simpler and more consistent than the Gauss system, e.g. concerning the fields in matter:

$$\mathbf{D} = \mathbf{E}+4\pi\mathbf{P} \quad \Leftrightarrow \quad \mathbf{D}^{\mathrm{H}} = \mathbf{E}^{\mathrm{H}}+\mathbf{P}^{\mathrm{H}} \quad \Leftrightarrow \quad \mathbf{D}^{\mathrm{SI}} = \epsilon_0\mathbf{E}^{\mathrm{SI}}+\mathbf{P}^{\mathrm{SI}},$$
$$\mathbf{B} = \mathbf{H}+4\pi\mathbf{M} \quad \Leftrightarrow \quad \mathbf{B}^{\mathrm{H}} = \mathbf{H}^{\mathrm{H}}+\mathbf{M}^{\mathrm{H}} \quad \Leftrightarrow \quad \mathbf{B}^{\mathrm{SI}} = \mu_0\mathbf{H}^{\mathrm{SI}}+\mu_0\mathbf{M}^{\mathrm{SI}}.$$

C.2.2 Systems in Electrodynamics with four Base Units

The reduction of electrical quantities to three mechanical quantities met with criticism towards the end of the 19th century. In the 1st edition of the book

Theorie der Electrizität, which for a long time could be considered the standard in the German-speaking area[2] Föppl [1894, p. 119] notes (translated from German):

A distinction was made between the dimensions of the electrostatic and the electromagnetic systems. Of course, there can only be one true dimension for each quantity, from which it immediately follows that at least one of those systems is incorrect. But they are probably both wrong.

The way out for three dimensions is the symmetric CGS system (i.e., Gaussian system). However, in this system [Föppl, 1894, § 41], $''\rho_{\text{free}}'' = \epsilon_r\epsilon_0\rho_f$ and $''\rho_{\text{true}}'' = \rho_f$ have the same dimension. Föppl [1894, p. 120] writes about this:

Thus, free and true quantities of electricity become quantities of the same kind, as do the electric force and the dielectric displacement. But all this is arbitrary and utterly unbelievable. True and free quantities of electricity must be kept completely separate everywhere, especially in terms of dimensions.

This demand requires four base units. From a physical standpoint, it is understandable that a base unit characteristic of electricity is desirable. In the law of gravity, we have the mass M, the size of which determines the force, as a base unit and the prefactor, the gravitational constant $G = 6.674 \times 10^{-11} \, \text{m}^3/\text{kg s}^2$ has dimensions. In the SI system, $\epsilon_0 = 1/4\pi k_C$ has dimensions and for $k_L = c$ there are also four fields in the vacuum that differ in dimensions. This contradicts the principle of parsimony (*Ockham's Razor*) according to which a theory should manage with as few assumptions and variables as possible.

Electrical and magnetic measurement system

It follows, in analogy to gravity, almost automatically to take the charge q as the fourth base unit. Together with $k_L = c$ this results in the non-rational *electrical measurement system*. Maxwell [1892, Art. 623] however only gives the dimensions of the fields, but avoids the calculation of the sizes, but carries these out with $k_C = 1$, i.e., he switches to electrostatic units.

With the pole strength p_m as the fourth unit and $k_L = c$, Maxwell obtains the (also non-rational) *magnetic system*, but p_m is calculated using $k_M = 1$, i.e., he switches to the electromagnetic system.

Access to a system with an electromagnetic base unit was not easy. Thus, Föppl [1894, §70] gives the dimensions of the electromagnetic quantities using the permittivity ($K \mathrel{\widehat{=}} \epsilon_r\epsilon_0$) for an electrical or the permeability ($\mu \mathrel{\widehat{=}} \mu_r\mu_0$) for a magnetic system and says:

---

[2] Föppl [1894, 1st ed.], Abraham [1904, 2nd ed.], Abraham-Becker [1930, 8th ed.] and Becker, Sauter [1973, 21th and last edition]

For each dimension, for the sake of easier overview and comparability, I will give two expressions, once considering K and the other time μ as the one size about whose dimension nothing could be decided so far.

With the relation $K\mu k_L^2 = 1$ valid in vacuum, Föppl would have made the decisive step further.

SI system

If you fix the SI system, as it is done here, to electrodynamics, i.e. to its representation, this is too short-sighted. It are the standards for the units, a total of 7 physical constants that determine these. If you make the "electrical system" rational ($k_C = 1/4\pi\epsilon_0$, $k_L = c$) and take MKSA units, i.e., you replace the charge with the current (C.2.8), then you have the SI system, in which the Coulomb force (C.2.5) in the denominator has the electric constant ϵ_0 (also called permittivity or dielectric constant of the vacuum). The magnetic constant μ_0 (also called permeability of the vacuum) fulfills the relation (C.2.7):

$$c = \frac{1}{\sqrt{\epsilon_0\mu_0}}. \tag{C.2.4}$$

The field constants were defined, see (C.2.10), using Ampère's force law (4.3.10). In Tab. C.5, some electrical units of the SI system are listed. It is surprising that in the SI system, despite knowledge of the STR, the asymmetry ($k_L = c$) between electrical and magnetic quantities, which was eliminated in the symmetric systems (Gauss, Heaviside-Lorentz), was reverted to. The two

Tab. C.5. SI units; $c = \{c\}$m/s with $\{c\} \approx 3 \times 10^8$, $\{\mu_0\} \approx 4\pi \times 10^{-7}$; $\{\epsilon_0\} \approx \dfrac{10^{-9}}{36\pi}$

μ_0	magnetic constant	$\mu_0 = \{\mu_0\}\,\mathrm{m\,kg\,s^{-2}\,A^{-2}}$	$(...)\,\mathrm{Vs/Am}$
$\epsilon_0 = 1/c^2\mu_0$	electric constant	$\epsilon_0 = \{\epsilon_0\}\,\mathrm{m^{-3}\,kg^{-1}\,s^4\,A^2}$	$(...)\,\mathrm{As/Vm}$
$\phi = \mathbf{E}\cdot\mathbf{l}$	volt	$V = \mathrm{m^2\,kg\,s^{-3}\,A^{-1}}$	$\mathrm{W/A}$
$\mathbf{E} = \mathbf{F}_c/Q$	V/m	$= \mathrm{m\,kg\,s^{-3}\,A^{-1}}$	
$Q = It$	coulomb	$C = \mathrm{sA}$	
$\mathbf{D} = \epsilon_r\epsilon_0\mathbf{E}$	C/m^2	$= \mathrm{m^{-2}\,s\,A}$	
$C = Q/\phi$	farad	$F = \mathrm{m^{-2}\,kg^{-1}\,s^4\,A^2}$	$\mathrm{C/V}$
$R = \phi/I$	ohm	$\Omega = \mathrm{m^2\,kg\,s^{-3}\,A^{-2}}$	$\mathrm{V/A}$
$G = 1/R$	siemens	$S = \mathrm{m^{-2}\,kg^{-1}\,s^3\,A^2}$	$1/\Omega$
$H = I/2\pi l$	A/m	$= \mathrm{m^{-1}\,A}$	
$\mathbf{B} = \mu_r\mu_0\mathbf{H}$	tesla	$T = \mathrm{kg\,s^{-2}\,A^{-1}}$	$\mathrm{Vs/m^2} = \mathrm{N/Am}$
$\Phi_B = Bl^2$	weber	$Wb = \mathrm{m^2\,kg\,s^{-2}\,A^{-1}}$	$\mathrm{Tm^2} = \mathrm{Vs}$
$L = \phi/\dot{I}$	henry	$H = \mathrm{m^2\,kg\,s^{-2}\,A^{-2}}$	$\mathrm{Wb/A}$

definitions, the magnetization $\mathbf{M}$ and magnetic polarization $\mathbf{J}$ may be considered as disadvantageous for SI, since their dimensions are different [Goldfarb, 2018]:

Sommerfeld convention: $\mathbf{B} = \mu_0(\mathbf{H}+\mathbf{M})$, Kennelly convention: $\mathbf{B} = \mu_0\mathbf{H}+\mathbf{J}$.

The electric and magnetic constants ϵ_0 and μ_0

In matter, the electric field is weakened by the factor ϵ_r and therefore the Coulomb force (1.2.1) between two charges embedded in a homogeneous, linear medium is correspondingly weaker:

$$\mathbf{F}_C = q\mathbf{E}, \qquad\qquad \mathbf{E} = -k_C\frac{q'}{\epsilon_r}\boldsymbol{\nabla}\frac{1}{|\mathbf{x}-\mathbf{x}'|}. \qquad (C.2.5)$$

If one introduces an electrical base unit, then the product $k_C qq'$ must yield a mechanical quantity, since on the left the force $\mathbf{F}_C$ stands, i.e., k_C is dimension dependent. It is obvious to set the proportionality factor for the strength of interaction in vacuum analogous to the medium, where it is $1/\epsilon_r$, with $k_C \sim 1/\epsilon_0$. If one opts for a rational system, then $1/4\pi$ is added to the proportionality factor, i.e., one has in the SI system $k_C = 1/4\pi\epsilon_0$.

In vacuum, the two constants k_C and k_L are sufficient, but in matter it was advantageous to introduce the auxiliary fields (5.2.17)

$$\mathbf{D} = \epsilon_r\epsilon_0\mathbf{E}, \qquad\qquad \mathbf{B} = \mu_r\mu_0\mathbf{H}. \qquad (C.2.6)$$

$\epsilon_r\epsilon_0$ is the dielectric constant and $\mu_r\mu_0$ is the permeability; here ϵ_r and μ_r are numbers, which are also referred to as relative permittivity (dielectric constant) and relative permeability. Corresponding to the dielectric constant of the vacuum ϵ_0, the constant μ_0 is introduced as "permeability of the vacuum". If one calculates the field energy (5.6.7) using (5.2.8'), then $k_L^2\epsilon_0\mu_0 = 1$ establishes the symmetry between electric and magnetic field energy:

$$u = \frac{1}{8\pi k_C\epsilon_0}\left[\mathbf{E}\cdot\mathbf{D}+(k_L^2\epsilon_0\mu_0)\mathbf{B}\cdot\mathbf{H}\right] \qquad \Rightarrow \qquad \mu_0 = 1/k_L^2\epsilon_0. \qquad (C.2.7)$$

μ_0 was then determined by the current (4.3.10) in two parallel wires. It should be pointed out again that ϵ_0 and μ_0 are not natural constants, but a consequence of the system of measurement. ϵ_0 is determined by k_C and μ_0 is also contributed by k_L.

C.2.3 On the Switch between Systems and Units

The definition of current strength

In an effort to trace all base units back to natural constants, the ampere has been redefined as part of the revision of the SI system since 2019, referring back to the elementary charge e_0. In addition, μ_0 is now determined from the constants h, e_0 and c set in the revision of the SI system with the experimental value of the fine structure constant α_f (see Tab. C.7, p. 645) [Goldfarb, 2018]:

$$\text{SI: } 1\,\text{A} = \frac{e_0 \times 10^{19}\,\text{s}^{-1}}{1.602\,176\,634}, \quad \mu_0 = \frac{2h}{ce_0^2}\alpha_f = \{\mu_0\}\frac{\text{N}}{\text{A}^2}, \quad \{\mu_0\} \cong 4\pi \times 10^{-7}. \quad \text{(C.2.8)}$$

μ_0 is therefore no longer exactly determined by the current of $1\,\text{A}$, but by an independent measurement of α_f. In electrodynamics, there are three older definitions of the ampere [Stille, 1961, p. 214]. In quadrant units, the absolute ampere was defined in 1881 at the 1st International Electricity Congress in Paris as a derived unit: $1\,\text{A} = 1$ V/Ω.

The *international ampere* was defined as the current that deposits 0.001118g of silver from a silver nitrate solution (established in 1898 in Germany; set at the International Conference in London in 1908).

The definition that was valid until recently was proposed by the *Comité International des Poids et Mesures* in 1946 [Stille, 1961, p. 221] and recognized by the IUPAP (The International Union of Pure and Applied Physics) in 1948:

An ampere is the strength of a constant electric current, which, flowing through two parallel conductors in a vacuum, placed 1 meter apart from each other, straight, infinitely long conductors of negligibly small, circular cross-section, would produce a force of 2×10^{-7} Newton per meter of conductor length.

The force that two parallel equal currents exert on each other, as sketched in Fig. C.1, is determined by (4.3.10). If this is given as $F = 2 \times 10^{-7}\,\text{N} = 0.02\,\text{dyn}$ for a length $l = d$, then the current strength is always equal to 1 ampere:

$$F = \frac{k_\text{C}}{c^2}\frac{2l\,I^2}{d} \quad \overset{d=l,\,K=0.02\text{dyn}}{\Longrightarrow} \quad \begin{aligned} I^\text{SI} &= \sqrt{4\pi/\mu_0}\,\sqrt{10^{-7}\text{N}} = 1\,\text{A}, \\ I^\text{emu} &= 0.1\sqrt{\text{dyn}} = 0.1\,\text{abA}, \\ I^\text{esu} &= 0.1c\sqrt{\text{dyn}} \approx 3 \times 10^9\,\text{statA}. \end{aligned} \quad \text{(C.2.9)}$$

The ampere therefore has different dimensions (numerical values), depending on the unit system [Cardarelli, 2003, p. 20–25]. In the old SI system, the mag-

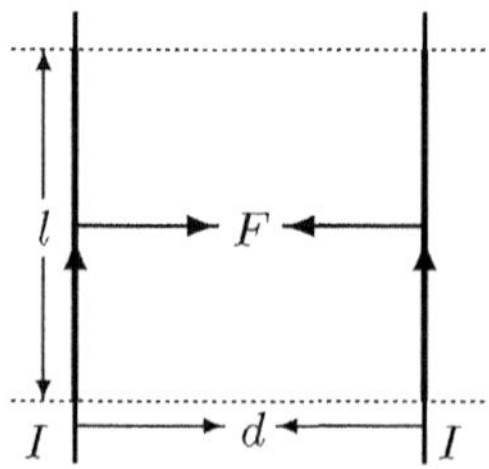

Fig. C.1. Force that two parallel current-carrying conductors exert on each other:
$d = l\,[= 1\,\text{m}]$, $F = 2 \times 10^{-7}\text{N} \Rightarrow I = 1\,\text{A}$.
If the currents are in the same direction, the conductors attract each other

netic constant is exactly given by (C.2.9). Using (C.2.4), the electric constant is also determined:

$$\boxed{\begin{aligned}
\text{SI}: \mu_0 &= \frac{4\pi F}{2I^2}\frac{d}{l} \overset{d=l}{\underset{F=2\times10^{-7}\mathrm{N}}{=}} 4\pi\times10^{-7}\frac{\mathrm{N}}{\mathrm{A}^2}, \\
\epsilon_0 &= \frac{1}{\mu_0\,c^2} = \frac{10^7}{4\pi\{c^2\}}\frac{\mathrm{A}^2\,\mathrm{s}^2}{\mathrm{m}^2\,\mathrm{N}} \approx \frac{1}{36\pi}\times10^{-9}\frac{\mathrm{A}^2\,\mathrm{s}^2}{\mathrm{m}^2\,\mathrm{N}}.
\end{aligned}}$$
$$\tag{C.2.10}$$

Tab. C.6 expresses the Gaussian quantities in terms of the SI quantities and thus provides a simple way to immediately rewrite any equation with electromagnetic quantities into the SI system.

Conversion of units

The starting point for the switch from Gaussian units, i.e., esu or emu, to SI units is the definition of the ampere (C.2.9):

$$1\,\mathrm{statA} = 1\,\sqrt{\mathrm{cm}^3\mathrm{g}}\,\mathrm{s}^{-2} = \frac{1\,\mathrm{A}}{10\{c\}} \approx \frac{1}{3}\times10^{-9}\,\mathrm{A}, \qquad c = \{c\}\,\mathrm{ms}^{-1},$$

$$1\,\mathrm{abA} = 1\,\sqrt{\mathrm{cmg}}\,\mathrm{s}^{-1} = 10\,\mathrm{A}. \tag{C.2.11}$$

With the help of these relations between the current intensities, the conversion factors can be given:

$$\frac{\mu_0}{4\pi} \overset{\mathrm{emu}}{=} 10^{-7}\frac{10^5\,\mathrm{dyn}}{0.01\,\mathrm{dyn}} = 1, \qquad 4\pi\epsilon_0 \overset{\mathrm{esu}}{=} \frac{10^7}{c^2}\frac{0.01c^2\mathrm{dyn}}{10^5\,\mathrm{dyn}} = 1. \tag{C.2.12}$$

The reverse way is simpler, by using the above cgs factor for the ampere when converting from SI to esu/emu. Thus, the SI system has four fields ($\mathbf{E}$, $\mathbf{D}$; $\mathbf{B}$, $\mathbf{H}$), which all have the same cgs dimension in the Gaussian system $\sqrt{\mathrm{cm}^{-1}\mathrm{gs}^{-1}}$. Using (C.2.12) we get:

$$\begin{aligned}
B: \quad & 1\,\mathrm{T} = \frac{1\,\mathrm{kg}}{\mathrm{s}^2\,\mathrm{A}}\sqrt{\frac{\mu_0}{4\pi}} = \frac{10^3\mathrm{gs}^{-2}}{0.1\sqrt{\mathrm{dyn}}} = 10^4\,\sqrt{\mathrm{cm}^{-1}\mathrm{gs}^{-1}} = 10^4\,\mathrm{G}, \\
H: \quad & \frac{1\,\mathrm{A}}{\mathrm{m}} = \frac{1\,\mathrm{A}}{\mathrm{m}\sqrt{4\pi\mu_0}} = \frac{0.1\sqrt{\mathrm{dyn}}}{4\pi10^2\mathrm{cm}} = \frac{10^{-3}}{4\pi}\sqrt{\mathrm{cm}^{-1}\mathrm{gs}^{-1}} = \frac{10^{-3}}{4\pi}\,\mathrm{Oe}, \\
J: \quad & 1\,\mathrm{T} = 1\,\mathrm{T}\sqrt{4\pi\mu_0} = 4\pi\times10^4\,\mathrm{G}, \\
m: \quad & 1\,\mathrm{Am}^2 = 1\,\mathrm{Am}^2\sqrt{\frac{4\pi}{\mu_0}} = 10^3\,\sqrt{\mathrm{cm}^5\mathrm{gs}^{-2}} = 10^3\,\mathrm{erg/G}.
\end{aligned}$$

The magnetic polarization for a magnetization of $1\,\mathrm{A/m}$ is given by $J^{\mathrm{SI}} = \{\mu_0\}\frac{\mathrm{N}}{\mathrm{mA}} = 4\pi\times10^{-7}\mathrm{T}$, since $\mathrm{N/A} = \mathrm{mT}$ [$=$ momentum/charge].

Physical constants

The physical quantities listed in Table C.7 refer to electrodynamics and atomic physics. There are relationships among the quantities, which are particularly important for estimations. Starting from a_B applies

Tab. C.6. Conversion Gauss $\leftrightharpoons$ SI. The factor 3 is actually $2.997\,924\,58$

Quantity	Relation		Conversion
Dipolemoment	$\mathbf{p}=\dfrac{1}{\sqrt{4\pi\epsilon_0}}\,\mathbf{p}^{\mathrm{SI}}$	$10^{18}\mathrm{Debye}$	$=1\sqrt{\mathrm{g\,cm^5\,s^{-1}}}\ \widehat{=}\ \tfrac{1}{3}\times 10^{-11}\,\mathrm{Cm}$
Charge	$Q=\dfrac{1}{\sqrt{4\pi\epsilon_0}}\,Q^{\mathrm{SI}}$	$1\,\mathrm{statC}$	$=1\sqrt{\mathrm{g\,cm^3\,s^{-1}}}\ \widehat{=}\ \tfrac{1}{3}\times 10^{-9}\,\mathrm{C}$
Polarization	$\mathbf{P}=\dfrac{1}{\sqrt{4\pi\epsilon_0}}\,\mathbf{P}^{\mathrm{SI}}$	$1\,\dfrac{\mathrm{statC}}{\mathrm{cm^2}}$	$=1\sqrt{\mathrm{g\,cm^{-1}\,s^{-1}}}\ \widehat{=}\ \tfrac{1}{3}\times 10^{-5}\,\dfrac{\mathrm{C}}{\mathrm{m^2}}$
Chargedensity	$\rho=\dfrac{1}{\sqrt{4\pi\epsilon_0}}\,\rho^{\mathrm{SI}}$	$1\,\dfrac{\mathrm{statC}}{\mathrm{cm^3}}$	$=1\sqrt{\mathrm{g\,cm^{-3}\,s^{-1}}}\ \widehat{=}\ \tfrac{1}{3}\times 10^{-3}\,\dfrac{\mathrm{C}}{\mathrm{m^3}}$
Current	$I=\dfrac{1}{\sqrt{4\pi\epsilon_0}}\,I^{\mathrm{SI}}$	$1\,\mathrm{statA}$	$=1\sqrt{\mathrm{g\,cm^3\,s^{-2}}}\ \widehat{=}\ \tfrac{1}{3}\times 10^{-9}\,\mathrm{A}$
Currentdensity	$\mathbf{j}=\dfrac{1}{\sqrt{4\pi\epsilon_0}}\,\mathbf{j}^{\mathrm{SI}}$	$1\,\dfrac{\mathrm{statA}}{\mathrm{cm^2}}$	$=1\sqrt{\mathrm{g\,cm^{-1}\,s^{-2}}}\ \widehat{=}\ \tfrac{1}{3}\times 10^{-5}\,\dfrac{\mathrm{A}}{\mathrm{m^2}}$
el. current	$\Phi_E=\sqrt{4\pi\epsilon_0}\,\Phi_E^{\mathrm{SI}}$	$1\,\mathrm{statV\,cm}$	$=1\sqrt{\mathrm{g\,cm^3\,s^{-1}}}\ \widehat{=}\ 3\times 10^0\,\mathrm{V\,m}$
Voltage	$\phi=\sqrt{4\pi\epsilon_0}\,\phi^{\mathrm{SI}}$	$1\,\mathrm{statV}$	$=1\sqrt{\mathrm{g\,cm}}\,\mathrm{s^{-1}}\ \widehat{=}\ 3\times 10^2\,\mathrm{V}$
el. Field	$\mathbf{E}=\sqrt{4\pi\epsilon_0}\,\mathbf{E}^{\mathrm{SI}}$	$1\,\dfrac{\mathrm{statV}}{\mathrm{cm}}$	$=1\sqrt{\mathrm{g\,cm^{-1}\,s^{-1}}}\ \widehat{=}\ 3\times 10^4\,\dfrac{\mathrm{V}}{\mathrm{m}}$
diel. Flux $[\Psi]$	$\Phi_D=\sqrt{\dfrac{4\pi}{\epsilon_0}}\,\Phi_D^{\mathrm{SI}}$	$1\,\mathrm{statC}$	$=1\sqrt{\mathrm{g\,cm^3\,s^{-1}}}\ \widehat{=}\ \dfrac{1}{12\pi}\times 10^{-9}\,\mathrm{C}$
diel. Field	$\mathbf{D}=\sqrt{\dfrac{4\pi}{\epsilon_0}}\,\mathbf{D}^{\mathrm{SI}}$	$1\,\dfrac{\mathrm{statC}}{\mathrm{cm^2}}$	$=1\sqrt{\mathrm{g\,cm^{-1}\,s^{-1}}}\ \widehat{=}\ \dfrac{1}{12\pi}\times 10^{-5}\,\dfrac{\mathrm{C}}{\mathrm{m^2}}$
Conductivity	$\sigma=\dfrac{1}{4\pi\epsilon_0}\,\sigma^{\mathrm{SI}}$	$\dfrac{1}{\mathrm{statO\,cm}}$	$=1\,\mathrm{s^{-1}}\qquad \widehat{=}\ \tfrac{1}{9}\times 10^{-9}\,\dfrac{\mathrm{S}}{\mathrm{m}}$
Capacitance	$C=\dfrac{1}{4\pi\epsilon_0}\,C^{\mathrm{SI}}$	$1\,\mathrm{statF}$	$=1\,\mathrm{cm}\qquad \widehat{=}\ \tfrac{1}{9}\times 10^{-11}\,\mathrm{F}$
Resistance	$R=4\pi\epsilon_0\,R^{\mathrm{SI}}$	$1\,\mathrm{statO}$	$=1\,\mathrm{cm^{-1}\,s}\qquad \widehat{=}\ 9\times 10^{11}\,\Omega$
Inductance	$L=4\pi\epsilon_0\,L^{\mathrm{SI}}$	$1\,\mathrm{statH}$	$=1\,\mathrm{cm^{-1}\,s^2}\qquad \widehat{=}\ 9\times 10^{11}\,\mathrm{H}$
mag. Flux	$\Phi_B=\sqrt{\dfrac{4\pi}{\mu_0}}\,\Phi_B^{\mathrm{SI}}$	$1\,\mathrm{Mx}$	$=1\sqrt{\mathrm{g\,cm^3\,s^{-1}}}\ \widehat{=}\ 1\times 10^{-8}\,\mathrm{Wb}$
Vector potential	$\mathbf{A}=\sqrt{\dfrac{4\pi}{\mu_0}}\,\mathbf{A}^{\mathrm{SI}}$	$1\,\dfrac{\mathrm{abV\,s}}{\mathrm{cm}}$	$=1\sqrt{\mathrm{g\,cm}}\,\mathrm{s^{-1}}\ \widehat{=}\ 1\times 10^{-6}\,\dfrac{\mathrm{V\,s}}{\mathrm{m}}$
Magnetic field	$\mathbf{B}=\sqrt{\dfrac{4\pi}{\mu_0}}\,\mathbf{B}^{\mathrm{SI}}$	$1\,\mathrm{G}$	$=1\sqrt{\mathrm{g\,cm^{-1}\,s^{-1}}}\ \widehat{=}\ 1\times 10^{-4}\,\mathrm{T}$
mag. Polarization	$\mathbf{J}=\dfrac{1}{\sqrt{4\pi\mu_0}}\,\mathbf{J}^{\mathrm{SI}}$	$1\,\mathrm{G}$	$=1\sqrt{\mathrm{g\,cm^{-1}\,s^{-1}}}\ \widehat{=}\ 4\pi\times 10^{-4}\,\mathrm{T}$
mag. Polestrength[a]	$p_B=\dfrac{1}{\sqrt{4\pi\mu_0}}\,p_B^{\mathrm{SI}}$	$1\,\mathrm{Unitpole}$	$=1\sqrt{\mathrm{g\,cm^3\,s^{-1}}}\ \widehat{=}\ 4\pi\times 10^{-8}\,\mathrm{Wb}$
mag. Potential	$\phi_M=\sqrt{4\pi\mu_0}\,\phi_M^{\mathrm{SI}}$	$1\,\mathrm{Gb}$	$=1\sqrt{\mathrm{g\,cm}}\,\mathrm{s^{-1}}\ \widehat{=}\ \dfrac{1}{4\pi}\times 10^1\,\mathrm{A}$
mag. auxiliary field	$\mathbf{H}=\sqrt{4\pi\mu_0}\,\mathbf{H}^{\mathrm{SI}}$	$1\,\mathrm{Oe}$	$=1\sqrt{\mathrm{g\,cm^{-1}\,s^{-1}}}\ \widehat{=}\ \dfrac{1}{4\pi}\times 10^3\,\dfrac{\mathrm{A}}{\mathrm{m}}$
mag. moment	$\mathbf{m}=\sqrt{\dfrac{\mu_0}{4\pi}}\,\mathbf{m}^{\mathrm{SI}}$	$1\,\mathrm{erg/G}$	$=1\sqrt{\mathrm{g\,cm^5\,s^{-1}}}\ \widehat{=}\ 1\times 10^{-3}\,\mathrm{A\,m^2}$
mag. polestrength	$p_m=\sqrt{\dfrac{\mu_0}{4\pi}}\,p_m^{\mathrm{SI}}$	$1\,\mathrm{dyn/G}$	$=1\sqrt{\mathrm{g\,cm^3\,s^{-1}}}\ \widehat{=}\ 1\times 10^{-1}\,\mathrm{Am}$
Magnetization	$\mathbf{M}=\sqrt{\dfrac{\mu_0}{4\pi}}\,\mathbf{M}^{\mathrm{SI}}$	$1\,\mathrm{erg/cm^3\,G}$	$=1\sqrt{\mathrm{g\,cm^{-1}\,s^{-1}}}\ \widehat{=}\ 1\times 10^3\,\dfrac{\mathrm{A}}{\mathrm{m}}$
Poynting vector	$\mathbf{S}=\mathbf{S}^{\mathrm{SI}}$	$1\,\mathrm{erg/cm^2\,s}$	$=1\,\mathrm{g\,s^{-3}}\qquad \widehat{=}\ 1\times 10^{-3}\,\mathrm{kg\,s^{-3}}$
electric/magnetic susceptibility	$\chi=\dfrac{1}{4\pi}\,\chi^{\mathrm{SI}}$		

[a] Semat, Katz [1958, (29-2)]: Conversion cgs-unit pole strength $\rightarrow$ Wb.

Tab. C.7. Atomic sizes: G: $k_C=1$, $k_L=1$,   SI: $k_C=1/4\pi\epsilon_0$, $k_L=c$

Speed of light in vacuum	$c = 2.997\,924\,58\ \times 10^{10}\,\mathrm{cm\,s^{-1}}$
Elementary charge	$e_0 = 4.803\,204\,40\ \times 10^{-10}\,\mathrm{statC}$
	$= 1.602\,176\,634\times 10^{-19}\,\mathrm{C}$
Planck's constant	$h = 6.626\,070\,15\ \times 10^{-34}\,\mathrm{J\,s}$
	$\hbar = h/2\pi = 1.054\,572\ \ \times 10^{-27}\,\mathrm{erg\,s}$
	$= 6.582\,119\ \ \ \times 10^{-16}\,\mathrm{eV\,s}$

Boltzmann constant	$k_B = 1.380\,649\times 10^{-16}\,\mathrm{erg/K}$
	$= 8.617\,343\times 10^{-5}\,\mathrm{eV/K}$
Electron mass	$m_e = 9.109\,382\times 10^{-28}\,\mathrm{g}$
	$= 0.510\,999\times 10^{6}\,\mathrm{eV}/c^2$
Fine-structure constant	$\alpha_f = k_C e_0^2/\hbar c = 1/137.036$
Impedance of free space	$Z_0 = 4\pi k_C/c = 376.73\,\Omega$

Compton wavelength of the electron	$\lambdabar_c = h/m_e c = 2.426\,31\times 10^{-10}\,\mathrm{cm}$
Bohr radius	$a_B = \hbar^2/k_C e_0^2 m_e = 5.291\,77\times 10^{-9}\,\mathrm{cm}$
Classical electron radius	$r_e = k_C e_0^2/m_e c^2 = 2.817\,94\times 10^{-13}\,\mathrm{cm}$

Rest energy of the electron	$m_e c^2 = 8.187\,10\times 10^{-7}\,\mathrm{erg}$
	$= 8.187\,10\times 10^{-14}\,\mathrm{J}$
	$= 5.109\,99\times 10^{5}\,\mathrm{eV}$
Bohr magneton	$\mu_B = k_L e_0 \hbar/2m_e c = 9.274\,01\times 10^{-21}\,\mathrm{erg/G}$
	$= 9.274\,01\times 10^{-24}\,\mathrm{J/T}$
	$= 5.788\,38\times 10^{-5}\,\mathrm{eV/T}$
Larmor frequency of the electron	$\omega_L = k_L e_0 B/2m_e c$
Cyclotron-frequency	$\omega_C = 2\omega_L$

$$\lambdabar_c = a_B \alpha_f \qquad \text{and} \qquad r_e = \lambdabar_c \alpha_f = a_B \alpha_f^2 . \qquad\qquad (\mathrm{C.2.13})$$

The energy levels of hydrogen-like atoms with atomic number Z are [Schwabl, 2007, (6.24')]

$$E_n = k_C^2 \frac{m_e Z^2 e_0^4}{2\hbar^2 n^2} = -k_C \frac{Z^2}{n^2}\frac{e_0^2}{2a_B} = -\frac{Z^2}{n^2} m_e c^2 \alpha_f^2 . \qquad\qquad (\mathrm{C.2.14})$$

C.3 Euclidean Metric with Imaginary Time

Minkowski [1907] introduced a four-dimensional notation in SRT and electrodynamics. The pseudo-Euclidean metric is packaged in an imaginary time axis which usually appears as x_4. We stick to x_0:

$$(\bar{x}_\mu) = (\mathrm{i}ct,\ \mathbf{x}), \qquad (\bar{\partial}_\mu) = \left(-\mathrm{i}\frac{\partial}{c\partial t},\ \boldsymbol{\nabla}\right) = -(\mathrm{i}\partial^0, (\partial^j)). \qquad (\mathrm{C.3.1})$$

Co- and contravariant indices are the same, which is why the superscript of the indices is omitted:

$$\bar{s}^2 = \bar{x}_\mu \bar{x}_\mu = \mathbf{x}^2 - c^2 t^2 \overset{(12.2.2')}{=} -g_{\mu\nu} x^\mu x^\nu. \tag{C.3.2}$$

In this notation, one automatically has the signature $(-1,1,1,1)$: space-like or east coast convention. It is used in many, especially older books and articles [Pauli, 1921; Landau, Lifshitz 2 , 1987; Greiner, 1991]. Tab. C.8 is based on the four-vectors of Tab. 13.2, p. 493, noting that the scalar products have the opposite sign.

Tab. C.8. Four-vectors in the metric with imaginary time axis $x_=ict$

Four-vector	Reference	Representation of the four-vector
Event	(12.2.1)	$(\bar{x}_\mu) = (ict, \mathbf{x})$
Gradient	(12.2.6)	$(\bar{\partial}_\mu) = (-i\partial_0, \boldsymbol{\nabla})$
Speed	(12.3.7)	$(\bar{u}_\mu) = (i\gamma, \gamma\boldsymbol{\beta})$
Momentum	(14.1.1)	$(\bar{p}_\mu) = (i\gamma mc, \boldsymbol{p})$
Acceleration	(14.1.6)	$(\bar{a}_\mu) = \left(i\dot{\gamma}(s), \dot{\gamma}(s)\boldsymbol{\beta} + \gamma\dot{\boldsymbol{\beta}}(s)\right)$
Current density	(13.1.4)	$(\bar{j}_\mu) = (ic\rho, \rho\mathbf{v})$
Potential	(13.1.10)	$(\bar{A}_\mu) = (i\phi/k_L, \mathbf{A})$
Force density	(14.1.25)	$(\bar{f}_\mu) = \rho\left(i\boldsymbol{\beta}\cdot\mathbf{E}, \mathbf{E} + k_L\boldsymbol{\beta}\times\mathbf{B}\right)$

$$\begin{pmatrix} \bar{T}_{00} & (\bar{T}_{0j}) \\ (\bar{T}_{i0}) & (\bar{T}_{ij}) \end{pmatrix} = \begin{pmatrix} -T^{00} & i(T^{0j}) \\ i(T^{i0}) & (T^{ij}) \end{pmatrix}, \quad \begin{pmatrix} \bar{F}_{00} & (\bar{F}_{0j}) \\ (\bar{F}_{i0}) & (\bar{F}_{ij}) \end{pmatrix} = -\begin{pmatrix} -F^{00} & i(F^{0j}) \\ i(F^{i0}) & (F^{ij}) \end{pmatrix}. \tag{C.3.3}$$

If one considers the stress-energy tensor from Minkowski [1908, (74)] one obtains

$$(T_M^{\mu\nu}) \overset{(14.1.31)}{=} \begin{pmatrix} -u_{\text{Field}} & -c\boldsymbol{p}_{\text{Field}}^T \\ -\mathbf{S}/c & (T_{mn}) \end{pmatrix} \;\rightarrow\; (\bar{T}_{M\mu\nu}) = \begin{pmatrix} u_{\text{Field}} & -ic\boldsymbol{p}_{\text{Field}}^T \\ -i\mathbf{S}/c & (T_{mn}) \end{pmatrix} \tag{C.3.4}$$

For the field tensors [Sommerfeld, 1952, (26.17), (26.14)] results, whereby in vacuum $H^{\mu\nu} = (1/\mu_0)F^{\mu\nu}$:

$$(F^{\kappa\lambda}) \overset{(13.1.16)}{=} k_S \begin{pmatrix} 0 & -\mathbf{E}^T/k_L \\ \mathbf{E}/k_L & -(\epsilon_{kln}B_n) \end{pmatrix} \;\rightarrow\; (\bar{F}_{\kappa\lambda}) = \begin{pmatrix} 0 & i\mathbf{E}^T/k_L \\ -i\mathbf{E}/k_L & (\epsilon_{kln}B_n) \end{pmatrix} \tag{C.3.5}$$

$$(H^{\kappa\lambda}) \overset{(13.2.1)}{=} k_S \begin{pmatrix} 0 & -k_L\mathbf{D}^T \\ k_L\mathbf{D} & -(\epsilon_{kln}H_n) \end{pmatrix} \;\rightarrow\; (\bar{H}_{\kappa\lambda}) = \begin{pmatrix} 0 & ik_L\mathbf{D}^T \\ -ik_L\mathbf{D} & (\epsilon_{kln}H_n) \end{pmatrix}.$$

Finally, the Maxwell equations ((13.1.24) and (13.2.2)):

$$F^{\nu\mu}{}_{,\nu} = k_S \frac{4\pi k_C}{ck_L} j^\mu, \qquad\qquad \bar{F}_{\nu\mu,\nu} = -\frac{4\pi k_C}{ck_L} \bar{j}_\mu, \tag{C.3.6}$$

$$H^{\nu\mu}{}_{,\nu} = k_S \frac{k_L}{c} 4\pi k_C \epsilon_0 j^\mu, \qquad \bar{H}_{\nu\mu,\nu} = -\frac{k_L}{c} 4\pi k_C \epsilon_0 \bar{j}_\mu, \qquad k_S \overset{(12.2.2')}{=} \pm 1.$$

Thus, essential equations are accessible for all systems.

References

Abraham M. *Theorie der Elektrizität: Einf"uhrung in die Maxwellsche Theorie der Elektrizität*, 2nd ed. Teubner, Leipzig (1904); 4th ed. (1912)

Abraham M., Becker R. *Theorie der Elektrizität 1*, 8th ed. Teubner, Leipzig (1930)

Becker R., Sauter F. *Theorie der Elektrizität 1*, 21th ed. Teubner, Stuttgart (1973)

Cardarelli F. *Encyclopaedia of Scientific Units, Weights and Measures: Their SI Equivalences and Origins* 1st ed. Springer London (2003)

Clausius R. *Ueber die verschiedenen Maasssysteme zur Messung electrischer und magnetischer Grössen*, Annalen der Physik und Chemie **252**, 529–551 (1882)

Fischer J. *Größen und Einheiten der Elektrizitätslehre* Springer Berlin (1961)

Föppl A. *Einführung in die Maxwellsche Theorie der Elektrizität*, Teubner, Leipzig (1894)

Forster C. et al. *Report of the Committee appointed for the purpose of constructing and issuing Practical Standards for use in Electrical Measurements* in Report of the 59. Meeting of the British Association for the Advancement of Science (BAAS), John Murray, London (1890), 41–44

Goldfarb R. B. *Electromagnetic Units, the Giorgi System, and the Revised International System of Units*, IEEE Magn. Lett. **9**, Article Sequence Number 1205905 (2018)

Greiner W. *Classical Electrodynamics*, 2nd ed. Springer (1991)

Jackson J. D. *Classical Electrodynamics*, 3rd ed., John Wiley & Sons Inc. (1998)

Landau L. D. and Lifshitz E. M. *The Classical Theory of Fields*, vol. 2, 4th ed. Butterworth-Heinemann, reprinted (1994)

Maxwell J. C. *A treatise on electricity and magnetism*, 2 volumes, 3rd ed. Clarendon Press Oxford (1892)

Minkowski H. *Die Grundgleichungen für die elektromagnetischen Vorgänge in bewegten Körpern*, Nachr. Königl. Ges. Wiss. Göttingen, 53–111 (1908)

Minkowski H. *Das Relativitätsprinzip*, Vortrag vor der Göttinger mathemat. Gesellschaft am 5. 11. 1907, Ann. Physik **47**, 927–938 (1915)

Pauli W. *Theory of Relativity*, Pergamon Press (1958) translated from *Relativitätstheorie*, Encyklopädie der Mathematischen Wissenschaften **V19**, Teubner (1921)

Schwabl F. *Quantum Mechanics*, 4th ed. Springer Berlin (2007)

Semat H., Katz R. *Physics*, Rinehart & Company Inc. New York (1958)

Sommerfeld A. *Electrodynamics*, Academic Press Inc., New York (1952)

Stille U. *Die Konstante c in der Elektrodynamik*, Phys. Blätter **13**, 14–22 (1957)

Stille U. *Messen und Rechnen in der Physik*, 2. Aufl. Springer Berlin (1961)

Thomson W. et al. *Second Report of the Committee for the Selection and Nomenclature of Dynamical and Electrical Units* in Report of the 44. Meeting of the BAAS, John Murray, London (1875)

Weber W. & Kohlrausch R. *Über die Elektrizitätsmenge welche bei galvanischen Strömen durch den Querschnitt der Kette fließt*, Pogg. Ann. XCIX, 10–25 (1856) or Ostwalds Klassiker vol. **142**, Akad. Verlagsges. Leipzig (1904)

Wheatstone C. et al. *Report of the Committee appointed to the British Association on Standards of Electrical Resistance* in Report of the 33. Meeting of the BAAS, John Murray, London (1864), p. 111–176

D

Compilation of Formulas for Electrodynamics

D.1 Equations of Electrodynamics in System-independent Form

Tab. 4.1. Prefactors for the unit systems of electrodynamics.

System	k_C	k_L	ϵ_0	$\mu_0 = \dfrac{1}{k_L^2 \epsilon_0}$	$k_r = k_C \epsilon_0$	$k_M = \dfrac{k_C}{k_L^2}$	$\dfrac{k_C}{ck_L}$	$\dfrac{4\pi k_C}{ck_L}$
Gauss	1	1	1	1	1	1	$\dfrac{1}{c}$	$\dfrac{4\pi}{c}$
Heaviside-Lorentz	$\dfrac{1}{4\pi}$	1	1	1	$\dfrac{1}{4\pi}$	$\dfrac{1}{4\pi}$	$\dfrac{1}{4\pi c}$	$\dfrac{1}{c}$
SI	$\dfrac{1}{4\pi\epsilon_0}$	c	ϵ_0	$\mu_0 = \dfrac{1}{c^2 \epsilon_0}$	$\dfrac{1}{4\pi}$	$\dfrac{\mu_0}{4\pi}$	$\dfrac{\mu_0}{4\pi}$	μ_0

Fundamental Equations of Electrodynamics

Microscopic Maxwell equations in free space

$$
\begin{array}{ll}
\text{(a)} \quad \boldsymbol{\nabla}\cdot\mathbf{E} = 4\pi k_C \rho & \text{(b)} \quad \boldsymbol{\nabla}\times\mathbf{E} + \dfrac{k_L}{c}\dot{\mathbf{B}} = 0 \\
\text{Gauss's law} & \text{law of induction} \\
\text{(c)} \quad k_L \boldsymbol{\nabla}\times\mathbf{B} - \dfrac{1}{c}\dot{\mathbf{E}} = \dfrac{4\pi k_C}{c}\mathbf{j} & \text{(d)} \quad \boldsymbol{\nabla}\cdot\mathbf{B} = 0 \\
\text{Ampère-Maxwell law} & \text{source-free magnetic field}
\end{array}
$$

$$(1.3.21)$$

Macroscopic Maxwell equations in matter

$$
\begin{array}{ll}
\text{(a)} \quad \boldsymbol{\nabla}\cdot\mathbf{D} = 4\pi k_r \rho_f & \text{(b)} \quad \boldsymbol{\nabla}\times\mathbf{E} + \dfrac{k_L}{c}\dot{\mathbf{B}} = 0 \\
\text{(c)} \quad \boldsymbol{\nabla}\times\mathbf{H} - \dfrac{k_L}{c}\dot{\mathbf{D}} = 4\pi k_r \dfrac{k_L}{c}\mathbf{j}_f & \text{(d)} \quad \boldsymbol{\nabla}\cdot\mathbf{B} = 0
\end{array}
$$

$$(5.2.16)$$

© The Editor(s) (if applicable) and The Author(s), under exclusive license
to Springer-Verlag GmbH, DE, part of Springer Nature 2025
D. Petrascheck, F. Schwabl, *Electrodynamics*, https://doi.org/10.1007/978-3-662-71502-4_18

Material or constitutive equations

$$\boxed{\begin{aligned} \mathbf{B} &= \mu_r\mu_0\mathbf{H} = (1+4\pi k_r\chi_m)\mu_0\mathbf{H} = \mu_0(\mathbf{H}+4\pi k_r\mathbf{M}) \\ \mathbf{D} &= \epsilon_r\epsilon_0\mathbf{E} = (1+4\pi k_r\chi_e)\epsilon_0\mathbf{E} = \epsilon_0\mathbf{E}+4\pi k_r\mathbf{P} \end{aligned}} \qquad (5.2.17)$$

$$\boxed{\mathbf{F} = q\big(\mathbf{E} + (k_L/c)\,\mathbf{v}\times\mathbf{B}\big)} \qquad \begin{array}{c}\text{Lorentz force:} \\ \text{Coulomb} + \text{Lorentz part}\end{array} \qquad (1.2.5)$$

$$\boxed{\dot{\rho}(\mathbf{x},t) + \boldsymbol{\nabla}\cdot\mathbf{j}(\mathbf{x},t) = 0} \qquad \text{Continuity equation} \qquad (1.1.3)$$

$$\boxed{\mathbf{j}_f = \sigma\mathbf{E}} \qquad \text{Ohm's law} \qquad (5.3.2)$$

Electrostatics

$$\boxed{\mathbf{E}\cdot\mathbf{n} = 4\pi k_C\sigma} \qquad \text{Induced surface charge on a metal surface} \qquad (2.3.3)$$

$$\boxed{C = Q/V = \epsilon_r A/4\pi k_C d} \qquad \text{Capacitance of the plate capacitor} \qquad (2.2.30)$$

$$\boxed{E_i^q = -k_C q\nabla_i\frac{1}{r}} \qquad \boxed{E_i^p = k_C p_j\nabla_i\nabla_j\frac{1}{r}} \qquad \boxed{E_i^Q = -\frac{k_C}{6}Q_{kl}\nabla_i\nabla_k\nabla_l\frac{1}{r}} \qquad (2.5.9)$$

Monopole,   dipole and   quadrupole field

$$U = \frac{k_C}{2}\int \mathrm{d}^3x\,\mathrm{d}^3x'\,\frac{\rho(\mathbf{x})\,\rho(\mathbf{x}')}{|\mathbf{x}-\mathbf{x}'|} = \frac{1}{2}\int \mathrm{d}^3x\,\rho(\mathbf{x})\,\phi(\mathbf{x}) = \frac{1}{8\pi k_C}\int \mathrm{d}^3x\,|E(\mathbf{x})|^2 \quad (2.4.3)$$

Energy of a charge distribution

$$\boxed{U(\mathbf{x}) = \int \mathrm{d}^3x'\,\rho(\mathbf{x}'-\mathbf{x})\phi^e(\mathbf{x}') = q\phi^e(\mathbf{x}) - \mathbf{p}\cdot\mathbf{E}^e(\mathbf{x}) - \frac{1}{6}Q_{ij}E_{i,j}^e(\mathbf{x})} \qquad (2.5.13)$$

Energy of a charge distribution in the external field

$$\boxed{\mathbf{K} = -\boldsymbol{\nabla}U} \qquad\qquad \text{Force on charge distribution} \qquad (2.5.17)$$

$$\boxed{\mathbf{N} = \mathbf{p}\times\mathbf{E}} \qquad\qquad \text{Torque acting on } \mathbf{p} \qquad (2.5.16)$$

$$G_D(\mathbf{x},\mathbf{x}') = 0\big|_{\mathbf{x}'\in\partial V} \qquad \text{Dirichlet boundary condition}$$

$$\phi(\mathbf{x}) = k_C\int_V \mathrm{d}^3x'\,G_D(\mathbf{x},\mathbf{x}')\rho(\mathbf{x}') - \frac{1}{4\pi}\oiint_{\partial V}\mathrm{d}\mathbf{S}'\cdot\phi(\mathbf{x}')\boldsymbol{\nabla}'G_D(\mathbf{x},\mathbf{x}') \qquad (3.1.6)$$

$$\mathbf{n}\cdot\boldsymbol{\nabla}'G_N(\mathbf{x},\mathbf{x}')\big|_{\mathbf{x}'\in\partial V} = -\frac{4\pi}{\partial V} \qquad \text{Neumann boundary condition}$$

$$\phi(\mathbf{x}) = k_C\int_V \mathrm{d}^3x'\,G_N(\mathbf{x},\mathbf{x}')\rho(\mathbf{x}') + \frac{1}{4\pi}\oiint_{\partial V}\mathrm{d}\mathbf{S}'\cdot G_N(\mathbf{x},\mathbf{x}')\boldsymbol{\nabla}'\phi + \langle\phi\rangle_{\partial V}. \qquad (3.1.7)$$

$$\phi(r, \vartheta, \varphi) = \sum_{l=0}^{\infty} \sum_{m=-l}^{l} (\alpha_{lm} r^l + \beta_{lm} r^{-l-1}) Y_{lm}(\vartheta, \varphi) \tag{3.2.38}$$

$$= \sum_{l=0}^{\infty} (a_l r^l + b_l r^{-l-1}) P_l(\cos\vartheta) \qquad \text{Solution of the Laplace equation} \tag{3.2.40}$$

$$\phi(r, \vartheta, \varphi) = k_C \sum_{l=0}^{\infty} \frac{4\pi}{2l+1} \frac{1}{r^{l+1}} \sum_{m=-l}^{l} Y_{lm}(\vartheta, \varphi) Q_{lm} \qquad \text{Spherical multipole expansion} \tag{3.3.10}$$

$$Q_{lm} = \int d^3x' \, \rho(\mathbf{x'}) \, r'^l \, Y_{lm}^*(\vartheta', \varphi') = (-1)^m \, Q_{lm}^*$$

$$\boxed{\phi(\mathbf{x}) = k_C \left\{ \frac{q}{r} - \mathbf{p} \cdot \boldsymbol{\nabla} \frac{1}{r} + \frac{1}{6} Q_{kl} \nabla_k \nabla_l \frac{1}{r} + \dots \right\}} \quad \text{Multipole expansion} \tag{2.5.11}$$

Boundary conditions

The symbols $\bigcirc$ and $\square$ denote the integration over an infinitesimal cylinder and an infinitesimal rectangle, respectively

Gauss law $\qquad \oiint_{\bigcirc} d\mathbf{S} \cdot \mathbf{D} = 4\pi k_r q_f = 0 \qquad\qquad \Rightarrow \mathbf{n} \cdot (\mathbf{D}_2 - \mathbf{D}_1) = 0, \quad \boldsymbol{\nabla} \cdot \mathbf{D} = 0 \quad (5.2.21)$

Induction law $\qquad \oint_{\square} d\mathbf{x} \cdot \mathbf{E} = -\frac{k_L}{c} \dot{\Phi}_B = 0 \qquad\qquad \Rightarrow \mathbf{n} \times (\mathbf{E}_2 - \mathbf{E}_1) = 0, \quad \boldsymbol{\nabla} \times \mathbf{E} = 0 \quad (5.2.23)$

Ampère-Maxwell $\quad \oint_{\square} d\mathbf{x} \cdot \mathbf{H} = \iint_{\square} d\mathbf{S} \cdot \frac{k_L}{c} [4\pi k_r \mathbf{j}_f + \dot{\mathbf{D}}] = 0 \Rightarrow \mathbf{n} \times (\mathbf{H}_2 - \mathbf{H}_1) = 0, \quad \boldsymbol{\nabla} \times \mathbf{H} = 0 \quad (5.2.25)$

Divergence less $\mathbf{B}$ $\quad \oiint_{\bigcirc} d\mathbf{S} \cdot \mathbf{B} = 0 \qquad\qquad\qquad \Rightarrow \mathbf{n} \cdot (\mathbf{B}_2 - \mathbf{B}_1) = 0, \quad \boldsymbol{\nabla} \cdot \mathbf{B} = 0. \quad (5.2.19)$

Energy and momentum

$$\boxed{u_{\text{field}}(\mathbf{x}, t) = \frac{1}{8\pi k_r} (\mathbf{E} \cdot \mathbf{D} + \mathbf{H} \cdot \mathbf{B})} \quad \text{Energy density of the electromagnetic field} \tag{5.6.4}$$

$$\boxed{\dot{u}_{\text{mech}} = \mathbf{j}_f \cdot \mathbf{E}} \quad \text{Power density as Joule heat} \tag{5.3.6}$$

$$\boxed{\mathbf{S} = \frac{c}{k_L} \frac{1}{4\pi k_r} \mathbf{E} \times \mathbf{H}} \quad \text{Energy flux density (Poynting vector)} \tag{5.6.3}$$

$$\boxed{\dot{u}_{\text{mech}} + \dot{u}_{\text{feld}} + \boldsymbol{\nabla} \cdot \mathbf{S} = 0} \quad \text{Poynting theorem (energy balance)} \tag{5.6.5}$$

$$\boxed{\dot{\boldsymbol{p}}_{\text{mech}} = \rho_f \mathbf{E} + \frac{k_L}{c} \mathbf{j}_f \times \mathbf{B}} \quad \text{Force density} \tag{5.6.8}$$

$$\boxed{\boldsymbol{p}_{\text{field}} = \frac{k_L}{c}\frac{1}{4\pi k_r}\,\mathbf{D}\times\mathbf{B} = \frac{\epsilon_r\mu_r}{c^2}\,\mathbf{S}}$$

Momentum density of the electromagnetic field   (5.6.15)

$$\boxed{T_{ij} = \frac{1}{4\pi k_r}\left[E_i D_j + H_i B_j - \frac{1}{2}(\mathbf{E}\cdot\mathbf{D} + \mathbf{H}\cdot\mathbf{B})\delta_{ij}\right]}$$

Maxwell's stress tensor   (5.6.11)

$$\boxed{\dot{p}_{\text{mech}\,i} + \dot{p}_{\text{field}\,i} = T_{ij,j}}$$

Balance equation for the momentum density   (5.6.12)

$$\boxed{\dot{\rho} + \boldsymbol{\nabla}\cdot\mathbf{j} = 0}\qquad \boxed{\dot{u}_g + \boldsymbol{\nabla}\cdot\mathbf{S} = 0}\qquad \boxed{\dot{p}_i + T_{ij,j} = 0}$$

Balance equations

Scalar and vector potential

$$\boxed{-\Delta\phi(\mathbf{x}) = 4\pi k_C\rho(\mathbf{x}),\qquad -\Delta\mathbf{A}(\mathbf{x}) = \frac{4\pi k_C}{ck_L}\mathbf{j}(\mathbf{x})}$$

Poisson equation   (2.1.5) (4.1.3)

$$\boxed{\square = \frac{1}{c^2}\frac{\partial^2}{\partial t^2} - \Delta}$$

d'Alembert-operator   (10.1.2)

$$\boxed{\begin{aligned}\square\phi &= 4\pi k_C\rho, \quad \phi(\mathbf{x},t) = k_C\!\int\! \mathrm{d}^3x'\,\frac{\rho(\mathbf{x}',t_r)}{|\mathbf{x}-\mathbf{x}'|},\quad t_r = t - \frac{|\mathbf{x}-\mathbf{x}'|}{c}\\[2mm] \square\mathbf{A} &= \frac{4\pi k_C}{ck_L}\mathbf{j}, \quad \mathbf{A}(\mathbf{x},t) = \frac{k_C}{ck_L}\!\int\!\mathrm{d}^3x'\,\frac{\mathbf{j}(\mathbf{x}',t_r)}{|\mathbf{x}-\mathbf{x}'|},\quad t_r = \begin{cases} 0 & \text{static}\\ t & \text{quasistatic}\\ t_r & \text{retarded}\end{cases}\end{aligned}}$$

$$\phi(\mathbf{x},t) = \frac{k_C q}{R(t_r)\,|J(t_r)|},\qquad\qquad J(t_r) = 1 - \mathbf{e}_R\cdot\boldsymbol{\beta}$$

Potentials of the point charge   (8.2.26)

$$\mathbf{A}(\mathbf{x},t) = \frac{1}{ck_L}\,\phi(\mathbf{x},t)\,\mathbf{v}(t_r) = \frac{\phi\boldsymbol{\beta}}{k_L}\qquad R(t_r) = |\mathbf{x}-\mathbf{s}(t_r)|,\quad t_r = t - R(t_r)/c.$$

Electromagnetic fields

$$\boxed{\mathbf{E} = -\boldsymbol{\nabla}\phi - (k_L/c)\dot{\mathbf{A}}}\qquad\qquad \boxed{\mathbf{B} = \boldsymbol{\nabla}\times\mathbf{A}}$$

Electric and magnetic field   (8.1.3)

$$\boxed{\boldsymbol{\nabla}\cdot\mathbf{A} + (1/ck_L)\dot{\phi} = 0}$$

Lorenz gauge   (8.1.8)

$$\boxed{\boldsymbol{\nabla}\cdot\mathbf{A} = 0}$$

Coulomb gauge   (8.2.48)

Magnetostatics

$$\iint_A \mathrm{da}\cdot\boldsymbol{\nabla}\times\mathbf{H} = 4\pi k_r \frac{k_L}{c}\iint_A \mathrm{da}\cdot\mathbf{j}_f \;\Leftrightarrow\; \boxed{\oint_{\partial F}\mathrm{dx}\cdot\mathbf{H} = 4\pi k_r\frac{k_L}{c}I}\;\begin{matrix}\text{Ampère's}\\\text{law}\end{matrix} \qquad (4.1.5)$$

$$\boxed{\mathbf{H} = \frac{k_C}{ck_L}I\int_C \mathrm{dx}'\times\frac{\mathbf{x}-\mathbf{x}'}{|\mathbf{x}-\mathbf{x}'|^3}}\qquad\text{Biot-Savart law}\qquad (4.1.9)$$

$$\boxed{\mathbf{m} = \frac{k_L}{2c}\int \mathrm{d}^3x'\,\mathbf{x}'\times\mathbf{j}}\qquad\begin{matrix}\text{Magnetic dipole moment}\\\text{plane loop }\mathbf{m} = IAk_L/c\end{matrix}\qquad (4.2.1)$$

$$\boxed{\mathbf{A} = \frac{k_C}{k_L^2}\frac{\mathbf{m}\times\mathbf{x}}{r^3}}\;\Rightarrow\;\boxed{\mathbf{B} = \frac{k_C}{k_L^2}\left[\left(\frac{3(\mathbf{m}\cdot\mathbf{x})\mathbf{x}}{r^5}-\frac{\mathbf{m}}{r^3}\right)+4\pi\mathbf{m}\delta^{(3)}(\mathbf{x})\right]}\qquad\begin{matrix}(4.2.10)\\(4.2.12)\end{matrix}$$

$$\boxed{U = -\mathbf{m}\cdot\mathbf{B}}\qquad\text{Energy of a dipole in the external field}\qquad (4.3.5)$$

$$\boxed{\mathbf{F} = -\boldsymbol{\nabla}U = \boldsymbol{\nabla}(\mathbf{m}\cdot\mathbf{B})}\qquad\text{Force on magnetic moment}\qquad (4.3.4)$$

$$\boxed{\mathbf{N} = \mathbf{m}\times\mathbf{B}}\qquad\text{Torque}\qquad (4.3.8)$$

$$\boxed{\rho_{\mathrm{M}} = -\boldsymbol{\nabla}\cdot\mathbf{M}}\quad(7.1.6)\quad\boxed{\mathbf{j}_{\mathrm{M}} = (c/k_L)\boldsymbol{\nabla}\times\mathbf{M}}\quad\begin{matrix}\text{Magnetic charge and}\\\text{magnetization}\end{matrix}\quad (7.1.3)$$

$$L_{ik} = \frac{k_C}{c^2}\mu_r\int \mathrm{d}^3x\,\mathrm{d}^3x'\frac{\mathbf{j}_i(\mathbf{x})\cdot\mathbf{j}_k(\mathbf{x}')}{I_iI_k|\mathbf{x}-\mathbf{x}'|}\quad\text{Self and mutual inductance}\qquad (7.2.13)$$

$$= \frac{k_C}{c^2}\mu_r\oint_{\partial A_i}\oint_{\partial A_k}\frac{\mathrm{dx}_i\cdot\mathrm{dx}_k}{|\mathbf{x}_i-\mathbf{x}_k|}\quad\text{für } i\neq k\quad\begin{matrix}\text{Neumann formula}\\\text{(mutual inductance)}\end{matrix}\qquad (7.2.19)$$

$$\boxed{U_m = \frac{1}{2}\sum_{i,k=1}^{n}L_{ik}\,I_iI_k}\quad\text{Energy of a (ferro-)magnet}\qquad (7.2.12)$$

$$\Phi_{B_{12}} = \oint_{\partial A_1}\mathrm{dx}_1\cdot\mathbf{A}_2(\mathbf{x}_1) = \frac{c}{k_L}L_{12}I_2\quad\begin{matrix}\text{Flow through loop 1}\\\text{from loop 2}\end{matrix}\qquad (7.2.19)$$

Radiation

$$\frac{\mathrm{e}^{ik|\mathbf{x}-\mathbf{x}'|}}{|\mathbf{x}-\mathbf{x}'|}\approx\frac{\mathrm{e}^{ikr}}{r}\mathrm{e}^{-i\mathbf{k}\cdot\mathbf{x}'}\qquad\text{Far zone, all multipoles, }\mathbf{k}=k\mathbf{e}_r\qquad (8.3.19)$$

$$\approx\frac{\mathrm{e}^{ikr}}{r}\left\{1+\frac{\mathbf{x}\cdot\mathbf{x}'}{r^2}(1-ikr)\right\}$$

1st term: electric dipole radiation
2nd term: magnetic dipole, electric quadrupole radiation
3rd term: far field

$$\boxed{\mathbf{A}(\mathbf{x},t) = \frac{k_C}{ck_L}\frac{\mathrm{e}^{ikr-i\omega t}}{r}\,\mathbf{j}(\mathbf{k}),}\quad \mathbf{j}(\mathbf{k}) = \int \mathrm{d}^3x'\,\mathbf{j}(\mathbf{x}')\,\mathrm{e}^{ik\mathbf{e}_r\cdot\mathbf{x}'},\;\text{Far field}\qquad(8.3.22)$$

$$\boxed{\mathbf{B}(\mathbf{x},t) = \boldsymbol{\nabla}\times\mathbf{A}} = ik\frac{k_C}{ck_L}\frac{\mathrm{e}^{ikr-i\omega t}}{r}\,\mathbf{e}_r\times\mathbf{j}(\mathbf{k}),\quad\text{Far field}\qquad(8.3.24)$$

$$\boxed{\mathbf{E}(\mathbf{x},t) = \frac{ic}{\omega}\boldsymbol{\nabla}\times\mathbf{B}} = ik\frac{k_C}{ck_L}\frac{\mathrm{e}^{ikr-i\omega t}}{r}\,\mathbf{e}_r\times\big(\mathbf{j}(\mathbf{k})\times\mathbf{e}_r\big)\qquad(8.3.25)$$

$$Z_0 = 4\pi k_C/c\qquad\qquad\text{Impedance of free space}\qquad(8.2.65')$$

$$\boxed{\langle\mathbf{S}\rangle = \frac{ck_L}{8\pi k_C}\,\mathbf{E}\times\mathbf{B}^* = \frac{1}{2Z_0}|\mathbf{E}|^2\mathbf{e}_r,}\quad\text{Average energy flux}\qquad(8.3.27)$$

$$\boxed{\frac{\mathrm{d}\langle P\rangle}{\mathrm{d}\Omega} = r^2\mathbf{e}_r\cdot\langle\mathbf{S}\rangle = \frac{r^2}{2Z_0}|\mathbf{E}|^2}\quad\text{Average power radiated in }\mathrm{d}\Omega\qquad(8.3.28)$$

$$\boxed{\mathbf{A}(\mathbf{x},t) = \frac{k_C}{ck_L}\frac{\mathrm{e}^{ikr-i\omega t}}{r}\mathbf{j}(0)}\quad\boxed{\frac{1}{c}\mathbf{j}(0) = -i\frac{\omega}{c}\mathbf{p}}\quad\begin{array}{l}\text{electric dipole radia}\\\text{tion near + far field}\end{array}\quad(8.4.5)$$

$$\boxed{\langle P\rangle = Z_0\frac{c^2k^4}{12\pi}|\mathbf{p}|^2}\quad\begin{array}{l}\text{Average emitted power of the}\\\text{electric dipole radiation}\end{array}\qquad(8.4.18)$$

Electromagnetic waves

$$\boxed{\mathbf{E} = \frac{E_+}{\sqrt{2}}\Big\{(1+r)\boldsymbol{\epsilon}_1 + i(1-r)\boldsymbol{\epsilon}_2\Big\}\mathrm{e}^{i(\mathbf{k}\cdot\mathbf{x}-\omega t)}}\quad\begin{array}{c}\text{elliptic polarization,}\\r=0\Rightarrow\text{circular}\end{array}\quad(10.2.3)$$

$$\boxed{\bar{\Box}\mathbf{E} = -\frac{4\pi k_C\mu_r\sigma}{c^2}\dot{\mathbf{E}}}\quad\text{p. 374}\quad\begin{cases}\epsilon_r\omega \gg 4\pi k_C\sigma & \text{bad conductor}\\\epsilon_r\omega \ll 4\pi k_C\sigma & \text{good conductor}\end{cases}$$

$$\begin{aligned}\mathbf{E} &= \big[\mathbf{E}_\perp(\mathbf{x}) + \mathbf{E}_z(\mathbf{x})\big]\mathrm{e}^{ikz-i\omega t}\\\mathbf{B} &= \big[\mathbf{B}_\perp(\mathbf{x}) + \mathbf{B}_z(\mathbf{x})\big]\mathrm{e}^{ikz-i\omega t}\end{aligned}\qquad\text{Waves in waveguides}\qquad(10.5.5)$$

1. TE modes: $\mathbf{E}_z = 0$
2. TM modes: $\mathbf{B}_z = 0$
3. TEM modes: $\mathbf{E}_z = \mathbf{B}_z = 0$

Special Theory of Relativity

$$\Lambda(\boldsymbol{\beta}, \boldsymbol{\alpha}) \equiv \Lambda(\mathbf{0}, \boldsymbol{\alpha})\,\Lambda(\boldsymbol{\beta}, \mathbf{0}) \qquad \text{Boost with } \boldsymbol{\beta} + \text{rotation by } \boldsymbol{\alpha} \tag{12.4.1}$$

$$\left.\begin{array}{l} ct' = \gamma\bigl(ct - \boldsymbol{\beta}\cdot\mathbf{x}\bigr) \\ \mathbf{x}'_\parallel = \gamma\bigl(\mathbf{x}_\parallel - \boldsymbol{\beta}ct\bigr) \\ \mathbf{x}'_\perp = \mathbf{x}_\perp \end{array}\right\} \mathbf{x}' = \mathbf{x} + (\gamma-1)(\hat{\boldsymbol{\beta}}\cdot\mathbf{x})\,\hat{\boldsymbol{\beta}} - \gamma\boldsymbol{\beta}ct \qquad \text{Boost, general direction: } \gamma = 1/\sqrt{1-\beta^2} \tag{12.4.2}$$

$$\Lambda(\boldsymbol{\beta}, \mathbf{0}) = \begin{pmatrix} \gamma & -\gamma\boldsymbol{\beta}^{\mathrm{T}} \\ -\gamma\boldsymbol{\beta} & \mathsf{E} + (\gamma-1)\hat{\boldsymbol{\beta}}\circ\hat{\boldsymbol{\beta}} \end{pmatrix} \qquad \begin{array}{l}\text{Boost, general direction} \\ \text{tensorial product: } \circ \equiv \otimes\end{array} \tag{12.4.3}$$

$$\boldsymbol{\beta}'' = \frac{\boldsymbol{\beta} + \boldsymbol{\beta}'_\parallel + \boldsymbol{\beta}'_\perp/\gamma}{1 + \boldsymbol{\beta}\cdot\boldsymbol{\beta}'} \qquad \begin{array}{l}\text{Addition of velocities} \\ \text{of arbitrary directions}\end{array} \tag{12.4.15}$$

1. $s^2 < 0$: vector spacelike: (b^μ) and (f^μ),
2. $s^2 = 0$: vector lightlike: (k^μ),
3. $s^2 > 0$: vector timelike: (z^μ), (v^μ).

$$l_0 = \gamma l \qquad\qquad\qquad \text{Lorentz-contraction} \tag{12.3.1}$$

$$\mathrm{d}s = c\mathrm{d}\tau = c\mathrm{d}t/\gamma \qquad\qquad \text{Proper time and its line element} \tag{12.3.2}$$

$$j^\mu{}_{,\mu} = 0 \qquad\qquad\qquad \text{continuity equation} \tag{13.1.7}$$

$$\partial^\nu\partial_\nu A^\mu = (4\pi k_{\mathrm{C}}/ck_{\mathrm{L}})j^\mu \qquad\qquad \text{wave equation} \tag{13.1.11}$$

$$A^\mu{}_{,\mu} = 0 \qquad\qquad\qquad \text{Lorenz gauge} \tag{13.1.12}$$

$$F^{\mu\nu} = \partial^\mu A^\nu - \partial^\nu A^\mu \qquad\qquad \text{field-strength tensor} \tag{13.1.14}$$

$$\tilde{F}^{\mu\nu} = (1/2)\,\epsilon^{\mu\nu\rho\sigma}\,F_{\rho\sigma} \qquad\qquad \text{dual field-strength tensor} \tag{13.1.20}$$

$$\epsilon_{0123} = 1 = -\epsilon^{0123} \qquad\qquad \text{(total) antisymmetric tensor} \tag{13.1.21}$$

$$F^{\nu\mu}{}_{,\nu} = (4\pi k_{\mathrm{C}}/ck_{\mathrm{L}})j^\mu \qquad \text{Inhomogeneous Maxwell equations} \tag{13.1.24}$$

$$\tilde{F}^{\nu\mu}{}_{,\nu} = 0 \qquad \text{Homogeneous Maxwell equations} \tag{13.1.25}$$

$$\bigl(F^{\mu\nu}\bigr) = \begin{pmatrix} 0 & -\mathbf{E}^{\mathrm{T}}/k_{\mathrm{L}} \\ \mathbf{E}/k_{\mathrm{L}} & -\epsilon_{ijk}B_k \end{pmatrix} \;(13.1.16) \qquad \bigl(\tilde{F}^{\mu\nu}\bigr) = \begin{pmatrix} 0 & \mathbf{B}^{\mathrm{T}} \\ -\mathbf{B} & -\epsilon_{ijk}E_k/k_{\mathrm{L}} \end{pmatrix} \tag{13.1.22}$$

$$\mathbf{E}' = \mathbf{E}_\parallel + \gamma(\mathbf{E}_\perp + k_{\mathrm{L}}\boldsymbol{\beta}\times\mathbf{B}) \qquad \text{Boost: transformation of } \mathbf{E} \tag{13.1.29}$$

$$\mathbf{B}' = \mathbf{B}_\parallel + \gamma(\mathbf{B}_\perp - \boldsymbol{\beta}\times\mathbf{E}/k_{\mathrm{L}}) \qquad \text{Boost: transformation of } \mathbf{B} \tag{13.1.30}$$

$$F^{\mu\nu}F_{\nu\mu} = 2(E^2/k_{\mathrm{L}}^2 - B^2) \qquad \tilde{F}^{\mu\nu}F_{\mu\nu} = 4\mathbf{E}\cdot\mathbf{B}/k_{\mathrm{L}} \qquad \text{Invariants} \tag{13.1.31}$$

Relativistic mechanics

$$\mathrm{p}^\mu = m v^\mu = m c u^\mu \quad \Leftrightarrow \quad \boxed{(\mathrm{p}^\mu) = m\gamma \begin{pmatrix} c \\ \boldsymbol{v} \end{pmatrix} = \begin{pmatrix} p^0 \\ \boldsymbol{p} \end{pmatrix}} \quad \text{Vector of momentum} \qquad (14.1.1)$$

$$\boxed{\mathrm{E} = \gamma m c^2} \qquad \Rightarrow \qquad \mathrm{p}^0 = \gamma m c = \mathrm{E}/c \qquad \text{Relativistic energy} \qquad (14.1.2)$$

$$\mathrm{p}^\mu \mathrm{p}_\mu = m^2 c^2 = \frac{\mathrm{E}^2}{c^2} - \boldsymbol{p}^2 \;\Rightarrow\; \mathrm{E} = c\sqrt{m^2 c^2 + \boldsymbol{p}^2} \quad \text{Relativistic energy} \quad (14.1.4)$$

$$\boxed{a^\mu = \ddot{x}^\mu(s) = \dot{u}^\mu(s)} = (\dot{\gamma},\, \dot{\gamma}\,\boldsymbol{\beta} + \gamma\,\dot{\boldsymbol{\beta}}) \qquad \text{Four-acceleration} \qquad (14.1.6)$$

$$m c \dot{u}^\mu = 0 \qquad \Rightarrow \qquad x^\mu(s) = x^\mu(0) + u^\mu s \qquad \text{Free particle} \qquad (14.1.9)$$

$$m c^2 a^\mu(s) = K^\mu \qquad\qquad\qquad \text{Equation of motion, p. 520}$$

$$K^0 = \mathbf{K} \cdot \boldsymbol{\beta} = \frac{1}{c}\frac{\mathrm{dE}}{\mathrm{d}\tau} \qquad \text{Work done by } \mathbf{K} \text{ on the particle} \times \frac{1}{c} \qquad (14.1.10)$$

$$P = -\frac{2k_C e^2 c}{3} a^\mu a_\mu = \frac{2k_C e^2}{3c}\gamma^6 \big[\dot{\beta}^2(t) - (\boldsymbol{\beta}\times\dot{\boldsymbol{\beta}})^2\big] \qquad \begin{aligned} &\beta = 0\text{: Larmor formula} \\ &\beta \neq 0\text{: Liénard formula} \end{aligned}$$
$$(14.1.13)$$

$$\boxed{(T^{\kappa\lambda}) = \begin{pmatrix} u_{\text{Feld}} & \mathbf{S}^{\mathsf{T}}/c \\ \mathbf{S}/c & -(\sigma_{kl}) \end{pmatrix}.} \qquad \text{Energy-momentum tensor} \qquad (14.1.33)$$

$$\sigma_{kl} = \frac{1}{4\pi k_C}\Big[E_k E_l + k_L^2 B_k B_l - \frac{1}{2}(E^2 + k_L^2 B^2)\delta_{kl}\Big]. \qquad (14.1.34)$$

$$u_{\text{Feld}} = \frac{1}{8\pi k_C}(E^2 + k_L^2 B^2), \qquad\qquad \mathbf{S} = c^2 \boldsymbol{p}_{\text{Feld}} = \frac{k_L c}{4\pi k_C}\mathbf{E}\times\mathbf{B},$$

$$\boxed{\mathrm{L} = -\frac{m c^2}{\gamma} - \frac{e k_L}{\gamma} u^\mu A_\mu} \qquad \text{Relativistic Lagrange function} \qquad (14.2.12)$$

$$\frac{\partial \mathrm{L}}{\partial \mathbf{x}} - \frac{\mathrm{d}}{\mathrm{d}t}\Big(\frac{\partial \mathrm{L}}{\partial \mathbf{v}}\Big) = 0 \quad \text{Euler-Lagrange equation} \qquad (14.2.2)$$

$$\dot{\boldsymbol{p}}(t) = e\big(\mathbf{E} + k_L \boldsymbol{\beta}\times\mathbf{B}\big) \quad \text{Euler-Lagrange equation} \qquad (14.2.14)$$

$$\boldsymbol{P} = \boldsymbol{p} + (e k_L/c)\mathbf{A} \qquad \text{Generalized momentum} \qquad (14.2.13)$$

$$\mathcal{H} = \boldsymbol{P}\cdot\mathbf{v} - \mathrm{L} \overset{(14.2.8)}{=} c\sqrt{\big(\boldsymbol{P} - (e k_L/c)\mathbf{A}\big)^2 + m^2 c^2} + e\phi \quad \text{Energy, p. 530}$$

$$\mathcal{L} = -\frac{k_L}{c} j_\mu A^\mu + \mathcal{L}_{\text{field}} \quad \mathcal{L}_{\text{field}} = \frac{-k_L^2}{16\pi k_C} F^{\mu\nu} F_{\mu\nu} \quad \text{Lagrange density} \quad (14.2.37)$$

Electromagnetic radiation field

$$\mathbf{A}(\mathbf{x},t) = \frac{c\sqrt{4\pi k_C}}{k_L\sqrt{2V}} \sum_{\mathbf{k}} \frac{\sqrt{\hbar}}{\sqrt{\omega_k}}\left(\mathrm{e}^{\mathrm{i}\mathbf{k}\cdot\mathbf{x}}\mathbf{A_k}(t) + \mathrm{e}^{-\mathrm{i}\mathbf{k}\cdot\mathbf{x}}A_{\mathbf{k}}^*(t)\right), \qquad (14.2.44)$$

$$\mathbf{A_k}(t) = \sum_{\lambda} \boldsymbol{\epsilon}_{\mathbf{k}\lambda} a_{\mathbf{k}\lambda}(t), \qquad a_{\mathbf{k}\lambda}(t) = \mathrm{e}^{-\mathrm{i}\omega_k t}a_{\mathbf{k}\lambda}, \qquad \omega_k = ck,$$

$$\mathrm{H} = \int \mathrm{d}^3x\, \frac{E^2+k_L^2 B^2}{8\pi k_C} = \frac{1}{2}\sum_{\mathbf{k},\lambda} \hbar\omega_{\mathbf{k}}(a_{\mathbf{k}\lambda}a_{\mathbf{k}\lambda}^* + a_{\mathbf{k}\lambda}^* a_{\mathbf{k}\lambda})\ \text{Field energy} \quad (14.2.47)$$

$$\boldsymbol{P}_{\mathrm{Feld}} = \frac{k_L}{4\pi c k_C}\int \mathrm{d}^3x\, \mathbf{E}\times\mathbf{B} = \sum_{\mathbf{k}\lambda} \hbar\mathbf{k}\, a_{\mathbf{k}\lambda}^* a_{\mathbf{k}\lambda} \quad \begin{array}{l}\text{Momentum of the}\\ \text{electromagnetic field}\end{array}$$

$$(14.2.49)$$

$$\boldsymbol{J}_{\mathrm{Feld}} = \boldsymbol{L} + \boldsymbol{S} = \frac{k_L}{4\pi c k_C}\int \mathrm{d}^3x\, \mathbf{x}\times(\mathbf{E}\times\mathbf{B}) \quad \begin{array}{l}\text{Angular momentum of}\\ \text{the electromagnetic field}\end{array} \quad (14.2.56)$$

$$\boldsymbol{S} = \frac{k_L}{4\pi c k_C}\int \mathrm{d}^3x\,(\mathbf{E}\times\mathbf{A}) \qquad \text{Spin} \qquad\qquad (14.2.59)$$

$$\boldsymbol{L} = \frac{k_L}{4\pi c k_C}\int \mathrm{d}^3x\, E_j(\mathbf{x}\times\boldsymbol{\nabla})A_j \qquad \text{Orbital angular momentum.}$$

$$(14.2.63)$$

D.2 Formulas of Electrodynamics in Gauss and SI

The fundamental equations of electrodynamics

Maxwell equations in free space

$$\text{SI:}\quad \begin{array}{ll}\text{(a)} & \boldsymbol{\nabla}\cdot\mathbf{E} = \dfrac{\rho}{\epsilon_0} \quad \text{(b)}\ \ \boldsymbol{\nabla}\times\mathbf{E} + \dot{\mathbf{B}} = 0\\[2mm] \text{(c)}\ \ \boldsymbol{\nabla}\times\mathbf{B} - \epsilon_0\mu_0\dot{\mathbf{E}} = \mu_0\mathbf{j} \quad \text{(d)} & \boldsymbol{\nabla}\cdot\mathbf{B} = 0\end{array}$$

$$\text{G:}\quad \begin{array}{ll}\text{(a)} & \boldsymbol{\nabla}\cdot\mathbf{E} = 4\pi\rho \quad \text{(b)}\ \ \boldsymbol{\nabla}\times\mathbf{E} + \dfrac{1}{c}\dot{\mathbf{B}} = 0\\[2mm] \text{(c)}\ \ \boldsymbol{\nabla}\times\mathbf{B} - \dfrac{1}{c}\dot{\mathbf{E}} = \dfrac{4\pi}{c}\mathbf{j} \quad \text{(d)} & \boldsymbol{\nabla}\cdot\mathbf{B} = 0\end{array} \qquad (1.3.21)$$

Maxwell equations in matter

$$\text{SI:}\quad \begin{array}{ll}\text{(a)} & \boldsymbol{\nabla}\cdot\mathbf{D} = \rho_f \quad \text{(b)}\ \ \boldsymbol{\nabla}\times\mathbf{E} + \dot{\mathbf{B}} = 0\\[2mm] \text{(c)}\ \ \boldsymbol{\nabla}\times\mathbf{H} - \dot{\mathbf{D}} = \mathbf{j}_f \quad \text{(d)} & \boldsymbol{\nabla}\cdot\mathbf{B} = 0\end{array}$$

$$\text{G:} \quad \begin{array}{ll} \text{(a)} \quad \nabla\cdot\mathbf{D} = 4\pi\rho_f & \text{(b)} \quad \nabla\times\mathbf{E} + \dfrac{1}{c}\dot{\mathbf{B}} = 0 \\[2ex] \text{(c)} \quad \nabla\times\mathbf{H} - \dfrac{1}{c}\dot{\mathbf{D}} = \dfrac{4\pi}{c}\mathbf{j}_f & \text{(d)} \quad \nabla\cdot\mathbf{B} = 0 \end{array} \tag{5.2.16}$$

Material or constitutive equations:

$$\text{SI:} \quad \begin{aligned} \mathbf{B} &= \mu_0\mu_r\mathbf{H} = \mu_0(1+\chi_m)\mathbf{H} = \mu_0(\mathbf{H}+\mathbf{M}) \\ \mathbf{D} &= \epsilon_0\epsilon_r\mathbf{E} = \epsilon_0(1+\chi_e)\mathbf{E} = \epsilon_0\mathbf{E}+\mathbf{P} \end{aligned}$$

$$\text{G:} \quad \begin{aligned} \mathbf{B} &= \mu_r\mathbf{H} = (1+4\pi\chi_m)\mathbf{H} = \mathbf{H}+4\pi\mathbf{M} \\ \mathbf{D} &= \epsilon_r\mathbf{E} = (1+4\pi\chi_e)\mathbf{E} = \mathbf{E}+4\pi\mathbf{P} \end{aligned} \tag{5.2.17}$$

Lorentz force (Coulomb + Lorentz part)

$$\text{SI:} \ \mathbf{F} = q\mathbf{E} + q\mathbf{v}\times\mathbf{B}, \qquad\qquad \text{G:} \ \mathbf{F} = q\mathbf{E} + \frac{q}{c}\mathbf{v}\times\mathbf{B}. \tag{1.2.5}$$

Continuity equation (balance equation of charge density)

$$\dot{\rho}(\mathbf{x},t) + \nabla\cdot\mathbf{j}(\mathbf{x},t) = 0. \tag{1.1.3}$$

Ohm's law: $\mathbf{j}_f^{\text{SI}} = \mathbf{j}_f\sqrt{4\pi\epsilon_0}$, $\sigma^{\text{SI}} = \sigma\,4\pi\epsilon_0$

$$\mathbf{j}_f = \sigma\mathbf{E}. \tag{5.3.2}$$

Electrostatics

Scalar potential: $\phi^{\text{SI}}(\mathbf{x}) = \phi(\mathbf{x})/\sqrt{4\pi\epsilon_0}$

$$\text{SI:} \ \phi(\mathbf{x}) = \frac{1}{4\pi\epsilon_0}\int d^3x'\, \frac{\rho(\mathbf{x}')}{|\mathbf{x}-\mathbf{x}'|}, \qquad \text{G:} \ \phi(\mathbf{x}) = \int d^3x'\, \frac{\rho(\mathbf{x}')}{|\mathbf{x}-\mathbf{x}'|}. \tag{2.1.8}$$

Electric field: $\mathbf{E}^{\text{SI}} = \mathbf{E}/\sqrt{4\pi\epsilon_0}$

$$\mathbf{E}(\mathbf{x}) = -\nabla\phi(\mathbf{x}). \tag{2.2.1}$$

For multipole fields (monopole, dipole, quadrupole, ...) this applies: $\mathbf{E}^{p\,\text{SI}} = \mathbf{E}^p/4\pi\epsilon_0$:

$$\text{SI:} \ \mathbf{E}^q = \frac{-q}{4\pi\epsilon_0}\nabla\frac{1}{r}, \qquad\qquad \text{G:} \ \mathbf{E}^q = -q\nabla\frac{1}{r}, \tag{2.5.4}$$

$$\text{G:} \ \mathbf{E}^p = (\mathbf{p}\cdot\nabla)\nabla\frac{1}{r}, \qquad\qquad \text{G:} \ \mathbf{E}^Q = -\frac{1}{6}(\nabla^T Q\nabla)\nabla\frac{1}{r}. \tag{2.5.9}$$

Induced surface charge on metal surface: $\sigma^{\text{SI}} = \sigma\sqrt{4\pi\epsilon_0}$

$$\text{SI:} \ \sigma = \epsilon_0\mathbf{E}\cdot\mathbf{n}, \qquad\qquad \text{G:} \ \sigma = \mathbf{E}\cdot\mathbf{n}/4\pi. \tag{2.3.3}$$

Capacitance of the plate capacitor: $C^{\text{SI}} = C\,4\pi\epsilon_0$

$$\text{SI: } C = Q/V = \epsilon_r\epsilon_0 A/d, \qquad\qquad \text{G: } C = Q/V = \epsilon_r A/4\pi d. \qquad (2.2.30)$$

Energy of a charge distribution

$$\text{SI: } U = \frac{1}{8\pi\epsilon_0}\int d^3x\, d^3x'\,\frac{\rho(\mathbf{x})\rho(\mathbf{x}')}{|\mathbf{x}-\mathbf{x}'|} = \frac{1}{2}\int d^3x\,\rho(\mathbf{x})\,\phi(\mathbf{x}) = \frac{\epsilon_0}{2}\int d^3x\,|E(\mathbf{x})|^2,$$

$$\text{G: } U = \frac{1}{2}\int d^3x\, d^3x'\,\frac{\rho(\mathbf{x})\rho(\mathbf{x}')}{|\mathbf{x}-\mathbf{x}'|} = \frac{1}{2}\int d^3x\,\rho(\mathbf{x})\,\phi(\mathbf{x}) = \frac{1}{8\pi}\int d^3x\,|E(\mathbf{x})|^2. \quad (2.4.3)$$

Force on charge distribution

$$\mathbf{K} = -\boldsymbol{\nabla} U. \qquad (2.5.17)$$

Torque acting on the dipole

$$\mathbf{N} = \mathbf{p}\times\mathbf{E}. \qquad (2.5.16)$$

Magnetostatics

Ampère's law in differential form:

$$\text{SI: } \boldsymbol{\nabla}\times\mathbf{B} = \mu_0\mathbf{j}, \qquad\qquad \text{G: } \boldsymbol{\nabla}\times\mathbf{B} = (4\pi/c)\mathbf{j}. \qquad (4.1.1a\text{–}b)$$

Vector potential: $\mathbf{A}^{\text{SI}} = \mathbf{A}\sqrt{\mu_0/4\pi}$

$$\text{SI: } \mathbf{A}(\mathbf{x}) = \frac{\mu_0}{4\pi}\int d^3x'\,\frac{\mathbf{j}(\mathbf{x}')}{|\mathbf{x}-\mathbf{x}'|}, \qquad \text{G: } \mathbf{A}(\mathbf{x}) = \frac{1}{c}\int d^3x'\,\frac{\mathbf{j}(\mathbf{x}')}{|\mathbf{x}-\mathbf{x}'|}. \qquad (4.1.4)$$

Biot-Savart law: $\mathbf{B}^{\text{SI}} = \mathbf{B}\sqrt{\mu_0/4\pi}$

$$\text{SI: } \mathbf{B} = \frac{\mu_0 I}{4\pi}\int_C d\mathbf{x}'\times\frac{\mathbf{x}-\mathbf{x}'}{|\mathbf{x}-\mathbf{x}'|^3}, \qquad \text{G: } \mathbf{B} = \frac{I}{c}\int_C d\mathbf{x}'\times\frac{\mathbf{x}-\mathbf{x}'}{|\mathbf{x}-\mathbf{x}'|^3}. \qquad (4.1.9)$$

Magnetic dipole moment: $\mathbf{m}^{\text{SI}} = \mathbf{m}\sqrt{4\pi/\mu_0}$

$$\text{SI: } \mathbf{m} = \frac{1}{2}\int d^3x'\,\mathbf{x}'\times\mathbf{j}(\mathbf{x}'), \qquad \text{G: } \mathbf{m} = \frac{1}{2c}\int d^3x'\,\mathbf{x}'\times\mathbf{j}(\mathbf{x}'). \qquad (4.2.1)$$

Vector potential and field of the magnetic dipole: $\mathbf{A}^{\text{SI}} = \mathbf{A}\sqrt{\mu_0/4\pi}$

$$\text{SI: } \mathbf{A} = \frac{\mu_0}{4\pi}\frac{\mathbf{m}\times\mathbf{x}}{r^3} \quad\Rightarrow\quad \text{SI: } \mathbf{B} = \frac{\mu_0}{4\pi}\left(\frac{3(\mathbf{m}\cdot\mathbf{x})\mathbf{x}}{r^5} - \frac{\mathbf{m}}{r^3}\right) + \mu_0\mathbf{m}\delta^{(3)}(\mathbf{x}),$$

$$\text{G: } \mathbf{A} = \frac{\mathbf{m}\times\mathbf{x}}{r^3} \quad(4.2.10)\quad\Rightarrow\quad \text{G: } \mathbf{B} = \frac{3(\mathbf{m}\cdot\mathbf{x})\mathbf{x}}{r^5} - \frac{\mathbf{m}}{r^3} + 4\pi\mathbf{m}\delta^{(3)}(\mathbf{x}). \quad (4.2.12)$$

Energy of the dipole in the external field

$$U = -\mathbf{m}\cdot\mathbf{B}. \tag{4.3.5}$$

Force on magnetic moment
$$\mathbf{K} = -\nabla U. \tag{4.3.4}$$

Torque on magnetic dipole
$$\mathbf{N} = \mathbf{m}\times\mathbf{B}. \tag{4.3.8}$$

Magnetic charge and magnetization current: $\rho_{\mathrm{M}}^{\mathrm{SI}} = \rho_{\mathrm{M}}\sqrt{4\pi/\mu_0};\ \mathbf{j}_{\mathrm{M}}^{\mathrm{SI}} = \mathbf{j}_{\mathrm{M}}\sqrt{4\pi\epsilon_0}$

$$\rho_{\mathrm{M}} = -\nabla\cdot\mathbf{M}, \tag{7.1.6}$$

$$\text{SI: } \mathbf{j}_{\mathrm{M}} = \nabla\times\mathbf{M}, \qquad\qquad \text{G: } \mathbf{j}_{\mathrm{M}} = c\nabla\times\mathbf{M}. \tag{7.1.3}$$

Neumann formula (mutual inductance $i \neq k$): $L_{ik}^{\mathrm{SI}} = L_{ik}/4\pi\epsilon_0$

$$\text{SI: } L_{ik} = \frac{\mu_0\mu_r}{4\pi}\oint_{C_i}\oint_{C_k}\frac{d\mathbf{x}_i\cdot d\mathbf{x}_k}{|\mathbf{x}_i-\mathbf{x}_k|}, \qquad \text{G: } L_{ik} = \frac{\mu_r}{c^2}\oint_{C_i}\oint_{C_k}\frac{d\mathbf{x}_i\cdot d\mathbf{x}_k}{|\mathbf{x}_i-\mathbf{x}_k|}. \tag{7.2.19}$$

Energy of a (ferro)magnet

$$U_m = \frac{1}{2}\sum_{i,k=1}^{n} L_{ik} I_i I_k. \tag{7.2.12}$$

Flow through loop 1 from loop 2: $\Phi_B^{\mathrm{SI}} = \Phi_B\sqrt{\mu_0/4\pi}$

$$\text{SI: } \Phi_{B_{12}} = \oint_{C_1}d\mathbf{x}_1\cdot\mathbf{A}_2(\mathbf{x}_1) = L_{12}I_2, \qquad \text{G: } \Phi_{B_{12}} = \oint_{C_1}d\mathbf{x}_1\cdot\mathbf{A}_2(\mathbf{x}_1) = cL_{12}I_2.$$

$$\tag{7.2.19}$$

Energy and momentum

Energy density of the electromagnetic field

$$\text{SI: } u_{\mathrm{field}}(\mathbf{x},t) = \frac{1}{2}\left(\mathbf{E}\cdot\mathbf{D}+\mathbf{H}\cdot\mathbf{B}\right), \quad \text{G: } u_{\mathrm{field}}(\mathbf{x},t) = \frac{1}{8\pi}\left(\mathbf{E}\cdot\mathbf{D}+\mathbf{H}\cdot\mathbf{B}\right). \tag{5.6.4}$$

Power density in the form of Joule heat
$$\dot{u}_{\mathrm{mech}} = \mathbf{j}_f\cdot\mathbf{E}. \tag{5.3.6}$$

Energy flux density (Poynting vector)

$$\text{SI: } \mathbf{S} = \mathbf{E}\times\mathbf{H}, \qquad\qquad \text{G: } \mathbf{S} = \frac{c}{4\pi}\mathbf{E}\times\mathbf{H}. \tag{5.6.3}$$

Poynting's theorem (energy balance)
$$\dot{u}_{\mathrm{mech}} + \dot{u}_{\mathrm{field}} + \nabla\cdot\mathbf{S} = 0 \tag{5.6.5}$$

Force density

$$\text{SI: } \dot{\boldsymbol{p}}_{\text{mech}} = \rho_f \mathbf{E} + \mathbf{j}_f \times \mathbf{B}, \qquad\qquad \text{G: } \dot{\boldsymbol{p}}_{\text{mech}} = \rho_f \mathbf{E} + \frac{1}{c}\mathbf{j}_f \times \mathbf{B}. \tag{5.6.8}$$

Momentum density of the electromagnetic field

$$\boldsymbol{p}_{\text{feld}} = \frac{\epsilon_r \mu_r}{c^2}\mathbf{S} \quad \Leftrightarrow \quad \text{SI: } \boldsymbol{p}_{\text{field}} = \mathbf{D}\times\mathbf{B}, \qquad \text{G: } \boldsymbol{p}_{\text{field}} = \frac{1}{4\pi c}\mathbf{D}\times\mathbf{B}. \tag{5.6.15}$$

Maxwell stress tensor

$$\text{SI: } T_{ij} = E_i D_j + H_i B_j - \frac{1}{2}(\mathbf{E}\cdot\mathbf{D} + \mathbf{H}\cdot\mathbf{B})\delta_{ij},$$

$$\text{G: } T_{ij} = \frac{1}{4\pi}\left[E_i D_j + H_i B_j - \frac{1}{2}(\mathbf{E}\cdot\mathbf{D} + \mathbf{H}\cdot\mathbf{B})\delta_{ij} \right]. \tag{5.6.11}$$

Balance equation for the momentum density
$$\dot{p}_{\text{mech}\,i} + \dot{p}_{\text{feld}\,i} = T_{ij,j}. \tag{5.6.12}$$

Radiation

Electric and magnetic field: $\mathbf{E}^{\text{SI}} = \mathbf{E}/\sqrt{4\pi\epsilon_0}$; $\mathbf{B}^{\text{SI}} = \mathbf{B}/\sqrt{\mu_0/4\pi}$

$$\text{SI: } \mathbf{E} = -\boldsymbol{\nabla}\phi - \dot{\mathbf{A}}, \qquad\qquad \text{G: } \mathbf{E} = -\boldsymbol{\nabla}\phi - (1/c)\dot{\mathbf{A}}, \tag{8.1.3}$$

$$\mathbf{B} = \boldsymbol{\nabla}\times\mathbf{A}.$$

Lorenz gauge

$$\text{SI: } \boldsymbol{\nabla}\cdot\mathbf{A} + (1/c^2)\dot{\phi} = 0, \qquad\qquad \text{G: } \boldsymbol{\nabla}\cdot\mathbf{A} + (1/c)\dot{\phi} = 0. \tag{8.1.8}$$

$$\boldsymbol{\nabla}\cdot\mathbf{A} = 0 \qquad\qquad\qquad\qquad \text{Coulomb gauge.} \tag{8.2.48}$$

Far field, all multipoles: $\mathbf{A}^{\text{SI}} = \mathbf{A}/\sqrt{\mu_0/4\pi}$; Fourier transform $\mathbf{j}(\mathbf{k})$ with $\mathbf{k} = k\mathbf{e}_r$

$$\text{SI: } \mathbf{A}(\mathbf{x},t) = \frac{\mu_0}{4\pi}\frac{e^{ikr-i\omega t}}{r}\mathbf{j}(\mathbf{k}), \qquad \text{G: } \mathbf{A}(\mathbf{x},t) = \frac{e^{ikr-i\omega t}}{r}\frac{1}{c}\mathbf{j}(\mathbf{k}). \tag{8.3.22}$$

Electric field, outside the power distribution: $\mathbf{E}^{\text{SI}} = \mathbf{E}/\sqrt{4\pi\epsilon_0}$

$$\text{SI: } \mathbf{E}(\mathbf{x},t) = (i/\omega)\boldsymbol{\nabla}\times\mathbf{B}, \qquad\qquad \text{G: } \mathbf{E}(\mathbf{x},t) = (ic/\omega)\boldsymbol{\nabla}\times\mathbf{B}. \tag{8.3.25}$$

Impedance of free space:

$$\text{SI: } Z_0 = \sqrt{\mu_0/\epsilon_0} = 377\,\Omega, \qquad \text{G: } Z_0 = 4\pi/c = 4.19\times10^{-10}\text{stat}\Omega. \tag{8.2.65'}$$

Average energy flux density:

$$\text{SI: } \langle \mathbf{S}\rangle = \frac{1}{2\mu_0}\mathbf{E}\times\mathbf{B}^* = \frac{|\mathbf{E}|^2}{2Z_0}\mathbf{e}_r, \qquad \text{G: } \langle \mathbf{S}\rangle = \frac{c}{8\pi}\mathbf{E}\times\mathbf{B}^* = \frac{|\mathbf{E}|^2}{2Z_0}\mathbf{e}_r. \tag{8.3.27}$$

Average power of the electric dipole radiation
$$\langle P\rangle = Z_0\frac{c^2 k^4}{12\pi}|\mathbf{p}|^2. \tag{8.4.18}$$

D.3 Vector Calculus

Vector products

$$\mathbf{a}\cdot(\mathbf{b}\times\mathbf{c}) =\mathbf{b}\cdot(\mathbf{c}\times\mathbf{a}) = \mathbf{c}\cdot(\mathbf{a}\times\mathbf{b}) \tag{A.1.55}$$

$$\mathbf{a}\times(\mathbf{b}\times\mathbf{c}) =(\mathbf{a}\cdot\mathbf{c})\mathbf{b}-(\mathbf{a}\cdot\mathbf{b})\mathbf{c} \tag{A.1.60}$$

$$(\mathbf{a}\times\mathbf{b})\cdot(\mathbf{c}\times\mathbf{d}) =(\mathbf{a}\cdot\mathbf{c})(\mathbf{b}\cdot\mathbf{d}) - (\mathbf{a}\cdot\mathbf{d})(\mathbf{b}\cdot\mathbf{c}) \tag{A.1.61}$$

$$\boldsymbol{\nabla}(\mathbf{a}\cdot\mathbf{b}) =(\mathbf{a}\cdot\boldsymbol{\nabla})\mathbf{b} + (\mathbf{b}\cdot\boldsymbol{\nabla})\mathbf{a} + \mathbf{a}\cdot(\boldsymbol{\nabla}\times\mathbf{b}) + \mathbf{b}\cdot(\boldsymbol{\nabla}\times\mathbf{a}) \tag{A.2.33}$$

$$\boldsymbol{\nabla}\cdot(\mathbf{a}\times\mathbf{b}) =\mathbf{b}\cdot(\boldsymbol{\nabla}\times\mathbf{a}) - \mathbf{a}\cdot(\boldsymbol{\nabla}\times\mathbf{b}) \tag{A.2.34}$$

$$\boldsymbol{\nabla}\times(\mathbf{a}\times\mathbf{b}) =\mathbf{a}(\boldsymbol{\nabla}\cdot\mathbf{b}) - \mathbf{b}(\boldsymbol{\nabla}\cdot\mathbf{a}) + (\mathbf{b}\cdot\boldsymbol{\nabla})\mathbf{a} - (\mathbf{a}\cdot\boldsymbol{\nabla})\mathbf{b} \tag{A.2.35}$$

$$\boldsymbol{\nabla}\times(\boldsymbol{\nabla}\times\mathbf{a}) =\boldsymbol{\nabla}(\boldsymbol{\nabla}\cdot\mathbf{a}) - \Delta\mathbf{a} \tag{A.2.38}$$

$$\boldsymbol{\nabla}(fg) = g\boldsymbol{\nabla}f+f\boldsymbol{\nabla}g \qquad\qquad \boldsymbol{\nabla}\times\mathbf{a}f = f\boldsymbol{\nabla}\times\mathbf{a} + (\boldsymbol{\nabla}f)\times\mathbf{a}$$

Coordinate systems

Cartesian coordinates

Del operator
$$\boldsymbol{\nabla} = \mathbf{e}_x\frac{\partial}{\partial x} + \mathbf{e}_y\frac{\partial}{\partial y} + \mathbf{e}_z\frac{\partial}{\partial z} \tag{A.2.10}$$

Divergence
$$\boldsymbol{\nabla}\cdot\mathbf{a} = \frac{\partial a_x}{\partial x} + \frac{\partial a_y}{\partial y} + \frac{\partial a_z}{\partial z} \tag{A.2.24}$$

Curl
$$\boldsymbol{\nabla}\times\mathbf{a} = \mathbf{e}_x\left[\frac{\partial a_z}{\partial y} - \frac{\partial a_y}{\partial z}\right] + \mathbf{e}_y\left[\frac{\partial a_x}{\partial z} - \frac{\partial a_z}{\partial x}\right] + \mathbf{e}_z\left[\frac{\partial a_y}{\partial x} - \frac{\partial a_x}{\partial y}\right] \tag{A.2.26}$$

Laplace operator
$$\Delta = \frac{\partial^2}{\partial x^2} + \frac{\partial^2}{\partial y^2} + \frac{\partial^2}{\partial z^2} \tag{A.2.29}$$

Cylindrical coordinates

$$
\begin{aligned}
x &=\varrho\cos\varphi & \mathbf{e}_\varrho &=\mathbf{e}_x\cos\varphi + \mathbf{e}_y\sin\varphi \\
y &=\varrho\sin\varphi & \mathbf{e}_\phi &= -\,\mathbf{e}_x\sin\varphi + \mathbf{e}_y\cos\varphi \\
z &=z & \mathbf{e}_z &=\mathbf{e}_z
\end{aligned}
\tag{A.3.8}
$$

Del operator
$$\boldsymbol{\nabla} = \mathbf{e}_\varrho\frac{\partial}{\partial\varrho} + \mathbf{e}_\varphi\frac{1}{\varrho}\frac{\partial}{\partial\varphi} + \mathbf{e}_z\frac{\partial}{\partial z} \tag{A.3.16}$$

Divergence
$$\boldsymbol{\nabla}\cdot\mathbf{a} = \frac{1}{\varrho}\frac{\partial\varrho a_\varrho}{\partial\varrho} + \frac{1}{\varrho}\frac{\partial a_\varphi}{\partial\varphi} + \frac{\partial a_z}{\partial z} \tag{A.3.18}$$

Curl
$$\boldsymbol{\nabla}\times\mathbf{a}=\mathbf{e}_\varrho\left[\frac{1}{\varrho}\frac{\partial a_z}{\partial\varphi} - \frac{\partial a_\varphi}{\partial z}\right] + \mathbf{e}_\varphi\left[\frac{\partial a_\varrho}{\partial z} - \frac{\partial a_z}{\partial\varrho}\right] + \mathbf{e}_z\frac{1}{\varrho}\left[\frac{\partial\varrho a_\varphi}{\partial\varrho} - \frac{\partial a_\varrho}{\partial\varphi}\right] \tag{A.3.19}$$

Laplace operator
$$\Delta = \frac{1}{\varrho}\frac{\partial}{\partial\varrho}\left(\varrho\frac{\partial}{\partial\varrho}\right) + \frac{1}{\varrho^2}\frac{\partial^2}{\partial\varphi^2} + \frac{\partial^2}{\partial z^2} \tag{A.3.20}$$

$$\text{Jacobian matrix} \quad \mathsf{J} = \frac{\partial(x\, y\, z)}{\partial(\varrho\, \varphi\, z)} = \begin{pmatrix} \cos\varphi & -\varrho\sin\varphi & 0 \\ \sin\varphi & \varrho\cos\varphi & 0 \\ 0 & 0 & 1 \end{pmatrix} \quad \det J = \varrho \qquad (A.3.10)$$

Volume element $\quad \mathrm{d}^3 x = \varrho\, \mathrm{d}\varrho \mathrm{d}\varphi \mathrm{d}z$

$$\text{Surface element} \quad \mathbf{da} = \mathbf{e}_\varrho\, \varrho\, \mathrm{d}\varphi \mathrm{d}z + \mathbf{e}_\varphi\, \mathrm{d}\varrho \mathrm{d}z + \mathbf{e}_z\, \varrho \mathrm{d}\varrho \mathrm{d}\varphi \qquad (A.3.12)$$

$$\text{Line element} \quad \mathbf{ds} = \mathbf{e}_\varrho\, \mathrm{d}\varrho + \mathbf{e}_\varphi\, \varrho \mathrm{d}\varphi + \mathbf{e}_z\, \mathrm{d}z \qquad (A.3.13)$$

$$\text{Delta function} \quad \delta^{(3)}(\mathbf{x}-\mathbf{x}') = \frac{1}{\varrho}\, \delta(\varrho-\varrho')\, \delta(\varphi-\varphi')\, \delta(z-z') \qquad (A.3.15)$$

Spherical coordinates

$$
\begin{aligned}
x &= r\cos\vartheta\cos\varphi \\
y &= r\cos\vartheta\sin\varphi \\
z &= r\sin\vartheta
\end{aligned}
\qquad
\begin{aligned}
\mathbf{e}_r &= \sin\vartheta(\mathbf{e}_x\cos\varphi + \mathbf{e}_y\sin\varphi) + \mathbf{e}_z\cos\vartheta \\
\mathbf{e}_\vartheta &= \cos\vartheta(\mathbf{e}_x\cos\varphi + \mathbf{e}_y\sin\varphi) - \mathbf{e}_z\sin\vartheta \\
\mathbf{e}_\varphi &= -\,\mathbf{e}_x\sin\varphi + \mathbf{e}_y\cos\varphi
\end{aligned}
\qquad (A.3.21)
$$

$$\text{Del operator} \quad \boldsymbol{\nabla} = \mathbf{e}_r \frac{\partial}{\partial r} + \mathbf{e}_\vartheta \frac{1}{r}\frac{\partial}{\partial\vartheta} + \mathbf{e}_\varphi \frac{1}{r\sin\vartheta}\frac{\partial}{\partial\varphi} \qquad (A.3.32)$$

$$\text{Divergence} \quad \boldsymbol{\nabla}\cdot\mathbf{a} = \frac{1}{r^2}\frac{\partial r^2 a_r}{\partial r} + \frac{1}{r\sin\vartheta}\frac{\partial \sin\vartheta a_\vartheta}{\partial\vartheta} + \frac{1}{r\sin\vartheta}\frac{\partial a_\varphi}{\partial\varphi} \qquad (A.3.34)$$

$$
\begin{aligned}
\text{Curl} \quad \boldsymbol{\nabla}\times\mathbf{a} = {}& \mathbf{e}_r \frac{1}{r\sin\vartheta}\Big[\frac{\partial \sin\vartheta a_\varphi}{\partial\vartheta} - \frac{\partial a_\vartheta}{\partial\varphi}\Big] \\
& + \mathbf{e}_\vartheta \frac{1}{r}\Big[\frac{1}{\sin\vartheta}\frac{\partial a_r}{\partial\varphi} - \frac{\partial r a_\varphi}{\partial r}\Big] + \mathbf{e}_\varphi \frac{1}{r}\Big[\frac{\partial r a_\vartheta}{\partial\varphi} - \frac{\partial a_\varphi}{\partial\vartheta}\Big] \quad (A.3.35)
\end{aligned}
$$

$$\text{Laplace operator} \quad \Delta = \frac{1}{r^2}\frac{\partial}{\partial r}r^2\frac{\partial}{\partial r} + \frac{1}{r^2\sin\vartheta}\frac{\partial}{\partial\vartheta}\sin\vartheta\frac{\partial}{\partial\vartheta} + \frac{1}{r^2\sin^2\vartheta}\frac{\partial^2}{\partial\varphi^2} \qquad (A.3.36)$$

$$\text{Jacobian matrix} \quad \mathsf{J} = \frac{\partial(x\, y\, z)}{\partial(r\, \vartheta\, \varphi)} = \begin{pmatrix} \sin\vartheta\cos\varphi & r\cos\vartheta\cos\varphi & -r\cos\vartheta\sin\varphi \\ \sin\vartheta\sin\varphi & r\cos\vartheta\sin\varphi & r\cos\vartheta\cos\varphi \\ \cos\vartheta & -r\sin\vartheta & 0 \end{pmatrix}$$

$$\det J = r^2\sin\vartheta \qquad (A.3.24)$$

Volume element $\quad \mathrm{d}^3 x = r^2\sin\vartheta\, \mathrm{d}r\, \mathrm{d}\vartheta\, \mathrm{d}\varphi \qquad (A.3.28)$

$$\text{Surface element} \quad \mathbf{da} = \mathbf{e}_r\, r^2\sin\vartheta\, \mathrm{d}\vartheta \mathrm{d}\varphi + \mathbf{e}_\vartheta\, r\sin\vartheta \mathrm{d}r \mathrm{d}\varphi + \mathbf{e}_\varphi\, r\, \mathrm{d}r \mathrm{d}\varphi$$

$$\text{Line element} \quad \mathbf{ds} = \mathbf{e}_r\, \mathrm{d}r + \mathbf{e}_\vartheta\, r\, \mathrm{d}\vartheta + \mathbf{e}_\varphi\, r\sin\vartheta\, \mathrm{d}\varphi \qquad (A.3.29)$$

$$\text{Delta function} \quad \delta^{(3)}(\mathbf{x}-\mathbf{x}') = \frac{1}{r^2\sin\vartheta}\, \delta(r-r')\, \delta(\vartheta-\vartheta')\, \delta(\varphi-\varphi') \qquad (A.3.31)$$

Index

Aberration, 454, 481
Abraham-Lorentz equation, 317
Absorption
    anomalous in the Laue case, 424
    in the Bragg case, 426
    in the Drude-Lorentz model, 188
Action integral, 528
Active resistance, 338
Addition theorem for spherical
    harmonics, 97, 616
Alfvén waves, 347
Ampère's circuital law, *see* Ampère's
    law
Ampère's force law, 145
Ampère's law, 17, 122
Ampère-Maxwell law, 16
    integral form, 15
Angular momentum of a charge
    distribution, 142
Angular momentum operator, 87
Anomalous dispersion, 191
Antenna
    dipole array, 307
    dipole group, 307, 323
    dipole line, 307, 323
    directional factor, 307
    radiation pattern, 308
Arago disc, 508
Azimuth angle, 579

Bar magnet, 227
Bilinear form, 563
Biot-Savart's law, 123, 124
Bohr magneton, 142, 645

Bohr radius, 645
Bohr-van Leeuwen theorem, 253
Boost, 447, 476
Borrmann fan, 428
Boundary conditions
    electric displacement field, 168
    electric field, 168
        charged layer, 41
        conductor, 47
Bragg case, 414
Bravais lattice, 403
Brewster angle, 361

Capacitance, 337
    coefficient, 64
Cauchy principal value, 606
Cauchy-Riemann differential equations,
    603
Cavity resonator, 379
Charge
    bound, 157
    free, 157
    true, 639
Charge conservation, 3
    under Lorentz transformation, 503
Charge density
    average, 6, 7
    continuous, 5
    magnetic, 220
    microscopic, 4
    of a point charge, 494
Charge invariance, 495
Cherenkov radiation, 283